D0161757

Color Imaging

Color Imaging

Fundamentals and Applications

Erik Reinhard
Erum Arif Khan
Ahmet Oğuz Akyüz
Garrett Johnson

A K Peters, Ltd.
Wellesley, Massachusetts

Editorial, Sales, and Customer Service Office

A K Peters, Ltd.
888 Worcester Street, Suite 230
Wellesley, MA 02482
www.akpeters.com

Library of Congress Cataloging-in-Publication Data

Reinhard, Erik, 1968–
Color imaging : fundamentals and applications / Erik Reinhard ... [et al.].
p. cm.
Includes bibliographical references and index.
ISBN: 978-1-56881-344-8 (alk. paper)
1. Computer vision. 2. Image processing. 3. Color display systems. 4. Color separation. I. Title.

TA1634.R45 2007
621.36'7--dc22

2007015704

Printed in India
12 11 10 09 08

10 9 8 7 6 5 4 3 2 1

Contents

Preface

Color is one of the most fascinating areas to study. Color forms an integral part of nature, and we humans are exposed to it every day. We all have an intuitive understanding of what color is, but by studying the underlying physics, chemistry, optics, and human visual perception, the true beauty and complexity of color can be appreciated—at least to some extent. Such understanding is not just important in these areas of research, but also for fields such as color reproduction, vision science, atmospheric modeling, image archiving, art, photography, and the like.

Many of these application areas are served very well by several specifically targeted books. These books do an excellent job of explaining in detail some aspect of color that happens to be most important for the target audience. This is understandable as our knowledge of color spans many disciplines and can therefore be difficult to fathom.

It is our opinion that in application areas of computer science and computer engineering, including such exciting fields as computer graphics, computer vision, high dynamic range imaging, image processing and game development, the role of color is not yet fully appreciated. We have come across several applications as well as research papers where color is added as an afterthought, and frequently wrongly too. The dreaded RGB color space, which is really a collection of loosely similar color spaces, is one of the culprits.

With this book, we hope to give a deep understanding of what color is, and where color comes from. We also aim to show how color can be used correctly in many different applications. Where appropriate, we include at the end of each chapter sections on applications that exploit the material covered. While the book is primarily aimed at computer-science and computer-engineering related areas, as mentioned above, it is suitable for any technically minded reader with an interest in color. In addition, the book can also be used as a text book serving a graduate-level course on color theory. In any case, we believe that to be useful in any engineering-related discipline, the theories should be presented in an intuitive manner, while also presenting all of the mathematics in a form that allows both a deeper understanding, as well as its implementation.

Most of the behavior of light and color can be demonstrated with simple experiments that can be replicated at home. To add to the appeal of this book, where possible, we show how to set-up such experiments that frequently require no more than ordinary household objects. For instance, the wave-like behavior of light is easily demonstrated with a laser pointer and a knife. Also, several visual illusions can be replicated at home. We have shied away from such simple experiments only when unavoidable.

The life cycle of images starts with either photography or rendering, and involves image processing, storage, and display. After the introduction of digital imaging, the imaging pipeline has remained essentially the same for more than two decades. The phosphors of conventional CRT devices are such that in the operating range of the human visual system only a small number of discernible intensity levels can be reproduced. As a result, there was never a need to capture and store images with a fidelity greater than can be displayed. Hence the immense legacy of eight-bit images.

High dynamic range display devices have effectively lifted this restriction, and this has caused a rethinking of the imaging pipeline. Image capturing techniques can and should record the full dynamic range of the scene, rather than just the restricted range that can be reproduced on older display devices. In this book, the vast majority of the photography was done in high dynamic range (HDR), with each photograph tone-mapped for reproduction on paper. In addition, high dynamic range imaging (HDRI) is integral to the writing of the text, with exceptions only made in specific places to highlight the differences between conventional imaging and HDRI. Thus, the book is as future-proof as we could possibly make it.

Acknowledgments

Numerous people have contributed to this book with their expertise and help. In particular, we would like to thank Eric van Stryland, Dean and Director of CREOL, who has given access to many optics labs, introduced us to his colleagues, and allowed us to photograph some of the exciting research undertaken at the School of Optics, University of Central Florida. Karen Louden, Curator and Director of Education of the Albin Polasek Museum, Winter Park, Florida, has given us free access to photograph in the Albin Polasek collection.

We have sourced many images from various researchers. In particular, we are grateful for the spectacular renderings given to us by Diego Gutierrez and his colleagues from the University of Zaragoza. The professional photographs donated by Kirt Witte (Savannah College of Art and Design) grace several pages, and we gratefully acknowledge his help. Several interesting weather phenomena were photographed by Timo Kunkel, and he has kindly allowed us to reproduce some of them. We also thank him for carefully proofreading an early draft of the manuscript.

We have had stimulating discussions with Karol Myszkowski, Grzegorz Krawczyk, Rafał Mantiuk, Kaleigh Smith, Edward Adelson and Yuanzhen Li, the results of which have become part of the chapter on tone reproduction. This chapter also benefitted from the source code of Yuanzhen Li's tone-reproduction operator, made available by Li and her colleagues, Edward Adelson and Lavanya Sharan. We are extremely grateful for the feedback received from Charles Poynton, which helped improve the manuscript throughout in both form and substance.

We have received a lot of help in various ways, both direct and indirect, from many people. In no particular order, we gratefully acknowledge the help from Janet Milliez, Vasile Rotar, Eric G Johnson, Claudiu Cirloganu, Kadi Bouatouch, Dani Lischinski, Ranaan Fattal, Alice Peters, Franz and Ineke Reinhard, Gordon Kindlmann, Sarah Creem-Regehr, Charles Hughes, Mark Colbert, Jared Johnson, Jaakko Konttinen, Veronica Sundstedt, Greg Ward, Mashhuda Glencross, Helge Seetzen, Mahdi Nezamabadi, Paul Debevec, Tim Cox, Jessie Evans, Michelle Ward, Denise Penrose, Tiffany Gasbarrini, Aaron Hertzmann, Kevin Suffern, Guoping Qiu, Graham Finlayson, Peter Shirley, Michael Ashikhmin, Wolfgang Heidrich, Karol Myszkowski, Grzegorz Krawczyk, Rafal Mantiuk, Kaleigh Smith, Majid Mirmehdi, Louis Silverstein, Mark Fairchild, Nan Schaller, Walt Bankes, Tom Troscianko, Heinrich Bülthoff, Roland Fleming, Bernard Riecke, Kate Devlin, David Ebert, Francisco Seron, Drew Hess, Gary McTaggart, Habib Zargarpour, Peter Hall, Maureen Stone, Holly Rushmeier, Narantuja Bujantogtoch, Margarita Bratkova, Tania Pouli, Ben Long, Native Visions Art Gallery (Winter Park, Florida), the faculty, staff, and students of the Munsell Color Science Laboratory, Lawrence Taplin, Ethan Montag, Roy Berns, Val Helmink, Colleen Desimone, Sheila Brady, Angus Taggart, Ron Brinkmann, Melissa Answeeney, Bryant Johnson, and Paul and Linda Johnson.

Part I
Principles

Chapter 1
Introduction

Color is a phenomenon that relates to the physics of light, chemistry of matter, geometric properties of objects, as well as to human visual perception and cognition. We may call sunlight yellow, and in this case we refer to a *property* of light. A car may be painted red, in which case the color red is an *attribute* of the car. When light enters the eye, a complex chain of events leads to the sensation of color, *a perceived quantity*. Finally, color may be remembered, associated with events, and reasoned about. These are *cognitive aspects* of color. Color means different things under different circumstances [814].

At the same time, it is clear that an understanding of color will involve each of these aspects. Thus, the study of color theory and its applications necessarily spans several different fields, including physics, chemistry, optics, radiometry, photometry, colorimetry, physiology, vision science, color appearance modeling, and image processing. As a result, what on the surface appears to be a relatively simple subject, turns out to have many hidden depths. Perhaps this is the reason that in practical applications found in computer science—computer vision, graphics and image processing—the use of color is often under-explored and misunderstood.

In our view, to understand color with sufficient depth, and to be able to apply this knowledge to your own area of interest, it is not enough to read the literature in any one specific discipline, be it computer graphics, computer vision, photography, art, etc. Instead, it is necessary to step outside one's own field in order to appreciate the subtleties and complexities of color.

The purpose of this book is therefore to explain color theory, its development and current state-of-the-art, as well as its practical use in engineering-oriented disciplines such as computer graphics, computer vision, photography, and film. Along the way, we delve into the physics of light, and its interaction with matter at the atomic level, such that the origins of color can be appreciated. We find that the intimate relationship between energy levels, orbital states, and electromagnetic waves helps to understand why diamonds shimmer, rubies are red, and the feathers of the blue jay are blue. Even before light enters the eye, a lot has already happened.

The complexities of color multiply when perception is taken into account. The human eye is not a simple light detector by any stretch of the imagination. Human vision is able to solve an inherently under-constrained problem: it tries to make sense out of a 3D world using optical projections that are two-dimensional. To reconstruct a three-dimensional world, the human visual system needs to make a great many assumptions about the structure of the world. It is quite remarkable how well this system works, given how difficult it is to find a computational solution that only partially replicates these achievements.

When these assumptions are violated, the human visual system can be fooled into perceiving the wrong thing. For instance, if a human face is lit from above, it is instantly recognizable. If the same face is lit from below, it is almost impossible to determine whose face it is. It can be argued that whenever an assumption is broken, a visual illusion emerges. Visual illusions are therefore important to learn about how the human visual system operates. At the same time, they are important, for instance in computer graphics, to understand which image features need to be rendered correctly and which ones can be approximated while maintaining realism.

Color theory is at the heart of this book. All other topics serve to underpin the importance of using color correctly in engineering applications. We find that too often color is taken for granted, and engineering solutions, particularly in computer graphics and computer vision, therefore appear suboptimal. To redress the balance, we provide chapters detailing all important issues governing color and its perception, along with many examples of applications.

We begin this book with a brief assessment of the roles color plays in different contexts, including nature and society.

1.1 Color in Nature

Living organisms are embedded in an environment with which they interact. To maximize survival, they must be in tune with this environment. Color plays an important role in three ways:

- Organisms may be colored by default without this giving them a specific advantage for survival. An example is the green color of most plants (Fig-

Figure 1.1. The chlorophyll in leaves causes most plants to be colored green.

ure 1.1), which is due to chlorophyll, a pigment that plays a role in photosynthesis (see Section 3.4.1).

- Color has evolved in many species in conjunction with the color vision of the same or other species, for instance for camouflage (Figure 1.2), for attracting partners (Figure 1.3), for attracting pollinators (Figure 1.4), or for appearing unappetizing to potential predators (Figure 1.5).

- Biochemical substances may be colored as a result of being optimized to serve a specific function unrelated to color. An example is hemoglobin which colors blood red due to its iron content. Such functional colors are normally found inside the body, rather than at the surface.

Plants reflect green and absorb all other colors of light. In particular, plants absorb red and blue light and use the energy gained to drive the production of carbohydrates. It has been postulated that these two colors are being absorbed as a result of two individual mechanisms that allow plants to photosynthesize with maximum efficiency under different lighting conditions [518, 769].

Figure 1.2. Many animals are colored similar to their environment to evade predators.

Figure 1.3. This peacock uses bright colors to attract a mate; Paignton Zoo, Devon, UK. (Photo by Brett Burridge (www.brettb.com).)

Figure 1.4. Many plant species grow brightly colored flowers to attract pollinators such as insects and bees; Rennes, France, June 2005.

Figure 1.5. These beetles have a metallic color, presumably to discourage predators.

Figure 1.6. This desert rose is colored to reflect light and thereby better control its temperature.

Color in plants also aids other functions such as the regulation of temperature. In arid climates plants frequently reflect light of all colors, thus appearing lightly colored such as the desert rose shown in Figure 1.6.

In humans, color vision is said to have co-evolved with the color of fruit [787, 944]. In industrialized Caucasians, color deficiencies occur relatively often, while color vision is better developed in people who work the land. Thus, on average, color vision is diminished in people who do not depend on it for survival [216].

Human skin color is largely due to pigments such as *eumelanin* and *phaeomelanin*. The former is brown to black, whereas the latter is yellow to reddish-brown. Deeper layers contain yellow carotenoids. Some color in human skin is derived from scattering, as well as the occurrence of blood vessels [520].

Light scattering in combination with melanin pigmentation is also the mechanism that determines eye color in humans and mammals. There appears to be a correlation between eye color and reactive skills. In the animal world, hunters who stalk their prey tend to have light eye colors, whereas hunters who obtain their prey in a reactive manner, such as birds that catch insects in flight, have dark eyes. This difference, as yet to be fully explained, extends to humankind where dark-eyed people tend to have faster reaction times to both visual and auditory stimuli than light-eyed people [968].

Figure 1.7. Color plays an important role in art. An example is this photograph of the Tybee Light House, taken by Kirt Witte, which won the 2006 International Color Award's Masters of Color Photography award in the abstract category for professional photographers (see also www.theothersavannah.com).

Figure 1.8. Statue by Albin Polasek; Albin Polasek Museum, Winter Park, FL, 2004.

1.2 Color in Society

The use of color by man to communicate is probably as old as mankind itself. Archaeologists have found colored materials, in particular different shades of ochre, at sites occupied some 300,000 years ago by *Homo erectus* [519]. Approximately 70,000 years ago, *Homo sapiens neanderthalensis* used ochre in burials, a tradition followed later by *Homo sapiens sapiens*.

Color, of course, remains an important means of communication in art [695, 1257, 1291] (Figure 1.7), even in cases where color is created by subtle reflections (Figure 1.8).

Color also plays an important role in religion, where the staining of church windows is used to impress churchgoers. The interior of a church reflects the times in which it was built, ranging from light and airy to dark and somber (Figure 1.9).

Figure 1.9. Color in different church interiors; Rennes, France, June 2005 (top); Mainau, Germany, July 2005 (bottom).

With modern technology, new uses of color have come into fashion. Important areas include the reproduction of color [509], lighting design [941], photography [697], and television and video [923]. In computer graphics, computer vision, and other engineering disciplines, color also plays a crucial role, but perhaps less so than it should. Throughout this book, examples of color use in these fields are provided.

1.3 In this Book

This book offers an in-depth treatment of various topics related to color. Our intention is to explain only a subset of color science, but to treat each of the topics we have chosen in depth. Our choice of topics is intended to be relevant to those who require a more-than-casual understanding of color for their work, which we envisage to be in computer graphics, computer vision, animation, photography, image processing, and related disciplines.

All use of color in any discipline is either explicitly or implicitly based on theories developed to describe the physics of light, as well as the perception of it by humans. We therefore offer in the first chapters a reasonably detailed account of light, its propagation through vacuum as well as other media, and its interaction with boundaries.

The physics of light is governed by Maxwell's equations, which form the basis of our thinking about the wave nature of light. Without these equations, there would not be any physical optics. Since almost everything in radiometry and geometric physics is derived from Maxwell's equations, we expect that areas of study further afield would also look very different without them. Fields affected include photometry, lighting design, computer graphics, and computer vision. As such, we begin this book with a detailed description of electromagnetic waves and show how radiometry, geometric optics, and much of computer graphics are derived from them.

While the wave nature of light constitutes a powerful model in understanding the properties of light, the theory of electromagnetic radiation is not able to explain all measurable light behavior. In particular, light sometimes behaves as particles, and this fact is not captured by electromagnetic theory. Hence, in Chapter 3, we briefly introduce quantum mechanics as well as molecular orbital theory, as these form the basic tools for understanding how light interacts with matter at the atomic scale. Such interaction is the cause of various behaviors such as absorption, emission, diffraction, and dispersion. These theories afford insight into questions such as "why is water pale blue" and "why is a ruby red and can sapphires be blue". Thus, Chapter 3 is largely concerned with providing answers to questions regarding the causes of color. Whereas Chapter 2 deals with the propagation of light, Chapter 3 is largely informed by the interaction of light with matter.

Chapter 4 provides a brief introduction to human vision. The optics of the eye, as well as the neurophysiology of what is collectively known as the human visual system, are described. The scientific literature on this topic is vast, and a full account of the neuronal processing of visual systems is beyond the scope of this book. However, it is clear that with the advent of more sophisticated techniques, the once relatively straightforward theories of color processing in the visual cortex, have progressed to be significantly less straightforward. This trend is continuing to this day. The early inferences made on the basis of single-cell recordings have been replaced with a vast amount of knowledge that is often contradictory, and every new study that becomes available poses intriguing new questions. On the whole, however, it appears that color is not processed as a separate image attribute, but is processed together with other attributes such as position, size, frequency, direction, and orientation.

Color can also be surmised from a perceptual point of view. Here, the human visual system is treated as a black box, with outputs that can be measured. In psychophysical tests, participants are set a task which must be completed in response to the presentation of visual stimuli. By correlating the task response to the stimuli that are presented, important conclusions regarding the human visual system may be drawn. Chapter 5 describes some of the findings from this field of study, as it pertains to theories of color. This includes visual illusions, adaptation, visual acuity, contrast sensitivity, and constancy.

In the following chapters, we build upon the fundamentals underlying color theory. Chapter 6 deals with radiometry and photometry, whereas Chapter 7 discusses colorimetry. Much research has been devoted to color spaces that are designed for different purposes. Chapter 8 introduces many of the currently-used color spaces and explains the strengths and weaknesses of each color space. The purpose of this chapter is to give transformations between existing color spaces and to enable the selection of an appropriate color space for specific tasks, realizing that each task may require a different color space.

Light sources, and their theoretical formulations (called illuminants), are discussed in Chapter 9. Chapter 10 introduces chromatic adaptation, showing that the perception of a colored object does not only depend on the object's reflectance, but also on its illumination and the state of adaptation of the observer. While colorimetry is sufficient for describing colors, an extended model is required to account for the environment in which the color is observed. Color appearance models take as input the color of a patch, as well as a parameterized description of the environment. These models then compute appearance correlates that describe perceived attributes of the color, given the environment. Color appearance models are presented in Chapter 11.

In Part III, the focus is on images, and in particular their capture and display. Much of this part of the book deals with the capture of high dynamic range images, as we feel that such images are gaining importance and may well become the de-facto norm in all applications that deal with images.

Chapter 12 deals with the capture of images and includes an in-depth description of the optical processes involved in image formation, as well as issues related to digital sensors. This chapter also includes sections on camera characterization, and more specialized capture techniques such as holography and light field data. Techniques for the capture of high dynamic range images are discussed in Chapter 13. The emphasis is on multiple exposure techniques, as these are currently most cost effective, requiring only a standard camera and appropriate software.

Display hardware is discussed in Chapter 14, including conventional and emerging display hardware. Here, the focus is on liquid crystal display devices, as these currently form the dominant display technology. Further, display calibration techniques are discussed.

Chapter 15 is devoted to a discussion on natural image statistics, a field important as a tool both to help understand the human visual system, and to help structure and improve image processing algorithms. This chapter also includes sections on techniques to measure the dynamic range of images, and discusses cross-media display technology, gamut mapping, gamma correction, and algorithms for correcting for light reflected off display devices. Color management for images is treated in Chapter 16, with a strong emphasis on ICC profiles. Finally, Chapter 17 presents current issues in tone reproduction, a collection of algorithms required to prepare a high dynamic range image for display on a conventional display device.

For each of the topics presented in the third part of the book, the emphasis is on color management, rather than spatial processing. As such, these chapters augment, rather than replace, current books on image processing.

The book concludes with a set of appendices, which are designed to help clarify the mathematics used throughout the book (vectors and matrices, trigonometry, and complex numbers), and to provide tables of units and constants for easy reference. We also refer to the DVD-ROM included with the book, which contains a large collection of images in high dynamic range format, as well as tonemapped versions of these images (in JPEG-HDR format for backward compatibility), included for experimentation. The DVD-ROM also contains a range of spectral functions, a metameric spectral image, as well as links to various resources on the Internet.

1.4 Further Reading

The history of color is described in Nassau's book [814], whereas some of the history of color science is collected in MacAdam's books [717, 720]. A historical overview of dyes and pigments is available in *Colors: The Story of Dyes and Pigments* [245]. An overview of color in art and science is presented in a collection of papers edited by Lamb and Bourriau [642]. Finally, a history of color order, including practical applications, is collected in Rolf Kuehni's *Color Space and its Divisions: Color Order from Antiquity to Present* [631].

Chapter 2
Physics of Light

Light travels through environments as electromagnetic energy that may interact with surfaces and volumes of matter. Ultimately, some of that light reaches the human eye which triggers a complicated chain of events leading to the perception, cognition, and understanding of these environments.

To understand the physical aspects of light, i.e., everything that happens to light before it reaches the eye, we have to study electromagnetism. To understand the various ways by which colored light may be formed, we also need to know a little about quantum mechanics and molecular orbital theory (discussed in Chapter 3). These topics are not particularly straightforward, but they are nonetheless well worth studying. They afford insight into the foundations of many fields related to color theory, such as optics, radiometry (and therefore photometry), as well as computer graphics.

The physical properties of light are well modeled by Maxwell's equations. We therefore begin this chapter with a brief discussion of Maxwell's equations. We then discuss various optical phenomena that may be explained by the theory of electromagnetic radiation. These include scattering, polarization, reflection, and refraction.

There are other optical phenomena that involve the interaction between light and materials at the atomic structure. Examples of these phenomena are diffraction, interference, and dispersion, each capable of separating light into different wavelengths—and are therefore perceived as producing different colors. With the exception of interference and diffraction, the explanation of these phenomena requires some insight into the chemistry of matter. We therefore defer their description until Chapter 3.

In addition to presenting electromagnetic theory, in this chapter we also introduce the concept of geometrical optics, which provides a simplified view of the theory of light. It gives rise to various applications, including ray tracing in optics. It is also the foundation for all image synthesis as practiced in the field of computer graphics. We show this by example in Section 2.10.

Thus, the purpose of this chapter is to present the theory of electromagnetic waves and to show how light propagates through different media and behaves near boundaries and obstacles. This behavior by itself gives rise to color. In Chapter 3, we explain how light interacts with matter at the atomic level, which gives rise to several further causes for color, including dispersion and absorption.

2.1 Electromagnetic Theory

Light may be modeled by a transverse electromagnetic (TEM) wave traveling through a medium. This suggests that there is an interaction between electric and magnetic fields and their sources of charge and current. A moving electric charge is the source of an electric field. At the same time, electric currents produce a magnetic field. The relationship between electric and magnetic fields are governed by Maxwell's equations which consist of four laws:

- Gauss' law for electric fields;
- Gauss' law for magnetic fields;
- Faraday's law;
- Ampere's circuital law.

We present each of these laws in integral form first, followed by their equivalent differential form. The integral form has a more intuitive meaning but is restricted to simple geometric cases, whereas the differential form is valid for any point in space where the vector fields are continuous.

There are several systems of units and dimensions used in Maxwell's equations, including Gaussian units, Heaviside-Lorentz units, electrostatic units, electromagnetic units, and SI units [379]. There is no specific reason to prefer one system over another. Since the SI system is favored in engineering-oriented disciplines, we present all equations in this system. In the SI system, the basic quantities are the meter (m) for length, the kilogram (kg) for mass, the second (s) for time, the ampere (A) for electric current, the kelvin (K) for thermodynamic temperature, the mole (mol) for amount of substance, and the candela (cd) for luminous intensity. (see Table D.3 in Appendix D).

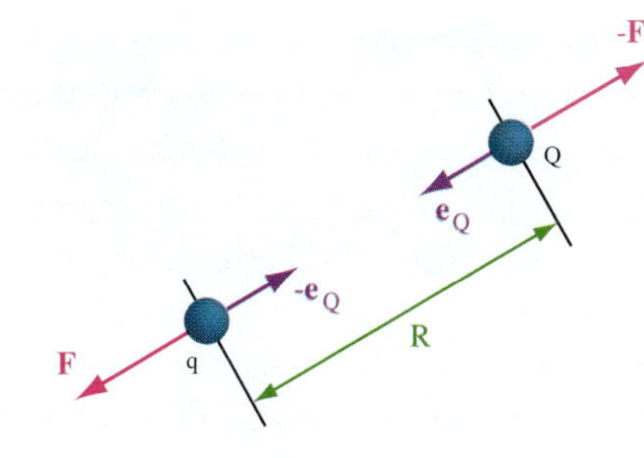

Figure 2.1. Two charges Q and q exert equal, but opposite force $\mathbf{F}$ upon each other (assuming the two charges have equal sign).

2.1.1 Electric Fields

Given a three-dimensional space, we may associate attributes with each point in this space. For instance, the temperature in a room may vary by location in the room. With heat tending to rise, the temperature near the ceiling is usually higher than near the floor.

Similarly, it is possible to associate other attributes to regions in space. If these attributes have physical meaning, we may speak of a *field*, which simply indicates that there exists a description of how these physical phenomena change with position in space. In a time-varying field, these phenomena also change with time.

For instance, we could place an electrical charge in space and a second charge some distance away. These two charges exert a force on each other dependent upon the magnitude of their respective charges and their distance. Thus, if we move the second charge around in space, the force exerted on it by the first charge changes (and vice versa). The two charges thus create a *force field*.

The force $\mathbf{F}$ that two charges Q and q exert on each other is given by Coulomb's law (see Figure 2.1):

$$\mathbf{F} = \frac{1}{4\pi\,\varepsilon_0}\,\frac{Qq}{R^2}\,\mathbf{e}_Q. \tag{2.1}$$

In this equation, R is the distance between the charges, and the constant $\varepsilon_0 = \frac{1}{36\pi} \times 10^{-9}$ (in Farad/meter) is called the *permittivity* of vacuum. The vector $\mathbf{e}_Q$ is a unit vector pointing from the position of one charge to the position of the other.

If we assume that Q is an arbitrary charge and that q is a unit charge, then we can compute the *electric field intensity* $\mathbf{E}$ by dividing the left- and right-hand

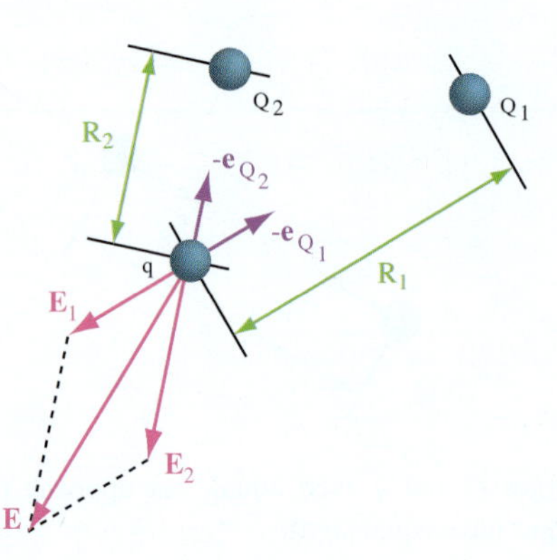

Figure 2.2. The electric field intensity **E** at the position of charge q is due to multiple charges located in space.

sides of Equation (2.1) by q:

$$\mathbf{E} = \frac{\mathbf{F}}{q} = \frac{Q}{4\pi\varepsilon_0 R^2}\,\mathbf{e}_Q. \tag{2.2}$$

If there are multiple charges present, then the electric field intensity at the position of the test charge is given by the sum of the individual field intensities (Figure 2.2):

$$\mathbf{E} = \sum_{i=1}^{N} \frac{Q_i}{4\pi\varepsilon_0 R_i^2}\,\mathbf{e}_{Q_i}. \tag{2.3}$$

Finally, if we scale the electric field intensity **E** by the permittivity of vacuum (ε_0), we obtain what is called the *electric flux density*, indicated by the vector **D** which points in the same direction as **E**:

$$\mathbf{D} = \varepsilon_0\,\mathbf{E} = \sum_{i=1}^{N} \frac{Q_i}{4\pi R_i^2}\,\mathbf{e}_{Q_i}. \tag{2.4}$$

For media other than vacuum, the permittivity will generally have a different value. In that case, we drop the subscript 0, so that, in general, we have

$$\mathbf{D} = \varepsilon\mathbf{E}. \tag{2.5}$$

This equation may be seen as relating the electric field intensity to the electric flux density, whereby the difference between the two vectors is in most cases determined by a constant unique to the specific medium. This constant, ε, is called the material's *permittivity* or *dielectric constant*. Equation (2.5) is one of three so-called *material equations*. The remaining two material equations will be discussed in Sections 2.1.3 and 2.1.8.

If, instead of a finite number of separate charges, we have a distribution of charges over space, the electric flux density is governed by Gauss' law for electric fields.

2.1.2 Gauss' Law for Electric Fields

In a static or time-varying electric field there is a distribution of charges. The relationship between an electric field and a charge distribution is quantified by Gauss' law for electric fields. It states that the total electric flux, $\mathbf{D} = \varepsilon\,\mathbf{E}$, emanating from a closed surface s is equal to the electric charge Q enclosed by that surface:[1]

$$\int_s \mathbf{D} \cdot \mathbf{n}\, ds = Q. \tag{2.6}$$

The integral is over the closed surface s and $\mathbf{n}$ denotes the outward facing surface normal. If the charge Q is distributed over the volume v according to a charge distribution function ρ (also known as *electric charge density*), we may rewrite Gauss' law as follows:

$$\int_s \mathbf{D} \cdot \mathbf{n}\, ds = \int_v \rho\, dv. \tag{2.7}$$

Thus, a distribution of charges over a volume gives rise to an electric field that may be measured over a surface that bounds that volume. In other words, the electric flux emanating from an enclosing surface is related to the charge contained by that surface.

2.1.3 Magnetic Fields

While charges may create an electric field, electric currents may create a magnetic field. Thus, analogous to electric flux, we may speak of magnetic flux that has the ability to exert a force on either a magnet or another electric current.

Given that an electric current is nothing more than a flow of moving charges, it is apparent that a magnetic field can only be produced from moving charges; stationary charges do not produce a magnetic field. Conversely, a magnetic field has the ability to exert a force on moving charged particles.

The *magnetic flux density* associated with a magnetic field is indicated by $\mathbf{B}$. A charged particle with a charge Q moving with velocity $\mathbf{v}$ through a magnetic field with a flux density of $\mathbf{B}$ is pulled by a force $\mathbf{F}$ which is given by

$$\mathbf{F} = Q\,\mathbf{v} \times \mathbf{B}. \tag{2.8}$$

---

[1] See also Appendix A which provides the fundamentals of vector algebra and includes further detail about the relationship between integrals over contours, surfaces, and volumes.

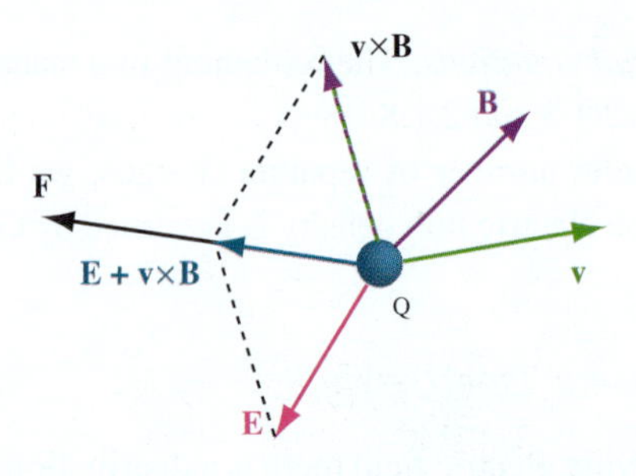

Figure 2.3. The Lorentz force equation: **F** is the sum of **E** and the cross product of the particle's velocity **v** and the magnetic flux density **B** at the position of the particle.

However, according to (2.2), a particle with charge Q is also pulled by a force equal to $F = Q\mathbf{E}$. The total force exerted on such a particle is given by the superposition of electric and magnetic forces, which is known as the Lorentz force equation (Figure 2.3):

$$\mathbf{F} = Q\,(\mathbf{E} + \mathbf{v} \times \mathbf{B})\,. \tag{2.9}$$

This equation thus provides a method to determine the motion of a charged particle as it moves through a combined electric and magnetic field.

Given the cross-product between the magnetic flux density and the velocity vector in the above equation, we deduce that the direction of the magnetic force is perpendicular to both the direction of movement and the direction of the magnetic flux. As such, this force has the ability to change the direction of the motion, but not the magnitude of the charged particle's velocity. It also does not change the energy associated with this particle.

On the other hand, the electric field exerts a force that is independent of the motion of the particle. As a result, energy may be transferred between the field and the charged particle.

The magnetic flux density **B** has a related quantity called the *magnetic vector*, indicated by **H**. The relationship between **B** and **H** is governed by a material constant μ called the *magnetic permeability*:

$$\mathbf{B} = \mu\mathbf{H}. \tag{2.10}$$

This equation is the second of three material equations and will be discussed further in Section 2.1.8.

2.1.4 Gauss' Law for Magnetic Fields

While electric fields are governed by Gauss' law for electric fields (the electric flux emanating from a closed surface depends on the charge present in the volume enclosed by that surface), magnetic fields behave somewhat differently. The total magnetic flux emanating from a closed surface bounding a magnetic field is equal to zero; this constitutes Gauss' law for magnetic fields:

$$\int_s \mathbf{B} \cdot \mathbf{n} \, ds = 0. \tag{2.11}$$

This equation implies that no free magnetic poles exist. As an example, the magnetic flux emanating from a magnetic dipole at its north pole is matched by the flux directed inward towards its south pole.

2.1.5 Faraday's Law

As shown earlier, currents produce a magnetic field. Conversely, a time-varying magnetic field is capable of producing a current. Faraday's law states that the electric field induced by a time-varying magnetic field is given by

$$\int_c \mathbf{E} \cdot \mathbf{n} \, dc = -\frac{d}{dt} \int_s \mathbf{B} \cdot \mathbf{n} \, ds. \tag{2.12}$$

Here, the left-hand side is an integral over the contour c that encloses an open surface s. The quantity integrated is the component of the electric field intensity $\mathbf{E}$ normal to the contour. The right-hand side integrates the normal component of $\mathbf{B}$ over the surface s. Note that the right-hand side integrates over an open surface, whereas the integral in (2.11) integrates over a closed surface.

2.1.6 Ampere's Circuital Law

Given a surface area s enclosed by a contour c, the magnetic flux density along this contour is related to the total current passing through area s. The total current is composed of two components, namely a current as a result of moving charged particles and a current related to changes in electric flux density. The latter current is also known as *displacement current.*

Moving charged particles may be characterized by the *current flux density* $\mathbf{j}$, which is given in ampere per square meter (A/m^2). If electric charges with a density of ρ are moving with a velocity $\mathbf{v}$, the current flux density $\mathbf{j}$ is given by

$$\mathbf{j} = \rho \, \mathbf{v}. \tag{2.13}$$

If the current flux density $\mathbf{j}$ is integrated over surface area, then we find the total charge passing through this surface per second (coulomb/second = ampere; C/s = A). Thus, the current resulting from a flow of charges is given by

$$\int_s \mathbf{j} \cdot \mathbf{n} \, ds. \tag{2.14}$$

The displacement current depends on the electric flux density. If we integrate the electric flux density over the same surface area, we obtain charge (coulomb; C):

$$\int_s \mathbf{D} \cdot \mathbf{n}\, ds. \tag{2.15}$$

If we differentiate this quantity by time, the result is a charge passing through surface s per second, i.e., current:

$$\frac{d}{dt} \int_s \mathbf{D} \cdot \mathbf{n}\, ds. \tag{2.16}$$

The units in (2.14) and (2.16) are now both in coulomb per second and are thus measures of current. Both types of current are related to the magnetic flux density according to Ampere's circuital law:

$$\int_c \mathbf{H} \cdot \mathbf{n}\, dc = \int_s \mathbf{j} \cdot \mathbf{n}\, ds + \frac{d}{dt} \int_s \mathbf{D} \cdot \mathbf{n} ds. \tag{2.17}$$

This law states that a time-varying magnetic field can be produced by the flow of charges (a current), as well as by a displacement current.

2.1.7 Maxwell's Equations

Both of Gauss' laws, Faraday's law, and Ampere's circuital law together form a set of equations that are collectively known as Maxwell's equations. For convenience we repeat them here:

$$\int_s \mathbf{D} \cdot \mathbf{n}\, ds = \int_v \rho\, dv; \qquad \text{Gauss' law for electric fields} \tag{2.18a}$$

$$\int_s \mathbf{B} \cdot \mathbf{n}\, ds = 0; \qquad \text{Gauss' law for magnetic fields} \tag{2.18b}$$

$$\int_c \mathbf{E} \cdot \mathbf{n}\, dc = -\frac{d}{dt} \int_s \mathbf{B} \cdot \mathbf{n}\, ds; \qquad \text{Faraday's law} \tag{2.18c}$$

$$\int_c \mathbf{H} \cdot \mathbf{n}\, dc = \int_s \mathbf{j} \cdot \mathbf{n}\, ds + \frac{d}{dt} \int_s \mathbf{D} \cdot \mathbf{n}\, ds. \qquad \text{Ampere's circuital law} \tag{2.18d}$$

The above equations are given in integral form. They may be rewritten in differential form, after which these equations hold for points in space where both electric and magnetic fields are continuous. This facilitates solving these four simultaneous equations.

Starting with Gauss' law for electric fields, we see that the left-hand side of (2.18a) is an integral over a surface, whereas the right-hand side is an integral over a volume. As we are interested in a form of Maxwell's equations that is valid for points in space, we would like to replace the left-hand side with an integral over the volume under consideration. This may be accomplished by applying Gauss' theorem (see Appendix A), yielding

$$\int_s \mathbf{D} \cdot \mathbf{n} \, ds = \int_v \nabla \cdot \mathbf{D} \, dv. \tag{2.19}$$

Combining this result with (2.18a) yields

$$\int_v \nabla \cdot \mathbf{D} \, dv = \int_v \rho \, dv, \tag{2.20}$$

and, therefore, in the limit when the volume v goes to zero, we have

$$\nabla \cdot \mathbf{D} = \rho. \tag{2.21}$$

A similar set of steps may be applied to Gauss' law for magnetic fields, and this will result in a similar differential form:

$$\nabla \cdot \mathbf{B} = 0. \tag{2.22}$$

Ampere's law, given in (2.18d), is stated in terms of a contour integral on the left-hand side and a surface integral on the right-hand side. Here, it is appropriate to apply Stokes' theorem to bring both sides into the same domain (see Appendix A):

$$\int_c \mathbf{H} \cdot \mathbf{n} \, dc = \int_s \nabla \times \mathbf{H} \cdot \mathbf{n} \, ds. \tag{2.23}$$

Substituting this result into (2.18d) yields a form where all integrals are over the same surface. In the limit when this surface area s goes to zero, Equation (2.18d) becomes

$$\nabla \times \mathbf{H} = \mathbf{j} + \frac{\partial \mathbf{D}}{\partial t}, \tag{2.24}$$

which is the desired differential form. Finally, Faraday's law may also be rewritten by applying Stokes' theorem applied to the left-hand side of (2.18c):

$$\int_c \mathbf{E} \cdot \mathbf{n} \, dc = \int_s \nabla \times \mathbf{E} \cdot \mathbf{n} \, ds = -\int_s \frac{\partial \mathbf{B}}{\partial t} \cdot \mathbf{n} \, ds. \tag{2.25}$$

Under the assumption that the area s becomes vanishingly small, this equation yields

$$\nabla \times \mathbf{E} = -\frac{\partial \mathbf{B}}{\partial t}. \tag{2.26}$$

Faraday's law and Ampere's law indicate that a time-varying magnetic field has the ability to generate an electric field, and that a time-varying electric field generates a magnetic field. Thus, time-varying electric and magnetic fields can generate each other. This property forms the basis for wave propagation, allowing electric and magnetic fields to propagate away from their source. As light can be considered to consist of waves propagating through space, Maxwell's equations are fundamental to all disciplines involved with the analysis, modeling, and synthesis of light.

2.1.8 Material Equations

The four laws that comprise Maxwell's equations are normally complemented with three *material equations* (also known as *constitutive relations*). Two of these were presented in Equations (2.5) and (2.10) and are repeated here for convenience:

$$\mathbf{D} = \varepsilon \mathbf{E}; \tag{2.27a}$$

$$\mathbf{B} = \mu \mathbf{H}. \tag{2.27b}$$

The third material equation relates the current flux density $\mathbf{j}$ to the electric field intensity $\mathbf{E}$ according to a constant σ and is known as *Ohm's law*. It is given here in differential form:

$$\mathbf{j} = \sigma \mathbf{E}. \tag{2.27c}$$

Material	σ	**Material**	σ
Good conductors			
Silver	6.17×10^{7}	Tungsten	1.82×10^{7}
Copper	5.8×10^{7}	Brass	1.5×10^{7}
Gold	4.1×10^{7}	Bronze	1.0×10^{7}
Aluminium	3.82×10^{7}	Iron	1.0×10^{7}
Poor conductors			
Water (fresh)	1.0×10^{-3}	Earth (dry)	1.0×10^{-3}
Water (sea)	4.0×10^{0}	Earth (wet)	3.0×10^{-2}
Insulators			
Diamond	2.0×10^{-13}	Porcelain	1.0×10^{-10}
Glass	1.0×10^{-12}	Quartz	1.0×10^{-17}
Polystyrene	1.0×10^{-16}	Rubber	1.0×10^{-15}

Table 2.1. Conductivity constants σ for several materials [532].

The constant σ is called the *specific conductivity*, and is determined by the medium. Values of σ for some materials are listed in Table 2.1. Materials with a conductivity significantly different from 0 are called *conductors*. Metals, such as gold, silver, and copper are good conductors. On the other hand, materials with low conductivity ($\sigma \approx 0$) are called *insulators* or *dielectrics* . Finally, for some materials, (*semiconductors*) the conductivity increases with increasing temperature. See also Section 3.3.3.

For most materials, the magnetic permeability μ will be close to 1. Some materials, however, have a permeability significantly different from 1, and these are then called magnetic.

The speed v at which light travels through a medium is related to the material constants as follows:

$$v = \frac{1}{\sqrt{\varepsilon \mu}}. \tag{2.28}$$

We will derive this result in Section 2.2.4. The symbol μ_0 is reserved for the permeability of vacuum. The value of μ_0 is related to both permittivity ε_0 and the speed of light c in vacuum as follows:

$$c = \frac{1}{\sqrt{\varepsilon_0 \mu_0}}. \tag{2.29}$$

Values for all three constants are given in Table 2.2. The permittivity and permeability of materials is normally given relative to those of vacuum:

$$\varepsilon = \varepsilon_0 \, \varepsilon_r; \tag{2.30a}$$

$$\mu = \mu_0 \, \mu_r. \tag{2.30b}$$

Values of ε_r and μ_r are given for several materials in Tables 2.3 and 2.4.

Normally, the three material constants, σ, ε, and μ, are independent of the field strengths. However, this is not always the case. For some materials these values also depend on past values of **E** or **B**. In this book, we will not consider such effects of *hysteresis*. Similarly, unless indicated otherwise, the material constants are considered to be *isotropic*, which means that their values do not change

Constant	Value	Unit
c	$3 \cdot 10^8$	m s^{-1}
ε_0	$\frac{1}{36\pi} \cdot 10^{-9}$	$\text{C}^2\,\text{s}^2\,\text{kg}^{-1}\,\text{m}^{-3} = Fm^{-1}$
μ_0	$4\pi \cdot 10^{-7}$	$\text{kg m C}^{-2} = Hm^{-1}$

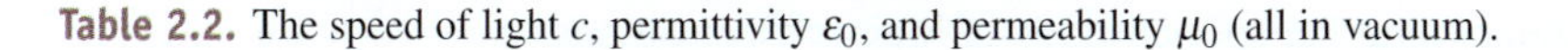

Table 2.2. The speed of light c, permittivity ε_0, and permeability μ_0 (all in vacuum).

Material	ε_r	Material	ε_r
Air	1.0006	Paper	$2-4$
Alcohol	25	Polystyrene	2.56
Earth (dry)	7	Porcelain	6
Earth (wet)	30	Quartz	3.8
Glass	$4-10$	Snow	3.3
Ice	4.2	Water (distilled)	81
Nylon	4	Water (sea)	70

Table 2.3. Dielectric constants ε_r for several materials [532].

Material	μ_r	Material	μ_r
Aluminium	1.000021	Nickel	600.0
Cobalt	250.0	Platinum	1.0003
Copper	0.99999	Silver	0.9999976
Gold	0.99996	Tungsten	1.00008
Iron	5000.0	Water	0.9999901

Table 2.4. Permeability constants μ_r for several materials [532].

with spatial orientation. Materials which do exhibit a variation in material constants with spatial orientation are called *anisotropic*.

2.2 Waves

Maxwell's equations form a set of simultaneous equations that are normally difficult to solve. In this section, we are concerned with finding solutions to Maxwell's equations, and to accomplish this, we may apply simplifying assumptions. The assumptions outlined in the previous section, that material constants are isotropic and independent of time, are the first simplifications. To get closer to an appropriate solution with which we can model light and optical phenomena, further simplifications are necessary.

In particular, a reasonable class of models that are solutions to Maxwell's equations is formed by time-harmonic plane waves. With these waves, we can explain optical phenomena such as polarization, reflection, and refraction. We first derive the wave equation, which enables the decoupling of Maxwell's equations, and therefore simplifies the solution. We then discuss plane waves, followed by time-harmonic fields, and time-harmonic plane waves. Each of these steps constitutes a further specialization of Maxwell's equations.

2.2.1 The Wave Equation

We assume that the source of an electromagnetic wave is sufficiently far away from a given region of interest. In this case, the region is called *source-free*. In such a region, the charge ρ and current distribution $\mathbf{j}$ will be 0, and as a result Maxwell's equations reduce to a simpler form:

$$\nabla \cdot \mathbf{D} = 0; \tag{2.31a}$$

$$\nabla \cdot \mathbf{B} = 0; \tag{2.31b}$$

$$\nabla \times \mathbf{E} = -\frac{\partial \mathbf{B}}{\partial t}; \tag{2.31c}$$

$$\nabla \times \mathbf{H} = \frac{\partial \mathbf{D}}{\partial t}. \tag{2.31d}$$

We may apply the material equations for $\mathbf{D}$ and $\mathbf{H}$ to yield a set of Maxwell's equations in $\mathbf{E}$ and $\mathbf{B}$ only:

$$\varepsilon \nabla \cdot \mathbf{E} = 0; \tag{2.32a}$$

$$\nabla \cdot \mathbf{B} = 0; \tag{2.32b}$$

$$\nabla \times \mathbf{E} = -\frac{\partial \mathbf{B}}{\partial t}; \tag{2.32c}$$

$$\frac{1}{\mu} \nabla \times \mathbf{B} = \varepsilon \frac{\partial \mathbf{E}}{\partial t}. \tag{2.32d}$$

This set of equations still expresses $\mathbf{E}$ in terms of $\mathbf{B}$ and $\mathbf{B}$ in terms of $\mathbf{E}$. Nevertheless, this result may be decoupled by applying the curl operator to (2.32c) (see Appendix A):

$$\nabla \times \nabla \times \mathbf{E} = -\frac{\partial}{\partial t} \nabla \times \mathbf{B}. \tag{2.33}$$

Substituting (2.32d) into this equation then yields

$$\nabla \times \nabla \times \mathbf{E} = -\mu \varepsilon \frac{\partial^2 \mathbf{E}}{\partial t^2}. \tag{2.34}$$

To simplify this equation, we may apply identity (A.23) from Appendix A:

$$\nabla(\nabla \cdot \mathbf{E}) - \nabla^2 \mathbf{E} = -\mu \varepsilon \frac{\partial^2 \mathbf{E}}{\partial t^2}. \tag{2.35}$$

From (2.32a), we know that we may set $\nabla \cdot \mathbf{E}$ to zero in (2.35) to yield the standard equation for wave motion of an electric field,

$$\nabla^2 \mathbf{E} - \mu \varepsilon \frac{\partial^2 \mathbf{E}}{\partial t^2} = 0. \tag{2.36}$$

For a magnetic field, a similar wave equation may be derived:

$$\nabla^2 \mathbf{B} - \mu\,\varepsilon \frac{\partial^2 \mathbf{B}}{\partial t^2} = 0. \tag{2.37}$$

The two wave equations do not depend on each other, thereby simplifying the solution of Maxwell's equations. This result is possible, because the charge and current distributions are zero in source-free regions, and, therefore, we could substitute $\nabla \cdot \mathbf{E} = 0$ in (2.35) to produce (2.36) (and similarly for (2.37)). Thus, both wave equations are valid for wave propagation problems in regions of space that do not generate radiation (i.e., they are source-free). Alternatively, we may assume that the source of the wave is sufficiently far away.

Under these conditions, we are therefore looking for solutions to the wave equations, rather than solutions to Maxwell's equations. One such solution is afforded by plane waves, which we discuss next.

2.2.2 Plane Waves

A plane wave may be thought of as an infinitely large plane traveling in a given direction. We will show that plane waves form a solution to the wave equations derived in the preceding section. For a position vector $\mathbf{r}(x,y,z)$ and unit direction vector $\mathbf{s} = (s_x, s_y, s_z)$, the solution will be of the form $\mathbf{E} = f(\mathbf{r} \cdot \mathbf{s}, t)$. A plane is then defined as

$$\mathbf{r} \cdot \mathbf{s} = \text{constant}. \tag{2.38}$$

For convenience, we choose a coordinate system such that one of the axes, say z, is aligned with the direction of propagation, i.e., the surface normal of the plane is $\mathbf{s} = (0,0,1)$. As a result, we have $\mathbf{r} \cdot \mathbf{s} = z$. Thus, if we consider a wave modeled by an infinitely large plane propagating through space in the direction of its normal $\mathbf{r} \cdot \mathbf{s}$, we are looking for a solution whereby the spatial derivatives in the plane are zero and the spatial derivatives along the surface normal are non-zero. The Laplacian operators in (2.36) and (2.37) then simplify to[2]

$$\nabla^2 \mathbf{E} = \frac{\partial^2 \mathbf{E}}{\partial z^2}. \tag{2.39}$$

Substituting into (2.36) yields

$$\frac{\partial^2 \mathbf{E}}{\partial z^2} - \mu\,\varepsilon \frac{\partial^2 \mathbf{E}}{\partial t^2} = 0. \tag{2.40}$$

[2]In this section we show results for **E** and note that similar results may be derived for **B**.

A general solution to this equation is given by

$$\mathbf{E} = \mathbf{E}_1\left(\mathbf{r}\cdot\mathbf{s} - \frac{1}{\sqrt{\mu\,\varepsilon}}t\right) + \mathbf{E}_2\left(\mathbf{r}\cdot\mathbf{s} + \frac{1}{\sqrt{\mu\,\varepsilon}}t\right) \tag{2.41a}$$

$$= \mathbf{E}_1\left(\mathbf{r}\cdot\mathbf{s} - vt\right) + \mathbf{E}_2\left(\mathbf{r}\cdot\mathbf{s} + vt\right), \tag{2.41b}$$

with $\mathbf{E}_1$ and $\mathbf{E}_2$ arbitrary functions.[3] The general solution to the wave equations therefore consists of two planes propagating into opposite directions ($+\mathbf{r}\cdot\mathbf{s}$ and $-\mathbf{r}\cdot\mathbf{s}$).

For simplicity, we consider only one of the planes and momentarily assume that (a partial) solution is given by a single plane traveling in the $+z$ direction:

$$\mathbf{E} = \mathbf{E}_1(\mathbf{r}\cdot\mathbf{s} - vt). \tag{2.42}$$

We will now show that for plane waves the vectors $\mathbf{E}$ and $\mathbf{H}$ are both perpendicular to the direction of propagation $\mathbf{s}$, and also that $\mathbf{E}$ and $\mathbf{H}$ are orthogonal to each other. We begin by differentiating the above expression with respect to time, which produces

$$\frac{\partial \mathbf{E}}{\partial t} = -v\mathbf{E}', \tag{2.43}$$

where the prime indicates differentiation with respect to the argument $\mathbf{r}\cdot\mathbf{s} - vt$ of $\mathbf{E}$. Next, we consider the curl of $\mathbf{E}$. The x-component of the curl of $\mathbf{E}$ is given by

$$(\nabla\times\mathbf{E})_x = \frac{\partial E_z}{\partial y} - \frac{\partial E_y}{\partial z} \tag{2.44a}$$

$$= E'_z s_y - E'_y s_z \tag{2.44b}$$

$$= \left(\mathbf{s}\times\mathbf{E}'\right)_x. \tag{2.44c}$$

The curl for the y- and z-components can be derived similarly. By substitution of (2.43) and (2.44) into Maxwell's equations (2.31) and applying the material equations (2.5) and (2.10), we find

$$\mathbf{s}\times\mathbf{H}' + v\varepsilon\mathbf{E}' = 0, \tag{2.45a}$$

$$\mathbf{s}\times\mathbf{E}' - v\mu\mathbf{H}' = 0. \tag{2.45b}$$

Using (2.29), we find that $v\varepsilon = \sqrt{\varepsilon/\mu}$ and $v\mu = \sqrt{\mu/\varepsilon}$. Integrating the previous equation then gives

$$\mathbf{E} = -\sqrt{\frac{\mu}{\varepsilon}}\mathbf{s}\times\mathbf{H}, \tag{2.46a}$$

$$\mathbf{H} = \sqrt{\frac{\varepsilon}{\mu}}\mathbf{s}\times\mathbf{E}. \tag{2.46b}$$

[3] We will derive expressions for these functions in the following sections.

To show that plane waves are *transversal*, i.e., both $\mathbf{E}$ and $\mathbf{H}$ are perpendicular to the direction of propagation $\mathbf{s}$, we take the dot product between $\mathbf{E}$ and $\mathbf{s}$ and $\mathbf{H}$ and $\mathbf{s}$:

$$\mathbf{E} \cdot \mathbf{s} = \left(-\sqrt{\frac{\mu}{\varepsilon}} \mathbf{s} \times \mathbf{H} \right) \cdot \mathbf{s}, \tag{2.47a}$$

$$\mathbf{H} \cdot \mathbf{s} = \left(\sqrt{\frac{\varepsilon}{\mu}} \mathbf{s} \times \mathbf{E} \right) \cdot \mathbf{s}. \tag{2.47b}$$

Using (A.6) in Appendix A, we find that $\mathbf{E} \cdot \mathbf{s} = \mathbf{H} \cdot \mathbf{s} = 0$, thereby proving that $\mathbf{E}$ and $\mathbf{H}$ are perpendicular to $\mathbf{s}$. Such waves are called *transversal electromagnetic (TEM) waves*.

These equations also imply that $\mathbf{s} = \mathbf{E} \times \mathbf{H}$. The vector $\mathbf{s}$ is generally known as the *Poynting vector* or *power density vector* and plays an important role in quantifying energy. This can be seen from the units employed in vectors $\mathbf{E}$ (V/m) and $\mathbf{H}$ (A/m), so that $\mathbf{s}$ is specified in watts per square meter $(V/m \cdot A/m = W/m^2)$. For planar waves in homogeneous media, the Poynting vector thus points in the direction of wave propagation, and its length is a measure of the energy flow induced by the electromagnetic field (discussed further in Section 2.4). In keeping with general practice, we will use the symbol $\mathbf{S}$ when referring to the Poynting vector in the remainder of this book.

2.2.3 Time-Harmonic Fields

A reasonable approach to modeling the oscillating wave behavior of light in free space is by using sinusoidal functions. Thus, we assume that the current and charge distributions, $\mathbf{j}$ and ρ, at the source vary with time as

$$a \cos(\omega t + \varphi), \tag{2.48}$$

where a is the amplitude and $\omega t + \varphi$ is the phase. The current and charge sources vary with time t as well as space $\mathbf{r}$ and are therefore written as $\mathbf{j}(\mathbf{r},t)$ and $\rho(\mathbf{r},t)$.

Using the results from Appendix C.4, we may separate these fields into a spatial component and a time-varying component:

$$\mathbf{j}(\mathbf{r},t) = \underline{\mathbf{j}}(\mathbf{r})\, e^{i\omega t}, \tag{2.49a}$$

$$\rho(\mathbf{r},t) = \underline{\rho}(\mathbf{r})\, e^{i\omega t}. \tag{2.49b}$$

Since Maxwell's equations are linear, this results in the following set of equations:

$$\nabla \cdot \underline{\mathbf{D}} e^{i\omega t} = \underline{\rho} e^{i\omega t}, \tag{2.50a}$$
$$\nabla \cdot \underline{\mathbf{B}} e^{i\omega t} = 0, \tag{2.50b}$$
$$\nabla \times \left(\underline{\mathbf{E}} e^{i\omega t}\right) = -i\omega \underline{\mathbf{B}} e^{i\omega t}, \tag{2.50c}$$
$$\nabla \times \left(\underline{\mathbf{H}} e^{i\omega t}\right) = i\omega \underline{\mathbf{D}} e^{i\omega t} + \underline{\mathbf{j}} e^{i\omega t}. \tag{2.50d}$$

Note that all the underlined quantities are (complex) functions of space $\mathbf{r}$ only. In addition, the time-dependent quantity $e^{i\omega t}$ cancels everywhere. For a homogeneous field, we may set $\underline{\mathbf{j}}$ and $\underline{\rho}$ to zero. For a homogeneous field in steady state we therefore obtain the following set of equations:

$$\nabla \cdot \underline{\mathbf{D}} = 0, \tag{2.51a}$$
$$\nabla \cdot \underline{\mathbf{B}} = 0, \tag{2.51b}$$
$$\nabla \times \underline{\mathbf{E}} = -i\omega \underline{\mathbf{B}}, \tag{2.51c}$$
$$\nabla \times \underline{\mathbf{H}} = i\omega \underline{\mathbf{D}}. \tag{2.51d}$$

By a procedure similar to the one shown in Section 2.2.1, wave equations for the electric field $\underline{\mathbf{E}}$ and the magnetic field $\underline{\mathbf{E}}$ may be derived:

$$\nabla^2 \underline{\mathbf{E}} - \omega^2 \mu \varepsilon \underline{\mathbf{E}} = 0, \tag{2.52a}$$
$$\nabla^2 \underline{\mathbf{B}} - \omega^2 \mu \varepsilon \underline{\mathbf{B}} = 0. \tag{2.52b}$$

2.2.4 Harmonic Plane Waves

In this section we will find a solution to the plane wave equations of harmonic waves in homogeneous media, i.e., a solution to Equations (2.52). We will start by assuming that the wave is traveling through a dielectric medium, i.e., the conductivity of the material is close to zero ($\sigma \approx 0$). Later in this section, we will consider conductive materials.

For general plane waves we assumed that the planes are traveling in the $+z$ and $-z$ directions. As both $\mathbf{E}$ and $\mathbf{H}$ (and thereby $\mathbf{D}$ and $\mathbf{B}$) are transversal to the direction or propagation, we find that

$$\frac{\partial \underline{\mathbf{E}}}{\partial x} = 0, \qquad \frac{\partial \underline{\mathbf{B}}}{\partial x} = 0, \tag{2.53a}$$
$$\frac{\partial \underline{\mathbf{E}}}{\partial y} = 0, \qquad \frac{\partial \underline{\mathbf{B}}}{\partial y} = 0. \tag{2.53b}$$

From Faraday's law, we find identities for the partial derivatives in the z directions for $\underline{\mathbf{E}}$:

$$-\frac{\partial \underline{\mathbf{E}}_y}{\partial z} = -i\,\omega\,\underline{\mathbf{B}}_x, \tag{2.54a}$$

$$\frac{\partial \underline{\mathbf{E}}_x}{\partial z} = -i\,\omega\,\underline{\mathbf{B}}_y, \tag{2.54b}$$

$$0 = -i\,\omega\,\underline{\mathbf{B}}_z. \tag{2.54c}$$

Similar solutions are found from Ampere's law for $\underline{\mathbf{B}}$:

$$-\frac{\partial \underline{\mathbf{B}}_y}{\partial z} = i\,\omega\,\varepsilon\,\mu\,\underline{\mathbf{E}}_x, \tag{2.54d}$$

$$\frac{\partial \underline{\mathbf{B}}_x}{\partial z} = i\,\omega\,\varepsilon\,\mu\,\underline{\mathbf{E}}_y, \tag{2.54e}$$

$$0 = i\,\omega\,\varepsilon\,\mu\,\underline{\mathbf{E}}_z. \tag{2.54f}$$

From these equations, we find that the components of $\underline{\mathbf{E}}$ and $\underline{\mathbf{B}}$ in the direction of propagation (z) are zero. Differentiation of (2.54b) and substitution from (2.54d) leads to

$$\frac{\partial^2 \underline{\mathbf{E}}_x}{\partial z^2} = -i\,\omega\frac{\partial \underline{\mathbf{B}}_y}{\partial z} = -\omega^2\,\mu\,\varepsilon\,\underline{\mathbf{E}}_x. \tag{2.55}$$

As we assume that the field is uniform, $\underline{\mathbf{E}}_x$ is a function of z only, and we may therefore replace the partial derivatives with ordinary derivatives, yielding the wave equation for harmonic plane waves ,

$$\frac{d^2 \underline{\mathbf{E}}_x}{dz^2} + \omega^2\,\mu\,\varepsilon\,\underline{\mathbf{E}}_x = 0. \tag{2.56}$$

Similar equations may be set up for $\underline{\mathbf{E}}_y$, $\underline{\mathbf{B}}_x$, and $\underline{\mathbf{B}}_y$. By letting

$$\beta^2 = \omega^2\,\mu\,\varepsilon, \tag{2.57}$$

a general solution for (2.56) is given by

$$\underline{\mathbf{E}}_x = \underline{\mathbf{E}}_m^+\,e^{-i\beta z} + \underline{\mathbf{E}}_m^-\,e^{i\beta z}. \tag{2.58}$$

For argument's sake, we will assume that the newly introduced constants $\underline{\mathbf{E}}_m^+$ and $\underline{\mathbf{E}}_m^-$ are real, and therefore we replace them with $\mathbf{E}_m^+$ and $\mathbf{E}_m^-$. The solution of the wave equation is then the real part of $\underline{\mathbf{E}}_x\,e^{i\omega t}$:

$$\begin{aligned} E_x(z,t) &= Re\{\underline{\mathbf{E}}_x\,e^{i\omega t}\} && (2.59\text{a}) \\ &= Re\{\mathbf{E}_m^+\,e^{i(\omega t - \beta z)} + \mathbf{E}_m^-\,e^{i(\omega t + \beta z)}\} && (2.59\text{b}) \\ &= \mathbf{E}_m^+\cos(\omega t - \beta z) + \mathbf{E}_m^-\cos(\omega t + \beta z). && (2.59\text{c}) \end{aligned}$$

Thus, the solution to the wave equation for harmonic plane waves in homogeneous media may be modeled by a pair of waves, one propagating in the $+z$ direction and the other traveling in the $-z$ direction.

Turning our attention to only one of the waves, for instance $\mathbf{E}_m^+ \cos(\omega t - \beta z)$, it is clear that by keeping to a single position z, the wave produces an oscillation with *angular frequency* ω. The *frequency* f of the wave and its period T may be derived from ω as follows:

$$f = \frac{\omega}{2\pi} = \frac{1}{T}. \tag{2.60}$$

At the same time, the wave $\mathbf{E}_m^+ \cos(\omega t - \beta z)$ travels through space in the positive z direction, which follows from the $-\beta z$ component of the wave's phase. The value of the cosine does not alter if we add or subtract multiples of 2π to the phase. Hence, we have

$$\omega t + \beta z = \omega t + \beta(z + \lambda) + 2\pi. \tag{2.61}$$

Solving for λ, which is called *wavelength*, and combining with (2.57) we get

$$\lambda = \frac{2\pi}{\beta} = \frac{2\pi}{\omega\sqrt{\mu\varepsilon}}. \tag{2.62}$$

With the help of Equations (2.29) and (2.60), we find the well-known result that the wavelength λ of a harmonic plane wave relates to its frequency by means of the speed of light:

$$\lambda = \frac{v_p}{f}. \tag{2.63}$$

The *phase velocity* v_p may be viewed as the speed with which the wave propagates. This velocity may be derived by setting the phase value to a constant:

$$\omega t - \beta z = C \quad \Rightarrow \quad z = \frac{\omega t}{\beta} - \frac{C}{\beta}. \tag{2.64}$$

For a wave traveling in the z direction through vacuum, the velocity of the wave equals the time derivative in z:

$$v_p = \frac{dz}{dt} = \frac{\omega}{\beta} = \frac{1}{\sqrt{\mu\varepsilon}}. \tag{2.65}$$

Thus, we have derived the result of (2.28). In vacuum, the phase velocity is $v_p = c$.

For conductive media, the derivation is somewhat different because the electric field intensity $\mathbf{j} = \sigma\mathbf{E}$ is now not zero. Thus, Ampere's law is given by

$$\nabla \times \mathbf{H} = \sigma\mathbf{E} + i\omega\varepsilon\mathbf{E}. \tag{2.66}$$

Rewriting this equation, we have

$$\nabla \times \mathbf{H} = i\omega \left(\varepsilon - \frac{i\sigma}{\omega} \right) \mathbf{E}. \tag{2.67}$$

The previous solution for dielectric materials may therefore be extended to include conductive media by substitution of ε by $\varepsilon - \dfrac{i\sigma}{\omega}$. As a result, the solution of (2.58) should be rewritten as

$$\underline{\mathbf{E}}_x = \underline{\mathbf{E}}_m^+ e^{-i\underline{\gamma} z} + \underline{\mathbf{E}}_m^- e^{i\underline{\gamma} z}, \tag{2.68}$$

where $\underline{\gamma}$ is complex:

$$\underline{\gamma} = i\omega \sqrt{\mu \left(\varepsilon - \frac{i\sigma}{\omega} \right)}. \tag{2.69}$$

The real and complex parts of $\underline{\gamma}$ are given by

$$\alpha = Re\{\underline{\gamma}\} = \frac{\omega\sqrt{\mu\varepsilon}}{\sqrt{2}} \sqrt{\sqrt{1 + \left(\frac{\sigma}{\omega\varepsilon} \right)^2} - 1}, \tag{2.70a}$$

$$\beta = Im\{\underline{\gamma}\} = \frac{\omega\sqrt{\mu\varepsilon}}{\sqrt{2}} \sqrt{\sqrt{1 + \left(\frac{\sigma}{\omega\varepsilon} \right)^2} + 1}. \tag{2.70b}$$

The complex form of the electric field is therefore given by (compare with (2.58))

$$\underline{\mathbf{E}}_x = \underline{\mathbf{E}}_m^+ e^{-\alpha z} e^{-i\beta z} + \underline{\mathbf{E}}_m^- e^{\alpha z} e^{i\beta z}. \tag{2.71}$$

As before, the real part of this complex solution is of interest. By splitting the complex amplitude $\underline{\mathbf{E}}_m^+$ into a real part and a phase part, i.e., $\underline{\mathbf{E}}_m^+ = E_m^+ e^{i\phi^+}$, and doing the same for $\underline{\mathbf{E}}_m^-$, the real part of the solution is given by

$$E_x(z,t) = E_m^+ e^{-\alpha z} \cos\left(\omega t - \beta z + \phi^+\right) \tag{2.72a}$$

$$+ E_m^- e^{\alpha z} \cos\left(\omega t + \beta z + \phi^-\right) \tag{2.72b}$$

It is easy to observe that this result is an extension of the dielectric case discussed earlier in this section. In particular, for dielectrics with $\sigma = 0$, the solution is real because $\alpha = 0$. In addition, for dielectrics the value of β is the same as before.

For a wave propagating along the $+z$-axis, we see that its amplitude is modulated by a factor $e^{-\alpha z}$. This means that its amplitude becomes smaller as the wave progresses through the material. A consequence of this is that waves can only penetrate a conductive medium by a short distance. As such, we note that conductive materials are generally not transparent.

Conversely, for dielectric materials with very low conductivity, the value of alpha tends to zero, so that the amplitude of waves propagating through dielectrics is not significantly attenuated. Hence, dielectric materials tend to be transparent.

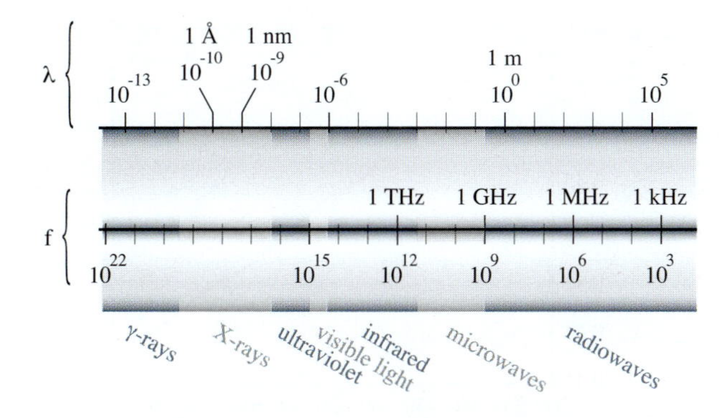

Figure 2.4. The electromagnetic spectrum.

2.2.5 The Electromagnetic Spectrum and Superposition

So far, we have discussed harmonic plane waves with a specific given wavelength. Such waves are called *monochromatic*. The wavelength can be nearly anything, ranging for instance from 10^{-13} m for gamma rays to 10^5 m for radio frequencies [447]. This range of wavelengths is known as the *electromagnetic spectrum* and is presented in Figure 2.4.

The range of wavelengths that the human eye is sensitive to is called the *visible spectrum* and ranges from roughly 380 nm to 780 nm. This range constitutes only a very small portion of the electromagnetic spectrum and coincides with the range of wavelengths that interacts sensibly with materials. Longer wavelengths have lower energies associated with them, and their interaction with materials is usually limited to the formation of heat. At wavelengths shorter than the visible range, the energy is strong enough to ionize atoms and permanently destroy molecules [815]. As such, only waves with wavelengths in the visible range interact with the atomic structure of matter in such a way that it produces color.

So far, our discussion has centered on monochromatic light; the shape of the wave was considered to be sinusoidal. In practice, this is rarely the case. According to the Fourier theorem, more complicated waves may be constructed by a superposition of sinusoidal waves of different wavelengths (or, equivalently, of different frequencies). This leads to the *superposition principle*, which is given here for a dielectric material (compare with (2.59c)):

$$E_x(z,t) = \int_{\omega=0}^{\infty} \left(E^{+}_{m,\omega} \cos(\omega t - \beta z) + E^{-}_{m,\omega} \cos(\omega t + \beta z) \right) d\omega. \tag{2.73}$$

In addition, radiators tend to emit light at many different wavelengths. In Figure 2.5, an example is shown of the wavelength composition of the light emitted by a tungsten radiator heated to 2000 K.

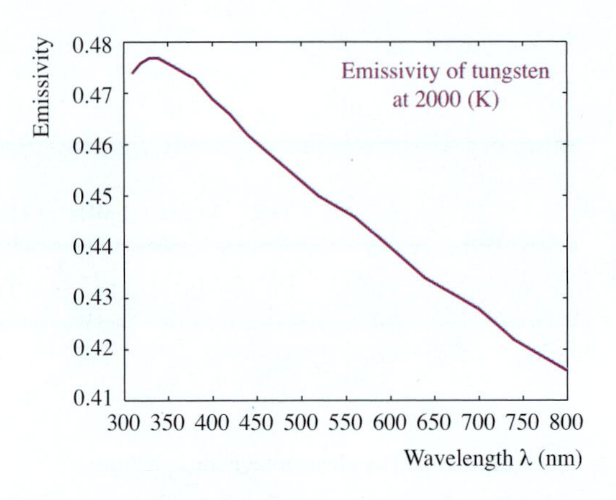

Figure 2.5. The relative contribution of each wavelength to the light emitted by a tungsten radiator at a temperature of 2000 *K* [653].

2.3 Polarization

We have already shown that harmonic waves are transversal: both $\underline{\mathbf{E}}$ and $\underline{\mathbf{H}}$ lie in a plane perpendicular to the direction of propagation. This still leaves some degrees of freedom. First, both vectors may be oriented in any direction in this plane (albeit with the caveat that they are orthogonal to one another). Further, the orientation of these vectors may change with time. Third, their magnitude may vary with time. In all, the time-dependent variation of $\underline{\mathbf{E}}$ and $\underline{\mathbf{H}}$ leads to *polarization*, as we will discuss in this section.

We continue to assume without loss of generality that a harmonic plane wave is traveling along the positive z-axis. This means that the vectors $\underline{\mathbf{E}}$ and $\underline{\mathbf{H}}$ may be decomposed into constituent components in the x- and y-directions:

$$\underline{\mathbf{E}} = \left(\underline{E}_x \, \mathbf{e}_x + \underline{E}_y \, \mathbf{e}_y\right) e^{-i\beta z}. \tag{2.74}$$

Here, $\mathbf{e}_x$ and $\mathbf{e}_y$ are unit normal vectors along the x- and y-axes. The complex amplitudes $\underline{E}_x$ and $\underline{E}_y$ are defined as

$$\underline{E}_x = |\underline{E}_x| \, e^{i\varphi_x}, \tag{2.75a}$$

$$\underline{E}_y = |\underline{E}_y| \, e^{i\varphi_y}. \tag{2.75b}$$

The phase angles are therefore given by φ_x and φ_y. For a given point in space $z = \mathbf{r} \cdot \mathbf{s}$, as time progresses the orientation and magnitude of the electric field intensity vector $\underline{\mathbf{E}}$ will generally vary. This can be seen by writing Equations (2.75) in their real form:

$$\frac{E_x}{|\underline{E}_x|} = \cos(\omega t - \beta z + \varphi_x), \tag{2.76a}$$

$$\frac{E_y}{|\underline{E}_y|} = \cos(\omega t - \beta z + \varphi_y). \tag{2.76b}$$

It is now possible to eliminate the component of the phase that is common to both of these equations, i.e., $\omega t - \beta z$, by rewriting them in the following form (using identity (B.7a); see Appendix B):

$$\frac{E_x}{|\underline{E}_x|} = \cos(\omega t - \beta z)\cos(\varphi_x) - \sin(\omega t - \beta z) sin(\varphi_x), \tag{2.77a}$$

$$\frac{E_y}{|\underline{E}_y|} = \cos(\omega t - \beta z)\cos(\varphi_y) - \sin(\omega t - \beta z) sin(\varphi_y). \tag{2.77b}$$

If we solve both equations for $\cos(\omega t - \beta z)$ and equate them, we get

$$\begin{aligned}\frac{E_x}{|\underline{E}_x|}\cos(\varphi_y) - \frac{E_y}{|\underline{E}_y|}\cos(\varphi_x) = {}& \sin(\omega t - \beta z) \\ & \times (\sin(\varphi_y)\cos(\varphi_x) - \cos(\varphi_y)\sin(\varphi_x)) \qquad (2.78a) \\ = {}& \sin(\omega t - \beta z)\sin(\varphi_y - \varphi_x). \qquad (2.78b)\end{aligned}$$

Repeating this, but now solving for $\sin(\omega t - \beta z)$ and equating the results, we find

$$\frac{E_x}{|\underline{E}_x|}\sin(\varphi_y) - \frac{E_y}{|\underline{E}_y|}\sin(\varphi_x) = \cos(\omega t - \beta z)\sin(\varphi_y - \varphi_x). \tag{2.79}$$

By squaring and adding these equations, we obtain

$$\left(\frac{E_x}{|\underline{E}_x|}\right)^2 + \left(\frac{E_y}{|\underline{E}_y|}\right)^2 - 2\frac{E_x}{|\underline{E}_x|}\frac{E_y}{|\underline{E}_y|}\cos(\varphi_y - \varphi_x) = \sin^2(\varphi_y - \varphi_x). \tag{2.80}$$

This equation shows that the vector $\mathbf{\underline{E}}$ rotates around the z-axis describing an ellipse. The wave is therefore *elliptically polarized*. The axes of the ellipse do not need to be aligned with the x- and y-axes, but could be oriented at an angle.

Two special cases exist; the first is when the phase angles φ_x and φ_y are separated by multiples of π:

$$\varphi_y - \varphi_x = m\pi \qquad (m = 0, \pm 1, \pm 2, \ldots). \tag{2.81}$$

For integer values of m, the sine operator is 0 and the cosine operator is either $+1$ or -1 dependent on whether m is even or odd. Therefore, Equation (2.80) reduces to

$$\left(\frac{E_x}{|\underline{E}_x|}\right)^2 + \left(\frac{E_y}{|\underline{E}_y|}\right)^2 = 2(-1)^m \frac{E_x}{|\underline{E}_x|}\frac{E_y}{|\underline{E}_y|}. \tag{2.82}$$

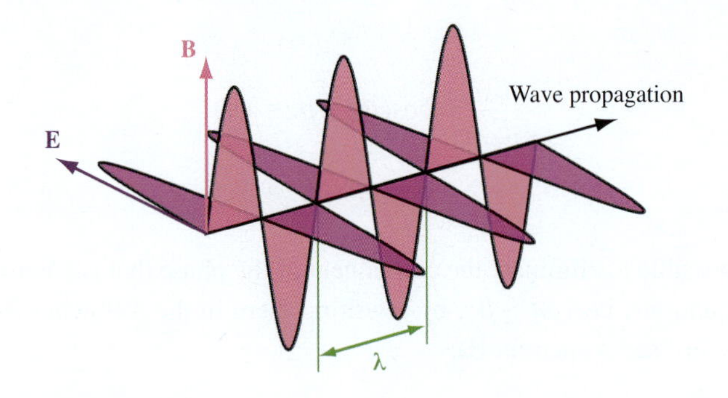

Figure 2.6. An electromagnetic wave with wavelength λ is defined by electric vector **E** and magnetic induction **B**, which are both orthogonal to the direction of propagation. In this case, both vectors are linearly polarized.

The general form of this equation is either $x^2 + y^2 = 2xy$ or $x^2 + y^2 = -2xy$. We are interested in the ratio between x and y, as this determines the level of eccentricity of the ellipse. We find this as follows:

$$x^2 + y^2 = 2xy. \tag{2.83a}$$

By dividing both sides by y^2 we get

$$\frac{x^2}{y^2} + 1 = 2\frac{x}{y}. \tag{2.83b}$$

Solving for the ratio x/y yields

$$\frac{x}{y} = 1. \tag{2.83c}$$

Solving $x^2 + y^2 = -2xy$ in a similar manner yields $x/y = -1$. We therefore find that the ratio between E_x and E_y is

$$\frac{E_x}{E_y} = (-1)^m \frac{\underline{E}_y}{\underline{E}_x}. \tag{2.84}$$

As such, the ratio between x- and y-components of $\underline{\mathbf{E}}$ are constant for fixed m. This means that instead of inscribing an ellipse, this vector oscillates along a line. Thus, when the phase angles φ_x and φ_y are in phase, the electric field intensity vector is *linearly polarized*, as shown in Figure 2.6. The same is then true for the magnetic vector $\underline{\mathbf{H}}$.

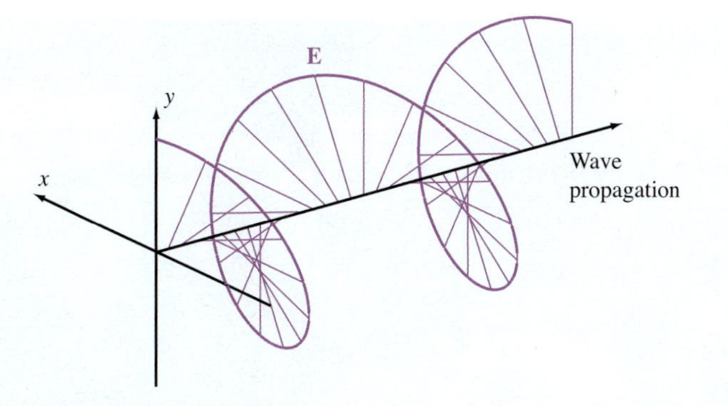

Figure 2.7. For a circularly polarized wave, the electric vector **E** rotates around the Poynting vector while propagating. Not shown is the magnetic vector, which also rotates around the Poynting vector while remaining orthogonal to **E**.

The second special case occurs when the amplitudes $|\underline{E}_x|$ and $|\underline{E}_y|$ are equal and the phase angles differ by either $\pi/2 \pm 2m\pi$ or $-\pi/2 \pm 2m\pi$. In this case, (2.80) reduces to

$$E_x^2 + E_y^2 = \pm|\underline{E}_x|^2. \tag{2.85}$$

This is the equation of a circle, and this type of polarization is therefore called *circular polarization*. If $\varphi_y - \varphi_x = \pi/2 \pm 2m\pi$ the wave is called a *right-handed circularly polarized wave*. Conversely, if $\varphi_y - \varphi_x = -\pi/2 \pm 2m\pi$ the wave is called *left-handed circularly polarized*. In either case, the field vectors inscribe a circle, as shown for **E** in Figure 2.7.

The causes of polarization include reflection of waves off surfaces or scattering by particles suspended in a medium. For instance, sunlight entering the Earth's atmosphere undergoes scattering by small particles, which causes the sky to be polarized.

Polarization can also be induced by employing polarization filters. These filters are frequently used in photography to reduce glare from reflecting surfaces. Such filters create linearly polarized light. As a consequence, a pair of such filters can be stacked such that together they block all light.

This is achieved if the two filters polarize light in orthogonal directions, as shown in the overlapping region of the two sheets in Figure 2.8. If the two filters are aligned, then linearly polarized light will emerge, as if only one filter were present. The amount of light E_e transmitted through the pair of polarizing filters

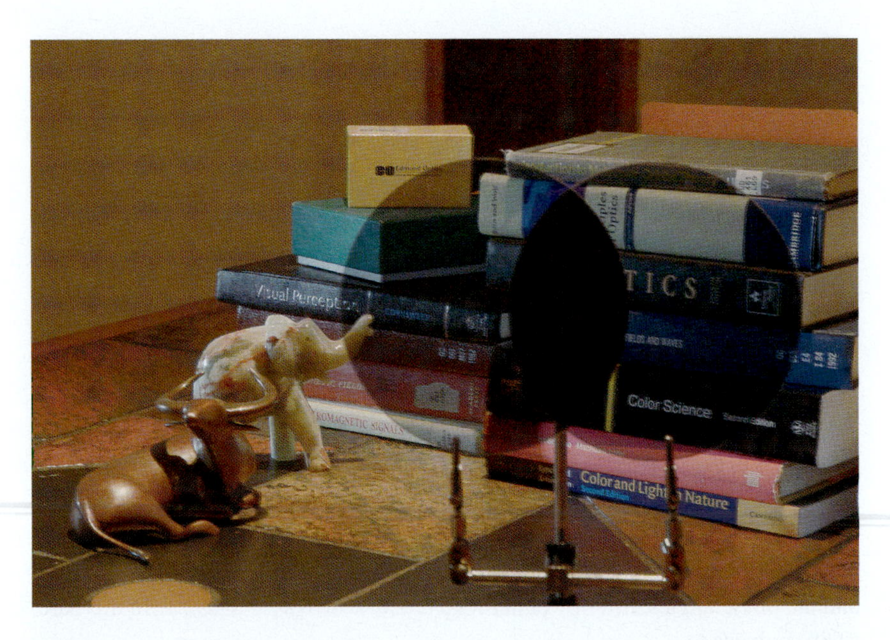

Figure 2.8. Two sheets of polarizing material are oriented such that together they block light, whereas each single sheet transmits light.

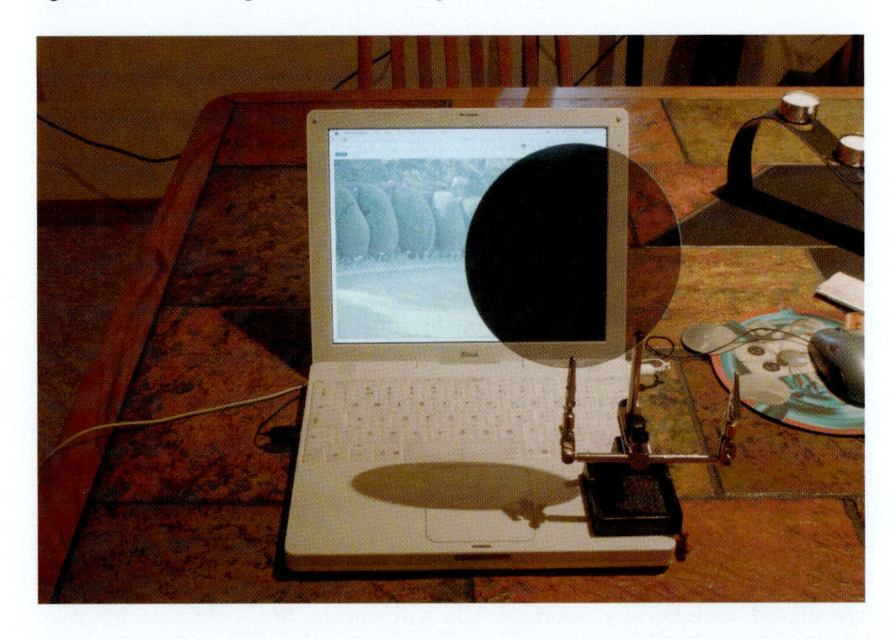

Figure 2.9. A polarizing sheet in front of an LCD screen can be oriented such that all light is blocked.

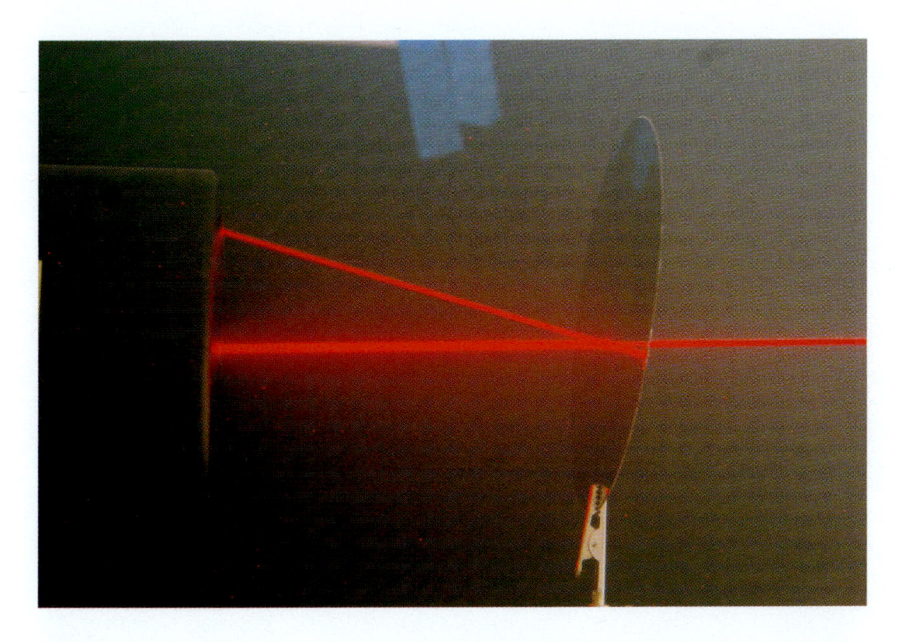

Figure 2.10. A polarizing sheet is oriented such that polarized laser light is transmitted.

is a function of the angle θ between the two polarizers and the amount of incident light $E_{e,0}$:

$$E_e = E_{e,0} \cos^2(\theta). \tag{2.86}$$

This relation is known as the Law of Malus [447, 730, 1128].

The same effect is achieved by placing a single polarizing filter in front of an LCD screen, as shown in Figure 2.9 (see also Section 14.2). Here, a backlight emits non-polarized light, which is first linearly polarized in one direction. Then the intensity of each pixel is adjusted by means of a second variable polarization in the orthogonal direction. Thus, the light that is transmitted through this sequence of filters is linearly polarized. As demonstrated in Figure 2.9, placing one further polarizing filter in front of the screen thus blocks the remainder of the light.

In addition, laser light is polarized. This can be shown by using a single polarization filter to block laser light. In Figure 2.10, a sheet of polarizing material is placed in the path of a laser. The sheet is oriented such that most of the light is transmitted. Some of the light is also reflected. By changing the orientation of the sheet, the light can be blocked, as shown in Figure 2.11. Here, only the reflecting component remains.

Polarization is extensively used in photography in the form of filters that can be attached to the camera. This procedure allows unwanted reflections to be re-

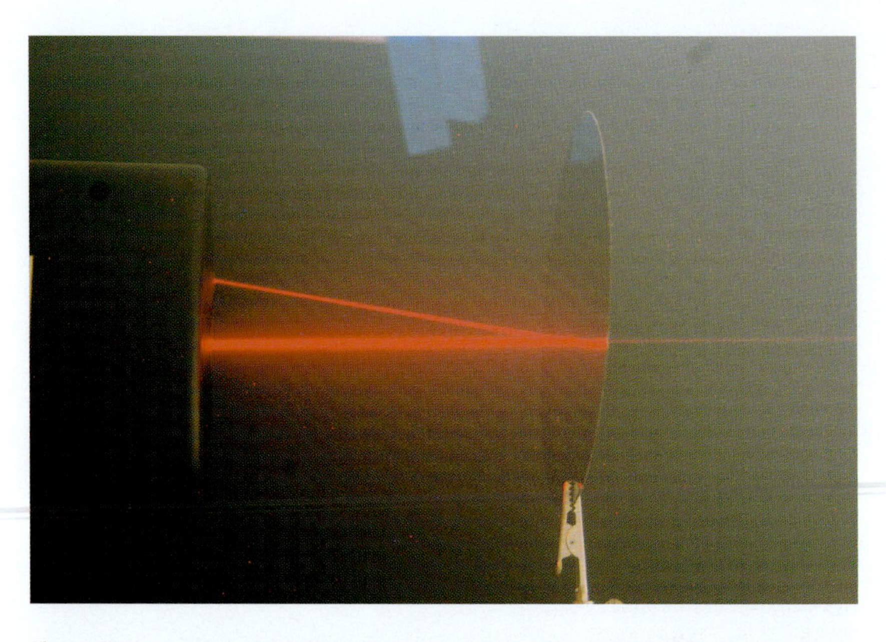

Figure 2.11. A polarizing sheet is oriented such that polarized laser light is blocked.

Figure 2.12. The LCD screen emits polarized white light, which undergoes further polarization upon reflection dependent on the amount of stress in the reflective object. Thus, the colorization of the polarized reflected light follows the stress patterns in the material.

moved from scenes. For instance the glint induced by reflections off water can be removed. It can also be used to darken the sky and improve its contrast.

In computer vision, polarization can be used to infer material properties using a sequence of images taken with a polarization filter oriented at different angles [1250]. This technique is based on the fact that specular materials partially polarize light.

Polarization also finds uses in material science, where analysis of the polarizing properties of a material provides information of its internal stresses. An example is shown in Figure 2.12 where the colored patterns on the CD case are due to stresses in the material induced during fabrication. They become visible by using the polarized light of an LCD screen.

2.4 Spectral Irradiance

In Section 2.2.2, we argued that the Poynting vector of a plane wave points in the direction of propagation. However, we did not discuss its magnitude. In general, the vectors **E** and **H** vary extremely rapidly in both direction and time. For harmonic plane waves, the magnitude of the Poynting vector may therefore be seen as giving the instantaneous energy density.

In the case of a linearly polarized harmonic plane wave, the **E** and **H** fields may be modeled as

$$\mathbf{E} = \mathbf{E}_0 \cos(\mathbf{r} \cdot \mathbf{s} - \omega t), \tag{2.87a}$$

$$\mathbf{H} = \mathbf{H}_0 \cos(\mathbf{r} \cdot \mathbf{s} - \omega t). \tag{2.87b}$$

The Poynting vector, given by $\mathbf{S} = \mathbf{E} \times \mathbf{H}$, is then

$$\mathbf{S} = \mathbf{E}_0 \times \mathbf{H}_0 \cos^2(\mathbf{r} \cdot \mathbf{s} - \omega t). \tag{2.88}$$

The instantaneous energy density thus varies with double the angular frequency (because the cosine is squared). At this rate of change, the Poynting vector does not constitute a practical measure of energy flow.

However, it is possible to average the magnitude of the Poynting vector over time. The time average $\langle f(t) \rangle_T$ of a function $f(t)$ over a time interval T is given by

$$\langle f(t) \rangle_T = \frac{1}{T} \int_{t-T/2}^{t+T/2} f(t)\, dt. \tag{2.89}$$

For a harmonic function $e^{i\omega t}$, such a time average becomes

$$\left\langle e^{i\omega t}\right\rangle_T = \frac{1}{T}\int_{t-T/2}^{t+T/2} e^{i\omega t}\,dt \tag{2.90a}$$

$$= \frac{1}{i\omega T} e^{i\omega t}\left(e^{i\omega T/2} - e^{-i\omega T/2}\right) \tag{2.90b}$$

$$= \frac{\sin\left(\dfrac{\omega T}{2}\right)}{\dfrac{\omega T}{2}} e^{i\omega t} \tag{2.90c}$$

$$= \operatorname{sinc}\left(\frac{\omega T}{2}\right) e^{i\omega t} \tag{2.90d}$$

$$= \operatorname{sinc}\left(\frac{\omega T}{2}\right)\left(\cos(\omega t) + i\sin(\omega t)\right). \tag{2.90e}$$

The sinc function tends to zero for suitably large time intervals, and the cosine and sine terms average to zero as well, i.e., $\langle\cos(\omega t)\rangle_T = \langle\sin(\omega t)\rangle_T = 0$.

However, the Poynting vector for a harmonic plane wave varies with $\cos^2$. It can be shown that the time average of such a function is $1/2$ for intervals of T chosen to be large with respect to the period τ of a single oscillation:

$$\left\langle \cos^2(\omega t)\right\rangle_T = \frac{1}{2} \qquad T \gg \tau. \tag{2.91}$$

Applying this result to the Poynting vector of (2.88), we find that the magnitude of the time-averaged Poynting vector is given by

$$\langle\|\mathbf{S}\|\rangle_T = \frac{1}{2}\|\mathbf{E}_0 \times \mathbf{H}_0\|. \tag{2.92}$$

To derive an expression for the time-averaged Poynting vector, we will first relate the magnitude of $\mathbf{E}$ to the magnitude of $\mathbf{H}$. From (2.46) we have that

$$\|\mathbf{E}\| = \sqrt{\frac{\mu}{\varepsilon}}\,\|\mathbf{s}\times\mathbf{H}\|, \tag{2.93a}$$

$$\|\mathbf{H}\| = \sqrt{\frac{\varepsilon}{\mu}}\,\|\mathbf{s}\times\mathbf{E}\|. \tag{2.93b}$$

The fact that all three vectors involved in the above equations are orthogonal means we may apply Equation (A.7), yielding

$$\|\mathbf{E}\| = \sqrt{\frac{\mu}{\varepsilon}}\,\|\mathbf{s}\|\,\|\mathbf{H}\|, \tag{2.93c}$$

$$\|\mathbf{H}\| = \sqrt{\frac{\varepsilon}{\mu}}\,\|\mathbf{s}\|\,\|\mathbf{E}\|. \tag{2.93d}$$

By solving both expressions for $\|\mathbf{s}\|$ and equating them, we find the following relation:

$$\sqrt{\mu}\,\|\mathbf{H}\| = \sqrt{\varepsilon}\,\|\mathbf{E}\|. \tag{2.94}$$

Using Equation (A.7) once more, the magnitude of the time-averaged Poynting vector follows from (2.92):

$$\langle\|\mathbf{S}\|\rangle_T = \frac{1}{2}\sqrt{\frac{\varepsilon_0}{\mu_0}}\|\mathbf{E}_0\|^2 \tag{2.95a}$$

$$= \frac{\varepsilon_0\,c}{2}\|\mathbf{E}_0\|^2. \tag{2.95b}$$

An important quantity proportional to this magnitude is called *irradiance*[4] E_e and is given by [447]

$$E_e = \frac{\varepsilon_0\,c}{2}\langle\|\mathbf{E}_0^2\|\rangle_T. \tag{2.96}$$

This quantity is also known as *radiant flux density* and is measured in W/m^2. Under the assumption that the material is dielectric, linear, homogeneous, and isotropic, this expression becomes

$$E_e = \frac{\varepsilon\,v}{2}\langle\|\mathbf{E}_0^2\|\rangle_T. \tag{2.97}$$

The irradiance is a measure of the amount of light illuminating a surface. It can be seen as the average energy reaching a surface unit per unit of time. Irradiance is one of the core concepts of the field of radiometry, and therefore lies at the heart of all color theory. It is discussed further in Chapter 6.

2.5 Reflection and Refraction

So far, we have discussed harmonic plane waves traveling in free space. For many applications in optics, computer graphics, and for the discussion of the causes of colored materials following in Chapter 3, the behavior of light at the boundary of two different media constitutes an important and interesting case. In the following sections, we discuss the direction of propagation as well as the strength of the reflected and refracted waves.

[4]In the physics literature, irradiance is often indicated with the symbol I. However, we are using the radiometric symbol E_e for consistency with later chapters.

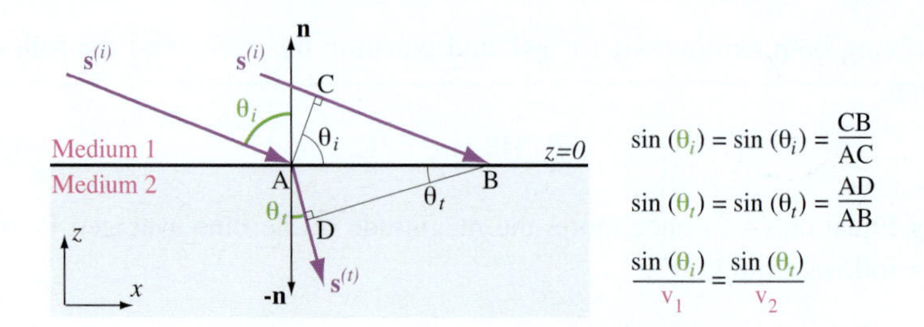

Figure 2.13. Geometry associated with refraction.

2.5.1 Direction of Propagation

For now, we assume that a harmonic plane wave is incident upon a planar boundary between two materials. At this boundary, the wave will be split into two waves, one which is propagated back into the medium and one that enters the second medium. The first of these two is thus *reflected*, whereas the second is *refracted* (or *transmitted*).

We will also assume that we have a coordinate system such that the boundary is located at the $z = 0$ plane, and that the incident plane travels in a direction indicated by $\mathbf{S}^{(i)}$. The direction vectors of the reflected and transmitted waves are named $\mathbf{S}^{(r)}$ and $\mathbf{S}^{(t)}$, respectively.

At any given location $\mathbf{r} = (x\ y\ 0)$ on the boundary, the time variation of each of the three fields will be identical:

$$t - \frac{\mathbf{r} \cdot \mathbf{S}^{(i)}}{v_1} = t - \frac{\mathbf{r} \cdot \mathbf{S}^{(r)}}{v_1} = t - \frac{\mathbf{r} \cdot \mathbf{S}^{(t)}}{v_2}. \tag{2.98}$$

As both the incident and the reflected wave propagate through the same medium, their velocities will be identical. However, the refracted wave will have a different velocity, since the medium has different permittivity and permeability constants. The above equalities should hold for any point on the boundary. In particular, the equalities hold for locations $r_1 = (1\ 0\ 0)$ and $r_2 = (0\ 1\ 0)$. These two locations give us the following set of equalities:

$$\frac{s_x^{(i)}}{v_1} = \frac{s_x^{(r)}}{v_1} = \frac{s_x^{(t)}}{v_2}, \tag{2.99a}$$

$$\frac{s_y^{(i)}}{v_1} = \frac{s_y^{(r)}}{v_1} = \frac{s_y^{(t)}}{v_2}. \tag{2.99b}$$

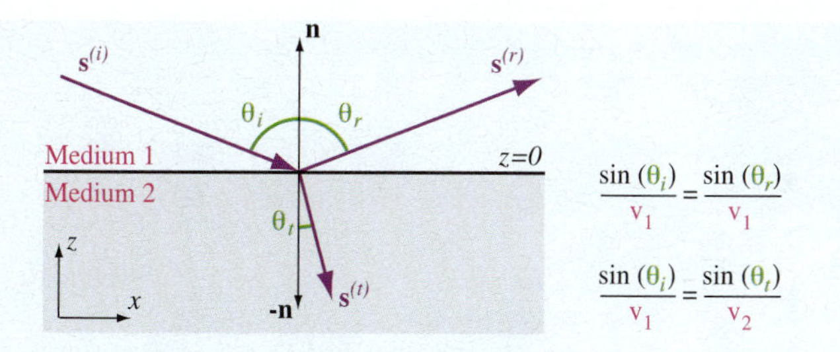

Figure 2.14. Geometry associated with reflection.

If we assume that the Poynting vectors for the incident, reflected, and transmitted waves lie in the x-z plane, then by referring to Figures 2.13 and 2.14, we have

$$\mathbf{S}^{(i)} = \begin{pmatrix} \sin(\theta_i) \\ 0 \\ \cos(\theta_i) \end{pmatrix}, \tag{2.100a}$$

$$\mathbf{S}^{(r)} = \begin{pmatrix} \sin(\theta_r) \\ 0 \\ \cos(\theta_r) \end{pmatrix}, \tag{2.100b}$$

$$\mathbf{S}^{(t)} = \begin{pmatrix} \sin(\theta_t) \\ 0 \\ \cos(\theta_t) \end{pmatrix}. \tag{2.100c}$$

The z-coordinates are positive for $\mathbf{S}^{(i)}$ and $\mathbf{S}^{(t)}$ and negative for $\mathbf{S}^{(r)}$. By combining (2.99) and (2.100) we find that

$$\frac{\sin(\theta_i)}{v_1} = \frac{\sin(\theta_r)}{v_1} = \frac{\sin(\theta_t)}{v_2}. \tag{2.101}$$

Since $\sin(\theta_i) = \sin(\theta_r)$ and the z-coordinates of $\mathbf{S}^{(i)}$ and $\mathbf{S}^{(r)}$ are of opposite sign, the angle of incidence and the angle of reflection are related by $\theta_r = \pi - \theta_i$. This relation is known as the *law of reflection*.

Since the speed of light in a given medium is related to the permittivity and permeability of the material according to (2.28), we may rewrite (2.101) as

$$\frac{\sin(\theta_i)}{\sin(\theta_t)} = \frac{v_1}{v_2} = \sqrt{\frac{\mu_2\,\varepsilon_2}{\mu_1\,\varepsilon_1}} = \frac{n_2}{n_1} = n. \tag{2.102}$$

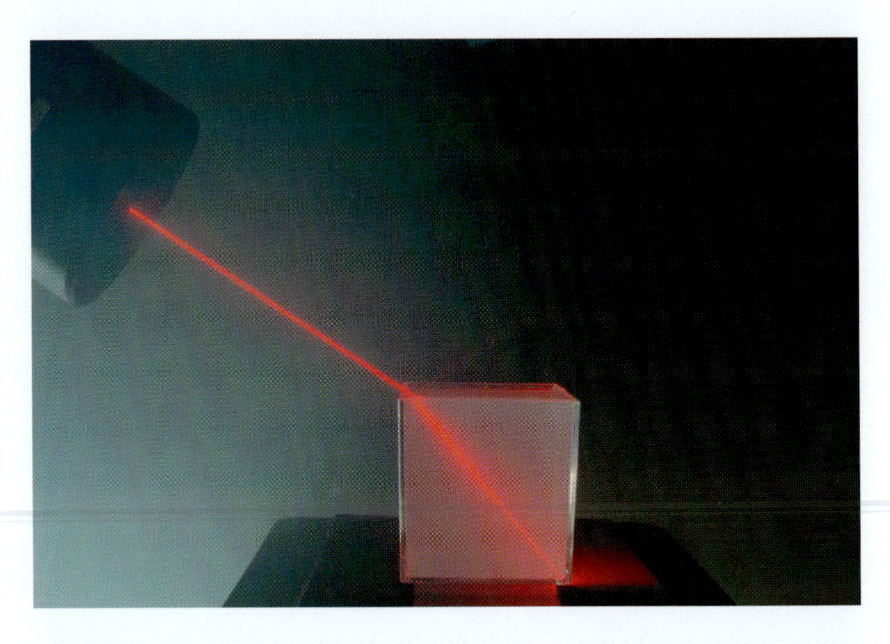

Figure 2.15. Refraction demonstrated by means of a laser beam making the transition from smoke-filled air to water.

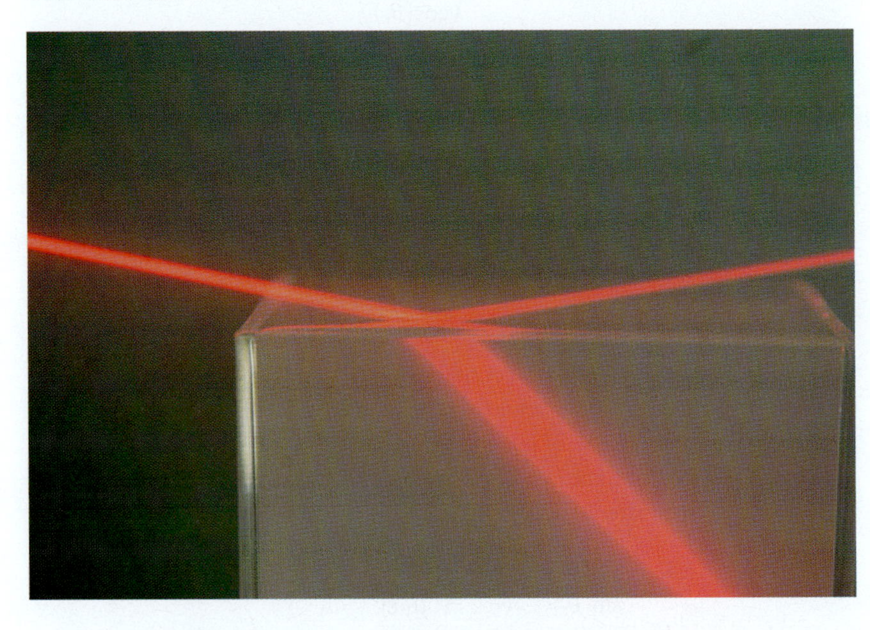

Figure 2.16. Laser light reflecting and refracting off an air-to-water boundary.

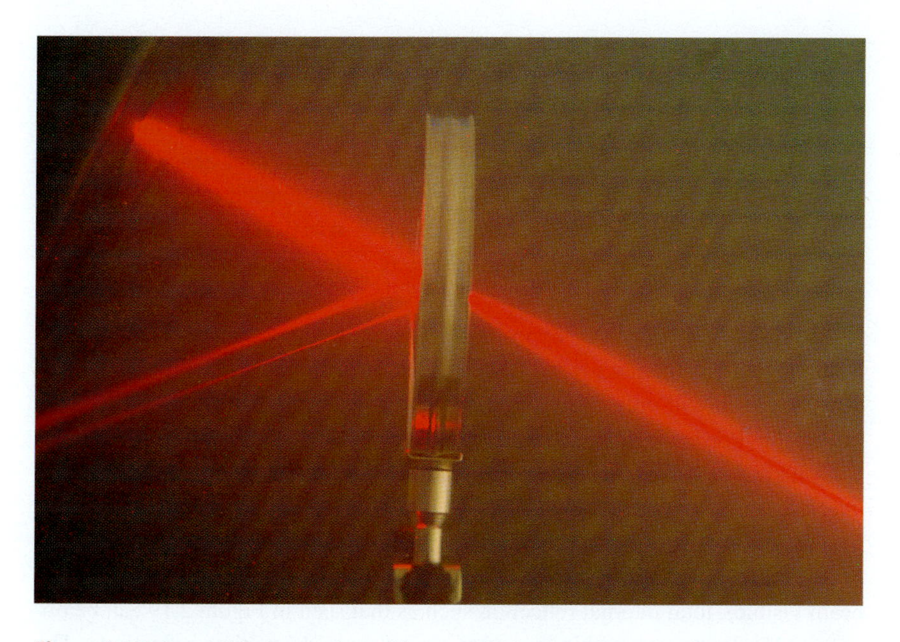

Figure 2.17. Laser light interacting with a magnifying lens. Note the displacement of the transmitted light, as well as the secondary reflections of front and back surfaces of the lens.

The values $n_1 = \sqrt{\mu_1 \varepsilon_1}$ and $n_2 = \sqrt{\mu_2 \varepsilon_2}$ are called the *absolute refractive indices* of the two media, whereas n is the *relative refractive index* for refraction from the first into the second medium. The relations given in (2.102) constitute *Snell's law*.[5]

An example of refraction is shown in Figure 2.15, where a laser was aimed at a tank filled with water. The laser is a typical consumer-grade device normally used as part of a light show in discotheques and clubs. A smoke machine was used to produce smoke and allow the laser light to scatter towards the camera. For the same reason a few drops of milk were added to the tank. Figure 2.16 shows a close-up with a shallower angle of incidence. This figure shows that light is both reflected and refracted.

A second example of Snell's law at work is shown in Figure 2.17 where a laser beam was aimed at a magnifying lens. As the beam was aimed at the center of the lens, the transmitted light is parallel to the incident light, albeit displaced. The displacement is due to the double refraction at either boundary of the lens. Similarly, the figure shows light being reflected on the left side of the lens. Two beams are visible here: one reflected off the front surface of the lens and the

[5]Named after the Dutch scientist Willebrord van Roijen Snell (1580–1626).

second beam reflected off of the back surface. These beams are predicted by a straightforward repeated application of Snell's law.

The refraction diagram shown in Figure 2.13 shows a wave traveling from an optically less dense medium into a more dense medium, and hence $\theta_t < \theta_i$. If we increase the angle of incidence until, in the limit, we reach $\theta_i = \pi/2$, then the wave is traveling parallel to the boundary and $\mathbf{S}^{(i)} \cdot \mathbf{n} = 0$. For this situation, there will be a maximum angle of refraction, $\theta_t = \theta_c$. This angle is called the *critical angle*.

A wave traveling in the opposite direction, i.e., approaching the boundary from the denser medium towards the less dense medium, refraction into the less dense medium will only occur if the angle of incidence is smaller than the critical angle. For angles of incidence larger than θ_c, the wave will not refract but will only reflect. This phenomenon is called *total reflection*, or sometimes *total internal reflection*, and is illustrated in Figure 2.18. It can be demonstrated by punching a hole in a plastic bottle and draining water through it. By aligning a laser with the opening in the bottle and adding a drop of milk (to make the laser beam visible), total internal reflection such as that seen in Figure 2.19 can be obtained. The light stays within the water stream through multiple reflections. The transmittance of light through strands of fiber, known as fiberoptics, works on the same principle (Figure 2.20) .

While Snell's law predicts the angles of reflection and transmission, it does not give insight into the amplitude of the reflected and refracted fields, **E** and **B**, respectively. To evaluate these amplitudes, we first need to look at the boundary conditions related to the interface between the two media, which are most easily evaluated by means of Maxwell's equations in integral form. The components of **D** and **B** normal to the boundary are assessed first, followed by the tangential components of these fields.[6]

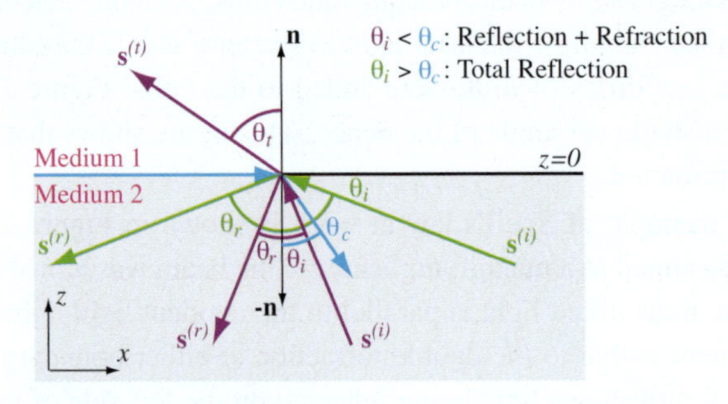

Figure 2.18. Geometry associated with total reflection.

[6]As $\mathbf{D} = \varepsilon\,\mathbf{E}$, the analysis is equivalent.

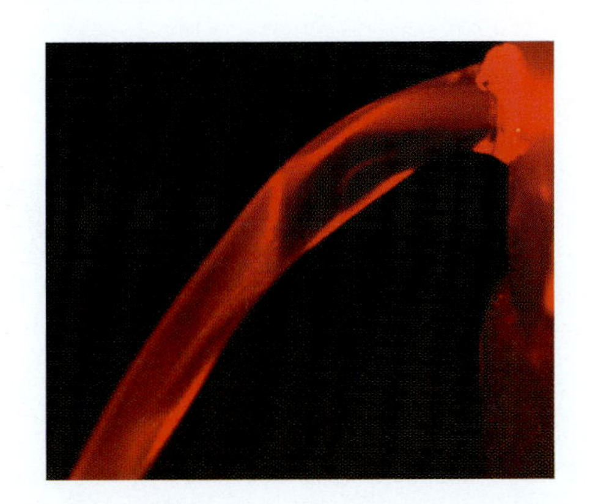

Figure 2.19. A stream of water exits a plastic bottle. Coming from the right, a laser beam enters the bottle and exits through the same hole as the water. Multiple reflections inside the stream of water are visible, essentially keeping the light trapped inside the stream.

Figure 2.20. Fiberoptic strands transmitting light with very little loss of energy due to total internal reflection inside the strands.

2.5.2 Strength of Reflected and Refracted Fields

We assume that the electric and magnetic fields across the boundary are not discontinuous, but merely have a sharp gradient. If we construct a pill box that in-

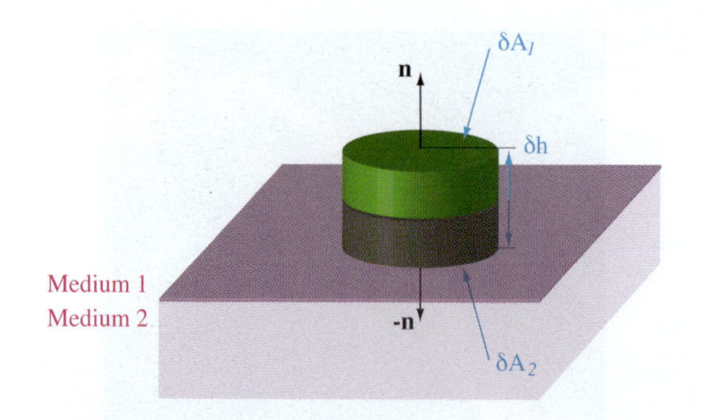

Figure 2.21. Pill box intersecting the boundary between two media.

tersects the boundary between two dielectric materials, as shown in Figure 2.21, then the fields gradually change from the top of the volume to the bottom, which is located on the other side of the boundary.

By Gauss' theorem (see Appendix A.5), the divergence of **B** integrated over the volume of the pill box is related to the integral of the normal component of **B**, when integrated over the surface of the pill box:

$$\int_v \nabla \cdot \mathbf{B}\, dv = \int_s \mathbf{B} \cdot \mathbf{n}\, ds = 0. \tag{2.103}$$

In the limit that the height of the pill box goes to zero, contribution of the cylindrical side of the pill box also goes to zero. If the areas of the top and bottom of the pill box (δA_1 and δA_2) are small and there is no flux or induction at the surface, then the above surface integral simplifies to

$$\mathbf{B}^{(1)} \cdot \mathbf{n} \delta A_1 - \mathbf{B}^{(2)} \cdot \mathbf{n} \delta A_2 = 0, \tag{2.104a}$$

$$\left(\mathbf{B}^{(1)} - \mathbf{B}^{(2)}\right) \cdot \mathbf{n} = 0, \tag{2.104b}$$

where $\delta A = \delta A_1 = \delta A_2$. This equation shows that the component of the magnetic induction **B** normal to the boundary between the two media is continuous across the media.

The normal component of the electric displacement **D** is analyzed similarly:

$$\int_v \mathbf{D}\, dv = \int_s \mathbf{D} \cdot \mathbf{n}\, ds = \int_v \rho\, dv. \tag{2.105}$$

In the limit that the pill box is shrunk to zero height, the charge density ρ, i.e., the charge per unit of volume, goes to infinity. For this reason, the concept of *surface*

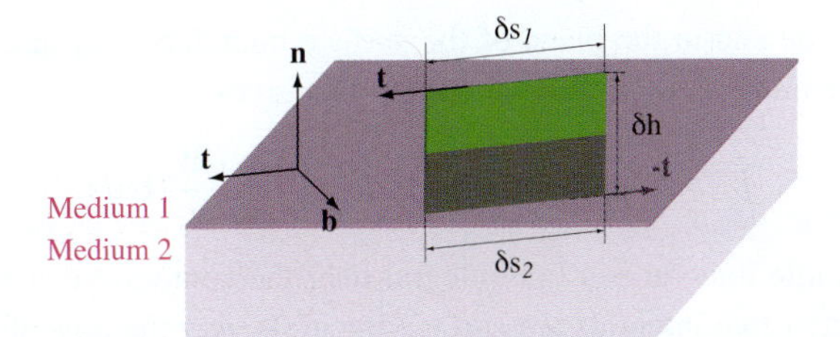

Figure 2.22. Plane intersecting the boundary between two media.

charge density $\hat{\rho}$ is introduced as follows:

$$\lim_{\delta h \to 0} \int_v \rho \, dv = \int_s \hat{\rho} \, dA. \tag{2.106}$$

The normal component of the electric displacement $\mathbf{D}$ may therefore be derived from

$$\left(\mathbf{D}^{(1)} - \mathbf{D}^{(2)} \right) \cdot \mathbf{n} = \hat{\rho}. \tag{2.107}$$

This means that upon crossing a boundary between two dielectrics, the surface charge density causes the normal component of $\mathbf{D}$ to be discontinuous and jump by $\hat{\rho}$.

To analyze the tangential components of $\mathbf{E}$ and $\mathbf{H}$, we can make use of Stokes' theorem (Appendix A.7), which relates line integrals to surface integrals. We place a plane perpendicular to the surface boundary that intersects the boundary, as shown in Figure 2.22. Note also that the incident, reflected and refracted Poynting vectors are assumed to lie in this plane, as shown in Figure 2.23.

In Figure 2.22, $\mathbf{n}$ is the normal of the boundary between the two dielectrics, $\mathbf{b}$ is the normal of the newly created plane, and $\mathbf{t} = \mathbf{b} \times \mathbf{n}$ is a vector that lies both

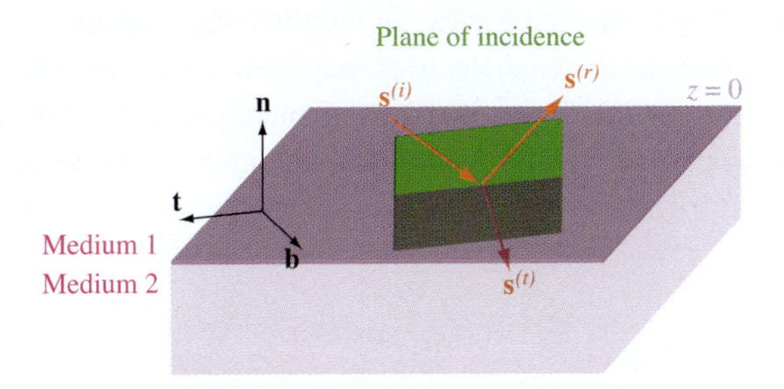

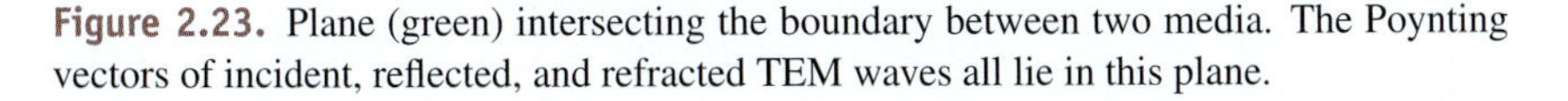
Figure 2.23. Plane (green) intersecting the boundary between two media. The Poynting vectors of incident, reflected, and refracted TEM waves all lie in this plane.

in the new plane and in the plane of the surface boundary. All three vectors are assumed to be unit length. Stokes' theorem then gives

$$\int_s \nabla \times \mathbf{E} \cdot \mathbf{b}\, ds = \int_c \mathbf{E} \cdot \mathbf{b}\, dc = -\int_s \frac{\partial \mathbf{B}}{\partial t} \cdot \mathbf{b}\, ds, \tag{2.108}$$

where the middle integral is a line integral over the contour c that surrounds the plane. The other two integrals are surface integrals over the area of the plane. If δs_1 and δs_2 are small, (2.108) simplifies as follows:

$$\mathbf{E}^{(2)} \cdot \mathbf{t}\, \delta s_2 - \mathbf{E}^{(1)} \cdot \mathbf{t}\, \delta s_1 = -\frac{\partial \mathbf{B}}{\partial t} \cdot \mathbf{b}\, \delta s\, \delta h. \tag{2.109}$$

In the limit that the height δ_h goes to zero, we get

$$\left(\mathbf{E}^{(2)} - \mathbf{E}^{(1)}\right) \cdot \mathbf{t}\, \delta s = 0. \tag{2.110}$$

By noting that $\mathbf{t} = \mathbf{b} \times \mathbf{n}$ and $-\mathbf{t} = -\mathbf{b} \times \mathbf{n}$, this equation can be rewritten as

$$\mathbf{b} \cdot \left(\mathbf{n} \times \left(\mathbf{E}^{(2)} - \mathbf{E}^{(1)}\right)\right) = 0. \tag{2.111}$$

This result indicates that at the boundary between two dielectrics, the tangential component of the electric vector $\mathbf{E}$ is continuous. A similar derivation can be made for the magnetic vector, which yields

$$\mathbf{n} \times \left(\mathbf{H}^{(2)} - \mathbf{H}^{(2)}\right) = \hat{\mathbf{j}}, \tag{2.112}$$

where $\hat{\mathbf{j}}$ is the *surface current density* (introduced for the same reason $\hat{\rho}$ was above). Thus, the tangential component of the magnetic vector $\mathbf{H}$ is discontinuous across a surface and jumps by $\hat{\mathbf{j}}$. Of course, in the absence of a surface current density, the tangential component of vector $\mathbf{H}$ becomes continuous.

The above results for the continuity and discontinuity of the various fields will now be used to help derive the *Fresnel equations*, which predict the amplitudes of the transmitted and reflected waves. To facilitate the derivation, the vector $\mathbf{E}$ at the boundary between the two dielectrics may be decomposed into a component parallel to the plane of incidence, and a component perpendicular to the plane of incidence, as indicated in Figure 2.24. The magnitudes of these components are termed $\underline{A}_{\parallel}$ and $\underline{A}_{\perp}$. The x and y components of the incident electric field are then given by

$$E_x^{(i)} = -\underline{A}_{\parallel} \cos(\Theta_i)\, e^{-i\tau_i}, \tag{2.113a}$$

$$E_y^{(i)} = \underline{A}_{\perp}\, e^{-i\tau_i}, \tag{2.113b}$$

$$E_z^{(i)} = \underline{A}_{\parallel} \sin(\Theta_i)\, e^{-i\tau_i}, \tag{2.113c}$$

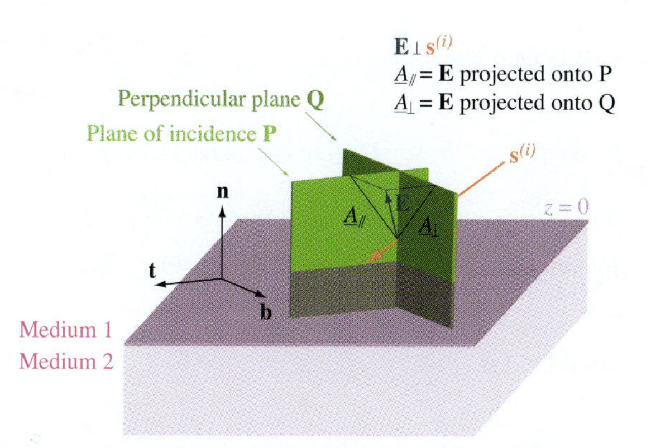

Figure 2.24. The electric vector **E** can be projected onto the plane of incidence as well as onto a perpendicular plane. The magnitudes of the projected vector are indicated by $A_{\|}$ and $A_{\perp}$.

with τ_i equal to the variable part of the wave function:

$$\tau_i = \omega\left(t - \frac{\mathbf{r}\cdot\mathbf{S}^{(i)}}{v_1}\right) = \omega\left(t - \frac{x\sin(\Theta_i) + y\cos(\Theta_i)}{v_1}\right). \tag{2.114}$$

As for dielectric materials, the magnetic permeability is close to one ($\mu = 1$), and therefore we can derive the components of the magnetic vector **H** from **E** by applying (2.46b):

$$\mathbf{H} = \sqrt{\varepsilon}\,\mathbf{S}\times\mathbf{E}, \tag{2.115}$$

so that we have

$$H_x^{(i)} = -\underline{A}_{\perp}\cos(\Theta_i)\sqrt{\varepsilon_1}\,,e^{-i\tau_i}, \tag{2.116a}$$

$$H_y^{(i)} = -\underline{A}_{\|}\sqrt{\varepsilon_1}\,e^{-i\tau_i}, \tag{2.116b}$$

$$H_z^{(i)} = \underline{A}_{\perp}\sin(\Theta_i)\sqrt{\varepsilon_1}\,e^{-i\tau_i}. \tag{2.116c}$$

With τ_r and τ_t equal to

$$\tau_r = \omega\left(t - \frac{x\sin(\Theta_r) + y\cos(\Theta_r)}{v_1}\right), \tag{2.117a}$$

$$\tau_t = \omega\left(t - \frac{x\sin(\Theta_t) + y\cos(\Theta_t)}{v_2}\right), \tag{2.117b}$$

and $\underline{R}$ and $\underline{T}$ the complex amplitudes of the reflected and refracted waves, expressions for the reflected and transmitted fields may be derived analogously.

The reflected field is then given by

$$E_x^{(r)} = -\underline{R}_{\|} \cos(\Theta_r)\, e^{-i\tau_r} \qquad H_x^{(r)} = -\underline{R}_{\perp} \cos(\Theta_r)\, \sqrt{\varepsilon_1}\,, e^{-i\tau_r}, \tag{2.118a}$$

$$E_y^{(r)} = \underline{R}_{\perp}\, e^{-i\tau_r} \qquad H_y^{(r)} = -\underline{R}_{\|}\, \sqrt{\varepsilon_1}\, e^{-i\tau_r}, \tag{2.118b}$$

$$E_z^{(r)} = \underline{R}_{\|} \sin(\Theta_r)\, e^{-i\tau_r} \qquad H_z^{(r)} = \underline{R}_{\perp} \sin(\Theta_r)\, \sqrt{\varepsilon_1}\, e^{-i\tau_r}, \tag{2.118c}$$

and the transmitted field is

$$E_x^{(t)} = -\underline{T}_{\|} \cos(\Theta_t)\, e^{-i\tau_t} \qquad H_x^{(t)} = -\underline{T}_{\perp} \cos(\Theta_t)\, \sqrt{\varepsilon_2}\,, e^{-i\tau_t}, \tag{2.119a}$$

$$E_y^{(t)} = \underline{T}_{\perp}\, e^{-i\tau_t} \qquad H_y^{(t)} = -\underline{T}_{\|}\, \sqrt{\varepsilon_2}\, e^{-i\tau_t}, \tag{2.119b}$$

$$E_z^{(t)} = \underline{T}_{\|} \sin(\Theta_t)\, e^{-i\tau_t} \qquad H_z^{(t)} = \underline{T}_{\perp} \sin(\Theta_t)\, \sqrt{\varepsilon_2}\, e^{-i\tau_t}. \tag{2.119c}$$

According to (2.111) and (2.112) and under the assumption that the surface current density is zero[7], the tangential components of **E** and **H** should be zero at the surface boundary. We may therefore write

$$E_x^{(i)} + E_x^{(r)} = E_x^{(t)} \qquad H_x^{(i)} + H_x^{(r)} = H_x^{(t)}, \tag{2.120a}$$

$$E_y^{(i)} + E_y^{(r)} = E_y^{(t)} \qquad H_y^{(i)} + H_y^{(r)} = H_y^{(t)}. \tag{2.120b}$$

By substitution of this equation, and using the fact that $\cos(\Theta_r) = \cos(\pi - \Theta_r) = -\cos(\Theta_r)$, we get

$$\cos(\Theta_i)\left(\underline{A}_{\|} - \underline{R}_{\|}\right) = \cos(\Theta_t)\, \underline{T}_{\|}, \tag{2.121a}$$

$$\underline{A}_{\|} + \underline{R}_{\|} = \sqrt{\frac{\varepsilon_2}{\varepsilon_1}}\, \underline{T}_{\|}, \tag{2.121b}$$

$$\cos(\Theta_i)\,(\underline{A}_{\perp} - \underline{R}_{\perp}) = \sqrt{\frac{\varepsilon_2}{\varepsilon_1}}\, \cos(\Theta_t)\, \underline{T}_{\perp}, \tag{2.121c}$$

$$\underline{A}_{\perp} + \underline{R}_{\perp} = \underline{T}_{\perp}. \tag{2.121d}$$

[7]This is a reasonable assumption for dielectric materials.

We use Maxwell's relation $n = \sqrt{\mu\,\varepsilon} = \sqrt{\varepsilon}$ and solve the above set of equations for the components of the reflected and transmitted waves, yielding

$$\underline{R}_{\parallel} = \underline{A}_{\parallel} \frac{n_2 \cos(\Theta_i) - n_1 \cos(\Theta_t)}{n_2 \cos(\Theta_i) + n_1 \cos(\Theta_t)}, \tag{2.122a}$$

$$\underline{R}_{\perp} = \underline{A}_{\perp} \frac{n_1 \cos(\Theta_i) - n_2 \cos(\Theta_t)}{n_1 \cos(\Theta_i) + n_2 \cos(\Theta_t)}, \tag{2.122b}$$

$$\underline{T}_{\parallel} = \underline{A}_{\parallel} \frac{2 n_1 \cos(\Theta_i)}{n_2 \cos(\Theta_i) + n_1 \cos(\Theta_t)}, \tag{2.122c}$$

$$\underline{T}_{\perp} = \underline{A}_{\perp} \frac{2 n_1 \cos(\Theta_i)}{n_1 \cos(\Theta_i) + n_2 \cos(\Theta_t)}. \tag{2.122d}$$

These equations are known as the *Fresnel equations*; they allow us to compute the *transmission coefficients* $t_{\parallel}$ and $t_{\perp}$ as follows:

$$t_{\parallel} = \frac{\underline{T}_{\parallel}}{\underline{A}_{\parallel}} = \frac{2 \cos(\Theta_i)}{\frac{n_2}{n_1} \cos(\Theta_i) + \cos(\Theta_t)}, \tag{2.123a}$$

$$t_{\perp} = \frac{\underline{T}_{\perp}}{\underline{A}_{\perp}} = \frac{2 \cos(\Theta_i)}{\cos(\Theta_i) + \frac{n_2}{n_1} \cos(\Theta_t)}. \tag{2.123b}$$

The *reflection coefficients* $r_{\parallel}$ and $r_{\perp}$ are computed similarly:

$$r_{\parallel} = \frac{\underline{R}_{\parallel}}{\underline{A}_{\parallel}} = \frac{\cos(\Theta_i) - \frac{n_1}{n_2} \cos(\Theta_t)}{\cos(\Theta_i) + \frac{n_1}{n_2} \cos(\Theta_t)}, \tag{2.124a}$$

$$r_{\perp} = \frac{\underline{R}_{\perp}}{\underline{A}_{\perp}} = \frac{\cos(\Theta_i) - \frac{n_2}{n_1} \cos(\Theta_t)}{\cos(\Theta_i) + \frac{n_2}{n_1} \cos(\Theta_t)}. \tag{2.124b}$$

We may rewrite both the reflection and the transmission coefficients in terms of the angle of incidence alone. This can be achieved by applying Snell's law (2.102), so that

$$\sin(\Theta_t) = \frac{n_1}{n_2} \sin(\Theta_i) \tag{2.125}$$

and therefore, using Equation (B.11), we have

$$\cos(\Theta_t) = \sqrt{1 - \left(\frac{n_1}{n_2}\right)^2 \sin^2(\Theta_i)} \tag{2.126}$$

By substitution into (2.123) and (2.124) expressions in n_1, n_2, and $\cos(\Theta_i)$ can be obtained. The reflection and transmission coefficients give the percentage of incident light that is reflected or refracted:

$$r = \frac{\|\mathbf{E}_{0,r}\|}{\|\mathbf{E}_{0,i}\|} \tag{2.127a}$$

$$t = \frac{\|\mathbf{E}_{0,t}\|}{\|\mathbf{E}_{0,i}\|} \tag{2.127b}$$

In the case of a wave incident upon a conducting (metal) surface, light is both reflected and refracted as above, although the refracted wave is strongly attenuated by the metal. This attenuation causes the material to appear opaque.

A water surface is an example of a reflecting and refracting boundary that occurs in nature. While all water reflects and refracts, on a wind-still day the

Figure 2.25. On a wind-still morning, the water surface is smooth enough to reflect light specularly on the macroscopic scale; Konstanz, Bodensee, Germany, June 2005.

boundary between air and water becomes smooth on the macroscopic scale, so that the effects discussed in this section may be observed with the naked eye, as shown in Figure 2.25.

2.5.3 Reflectance and Transmittance

If we assume that a beam of light is incident upon a dielectric surface at a given angle Θ_i, and that the cross-section of the beam has an area A, then the incident energy per unit of time $P_{e,i}$ is given by

$$P_{e,i} = E_{e,i} A \cos(\Theta_i), \tag{2.128}$$

where $E_{e,i}$ is the radiant flux density of Equation (2.96). Energy per unit of time is known as *radiant flux* and is discussed further in Section 6.2.2. Similarly, we can compute the reflected and transmitted flux ($P_{e,r}$ and $P_{e,t}$) with

$$P_{e,r} = E_{e,r} A \cos(\Theta_r), \tag{2.129a}$$

$$P_{e,t} = E_{e,t} A \cos(\Theta_t). \tag{2.129b}$$

The *reflectance* R is defined as the ratio of reflected to incident flux:

$$R = \frac{P_{e,r}}{P_{e,i}} = \frac{E_{e,r} A \cos(\Theta_r)}{E_{e,i} A \cos(\Theta_i)}. \tag{2.130}$$

Given that the angle of incidence and the angle of reflectance are identical for specular reflection, this expression simplifies to

$$R = \frac{E_{e,r}}{E_{e,i}}. \tag{2.131}$$

Using (2.96), this expression can be rewritten as

$$R = \frac{\frac{\varepsilon_r v_r}{2} \|\mathbf{E}_{0,r}^2\|}{\frac{\varepsilon_i v_i}{2} \|\mathbf{E}_{0,i}^2\|}. \tag{2.132}$$

As the incident and reflected beams are in the same medium, we have $\varepsilon_i = \varepsilon_r$ and $v_i = v_r$, so that the reflectance becomes

$$R = \left(\frac{\|\mathbf{E}_{0,r}\|}{\|\mathbf{E}_{0,i}\|} \right)^2 = r^2. \tag{2.133}$$

The reflectance can be split into its component forms, leading to

$$R_{\parallel} = r_{\parallel}^2, \tag{2.134a}$$

$$R_{\perp} = r_{\perp}^2. \tag{2.134b}$$

The *transmittance* T is defined as the ratio between incident flux to transmitted flux:

$$T = \frac{P_{e,i}}{P_{e,t}} = \frac{E_{e,i} A \cos(\Theta_i)}{E_{e,t} A \cos(\Theta_t)}. \tag{2.135}$$

Through similar arguments, the transmittance can be derived to be

$$T = \frac{n_2 \cos(\Theta_t)}{n_1 \cos(\Theta_i)} t^2, \tag{2.136a}$$

$$T_{\parallel} = \frac{n_2 \cos(\Theta_t)}{n_1 \cos(\Theta_i)} t_{\parallel}^2, \tag{2.136b}$$

$$T_{\perp} = \frac{n_2 \cos(\Theta_t)}{n_1 \cos(\Theta_i)} t_{\perp}^2. \tag{2.136c}$$

Under the assumption that no absorbtion takes place, all light is either reflected or transmitted, so that we have

$$R + T = 1, \tag{2.137a}$$

$$R_{\parallel} + T_{\parallel} = 1, \tag{2.137b}$$

$$R_{\perp} + T_{\perp} = 1. \tag{2.137c}$$

An especially important case arises when the angle of incidence is perpendicular to the boundary between the two media. In this case Θ_i becomes 0, and therefore Θ_t also approaches 0. In that case, the distinction between parallel and perpendicular components vanishes, so that $R_{\parallel} = R_{\perp} = R$ and $T_{\parallel} = T_{\perp} = T$. The resulting reflectance and transmittance is then [447]

$$R = \left(\frac{n_2 - n_1}{n_2 + n_1} \right)^2, \tag{2.138a}$$

$$T = \frac{4 n_1 n_2}{(n_1 + n_2)^2}. \tag{2.138b}$$

Finally, in the case that the surrounding medium is air, the index of refraction may be taken to approximate $n_1 \approx 1$, so that the reflectance at normal angles is

$$R \approx \left(\frac{n_2 - 1}{n_2 + 1} \right)^2, \tag{2.139a}$$

$$T \approx \frac{4 n_2}{(n_2 + 1)^2}. \tag{2.139b}$$

For instance, for glass having an index of refraction of around 1.5, this means that approximately 4% of light is reflected. This can become a problem in lens design, where compound lenses are made of several air-glass interfaces. The amount of light reflected by a dielectric can be reduced by applying a thin coat of anti-reflective material. This is discussed further in Section 12.6.

Figure 2.26. A calcite crystal causes birefringence which depends on the orientation of the crystal.

2.6 Birefringence

Some materials have an index of refraction that depends on the polarization of the light. This means that unpolarized light can be split into two or more polarized light paths. This effect is called *birefringence* or *double refraction* and is demonstrated in Figure 2.26 by means of a calcite crystal (also known as *iceland spar*). A second example of a birefringent material is the cornea of the human eye (see Section 4.2.2) [755].

To see that light becomes linearly polarized into two orthogonal directions, Figure 2.27 contains photographs of the same calcite crystal placed upon a detail of Figure 1.7. The light is linearly polarized, which is demonstrated by placing a linear polarization filter on top of the crystal. This allows one of the two linearly polarized light directions to pass through, while blocking the other. As a consequence, slowly rotating either the filter or the crystal, will suddenly change the path selected to pass through the polarizer. The combination of linear polarizers and liquid birefringent materials, which have polarizing behavior that can be controlled by applying an electric field, forms the basis of liquid crystal displays (LCDs), which are discussed further in Section 14.2.

Molecules with an elongated shape tend to give rise to birefringence. Such rod-like molecules are called *calamitic*. To create birefringence in bulk, these molecules must be aligned, so that the material itself will exhibit birefringence. The vector pointing along the long axis of the molecules is called the *director*. In that case, the relative permittivity (i.e., the dielectric constant) of the material

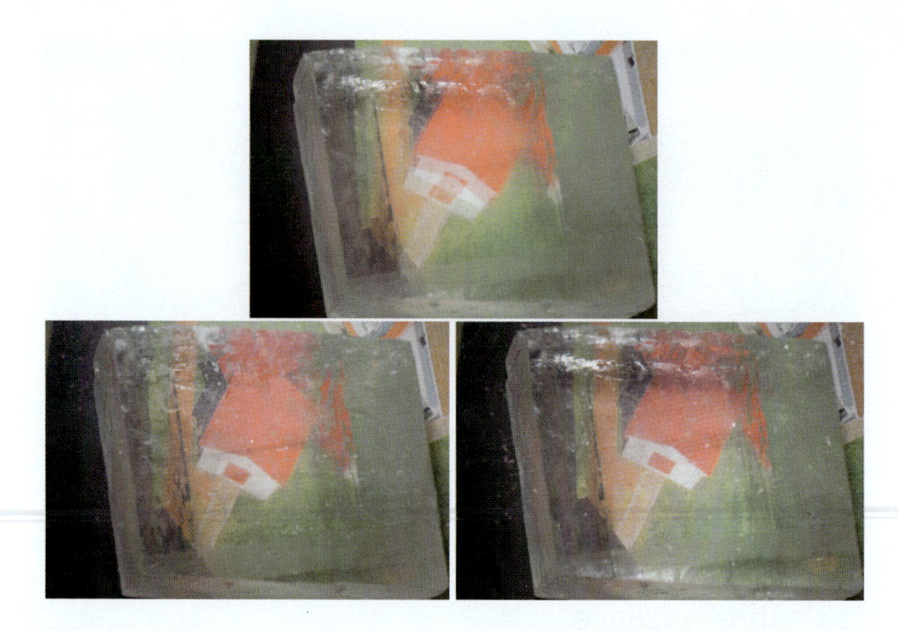

Figure 2.27. The block of birefringent calcite is placed on a print of Figure 1.7 (top), showing double refraction. The bottom two images are photographs of the same configuration, except that now a linear polarizing filter is placed on top of the crystal. Dependent on the orientation (which is the difference between the bottom two photographs), one or the other refraction remains. A polarizer used in this manner is called an *analyzer*.

is dependent on direction, yielding $\varepsilon_{\parallel}$ for directions parallel to the director and $\varepsilon_{\perp}$ for all directions orthogonal to the director. The dielectric anisotropy is then given by $\Delta\varepsilon$:

$$\Delta\varepsilon = \varepsilon_{\parallel} - \varepsilon_{\perp}. \tag{2.140}$$

As the index of refraction of a material depends on its permittivity, in the direction of the director the material will have an index of refraction of $n_{\parallel}$, whereas in orthogonal directions it is given by $n_{\perp}$. The optical anisotropy Δn is then

$$\Delta n = n_{\parallel} - n_{\perp}. \tag{2.141}$$

We now consider the amplitude of the electric field intensity in two orthogonal directions x and y, as given in (2.76), with x aligned with the director. Recall that the distance z in this equation is only a convenient notation representing a point $\mathbf{r}$ in space under a plane wave traveling in direction $\mathbf{S}$. In other words, in the more general case we have $\beta z = \beta \mathbf{r} \cdot \mathbf{S}$. With β given by

$$\beta = \frac{\omega}{c} = \frac{2\pi n}{\lambda}, \tag{2.142}$$

the wave vector **k** is given by

$$\mathbf{k} = \frac{2\pi n}{\lambda}\mathbf{S}. \tag{2.143}$$

Thus, the wave vector **k** is nothing more than **S** scaled by an amount dependent on wavelength λ and index of refraction n.

With indices of refraction that are different for the x and y directions, (2.76) becomes

$$\frac{E_x}{|\underline{E}_x|} = \cos(\omega t - \mathbf{k}_{\|}\mathbf{r} + \varphi_x), \tag{2.144a}$$

$$\frac{E_y}{|\underline{E}_y|} = \cos(\omega t - \mathbf{k}_{\perp}\mathbf{r} + \varphi_y), \tag{2.144b}$$

with

$$\mathbf{k}_{\|} = \frac{2\pi n_{\|}}{\lambda}\mathbf{S}, \tag{2.145a}$$

$$\mathbf{k}_{\perp} = \frac{2\pi n_{\perp}}{\lambda}\mathbf{S}. \tag{2.145b}$$

In the case that the director makes an angle, say ψ, with our choice of coordinate system, a further scaling of the electric field magnitudes is required:

$$\frac{E_x}{|\underline{E}_x|} = \cos(\psi)\ \cos(\omega t - \mathbf{k}_{\|}\mathbf{r} + \varphi_x), \tag{2.146a}$$

$$\frac{E_y}{|\underline{E}_y|} = \sin(\psi)\ \cos(\omega t - \mathbf{k}_{\perp}\mathbf{r} + \varphi_y). \tag{2.146b}$$

For computations involving liquid crystals, it is often convenient to remove all components that do not involve **r**, and add them back in afterwards. Thus, we may analyze the complex phasors P_x and P_y:

$$P_x = E_x \cos(\psi) \exp\left(-i\mathbf{k}_{\|}\mathbf{r}\right), \tag{2.147a}$$

$$P_y = E_y \sin(\psi) \exp\left(-i\mathbf{k}_{\perp}\mathbf{r}\right), \tag{2.147b}$$

whereby we momentarily ignore the factors $\exp(i\omega t)$ and $\exp(i\varphi)$. Absorbing the scaling according to ψ into the constants E_x and E_y and placing the two phasors in a column vector, we obtain the so-called *Jones vector* **J**:

$$\mathbf{J} = \begin{bmatrix} J_x \\ J_y \end{bmatrix} = \begin{bmatrix} E_{x,\psi} \exp\left(-i\mathbf{k}_{\|}\mathbf{r}\right) \\ E_{y,\psi} \exp\left(-i\mathbf{k}_{\perp}\mathbf{r}\right) \end{bmatrix}. \tag{2.148}$$

The use of the Jones vector will return in the analysis of liquid crystals in Section 14.2.

2.7 Interference and Diffraction

The emphasis of this chapter is on the wave nature of light. One of the most compelling demonstrations of wave behavior is afforded by two closely related phenomena, interference and diffraction.

We are modeling light with time-harmonic electromagnetic waves. The superposition principle discussed in Section 2.2.5 shows that multiple waves may be summed to determine a compound solution at any given position in space. For waves with random phases, the result yields an average intensity that does not vary appreciably with space, except for the attenuation caused by the medium of propagation. This assumption is, for instance, made in geometrical optics, which will be discussed in Section 2.9.

2.7.1 Interference

If the phase function of each of the waves is correlated, we speak of *coherent light*. When such waves are superposed, the result shows peaks and troughs as a result of the sinusoidal variation of both electric and magnetic vectors, as shown in Figure 2.28. If the two waves are out of phase by π, then the two waves exactly cancel. Thus, dependent on the phase difference at a given point in space, the two waves may reinforce each other or cancel each other. It is said that coherent waves *interfere*, and the resulting pattern is called an *interference pattern*.

The irradiance at a point in space is given by (2.96), where $\left\langle \|\mathbf{E}_0^2\| \right\rangle_T$ is the time average of the magnitude of the electric field vector. Applying the superposition principle for a pair of coherent monochromatic time-harmonic plane waves with

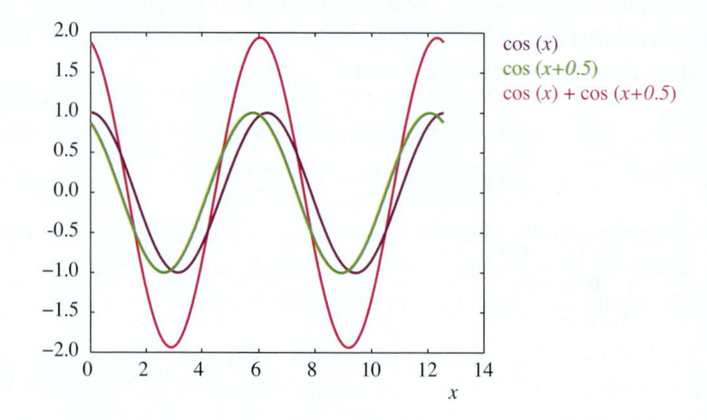

Figure 2.28. Two waves superposed forming a new wave, in this case with a larger amplitude than either of the constituent waves.

electric field vectors $\mathbf{E}_1$ and $\mathbf{E}_2$, the irradiance becomes

$$E_e = c\,\varepsilon_0 \left\langle (\mathbf{E}_1 + \mathbf{E}_2)^2 \right\rangle_T. \tag{2.149}$$

Since the electric vector is now the sum of two components, the square of its magnitude is given by

$$(\mathbf{E_1} + \mathbf{E_2})^2 = \mathbf{E}_1^2 + \mathbf{E}_2^2 + 2\,\mathbf{E}_1 \cdot \mathbf{E_2}. \tag{2.150}$$

The implications for the irradiance are that this quantity is composed of three terms, namely the irradiance due to either wave as well as a cross term:

$$E_e = \varepsilon_0 c \left(\left\langle \mathbf{E}_1^2 \right\rangle_T + \left\langle \mathbf{E}_2^2 \right\rangle_T + 2 \left\langle \mathbf{E}_1 \cdot \mathbf{E}_2 \right\rangle_T \right). \tag{2.151}$$

The cross term $2\,\langle \mathbf{E}_1 \cdot \mathbf{E}_2 \rangle_T$, known as the *interference term*, explains the patterns that may be observed when two coherent bundles of light are superposed.

We assume that the electric vectors of the two waves are modeled by

$$\mathbf{E}_1 (\mathbf{r},t) = \mathbf{a}_1 \cos\left(\mathbf{r} \cdot \mathbf{S}_1 - \omega t + \varphi_1\right), \tag{2.152a}$$

$$\mathbf{E}_2 (\mathbf{r},t) = \mathbf{a}_2 \cos\left(\mathbf{r} \cdot \mathbf{S}_2 - \omega t + \varphi_2\right), \tag{2.152b}$$

where $\mathbf{a}_i$ are the amplitudes and φ_i are the initial phases of the two (linearly polarized) waves. The interference term is then

$$\begin{aligned} \mathbf{a}_1 \cdot \mathbf{a}_2 & \left[\cos(\mathbf{r} \cdot \mathbf{S}_1 + \varphi_1) \cos(\omega t) + \sin(\mathbf{r} \cdot \mathbf{S}_1 + \varphi_1) \sin(\omega t)\right] \\ & \times \left[\cos(\mathbf{r} \cdot \mathbf{S}_2 + \varphi_2) \cos(\omega t) + \sin(\mathbf{r} \cdot \mathbf{S}_2 + \varphi_2) \sin(\omega t)\right]. \end{aligned} \tag{2.153}$$

It can be shown that the time-averaged quantity of the interference term becomes

$$2\,\langle \mathbf{E}_1 \cdot \mathbf{E}_2 \rangle_T = \mathbf{a}_1 \cdot \mathbf{a}_2 \cos\left(\mathbf{r} \cdot \mathbf{S}_1 + \varphi_1 - \mathbf{r} \cdot \mathbf{S}_2 - \varphi_2\right). \tag{2.154}$$

The interference term can then be rewritten in the following form to highlight the fact that its oscillations depend on the phase difference between the two waves:

$$2\,\langle \mathbf{E}_1 \cdot \mathbf{E}_2 \rangle_T = \mathbf{a}_1 \cdot \mathbf{a}_2 \cos\left(\delta\right), \tag{2.155}$$

where the phase difference δ equals $\mathbf{r} \cdot \mathbf{S}_1 + \varphi_1 - \mathbf{r} \cdot \mathbf{S}_2 - \varphi_2$. This result shows that the phase difference is due to a difference in initial phase, $\varphi_1 - \varphi_2$, as well as a difference in path length between the origins of the two waves and the point of interest $\mathbf{r}$ (Figure 2.29).

The irradiance at $\mathbf{r}$ is maximum when $\cos(\delta) = 1$ and occurs when the phase difference is a multiple of 2π:

$$\delta = 0, \pm 2\pi, \pm 4\pi, \ldots. \tag{2.156}$$

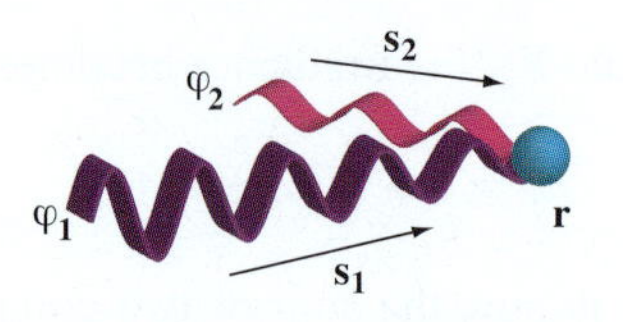

Figure 2.29. The phase at point **r** is determined by the phases at the origins of the two waves, as well as the distance traveled.

In this situation, we have *total constructive interference*. The irradiance attains a minimum when the phase difference is

$$\delta = \pm\pi, \pm 3\pi, \pm 5\pi, \ldots. \tag{2.157}$$

In this case, there is *total destructive interference*.

So far we are assuming that the waves are planar and that they correspond to a source that is infinitely far away. This, of course, is rarely the case although many interesting theoretical results may be derived in this way. However, if the source of radiation is not infinitely far away, it is better to assume a spherical wavefront, i.e., a wave originating at a point in space and traveling outward in all directions, as shown in Figure 2.30. Such a wavefront may be modeled by

$$\mathbf{E}(r,t) = \mathbf{a}(r) \exp\left[i\left(\frac{2\pi r}{\lambda} - \omega t + \varphi\right)\right]. \tag{2.158a}$$

Here, r is the distance between the point of interest and the source of the wavefront, i.e., the radius. A pair of spherical wavefronts may show interference just as planar waves would. The phase difference δ of two spherical wavefronts $\mathbf{E}_1$

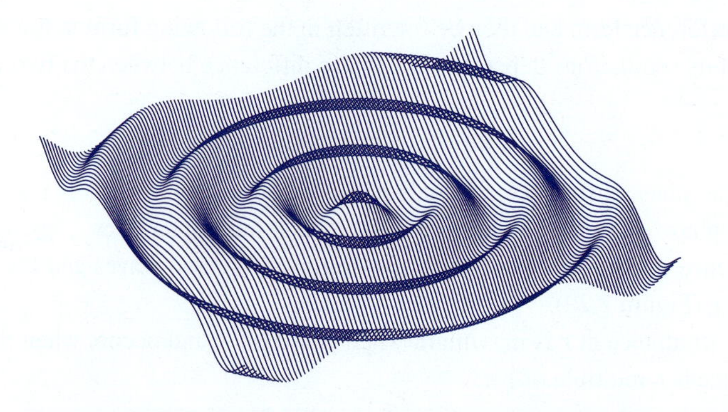

Figure 2.30. Radial wavefront due to a point source.

and $\mathbf{E}_2$ is then

$$\delta = \frac{2\pi}{\lambda}(r_1 - r_2) + (\varphi_1 - \varphi_2). \tag{2.159}$$

If for simplicity we assume the amplitudes $\mathbf{a}_1$ and $\mathbf{a}_2$ to be identical and constant over the region of interest, it can be shown that the irradiance depends on the phase difference as follows [447]:

$$E_e = 4E_{e,0}\cos^2\left(\frac{\delta}{2}\right), \tag{2.160}$$

where $E_{e,0}$ is the irradiance due to either source (which in this case are identical). Maxima in the irradiance E_e occur at intervals,

$$\delta = 0, \pm 2\pi, \pm 4\pi, \ldots, \tag{2.161}$$

and minima occur at

$$\delta = \pm\pi, \pm 3\pi, \pm 5\pi, \ldots. \tag{2.162}$$

Thus, maxima occur for integer multiples of 2π, i.e., when

$$2\pi m = \frac{2\pi}{\lambda}(r_1 - r_2) + (\varphi_1 - \varphi_2) \qquad m = 0, \pm 1, \pm 2, \ldots. \tag{2.163}$$

This expression can be simplified to

$$r_1 - r_2 = \frac{\lambda(\varphi_1 - \varphi_2)}{2\pi} - m\lambda. \tag{2.164}$$

If the two sources are emitting light in phase, then $\varphi_1 = \varphi_2$, and we obtain

$$r_2 - r_1 = m\lambda. \tag{2.165}$$

This result indicates that the location of peaks and troughs in the interference pattern depends on the wavelength of the two sources, which in our simplified analysis are assumed to be in phase and emit light of the same wavelengths. Thus, if the phase difference between two bundles of light can be coherently manipulated by fractions of a wavelength, we can expect white light to be broken up into its constituent parts.

Further, a thin layer which reflects off its front surface as well as its back surface, can create interference patterns. This occurs because the path length from the light source to the human observer is slightly shorter for the light reflected off the front surface than it is for the light reflected off the back surface. As a result, according to (2.164) this results in small phase differences. If the spacing of the reflecting layers causes a phase difference in the order of a visible wavelength, this is seen as interference color. As shown in Figure 2.31, the viewing

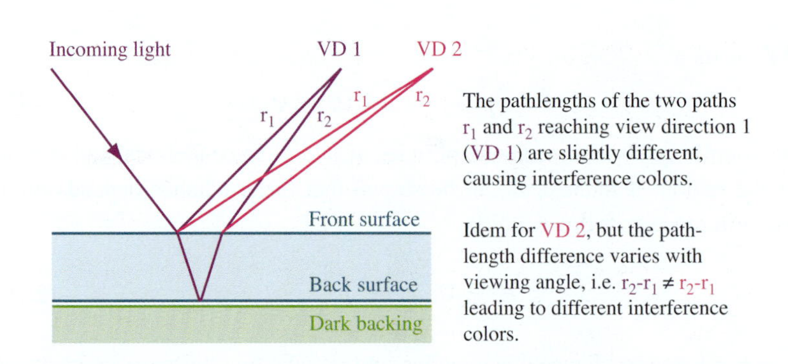
Incoming light
VD 1
VD 2
r1
r2
r1
r2
Front surface
Back surface
Dark backing
The pathlengths of the two paths r1 and r2 reaching view direction 1 (VD 1) are slightly different, causing interference colors.
Idem for VD 2, but the path-length difference varies with viewing angle, i.e. r2-r1 ≠ r2-r1 leading to different interference colors.

Figure 2.31. Interference colors, which depend on viewing angle, are caused by a semi-transparent layer creating two thinly spaced reflective layers. The path lengths of light reflected off each surface are slightly different, causing interference colors.

angle has an impact on the relative path lengths, and therefore one can expect the color of the surface to vary with viewing angle. The dark substrate serves to create brighter and more brilliant colors. In practice, this effect is not limited to a single transparent layer, but further reflections can be created by a stack of layers.

Figure 2.32. This *Morpho zephyritis* shows iridescent blue colors as a result of interference in its wings.

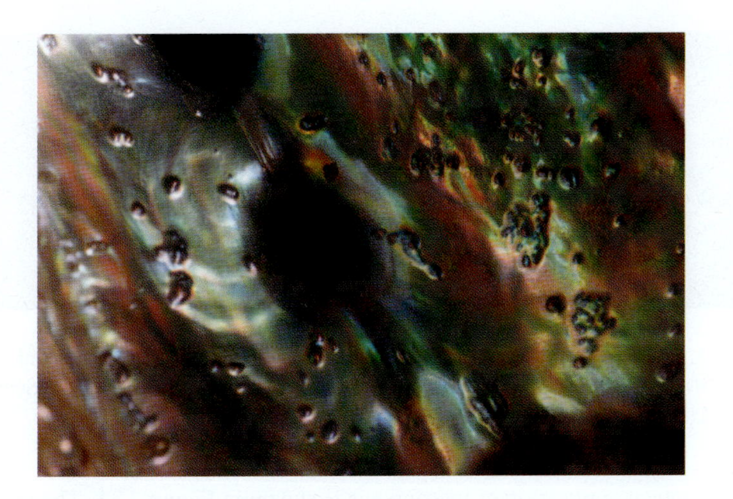

Figure 2.33. Iridescence in sea shells. Shown here is a close-up of the inside of an abalone shell.

Recently, display devices have been proposed which are based on this principle, as discussed in Section 14.10.

Interference colors occur in nature, for instance, in the wings of certain species of butterfly such as the *Morpho zephyritis* shown in Figure 2.32. Colors resulting from interference at wavelength scales are usually termed *iridescent*. Multiple colors may be present and normally vary with viewing angle. In elytra, the wing-cases of beetles (see Figure 1.5), as well as in butterfly wings, the layers causing interference are backed by a dark layer, usually made of melanin, which strengthens the color. Iridescence is also present in many birds, including the peacock (Figure 1.3) and hummingbirds, as well as sea shells such as abalones and oysters (Figure 2.33).

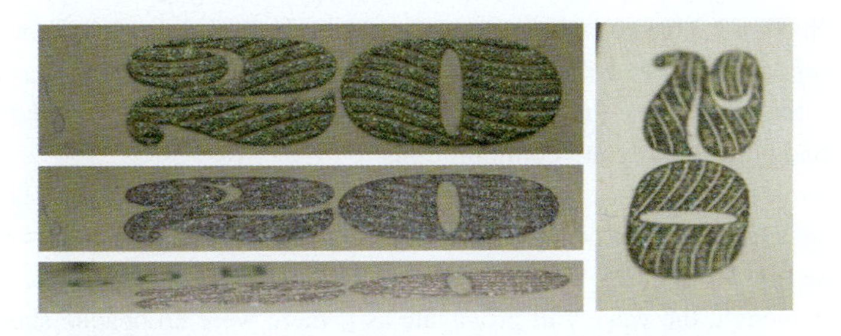

Figure 2.34. The number 20 from a US $20 bill photographed under different viewing angles. The particles in the ink are aligned to produce a distinct change in color with viewing angle.

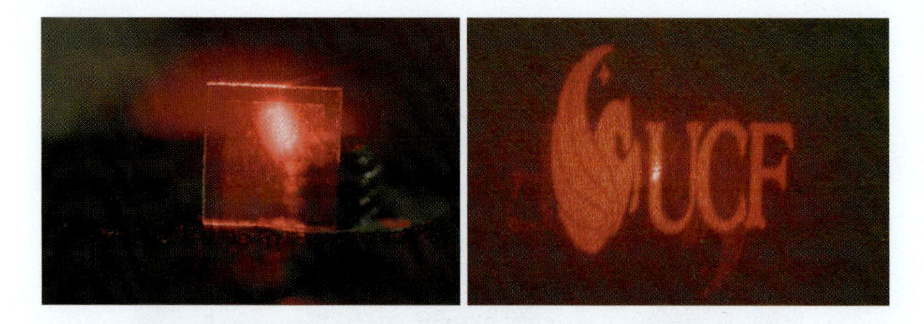

Figure 2.35. An interference pattern is etched into a small piece of plastic, which only reveals itself when hit by a laser beam. (Etching courtesy of Eric G. Johnson, School of Optics, University of Central Florida.)

A practical application of interference is shown in Figure 2.34. The ink used to print the number 20 on US currency contains small particles that produce interference colors [447]. The color changes with viewing angle due to the structured orientation of the particles. A second application of interference in optics is shown in Figure 2.35. Here, a small piece of plastic is etched with a pattern that does not appear to contain information (left side of figure). When a laser is aimed at it, an entirely different pattern is formed, as seen in the same figure (right). Finally, an algorithm for simulating iridescence was recently proposed [1107]. It allows the simulation of, for instance, butterfly wings.

2.7.2 Diffraction

We have thus far discussed interference in the form of a pair of hypothetical light sources that emit light of the same wavelength and are in phase. In this section we discuss diffraction, which is the behavior of light in the presence of obstacles. Diffraction may cause interference patterns.

In Section 2.5 the behavior of light near a boundary was discussed. In the case of a boundary between a dielectric and a metal, the wave will not propagate very far into the metal. The theory of diffraction deals with opaque boundaries with edges and holes in different configurations.

If we assume that we have a thin opaque sheet of material with a hole in it, then the interaction of the wave with the sheet away from the hole will be as discussed in Section 2.5. In the center of the hole, and assuming that the hole is large enough, the wave will propagate as if there were no opaque sheet there. However, near the edge of the hole, diffraction occurs. Here the wave interacts with the material, causing new wavefronts. A detailed treatment of this interaction would involve atomic theory (touched upon in Chapter 3) and is beyond the scope

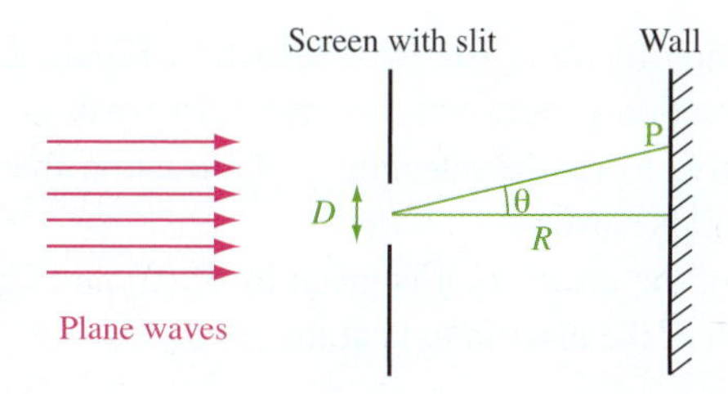

Figure 2.36. Light passing through a slit is projected onto a wall. The distribution of irradiance values is observed for points **P** along the wall.

of this book. We briefly outline the behavior of light near edges using a simplified wave theory of light.

For a sheet of material with a hole, the visible manifestation of diffraction depends on the size of the hole in relation to the wavelength of the incident light. If the hole is very large, the sheet casts a shadow and the result is much like everyday experience predicts. When the hole becomes small, the edge of the hole will diffract light and cause interference patterns.

For a rectangular hole, i.e., a *slit*, we are interested in the irradiance distribution as a result of diffraction. If the distance R between the slit and the line of observation on a wall is large in comparison with the size of the slit ($R \gg D$, see Figure 2.36), we speak of *Fraunhofer diffraction*. It can be shown that the irradiance distribution along the wall as a function of angle θ is given by [447]

$$E_e(\theta) = E_e(0)\,\mathrm{sinc}^2\left(\frac{D\pi}{\lambda}\sin(\theta)\right). \tag{2.166}$$

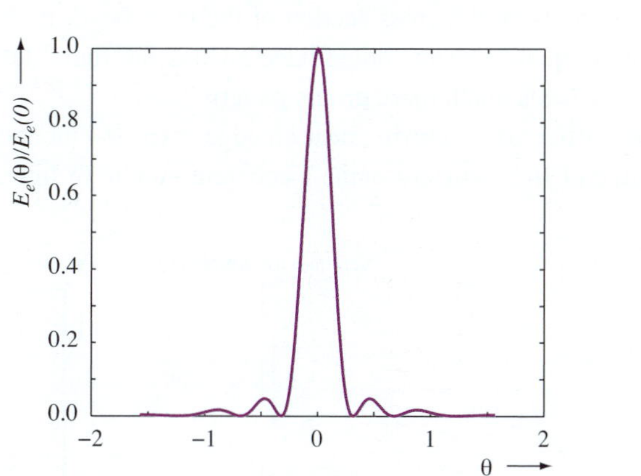

Figure 2.37. Irradiance as function of eccentricity angle θ at a distance R from the slit. The size of the slit was chosen to be $D = 10\lambda/\pi$.

A plot of this function for $D = 10\lambda/\pi$ is shown in Figure 2.37. The equation shows that the diffraction pattern will become very weak as the size of the slit becomes large relative to the wavelength λ. Diffraction therefore occurs most strongly when the slit is narrow.

The irradiance of the center spot is given by $E_e(0)$ and may be expressed in terms of the strength of the electric field at the slit Ξ_0:

$$E_e(0) = \frac{1}{2}\left(\frac{\Xi_0 D}{R}\right)^2. \tag{2.167}$$

Fraunhofer diffraction can also be shown for other configurations. An often-seen demonstration is that of a double slit. Here, two parallel slits each cause an interference pattern; these patterns are superposed. The irradiance is then

$$E_e(\theta) = E_e(0)\left(\frac{\sin^2\left(\frac{D\pi}{\lambda}\sin(\theta)\right)}{\left(\frac{D\pi}{\lambda}\sin(\theta)\right)^2}\right)\cos^2\left(\frac{a\pi}{\lambda}\right), \tag{2.168}$$

with a the separation distance between the two slits, as shown in Figure 2.38.

Thus, a diffraction effect may be obtained by aiming a coherent bundle of light at a single or double slit. Such an experiment can be readily devised from a laser pointer, a comb, and pieces of cardboard. By using the cardboard to mask off all the gaps in the comb bar, a single or double slit can be fabricated (Figure 2.39). The projection of the light on a wall is as shown in Figure 2.40. The horizontal bright-and-dark pattern is caused by interference of the light passing through the slits. For comparison, the cross section of the laser beam is shown in the same figure. A good quality beam should have a Gaussian radial fall-off. The cheap laser used here has a much more erratic pattern.

To show diffraction occurring near an edge, a (near-) monochromatic laser is a good source of light, since it emits a coherent bundle of light. By shining this

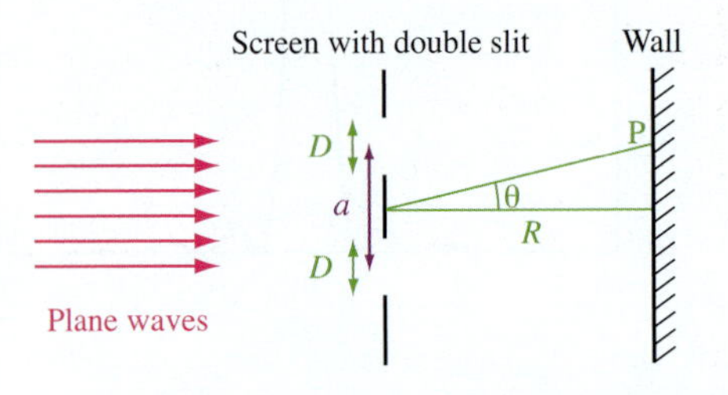

Figure 2.38. Light passing through a double slit is projected onto a wall.

Figure 2.39. With a source of laser light, a comb, and two bits of cardboard, a setup may be created to demonstrate interference patterns. The cardboard is used to mask off most of the comb such that only one or two slits are available to let light through. This is the setup used to demonstrate single- and double-slit diffraction in Figure 2.40.

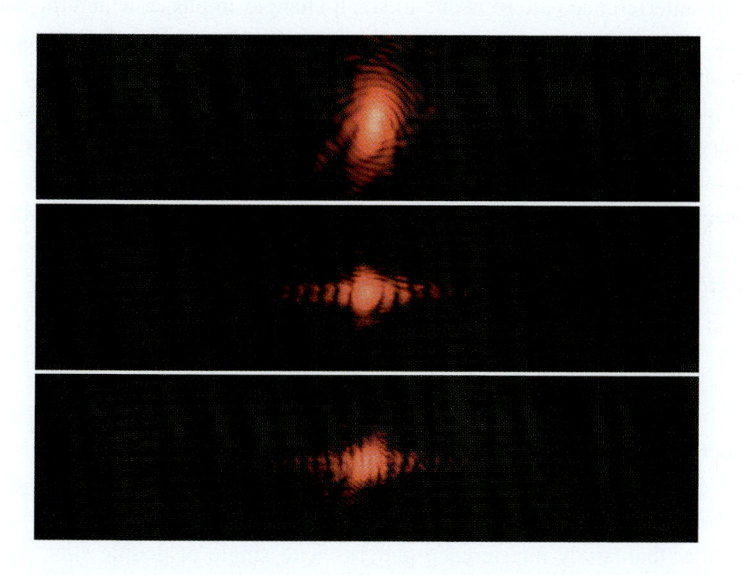

Figure 2.40. Interference patterns created by passing laser light through a single slit (middle) and a double slit (bottom), as shown in Figure 2.39. The laser's cross section is shown at the top.

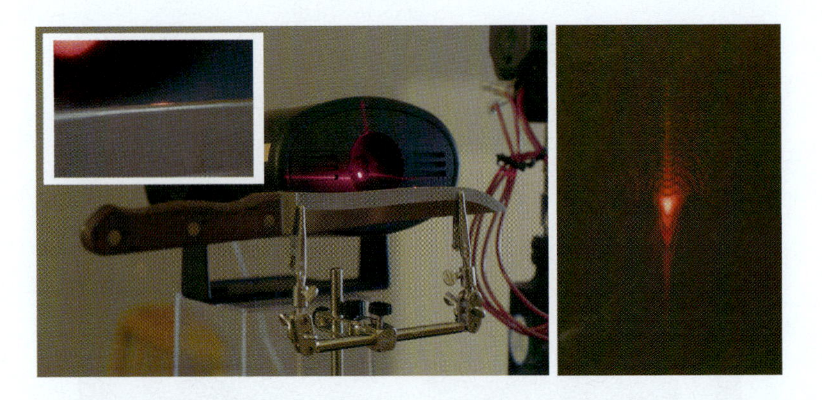

Figure 2.41. A laser is shone just over a knife's edge (left). This causes an interference pattern (right).

source over the edge of a knife, as shown in Figure 2.41, a diffraction pattern can be created. The right panel in this figure shows the pattern of light created on a wall several meters away from the laser and the knife. Below the center spot there is a rapid fall-off, whereas above the center spot a series of additional spots are visible.

This interference pattern is due to small changes in phase, which themselves are due to the light interacting with the knife's edge in a manner dependent on its minimum distance to the edge.

While pointing a laser at the edge of a knife is not a common everyday occurrence, light streaking past an edge is very common. The above experiment shows that shadow edges in practice are never completely sharp. This is in part because light sources are never infinitely far away, nor infinitely small, although even in such an idealized case, the diffraction around the edge will cause some interference and therefore softening of the edge.

2.7.3 Diffraction of Polychromatic Light

It can be seen that the interference patterns caused by edges, slits, and apertures of other shapes are dependent on the size of the aperture in relation to the wavelength. In the case of an edge, there is a dependency between wavelength and closest distance between the wavefront and the edge.

Rather than study interference patterns with single or double slits, we may construct a large number of parallel slits. Such an array of diffracting obstacles is called a *diffraction grating*. Another way of creating a diffraction grating is by creating parallel grooves in a transparent sheet of material. If the grooves are

Figure 2.42. A diffraction grating lit by a D65 daylight simulator (left). The rainbow patterns on both the reflected (middle) and transmitted sides (right) are shown.

close enough together ($D \approx \lambda$), light is diffracted into different directions based on its wavelength.

Polychromatic light sources consist of wavelengths distributed over the entire visible spectrum. Most common light sources, including daylight, are of this variety. Passing such light through a diffraction grating causes a rainbow pattern, as observed in Figure 2.42. As the sheet shown in this figure both transmits and reflects light, the rainbow pattern is seen on both sides.

Figure 2.43. A DVD lit by a D65 daylight simulator. The small pits encoding the data are of the order of a wavelength, and therefore cause a diffraction pattern resulting in a set of rainbow colors.

A more common occurrence of color caused by diffraction is shown in Figure 2.43, where a white light source is shone at a DVD. The pits encoding binary information in the reflecting layer are of the order of a wavelength and, therefore, cause a rainbow pattern.

2.8 Scattering

Many optical phenomena may be described as scattering, although it is often more convenient to use specialized theories. For instance, we have described reflection and refraction of electro-magnetic waves at a boundary between two dielectrics with different indices of refraction in Section 2.5. Similarly diffraction and interference can be seen as special cases of scattering [111].

Scattering is an interaction of light with particles such as single atoms, molecules, or clusters of molecules. Light can be absorbed by such particles and almost instantaneously be re-emitted. This repeated absorption and re-emission of light is called *scattering*. The direction of scattering may be different from the direction of the incident light, leading for instance to the translucent appearance of some materials as shown in Figure 2.44. Scattering events are called *inelastic* if the wavelength of the re-emitted light is similar to the wavelength of the incident light. If the wavelengths alter significantly, the scattering is called *elastic*.

Figure 2.44. This statue scatters light internally, resulting in a translucent appearance.

While a full account of scattering theory would be too involved for a book on color theory, in the following subsections we offer a simplified explanation of different types of scattering. More detailed accounts of light scattering can be found in "Light Scattering by Small Particles" [499], "Dynamic Light Scattering" [88], "Scattering Theory of Waves and Particles" [832] and "Electromagnetic Scattering in Disperse Media" [54].

2.8.1 Rayleigh Scattering

Consider a linearly polarized electric dipole with charge q. The magnitude of its moment p as a function of time t is then described by

$$p(t) = q\,d\,\cos(\omega t), \tag{2.169}$$

where d is the initial maximum distance between the positive and negative charges. The oscillation of this dipole gives rise to an electro-magnetic field with a relatively complicated structure near the dipole. At some distance away from the dipole, the field has a simpler structure whereby the **E** and **H** fields are transverse, perpendicular, and in phase [447]. The magnitude of the electric vector is given by

$$\|\mathbf{E}\| = \frac{qd\pi\sin(\theta)}{\lambda^2\varepsilon_0}\,\frac{\cos(\mathbf{r}\cdot\mathbf{s} - \omega t)}{r}, \tag{2.170}$$

where position $\mathbf{r}$ is a distance r away from the dipole. The irradiance E_e associated with this field is obtained by applying (2.96) and noting that the squared cosine term becomes 1/2 after time-averaging (see Section 2.4):

$$E_e = \frac{q^2 d^2 \pi^2 c \sin^2(\theta)}{2\lambda^4 r^2}. \tag{2.171}$$

This equation shows that a dipole oscillating with an angular frequency ω, generates an electro-magnetic field with an irradiance that falls off with the square of the distance and, as such, obeys the *inverse-square law*. We therefore speak of a *dipole radiator*.

In addition, the irradiance depends inversely on the fourth power of wavelength. Thus, the higher the frequency of the oscillations (smaller λ), the stronger the resulting field.

Dipole radiation can be induced in small particles in the presence of an electromagnetic field. Such particles thus scatter light at the blue end of the visible spectrum more than light with longer wavelengths. If these scattering particles are spaced in a medium at distances of at least one wavelength (in the visible region), then the scattered waves are essentially independent. Light passing through such

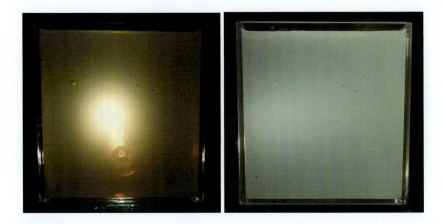

Figure 2.45. Rayleigh scattering demonstrated with a tank of water and milk. On the left, a D65 daylight simulator is placed behind the tank. The scattering causes a color that tends towards orange. The image on the right shows the same tank and contents lit from the side by the same light source. The substance now has a bluish tinge.

a medium will be absorbed and re-emitted in all directions. As a result, waves are independent from each other in all directions, except forward. Light scattered in the forward direction will generally maintain its phase coherence. This type of scattering is called *Rayleigh scattering*.

Rayleigh scattering can be demonstrated by shining light through a tank of water which contains a small amount of milk. The milk solution acts as a scattering medium. The light passing through in the forward direction will take on an orange tint, since longer wavelengths are scattered less than shorter wavelengths. Light scattered towards the side will have a blue tinge, as shown in Figure 2.45.

Another example of Rayleigh scattering is the blue color of some people's eyes (Figure 2.46). Both blue skies and orange and red sunrises and sunsets are explained by Rayleigh scattering as well. The latter example is discussed further

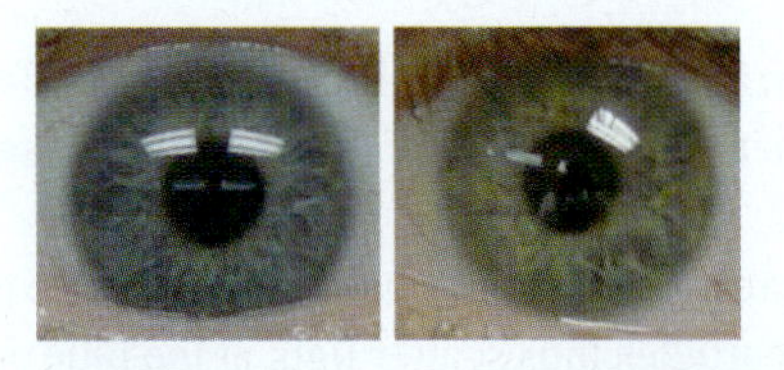

Figure 2.46. The blue color in the iris shown on the left is due to Rayleigh scattering. Green eyes, as shown on the right, occur due to a combination of Rayleigh scattering and staining due to melanin. (Photographs adapted from [667].)

Figure 2.47. The blue coloring of the feathers of this blue jay are caused by Rayleigh scattering; Yosemite National Park, California, 1999.

in Section 2.11. Rayleigh scattering is rare in the plant world, but surprisingly common among animals. It is for instance the cause of the blue color in blue jays (Figure 2.47).

2.8.2 Mie Scattering

While Rayleigh scattering affords a relatively simple model for small scattering particles, for media containing particles larger than about a wavelength this model is not appropriate. Instead, the more complicated Mie scattering model may be used [111, 112, 436, 779]. The complexity of this model is such that it is beyond the scope of this book to give a detailed mathematical account.

This complexity arises as follows. The model assumes that the particle is spherical, with a planar wave-front scattering off of it. The wave equations (2.36) and (2.37) are most conveniently restated in spherical coordinates. However, the plane wave is most naturally represented in Cartesian coordinates. Converting the plane wave representation to spherical coordinates is somewhat involved. Conversely, expressing the spherical boundary conditions in a Cartesian coordinate system is similarly involved [1233].

Figure 2.48. Large particles, such as water droplets, cause light to be scattered in a largely wavelength-independent manner. Hence, clouds appear a neutral gray or white; A Coruña, Spain, August 2005.

Figure 2.49. Early morning fog causing Mie scattering; Casselberry, FL, 2004.

Figure 2.50. Smoke machines used for visual effect at concerts and in clubs produce fine droplets that remain suspended in air, causing Mie scattering.

Mie scattering is only weakly dependent on wavelength and may scatter red and green wavelengths in specific directions [815]. This is called *polychroism* and occurs when the particles are larger than those causing Rayleigh scattering, but still relatively small. For large particles, the scattering is largely independent of wavelength, causing white light to remain white after scattering. As an example, Mie scattering causes clouds to appear white or gray, as shown in Figure 2.48. Mist and fog are other examples of Mie scattering (Figure 2.49), including artificially created fog such as that used at concerts (Figure 2.50).

A further natural diffraction-related phenomenon explained by Mie's theory, is the back scattering that occurs in clouds near the anti-solar point. A viewer observing clouds with the sun directly behind, most commonly observed from an airplane, may see a circular pattern called a *glory*. Dependent on the distance between the observer and the clouds, the center of the glory will be occupied by the shadow of the airplane. Figure 2.51 shows an example of a glory. Here the distance between the clouds and the plane is large, and therefore the shadow is negligibly small.

Figure 2.51. A glory; New Zealand, 2002. (Photograph courtesy of Timo Kunkel (www.timo-kunkel.com).)

2.9 Geometrical Optics

The wavelength of light in the visible range is normally very small (on the order of 10^{-7} meters) in relation to the size of the optical features of interest, which include surface boundaries, volumes of space, and lenses. In these cases, to a first approximation, it is possible to model the behavior of wavefronts by neglecting wavelengths. Thus, in the limit that λ goes to zero, light may be modeled in terms of geometrical behavior. When this is done, we speak of *geometrical optics*.

In this section, we assume that regions are free of currents and charges, i.e. $\mathbf{j} = 0$ and $\rho = 0$. The general form of the fields considered here is

$$\underline{\mathbf{E}} = \underline{e}(\mathbf{r}) e^{i\beta S(\mathbf{r})}, \tag{2.172a}$$

$$\underline{\mathbf{H}} = \underline{h}(\mathbf{r}) e^{i\beta S(\mathbf{r})}, \tag{2.172b}$$

where $S(\mathbf{r})$ is a real scalar function of space $\mathbf{r}$. The function $S(\mathbf{r})$ is known as the *optical path* or the *eikonal*. Vectors $\underline{e}(\mathbf{r})$ and $\underline{h}(\mathbf{r})$ are complex vector valued functions of space. Maxwell's equations for this particular solution can then be rewritten using the identities given in Section A.9 as

$$(\nabla S) \times \mathbf{h} + \varepsilon \mathbf{e} = -\frac{1}{i\beta} \nabla \times \mathbf{h}, \tag{2.173a}$$

$$(\nabla S) \times \mathbf{e} - \mu \mathbf{h} = -\frac{1}{i\beta} \nabla \times \mathbf{e}, \tag{2.173b}$$

$$\mathbf{e} \cdot (\nabla S) = -\frac{1}{i\beta} (\mathbf{e} \cdot \nabla \ln \varepsilon + \nabla \cdot \mathbf{e}), \tag{2.173c}$$

$$\mathbf{h} \cdot (\nabla S) = -\frac{1}{i\beta} (\mathbf{h} \cdot \nabla \ln \mu + \nabla \cdot \mathbf{h}). \tag{2.173d}$$

In the limit that the wavelength goes to zero, the value of β goes to infinity (because $\beta = \frac{2\pi}{\lambda}$). As a result Maxwell's equations simplify to

$$(\nabla S) \times \mathbf{h} + \varepsilon \mathbf{e} = 0, \tag{2.174a}$$

$$(\nabla S) \times \mathbf{e} - \mu \mathbf{h} = 0, \tag{2.174b}$$

$$\mathbf{e} \cdot (\nabla S) = 0, \tag{2.174c}$$

$$\mathbf{h} \cdot (\nabla S) = 0. \tag{2.174d}$$

If $\mathbf{h}$ from (2.174b) is substituted into (2.174a), the latter equation yields

$$\frac{1}{\mu} \left((\mathbf{e} \cdot (\nabla S)) \nabla S - \mathbf{e} (\nabla S)^2 \right) + \varepsilon \mathbf{e} = 0. \tag{2.175}$$

Substituting (2.174c) into this equation gives

$$(\nabla S)^2 = \mu \varepsilon, \tag{2.176a}$$

$$(\nabla S)^2 = n^2, \tag{2.176b}$$

or equivalently

$$\left(\frac{\partial S}{\partial x}\right)^2 + \left(\frac{\partial S}{\partial y}\right)^2 + \left(\frac{\partial S}{\partial z}\right)^2 = n^2(x, y, z). \tag{2.177}$$

This equation is known as the *eikonal equation*. The gradient of S can be seen as the normal vector of the wavefront. It can be shown that the average Poynting vector $\mathbf{S}$ is in the direction of the wavefront:

$$\mathbf{S} = \frac{\nabla S}{n}. \tag{2.178}$$

2.9.1 Fermat's Principle

The eikonal equation provides a description of geometrical optics. An alternative description is afforded by *Fermat's principle*, which states that light follows a trajectory (a ray) such that the optical path length is an extremum. The optical path length is defined as

$$\int_a^b n \, ds, \tag{2.179}$$

where ds is an element of arc length, and n is the index of refraction. The path has fixed end points a and b. Minimizing this integral can be achieved using the calculus of variations [736, 1082] resulting in

$$\frac{d}{d\mathbf{l}} \left(n(\mathbf{r}) \frac{d\mathbf{r}}{d\mathbf{l}} \right) = \nabla n(\mathbf{r}). \tag{2.180}$$

This equation is known as the *ray equation* [466, 617, 650, 1081] and is valid for inhomogeneous isotropic media that are stationary over time. This equation expresses the fact that at every point in the medium, the tangent and the normal vector associated with the ray path span a plane called the *osculating plane*, and the gradient of the refractive index must lie in this plane.

A consequence of Fermat's principle is that in a homogeneous medium where $n(\mathbf{r})$ is constant, light travels in a straight line. In inhomogeneous media, however, light may travel along curved arcs according to the above ray equation. This happens, for instance, in the atmosphere, as shown in Section 2.10.4. Finally, Snell's law for reflection and refraction can be derived from Fermat's principle (Section 2.5.1).

2.9.2 The Cosine Law of Illumination

The cosine law of illumination states that a surface can receive radiation only in proportion to its area projected in the direction of the light source. If the direction of the light source is given by $\mathbf{L}$ and the surface normal is $\mathbf{N}$, then the cosine of the *angle of incidence* θ is given by their dot product:

$$\cos(\theta) = \mathbf{N} \cdot \mathbf{L}. \tag{2.181}$$

The projected area is then

$$A' = A\, \mathbf{N} \cdot \mathbf{L} = A\, \cos(\theta), \tag{2.182}$$

as illustrated in Figure 2.52. In this figure, a large surface is angled away from the light source, whereas a smaller surface is aimed directly at the light source. The

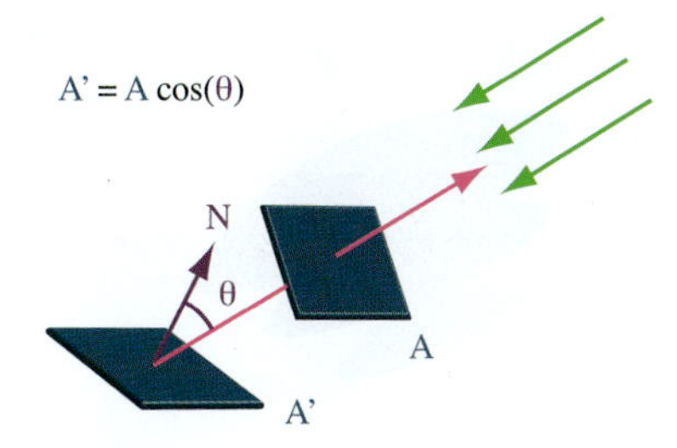

Figure 2.52. A surface can only receive radiation proportional to its projected area in the direction of the light source.

projected area of the larger surface in the direction of the light source is identical to the projected area of the smaller surface, and therefore both surfaces receive the same amount of radiation.

A similar law called the *cosine law of emission* states that radiation emitted from *iso-radiant* or *Lambertian* surfaces [8] decreases with a factor of $cos(\theta)$, where θ is the *angle of exitance*. An example is shown in Figure 2.53, where the radiation emitted from a point decreases as the angle of exitance increases.

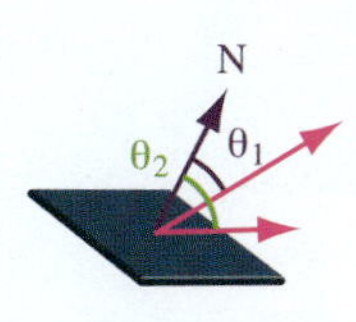

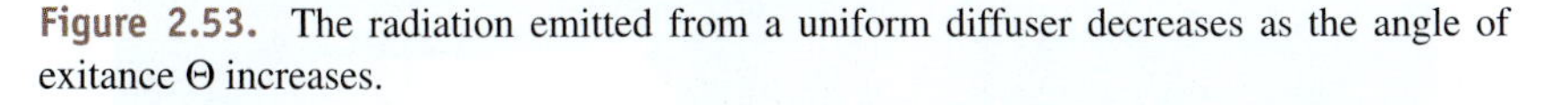

Figure 2.53. The radiation emitted from a uniform diffuser decreases as the angle of exitance Θ increases.

Surfaces that obey the cosine law of emission appear equally light regardless of the viewing direction. Although no perfect Lambertian surfaces exist in practice, many surfaces are well approximated by this law. These are called *matte* or *diffuse* surfaces.

2.9.3 The Inverse Square Law

In Section 2.1.1, we saw that according to Coulomb's law two charges exert a force on each other that diminishes with the square of their distance. The cause

[8] Strictly speaking, these surfaces are iso-radiant. Although a surface might be Lambertian at each point individually, there may be spatial variation in the amount that is emitted. Iso-radiance requires uniformity across the entire surface [839].

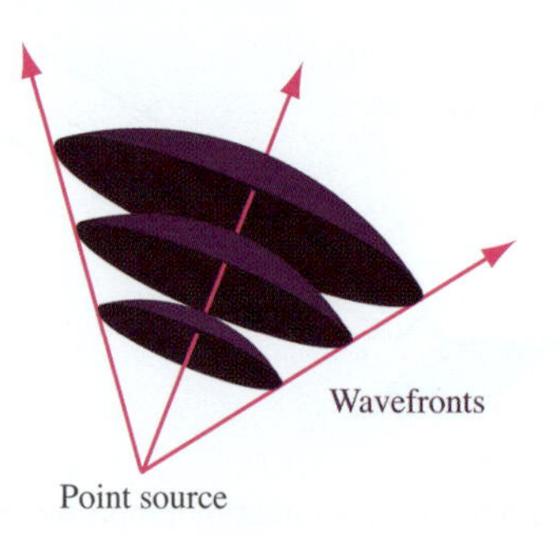

Figure 2.54. Wavefronts diverging spherically from a point source.

of this relation lies in a more general law which stems from geometrical considerations; this more general law is called the *inverse square law*[9].

Consider a point source that emits light. This light propagates through space as spherical wavefronts. The farther the wave has traveled, the greater the surface area spanned by the wave. The energy carried by the wave is thus distributed over increasingly large areas, and therefore the energy density is reduced (Figure 2.54).

Figure 2.55. The objects farther from the light source (a 25W omni-directional incandescent light) are dimmer than the objects closer to it due to the inverse square law.

[9]This law applies to other modalities as well, including gravity and sound.

The inverse square law states that the surface density of radiant energy emitted by a point source decreases with the squared distance d between the surface and the source:

$$E_e \propto \frac{1}{d^2}. \tag{2.183}$$

An example is shown in Figure 2.55 where the mugs farther from the light source are dimmer than the objects nearer to it.

As a general rule of thumb, if the distance of an observer to a source is greater than 10 times the largest dimension of the source, it can be approximated as a point source with respect to the observer [397]. The error associated with this approximation is important in radiometry and will be further discussed in Section 6.2.11.

2.9.4 Bouguer, Beer, and Lambert's Laws

Light traveling through a dielectric medium will occasionally collide with a particle and either undergo scattering, or be absorbed and turned into a small amount of heat. Both phenomena are discussed in detail in the following chapter. However, the effect of either result of the collision is that a photon has a probability of interacting with the medium for every unit of distance traveled.

Hence, the irradiance after traveling a distance Δs through the dielectric material will be attenuated by some amount:

$$E_e\left(s+\Delta s\right) = E_e\left(s\right)\,\left(1-\sigma_a \Delta s\right). \tag{2.184}$$

Here, σ_a is a wavelength-dependent quantity modeling the attenuation induced by the medium. It is known as the *spectral absorbtivity*. Rewriting this equation and taking the limit $\Delta s \rightarrow 0$, we find

$$\lim_{\Delta s \rightarrow 0} \frac{E_e\left(s+\Delta s\right) - E_e\left(s\right)}{\Delta s} = -\sigma_a\, E_e\left(s\right), \tag{2.185a}$$

$$\frac{d\,E_e\left(s\right)}{ds} = -\sigma_a\, E_e\left(s\right). \tag{2.185b}$$

Assuming that the medium is homogeneous and isotropic, this equation can be solved to yield

$$E_e\left(s\right) = E_e(0)\, e^{-\sigma_a s}. \tag{2.186}$$

Thus, the attenuation in an absorbing dielectric material is exponential. This result is known as *Bouguer's law*. The ratio $E_e(s)/E_e(0)$ is called the transmittance T:

$$T = \frac{E_e(s)}{E_e(0)} \tag{2.187a}$$

$$= e^{-\sigma_a s}. \tag{2.187b}$$

A related quantity is absorbance A, written as

$$A = -\ln \frac{E_e(s)}{E_e(0)} \tag{2.188a}$$

$$= \sigma_a s. \tag{2.188b}$$

In spectroscopy and spectrophotometry, transmittance and absorbance are often expressed in powers of 10. The conversion between the two forms is a factor of $\ln(10) \approx 2.3$:

$$T = 10^{-\sigma_a s/\ln(10)}, \tag{2.189a}$$

$$A = \sigma_a s/\ln(10). \tag{2.189b}$$

The attenuation σ_a can be split into a material- dependent constant ε, which is called either molar absorbtivity, extinction coefficient, or the concentration of the absorbing material c. The absorbance is then given as

$$A = \varepsilon c s, \tag{2.190}$$

and the transmittance is

$$T = 10^{-\varepsilon c s}. \tag{2.191}$$

This relation is known as Beer's law. Here, we have merged the $\ln(10)$ factor with the constant ε. As both concentration c and pathlength s have the same effect on absorbance and transmittance, it is clear that a given percent change in absorbance or transmittance can be effected by a change in pathlength, or a change in concentration, or a combination of both.

Beer's law is valid for filters and other materials where the concentration c is low to moderate. For higher concentrations, the attenuation will deviate from that predicted by Beer's law. In addition, if scattering particles are present in the medium, Beer's law is not appropriate. This is further discussed in Section 3.4.6.

Finally, if the pathlength is changed by a factor of k_1 from s to $k_1 s$, then the transmittance changes from T to T' as follows:

$$T' = T^{k_1}, \tag{2.192}$$

known as Lambert's law. Similarly, if the concentration is changed from c to $k_2 c$, the transmittance goes from T to T'':

$$T'' = T^{k_2}. \tag{2.193}$$

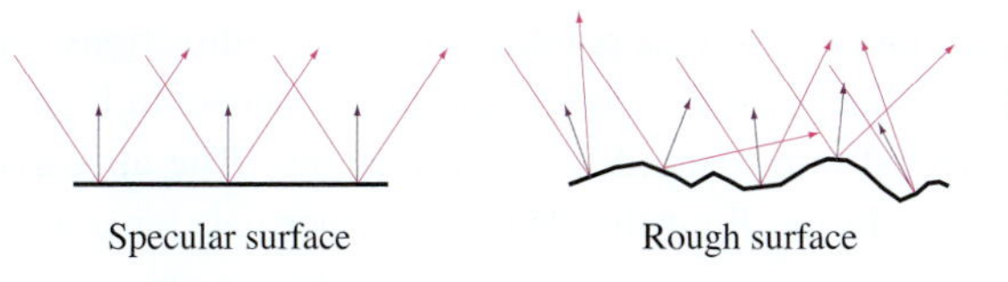

Figure 2.56. A perfect specular reflector (left) and a rough surface (right). Light incident from a single direction is reflected into a single direction for specular surfaces, whereas rough surfaces reflect light into different directions.

2.9.5 Surface Reflectance

We have shown that when a plane wave is incident upon a boundary between two dielectrics, two new waves are created. These new waves travel in the reflected direction and the transmitted direction. For a boundary between a dielectric and a conductive object, such as a metal, the refracted wave is dampened and, therefore, travels into the material for a short distance. Reflected and refracted waves traveling in directions as predicted by Snell's law are called *specular*.

There are several aspects of material appearance that are not explained by this model. For instance, it is not clear why shining a white light at a tomato produces a red color; nor are diffuse, glossy, or translucent materials explained. While the color of an object depends on the atomic structure of the material (this is the topic of Chapter 3), non-specular surface reflectance and transmittance are the cause of microscopic variations in surface orientation.

Thus, the occurrence of different levels of shininess depends on surface features that are larger than single atoms and molecules, but still below the resolving power of the human visual system. Specular reflection can, therefore, only occur when the boundary is perfectly smooth at the microscopic level, i.e., the surface normal is constant over small surface areas. At the macroscopic level, the surface normal may still vary; this is then interpreted by the human visual system as a curving but smooth surface.

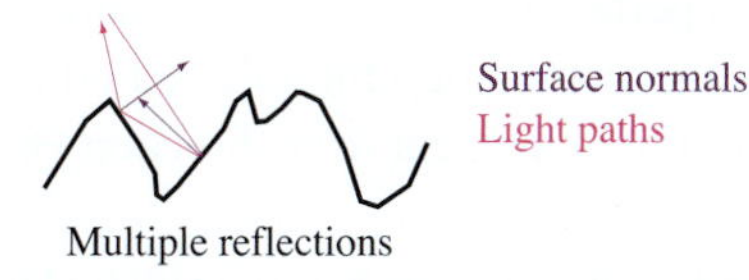

Figure 2.57. Self-occlusion occurs when reflected rays are incident upon the same surface and cause a secondary reflection.

A rough surface has surface normals pointing in different directions. Since at each point of the surface, light is still reflected specularly, shown in Figure 2.56, a beam of light is reflected into many different directions. The appearance of such an object is glossy. In the limit that the surface normals have a uniform distribution over all possible directions facing away from the object, light will be reflected in all directions equally. Such surfaces are perfectly diffuse and are called *Lambertian* surfaces.

The distribution of surface undulations can be such that reflected rays are incident upon the same surface before being reflected again away from the surface. Such *self-occlusion* is shown in Figure 2.57.

2.9.6 Micro-Facets

The most direct way to model the macroscopic reflectance behavior of surfaces is to treat a surface as a large collection of very small flat surfaces, called *micro-facets*. It is generally assumed that each facet is perfectly specular, as outlined in the preceding section. The orientation of each facet could be explicitly modeled, but it is more efficient to use a statistical model to describe the distribution of orientations. An example is the first micro-facet model, now named the Torrance-Sparrow model [1140].

Blinn proposed to model the distribution of orientations $D(\Theta_\mathrm{i}, \Theta_\mathrm{o})$ as an exponential function of the cosine between the half-angle and the aggregated surface normal [105]. The half-angle vector Θ_h is taken between the angle of incidence Θ_i and the angle of reflection Θ_o:

$$D(\Theta_\mathrm{i}, \Theta_\mathrm{o}) = (\Theta_\mathrm{h} \cdot \mathbf{n})^e \tag{2.194a}$$

$$\Theta_\mathrm{h} = \frac{\Theta_\mathrm{i} + \Theta_\mathrm{o}}{||\Theta_\mathrm{i} + \Theta_\mathrm{o}||}. \tag{2.194b}$$

The exponent e is a user parameter.

Different micro-facet models make different assumptions, for instance, on the existence of self-occlusion. Another common assumption is that micro-facets form a regular grid of V-shaped grooves. One model, namely Oren-Nayar's, assumes that each facet is a perfectly diffuse reflector, rather than a specular reflector [859]. In this model, the amount of light reflected off a surface in a particular direction Θ_o as function of light incident from direction Θ_i is given by

$$f_r(\Theta_\mathrm{i}, \Theta_\mathrm{o}) = \frac{\rho}{\pi} \left(A + B \max\left(0, \cos(\phi_\mathrm{i} - \phi_\mathrm{o}) \sin(\alpha) \tan(\beta)\right)\right), \tag{2.195a}$$

$$A = 1 - \frac{\sigma^2}{2\sigma^2 + 0.66}, \tag{2.195b}$$

$$B = \frac{0.45\sigma^2}{\sigma^2 + 0.09}, \tag{2.195c}$$

$$\alpha = \max(\theta_\text{i}, \theta_\text{o}), \tag{2.195d}$$

$$\beta = \min(\theta_\text{i}, \theta_\text{o}). \tag{2.195e}$$

In this set of equations, σ is a free parameter modeling the roughness of the surface. The factor ρ is the (diffuse) reflectance factor of each facet. The directions of incidence and reflection (Θ_i and Θ_o) are decomposed into polar coordinates, where polar angle θ is the elevation above the horizon and azimuth angle ϕ is in the plane of the facet:

$$\Theta = (\theta, \phi). \tag{2.196}$$

The above micro-facet models are *isotropic*, i.e., they are radially symmetric. This is explicitly visible in the Oren-Nayar model where the result depends only on the difference between ϕ_i and ϕ_o, but not on the actual values of these two angles. This precludes modeling of *anisotropic materials*, where the amount reflected depends on the angles themselves. A characteristic feature of anisotropic materials is, therefore, that rotation around the surface normal may alter the amount of light reflected. An often-quoted example of an anisotropic material is brushed aluminium, although it is also seen at a much larger scale for example in pressed hay as shown in Figure 2.58. Several micro-facet models that take anisotropy into account are available [45, 46, 702, 921, 1005, 1212].

Figure 2.58. Pressing hay into rolls, as shown here, produces an anisotropic effect on the right side due to the circular orientation of each of the grass leaves; Rennes, France, June 2005.

2.9.7 Bi-Directional Reflectance Distribution Functions

A further abstraction in the modeling of surface reflectance is afforded by *bi-directional reflectance distribution functions*, or BRDFs for short [838]. The micro-facet models presented in the preceding section can be seen as a specific class of BRDF. In general, BRDFs are functions that return for each possible angle of incidence and angle of reflectance, the fraction of light that is reflected at the surface.

For BRDFs to be plausible models of physical reality, at the very least they should be reciprocal as well as energy conserving. Reciprocal means that if the angle of incidence and reflectance are reversed, the function returns the same value:

$$f_r(\Theta_\mathrm{i}, \Theta_\mathrm{o}) = f_r(\Theta_\mathrm{o}, \Theta_\mathrm{i}). \tag{2.197}$$

The energy-conserving requirement refers to the fact that light incident upon a surface is either absorbed, transmitted, or reflected. The sum of these three components can not be larger than the incident amount. Light emitted by self-luminous surfaces is modeled separately and does not form part of a BRDF.

For computational convenience, the reflection of light may be classified into four broad categories: diffuse, glossy specular, perfect specular, and retro-reflective, although most surfaces exhibit a combination of all four types [895].

Fully specular BRDFs behave like the reflection detailed in Section 2.5, i.e., the Poynting vector of incident light is mirrored in the surface normal to determine the outgoing direction [666]:

$$f_r(\Theta_\mathrm{i}, \Theta_\mathrm{o}) = \frac{\delta(\theta_\mathrm{o} - \theta_\mathrm{i})\; \delta(\phi_\mathrm{o} - \phi_r - \pi)}{\sin(\theta_r)\; \cos(\theta_r)}, \tag{2.198}$$

where we have used (2.196), and $\delta()$ is the Kronecker delta function. At the other extreme, *Lambertian* surface models assume that light incident from all different directions is reflected equally [645]:

$$f_r(\Theta_\mathrm{i}, \Theta_\mathrm{o}) = \frac{1}{\pi}. \tag{2.199}$$

Although this is not a physically plausible model, it is a reasonable approximation for several materials, including matte paint.

An early model of glossy reflection was presented by Phong [897] and is given here in modified form with the coordinate system chosen such that x and y are in the plane of the surface and the z-coordinate is along the surface normal [895]:

$$f_r(\Theta_\mathrm{i}, \Theta_\mathrm{o}) = \left(\Theta_\mathrm{i} \cdot (-\Theta_{ox}, -\Theta_{oy}, \Theta_{oz})^T\right)^e. \tag{2.200}$$

Here, the outgoing direction Θ_o is scaled first by $(-1,-1,1)^T$, thus mirroring this vector in the surface normal. It is now possible to make this scaling a user parameter, allowing the modeling of various types of off-specular reflection:

$$f_r(\Theta_i, \Theta_o, \mathbf{s}) = \left(\Theta_i \cdot \left(\mathbf{s}\,\Theta_o^T\right)\right)^e. \tag{2.201}$$

While this BRDF could be used to model glossy materials, there is a further refinement that can be made by realizing that this function specifies a single lobe around the angle of incidence. More complicated BRDFs which are still physically plausible can be created by modeling multiple lobes and summing them:

$$f_r(\Theta_i, \Theta_o) = \frac{\rho_d}{\pi} + \sum_{i=1}^{n} \left(\Theta_i \cdot \left(\mathbf{s_i}\,\Theta_o^T\right)\right)^{e_i}. \tag{2.202}$$

The term $\frac{\rho_d}{\pi}$ models a diffuse component. Each of the n lobes is modeled by a different scaling vector $\mathbf{s_i}$ as well as a different exponent e_i. The result is called the *Lafortune model* [637] .

An empirical anisotropic reflectance model takes into account the orientation of the surface. The amount of light reflected into Θ_o depends on the angle of rotation ϕ_n around the surface normal. Defining directional vectors for the incoming and outgoing directions ($\mathbf{v}_{\Theta_i}$ and $\mathbf{v}_{\Theta_o}$), the half-vector $\mathbf{v}_h$ is defined by

$$\mathbf{v}_h = \frac{\mathbf{v}_{\Theta_i} + \mathbf{v}_{\Theta_o}}{\|\mathbf{v}_{\Theta_i} + \mathbf{v}_{\Theta_o}\|}. \tag{2.203}$$

The angle Θ_{nh} is between the half-vector and the surface normal. Ward's anisotropic BRDF is then given by [269, 1200, 1212]

$$f_r(\Theta_i, \Theta_o, \mathbf{s}) = \frac{\rho_d}{\pi} + \frac{\rho_s}{\sqrt{\cos(\theta_i)\,\cos(\theta_o)}}\,\frac{A}{4\pi\,\sigma_x\,\sigma_y}, \tag{2.204a}$$

where

$$A = \exp\left(-\tan\left(\Theta_{nh}\left(\cos^2\left(\frac{\phi_n}{\sigma_x^2}\right) + \sin^2\left(\frac{\phi_n}{\sigma_y^2}\right)\right)\right)\right). \tag{2.204b}$$

The specular and diffuse components are modeled by ρ_s and ρ_d. Their sum will be less than 1. The function is parameterized in σ_x and σ_y, each of which are typically less than 0.2. These two parameters quantify the level of anisotropy in the model. If they are given identical values, i.e., $\sigma = \sigma_x = \sigma_y$, the model simplifies to an isotropic model:

$$f_r(\Theta_i, \Theta_o, \mathbf{s}) = \frac{\rho_d}{\pi} + \frac{\rho_s}{\sqrt{\cos(\theta_i)\,\cos(\theta_o)}}\,\frac{\exp\left(-\tan\left(\frac{\Theta_{nh}}{\sigma^2}\right)\right)}{4\pi\,\sigma^2}. \tag{2.205}$$

Material	ρ_d	ρ_s	σ_x	σ_y
Lightly brushed aluminium	0.15	0.19	0.088	0.13
Rolled aluminium	0.10	0.21	0.04	0.09
Rolled brass	0.10	0.33	0.05	0.16
Enamel finished metal	0.25	0.047	0.080	0.096

Table 2.5. Example parameter settings for Ward's model of anisotropic reflection (After [1212]).

Parameters measured for several materials are listed in Table 2.5.

Although BRDFs are typically modeled as the ratio of incident to reflected light and depend on incoming and outgoing directions, they can be extended to be wavelength dependent, i.e., $f_r(\Theta_\mathrm{i}, \Theta_\mathrm{o}, \lambda)$. In addition, polarization can be accounted for in formulations related to BRDFs; these formulations are then called *bi-directional surface scattering reflectance distribution functions* (BSS-RDF) [36].

2.10 Application: Image Synthesis

Light travels normally in straight lines until it reaches a boundary. When such a boundary is encountered, BRDFs describe the aggregate behavior of a differential surface area, which may be thought of consisting of a large number of differently oriented boundaries.

These two observations together may be used to compute the distribution of light over a scene. Within a closed environment that contains both light sources as well as reflective materials (such as a room, or approximately any outdoor scene), light is at *equilibrium*, i.e., the amount of light emitted by all the light sources together equals the amount of light absorbed by all the surfaces. Equilibrium is reached after a very short period of time after switching on a light source.

For any point **P** on a surface, we already know how it reflects light; this is given by the BRDF. To compute how the ability to reflect light contributes to the light incident upon another point in the scene, we need to compute from where in the scene light arrives at **P**. In other words, to compute how much light is reflected in a given Θ_o direction, we need to integrate over all angles of incidence[10]:

$$L_\mathrm{o}(\mathbf{P}, \Theta_\mathrm{o}) = \int_{\Omega_\mathrm{i}} f_r(\mathbf{P}, \Theta_\mathrm{i}, \Theta_\mathrm{o})\, L_\mathrm{i}(\mathbf{P}', \Theta_\mathrm{i}) \cos(\theta_\mathrm{i})\, d\omega_\mathrm{i}, \tag{2.206}$$

[10]We follow the notation used by Arjan Kok in his PhD thesis [603]. We use the non-descript term "light" here, indicated with the symbol L, since the appropriate radiometric term, radiance, is not defined yet. Its definition can be found in Section 6.2.7.

where $\mathbf{P}'$ are points on other surfaces in the scene that contribute to the incident light. The total amount of light traveling from point $\mathbf{P}$ in direction Θ_o is then simply the sum of all reflected light into that direction plus the light emitted at surface point $\mathbf{P}$. The latter value is only non-zero for self-luminous surfaces. The equation governing light transport in a scene is known as the *rendering equation* and is given by [568]

$$L_o(\mathbf{P},\Theta_o) = L_e(\mathbf{P},\Theta_o) + \int_{\Omega_i} f_r(\mathbf{P},\Theta_i,\Theta_o) L_i(\mathbf{P}',\Theta_i) \cos(\theta_i) d\omega_i. \qquad (2.207)$$

Although not explicitly shown, the terms in this integral, specifically L_o, L_i, L_e, and f_r, are wavelength-dependent. In later chapters, it is shown that human vision encodes color as a triplet of values, and therefore renderers can be optimized with little harm to sample three color bands, rather than many separate wavelengths. This does reduce the repertoire of visual phenomena that can be rendered, though. For instance, dispersion that occurs due to a wavelength-dependent index of refraction can only be modeled within a spectral renderer.

Evaluation of (2.207) is difficult in practice for several reasons. The quantity computed also appears as a factor under the integral. This makes (2.207) a Fredholm integral of the second kind, a type of equation that needs to be solved numerically for all but the simplest integration domains.

However, in practice, the domain of integration is over a hemisphere placed over point $\mathbf{P}$ which is visible by a good part of the scene. Thus, numerical integration involves sampling the environment to compute the factor $L_i(\mathbf{P}',\Theta_i)$ for a large set of directions Θ_i and then multiplying by the BRDF f_r.

Since most of the environment contributes some quantity of light to the reflection in point $\mathbf{P}$, rendering algorithms that compute the light interaction between a point in the scene with every other point in the scene are called *global illumination* algorithms. Global illumination solutions differ in the way that the rendering equation is approximated, where each affords a different trade-off between rendering accuracy and computational cost [370, 895, 1043].

Common techniques include Monte Carlo sampling [271], ray tracing [370, 1042, 1102, 1242], radiosity [380, 1046], and photon mapping [549]. None of these techniques currently run in real time, perhaps with the exception of ray tracing, which is on the brink of achieving interactive rates [875, 876, 1192, 1193].

2.10.1 Ray Tracing

The ray-tracing algorithm simulates the flux of light by recursively tracing rays through an environment [1242]. A view point and view plane with associated parameters is specified within the 3D environment that is to be rendered. This serves to specify what will be visible in the image, much like holding a photo camera in a real environment determines the composition of a photograph.

Rather than start at the light sources and follow the path that photons take through the scene until they hit the view plane, the process is usually reversed and light is traced backwards from the view point through the view plane into the scene. This optimization helps reduce the number of rays that needs to be traced, as there is no guarantee that any ray started at a light source will ever reach the view plane.

When a ray hits a surface at a point **P**, called the *intersection point*, the rendering equation can be evaluated by sampling the environment with further (secondary) rays. This sampling is potentially costly, as most of the environment would have to be sampled to determine which parts of the scene illuminated the intersection point.

To reduce the cost of sampling the environment, it is noted that some parts of the scene contribute more to the illumination of a point than others. In particular, nearby light sources that are directly visible from **P** are likely to contribute the bulk of illumination. Similarly, unless the material is Lambertian, light arriving from the reflected direction is more likely to contribute than that from other directions. By splitting the integral in (2.207) into separate components, an efficient approximate solution can be found:

$$\begin{aligned} L_o(\mathbf{P},\Theta_o) = {} & L_e(\mathbf{P},\Theta_o) \\ & + \sum_S \int_{\mathbf{P}_S \in S} v(\mathbf{P},\mathbf{P}_S)\; f_r^{\text{diff}}(\mathbf{P})\; L_e\left(\mathbf{P}_S,\Theta_o'\right)\; \cos(\theta_S)\, d\omega_S \\ & + \int_{\Theta_R \in \Omega_R} f_r^{\text{spec}}(\mathbf{P},\Theta_R,\Theta_o)\; L_o(\mathbf{P}_R,\Theta_R)\; \cos(\theta_R)\, d\omega_R \\ & + \rho_d(\mathbf{P})\; L_a(\mathbf{P}). \end{aligned} \tag{2.208}$$

In this equation, the first term signifies the light emitted from point **P**, as before. The second term samples all the light sources directly. In it, $v(\mathbf{P},\mathbf{P}_S)$ is the *visibility term* which equals 1 if there is a direct line of sight between **P** and the sample point $\mathbf{P}_S$ on the light source. The visibility term is 0 otherwise. In (2.208), the third term samples a small set of directions centered around the reflected direction to account for specular and possibly glossy reflection. All other light directions are not sampled, but are approximated with a constant, the fourth term in (2.208). This last term is named the *ambient term*. It should also be noted that the BRDF is split into a specular and diffuse component (f_r^{spec} and f_r^{diff}) for the sake of convenience.

Figure 2.59. Diffuse inter-reflection between the rocks creates an intense orange glow; Bryce Canyon, Southern Utah. (Image courtesy of the National Park Service (http://www.nps.gov/archive/brca/photogalleryarchive.htm).)

2.10.2 Radiosity

While ray tracing is able to produce serviceable images in many applications, the approximations are such that many light paths are not sampled. For instance, light reflected from a Lambertian surface incident upon a second Lambertian surface, i.e., *diffuse inter-reflection*, is not accounted for. Instead, in ray tracing this component is approximated with a single constant, the fourth term in (2.208). An example of diffuse inter-reflection in nature is shown in Figure 2.59.

Under the assumption that an environment consists of Lambertian surfaces, BRDFs for all surfaces are simplified to be constant:

$$f_r(\mathbf{P}) = \frac{\rho_d(\mathbf{P})}{\pi}. \tag{2.209}$$

The rendering equation 2.207 then simplifies to become the *radiosity equation* [200]:

$$\begin{aligned} L(\mathbf{P}) = {} & L_e(\mathbf{P}) \\ & + \frac{\rho_d(\mathbf{P})}{\pi} \int_{\text{all }\mathbf{P}'} L(\mathbf{P}') \, \frac{\cos(\Theta_\mathrm{i}) \cos(\Theta'_\mathrm{o})}{\|\mathbf{P}' - \mathbf{P}\|^2} \, v(\mathbf{P}, \mathbf{P}') \, dA', \end{aligned} \tag{2.210}$$

where the integration is over differential surface areas dA'. To make the solution to this equation tractable, the scene is usually subdivided into small patches A_i, where the energy associated with each patch is assumed to be constant. For a patch i this yields the following equation:

$$L_i = L_{e,i} + \frac{\rho_{d,i}}{\pi} \sum_j \frac{L_j}{A_i} \int_{A_i} \int_{A_j} \frac{\cos(\Theta_i)\cos(\Theta_j)}{\pi r^2} \delta_{ij} dA_j dA_i. \quad (2.211)$$

In this equation, r is the distance between the patches i and j and δ_{ij} gives the mutual visibility between the delta areas of the patches i and j. This equation can be rewritten as

$$L_i = L_{e,i} + \rho_{d,i} \sum_j L_j f_{i \to j}, \quad (2.212a)$$

$$f_{i \to j} = \frac{1}{A_j} \int_{A_i} \int_{A_j} \frac{\cos(\Theta_i)\cos(\Theta_j)}{\pi r^2} \delta_{ij} dA_j dA_i. \quad (2.212b)$$

Here, the *form factor* $f_{i \to j}$ is the fraction of power leaving patch i that arrives at patch j. Form factors depend solely on the geometry of the environment, i.e., the size and the shape of the elements and their orientations relative to each other. Therefore, the radiosity method is inherently view independent. It is normally used as a preprocessing step to compute the diffuse light distribution over a scene and is followed by a rendering step that produces the final image (or sequence of images).

2.10.3 Photon Mapping

Light reflected off curved surfaces incident upon a Lambertian surface, or light transmitted through a transparent object and incident upon a Lambertian surface, are not accounted for in ray tracing and radiosity. These light paths create *caustics*, patterns of light and dark, as for instance cast by the transparent objects in Figure 2.60.

In ray tracing, where light paths are traced backwards, it is not possible to determine in which direction to sample a light source, if a transparent object is located along the line of sight between an intersection point and the light source. The refraction caused by the transparent object would veer the ray off course and, in all likelihood, it would miss the light source.

Thus, to properly account for caustics, rays should be traced from the light source into the scene. However, as argued before, the probability that any of these rays would hit the view plane is remote at best. A useful intermediate solution is to begin by tracing rays/photons from the light source into the scene and deposit

Figure 2.60. Caustics cast onto the table by various objects.

energy whenever such rays hit a Lambertian surface. The data structure used to store the energy of photons is called a *photon map* [549].

As light is properly refracted and reflected during this initial pass, the photon map accurately encodes the caustics caused by transparent and specular objects.

During a second pass, conventional ray tracing may be used in a slightly modified form. Whenever a Lambertian surface is intersected, rather than continue as normal, the photon map may be sampled instead. Variations are possible where rather than storing and sampling energy in a photon map, angles of incidence are recorded. In the second pass, rather than aiming rays directly at the light sources, the values in the photon map are read to determine secondary ray directions. In either solution, caustics can be rendered. In addition, photon mapping may be extended to simulate the atmosphere as detailed in Section 2.11.

2.10.4 Rendering Inhomogeneous Media

The earth's atmosphere typically has a varying index of refraction. This leads to curved ray paths as discussed in Section 2.9.1. The density of air is determined by

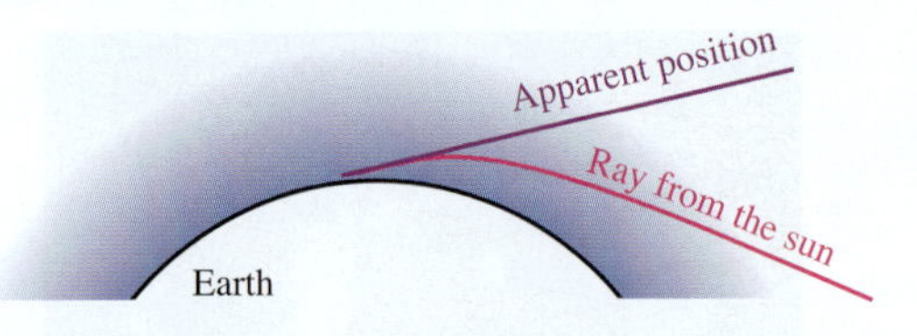

Figure 2.61. The varying density of the atmosphere causes the sun to be in a different position than where it appears to be.

its temperature—the higher the temperature, the lower its density. The index of refraction in turn depends on the density of the medium. Within a medium with a smoothly varying refractive index, rays bend towards the area with a greater index of refraction [112]. For the earth's atmosphere, this has the implication that when the sun is seen just above the horizon, it is in fact just below the horizon, as shown in Figure 2.61.

In the atmosphere, the density of air is mediated by both pressure and temperature, with both normally being highest near the earth's surface. However, local variations in temperature can bend rays in unexpected directions, yielding effects such as mirages and fata morganas [783]. The inhomogeneous density of air results in a spatially varying index of refraction.

To bend rays along their path, after advancing each ray by a small distance, its direction $\mathbf{l}$ needs to be recomputed as a function of the local gradient of the refractive index $\nabla n(\mathbf{r})$. This can be achieved by evaluating the ray equation (2.180).

If sunlight heats the ground, then close to the ground the index of refraction changes more rapidly than higher above the ground. If an observer looks at a distant object, then this may give rise to multiple paths that light rays may take between the object and the observer. As an example, Figure 2.62 shows how a curved path close to the ground plane may carry light between the roof and

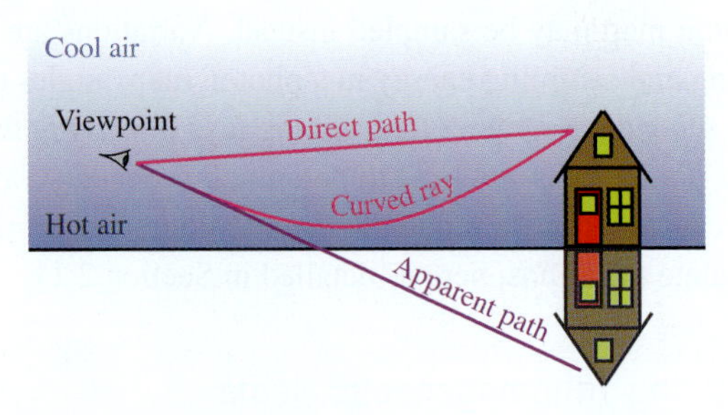

Figure 2.62. The roof of the building is visible along a direct path, as well as along a curved path. The latter gives the impression of the building being reflected in the ground plane.

Figure 2.63. Rendering of an inferior mirage. (Image courtesy of Diego Gutierrez, Adolfo Munoz, Oscar Anson, and Francisco Seron; Advanced Computer Graphics Group (GIGA), University of Zaragoza, Spain [411].)

the observer, whereas a straighter path farther from the ground also carries light between the roof and the observer. The effect is that the building is visible both upright and inverted. This is known as a *mirage*.

Dependent on atmospheric conditions, layers of colder and warmer air sometimes alternate, leading to different types of mirages. The configuration shown in Figure 2.62 leads to an *inferior mirage*. A rendered example is shown in Figure 2.63. An inversion layer may cause the temperature at some altitude to be higher than at ground level, and this gives rise to a *superior mirage*, as rendered in Figure 2.64. Finally, if several such inversion layers are present, observers may see a *fata morgana*, as shown in Figure 2.65. Further atmospherical phenomena, including rainbows, halos, and scattering, are discussed in Section 2.11.

Figure 2.64. Rendering of a superior mirage. (Image courtesy of Diego Gutierrez, Adolfo Munoz, Oscar Anson, and Francisco Seron; Advanced Computer Graphics Group (GIGA), University of Zaragoza, Spain [411].)

Figure 2.65. Rendering of a fata morgana. (Image courtesy of Diego Gutierrez, Adolfo Munoz, Oscar Anson, and Francisco Seron; Advanced Computer Graphics Group (GIGA), University of Zaragoza, Spain [411].)

2.10.5 Optimizations

Many optimization techniques have been applied to make rendering algorithms faster, including spatial subdivision techniques, which essentially sort the objects in space allowing intersection points to be computed without testing each ray against all objects [437]. Other strategies include distributed and parallel processing, allowing multiple processing units to partake in the computations [161].

As sampling is a core technique in rendering algorithms, optimizing sampling strategies has received a great deal of attention. In particular, from (2.207) we have that each point in the scene contributes some amount of light to be reflected in a direction of our choosing. Using pre-processing techniques, sampling may be directed towards those parts of the scene that contribute most to the final result. Such techniques are collectively called *importance sampling*. An example is Ostromoukhov's technique, which is based on Penrose tiling [863].

At the same time, the BRDF kernel will weight these contributions. It is therefore reasonable to only sample directions that will be weighted with a large factor. Sampling techniques which take both the environment as well as the BRDF into account to generate sampling directions should allow accurate results with minimal loss of accuracy [655].

2.11 Application: Modeling the Atmosphere

Every day the Earth's atmosphere produces beautiful displays of color, ranging from blue skies, clouds, orange and red sunrises, and sunsets to rarer phenom-

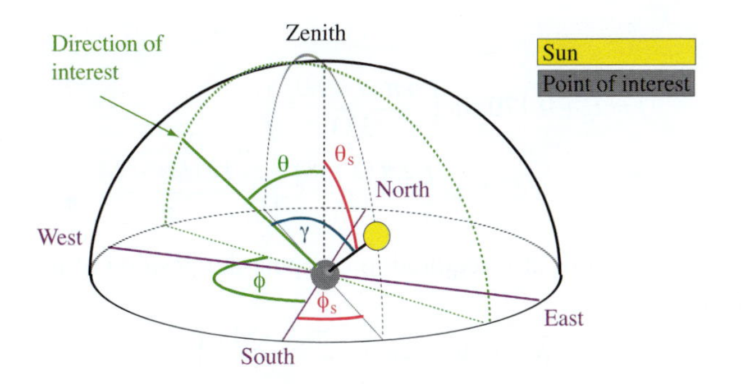

Figure 2.66. The geometry related to computing the position of the sun in the sky.

ena such as rainbows, halos, and the northern and southern lights. Each of these phenomena can be explained [716, 783], and most have been modeled and rendered. Examples include simulation of night scenes [1124], rendering of northern and southern lights [59], and rendering of mirages and fata-morganas [410, 411]. In this section, we begin by briefly explaining several atmospheric phenomena and then show how some of them may be incorporated into an image synthesis algorithm.

2.11.1 The Sun

As the predominant light source, the interaction of the sun with the atmosphere is the most important feature giving rise to sky coloring, as well as the sun's spectral distribution as seen at the surface of the Earth. The atmospheric conditions as well as the length of the trajectory of the sun's light through the atmosphere determine the range of colors that can be observed. To model atmospheric phenomena, it is therefore important to know the position of the sun in the sky, as well as the spectral distribution of the sun's light before it reaches the atmosphere.

The position of the sun is determined by the set of polar coordinates (θ_s, ϕ_s). These, as well as several other angles used in the following section are depicted in Figure 2.66. The position of the sun may be determined from solar time t which itself is given in terms of the Julian date J,[11] the longitude (in radians), the standard meridian M_S for the time zone (in radians), and the standard time (in decimal hours):

[11]The Julian date is an integer in the range $[1, 365]$ indicating the day of the year.

$$t = t_s + 0.170 \sin\left(\frac{4\pi(J-80)}{373}\right) \tag{2.213}$$

$$-0.129 \sin\left(\frac{2\pi(J-8)}{355}\right) + \frac{12(M_S - L)}{\pi}. \tag{2.214}$$

The solar declination δ, also required to compute the position of the sun, is approximated by

$$\delta = 0.4093 \sin\left(\frac{2\pi(J-81)}{368}\right). \tag{2.215}$$

With the help of the site latitude l (in radians), the angle in the sky of the sun is then given by (see also Figure 2.66)

$$\theta_s = \frac{\pi}{2} - sin^{-1}\left(\sin(l)\sin(\delta) - \cos(l)\cos(\delta)\cos\left(\frac{\pi t}{12}\right)\right), \tag{2.216}$$

$$\phi_s = \tan^{-1}\left(\frac{-\cos(\delta)\sin\left(\frac{\pi t}{12}\right)}{\cos(l)\sin(\delta) - \sin(l)\cos(\delta)\cos\left(\frac{\pi t}{12}\right)}\right). \tag{2.217}$$

The second characteristic of the sun required for accurate modeling of the atmosphere is its spectral distribution. Outside the atmosphere, the spectral distribution of the solar irradiance is given in Figure 2.67 [403, 404]. At the Earth's

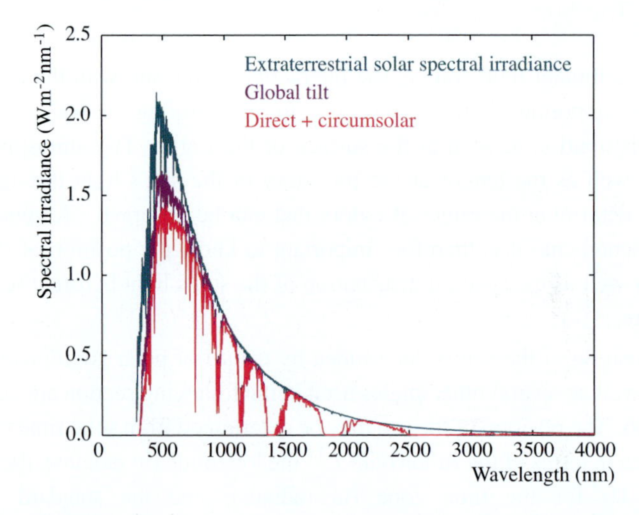

Figure 2.67. The extraterrestrial solar irradiance [403, 404], as well as two terrestrial conditions, as specified by the American Society for Testing and Materials (ASTM) in standard G173-03 [48].

surface, this distribution is altered due to interaction with the atmosphere [1037]. The American Society for Testing and Materials (ASTM) has defined two reference spectra using conditions representative of average conditions encountered in mainland USA [48]; these are plotted in Figure 2.67. These terrestrial standards make assumptions on the atmosphere, including the use of the 1976 U.S. Standard Atmosphere [1166] and a surface spectral reflectivity as produced by light soil [486].

2.11.2 The Atmosphere's Color

The most easily observed aspects of the atmosphere include the color of the sky and clouds. The sky appears blue during the day and colors orange to red during sunsets and sunrises. The range of colors the sky exhibits during the course of a day are all explained by Rayleigh scattering of small particles in the upper atmosphere.

As the upper atmosphere scatters light, the amount of light scattered is determined by the distance that light travels through the atmosphere. The longer the pathlength, the more scattering can occur. During the day, when the sun is more or less overhead, this pathlength is relatively short, and the sun therefore looks pale yellow. At sunrise and sunset, the pathlength is much longer, and thus most of the blue has been scattered out before the direct sunlight reaches the observer.

Figure 2.68. A sunset. The distance traveled by sunlight through the thin upper atmosphere causes much of the light at blue wavelengths to be scattered out, leaving the red part of the spectrum. This is an example of Rayleigh scattering; Clevedon, UK, August 2007.

Figure 2.69. The blue sky is caused by Rayleigh scattering in the thin upper atmosphere; St. Malo, France, June 2005.

This colors the sun orange and red, as shown in Figure 2.68. For a model for rendering twilight phenomena the reader is referred to a paper by Haber [413].

The color of the sky at mid-day is also due to Rayleigh scattering. While most wavelengths travel through the atmosphere undisturbed, it is mostly the blue that is scattered, as shown in Figure 2.69.

The atmosphere contains many particles that are relatively large. These particles give rise to Mie scattering , leading, for instance, to haze, a whitening of the sky's color. Finally, the scattering of light causes objects in the distance to appear desaturated and shifted towards the blue. This phenomenon is known as *aerial perspective*.

Haze may be described with a heuristic parameter called *turbidity* [760]. To determine an appropriate value for turbidity on the basis of a description of the atmosphere or the meteorological distance,[12] Table 2.6 may be consulted (after [925][13]).

With the position of the sun known, and a value for turbidity specified, it is now possible to compute aerial perspective. When light is reflected off a distant object, along its path towards the observer it may be attenuated. In addition, the

[12]The meteorological distance is defined as the distance under daylight conditions at which the apparent contrast between a black target and the horizon is just noticeable. It roughly corresponds to the most distant feature that can be distinguished.

[13]The following derivation to compute the attenuation along a given path through the sky is also after Preetham, Shirley, and Smits [925].

Atmosphere	R_m (in km)	Turbidity T
pure air	256	1.0
very clear	55	2.0
clear	15	3.8
light haze	6	8.0
haze	3	24.0
thin fog	1	60.0

Table 2.6. Turbidity T as function of the type of atmosphere. Values are approximate.

atmosphere scatters light at all positions in space, and in particular along this path, some light may be scattered into the direction of the observer. This is called *in-scattering*. The light arriving at the observer is therefore a function of the in-scattered light L_{in} and the attenuated light L_0:

$$L = \tau L_0 + L_{\text{in}}, \tag{2.218}$$

where L is the amount of light arriving at the observer and τ is the extinction factor. Both the in-scattering term and the extinction factor depend on the scattering medium and are, therefore, subject to both Rayleigh scattering for molecules and Mie scattering for larger particles. The extinction factor (as well as L_{in}) depends on the scattering coefficients β_m and β_h for molecules and haze respectively:

$$\tau(s) = \exp\left(-\int_0^s \beta_m(h(x))\right) \exp\left(-\int_0^s \beta_h(h(x))\right). \tag{2.219}$$

The function $h(x)$ denotes the height of each point x along the path between 0 and s and is given by $h(x) = h_0 + x\cos(\Theta)$. The extinction factor is thus an integral over the path taken by the light through the atmosphere. As the scattering function depends on the density of the particles, and therefore on the height h above the Earth's surface, the scattering function for both β_m and β_h is given by

$$\beta(h) = \beta^0 \exp(-\alpha h), \tag{2.220}$$

where β^0 is the scattering coefficient at the Earth's surface and $\alpha = 1.3$ is the exponential decay constant. The value of β^0 can be computed as a function of angle θ (see Figure 2.66), or can be integrated over all angles. These expressions for the Rayleigh scattering part are given by β_m^0:

$$\beta_m^0(\theta) = \frac{\pi^2 \left(n^2-1\right)^2}{2N\lambda^4} \left(\frac{6+3p_n}{6-7p_n}\right) \cos\left(1+\cos^2(\theta)\right), \tag{2.221}$$

$$\beta_m^0 = \frac{8\pi^3 \left(n^2-1\right)^2}{3N\lambda^4} \left(\frac{6+3p_n}{6-7p_n}\right). \tag{2.222}$$

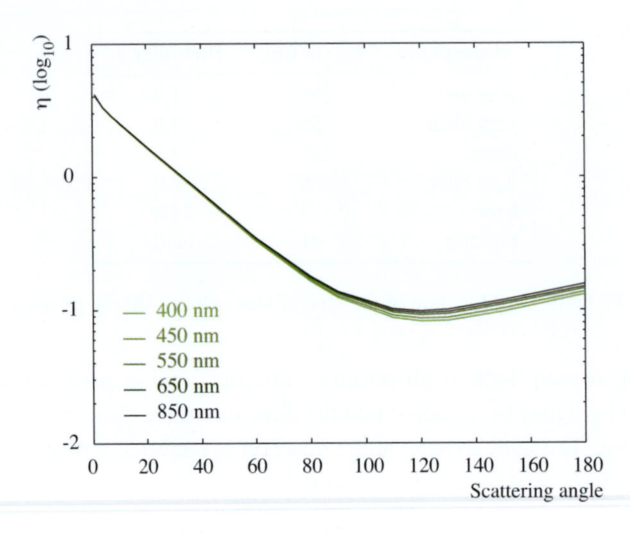

Figure 2.70. Values for η given different wavelengths and scattering angles. These values are valid for a value of 4 for Junge's constant v.

Here, $n = 1.0003$ is the index of refraction of air, $N = 2.545^{25}$ is the number of molecules per unit volume, and $p_n = 0.035$ is a de-polarization factor. The scattering coefficients for haze are due to Mie scattering and are given by

$$\beta_h^0(\theta) = 0.434\, c \left(\frac{2\pi}{\lambda}\right)^{v-2} \frac{\eta(\theta,\lambda)}{2}, \tag{2.223}$$

$$\beta_h^0 = 0.434\, c \left(\frac{2\pi}{\lambda}\right)^{v-2} K(\lambda), \tag{2.224}$$

where v is Junge's exponent which takes a value of 4 for sky modeling. The concentration factor c depends on the turbidity T as follows:

$$c = (0.6544\,T - 0.6510) \times 10^{-16}. \tag{2.225}$$

Curves for $\eta(\theta,\lambda)$ and K are plotted in Figures 2.70 and 2.71.

With the scattering coefficients for Rayleigh and Mie scattering defined, the final expression of the extinction coefficient can be derived. As the height h varies along the path as indicated above, the integrals in (2.219) can be solved, and this equation can be rewritten as follows:

$$\tau(s) = \exp\left(-a_m\left(b_m - u_m(s)\right)\right)\, \exp\left(-a_h\left(b_h - u_h(s)\right)\right), \tag{2.226}$$

$$u(s) = \exp\left(-\alpha(h_0 + x\cos(\theta))\right), \tag{2.227}$$

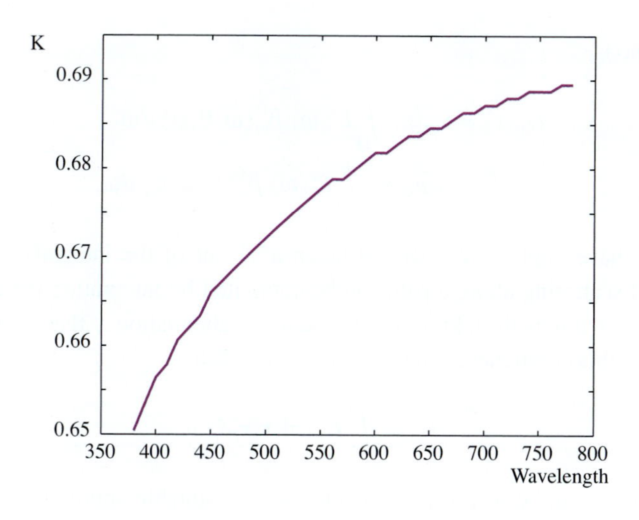

Figure 2.71. Values for K as function of wavelength λ.

$$a = -\frac{\beta^0}{\alpha \cos(\theta)}, \tag{2.228}$$

$$b = \exp(-\alpha h_0). \tag{2.229}$$

This expression thus forms the first component of the computation of atmospheric attenuation and can be plugged into (2.218). The second component involves the computation of the light that is scattered into the path. At any given point along the path, some light is scattered in, and this quantity thus needs to be integrated over the length of the path.

We first compute the light scattered in at a specific point. This quantity depends on the scattering functions β_m^0 and β_h^0 at sea level, as well as altitude h. We can thus write expressions for the angular scattering functions in a manner similar to (2.220):

$$\beta(\omega, \theta, \phi, h(x)) = \beta^0(\omega, \theta, \phi)\, u(x). \tag{2.230}$$

The function $\beta(\omega, \theta, \phi, h(x))$ can be seen as a phase function in that it describes how much of the light coming from ω is scattered into the direction of interest (θ, ϕ). The total amount of light from ω scattering into this direction is then $L^s(\omega)\,\beta(\omega, \theta, \phi, h(x))$. The total amount of light scattered at position x is given by the integral over all directions ω (repeated for both Rayleigh and Mie scattering components):

$$L_{\text{in}}(x) = u_m(x) \int_{\Omega} L^s(\omega)\, \beta_m^0(\omega, \theta, \phi)\, d\omega + u_h(x) \int_{\Omega} L^s(\omega)\, \beta_h^0(\omega, \theta, \phi)\, d\omega, \tag{2.231}$$

where we have applied (2.230) and taken $u(x)$ out of the integrals. The total amount of scattering along a path can be computed by integrating the above expression over this path, taking into account the attenuation τ that occurs at the same time (this component can be taken from (2.226)):

$$L_{\text{in}} = \int_0^s L_{\text{in}}(x)\ \tau(s) dx. \tag{2.232}$$

This equation can be solved numerically, or with suitable approximations it can be simplified to allow analytical solutions, as shown by Preetham et al [925]. This then also allows (2.218) and hence the attenuation along a ray path due to both molecular Rayleigh scattering as well as Mie scattering due to larger haze particles to be computed.

2.11.3 Rainbows

During a rain shower, droplets of water are suspended in air. If a directional light source, such as the sun, illuminates a volume with water droplets, each droplet

Figure 2.72. A rainbow; Millford Sound, New Zealand, 2001. (Photo courtesy of Timo Kunkel (www.timo-kunkel.com).)

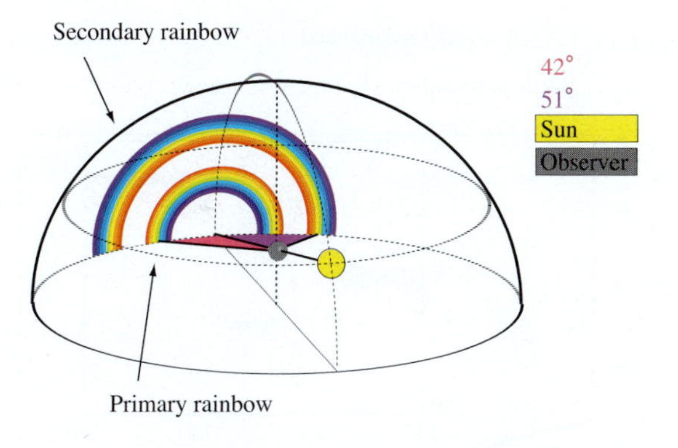

Figure 2.73. Geometry associated with rainbows.

reflects and refracts light. The refraction causes dispersion of the light which depends on wavelength. This gives rise to a rainbow (Figure 2.72). The rainbow is due to a single internal reflection inside the water droplet. If a second reflection occurs, this gives rise to a weaker secondary rainbow. The location of these rainbows with respect to the sun and the observer is diagrammed in Figure 2.73. The

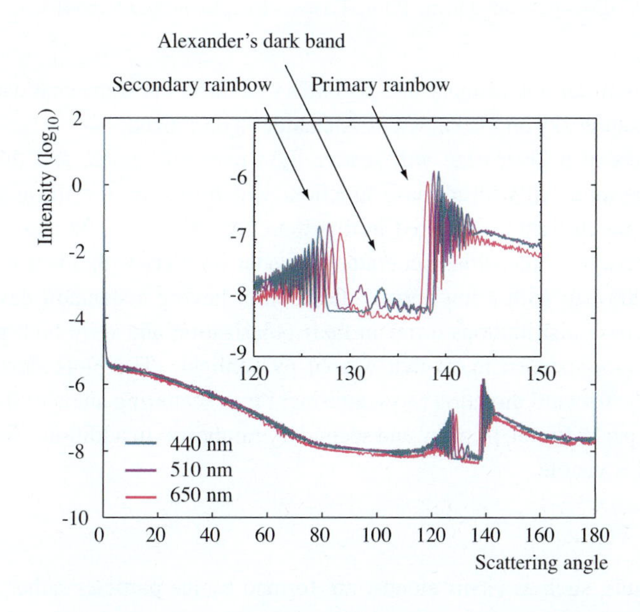

Figure 2.74. Intensity as function of scattering angle for water droplets of 1000 μm (std. dev. 120 μm) suspended in air and lit by sunlight. These plots were computed for perpendicular polarization. (Data generated using Philip Laven's Mieplot program [654].)

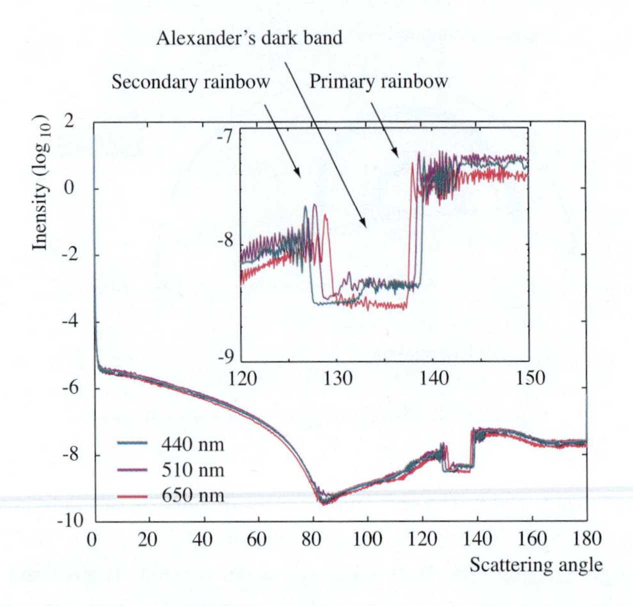

Figure 2.75. Intensity as function of scattering angle for water droplets of 1000μm (std. dev. 120μm) suspended in air and lit by sunlight. These plots were computed for parallel polarization. (Data generated using Philip Laven's Mieplot program [654].)

region in between the primary and secondary rainbows is somewhat darker than the surrounding sky and is known as Alexander's dark band.

Droplets of a given size will scatter light over all angles, and this behavior may be modeled with a phase function, which records for all outgoing angles, how much light is scattered in that direction. Figures 2.74 and 2.75 show the distribution of intensities generated by water droplets with a size distributed around 1000 μm with a log normal distribution having a standard deviation of 10 μm. These distributions differ in their polarization and were both generated for droplets suspended in air that was lit by sunlight. The plots show marked peaks in the forward direction (no scattering, i.e., a scattering angle of 0 degrees), as well as peaks for the primary and secondary rainbows. In addition, Alexander's dark band is visible.

2.11.4 Halos

Some clouds, such as cirrus clouds, are formed by ice particles rather than rain droplets. These ice particles are usually assumed to be shaped like a hexagonal slab with a varying height-to-width ratio. Crystals that are flat are called plates, whereas crystals that are elongated are called pencils. Under wind-still conditions,

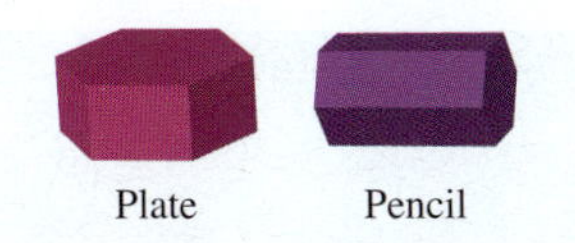

Figure 2.76. Ice crystals formed in cirrus clouds are usually assumed to be either plates or pencils.

these crystals are suspended in air in approximately the orientation as shown in Figure 2.76.

Sunlight penetrating cirrus clouds will interact with ice particles [387]. As the normal laws of reflection and refraction apply when light enters these crystals, light will be scattered into preferential directions determined by the geometry of these crystals. These (multiple) scattering events lead to the formation of halos. The halos occurring most often are located at an angle of 22° around the sun. A weaker halo is sometimes seen at 46°. Further halo effects, such as sun-dogs (parhely) and the upper tangent arc are also well known. Their position in relation to the observer and the sun is diagrammed in Figure 2.77. A photograph of a halo as well as a sun-dog is shown in Figure 2.78. The 46° halo is shown in Figure 2.79.

The scattering behavior of hexagonal ice particles can be modeled with a phase function, i.e. a function that models for every direction how much light is scattered in that direction [469, 691]. Phase functions indicate fractions of light being scattered in specific directions and therefore integrate to one. Such phase functions have been measured for halo-forming cirrus clouds, and are plotted in Figure 2.80.

The assumption that all ice crystals are either plates or pencils is an oversimplification. In particular, the collection and analysis of samples of ice crystals has revealed that ice crystals in halo-forming clouds are frequently hollow, and also

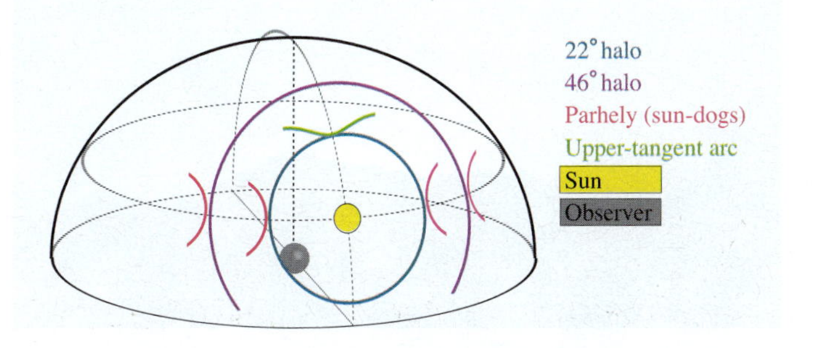

Figure 2.77. Different types of halos and their position in relation to the observer and the sun (after [377]).

Figure 2.78. Halo around the sun. Toward the lower left a second refraction is visible; Southern Utah, 2001.

have more complicated shapes than simple hexagonal forms [996]. As a result, the phase function shown in Figure 2.80, which is based on a simplified model, suggests that the 22° and in particular the 46° are more common than they are in practice. See also Sassen and Ulanowski for a discussion [997, 1161].

Figure 2.79. The 46° halo, photographed with a wide-angle lens; Wanaka, New Zealand, 2001. (Photo courtesy of Timo Kunkel (www.timo-kunkel.com).)

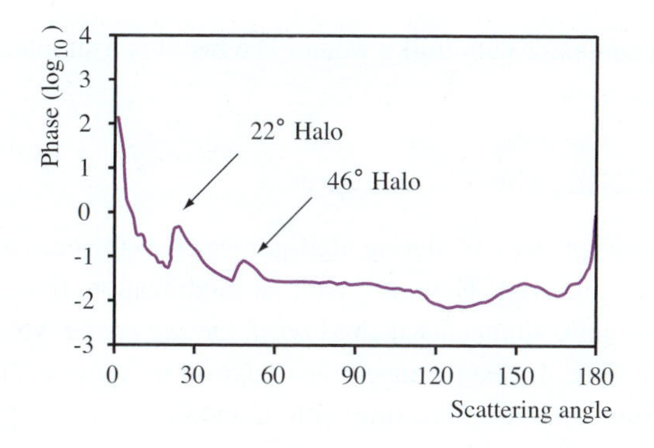

Figure 2.80. Phase function for hexagonal ice crystals (after [963]).

2.11.5 The Green Flash

At the last few seconds of sunset, sometimes the sun may be colored green, a phenomenon known as the *green flash*. It results from a combination of Rayleigh scattering and dispersion. While rarely seen in nature, predominantly because this phenomenon only lasts for a brief moment, this effect can be reproduced with a projector, a mirror, and a pan of water with a few drops of milk suspended [109]. With such a set of tools, it is possible to construct a dispersing prism made of a material which exhibits Rayleigh scattering (water with some milk). In Figure 2.81, we show a similar demonstration, created by shining a beam of white light through an ordinary dispersing prism, followed by passing the dispersed

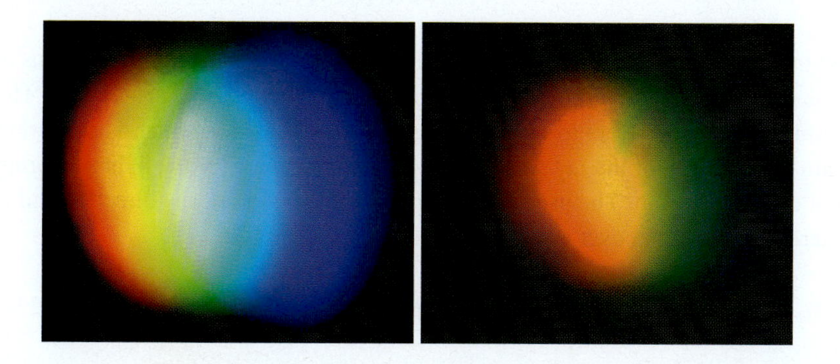

Figure 2.81. On the left, white light is passed through a dispersing prism as well as a tank of clear water. Normal dispersion is observed. On the right, a few drops of milk are added to the water. The blue component of the dispersed light is scattered laterally, so that a red disk with a green rim remains.

light through a container with milky water. The result is a simulation of a sunset with added dispersion.

2.11.6 Rendering the Atmosphere

To render the atmosphere, a rendering algorithm such as the ones outlined in Section 2.10 can be employed. However, without modification, these algorithms do not model important components such as *participating media*, volumes of matter interacting with light. Instead, conventional algorithms assume that light travels through a medium without interacting with it, and only changes paths at surface boundaries.

This is not sufficient to model the atmosphere. Instead, the *radiative transfer equation* can be evaluated for every point along a ray path. Thus, tracing a ray involves advancing the ray by a short distance, evaluating the radiative transfer equation to determine the change in light due to the particles present in a small volume around the point of interest. The ray is then advanced a short distance again, and the process is repeated. At surface boundaries, light is reflected and refracted as normal.

The radiative transfer equation takes into account how much light is emitted, scattered, and absorbed at a point $\mathbf{P}$ on a ray aimed into direction Θ:

$$\begin{aligned}\frac{\partial L(\mathbf{P},\Theta)}{d\mathbf{P}} = \; & \alpha(\mathbf{P})\, L_e(\mathbf{P},\Theta) \\ & + \alpha(\mathbf{P})\, L_i(\mathbf{P},\Theta) \\ & + \sigma(\mathbf{P})\, L(\mathbf{P},\Theta) \\ & + \alpha(\mathbf{P})\, L(\mathbf{P},\Theta).\end{aligned} \tag{2.233}$$

With α the absorption coefficient and σ the scattering coefficient, the four terms in this equation model emission, in-scattering, out-scattering, and absorption. The term in-scattering refers to light coming from any direction which is scattering into the direction of the ray path. Out-scattering accounts for the light scattering out of the ray's path. The in-scattered light L_i can come from any direction and is therefore an integral over a full sphere Ω:

$$L_i(\mathbf{P},\Theta) = \int_{\Omega} p(\mathbf{P},\Theta,\Theta')\, L(\mathbf{P},\Theta')\, d\Theta'. \tag{2.234}$$

Thus, the light coming from each direction is weighted by a phase function p that models the scattering properties of the volume, i.e., the directionality of the scattering. As desired, this function can be chosen to account for rainbows [963], halos [371, 372, 377, 541, 542], as well as clouds. In addition, a specific form of

Figure 2.82. Rendering of participating media. The left image shows distortions caused by spatially varying temperature. The middle image shows a 3D turbulence field simulating smoke. An anisotropic medium is modeled in the rightmost image. (Images courtesy of Diego Gutierrez, Adolfo Munoz, Oscar Anson, and Francisco Seron; Advanced Computer Graphics Group (GIGA), University of Zaragoza, Spain [411].)

this function was shown in Section 2.11.2 that accounts for the coloring of the sky. Section 3.8.2 discusses an alternative phase function for rendering flames.

In the above, it is implied that each function is also dependent on wavelength. The scattering is inelastic, i.e., scattering does not change a particle's wavelength. To account for elastic scattering, and thus model fluorescence and phosphorescence, an extra term can be added to the radiative transfer equation :

$$\int_{\Omega}\int_{\lambda} \alpha_{\lambda_i}(\mathbf{P})\, f_{\lambda_i \to \lambda}(\mathbf{P})\, L_{\lambda_i}(\mathbf{P},\Theta')\, \frac{p_{\lambda}(\mathbf{P},\Theta',\Theta)}{4\pi}\, d\Theta'\, d\lambda_i. \tag{2.235}$$

Here, f is a probability distribution function that models the likelihood of a photon at wavelength λ_i being re-emitted at wavelength λ. Simulations of inelastic scattering are shown in Figure 2.82.

2.12 Summary

Light is modeled as harmonic electromagnetic waves. The behavior of electric and magnetic fields gives rise to the propagation of light through free space. The electric and magnetic vectors are orthogonal to the direction of propagation, which is indicated by the Poynting vector. The magnitude of the Poynting vector is commensurate with the amount of energy carried by the wave.

While propagating, the electric and magnetic vectors normally rotate around the Poynting vector, inscribing an ellipse. When the ellipse degenerates either into a circle or a line, the light is said to be polarized.

In homogeneous isotropic materials, light propagates by the same principle, albeit at a different velocity. If light is modeled as a superposition of harmonic plane waves, then each of the constituent components oscillates with a given frequency and associated wavelength. Visible light consists of a very narrow range of wavelengths—roughly between 380 and 780 nm.

At surface boundaries, light is reflected and refracted. The angle of reflection and refraction are given by Snell's law, whereas the amount that is reflected and refracted is modeled by the Fresnel equations.

2.13 Further Reading

There are many texts on the physics of light and optics. The classic texts are Born and Wolf, *Principles of Optics* [112] as well as Hecht, *Optics* [447]. For the preparation of this chapter, we also used Iskander, *Electromagnetic Fields and Waves* [532] and the *Handbook of Optics* [73]. Other sources are [379, 428, 436, 736, 1016]. There are many books on computer graphics, of which we can list only some [23, 271, 370, 895, 1043]. Several books on weather phenomena are worth mentioning [387, 716, 783]. Williamson and Cummins, *Light and Color in Nature and Art* [1247] discusses many of the topics of this chapter, including modeling the atmosphere.

Chapter 3
Chemistry of Matter

In the previous chapter, light was treated exclusively as a wave phenomenon. This model is suitable for explaining reflection, refraction, and polarization. However, the wave model alone is not able to explain all phenomena of light. It is therefore necessary to extend the theory to include the notion of particles. This leads to the concept of wave-particle duality: light has behaviors commensurate with both wave models as well as particle models.

It was postulated by de Broglie that not only light, but all matter exhibits both wave-like and particle-like behaviors. This includes protons, neutrons and electrons that form atomic and molecular structures.

Thus, a model is needed to account for both wave and particle behaviors, and such a model is afforded by Schrödinger's equation. This equation can be used to show that particles can only have specific energies, i.e., energy levels of particles are generally quantized. The study of particles at the microscopic level is therefore termed *quantum mechanics* [179, 683, 1109].

This chapter first briefly reviews classical physics which forms the basis of quantum mechanics. Then, quantum mechanics and molecular orbital theory is explained. These theories are necessary to understand how light interacts with matter. In particular, we discuss the close interaction of energy states of electrons orbiting atomic nuclei and light of particular wavelengths being emitted, absorbed, or reflected.

The material presented in this chapter, in conjunction with electromagnetic theory, also affords insight into the behavior of light when it interacts with matter which itself has structure at the scale of single wavelengths.

3.1 Classical Physics

As quantum mechanics relies heavily on classical physics, we start this chapter with a brief overview of classical physics. We ignore magnetic interactions for the moment to simplify the discussion.

A particle traveling through space has various inter-related attributes such as mass m, (linear) velocity v, and momentum p:

$$p = mv. \tag{3.1}$$

Along with linear momentum, a particle may possess angular momentum L if it rotates around a point[1]. If the point of rotation is a distance r away from the particle, the angular momentum is defined as (see also Figure 3.1)

$$L = mvr. \tag{3.2}$$

The momentum of a particular will change over time if a force is applied to it. According to Newton's second law of mechanics, a constant force F gives rise to a linear change of momentum with time:

$$F = \frac{d(mv)}{dt} = \frac{dp}{dt}. \tag{3.3}$$

For a particle with constant mass m, this law may be rewritten to yield

$$F = m\frac{dv}{dt} = ma, \tag{3.4}$$

where a is the acceleration of the particle. In addition to mechanical forces, particles may be subject to forces induced by charges, as predicted by Coulomb's law and given in Equation (2.1).

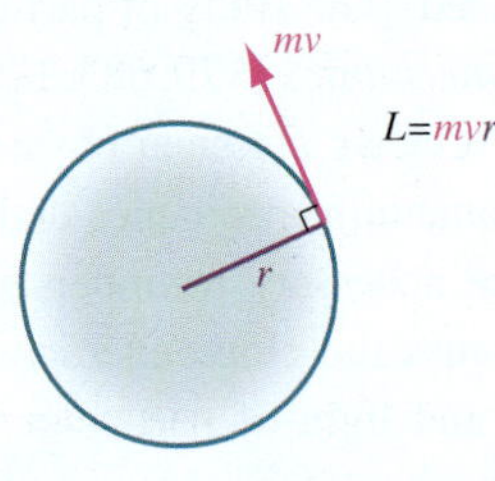

Figure 3.1. Angular momentum.

[1]In this chapter, the quantity L is reserved for angular momentum. However, this is also the common symbol used for luminance. In the remainder of this book, L is therefore taken to mean luminance.

The total energy E of a particle may be decomposed into a kinetic energy T, as well as a potential energy U:

$$E = T + U. \tag{3.5}$$

Given the above, the kinetic energy is given by

$$T = \frac{mv^2}{2} = \frac{p^2}{2m}. \tag{3.6}$$

Thus, the kinetic energy of a particle may be expressed in terms of its linear momentum p. The potential energy of a charged particle in the electrostatic field of another particle a distance r away is given by

$$U = \frac{Q_1 Q_2}{r}. \tag{3.7}$$

In the following, we will assume that, for a given system of particles, the total energy remains constant. In that case, the Hamiltonian H of the system equals its total energy:

$$H = E. \tag{3.8}$$

If we assume that heavy particles, such as protons, are stationary, then the kinetic energy is due to the electrons only. For instance, the Hamiltonian of a hydrogen atom (which consists of one proton and one electron) becomes

$$H = T + U \tag{3.9a}$$

$$= \frac{p^2}{2m} - \frac{e^2}{r}. \tag{3.9b}$$

In this equation, e is the unit charge, which is positive for the proton and negative for the electron, and r is the distance between the proton and the electron. For atoms with multiple protons and electrons, the Hamiltonian becomes more complex as it will include a term for each pair of particles. As an example, the Hamiltonian for a helium atom with two protons and two electrons would be

$$H = \frac{p_1^2}{2m} + \frac{p_2^2}{2m} - \frac{2e^2}{r_1} - \frac{2e^2}{r_2} + \frac{e^2}{r_{12}}. \tag{3.10}$$

For complex configurations, the Hamiltonian can not be evaluated directly, and one would have to resort to approximations to estimate the total energy.

The total energy of a system is an indication of its stability. For instance, if an electron with kinetic energy T approaches a proton, the system's potential energy decreases. However, the total energy of the system remains constant, and therefore the kinetic energy must increase. For the electron and the proton to remain together, this excess kinetic energy must be lost. Otherwise, the kinetic energy of the electron can become so large as to overcome the attractive force of the proton. In that case, the electron would escape. Stability of a system thus implies that the negative potential energy is greater than the positive kinetic energy. A consequence is that the total energy of the system has to be negative for the system to be stable.

While many phenomena may be adequately explained by classical physics, at the microscopic level there are discrepancies between effects that can be measured and those that are predicted by the theory. This has necessitated a new theory.

3.2 Quantum Mechanics

In 1905, Einstein postulated that energy and mass are related by his famous equation [279]:

$$E = mc^2. \tag{3.11}$$

Here, E is the relativistic energy of a particle, and m is the relativistic mass of a particle. The constant c is, as before, the speed of light in vacuum. One of the implications of this equation is that the mass of a slow-moving particle is lower than the mass of the same particle traveling at a higher speed.

For now, we will assume that a light wave is composed of photons[2]. In that case, the momentum of a photon traveling through a vacuum may be defined as

$$p = mc. \tag{3.12}$$

Here, p is the relativistic momentum of the photon. The relativistic energy of the photon is then given by

$$E = pc. \tag{3.13}$$

For a given particle, such as an electron or a photon, Max Planck realized that oscillations occur not at all frequencies, but only at specific frequencies. This implies that the energy carried by a particle can only take on discrete values and may be expressed in terms of frequency f:

$$E = nhf, \qquad n = 0, 1, 2, 3, \ldots. \tag{3.14}$$

[2]This hints at the well-known wave-particle duality, which will be described in Section 3.2.1.

Here, n is an integer value, and h is a constant of proportionality, known as Planck's constant. Its value is

$$h = 6.626176 \times 10^{-34} \text{ J s}. \tag{3.15}$$

Thus, quantum theory derives its name from the quantization of energy levels.

3.2.1 Wave-Particle Duality

Light was explained as a wave phenomenon in the previous chapter. However, several experiments demonstrate that the behavior of light can not be explained solely by waves. Thus, the notion emerged that light is both. It loosely behaves as a wave when traveling through free space and as a particle when it interacts with matter.

This can be adequately demonstrated for light. The particles are called *quanta*, except when the particles have frequencies in the visible range. Then, these particles are called *photons*.

Furthermore, the same wave-particle duality can be demonstrated for larger particles such as electrons. If any particle can act as a wave, then associated with it, there should be a wavelength. From (2.63), we have a relation between the wavelength λ, the velocity of the wave v, and the frequency f. At the same time, from (3.14) we have $E = hf$, and from (3.13) the energy of a photon is given as $E = pc$. Combining these three equations, we get the de Broglie's postulate:

$$\lambda = \frac{h}{p}, \tag{3.16}$$

where h is Planck's constant. Thus, the wavelength of a wave/particle is related to its momentum. The significance of this equation is that it directly links a quantity associated with particles, namely their momentum, to a quantity λ associated with waves.

3.2.2 Bohr's Atom Model

In this section, we consider a very simple model of atoms, which we will refine in later sections. Here, we assume that an atom consists of heavy particles such as protons and neutrons forming a nucleus, with lighter particles orbiting around the nucleus at some distance. This is known as Bohr's model of the atom.

Protons have a positive charge, indicated with $+e$, whereas electrons have a negative charge of $-e$. Neutrons have no charge. The number of protons in a nucleus determines the atom's *atomic number*, which is given the symbol Z.

Normally, the number of protons and neutrons in a nucleus are identical. If these numbers are different, then the atom is called an *isotope*.

The number of electrons surrounding the nucleus determines the charge of the atom. If the number of electrons equals the number of protons in the nucleus, then the total charge is zero. If the number of electrons and protons is not matched, the charge is non-zero, and the atom is called an *ion*. If the charge is positive due to a lack of electrons, the atom is called a *cation*. An excess of electrons causes the atom to be negatively charged, and the atom is then called an *anion*.

We consider an atom with only a single proton and electron, i.e., a hydrogen atom. Since the electron is in orbit around the proton, it neither gains nor loses energy. If this were not the case, the electron would gravitate towards the proton or spiral away from the proton.

The orbits of electrons are for now assumed to be circular. The angular momentum L of an electron can then be derived from de Broglie's postulate. The length of the orbit is related to its radius r by a factor of 2π. If an electron is not gaining or losing energy, then, after each orbit, it will return to the same position. Ascribing a wave to this electron, the circumference should be an integral number of wavelengths, i.e.,

$$2\pi r = n\lambda, \qquad n = 1, 2, 3, \ldots. \tag{3.17}$$

Substituting (3.16) then gives

$$2\pi r = \frac{nh}{p} = \frac{nh}{mv}, \qquad n = 1, 2, 3, \ldots. \tag{3.18}$$

Rearranging terms gives

$$mvr = \frac{nh}{2\pi} = L, \qquad n = 1, 2, 3, \ldots. \tag{3.19}$$

Thus, the angular momentum is considered quantized and non-zero (because $n > 0$). As an aside, the quantity $\frac{h}{2\pi}$ occurs frequently; it has been given its own symbol $\hbar = \frac{h}{2\pi}$. The expression for mvr allows us to solve for r, i.e., we can compute the radius at which the electron is orbiting. However, the result depends on its velocity, which is unknown.

To arrive at an expression for the radius, we proceed as follows. For a proton with an electron in orbit, the Coulomb force is given by

$$F = \frac{e^2}{r^2}. \tag{3.20}$$

From Newton's second law, we know that this equals ma. The acceleration as a result of the Coulomb force can be computed from the change in velocity after the electron has rotated around the proton once. The magnitude of the velocity is given by

$$v = \frac{2\pi r}{T}, \tag{3.21}$$

where T is the time required for one orbit. The change in magnitude Δv for a single orbit is $\Delta v = 2\pi v$. The acceleration is then simply $a = \Delta v/T = 2\pi v/T$. Using (3.21), this can be written as

$$a = \frac{v^2}{r}. \tag{3.22}$$

We now use (3.20) and find the following relation:

$$\frac{e^2}{r^2} = \frac{mv^2}{r}. \tag{3.23}$$

To eliminate the unknown velocity, we use the expression for angular momentum L to solve for v:

$$v = \frac{nh}{2\pi mr}. \tag{3.24}$$

Substituting into (3.23), we find

$$r = \frac{n^2 h^2}{4\pi^2 m e^2}, \qquad n = 1,2,3,\ldots. \tag{3.25}$$

As a result of quantizing the angular momentum, we thus find that the distance between the electron and the proton is quantized. Since all quantities are known, the radius is a direct function of quantization level n. The resulting orbits are called *Bohr orbits*.

Given these orbits, we can now compute the energy of the electron in terms of its orbit. The total energy of a hydrogen atom is given by

$$E = T + U = \frac{mv^2}{2} - \frac{e^2}{r} \tag{3.26a}$$

$$= \frac{p^2}{2m} - \frac{e^2}{r}. \tag{3.26b}$$

Using the equations above, this may be restated as

$$E = \frac{e^2}{2r} - \frac{e^2}{r} = -\frac{e^2}{2r}. \tag{3.27}$$

Substituting (3.25), the total energy associated with a hydrogen atom is given as

$$E = -\frac{2\pi^2 m e^4}{h^2}\frac{1}{n^2}, \qquad n = 1,2,3,\ldots. \tag{3.28}$$

The ground state of an atom occurs for $n = 1$. In this state, the energy is most negative, and the atom is therefore most stable. For larger values of n, we speak of excited states. As the energy for excited states is less negative than for the ground state, these states are also less stable. The energy associated with the ground state is the negative of the ionization potential, which is the energy required to remove an electron. Thus, the quantization of the angular momentum has led to the quantization of possible energy states of the atom.

Now, an electron may jump from one state to another, thereby changing the total energy of the atom. If changing from state n_1 to n_2, the corresponding change in total energy is

$$\Delta E = -\frac{2\pi^2 m e^4}{h^2}\left(\frac{1}{n_2^2} - \frac{1}{n_1^2}\right). \tag{3.29}$$

If this energy change is due to a jump to a lower excited state, i.e., $n_2 < n_1$, then this energy must have been dissipated in some manner, for instance by emission of a photon. The energy of the photon must equal the change in energy of the atom. We can now determine the frequency f of the photon, given that its energy equals $\Delta E = hf$. Solving for f yields an expression for the frequency of the photon:

$$f = -\frac{2\pi^2 m e^2}{h^3}\left(\frac{1}{n_2^2} - \frac{1}{n_1^2}\right). \tag{3.30}$$

With the frequency, as well as the medium through which the photon is traveling known, the wavelength is also determined according to (2.63). As a result, the change in energy state of an electron from a higher excited state to a lower excited state has caused the emission of a photon with a wavelength determined by the quantitative change in the atom's energy. This change is quantized according to the equations given above, and thus the color of the light emitted is quantized as well.

Conversely, a photon may be absorbed by an electron. The extra energy imparted onto the electron is used to bring it to a higher excited state. This absorption can only happen if the energy of the photon equals the amount of energy required to make the electron jump between states of excitation. Thus, a photon of the right wavelength may be absorbed, whereas photons of longer or shorter wavelengths may pass through the atomic structure, or alternatively get deflected.

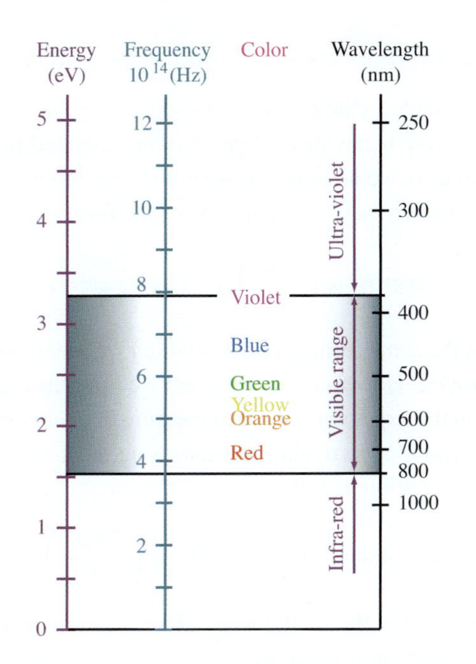

Figure 3.2. Energy diagram showing four different scales to represent energy.

The absorption of photons with specific wavelengths is thus governed by the atomic structure of the matter with which the photon interacts. Different atoms and molecules afford different possible states of excitation and will, therefore, absorb photons with different wavelengths. Photons with other wavelengths are not absorbed and cause reflected and transmitted light of different colors to appear.

As indicated above, the energy of excited states increases with increasing n, i.e., becomes less negative. For $E = 0$, the electron is removed and the atom is left in an *ionized state*. The electron is then said to be in *free state* or *unbound state*. If more energy is added to the atom, if will be in the form of kinetic energy. This energy may be added in any amount and is not limited to specific quantized amounts.

Finally, the change of energy states can be conveniently represented in an energy diagram, as depicted in Figure 3.2. In addition to color, wavelength, and frequency, a new unit of energy is introduced here, the electronvolt (eV; see also Table D.2). This is a convenient unit to express the quantized energy of an electron, atom, or molecule. The conversion between wavelength λ and energy ΔE (if specified in electron volts (eV)) is given by

$$\lambda \Delta E = 1239.9. \tag{3.31}$$

3.2.3 The Time-Independent Schrödinger Equation

As particles are known to exhibit wave behavior, it may be assumed that a solution describing the wave behavior for light may be extended to include the wave behavior of particles. If we associate a wave with the motion of an electron, its solution in one dimension may be taken to be of the form

$$\psi = A \sin\left(\frac{2\pi z}{\lambda}\right) + B \cos\left(\frac{2\pi z}{\lambda}\right). \tag{3.32}$$

This equation is of the same form as (2.59c), which we presented as a solution for harmonic plane waves. However, we have removed the time-dependent component. In addition, in this equation ψ is not a measure of energy, but the probability of finding the electron at a particular position along the z-axis[3]. We may substitute (3.16) into this equation as follows:

$$\psi = A \sin\left(\frac{2\pi p z}{h}\right) + B \cos\left(\frac{2\pi p z}{h}\right). \tag{3.33}$$

Under the assumption that the total energy of an electron remains constant, the Hamiltonian for an electron is given by

$$H = E = \frac{p^2}{2m} + U. \tag{3.34}$$

The equalities remain intact if all terms are multiplied by ψ. This yields the time-independent Schrödinger equation:

$$H\,\psi = E\,\psi. \tag{3.35}$$

In this formulation, the energy E of an electron is called the eigenvalue and ψ is called the *eigenfunction*. The eigenfunctions correspond to orbitals. Thus, an electron in orbital ψ has energy E. Equation (3.34) also yields

$$\frac{p^2\,\psi}{2m} = (E - U)\,\psi. \tag{3.36}$$

To derive an expression for the Hamiltonian operator, in this case, we begin by differentiating ψ twice:

$$\frac{\partial^2 \psi}{\partial z^2} = -\frac{4\pi^2 p^2}{h^2}\left(A \sin\left(\frac{2\pi p z}{h}\right) + B \cos\left(\frac{2\pi p z}{h}\right)\right) \tag{3.37a}$$

$$= -\frac{4\pi^2}{h^2}\,p^2\,\psi. \tag{3.37b}$$

[3]This expression will be extended to three dimensions later in this section.

Rearranging terms and dividing left and right by $2m$ yields

$$\frac{p^2 \psi}{2m} = -\frac{h^2}{8\pi^2 m} \frac{\partial^2 \psi}{\partial z^2}, \tag{3.38}$$

and, therefore,

$$-\frac{h^2}{8\pi^2 m} \frac{\partial^2 \psi}{\partial z^2} = (E - U)\,\psi. \tag{3.39}$$

Thus, the Hamiltonian is now given in operator form as follows:

$$H = -\frac{h^2}{8\pi^2 m} \frac{\partial^2 \psi}{\partial z^2} + U. \tag{3.40}$$

This operator may be extended to three dimensions by replacing the partial derivative in the z-direction with the Laplacian operator, as given in (A.19). The Hamiltonian operator is then formulated as

$$H = -\frac{h^2}{8\pi^2 m} \nabla^2 + U. \tag{3.41}$$

3.2.4 Quantum Numbers

The integer numbers n which characterize energy states of bound electrons are called *principal quantum numbers*. The atomic structure may be further described by additional quantum numbers.

As argued in Section 3.2.2, the angular momentum L is quantized. Empirically, it has been found that the square of the angular momentum can only have specific values:

$$l(l+1)\hbar^2, \qquad l = 0, 1, 2, \ldots, n-1. \tag{3.42}$$

The integer l is called the *angular momentum quantum number*, and its value, describing the rotation of an electron around a nucleus, is strictly less than n. This implies that for higher energy states, i.e., larger values of n, there are more possible angular momentum states. The *magnetic quantum number* m can have both positive and negative integral values between $-l$ and l. It describes the projection of the angular momentum of an electron onto an axis determined by an external magnetic field [370]. The fourth and final quantum number is the *spin quantum number* with values of either $+\frac{1}{2}$ or $-\frac{1}{2}$; it is a number describing the rotation of an electron around itself.

An atom has a nucleus surrounded by an electron cloud. The electrons are arranged in *orbits* (also called *shells*), with each orbit containing between zero

l	0	1	2	3	4	5	...
Symbol	s	p	d	f	g	h	...

Table 3.1. Symbols associated with values of the angular momentum quantum number l.

and two electrons. If there are two electrons in an orbit, then these will have opposite spin.

By convention, orbital states are indicated with a value for n, followed by a letter for l. The letter scheme is as indicated in Table 3.1. The first four letters stand for sharp, principal, diffuse, and fundamental.The letters following these four are arranged alphabetically [768]. Lowercase letters are used for single electrons, whereas the *total angular quantum number L* of all the electrons in an atom, taking values 0, 1, 2, 3, 4, 5, are given the uppercase letters S, P, D, F, G, H.

It is not possible for any two electrons in the same atom to have the same set of quantum numbers. This is known as *Pauli's exclusion principle* [750, 886]. Atoms are formed by filling orbitals in an order dictated by the lowest available energy states.

The lowest energy orbit has principal quantum number 1 and has a single atomic orbital, which is referred to as the *s-orbital*. If this orbital has one electron, then the electronic configuration corresponds to hydrogen (H), whereas the

n	l	m	s	**Designation**	**Number of electrons**
1	0	0	$+\frac{1}{2},-\frac{1}{2}$	1s	2
2	0	0	$+\frac{1}{2},-\frac{1}{2}$	2s	2
2	1	-1	$+\frac{1}{2},-\frac{1}{2}$		
2	1	0	$+\frac{1}{2},-\frac{1}{2}$	2p	6
2	1	1	$+\frac{1}{2},-\frac{1}{2}$		
3	0	0	$+\frac{1}{2},-\frac{1}{2}$	3s	2
3	1	-1	$+\frac{1}{2},-\frac{1}{2}$		
3	1	0	$+\frac{1}{2},-\frac{1}{2}$	3p	6
3	1	1	$+\frac{1}{2},-\frac{1}{2}$		
3	2	-2	$+\frac{1}{2},-\frac{1}{2}$		
3	2	-1	$+\frac{1}{2},-\frac{1}{2}$		
3	2	0	$+\frac{1}{2},-\frac{1}{2}$	3d	10
3	2	1	$+\frac{1}{2},-\frac{1}{2}$		
3	2	2	$+\frac{1}{2},-\frac{1}{2}$		

Table 3.2. Quantum numbers associated with the lowest electron orbitals.

Atom		Configuration	
Lithium	Li	$1s^2 2s^1$	[He] $2s^1$
Beryllium	Be	$1s^2 2s^2$	[He] $2s^2$
Boron	B	$1s^2 2s^2 2p^1$	[He] $2s^2 2p^1$
Carbon	C	$1s^2 2s^2 2p^2$	[He] $2s^2 2p^2$
Nitrogen	N	$1s^2 2s^2 2p^3$	[He] $2s^2 2p^3$
Oxygen	O	$1s^2 2s^2 2p^4$	[He] $2s^2 2p^4$
Fluorine	F	$1s^2 2s^2 2p^5$	[He] $2s^2 2p^5$
Neon	Ne	$1s^2 2s^2 2p^6$	[He] $2s^2 2p^6$

Table 3.3. Electron configurations for atoms with electrons in the first two shells.

two-electron case corresponds to helium (He). Their electronic configurations are indicated with $1s^1$ and $1s^2$, respectively. The notation names the principal quantum number first (1), then the orbital (s), and the superscript indicates the number of electrons in this orbital.

For a principal quantum number of 2, the next lowest shell contains one s-orbital and three p-orbitals. All four orbitals have the same energy, and each can hold up to two electrons. The p-orbitals are differentiated by a subscript: p_x, p_y, and p_z. An electron in one of the 3p orbital indicates that $n = 3$ and $l = 1$. The quantum numbers of several of the lowest electron orbitals are given in Table 3.2. The electron configurations of atoms with electrons in the first two shells are given in Table 3.3. As the first shell is always completely filled, an abbreviated notation exists, listing the name of the atom rather than its corresponding electronic configuration. The electronic configuration of Lithium is $1s^2 2s^1$ and can therefore be abbreviated to [He] $2s^1$.

Atoms with principal quantum number 3 have three shells, with the first two shells completely filled with electrons. The third shell has one s-orbital, three p-orbitals, and five d-orbitals. Atoms with empty d-orbitals are listed in Table 3.4. The d-orbitals are indicated with the following subscripts: d_{xy}, d_{xz}, d_{yz}, $d_{x^2-y^2}$, and d_{z^2}. Each orbital takes at most two electrons, for a maximum of 10 electrons in the d-orbitals. The atoms which have partially filled d-orbitals are called *transition metals*. Their electronic configuration is given in Table 3.5. Note that the energy levels of the 4s orbital and the 3d orbitals are very similar, causing the 4s orbital usually to be filled before the 3d orbitals.

The *inert gases* all have their s and p-orbitals fully filled. Their abbreviated configurations are given in Table 3.6.

All atoms can be arranged according to their electronic states and chemical behavior, such as in the periodic table of elements (Figure 3.3). The rows in this table are called *periods*, whereas the columns are called *groups*.

Atom		Configuration	
Sodium	Na	$1s^2 2s^2 2p^6 3s^1$	[Ne] $3s^1$
Magnesium	Mg	$1s^2 2s^2 2p^6 3s^2$	[Ne] $3s^2$
Aluminium	Al	$1s^2 2s^2 2p^6 3s^2 3p^1$	[Ne] $3s^2 3p^1$
Silicon	Si	$1s^2 2s^2 2p^6 3s^2 3p^2$	[Ne] $3s^2 3p^2$
Phosphorus	P	$1s^2 2s^2 2p^6 3s^2 3p^3$	[Ne] $3s^2 3p^3$
Sulphur	S	$1s^2 2s^2 2p^6 3s^2 3p^4$	[Ne] $3s^2 3p^4$
Chlorine	Cl	$1s^2 2s^2 2p^6 3s^2 3p^5$	[Ne] $3s^2 3p^5$
Argon	Ar	$1s^2 2s^2 2p^6 3s^2 3p^6$	[Ne] $3s^2 3p^6$

Table 3.4. Electron configurations for atoms with electrons in the first three shells (excluding d-orbitals).

Atom		Configuration	
Scandium	Sc	$1s^2 2s^2 2p^6 3s^2 3p^6 3d^1 4s^2$	[Ar] $3d^1 4s^2$
Titanium	Ti	$1s^2 2s^2 2p^6 3s^2 3p^6 3d^2 4s^2$	[Ar] $3d^2 4s^2$
Vanadium	V	$1s^2 2s^2 2p^6 3s^2 3p^6 3d^3 4s^2$	[Ar] $3d^3 4s^2$
Chromium	Cr	$1s^2 2s^2 2p^6 3s^2 3p^6 3d^5 4s^1$	[Ar] $3d^5 4s^1$
Manganese	Mn	$1s^2 2s^2 2p^6 3s^2 3p^6 3d^5 4s^2$	[Ar] $3d^5 4s^2$
Iron	Fe	$1s^2 2s^2 2p^6 3s^2 3p^6 3d^6 4s^2$	[Ar] $3d^6 4s^2$
Cobalt	Co	$1s^2 2s^2 2p^6 3s^2 3p^6 3d^7 4s^2$	[Ar] $3d^7 4s^2$
Nickel	Ni	$1s^2 2s^2 2p^6 3s^2 3p^6 3d^8 4s^2$	[Ar] $3d^8 4s^2$
Copper	Cu	$1s^2 2s^2 2p^6 3s^2 3p^6 3d^{10} 4s^1$	[Ar] $3d^{10} 4s^1$
Zinc	Zn	$1s^2 2s^2 2p^6 3s^2 3p^6 3d^{10} 4s^2$	[Ar] $3d^{10} 4s^2$

Table 3.5. Electron configurations for the 3d transition metals.

Atom		Abbreviated configuration
Helium	He	$1s^2$
Neon	Ne	[He] $2s^2 2p^6$
Argon	Ar	[Ne] $3s^2 3p^6$
Krypton	Kr	[Ar] $3d^{10} 4s^2 4p^6$
Xenon	Xe	[Kr] $4d^{10} 5s^2 5p^6$

Table 3.6. Electron configurations of inert gases.

$\mathrm{I_a}$	$\mathrm{II_a}$	$\mathrm{III_b}$	$\mathrm{IV_b}$	$\mathrm{V_b}$	$\mathrm{VI_b}$	$\mathrm{VII_b}$	VIII			$\mathrm{I_b}$	$\mathrm{II_b}$	$\mathrm{III_a}$	$\mathrm{IV_a}$	$\mathrm{V_a}$	$\mathrm{VI_a}$	$\mathrm{VII_a}$	O
1 H 1.0079 Hydrogen Colorless																	2 He 4.0026 Helium Colorless
3 Li 6.941 Lithium Silvery white gray	4 Be 9.01218 Beryllium Lead gray											5 B 10.81 Boron Black	6 C 12.011 Carbon Black Colorless	7 N 14.0067 Nitrogen Colorless	8 O 15.9994 Oxygen Pale blue	9 F 19.9984 Fluorine Pale yellow	10 Ne 20.179 Neon Colorless
11 Na 22.98977 Sodium Silvery white	12 Mg 24.305 Magnesium Silvery white											13 Al 26.98154 Aluminium Silvery	14 Si 28.086 Silicon Dark gray, bluish tinge	15 P 30.97376 Phosphorus Colorless/red silvery white	16 S 32.06 Sulfur Lemon yellow	17 Cl 35.453 Chlorine Yellowish green	18 Ar 39.948 Argon Colorless
19 K 39.098 Potassium Silvery white	20 Ca 40.08 Calcium Silvery white	21 Sc 44.9559 Scandium Silvery white	22 Ti 47.90 Titanium Silvery metallic	23 V 50.9414 Vanadium Gray metallic	24 Cr 51.996 Chromium Silvery metallic	25 Mn 54.9380 Manganese Silvery metallic	26 Fe 55.847 Iron Metallic, grayish tinge	27 Co 58.9332 Cobalt Metallic, grayish tinge	28 Ni 58.70 Nickel Metallic, silvery tinge	29 Cu 63.546 Copper Copper, metallic	30 Zn 65.38 Zinc Bluish, pale gray	31 Ga 69.72 Gallium Silvery white	32 Ge 72.59 Germanium Grayish white	33 As 74.9216 Arsenic Metallic gray	34 Se 78.96 Selenium Metallic gray	35 Br 79.904 Bromine Red-brown, metallic	36 Kr 83.80 Krypton Colorless
37 Rb 85.4648 Rubidium Silvery white	38 Sr 87.62 Strontium Silvery white	39 Y 88.9059 Yttrium Silvery white	40 Zr 91.22 Zirconium Silvery white	41 Nb 92.9064 Niobium Gray metallic	42 Mo 95.94 Molybdenum Gray metallic	43 Tc 98.9062 Technetium Silvery gray metallic	44 Ru 101.07 Ruthenium Silvery white metallic	45 Rh 102.9055 Rhodium Silvery white metallic	46 Pd 106.4 Palladium Silvery white metallic	47 Ag 107.868 Silver Silver	48 Cd 112.40 Cadmium Silvery gray metallic	49 In 114.82 Indium Silvery gray	50 Sn 118.69 Tin Silvery gray	51 Sb 121.75 Antimony Silvery gray	52 Te 127.60 Tellurium Silvery gray	53 I 126.9045 Iodine Violet-dark gray	54 Xe 131.30 Xenon Colorless
55 Cs 132.9054 Caesium Silvery gold	56 Ba 137.34 Barium Silvery white	71 Lu 174.97 Lutetium Silvery white	72 Hf 178.49 Hafnium Gray steel	73 Ta 180.9479 Tantalum Gray blue	74 W 183.85 Tungsten Grayish white	75 Re 186.207 Rhenium Grayish white	76 Os 190.2 Osmium Bluish gray	77 Ir 192.22 Iridium Silvery white	78 Pt 195.09 Platinum Grayish white	79 Au 196.9665 Gold Gold	80 Hg 200.59 Mercury Silvery white	81 Tl 204.37 Thallium Silvery white	82 Pb 207.2 Lead Bluish white	83 Bi 208.9804 Bismuth Reddish white	84 Po 210 Polonium Silvery	85 At 211 Astatine Metallic	86 Rn 222 Radon Colorless
87 Fr 223 Francium Metallic	88 Ra 226.0254 Radium Metallic	103 Lr 257 Lawrencium Unknown	104 Rf 257 Rutherfordium Unknown	105 Db 264 Dubnium Unknown	106 Sg 266 Seaborgium Unknown	107 Bh 264 Bohrium Unknown	108 Hs 269 Hassium Unknown	109 Mt 268 Meitnerium Unknown	110 Ds 281 Darmstadtium Unknown	111 Rg 272 Roentgenium Unknown	112 Uub 285 Ununbium Unknown						

Legend:

Atomic number Symbol Atomic weight Name Color

Color coding:

Metals Transition Metals Metalloids Non-Metals Inert Gases

Lanthanide series	57 La 138.9055 Lanthanum Silvery white	58 Ce 140.12 Cerium Silvery white	59 Pr 140.9077 Praseodymium Silvery white yellowish	60 Nd 144.24 Neodymium Silvery white yellowish	61 Pm 147 Promethium Metallic	62 Sm 150.4 Samarium Silvery white	63 Eu 151.96 Europium Silvery white	64 Gd 157.25 Gadolinium Silvery white	65 Tb 158.9254 Terbium Silvery white	66 Dy 162.50 Dysprosium Silvery white	67 Ho 164.9304 Holmium Silvery white	68 Er 167.26 Erbium Silvery white	69 Tm 168.9342 Thulium Silvery white	70 Yb 173.04 Ytterbium
Actinide series	89 Ac 227 Actinium Silvery	90 Th 232.0381 Thorium Silvery white	91 Pa 231.0359 Protactinium Silvery metallic	92 U 238.029 Uranium Metallic gray	93 Np 237.0482 Neptunium Silvery metallic	94 Pu 239.11 Plutonium Silvery white	95 Am 241 Americium Silvery white	96 Cm 244 Curium Silver	97 Bk 249 Berkelium Unknown	98 Cf 252 Californium Unknown	99 Es 253 Einsteinium Unknown	100 Fm 254 Fermium Unknown	101 Md 256 Mendelevium Unknown	102 No 254 Nobelium Unknown

Figure 3.3. The periodic table of elements.

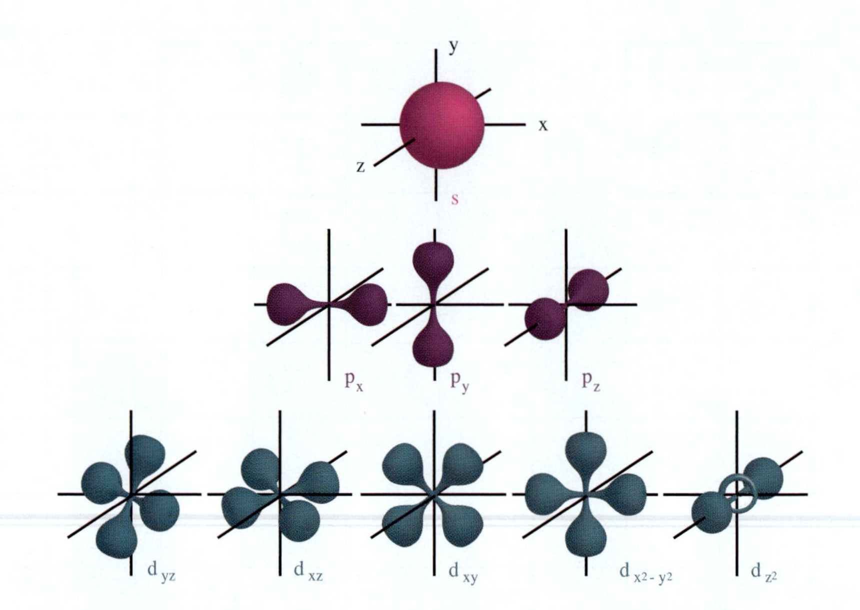

Figure 3.4. A schematic of s-, p-, and d-orbitals. Note that this is a representation of a probability density function, and that therefore the boundary is not sharp in reality, nor is the shape accurate.

3.2.5 Orbital Shapes and Bonds

In Section 3.2.3, the time independent Schrödinger equation was presented. A solution to this equation is a *wave function* ψ . Squaring and normalizing ψ yields a function which indicates the electron cloud density. It is therefore a function of the probability of finding an electron in any particular position around a nucleus.

For s orbitals, the electron cloud density is spherical, with the center coinciding with the nucleus. There are three p-orbitals in the 2p shell, which are each dumbbell-shaped and orthogonal to each other. The p-orbitals are aligned with the *x*-, *y*-, and *z*-axes. The d_{xy}-, d_{xz}-, and d_{yz}-orbitals have lobes that lie in between the axes, whereas the $d_{x^2-y^2}$- and d_{z^2}-lobes are once more axis aligned [1128]. A schematic of s-, p-, and d-orbitals is given in Figure 3.4.

When a pair of hydrogen atoms form a hydrogen molecule, their 1s orbitals may overlap forming a bond between the two atoms. If the wave-functions associated with both orbitals have the same sign, then we speak of *bonding orbitals*. If the signs are opposite, the electrons in the two orbitals repel each other. This is called *anti-bonding* . The combined 1s orbitals are termed *σ-orbitals* and their associated bonds are called *σ-bonds* and *σ^*-anti-bonds*.

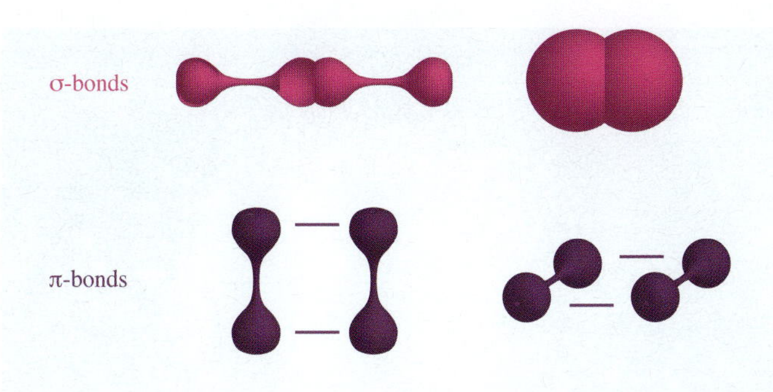

Figure 3.5. A schematic of bonding between s- and p-orbitals.

The three 2p-orbitals also form bonds, see Figure 3.5. In particular, the p_x-orbitals of two atoms may directly overlap, forming a strong $2p\sigma$-bond or $2p\sigma^*$-anti-bond. The two $2p_y$- and $2p_z$-orbitals can also bond, forming weaker *π-bonds* and *π^*-anti-bonds*, dependent on whether the wave-functions have identical or opposite signs.

Molecules can bond more than once, i.e., double or triple bonding can occur. For instance, oxygen (O_2) produces a double bond—one $2p\sigma$-bond due to the p_x-orbital and a $2p\pi$-bond due to either the p_y- or p_z-orbital. Nitrogen N_2, which has a $2s^2 2p^3$ configuration, forms a triple bond involving p_x-, p_y-, and p_z-orbitals, as shown in Figure 3.6.

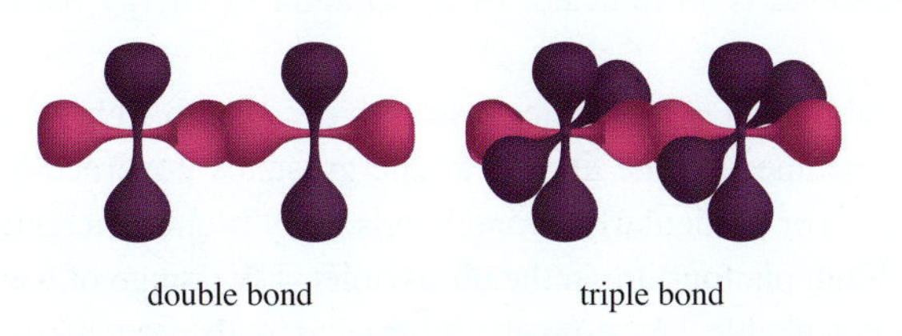

Figure 3.6. A schematic of double and triple bonding.

Each bonding between atoms creates an energy level lower than the sum of the energies of the orbitals associated in the bonding. Similarly, anti-bonding involves a higher energy level than two unbonded orbitals. As such, we may expect that molecules with orbitals when bonding with orbitals in other atoms, absorb photons with different energies than single atoms. Thus, much of the richness and variety of color in substances is due to the formation of bonds in molecules, or to the presence of atoms and ions inside a crystal lattice consisting of different types of atoms.

Figure 3.7. The strong ionic bonding between Na^+ and Cl^- causes absorption at wavelengths outside the visible range. As a result, kitchen salt is colorless.

3.3 Atoms and Ions

In atoms, the energy states that electrons can attain are determined by the set of allowable transitions between orbitals. In molecules, which consist of multiple atoms, the electrons are involved in the bonding between individual atoms. The strength of the bond is an indicator of the amount of energy required to free an electron.

Thus, the ability of a molecule to selectively absorb light of specific wavelengths, is determined by the allowable energy states governed by the bonding between atoms. For particularly strong bonds, only highly energetic photons can be absorbed. Such photons are in the ultra-violet (UV) range of wavelengths, and are therefore not visible. As a result, substances with particularly strong bonds tend to appear colorless.

An example is the ionic bonding in salt (Figure 3.7), which consists of a sodium atom and a chlorine atom. The sodium atom donates the electron in its outermost shell to the chlorine atom. As a result, the two atoms become sodium and chloride ions, Na^+ and Cl^-, each having a fully filled outer orbit, so that all electrons are paired. Having all electrons paired creates a very stable bonding, so that only highly energetic photons can be absorbed.

In contrast to ionic bonding, there is the situation where an electron in the outer shell is shared equally by two atoms. Such covalent bonding can also produce very strong bonds between atoms, as evidenced by diamond. The strength of the bonds in this cubic form of carbon, resulting from the pairing of electrons, causes diamond to be colorless.

3.3.1 Color from 3d Transition Metals

In many inorganic compounds, the bonding is neither fully ionic, nor fully covalent. An example is corundum, Al_2O_3, which has all its electrons paired off in a configuration which is about 60% ionic and 40% covalent. The strength of the bonding again causes the material in its pure form to be colorless (it is then known as white sapphire).

However, if the material contains about 1% chromium sesquioxide, Cr_2O_3, the material becomes red and is then known as ruby. This color is a result of unpaired electrons that can be excited to various states making transitions that correspond to a release of radiation in the visible range, i.e., fluorescence.

It is the 3d-orbitals and, in particular their shape, that are involved in the formation of color. In a single atom, each of the five different 3d-orbitals has the same energy. However, the d-orbitals can be classified in two groups. These are the d_{xy}-, d_{xz}-, and d_{yz}-orbitals, which have lobes lying in between the axes, and the $d_{x^2-y^2}$- and d_{z^2}-orbitals, which are axis aligned. The former group is collectively called the t_{2g} or t_2 set, whereas the latter group is called the e_g or e set [1128].

If the atom is located in a crystal, then there are nearby electrons which repel the electrons in the d-orbitals. This changes the shape of the d-orbitals by an amount that depends on their orientation. The energy associated with some of the d-orbitals increases more than others in the presence of a surrounding crystal lattice. This phenomenon is called *crystal field splitting* or *ligand field splitting* [1128]. Of course, the shape of the surrounding crystal lattice, as well as the distance to the transition metal, also play a role in which d-orbitals are raised in energy, and by how much.

The energy levels associated with the d-orbitals in a transition metal lodged in a crystal lattice determine which photons may be absorbed, and thereby determine the absorption spectrum of the material.

If ruby is illuminated with white light, then electrons in Cr^{3+} ions are excited from their ground state E_0 to two upper levels, E_2 and E_3, leading to two broad absorption spectra. The transition from E_0 to E_2 corresponds to a range of wavelengths centered around 556 nm (yellow/green). The transition from E_0 to E_3 corresponds to a range of wavelengths centered around 407 nm (violet). The transition from E_0 to E_1 is not within the visible range of wavelengths. The light that is not absorbed is predominantly red, i.e., the color of ruby [1128].

A second example is given by emerald. The shape of the crystal lattice is the same, albeit somewhat larger. The impurities are also formed by Cr^{3+} ions. This results in an energy diagram which is essentially the same, although the upper energy levels are closer to the ground state. This shifts the absorption spectrum towards red, yielding the green color with a hint of blue typically seen in emeralds.

The impurities that give ruby and emerald their color (Cr_2O_3 in this case), are called *transition metal impurities*. The resulting colors are called *allochromatic transition metal colors* . Minerals may be colored with chromium, cobalt, iron, manganese, nickel, samarium, and vanadium impurities, and, in general, the transition metals listed in Table 3.5.

If the transition element is also the main compound, we speak of *idiochromatic transition metal colors*. Elements involved in idiochromatic transition metals are chromium, manganese, iron, cobalt, nickel, and copper. Green, for instance, can be formed in several ways in crystals, including chrome green (Cr_2O_3), manganese (II) oxide (MnO), melantite (Fe_2O_3), cobalt (II) oxide (CoO), bunsenite (NiO), and malachite ($Cu_2(CO_3)(OH)_2$). An example of a red colorization is rhodochrosite ($MnCO_3$), whereas blue can be formed for instance by Thenard's blue (Al_2CoO_4), azurite ($Cu_3(CO_3)_2(OH)_2$), and turquoise ($CuAl_6(PO_4)(OH)_8$) [815].

3.3.2 Charge Transfer Color

If multiple transition metals are present as impurities, it may become possible for an electron to move from one transition metal ion to another and absorb a photon in the process. As an example, corundum with 1% of titanium or iron does not have color (titanium) or a very pale yellow color (iron). However, if both impurities are present, *inter-valence charge transfer* can occur:

$$Fe^{2+} + Ti^{4+} \rightarrow Fe^{3+} + Ti^{3+} \tag{3.43}$$

Here, an electron has moved from the iron ion to the titanium ion. The difference in energy between the two states is 2.11 eV. Thus, absorption of photons with energies of 2.11 eV is possible, resulting in a deep blue color. Corundum containing both titanium and iron is known as *blue sapphire*.

The crystal structure of corundum is such that the above charge transfer occurs for ions that are 2.65 Å apart.[4] However, the crystal lattice also allows Ti and Fe ions at a distance of 2.79 Å, resulting in a weaker second absorption which creates a blue-green color. These two configurations occur at right angles, which means

[4]The Ångstrom is a measure of distance, and equals one tenth of a nanometer.

that the blue and blue-green colors are seen dependent on the orientation of the crystal. Blue sapphire thus exhibits dichroism.

Other examples of colors caused by charge transfer include many pigments, such as Prussian blue, as well as yellow, brown, and black colors seen in rocks and minerals that contain iron.

3.3.3 Metals and Semiconductors

Metals are characterized by high conductivity, which increases with temperature. The opposite of conductivity is resistivity, which is therefore low in metals. Dielectrics, on the other hand, have high resistivity, and thus do not conduct electricity very well. Materials that fall in between these extremes are called *semiconductors* and are characterized by a decrease in conductivity when the temperature is raised.

To explain the properties of metals and semiconductors, one may resort to *band theory*, which can be seen as a special case of *molecular orbital theory*. To begin, we consider a pair of hydrogen atoms approaching each other from infinity (after [815]). Each atom has an electron in the 1s orbital. If these two electrons have parallel spins, they will repulse more and more as they approach, creating a higher energy. If the two electrons have opposite spin, then the energy will fall as the two atoms approach. In this case, as the atoms move very close together, the nuclei will start to repulse each other and will counteract the attraction between the two opposite spinning electrons.

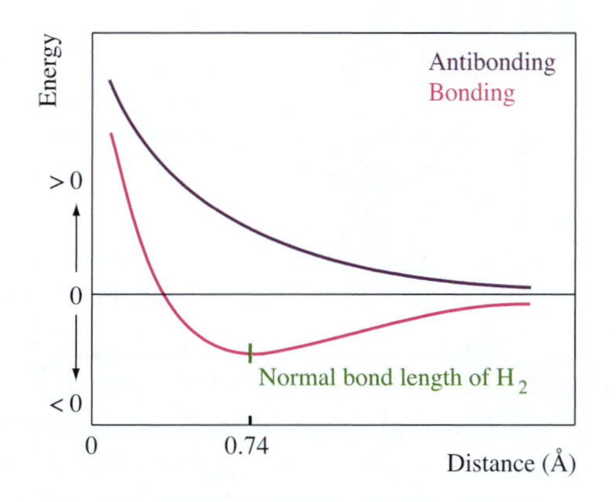

Figure 3.8. Morse curve showing energy as a function of distance. The inter-atomic distance is normally measured in Ångstrom, which is one-tenth of a nanometer (see also Figure 2.4).

Figure 3.8 shows the energy of the ensemble of two hydrogen atoms as function of the distance between the two atoms. The top curve is for two electrons having parallel spin, whereas the bottom curve is for electrons having opposite spin. These types of plot are called *Morse curves* or *potential energy diagrams*. For hydrogen with electrons in opposite spin, the distance between atoms is 0.74 Å.

The molecular hydrogen orbitals in one H_2 molecule can accommodate the same number of electrons in two orbitals as two atomic hydrogen orbitals in one orbital each. Extending this approach by involving N hydrogen atoms, the number of orbitals and energy levels increases by N. These energy levels will be very close together (and depend on the inter-atoms distance between each pair of hydrogen atoms), such that individual levels can not be distinguished. Thus, a large number of hydrogen atoms in close proximity creates an energy band.

Typically, electrons in partially filled outer orbitals form bands, whereas fully filled inner orbitals are not involved in bands unless their inter-atomic distance becomes very small. Unfilled shells can form bands at much larger distances, as shown in Figure 3.9.

For some materials, two or more of the outer orbitals, such as the 2s- and 2p-orbitals of lithium, each broaden into two bands [781]. For metals, these bands may overlap, causing interactions between electrons in different bands. This aids conductivity in metals.

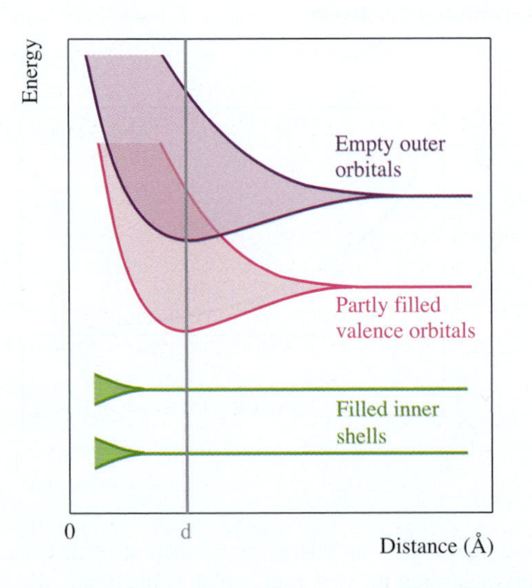

Figure 3.9. Bands formed by clusters of atoms (after [815]).

The band structure of metals has the effect that light is strongly absorbed at many different wavelengths. Refracted light usually travels into the metal for less than a single wavelength (several hundred atoms), before it is completely absorbed. This absorption creates a current through the metal near the surface, which in turn produces new photons which appear as reflected light. As a consequence, metals strongly reflect light.

Within the band structure, some transitions in energy levels are favored over others. While the precise mechanism is currently poorly understood [815], the limitations of state transitions give rise to a wavelength dependent reflectivity as a function of the atomic structure of different metals. For instance, silver reflects light over a broader range of wavelengths than gold, which reflects less light at shorter wavelengths.

In non-conducting materials, bands formed by different orbitals may repel, as in the case of the 2s- and 2p-electrons in diamond. This causes a gap to occur between the 2s- and 2p-bands, which for diamond are well separated at the carbon bond distance of 1.54 Å. The lower energy band is called the *valence band* and the upper band the *conduction band*. As the conduction band is at a higher energy level than the most energetic electrons in the material,[5] the conduction band is empty, and diamond therefore acts as an insulator [815].

The different atomic structures of materials thus give rise to different interactions between bands, as well as different sizes of the energy gap between bands. If the bands overlap in energy level, then we speak of metals. If there is a large gap between bands, such as the 5.4 eV band gap of pure diamond, the material is an insulator. If the gap is relatively small, then we speak of semiconductors—a class of materials that plays a crucial role in the manufacture of transistors and chips.

The energy, and therefore the wavelength, of light that can be absorbed is dependent on the size of the band gap. Light needs to have sufficient energy to excite an electron such that it can jump from the valence band into the conduction band. For diamond the band gap is so large that no visible light can be absorbed and, hence, the diamond appears colorless.

For a medium band-gap semiconductor, such as the pigment cadmium yellow (CdS), the band gap is 2.6 eV, and this is small enough to absorb some light in the violet and blue range. The pigment vermillion (HgS) has a yet smaller band gap of 2.0 eV, allowing the absorption of most visible wavelengths except red. Band gaps below 1.77 eV allow absorption of all visible wavelengths, and materials with small band gaps therefore appear black.

It is possible to add impurities to pure semiconductors. This involves replacing a small amount of semiconductor material with a different substance. Such

[5]The energy of the most energetic electrons in a material is called the *Fermi energy* or *Fermi level*.

impurities are called *activators* or *dopants*. For instance, adding a small concentration of nitrogen (N) to diamond will introduce extra electrons that form a new energy level within the band gap, called a *donor level*. As a result, the band gap is effectively split into two smaller gaps, and now light in the blue and violet range can be absorbed. Thus, diamond containing some nitrogen impurities will have a yellow color.

Similarly, impurities may be added that create an electron deficiency, and therefore they may insert an *acceptor level* within the band gap. Wherever an electron deficiency exists, we speak of a *hole*. An example is the addition of boron which, when added to diamond, causes a blue coloration.

3.4 Molecules

Molecules consist of collections of atoms that form bonds. Collections of atoms in a single molecule can be of a particular form and can occur in different molecules. Such collections are called *groups*. Some groups form bonds with energy levels enabling the absorption of photons within the visible range of wavelengths. In that case, the group is said to be color-bearing.

Other groups may bring about shifts in intensity and/or wavelength. Groups that increase or decrease intensity are referred to as *hyperchromic* and *hypochromic*, respectively. Substances that aid in color shifts are called *bathochromic* if the shift is towards the red end of the spectrum and *hypsochromic* if the shift is towards the blue end.

In (organic) molecules, color-bearing groups of atoms are called *chromophores*. Examples of chromophores include carbon-carbon double bonds as well as nitrogen double bonds. The molecules containing such chromophores are called *chromogens*, although they frequently have other groups as well that are involved in color formation. These extra groups are called *auxochromes* [815, 1128]. The auxochromes shift as well as strengthen the color, i.e., they are both bathochromic and usually hypochromic. Auxochromes include NH_2, NO_2, and CH_3 structures as well as NHR, and NR_2, where R is a more complex organic group.

Several important classes of organic molecules exists, and these are briefly discussed in the following sections.

3.4.1 Porphyrins

Carbon atoms often form double bonds, yielding an absorption spectrum that peaks outside the visible range. However, chains of carbon atoms that alternate

Aldehyde	Chain length	λ_{max}(in nm)
CH_3-(CH=CH)-CH_3	1	220
CH_3-$(CH=CH)_2$-CH_3	2	270
CH_3-$(CH=CH)_3$-CH_3	3	312
CH_3-$(CH=CH)_4$-CH_3	4	343
CH_3-$(CH=CH)_5$-CH_3	5	370
CH_3-$(CH=CH)_6$-CH_3	6	393
CH_3-$(CH=CH)_7$-CH_3	7	415

Table 3.7. Peak absorption wavelength as a function of the number of conjugated double bonds (after [1099]).

between single and double bonds create absorption spectra that are shifted toward the visible range. Such chains are called *conjugated* double bonds. For instance, a chain of six such double bonds:

$$CH_3 - (CH = CH)_6 - CH_3 \tag{3.44}$$

reaches the blue end of the spectrum. Molecules containing such a group therefore appear yellow. Examples are α- and β carotene, which appear in carrots. The number of conjugated double bonds determines where the peak in the absorption spectrum lies. The π-electrons in such chains are mobile across the chain. Adding further double bonds increases their mobility and, thereby, the number of resonant configurations. This lowers the energy level of the lowest excited level and, thus, shifts the absorption spectrum further into the visible range [767]. Table 3.7 lists the peak absorption wavelength for several aldehydes.

Conjugated bonds also appear in more elaborate molecular structures, for instance, in ones where bonds circle a single metal atom. Such structures are called porphyrins, with chlorophyll being one of the most important naturally occurring one, giving plants and algae their green color (Figure 1.1). The chlorophyll molecule has a magnesium atom at its center.

Iron is the center of a different porphyrin, called *heme*, which in many species is part of the hemoglobin molecule. This molecule is responsible for transporting iron through the bloodstream. Heme is responsible for the color of blood. However, this color is derived from π to $\pi*$ transitions, rather than from the presence of an iron atom.

3.4.2 Photochromic Compounds

Some groups of molecules change color under the influence of light. If such a change is reversible, then the molecule is said to be *photochromic*. In particular, some molecules turn colorless when illuminated, a process called called *bleaching*.

The most important photochromic molecules are the ones that transduce light in the visual system. For instance, the molecules present in the rods, called rhodopsin, have a red-purple color, changing to colorless under the influence of light. Under illumination, rhodopsin transforms via several intermediate forms, to a range of bleached molecules called *metarhodopsins*. In the dark, these molecules revert back to rhodopsin, through a range of intermediate forms.

The chromophore in rhodopsin is called 11-*cis*-retinal. It is connected to an opsin through the amino acid lycene. The opsin acts as an auxochrome and causes a bathochromic shift by raising the peak of the absorption spectrum of 11-*cis*-retinal from 370 nm to 570 nm.

Under medium light intensities, the incidence of a photon causes the 11-*cis*-retinal to change to the trans form, called all-*trans*-retinal. The reaction that effects this change also causes a chain of events that leads to the rod emitting a signal, thereby mediating vision. The configuration of all-*trans*-retinal connected to the opsin via lycene is called metarhodopsin.

In cones, the retinal is believed to be connected to different opsins, which cause different bathochromic shifts. This creates different peak sensitivities in the three cone types, thereby enabling color vision.

Under low light levels, the *trans*-retinal may become fully dislodged from its opsin and will be transformed back to its *cis* form by means of enzymes located in the eye. Under yet lower light conditions, the *trans*-retinal may leave the eye altogether. It is then carried through the bloodstream to the liver, where it is regenerated into the *cis*-form. This complex sequence of events under low light conditions causes dark adaptation to progress slowly over a long period of time [1128] (see Section 4.3.2).

Another application of a photochromic material is found in sunglasses that change to a darker state under UV illumination. Hence, outside in the sun, they will be darker than indoors.

3.4.3 Colorants

Color may be given to objects by means of *colorants*, which is the general term used for substances that produce color. A distinction is normally made between *pigments* and *dyes*, as seen in the definition of pigments offered by the Color Pigments Manufacturers Association (CPMA) [678]:

> Pigments are colored, black, white or fluorescent particulate organic and inorganic solids which are usually insoluble in, and essentially physically

> and chemically unaffected by, the vehicle or substrate in which they are incorporated. They alter appearance by selective absorption and/or by scattering of light.
>
> Pigments are usually dispersed in vehicles or substrates for application, as for instance in inks, paints, plastics or other polymeric materials. Pigments retain a crystal or particulate structure throughout the coloration process.
>
> As a result of the physical and chemical characteristics of pigments, pigments and dyes differ in their application; when a dye is applied, it penetrates the substrate in a soluble form after which it may or may not become insoluble. When a pigment is used to color or opacify a substrate, the finely divided insoluble solid remains throughout the coloration process.

If after application the substrate evaporates, leaving only the pigment particles behind, we speak of an *ink*; an example is watercolor. In other cases, the substrate may harden and form a protective coating, as for instance in oil paints.

Dyes, on the other hand, are usually dissolved in a solvent [1247]. The dye molecules do not stay together, but mix throughout the solvent. These molecules absorb light selectively, which gives the solution its color. In between dyes and pigments is a third category of colorants, called *lakes*. These are essentially pigments in which the color of the particles is created by dyeing. Before dyeing, the particles, which are typically made of aluminium oxide, are colorless and translucent.

Almost all colorants can be synthetically produced, with the exception of chlorophyll, which is used to color, for instance, soap and chewing gum green. For yellow, orange, and red colorants, a single absorption band in the blue-violet-green region is required. A single absorption band in the red-orange-yellow region creates blue and violet colors. However, to produce green, two absorption bands in the red and violet parts of the spectrum are required, which presents a difficult problem for designers of organic colorants [815]. Using a mixture of yellow and blue colorants tends to produce less saturated colors than a single green colorant and may cause a change in color when one of the two colorants fades faster than the other.

There are many ways to specify a colorant. One is the *color index*, which is produced by the Society of Dyers and Colourists and the American Association of Textile Chemists and Colorists, and contains some 8000 dyes and pigments [1067]. In addition, pigments and dyes can have many designations, given by different associations and agencies. For instance color index C.I. 14700, which was formerly used to color food, is also known as D & C Red No 4 by the U.S. Food and Drug Administration, as well as Food Red 1 and Ponceau SX [815]. Some substances are neither dyes or pigments, such as substances used to color glasses, glazes, and enamels.

3.4.4 Dyes

As dye molecules are fully dissolved, the interaction of light with dyes is relatively straightforward and follows Beer's Law (see (2.191)). This means that the transmittance of a material depends directly on its concentration of dye molecules. Similarly, the amount of light absorbed depends on the thickness of the layer, assuming that the dye is applied in layers.

If a dye is constructed by mixing multiple types of dye, then the wavelength-dependent transmittance of the mixtures can be inferred from the transmittance functions T_i of each dye i separately by simply multiplying the transmittance functions together for each wavelength:

$$T_{\text{mix}}(\lambda) = \prod_{i=0}^{n} T_i(\lambda). \tag{3.45}$$

Changing the concentration of a colorant may alter its perceived color. Several mechanisms contribute to this, including scattering, which for instance causes the white head on a glass of beer. A second mechanism is *dye dichroism*,[6] as illustrated in Figure 3.10. The two curves in this figure show absorption in the same region, which is partly in the visible range. The low concentration absorbs a small amount in the violet region, causing a yellow color. As the concentration is increased, a more significant portion of the visible spectrum is absorbed, including blue and green, yielding an orange to red color. This effect would not occur

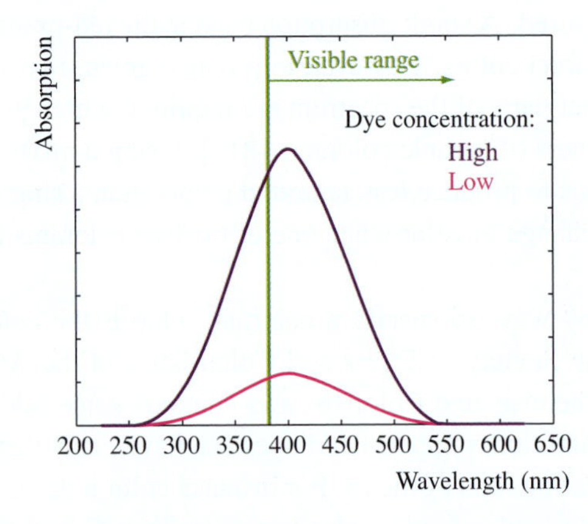

Figure 3.10. Dye dichroism as a result of varying concentrations of dye in a solution.

[6]This phenomenon is to be distinguished from the orientation-sensitive dichroism discussed in Section 3.7.

Figure 3.11. Undiluted yellow food coloring, containing E102, E110, and E124, is red (left), whereas diluting this substance with water produces a yellow color (right).

if the absorption spectrum were entirely located in the visible region. Some yellow food colorings show this dependency of color on concentration, as shown in Figure 3.11. Dichroic dyes find application in the construction of certain LCD display devices [537].

A related phenomenon is outlined in Figure 3.12, where all absorption takes place in the visible spectrum. At low concentrations, the dent around 600 nm causes the dye to look yellow. Increasing the concentration makes this dent relatively unimportant, as yellow wavelengths are now mostly absorbed. As a result, the lack of absorption in the violet range causes the material to appear violet.

Assume that a surface is painted with a dye where the chromogens are dissolved at a concentration c, and that the layer has a thickness of h. The painted layer can then be modeled using Beer's law (Section 2.9.4). Dyes can thus be characterized by their thickness as well as the concentration of their chromogens. As dyes do not have a particulate structure, more complex phenomena, such as scattering, do not need to be taken into account. This is in contrast to pigmented paints, which are discussed in Section 3.4.6.

3.4.5 Bleaching

The chromogens of organic dyes can be split by chemical reactions. The active ingredients involved in these reactions are usually either chlorine or hydrogen

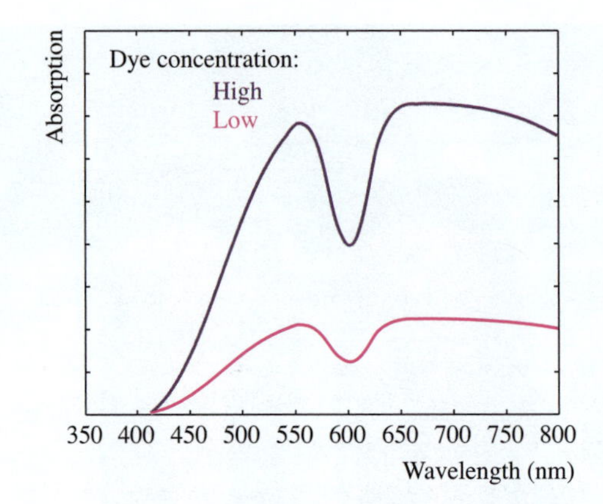

Figure 3.12. A change in concentration yields a different perceived color.

peroxide—substances used to bleach fabrics, paper, and hair. In each case, the reaction breaks the π component of a double bond, leaving only a single σ-bond, or it may even break both bonds.

As a result, the reaction causes a hypsochromic shift towards the blue end of the spectrum and may move the peak of the absorption spectrum into the ultraviolet region. As the peak of absorption is then outside the visible region, the result of chemical bleaching is a loss of color.

3.4.6 Pigments

As pigments are insoluble particles, their index of refraction is usually different from the material they are suspended in. As a result, pigments can be used both to give color to a material and to make the material more opaque. To maximize the *hiding power* of a pigment, the difference between the index of refraction of the pigment and its carrier should be as large as possible. This then creates particles that mostly reflect and scatter light, rather than transmit it. Pigments in white paints, for instance, usually have a refractive index greater than 2.0.

A paint usually contains a carrier, such as linseed oil in the case of artists' oil paints, with pigment particles suspended in it. A full account of light interaction with a layer of paint would involve scattering, which could be modeled as Mie scattering. However, this approach would be too complicated [629], and a simpler theory is required to model how light is absorbed and scattered in a pigmented paint. In this case, Beer's law is not appropriate either, as this empirical law does not assume that scattering occurs in the medium.

Instead, Kubelka-Munk theory is normally applied to describe light behavior in a layer of paint [628]. Light traveling through such a layer has at every point

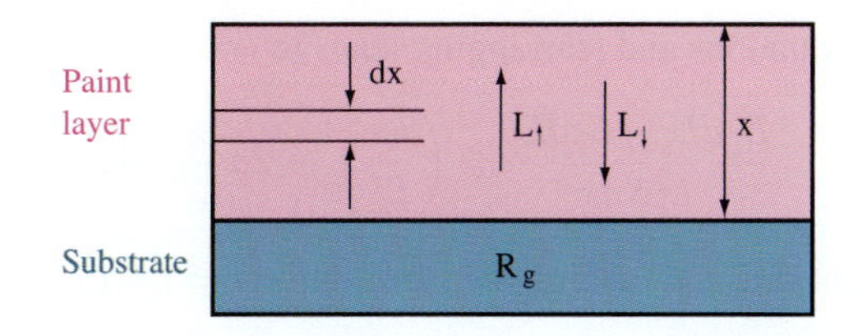

Figure 3.13. The geometry associated with Kubelka-Munk theory.

a probability of being scattered. The simplifying assumption is made that light scatters only into the direction from which it came. Thus, for light penetrating the layer in the forward direction, a scattering event would cause some fraction of this light to be reflected back towards the surface of the layer. If light traveling towards the surface undergoes a further scattering event, it will be scattered in a direction away from the surface. Lateral scattering is not accounted for.

The geometry associated with Kubelka-Munk theory is shown in Figure 3.13. Assume that the layer of paint has a uniform thickness of x. With S the scattering component and K the absorption, in a thin sublayer of thickness dx, the light traveling downward is attenuated by an amount equal to [412]

$$(K+S)\ L_{\downarrow}\, dx. \tag{3.46}$$

The attenuation induced by the same sublayer for light traveling upward is similarly

$$(K+S)\ L_{\uparrow}\, dx. \tag{3.47}$$

Since we assume that the light that is scattered is always scattered in the opposite direction, the attenuation in each direction is modified by $-SLdx$. Hence, the total change for light traveling in either direction due to a sublayer of thickness dx is

$$dL_{\downarrow} = (K+S)\, L_{\downarrow}\, dx - S L_{\uparrow}\, dx, \tag{3.48a}$$
$$-dL_{\uparrow} = (K+S)\, L_{\uparrow}\, dx - S L_{\downarrow}\, dx. \tag{3.48b}$$

This pair of equations can be solved as follows [412]. First, we rearrange terms and substitute $a = 1 + K/S$:

$$\frac{dL_{\downarrow}}{S dx} = a L_{\downarrow} - L_{\uparrow}, \tag{3.49a}$$
$$\frac{-dL_{\uparrow}}{S dx} = a L_{\uparrow} - L_{\downarrow}. \tag{3.49b}$$

Summing these equations and rearranging leads to

$$\frac{L_\downarrow dL_\uparrow - L_\uparrow dL_\downarrow}{L_\downarrow^2 S dx} = \left(\frac{L_\uparrow}{L_\downarrow}\right)^2 - 2a\frac{L_\uparrow}{L_\downarrow} + 1. \tag{3.50}$$

Using the quotient rule and making the substitution $r = L_\uparrow / L_\downarrow$, we get

$$\frac{dr}{S dx} = r^2 - 2ar + 1. \tag{3.51}$$

Rearranging and integrating yields

$$\int \frac{dr}{r^2 - 2ar + 1} = S \int dx. \tag{3.52}$$

For a thickness of 0, the value of r will be equal to the reflectance of the underlying material, R_g. For thicknesses larger than 0, the reflectance of the paint plus its substrate is R. Hence, the integration limits of the left-hand side of the above equation are R_g and R, allowing us to carry out the integration. Using $b = \sqrt{a^2 - 1}$, we have

$$\int_{R_g}^{R} \frac{dr}{r^2 - 2ar + 1} = \frac{1}{2b} \int_{R_g}^{R} \frac{dr}{r - (a+b)} - \frac{1}{2b} \int_{R_g} R \frac{dr}{r - (a-b)} \tag{3.53a}$$

$$= \frac{1}{2b} \ln \left(\frac{(R - a - b)(R_g - a + b)}{(R - a + b)(R_g - a - b)} \right), \tag{3.53b}$$

and therefore from (3.52)

$$= St, \tag{3.53c}$$

where t is the thickness of the layer of paint. For a layer of paint with a hypothetical thickness of infinity, the hiding of the substrate is complete. In this case, the value of R_g can be set to any convenient value, such as $R_g = 0$. A further rearrangement gives

$$\lim_{t \to \infty} \frac{(R - a - b)(-a + b)}{\exp(2Stb)} = (R - a + b)(-a - b), \tag{3.54a}$$

$$0 = (R - a + b)(-a - b). \tag{3.54b}$$

Hence, for complete hiding, the reflectance of a paint is given by

$$R = \frac{1}{a + \sqrt{a^2 - 1}} \tag{3.55a}$$

$$= \frac{1}{1 + \frac{K}{S} + \sqrt{\left(1 + \frac{K}{S}\right)^2 - 1}}. \tag{3.55b}$$

This last equation is the Kubelka-Munk equation, which expresses the reflectance of a layer of paint given absorption and scattering coefficients, K and S, respectively. Equivalent forms of this equation are

$$R = 1 + \frac{K}{S} - \sqrt{\left(\frac{K}{S}\right)^2 + 2\frac{K}{S}}, \tag{3.56a}$$

$$\frac{K}{S} = \frac{(1-R)^2}{2R}. \tag{3.56b}$$

These equations are valid for pigmented layers of paint where the pigment is dispersed homogeneously throughout the carrier and the pigment particles themselves all have equal density.

It is assumed here that both absorption and scattering coefficients are wavelength-dependent functions. Their absolute units are unimportant, as only their ratio is ever used. This simplifies measuring K and S values substantially. Consider, for instance, that the spectral reflectance function $R(\lambda)$ of a paint is given. It is then possible to simply set $S(\lambda) = 1$, and compute $K(\lambda)$ using the Kubelka-Munk equation (3.56b).

A paint containing several different pigments with absorption and scattering coefficients $K_i(\lambda)$ and $S_i(\lambda)$ yields absorption and scattering coefficients K_M and S_M for the mixture as follows:

$$K_M(\lambda) = \sum_{i=1}^{n} K_i(\lambda)\, c_i, \tag{3.57a}$$

$$S_M(\lambda) = \sum_{i=1}^{n} S_i(\lambda)\, c_i. \tag{3.57b}$$

Here, c_i is the concentration of the ith pigment. Once the K and S functions for a given pigment are known, the K and S values for paints containing this pigment can be derived on the basis of the Kubelka-Munk equation and the linear superposition of the known and unknown pigments.

One of the areas where the Kubelka-Munk theory is useful is in predicting the difference between organic and inorganic paint mixtures. A mixture of white with an inorganic pigment tends to appear more gray than a mixture of white with an organic pigment at comparable concentrations. Haase and Meyer show an example where a mixture of titanium dioxide (white) with cadmium red (inorganic) is compared with a mixture of titanium dioxide with naphthol red (organic) [412]. Both reds appear similar when applied directly. However, when mixed with titanium dioxide, the cadmium red looks more gray.

The Kubelka-Munk theory has several applications, including color matching. In this case, color matching refers to the process of taking a given color and mixing a paint to match that color. It also has uses in computer graphics. The discussion of both applications is deferred until Section 8.13, because the applications require knowledge of color spaces and color difference metrics, which are not explained until Chapter 8.

3.4.7 Water and Ice

Water and ice consist of H_2O molecules. Water molecules can vibrate in three different ways, as shown in Figure 3.14. These vibrations are designated v_1, v_2, and v_3, and they are associated with wavelengths in the infrared range (2755 nm, 6199 nm, and 2638 nm, or 0.45 eV, 0.20 eV, and 0.47 eV). These vibrations can have harmonic overtones, i.e., linear combinations are possible, so that radiation due to vibration is also possible at energies associated with, for example, $v_1 + v_2$, $2v_2 + v_3$, etc.

It is rare that any of these overtones reach the visible spectrum in isolated water molecules. However, in both water and ice, H_2O molecules form hydrogen bonds between molecules causing stronger bonding and associated absorption nearer the visible spectrum. This typically leads to some absorption in the red part of the visible spectrum [218]. As a result, clear water has a pale blue color [1056], which can be observed, for instance, against a sandy sea floor as shown in Figure 3.15.

It should be noted that the blue color of water is typically not due to the reflection of the blue sky, except perhaps at grazing angles. A polarizer can be used to eliminate reflections, revealing that water is blue by itself.

Water and ice frequently contain contaminations, such as algae, which may color the water blue-green to green. Water in glacier-fed lakes, for instance, contains silt giving it its characteristic turquoise color, which is due to scattering events (Figure 3.16) [1, 716]. As water is rarely ever pure, there is great variability of its color [108, 110, 905, 1055]. For uncontaminated water, absorption spectra have been measured [917, 1068] and are shown in Figure 3.17.

When water stains other materials, such as paper, fabric or rock, the effect is almost always that the material appears darker [39, 1156]. For dry porous materi-

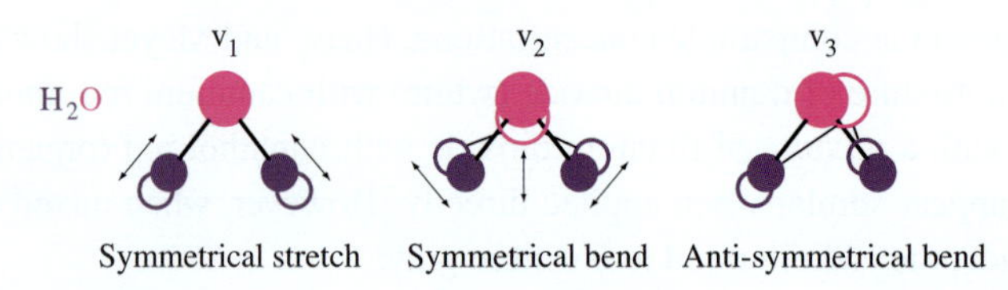

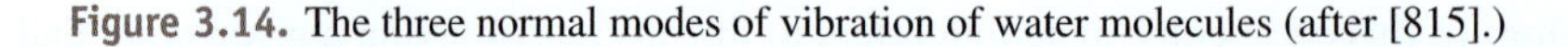

Figure 3.14. The three normal modes of vibration of water molecules (after [815].)

Figure 3.15. The pale blue color of water can be observed against sand, as seen here in this aerial shot of a coast line; Grand Bahama, 2004.

Figure 3.16. The silt in this glacier-fed lake colors the water turquoise; Alberta, Canada, July 2001. (Image courtesy of Greg Ward.)

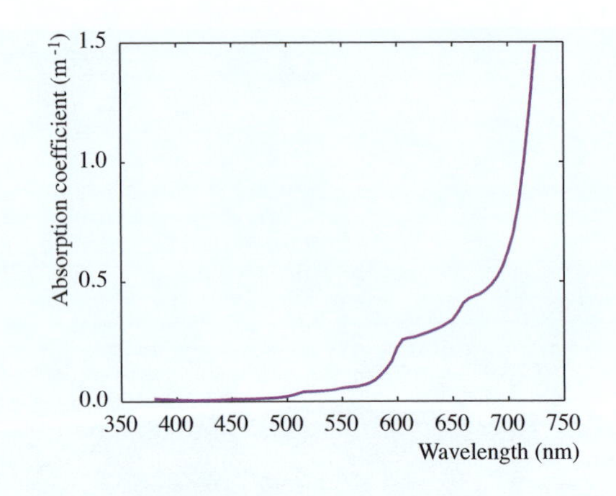

Figure 3.17. Absorption spectrum for clear water [917].

als, the index of refraction is high, for instance 2.0. The amount of light reflected depends on the ratio of the two indices of refraction. Going from air to dry porous material in our example, this ratio would be 1.0 / 2.0. Wet material on the other hand is coated with a thin layer of water with an index of refraction of around 1.33. The above ratio is then reduced to 1.0/1.33, and therefore less light is reflected (Figure 3.18) [716].

For non-porous materials, water forms a thicker layer. Light penetrating this layer is partially reflected by the underlying material. This reflected light under-

Figure 3.18. Water seeping down this rock changes the appearance of the material.

goes further reflections and refractions at the water-to-air boundary. Some total internal reflection occurs, so that less light emerges [668]. A computational model to synthesize imagery of wet materials was described by Jensen et al [548].

3.4.8 Glass

Glass is formed by melting and cooling. The cooling occurs sufficiently rapidly so that crystallization does not occur. The material is said to be in a *glassy* or *vitreous* state. Different types of glass are used for different applications. For instance, mercury vapor lamp jackets are made of *fused silica* (SiO_2). Ordinary *crown glass* (soda-flint-lime) contains SiO_2 as well as Na_2O and CaO and is used for window panes and bottles. Ovenware is made from *borosilicate crown* and contains SiO_2, B_2O_3, Na_2O, and Al_2O_3. Finally, crystal tableware is made out of *flint*, containing SiO_2, PbO, and K_2O.

Typical crown glass has a green tinge, as can be observed by looking at a window pane from the side, or, as shown in Figure 3.19, by observing the side of a block of glass. This colorization is due to the use of iron in the manufacture of glass. It is possible to minimize the amount of green ferrous iron (Fe^{2+}) by means of *chemical decolorizing* or *physical decolorizing*. Adding some manganese oxide (MnO_2), also known as *glassmaker's soap*, achieves both. First, physical decolorizing occurs because manganese oxide has a purple color, offset-

Figure 3.19. The top of this block of glass shows a green tinge.

ting the green of the ferrous iron. Second, chemical decolorizing is achieved by the following reaction:

$$Mn^{4+} + 2Fe^{2+} \rightarrow Mn^{2+} + 2Fe^{3+}. \tag{3.58}$$

The resulting ferric iron (Fe^{3+}) is colorless.

Color may be added to glass by adding small quantities of other materials. The mechanisms by which color is formed include ligand field effects, such as occurs when adding transition metals to the molten glass. Many allochromatic colors can be created by adding a variety of irons, including Ti^{3+}, Mn^{2+}, Mn^{3+}, Fe^{2+}, Fe^{3+}, and Co.

By adding iron sulfates, yellow or brown charge-transfer colors may be created, as is the case in the well-known beer-bottle brown. Adding gold or silver to the molten glass will produce metal scattering. Gold produces a deep red or purple color (purple of Cassius), whereas silver creates a yellow color. The addition of copper also produces red, albeit not as intense or deep as gold. Finally, adding semiconductors to the glass, such as cadmium sulfide-selenide ($CdS \cdot CdSe$) can produce yellow, orange, red, and black [815].

The archetypical use of colored glass is in the stained-glass windows in churches and cathedrals, as shown in Figure 3.20.

Figure 3.20. Colored glass in stained-glass windows; Rennes, France, June 2005.

3.5 Sources of Radiation

Atoms and molecules consist of nuclei with clouds of electrons circling around the nuclei. By adding thermal energy (heat), molecules and atoms may be set into vibration. Such vibrations and oscillations may cause energy to be released. It can be shown that all atoms and molecules radiate energy, unless their temperature is reduced to absolute zero ($0K$). Thus, the amount of radiation is dependent on temperature, and for certain temperatures, the radiation may be in the visible spectrum.

Thermal radiation is energy emitted by all objects as a result of motion by the objects' atoms. A body may absorb energy and radiate energy. If its temperature remains constant, the amount of absorption and radiation is the same, and the body is said to be at *equilibrium*. Absorption and emission may be modeled with a wavelength-dependent absorption coefficient α_λ and a wavelength-dependent emission σ_λ.

The absorption coefficient is the unitless fraction of incident radiant energy that is absorbed. The emission coefficient is the energy per unit time per unit area that is emitted by a body at each given wavelength. Both coefficients depend on the nature of the body, as well as wavelength.

If we were to heat a chamber, then its interior walls would absorb and emit light. At each wavelength independently, the absorption and emission would be equal if the temperature were to remain constant. This is formalized in *Kirchhoff's Radiation Law*:

$$\frac{\alpha_\lambda}{\sigma_\lambda} = M_{e\lambda}, \tag{3.59}$$

where $M_{e\lambda}$ is known as the *spectral radiant exitance*,[7] and is measured in $W\mathrm{m}^{-2}$. This function depends on temperature T and wavelength λ, but not on the material of the cavity walls, nor on the cavity's spatial configuration. This fact was empirically observed as early as 1792 by ceramist Thomas Wedgwood who noted that a kiln fired to a certain temperature would glow the same color as the pottery inside [447].

3.5.1 Blackbody Radiation

If an object is a perfect absorber, i.e., all incident radiation is absorbed, then the fraction α_λ equals 1 for all wavelengths λ. Such an object appears black and is

[7]This is a radiometric term and will be explained along with other radiometric terms in Chapter 6.

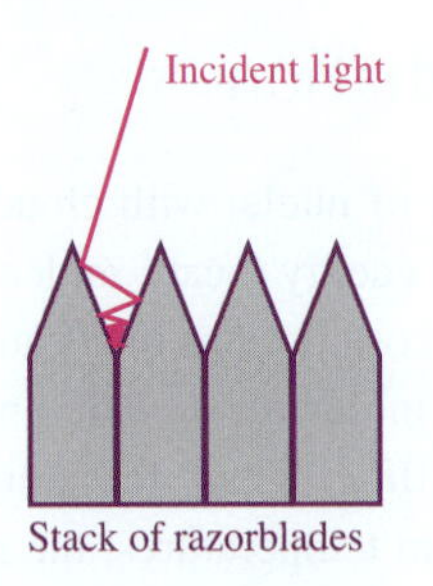

Figure 3.21. The geometry of a stack of razor blades is such that it can absorb most light, and thus simulate a blackbody.

called a *blackbody radiator*. Since no materials are perfect absorbers, it is not possible to directly observe a perfectly black material.

However, it is possible to construct a cavity, as mentioned above, and create a small hole in one of the walls. Nearly all of the energy entering the cavity from outside is then absorbed after one or more reflections inside the cavity. Only a tiny fraction of this light is reflected towards the hole and subsequently leaves the cavity. The hole, therefore, appears nearly black and could be regarded as a close approximation to a blackbody radiator.

As an aside, an alternative way to construct an object resembling a blackbody radiator is by stacking a number of old-fashioned razor-blades. Although each blade is a good reflector ($\alpha_\lambda \ll 1$), the spatial configuration is such that light is funneled into the grooves formed by the stack of blades and is, therefore, not able to escape, as shown in Figure 3.21.

A stack of razor blades is shown in Figure 3.22, with the light source around 20 cm above the grooves [1247].[8] The grooves absorb most of the light. At different angles, the grooves do not behave as a blackbody, as shown in Figure 3.23.

Returning to a cavity acting as a blackbody radiator, the walls absorb light entering the cavity through a hole. If this cavity is heated, the atoms of the walls vibrate more vigorously. As a result they each radiate more energy. The interaction of atoms with neighboring atoms causes each to be capable of radiating over a broad spectrum, rather than the narrow bandwidth spectra associated with the excited states of individual electrons. Nonetheless, the broad spectrum generated by a heated cavity shows a peak at a wavelength determined by the temperature of the cavity. *Wien's displacement law* provides a formula relating the peak wave-

[8]This demonstration was given by Alain Fournier at his keynote speech for the 6th Eurographics Workshop on Rendering, held in Dublin in 1995.

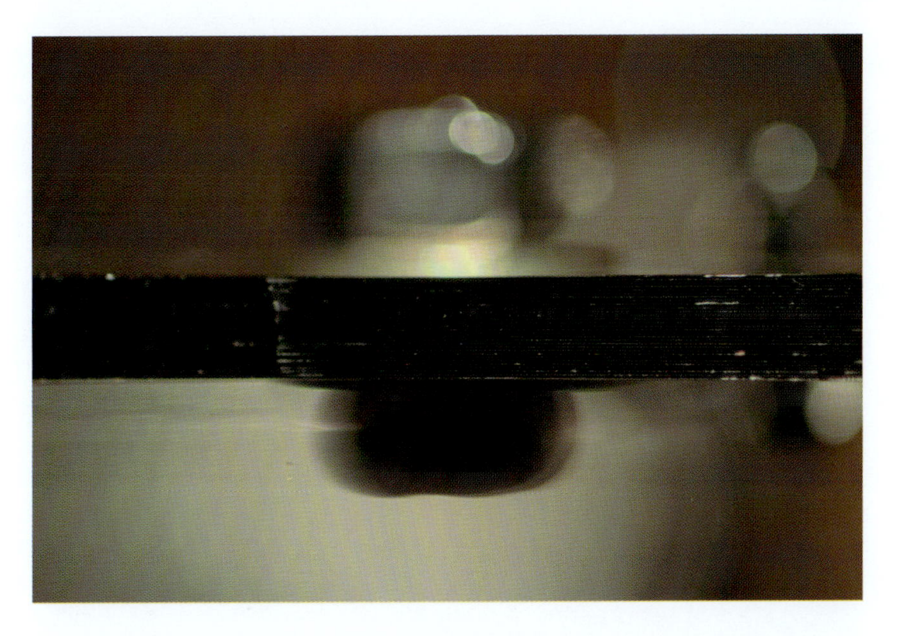

Figure 3.22. A stack of 30 razor blades approximating a blackbody radiator.

Figure 3.23. If the light is incident at a shallow angle, and the viewpoint is also chosen at a shallow angle, the grooves may reflect rather than absorb.

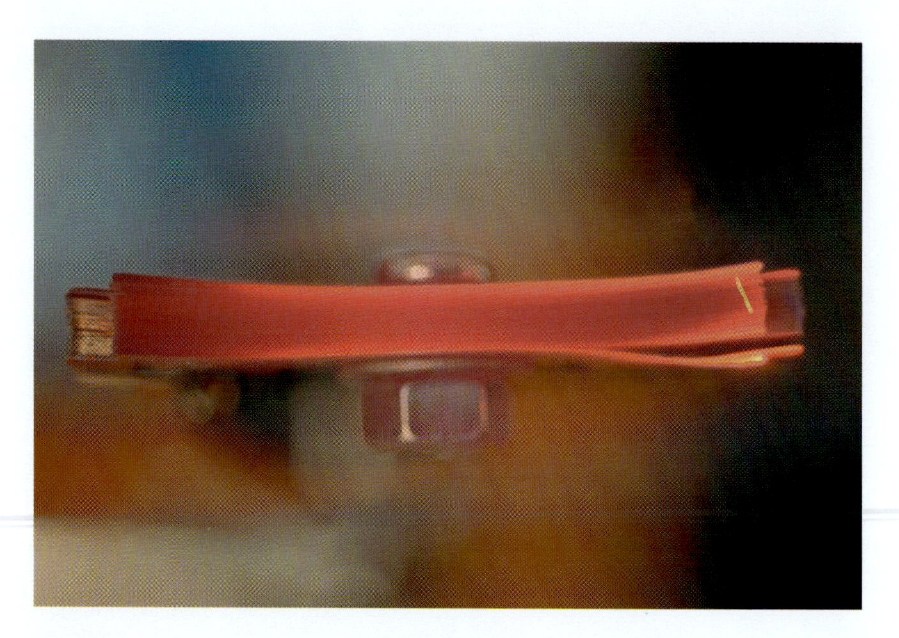

Figure 3.24. The stack of razor blades from Figure 3.22 is heated with a butane fueled torch.

length $\lambda_{\max}$ to the temperature of the cavity:

$$\lambda_{\max} T = \frac{hc}{4.965114 k_B} \tag{3.60a}$$

$$= 2.897791 \times 10^{-3}, \tag{3.60b}$$

where h is Planck's constant, c is the velocity of light, and k_B is Boltzmann's constant (see Appendix D). The right-hand side is given in (mK). This equation shows that the peak wavelength decreases with increasing temperature, a fact that can be empirically verified. For instance, iron begins to radiate dull red at around 700 °C. At melting point (1500 °C), the iron glows yellow. The temperature of the sun, which is normally perceived as white, is around 5700 °C. Even hotter stars appear bluish-white [815]. An example of thermal radiation is shown in Figure 3.24.

For a blackbody radiator, *Planck's radiation law* provides a functional form for the *spectral density* at each wavelength as function of the temperature [1262]:

$$\mu_{e\lambda} = \frac{8\pi hc}{\lambda^5} \left(\exp\left(\frac{hc}{\lambda k_B T} \right) - 1 \right)^{-1}. \tag{3.61}$$

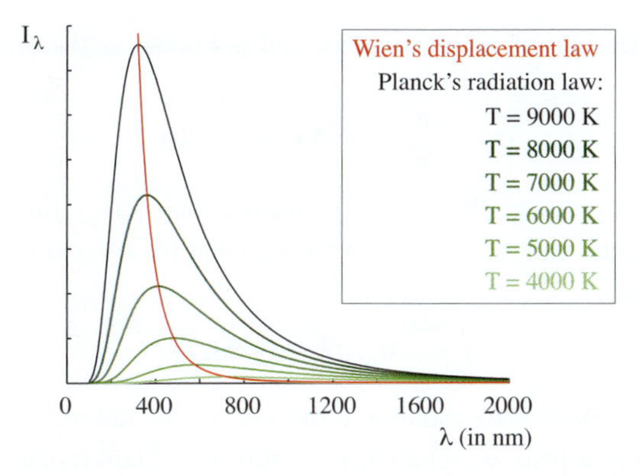

Figure 3.25. The emission spectrum for blackbody radiators heated to different temperatures.

The *spectral radiant exitance* $M_{e\lambda}$ as function of wavelength and temperature is:

$$M_{e\lambda} = \frac{c\mu_{e\lambda}}{4} \tag{3.62a}$$

$$= \frac{2\pi hc^2}{\lambda^5} \left(\exp\left(\frac{hc}{\lambda k_B T} \right) - 1 \right)^{-1}. \tag{3.62b}$$

This function is plotted for various values of T in Figure 3.25. The same figure also shows in red the curve determined by Wien's displacement law. Note that the range of wavelengths shown is much larger than the visible range. Within the visible range, each of the curves is relatively flat, with the exception of very high and very low temperatures. This means that for intermediate temperatures, blackbodies radiate a broad spectrum, thereby appearing predominantly white with perhaps a tinge of blue for higher temperatures and red for lower temperatures.

Planck's radiation law for blackbody radiators gives a spectral distribution of energy with its peak located at λ_{max}. Wien's displacement law predicts that this peak shifts if the temperature is increased. Heating an object not only increases the amount of light that is emitted, but also changes the apparent color. It is therefore natural to refer to a given spectral distribution of a blackbody radiator by its *color temperature*, which is given in degrees Kelvin.

The area under each of the curves in Figure 3.25 has special significance and represents the *total radiant exitance* of the blackbody radiator. The radiant exitance M_e can be computed by integrating (3.62) over all wavelengths:

$$M_e = \int_0^\infty M_{e\lambda} \, d\lambda = \sigma T^4. \tag{3.63}$$

Here, σ is a constant of proportionality and is known as the *Stefan-Boltzmann constant* :

$$\sigma = \frac{2\pi^5 k^4}{15h^3c^2} = 5.67032 \times 10^{-8}. \tag{3.64}$$

This constant is given in ($W\mathrm{m}^{-2}\mathrm{K}^{-4}$). Rather than integrating over all wavelengths, the radiant exitance may be approximated by integrating over all wavelengths up to $\lambda_{\max}$:

$$\int_0^{\lambda_{\max}} M_{e\lambda}\, d\lambda \approx 0.25\sigma T^4. \tag{3.65}$$

An example of a blackbody radiator is the sun, which radiates at approximately 5700 K, yielding a peak wavelength near 500 nm. Stars radiate between 5000 K and 11,000 K. Incandescent lamps have a temperature of around 3000 K. Human skin also radiates, but only weakly and at a temperature outside the visible spectrum (around 10,000 nm) [1247].

3.5.2 Incandescence

Objects that are heated radiate energy. Some of the energy is radiated in the visible spectrum, dependent on the object's temperature. Blackbodies are a special case of radiating bodies, in that they absorb all incident light. In the more general case, any radiation occurring as a result of heating objects is referred to as *incandescence*.

Candles, fires, tungsten filaments, hot coal, and iron are all examples of incandescent light sources. The heat of a candle flame, for instance, melts the wax which subsequently flows upward in the wick by means of capillary action and is then vaporized. Such a flame emits yellow light near the top and blue light near the bottom.

The vaporized wax is the fuel that can react with oxygen. The center of the flame contains a limited amount of oxygen, which diffuses in from outside the flame. The reaction in the center occurs at around 800 °C to 1000 °C [815]:

$$C_{17}H_{35}CO_2H + 11O_2 \rightarrow 9H_20 + 5CO_2 + 5CO + 8C + 9H_2. \tag{3.66}$$

Near the outside of the flame where more oxygen is available, further reactions occur at the higher temperature of between 1000 °C to 1200 °C, yielding yellow light:

$$2CO + O_2 \rightarrow 2CO_2, \tag{3.67a}$$

$$2C + 2O_2 \rightarrow 2CO_2, \tag{3.67b}$$

$$2H_2 + O_2 \rightarrow 2H_20. \tag{3.67c}$$

Figure 3.26. The flame of a lighter. The lower part of the flame contains a pre-mixed combination of air and fuel, causing total combustion which leads to a blue flame. The fuel in the top part of the flame is only partially combusted, leading to soot as well as a yellow flame.

Thus, the above reactions produce energy in the form of light, as well as water and carbon dioxide. A candle flame is classified as a *diffusion flame*, because oxygen diffuses inward before it reacts with the fuel.

Near the bottom of the flame, a different situation arises, since before ignition the fuel is mixed with oxygen. Such *premixed flames* appear blue and burn cleanly since no soot particles are formed in the process.

Both premixed and diffusion flames occur in candle light, but they are generally easier to observe in the flame of a cigarette lighter, as shown in Figure 3.26. The lower temperatures in the center of a flame due to a lack of oxygen in the flame's interior can be demonstrated by lowering a wire mesh into the flame. The flame will burn underneath the mesh, but not above it. The mesh therefore allows us to observe a cross-section of the flame, as shown in Figure 3.27. The dark center is easily observed through the mesh.

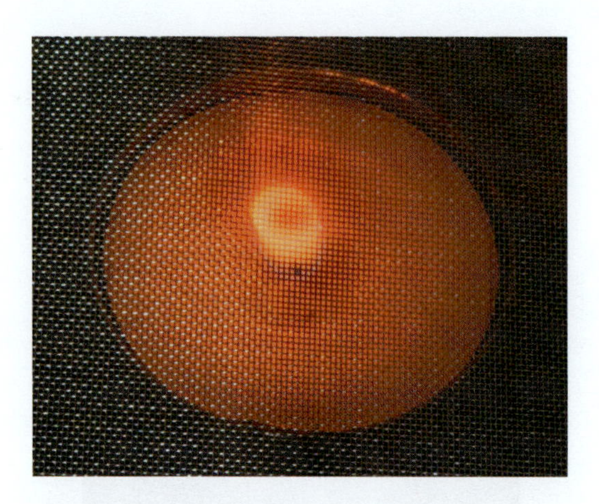

Figure 3.27. The emission of light from an open flame occurs near the surface of the flame. The interior lacks sufficient oxygen to burn. That a flame is hollow is demonstrated here by lowering a mesh into the flame of a candle, revealing a dark center. The flame emits light underneath the mesh, but not above.

Incandescence is also the mechanism employed in tungsten filaments, as shown in Figure 3.28. Here, an electric current is sent through a coil. The coil's resistance causes electric energy to be converted into heat, which causes the tungsten to emit light. The temperature of filament light bulbs is around 2500 °C. Incandescence is also found in old-fashioned vacuum tubes, as shown in Figure 3.29.

Figure 3.28. An incandescent tungsten filament.

Figure 3.29. Incandescence in a vacuum tube.

Other forms of incandescent light are found in pyrotechnics, for instance when magnesium powder is burned. This reaction produces very high temperatures, leading to very bright white light, which may be colored by adding nitrates, chlorates, or perchlorates. This reaction is also used in flares and tracer bullets. A final example of incandescence is the photographic flashlight. The bulb of a flashlight contains shredded zirconium metal and oxygen. On ignition, this produces molten zirconium and zirconium oxide which radiates at a color temperature of around 4000 °C [815].

Incandescent light is characterized by a lack of polarization. In addition, the phase of each emitted electro-magnetic wave is random. Light is generally emitted equally in all directions. The temperature of an incandescent light source determines its color, similar to blackbody radiation. With temperature, the radiation progresses from black to red, orange, yellow, white, and finally blue-white.

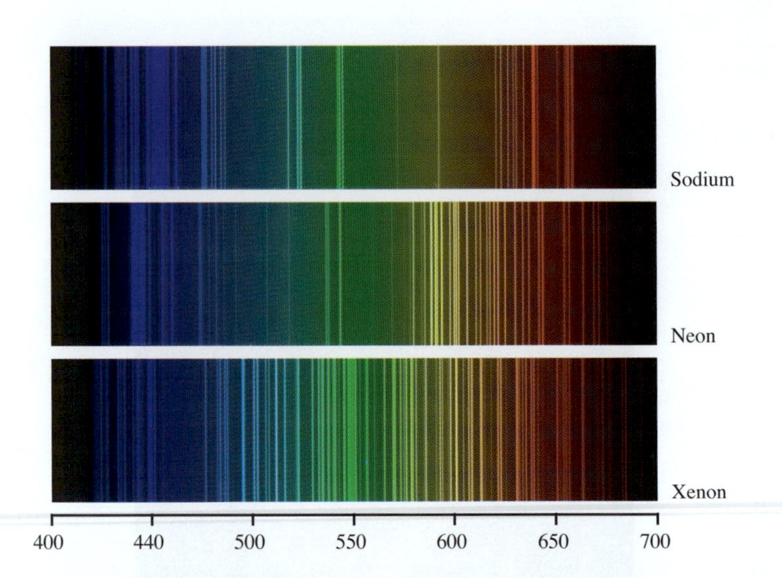

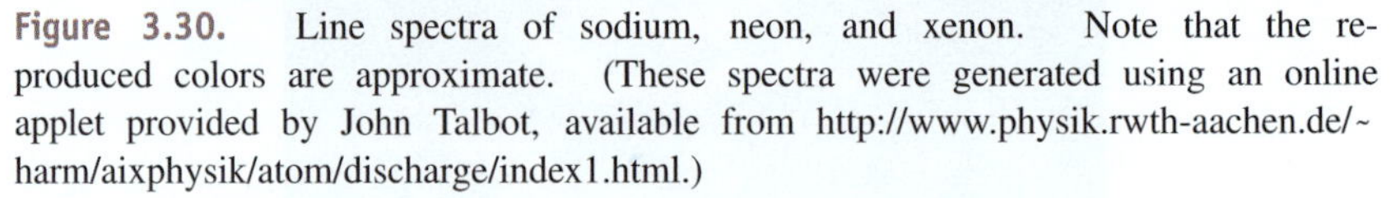

Figure 3.30. Line spectra of sodium, neon, and xenon. Note that the reproduced colors are approximate. (These spectra were generated using an online applet provided by John Talbot, available from http://www.physik.rwth-aachen.de/~harm/aixphysik/atom/discharge/index1.html.)

3.5.3 Gas Excitation

Radiant energy may be produced by sending an electric current through a tube filled with inert gas. This may be accomplished by two electrodes at either end of the tube providing an electric field. Electrons accelerate through the tube and collide with the gas molecules. The kinetic energy of the electrons is then absorbed by the molecules, yielding electrically excited molecules, atoms, and ions, as well as electrons. Of these, the charged particles (ions and electrons) are subjected to the electric field and accelerate themselves, producing further collisions.

When excited molecules and atoms return to their original state, photons are emitted. The number of allowable electric states is limited, and therefore *line spectra* are produced, as shown in Figure 3.30. Such spectra are characterized by narrow emission lines, superposed with a continuous broad spectrum. Joseph von Fraunhofer mapped the wavelengths of a large number of spectral lines, which were found to be associated with specific elements. The lines were assigned letters and are collectively called *Fraunhofer lines*. A collection of these are listed in Table 3.8. Fraunhofer lines are often used to characterize the dispersive properties of dielectric materials (see for instance Section 12.3.6).

Designation	Element	Wavelength	Designation	Element	Wavelength
T	Fe	302.108	F	H	486.134
P	Ti^+	336.112	D_2	Na	588.995
N	Fe	358.121	D_1	Na	589.592
L	Fe	382.044	C	H	656.281
K	Ca^+	393.368	B	O_2	686.719
H	Ca^+	396.847	A	O_2	759.370
G	Ca	430.774	Z	O_2	822.696
G	Fe	430.790			

Table 3.8. A selection of Fraunhofer lines.

The spectrum emitted by a gas discharge lamp is primarily determined by the inert gas used. Only inert gases are chosen since, otherwise, the gas may react with the electrodes and break the lamp. Inert gases include neon, argon, helium, krypton, and xenon (see Table 3.6).

Neon, as used in neon signs, gives a red color, as does neon with some argon added. Argon by itself, or argon with added mercury, produces blue light. Yellow and green light may be created by using colored glass tubes. Less frequently

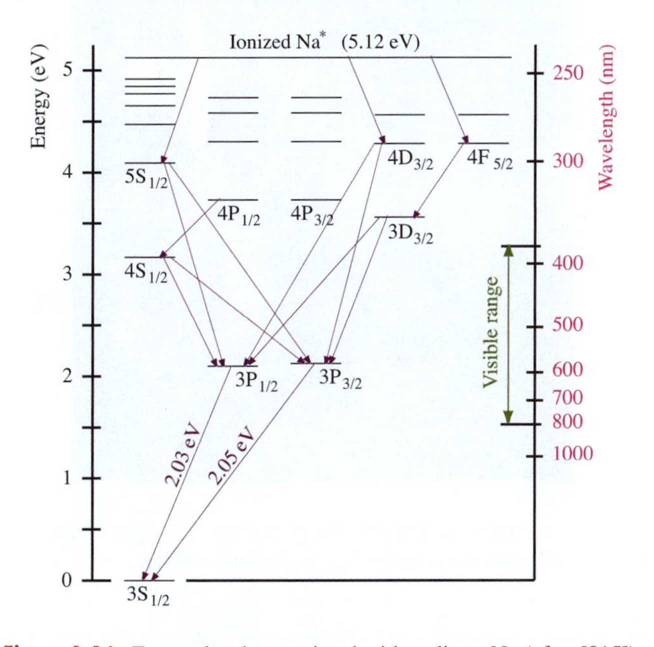

Figure 3.31. Energy levels associated with sodium, Na (after [815]).

used are helium, krypton, and xenon because they are either not producing sufficient intensity, or they are too costly to produce. These gases produce yellow, pale lavender, and blue, respectively. Further colors may be produced by adding phosphor powders or by using phosphorescent glass tubes.

Sodium vapor lamps operate on a similar principle, although sodium is a solid metal at room temperature. To produce sodium vapor, neon is added which produces the initial discharge, as well as heat.[9] The heat then vaporizes the sodium (Na) which is then ionized due to collisions with electrons:

$$Na \rightarrow Na^{+} + e^{-}. \tag{3.68}$$

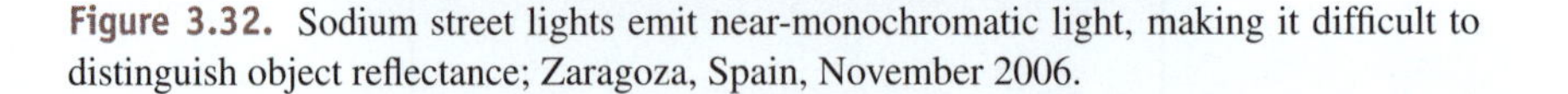

Figure 3.32. Sodium street lights emit near-monochromatic light, making it difficult to distinguish object reflectance; Zaragoza, Spain, November 2006.

[9]A sodium vapor lamp emits pink light for the first couple of minutes after being switched on as a result of the neon gas discharge.

The ion may later recombine with an electron and thereby return to a lower excited energy level, before returning to a yet lower level and emitting a photon:

$$\mathrm{Na}^+ + e^- \to \mathrm{Na}^*, \tag{3.69a}$$

$$\mathrm{Na}^* \to \mathrm{Na} + \text{photon}. \tag{3.69b}$$

There are several different energy states the sodium may reach, as shown in Figure 3.31. Many of the energy states are outside the visible range and, therefore, produce radiation that is not detected by the human eye. Around half of the light emitted is due to a doublet of states at 2.03 eV and 2.05 eV, which both correspond to yellow light ($\lambda \approx 589.6$ and 589.0 nm). The strong yellow color makes sodium vapor lamps less suitable for a range of applications, as it is difficult to determine the color of objects under sodium lighting. Sodium lamps are predominantly used in street lighting (Figure 3.32 and Section 9.3).

Mercury vapor lamps produce a bluish overall spectrum by a mechanism similar to sodium vapor lamps. Their principle emission lines in the visible region occur at around 405, 436, 546, and 578 nm. The color of mercury vapor lamps can be improved by coating the glass envelope with a phosphor. Such phosphor-coated mercury vapor lamps are in wide use today.

3.5.4 Fluorescence

In most cases, when a photon is absorbed by an atom, an electron jumps to a higher state and immediately returns to its ground state, whereby a new photon is emitted of the same wavelength as the photon that was absorbed. Some materials, however, allow electrons to return to their ground state via one or more intermediate energy states. This creates state transitions during which photons of different wavelengths are emitted. The wavelength of the emitted photon is determined by the difference in energy levels of the two states involved in the transition. This means that new photons are emitted at longer wavelengths, a phenomenon known as *Stokes' law*. The process of absorption and re-emission at a different wavelength is called *fluorescence*.

As indicated in the preceding section, the inside of a mercury vapor lamp may be coated with a phosphor, creating a fluorescent light source. An example of a fluorescent light source is shown in Figure 3.33. The mercury vapor emits light, predominantly in the ultra-violet region (253.7 nm), which excites the phosphor. The phosphor in turn re-radiates light at a longer wavelength. The color of a fluorescent tube is determined by the type of phosphor used. For instance, magnesium tungstate creates an emission spectrum with a maximum emittance at 480 nm. Maximum emittance at longer wavelengths is possible with, for instance,

Figure 3.33. A fluorescent light source.

calcium halophosphate (590 nm), strontium magnesium phosphate (620 nm), or magnesium fluoro-germanate (660 nm) [1262].

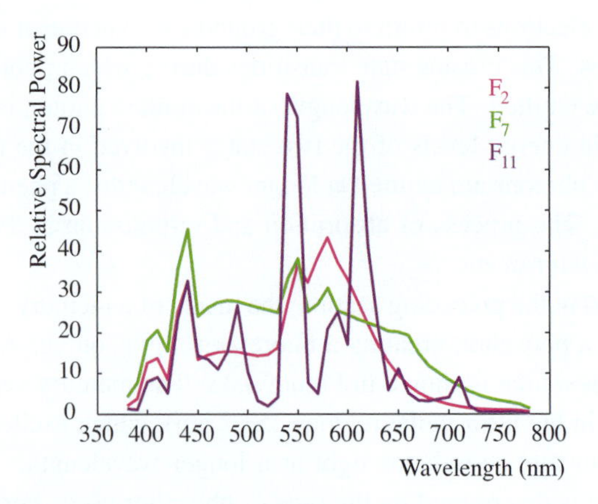

Figure 3.34. Relative spectral power distributions for CIE F_2, F_7 and F_{11} fluorescent light sources (data from [509]).

Figure 3.35. This set-up demonstrates frequency up-conversion. The small transparent panel is coated with phosphors that emit at blue wavelengths when subjected to near-infrared light. (Demonstration kindly provided by Janet Milliez, School of Optics, University of Central Florida.)

Fluorescent lamps can approximate daylight, although they are found to be unsuitable for critical color-matching tasks, as their spectral power distribution (shown for typical fluorescent lights in Figure 3.34) deviates too much from average daylight.

It is possible to construct phosphors that fluoresce in the visible range if excited by near-infrared electromagnetic radiation [866, 896]. In this case, the re-emitted light is of a shorter wavelength than the incident light. This is effectively the opposite of fluorescence, where re-emission takes place at longer wavelengths. The frequency of light being the reciprocal of its wavelength, this process is termed *frequency up-conversion*. Recent advances in up-conversion materials show promise for the construction of emissive display devices with a remarkable brightness, paired with a wide color gamut which includes very saturated colors [936, 937].

A demonstration of a small, yet very bright, vector display device is shown in Figure 3.35. Here, near-infrared light is directed in a lissajous pattern toward a phosphor-coated piece of plastic which is designed to emit blue light. Phosphors emitting at red and green wavelengths are also possible, as shown in Figure 3.36. The blue phosphor in this figure has a secondary emission at red wavelengths and, therefore, appears somewhat purple. Nonetheless, this is a potentially suitable technology for constructing versatile display devices with low weight and depth [936, 937].

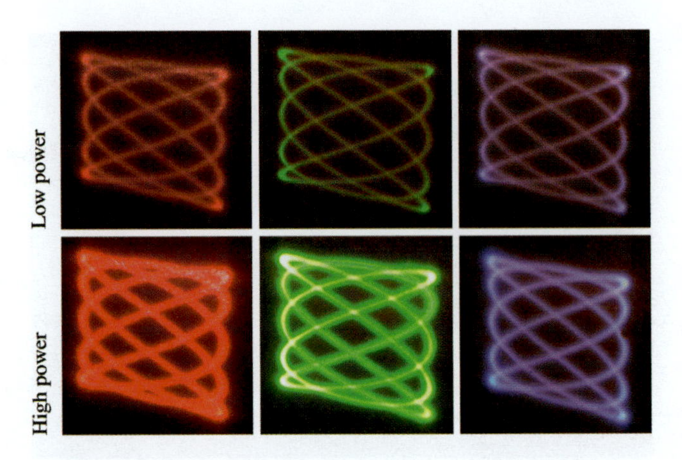

Figure 3.36. Red, green, and blue phosphors shown in bright and dim variants. The substrate was opaque. (Demonstration kindly provided by Janet Milliez, School of Optics, University of Central Florida.)

3.5.5 Phosphorescence

For some substances, electrons may be excited into a higher energy state, and they may alternate between two meta-stable states before returning to the ground state. These alternate states may be reached by further absorption of radiant energy. If the eventual return to the ground state emits a photon, it will be at a later time than the initial absorption. Materials exhibiting this behavior are called *phosphorescent*.

If the source of radiation that caused the transition between states is switched off, there will be a delay before all atoms in the substance have returned to their ground states. The decay time can be measured by noting the time necessary before the emitted radiant power is reduced to a fraction of e^{-1} of its original value.

Fluorescent materials generally have a very short delay time, on the order of 10^{-8} seconds. Phosphorescent materials have decay times that range from 10^{-3} seconds to many days [1262]. The decrease in energy is typically exponential. However, when an electron absorbs enough energy to ionize the atom, the electron will follow a power-law decay. The radiance L emitted by a phosphor as function of time t may be modeled with an empirical formula [675]:

$$L = \frac{1}{b\left(\frac{1}{\sqrt{L_o\, b}} + t\right)^2}, \tag{3.70}$$

Figure 3.37. The hands of this pocket watch are painted with a phosphorescent paint. Before this photograph was taken, the watch was lit for around a minute by a D65 daylight simulator. The photograph was taken in low light conditions to show the phosphorescent behavior of the paint.

where L_0 is the radiance emitted at time $t = 0$, and b is a material dependent constant.

Examples of phosphorescent materials are the paint used to paint the hands and dials of watches, as shown in Figure 3.37, and the light emitted by cathode ray tubes (Figure 3.38).

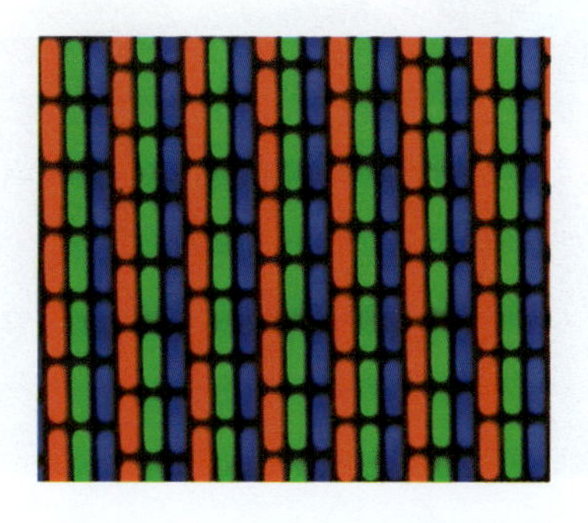

Figure 3.38. Red, green, and blue dots of a phosphorescent cathode ray tube. At a distance, the human visual system fuses the separate dots and the result is perceived as a uniform white.

3.5.6 Luminescence

Fluorescence and phosphorescence discussed in Sections 3.5.4 and 3.5.5 are two examples of a general principle, called *luminescence*. The term luminescence refers to all production of light that is not caused by thermal radiation. This leaves many forms of luminescence [815]:

Photoluminescence is luminescence induced by visible light or ultraviolet light. Examples include day-glo paint, which contains organic dyes that fluoresce causing an appearance brighter than white, and fluorescence brighteners used in laundry detergent.

Radioluminescence in certain substances is caused by radiation of energetic particles such as alpha, beta, or gamma rays. For example, radioactive materials can be mixed with paint to create a self-luminous substance, used, for instance, in dials of watches.

Electroluminescence occurs due to the application of an electric field or current. Examples include gas discharge lamps as discussed in Section 3.5.3, as well as the plasma technology outlined in Section 14.4.

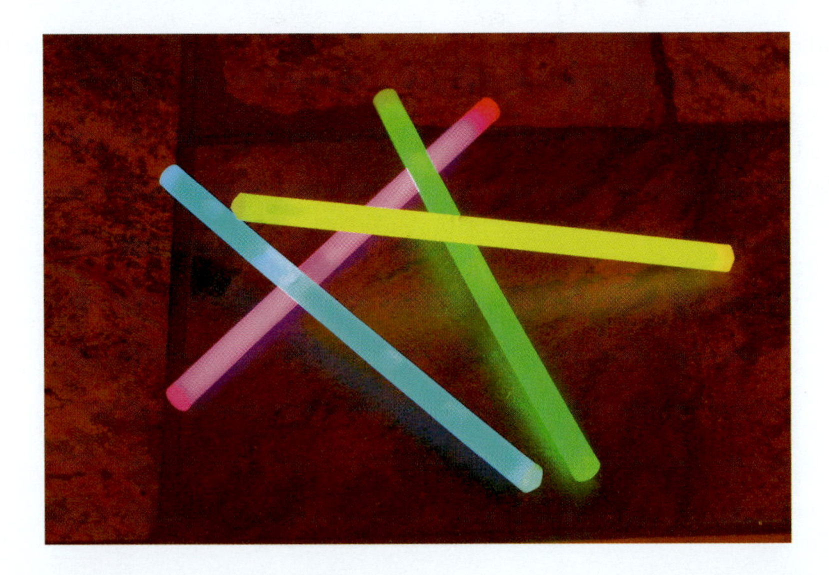

Figure 3.39. Glow sticks emit light by chemiluminescence. The yellow, blue and green glow sticks are enclosed in clear plastic holders, while the pink color occurs in part due to the red plastic of the tube.

Triboluminescence is produced by mechanical force, such as grinding crystals. It can be produced at home by crushing hard candy, in particular wintergreen lifesavers.

Chemiluminescence is created by chemical energy, as for instance used in glow sticks (Figure 3.39). Glow sticks can be frozen to halt the reaction that will resume afterwards once the sticks reach room temperature.

Bioluminescence is a form of chemiluminescence produced by biological mechanisms such as glow worms and fireflies. Also many (deep-)sea creatures exhibit bioluminescence, including species of jellyfish and angler fish. Some algae have a triboluminescent form, which makes them glow when disturbed. This creates the "burning of the sea" effect.

Further distinctions between different forms of luminescence may be made [815], which are beyond the scope of this book.

3.5.7 Light-Emitting Diodes

Semiconductors, as discussed in Section 3.3.3 can be doped with donors as well as acceptors, and, in either case, this may change the color of the semiconductor. In addition, donor impurities may increase the material's conductivity at high temperatures. Acceptors increase conductivity at all temperatures.

If a semiconductor such as silicon is doped with phosphorus, a donor, the result is called *n-type silicon* because there exist negatively charged electrons in the conduction band. Doping silicon with boron, an acceptor, results in positively charged holes in the valence band, and such material is called *p-type silicon*. Placing two such materials side-by-side creates a p-n contact region (a *p-n junction*) where all donors and acceptors are ionized. The Fermi level is now just above the acceptor level in the p-region and just below the donor level in the n-region.

By applying a voltage to this configuration, the Fermi level will jump to a different level. Electrons in the conduction band will then be attracted to the holes in the valence band, and their recombination will result in the emission of photons. In silicon, the band-gap energy is at 1.1 eV, and this is sufficient to produce photons at infrared wavelengths. This process is called *injection electroluminescence*. Other semiconductors exist with band gaps that correspond to visible light with different colors, such as $GaAs_{1-x}P_x$ (red), AlP (green) and ZnS (blue) [815].

Light sources based on this principle are called *light-emitting diodes*, or LEDs for short [479], and are discussed further in Section 9.8. An example of a set of red LEDs is shown in Figure 3.40.

Figure 3.40. A set of red LEDs.

The principle used by LEDs can be reversed, such that light absorbed at a p-n junction creates a voltage, a principle used by solar panels (Figure 3.41).

Figure 3.41. Solar panels use the same semiconductor technology as LEDs, albeit in reverse by absorbing light and thus producing an electric current.

3.5.8 Organic Light-Emitting Diodes

Light-emitting devices can be constructed from semiconductor materials, as discussed above. Recently, there has been considerable interest in light-emitting devices built from organic molecules. Such devices are known as *organic light-emitting diodes* (OLEDs) [1117]. This approach has the advantage of potentially high efficiency and ease of manufacturing. In particular, there are developments towards modifying standard printing techniques such as screen printing [96, 97, 663, 870] and ink jet printing [431] to produce OLED displays [802].

OLEDs can be configured to emit light over the entire visible spectrum, for instance by altering the driving voltage, or interference effects in multi-layer OLEDs [385]. While high efficiency can currently be combined with a high device stability for red and green emitters [79, 181, 852, 1148, 1224], stability, efficiency, high color saturation, and a long lifetime are still problematic for blue emitters [383, 1234].

Organic LED technology may be coupled with inorganic nanocrystals, for instance made of cadmium-selenium (CdSe) [199]. These crystals afford narrower emission spectra, i.e., more saturated colors, than conventional OLEDs which typically emit light over a 50–100 nm range [1150].

OLEDs hold great promise for the future, in particular for the production of efficient and bright display devices, as discussed in Section 14.6. An intermediate solution is the use of hybrid TFT-OLED technology to overcome current device limitations [1234].

3.5.9 Lasers

As we have seen in Section 3.5.4, an atom may absorb energy in the form of a photon, causing an electron to jump to a higher energy level, or equivalently, a different orbital. When the electron jumps back to its ground state, possibly via one or more intermediate energy levels, new photons may be emitted, leading to fluorescence. If energy is added to the atom in the form of kinetic energy (heat), then electrons also jump to higher energy levels. Upon return to the ground state, emission of photons in the visible range may also occur, leading to incandescence.

We will assume that there is a population of N_1 atoms at a given lower energy level E_1, and a population of N_2 atoms in a higher (excited) energy level E_2. In a material with vast quantities of atoms, the relaxation to lower energy levels occurs at an exponential rate, determined by the material itself. With t indicating time, and γ_2 the *spontaneous decay rate*, the size of the population of atoms at the higher energy level is given by [1045]:

$$N_2(t) = N_2(0)\, e^{-\gamma_2 t} \tag{3.71a}$$

$$= N_2(0)\, e^{-t/\tau_2}. \tag{3.71b}$$

The reciprocal of γ_2 is given by $\tau_2 = 1/\gamma_2$ and signifies the lifetime of the upper level E_2. The above decay rate is independent of which lower energy levels are reached. The decay rate from level E_2 to level E_1 can be modeled with a similar equation, except that γ_2 is replaced by a different constant γ_{21}. For convenience, we may write this equation as a differential equation, which models the spontaneous emission rate:

$$\frac{dN_2(t)}{dt} = -\gamma_{21} N_2(t). \tag{3.72}$$

So far we have not discussed how atoms end up in a higher energy state, other than that photons at the right wavelengths may be caught. Normally, this happens irregularly dependent on the temperature of the material and the amount of incident light. However, it is possible to force atoms into higher energy levels, for instance, by flashing the material with a short pulse of light that is tuned to the transition frequency of interest. This process is called *pumping*.

After such a short pulse is applied, the population of atoms in the higher energy state can be modeled by the following atomic rate equation:

$$\frac{dN_2(t)}{dt} = K n(t) N_1(t). \tag{3.73}$$

Here, $n(t)$ is the photon density of the applied signal measured in number of photons per unit volume, or equivalently as the electromagnetic energy density divided by $\hbar\omega$. This is the process of *stimulated absorption*, i.e., the application of a short pulse of light forces atoms that are initially in the lower energy state E_1 into the higher energy state E_2.

Similarly, the same pulse forces atoms that are initially in the higher energy state E_2 into the lower energy state E_1, thereby emitting photons. This process is called *stimulated emission* and is modeled by the following atomic rate equation:

$$\frac{dN_2(t)}{dt} = -K n(t) N_2(t). \tag{3.74}$$

Note that the constant of proportionality K in the above two equation is identical, but of opposite sign. This constant will be largest if the pulse of light that is applied is of the same energy (frequency/wavelength) as the transition energy between levels E_1 and E_2 of the material.

The total rate equation when pumping is applied to a volume of material is then the superposition of the above three equations:

$$\frac{dN_2(t)}{dt} = K\,n(t)\,(N_1(t) - N_2(t)) - \gamma_{21} N_2(t). \tag{3.75}$$

Stimulated emission produces photons with energies that are able to force stimulated absorption or emission in other atoms. As a result, stimulated emission causes amplification of the initial pulse of light (hence the acronym *laser*, which stands for *light amplification by stimulated emission of radiation*). Whether a population of atoms, i.e., a volume of material, amplifies light or absorbs light depends on the ratio of atoms in the lower and higher energy states, i.e., N_2/N_1. In essence, if we can force a majority of the population of atoms into the higher energy state, then the volume of material will act as an amplifier; otherwise, the material will absorb energy.

This can also be seen from (3.75), where the difference between N_1 and N_2 is on the right-hand side of the equation. If this difference is positive, then the material absorbs energy. If this difference is negative, i.e., when the material is in a condition known as a *population inversion*, then more state transitions are from E_2 to E_1, leading to emission of additional photons which cause further stimulated emissions. Under this condition, the initial pulse of light is amplified.

For a material that is in thermal equilibrium at temperature T, the ratio N_2/N_1 adheres to a fundamental law of thermodynamics, known as *Boltzmann's Principle*:

$$\frac{N_2}{N_1} = \exp\left(-\frac{E_2 - E_1}{kT}\right). \tag{3.76}$$

This ratio can also be expressed as a difference:

$$N_1 - N_2 = N_1\left(1 - \exp\left(-\frac{\hbar\,\omega}{kT}\right)\right). \tag{3.77}$$

This equation shows that for a material in thermal equilibrium, the majority of the population is in the lower energy level, and therefore light is absorbed. Thus, to create laser light, a pumping mechanism may be used to move the majority of the population into a non-equilibrium state.

When such a pumping process is applied, each atom may react to it by changing energy level. Since each atom responds in the same manner, the resulting amplification is *coherent* in both phase and amplitude. Thus, the output of a laser is an amplified reproduction of the input (pumping) signal—at least for a narrow band around the wavelength associated with the transition energy $E_2 - E_1$. However, some noise will be introduced due to spontaneous emission and absorption.

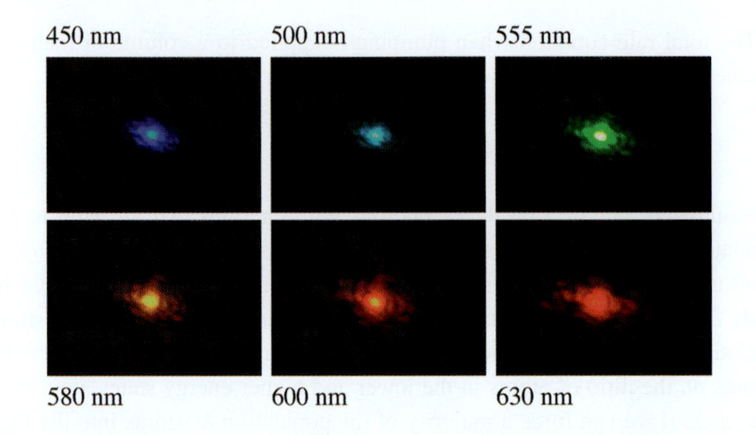

Figure 3.42. Tunable laser photographed at various wavelengths. (Photographs created with the expert help of Claudiu Cirloganu, School of Optics, University of Central Florida.)

The components of a practical laser include:

A laser medium, which is the collection of atoms, molecules, or ions that will mediate amplification.

A pumping process, which will excite the laser medium into a higher energy level, causing a population inversion.

Optical elements, which will mirror a beam of light to pass through the laser medium one or more times. If the beam of light passes through once, we speak of a laser amplifier, otherwise of a laser oscillator.

Lasers emit light at a narrow band of wavelengths, and therefore produce near monochromatic light. In addition, as outlined above, this light is typically coherent, i.e., in phase and amplitude, making laser light very different from most light sources normally encountered. Most lasers are set to one specific wavelength, although high-end lasers can be made to be tunable. For such lasers, it is possible to dial in the wavelength at which they will emit monochromatic light (Figure 3.42).

3.6 Polarization in Dielectric Materials

In dielectric materials, electrons are bound to the atomic structure, i.e., conduction of free electrons such as that which occurs in metals is absent. In the presence of an electromagnetic field, electrons remain bound, but may be displaced. In such cases, the electrons may be thought of as orbiting around a point that is some distance away from the nucleus. As a result, the negative charge of the electrons is slightly offset with respect to the positive charge of the nucleus. An atom under the influence of an electromagnetic field may thus form an *electric dipole*. The atom is said to be polarized under these conditions.

Some dielectrics are known as polar substances, where the molecular structure is such that dipoles exist even in the absence of an external electromagnetic field. This is for instance the case with water. However, for polar substances, the orientation of each molecule is random, and therefore the net polarization of the material in bulk is zero. When an electric field is applied, the forces acting upon the molecules will align them with the field, and the material will become polarized. This phenomenon is called *orientational polarization*.

Some materials consist of ions that are bound together. As an example, sodium chloride NaCl consists of positive and negative ions. Application of an external electric field will separate the ions and thus form electric dipoles. The result is called *ionic polarization*.

A polarized atom may be modeled as a positive point charge $+Q$ and a negative point charge $-Q$ separated by a distance $\mathbf{r}$. The *dipole moment* $\mathbf{p}$ is then defined as

$$\mathbf{p} = Q\mathbf{r}. \tag{3.78}$$

It would be difficult to model the interaction of an electromagnetic wave with a material if the dipole moments of each individual atom or molecule need to be considered. It is more convenient to compute the average behavior of the material at the microscopic scale. The *electric polarization* $\mathbf{P}$ of a material is then given by

$$\mathbf{P} = NQ\mathbf{r}_a = N\mathbf{p}, \tag{3.79}$$

where N is the number of atoms per unit volume and $\mathbf{r}_a$ is the average separation between positive and negative charges in atoms. It is now possible to model the electric displacement $\mathbf{D}$ as an additive quantity rather than a multiplicative quantity as shown in (2.5):

$$\mathbf{D} = \mathbf{E} + 4\pi\mathbf{P}. \tag{3.80}$$

For suitably weak fields, the electric polarization can be taken to be proportional to the magnitude of the electric field, i.e., $\mathbf{P} \propto \mathbf{E}$. The constant of proportionality is then called the *dielectric susceptibility* η:

$$\mathbf{P} = \eta\,\mathbf{E}. \tag{3.81}$$

In this case, by comparing (2.5) and (3.81), we find a relation between the dielectric susceptibility and the material constant ε:

$$\varepsilon = 1 + 4\pi\eta. \tag{3.82}$$

3.6.1 The Lorentz-Lorenz Formula

Since an atom consists of a heavy nucleus surrounded by electrons orbiting at specific distances, we can expect that the impact of an electromagnetic field on a single atom yields somewhat different mathematics than the impact on a volume of atoms or molecules. Thus, the *effective field*, $\mathbf{E}'$ and $\mathbf{H}'$, is acting on an individual atom, whereas the *mean* or *observed field*, $\mathbf{E}$ and $\mathbf{H}$, is averaged over a large number of atoms. It can be shown that the effective and mean fields can be expressed in terms of each other by solving a Poisson equation over a uniformly charged sphere, yielding [112]

$$\mathbf{E}' = \mathbf{E} + \frac{4\pi}{3}\mathbf{P}. \tag{3.83}$$

We assume that the electric dipole moment of an atom or molecule is proportional to the strength of the electric field, and hence

$$\mathbf{p} = \alpha \mathbf{E}'. \tag{3.84}$$

The constant α is called the *mean polarizability*. With N atoms per unit volume, as given in (3.79), the electric polarization can be rewritten as

$$\mathbf{P} = N\mathbf{p} = N\alpha\mathbf{E}'. \tag{3.85}$$

Combining this with (3.83), we find

$$\mathbf{P} = \frac{N\alpha}{1 - \frac{4\pi}{3}N\alpha}\mathbf{E}. \tag{3.86}$$

We therefore have the following expression for the dielectric susceptibility:

$$\eta = \frac{N\alpha}{1 - \frac{4\pi}{3}N\alpha}. \tag{3.87}$$

Substitution of this result into (3.82) yields an expression for the permittivity ε:

$$\varepsilon = \frac{1 + \frac{8\pi}{3}N\alpha}{1 - \frac{4\pi}{3}N\alpha}. \tag{3.88}$$

With $n = \sqrt{\varepsilon}$, the mean polarizability can be expressed in terms of the index of refraction as follows:

$$\alpha = \frac{3}{4\pi N}\frac{n^2 - 1}{n^2 + 2}. \tag{3.89}$$

In the context of refractivity, this relation is called the *Lorentz-Lorenz* formula, and is named after two scientists who discovered this relation independently and almost at the same time. When written in the following form,

$$4\pi\alpha N = 3\,\frac{\varepsilon - 1}{\varepsilon + 2}, \tag{3.90}$$

and used in the context of dielectric materials, this equation is also known as the *Clausius-Mossotti equation* [422,829]. Finally, in the context of conductivity, this equation is known as *Maxwell's formula* [739]. The importance of these formulae lies in the fact that they relate Maxwell's equations with the atomic theory of matter.

3.6.2 Dispersion

Newton's famous experiment, in which light is refracted through a prism to reveal a set of rainbow colors, cannot be explained with the machinery described so far. In this experiment, the different wavelengths impinging upon a dispersing prism are refracted by different amounts. This suggests that the index of refraction is not a simple constant related to the permittivity and permeability of the material, but that it is dependent on wavelength λ (or equivalently, angular frequency ω).

It is indeed possible to derive a wavelength-dependent model for the index of refraction to account for the wavelength-dependent nature of certain dielectric materials.

We begin by assuming that electrons move with a velocity much smaller than the speed of light. The Lorentz force acting on electrons, as given in (2.9), then simplifies since the second term may be neglected. With a mass m, the position of the electron, as a function of forces acted upon it, is modeled by the motion equation:

$$m\,\frac{\partial^2 \mathbf{r}}{\partial t^2} + q\,\mathbf{r} = e\,\mathbf{E}'. \tag{3.91}$$

If the electric field acting upon the electron is given by

$$\mathbf{E}' = \mathbf{E}'_0\, e^{-i\omega t}, \tag{3.92}$$

then a solution in terms of $\mathbf{r}$ is given by [112]

$$\mathbf{r} = \frac{e\,\mathbf{E}'}{m\left(\omega_0^2 - \omega^2\right)}. \tag{3.93}$$

Here, the quantity

$$\omega_0 = \sqrt{\frac{q}{m}}$$

is the *resonance* or *absorption frequency*. Thus, the angular frequency ω of the incident field $\mathbf{E}'$ causes the electron to oscillate with the same angular frequency.

The electron contributes a dipole moment $\mathbf{p} = e\,\mathbf{r}$ to the total polarization $\mathbf{P}$. Under the assumption that the influence of the nuclei can be neglected, we therefore have for N atoms, each with a single electron with resonance frequency ω_0, a total polarization

$$\mathbf{P} = N e\,\mathbf{r} = N \frac{e^2}{m} \frac{\mathbf{E}'}{\left(\omega_0^2 - \omega^2\right)}. \tag{3.94}$$

However, if this equation is compared with (3.85), it can be seen that the mean polarizability α is related to the angular frequency ω:

$$N\alpha = N \frac{e^2}{m\left(\omega_0^2 - \omega^2\right)}. \tag{3.95}$$

We can now use the Lorentz-Lorenz formula (3.89) to relate the index of refraction to the angular frequency:

$$\frac{n^2 - 1}{n^2 + 2} = \frac{4\pi}{3} \frac{N e^2}{m\left(\omega_0^2 - \omega^2\right)}. \tag{3.96}$$

It can be seen that the right-hand side of this equation tends to infinity for angular frequencies ω approaching the resonance frequency ω_0. This is a consequence of the formulation of the motion equation (3.91), which does not include a damping term that models a resisting force. If this term were added, the position of an electron would be modeled by

$$m \frac{\partial^2 \mathbf{r}}{\partial t^2} + g \frac{\partial \mathbf{r}}{\partial t} + q\,\mathbf{r} = e\,\mathbf{E}', \tag{3.97}$$

and the solution would be

$$\mathbf{r} = \frac{e\,\mathbf{E}'}{m\left(\omega_0^2 - \omega^2\right) - i\,\omega\,g}. \tag{3.98}$$

As a result, the polarization P becomes a complex quantity, as does the index of refraction n. The imaginary part of the index of refraction models the absorption of the dielectric and is known as the *extinction coefficient*.

Materials normally have several resonance frequencies. Equation (3.96) may therefore be extended as follows:

$$\frac{n^2 - 1}{n^2 + 2} = \frac{4\pi e^2}{3m} \sum_k \frac{N_k}{\omega_k^2 - \omega^2}. \tag{3.99}$$

In this equation, N_k represents the number of electrons that correspond to resonance frequency ω_k. While we omit the derivation (see [112]), this equation may be rewritten for gasses as well as for liquids and solids. The derivation proceeds differently for gasses than for substances where n differs significantly from unity. In either case, using the correspondence between angular frequency and wavelength, $\omega/2\pi = c/\lambda$, the above equation becomes

$$n^2 - 1 = a + \sum_k \frac{b_k}{\lambda^2 - \lambda_k^2}. \tag{3.100}$$

The constants a and b_k are given by

$$a = \frac{1}{c^2} \sum_k \rho_k \lambda_k^2, \tag{3.101a}$$

$$b_k = \frac{1}{c^2} \rho_k \lambda^4, \tag{3.101b}$$

$$\rho_k = \frac{e^2 N_k}{\pi m}. \tag{3.101c}$$

The above equation is known as *Sellmeier's dispersion formula* and the constants a and b_k are called *Sellmeier's constants*. The refractive behavior of materials can normally be modeled to a reasonable accuracy with a limited number of constants. In many cases, taking into account around two resonant frequencies suffices—especially in rendering applications.

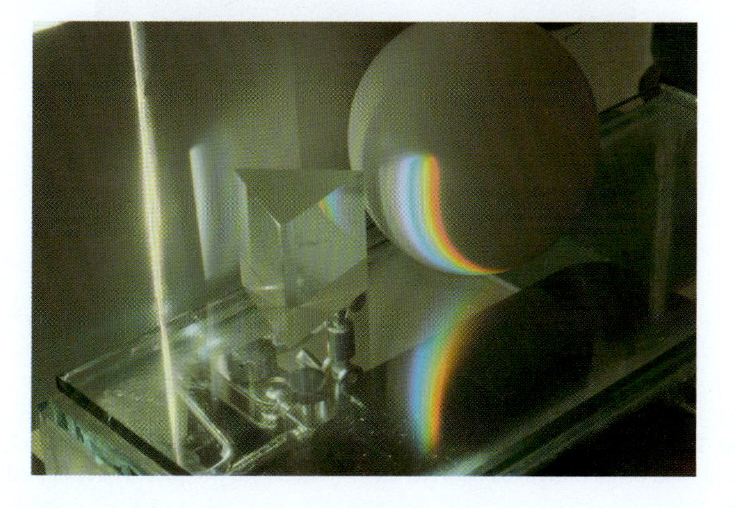

Figure 3.43. Dispersion through a prism. The slit is visible towards the left. The dispersed light is projected onto a sphere for the purpose of visualization.

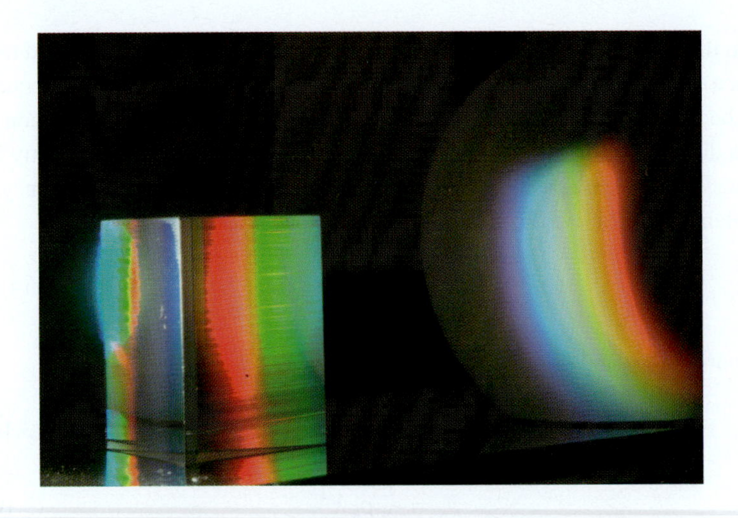

Figure 3.44. Dispersion through a prism. The light source is a Gretag Macbeth D65 daylight simulator, which produces light similar to that seen under an overcast sky.

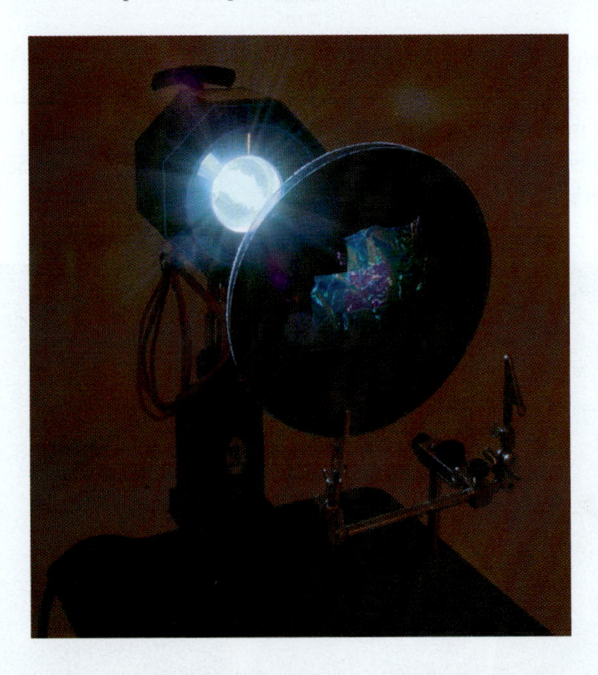

Figure 3.45. A Gretag Macbeth D65 daylight simulator lights two sheets of polarizing material from behind. Wedged between the two sheets are two pieces of crumpled cellophane.

An example of dispersion is shown in Figure 3.43 and Figure 3.44. Here, a D65 daylight simulator is used to shine light through a slit formed by two cardboard panels a short distance apart. The light is refracted and dispersed by a 35 mm dispersing prism. The dispersed light forms a rainbow pattern, which is reflected by the diffuse white sphere.

Figure 3.46. Wavelength-dependent polarization shown by wedging pieces of crumpled cellophane between polarizing sheets oriented at right angles.

Figure 3.47. Two polarizing sheets are aligned. The wedged pieces of cellophane produce complementary colors to those shown in Figure 3.46.

Another example, showing the relationship between polarization and wavelength, is afforded by crumpled cellophane wedged between two linear polarizers [447]. The setup of such an experiment is shown in Figure 3.45. Dependent on its thickness, the cellophane produces different colors of the rainbow by rotating the **E** field in a wavelength-dependent manner. Between crossed polarizing sheets, the results are as shown in Figure 3.46. Complementary colors are produced if the polarizing sheets are aligned, as shown in Figure 3.47.

3.7 Dichroism

There exist at least three phenomena which are each called *dichroism*. In this section, we briefly describe each for the purpose clarifying the differences. In each case, dichroism refers to the ability of a material or substance to attain two different colors. However, the mechanism by which this two-coloredness occurs is different.

First, dichroism can be related to polarization. In Section 2.3, we described how birefringence may occur in certain minerals. In that case, the index of refraction was dependent on the polarization of light. For certain materials, the wavelength composition of transmitted light may depend on polarization as well. Ruby crystals, for instance, have octahedral structures. If such crystals are lit by polarized light, it is possible to change the color of the ruby from purple-red to orange-red.

The absorption of polarized light by rods in certain species of fish has been shown to differ by the angle of polarization [248]. Thus, at least some fish retinas show dichroism of this form, particularly if the retina is unbleached.

A second effect occurs in other minerals, where the color of the transmitted light is determined by the angle at which light strikes the stone. As such, the color of transmitted light depends on the angle it makes with the optical axis of the material. This effect is also known as dichroism, but it should be distinguished from the case where the wavelength composition of transmitted light is dependent on polarization. Iolite, a relatively common gem stone, exhibits this form of dichroism, as shown in Figure 3.48. Some materials show different colors along three or more optical axes, yielding an effect known as *pleochroism*.

Thirdly, dye dichroism refers to the different colors that dye solutions may obtain as function of the concentration of the dye, as discussed in Section 3.4.3.

Finally, *Dichroic filters* and *dichroic mirrors* are based on interference effects. Such devices selectively reflect part of the visible spectrum while transmitting the remainder. They are built by coating a glass substrate with very thin layers of

Figure 3.48. The stone in this pendant is made of iolite, a dichroic material. As a result, viewing this stone from different angles reveals a different set of colors; in this case the color varies between purple and yellowish-gray.

optical coatings. It is possible, for instance, to build a beam splitter by coating several prisms with a dichroic optical material, and thus split light into red, green, and blue components. Such beam splitters are used in some cameras that have separate sensors for the three color channels. They are also commonly used in LCD projectors (Section 14.12) as well as LCoS projectors (Section 14.14).

3.8 Application: Modeling of Fire and Flames

The chemical reactions occurring in a typical combustion process were discussed in Section 3.5.2. Combustion can be classified into two groups, which are detonations and deflagrations. The former are explosive reactions, whereas the latter are slow reactions visible as flames and fire. During a deflagration the chemical reactions cause an expansion of the fuel being burned, and create heat as well. Both phenomena give rise to turbulence.

Modeling a flame can be achieved by finding a solution to the Navier-Stokes equations for flow. Given a 3D grid with each grid point defining a velocity vector, each iteration simulates a time step, updating the velocity field, and thereby the flow. These equations can be used for both fluid flow as well as gas flow. There exist several variations of the Navier-Stokes equations, which model different aspects of the fluid/gas. For flames, viscosity is unimportant, and therefore the equations for incompressible flow are appropriate.

Considering a velocity field with each discrete position holding a velocity $\mathbf{u}$, then incompressibility can be enforced by ensuring that the divergence is zero:

$$\nabla \cdot \mathbf{u} = 0. \tag{3.102}$$

Then, at time t the velocities are computed from the velocities of the previous time step $t-1$ using:

$$\mathbf{u}_t = -(\mathbf{u}_{t-1} \cdot \nabla)\,\mathbf{u}_{t-1} - \frac{\nabla p}{\rho} + \mathbf{f}, \tag{3.103}$$

where p is the pressure, ρ is the density, and $\mathbf{f}$ models any external forces affecting the flow. These quantities are defined at each grid point, and may or may not themselves be updated for each time step. The above two equations together are termed the Navier-Stokes equations for incompressible flow. Solutions to this equation can be obtained for instance using semi-Lagrangian methods [307, 1079].

Unfortunately, the gas expansion discussed above cannot be taken into account with this model (it would require the more complex equations for compressible flow). However, the Navier-Stokes equations for incompressible flow can still be used, for instance by modeling the reaction zone where fuel is burnt as a thin surface in 3D space, rather than a volume [741]. Recall from Section 3.5.2 that a deflagration requires oxygen which diffuses in from the surrounding air, leaving a core within the flame where no reaction occurs. Thus, the reaction zone is reasonably well modeled with an implicit surface.

3.8.1 Modeling a Flame

Nguyen et al. use a dynamic implicit surface to track a flame over time. Their method is explained in some detail in this and the following subsection. They use separate sets of equations to model the distribution of fuel and the flow of hot gaseous products [835]. The updates to the two sets of Navier-Stokes equations for each time step are coupled, however, enabling the conservation of mass and momentum. To model turbulence, the Navier-Stokes equations are augmented with a vorticity confinement term [307, 1084].

The implicit surface separates the gaseous fuel from the hot gaseous products and the air around the flame. In a system where fuel is injected over a surface with area A_f, the fuel, which is premixed at first, will begin to heat up, and emit light. In this stage, the color will be blue. Given a flame speed S indicating the rate with which fuel is burned (a fuel-dependent property [1155]), the surface area of the blue core can be estimated to be A_S using

$$v_f A_f = S A_S, \tag{3.104}$$

where v_f is the speed of fuel injection. Thus, the size of the blue core depends on the speed of injection in this model.

The reactions occurring in the flame, including the blue core, cause expansion which is normally modeled as the ratio of densities ρ_f/ρ_h, where ρ_f is the density of the fuel, and ρ_h is the density of the hot gaseous products. Across the dynamic implicit surface modeling the thin flame, the expansion of the gas is modeled by requiring conservation of mass and momentum, leading to the following equations:

$$\rho_h (V_h - D) = \rho_f (V_h - D), \tag{3.105a}$$

$$\rho_h (V_h - D)^2 + p_h = \rho_f (V_h - D)^2 + p_f. \tag{3.105b}$$

Here, p_f and p_h are pressures, and V_f and V_h are normal velocities of the fuel and hot gaseous products. The speed of the implicit surface in its normal direction is given by

$$D = V_f - S. \tag{3.106}$$

For the special case of solid fuels, the pressure will be zero; with the density and normal velocity of the solid fuel given by ρ_s and V_s, we have

$$\rho_f (V_f - D) = \rho_s (V_s - D), \tag{3.107}$$

and therefore

$$V_f = V_s + \left(\frac{\rho_s}{\rho_f} - 1 \right) S. \tag{3.108}$$

Hence, the velocity of the gasified fuel in its normal direction at the boundary between its solid and gaseous state is related to the velocity of the solid fuel plus an expansion-related correction.

The dynamic implicit surface indicating the boundary of the blue core moves at each time step with a velocity given by **w**:

$$\mathbf{w} = \mathbf{u}_f + S\mathbf{n}, \tag{3.109}$$

where the velocity of the gaseous fuel is $\mathbf{u}_f$ and $S\mathbf{n}$ is a term describing the velocity induced by the conversion of solid fuel to gas. At each grid point, the surface normal **n** is determined by

$$\mathbf{n} = \frac{\nabla \phi}{|\nabla \phi|}, \tag{3.110}$$

where ϕ defines the region of the blue core. Its value is positive where fuel is available, zero at the boundary, and negative elsewhere. The motion of the implicit surface ϕ is then defined as

$$\phi_t = -\mathbf{w} \cdot \nabla \phi. \tag{3.111}$$

This equation can be solved at each grid point, after which an occasional conditioning step is inserted to keep the implicit surface ϕ well-behaved [835, 1027, 1146].

The implicit surface indicates the boundary between the fuel and hot gaseous products. These two types of gas have their own velocity fields, $\mathbf{u}_f$ and $\mathbf{u}_h$, which are both independently updated with the Navier-Stokes equations above. As these velocity fields are valid on opposite sides of the implicit surface, at this boundary care must be taken to update the velocities appropriately. For instance, when computing velocities $\mathbf{u}_h$ for hot gaseous products near the implicit surface, values on the other side of the boundary must be interpolated with the aid of the normal velocity of the fuel V_f, which is computed with

$$V_f = \mathbf{u}_f \cdot \mathbf{n}. \tag{3.112}$$

The corresponding normal velocity for the hot gaseous products can be computed with

$$V_h^G = V_f + \left(\frac{\rho_f}{\rho_h} - 1 \right) S \tag{3.113}$$

The superscript G indicates that this is a ghost value, i.e., it is extrapolated for a region where no hot gaseous fuel is present. The velocity of the hot gaseous products on the opposite side of the implicit surface can now be computed with

$$\mathbf{u}_h^G = V_h^G \mathbf{n} + \mathbf{u}_f - \left(\mathbf{u}_f \cdot \mathbf{n} \right) \mathbf{n}. \tag{3.114}$$

This mechanism allows the Navier-Stokes equations to be computed throughout the volume, including the boundary of the flame.

The final component of the flame model involves external forces $\mathbf{f}$ which are applied to the velocity field. These come in two forms. The first external force is a dependency on temperature, which is defined as a scalar field T (discussed below) and leads to buoyancy. The temperature-dependent force is defined as

$$\mathbf{f}_T = \alpha \left(T - T_{\text{air}} \right) \mathbf{z}, \tag{3.115}$$

where α is a user-defined constant, and $\mathbf{z} = (001)^T$ is the up vector.

The second external force is termed *vorticity confinement* and constitutes a correction to add turbulence and swirling, which is otherwise dampened as a result of the low resolution of the velocity field that is normally used in rendering applications [1084]. Vorticity confinement $\mathbf{f}_C$ is defined as

$$\mathbf{f}_C = \varepsilon h \left(\mathbf{n}_\omega \times \omega\right). \tag{3.116}$$

The amount of vorticity is determined by ε, with $\varepsilon > 0$, and h is the spacing between the grid points. The vorticity ω and normalized vorticity gradients $\mathbf{n}_\omega$ are computed with

$$\omega = \nabla \times \mathbf{u}, \tag{3.117a}$$

$$\mathbf{n}_\omega = \frac{\nabla |\omega|}{|\nabla |\omega||}. \tag{3.117b}$$

As fuel is converted from solid to gas, ignites, and moves through the blue core out into the area where the flame is categorized as a diffusion flame (see Section 3.5.2), the temperature follows a profile which rises at first and then cools down again. To model this both over space and over time, it is useful to track fuel particles during their life cycle, as follows:

$$Y_t = -\left(\mathbf{u} \cdot \nabla\right) Y - 1. \tag{3.118}$$

This equation is solved once more with the aid of semi-Lagrangian fluid solvers. With suitably chosen boundary conditions, the value $1 - Y$ can be taken to indicate the time elapsed since a particle left the blue core. The temperature T is then linearly interpolated between the ignition temperature T_{ignition} and the maximum temperature $T_{\max}$, which occurs some time after leaving the blue core. In the temperature fall-off region, with the help of a cooling constant c_T, the temperature can be computed with [835]

$$T_t = -\left(\mathbf{u} \cdot \nabla\right) T - c_T \left(\frac{T - T_{\text{air}}}{T_{\max} - T_{\text{air}}}\right)^4. \tag{3.119}$$

Finally, the density ρ_t at time step t is computed similarly, using

$$\rho_t = -\left(\mathbf{u} \cdot \nabla\right) \rho. \tag{3.120}$$

3.8.2 Rendering a Flame

While this procedure models the flame and the progression of its shape over time, the rendering can be achieved by a ray-marching algorithm, visiting each grid element along its path. The temperature recorded at each grid point can be used to compute the amount of blackbody radiation (see Section 3.5.1). The technique for rendering a flame bears strong resemblance to the atmospheric rendering technique discussed in Section 2.11.6.

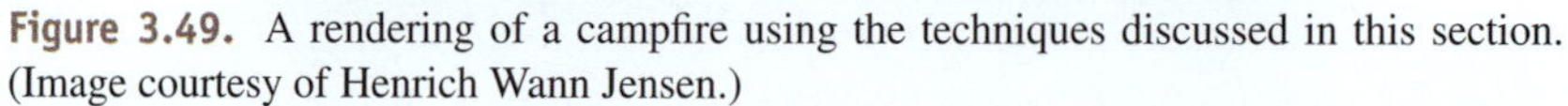

Figure 3.49. A rendering of a campfire using the techniques discussed in this section. (Image courtesy of Henrich Wann Jensen.)

However, in this case, it is appropriate to use a more general phase function to account for scattering events. To model the spherical distribution of scattered light, the Henyey-Greenstein phase function p can be used, which only depends on incoming and outgoing directions Θ and Θ' [460]:

$$p(\Theta, \Theta') = \frac{1 - g^2}{4\pi \left(1 + g^2 - 2g\,\Theta \cdot \Theta'\right)^{1.5}}. \tag{3.121}$$

This function is then used in (2.234), which models in-scattering. Accounting for in-scattering, out-scattering, as well as light emitted along each ray segment, we use (2.233). However, light emission for modeling flames is governed by Planck's radiation law, so that $L_{e\lambda}$ equals the spectral radiant exitance $M_{e\lambda}$ given in (3.62).

An example result obtained with this combined modeling and rendering technique is shown in Figure 3.49, showing a campfire where a cylindrical log emits fuel which then catches fire.

3.9 Further Reading

A more elaborate discussion on the sources of color can be found in Nassau [815] and Tilley [1128]. Wyszecki and Stiles is a must-have reference work, containing a vast amount of data [1262]. Several books on quantum mechanics were consulted for the preparation of this chapter [179, 683, 1109]. Williamson and Cummins also covers some of the material in this chapter [1247].

Chapter 4
Human Vision

In the preceding chapters, light and its interaction with matter was discussed. One particularly important interaction occurs when light enters the human eye. Light falling on the retina triggers a most remarkable sequence of events. In this and the following chapter, we will focus on this chain of events, insofar the current state of knowledge allows us. It should be noted that much of the neurological underpinnings of color still forms an active area of research, and knowledge of higher-level color processing in the human brain therefore remains relatively sparse [496]. In addition, the more we learn about the functional organization of the visual cortex, the more complex the human visual system turns out to be. Nonetheless, we consider this topic the other half of essential background information, necessary to understand and effectively use theories of color.

In this chapter, we primarily deal with the substrate that forms the human visual system: the eye, the retina, and neural circuitry in the brain. Much of the information collected in this chapter stems from studies of Old-World monkeys [227, 1032]. It has been shown that their photoreceptors have the same spectral tuning as human photoreceptors [78, 1009, 1010], as well as a similar retinal layout [972]. In addition, cell types and circuits found in the macaque visual system are very similar to those in the human retina [224–226, 604, 606, 608, 892].

Both anatomy and histology use terms and descriptions for structures relative to the orientation of the body. Figure 4.1 shows the planes of reference used to describe where structures are. These planes are [329]:

The midsagittal plane, which divides the head into a left and right half. Objects located closer to this plane are called *medial* and those that are further away are called *lateral.*

The coronal plane divides the head into an *anterior* and *posterior* region.

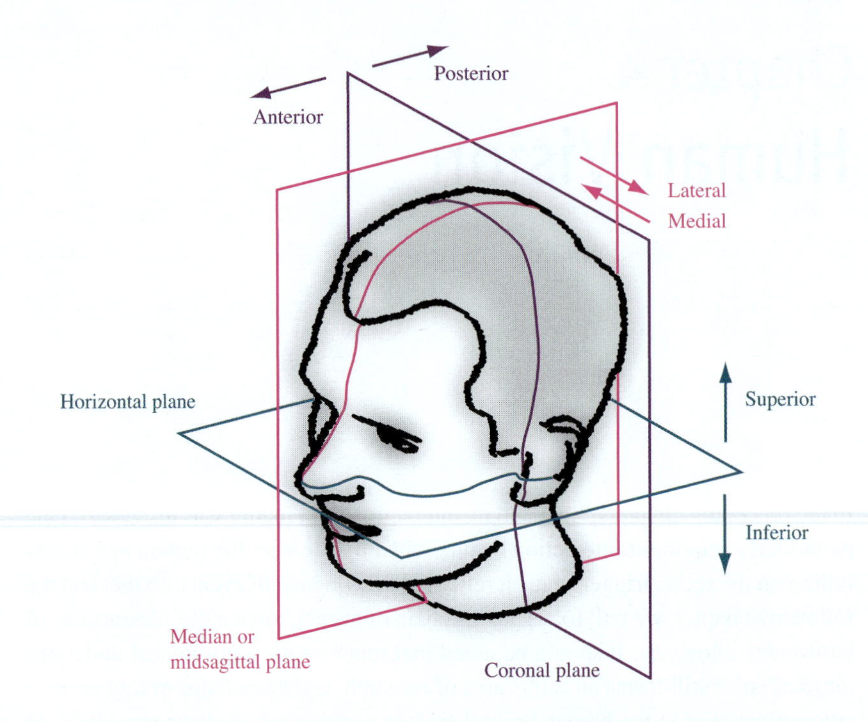

Figure 4.1. Terminology used to indicate anatomic planes of reference.

The horizontal plane divides the head into a *superior* and *inferior* region. These regions are also called *cranial* and *caudal*.

One further term of reference is *deep*, indicating distance from the surface of the body.

4.1 Osteology of the Skull

The skull consists of an upper part, called *cranium* or *neurocranium*, and a lower part, called *facial skeleton* or *viscerocranium*. The cranium itself consists of the *cranial vault* and the *cranial base*.

The skull contains four cavities, each with a different function:

The cranial cavity houses the brain.

The nasal cavity serves in respiration and olfaction.

The orbits house the eyes.

The oral cavity forms the start of the gastrointestinal tract.

The two orbital cavities lie between the cranium and the facial skeleton, and are separated by the nasal cavity. They serve as sockets to hold the eyes as well as the adnexa, and they transmit nerves and blood vessels that supply the face around the orbit [329].

The shape of each orbit is roughly a quadrilateral pyramid with the apex forming the optic canal (which provides passage for the optic nerve and the ophthalmic artery between the orbits and the cranium). Their walls are approximately triangular, except for the medial wall which is oblong. The medial walls are parallel to the midsagittal plane, whereas the lateral walls are angled at 45 degrees to this plane. The height, width, and depth of the orbits are around 40 mm each, with a volume of 30 ml.

4.2 Anatomy of the Eye

The eyes are used for photoreception and for communication of the resulting action potentials to the brain. A diagram of a horizontal cross-section is shown in Figure 4.2. The eyes are located in the anterior portion of the orbit, closer to the lateral wall than the medial wall, and closer to the roof than the floor of the orbit.

The shape of the eye is approximately that of two intersecting spheres, the smaller one forming the cornea and the larger one the sclera. The cornea has a radius of 7.8 mm and forms one-sixth of the circumference of the eye. The sclera accounts for the rest of the circumference, forming a sphere with a radius of 11.5 mm. The axial length of the globe is on average 24 mm.

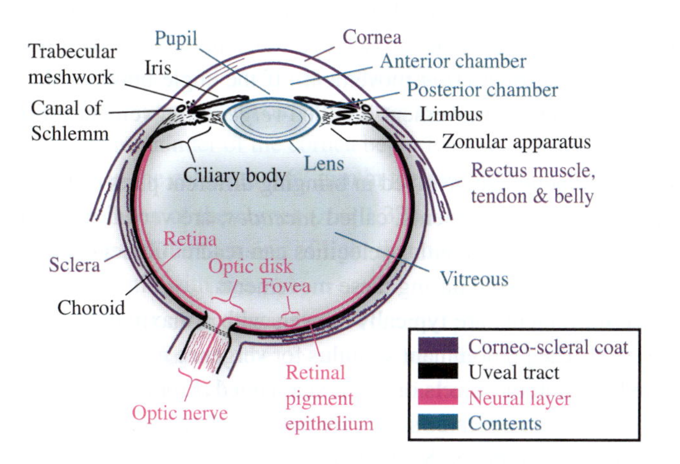

Figure 4.2. Diagram of a horizontal cross-section of the human eye.

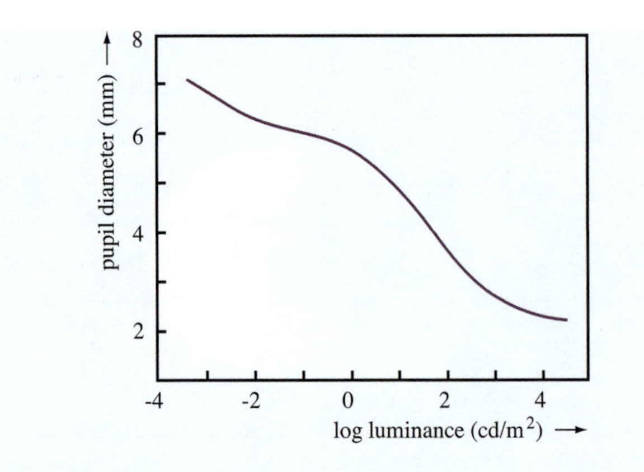

Figure 4.6. Entrance pupil diameter as a function of ambient illumination for young adult observers (after [305]). The photometric unit of candela per square meter is formally introduced in Chapter 6.

including age, accommodation, drugs, and emotion [166]. The range of pupillary constriction affords a reduction in retinal illumination by at most a factor of seven (0.82 log units [118]) to ten [1262].

4.2.5 The Lens

Posterior to the iris and the pupil lies the lens [236]. It is suspended between the aqueous humor and the vitreous body and connects to lens zonules (suspensory ligament), which in turn connect to the ciliary body. The latter contains the ciliary muscle, which together with the elasticity of the lens itself is capable of changing the refractive power of the lens, a process called *accommodation*. This process allows the eye to focus on objects both near and far. The ability to accommodate diminishes with age, which is why many older people need reading glasses.

The transparency of the lens is mediated by its shape, internal structure, and biochemistry. It receives nourishment from the aqueous humor and the vitreous body, thereby allowing the lens to be avascular. The curvature of the lens anteriorly has a radius ranging from 8 to 14 mm, whereas the curvature posteriorly has a radius between 4.5 and 7.5 mm. The lens has a diameter of around 10 mm and an axial length of 4 mm, although the lens continues to grow at a rate of 0.0023 mm per year.

The lens consists of three layers: the capsule, the lens epithelium, and the lens fibers. The capsule, which envelopes the lens, is elastic and shapes the lens to be more rounded when not under tension by the ciliary muscle.

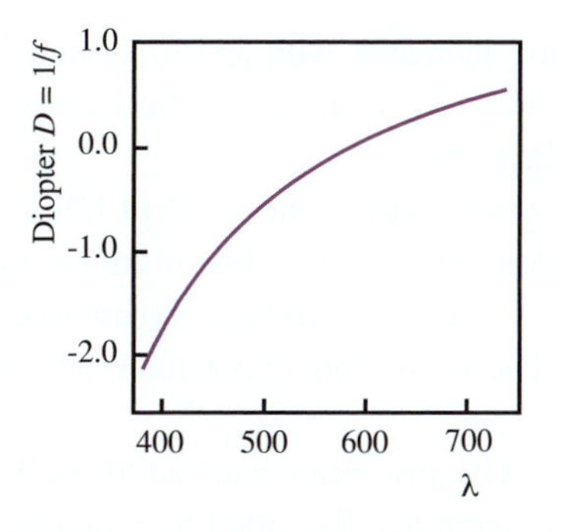

Figure 4.7. Chromatic aberration in the human eye, measured in diopters D, as function of wavelength λ (after [80]).

The curvature of the lens is not the same in horizontal and vertical directions, leading to *astigmatism* (the phenomenon that horizontal and vertical lines have different focal lengths). A consequence of astigmatism is that, at the retina, horizontal edges will consistently show colored fringes, *chromatic aberration*, whereas vertical edges will show differently colored fringes [477]. Both Young and Helmholtz have demonstrated chromatic aberration in their own eyes [164, 165], and this effect was later measured by Wald and Griffin [1191] and Bedford and Wyszecki [80]. The amount of chromatic aberration in the human eye is measured in diopters, i.e., the reciprocal of focal length, as shown in Figure 4.7. This effect is believed to be corrected with later neural mechanisms and possibly also by waveguiding due to the packing of photoreceptors, as discussed in Section 4.2.7.

Other than accommodation, a second function of the lens is to prevent optical radiation between 295 and 400 nm from reaching the retina [344]. This filtering is required to protect the retina from damage by ultra-violet (UV) radiation. With age, the lens yellows and also becomes more opaque, which increases the amount of UV filtering that occurs. The yellowing is due to photochemical reactions rather than biological aging [254, 255, 344, 1223] and causes changes in color perception with age. The decrease in light transmittance is in part due to disturbances in the regular packing of crystallins, which in turn induces increased scattering [329], thus yielding a more opaque lens. This is known as a *cataract* (see also Section 5.11).

Finally, the lens is a source of fluorescence (Section 3.5.4), in that incident UV radiation is absorbed by the lens and re-emitted in the visible range. This phenomenon increases with age. For a healthy lens of a 30-year old, the ratio between incident (sky) light and the amount of produced fluorescence is around

0.002 [1222]. This ratio increases with age to around 0.017 (noticeable) for a 60-year old and 0.121 for an 80-year old. At these ratios, the occurrence of fluorescence is seen as veiling glare.

Dependent on the wavelength of the incident UV light, the range and peak wavelength of the fluorescence varies. For instance, incident light at 360 nm produces fluorescence with a range between 380 nm to over 500 nm, with a peak at 440 nm [672, 1302]. The wavelength of 360 nm is near the absorption maximum of the lens.

In the aging lens, the UV absorption band additionally broadens into the blue, suggesting that human vision may be aided by employing filters that block UV and short wavelengths in the visible range. This would reduce both veiling glare as well as fluorescence [657, 1297].

4.2.6 The Ocular Media

In addition to accommodation, the ciliary body is responsible for the production of the aqueous humor, the fluid that fills the anterior and posterior chambers. Fluid secreted by the ciliary body first enters the posterior chamber, which communicates with the anterior chamber through the pupil. The flow continues towards the trabecular meshwork and the canal of Schlemm (Figure 4.2). Replenishment of the aqueous humor occurs at a rate of 4 μl per minute. The anterior chamber holds around 250 μl of fluid.

The functions of the aqueous humor are to nourish the cornea and the lens, drain toxic metabolic waste, and maintain pressure on the cornea and lens. The latter is achieved by balancing the production and drainage of the fluid and is necessary for the normal functioning of the eye.

The vitreous cavity makes up about two-thirds of the volume of the eye. The contents are called the *vitreous body* or *vitreous* and weighs 3.9 grams. It consists of a transparent viscoelastic gel with an index of refraction of 1.337. The vitreous body is largely acellular.

Medium	n
Cornea	1.376
Aqueous humor	1.336
Lens (center)	1.406
Lens (outer layers)	1.386
Vitreous body	1.337

Table 4.1. Indices of refraction for the different ocular media.

Image formation on the retina is thus mediated by the cornea, the aqueous humor, the lens, and the vitreous body. Their respective refractive indices are summarized in Table 4.1 [447]. Together, they are able to focus light on the retina. However, the ocular media is not a perfect transmitter. Light losses occur due to Fresnel reflection, absorption, and scattering [1262]. Fresnel reflection is generally less than 3 to 4 percent [166]. The optical system is optimized for use in open air. For instance, in underwater environments, the index of refraction from water to cornea is different, and vision appears blurred as a result [208].

Further, the index of refraction of the ocular media is dependent on wavelength, causing edges projected on the retina to show colored fringes, known as *chromatic dispersion* [50].

Aberrations are strongest at the first refractive surface, the anterior corneal surface. The aberrations occurring here only slightly increase with age and are largely corrected by subsequent ocular surfaces in young individuals. However, the corrective power of later refractive boundaries in the ocular media decreases with age, so that the total aberration at the retina increases with age [43]. For instance, the continued growth of the lens causes aberration of the lens to increase with age [367].

Absorption of short wavelengths occurs at the cornea. This may disrupt surface tissue, leading to *photokeratitis* (snow blindness or welder's flash) [166]. The damage threshold is lowest (0.4 J/cm^2) for light at 270 nm [904]. Further absorption at short wavelengths occurs in the center of the lens [772]. This absorption increases with age. Finally, anterior to the outer segments of the cones in the fovea lies a layer of yellow pigment (xanthophyll), which absorbs light at short wavelengths [1065]. The average transmittance of the human eye is plotted in Figure 4.8.

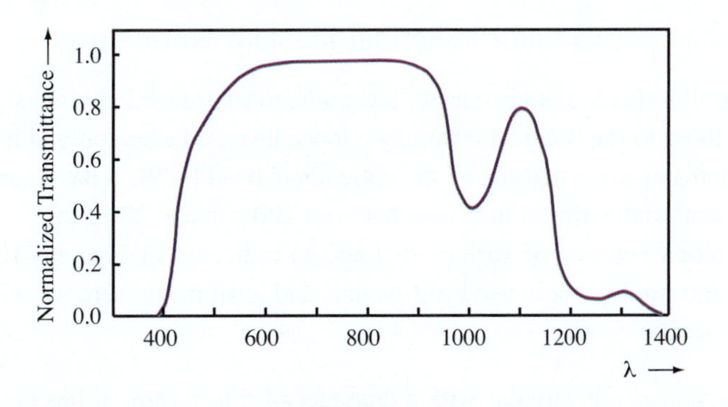

Figure 4.8. Wavelength-dependent transmittance of the ocular media (after [351]).

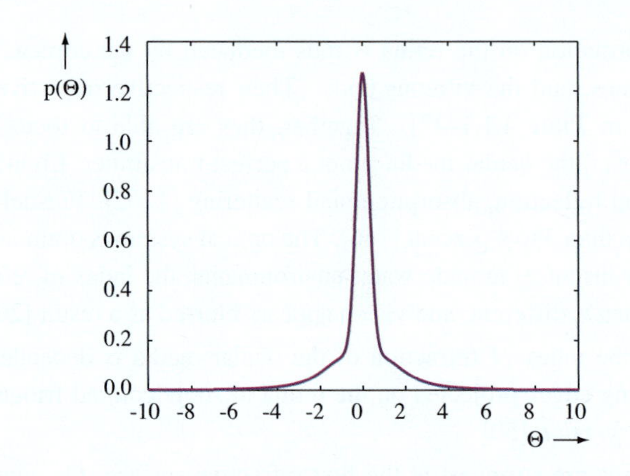

Figure 4.9. Point spread function for ocular light scatter (after [75]).

Some light is scattered in the eye. This means that if the eye is exposed to a point light source, a small area of the retina is illuminated, rather than a single point. Ocular light scatter is also known as *veiling glare*. The intensity fall-off as a function of radial angle (in arc minutes) can be modeled by a point spread function $p(\Theta)$ [75]:

$$p(\Theta) = \begin{cases} 1.13\, e^{-5.52\Theta^2} + \dfrac{1.42}{9.85 + \Theta^{3.3}}, & |\Theta| \leq 7', \\ 9.35 \times 10^{-6}\,(7.8 + \Theta)^{-2.5}, & |\Theta| > 7'. \end{cases} \tag{4.2}$$

This function is plotted in Figure 4.9, and following Baxter et al. [75], it consists of data collected for visual angles less than 5 arc minutes [615] superposed with data for angles larger than 9 arc minutes [336, 402, 1185].

4.2.7 The Retina and the Retinal Pigment Epithelium

The (neural) retina is a neurosensory layer able to transduce light into signals and transmit those to the brain. It is largely a loose layer, attached only at its anterior termination and at the margins of the optic nerve head [329]. It has a surface area of 1250 mm^2 and varies in thickness between 100 μm and 230 μm.

The retina consists of various regions, as indicated in Figure 4.10. While different terminology is in use by clinicians and anatomists, here we will use the anatomic terms:

The area centralis is circular with a diameter of 5 to 6 mm. It lies between the superior and inferior temporal arteries.

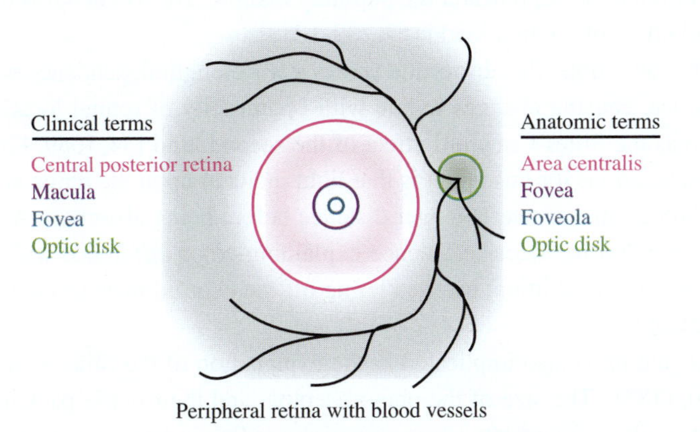

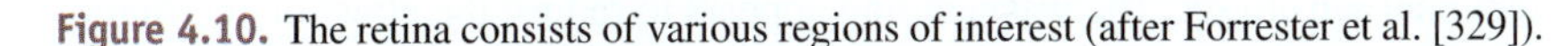

Figure 4.10. The retina consists of various regions of interest (after Forrester et al. [329]).

The fovea is 1.5 mm in diameter and lies 3 mm lateral to the optic disk. It contains a layer of xanthophyll pigment which gives it a yellowish color (hence the clinical term *macula lutea*).

The foveola is a 0.35 mm-diameter zone which consists of a depression surrounded by a rim. This area is avascular. The visual acuity of the retina (resolving power) is highest in this region.

The optic disk contains no retinal layer or photoreceptors. It is known as the blind spot. It lies 3 mm medial to the center of the fovea.

The peripheral retina is the remainder of the retina.

The histology of the neurosensory retina is further discussed in Section 4.3.

The retina is backed by the retinal pigment epithelium, which provides adhesion and nourishment to the retina. It is also heavily pigmented with melanin, which aids in the absorption of light. This reduces light scatter in the eye and thus improves image resolution.

A peculiarity of the retina is that its BRDF is not Lambertian, which implies that its ability to absorb light depends on the direction of incidence. As a consequence of the positioning of the retina with respect to the pupil, light reaching the periphery of the retina will be incident at an angle, where it is less efficiently absorbed than in the center of the retina.

Thus, light entering the pupil through the center and reaching a given position on the retina will be absorbed better than light reaching the same location after passing through the pupil nearer the pupillary margin. This is known as the Stiles-Crawford effect of the first kind [167, 789, 1088].

At the same time, the absorption shows a wavelength dependence with angle of incidence, causing changes in hue with eccentricity of retinal location. This is known as the Stiles-Crawford effect of the second kind [34, 1089, 1262]. One of the explanations for this effect is that light incident upon the retina at an angle passes through matter over a longer distance before being absorbed. As a result, more attenuation has occurred, hence explaining color difference with angle of incidence [85]. In addition, backscattering off the choroid may account for some of this effect [85].

Waveguiding is also implicated in the explanation of the Stiles-Crawford effects [40, 1183]. The size of the photoreceptors and their dense packing creates an effect similar to fiber-optics (more on waveguiding in Section 4.7), creating directional sensitivity. The waveguiding appears to reduce the effect of aberrations in the ocular media [1183].

4.3 The Retina

When light enters the eye, it passes through the ocular media, and then passes through several layers of cells before being absorbed by the photoreceptors. The photoreceptors are thus at the very back of the retina [803], in close contact with the retinal pigment epithelium, which provides the photoreceptors with *retinal* (vitamin A). This molecule is essential in photo-transduction.

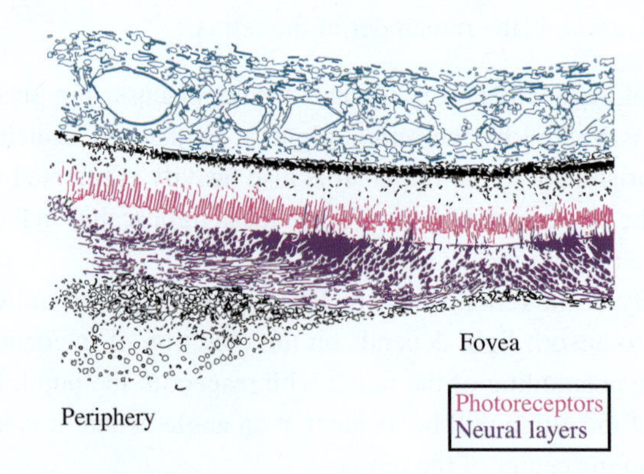

Figure 4.11. Neural cell layers are pushed aside in the fovea to allow more light to reach the photoreceptors.

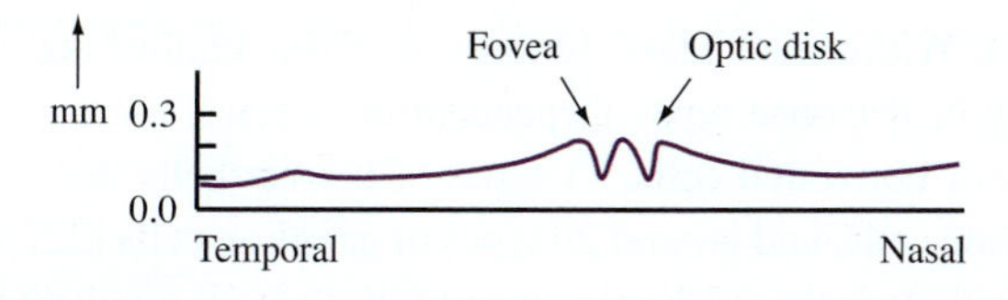

Figure 4.12. The thickness of the retinal layers varies with position (after [329]).

The layers of cells preceding the retina are varied in thickness, and, in particular in the fovea, light passes through a much thinner layer before reaching the photoreceptors, as shown in Figure 4.11. A plot of the variation in thickness of the retina is given in Figure 4.12.

The neurosensory retinal layer transduces light into neural signals that are then transmitted to the brain for further processing. To this end, the retina consists of several types of cells, including photoreceptors, bipolar, ganglion, amacrine, and

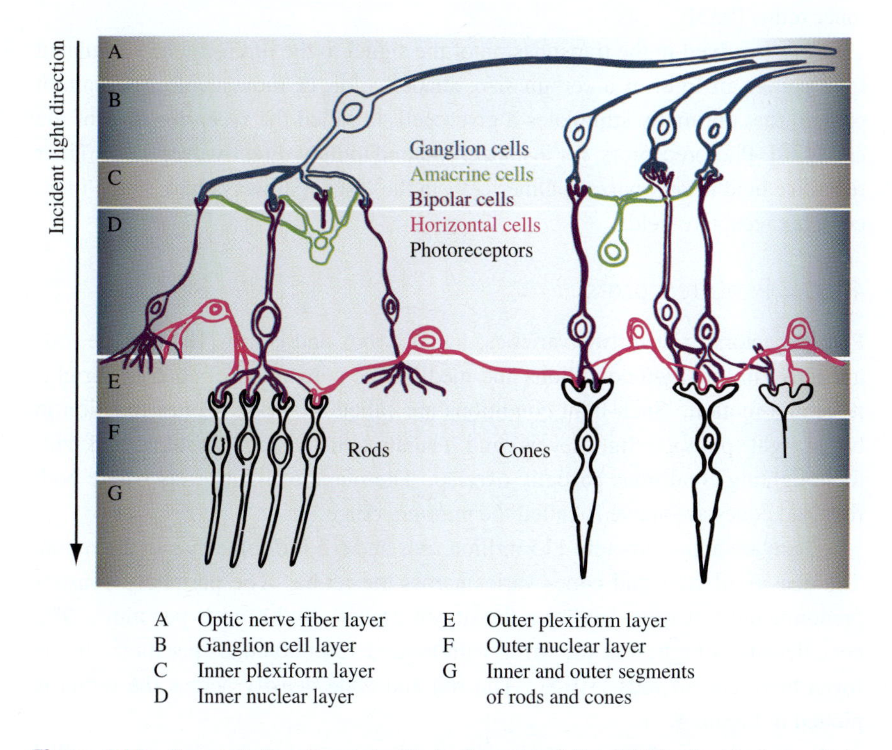

Figure 4.13. Connectivity of neural cells in the retina (after [329]). In this figure, light comes from the top, passes through the layers of ganglion, amacrine, bipolar, and half cells before reaching the light sensitive photoreceptors.

half cells [262]. Within each class, further sub-classifications are possible based on morphology or response type. Dependent on species, there are between one and four types of horizontal cells, 11 types of bipolar cells, between 22 and 40 types of amacrine cells, and around 20 types of ganglion cells [227, 609, 723, 749, 1177, 1217]. Although the predominant cell types are all involved in neural processing of visual signals, the retina also contains glial cells, vascular endothelium, pericytes, and microglia.

The photoreceptors, bipolars, and ganglion cells provide vertical connectivity [915], whereas the horizontal and amacrine cells provide lateral connectivity. A schematic of the cell connectivity of neural cells in the retina is given in Figure 4.13. These cells are organized into three cellular layers, indicated in the figure with blue backgrounds, and two synaptic layers where cells connect. The latter are called the *inner* and *outer plexiform layers* [262]. Each of the cell types occurring in the retina is discussed in greater detail in the following sections. Visual processing typically propagates from the outer layers towards the inner retina [935].

Cells involved in the transmission of the signal to the brain can be stimulated by a pattern of light of a certain size, shape, color, or movement. The pattern of light that optimally stimulates a given cell, is called the *receptive field* of the cell [430]. Photoreceptors, for instance, respond to light directly over them. Their receptive field is very narrow. Other cells in the visual pathways have much more complex receptive fields.

4.3.1 Photoreceptors

Photoreceptors exist in two varieties, namely *rods* and *cones* [1015]. The rods are active in low light conditions and mediate the perception of contrast, brightness, and motion. Such light conditions are called *scotopic*. Cones function in bright light (photopic light levels) and mediate color vision. Scotopic and photopic lighting conditions partially overlap. The range of light levels where both rods and cones are active is called the *mesopic* range.

There are approximately 115 million rods and 6.5 million cones in the retina. The density of rods and cones varies across the retina. The periphery consists predominantly of rods, having a density of around 30 000 rods per mm^2. The cone density is highest in the fovea with around 150 000 cones per mm^2. In the fovea there are no rods [1015]. The rod and cone density across the retina is plotted in Figure 4.14.

Rods and cones form the beginning of a sequence of processing stages called *pathways*. Pathways are sequences of cells that are interconnected. For instance, photoreceptors are connected to bipolar cells as well as to half cells. Bipolars are

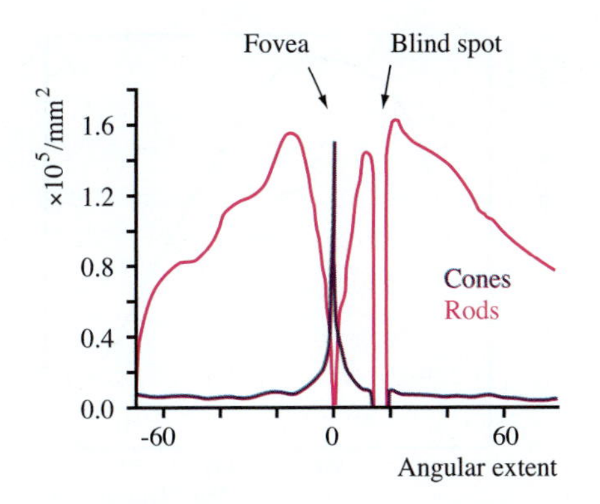

Figure 4.14. The density of rods and cones in the retina (after [901]).

also interconnected by a layer of amacrine cells before connecting to the ganglion cells. The ganglion cells in turn carry the signal away from the eye and into the lateral geniculate nucleus (LGN), a small area in the brain that in turn connects to the visual cortex.

The interconnections at each stage along the visual pathways are complex and varied. In addition to feed-forward processing, all stages and neural layers have feedback loops. There also exists feedback from the brain back into the retina [347, 482].

A full understanding of the visual processing occurring in the retina remains beyond our abilities. Nonetheless, more is known about the retina than most other centers of processing in the brain. In the remainder of this chapter, we give a brief overview of some of the peculiarities that are known about the human visual system, as this is the substrate of human (color) vision.

4.3.2 Rods

Under favorable circumstances the human visual system can detect a single photon [446, 1018, 1205], a feature attributed to the sensitivity of the rods, and the pooling occurring in the rod pathway. Rods have a peak density in a ring 5 mm from the center of the fovea [862] (see also Figure 4.14). Rods are packed in a hexagonal pattern, separating cones from each other.

Much is known about the molecular steps involved in light transduction in both rods and cones [289, 928]. A simplified overview is given here. Rods contain a visual pigment called *rhodopsin*, which has a peak spectral sensitivity at $\lambda_{\max} = 496$ nm (see Figure 4.15). Whenever a rhodopsin molecule absorbs a photon, a chemical reaction occurs that bleaches the molecule, leaving it in a state that does

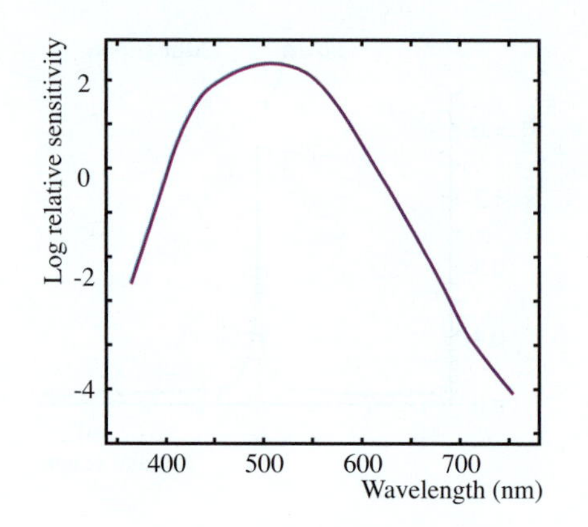

Figure 4.15. The log relative sensitivity of the rods.

not allow it to absorb another photon. This reduces the sensitivity of the rod. After some period of time, the process is reversed, and the rhodopsin is regenerated.

The regeneration process is relatively slow, although not as slow as the process of dark adaptation. Dark adaptation can be measured psychophysically as the time it takes to regain sensitivity after entering a dark room. While dark adaptation may take at least 30 minutes, rhodopsin regeneration requires around 5 minutes. Some of the reasons for the duration of dark adaptation are outlined in Section 3.4.2.

The rate of regeneration is related to the amount of rhodopsin present. If $p \in [0,1]$ is the fraction of unbleached rhodopsin, the rate of change dp/dt can be described by [118]:

$$\frac{dp}{dt} = \frac{1-p}{400}, \tag{4.3}$$

where 400 is a time constant. A solution to this differential equation is given by

$$p = 1 - \exp\left(\frac{-t}{400}\right). \tag{4.4}$$

This behavior was indeed observed [987]. The bleaching process follows a similar rate of change [118]:

$$\frac{dp}{dt} = \frac{-L_v\, p}{400\, L_0}. \tag{4.5}$$

Here, L_v is the intensity of the light causing the bleaching, and L_0 is the intensity that bleaches 1/400 of the rhodopsin per second. The solution to this differential equation is

$$p = \exp\left(\frac{-L_v\, t}{400\, L_0}\right). \tag{4.6}$$

The combined effect of bleaching and generation is then given by

$$\frac{dp}{dt} = \frac{1}{400}\left(1 - p - \frac{L_v\, p}{L_0}\right). \tag{4.7}$$

As bleaching and regeneration occur concurrently, exposure to a constant amount of light will cause the amount of rhodopsin to converge to a steady state. Once such a steady state is reached, the rate of change by necessity becomes zero. As a result, (4.7) gives a relation between the fraction of unbleached rhodopsin p and the steady state illumination L_v:

$$1 - p = \frac{L_v}{L_v + L_0}. \tag{4.8}$$

The value of L_0 is known as the *half-bleach constant*. This equation shows that for very small values of $L_v \ll L_0$, the relation between L_v and p is nearly linear. For very large values of L_v, the fraction of unbleached rhodopsin asymptotically approaches zero. Centered around L_0, the range of intensities for which p is between 0.01 and 0.99 is approximately 2 log units (i.e., $L_{p=0.01}/L_{p=0.99} \approx 100$).

4.3.3 Cones

The bleaching and regeneration processes known to occur in rhodopsin also occur in the opsins (visual pigments) associated with cones. The rate of change of opsins p as function of intensity L_v is given by [118]

$$\frac{dp}{dt} = \frac{1}{120}\left(1 - p - \frac{L_v\, p}{L_0}\right). \tag{4.9}$$

As only the constant 120 is different from (4.7), the steady-state behavior for opsins is the same as for rhodopsin and is given by (4.8).

The cones come in three types which have peak sensitivities to different wavelengths. These three types can be broadly classified as sensitive to long, medium, and short wavelengths (abbreviated as L, M, and S cones), as shown in Figure 4.16.[1] The peak sensitivities λ_{max} lie around 440 nm, 545 nm, and 565 nm, respectively [1036]. With three different cone types, human color vision is said to be *trichromatic*.

Figure 4.16 also shows that each cone type is sensitive to a wide range of wavelengths, albeit in different relative proportions. The signal produced by a photoreceptor is, therefore, proportional to the mix of wavelengths incident upon

[1]Note that the curves shown here are not corrected for scattering and absorption of the ocular media.

$\log L_v$	-0.82	0	1	2	3	4	5	6
A	7.18	6.56	5.18	3.09	1.87	1.31	1.09	1.00

Table 4.2. Pupillary area as a function of luminance $\log L_v$ (after [656]).

Upon stimulation, photoreceptors produce a graded potential which in turn triggers a response in subsequent cell layers (the bipolar cells and amacrine cells). The dynamic range of the output is limited, as in all neurons [355], and it can be modeled for brief flashes of light with either a modified Michaelis-Menten function V_{MM}, or an exponential function $V_{\exp}$:

$$V_{\mathrm{MM}}(L_v) = \frac{j L_v^n}{L_v^n + \sigma^n}, \tag{4.11}$$

$$V_{\exp}(L_v) = j\left(1 - 2^{-L_v^n/\sigma^n}\right). \tag{4.12}$$

Here, L_v is the luminance (photometrically weighted radiance, see Chapter 6), j is the maximum photocurrent, σ is the semi-saturation constant, and n is an exponent which normally has a value somewhat below 1 [77, 483, 1169]. In particular, Boynton and Whitten estimate the exponent to be 0.7 [117].

The semi-saturation constant determines which value of L_v is mapped to half the maximum output. The photoreceptor output is thus modeled with a function that is linear for small values and, for large values, asymptotically reaches the maximum output j. However, photoreceptors, and in particular cones, are able to adjust their gain, i.e., σ [355].

Equation (4.11) models the non-linearity of the cone response under the assumptions that the pupil is fixed and no bleaching occurs. Taking these effects into account, the cone response may be modeled by [118]:

$$V_{\mathrm{MM}}(L_v) = \frac{j\,(L_v A p)^n}{L_v A p^n + \sigma^n}, \tag{4.13}$$

where p is the fraction of unbleached opsins derived from (4.8) and A is the relative pupil area, which is given as function of L_v in Table 4.2. This equation applies to a steady-state situation where L_v is kept constant.

A recent and more complete computational model of cone response, including photo-transduction, calcium feedback, and the transduction occurring in the cone inner segment is described by van Hateren (see Section 4.8) [434].

4.3.4 The Photoreceptor Mosaic

The S-cones are sparse in comparison with the L- and M-cones, and become even sparser in the fovea. They are absent in a central retinal disk (the foveola) with

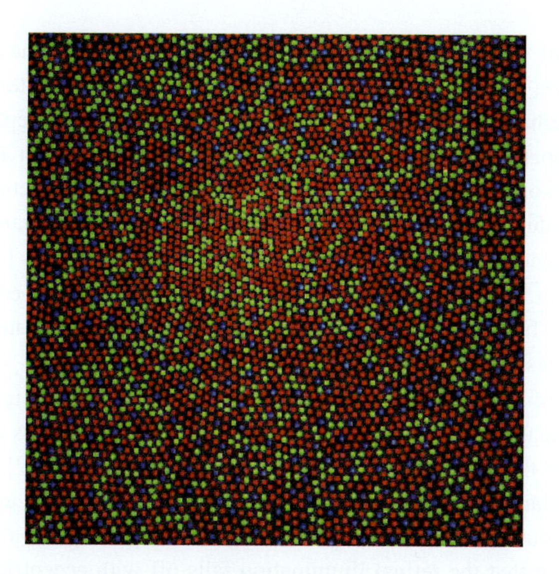

Figure 4.19. Photoreceptor mosaic. (From Mark D Fairchild, *Color Appearance Models*, 2nd edition, Wiley, Chichester, UK, 2005.)

a diameter of 100 μm (0.34 deg) [220]. This means that even in humans with normal color vision, the fovea is *tritanopic* (loss of S-cone function) for small and brief targets [610, 1248].

With the peak response of L- and M-cones being only 30 nm apart, they exhibit large amounts of overlap in spectral sensitivity, thus contributing more or less evenly to a luminance signal. The S-cones, on the other hand, have a very different peak sensitivity and are a poor representation for a luminance signal [861]. The central part of the fovea, therefore, appears to be geared towards maintaining high spatial acuity rather than high contrast color vision [352].

An alternative explanation for the absence of S-cones in the center of the fovea is possible, however. For instance, it can be argued that this absence of S-cones in the fovea helps to counteract the effects of Rayleigh scattering in the ocular media [1036].

The S-cones in the periphery are packed at regular distances in non-random order [1245]. The packing of L- and M-cones is essentially random. A schematic of the packing of cones in the retina is given in Figure 4.19.

The ratio between cone types in the human retina varies considerably between individuals [785, 977]. For instance, the measured ratio of L- and M-cones was 3.79:1 for one individual and 1.15:1 for another [1245]. This difference has only minor influence on the perception of color [121, 911], which may be due to a combination of various factors. First, the statistics of natural scenes are such that high frequency, high contrast signals are rare. Further, the scatter occurring in the ocular media may help to prevent aliasing. Lastly, post-receptoral processing may compensate for the idiosyncrasies of the photoreceptor mosaic [824]. Nonetheless, some variability in photopic spectral sensitivity remains between healthy individuals, which is largely attributed to the random arrangement of L- and M-cones and the variability in ratio between these cone types [227, 670].

The ratio of the combined population of L- and M-cones to S-cones is around 100:1 [1205]. If we assume that the L/M ratio is 2:1, then the ratio between all three cone types is given by L:M:S = 0.66 : 0.33 : 0.01.

Finally, the packing density of cones in the fovea varies significantly between individuals. Measurements have shown individuals with as few as 98,200 cones/mm^2 in the fovea up to around 324,100 cones/mm^2 [219]. Outside the fovea, the variance in receptor density is much lower. It can be argued that the cone density outside the fovea is just enough to keep the photon flux per cone constant, given that the retinal illumination falls off with eccentricity. Between $\phi = 1^\circ$ and $\phi = 20^\circ$ degrees of eccentricity, the cone density $d(\phi)$ can be modeled by [1157]

$$d(\phi) = 50{,}000 \left(\frac{\phi}{300} \right)^{-2/3}. \tag{4.14}$$

Finally, the packing density is slightly higher towards the horizontal meridian as well as towards the nasal side [219].

4.3.5 Bipolar Cells

Bipolar cells take their primary input from the photoreceptors. They are differentiated into different types according to their response to the neurotransmitter glutamate received from the photoreceptors. For instance, they may be tuned to faster or slower fluctuations in the visual signal [609]. The neuronal chemistry of bipolar cells (and all other cells in the brain) determines their function [457].

Some bipolars activate ON pathways, i.e., they detect light against a darker background. Other bipolars respond to dark spots against a light background and start an OFF pathway. ON-bipolars and OFF-bipolars work in parallel and may take their input from the same cones. The parallel processing of ON and OFF signals are continued later in the visual pathway; bipolar ON and OFF cells connect to separate ON and OFF ganglion cells.

Having separate ON and OFF pathways implies that the sensation of light and dark is mediated by separate processes. On the basis of psychophysical experiments this was already postulated over a century ago by Hering [461].

In the fovea, L- and M-cones each connect to a single *midget* bipolar cell. As a consequence, midget bipolar cells carry chromatic information. However, the fovea also contains *diffuse* bipolar cells that connect to different cone types, and thus carry achromatic information [1217]. Diffuse bipolar cells have a center-surround organization [223] (see Section 4.3.6).

The S-cones form the start of a separate ON pathway. Separate blue-cone bipolar cells connecting to blue cones have been identified [612], which mediate the S-cone ON pathway. In addition, each S-cone connects to a single midget bipolar cell [597, 598], which potentially forms the substrate of a blue OFF signal [227].

4.3.6 Horizontal Cells

In addition to taking input from photoreceptors, bipolar cells are interconnected laterally by horizontal cells. These cells have very wide receptive fields, i.e., they connect to a large number of bipolars and photoreceptors. At the same time, the plasma membranes of horizontal cells interconnect, thereby widening their receptive field even further.

Horizontal cells connect to all cones within their dendritic field (i.e., receptive field). As different cone types are likely to be available within a horizontal cell's receptive field, this implies that horizontal cells are not necessarily chromatically selective [1217]. However, two dominant horizontal cell types have been found, H1 and H2, whereby the former takes input preferentially (but not exclusively) from L- and M-cones, whereas the latter preferentially connects to S-cones [223].

Horizontal cells provide an inhibitory *opponent* signal to the bipolar cells, either through direct connection, or through a proposed feedback mechanism involving photoreceptors [609]. If the area surrounding the receptive field of a bipolar cell is different from the bipolar's receptive field itself, the signal is amplified further due to the effect of the horizontal cells. Thus, the receptive field of a bipolar cell is indirectly modified by activity of neighboring photoreceptors and bipolar cells.

The response to light of bipolars due to their connectivity with photoreceptors is called the *center*, whereas the modulation by a wider receptive field due to horizontal cell activity, is called the *surround*. Typically, an ON-bipolar is augmented by an OFF-surround, and an OFF-bipolar has an ON-surround—hence the term opponent signal.

The activity of horizontal cells and, therefore, indirectly photoreceptors and bipolar cells, is modulated by further feedback from signals originating in the inner plexiform layer. This allows photoreceptors to neuronally adapt to light, in addition to the non-linear response mechanisms outlined in Section 4.3.3. This feedback mechanism also affords the possibility of color coding the response of bipolar cells [609]. As a result, the notion that retinal processing of signals is strictly feed-forward is false, as even the photoreceptors appear to receive feedback from later stages of processing.

4.3.7 Ganglion Cells

Ganglion cells form the connectivity between the retina and the rest of the brain. Their input, taken from both bipolar cells and amacrine cells, is relayed to the lateral geniculate nucleus through the ganglion's axons which form the optic nerve bundle.

The predominant ganglion cell types in the human retina are ON-center and OFF-center [430]. An ON-center ganglion cell is activated when a spot of light falls on the center of its receptive field. The opposite happens when light falls onto the periphery of its receptive field. For OFF-center ganglion cells, the activation pattern is reversed with activation occurring if light falls on the periphery; illumination of the center of its receptive field causing inhibition. Thus, the receptive fields of these ganglion types are concentric, with center-surround organization (ON-center, OFF-surround and vice versa) [632].

In the fovea, the receptive fields of ganglion cells are much smaller than in the rest of the retina. Here, *midget* ganglion cells connect to midget bipolar cells in a one-to-one correspondence. Each red and green cone in the fovea (but not blue cone) connects to one ON and one OFF midget bipolar cell, and these are in turn connected to an ON and OFF ganglion cell, thereby continuing the ON and OFF pathways. This allows cones in the fovea to transmit dark-on-light and, separately, light-on-dark information at very high spatial resolution [609].

Ganglion cells respond to colored stimuli in one of two ways: achromatically or in color-opponent fashion. A particular ganglion cell type, the *parasol* ganglion cell, is the substrate for carrying achromatic information, i.e., information about the sum of the L and M channels.

The classic color opponent pattern arises when a ganglion cell is stimulated by long wavelength light (red) and inhibited by short wavelength light (green) [1190]. Such red/green color opponency is counter-balanced by the existence of a green/red color opponent ganglion cell type.[2] The information about the difference between L- and M-cone excitations is carried by *midget ganglion* cells [223].

However, it should be noted that this color opponency only arises in the fovea where the size of the center is reduced to a single cone type. In the periphery, the

[2] Note that although we call this opponent process "red/green," this process corresponds in practice to a color axis that goes from pinkish-red to cyan [209], see Section 4.6.

receptive field size gradually increases with eccentricity, and here midget ganglion cells receive input from L- and M-cones for both center and surround. This is consistent with psychophysical evidence that red/green sensitivity declines with eccentricity [223, 801].

Small-bistratified ganglion cells form a further cell type that carry blue-ON/yellow-OFF signals [221].[3] They are not spatially opponent, in that their center and surrounds have nearly the same size. The ON-signal is derived from S-cones (through the blue-cone bipolars), whereas the surround is derived from L- and M-cones (by means of diffuse bipolar cells). The bistratifed ganglion cells thus carry an S-(L+M) signal.

Several other ganglion cell types with complex receptive fields exist in the retina, for instance, ones that are selective for direction and motion [60, 62].

4.3.8 Ganglion Cell Classifications

A further classification of ganglion cells is given by Gouras [382], who subdivides these cells into *tonic* and *phasic* types, which exhibit different responses to light stimuli. Tonic cells respond to light in a steady manner. They have very small receptive fields and are color-specific. These cells project to the parvocellular layer of the Lateral Geniculate Nucleus (LGN; see Section 4.4) and are therefore part of the *parvocellular pathway*. Their optimal response is for high-contrast signals with high spatial resolution.

Tonic cells can have red, green, or blue centers (both ON and OFF), corresponding to the three cone types. Their surround is also spectrally selective, although drawn from a different cone type than the center. Blue OFF-center types are extremely rare [788].

The phasic ganglion cells have larger receptive fields and respond transiently to either the onset or the offset of a stimulus. For small spots, the transient discharge is followed by a smaller sustained discharge. Large spots of light only produce the initial transient. The transient behavior is caused by the inhibiting surround response, which is delayed with respect to the center response.

Phasic cells are generally achromatic and project to the magnocellular layer of the LGN. Changes in luminance are detected, but color changes are not [221]. Phasic cells are more sensitive to low contrast signals covering larger areas [826].

There are further color-opponent ganglion cell types. For instance, cells exist with a color-opponent center, but with no surround. Such cells exist for blue-yellow color opponency [221].

[3] Note that although this dimension is referred to as blue/yellow, it encodes colors that are perceived as violet/lime [257] (see Section 4.6).

Ganglion cells are tuned to spatial frequency, i.e., their sensitivity to spatially varying patterns is highest for certain spatial frequencies of the pattern. The frequency response of ganglion cells can be characterized by a grating composed of a sine wave oriented in some direction. The contrast of the sine wave is measured as the difference between the lightest and darkest parts of the wave. By introducing such a grating into the receptive field of the ganglion cell, and reducing the contrast until the cell just barely responds, the threshold of visibility can be deduced [826].

By repeating this process for sine waves of different spatial frequencies, a threshold curve can be created. Such curves, known as contrast-sensitivity functions (CSFs), show a peak at a given frequency [284]. This is the frequency to which the ganglion cell is most responsive.

In addition to peaked-contrast sensitivity, ganglion cells may be characterized by the sensitivity to motion within their receptive fields (called *hyper-acuity*). The smallest detectable displacements are frequently much smaller than might be expected given the size of the center receptive field, or the peak contrast sensitivity [1031].

A useful feature of both contrast sensitivity and hyper-acuity measures is that they apply to ganglion cells as well as human psychophysical performance. This leads to the conclusion that organisms cannot detect visual stimuli that are not detected by ganglion cells [826].

Ganglion receptive fields can be reasonably well modeled by the subtraction of two Gaussian profiles. The center is modeled by a positive Gaussian function with a smaller radius and larger amplitude than the surround. The Gaussian profile modeling the surround is subtracted from the center profile. This is the so-called *Difference of Gaussians* or DoG model.

4.3.9 Amacrine Cells

Amacrine cells provide lateral connectivity between bipolar cells and ganglion cells, as well as vertical connections between ON and OFF visual pathways. In addition, the rod and cone pathways are interconnected through amacrine cells, as discussed in Section 4.3.10.

There is a large number of different amacrine cell types which appear to be involved in adaptation processes that correct for changes in the environment. For instance, the spectral composition of daylight during a sunrise and a sunset is very different from mid-day. To feed the brain with an optimal signal during changing light conditions requires a light-driven modulatory system that is perhaps mediated by amacrine cells.

As an example, feedback is provided to the horizontal cells. For instance, a specific amacrine cell releases dopamine when stimulated with a flashing light. This neurotransmitter in turn decouples horizontal cells, thereby effectively reducing their receptive field size [609].

Another example is given by the *somatostatin-immunoreactive* amacrine cells [961, 988] which have cell bodies and dendrites largely located in the lower part of the retina. Their axons project into the upper-half of the retina. Under normal viewing conditions light from the sky is projected predominantly onto the lower-half of the retina, whereas reflections off the ground are projected onto the upper-half of the retina. Somatostatin-immunoreactive amacrine cells may therefore be involved in adaptation to the tremendous variation in light received from different parts of the visual field [1217].

4.3.10 The Rod Visual Pathway

Rods project to just one type of bipolar cell—the rod bipolar. Large numbers of rods project to a single rod bipolar cell. These bipolars do not directly connect to ganglion cells, but synapse into amacrine cells instead [605, 607], thus diverging the signal. There are two types of amacrine cells dominant in the rod visual pathway, with designated names *AII* and *A17*.

The AII amacrines link the rod and cone pathways, such that the rod signals can also use the cone bipolars to reach the ganglion cells. This type of amacrine is mostly post-synaptic to rod bipolar cells and synapses into ganglion cells. They are small-field, which means that their lateral spread is limited. Nonetheless, AII amacrines are also directly connected to each other, much like horizontal cells are interconnected. This widens their area of influence. In bright light conditions, when the narrow-field cone pathways need to maintain their spatial resolution, a different type of amacrine cell releases a neurotransmitter that uncouples the AII amacrines, thereby narrowing their receptive field and minimizing the interference of the rod pathway.

On the other hand, the A17 amacrine cells are wide-field and typically receive input from approximately 1000 bipolar cells, making them very sensitive to light. The A17 does not synapse into other amacrine or ganglion cells. Its large receptive field appears to be used to integrate the signal, helping to set sensitivity levels [825].

The significant convergence of rods into bipolars and the divergence of the signal through amacrine cells is thought to increase light sensitivity of the rod pathway. It achieves this very high sensitivity at the cost of spatial resolution, which is reduced by the pooling of signals.

Finally, in the mesopic range, rod vision has a (relatively small) effect on color vision [599, 1077, 1078].

4.4 The Lateral Geniculate Nucleus

The signal leaving the retina projects to several parts of the brain, including the superior colliculus and the lateral geniculate nucleus.[4] Of interest to color vision are the pathways that go through the lateral geniculate nuclei (LGN). These are located on either side of the thalamus and relay the signal from the eye to the primary visual cortex.

The optic nerve leaving each eye is split, with the nasal part crossing over (Figure 4.20). This means that each LGN receives input from both eyes. The left LGN is responsible for processing the visual input of the left half of each eye, and the right LGN processes the right half of the visual field. The part where the nerve bundles cross over is called the *optic chiasm*.

There exist three pathways that project from the retina into each LGN and from there into the cortex. These pathways are collectively known as the *retino-geniculo-cortical* pathway. Each LGN contains several layers, and each layer receives input from one eye only. In addition, each layer is classified by cell type and receives input from predominantly one ganglion cell type. Thus, signals that are processed separately in the retina remain separate in the LGN (Figure 4.21). There are three types of layers, with the third mediating output from the blue cones being a relatively recent discovery [157, 257, 458].

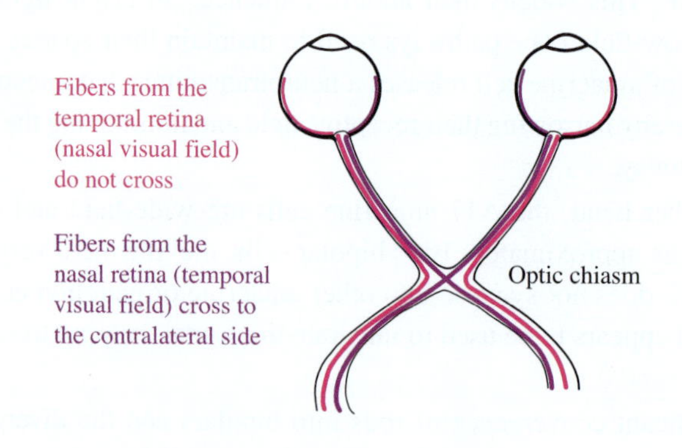
Fibers from the temporal retina (nasal visual field) do not cross
Fibers from the nasal retina (temporal visual field) cross to the contralateral side
Optic chiasm

Figure 4.20. The optic chiasm.

[4]The term *geniculate* refers to its structure and means knee-shaped.

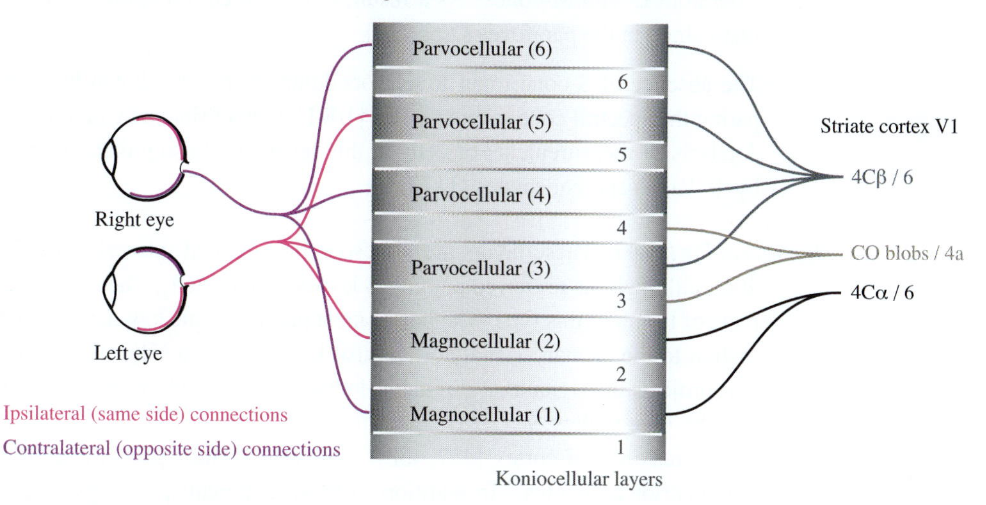

Figure 4.21. A diagram of the LGN showing both its main inputs and main outputs, which project to the striate cortex.

The three types of layers are [257][5]:

Magnocellular layers. These layers receive input from parasol ganglion cells and are therefore responsible for carrying the achromatic L+M signal [227, 660]. These cells have a spatial center-surround configuration, where ON-center cells receive excitatory input from both L- and M-cones in the center and inhibitory input from both L- and M-cones in the surround. OFF-center cells have opposite polarity [352].

Parvocellular layers. Receiving input from midget ganglion cells, these layers are responsible for carrying the L-M signal, i.e., they mediate red-green color opponency [673,891]. Wiesel and Hubel first proposed the possibility of neural circuitry whereby the center of the receptive field was excited by one cone type, and the surround by another [1244]. Evidence for such organization has been confirmed, for instance, in the red-green opponent ganglion cells [661] and the LGN [746, 946, 947]. However, it should be noted that the dendritic field sizes of midget ganglion cells in the parafovea increase in size, and as a result both their centers and surrounds take input

[5]Note that the following description of these three layers in the LGN is simplified and not necessarily universally accepted. However, for the purpose of this book, the current description is sufficiently complete.

from both L- and M-cones. As a result, red-green color opponency is not maintained in the parafovea [222, 223].

The absence of S-cone input to parvocellular layers was shown by comparing the spectral cone response data [1091] with a Principal Components Analysis of measurements of cells in this layer [976] using data obtained by de Valois and colleagues [1172].

Koniocellular layers. These layers are also sometimes called intercalated [322, 406], although the preferred naming is koniocellular [564]. There are three pairs of layers of this type. The dorsal most pair transmits low-acuity information to the primary visual cortex (also known as area V1). The middle pair projects the signal that originates in bistratified ganglion cells and carries the blue yellow S-(L+M) signal to the cytochrome oxydase blobs of the primary visual cortex [257, 458]. The ventral-most pair is related to the superior colliculus. In addition to these dominant pathways, koniocellular neurons project to parts of the extrastriate cortex, bypassing V1 altogether [458, 973, 1052, 1286].

4.5 The Visual Cortex

The visual cortex is the primary location for processing visual signals in the brain. It contains a large number of separate functional units that are interconnected through both feed-forward and feedback connections. These cortical areas include:

V1. This is the principle part of the visual cortex that receives input from the LGN. In an older classification scheme, this area is also known as Brodmann area 17 [138, 710]. It is also known as the primary visual cortex or the striate cortex.

V2. Area V2, after staining with cytochrome oxydase, reveals thin stripes, thick stripes, and pale stripes, each receiving input from different areas in V1.

V3. This is an area thought to be involved in the integration of signals from different pathways.

V4. This area is thought to play an important role in the processing of color information.

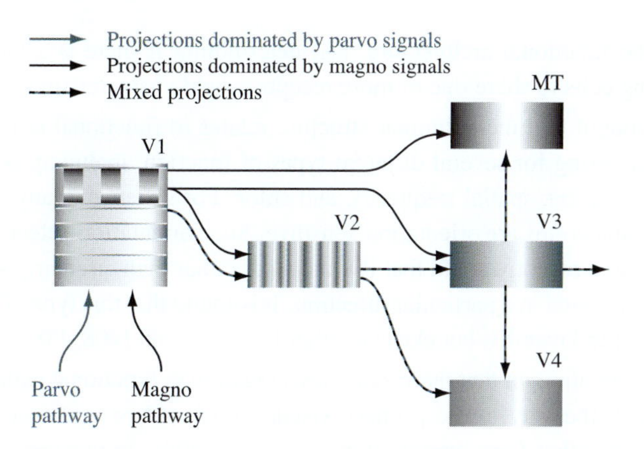

Figure 4.22. The main feed-forward projections between several cortical areas (after [592]).

MT. Area MT is also known as area V5, or the middle temporal area. Cells in this area tend to be strongly sensitive to motion. It predominantly receives input from the geniculate magno system.

Many other areas exist, but most is known about areas V1, V2, and MT [563]. The main feed-forward projections between these areas are diagrammed in Figure 4.22. Color-selective cells have been found in several areas, including V1, V2, V3, and in particular V4 [206,624,705,1113,1289,1290]. However, it should be noted that the localization of V4 as a center for color processing is still under debate, as some studies suggest a different organization [282, 407, 1139]. In the following we briefly discuss some of the cortical areas relevant to color processing.

4.5.1 Area V1

Area V1 is the main entry point to the visual cortex for signals from the lateral geniculate cortex. It consists of six layers, with some layers divided into sublamina [138, 710]. In addition, several layers, including layers two and three, have columnar structures called *blobs* or *patches*. The areas in between are called *inter-blobs* or *inter-patches*. The separation between patches and inter-patches is based on a histochemical technique called cytochrome oxydase (CO) that differentiates cells based on metabolism, revealing structure in layers two and three of the primary visual cortex.

One of the challenges still remaining is to determine what the function might be of cells located in different layers. This problem may be interpreted as unraveling the functional architecture, i.e., determining if there is a tendency for neighboring cells to share one or more receptive-field characteristics [493].

It is thought that the columnar structure relates to functional structure, with proposals existing for several different types of function, including ocular dominance, orientation, spatial frequency, and color. For instance, many cells in the primary visual cortex are orientation-sensitive. Such orientation-selective cells respond best to a receptive field that is elongated, rather than circularly symmetric, and that is oriented in a particular direction. It is found that this type of selectivity is prominent in layer 4B, but exists in other layers as well [408, 438, 964].

It was long thought that there exists a segregation in functional pathways from the retina all the way to the primary visual cortex and beyond, separating the processing of color, form, and motion/stereo [287, 694]. In this tripartite model, area V1 transforms the three streams which are then projected to area V2 [692–694].

The optic radiations originating in the lateral geniculate nucleus synapse into different layers and into the patches of layers two and three, as diagrammed in the left part of Figure 4.23. Here, it is seen that different layers in the LGN project to different parts of area V1. This appears to provide evidence that the different pathways originating in the retina remain functionally segregated in V1.

However, the intra-cortical circuitry, i.e., the interconnectivity within area V1 after signals from the LGN have been projected, suggests that the pathways do

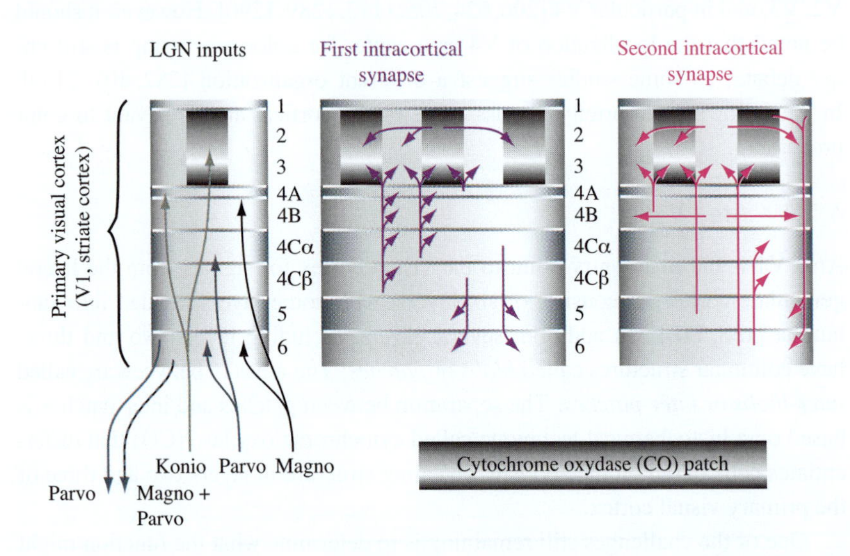

Figure 4.23. Area V1 (primary visual cortex) and its intracortical circuitry (after [1051]).

not remain separate, but are combined instead [148, 636, 1051, 1265, 1281]. A summary of intra-cortical connectivity is shown in Figure 4.23. This finding was corroborated by studies that revealed that orientation-selective cells can be tuned to color-selective as well [209, 491, 669, 674, 1127, 1181, 1189].

A further complicating factor is that area V1 has both feed-forward and feed-back connections from many other functional areas in the brain [1008]:

- Feed-forward projections from other areas to V1 originate in: pulvinar, LGN, claustrum, nucleus paracentralis, raphe system, locus coeruleus, and the nucleus basalis of Meynert [6, 104, 260, 384, 459, 635, 844, 889, 957].

- Feedback projections from other areas to V1 originate in: V2, V3, V4, MT, MST, frontal eye field (FEF), LIP, and the inferotemporal cortex [69, 889, 970, 1041, 1108, 1164, 1165].

- Area V1 projects forward to V2, V3, MT, MST, and FEF [115, 323, 693, 709, 753, 1041, 1164, 1165].

- Area V1 extends backward projections to SC, LGN, pulvinar, and pons [333, 334, 384, 409, 709].

For the purpose of modeling the micro-circuitry of area V1, as well as understanding the function of V1, these feedback mechanisms are likely to play roles that are at least as important as the geniculate input [747].

While the color processing in the LGN occurs along dark-light, red-green, and yellow-blue axes, this is not so in area V1. LGN projections are recombined in both linear and non-linear ways, resulting in color sensitivity that is more sharply tuned to specific ranges of color, not necessarily aligned with the color directions seen in the LGN [209, 419, 1175]. In particular, clusters of cells have been identified that respond to purple [1189]. Psychophysical studies have confirmed the likelihood of a color transform from the LGN to the striate cortex [1174, 1176]. There is a possibility that these non-linear recombinations underlie the human perception of color categories. It is also possible that separate color transformations in V1 contribute to the perception of color boundaries and colored regions [1030]. A proposed three-stage color model, accounting for phenomena measured through neurophysiological experiment as well as psychophysics, is discussed in Section 4.6.

In summary, many cells in area V1 respond to multiple stimuli, such as color, position, orientation, and direction. The notion that pathways for the processing of color, shape, and the like remain separate is unlikely to be true. In addition, area V1 is one of many areas where color processing takes place, thus moving gradually from the color-opponent mechanisms found in the LGN to mechanisms that are more closely aligned with color perception.

4.5.2 Area V2

Area V2 is characterized by thick, thin, and pale stripes that are revealed under cytochrome oxydase staining. There is some degree of functional segregation along these histochemically defined zones, although many cells are found that are selective for multiple stimulus attributes [592]. The thick and pale stripes have a higher degree of orientation selectivity than the thin stripes [726]. The thin stripes have a tendency towards color selectivity and are found to be sensitive to wavelength composition [800]. Nonetheless, the precise nature of functional segregation in area V2 is still under dispute [1051] and significant cross-talk between the processing of various stimulus attributes exists.

The pathway between V1 and V2 is currently known to be bipartite [1050], with patches in V1 connecting to thin stripes and inter-patches connecting to both thick and pale stripes in V2 [1051]. These projections are well segregated [490].

There is also significant feedback from V2 to V1 [971], although the organization of these feedback projections is only beginning to emerge [38, 726, 1044, 1251]. The effect of removing these feedback connections produces only subtle changes in the responses of V1 neurons [511, 993]. It is possible that these feedback projections alter the spatial extent of surround inhibitory circuits [1033], increase the saliency of stimuli [510], or possibly aid in disambiguation [854].

4.5.3 Area V3

Current techniques have not revealed anatomically or functionally distinct compartments within area V3 [592], although a columnar organization has been reported [4]. In addition, little is known about the physiology of neurons in this area [308]. V3 may play a role in the integration of different stimulus attributes, aiding in visual perception. It has also been suggested that this area is involved in the analysis of three-dimensional form [4].

The selectivity of neurons in V3 for different stimulus attributes is similar to those in V2, except that a higher sensitivity to direction was found (40% of the cells in V3 versus 20% of the cells in V2 [592]). Orientation selectivity is found in around 85% of the cells, selectivity for size is found in approximately 25%, and color sensitivity in approximately 50% of the cells in both V2 and V3 [592]. There is no correlation between selectivity for different attributes, suggesting that different attributes are not processed along separate pathways in V3 (or in V2).

4.5.4 Area V4

Area V4 has been the subject of significant controversy. It has been implicated as a color center on the basis of a rare condition called *achromatopsia*. This is a deficiency that causes the world to appear colorless and is also known as *cerebral color blindness*. This deficiency has been related to lesions in a particular part of the brain, known as V4. As a result, this area may be involved in computations relating to the perception of color [71, 230, 887, 1289]. It should be noted, however, that this viewpoint is challenged by several studies [282, 407, 1139] and, as such, the question of whether V4 can be deemed a color center remains open [392].

4.5.5 Area MT

Area MT is known to be involved in the perception of motion. It responds more strongly to moving versus stationary stimuli. Direction-selective populations of neurons have been found in both humans [498, 1138] and macaque monkeys [135, 754, 991]. This area is predominantly sensitive to achromatic stimuli.

4.5.6 Retinotopic Maps

The optics of the eye project an image onto the retina, where neighboring photoreceptors image neighboring portions of the retinal image. The signal that is transmitted towards the visual cortex generally maintains this topography. In the LGN, the spatial localization of image features is preserved, and the mapping of the retinal image to neurons in this area is therefore topographic. The term *retinotopic* is reserved for this mapping.

In the primary visual cortex, this mapping is preserved, although the fovea is represented in a relatively large portion of area V1 and beyond. Although neighboring areas in the retinal image are still represented by neighboring clusters of neurons, in V1 the mapping has undergone a log-polar transformation. This means that along one dimension, the log distance from the fovea is mapped, and the other dimension represents the polar angle with respect to the (arbitrarily chosen) horizontal axis of the retinal image (Figure 4.24). Retinotopic mapping exists in area V1 as well as areas beyond V1, although further away from V1 the precision of this organization declines [393, 394, 432, 677].

The logarithmic relationship between retinal distance to the fovea and distance measured along the surface of the primary visual cortex results in a large area of the cortex being devoted to processing signals stemming from the fovea. This is known as *cortical magnification* [268, 1017].

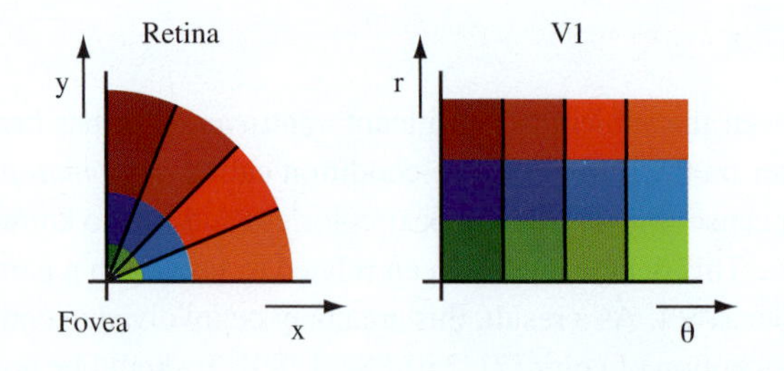

Figure 4.24. The retinal image is laid out in the cortex according to a log-polar mapping (after [392]).

4.5.7 Summary

As the simple model of segregated pathways through the modules in the visual cortex is being refined, it is too early to attribute specific functionality to separate regions in the visual cortex. The exact processing that takes place is yet to be unraveled, and it is therefore not appropriate to assign a substrate to the higher levels of functioning of the human visual system. This is the case for the primary visual cortex, as well as for processing elsewhere in the visual cortex.

Dependent on the techniques used to measure functional aspects of the visual cortex, it is possible to arrive at different conclusions. For instance, there are those that argue in favor of segregated pathways for color, motion, form, etc. Others have argued in favor of a distributed processing of features, where functional areas in the visual cortex are sensitive to combinations of motion, form, direction, orientation, color, and the like.

On perceptual grounds, it is also not possible to arrive at a single conclusion. While some perceptual attributes appear to be dissociated from others, this is not always strictly true [1202]. For instance, the apparent velocity of an object is related to its contrast and color [159], and high spatial and temporal frequencies appear desaturated [908].

One difficulty for determining the functional architecture of the visual cortex is due to the limitations of the currently available methodologies. Most use point methods, i.e., record responses from single neurons by means of single-cell recordings or tracer micro-injections. Given the number of neurons present, and their vast interconnection network, the problems involved are obvious [494]. Nonetheless, there is promise for the future, in that new techniques are becoming available that allow large populations of cells to be imaged at cell resolution [846].

In the following chapter, we therefore turn our attention to an altogether different strategy to learn about the (color) processing that takes place in the human visual system, namely psychophysics. Here, the human visual system is treated as a black box, which reveals some of its functionality through measuring task response.

4.6 A Multi-Stage Color Model

The wavelengths making up the visible spectrum can be arranged on a circle. The 0°, 90°, 180° and 270° angles are chosen to coincide with stimuli that isolate responses from the two types of opponent neuron in the LGN [209], as shown in Table 4.3.

Angle	LGN axis	Perceptual color
0°	L-M	"pinkish-red"
90°	S-(L+M)	"violet"
180°	M-L	"cyan"
270°	-S+(L+M)	"greenish-yellow"

Table 4.3. Major axes in MBDKL color space.

The relative strength of each color is adjusted such that all colors appear equally bright. In particular, the contrasts along the L-M axis are 8% for the L-cones and 16% for the M-cones, whereas the S-cone contrast is 83% [1175]. Contrasts for intermediate angles are interpolated. The resulting colors are called *iso-luminant*. This configuration is known as the MBDKL color space, named after Macleod, Boynton, Derrington, Krauskopf, and Lennie [250, 722]. Computational details of this color space are given in Section 8.6.3.

Although its axes are commonly referred to as red-green and yellow-blue, the actual colors they encode are as given above. The axes of the color space used by the LGN therefore do not coincide with the perceptual dimensions of red, green, blue, and yellow, which are rotated in the MBDKL color space. It was postulated by Russell and Karen de Valois that the red, green, yellow, and blue perceptual color systems are formed from the LGN output by adding or subtracting the S signal from the L and M opponent cells [1171].

The multi-stage color model proposed by Russell and Karen de Valois is composed of four separate stages. They will be discussed in the following sections [1171]:

1. Three cone types with three to five pigments;

2. Cone opponency;

3. Perceptual opponency;
4. Color selectivity in complex cells.

4.6.1 First Stage

The three-cone model is currently widely accepted, each with a single cone pigment. However, it is possible that there exist two variants of the L-cone pigment, with a peak sensitivity spaced a few nanometers apart, and possibly a similar pair of M-cone pigments [823]. The ratio of L:M:S cones is taken to be 10:5:1. The S-cones are assumed to be regularly spaced, whereas the L- and M-cones are clustered together.

4.6.2 Second Stage

The second stage is postulated to occur in the retina and is involved with the formation of spectrally opponent and non-opponent signals. The horizontal cells in this model sum their inputs, which are taken from several cones in a spatial region, calculating a L+M+S signal. At the bipolar cell level, the midget bipolars compute L - (L+M+S), M - (L+M+S), and S - (L+M+S) signals. This notation indicates that the center is formed by either L, M, or S, and the spatial surround is (L+M+S). For brevity, these signals are denoted M_o, L_o and S_o.

In addition, the opposite signals are also computed: -L + (L+M+S), -M + (L+M+S), and -S + (L+M+S), assuming that the spatial surround is formed by indiscriminately summing over all L-, M-, and S-cones within a local neighborhood.[6] These signals are the start of the parvocellular pathway. These signals are also written as $-L_o$, $-M_o$, and $-S_o$.

The center and the surround have more or less the same strength. Taking into account the cone ratios, for L_o, the model thus weights 16L for the center against a surround of 10L + 5M + S:

$$\begin{aligned} L_o &= 16L - (10L + 5M + S) && (4.15a) \\ &= 6L - 5M - S. && (4.15b) \end{aligned}$$

In the discrete model, diffuse bipolars are weighted as 6L - 5M. In either case, the result is close to a conventional "red-green" color opponent cell. The same is true for M_o and S_o, which end up being weighted as 11M - 10L - S and 15S - 10L - 5M, respectively.

[6] An alternative model is also presented, whereby the receptive fields are modeled as L-M, -L+M, M-L, and -M+L. This model assumes that the surround consists of only L and M inputs, and is therefore termed "discrete."

The diffuse bipolars appear to receive input from a small number of L- and M-cones, as well as a surround input derived from horizontal cells. The signals computed by diffuse bipolar cells is therefore designated L+M - (L+M+S) and -L-M + (L+M+S), forming the start of the magnocellular pathway.

4.6.3 Third Stage

In the third stage, perceptual opponency is modeled. The output of the diffuse bipolars, through parasol ganglion cells and the magnocellular layer of the LGN, is assumed to continue separately and to form the photopic luminosity function (see Section 6.1).

The six parvocellular LGN types, L_o, M_o, S_o, $-L_o$, $-M_o$, and $-S_o$ undergo further transformations as the third stage of the multi-stage color model. LGN cells respond to both luminance and color variations, but with different receptive fields [1170]. For instance, an L_o cell has an excitatory center and inhibitory surround with respect to an increase in luminance. The reason for this behavior is that both L- and M-cones respond similarly to changes in luminance, thus driving the center and surround in opposite directions.

However, at the same time, this cell has a uniformly excitatory response to an equiluminant change in color if the shift is towards longer wavelengths. Here, the L- and M-cones are driven in opposite directions, yielding a synergistic response at an L_o cell. Cells at the second stage therefore confound luminance and chromatic information. Similar arguments can be made for the five other cell types in the LGN. An $-M_o$, for instance, has an inhibitory center and excitatory surround for luminance changes, but a uniform excitatory response to a chromatic change towards longer wavelengths.

It is, however, possible to separate color and luminance information by recombining the outputs of second stage cells in specific ways. For instance, summing L_o and $-M_o$ at the third stage results in cancellation of sensitivity to luminance, yielding only a response to chromatic changes. Summing L_o and $-M_o$, on the other hand, leads to canceling of chromatic responsivity and extracts the luminance. Hence, L_o - M_o and M_o - L_o form chromatic signals, sensitive to changes from red to green, and from green to red. Similarly, L_o + M_o and M_o + L_o are spatially opponent luminance signals.

In this model, the third stage is dominated by L_o and M_o signals, contributing to both the perceptual red-green and yellow-blue systems. It is thus postulated, for instance, that the M-cones are the main contributors to the blue signal. The S_o signal is then used to modulate this dominant L/M pathway to retrieve the perceptual red, green, yellow, and blue color axes. In particular, these color systems

Signal 1	Signal 2	Sum	Result
L_O - M_O	$+S_O$	$L_O - M_O + S_O$	Red
L_O - M_O	$-S_O$	$L_O - M_O - S_O$	Yellow
M_O - L_O	$+S_O$	$M_O - L_O + S_O$	Blue
M_O - L_O	$-S_O$	$M_O - L_O - S_O$	Green
$L_O + M_O$	$+S_O$	$L_O + M_O + S_O$	Light
$-L_O - M_O$	$-S_O$	$-L_O - M_O - S_O$	Dark

Table 4.4. Recombination of L_O, M_O, and S_O signals.

can be computed by adding or subtracting S_O from the L_O - M_O and M_O - L_O cells. The four possible combinations are listed in Table 4.4

It is assumed that the contribution from S_O is given different weights for the four color systems. To match hue scaling experiments [1174], the S_O is given a weight of 1.0 for the perceptual red system, 2.0 for green, 2.0 for blue, and 3.5 for the yellow system. The contributions of the three cone types to each of these four perceptual color axes is given in Table 4.5.

Perceptual color axes	Stage-3 weights	Cone contributions
Red	$10L_O - 5M_O + S_O$	100L - 110M + 10S
Green	$10L_O - 5M_O + 2S_O$	90L - 115M + 25S
Yellow	$-10L_O + 5M_O + 3.5S_O$	-145L + 87.5M + 57.5S
Blue	$-10L_O + 5M_O + 2S_O$	-130L + 95M + 35S

Table 4.5. Cone contributions to each of the four perceptual color axes. The weights given to S_O are 2 for all four axes in the original model, but have been adjusted here to account for more recent hue-scaling results [1174].

4.6.4 Fourth Stage

It has been shown that most simple cells produce a half-wave rectified output. The fourth stage produces such an output. Complex cells, located beyond V1 can then be modeled by summing together two half-wave rectified simple-cell responses.

4.6.5 Neurophysiological Experiments

Using the MBDKL color space as a basis, neurophysiological and psychophysical experiments have been carried out that support this view [209, 1174–1176]. The color responses measured in parvocellular LGN cells of macaque monkeys are shown in Figure 4.25. The LGN responses as function of color angle are well

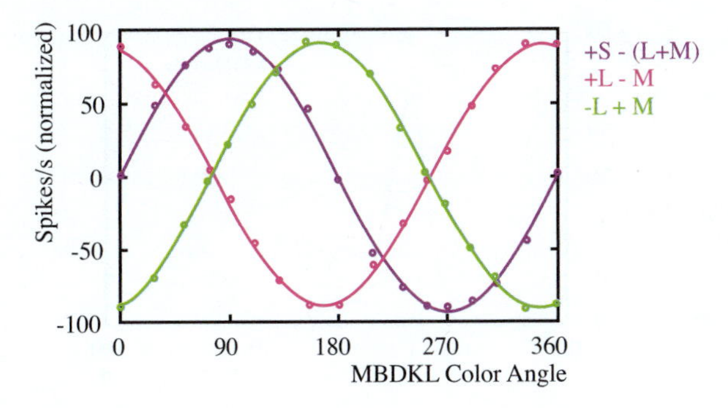

Figure 4.25. Sinusoidal fit to color responses of various LGN cells (after [1175]).

modeled by sinusoidals, indicating that the input transmitted from the photoreceptors is linearly recombined in the LGN.

The optimal tuning of LM opponent cells is along the 0°-180° axis in MBDKL space. The M center cells (+M-L and -L+M) on average have their maximum response at 180° and 0°. The L center cells (+L-M and -L+M) peak slightly off-axis at approximately 345° and 165°. The +S-(L+M) cells peak at 90%. The LGN cells, in general, respond strongly to colors along the 0°-180° and 90°-270° axes and show little response to other colors, as shown in Figure 4.26.

This figure also shows the color tuning of a set of cortical cells. Cortical cells are tuned to a variety of different colors, albeit with dips around the 0°-180° and 90°-270° axes [1175]. Other studies have found a similarly wide variety of tunings in V1 [669, 1127], although other experiments have shown cell types that respond discretely to red, green, yellow, and blue [1147, 1180].

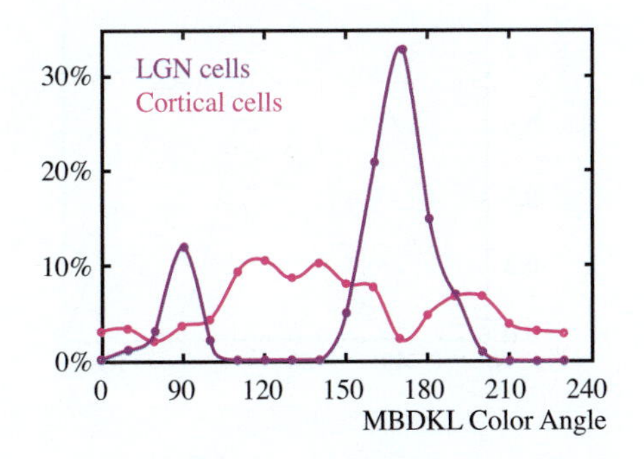

Figure 4.26. The tuning distribution of 100 LGN cells and 314 striate cortex cells. Opposite phases are combined, i.e., 0° and 180° are both mapped to 180° (after [1175]).

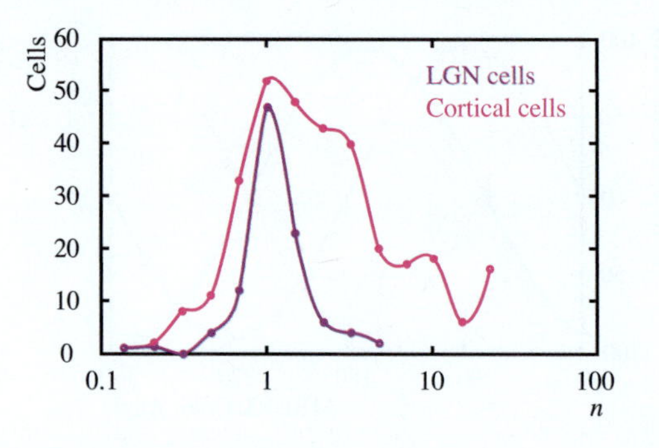

Figure 4.27. The distribution of exponents for cells in the LGN and in the striate cortex. The median values for LGN and V1 cells are 1.08 and 1.90 (after [1175]).

While the LGN cells respond linearly to their inputs, $V1$ cells exhibit a range of non-linearities. Such non-linearities can be modeled with an exponentiated sine function in the MBDKL space:

$$R = A \sin^n (C - \phi). \tag{4.16}$$

Here, the response R is modeled as a function of gain A, the color axis C under investigation, and phase ϕ. The exponent n models the degree of non-linearity. For a group of LGN and V1 cells, the distribution of exponents that best fits the individual cells' responses are plotted (Figure 4.27). The median value found for V1 cells is 1.90, indicating that nearly one half of the cells exhibit significant non-linear behavior.

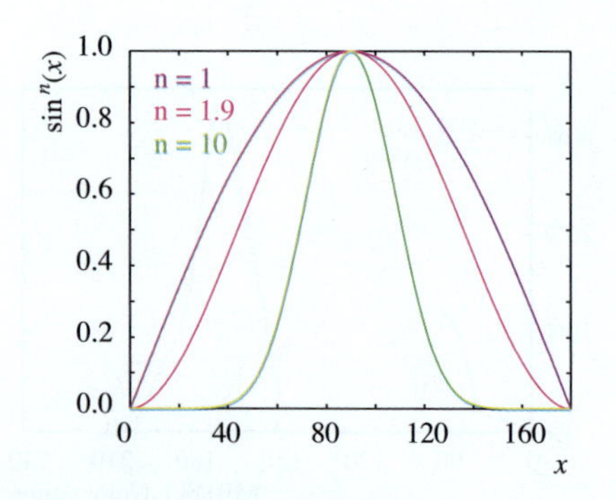

Figure 4.28. A sine function exponentiated with different values of n. As can be seen here, larger exponents cause the function to be more peaked, i.e., tuned to a narrower set of wavelengths.

The interpretation of an exponent n that is larger than 1 is that such responses are more sharply tuned, i.e., the responses are limited to a smaller range of wavelengths. This can be seen in Figure 4.28 which plots sine functions exponentiated with different values of n.

4.6.6 Psychophysical Experiments

A further experiment, this time psychophysical, compared responses for large stimuli (subtending 2° of visual angle) against small stimuli (0.1° of visual angle) both presented to the foveal region [1176]. As S-cones are absent in the fovea, the small stimuli activate only L- and M-cones, whereas the large stimuli activate all three cone types. The result of the 2° color-scaling experiment is reproduced in Figure 4.29.

Once more, the peaks in the graphs for red, green, blue, and yellow color regions do not coincide with the LGN cell peaks, but are rotated. Second, the shape of the graphs is non-sinusoidal, but the regions are better modeled with an exponentiated sine curve (4.16). The exponents found for red, green, blue, and yellow are, respectively, 1, 1.7, 1.8, and 2.9.

For small stimuli presented to the fovea, the S-cones will contribute little or nothing to the perception of color [1176]. As a result, stimuli around the 90° and 270° angles are frequently not noticed. The color scaling results for small stimuli are reproduced in Figure 4.30. Here, stimuli in the 90° to 270° range were perceived almost equally as green and blue. The range between 270° and 450° was seen as red with a small yellow component added in.

The absence of input from the S-cones has therefore resulted in only two regions with more or less uniform color appearance. The contribution of the S system is summarized in Table 4.6. It can be seen that the S-cone signals make

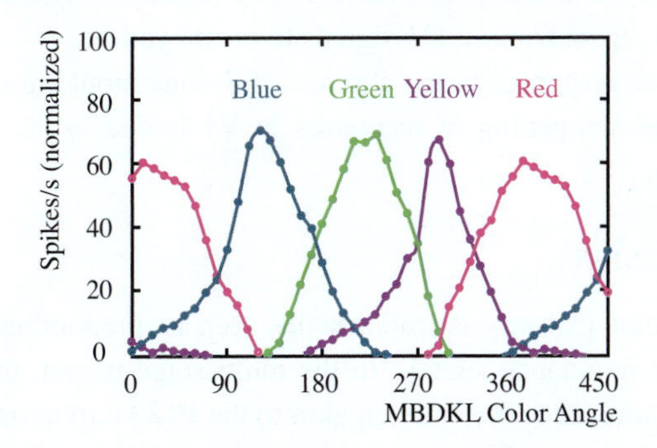

Figure 4.29. Color-scaling results for 2° stimuli (after [1176]).

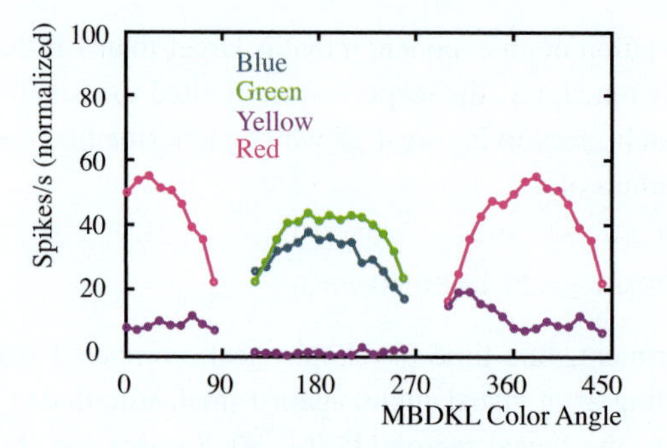

Figure 4.30. Color scaling results for 0.1° stimuli. Near-zero responses, as well as data around 90° and 270°, are not shown (after [1176]).

MBDKL angle		Perceptual color	S-cone contribution
90°-	180°	Green	Decrease
90°-	180°	Blue	Increase
180°-	270°	Green	Increase
180°-	270°	Blue	Decrease
0°-	90°	Red	Slight increase
270°-	360°	Red	Slight decrease
0°-	90°	Yellow	Decrease
270°-	360°	Yellow	Increase

Table 4.6. S-cone contribution to color perception.

significant contributions to the perception of both green and blue. This is consistent with the possibility that in V1, S signals are added to the +M-L cells to form blue and subtracted from the -M+L cells to form green. Similarly, these findings are in accord with a model which adds S-cone signals to +L-M signals to form red and subtracts them from -L+M signals to create yellow.

Moreover, the responses in the absence of S-cone input are not peaked. This suggests that the sharpening of responses in V1 is due to the contribution the S-cone pathway.

4.6.7 Discussion

The magnocellular pathway is traditionally seen as predominantly responsible for carrying the luminance signal. In the multi-stage model, this pathway carries essentially luminance information akin to the $V(\lambda)$ curve (introduced in Section 6.1). However, there also exists evidence that the parvocellular pathway may carry achromatic information which is multiplexed with color information [1170,

1173]. The multi-stage color model is able to model this and then separate the chromatic and achromatic information at the cortical stage.

In this model, the L- and M-cones provide the dominant input to both the red-green and yellow-blue systems. The S-cones modulate this dominant pathway in the third stage to construct perceptual color axes. It suggests that the dominant input to the blue system is formed by the M-cones, not the S-cones. This feature is in accord with measurements of tritanopes (see Section 5.11) who are missing or have non-functional S-cones. They typically see all wavelengths below 570 nm as blue or partially blue. However, S-cones do not respond to wavelengths above around 520 nm. The gap between 520 and 570 nm where blue is detected cannot be explained by conventional color models, but can be explained with the multi-stage model.

Finally, the rotation to perceptual color axes is in accord with both hue-cancellation experiments [515], as well as cross-cultural color-naming research [87, 449].

4.7 Alternative Theory of Color Vision

This chapter so far has focused on generally accepted theories of color vision. However, recently an intriguing alternative theory has emerged which can be viewed as complementary to known facts about human vision. This theory deals with the transduction of photons into electrical signals in the photoreceptors.

It can be argued that there are several unusual aspects in current theories and measurements. First, the absence of blue cones in the fovea does not appear to impede color perception in the foveal region. Second, the seemingly random distribution of L- and M-cones does not appear to have a marked effect on color vision. This is odd, if it is assumed that each of the three cone types is characterized by the presence of a single type of photopigment.

Moreover, recent studies have for the first time been able to probe individual cones by correcting the light scatter in the ocular media [476]. By stimulating individual cones with monochromatic light at wavelengths of 600, 550, and 500 nm, the expectation is that dependent on the cone type, the corresponding sensation of red, green, or blue will be elicited. However, this did not turn out to be the case, and it was concluded that the diversity in cone response could not be completely explained by L- and M-cone excitation. Instead, single cones appear to be capable of producing more than one sensation.

A further counter-intuitive property of color vision that cannot be explained by the three-receptor model was first reported by Tyndall [1158] and discussed by Mollon and Estévez [786]. It was observed that wavelength discrimination at around 455 nm improves when a white desaturating light is added. For a three-cone mechanism, any wavelength discrimination must be reduced if white light is added to the stimulus, and therefore Tyndall's discovery is not well-explained by current theories of color vision.

There are several other anomalies in current models of color vision [771]. Rather than listing them all, we mention one further remarkable observation. It is possible to mix red and green light to form yellow light. It is also possible to construct monochromatic yellow light. These two yellows are perceived the same, i.e., they are *metameric* (see Section 7.2.1). However, when a vertical yellow bar is moved horizontally across the visual field, the monochromatic yellow remains yellow, whereas the yellow color mixture is broken up into a red leading edge, a green trailing edge, and a yellow middle region. This phenomenon was noted first by Ives in 1918 [539].

In a three-cone model, the explanation of this result is problematic, even if different time constants are assumed for the different cone types. In particular, it would be expected that the moving color mixture can be broken up into red and green edges. However, it would be expected that the monochromatic yellow bar would also break up into a red leading edge and a green trailing edge, on the basis that the quantal catches for each cone type should be identical for both stimuli. This is contrary to the experimental result.

Although the three-cone model is able to explain many visual phenomena and is a useful starting point for most applications in engineering-related disciplines, there are sufficiently many observations which cannot be explained by this model. There is therefore room for alternate hypotheses which may explain the same phenomena, but in addition explain some of the cases where the current model breaks down.

One such model considers the shape of the photoreceptors. In particular, the cone shape of the receptors that mediate color vision may be important. In Section 2.5, we showed that a tube with a different index of refraction than its surrounding area may conduct light as a result of total internal reflection. If the cross-section of the tube is made very small, of the order of a single wavelength of light, the tube becomes a waveguide that is characterized by the waveguide parameter V [1064]:

$$V = \frac{\pi d}{\lambda} \sqrt{n_1^2 - n_2^2}. \tag{4.17}$$

Here, d is the diameter of the tube, n_1 and n_2 are the refractive indices of the tube and its surround, and λ is the wavelength for which the waveguide parameter is computed. The waveguide parameter V can be used to determine if the tube is wide enough for geometric optics to be a valid model, or if waveguide effects play a role. The latter would be the case for smaller values of V. For very small values of V, light cannot be confined to the interior of the tube, and light will radiate away.

As V depends on both the diameter of the tube, as well as the wavelength, it is clear that given a specific tube diameter, longer wavelengths cannot be propagated through the tube, whereas short wavelengths can. Replacing the tube with a tapered cylinder, at the wide entrance many different wavelengths can be propagated, whereas moving towards the thin end of the taper, only shorter and shorter wavelengths can be propagated.

If the wavelength composition along the length of the tapered cylinder could be read out, then the waveguiding principle can be used to construct a spectrometer. It turns out that the size and shape of the cones in the retina are such that waveguiding could occur. It is therefore possible that cones act as waveguides, irrespective of which photopigment, or combination of photopigments, is present in each.

This model of color vision, however, is currently not widely accepted. Although it appears to explain several phenomena that current models of human color vision cannot explain, significant further research would be required to determine if cones in the human retina use waveguiding to mediate color vision. So far, the model is commensurate with existing literature. However, direct verification of this model has yet to occur.

4.8 Application: Modeling a Human Retina

All displays and all rendering algorithms produce stimuli that are ultimately intended to be viewed by humans. A relatively complete model of low-level human vision may therefore help in the assessment of how images are actually perceived. Such a model may, for instance, help assess the retinal image quality of images displayed on new hardware before the hardware is built [243]. For rendering algorithms, such a model may help assess where shortcuts in the rendering pipeline may be achieved.

Deering presents a model taking into account the eye optics, the photoreceptor mosaic, as well as the transduction of photons by the cones (the rods are excluded, so that the model is suitable for photopic light conditions only) [243]. As part of the simulation, which covers much of the material discussed in this chapter, a new algorithm for synthesizing artificial retinas is developed, placing and packing cones according to some simple rules. A new retina is started with just seven cones; a center cone surrounded by six hexagonally packed cones. In each following iteration, a ring of new cones is placed outside this cluster. Based on the location where cones are created, they have a larger or smaller anatomical target radius. The cone centers are then migrated inward.

This migration is governed by two rules, the first being adherence to each cone's target radius and the second a tendency to move inward as far as possible. For a pair of cones with positions $\mathbf{p}_i$ and target radii r_i, the normalized distance d is given by

$$d(i,j) = \frac{|\mathbf{p}_i - \mathbf{p}_j|}{r_i + r_j}. \tag{4.18}$$

Cones i and j are defined as neighbors if $d(i,j) \geq 1.5$.

The cone positions are then iteratively updated using the cone force equation:

$$\mathbf{p}_i' = \mathbf{p}_i + k_1\,\mathbf{n} + k_2\,\mathbf{r} + k_3 \sum_{j|d(i,j)\geq 1.5} s(d(i,j))\,\frac{\mathbf{p}_i - \mathbf{p}_j}{|\mathbf{p}_i - \mathbf{p}_j|}. \tag{4.19}$$

Here, $\mathbf{n}$ is a vector pointing from $\mathbf{p}_i$ to the center of the retina, and $\mathbf{r}$ is a random vector such that $|\mathbf{r}| \in [0,1]$. The factors k_1, k_2, and k_3 are constant, and s is an interpolation function given by

$$s(d) = \begin{cases} 0, & d > 1, \\ 3.2\,(1-d), & 0.75 < d \leq 1, \\ 0.266\,(3.75-d), & 0 \leq d \leq 0.75. \end{cases} \tag{4.20}$$

These migration steps are executed between 25 and 41 times to let the cones settle into a pattern. The upper bound of migration iterations leads to a retinal mosaic that is mixed regular and irregular in a manner similar to real retinas. With this parameterization, each cone has on average 6.25 neighbors, which is a number also found in human retinas. Outside the fovea, fewer migration iterations are carried out to simulate that there the retina is less regularly packed.

Once the current ring of cones has finished migrating, the next iteration starts by surrounding the current cluster with a new ring of cones, which in turn will migrate inwards.

It would be possible to model the response of each cone with any of a number of models. A recent model that predicts a wide range of measurements was presented by van Hateren [434]. It is based on state-of-the-art knowledge of the various stages involved in photo-transduction. For a full account of this model, the reader is encouraged to track down the reference [434], which is freely available on the Journal of Vision website. As the details of the model are beyond the scope of this book, we only give an overview of this model in the remainder of this section.

The basic building blocks are temporal low-pass filters, static (non-) linear filters, and subtractive and divisive feedback loops. Each of these building blocks forms a plausible model of some aspect of cone functionality, whether cellular or molecular. A low-pass filter transforms an input $x(t)$ to an output $y(t)$ with

$$y(t) = \int_{-\infty}^{\infty} h(t-s)\,x(s)\,ds, \tag{4.21a}$$

$$h(t) = \begin{cases} \frac{1}{\tau}\exp\left(\frac{-t}{\tau}\right), & t \geq 0 \\ 0 & t < 0 \end{cases} \tag{4.21b}$$

This equation can also be written in differential form:

$$\tau \frac{dy(t)}{dt} = x(t) - y(t). \tag{4.22}$$

In both forms, τ is a time constant.

In Figure 4.31, the cascading of processing functions is indicated with a control flow diagram. The four main blocks are the photo-transduction stage, a calcium feedback loop, processing occurring in the inner segment, and finally the

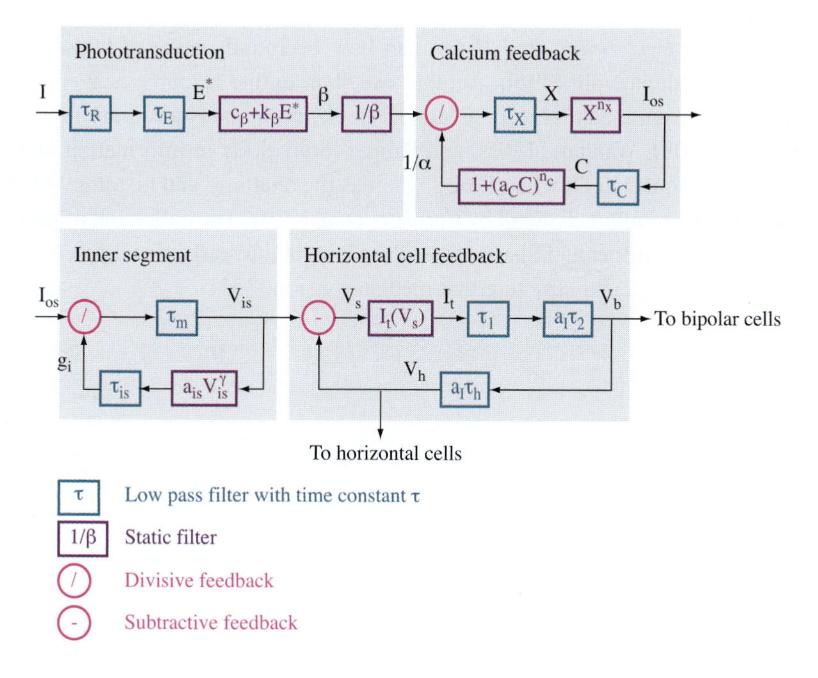

Figure 4.31. A model of cone response (after [434]).

horizontal cell feedback. While the constants and functions indicated in this flow chart are detailed in van Hateren's paper [434], it is included here to show the sort of complexity that is required to model only the very first stage of human vision, the process of photoreception in a single cone.

Compare this complexity with the simple sigmoidal response functions (Michaelis-Menten or Naka-Rushton) discussed in Section 4.3.3. These models treat photoreceptors as black boxes with an input and an output, whereas the present model incorporates most of the separate molecular and cellular components that define cone functionality. In addition, it takes temporal aspects into account (except slow processes such as bleaching).

No matter which model is preferred, it is now known that most of the sensitivity regulation in the outer retina of primates is already present in the horizontal cells and can be measured there [1057]. This is necessarily so, and is consistent with the model discussed here.

It makes the model useful for applications in visual computing. For instance, it has been shown to be capable of encoding high dynamic range video [435].

4.9 Further Reading

Hubel's book *Eye, Brain, and Vision* can bow be found online at http://neuro.med.harvard.edu/site/dh/ [496]. Another excellent online resource is Webvision at http://webvision.med.utah.edu. Classical books on human vision are those by Cornsweet [208], Wandell [1205], and Palmer [868]. Lots of information on the retina can be found in Dowling [262], whereas the anatomy and histology of the eye are described in much detail by Forrester et al. [329]. The collection of papers edited by Gegenfurtner and Sharpe gives deep insight into early visual processing, including the molecular structure that mediates vision [353].

Chapter 5
Perception

So far, we have described components of the human visual system (HVS) and some of what is known about their interconnectivity. Long before anything was known about the components of the human visual system—the substrate that mediates vision—inferences about how humans make sense of the world were based on psychophysical experiments.

In a typical psychophysical experiment, a participant is shown a stimulus and is given a task. By varying aspects of the stimulus, the task performance can be related to the variation in the stimulus. For instance, it is possible to ask a participant if a patch of light can be distinguished from its background. By varying the intensity of the patch between trials, the threshold at which the patch becomes visible can be determined.

In a sense, psychophysical experiments treat the human visual system as a black box. The input is the stimulus, and the output is the task response given by the observer. With suitably designed psychophysical experiments, many aspects of human vision can be inferred. These types of experiments can be designed to ask questions at a much higher level than neurophysiology currently is able to address. In particular, many experiments exist that answer questions related to human visual perception.

Although many different definitions of human visual perception exist, we will refer to *human perception* as the process of acquiring and interpreting sensory information. *Human visual perception* is then the process of acquiring and interpreting visual information. Thus, perception is related to relatively low-level aspects of human vision.

High-level processing of (visual) information is the domain of *cognition* and involves the intelligent processing of information, as well as memory, reasoning, attention, thought, and emotion. In general, the higher-level processes are much more difficult to understand through psychophysical means than the lower-level perception attributes of human vision. As such, most of this chapter will focus on perception, rather than cognition.

One of the cornerstones of color perception is the triple of trichromatic, color-opponent, and dual-process theories. After giving general definitions and discussing some of the problems faced by the human visual system, we therefore discuss these early in this chapter.

The trichromatic, opponent-color and dual process theories are able to explain several aspects of human vision. However, there are many more peculiarities of vision that they do not explain. Thus, dependent on the application, further refinements and additions to these models will have to be considered. Rather than present a comprehensive catalogue of visual illusions, we show a more-or-less representative selection. It will be clear that human vision has many complexities that may have to be taken into account when developing visual applications. In addition, visual illusions play an important role in understanding the HVS [272]. After all, knowing when visual processing breaks down allows us to learn about the operation of the HVS.

In the remainder of this chapter, we then discuss some of the mechanisms of early vision that have been uncovered that help to explain some of these visual illusions. These mechanisms, which include various forms of adaptation, sensitivity, and constancy, can be used as building blocks in engineering solutions. As an example, models of lightness and brightness perception have been used as part of tone-reproduction operators (which are the topic of Chapter 17), and retinex theory has been used in several different applications. Chromatic adaptation (Chapter 10) is important in white-balancing applications in photography.

We conclude this chapter with a discussion of higher-level processing involving color naming, as well as color deficiencies and tests for them.

5.1 Lightness, Brightness, and Related Definitions

Before we are able to discuss human vision in greater detail, it is helpful to introduce several definitions related to human visual perception. The definitions we have chosen here, are following Wyszecki and Stiles [1262]:

Light is that aspect of radiant energy of which a human observer is aware through the visual sensations that arise from the stimulation of the retina of the eye by radiant energy.

Color is that aspect of visual perception by which an observer may distinguish differences between two structure-free fields of view of the same size and shape, such as may be caused by differences in the spectral decomposition of the radiant energy concerned in the observation.

Hue is the attribute of the perception of a color denoted by blue, green, yellow, red, purple, and so on.

Unique hues are hues that cannot be further described by the use of the hue name other than its own. There are four unique hues, each of which shows no perceptual similarity to any of the others. They are red, green, yellow, and blue.

Brightness is the attribute of visual sensation according to which a given visual stimulus appears to be more or less intense, or, according to which the area in which the visual stimulus is presented appears to emit more or less light. Brightness ranges from "dim" to "bright."

Lightness is the attribute of visual sensation according to which the area in which the visual stimulus is presented appears to emit more or less light in proportion to that emitted by a similarly illuminated area perceived as the "white" stimulus. It is sometimes referred to as *relative brightness.* Lightness ranges from "dark" to "light".

Chromaticness is the attribute of visual sensation according to which the (perceived) color of an area appears to be more or less chromatic. It is also known as *colorfulness.*

Chroma is the attribute of visual sensation that permits a judgment to be made of the degree to which a chromatic stimulus differs from an achromatic stimulus of the same brightness.

Saturation is the attribute of visual sensation that permits a judgment to be made of the degree to which a chromatic stimulus differs from an achromatic stimulus regardless of their brightness.

In addition to these basic terms, it should be noted that the above definition of color refers to a perceived quantity. It is therefore also sometimes referred to as *perceived color.*

Further specifications of color are as given below [1262]:

Related colors are perceived to belong to an area seen in relation to other colors.

Unrelated colors are perceived to belong to an area seen in isolation from other colors.

Chromatic colors are perceived as possessing hue.

Achromatic colors are perceived as devoid of hue.

The four unique hues have specific properties, in that unique red and unique green are perceived to be neither yellow or blue. Similarly, unique yellow and unique blue are perceived to be neither red or green. The hueness of a color stimulus is sometimes described as a combination of two unique hues. For instance, orange can be described as reddish-yellow or yellowish-red [1262].

The terms brightness and lightness are often confused with each other, and with luminance and illuminance (which are photometric terms introduced in Section 6.2). Wyszecki and Stiles' definition given above will be the one that we use as well, but it should be noted that in some fields these terms are used to describe other quantities. For instance, brightness is sometimes defined as perceived luminance and lightness as perceived reflectance [13].[1] In those cases, the term *surface lightness* could be used [140]. Finally, brightness is also known as *luminosity*.

5.2 Reflectance and Illumination

The human visual system solves a remarkably difficult problem, namely to make sense of the environment. The carrier of information is light, which falls on the retina and is then processed. In Section 2.10, the process of image synthesis is outlined. This process is a computational model of how light interacts with the environment to ultimately form an image on either a camera sensor or the retina.

In particular, the rendering equation (2.207) provides a relatively complete model of light interaction. For a given point on a surface, this equation computes how much light is reflected into a direction of choice, given the light incident upon this point from all surface points in the scene. Thus, a single evaluation of this equation requires the full geometry of the scene as well as all light incident upon a single point, arriving from all other points.

While this is a computationally expensive equation, the human visual system has evolved to evaluate the inverse of this equation. Thus, starting with light

[1] A lively debate on lightness and brightness perception is available online, at www.psych.ndsu.nodak.edu/mccourt/CV/Hyperlinkedpapers/YorkCD/.

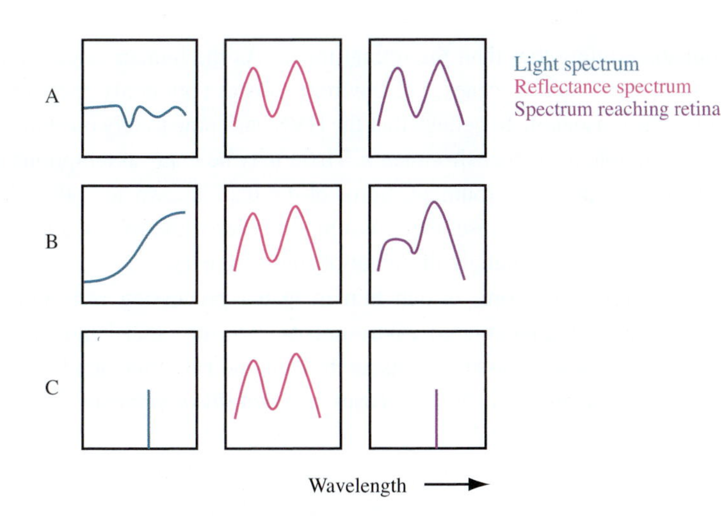

Figure 5.1. The light spectrum and the surface reflectance together determine the spectrum of light reaching our eyes (after [868]).

falling on the retina, an impression of the distribution of light in the environment is derived. In addition, the placement and shape of objects in the environment is deduced.

One of the tasks the human visual system appears to carry out is the computation of surface reflectance, i.e., the illumination and surface-reflectance components are disentangled. We note, however, that this is not a universally held opinion. There are still many uncertainties and open questions related to what the human visual system perceives, and how.

However, at some point in the chain of processing, the human visual system infers 3D shape from 2D retinal input. This is an under-constrained problem. First, the environment has three dimensions, whereas the retinal input has only two. Second, Equation (2.207) is not invertible.

An example of the complexities involved are shown in Figure 5.1. Here, the same surface (middle column) is illuminated with three different light sources (left column). The resulting spectral composition falling on the retina (right column) is different in each case. Yet, under suitable conditions, the human visual system is able to infer the reflectance spectrum of the surface to a reasonable level of accuracy. This allows the illuminant to be discounted and objects in environments to be distinguished. Ignoring the illumination is known as *color constancy* [42, 120] and is discussed in more detail in Section 5.9.

Any under-constrained (computational) system requires additional information to allow a solution to be approximated. This additional information has to come from something other than the retinal input. As the human visual system has evolved to view natural scenes, i.e., environments as commonly encountered in daily life, it is reasonable to believe that the HVS has come to rely on statistical regularities available in such environments. There may be many assumptions that the HVS makes about its environment. Some of these are known and others have yet to be discovered. In any case, these assumptions serve the role of additional information that helps disentangle illumination and geometry.

The key insight, therefore, is that human visual processing is beautifully matched to commonly encountered environments. On the other hand, if some of these known or unknown assumptions are broken, the HVS may not be able to infer its environment correctly. In those cases, a *visual illusion* may result.

5.3 Models of Color Processing

The perception of color is mediated by many different processes occurring along the visual pathways. Before any deep insights were gained into the inner workings of the human visual system, several models were developed to account for some of the peculiarities of human color vision. These include trichromatic theory, postulating three types of light transducer, and opponent-process theory, which better predicts certain forms of color blindness. Both theories are now held to be true, and are united into the dual-process theory. Each of these are discussed in the following sections. Note that these theories are in many ways the perceptual equivalents of the processing in the early visual system, as inferred from the neurophysiology discussed in the preceding chapter.

5.3.1 Trichromatic Theory

Trichromatic theory was introduced by George Palmer in 1777 [867] and independently proposed by Sir Thomas Young in 1802 [1283]. It was extended in 1855 by James Clerk Maxwell [756] and then by Hermann von Helmholtz [453] in 1867. It is now known as the *Young-Helmholtz trichromatic theory*.

In particular, Helmholtz determined that new colors can be formed by mixtures of other colors. His color-matching experiments involved a test color and a set of light sources that can be blended in various amounts. A participant in a color-matching experiment can adjust the strength of each light source in the set to visually match the test color. Helmholtz found that the minimum number of light sources required to match any given test color was three. Thus, trichromatic theory predicts that any perceivable color can be matched with a mixture of three primary colors.

Hence, it was surmised that the human visual system has three different mechanisms to detect combinations of wavelengths. It was confirmed many years later that the human visual system indeed has three types of photoreceptors that are active under photopic conditions, as outlined in Section 4.3.3.

5.3.2 Opponent-Process Theory

While the trichromatic theory for color vision successfully predicts the outcome of color-matching experiments, it fails to give a full account of human color vision. In particular, it does not account for the specific anomalies related to color blindness. In color blind individuals, pairs of colors are lost simultaneously. In addition, these pairs are always either red and green, or yellow and blue.

Second, the subjective experience of color is not explained by the trichromatic theory. Although the color mixing of three colors to match a test color may explain why purple appears as a mixture of red and blue, the subjective experience of a mixture of red and green is yellow, rather than the predicted reddish-green.

To account for these unexplained phenomena, the *opponent-process theory* was proposed [462]; it was later validated by Jameson and Hurvich [516]. Here, primary colors are recombined into opponent processes that pair colors in a mutually exclusive manner. In particular, reds and greens are paired, and so are yellows and blues. The red-green channel encodes all shades of red and green in one dimension, with the reds encoded as positive values and the greens as negative values.

Thus, the red-green channel progresses from highly saturated greens to desaturated greens, to gray, to desaturated reds, and finally to saturated reds. A second channel encodes yellows against blues in a similar manner. The third channel is non-opponent and encodes achromatic responses, from black to white.

A consequence of this encoding is that some pairs of colors are perceived as a mixture of two colors, whereas others are not. For example, the subjective experiences of a mixture of red and blue is purple, and that of red and yellow is orange. On the other hand, mixtures along each of the dimensions defined by the color-opponent theory are not perceived together. Thus, reddish-green and yellowish-blue are not among our subjective color experiences (see also Section 4.3.7).

Color-opponent theory successfully predicts that disorders affecting one of the neural encoding mechanisms causes colors to be lost in pairs. This is borne out in practice, with color blindness affecting either the red-green channel, or the yellow-blue channel (see Section 5.11).

Finally, each of the two opponent channels have a point that is perceived as neutral grey. The neutral point in the red-green channel is perceived as neither red or green. A similar neutral point lies between yellow and blue. Prolonged

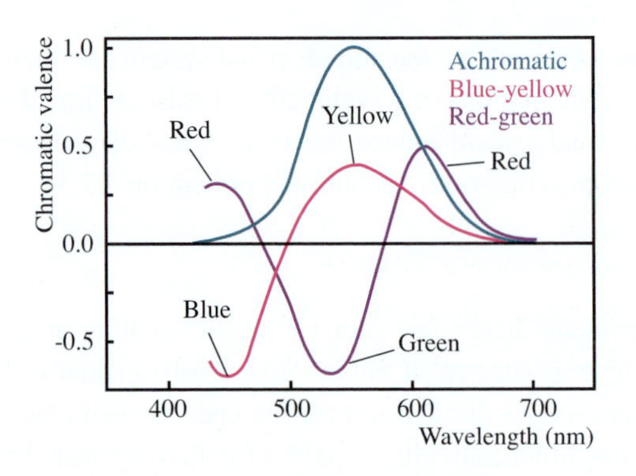

Figure 5.2. The spectral response of the red-green and yellow-blue opponent processes. Also shown is the spectral response of the achromatic channel (after [516]).

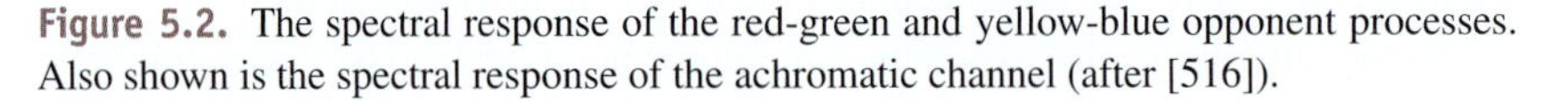

exposure to colored stimuli causes adaptation to occur, resulting in a shift that places the neutral gray at a different value. Hence, opponent processing may account for simultaneous color contrast as well as color after-images.

In addition, each channel responds to one of two colors in a mutually exclusive manner. This affords an opportunity to devise a psychophysical experiment that determines the spectral response function of the color-opponent mechanisms.

A monochromatic test stimulus of a given wavelength may be presented to an observer, who is given control over red, green, yellow, and blue light sources that can be mixed with the test stimulus. The aim of the experiment is to adjust one of the red and green, and one of the yellow and blue light sources, to yield a light mixture that is perceived as achromatic. By recording the amount of red, green, yellow, and blue that needs to be added to the test stimulus for each wavelength, the spectral distribution of each opponent process is found, as shown in Figure 5.2. The positive and negative designations are appointed arbitrarily.

5.3.3 Dual-Process Theory

While the trichromatic theory predicts the results of color-matching experiments, and the opponent-process theory predicts the pairwise loss of color in color blindness and the subjective experiences of mixtures of colors, neither theory gives a full account of all color phenomena. In particular, the phenomena explained by one theory are not explained by the other, and vice versa. As such, there is value in both theories, and in practice they can be unified into a single model of human visual processing.

Such a unified approach is afforded by the *dual-process theory*. The precise, testable form of this theory was developed by Leo Hurvich and Dorothea Jameson in 1957 [516]; they suggested two sequential stages of color processing. The first step includes a trichromatic stage, in which light entering the eye is converted into three values determined by the sensitivity of the photoreceptors in the eye. The second step is an opponent-process stage which encodes the three outputs from the previous step and thus forms color opponency.

Although the trichromatic, opponent-color, and dual-process theories were developed on the basis of psychophysical results, they are confirmed by the neurophysiological results presented in the preceding chapter. Computational implementations of these models are discussed in Chapter 8. Although they constitute a useful set of color models, they do not form a complete model of human color vision. There are several levels of visual processing that influence the perception of color, as discussed in the remainder of this chapter.

5.4 Visual Illusions

When the human visual system is faced with a stimulus that is in some way anomalous, for instance, because it is designed to deviate from natural images in a specific way, then it may misinterpret the stimulus. For instance, a static display may be perceived as in motion. The specific combination of color, contrast, luminance, and geometry, or a subset of these, are required to create a visual illusion. This serves to show that human visual processing is inherently non-linear, and that different attributes of a stimulus interact in fascinating and unexpected ways.

Arguably, one of the tasks of the human visual system is to make sense of the world. When we take a white object into a sunlit environment, we are able to determine that the object is white and, at the same time, establish that it is in a bright environment. Taking that object subsequently into a dark room, we are still able to detect that the object is white and that it is now placed in a dim environment. In effect, we are sensitive to relative luminances, but at the same time are able to detect absolute values. A different way of looking at this is to say that the human visual system is able to separate reflectance from illumination.

If the human visual system acted as a linear light meter, we would never be able to accomplish this. Taking the object into the dark room would simply mean that the measured luminance values would be lower. Thus, if the human visual system does not act as a simple photometer, then it becomes interesting to ask what it is that the human visual system computes.

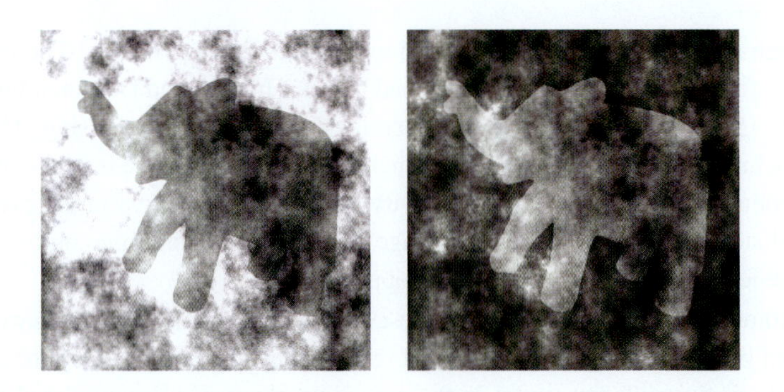

Figure 5.3. The elephants in both panels are identical, yet they appear to be darker on the left and lighter on the right.

For the perception of lightness, it is now accepted that the retinal input is split into separate components that are processed more or less independently [602]. However, there are two proposals as to how such a decomposition may occur, namely into frameworks and into layers [366].

For frameworks, the retinal image is thought to be decomposed into contiguous regions of illumination. Within each region, the highest intensity serves as an anchor (see Section 5.8.5), and all other values within a region are relative to this anchor.

The alternative theory presumes that the retinal image is split into overlapping layers that are each treated separately. The image is then assumed to consist of an illumination pattern that is projected onto a pattern of surface reflectances. Of course, this would enable the human visual system to reverse engineer reflectance from illumination by a process known as *inverse optics* [742]. There are many real-world examples where reflectance and illumination are perceived simultaneously. An example is a white house reflected in a black car [366]. The intensity of the reflected pattern is gray; yet, neither the house nor the car appear gray. Instead, the image is perceptually split into two separate layers.

Another example where an image is split into separate layers is shown in Figure 5.3 [35]. Here both images show an elephant with identical pixels. Only the surround is changed, namely lightened in the left image and darkened in the right image. Both images, however, still appear to be composed of a cloudy pattern with a dark elephant on the left and a light elephant on the right. The result of this apparent layered processing is that the human visual system has not perceived the two elephants as identical.

Thus, the accuracy with which the human visual system is able to disentangle reflectance from illumination is not perfect, and this is one source of visual illusions. Visual illusions can be seen as systematic errors made by the visual system, and they constitute a "signature of the software used by the brain" [365,366]. Many of the errors made by the human visual system can be explained by a model of frameworks [13, 363]. Likewise, a layers model can also be expanded to include an account for errors [35, 366]. It is currently not known which model is more plausible, or even if the two models can be unified.

In any case, visual illusions are important to understand aspects of human visual processing. In addition, some of them have been directly applied in some applications. For instance, counter-shading is a well-known technique used by artists to increase the apparent contrast of images and painting (see Section 17.9). The Cornsweet-Craik-O'Brien illusion, introduced in Section 5.4.4, has been used to assess tone-reproduction operators (Section 17.11.4).

5.4.1 The Ponzo Illusion

The *Ponzo illusion* can be created by placing two identical bars over parallel lines that recede into the distance [916]. Although the two bars are of identical width, they appear to be of different size, as shown in Figure 5.4. One explanation for this effect is that the top bar appears to be further away from the viewer due to the presence of converging lines which are often associated with distance [134]. As objects that are further away also tend to be smaller, the human visual system interprets the top bar as being both further away and larger.

This is an example of *size constancy*, which is the ability of the human visual system to perceive objects to be of the same size, independent of their distance, and therefore the size of the projection of the object onto the retina (Figure 5.5).

Different forms of constancy play an important role in human vision. Variations of it occur in many different image attributes, such as lightness constancy, color constancy, and contrast constancy, which are discussed in the following section, as well as in Section 5.8 (lightness constancy) and Section 5.9 (color constancy).

Figure 5.4. The Ponzo illusion. Two identical colored horizontal bars appear to be of different length.

Figure 5.5. An example of size constancy. The menhirs in this photograph are all approximately the same size; Parque Escultórico de la Torre de Hécules, A Coruña, Spain, August 2005.

5.4.2 Simultaneous Contrast and Contrast Constancy

A uniform gray patch will appear darker or lighter depending on the environment it is in. We have shown in Section 5.4 that the human visual system does not act as a linear light meter, but rather it must balance the detection of relative values, which arguably is required for effective object recognition, with the detection of absolute values, which is necessary to determine the illumination of an environment.

While the human visual system gets it right in most naturally occurring environments, in degenerate cases such as presented in Figure 5.6, the stimulus is simple enough to see that this balance between absolute and relative values is tenuous. Here, the two patches are identical, but their surrounds are not. As a result, the two patches appear somewhat different.

The traditional explanation of simultaneous contrast is that the edges between patch and surround are detected by antagonistic center-surround receptive fields (Section 4.3.4), after which edge information is propagated into the interior of the patch [207, 396, 869]. The propagation of values from an edge to an interior of a patch is called *filling-in* and is discussed further in Section 5.8.3.

The simultaneous-contrast illusion demonstrates that we cannot simply take a colored patch and derive from it the visual percept that it generates. Instead, we have to take the viewing environment into account. The sub-domain in color

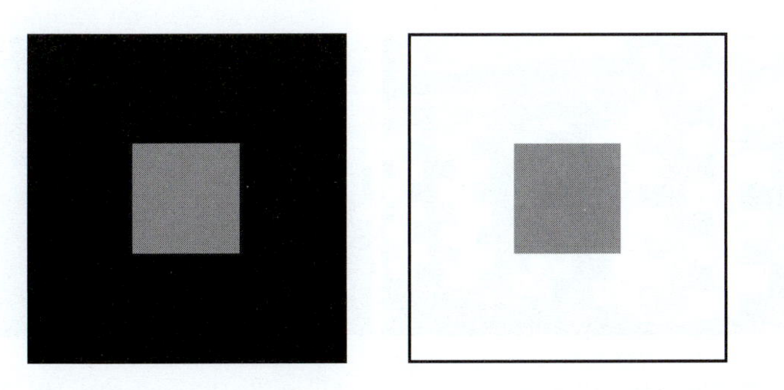

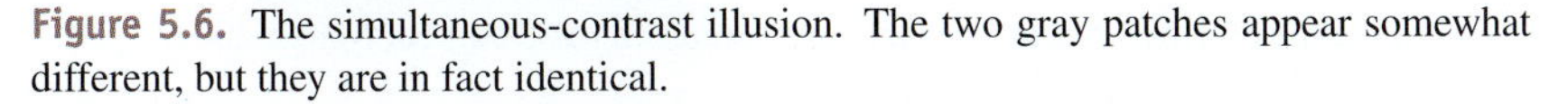

Figure 5.6. The simultaneous-contrast illusion. The two gray patches appear somewhat different, but they are in fact identical.

science that deals with this is called color-appearance modeling, and is discussed in detail in Chapter 11.

Figure 5.7 shows two checkerboard patterns each surrounded by a larger checkerboard pattern with either a higher or a lower contrast. Although the two center panels have the same contrast, the one on the right appears to have higher contrast. This phenomenon is known as *contrast constancy* or *contrast contrast* [180, 1069]. Compare this illusion to the one shown in Figure 5.3.

Figure 5.7. Contrast constancy. The center part of both checkerboard patterns has the same contrast (after [180]).

5.4.3 The Café-Wall Illusion

The *café-wall illusion* was named after the place where it was first observed and, consequently, described: on the wall of a café in Bristol [388,389]. In this illusion, tiles in different rows are offset by some amount. Between the tiles, there is mortar of a color that is in between the dark and the light tiles. The presence of the mortar is essential, as having no lines between the tiles would not create an illusion. The café-wall illusion produces lines that are perceived as curvy, whereas the lines are actually perfectly straight, as shown in Figure 5.8.

Figure 5.8. The café wall at the bottom of St. Michael's Hill in Bristol, UK. The mortar between the tiles forms straight lines that are perceived to be curved; March 2005.

5.4.4 The Cornsweet-Craik-O'Brien Illusion

The *Cornsweet-Craik-O'Brien illusion* [208, 842, 1038] is shown in Figure 5.9. This illusion consists of two identical gray patches separated by two non-linear ramps of opposite polarity. A sharp gradient is present between the two ramps, resulting in a luminance profile.

The shape of the ramps is chosen such that they are interpreted by the human visual system as constant luminance. The C1-continuity between the ramps and the constant sections ensures that the ramps cannot be differentiated from the constant-luminance sections. It may be argued that low level center-surround processes, such as those found in the retina [262, 632, 742, 868], are the cause for this failure to detect the ramps. These processes are tuned to sharp discontinuities rather than slow gradients. The discontinuity between the left half and the right half of this illusion is therefore the only feature that is detected by low level visual processing. Also, note that center-surround processing of an actual step function would produce a similar signal.

This visual illusion is implicitly exploited by many artists who apply counter-shading to their drawings and paintings. A computational simulation for the pur-

Figure 5.9. The Cornsweet-Craik-O'Brien illusion. The left and right quarters of this image have the same value, whereas the middle half contains two smooth ramps, separated by a sharp jump. This configuration is perceived as a step function, with the left half and right half of unequal brightness (after [208]).

pose of contrast enhancement is outlined in Section 17.9. In addition, this model was used as a stimulus in a validation study to assess the performance of tone-reproduction operators, as discussed in Section 17.11.4.

5.4.5 The Hermann Grid and the Scintillating Grid

The *Hermann grid*, shown in Figure 5.10 (left) consists of a regular array of black blocks which are separated by a white background [130, 463, 1076, 1145]. In the periphery of vision, the crossings of the white lines will exhibit dark patches. The classic explanation of this effect is that center-surround mechanisms have much smaller receptive fields in the fovea than in the periphery. Thus, if the stimulus is viewed from an appropriate distance, the black patches only occur in the periphery of vision, where the surround overlaps with both the white bands and the black squares. When focused on a crossing of white bands, the center-surround mechanisms in the fovea are small enough to fit entirely within the white bands and, thereforeb do not overlap with the black blocks. The differential output of the center and surrounds in the fovea and the periphery are therefore different.

While this explanation may seem reasonable, recently modifications to the Hermann grid have made it possible to make the illusion disappear. As shown in Figure 5.10 (right), by replacing the straight edges of the black blocks with wavy patterns, the Hermann illusion vanishes. As the width between the blocks and the size of the blocks is not altered, the classic explanation for the Hermann grid cannot be maintained [354].

A stronger version of the Hermann grid illusion is known as the *scintillation illusion* [1011, 1012, 1178]. Figure 5.11 shows the scintillation illusion in two

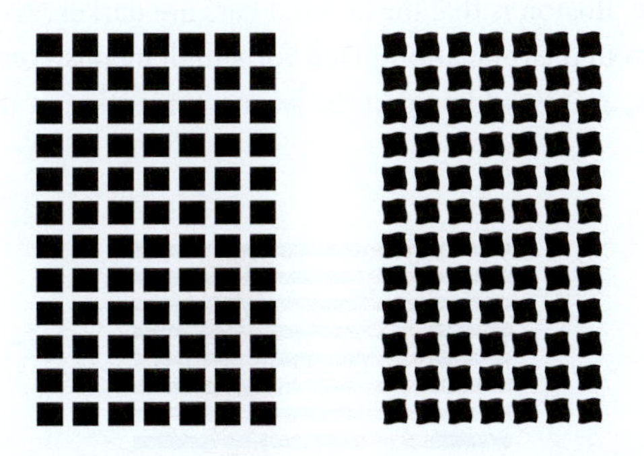

Figure 5.10. The Hermann grid (left), showing phantom dark blobs on the intersections of the white bands. The classical explanation of the Hermann grid can be invalidated by creating curved sides to each block, which makes the illusion disappear (right).

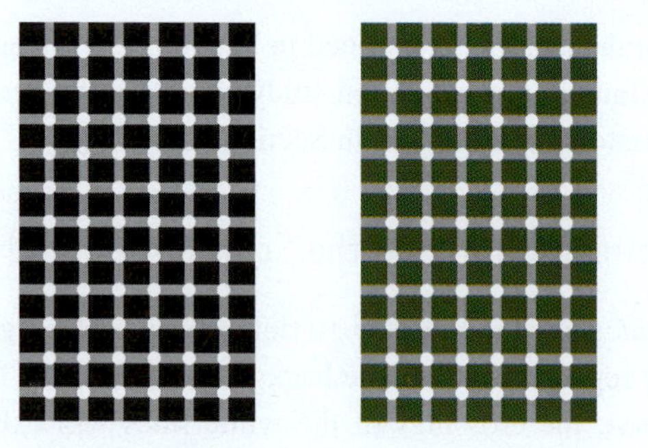

Figure 5.11. By placing white dots at the cross sections of a Hermann grid, the scintillation effect can be induced. The dark blobs take on the same color as the blocks.

versions. Note that the black boxes give rise to black illusory patches, whereas the green boxes give rise to green illusory patches.

5.4.6 The Munker-White Illusion

The *Munker-White illusion* consists of horizontal black and white stripes. If, in one vertical area, all the black stripes are replaced with stripes of a different color, and, in another vertical area, all the white stripes are replaced by the same color, then these two colored areas appear to be of different intensity, as shown in Figure 5.12 [1241]. This illusion shows that the apparent lightness of a surface can be affected by the spatial configuration in which it appears.

In comparison with the simultaneous-contrast illusion discussed in Section 5.4.2, the perceived lightness values are opposite. Here, in a predominantly light surround, the colored bars appear lighter rather than darker. In a predominantly dark surround, the illusion is that the colored bars are darker. As a consequence, a simple center-surround explanation given for simultaneous contrast does not apply to this illusion, as it would predict the wrong perception of these patches [37].

Figure 5.12. The Munker-White illusion.

This has resulted in several possible explanations of the Munker-White illusion. In the first, it is argued that a more complex center-surround configuration can still give rise to this illusion [102, 325, 799]. Second, the illusion may be the result of grouping and segmentation. It is typically assumed in such models that T-junctions are detected by the human visual system, which are then interpreted in a specific way.

For instance, T-junctions may be thought of as providing information regarding occlusions of objects. In a T-shaped edge, the surface above the horizontal bar occludes the objects separated by the stem of the T-shape. Thus, T-shaped edges allow surfaces to be ranked in order of "belongingness": the objects either side of the stem of the T belong together more than they belong to the surface above the horizontal bar [363]. In White's illusion, the colored bars therefore belong more to the bars to which they are superposed than to the ones to which they are adjacent [1129, 1288].

The anchoring theory proposed by Alan Gilchrist also provides an explanation of White's illusion [37, 363]. This theory is discussed further in Section 5.8.5. An extensive review of the Munker-White illusion is presented by Anderson [37].

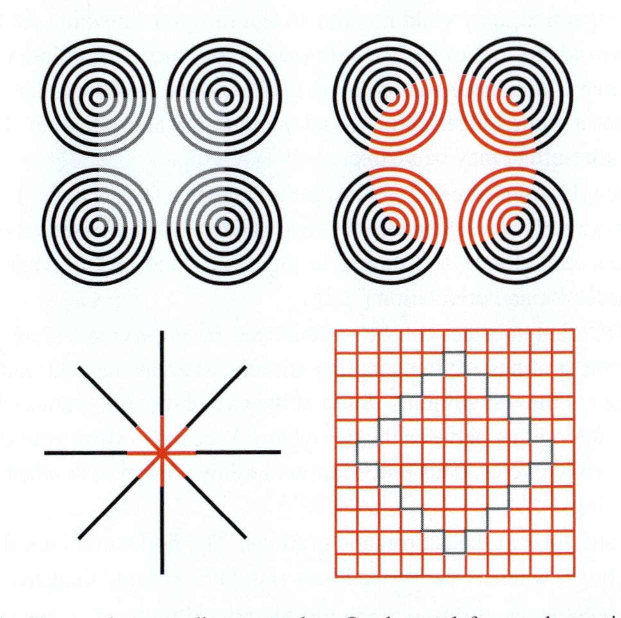

Figure 5.13. Neon color-spreading examples. On the top left, an achromatic version is shown, which is also known as *neon brightness spreading*. The top-right figure is due to Varin [1179], whereas the bottom-left illusion was developed by van Tuijl [1151]. The bottom right illusion is from [127].

5.4.7 Neon Color Spreading

The *neon color-spreading* illusion typically occurs when an incomplete black line pattern is completed with lines of a different color. The colored parts of the figure then dilute into the white surrounding areas, as seen in Figure 5.13. The color of the red bars in this figure spreads into its neighboring surrounding area, but is importantly halted by an illusory contour.

This illusion was first discovered by Varin [1179] and later independently rediscovered by van Tuijl [1151]. The bottom-left figure is a variation of the *Ehrenstein figure*, showing that a very minimal configuration can still result in the neon color spreading effect [277, 278].

Discussion of this illusion can be found in various places [593, 942, 943]. See also Bressan et al. for an overview of conditions under which the neon color spreading effect is likely to occur [127].

5.4.8 Color After-Effects

Staring at a field of a particular color for a period of time, and then focusing on a white background, may yield an after image of opposite color. As the human visual system adapts to prevailing conditions, for instance, the prolonged presence of a particular color, once that color is removed, the HVS remains adapted to this color for a while longer before re-adapting to the new situation. During this period, an after-effect may be visible.

Interestingly, some after-effects last for more than a few seconds. In particular, *contingent after-effects* may endure for several days. Such after-effects depend on two stimulus variables. For instance, in the case of the McCollough effect, the variables are color and orientation [762].

The McCollough effect can be experienced by observing Figure 5.14. The two chromatic gratings are the adapting stimuli, whereas the achromatic grating at the bottom is the test stimulus. Stare at the two chromatic gratings for at least 10 minutes, alternating between the two every 15 seconds. Move your eyes slowly around the white circles. This procedure will allow your eyes to adapt to the two colored gratings.

Afterward, look at the achromatic gratings. The horizontal bars should have a greenish tinge, whereas the vertical bars should be colored magenta. These are thus the complements of the colors in the adapting stimuli.

The induction protocol that must be followed to create the McCollough effect includes a uniformly oriented pattern of a particular color, which may be alternated with orthogonally oriented patterns of a complementary color. After

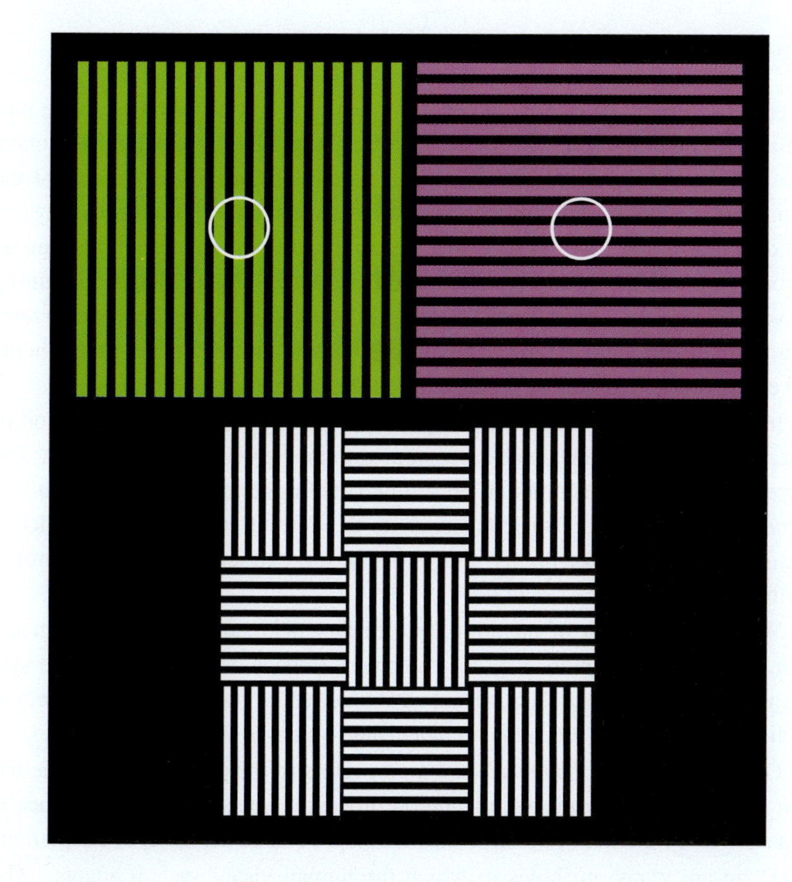

Figure 5.14. The McCollough effect.

induction, observing achromatic versions of these patterns, a faint tinge of a complementary color can be seen [139].

It should be noted that the McCollough effect can be extremely long lasting, making it special among visual illusions. The 10-minute adaptation time recommended above was found to be sufficient to make the after effect last for 24 hours [557]. It is therefore known an a *contingent after-effect*.

As discussed in Section 4.2.5, as a result of astigmatism, the optics of the eye produces colored fringes whereby the color depends on the orientation of the edges. One theory proposes that the McCollough effect is a manifestation of the neural circuitry that corrects for some of the artifacts, namely chromatic aberration, introduced by the eye's optics [139]. Overviews of the McCollough effect are published by Stromeyer [1100] and McCollough [763].

5.5 Adaptation and Sensitivity

When observers move through an environment, they get accustomed to the particular lighting conditions of the environment. According to Goethe, "Prisoners who have been long confined in darkness acquire so great susceptibility of the retina, that even in the dark (probably a darkness very slightly illumined) they can still distinguish objects." [373]. Under normal conditions, changes in lighting by factors of up to 1000 go largely unnoticed [63]. Much larger changes in lighting are very noticeable. For instance, entering a cinema during the day leads to temporary blindness; sight is slowly regained after some time. Leaving the cinema and entering daylight may be a painfully bright experience as well.

In both cases, the human visual system adapts, so that after some period of time, the vision system becomes accustomed to the new lighting conditions. The process of getting accustomed to prevailing (lighting) conditions is called *adaptation*. It regulates the visual sensitivity to the current environment in order to maximize the ability for higher-level processes to extract information from the environment.

Within the concept of adaption, two further distinctions need to be made. First, *steady-state adaptation* refers to the performance of the human visual system when fully adapted. Thus, visual attributes can be measured with respect to the lighting conditions when the human visual system is at equilibrium.

On the other hand, *temporal adaptation* refers to the performance of the human visual system during the adaptation process. Thus, visual performance is measured as a function of the time that has elapsed since the onset of adaptation.

There are many attributes to which the human visual system adapts. The two specific examples given above are dark adaptation and light adaptation for entering and leaving the cinema, respectively. There are many other examples of situations where the human visual system adapts, including chromatic adaptation which allows the human visual system to adapt to lighting of a particular color. An example in which this process causes unexpected behavior are the color aftereffects discussed in Section 5.4.8.

Light adaptation, dark adaptation, and chromatic adaptation are important effects for the study of color and are discussed in the following sections. We also discuss sensitivity, as sensitivity and adaptation are closely related.

5.5.1 Dark Adaptation

The human visual system functions over a very large range of illumination. While the range of light that can be processed and interpreted simultaneously is limited,

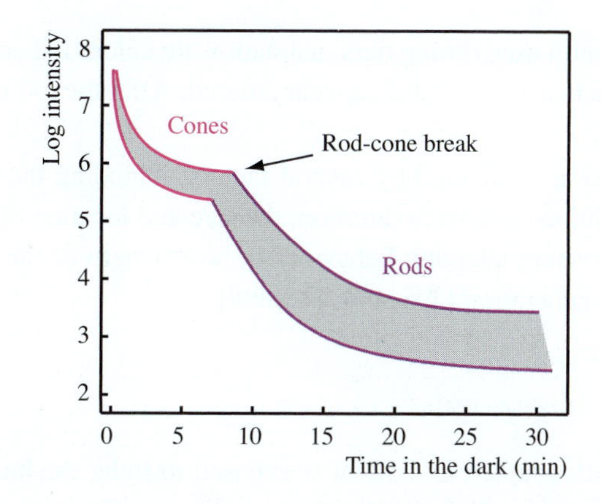

Figure 5.15. The time course of dark adaptation (after [443]).

over the course of time the human visual system adapts to a much larger range of illumination. These processes are not instantaneous. The time course of adaptation for the transition between a light environment and a dark environment is called *dark adaptation*.

There are several mechanisms at play during dark adaptation, including dilation of the pupil, photochemical activity in the retina, and neural processes in the remainder of the visual system. While the pupil rapidly adjusts to new lighting conditions, the purpose of constriction is to improve the optical resolution of the stimulus falling on the retina, rather than a mechanism to stabilize image luminance [150, 151, 1238].

Dark adaptation can be measured by letting an observer gaze at a pre-adapting light for approximately five minutes before placing the observer in a dark room and increasing the intensity of a test stimulus until it is detected by the observer [53]. The intensity of this threshold stimulus can be plotted as a function of time, as shown in Figure 5.15. The adaptation curve followed by individual observers varies somewhat, but tends to lie within the light gray region in this figure.

The curve consists of two separate regions that are marked by a break. During the first few minutes of dark adaptation, the cones gain in sensitivity and dominate the adaptation curve. The threshold level at the end of the cone adaptation curve is called the *cone threshold*. The rod system is slower to adapt, and only begins to dominate the adaptation process after the cones are at equilibrium. The curve followed for the rods ends with a maximum sensitivity after around 40 minutes. The threshold reached at that point is called the *rod threshold* and has a value of around 10^{-5} cd/m^2.

If the test stimuli used during dark adaptation are colored, then before the rod-cone break is reached, these stimuli appear colored. After the rod-cone break, they appear colorless.

Dark adaptation is affected by several factors, including the intensity of the pre-adapting light, the exposure duration, the size and location of the spot on the retina used to measure adapting behavior, the wavelength of the threshold light, and rhodopsin regeneration [208, 444, 445, 899].

5.5.2 Light Adaptation

When a fully dark-adapted individual is exposed to light, the human visual system adapts to the new and brighter lighting conditions. This process is called *light adaptation*. It can be measured with *increment threshold* experiments, whereby an observer is exposed to test stimuli presented on a background with a given intensity. The test stimulus is then increased until it is detected. The results of such an experiment are plotted in Figure 5.16. On the horizontal axis the background intensity is plotted, and the vertical axis plots the threshold intensity of the test stimulus. For this reason, curves as shown in this plot are called *threshold-versus-intensity* (or TVI) curves .

In this particular plot, the curve consists of two branches; the lower branch is caused by the rod system and the upper branch is due to the cone system. TVI curves as a function of the adapting luminance L_A can be modeled for cones

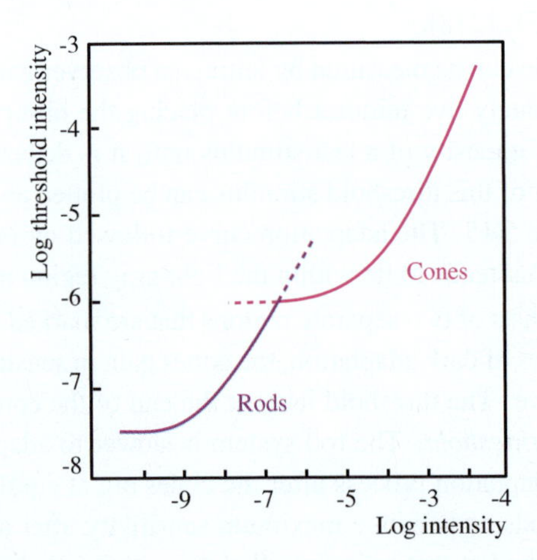

Figure 5.16. The time course of light adaptation (after [236]).

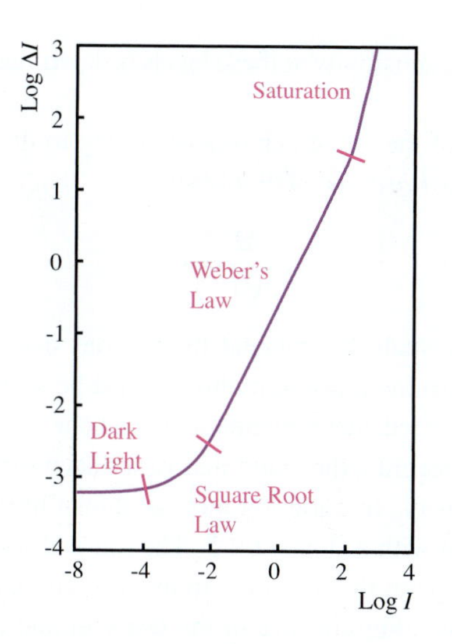

Figure 5.17. A schematic threshold-versus-intensity curve for a single mechanism. The background intensity I is plotted against the difference threshold ΔI. Four different sections can be discerned (after [16]).

by [310]

$$\log(\mathrm{TVI}(L_A)) = \begin{cases} -0.72 & \text{if } \log(L_A) \leq -2.6, \\ \log(L_A) - 1.255 & \text{if } \log(L_A) \geq 1.9, \\ (0.249 \log(L_A) + 0.65)^{2.7} - 0.72 & \text{otherwise.} \end{cases} \tag{5.1}$$

For rods, the TVI curve can be approximated by [310]

$$\log(\mathrm{TVI}(L_A)) = \begin{cases} -2.86 & \text{if } \log(L_A) \leq -3.94, \\ \log(L_A) - 0.395 & \text{if } \log(L_A) \geq -1.44, \\ (0.405 \log(L_A) + 1.6)^{2.18} - 2.86 & \text{otherwise.} \end{cases} \tag{5.2}$$

Under suitable experimental conditions, it is possible to isolate either the rod or the cone system, and then a single curve is obtained [16, 63]. Such a curve, plotting the difference threshold ΔI against the background intensity I, is shown in Figure 5.17.

This curve shows four different sections. At the bottom left, a linear section characterizes the human visual system's behavior at the *dark-light* level. Increasing the background intensity does not initially bring about an increase in detection threshold, because the sensitivity at these levels is determined by neural noise (the dark light).

The second part of the curve behaves according to the *square root law*, also known as the *de Vries-Rose law* [979, 1188]:

$$\frac{\Delta I}{\sqrt{I}} = k. \tag{5.3}$$

Here, light levels are such that quantal fluctuations determine the threshold of vision. Thus, to determine a test stimulus, it must be sufficiently different from the background to exceed these fluctuations. Under such conditions, an ideal light detector would record a threshold that varies according to the square root of the background intensity. In a log-log plot, as shown in Figure 5.17, this would yield a curve segment with a slope of 0.5. The rod pathway typically has a slope of 0.6 [417]. The shape of the transition to the next curve segment is determined by several factors, including the size of the test spot and the duration of the test stimulus.

A large part of the threshold-versus-intensity curve has a slope of around 1. This means that the intensity threshold ΔI is proportional to the background intensity I [114, 1225, 1226]. This relation is known as *Weber's law* [100, 900]:

$$\frac{\Delta I}{I} = k. \tag{5.4}$$

The constant k is the *Weber constant* or the *Weber fraction*. The intensity threshold ΔI is also known as the *just noticeable difference* (JND) . To account for neural noise, the dark light may be modeled as a small constant I_0 that is added to the background intensity:

$$\frac{\Delta I}{I + I_0} = k. \tag{5.5}$$

The last part of the curve in Figure 5.17 steepens, indicating that at those background levels, very large intensity differences become undetectable. Thus, the system saturates. For the rod system, this occurs at around 6 cd/m^2 for a natural pupil in daylight [63].

The cone system does not saturate, but shows a constant difference threshold over a very large range of background intensities [133].

5.5.3 Supra-Threshold Sensitivity

Weber's law, as discussed in the preceding section, relates stimulus detection at threshold to the magnitude of the background. It is therefore valid to detect increments, but not directly applicable to supra-threshold stimuli, i.e., stimuli that are well above the detection threshold. To yield a relation between the magnitude of the stimulus and the strength of the evoked sensation, it is tempting to integrate Weber's law, after noting that the Weber fraction can be interpreted as a constant times the differential change in perception $d\psi$, i.e., $k = k'\,d\psi$. Restating Weber's law and integrating yields

$$d\psi = \frac{\Delta I}{k' I}, \tag{5.6a}$$

$$\psi = k' \ln(I) + C. \tag{5.6b}$$

For convenience, we set the arbitrary constant of integration C to $-k' \ln(I_0)$, where I_0 is a threshold stimulus. The result is *Fechner's law* and relates the perceived intensity ψ of a patch to its radiometric value I:

$$\psi = k' \ln\left(\frac{I}{I_0}\right). \tag{5.7}$$

According to this equation, the magnitude of the sensation has a logarithmic relation to the size of the stimulus. This law makes the erroneous assumption that threshold discriminations, which are a measure of uncertainty and variability in sensory discrimination, are integrable to give a relation between stimulus and sensation.

Thus, currently Fechner's law has largely been abandoned in favor of a different law that holds for many modalities, including brightness, lightness, smell, taste, temperature, and many others. It goes by the name of *Stevens' power law* and relates stimulus I to subjective magnitude ψ as follows [1086]:

$$\psi = k(I - I_0)^n, \tag{5.8}$$

where the exponent n varies according to which of the many sensations is measured. For example, the exponent is 0.33 for brightness,[2] 1.2 for lightness, and 3.5 for electric shock.

5.5.4 Contrast Sensitivity

The threshold of visibility depends on many factors. As suggested by Weber's law, the visibility of a test stimulus depends on the intensity of the background. In addition, the ability to discern stimuli from their background depends on the spatial frequency of the stimuli. A convenient way to adjust the spatial frequency of a stimulus is by employing a grating, which is a pattern that alternates regularly over space.

[2] Assuming a 5° target in the dark.

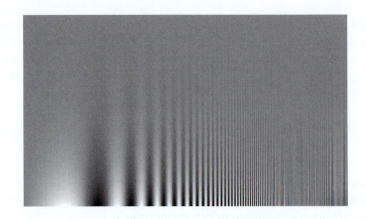

Figure 5.18. Campbell-Robson contrast-sensitivity chart [152] (Figure courtesy of Izumi Ohzawa, Osaka University with permission).

A sine function is an example of a grating. Such a function can be created at many spatial frequencies, as well as different amplitudes. It is thus possible to construct a 2D function where the spatial frequency is increased along the x-axis and the amplitude is varied along the y-axis:

$$I(x,y) = \log(y/y_{\max})\,\frac{1+\sin\left(\exp(x/x_{\max})\right)}{2}, \tag{5.9}$$

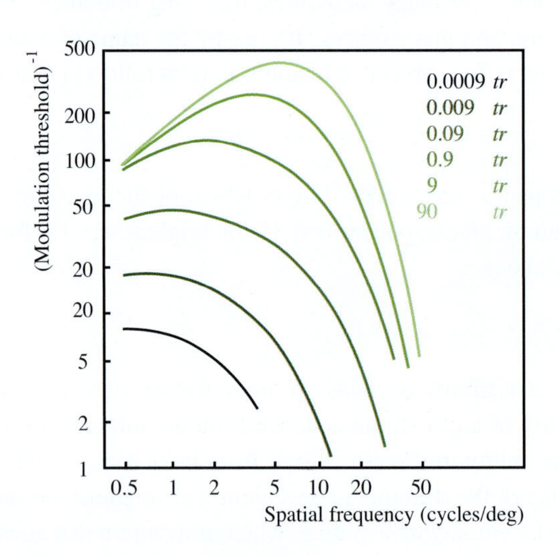

Figure 5.19. Contrast sensitivity for different levels of retinal illumination. (Data from van Ness [827]; figure redrawn from [646].)

where x_{max} and y_{max} are the maximum display dimensions. The resulting image is shown in Figure 5.18, which is called a *Campbell-Robson* contrast-sensitivity chart. In this chart, it is possible to trace from left to right where the sine grating transitions into a uniform gray. In the uniform gray area there is still contrast, but it falls below the threshold of visibility. The trace forms an inverted U-shape, which delineates the contrast sensitivity as a function of spatial frequency.

Contrast sensitivity is, in addition, dependent on the amount of light falling on the retina. This amount itself is modulated by both the environment as well as the pupil size. For this reason, the preferred measure for retinal illuminance is the *troland*, which is luminance multiplied by the pupillary area.[3] The contrast sensitivity for different levels of retinal illuminance are plotted in Figure 5.19.

Finally, the sensitivity to contrast may be altered in individuals with certain diseases such as multiple sclerosis and disorders such as cataracts and amblyopia [41].

5.5.5 Chromatic Adaptation

The TVI curves explained in Section 5.5.2 are applicable to both the rod and the cone systems. As there are three types of cones with peak sensitivities at different wavelengths, it is reasonable to assume that each cone type follows the TVI curve independently. This means that if an environment has a strong color cast, usually because the illuminant is colored, each of the three cone types will adapt separately and, simultaneously, have a different difference threshold ΔI.

The different levels of adaptation of the three cone types cause the human visual system to be adapted to the dominant color of the environment. This mechanism is called *chromatic adaptation*. As normally the dominant color of the environment is caused by the color of the light source, the human visual system thus becomes adapted to the prevailing lighting conditions. This property of the HVS is known as *discounting the illuminant*. The independence of the three receptor types thus creates an automatic *white balancing* of the HVS.

Figure 5.20 shows an example of chromatic adaptation. By staring at the cross in the panel at the top for a few minutes, the retina becomes adapted to this bipartite field. The image on the bottom is processed to have a similar color cast in both the left- and right-half to that of the bipartite field. By observing the image after adaptation to the top panel, the image will look normal, because the HVS will discount the color casts present in the image.

If the spectral sensitivities of the three receptors at one state of adaptation are given by $\bar{l}_1(\lambda)$, $\bar{m}_1(\lambda)$, and $\bar{s}_1(\lambda)$, and by $\bar{l}_2(\lambda)$, $\bar{m}_2(\lambda)$, and $\bar{s}_2(\lambda)$ for a different

[3] Luminance is a photometric term, explained in Section 6.2.7.

state of adaptation, then these spectral sensitivities are related as follows:

$$\bar{l}_1(\lambda) = k_l \bar{l}_2(\lambda), \tag{5.10a}$$

$$\bar{m}_1(\lambda) = k_m \bar{m}_2(\lambda), \tag{5.10b}$$

$$\bar{s}_1(\lambda) = k_s \bar{s}_2(\lambda). \tag{5.10c}$$

These relations are known as the *von Kries coefficient law* [622]. The coefficients k_l, k_m, and k_s are inversely related to the relative strengths of activation [544]. Von Kries chromatic adaptation is further discussed in Chapter 10.

Figure 5.20. An example of chromatic adaptation. Stare for 20 seconds at the small cross in the top panel. Then focus on the cross in the bottom panel. The image should look normal; Rathaus Konstanz, Germany, June 2005.

5.5.6 Spatial Summation

The human visual system has the ability to sum light falling on neighboring parts of the retina. This implies that there is a relation between the threshold intensity ΔI and the size of the stimulus A. It has been found that if the stimulus area is doubled, the value of the threshold is halved:

$$I A^n = k, \tag{5.11}$$

where the exponent n describes the degree of summation. If summation occurs fully, then n equals 1. Otherwise, n lies between 0 and 1. This relation is known as *Ricco's law* [236, 329, 570].

5.6 Visual Acuity

The photoreceptor mosaic is a discrete array of light receptors. This means that the spatial resolution of detail that can be processed by the HVS is limited. The spatial resolving power of the human visual system is referred to as *visual acuity*.

There are several factors limiting visual acuity, including the photoreceptor density and diffraction and abberations in the eye [1054]. In addition, there are several other factors affecting visual acuity, which include illumination, contrast, color, and the location of the retina under consideration.

Based on cone spacing alone, the resolving power of the retina is around 60 cycles per degree of visual angle [219]. This is higher than can be achieved with clinical techniques to infer visual resolving power, as no correction for the optics of the eye and neural processing was applied.

5.6.1 Factors Determining Visual Acuity

For very young children, visual acuity dramatically depends on age [492]. Figure 5.21 shows a computer simulation of how visual acuity develops over the first few months of life. The object is assumed to be 24 inches away. Note that both visual acuity and color vision are not well developed at birth.

There are at least seven factors that have an impact on visual acuity for adult humans in addition to the hard limits imposed by the photoreceptor array. These factors are refractive errors in the optics of the eye, the size of the pupil, the amount of illumination, the time the target was exposed, the area of the retina involved in target detection, the state of adaptation of the eye, and the presence of eye movement.

When a point light is present in the visual field, a thin beam of light is projected onto the retina. The optics of the eye will spread the beam somewhat.

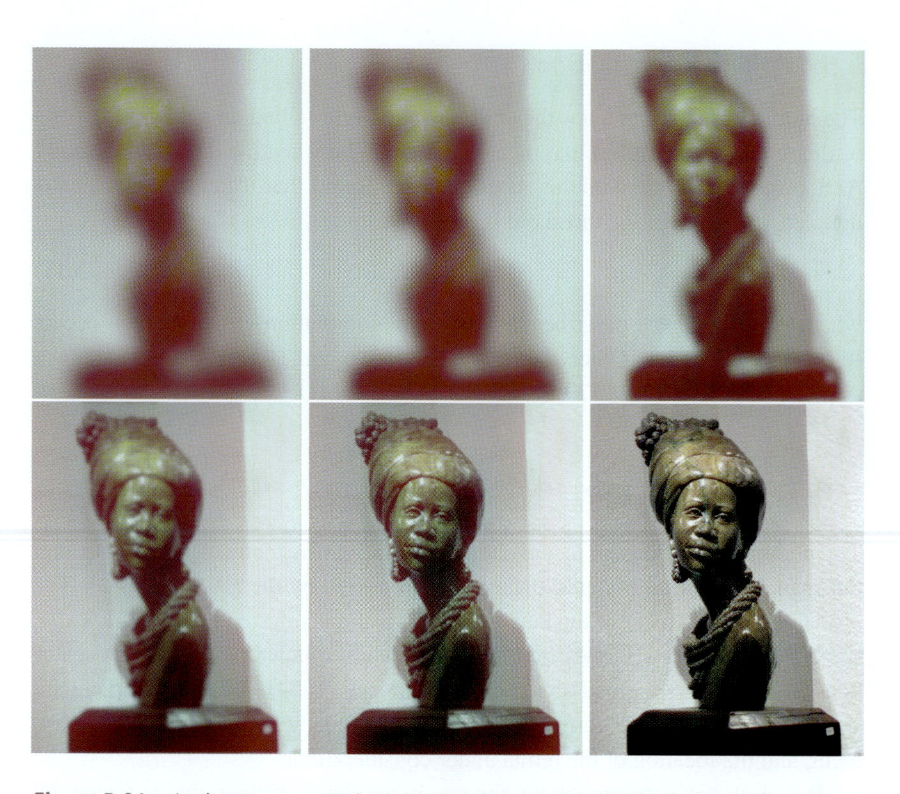

Figure 5.21. An image processed to simulate human vision after 0, 4, and 8 weeks of development (top row), and 3 and 6 months of development followed by adult vision (bottom row). The simulation is from tinyeyes.com. Statue by James Tandi; photographed at Native Visions Art Gallery, Orlando, FL.

The distribution of light on the retina due to a point source is characterized by a point-spread function (PSF).

Errors in the optics of the eye cause light to not be sharply focused on the retina; these errors create some amount of defocus. The PSF will overlap with a larger portion of the retina, leading to a blurred view of the world. Some causes of blur are *myopia* (short-sightedness) and *hyperopia* (long-sightedness). A myopic eye causes the light to be sharply focused somewhere in front of the retina, whereas a hyperopic eye focuses the image behind the retina.

The pupil determines how much light enters the eye. A larger pupil causes more light to enter. More light will allow more photons to stimulate the retina and also reduces diffraction effects. However, abberation will increase. Reducing the pupil size, on the other hand, will reduce abberation effects, but increase

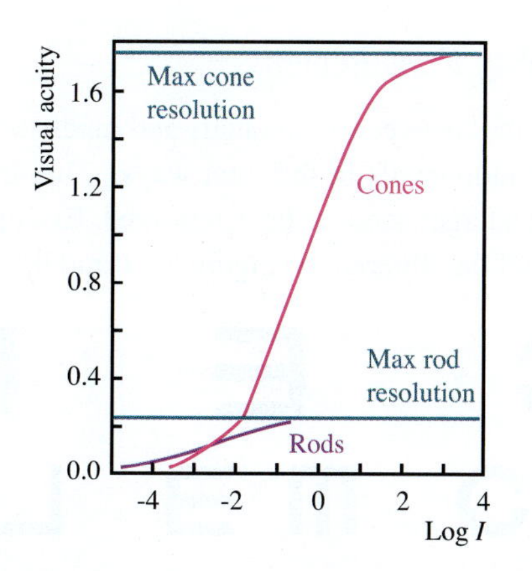

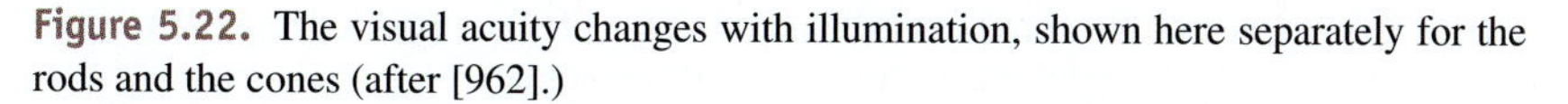

Figure 5.22. The visual acuity changes with illumination, shown here separately for the rods and the cones (after [962].)

diffraction [51]. Thus, the size of the pupil has a direct effect on the point-spread function of the projected light.

The amount of background illumination has a direct impact on visual acuity. For recognition tasks, the relationship between the two is given in Figure 5.22.

The resolving power is not uniformly distributed over the retina, but decreases with distance to the fovea. The relationship is shown in Figure 5.23.

Under photopic lighting conditions, the best visual acuity is achieved for test stimuli with the same intensity as the intensity to which the eye is adapted. Under scotopic conditions, the visual acuity is much lower and is determined by the AII amacrine cell [606, 1218].

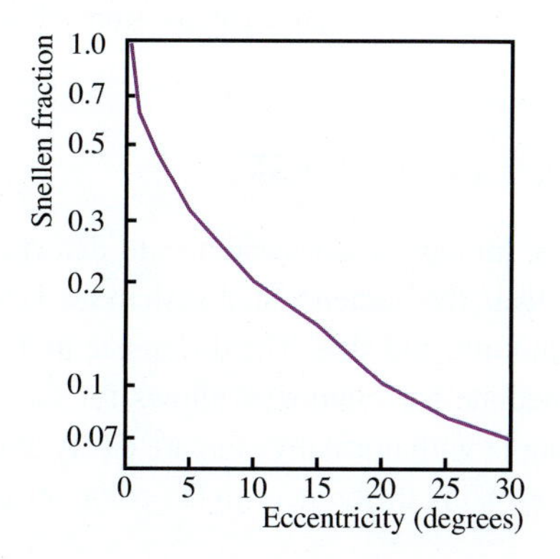

Figure 5.23. Visual acuity as function of retinal position, measured as the eccentricity from the fovea (after [1239]).

5.6.2 Visual Acuity Measurements

The visual acuity differs between individuals and tends to degrade with age. It may be measured in individuals in different ways. The first is *target detection* where the location of a target needs to be determined. Examples are Landolt rings (Figure 5.24 (left)) and the illiterate E (Figure 5.24(middle)).

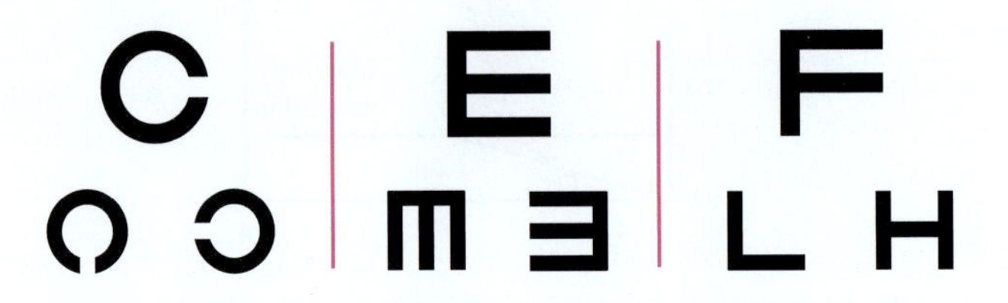

Figure 5.24. Landolt rings (left) and illiterate Es (middle) are used for target detection, whereas Snellen charts (right) are useful for target recognition.

Target recognition tasks are also frequently used to measure an individual's visual acuity. The Snellen chart is a good example (Figure 5.24 (right)).

Finally, *target localization* is a separate technique whereby a small displacement of an object, such as a discontinuity in a contour, can be detected (Figure 5.25). If visual acuity is measured in this way, it is called *Vernier* acuity.

Figure 5.25. Vernier acuity can be measured by having observers locate small displacements. In both panels, the displacements are increased from left to right.

5.7 Simultaneous Contrast

One of the tasks of the human visual system is to detect objects in its environment. To perform this task, the human visual system uses cues such as brightness, color, texture, shape, motion, and size. The difference in these cues, coming from objects and their immediate surroundings, allows for the detection of these objects. For example, people with normal vision are easily able to detect a red apple against green foliage, by using differences in the color cue coming from the apple and its surround.[4]

[4]A relation between the need to distinguish certain tropical fruits and the evolution of primate vision has been proposed. An overview of arguments for and against this hypothesis is given by

Figure 5.26. An example of simultaneous contrast. The central circle on the left appears larger than the one on the right, even though they are the same size (after [868]).

The human visual system enhances such differences in cues, allowing for easier detection of objects. Take the identical inner circles in Figure 5.26, for example. In this artificial scenario, the circle on the left is surrounded by smaller circles, and the one on the right is surrounded by larger circles. The human visual system enhances the difference between the sizes of the central circles and their surrounds, making the circle on the left appear larger and the one on the right appear smaller. As a result, the two central circles appear to have different sizes. The enhancement of the difference between cues from the object and its surround is termed *simultaneous contrast*.

Similarly, a gray region appears brighter when viewed on a darker surround, and darker when viewed on a brighter surround [171]. This type of simultaneous contrast is called *simultaneous lightness contrast*. *Simultaneous color contrast* refers to the perceived enhancement of differences in the color cue. Both these types of simultaneous contrast are discussed in the following sections.

5.7.1 Simultaneous Lightness Contrast

An example of simultaneous lightness contrast is shown in Figure 5.27, where the ring of uniform gray appears to be composed of four different shades of gray. The ring, or stimulus, appears darker if its surround is lighter, and lighter if its surround is darker. If the ring were cut out and removed from its surround, it would appear a uniform gray.

Mach bands form another well-known effect of simultaneous lightness contrast. These are named after their discoverer, Ernst Mach. Mach bands are the perception of bands along C1-discontinuities in images. These bands accentuate the difference across such edges, as illustrated in Figure 5.28. There is a perception of bands on either side of the edge. The band on the left of the edge appears brighter than the region on the left, and the band on the right appears darker than

Regan et al [944].

Figure 5.27. A gray ring of uniform lightness appears to have four sections of different lightness. The perceived lightness of each section is influenced by the lightness of the surround (after [868]).

the region on the right. The graph shows the difference between the perceived and the actual lightness across the edge.

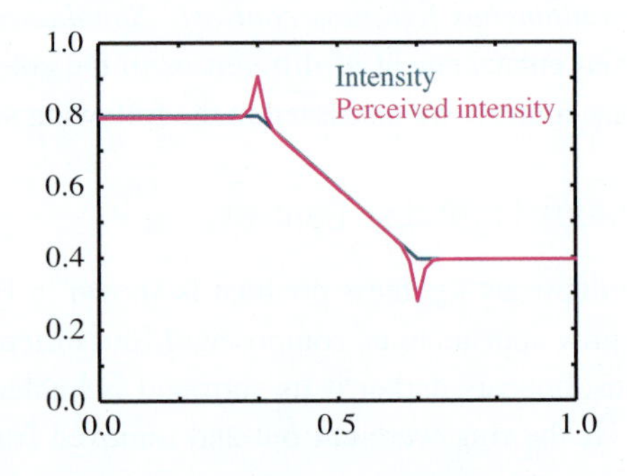

Figure 5.28. Bands are perceived on either side of the edge shown above. The presence of these bands enhances the physical difference across the edge.

5.7.2 Simultaneous Color Contrast

Similar to color after-effects, which involve the perception of complementary hues over time, simultaneous color contrast involves the perception of complementary hues across space. Simultaneous color contrast is also known as *induced*

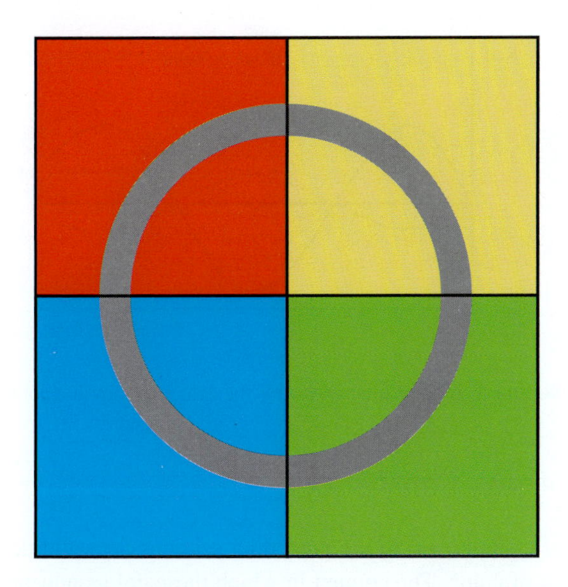

Figure 5.29. The uniformly colored ring is perceived to have different colors, determined by the background color behind the ring (after [868]).

color. As with simultaneous lightness contrast, induced color may be demonstrated by placing a uniform color against different backgrounds, so that the color appears different. Figure 5.29 demonstrates this effect. The perceived color of the uniform gray ring is determined by the background behind it. In the case of a blue background, the ring appears to have the hue of the complementary hue (yellow), while against the yellow background, the ring appears blue.

The magnitude of the perceived contrast effect may be measured by using a nulling method [614]. In this method, the observer is asked to fixate on a white disk. If its surround is set to green, for example, the white disk appears to be colored by the opposite hue, which is red in this case. The perceived red color of the disk may be neutralized by adding a little of the opposite color (green) to the disk. The amount of green light required to make the disk appear white again is the measure of the contrast effect.

When this experiment is repeated for different colors, the observer is required to keep a standard white in mind. This may be avoided by modulating the color of the surround from one color to its complementary hue. In this version, the observer simply has to keep the appearance of the disk steady as the color of the surround is modulated from say, red to green. Figure 5.30 illustrates the result of such a measurement.

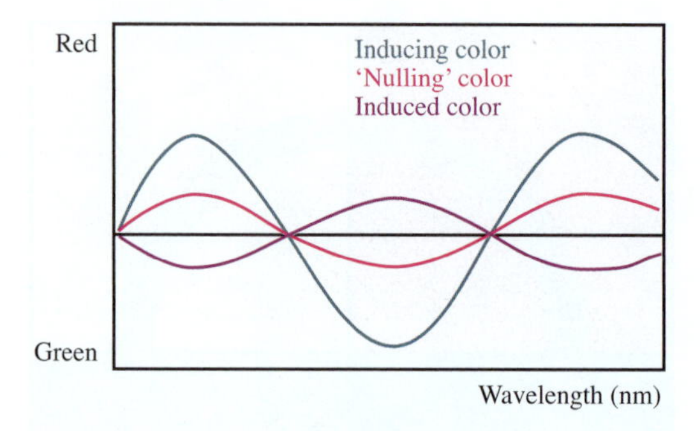

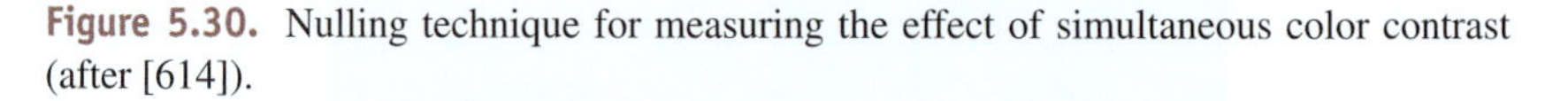

Figure 5.30. Nulling technique for measuring the effect of simultaneous color contrast (after [614]).

5.8 Lightness Constancy

The color theories described in the preceding sections are valid descriptions for isolated patches of color. The context in which these colors appear are therefore ignored. Nevertheless, this context plays an important role in the perception of color. Our color perception is affected by factors such as the geometry of the environment and its lighting conditions.

An example is shown in Figure 5.31. Here, the colors of the left and right half of the light house are different when seen in isolation. When observed in context, it is clear that the surface reflectance of the light house remains constant, but the illumination changes. The complexities involved in disentangling reflectance from illumination were outlined in Section 5.2. Figure 5.31 shows that the HVS effortlessly separates surface reflectance from illumination. This particular feature is called *constancy*. It enables the perception of objects in scenes, independent of illumination.

Formally, constancy may be defined as the ability to perceive the reflectance properties of surfaces despite changes in lighting and other viewing conditions [868]. It is a feature of the human visual system that applies to several different modalities. In particular, *lightness constancy* applies to achromatic stimuli and the perception of black and white. An example is given in Figure 5.32. The perceived difference in the two checks marked A and B are due to some of the inferences the human visual system makes.

First, local contrast suggests that check A is darker than average, whereas check B is interpreted as lighter than average. As this assumption is based on local contrast only, the presence of the shadow does not alter this logic. Second, high frequency edges are typically interpreted as belonging to textures, whereas low frequency content may be due to shadows, and is therefore ignored.

Figure 5.31. The left and right half of the light house each stimulate the photoreceptors differently, as demonstrated by the insets. However, the wall of the light house is still perceived as one continuous surface. The difference in illumination is discounted by the human visual system, as long as the context is present; Light house at Ponce's Inlet, FL, June 2005.

These inferences normally help in reconstructing 3D information from retinal input, and are the cause of lightness constancy. Assume, for instance, that the human visual system operated as a linear light meter. In that case, we would not be able to distinguish a white surface in a dark environment from a black surface in a very light environment [13]. However, humans can normally accomplish this task, which is a desirable property enabling us to make inferences about the environment. There exist several theories to explain different forms of constancy, as discussed in the following sections.

5.8.1 Adaptational Theories

One of the mechanisms that partially aid color and lightness constancy is adaptation. If it is assumed that the lighting in an environment is constant, then adaptation to this lighting will subdue the active perception of it. Therefore, the lighting in the scene is automatically discarded by retinal mechanisms [517, 622].

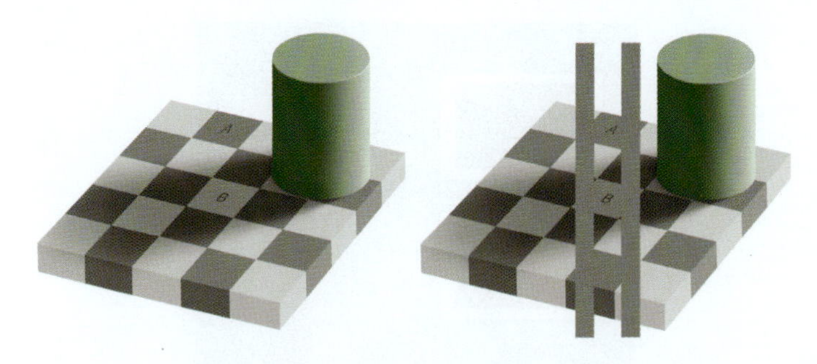

Figure 5.32. The check shadow illusion. Although the two squares marked A and B have the same luminance, they are perceived differently due to the presence of a shadow (left). If the illusion of a shadow is removed, then the two checks appear the same (right). (Images courtesy of Edward H. Adelson, (http://web.mit.edu/persci/people/adelson/checkershadow_illusion.html).)

Though adaptation may aid constancy, it cannot give a full account of color and lightness constancy. Adaptation theories are only valid to explain global changes in lighting conditions; they do not address local conditions such as the change in lighting due to shadows in the same environment. In particular, the simultaneous effects in Figure 5.31 and 5.32 are not well explained by adaptation theories.

5.8.2 Relational Theories

It is a well-known fact that the human visual system is largely sensitive to differences, contrasts, and changes over time. Sensitivity to absolute values is far less dominant than sensitivity to relative values. As relative values can be inferred from the retinal input only, no prior knowledge is required, nor are absolute values important. In addition, the ratio of intensity values received by the retina remains invariant under changing lighting conditions. This invariance is the basis for *relational theories*, which propose that relative values convey information regarding surface reflectance [331].

This notion was first introduced by Hering and formulated in a testable form by Wallach [1196], who suggests that perceived lightness is determined by ratios of light entering the eyes. In Wallach's experiment, a white screen receives light from two different projectors, such that the light from one of the projectors creates a disk on the surface, and the second projector displays a ring around that disk (see Figure 5.33). The disk and the ring are displayed at different intensities.

This set-up is then duplicated, albeit with the ring displayed at a different intensity than the first ring. The intensity of the disk is controllable by the observer.

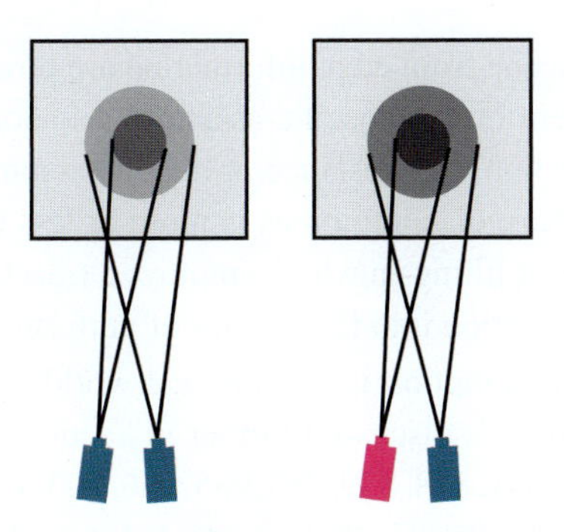

Figure 5.33. The experimental setup of Wallach's experiment. Subjects were asked to modify the light of the circle on the right so that it was perceived to have the same reflectance as the circle on the left (after [1196]).

Participants in this experiment are then asked to adjust the intensity of the second disk such that its reflectance is perceived to be identical to the first disk.

If participants are able to perceive absolute values, then the values of the two disks should be matched. If participants are only able to perceive relative values, then the ratios of the values of the disk and ring should be matched. Wallach's experiment revealed that participants tended to match ratios rather than absolute values. This gives weight to the hypothesis that ratios of light entering the eye are important in the perception of lightness.

5.8.3 Filling-In

Given that relative contrast is important to human vision, the question arises at which spatial scale such computations may occur. In particular, such computations could be carried out on either a local or global scale. Visual phenomena, such as the Cornsweet-Craik-O'Brien illusion imply that this computation may be local (see Section 5.4.4).

In this illusion, the left and right quarters of the image have identical values. Nevertheless, these parts of the image are perceived to be different due to the presence of an edge. This edge is easily perceived, but the gradual ramps directly adjoining the edge contain no C_1-discontinuities, and they are therefore not noticed. As a result, the change in value due to the edge is propagated along the surface; hence, the perceived intensity difference between left and right sides of this illusion.

The propagation of edge information to the interior of surfaces is called *filling-in* [280, 395, 616, 1197]. This process is under-constrained, since many different ways to reconstruct a scene from edge information are possible. It is thought that the human visual system makes specific assumptions about the world to enable a reasonable reconstruction. For instance, it may be assumed that the luminance distribution between sharp discontinuities is more or less uniform. This leads to a computational model of filling-in which employs a (reaction-) diffusion scheme to propagate edge information into the interior of uniform regions [830].

A different useful assumption is that the real world exhibits the well-known $1/f$ image statistic, which is discussed further in Section 15.1 [52, 55, 58, 82, 83, 144, 172, 259, 313, 314, 651, 878, 879, 983, 985, 986, 1001]. Given that low spatial frequencies are attenuated in the edge signal, but are not completely absent, a signal may be reconstructed by boosting low frequencies according to this $1/f$ statistic [228].

5.8.4 Retinex Theory

The filling-in procedure described in the preceding section provides a model for the global perception of intensities derived from edge information. The contrast ratio produced by the edge is then a cue for the relative lightness level at a global scale.

An alternative model for the propagation of locally computed contrast ratios is afforded by the retinex model [647]. In smooth regions, local contrast is small enough to fall below the detection threshold, and thus smooth gradients are discounted. Contrast ratios are then propagated over large regions by multiplication of all the ratios that are above threshold.

This model may be tested by the construction of an appropriate psychophysical test. To this end, Land and McCann constructed 2D displays of uniform rectangular patches. These displays are called *mini-worlds of Mondriaan*. These displays are lit with a light source creating a spatially slowly varying illumination. In this set-up, regions of Mondriaans with low surface reflectance could possibly reflect more light than regions with higher reflectance. Figure 5.34 shows a simplified Mondriaan along with the computations that Land and McCann suggested may be taking place in the HVS.

Through these informal experiments, they were able to show how humans are capable of perceiving relative reflectances of surfaces, despite slowly varying scene lighting and regardless of the absolute values of light in the scene. They proposed the retinex theory, which suggests that the HVS has a densely connected structure of systems. This structure computes products of ratios as described above. Retinex theory is used in several applications, including shadow removal from images, which is discussed in Section 5.12.

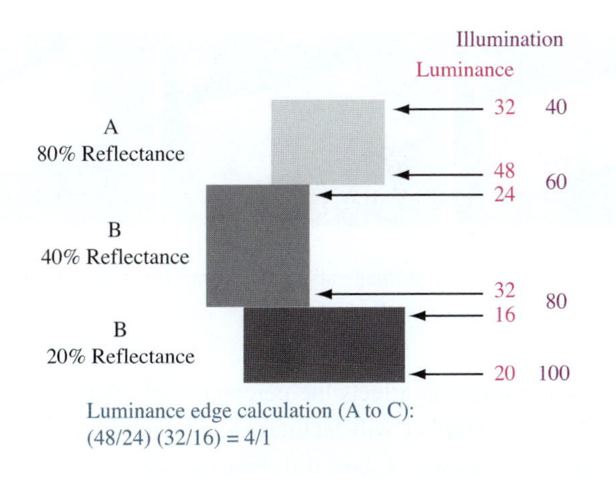

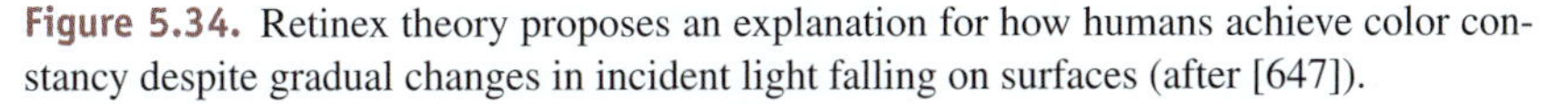

Figure 5.34. Retinex theory proposes an explanation for how humans achieve color constancy despite gradual changes in incident light falling on surfaces (after [647]).

5.8.5 Anchoring

While human vision is predominantly driven by relative values, the absolute impression of how light or dark an environment, or part thereof is, also plays an important role. The Retinex theory offers no model for how humans are able to perceive absolute values. To compute absolute values given the relative values from the Retinex theory, several heuristics may be applied. Finding the absolute values from relative values is termed the *scaling problem*.

An example of a heuristic that may be involved includes the *anchoring heuristic*, which maps the surface with the highest reflectance to white [679]. Evidence for this heuristic can be demonstrated by showing a range of gray values between black and dark gray. The lightest of these will be perceived as white, whereas the black will be perceived as gray [158]. A related example is shown in Figure 5.35, where in the left image, a piece of paper is shown that appears white. When successively whiter sheets are added, the original piece of paper will appear less and less white. This is known as the *Gelb effect* and is an example of a situation where lightness constancy breaks.

In addition to a tendency for the highest luminance to appear white, there is a tendency for the largest surface area to appear white [363]. If the largest area also has the highest luminance, this becomes a stable anchor which is mapped to white. However, when the largest area is dark, the highest luminance will be perceived as self-luminous.

In all likelihood, a combination of both heuristics is at play when anchoring white to a surface in an environment. In addition, the number of different sur-

Figure 5.35. The paper in the left image appears less and less white when whiter paper is added to the scene.

faces and patches in a region affects the perception of white. This is known as *articulation*. A higher number will facilitate local anchoring (see Figure 5.36). Finally, *insulation* is a measure of how different surfaces or patches in a scene are grouped together [363]. The patches surrounding the test square in Figure 5.36, for instance, may be considered a local framework, and the rest of the page may be considered a global framework.

In general, frameworks can be viewed as regions of common illumination. In addition, proximity is an important grouping factor. The latter essentially means that nearby regions are more likely to be in the same framework than distant regions.

Another anchoring rule may be that the average luminance in an environment is mapped to middle gray [456]. Similar rules are used frequently in photography, as well as in tone reproduction [950] (see Chapter 17). However, experimental evidence suggests that anchoring to highest and lowest luminance as well as largest area provides a better model of human visual perception [362, 679]. A computational model of lightness constancy was implemented for the purpose of tone reproduction, as discussed in Section 17.8.

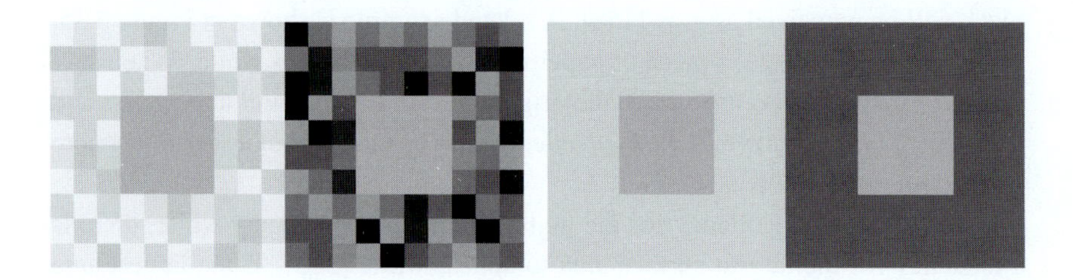

Figure 5.36. The patches in the centers of all four groups have the same reflectance. The right pair of patches look somewhat different because of their surrounds. If the surrounds are made of a larger number of patches, the effect of lightness difference is enhanced (left pair), even though the average intensity of light reaching the eyes from the surrounds in both groups is the same (after [13]).

Once an anchor is found, all other surface reflectances can then be derived from the lightest surface by relative Retinex-like computations [647].

5.8.6 Shadow and Reflectance Edges

Images contain edges, usually at different spatial frequencies. The sharpness of an edge may be a cue to the visual system as to what type of edge it is. In particular, edge sharpness may help in determining if the edge signals a change in surface reflectance or in illumination. The Retinex theory, however, does not distinguish between different types of edges.

To determine whether it matters that an edge is of a particular type, an elegant experiment may be devised where participants match a test patch against a set of reference patches [364]. The test patch is placed in the shade, whereas the reference patches are directly illuminated by a light source. Thus, there exists a shadow edge between the test and reference patches, as shown in Figure 5.37.

Observers participated in the matching experiment twice. During one set of the trials, a mask was placed in front of the observer such that it was not possible to determine the nature of the edge. In the other set of trials, the mask was removed, and the observers were made aware that the edge was a shadow edge.

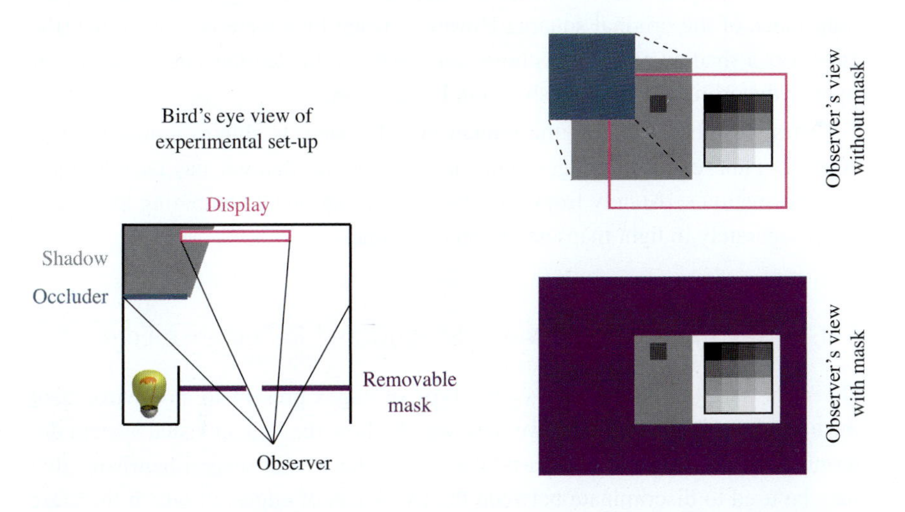

Figure 5.37. The set-up of Gilchrist's experiment to determine the importance of shadow-edge identification for constancy. Subjects were asked to select from the squares on the right, the one that best matched the reflectance of the square on the left. They performed the experiment twice. Once without the mask (top right), so they knew the edge in the center was a shadow edge, and once with the mask (bottom right), so they did not know what type of edge it was (after [364]).

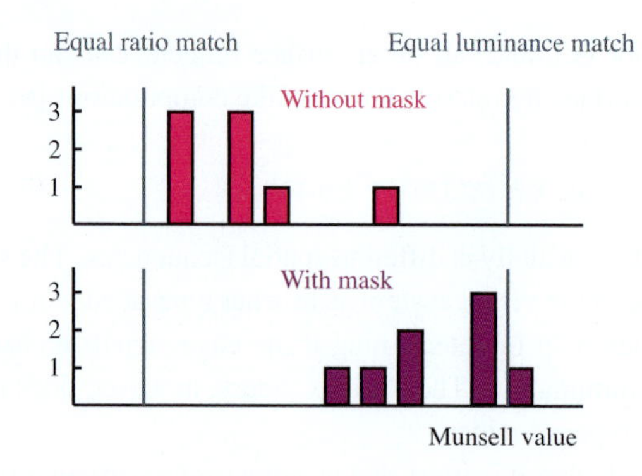

Figure 5.38. The results from Gilchrist's experiment. When subjects were aware that the edge in the middle was a shadow edge, they were able to ignore it and select a square such that its reflectance was almost identical to the original patch. When this information was not available, they chose squares that had the same amount of light reflected off their surfaces as the original square. Thus, lightness constancy failed when the subjects were unaware that an edge was a shadow edge (after [364]).

When the subjects knew that the edge was due to a change in incident light, they chose squares from the candidate squares that were very close to the actual reflectance of the original square. However, when they were not aware that the edge was a shadow edge, they chose squares that reflected the same intensity of light as the original square, as shown in Figure 5.38.

These results imply that the human visual system is able to ignore shadow edges, but not reflectance edges. This, in turn, implies that we may encode lighting information separately from reflectance information by computing local contrasts separately in light maps and reflectance maps.

5.8.7 Distinguishing between Shadow and Reflectance Edges

Given the fact that knowledge of the type of edges present in an environment mediates lightness perception, we may wonder how the human visual system determines that an edge is of a particular type. There exist several heuristics that may be used to discriminate between the two kinds of edges, of which three are discussed here

In the absence of any other distinguishing information, the HVS may decide on the nature of an edge based on how sharp it is. Fuzzy edges are typically due to changes in lighting rather than reflectance. This is the reason why the sharp edges in Wallach's experiment discussed earlier were interpreted as reflectance edges.

Sometimes, a sharp edge may be perceived as a luminance edge. This is due to another heuristic employed by the HVS. If depth information available to the HVS shows that two regions are not coplanar, the edge dividing them, even if it is sharp, is interpreted as a shadow edge. Such edges explain the different amounts of light falling on the different planes of an object, due to its shape and position.

The Retinex theory implies that the HVS keeps track of ratios across edges. If these edges are caused by changes in reflectance, then the maximum ratio is obtained if the edge is adjoined by a black and a white surface. A white surface typically does not reflect more than 90 percent of the incident light, whereas black reflects no less than about 10 percent of the incident light [868]. Therefore, if a ratio across an edge is more than 9:1, then the edge is most likely caused by changes in lighting.

5.9 Color Constancy

The chromatic equivalent of lightness constancy is called *color constancy*. Color constancy is not studied in as much detail as lightness constancy, and consequently, less is known about how color constancy is achieved by humans. To achieve color constancy, on the basis of the photoreceptor input, reflectance and illuminance need to be disentangled. With three types of photoreceptor, there are six unknowns corresponding to the reflectance and illuminance for each of the three response functions.

As with lightness constancy, it is assumed that chromatic adaptation plays a partial role in discounting effects such as the chromatic content of the lighting in the scene [1227, 1300]. However, the higher-dimensionality of the problem affords additional assumptions to be made about the environment and, thus, aids in the interpretation of the environment.

5.9.1 Color Constraints

The environment around us typically contains certain unvarying features which make it easier to use heuristics to infer information from the scene. One of these is consistency of light within a single scene. Usually, scenes are lit by a specific spectrum of light from a single type of light source. This implies that the intensity of the light may vary across a scene, but not its spectrum [440]. A recent proposal exploits this by assuming that the illumination can be modeled by blackbody radiation, using Planck's law to solve the color constancy problem [583].

Another important constraint is that there is only a limited number of different types of light spectra in typical scenes. The spectrum of light in an outdoor environment is different depending on whether there are clouds, what time of day it is, etc. However, these changes are relatively minor. For indoor environments, a similar argument can be made. Here, there are several different kinds of light sources, but their variation is not significant.

Finally, only a limited number of different surface reflectances are typically encountered [233]. Thus, the limited variability in lighting and surface reflectance may be exploited by the human visual system to achieve color constancy.

5.9.2 Shadow and Reflectance Edges

Color is a useful attribute for discriminating between shadow and reflectance edges [868]. An interesting difference to note between shadow and reflectance edges is their change in spectrum across the edges. Across a shadow edge, there is typically a sharp change in intensity, but not in the shape of the spectrum. Across a reflectance edge, on the other hand, there is typically a change in the shape of the spectrum (see Figure 5.39).

Thus, the absence or presence of changes in spectral composition across an edge may be exploited to determine the edge's type [982]. Given knowledge of the intensity at two different wavelengths at either side of the edge, the type of edge may be inferred. The two known wavelengths on either side define both slopes as well as spectral cross-points.

Figure 5.40 shows the four possible configurations. The diagram at the top left illustrates the case in which the measured intensity of wavelength 1 has increased across the edge while intensity of wavelength 2 has decreased. Since there is

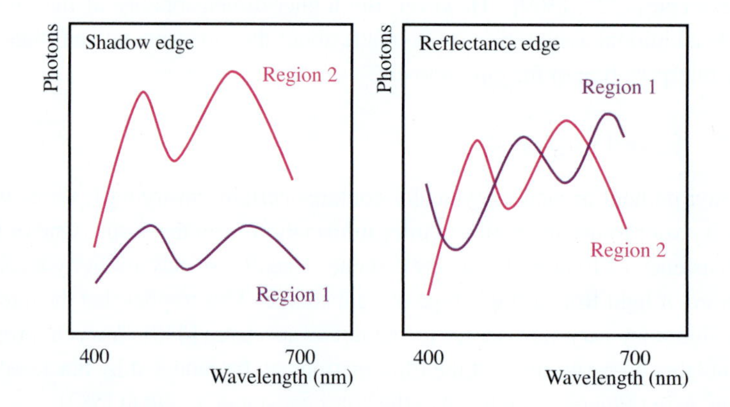

Figure 5.39. Possible changes in spectra across shadow and reflectance edges (after [868]).

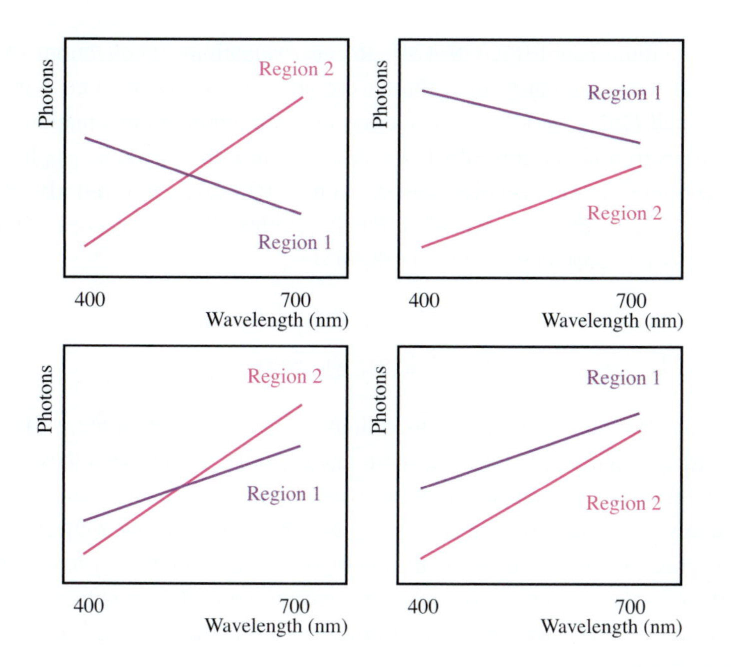

Figure 5.40. The four different cases in the edge-classification algorithm proposed by Rubin and Richards [982].

incomplete information about the spectrum, it is approximated with a line joining the two known measurements. In such a situation, the slopes intersect, indicating that the spectra change not only in intensity across the edge, but also in shape. This spectral crosspoint indicates the presence of a reflectance edge.

If the shape of the spectrum has in fact changed across the edge, then the slopes of the spectra on either side of the edge may also be of different signs, as shown in the diagram on the top right. Opposite slope signs also indicate the presence of reflectance edges.

Thus, if either (or both) of these features are present across an edge, as in all cases except for the diagram at the bottom right, the edge is classified as a reflectance edge. If neither are present, as in the diagram at the bottom right, then the shape of the spectra is assumed to be similar enough on either side of the edge to classify it as a shadow edge.

5.9.3 Applications of Color Constancy

Algorithms modeling color constancy attempt to retrieve the perceived surface color of an object projected onto the photoreceptors or camera sensor, independent of the illuminant [317, 319, 474]. Recent evaluations of collections of computational color-constancy algorithms were presented by Barnard et al [66] and Hordley and Finlayson [487]. Such algorithms are important in situations where the illumination is not controlled, for instance, in remote-sensing applications, guiding robots, as well as color reproduction [1203, 1204]. It also finds practical application in matching paints [474]. Finally, Retinex theory has been adapted to solve the color constancy problem [738, 965].

5.10 Category-Based Processing

Humans tend to classify colors into various categories. Each of these categories is accorded a name, and all the colors in that category, even though they generate different response patterns in the retina, may be called that name. Names of some commonly used categories include red, green, blue, and yellow. Objects such as leaves, grass, and unripe fruit are all called green, even though they have different reflectance spectra. This categorization may be viewed as dividing the three-dimensional color space into several different subsets and giving each subset a different name.

Given our ability to categorize color, several questions arise. When we observe a color, why do we immediately classify it as a specific category and give it a label? Why have we divided the space of colors into various regions with labels? Is it because we were taught these labels in our childhood and, therefore, learned to classify all colors according to these labels? Or is there some physiological reason for these specific categories? Various experiments have been conducted to study the nature of color categorization, which indicate that the physiology of the human visual system determines to a large extent how we categorize color. We will discuss these in the remainder of this section.

5.10.1 Color Naming

Before any cross-cultural research had been performed, color naming was thought to be an example of *cultural relativism*, which is the theory that a person's culture influences his or her thoughts and experiences. It implies that two people from different cultures think about the world differently. In linguistics, this hypothesis is named the Sapir-Whorf hypothesis after two linguists who made it famous, Edward Sapir and Benjamin Lee Whorf.

In terms of color naming, cultural relativism implies that a name for a particular kind of color emerges in a language as its need develops in the society speaking that language. One example of this theory is that the Inuit, whose surroundings are predominantly covered with snow, and who thus have a greater need to distinguish between different kinds of snow, have more than a dozen names for it. Cultural relativists further believed that this was the only factor influencing the naming of colors in different languages, and that human physiology played no part in determining it. Hence, there would be no correlation in the division of color space into various categories across different languages.

The first cross-cultural research to study the nature of color categorization was performed by Brent Berlin and Paul Kay in 1969 [87]. This work was prompted by their intuitive experience in several languages of three unrelated stocks. They felt that color words translated too easily among various pairs of unrelated languages for the relativity thesis to be valid.

Basic color terms are names of colors with the following properties. First, they are single lexical terms such as red, green, or pink; not light-green or greenish-blue. Second, they refer to the color of objects, rather than objects themselves. Therefore, colors such as bronze and gold are not valid color terms. Third, the colors are applicable to a wide range of objects, so that colors like blonde are not valid. Fourth, the colors are in frequent use.

Berlin and Kay use an array of Munsell color chips (see Section 8.9.1) to determine the references of basic color terms of a language. For each basic color term, participants in the experiment are asked to specify the best example of the color, as well as the region of chips on the array that can be called by the same name as the basic color term. Figure 5.41 shows a copy of the Munsell Chart that was shown to the participants.

Berlin and Kay experimentally examined 20 languages directly and 78 more through literature reviews. They discovered two interesting trends. First, there

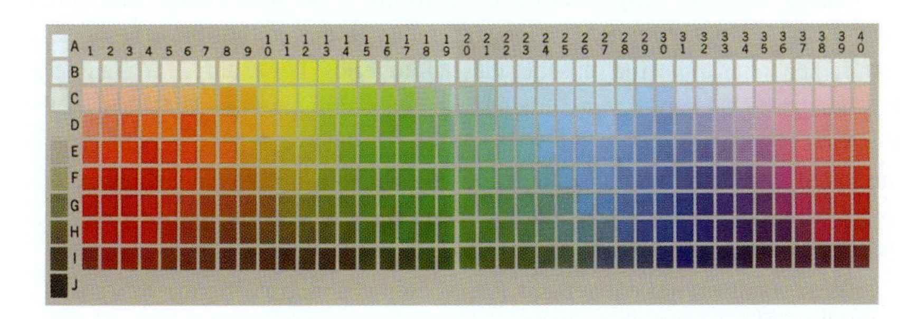

Figure 5.41. An illustration of the Munsell Chart that was shown to participants in the work by Berlin and Kay [87]. (See also the World Color Survey, http://www.icsi.berkeley.edu/wcs/.)

Type	Basic terms	White	Black	Red	Green	Yellow	Blue	Brown	Pink	Purple	Orange	Gray
1	2	●	●	○	○	○	○	○	○	○	○	○
2	3	●	●	●	○	○	○	○	○	○	○	○
3	4	●	●	●	●	○	○	○	○	○	○	○
4	4	●	●	●	○	●	○	○	○	○	○	○
5	5	●	●	●	●	●	○	○	○	○	○	○
6	6	●	●	●	●	●	●	○	○	○	○	○
7	7	●	●	●	●	●	●	●	○	○	○	○
8	8	●	●	●	●	●	●	●	●	○	○	○
9	8	●	●	●	●	●	●	●	○	●	○	○
10	8	●	●	●	●	●	●	●	○	○	●	○
11	8	●	●	●	●	●	●	●	○	○	○	●
12	9	●	●	●	●	●	●	●	●	●	○	○
13	9	●	●	●	●	●	●	●	●	○	●	○
14	9	●	●	●	●	●	●	●	●	○	○	●
15	9	●	●	●	●	●	●	●	○	●	●	○
16	9	●	●	●	●	●	●	●	○	●	○	●
17	9	●	●	●	●	●	●	●	○	○	●	●
18	10	●	●	●	●	●	●	●	●	●	●	○
19	10	●	●	●	●	●	●	●	●	●	○	●
20	10	●	●	●	●	●	●	●	●	●	○	●
21	10	●	●	●	●	●	●	●	●	○	●	●
22	11	●	●	●	●	●	●	●	●	●	●	●

Table 5.1. The 22 possible subsets of the 11 basic color terms.

appear to be just 11 universal basic color terms.[5] This small number is due to the fact that color categories are shared by languages. The 11 basic colors are: white, black, red, green, yellow, blue, brown, purple, pink, orange, and gray.

Second, there are only 22 different subsets of these basic color terms from a total of $2^{16} - 1$ possibilities. If a language does not have all of these basic color terms, then it has a specific subset of these color terms. For example, if a language has only two basic color terms, then these color terms are the equivalent of the English color names white and black. Table 5.1 shows the possible subsets that are found across all these languages. Based on this result, it is hypothesized that as languages develop, they acquire the basic color terms in a specific order that can be termed evolutionary.

These results indicate that there are physiological reasons for the very specific evolution of languages in terms of color. It is worth noting that colors important in Hering's opponent theory (red, green, blue, yellow, white, and black) are also

[5]This number was increased to 16 by later studies [584].

some of the first colors to be recognized in languages as separate categories. On the basis of these results, the Sapir-Whorf hypothesis has been severely undermined. There is now concrete evidence for the existence of semantic universals in the domain of color vocabulary.

Research on color naming continues to this day [391, 945], and the reader is encouraged to visit the World Color Survey at www.icsi.berkeley.edu/wcs/.

5.10.2 Focal Colors and Prototypes

In the course of their research, Berlin and Kay also discovered the existence of best examples, or *focal colors*, in color categories. To find out which basic color terms best describe any particular color, they use two different approaches. In one approach, they obtain information about the boundary of color terms by showing participants various colors and asking them to point out those colors that lie on the boundary of a specified color term. In the second approach, they show colors and ask participants to identify the colors that best describe a specified color term. Participants invariably give faster and more consistent responses in the second approach. This implies that color categories are structured around focal colors, rather than around colors at the boundaries of color terms.

This idea received more support from several experiments conducted by Eleanor Rosche and colleagues [450, 978]. She performed a number of experiments with members of the Dani tribe in New Guinea as her participants. The Dani have just two basic color terms in their language. In one experiment, the Dani are shown different colors and later asked to recognize them from a set of colors. The Dani point out focal colors faster and more reliably than those lying at the boundaries of color categories. In a similar experiment, the Dani's learning ability is tested in terms of colors. They learn focal colors more quicker and more accurately than colors at the boundaries of categories. These and other results form mounting evidence that focal colors play a crucial role in our internal representation of color categories.

In light of these discoveries, Kay and McDaniel [584] propose a model for color naming which is based on fuzzy set theory. They argue that since people are able to find and remember focal colors more easily than boundary colors, focal colors are better members of their color categories. Thus, each category can be seen as a set, such that membership of a particular color is defined by a number ranging from 0 to 1, where 1 indicates that the color is a focal color, and 0 indicates that the color does not belong to that specific color category. A color may be a member in more than one category. Bluish-green for instance, would be a member of both blue and green categories. The membership of such a color in these categories is represented by fractions between 0 and 1.

This model predicts that it is easier to reliably detect and remember focal colors. A focal color clearly belongs to one category, in which its membership value is high. For colors on the boundary, belonging to more than one category, a decision has to be made regarding which category they belong to. This could possibly make the task of detection and recognition less reliable.

5.11 Color Anomalies

Individuals with normal color vision are called *trichromats*, referring to the triplet of cone types present in a normal healthy retina. About 8% of all males, however, suffer from some color vision deficiency [128]. Common visual anomalies affecting color vision fall in two dominant classes. The first class is *anomalous trichromacy* and is characterized by the presence of one or more photopigments that have a peak sensitivity different from normal opsins [30–33].

There are three different types of anomalous trichromats. In a protanomalous trichromat, the absorption spectrum of the L-Cone (erythrolabe) is displaced. In a deuteranomalous trichromat, the absorption spectrum of the M-Cone (chlorolabe) is displaced. Finally, in a tritanomalous trichromat, the absorption spectrum of the S-Cone (cyanolabe) is displaced. These displacements of cone photopigments from their optimal positions result in deficient color discrimination [1019].

Apart from displaced absorption spectra, anomalous trichromats may have weaker absorption spectra, leading to an anomaly called *color weakness*. Such individuals require a more intense stimulus to perceive the same color.

The second class of anomalies is called *dichromacy*. Dichromatic individuals perceive the world as mixtures of only two basic colors. This anomaly is also related to the photopigments of the cones in the eye. A protanope is missing erythrolabe (L-cones are affected), a deuteranope is missing chlorolabe (affecting the M-cones), and a tritanope is missing cyanolabe (affecting the S-cones). An overview is given in Table 5.2.

The terms *protan*, *deutan*, and *tritan* refer to the affected photo-pigment. Protans and deutans tend to confuse reds and greens. These defects are usu-

	Red-green defects		**Yellow-blue defects**
Photopigment status	Protan defect	Deutan defect	Tritan defect
Dichromat	Protanope	Deuteranope	Tritanope
Anomalous trichromat	Protanomalous trichromat	Deuteranomalous trichromat	Tritanomalous trichromat

Table 5.2. A table listing the different types of dichromacies and anomalous trichromacies.

ally, but not always, inherited. Tritans tend to confuse blues and yellows, a defect that is almost always acquired. Figure 5.42 and Figure 5.43 show images as they appear to a normal trichromat, as well as how they might appear to the different types of dichromats. These images were created with Vischeck, a freely available program to treat images so that they appear as they would to people with different color anomalies. Computer simulation of color deficiencies is further discussed by Brettel et al [129].

Achromotopsias are rare anomalies in which the affected person behaves largely as a monochromat [3, 910]. As the name implies, monochromats see the world as shades of a single color. The most common form of achromatopsia is rod monochromacy, in which individuals have no cones. Rod monochromats do not show the Purkinje effect (see Section 6.1), have achromatic vision, low visual acuity, high sensitivity to bright lights, nystagmus (involuntary eye movement), and macular dystrophy. It is a color deficiency that occurs in about 1 in 30,000 [236, 329].

Figure 5.42. An image as it would be perceived by normal trichromats (top left), protanopes (top right), deuteranopes (bottom left), and tritanopes (bottom right); Insel Mainau, Germany, August 2005. (Images created with the freely available Vischeck program (www.vischeck.com)).

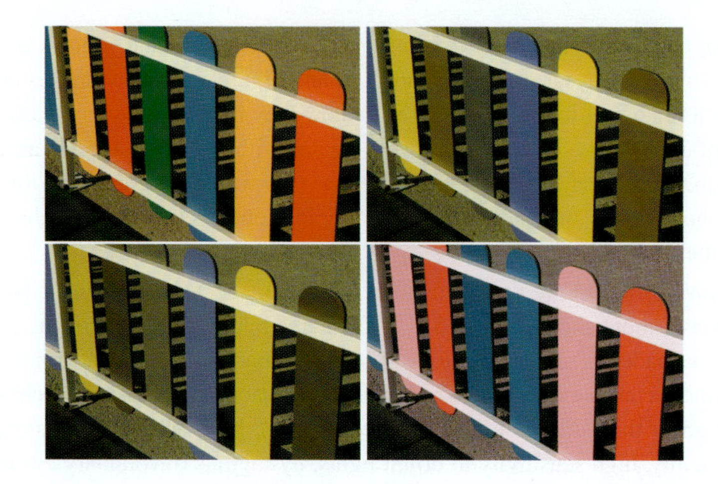

Figure 5.43. An image as it would be perceived by normal trichromats (top left), protanopes (top right), deuteranopes (bottom left), and tritanopes (bottom right); Zaragoza, Spain, November 2006. (Images created using Vischeck (www.vischeck.com).)

Another form of achromatopsia is cone monochromacy, which occurs in about 1 in one-hundred million humans. Here, the discrimination of colored lights is lost if their luminances are equalized. Visual acuity is unaffected. Although the name suggests that cone monochromats have only one type of functioning cone, experiments have shown that it is more likely of cortical origin [316, 329, 361].

Further, *chromatopsias* are color anomalies that may not be termed color defects, as they do not result in decreased ability to discriminate colors. They simply present a distortion of color vision, much like looking through a colored filter. As an example, such anomalies may be caused temporarily due to cataract extraction. A particular kind of cataract, known as a nuclear cataract, behaves like a yellow filter, thereby absorbing blue light. Since the individual with the cataract has grown accustomed to having less blue light incident on the retina, he or she becomes extremely sensitive to blue stimuli over time. When the cataract is finally removed and the individual is exposed to a "bluer" spectrum, he or she usually reports a blue cast on the whole scene. As might be expected, this perception fades away with time.

5.11.1 Spectral Sensitivity

There are various ways in which the differences in color perception by people with color anomalies may be visualized and compared with the perception of normal trichromats.

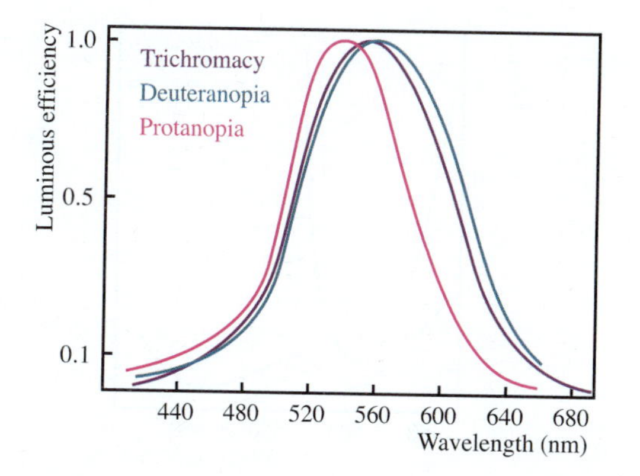

Figure 5.44. Achromatic response function for normal trichromats, protanopes, and deuteranopes (after [516]).

One such mechanism was introduced in Section 5.3.3 and quantifies an observer's spectral sensitivity through a nulling method [516]. When the experiment is performed by an observer with a color anomaly, the different spectral sensitivity of this observer may be compared with that of a normal observer, resulting in the achromatic response functions shown in Figure 5.44. The protanope's response function is significantly different from the trichromat's and deuteranope's response function. Figure 5.45 shows the theoretical chromatic and achromatic responses of a protanope and a deuteranope. In this case, there is no perception of reds and greens, leading to only a single chromatic response.

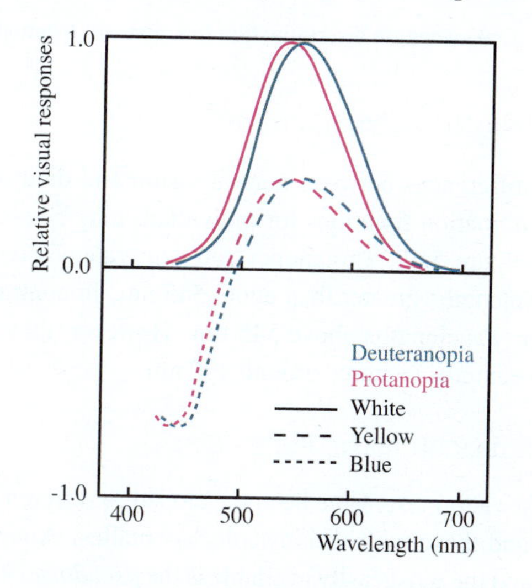

Figure 5.45. Spectral response for dichromats with red-green defect (after [516]).

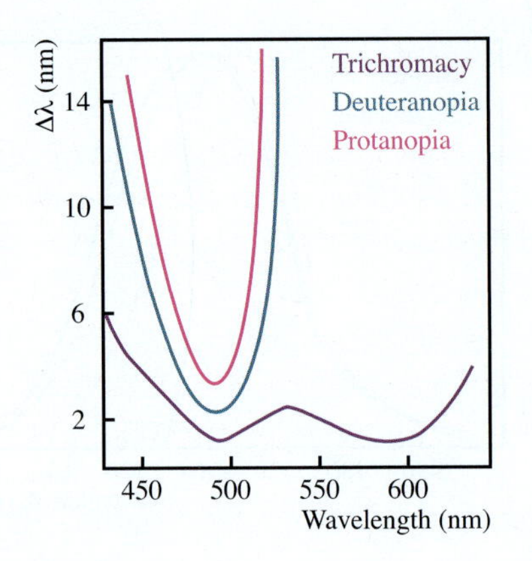

Figure 5.46. Wavelength-discrimination functions for protanopia and deuteranopia [903].

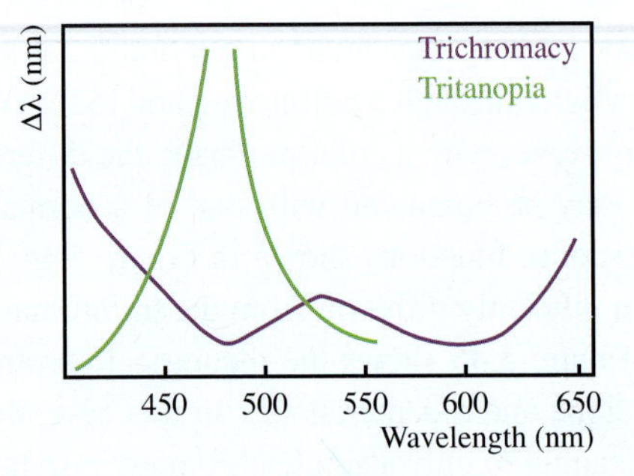

Figure 5.47. Wavelength discrimination functions for tritanopia [1254].

5.11.2 Wavelength Discrimination

To visualize the differences between normal vision and dichromatic vision, the wavelength-discrimination functions for individuals may be plotted, as shown in Figure 5.46 and Figure 5.47. Protanopes and deuteranopes respond like monochromats for wavelengths greater than about 545 nm. In contrast, tritanopes discriminate well for wavelengths above 545 nm. However, they behave as monochromats for wavelengths centered around 490 nm.

5.11.3 Pseudoisochromatic Plate Tests

A variety of color vision tests have been developed to differentiate between normal trichromats and those with various color anomalies. Among these, the simplest to perform and the most easily available is the *pseudoisochromatic plate test*, also known as the *Ishihara test* [531] . This test consists of a booklet of different plates. Each plate is drawn such that its pattern is only detectable by observers with normal color vision.

The Ishihara test set exists in two variations, one consisting of 24 plates and one consisting of 34 plates. If administering an Ishihara test, it is recommended to use plates from the larger set, as the 24-plate set contains relatively few reliable plates. Both sets contain two groups of plates. The first group is suitable for numerates, containing numerals painted in a different color from the background. The second group is suitable for innumerates, contains snaking patterns, but is rarely used because the results are much more difficult to assess, and these plates are less reliable.

The plates containing numerals can be classified into four colorimetrically different groups:

Transformation plates cause anomalous color observers to give different answers than those with normal vision. Plates 2–9 are in this group (of the 34-plate set).

Disappearing digit plates, also known as vanishing plates contain numerals that can only be observed by those with normal color vision. For color-deficient viewers, these plates appear as a random dot pattern. These are Plates 10–17.

Hidden digit plates are designed such that anomalous observers will report a numeral, whereas normal observers do not see any pattern. Plates 18–21 are in this group.

Qualitative plates are designed to assess the severity of color blindness and separate protan from deutan color perception. These are Plates 22–25.

Of these four types, only the transformation and disappearing plates are reliable and should be used to test for color blindness. Figure 5.48 shows examples of all four plate types.

It should be noted that the Vischeck simulation applied to the hidden-digit plate does not reveal the number that was intended. Hidden-digit plates are typically not very reliable and, in addition, the Vischeck algorithm may not be optimally matched to this particular stimulus. It is included for completeness.

A typical color plate test procedure begins by showing the observer Plate 1 (see Figure 5.49), which is a demonstration plate. If the observer does not give the correct answer for this plate, then the remainder of the test should not be administered.

To interpret the results, the false positives and false negatives should be taken into consideration. The false-positive rate varies significantly with age, and to a

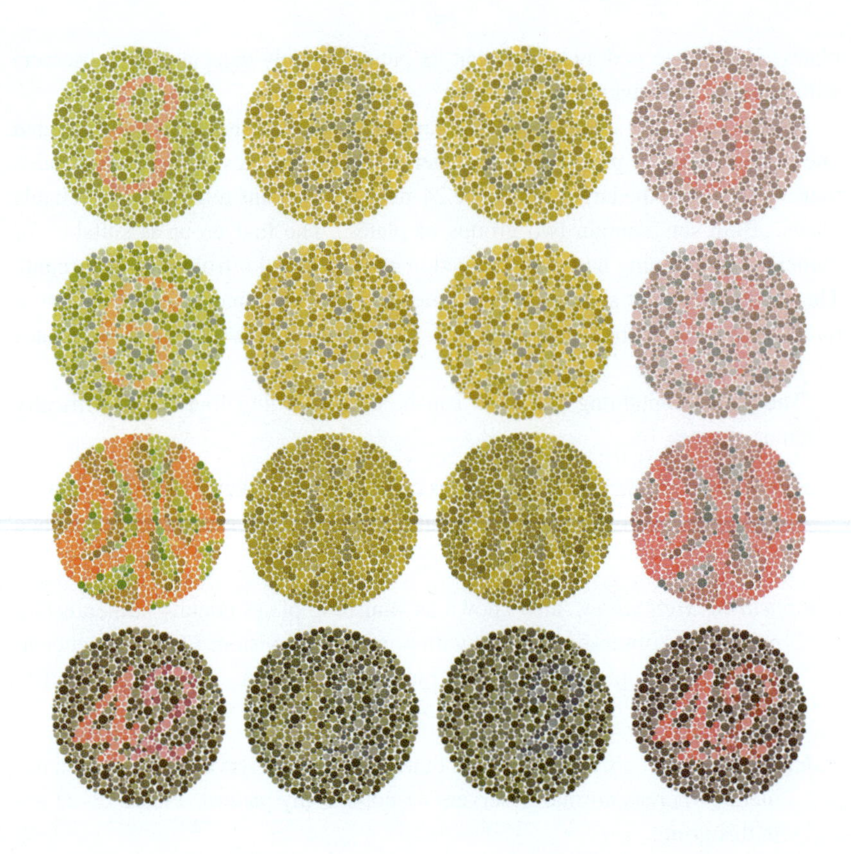

Figure 5.48. Ishihara test plates from the 24-set to determine color blindness. From left to right: Ishihara plate, same plate as seen by a protanope, a deuteranope, and a tritanope. Top to bottom: a transformation plate, a disappearing plate, a hidden-digit plate, and a qualitative plate. Note that due to color reproduction issues as well as viewing conditions, these reproductions are not suitable to determine color blindness. (Test plates from [531]. Color anomalies simulated using Vischeck (www.vischeck.com).)

lesser extent with education [638]. The false-negative rate depends on the type of anomalous vision, but tends to be low. The question then is where to place the cut-off, i.e., how many errors before we can decide that someone is color normal?

The answer depends on many factors. For a male population, a good balance between false positives and false negatives is achieved by placing the cut-off between four and six errors, provided Plates 2–17 are used (in particular excluding the hidden-digit plates).

For females, the occurrence of anomalous color vision is lower, meaning the cut-off point would have to be chosen differently, and even then the interpretation

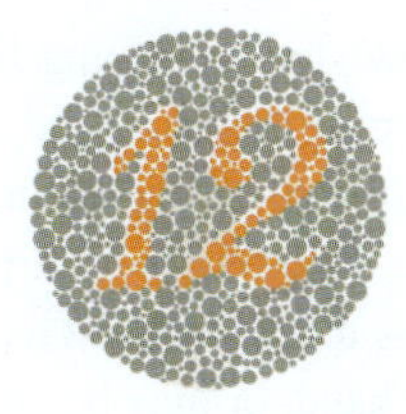

Figure 5.49. Demonstration test. This numeral should be visible by both color blind observers as well as color normal observers. (Test plate from [531].)

of the results would be more complicated. In addition, as color deficiencies in females may have a somewhat different nature, the Ishihara test is not recommended as a stand-alone test, but should always be augmented with other tests.

If the hidden-digit plates are included, then the interpretation of the results also becomes much more difficult. The results obtained with this group of plates varies significantly with age and with degree of anomaly.

In summary, the Ishihara test is suitable for quick and convenient screening of individuals. For any one suspected to have anomalous color vision, further tests should always be carried out, as failing the Ishihara test does not automatically mean that the observer is color deficient. In addition, the Ishihara test is not suitable for determining the degree of color deficiency.

5.12 Application: Shadow Removal from Images

Retinex theory, introduced in Section 5.8.4, is the basis for several applications (see [332] and [340] for implementations). For instance, it has been used successfully for dynamic range reduction [553, 775], the topic of Chapter 17. Retinex theory has also been used in algorithms addressing the issues of color correction [64, 553, 737], gamut mapping [759], and image enhancement [774].

In this section, we will discuss an algorithm that employs Retinex theory to remove shadows from images [320]. Shadow removal in an image is useful for a variety of tasks. Several tasks, such as image segmentation and object tracking and recognition, might perform better on an image from which the effects of scene lighting have been removed. Often these algorithms are fooled by the presence of shadow edges which they mistakenly respond to as reflectance edges. If shadows are removed first, then image data will conform more closely to assumptions made by these algorithms. Removing lighting effects from an image may also be considered as the first step in scene relighting, a task in which images of a scene are re-rendered with modified lighting.

Retinex theory explains how slowly varying lighting may be discounted by the human visual system and may, therefore, be used to remove some lighting effects from an image. The algorithm presented in this section adapts this theory so that it can be used to remove the effects of sharp variations in lighting as well. Location of shadow edges needs to be specified as input into the algorithm. This information may be marked by hand, or may be computed automatically in some cases, by using existing techniques [317].

In the following, we first discuss an algorithm for Retinex theory (Section 5.12.1) and then detail the changes that are made to this algorithm to enable the removal of shadows (Section 5.12.2).

5.12.1 Retinex Algorithm

Given an image as input, the general concept of Retinex theory, as discussed in Section 5.8.4, may be implemented in several different ways [340, 553, 647–649, 737]. Land's path-based algorithm is one such approach that lends itself easily to adaptation, so that shadows may be removed [647].

Land proposed that a region's reflectance, relative to the region in the scene with the highest reflectance, may be obtained by computing the product of ratios along a path starting from the region with the highest reflectance and ending at the region whose relative reflectance is being computed. For instance, if the path starts from pixel p_1, goes through pixels p_2, p_3, and p_4, and ends at p_5, then the relative reflectance $\Theta_{1,5}$ at p_5 with respect to p_1 may be computed in the following manner:

$$\Theta_{1,5} = \frac{p_5}{p_1} = \frac{p_2}{p_1}\,\frac{p_3}{p_2}\,\frac{p_4}{p_3}\,\frac{p_5}{p_4}. \tag{5.12}$$

If products of ratios are computed in log space, ratios become subtractions and products become summations. Therefore, the relative reflectance becomes

$$\Theta_{i,j} = \exp\left(\sum_{k \in P, k<j} \log p_{k+1} - \log p_k\right), \tag{5.13}$$

where $\Theta_{i,j}$ is the reflectance of pixel j with respect to pixel i and P is the path from pixel i to pixel j. This computation requires knowledge about the location of the pixel with the highest reflectance. Since the value at a pixel is affected by its reflectance, as well as the intensity of incident light, it cannot be used to determine the pixel with the highest reflectance. Such a pixel may only be found with an exhaustive evaluation of all possible sequential products, comparing the results and selecting the lowest value. This value corresponds to the ratio of pixels with the highest and lowest reflectance in the scene.

Land suggested that an approximation to the above procedure may be obtained without an exhaustive search by computing products of ratios along several random paths ending at the pixel whose relative reflectance is being computed. Each path is assumed to start from a pixel with a 100% reflectance. Since this is obviously incorrect for most randomly selected starting pixels, often the product of ratios becomes greater than 1 somewhere along the path. This occurs every time the starting pixel's reflectance is lower than that of the current pixel on the path. In such cases, the ratio computation is set to 1 and started afresh from the current pixel. This pixel is now considered the new starting pixel of the path.

If several such paths are computed for each pixel, they produce different relative reflectances, since the starting pixel for each reflectance computation is different. However, averaging these relative reflectances gives a reasonable estimate of the final relative reflectance of each pixel:

$$\rho_j = \frac{1}{N}\sum_{i=1}^{N} \Theta_{i,j}, \tag{5.14}$$

where ρ_j is the final relative reflectance of pixel j and N is the number of paths used in the computation.

The ratios computed between different adjacent pixels may give some indication of whether the pixels are on opposite sides of an edge or belong to the same uniform region. If the pixels have similar values, then the ratio will be close to 1. In log space, this is equivalent to a value close to 0. Log ratios significantly different from 0 will mean that an edge exists between the pixels. To eliminate the effects of slowly varying lighting, a threshold t may be applied to the ratios. Those ratios that are close to 0 will be set to 0, as shown below:

$$\Theta_{i,j} = \sum_{k \in P, k < j} T(dp_k), \tag{5.15a}$$

$$dp_k = log p_{k+1} - log p_k, \tag{5.15b}$$

$$T(dp_k) = \begin{cases} 0 & \text{if } |dp_k| < t, \\ dp_k & \text{otherwise.} \end{cases} \tag{5.15c}$$

5.12.2 Shadow Edge Removal

So far, the algorithm for Retinex will only discount small changes in illumination. If shadow edges are to be removed using the same technique, a higher value for the threshold t has to be set. However, a higher value for t not only discounts shadow edges, but some reflectance edges as well, giving rise to erroneous relative reflectances. Therefore, some mechanism is required to distinguish between shadow and reflectance edges, so that the algorithm can react differently to them.

A classification of edges can be given as input to the Retinex algorithm. This can be done by providing a hand-marked binary image S as input, in which the shadow edges are marked white and the rest of the image is black. Alternatively, the binary image may be computed automatically in some cases, using existing techniques [317]. If a shadow edge is encountered, then the ratio at that point can be set to 0 in log space. The function T now becomes

$$T(dp_k) = \begin{cases} 0 & \text{if } |dp_k| < t \text{ or } S_k = 1, \\ dp_k & \text{otherwise.} \end{cases} \tag{5.16}$$

By incorporating this change into the Retinex algorithm, sharp changes in lighting will also be discounted, producing a representation of the scene approximately devoid of lighting effects [320].

5.13 Application: Graphical Design

Understanding the different forms of color deficiency is helpful in graphical design tasks, including the design of logos, brochures, and web pages. In this section, we briefly list some of the design rules that may help to create safe color combinations, i.e., designs that can be distinguished by both color normal and color-deficient observers.[6]

There are certain color combinations that are preferred for the color of text printed on a background and there are others that should be avoided, for various reasons. In particular, black letters on a white background are always good. Black letters on other backgrounds should mostly be avoided. In particular, black on red or green backgrounds will be seen by some as black on black. On the other hand, black on a yellow background is typically a good combination.

White lettering on a black background is also always acceptable. However, other colors on a black background should be avoided. Blue tends to be a better color than others, but it should not be used on a black background for fine detail, since visual acuity for blue colors is worse in all types of color vision.

Mixed backgrounds should be avoided, as masking effects will make text difficult to read for anyone, but especially for color-deficient viewers.

Saturation also plays a role in graphical design for color-deficient viewers. In general, more saturated colors will be better. However, for green dichromats, red will be seen as yellow, and may therefore largely vanish against a white background. To accommodate green dichromats, it would be best to ensure that the

[6] See www.firelilly.com/opinions/color.html.

red is displayed against a background that would yield good contrast with yellow as well. Making the text large and bold will also be beneficial.

A saturated red right next to a saturated blue is a special case. The wavelengths of these colors are at opposite ends of the visible spectrum, and this causes the focal length for lenses (including the ones in human eyes) to be distinctly different for these colors. This may leave the impression that the text is not stationary, but moves around with eye movements. This is due to continuous refocusing of the eye for this color combination, even though the eye-object distance is identical. This effect occurs for color-normal observers.

Figure 5.50. From left to right, top to bottom: input image, simulation of a deuteranope, daltonized input image, and a simulation of how a deuteranope may perceive the daltonized input image; Insel Mainau, Germany, August 2005. (Simulations created using Vischeck (www.vischeck.com/daltonize).)

While the above ought to help with choosing text and backgrounds, it is not so easy to adjust images so that they can be viewable for all web users. However, it is possible to adjust images so that color-deficient observers find it easier to recognize objects in them. One approach is to increase the red and green contrast in the image, as many color deficient viewers have residual red-green sensitivity. Second, red-green variations can be mapped to other image attributes, such as the yellow-blue channel, or luminance. The combination of these image-processing techniques is called *daltonization* [261]. An example of such a simulation is shown in Figure 5.50.

5.14 Application: Telling Humans and Computers Apart

Visual illusions afford the possibility to detect the difference between humans and computers, for instance in Internet-based applications where a webpage should be visible to humans, but not to, say, malicious web crawlers. A commonly used technique is image-based, whereby the viewer is presented an image of some text and is asked to type in the text they observe [18, 19]. Geometric distortions and random patterns can be added to the image to make it more difficult for a computer to decipher the text. However, such systems can be defeated by clever object-recognition algorithms [790].

Gavin Brelstaff and Francesca Chessa have argued that it may be possible to use human visual illusions, which may be designed such that the complicated human visual machinery is required to decipher a message encoded within the display [125]. Given the huge complexity of the human visual system, and the fact that much of it has not yet been explored, it may be very difficult for computational models to defeat such approaches.

5.15 Further Reading

Margaret Livingstone's book is a good source linking human vision with art [695], as is Semir Zeki's book [1290]. Good general resources for the material discussed in this chapter are Davson [236], Palmer [868], and Marr [742]. Further reading on lightness perception can be found online at

- http://www-psych.rutgers.edu/~alan/theory3/ [363]

Interesting web sites with further optical illusions can be found at:

- http://www.echalk.co.uk/amusements/OpticalIllusions/illusions.htm
- http://www.michaelbach.de/ot/
- http://www.ritsumei.ac.jp/ akitaoka/index-e.html
- http://www.viperlib.org/

Color deficiencies are described in Donald McIntyre's *Colour Blindness: Causes and Effects* [766].

Many of the plots in this book were derived from data publicly available on the Internet. In particular, we refer to the Munsell Color Science Laboratory (http://www.cis.rit.edu/mcsl/online/) and the CVRL color and vision database (http://www.cvrl.org), which comes courtesy of the Institute of Ophthalmology, University College London.

Part II
Color Models

Chapter 6
Radiometry and Photometry

Photons are the carriers of optical information. Light, emitted by display devices, light sources, or the sun, reaches our eyes, usually after reflection off objects in a scene. After transduction, a signal is transmitted to the brain which is interpreted, and dependent on the circumstances, this may provoke a response from the observer.

Since light is the carrier of visual information, a fundamental activity related to light is its measurement. Light measurement exists in two forms. The first, *radiometry*, is the science of measuring all optical radiation whose wavelengths lie between approximately 10 nm and 10^5 nm (see Figure 2.4). This region includes ultraviolet, visible, and infrared radiation. We only consider *optical* radiation, i.e., radiation that obeys the principles of optics.

Whereas radiometry considers many more wavelengths than those to which the human visual system is sensitive, the second form of light measurement, *photometry*, is concerned only with radiation in the visible range. The quantities derived in radiometry and photometry are closely related: photometry is essentially radiometry weighted by the sensitivity of the human eye. As such, radiometry measures physical quantities and photometry measures psycho-physical quantities.

In this chapter, we first revisit the topic of human wavelength sensitivity, enabling us to refine our explanation of radiometry and photometry. We then define several important quantities relevant to both disciplines. We introduce computational models that attempt to compute perceptual quantities from their physical counterparts. Thus, this chapter affords an insight into radiation measurements that are important to understand the color measurements discussed in the next chapter.

6.1 The Sensitivity of the Human Eye

The human eye is only sensitive to electromagnetic radiation between approximately 380 to 780 nm. Sensitivity peaks at $\lambda = 555$ nm and tails off toward the minimum and maximum wavelengths. Visual sensitivity varies slightly between individuals, and depends upon factors such as the illumination of the viewing environment, the extent of the viewing field, and the observer's adaptation level [183].

Under controlled viewing conditions, the sensitivity of the human eye can be measured with sufficient accuracy. Such measurements were performed in the early 1900s [358, 360]. Here, the field of view was fixed at 2° (approximately the angular extent of a thumb nail when viewed at arm's length) and the luminance of the viewing field was set to at least $8\,\text{cd}/\text{mm}^2$, i.e., the lighting conditions are in the photopic range. The accepted international values are based on five different sets of measurements [1199] and form the basis of the *CIE 1924 2° photopic luminous efficiency curve* $V(\lambda)$ (see Figure 6.1).

$V(\lambda)$ was modified by the CIE in 1988 to better account for human visual sensitivity below 460nm. This modified curve is denoted by $V_M(\lambda)$, and is considered supplemental to $V(\lambda)$ rather than a replacement (Figure 6.1) [189].

For scotopic vision a separate curve was standardized by the CIE in 1951 [182, 212, 1194] and is called the *CIE scotopic luminous efficiency curve*, denoted by $V'(\lambda)$. This curve is plotted in Figure 6.2 in both linear and logarithmic scales. The sensitivity curves are tabulated in Appendix E.

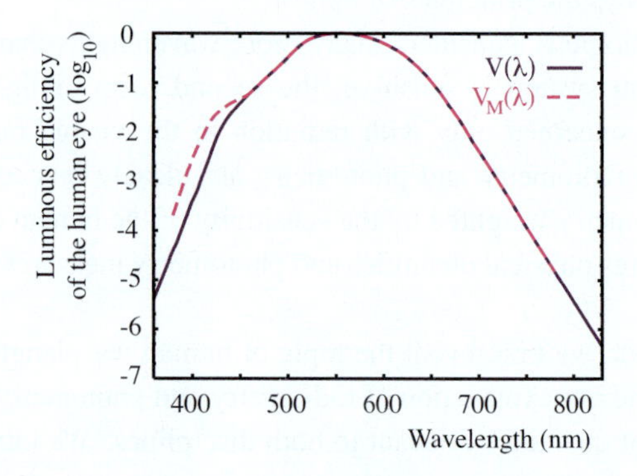

Figure 6.1. $V(\lambda)$ and $V_M(\lambda)$.

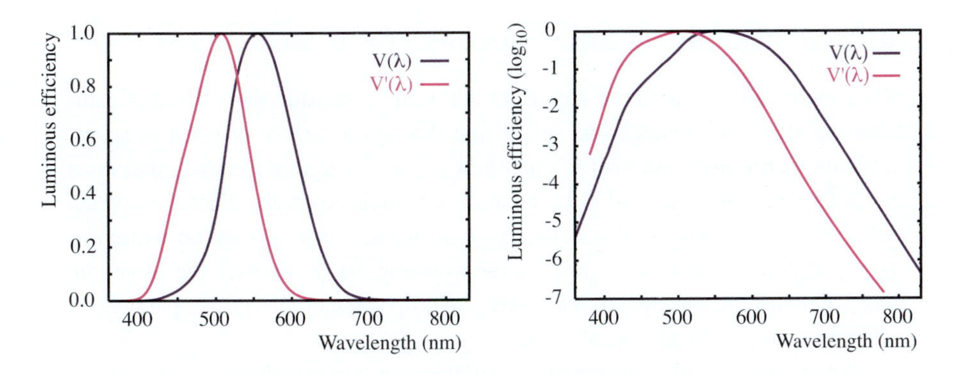

Figure 6.2. Photopic $V(\lambda)$ and scotopic $V'(\lambda)$ luminous efficiency curves.

As mentioned in the introduction of this chapter, photometric quantities can be obtained by spectrally weighting their radiometric counterparts. The weighting function is given by the $V(\lambda)$ curve.

Under photopic lighting conditions the L-, M-, and S-cones mediate vision. Each cone type has a different spectral sensitivity and population (see Sections 4.3.3 and 4.3.4). Therefore, $V(\lambda)$ approximates the weighted sum of the sensitivities of each type and takes into account their relative populations [302]. Conversely, as only rods are active under scotopic lighting conditions (see Section 4.3.2), $V'(\lambda)$ models the sensitivity of the rods to light at different wavelengths.

Under mesopic lighting conditions, the relative contributions of rods and cones vary with illumination level. As a result, a single curve would not represent human visual sensitivity appropriately; a family of curves would be a more likely solution [567,611,1199]. Nevertheless, no curve has officially been recommended by the CIE.

As shown in Figure 6.2, the shapes of the luminous efficiency curves for photopic and scotopic vision are similar. However, the peak of the scotopic curve is shifted by approximately 50 nm toward the shorter wavelengths (i.e., toward the blue). Thus, a change in illumination causes a shift in the relative intensity of colors. For instance, given a red and a blue object with identical surface reflectance under ordinary illumination, the red one will appear darker than the blue object as the illumination is reduced to the scotopic level. This phenomenon is called the *Purkinje effect* or the *Purkinje shift* [930, 1199].

6.2 Radiometric and Photometric Quantities

Radiometric and photometric quantities are used to measure the amount of radiation incident on, exitant from, or passing through a surface or point in space. All radiometric and photometric quantities are wavelength dependent, and most of them have positional and directional dependencies as well. Therefore, when referring to a quantity Q that depends on all these factors it may be written as $Q(x,y,\theta,\phi,\lambda)$, where (x,y) denotes the position, (θ,ϕ) denotes the direction, and λ denotes the wavelength. However, we drop some or all of these parameters when this does not lead to confusion.

We assume that the radiometric quantities we begin with are spectrally unweighted (or have an equal weight applied to all wavelengths). Then, any given photometric quantity $Q_v(\lambda)$ may be derived from its radiometric counterpart $Q_e(\lambda)$ by

$$Q_v = K_m \int_{380}^{780} Q_e(\lambda)\, V(\lambda)\, d\lambda, \tag{6.1}$$

where the e subscript (for energy) is used for radiometric quantities and the v subscript (for visual) is used for photometric quantities. These may also be dropped when the meaning is clear. The result of integration is normalized by a constant $K_m = 683$ lm/W to obtain the photometric units at an absolute scale. The unit lm/W (lumens per watt) affords a conversion from the fundamental radiometric unit watt to the photometric unit lumen.

A similar computation can be carried out for values in the scotopic range. Here, $V(\lambda)$ is replaced with $V'(\lambda)$ and K_m is replaced with $K'_m = 1700$ lm/W resulting in

$$Q'_v = K'_m \int_{380}^{780} Q_e(\lambda)\, V'(\lambda)\, d\lambda. \tag{6.2}$$

To calculate the value of a photometric quantity for only a single wavelength λ_0, the computation reduces to

$$Q_v(\lambda_0) = K_m\, Q_e(\lambda_0)\, V(\lambda_0). \tag{6.3}$$

The most commonly used radiometric and photometric quantities are listed in Table 6.1; they are explained in the following sections.

6.2.1 Radiant and Luminous Energy

To measure the total energy in a beam of radiation, a quantity called *radiant energy* is used. Radiant energy is denoted by Q_e and measured in *joules* (J).

Similarly, to measure the total energy in a beam of radiation as perceived by humans *luminous energy* is used. Thus, luminous energy is zero outside the

Radiometric quantity	Unit	Photometric quantity	Unit
Radiant energy (Q_e)	J (joule)	Luminous energy (Q_v)	$\mathrm{lm \cdot s}$
Radiant flux (P_e)	$\mathrm{J \cdot s^{-1}} = \mathrm{W}$ (watt)	Luminous flux (P_v or F_v)	lm (lumen)
Radiant exitance (M_e)	$\mathrm{W \cdot m^{-2}}$	Luminous exitance (M_v)	$\mathrm{lm \cdot m^{-2}} = \mathrm{lx}$ (lux)
Irradiance (E_e)	$\mathrm{W \cdot m^{-2}}$	Illuminance (E_v)	$\mathrm{lm \cdot m^{-2}} = \mathrm{lx}$ (lux)
Radiant intensity (I_e)	$\mathrm{W \cdot sr^{-1}}$	Luminous intensity (I_v)	$\mathrm{lm \cdot sr^{-1}} = \mathrm{cd}$ (candela)
Radiance (L_e)	$\mathrm{W \cdot m^{-2} \cdot sr^{-1}}$	Luminance (L_v)	$\mathrm{lm \cdot m^{-2} \cdot sr^{-1}} = \mathrm{cd \cdot m^{-2}}$ (nit)

Table 6.1. Radiometric and photometric quantities.

380 – 780 nm range. Luminous energy may be computed from radiant energy using (6.1). It is denoted by Q_v and measured in lumen seconds (lm s). The lumen is the base unit of photometry and is explained in the next section. Radiant and luminous energy are used in applications where the aim is to measure the total dosage of radiation emitted from a source.

6.2.2 Radiant and Luminous Flux

As radiation may be variable over time, an important quantity is the rate of change of radiation. This is called *radiant flux*. The symbol used for radiant flux is P_e and it is defined as the time derivative of radiant energy:

$$P_e = \frac{dQ_e}{dt}. \tag{6.4}$$

With Q_e expressed in joules and t in seconds, radiant flux P_e is given in watts ($1\ \mathrm{W} = 1\ \mathrm{J/s}$). The photometric counterpart of radiant flux is called *luminous flux*:

$$P_v = \frac{dQ_v}{dt}. \tag{6.5}$$

The same quantity may be derived from radiant flux:

$$P_v = K_m \int_{380}^{780} P_e(\lambda)\, V(\lambda)\, d\lambda. \tag{6.6}$$

The unit of luminous flux is the *lumen* (lm), which is the base unit of photometry. Luminous flux is a measure of the capacity of radiant flux to invoke visual sensations and is equal to $1/683$ watt of radiant power at a frequency of 540×10^{12} Hz, which corresponds to a wavelength of about 555 nm.

6.2.3 Radiant and Luminous Intensity

Radiant flux emitted from an ideal point source is not constrained as a function of direction. As an emitter may radiate more in some directions than others, a further unit of measurement is used. *Radiant intensity* may be used to quantify the amount of flux emitted in a given direction. It is defined as the amount of radiant energy emitted per second into a differential solid angle (solid angle is defined in Appendix B.6). Therefore the radiant intensity I_e is given by

$$I_e = \frac{dP_e}{d\omega} \tag{6.7a}$$

$$= \frac{d^2\, Q_e}{dt\, d\omega}. \tag{6.7b}$$

The unit of radiant intensity is watts per steradian (W/sr). If the radiant intensity of a point source is the same in all directions, the source is said to be *uniform* or *isotropic*. For a uniform source with radiant intensity I_e, the total radiant flux equals

$$P_e = \omega\, I_e = 4\pi\, I. \tag{6.8}$$

For instance, if $I_e = 1$ W/sr then $P_e \approx 12.56$ W. A related quantity to radiant intensity is the *average radiant intensity* which is defined as the ratio of the total flux emitted by a source to the solid angle of emittance which is usually 4π (i.e., the entire sphere surrounding the source).

The luminous intensity is the photometric counterpart of radiant intensity and may be computed from luminous flux:

$$I_v = \frac{dP_v}{d\omega} \tag{6.9a}$$

$$= \frac{d^2\, Q_v}{dt\, d\omega}, \tag{6.9b}$$

or may be derived from radiant intensity:

$$I_v = K_m \int_{380}^{780} I_e(\lambda)\, V(\lambda)\, d\lambda. \tag{6.10}$$

The unit of luminous intensity is lumens per steradian (lm/sr) which is normally called *candela* (cd). The candela is defined as the amount of luminous intensity in a given direction of a source that emits monochromatic radiation of frequency 540×10^{12} Hz and that has a radiant intensity in that direction of $1/683\,\mathrm{W\,sr}^{-1}$.

Luminous intensity may be computed for any source (i.e., both point and extended), but in general it is only meaningful for point sources [839]. For extended sources, *radiance* and *luminance* are appropriate. They are defined in the following sections.

The *average luminous intensity*, which is the photometric equivalent of average radiant intensity, is also referred to as the *mean spherical candle power* (MSCP) [1199].

6.2.4 Irradiance and Illuminance

When optical radiation reaches a surface, that surface is irradiated or illuminated. The radiometric irradiance and the photometric illuminance are measures of how intensely irradiated or illuminated a point on a surface is.

The irradiance E_e is defined as the ratio of the radiant flux striking a differential surface area dA (i.e., radiant energy per second per surface area):

$$E_e = \frac{dP_e}{dA}. \tag{6.11}$$

As irradiance does not have directional dependency, to compute irradiance at a point, the radiation arriving from all incident directions must be included. For a differential surface area of an opaque surface, this requires integrating over a hemisphere covering the area of interest. Irradiance is measured in watts per square meter (W/m^2).

The illuminance E_v at a point is defined as the luminous flux per differential area:

$$E_v = \frac{dP_v}{dA}. \tag{6.12}$$

It may also be derived directly from irradiance E_e:

$$E_v = K_m \int_{380}^{780} E_e(\lambda)\, V(\lambda)\, d\lambda. \tag{6.13}$$

The unit of illuminance is lumens per square meter (lm/m^2), which is also called *lux*.

Irradiance and illuminance are measures of the density of incident radiation. To define the surface density of exitant radiation leaving a point, the quantities called *radiant exitance* and *luminous exitance* are used. They have the same units as irradiance and illuminance.

In the following sections we demonstrate the computation of illuminance with two examples. In the first example, the surface lit by a light source is assumed to be differential. Therefore, the intensity of the light source can be regarded as the same for all directions spanning the surface. In the second example, a larger surface is considered where this assumption cannot be made.

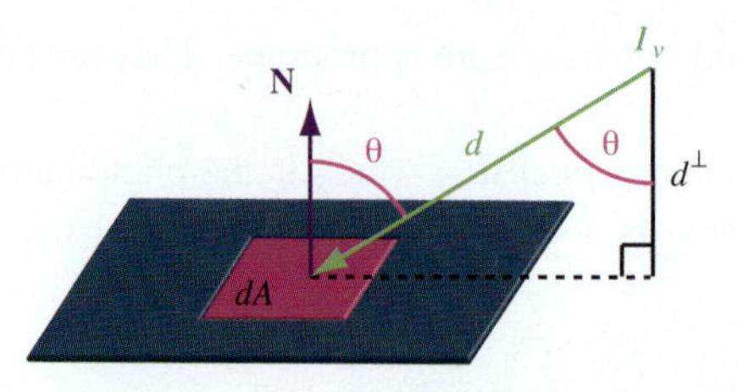

Figure 6.3. A differential surface area lit by a point light with intensity I_v.

6.2.5 Illuminance on a Differential Surface Area

The configuration under consideration is depicted in Figure 6.3. As the light strikes a differential surface area, we can assume that all points contained in this area are exposed to the same luminous intensity. The flux dP_v emitted from the light source that reaches the surface element dA is then

$$dP_v = I_v \, \frac{dA \, \cos(\theta)}{d^2}, \tag{6.14}$$

where I_v is the luminous intensity of the source in the direction of the surface, θ is the angle of incidence, and d is the distance between the source and the surface. Note that $dA \, \cos(\theta)/d^2$ is the solid angle spanned by the surface with respect to the source and is obtained by applying both the cosine law and the inverse square law (Sections 2.9.2 and 2.9.3). The illuminance is then given by

$$E_v = \frac{dP_v}{dA} \tag{6.15a}$$

$$= I_v \, \frac{\cos(\theta)}{d^2}. \tag{6.15b}$$

This equation may be written in a different form which will help the computation of illuminance over large surface areas. Since d equals $d^\perp/\cos(\theta)$, where $d^\perp$ is the distance between the surface and the point source along the surface's normal, substituting d in (6.15b) gives

$$E_v = I_v \, \frac{\cos^3(\theta)}{(d^\perp)^2}. \tag{6.16}$$

This equation is known as the *cosine-cubed law*. In this formulation, each point on a surface lit by a uniform light source receives illumination that varies with θ, since $d^\perp$ is the same for all points on a large flat surface. For non-uniform light sources, the illumination changes with θ and I_θ (the intensity of the source in the given direction θ).

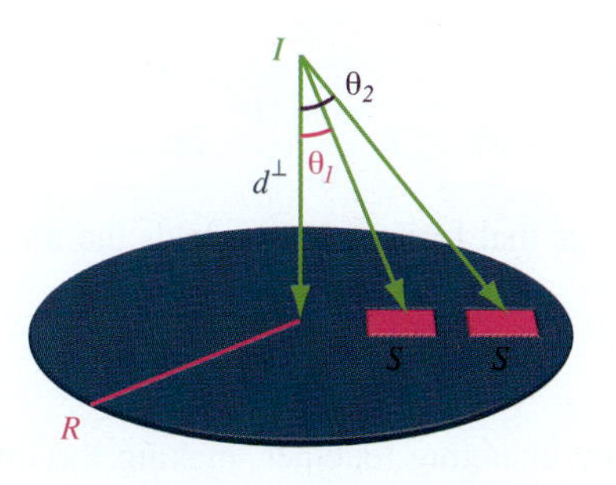

Figure 6.4. A large surface is illuminated by a point source.

6.2.6 Illuminance on Large Surfaces

The equations of the previous section can be extended to compute the average illumination of any surface, for instance by computing the illuminance at each point and then averaging the results. Consider the surface and the point light in Figure 6.4.

For simplicity, we assume that the surface is disk-shaped and has a radius R. We also assume that a point source of luminous intensity I_v is located directly above the center of the disc. Under this configuration, the cosine-cubed law suggests that all points lying on a ring with a given radius will have the same illuminance (Figure 6.5). The area of a ring of radius r is equal to $2\pi r\, dr$, where dr is the differential increase in radius across the ring. The total illuminance over this

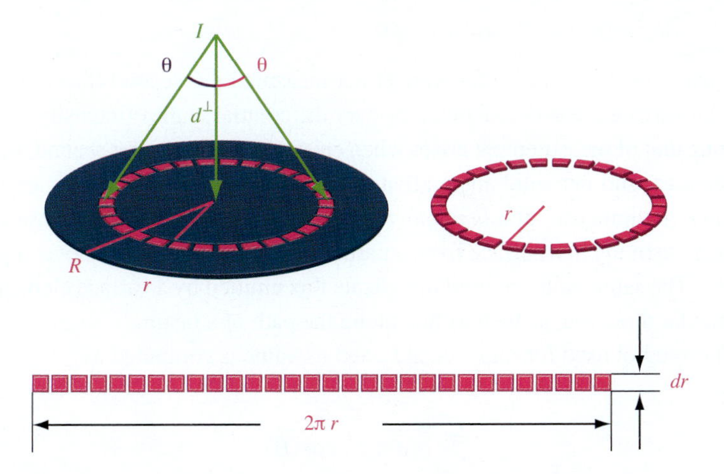

Figure 6.5. Top: Each differential area on a ring has the same angle of incidence θ and, therefore, has the same illuminance. Bottom: The area calculation of the ring is shown.

ring is then

$$P_{v,r} = I_v \frac{\cos^3(\theta)}{(d^\perp)^2} 2\pi r \, dr. \tag{6.17}$$

Integrating over all rings that form the disk yields the total illuminance over the surface:

$$P_v = \int_0^R I_v \frac{\cos^3(\theta)}{(d^\perp)^2} 2\pi r \, dr. \tag{6.18}$$

Here, both r and θ are changing together, making the integration problematic. This can be avoided by rewriting r in terms of θ as $r = d^\perp \tan(\theta)$. Then $dr = d^\perp \sec^2(\theta)\, d\theta$ and the upper limit of the integral becomes $\tan^{-1}(R/d^\perp)$:

$$P_v = 2\pi I_v \int_0^{\tan^{-1}(R/d^\perp)} \sin(\theta)\, d\theta \tag{6.19a}$$

$$= 2\pi I_v \left(1 - \frac{d^\perp}{\sqrt{R^2 + (d^\perp)^2}}\right). \tag{6.19b}$$

The average illumination $\bar{E}_v$ is calculated by dividing P_v by the disc's surface area:

$$\bar{E}_v = \frac{2 I_v}{R^2} \left(1 - \frac{d^\perp}{\sqrt{R^2 + (d^\perp)^2}}\right). \tag{6.20}$$

It can be verified that as R becomes small compared to $d^\perp$ (i.e., $R/d^\perp \to 0$), this equation reduces to $\bar{E}_v = I_v/(d^\perp)^2$.

6.2.7 Radiance and Luminance

The quantities discussed so far were either measured per second (flux), over differential surface areas (irradiance), or over differential angles (intensity). An interesting unit of measurement arises when energy is measured per second, per unit surface area, and per solid angle all at the same time. Such quantities are called *radiance* (radiometric) and *luminance* (photometric) and allow the measurement of energy arriving at a surface from a particular direction integrated over a period of time. The same units are used to indicate flux emitted by a surface element into a particular direction, as well as flux along the path of a beam.

The symbol used for radiance is L_e and its value is computed as

$$L_e = \frac{d^2 P_e}{d\omega \, dA \cos(\theta)}, \tag{6.21}$$

where θ is the angle of incidence or exitance (see also Figure 6.6). The unit of radiance is watts per steradian per square meter ($\mathrm{W\,sr^{-1}\,m^{-2}}$).

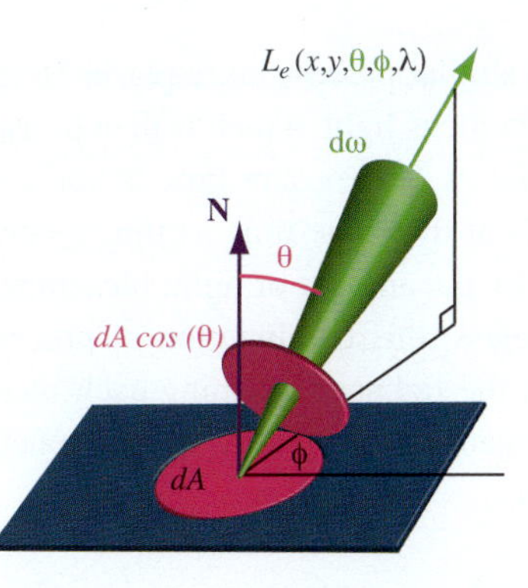

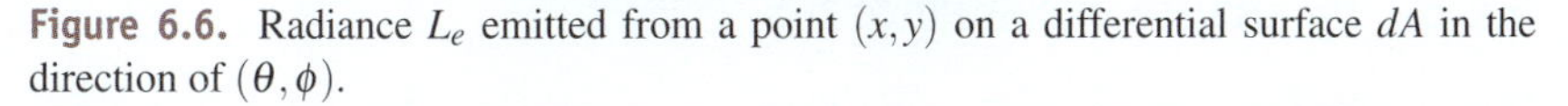

Figure 6.6. Radiance L_e emitted from a point (x, y) on a differential surface dA in the direction of (θ, ϕ).

The radiance L_e emitted from a differential area is directly related to its radiant intensity I_e:

$$L_e = \frac{dI_e}{dA \cos(\theta)}, \tag{6.22}$$

where θ is the exitant angle.

Luminance L_v is the photometric counterpart of radiance. It may be computed from luminous flux P_v as

$$L_v = \frac{d^2 P_v}{d\omega dA \, \cos(\theta)}. \tag{6.23}$$

Alternatively, luminance may be derived from radiance:

$$L_v = K_m \int_{380}^{780} L_e(\lambda)\, V(\lambda)\, d\lambda. \tag{6.24}$$

Luminance is measured in candela per square meter ($\mathrm{cd\,m^{-2}}$). The *nit* is also sometimes used instead of candela per square meter, although this term is now obsolete.

It is easy to see that a camera records spectrally weighted radiance. Light arriving from many directions is focused by the lens, measured over a given exposure time, and for each pixel individually recorded over a small surface area. Thus, the charges recorded in the sensor, between the time that the shutter is opened and closed, represents radiance values. The firmware in the camera then processes the data, so that we can not be certain that the values retrieved from the camera still correspond to radiances (although most cameras now export a raw format, designed to bypass much of the in-camera processing).

In the human eye a similar process takes place. Here the refractive indices of the various ocular media focus light, which is then transduced by the photoreceptors. The human equivalent of exposure time is not so easily defined. Photoreceptors transmit signals at the same time as they receive photons. The amount of signal is modulated by the amount of light, bleaching, and by neural feedback mechanisms. Nevertheless, this combination of processes causes the signal to be commensurate with the radiance reaching each receptor. Subsequent visual processing stages then cause the signal to be photometrically weighted, so that luminance values result.

6.2.8 Invariance of Radiance

If we assume that energy propagates between an emitter and a surface through a medium that does not absorb any energy, then this medium is called *lossless*. In such media, it is possible to measure the emitted radiance from a source by only measuring the incident radiance on a detector.

In Figure 6.7, two surface elements are placed in arbitrary orientations; the distance between them is also arbitrary. We assume that the surfaces are small enough such that the radiance emitted and received by any one point on either surface is the same for all other points on the same surface.

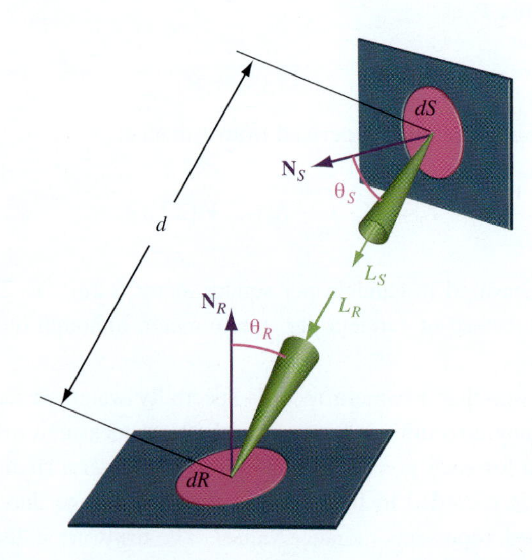

Figure 6.7. A differential surface dR is radiated by a differential source dS located at a distance d from the surface. In a lossless medium, the radiance L_S emitted from the source is equal to the radiance L_R received by the surface.

If the radiance emitted from the surface with surface area dS is equal to L_S, then the total flux dP_S that is emitted in the direction of dR equals

$$dP_S = L_S \cos(\theta_S)\, dS\, d\omega_R, \tag{6.25}$$

where $d\omega_R$ is the solid angle subtended by dR at the center of dS. The differential solid angle $d\omega_R$ can be rewritten as follows:

$$d\omega_R = \frac{\cos(\theta_R)\, dR}{d^2}, \tag{6.26}$$

where d is the distance between the center of the surfaces. Substituting $d\omega_R$ into the previous equation we obtain

$$dP_S = \frac{L_S \cos(\theta_S) \cos(\theta_R)\, dS\, dR}{d^2}. \tag{6.27}$$

If the radiance received by any point on the surface with area dR is equal to L_R, then the total flux received from the direction of dS is

$$dP_R = L_R \cos(\theta_R)\, dR\, d\omega_S. \tag{6.28}$$

This expression can be rewritten as

$$dP_R = \frac{L_R \cos(\theta_R) \cos(\theta_S)\, dR\, dS}{d^2}. \tag{6.29}$$

In a lossless medium, the flux emitted from dS in the direction of dR must be equal to the flux received by dR from the direction of dS (see Figure 6.8):

$$\frac{L_S \cos(\theta_R) \cos(\theta_S)\, dR\, dS}{d^2} = \frac{L_R \cos(\theta_R) \cos(\theta_S)\, dR\, dS}{d^2}, \tag{6.30}$$

and hence

$$L_S = L_R. \tag{6.31}$$

This result expresses the *invariance of radiance*. Since invariance of radiance is independent of the points chosen on dS and dR (their centers were arbitrarily chosen), it holds between any pair of points on both surfaces. Furthermore, the equation is also independent of d, which means that for any pair of points along a ray connecting dS to dR, the radiance is the same. The significance of this result lies in the fact that by measuring the *incident* radiance on a receiver, such as a detector, the radiance *emitted* from a source can be inferred.

So far we have assumed that the intervening medium is lossless, and that there is no optical system between the source and the receiver. However, most

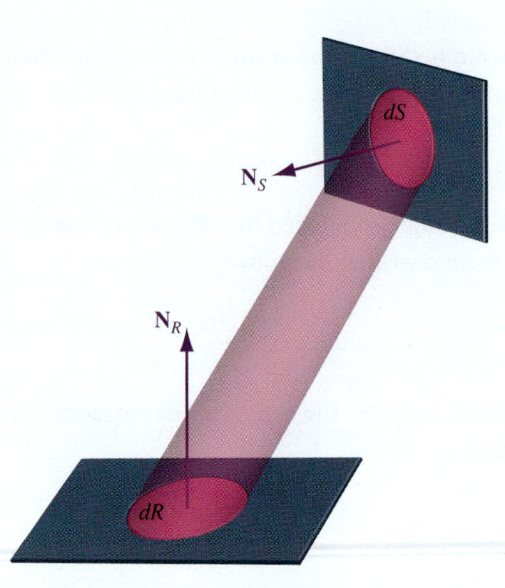

Figure 6.8. The flux emitted by dS in the direction of dR is equal to the flux received by dR from the direction of dS.

measurement instruments have lenses to refract the incident radiation towards a detector.

At a differential surface element dA of the lens, a beam of radiation is refracted according to Snell's law, given in (2.102). The differential flux on either side of the surface of the lens is given by

$$dP_1 = L_1 \, dA \, cos(\theta_1) \, d\theta_1 \, d\phi, \tag{6.32a}$$

$$dP_2 = L_2 \, dA \, cos(\theta_2) \, d\theta_2 \, d\phi. \tag{6.32b}$$

In a lossless medium, both quantities will be identical ($dP_1 = dP_2$). Combining this result with (2.102), we find

$$\frac{L_1}{n_1^2} = \frac{L_2}{n_2^2}. \tag{6.33}$$

The quantity L/n_i^2 is called the *basic radiance*, and it is always preserved in a lossless medium [839]. Once light exits the lens and returns to the medium with refractive index n_1, the radiance also returns to its old value.

In practice, radiance is not fully invariant due to energy losses during transmission, refraction, and reflection. The fraction of initial radiance L_S that is successfully propagated to a receiver L_R is called the *propagance* τ of the entire path of radiation:

$$\tau = \frac{L_S}{L_R}. \tag{6.34}$$

The value of propagance of an optical path is due to all attenuation caused by absorption, scattering, reflection, and refraction.

6.2.9 Irradiance from Radiance

Given a source and a receiver that face each other (Figure 6.7), it is possible to compute the irradiance on a receiver as function of the radiance emitted by the source. The flux dP_R received by dR from dS is given by (6.29). Since the irradiance is equal to the incident flux per differential surface area, the irradiance E_R on dR is equal to dP_R/dR, and hence

$$E_R = \frac{L_S \cos(\theta_R) \cos(\theta_S) \, dS}{d^2}, \tag{6.35}$$

where we have applied (6.31). In this equation, the factor $\cos(\theta_S) \, dS/d^2$ is the solid angle subtended by dS, and hence (6.35) may be rewritten as follows:

$$E_R = L_S \cos(\theta_R) \, d\omega_S. \tag{6.36}$$

An interesting result is obtained if both the source dS and the receiver dR lie on the interior surface of a sphere. In that case, we have,

$$\theta_R = \theta_S = \theta, \tag{6.37a}$$

$$d = 2r \cos(\theta), \tag{6.37b}$$

where r is the radius of the sphere as shown in Figure 6.9. Substituting these into (6.35), the irradiance on dR is given by

$$E_R = \frac{L_S \, dS}{4r^2}. \tag{6.38}$$

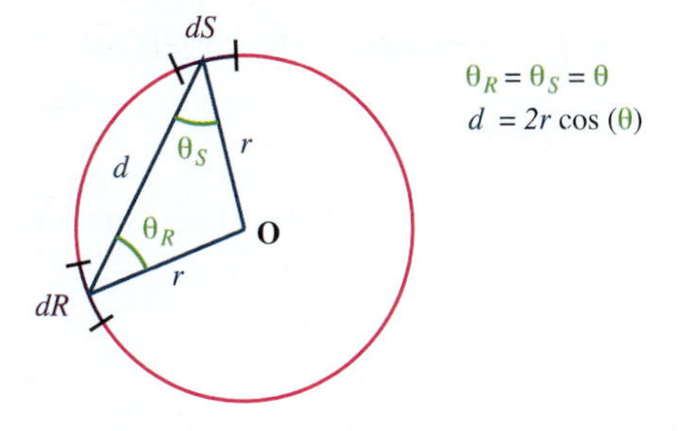

Figure 6.9. The two differential patches dS and dR lie on a sphere. In this case, the irradiance of dR due to emission from dS is independent of the distance d and both of the angles θ_S and θ_R.

Note that this result is independent of both θ and d. Therefore, all the points on a sphere are equally irradiated from any other point on the sphere. This is a particularly convenient configuration which may be exploited to measure both power as well as irradiance (or their photometric equivalents) using a device called the *integrating sphere* (see Section 6.8.2).

6.2.10 Iso-Radiant Surfaces

As discussed in Section 2.9.2, the radiation emitted from any point of an iso-radiant surface decreases with the cosine of the angle between the surface normal and the direction of emission. In other words, for an iso-radiant surface the radiant intensity of each point (i.e., differential surface element) along the direction θ is equal to

$$dI = I_0 \cos(\theta), \tag{6.39}$$

where I_0 is the radiant intensity in the direction normal to the surface. For such surfaces, the radiance emitted from any point in the direction θ is given by

$$L = \frac{dI}{dA \cos(\theta)} = \frac{I_0}{dA}. \tag{6.40}$$

As such, radiance (and similarly luminance) emitted from an iso-radiant surface is the same irrespective of the outgoing direction. For such surfaces, the flux dP emitted per differential area may be conveniently related to the radiance of the surface:

$$
\begin{aligned}
dP &= \int_0^{\pi/2} \int_0^{2\pi} L dA \cos(\theta)\, d\omega && \text{(6.41a)} \\
&= \int_0^{\pi/2} \int_0^{2\pi} L dA \cos(\theta) \sin(\theta)\, d\phi\, d\theta && \text{(6.41b)} \\
&= \int_0^{\pi/2} 2\pi L dA \cos(\theta) \sin(\theta)\, d\theta && \text{(6.41c)} \\
&= \pi L dA. && \text{(6.41d)}
\end{aligned}
$$

The total flux emitted from the entire surface is then computed as

$$
\begin{aligned}
P &= \int_A dP && \text{(6.42a)} \\
&= \int_A \pi L dA && \text{(6.42b)} \\
&= \pi L A. && \text{(6.42c)}
\end{aligned}
$$

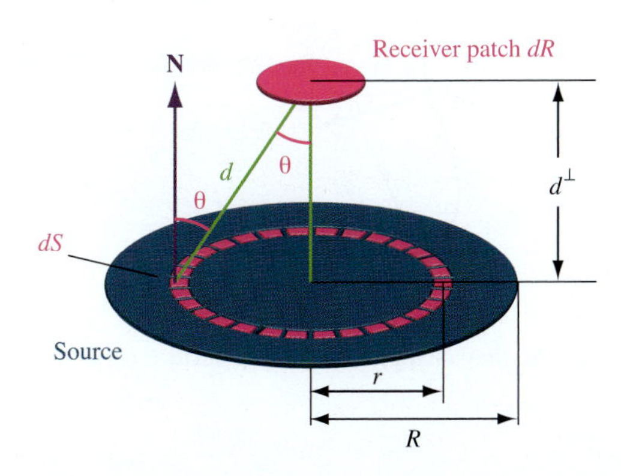

Figure 6.10. The differential surface patch dR is illuminated by a flat circular extended source. A differential ring on the source surface is shown, and the radiance L emitted from a differential area dS of this ring is shown to strike dR.

6.2.11 Error Associated with the Point-Source Assumption

While computations with point sources are convenient, in practice, light sources are extended. In Section 2.9.3, we gave a rule of thumb stating that if the distance to the source is at least ten times the size of the largest dimension of the source, this source can be approximated as a point source. The error associated with this rule is analyzed here.

We begin by considering an extended circular source of constant radiance L illuminating a differential surface area dR which is $d^{\perp}$ units apart from the source (see Figure 6.10). The source and receiver are facing each other, i.e., their surface normals point in opposite directions.

The amount of flux received by dR from each differential element dS of the source is equal to

$$d^2 P = \frac{L \cos^2(\theta)\, dR\, dS}{d^2}. \tag{6.43}$$

As θ is the same for all points on a ring, we can rewrite the total flux received from each ring as

$$d^2 P_{\text{ring}} = 2\pi r\, dr\, \frac{L \cos^2(\theta)\, dR}{d^2}, \tag{6.44}$$

and, therefore, the flux received by the receiver dR from the extended source is equal to

$$dP = \int_0^R 2\pi r\, dr\, \frac{L \cos^2(\theta)\, dR}{d^2}. \tag{6.45}$$

This form is difficult to integrate since both d and θ are changing together for each differential band. Making similar substitutions as outlined in Section 6.2.6

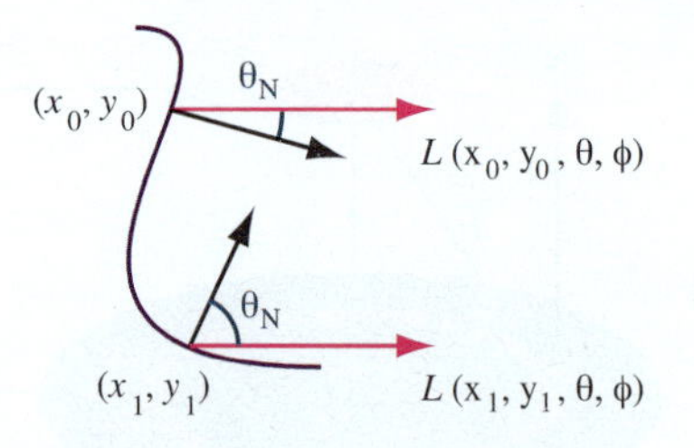

Figure 6.11. The angle θ_N may change based on the surface curvature whereas θ is fixed.

yields

$$P = \int_0^{\tan^{-1}(R/d^{\perp})} 2\pi L\, \sin(\theta)\, \cos(\theta)\, d\theta\, dR \tag{6.46a}$$

$$= \pi L\, dR \left(\frac{R^2}{R^2 + (d^{\perp})^2} \right), \tag{6.46b}$$

and the irradiance E is equal to

$$E = \frac{dP}{dR} = \pi L \left(\frac{R^2}{R^2 + (d^{\perp})^2} \right). \tag{6.47}$$

In the general case, the radiant intensity of an extended source in the direction of (θ, ϕ) is defined as

$$I(\theta, \phi) = \int_S L(\theta, \phi)\, \cos(\theta_N)\, dS, \tag{6.48}$$

where $\int_S$ indicates that integral is computed over the entire surface. Also note that θ_N is the angle between the surface normal and the direction of emission. The angle θ_N is not necessarily equal to θ as illustrated in Figure 6.11.

For an extended circular source (Figure 6.10), the intensity in the direction of the surface normal is given by

$$I = \pi R^2 L. \tag{6.49}$$

By substituting this result into (6.47) we obtain the irradiance E_a on the differential surface dR in terms of the intensity of the source:

$$E = \frac{I}{R^2 + (d^{\perp})^2}. \tag{6.50}$$

If the source is very small $(R \to 0)$ or if the distance from the source is very large $(d^{\perp} \to \infty)$ (e.g., starlight), the approximate irradiance E_a is given by

$$E_a \approx \frac{I}{(d^{\perp})^2}. \tag{6.51}$$

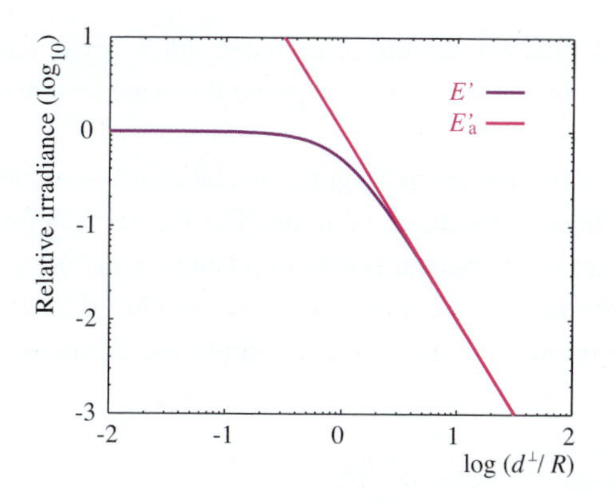

Figure 6.12. The relative irradiance on a differential surface due to an extended source. Both exact values E' and the point-source approximation E'_a are shown.

Therefore, the error associated with treating an extended source as a point source stems from ignoring the R^2 in the denominator. We can see this by dividing both E and E_a by R^2 so that

$$E' = \frac{I}{1 + (d^{\perp}/R)^2}, \tag{6.52a}$$

$$E'_a \approx \frac{I}{(d^{\perp}/R)^2}. \tag{6.52b}$$

If we plot both E' and E'_a for different ratios of $d^{\perp}$ and R (Figure 6.12), we see that the two curves almost overlap when the distance between the source and the receiver is at least ten times the radius of the source. At this distance, the error drops below 1%, and it reduces further for longer distances.

6.3 The Efficacy of Optical Radiation

As evidenced by the $V(\lambda)$ curve, the human visual system has a different sensitivity to each wavelength. Wavelengths close to the peak sensitivity (555 nm) invoke a greater visual sensation than other wavelengths. Thus, radiation composed of wavelengths around the HVS' peak sensitivity have a higher luminous efficiency than other radiation.

The ability to invoke a visual sensation can be expressed in terms of efficacy as well as efficiency. The *efficacy* of radiation is the ratio of its luminous power to the radiant (or electrical) power that produced it. Thus, efficacy is measured in lumens per watt (lm/W). On the other hand, *efficiency* is a unitless term that expresses the ratio of the efficacy at a particular wavelength to the efficacy at 555 nm [1090].

Two types of efficacies are distinguished: the efficacy of electromagnetic radiation and the efficacy of lighting systems. The former is defined as the ratio of the luminous power to the radiant power in a beam of radiation, whereas the latter is defined as the ratio of the luminous power produced to the electrical power consumed in a lighting system. These concepts are discussed in the following sections.

6.3.1 Luminous Efficacy of Radiation

The luminous efficacy of radiation K_r is defined as the quotient of the luminous flux and the radiant flux in a beam of electromagnetic radiation:

$$K_r = \frac{P_v}{P_e}. \tag{6.53}$$

It is measured in lm/W. When a beam of radiation is composed of a single wavelength λ_0, its efficacy $K_r(\lambda_0)$ is equal to the value of the luminous sensitivity curve for that wavelength times a normalization constant. For instance, for photopic vision the luminous efficacy at a single wavelength λ_0 is

$$K_r(\lambda_0) = \frac{683\, P_e(\lambda_0)\, V(\lambda_0)}{P_e(\lambda_0)} = 683\, V(\lambda_0) \quad \text{lm/W} \tag{6.54}$$

The luminous efficacy curve for photopic vision may be obtained by plotting $K_r(\lambda)$ for all wavelengths in the visible spectrum. A similar curve $K_r'(\lambda)$ can be created from $V'(\lambda)$ for scotopic conditions. Both curves are plotted in Figure 6.13. Whereas $V(\lambda)$ and $V'(\lambda)$ give relative information, $K_r(\lambda)$ and $K_r'(\lambda)$ provide absolute values of sensitivity.

For polychromatic light, the luminous efficacy may be computed by

$$K_r = \frac{683 \int_{380}^{780} P_v(\lambda)\, V(\lambda)\, d\lambda}{\int_0^\infty P_e(\lambda)\, d\lambda}. \tag{6.55}$$

The luminous efficacies of ultraviolet and infrared radiation, as well as of other types of radiation outside the visible spectrum, are zero. The maximum possible luminous efficacy for photopic vision is equal to 683 lm/W, and it is only attainable if all the radiation is concentrated at 555 nm. As an aside, the efficiency of such radiation is unity. If radiation is uniformly distributed at all wavelengths in the visible range and zero everywhere else, its efficiency is around 0.4 [1199]. The luminous efficacies of commonly encountered types of radiation are given in Table 6.2.

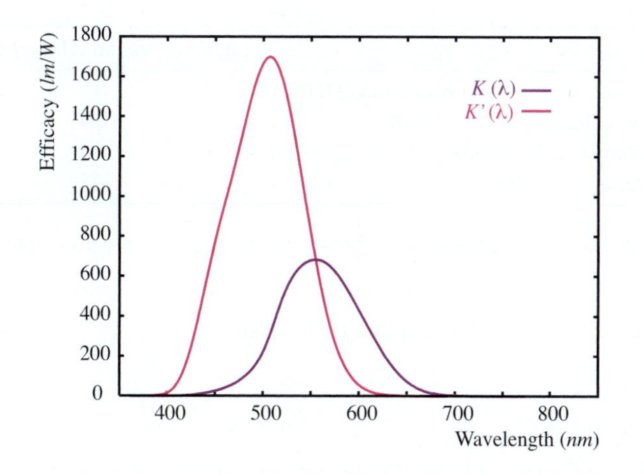

Figure 6.13. Luminous efficacy curves for photopic ($K(\lambda)$) and scotopic ($K'(\lambda)$) vision.

Source	Luminous efficacy (lm/W)
Monochromatic light (555 nm)	683
White light, constant over visible spectrum	220
Blue-sky light	125–140
Direct-beam sunlight, midday	90–120
Direct-beam sunlight, extraterrestrial	99.3
Direct-beam sunlight, sunrise/sunset	50–90
Overcast sky	103–115
Uncoiled tungsten wire at its melting point	53
Radiation from typical tungsten filament lamp	15
Phosphorescence of cool white fluorescent lamp	348

Table 6.2. Luminous efficacies of commonly encountered light sources (from [761]).

6.3.2 Luminous Efficacy of Lighting Systems

The efficacy of electromagnetic radiation does not give insight into how much power needs to be supplied to a lighting system for it to create that radiation. For instance, a monochromatic laser beam emitting at 555 nm has high efficiency, even if a large amount of electrical power is consumed in the operation of such a laser. Thus, a different measure is required to compute the overall efficacies of lighting systems.

The luminous efficacy of lighting systems K_s is defined as the ratio of the total luminous flux P_v produced by a lighting system (i.e., lamp, laser, etc.) and

Source	Lighting system efficacy (lm/W)
Standard incandescent filament lamp, 2700K	17
Tungsten-halogen (linear), 2950K	20
Fluorescent T12 4ft, 3500K	82
Mercury vapor, 6800K	45

Table 6.3. Luminous efficacies of lighting system for commonly used light sources.

the total electrical power P_{el} supplied to the system:

$$K_s = \frac{P_v}{P_{\mathrm{el}}}. \tag{6.56}$$

The efficacies of commonly used light sources are given in Table 6.3.

6.4 Luminance, Brightness, and Contrast

While luminance is a psycho-physical quantity relating radiance to the spectral sensitivity of the human visual system, it is not always a good measure of how the stimulus that caused this quantity of luminance is perceived. The perception of a certain amount of luminance, whether emitted by a source, or reflected off a surface, depends on environmental and physiological conditions, and the perception of the same absolute amount of luminance may thus vary with viewing conditions. For example, the headlights of a car appear brighter during the night than during the day, even though they are the same lights (see Figure 6.14).

To describe the perception of how much light is emitted, brightness, as defined in Section 5.1, can be used. Although brightness depends on external factors,

Figure 6.14. The car lights are on in both images, but they appear much brighter in the dark environment. (This image was published in Erik Reinhard, Greg Ward, Sumanta Pattanaik, and Paul Debevec, “High Dynamic Range Imaging, Acquisition, Display, and Image-Based Lighting”, © Elsevier 2006.)

such as the appearance of surrounding areas, in many cases it is proportional to luminance. Under appropriate conditions, the relation between brightness B and luminance L can be approximated with Stevens' power law (Section 5.5.3) [1262]:

$$B = a L^{1/3} - B_0. \tag{6.57}$$

Here, a and B_0 are constants depending on the viewing conditions. These conditions include the state of adaptation of the observer as well as the background and spatial structure of the viewed object [865].

A fundamental quantity which is closely related to luminance and brightness is *contrast*. Essentially, two types of contrast may be identified: *physical contrast* and *perceived contrast*.

Physical contrast, also known as *luminance contrast*, is an objective term used to quantify luminance differences between two achromatic patches.[1] It may be defined in several ways depending on the spatial configuration of patches. For instance, the contrast C between a foreground patch superimposed on a background patch is characterized by

$$C = \left| \frac{L_f - L_b}{L_b} \right|, \tag{6.58}$$

where L_f and L_b are the luminances of the foreground and background patches. If two patches L_1 and L_2 reside side by side forming a bi-partite field, their contrast may be computed by

$$C = \frac{|L_1 - L_2|}{\max(L_1, L_2)}. \tag{6.59}$$

Finally, for periodic patterns, such as sinusoidal gratings, the contrast may be computed by

$$C = \frac{L_{\max} - L_{\min}}{L_{\max} + L_{\min}}, \tag{6.60}$$

where L_{max} and L_{min} are the maximum and the minimum luminances in the grating. This type of contrast is also called *Michelson contrast* or modulation .

For pairs of diffuse surfaces, the luminance contrast depends solely on their reflectances [941]. Thus, even if the illumination on these surfaces is altered, the luminance will change proportionally, so that the luminance contrast will remain the same.

The second type of contrast is perceived contrast, also known as *brightness contrast*. It is the apparent difference between the brightnesses of two achromatic patches. Since brightness does not only depend on luminance, brightness contrast does not strictly depend on luminance contrast either. In general, brightness

[1]Chromatic patches might also induce color contrast; this is discussed in Chapter 5.

contrast increases with luminance, and this phenomenon is called the Stevens effect [302]. For instance, a black-and-white photograph appears less rich in contrast if viewed indoors compared to viewing it outdoors. Outdoors, blacks appear more black and whites appear more white. The luminance contrast is the same in both cases, unless the photograph is held at angles that cause highlights.

6.5 Optical Detectors

The various radiometric and photometric terms explained so far provide quantities that indicate how much light is emitted, transmitted, or received. We now discuss the devices that are used to measure light and how these quantities may be measured in practice.

The basic function of an *optical detector* is to measure optical radiation. When a photon strikes the surface of a detector, the detector undergoes a change of state.

A photon striking a material may free one of its electrons if the energy of the photon is above a certain threshold, a process called *photo-electricity*. The energy of the released electron is equal to the energy imparted on it by the photon minus the energy it required to depart from its orbit. Thus, by measuring the energy of a released electron, one can find the energy of the incident photon.

Most optical detectors work on this principle, although they vary in construction. Three main types of optical detectors may be distinguished: *photo-emissive*, *photo-conductive*, and *photo-voltaic* detectors. Each detector has a relationship between the incident radiation, and the output signal produced. This relationship is characterized by a detector's responsivity.

6.5.1 Responsivity

The output signal of an optical detector may be expressed in several ways, such as a change in the voltage, current, or conductivity. If we denote the output signal by S and the incident flux by P, the *flux responsivity* R_P of a detector is given by

$$R_P(\lambda, \lambda_0) = \frac{S(\lambda, \lambda_0)}{P(\lambda)}, \tag{6.61}$$

where λ_0 is the wavelength to which the detector is most sensitive. Thus, the responsivity may change as a function of position, direction, and wavelength. Furthermore, the responsivity may change as a function of incident radiation. A detector is said to have a *linear* response if its flux responsivity does not change with the amount of incident radiation, assuming all other factors are kept constant. Linearity is a desired property of optical detectors and often satisfied within the wavelength range of the detector [839].

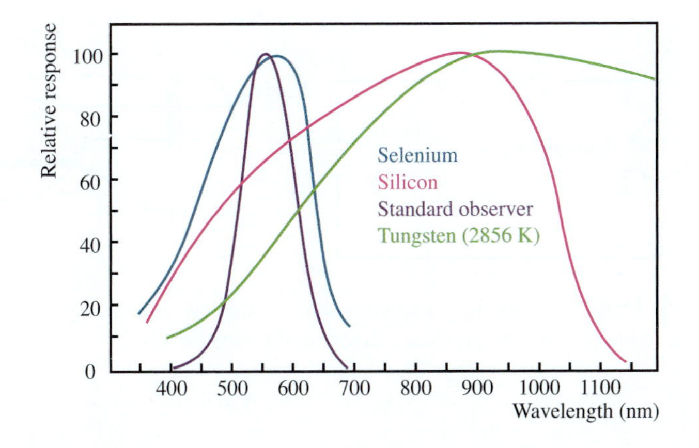

Figure 6.15. The spectral response of various materials used in detectors, as well as the responsivity of the human eye (after [1090]).

In Figure 6.15, the responsivities of two commonly used detector materials, selenium and silicon, are plotted together with the sensitivity of the human eye (i.e., $V(\lambda)$). As can be seen, there are significant differences between the responsivities of different materials and the human eye. In practice, no material has precisely the same responsivity as the human eye. Therefore, to make photometric measurements with optical detectors, specially designed light filters called *photopic correction filters* are used to match the responsivity of detectors to the sensitivity of the human eye.

6.5.2 Photo-Emissive Detectors

Photo-emissive detectors are composed of a vacuum tube and an open circuit through which no current passes in the absence of light (see Figure 6.16, left). The vacuum tube comprises a positively charged anode and a negatively charged cathode. The cathode contains a suitable material that releases electrons when exposed to radiation. A diagram of a vacuum tube is shown in Figure 6.16 (right). The envelope of the vacuum tube is made of glass which lets light in. The envelope is attached with earthed-metal guard rings to prevent leakage of electrons over the glass surface. In some cases the tube is filled with a noble gas, such as argon, to enable a higher current between anode and cathode [1199].

Usually, the cathode contains alkali metals which release their single valence band electrons easily when exposed to light. Therefore, when light strikes the cathode it releases some of its electrons. These free electrons are collected at the

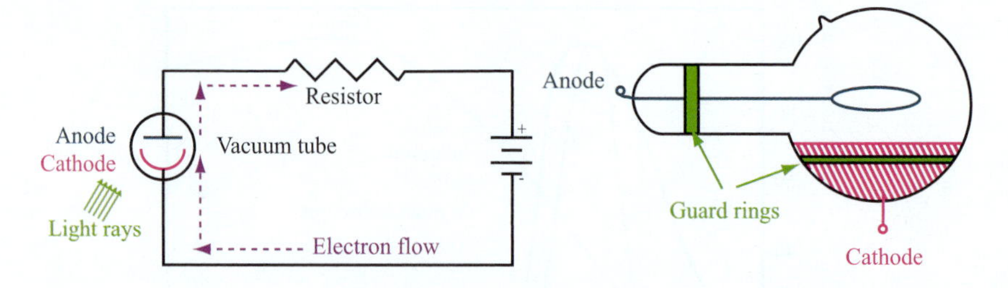

Figure 6.16. Photo-detection by photo-emission. On the left, a photo-emissive circuit is shown. The figure on the right presents a diagram of a vacuum tube.

anode due to its positive voltage, resulting in a continuous flow of current. By measuring this current, the flux incident upon the detector can be inferred.

Vacuum tubes by themselves are rarely used in practice. However, a variant called the *photo-multiplier tube* (PMT) is used in applications requiring high sensitivity [941]. The difference between a PMT and a classical vacuum tube is that there are several other electrodes (called dynodes) between the anode and the cathode. When an electron released from the cathode hits the first dynode, more electrons are released. Each subsequent dynode further amplifies the signal, until the anode collects a large number of electrons, yielding a sensitive measuring device. One disadvantage of PMTs is that they require stable and very high voltages to operate.

6.5.3 Photo-Conductive Detectors

Photo-conductive detectors are built from semiconductor materials such as silicon. Normally semiconductors have high resistance and, therefore, low conductivity. In a circuit, as shown in Figure 6.17, no current passes in the absence of light.

However, when exposed to light, the photo-electric effect causes some of the valence electrons in the semiconductor to be freed. With an increasing number of free electrons, the conductivity of the material increases (i.e., the resistance decreases), which results in a higher electrical current flowing through the circuit. Thus, by measuring the change in the current, one can measure the amount of light falling on the detector.

Historically, the most common materials used in photo-conductive detectors are cadmium sulfide (CdS), cadmium selenide (CdSe), and lead sulfide (PbS), whereas currently silicon (Si) is widely used [1024].

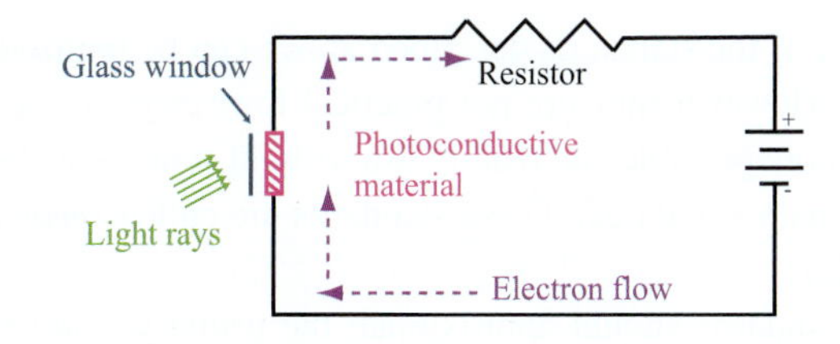

Figure 6.17. Photo-detection by photo-conduction.

6.6 Light Standards

The efforts to quantify light and define certain light sources as standards dates back to the early 1800s. First, flames of certain candles were used as standards, but these were not reliable to the accuracy needed by photometric measurements. Then, more controlled flame lamps such as the Carcel lamp, the pentane lamp, and the Hefner lamp replaced the standard candles during the nineteenth century [1199]. Although they are more reliable and accurately reproducible than candles, flame lamps also do not afford the desired accuracy. After their invention, incandescent filament lamps were used as standards. However, they also lack the desired accuracy and stability.

Proposals were made from time to time to use the radiation emitted from a heated metal of a specified surface area to be used as standard. Different types of metals were tested,[2] and platinum at its melting temperature was found to be the most appropriate. Even platinum was not perfectly reliable due to the difficulties of preventing contamination of pure platinum. Later this standard was revamped, and a full radiator[3] immersed in pure molten platinum at its melting temperature was used as the primary standard.

Although this latest standard served reasonably well in photometric measurements, it has difficulties in construction. It has been a goal for a long time to relate the unit of luminous intensity, the candela, to the optical watt [242]. This desire was realized in 1979 when a new definition of candela was adopted.

The candela is now defined as the amount of luminous intensity in a given direction of a source that emits monochromatic radiation of frequency 540×10^{12} Hz and that has a radiant intensity in that direction of $1/683$ W/sr. This definition is been adopted as the primary standard of light.

---

[2]Suggested radiators were an iron vessel containing boiling zinc, a coil of platinum wire heated to the temperature of a hydrogen flame, and burning magnesium [1199].

[3]Although it is not possible to obtain a full (blackbody) radiator in nature, very close approximations can be made with a thin hollow cylinder with absorbing walls.

Of course, all of the standards described above can be realized under laboratory conditions. However they are not practical for everyday use. For practical measurements, more portable and readily usable light sources are created by comparison with primary standards. These standards are called *secondary standards* (or sub-standards).

Secondary standards should approximate the primary standards to the adequate degree of accuracy. In most cases, the hierarchy is deepened one more level to produce what is called *working standards* by comparing against secondary standards. The accuracy of working standards should be adequate enough to be used in ordinary photometric work [1199].

6.7 Detector Standards

Light standards discussed in the previous section are used to carry out photometric measurements by comparing test lamps against them (see also Section 6.9.1). However, today, such comparisons are rarely made, mainly because of the practical difficulties involved and low reproducibility of results. Instead, calibrated optical detectors are used for both radiometric and photometric measurements.

The primary standard that underpins the calibrated detectors is the *electrical substitution radiometer*[4] (ESR) that operates at cryogenic temperatures (about $5\text{K} = -272^\circ\text{C}$). The ESR works on the principle of matching the temperature change caused by incident radiant flux (radiant power) with the temperature change due to applied electrical power. When an equilibrium is achieved, the radiant flux is computed from the equivalent electrical power. A simplified illustration of the ESR is shown in Figure 6.18.

When the shutter is open, the radiant flux enters the receiving cone and increases the temperature of the heat sink whose initial temperature is T_0. Then, the shutter is closed and the temperature of the heat sink is maintained by increasing the electrical current through the heater with resistance R. If we denote the current that is just sufficient to maintain the temperature of the heat sink by i, the electrical power P consumed at the heater and hence the optical power P_e will be equal to

$$P_e = P = i^2 R. \tag{6.62}$$

Due to its low operating temperature, very accurate measurements can be performed with a cryogenic ESR. However, the main purpose of an ESR is to calibrate secondary standards or *transfer standard detectors* which in turn are used to calibrate working standards [877].

[4] Also known as *absolute cryogenic radiometer* or *electrically calibrated radiometer*.

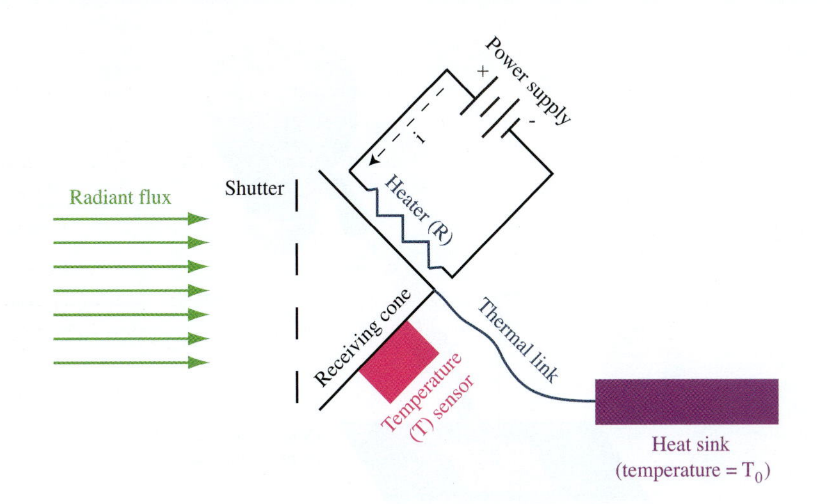

Figure 6.18. An illustration of the ESR, simplified to bring out its important components. (This illustration is based on the drawing in NIST Technical Note 1421 [877].)

6.8 Measurement of Optical Radiation

We discuss some of the equipment that may be used to measure various radiometric and photometric quantities. These include the integrating sphere, gonioradiometers, photometers, and spectroradiometers. We begin with the measurement equation, which lies at the heart of the operation of these measurement devices.

6.8.1 The Measurement Equation

The measurement equation relates the exitant radiation from a surface to the output signal of a radiometric instrument. All radiometric quantities can be measured by proper manipulation of the measurement equation.

The element of radiant flux incident on the detector surface at the differential area dA centered around point (x,y), and within a solid angle $d\omega$ centered around direction (θ,ϕ), is given by (Figure 6.19)

$$dP(x,y,\theta,\phi,\lambda) = L(x,y,\theta,\phi,\lambda)\, dA\, \cos(\theta)\, d\omega\, d\lambda. \tag{6.63}$$

If the responsivity of the detector is R_P, each element of radiant flux produces in the detector a differential output signal dS equal to

$$dS(x,y,\theta,\phi,\lambda,\lambda_0) = R_P(x,y,\theta,\phi,\lambda,\lambda_0)\, dP(x,y,\theta,\phi,\lambda). \tag{6.64}$$

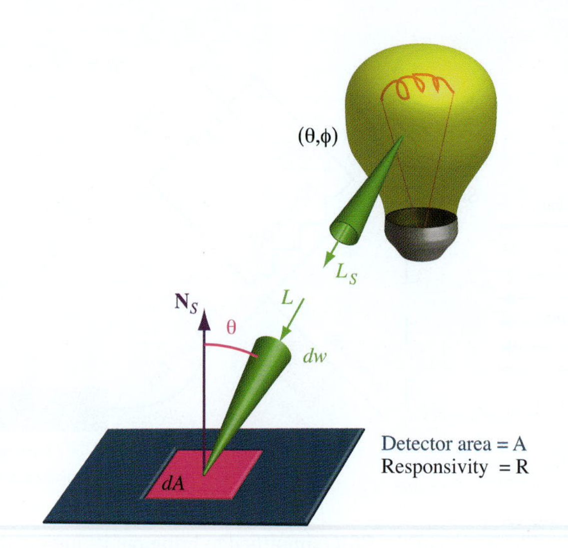

Figure 6.19. The detector with responsivity R and total surface area A is configured to measure the radiance emitted from the light source. For simplicity, the optical system that typically sits in front of the detector is not shown.

Here, λ_0 is the wavelength at which the detector is designed to operate. There is a range of wavelengths $\Delta\lambda$ around λ_0 for which the detector has a significant non-zero response. The total signal S produced by the detector is equal to the sum of all differential signals dS, yielding the *measurement equation*:

$$S(A,\omega,\Delta\lambda,\lambda_0) = \int_{\Delta\lambda}\int_A\int_\omega dS(x,y,\theta,\phi,\lambda,\lambda_0)\, d\omega\, dA\, d\lambda \tag{6.65a}$$

$$= \int_{\Delta\lambda}\int_A\int_\omega R_P(x,y,\theta,\phi,\lambda,\lambda_0)\, L \cos(\theta)\, d\omega\, dA\, d\lambda, \tag{6.65b}$$

where ω is the solid angle subtended by the source being measured at point (x,y). As an aside, it should be noted that ω may change based on the position (x,y) on the detector surface, and, therefore, ω is a function of position (x,y).

Equation (6.65b) is valid for the configuration given in Figure 6.19. In this form, we are measuring the incident radiance on the detector rather than the emitted radiance from the source. In a lossless medium, these would be equal due to the invariance of radiance. In practice, however, the radiance may be attenuated by the intervening medium and due to losses during refraction by the optical system. As discussed in Section 6.2.8, the incident radiance L is related to the source radiance L_S via the propagance τ of the optical path:

$$L(x,y,\theta,\phi,\lambda) = L_S(x,y,\theta,\phi,\lambda)\, \tau(x,y,\theta,\phi,\lambda). \tag{6.66}$$

Consequently, the final form of the measurement equation becomes

$$S(A,w,\Delta\lambda,\lambda_0) = \int_{\Delta\lambda}\int_{A}\int_{\omega} R_P L_S \tau \cos(\theta)\, d\omega\, dA\, d\lambda. \tag{6.67}$$

In this equation, L_S is the radiance to be measured, and S is the output signal of the measurement instrument. If the instrument is calibrated, then S equals L_S. However, if the device is uncalibrated, a second measurement should be made whereby the light source is replaced with a standard lamp with known radiance. For a standard lamp, the measurement equation is

$$S^{\mathrm{s}}(A,w,\Delta\lambda,\lambda_0) = \int_{\Delta\lambda}\int_{A}\int_{\omega^{\mathrm{s}}} R_P L_S^{\mathrm{s}} \tau^{\mathrm{s}} \cos(\theta)\, d\omega\, dA\, d\lambda. \tag{6.68}$$

Here, L_S^{s} is the known radiance of the standard lamp. The known quantities are the results of the two measurements S and S^{s} and the radiance L_S^{s} of the standard lamp. To solve for L_S using (6.67) and (6.68), we assume that the values of L_S and L_S^{s} are constant over the integration intervals A, w, w^{s}, and $\Delta\lambda$. In practice, the aperture size of the measuring device is about 1°, and the response is sufficiently flat over $\Delta\lambda$, making this assumption valid [839]. As a result, L_S and L_S^{s} may be taken out of their integrals in (6.67) and (6.68).

In addition, we will assume that τ equals τ^{s} and ω equals ω^{s}. The integrals in these equations are then identical, so that combining these two equations yields a solution for L_S:

$$L_S = S\left(\frac{L_S^{\mathrm{s}}}{S^{\mathrm{s}}}\right). \tag{6.69}$$

6.8.2 The Integrating Sphere

An apparatus commonly used for radiant and luminous power as well as irradiance and illuminance measurements is the *integrating sphere*, also known as *Ulbricht sphere* or *averaging sphere*. The integrating sphere is a hollow sphere with uniform diffusely reflecting inner walls (Figure 6.20). Light entering the entrance port or diverging from a lamp inside the sphere is diffusely reflected several times until it strikes the detector. The detector is protected from direct light with a screen, which ensures that radiation is reflected at least once before impinging upon the detector.

Of the two integrating spheres shown in Figure 6.20, the sphere on the left is appropriate to measure the radiant power of the light source inside, whereas the sphere on the right is designed to measure the irradiance at the entrance port.

The usefulness of the integrating sphere lies in the fact that illumination due to reflected light at any point on the surface of the sphere is the same independent of

which can be rewritten as

$$dP_{i,1,2} = \frac{dR}{4r^2}\frac{\rho P}{\pi} + \frac{dR}{4r^2}\frac{\rho^2 P}{\pi}. \tag{6.82}$$

After an infinite number of reflections, the flux received by dR can be written as

$$dP = dP_{i,1,2,\ldots} = \frac{dR\,P}{4\pi r^2}(\rho + \rho^2 + \ldots). \tag{6.83}$$

Since $0 \leq \rho < 1$, this expression can be further simplified:

$$dP = \frac{dR\,P}{4\pi r^2}\frac{\rho}{1-\rho}. \tag{6.84}$$

And, finally, the irradiance (or the illuminance) on dR is equal to

$$E = \frac{dP}{dR} = \frac{P}{4\pi r^2}\frac{\rho}{1-\rho}, \tag{6.85}$$

which is independent of the location of the light source and directly proportional to its power. Thus, by measuring the irradiance or illuminance due to reflected light (i.e., by means of a detector), it is possible to compute the total flux of the light source. The first use of an integrating sphere in this manner (i.e., as a photometer) was made by Ulbricht [1162].

If used for irradiance measurements, rather than the total flux of the source, the irradiance at the entrance port may be computed from the incident flux and the area of the entrance port.

So far, we assumed a perfect sphere with perfectly diffusing walls. Unfortunately, the presence of the light source, the screen, the entrance port, and the detector cause deviations from perfect behavior. This problem may be mitigated by constructing the sphere as large as possible. For flux measurements, it is not uncommon to use integrating spheres with a diameter of 5 m [841]. The distance of the screen to the light source is also important. It is found that optimum results are achieved with a screen distance of $0.4\,r$ [1121].

Integrating spheres used for irradiance measurements are much smaller, often in the order of a few centimeters. The main reason to use an integrating sphere for measuring irradiance is to enable an accurate cosine response. In the absence of an integrating sphere, radiation arriving from wide angles tends to reflect more of the detector surface, causing an inaccurate measurement.

The integrating sphere has proven to be useful for purposes other than flux or irradiance measurements. For instance, if the detector is removed, radiance emitted from the exit port will be uniform in all directions. Therefore, it can serve as a uniform diffuser which enables its use for accurate camera calibration.

6.8.3 Measurement of Radiant and Luminous Flux

Since radiant and luminous flux are measures of the total radiation emitted from a source in all directions, the preferred technique to measure these quantities is by use of an integrating sphere. Dependent on the geometry and the emission characteristics, the source may be placed at the entrance port or inside the sphere. For instance, if the emission is limited to a hemispherical solid angle, such as most LEDs and spot lights, the source is placed at the entrance port facing toward the center of the sphere. For sources emitting over wider solid angles, including incandescent and fluorescent lamps, the source must be placed inside the sphere.

The total flux of a test light source may be measured with one of the following four techniques:

1. the direct measurement method;
2. the substitution method;
3. the simultaneous method;
4. the auxiliary lamp method.

In the direct measurement method, the detector is replaced with a spectroradiometer to obtain the spectral radiant flux. This method has several benefits, in that not only the total power of the lamp is obtained but also important color properties such as chromaticity coordinates (Section 7.5.1), the source's correlated color temperature (Section 9.2.3), and its color-rendering index (Section 9.3).

In the other three methods, a standard lamp and the test lamp are measured alternatively. In this case, the ratio of the photo-detector signals is assumed to be directly proportional to the ratio of the power of each lamp:

$$P_T = P_S \left(\frac{S_T}{S_S} \right), \tag{6.86}$$

where S denotes the output signal of the photo-detector and the subscripts S and T identify the standard and the test lamps.

In the substitution method, the lamps are placed inside the sphere one by one, preferably in the center of the sphere [1199]. In the simultaneous method, both lamps are in the sphere together (symmetrically along the vertical axis), although only one lamp is lit for each measurement. An additional screen is placed between the lamps to prevent the opaque parts of the lamps absorbing direct light from each other. In the auxiliary lamp method, a third lamp is placed inside the sphere to control the self-absorption of light emitted by each lamp itself.

6.8.4 Measurement of Radiant and Luminous Intensity

Radiant and luminous intensity are measures of the directional flux distribution of a light source. Therefore, to measure these quantities a means of taking an adequate number of measurements from various angles around the source is necessary.

An instrument that is used for this purpose is called the *goniophotometer* (and similarly *gonioradiometer* for radiometric measurements). A goniophotometer allows rotation of an optical detector over 360° around a fixed light source both in vertical and horizontal directions. The frequency of measurements depends on the directional characteristics of the light source. For instance, if the light source is fairly uniform, 36 measurements each separated by 10° in the horizontal plane may be sufficient [1199]. More measurements are likely to be needed around the vertical plane, as most sources manifest a greater variation along their vertical axis.

The movement of the detector follows a spherical path such that the distance between the measured source and the photo-detector is unaltered between measurements. Also, the light source should be entirely within the field of view of the detector. All additional areas that are in the field of view must be blackened so that only the source radiation is measured.

6.8.5 Measurement of Irradiance and Illuminance

Irradiance and illuminance represent the surface density of the incident electromagnetic radiation. In most cases, radiation arrives from a hemispherical solid angle surrounding the surface to be measured.

The chief practical consideration for measuring irradiance is to account for the cosine law (Section 2.9.2). According to the cosine law, a surface can only receive radiation proportional to its area projected in the direction of propagation. If the angle between the surface normal and the incident light rays is θ, the projected area of the surface in the direction of propagation is equal to

$$A' = A \cos(\theta). \tag{6.87}$$

The difficulty to satisfy the cosine law arises from the fact that as the light rays arrive at the surface from grazing angles, they also tend to reflect more off the surface (see Section 2.5.2).[5] Without accounting for this effect, irradiance and illuminance values will be underestimated.

[5] This is a different way of saying that near grazing-angles paints tend to be more reflective than a Lambertian surface.

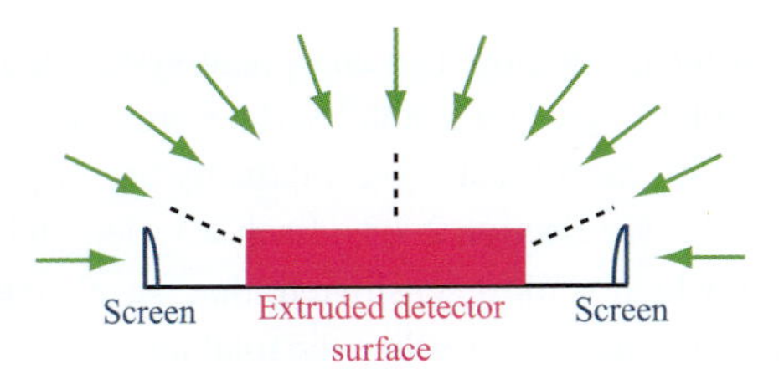

Figure 6.21. An extruded detector surface. Paths of three of the light rays are depicted with dashed lines. The rays incident from narrow angles are received by the top of the detector, while the rays incident from wide angles are received by the edges. Screening prevents contribution of light from very wide angles.

The solution to this problem is called *cosine correction* and can be achieved in several ways. Most detectors have highly diffuse covers to minimize the reflectance of the detector. In addition, the detector surface may be extruded to enable light entrance through the edges of the detector as well. Such a detector is depicted in Figure 6.21. In this case, light arriving from the horizontal direction ($\theta = 90°$) should be blocked with proper screening. A second alternative is to use an integrating sphere to average out light arriving from a very wide solid angle.

6.8.6 Measurement of Radiance and Luminance

Radiance and luminance are best measured with a *spectroradiometer*, which can also reveal the wavelength distribution of electromagnetic radiation. Luminance may be computed from radiance with built-in circuitry.

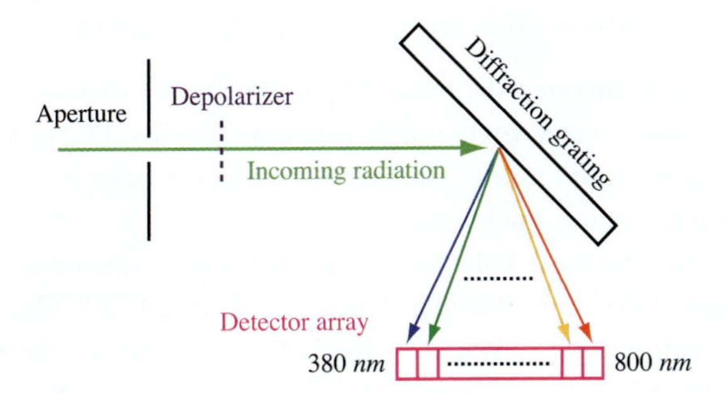

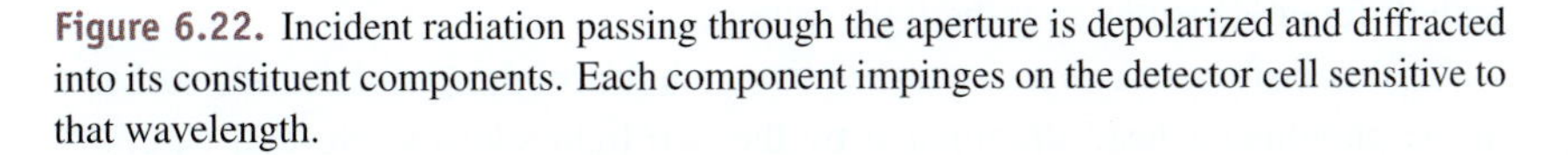

Figure 6.22. Incident radiation passing through the aperture is depolarized and diffracted into its constituent components. Each component impinges on the detector cell sensitive to that wavelength.

A spectroradiometer is composed of an optical system that admits radiation from a very narrow solid angle and a small surface area. The admitted radiation is then diffracted into its constituent wavelengths by use of a diffraction grating as shown in Figure 6.22 (see Section 2.7.3). Each wavelength is directed to a cell on the detector array which is most sensitive to that wavelength. The number of cells on the detector array may be less than the total number of wavelengths (e.g., a cell for every four wavelengths). This then becomes the wavelength resolution of the instrument.

6.9 Visual Photometry

Photometric measurements may also be carried out using visual comparison performed by human observers. Using human observers in photometric measurements was predominantly the method of choice until the 1940s. Such methods, however, suffer from several drawbacks. For instance, results obtained from one set of measurements can deviate from the results obtained from another, even for the same group of observers. Thus, beside the inconvenience of an elaborate set-up, visual photometry has the disadvantages of poor reproducibility and low precision [185].

A further disadvantage is that humans can only compare lights in a subjective sense. For instance, one can distinguish the brighter of the two sources, but one cannot quantitatively say how much brighter that source is. In general, visual photometry is impractical for daily use. However, a faster and more convenient method for measuring color matching functions is outlined in Section 7.7 that largely overcomes the aforementioned disadvantages.

6.9.1 Visual Photometry using Brightness Matches

The fundamental technique used in visual photometry is matching the brightness of two surfaces illuminated by two light sources. Usually, one of the sources is taken as a standard with known characteristics, and the other is a test source for which the measurement is made.

The surfaces lit by these light sources are presented to the observer by means of an instrument called the *photometer head* (see Figure 6.23). The observer can adjust the amount of light reflected by the surfaces by moving the light sources closer to or farther from the photometer head, or by placing neutral density filters with different densities over the light sources.

In this system, observers are asked to match the brightness of the two surfaces in the photometer head illuminated by the two light sources. Several carefully

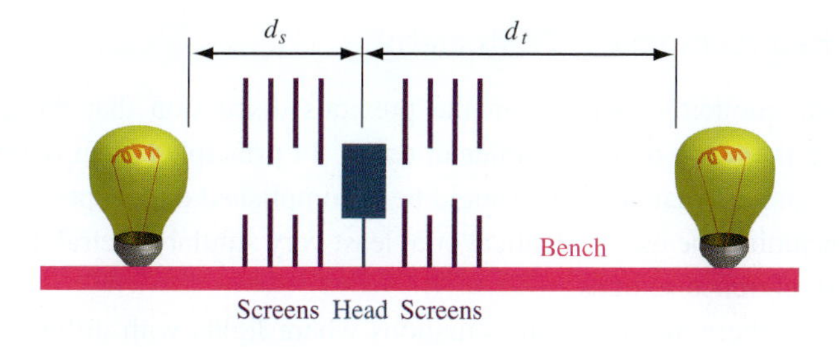

Figure 6.23. A photometer based on the inverse square law.

placed screens are used to block stray light so that light enters the photometer head only from the light sources. One of the sources is a known standard whose intensity is previously established, and the other source is a test source whose intensity is to be measured. The observer is able to adjust the brightness of the illuminated surfaces by sliding the lights towards and away from the photometer head, effectively using the inverse-square law (Section 2.9.3)

The brightness match is obtained when the illumination of both surfaces inside the photometer head appear the same:

$$\frac{I_s \rho_s}{d_s^2} = \frac{I_t \rho_t}{d_t^2}, \tag{6.88}$$

where I_s and I_t are the intensities of the standard and the test sources, ρ_s and ρ_t are the reflectances of the illuminated surfaces, and d_s and d_t are the distances of the sources from the surfaces. The measured term I_t may then be computed with

$$I_t = I_s \frac{\rho_s}{\rho_t} \left(\frac{d_t}{d_s} \right)^2, \tag{6.89}$$

or, if the surfaces have the same reflectance,

$$I_t = I_s \left(\frac{d_t}{d_s} \right)^2. \tag{6.90}$$

Although the photometer in this example is designed to measure luminous intensity, other photometric quantities may be measured similarly with different configurations of the instrument. For instance, since luminance is intensity per unit area, luminance at any point of the light source may be measured by admitting light from a smaller portion of the source.

It should be noted that there are numerous instruments and measurement techniques used in visual photometry. An excellent reference is Walsh [1199].

6.9.2 Heterochromatic Photometry

It has been implicitly assumed in the previous discussion that the compared sources are the same or very similar in color. In principle, when observers are making brightness matches, this should be accomplished under specific test conditions, including the use of identical or at least very similar spectral distributions of test and reference sources.

However, there are numerous situations where lights with different spectral distributions need to be compared. Even for two metameric light sources, their spectral distribution is unlikely to be identical. Therefore when making photometric measurements, techniques to mitigate the effects of color differences should be considered. Visual photometry carried out with stimuli of different spectral composition is called *heterochromatic photometry*.

The first solution is comparing lights of similar colors. Even though the spectral compositions of the lights may not be identical, if the differences are subtle these might not be perceived by the eye and may not harm the measurement. For instance, with a difference of 100° K in color temperature, measurements can be performed with an accuracy of about 1 percent [1199] (color temperature is discussed in Sections 3.5.1 and 9.2). When the color differences are larger, the gap in color between the sources may be divided into steps, and at each step an intermediate auxiliary lamp may be used. Then each lamp is compared with the next one, forming a link between the original sources that are compared.

Another technique is called the *compensation method* or the *mixture method*. In this method, light emitted from one of the sources is mixed with light from an auxiliary source of a different color forming a mixture which is similar in color to the other source. For instance, by adding a small amount of blue light to light emitted from a standard source of low color temperature, its color temperature may be increased to a level where it becomes comparable to the test source. As long as the intensity of the added blue light is known, the intensity of the test source can be calculated.

A fundamental but more involved method splits the radiation emitted from each light into its spectral components that are monochromatic and compares each component separately to find the intensity of each. These intensities are then summed to find the total intensity of the test source. This process may be carried out with spectrophotometers. In this method, measurements should be made with care, since errors introduced for each component will also add up in the final result.

It is possible to use color filters to match the spectral distributions of the compared lights. If a specially designed color filter is placed in front of one of the lamps, the spectral distribution of the light passing through this filter will match that of the other light with which the comparison is made. Then, classical methods of homochromatic photometry may be used to measure the intensity of the transmitted light. Of course the transmission characteristics of the filter should be known (or measured with some other method) to adjust the measured intensity accordingly.

In the case of large chromatic differences between source and test patches, it is still possible to carry out brightness matches with good accuracy, for instance, with the flicker photometer method. An instrument called the *flicker photometer* is used for the purpose of presenting two lights on a surface in alternation at a minimum frequency such that when the brightness of the two lights match, the perception of flicker will disappear [1199]. The method relies on the fact that the temporal resolving power of the human eye is lower for colored stimuli than for achromatic stimuli. This means that there exist frequencies for which chromatic differences between the alternating patches is not perceived. By adjusting the intensity of one of the patches, a brightness match occurs if the perceived achromatic differences disappear as well. Typical frequencies are between 10 Hz and 40 Hz [1262].

While a heterochromatic flicker-photometry experiment can be constructed with a physical device whereby synchronized shutters are opened and closed in opposite phase, an accurate experiment carried out on a display device is complicated. Computers and graphics cards are getting ever faster, but are rarely able to guarantee a fixed and stable frame-rate. Drift and fluctuations are likely to occur, and they affect the accuracy of the measurement [591]. A general problem with heterochromatic flicker photometry is that the frequencies required to measure brightness matches may induce photosensitive epileptic seizures in some individuals, making the technique usable only in controlled laboratory experiments [530].

Brightness matching, in the case of large chromatic differences between source and test patches, moves into the realm of color matching, which is discussed in the following chapter on colorimetry. There, we will also introduce a recent technique to improve the accuracy of visual brightness matching that can be performed on computer displays and show how it can be used to derive observer-specific iso-luminant color maps (Section 7.7.1).

6.10 Application: Measuring Materials

Measurements of object reflectance and light emission are important in many different fields, for instance in lighting design. In computer graphics, accurate image synthesis can only be achieved if modeling, rendering, and display are all based on accurate data and measurements. Even the best renderer is still a garbage-in-garbage-out system, meaning that if the modeling is not accurate, the resulting image will not be accurate.

Modeling of an environment requires the acquisition of geometry, the emission spectra of light sources, as well as the measurement of materials. In this section, we consider the problem of measuring materials. Opaque spatially homogeneous materials can be modeled with bi-directional reflectance distribution functions (BRDFs), which were introduced in Section 2.9.7. Such functions relate the amount of light reflected in a direction of interest to the amount of light incident upon a point on a surface. The relation is given as a ratio and may be defined as a function of wavelength.

With incident and exitant direction defined as two-dimensional polar coordinates, a BRDF is a five-dimensional function. For isotropic BRDFs the dimensionality is reduced to four, as rotation of the surface around the surface normal at the point of interest does not alter the ratio of reflected light. If the BRDF is, in addition, measured with tristimulus values rather than a full spectral set of samples, the wavelength dependence is removed, and the dimensionality is reduced to three dimensions.

This means that to measure an isotropic BRDF for the purpose of using it in a renderer requires a material sample to be measured along three dimensions. The traditional way to measure a BRDF is with the use of a *gonioreflectometer*.[6] A flat sample containing the material to be measured is placed in a fixed position. A calibrated point light source can move along one direction over the hemisphere placed over the sample material. The detector can independently move over the entire hemisphere. Together, they span the three degrees of freedom required to sample all possible combinations of incident light and measurement angles, as shown in Figure 6.24 [744]. Anisotropic BRDFs can be measured by giving the position of the light source a second degree of freedom.

Every measurement, thus, provides a data point for one angle of incidence paired with one angle of reflectance. A similar result can be obtained by placing the detector in a fixed position and rotating the sample over two degrees of freedom. The light source is then still moved over the same path as before.

In either set-up, the number of samples needs to be large to get a dense sampling of the BRDF. Such a dense sampling is required in the absence of further knowledge about the sort of material being measured. This makes the acquisition of a full BRDF long-winded, even if the gantry maneuvering the detector and the

[6]See the Cornell Light Measurement Laboratory (http://www.graphics.cornell.edu/research/measure/) and the Nist reference reflectometer at the Starr facility (http://physics.nist.gov/).

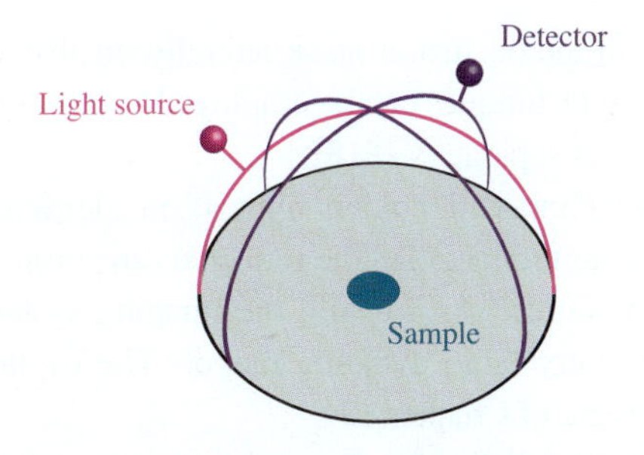

Figure 6.24. In a gonioreflectometer, the light source and detector have one and two degrees of freedom to move over the hemisphere placed over the sample to be measured. This configuration enables the acquisition of isotropic BRDFs.

light source is under computer control. If specific materials are being measured, then it may be possible to measure only a subset of all angles of incidence and reflectance.

A different approach to make the acquisition procedure tractable is to capture many samples at once, usually by employing a digital camera. Two of the degrees of freedom in the measurement device are then replaced by the two image dimensions. For flat samples this may be achieved by placing a hemispherical mirror over the sample and photographing it through a fish-eye lens [1212]. A perhaps simpler hardware set-up may be achieved by requiring the sample to be curved. Spherical [744, 745, 751, 752] or cylindrical samples [701] are simplest to capture, since a single photograph will capture many different orientations of the surface normal with respect to the light source. The requirement of simple shapes exists because this allows an analytic model to infer the surface orientation at each pixel.

More complex material samples can also be used, as long as they are convex to avoid inter-reflections within the sample material. In that case, the camera may be augmented with a laser scanner device to capture the 3D geometry of the sample [744]. From the geometry, the surface normals at each pixel can be inferred; these are, in turn, used to relate pixel intensities to angles of incidence and angles of reflectance.

For concave objects one may expect significant inter-reflections that will render the material measurement inaccurate. One would have to separate the pixel values into direct and indirect components. While to our knowledge this has not been achieved in the context of material measurements, separation of direct and indirect components is possible by placing an opaque surface with small regularly spaced holes (an occlusion mask) in front of the light source. After taking multiple captures with the occlusion mask laterally displaced by small amounts, image post-processing techniques can be employed to infer the direct and indirect illumination on surfaces separately [818].

BRDFs can reflect almost all incident light. If, in addition, the BRDF is highly specular, then the amount reflected in the reflected direction is only limited by the intensity of the light source. As a result, the imaging system should be able to capture a potentially very large dynamic range. The capture of high dynamic range images is the topic of Chapter 12.

The number of materials that can be represented by a BRDF, although large, is still limited as it assumes that all surface points have identical reflectance properties. It is possible to extend the measurement of BRDFs to account for spatially varying reflectance. Such spatially varying reflectance functions are called *bidirectional texture functions* (BTFs). They can be measured with the aid of a robot arm, a light source, and a digital camera using flat samples [231, 232].[7]

In the case of BRDFs, and especially BTFs, the number of samples required to represent a material is very large. It is therefore desirable to fit this quantity of data to analytic BRDF models. This can be achieved in several different ways. Some of the resulting models are discussed in Section 2.9.7. For further information, we refer the interested reader to the relevant references [231, 582, 601, 637, 639, 640, 671, 744, 764, 998, 1013, 1212, 1240, 1285].

6.11 Further Reading

Radiometry and photometry are explained in Wyszecki and Stiles' *Color Science: Concepts and Methods, Quantitative Data and Formulae* [1262]. There are several dedicated books, including *Introduction to Radiometry and Photometry* by McCluney [761], *Optical Radiation Measurements* by Grum and Becherer [397], *Handbook of Applied Photometry* by DeCusatis [242], and *Photometry and Radiometry for Engineers* by Stimson [1090].

[7] A collection of BTFs and BRDFs obtained with this technique is available from the Columbia-Utrecht reflectance and texture database at http://www.cs.columbia.edu/CAVE/software/curet/index.php.

Chapter 7
Colorimetry

Color is a purely psychological phenomenon; that is to say, without a human observer there could be no color. Though perceptual in nature, that does not mean that color cannot be well measured, quantified, and communicated. *Colorimetry* is the science of color measurement and description. To measure color, we must take into account the physical properties of the environment including material and light sources, as well as the physiological properties of the human visual system.

Since color is ultimately a human response, the science of color measurement should begin with human observations. As we learned in Section 4.3.1, the human visual system contains two types of photodetectors; rods and cones. The three types of cones are primarily responsible for color vision. The simplest form of color measurement might be to directly record the cone responses for any given stimulus. While this might create a direct physiological measurement, the potentially invasive procedure, not to mention the variation between any two individuals, probably outweighs its usefulness. Another method for color measurement that correlates with the human cone responses is necessary, namely colorimetry.

This chapter outlines the experimental procedures, as well as statistical models, used to measure and specify colors. We will start with an explanation of Grassmann's laws of additive color matching. Using these laws, we will build a visual color-matching experiment, as well as a visual colorimeter. Historical color-matching experiments and their results will be discussed. These results have resulted in several standardized "observers," which can be thought of as statistical approximations to actual human observers. Color measurements can be specified through integration of these standard observers, as discussed in Section 7.5. Finally, the practical applications of colorimetry and the standard observers will be discussed.

It is very important to stress that although colorimetry is based upon the human visual system, it is not a technique for describing how colors actually appear to an observer. It is purely a method for accurately quantifying color measurements and describing when two colors "match" in appearance. The ultimate visual response to any stimulus requires much more information about the viewing conditions, including (but not limited to) size, shape, and structure of the stimulus itself, color of surround, state of adaptation, and observer experience. The science of describing what a stimulus actually looks like is called color appearance modeling and is described in detail in Chapter 11.

7.1 Grassmann's Laws

Before we can begin any discussion of colorimetry, or visual color matching, it is important to make a few assumptions. Wyszecki and Stiles sum up the experimental laws of color matching as those of trichromatic generalization [1262]. The trichromatic generalization states that since humans have three fundamental cone types, any color stimulus can be matched completely with an additive mixture of three appropriately modulated fixed primary-color sources. We see this generalization in practice daily, as most television and desktop monitors utilize only three primaries (typically red, green, and blue) to recreate most of the colors of the world. In practice, there are limitations to the total number of colors that can reproduced with a given set of primaries, known as the reproducible gamut of the display device. More details on display devices can be found in Chapter 14.

The trichromatic generalization allows us to make color matches between any given stimulus and an additive mixture of other color stimuli. These additive color matches follow the general rules of algebra. The mathematical rules of color matching were described by Grassmann in 1853 and are known as *Grassmann's laws of additive color matching* [386]. These laws, like typical algebraic expressions, can be explicitly broken down into four rules [1262]:

Symmetry Law. If color stimulus **A** matches color stimulus **B** then **B** matches **A**. Although this may seem obvious, it is included to show the similarities between Grassmann's laws, and generalized algebra.

Transitive Law. If color stimulus **A** matches **B**, and color stimulus **B** matches **C**, then color stimulus **A** matches **C**. Again, this should seem obvious.

Proportionality Law. If color stimulus **A** matches **B**, then $\alpha\mathbf{A}$ matches $\alpha\mathbf{B}$, where α is any positive scaling factor. It is important to stress, that for the proportionality law to hold, only the radiant power of color stimulus can change;

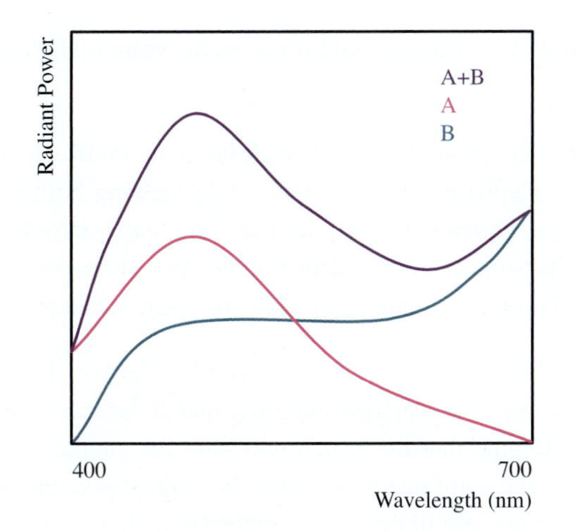

Figure 7.1. Additive color mixing of the spectral power distributions of two color stimuli **A** and **B**.

the overall spectral power distribution cannot change. This can be important when designing color-matching experiments. For instance, if the three primaries are tungsten light bulbs, the radiant power cannot be changed by decreasing or increasing the current to the light source as this will also change the fundamental spectral power distribution of the lights.

Additivity Law. The additivity law is the most important law for generalized color matching and forms the basis for colorimetry as a whole. Essentially, if there are four color stimuli **A**, **B**, **C**, and **D**, and if two of the color matches are true, i.e., **A** matches **B** and **C** matches **D**, $\mathbf{A}+\mathbf{C}$ matches $\mathbf{B}+\mathbf{D}$, then it follows that $\mathbf{A}+\mathbf{D}$ matches $\mathbf{B}+\mathbf{C}$.

When we talk about additive color mixing, in these particular examples, we are essentially talking about adding the spectral power distributions of radiant light. That is to say, we are assuming there are no interactions between wavelengths, and we simply add the power of two stimuli or sources at any given wavelength interval. An example of this is shown in Figure 7.1.

It is important to understand that additive color mixing is generally only valid for adding light itself. It should not be assumed that material properties such as transmittance and reflectance will behave in a linear manner. Wyszecki and Stiles also stress that there are three fundamental considerations that are ignored regarding Grassmann's laws [1261]:

1. *The dependence on the observational conditions on a match.* If two stimuli match under one given viewing condition, there is no reason to assume that

the match will continue to hold if one of the stimuli is placed in a different condition.

2. *Observer adaptation state.* Although the match will generally hold if the observer's adaptation state changes while viewing both stimuli, an effect known as *persistence of color matches*, the match may break down when viewing the stimuli independently under two different adaptation states. More details on chromatic adaptation are given in Chapter 10.

3. *The dependence on the match for a given observer.* If two stimuli (with disparate spectral power distributions) match for one person, there is no reason to assume that the match will hold for another person. The variability between color matches among the general population is surprisingly large. When we add in the color deficient population, it should be obvious to the casual observer that what they consider to be a color match will not necessarily match for all (or any!) other people.

Grassmann's laws provide the framework for developing a robust method for measuring and communicating colors. Remember that an ideal method for measuring colors would be to directly measure the human cone responses for any given stimulus. Unfortunately this is not a feasible method for specifying color. If we had the spectral sensitivities of the cone photoreceptors, as described in Section 4.3.3, it would be possible to estimate the response through mathematical integration of the color stimulus and the cones. It is only within the last several decades that we have had a solid understanding of what the cone responses actually are. So, historically, we have not been able to use the cones as a method for describing color measurements.

Using Grassmann's laws of additive color mixing, researchers in the early 1920s were able to estimate the cone responses in a series of color-matching experiments. Section 7.2 describes those historical experiments and how they were used to generate computational systems of colorimetry.

7.2 Visual Color Matching

If we knew the cone responses, it would be easy to specify color matches in terms of integrated responses to radiant power stimuli. Thus, for each of the three cone types, the radiant power of the stimulus is multiplied on a wavelength-by-wavelength basis with the cone spectral sensitivity, and then the response is integrated across all wavelengths. This concept is illustrated in (7.1), where $\Phi(\lambda)$

is the spectral power distribution of the stimulus, and $L(\lambda)$, $M(\lambda)$, and $S(\lambda)$ are the relative cone spectral sensitivities:

$$\mathbf{L}_1 = \int_\lambda \Phi(\lambda) L(\lambda)\, d\lambda, \tag{7.1a}$$

$$\mathbf{M}_1 = \int_\lambda \Phi(\lambda) M(\lambda)\, d\lambda, \tag{7.1b}$$

$$\mathbf{S}_1 = \int_\lambda \Phi(\lambda) S(\lambda)\, d\lambda. \tag{7.1c}$$

From these equations, we can state that if the relative integrated cone responses $\mathbf{L}_1$, $\mathbf{M}_1$, and $\mathbf{S}_1$ for one stimulus equal those for a second, $\mathbf{L}_2$, $\mathbf{M}_2$, and $\mathbf{S}_2$, then the two color stimuli must, by definition, match.

The integrated **LMS** cone responses can be thought of as *tristimulus values*. We can now state that two color stimuli match when their integrated tristimulus values are identical. Knowing the cone spectral-power distributions would allow us to perform these tristimulus calculations to specify color measurements. Unfortunately, it is only recently that we know how to obtain cone sensitivities. Historically, they have been very difficult to measure. If the cone sensitivities were idealized, such as those shown in Figure 7.2, then they would be easy to measure. We could just use a monochrometer to scan the visible wavelength range and record the relative response to each wavelength. The real cone spectral sensitivities, as discussed in Section 4.3.3 and shown again in Figure 7.3 have a high degree of overlap. This overlap means that it is impossible to isolate the visual response to a single cone excitation and impossible to experimentally measure the individual cone sensitivities. Other techniques for estimating the cone responses are necessary.

At this point, one might question why our eyes have evolved in such a way that our photodetectors are so highly overlapping. From an information-processing point of view, this suggests that the **LMS** cone signals are highly correlated and not data-efficient. This is true, although processing at the ganglion level does partially de-correlate the cone signals into opponent colors before transmission to the brain. The high degree of cone overlap actually provides the human visual system with a very high sensitivity to wavelength differences by examining the relative **LMS** signals with each other.

Imagine two monochromatic lights with the same radiant power at 540 nm and 560 nm. If these lights are viewed using the idealized cone spectral sensitivities shown in Figure 7.2, only the middle sensitive cone will be excited. The integrated

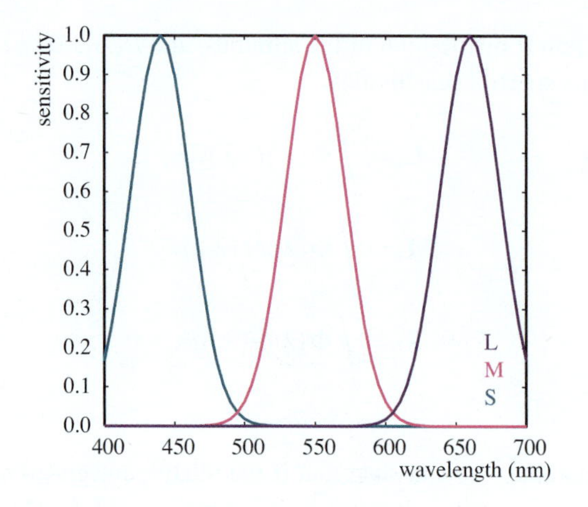

Figure 7.2. Idealized cone spectral sensitivities. The cones evenly sample the visible wavelengths and have minimal overlap.

response to both the lights will be the same, and the observer will not be able to distinguish the color difference between the two light sources.

Now consider the same two light sources, and the real cone spectral sensitivities as shown in Figure 7.3. All three cones will respond to the 540 nm light, while the **L** and **M** cones will respond to the 560 nm light. The relative **LMS** responses to the two lights will be very different, however. By comparing the

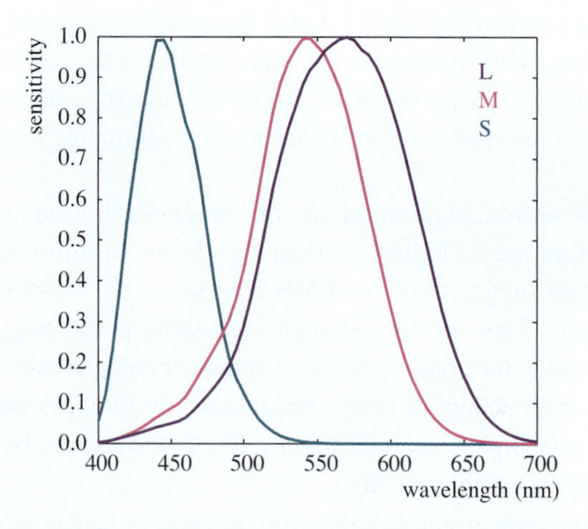

Figure 7.3. Actual cone spectral sensitivities. Notice the high degree of overlap (correlation) between the three cone types. (From [1091].)

relative responses, it is then possible for the visual system to distinguish the two color stimuli as being different. This behavior is discussed in further detail in Section 7.2.1.

7.2.1 Univariance and Metamerism

An important consideration for colorimetry, and generating color matches in general, should be immediately obvious when viewing (7.1). The act of integrating the cone responses suggests that our visual system does not act as a wavelength detector. Rather, our photoreceptors act as approximate linear integrators. That means that two color stimuli that have distinctly different spectral power distributions can elicit identical cone responses. In the example given in the previous section, we saw that two different light sources can cause identical cone responses. This is known as the *principle of univariance.*

Another example of this can be seen in Figure 7.4. Two monochromatic light sources with the same radiant power, at approximately 420 nm and 475 nm have the same integrated response from our short wavelength **S** cones. The relative responses of the **L** and **M** cones is different, so our visual system is still able to detect the difference between these two light sources in this example.

The principle of univariance, and the fact that our cones act as photo-integrators, allows two color stimuli that have very different spectral-wavelength properties to appear to match to a human observer. This is known as *metamerism* and allows for a wide range of color-reproduction devices. Essentially, we can

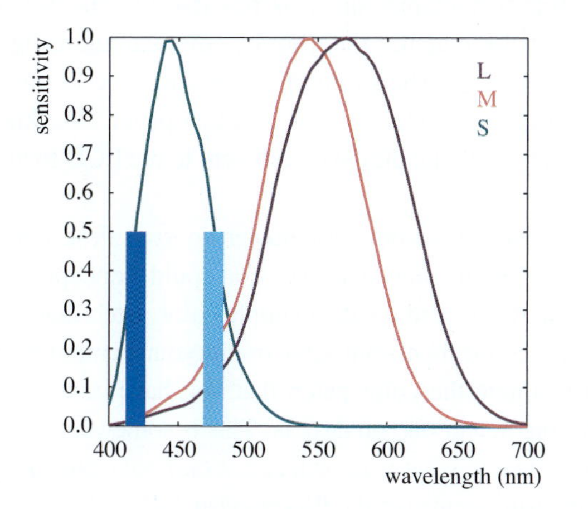

Figure 7.4. Demonstration of the principle of univariance. Two monochromatic light sources at different wavelengths integrate to the same **S** response.

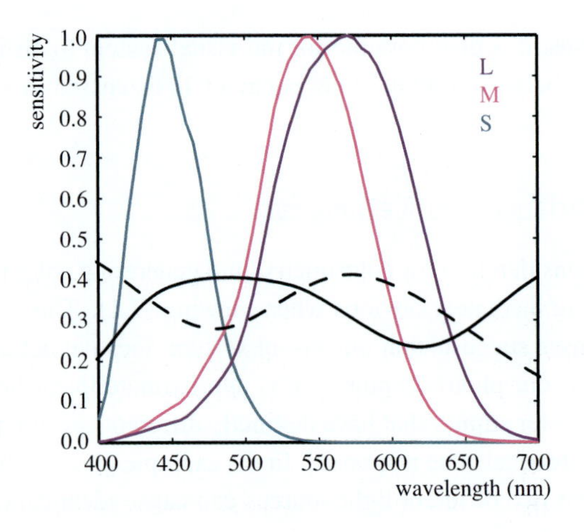

Figure 7.5. Iconic example of metamerism. The two color stimuli, represented by the solid and dashed black lines, integrate to the same **LMS** signals and appear to match, despite have markedly different spectral power distributions.

now conceive of creating a color match between any given spectral power distribution by generating a differing spectral power distribution that integrates to the same **LMS** cone responses. An iconic example of this is shown in Figure 7.5.

7.2.2 A Device for Visual Color Matching

Using this principle of metamerism, it is possible to conceive of a device that allows us to create color matches for arbitrary color stimuli using a small number of known color *primaries*. The simplest concept may be to think of using your computer and a display device, such as an LCD panel. We know that we can adjust the red, green, and blue digital counts sent to the LCD to generate arbitrary colors.

Suppose you have a favorite color flower on your desk that you want your friend who is in a different country to see. You could, conceptually, open up your favorite graphics software and create a simple color patch that matches the color of the flower. You could then send your friend your computer and LCD screen and have them examine the color patch that you have created. They will then know what color the flower on you desk is. This is somewhat inconvenient, but in essence you have created what is known as a *visual colorimeter*; a device used to generate color measurements (of the flower) visually.

If your friend has the identical computer and LCD panel, it may be possible to simply send them the graphics file that contains the correct red, green, and blue values so that they can open that file on their own system. If the manufacturer of the computer and LCD has excellent quality control, then it is possible that your friend will see the same color that matches the flower on your desk. Of course, because color perception depends on many factors other than just the display itself, such as whether the computer is viewed in the dark or in a brightly lit office, there is no way to guarantee that your friend is seeing the same color.

Now we can imagine a device specifically designed to generate these types of visual color matches. It could be as simple as a box with three light sources and a screen. We can generate color matches by adjusting the relative amounts of those three primaries and then send those coordinates to anyone else in the world who has the same device. If the device is viewed in identical conditions, then we can be relatively satisfied that when they dial in the same relative amounts of the primaries, they will get a color match. With this concept, it is possible to *measure* the color of any stimulus in the world simply by dialing in a color match. This form of color measurement does not need to have any knowledge of the actual cone spectral sensitivities, as long as we create the color matches with a consistent set of primaries.

There are obvious practical limitations to this type of color measurement. For one, it is quite inconvenient to have to create visual matches for all measurement needs. There is also the problem with the difference between matches created by different people. What matches for one person will quite possibly not match for another person.

If we instead perform this type of visual color matching for a wide variety of colored stimuli and a wide variety of observers, it is possible to generate an "average" color-matching data set. Experimentally, this can be performed using a set-up similar to that shown in Figure 7.6.

An observer looks at an opening in a wall and sees a circular patch with two colors shown in a bipartite field. His task is to adjust the colors of one of the fields such that it matches the other. Behind the wall are three light-source primaries on one side and a single monochromatic light source on the other, all pointing at a white screen. The observer can adjust the relative amounts of the three primaries (without altering the spectral power distribution) to create a match for the monochromatic light. He can then repeat this procedure across all wavelengths in the visible spectrum.

In practice, as shown in Figure 7.6, it is necessary to also have adjustable primaries in the reference field. This is because monochromatic light appears very saturated to humans, and it is difficult to match with three broad primaries. Adding light from the broad primaries to the monochromatic light will desaturate the appearance. In essence, using Grassmann's laws as described in Section 7.1,

The spectral tristimulus values indicate the amount of each of the three primaries (hence, *tri*stimulus) necessary to generate a color match, for any given spectral wavelength. For instance, at 440 nm, we would need approximately -0.0026 units of **R**, 0.0015 units of **G** and 0.3123 units of **B**. These curves can also be thought of as *color-matching functions.* We can can determine the amount of each primary necessary to create a color match by integrating the spectral power distribution of the color stimuli with the color-matching functions. These integrated values can then be thought of as *tristimulus values.* (Note, we no longer need to refer to them as spectral.) The generic form of this calculation is:

$$\mathbf{R} = \int_{\lambda} \Phi(\lambda) \cdot \bar{r}(\lambda)\, d\lambda, \tag{7.3a}$$

$$\mathbf{G} = \int_{\lambda} \Phi(\lambda) \cdot \bar{g}(\lambda)\, d\lambda, \tag{7.3b}$$

$$\mathbf{B} = \int_{\lambda} \Phi(\lambda) \cdot \bar{b}(\lambda)\, d\lambda. \tag{7.3c}$$

In these equations, **R**, **G**, and **B** are the tristimulus values, Φ is the color-stimuli spectral power distribution, and $\bar{r}(\lambda)$, $\bar{g}(\lambda)$, $\bar{b}(\lambda)$ are the CIE *rgb* color-matching functions shown in Figure 7.7. Using these tristimulus values to measure color, it is now possible to determine if two stimuli match. If their **RGB** values are the same, then the colors match. We can eliminate the need for visual observation by calculating these relationships directly. The generalized relationship for a color match is shown in (7.4). It is important to stress that all three tristimulus values must be the same, in order to guarantee a color match.

$$\int_{\lambda} \Phi_1(\lambda) \cdot \bar{r}(\lambda)\, d\lambda = \int_{\lambda} \Phi_2(\lambda) \cdot \bar{r}(\lambda)\, d\lambda, \tag{7.4a}$$

$$\int_{\lambda} \Phi_1(\lambda) \cdot \bar{g}(\lambda)\, d\lambda = \int_{\lambda} \Phi_2(\lambda) \cdot \bar{g}(\lambda)\, d\lambda, \tag{7.4b}$$

$$\int_{\lambda} \Phi_1(\lambda) \cdot \bar{b}(\lambda)\, d\lambda = \int_{\lambda} \Phi_2(\lambda) \cdot \bar{b}(\lambda)\, d\lambda. \tag{7.4c}$$

We now have a method for calculating when colors match, or we can measure a color, based upon the spectral power distribution of the stimuli. It should be possible to communicate the **RGB** tristimulus values to our friend and have them generate a color match using their three monochromatic primaries at 435.8, 546.1, and 700 nm. If they do not have access to those primaries, they can calculate the appropriate linear transform (3×3 linear matrix) necessary to generate a match with the primaries that they do have. One such transformation, to an imaginary primary set, is discussed in Section 7.4.

7.4 CIE 1931 and 1964 Standard Observers

Using Grassmann's laws of additive color mixing, as well as the transformation of primaries, the CIE was able to calculate a set of color-matching functions based upon the 17 observers from the Wright and Guild experiments. These color-matching functions, $\bar{r}(\lambda)$, $\bar{g}(\lambda)$, and $\bar{b}(\lambda)$, were based upon a specific set of real primaries: monochromatic lights at 435.8, 546.1, and 700 nm.

These primaries were chosen because the first two were spectrum lines from a mercury lamp, while the 700 nm light was chosen so as to isolate the **L** cone as much as possible, using the knowledge of the visual system available at the time [93]. In 1931 numerical integration techniques were difficult and expensive to perform, and integration with both positive and negative numbers (as necessary using the CIE **RGB** color-matching functions) were even more difficult. Additionally, the CIE desired to combine the system of colorimetry with the 1924 CIE system of photometry, as discussed in Chapter 6. Having a fourth integrating function, CIE $V(\lambda)$ would require more computation than what was desired.

The CIE set out to define a transform to a new set of primaries with two specific goals: having an all positive response and having one of the primaries, or color-matching functions, be the photopic luminance response function, CIE $V(\lambda)$. More details on the derivation of this transformation can be found in Fairman [304].

This linear transformation from RGB to an all positive set of color-matching functions required a set of imaginary primaries. That is to say, in order to successfully reproduce all colors in the visible spectrum, we need to use lights that cannot be physically realized. The color-matching functions settled upon were referred to as $\bar{x}(\lambda)$, $\bar{y}(\lambda)$,and $\bar{z}(\lambda)$, and the integrated tristimulus values are CIE XYZ so as to not be confused by realizable RGB tristimulus. Using imaginary primaries means that it is impossible to create a device to actually recreate the color matches, as we could using the three monochromatic primaries described in Section 7.3. However, this just means that we use this new system of colorimetry strictly in a computational manner. The linear transformation from CIE RGB to XYZ is shown in Equation 7.5:

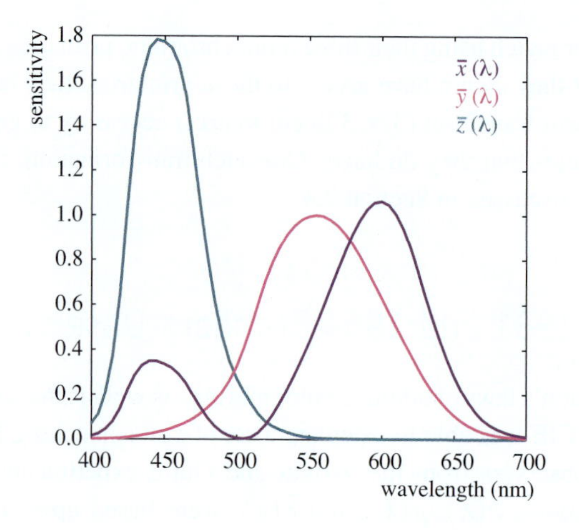

Figure 7.8. The CIE 1931 $\bar{x}(\lambda)$, $\bar{y}(\lambda)$, and $\bar{z}(\lambda)$ color-matching functions.

$$\begin{bmatrix} X \\ Y \\ Z \end{bmatrix} = \frac{1}{0.17697} \begin{bmatrix} 0.4900 & 0.3100 & 0.2000 \\ 0.17697 & 0.81240 & 0.01063 \\ 0.0000 & 0.0100 & 0.9900 \end{bmatrix} \cdot \begin{bmatrix} R \\ G \\ B \end{bmatrix}. \tag{7.5}$$

The scaling factor $\frac{1}{0.17697}$ shown in this equation was used to normalize the functions to the same units as CIE $V(\lambda)$. Applying (7.5) to the $\bar{r}(\lambda)$, $\bar{g}(\lambda)$, and $\bar{b}(\lambda)$ color-matching functions gives us the $\bar{x}(\lambda)$, $\bar{y}(\lambda)$, and $\bar{z}(\lambda)$ color-matching functions:

The very first thing to notice about these color-matching functions is that, indeed, they are all positive. It should also be immediately obvious that these functions do not look like the human cone spectral sensitivities, although we can use them to calculate color matches in much the same way as we could with cones. The $\bar{y}(\lambda)$ color-matching function should look almost identical to the 1924 photometric standard observer. These color-matching functions have come to be known as the *CIE 1931 standard observer*, or similarly the *CIE* $2°$*-observer*; they are meant to represent the color-matching results of the average human population [186].

The term $2°$-observer comes from the fact that both Wright and Guild used a bipartite field that subtended approximately two degrees of visual angle, and as such all the color matches were made using just the fovea. As we learned in Section 4.3.1, this meant that the color matches were performed using only the cones while eliminating any rod contribution. This was vitally important, because

Grassmann's laws of additive color mixing can break down when there are both rod and cone contributions to color mixing.

What this means, is that the CIE 1931 standard observer is designed to predict color matches of very small stimuli (think a thumbnail at arm's length). Another important consideration is that the average human response used for color measurements was taken from a pool of 17 observers, over 75 years ago! Despite the age, and apparent limitations of the 1931 standard observer, it is still used with great success today and is still the foundation of modern colorimetry.

Not willing to rest on their laurels with the introduction of the 1931 standard observer, the CIE continued to encourage active experimentation to both validate the existing color-matching functions and to test the use of them for larger color patches. Through observations lead by Stiles at the National Physics Laboratory in the UK, with ten observers using both 2° and 10°, it was determined that the 1931 standard observer was appropriate for the small-field color matches [1262].

There were discrepancies between the small field results and the large field results, and so further testing was required. Stiles and Burch proceeded to measure an additional 49 observers using the 10°-bipartite field [1087]. The Stiles and Burch data was collected at a relatively high luminance level, in an attempt to minimize the contribution of the rods to the color matches. Some computational techniques were also utilized to eliminate any remaining rod contribution [93].

Around the same time, Speranskaya measured the color-matching functions for 27 observers, also using a 10° viewing field [1074]. This data was measured at a lower luminance level, and so the contributions of the rods were thought to be higher. The CIE computationally removed the rod contribution and combined these two data sets. A transformation to imaginary primaries, very similar to that performed to generate the 1931 standard observer, followed. What resulted were the *1964 CIE supplementary standard observer* [186, 1262].

Like the 1931 standard observer, the 1964 standard observer is also commonly referred to as the 10°-*standard observer*. The color-matching functions for the 10°observer are expressed as: $\bar{x}_{10}(\lambda)$, $\bar{y}_{10}(\lambda)$, and $\bar{z}_{10}(\lambda)$. These functions are shown in Figure 7.9. For comparisons to the 1931 standard observer, both sets of color-matching functions are shown in Figure 7.10.

The CIE recommends the use of the 1964 color-matching functions for all color stimuli that are larger than 4° of visual angle. It is important to emphasize, as shown in Figure 7.10, that the $\bar{y}_{10}(\lambda)$ color-matching function does not equal the photometric observer ($\bar{y}(\lambda)$ or $V(\lambda)$), and that calculations using the 1964 standard observer do not directly translate into luminance measurements.

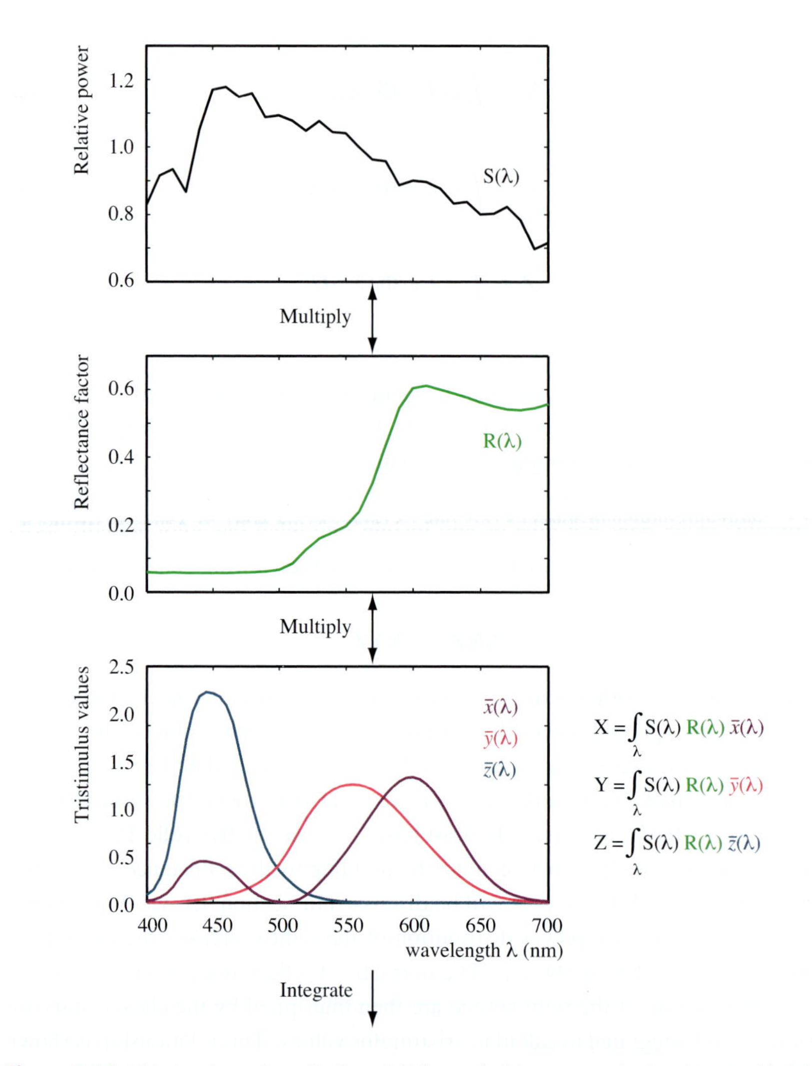

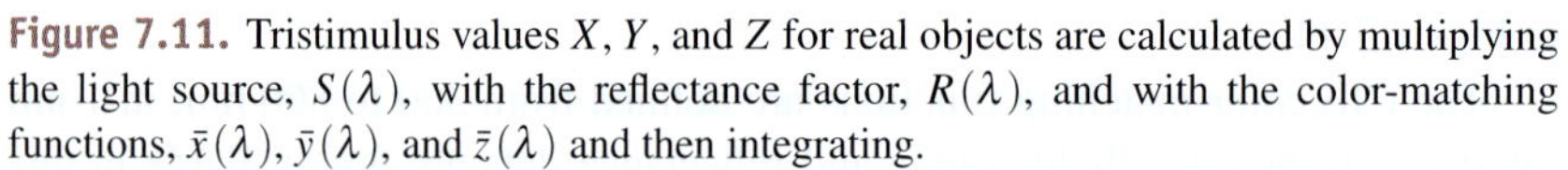

Figure 7.11. Tristimulus values X, Y, and Z for real objects are calculated by multiplying the light source, $S(\lambda)$, with the reflectance factor, $R(\lambda)$, and with the color-matching functions, $\bar{x}(\lambda)$, $\bar{y}(\lambda)$, and $\bar{z}(\lambda)$ and then integrating.

These calculations are

$$X = k\sum_{\lambda}\bar{x}(\lambda)\,S(\lambda)\,R(\lambda)\,\Delta\lambda, \tag{7.8a}$$

$$Y = k\sum_{\lambda}\bar{y}(\lambda)\,S(\lambda)\,R(\lambda)\,\Delta\lambda, \tag{7.8b}$$

$$Z = k \sum_{\lambda} \bar{z}(\lambda) S(\lambda) R(\lambda) \Delta\lambda \tag{7.8c}$$

$$k = \frac{100}{\sum_{\lambda} \bar{y}(\lambda) S(\lambda) \Delta\lambda.} \tag{7.8d}$$

In these equations, we see that the normalization constant k is designed to assure that calculations for a perfect reflecting diffuser will always result in a *luminance factor* of $Y = 100$. The luminance factor can be defined as the ratio of the luminance of an object to that of a perfect reflecting diffuser, when viewed under the same geometry and lighting conditions. When we calculate relative tristimulus values, as we have above, the CIE Y tristimulus value always corresponds to the luminance factor. In (7.8), $\Delta\lambda$ is the sampled wavelength interval, typically 5, 10, or 20 nm. In practice, when normalizing the tristimulus values, the sampling interval gets removed from the calculations due to its presence in the k calculation.

Suppose we do not want to measure reflecting objects, but rather actual light sources or self illuminating colors (such as a CRT display) that have physically meaningful units. To do this, we must revisit the photometric luminous efficiency function, $V(\lambda)$. Recall that a *lumen* is a unit of luminous flux defined to be 1/683 of a watt at 555 nm, the peak of $V(\lambda)$. Thus, there are 683 lm/W at 555 nm, and there is a relationship between lumens and the weights of the $V(\lambda)$ photometric observer.

When measuring self-illuminating materials, we typically measure either the spectral radiance ($Wm^{-2}sr^{-1}$) or irradiance (Wm^{-2}). Using (7.9), we can calculate either luminance units (cd m^{-2}) or illuminance units (lux, or lm m^{-2})[1]. In these equations, we will use $\Phi(\lambda)$ to be either radiance or irradiance measurements. It is important, for these calculations, to include the wavelength-sampling interval, $\Delta\lambda$, as there is no normalization constant k:

$$X = 683 \sum_{\lambda} \bar{x}(\lambda) \Phi(\lambda) \Delta\lambda, \tag{7.9a}$$

$$Y = 683 \sum_{\lambda} \bar{y}(\lambda) \Phi(\lambda) \Delta\lambda, \tag{7.9b}$$

$$Z = 683 \sum_{\lambda} \bar{z}(\lambda) \Phi(\lambda) \Delta\lambda. \tag{7.9c}$$

7.5.1 Chromaticity Coordinates

The CIE standard observers allow us to calculate consistent color measurements and also determine when two colors match each other. Although these coordinates

[1] See appendix D for an overview of SI units as well as derived units.

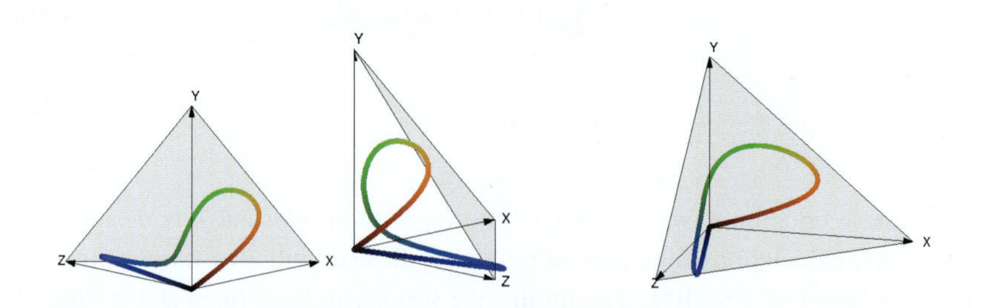

Figure 7.12. The visible spectrum displayed in CIE XYZ tristimulus space, seen from different vantage points. This can also be considered a three-dimensional plot of the 2° color-matching functions. It's projection onto a 2D plane produces the more familiar chromaticity diagrams.

do not represent color appearances, we can use them to represent a stimulus' location in a three-dimensional *color space*. Each of the axes in this space represent the imaginary X, Y, and Z primaries, and a color stimulus' location in the space is its integrated tristimulus values. We can plot the location of the visible spectrum to get an idea of the behavior and shape of the CIE XYZ space. Figure 7.12 shows the spectral colors, which are also the color-matching functions of the 1931 standard observer, plotted in XYZ tristimulus space.

Although we cannot ascertain the appearance of any color based upon its tristimulus values, we show the spectral lines in Figure 7.12 in color. While not entirely appropriate and done mostly for illustrative purposes, the monochromatic colors of the spectrum do not drastically change as a result of viewing conditions and chromatic adaptation. In essence, we are assuming that the colors drawn are those of the individual wavelengths when viewed in isolation, otherwise known as *unrelated colors*.

It is difficult to visualize the shape of the spectral colors in the three-dimensional space. It is also fairly difficult to determine where any given color lies in the space. Often we are interested in knowing and specifying the general region of space a color occupies, but in an easy to understand manner. We know, by definition, that for the 1931 standard observer, the Y tristimulus value is a measure of luminance, which is directly related to our perception of lightness.

The other two tristimulus values do not have as easily interpretable meanings. By performing a projection from three to two dimensions, we can generate a space that approximates *chromatic* information: information that is independent of luminance. This two-dimensional projection is called a chromaticity diagram and is obtained by performing a type of perspective transformation that normalizes the tristimulus values and removes luminance information.

We can think of chromaticity coordinates as ratios of tristimulus values that do not contain any magnitude information. This normalization is shown in (7.10). Chromaticity coordinates are generally expressed using lowercase letters, while we use uppercase letters for tristimulus values.

$$x = \frac{X}{X+Y+Z}, \tag{7.10a}$$

$$y = \frac{Y}{X+Y+Z}, \tag{7.10b}$$

$$z = \frac{Z}{X+Y+Z}, \tag{7.10c}$$

$$x+y+z = 1. \tag{7.10d}$$

It should be obvious that the projection of three-dimensions of information down to two always results in the loss of information. For chromaticity coordinates, the ratio normalization essentially throws away the luminance information. It should also be obvious that because the chromaticity coordinates always sum to 1, if we know two of them we can easily determine the third. On their own, chromaticity coordinates do not possess enough information to adequately describe a color measurement or a color match. Traditionally, we can use two coordinates (usually x and y) along with one of the tristimulus values to fully describe a color. Since the Y tristimulus value describes luminance, you will often see colors described with their Yxy values. We can calculate tristimulus values back from chromaticities using (7.11).

$$X = \frac{x}{y}Y, \tag{7.11a}$$

$$Z = \frac{z}{y}Y. \tag{7.11b}$$

Plotting the spectral colors in the chromaticity diagram results in the familiar horseshoe-shaped curve. These spectral locations are referred to as the spectrum locus, and they represent the boundary of physically realizable colors. The spectrum locus for the 1931 standard observer is shown in Figure 7.13. Once again, in this figure we plot the spectral wavelengths in color. While this is somewhat appropriate for the monochromatic lights (again, when viewed in isolation) it is not appropriate to draw colors inside the spectral locus.

As we have emphasized in this chapter, colorimetry is not designed to predict how colors appear, but rather just provide a means for measuring and describing color matches. Chromaticity diagrams have an even more limited description

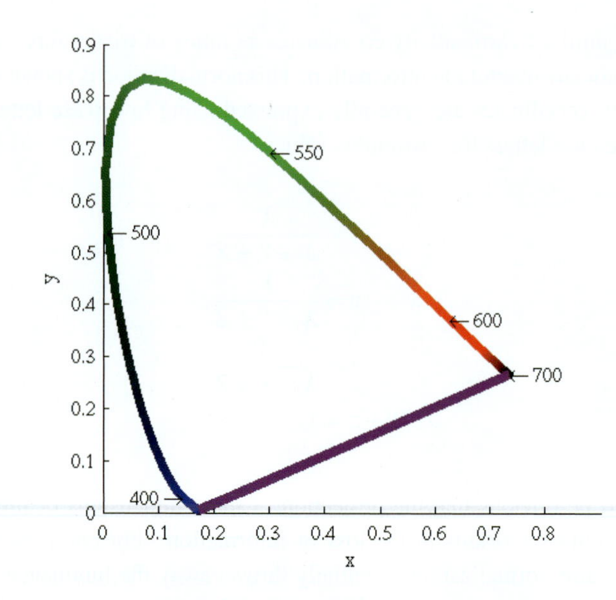

Figure 7.13. The visible spectrum displayed in the CIE xy chromaticity diagram.

of color, as they lack any luminance information. Depending on viewing conditions, luminance levels, and state of adaptation, it is possible to make almost any location in the chromaticity diagram appear either black or white (or any other color!). This is evidenced in Figure 7.14, where the chromaticities of several CIE standard illuminants are plotted in a chromaticity diagram. Despite the variety of chromaticity locations, real light sources at these same coordinates would generally appear white.

The "purple" line plotted at the bottom of the horseshoe-shaped spectrum locus in Figure 7.13 does not exist as a pure spectral color, but rather as a combination between the short and long wavelength stimuli. Appropriately enough, this is referred to as the *purple line* or *purple boundary*.

As chromaticities are simply normalized tristimulus values, it is possible to specify the range of colors that can be generated by two or more primaries, when these primaries are represented in a chromaticity diagram. For instance, for a two-primary system, we draw a straight line in the chromaticity diagram connecting the two primaries together. The system can then reproduce any of the chromaticities on that line, by changing the relative luminances of the two primaries.

For a three-primary system, we draw lines in the chromaticity diagram that connect the three primaries together, with the range of reproducible chromaticity coordinates lying inside the triangle bounded by those lines. This is illustrated in

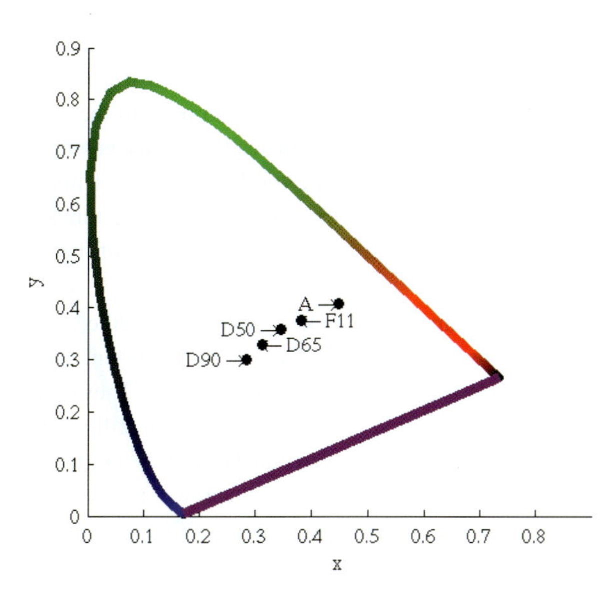

Figure 7.14. The location of various CIE standard illuminants in the 1931 *xy* chromaticity diagram.

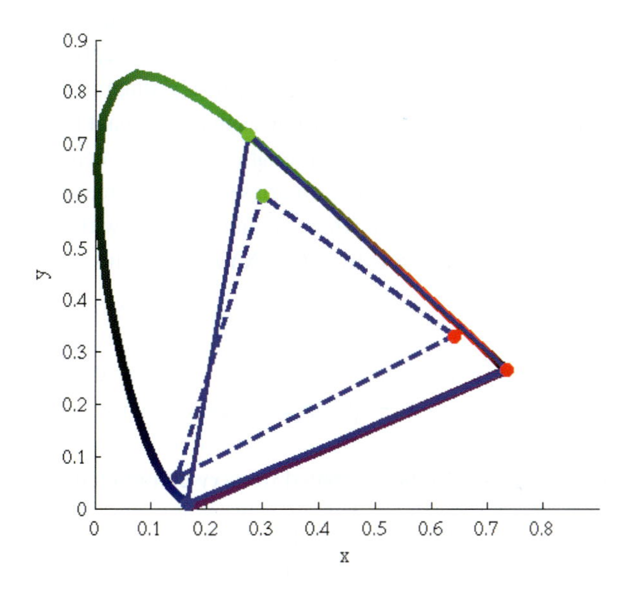

Figure 7.15. The chromaticity boundaries of the CIE RGB primaries at 435.8, 546.1, and 700 nm (solid) and a typical HDTV(dashed).

that is not tied to any specific display device or specific set of primaries. Most readers have probably encountered the situation where they create or examine an image on one computer display and then look at the image on a second computer display, only to find out that the image looks markedly different. Despite the fact that the digital file contains the exact same red, green, and blue values, each computer interprets these data differently. The images are assumed to be *device-dependent*, essentially measurements tied to the specific device on which they were created. Imagine if other measurements in the world were based in such a manner, for instance a unit of length based upon each individual foot size!

By removing the display device from the picture, we can specify color in a meaningful way, based upon the imaginary but well-defined XYZ primaries. This is similar to basing the measurement of length on an imaginary, but well-defined, *golden foot*. We can then transform our colorimetric units into any other unit, or any other display device, much like we can convert the foot to the meter, fathom, or furlong. This is the basis of modern color management.

So how can one take advantage of the CIE system of colorimetry? When creating digital images, we can create them directly in tristimulus space, though due to the imaginary nature of the space this may prove difficult. For computer-graphics renderings that rely on physics, there are several ways to directly take advantage of colorimetry. If, in the rendering process, all calculations are performed in the spectral domain, including light source propagation as well as material and light source interactions, then it is easy to just integrate the resulting spectral radiance values of the scene with the CIE $\bar{x}(\lambda)$, $\bar{y}(\lambda)$, and $\bar{z}(\lambda)$ color-matching functions, and integrate as shown in (7.6). We will then have perceptually accurate measurements of the rendered image.

But how do we display colorimetric data? For that, it is necessary to apply another *transformation of primaries*.

7.6.1 Transforming Primaries

Recall from Section 7.3, that the original color-matching data from Wright and Guild were transformed to a common set of primaries before being averaged together. This transformation utilized Grassmann's laws to express one set of spectral tristimulus values in terms of another. This was accomplished using a linear transform, which for a three-primary display is represented by a 3×3 matrix transform. For additive color systems that obey Grassmann's laws, the calculation of this transform is very straightforward.

Desktop display systems, such as CRTs, behave as additive systems and, as such, calculating the transformation from XYZ tristimulus values into the native

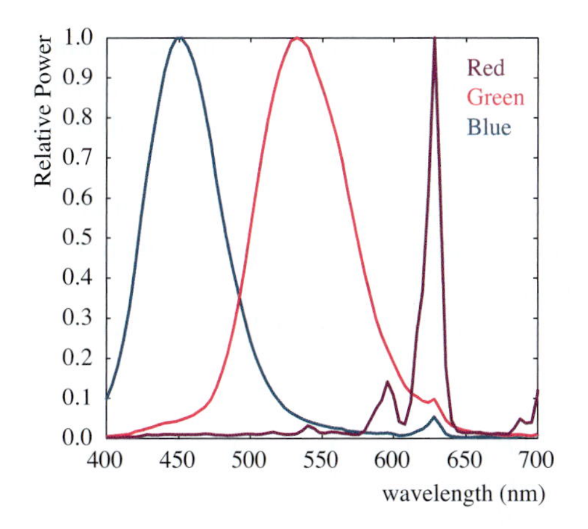

Figure 7.17. The relative spectral power distribution of a representative CRT display.

primaries of the display is relatively easy. This is known as *display characterization*; more details can be found in Section 14.18, as well as in the works by Berns [93] and Day [237]. LCDs are more complicated, as they have a very bright black level, and also have different spectral characteristics dependent on intensity. The following discussion is therefore only approximate for LCD displays.

It is important to stress that Grassmann's laws assume a completely linear system, which most displays are inherently not. When characterizing a display, it is very important to first remove the nonlinear components, often referred to as *gamma*, before applying the colorimetric transformation. We will ignore the nonlinear component in this section, though further details are presented in Chapter 14.

A typical CRT display has three additive primaries; red, green, and blue. An example of such a display is shown in Figure 7.17. Due to the phosphors of a typical CRT display, there are relatively broadband blue and green primaries and a relatively sharp red primary. If we would like to figure out the transformation from CIE XYZ tristimulus values into the units of the display device, we need to first measure the XYZ values of the three primaries of the CRT. We can then generate a transformation that will take us from display RGB values into XYZ tristimulus values. This is accomplished using (7.15):

$$\begin{bmatrix} X \\ Y \\ Z \end{bmatrix} = \begin{bmatrix} X_{\text{red}} & X_{\text{green}} & X_{\text{blue}} \\ Y_{\text{red}} & Y_{\text{green}} & Y_{\text{blue}} \\ Z_{\text{red}} & Z_{\text{green}} & Z_{\text{blue}} \end{bmatrix} \begin{bmatrix} R \\ G \\ B \end{bmatrix}. \tag{7.15}$$

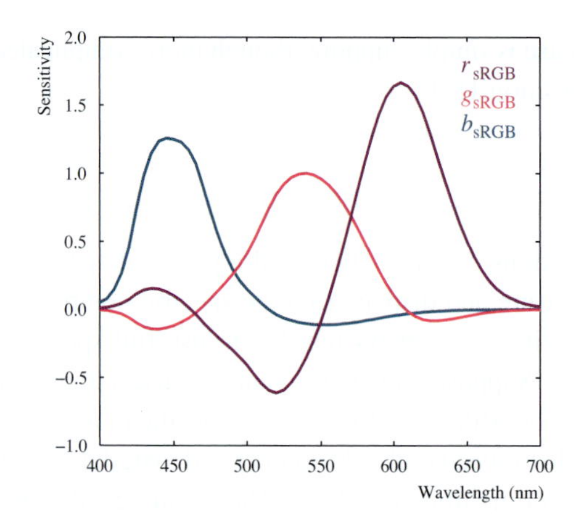

Figure 7.18. The color-matching functions corresponding to the sRGB primaries.

transform from the desired primaries. The special case, where the spectral sensitivities of the capture device are a direct linear transform from CIE XYZ tristimulus values, means that the device satisfies the *Luther-Ives* condition and is *colorimetric* [538, 715].

7.6.3 XYZ to Spectrum Conversion

For some applications, including rendering if a mixture of spectral and tristimulus data is to be used within the same application [416], it may be necessary to convert a tristimulus value XYZ to a plausible spectral representation $\Phi(\lambda)$. This amounts to inverting (7.6), yielding infinitely many solutions, which are each metamers of each other.

To solve this problem, additional constraints must be imposed. The simplest approach is to introduce three linearly independent basis functions $f_1(\lambda)$, $f_2(\lambda)$, and $f_3(\lambda)$. The derived spectrum is then a linear combination of these basis functions:

$$\Phi(\lambda) = \sum_{j=1}^{3} x_j f_j(\lambda), \tag{7.18}$$

where x_j are three coefficients. The values for these coefficients can be found by substituting this equation into (7.6), giving

$$X = \int_\lambda \bar{x}(\lambda) \sum_{j=1}^{3} x_j f_j(\lambda)\, d\lambda, \tag{7.19a}$$

$$Y = \int_\lambda \bar{y}(\lambda) \sum_{j=1}^{3} x_j f_j(\lambda)\, d\lambda, \tag{7.19b}$$

$$Z = \int_\lambda \bar{z}(\lambda) \sum_{j=1}^{3} x_j f_j(\lambda)\, d\lambda. \tag{7.19c}$$

Making the following substitutions:

$$c_0 = X, \tag{7.20a}$$

$$c_1 = Y, \tag{7.20b}$$

$$c_2 = Z, \tag{7.20c}$$

$$\bar{c}_0 = \bar{x}, \tag{7.20d}$$

$$\bar{c}_1 = \bar{y}, \tag{7.20e}$$

$$\bar{c}_2 = \bar{z}, \tag{7.20f}$$

and rearranging terms in (7.19), we have

$$c_i = \sum_{j=1}^{3} \int_\lambda \bar{c}_i(\lambda)\, f_j(\lambda)\, d\lambda\; x_j. \tag{7.21}$$

With the color-matching functions $\bar{c}_i(\lambda)$, the basis functions $f_j(\lambda)$, and the XYZ tristimulus value c_i known, the coefficients x_j can be computed. Plugging these coefficients into (7.18) yields the desired spectral representation of the input XYZ tristimulus value c_i.

However, the choice of basis functions $f_j(\lambda)$ is important, as it will have a profound impact on the shape of the resulting spectrum. For computational simplicity, a set of delta functions could be chosen [369]:

$$f_j(\lambda) = \delta(\lambda - \lambda_j) = \begin{cases} 1 & \text{for } \lambda = \lambda_i, \\ 0 & \text{for } \lambda \neq \lambda_i. \end{cases} \tag{7.22}$$

The coefficients x_j can then be derived from (7.21), which simplifies to

$$c_i = \sum_{j=1}^{3} \bar{c}_i(\lambda_j)\, x_j. \tag{7.23}$$

To achieve a numerically stable solution, the values for λ_j can be chosen to coincide with the peaks in the color-matching functions $\bar{c}_i$, which are $\lambda_1 = 590$ nm,

$\lambda_2 = 560$ nm, and $\lambda_3 = 440$ nm. Of course, with this approach the resulting spectrum will only have three non-zero values. This is far removed from typical spectra, which tend to be much smoother.

Alternatively, the following box functions could be used as basis functions [414, 415]:

$$f_j(\lambda) = \begin{cases} 1 & \text{for } \lambda_i < \lambda < \lambda_{i+1}, \\ 0 & \text{otherwise.} \end{cases} \tag{7.24}$$

The boundaries of the intervals are given by $\lambda_1 = 400$ nm, $\lambda_2 = 500$ nm, $\lambda_3 = 600$ nm, and $\lambda_4 = 700$ nm. This approach yields spectra which vary abruptly at the interval boundaries.

For use in rendering, it is desirable to create spectra which are relatively smooth, especially if they represent reflectance functions of materials [1063]. To enforce a smoothness constraint, the first three Fourier functions can be chosen as basis functions [265, 1204]:

$$f_1(\lambda) = 1, \tag{7.25a}$$

$$f_2(\lambda) = \cos\left(2\pi \frac{\lambda - \lambda_{\min}}{\lambda_{\max} - \lambda_{\min}}\right), \tag{7.25b}$$

$$f_3(\lambda) = \sin\left(2\pi \frac{\lambda - \lambda_{\min}}{\lambda_{\max} - \lambda_{\min}}\right). \tag{7.25c}$$

This approach leads to smooth spectra, but there is a significant probability that parts of the spectra are negative, and they can therefore not be used. This tends to happen most for highly saturated input colors. Although the negative parts of any spectrum could be reset to zero, this would introduce undesirable errors.

An arguably better approach would be to parameterize the basis functions to make them adaptable to some characteristic of the tristimulus value that is to be converted to a spectrum. In particular, it would be desirable to create smooth spectra for desaturated colors, whereas saturated colors would yield more spiked spectra. As an example, Gaussian basis functions may be parameterized as follows [1106]:

$$f_j(\lambda) = \exp\left(-\ln 2 \left(\frac{2(\lambda - \lambda_j)}{w_j}\right)^2\right). \tag{7.26}$$

Here, λ_j indicates the center position of the jth Gaussian, and w_j is its width. The values λ_j are $\lambda_1 = 680$ nm, $\lambda_2 = 550$ nm, and $\lambda_3 = 420$ nm.

For a tristimulus value given in, for instance, the sRGB color space, a rough indication of the degree of saturation is obtained by the following pair of values:

$$s_1 = \frac{|sR - sG|}{sR + sG}, \tag{7.27a}$$

$$s_2 = \frac{|sB - sG|}{sB + sG}. \tag{7.27b}$$

We can now compute the widths of the Gaussian basis functions by linearly interpolating between the user-defined minimum and maximum Gaussian widths $w_{\min} = 10$ nm, and $w_{\max} = 150$ nm:

$$w_1 = s_1\, w_{\min} + (1 - s_1)\, w_{\max}, \tag{7.28a}$$

$$w_2 = s_2\, w_{\min} + (1 - s_2)\, w_{\max}, \tag{7.28b}$$

$$w_3 = \min(w_1, w_2). \tag{7.28c}$$

This approach will produce smooth spectra for desaturated colors and progressively more peaked spectra for saturated colors. As there are infinitely many solutions to the conversion between XYZ tristimulus values and spectral representations, this appears to be a reasonable solution, in keeping with many spectra found in nature.

7.6.4 Beginnings of Color Management

We have now seen how it is possible to measure and specify color in a device-independent manner. That is to say, we can describe color in a way that is not tied to a specific input or output display device. This specification lies at the heart of all modern color-management systems. The goal of a color-management system is to have a color, or image, specified in such a way that we can reproduce it with accurate results on any other device. This allows us to see an image on our computer monitor, print the image, or send the image to our friends across the country, and still have all the images match in appearance.

We can do this by calculating the transformation of primaries from one device to all the others, as described earlier in this section. This, of course, assumes all the devices are linear and additive, which is most definitely not the case for printers. The other technique is to calculate the *colorimetric characterization* for each device, both on the input side and the output side, from their own respective primaries into CIE XYZ tristimulus space. From this space, which is no longer tied to any particular device, we can then transform back out into any other device. The general flowchart for such a color management system is shown in Figure 7.19.

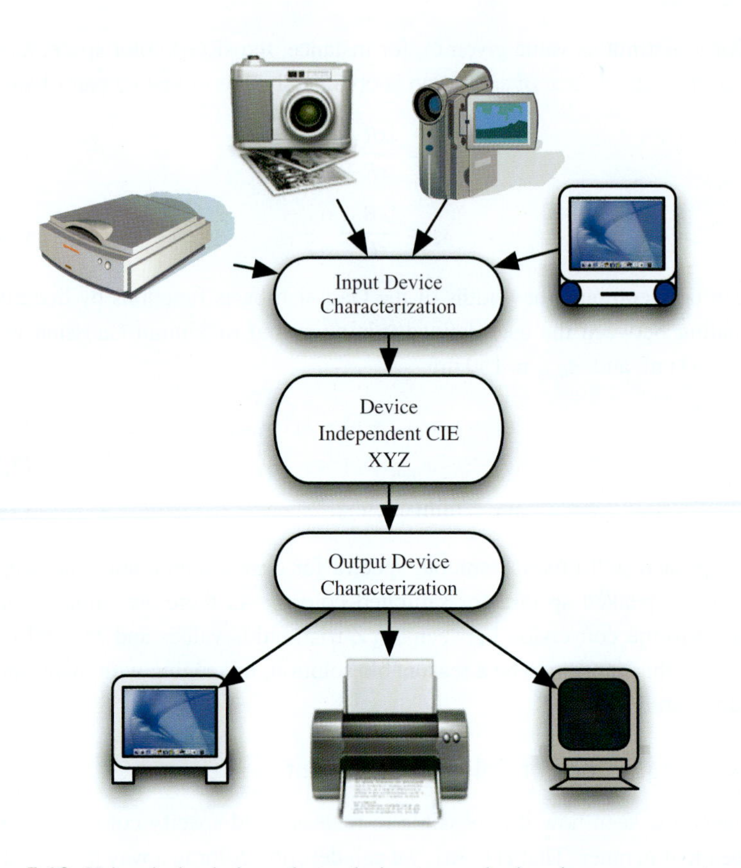

Figure 7.19. Using device-independent colorimetry as a basic color management process.

While such techniques do lie at the heart of most modern color-management systems, Figure 7.19 represents an over-simplification. Recall that colorimetry is only designed to measure colors and determine if two colors match, but not specify what two colors look like. This color match may break down when certain aspects of the viewing conditions change, such as overall amounts or colors of the lighting. In these situations, a colorimetric match between input and output devices does not guarantee an appearance match. In fact, more often than not, an appearance mis-match is to be expected. As an example, displaying the same image on two different displays will yield a different percept even if the primaries are identical, but only the white points do not match.

In order to obtain more accurate color reproduction across a wide variety of media, it is necessary to employ chromatic-adaptation models or a full color appearance model, as discussed in Chapter 11.

7.7 Application: Iso-Luminant Color Maps

The field of visualization is concerned with the mapping of data to pixel values in such a way that meaningful conclusions about the data can be drawn. There are several forms of visualization, including medical visualization, scientific visualization, and information visualization. The origins of the data are therefore varied. However, in general, the data have many different attributes that need to be mapped to pixel values. This means that high-dimensional attribute spaces are mapped to three-dimensional color spaces.

It is frequently desirable to visualize several attributes of the data simultaneously. There are many different ways to accomplish this, but we focus on the sub-problem of mapping data to color values. As luminance plays an extremely important role in human vision, and in particular in the perception of image structure and surface shape [868], it requires careful consideration when designing visualization algorithms.

For instance, it may be possible to map one attribute of the data to a range of luminance values, and a second attribute of the data to chromaticity values, leading to a bivariate color map. In such an approach it would be crucial to design a color map in which luminance variations are absent, i.e., an iso-luminant color map. In the case of univariate color maps where only one attribute is mapped to a color scale, iso-luminant color maps are also desirable, as they help avoid errors in interpretation as a result of perceptual effects such as simultaneous contrast [1215] (see Section 5.7). Similar advantages apply to the use of univariate color maps in combination with surface shading [326].

The design of an iso-luminant color map requires at least three problems to be solved. First, the display device used for the visualization is often uncalibrated, with unknown primaries, white point, and gamma function. Variations between display devices can be considerable [370]. Second, the lighting conditions in the viewing environment are typically unknown and uncontrolled [251,1216], leading to reflections off the display and an unknown state of adaptation of the viewer. Third, the spectral sensitivity in humans may vary between individuals, and may even change with age, for instance as the result of the build-up of yellow pigments in the ocular media [1262].

To overcome these problems, it is possible to design an iso-luminant color map specifically for a particular combination of monitor, display environment, and observer. This can be achieved by modifying an existing color-matching experiment, such that it can be conveniently and quickly executed by any observer [591]. Unlike most research on the design and application of color maps in visualization [86, 958, 959, 966, 1215], this technique does not assume calibrated displays and/or viewing environments.

7.7.1 Color Matching

The construction of an iso-luminant color map requires matching a set of colors to a single achromatic luminance value. This is similar to the problem of measuring the luminous efficiency function $V(\lambda)$, which is normally achieved by means of a color-matching experiment. However, in our case, the color-matching experiment takes place in the device-dependent color space of the monitor that is to be used for visualization.

Color-matching experiments either present stimuli in spatial juxtaposition, or exploit differences in temporal resolution between chromatic and achromatic perception [118, 1143, 1262]. Temporal approaches such as flicker photometry show stimuli that rapidly alternate between achromatic and chromatic values, whereby the observer can adjust the achromatic signal until the image appears stable. Although accurate results can be achieved, the flicker frequency should be around 15 Hz to 20 Hz, which is in a range that is difficult to reproduce accurately on typical display devices. In addition, this frequency may induce epileptic seizures in some individuals. This makes flicker photometry unattractive for use as a general technique to be applied outside carefully controlled laboratory conditions.

An alternative is to display two patches side by side in a bipartite field, and the observer can adjust the intensity of one of the patches to match the apparent brightness of the other patch. If both patches are chromatic, but of different wavelengths, the resulting technique is called heterochromatic brightness matching.

This task is rather difficult, making it problematic to obtain accurate results. In particular, this method suffers from a phenomenon called *additivity failure*, described as follows [118]. First, two chromatic patches are individually matched to an identical white field by adjusting the intensity of the chromatic patches. If these two chromatic patches are then super-posed, and the white field is doubled in intensity, then the combined colors should match the new white field.

However, in practice, the combined field will appear dimmer than the doubled white field. In addition, saturated colors will appear to glow with a brightness that is not in proportion to their actual luminance. The latter is known as the Helmholtz-Kohlrausch effect [302, 1262] (see also Section 11.2.2).

Related to direct heterochromatic brightness matching is the minimally distinct border technique. Here, the two fields are precisely juxtaposed, i.e., the fields are not separated in space. One of the fields can now be adjusted until the border between them is minimally distinct to the observer. The results of this method tend to be similar to heterochromatic brightness matching, although it solves the problem of additivity failure, and therefore does not suffer from the Helmholtz-Kohlrausch effect [118, 1262]. It is therefore possible to regard the minimally distinct border technique as a luminance matching task, making it an interesting technique for the construction of iso-luminant color maps.

However, finding the minimally distinct border between two patches is still not a simple task. Especially if the patches are very different in spectral composition, the border between them will remain visible. As a result, the matches obtained with this method may show relatively large variance.

7.7.2 Color Matching Using Face Stimuli

The lack of accuracy of the minimally distinct border technique is a problem if it is to be used in visualization tasks. A luminance-matching task with higher accuracy may be constructed by considering human faces [591].

Humans are generally very good at distinguishing and recognizing faces, possibly because the brain contains dedicated circuitry for face recognition [94, 281, 350]. However, at the same time, face recognition is much more difficult if the main lighting is not coming from above [934], which was confirmed in an fMRI study [356]. In addition, incorrect luminance levels are more easily detected in faces than in other objects [1101]. Given that luminance mediates face perception in thresholded face images [160], this can be exploited by constructing a test pattern as shown in Figure 7.20.

This pattern was constructed by thresholding a photograph of a human face which was lit by a strong directional light source. The pattern on the left in this

Figure 7.20. The two-tone pattern on the left is easily recognized as a human face, whereas the image on the right is not. (Gordon Kindlmann, Erik Reinhard, and Sarah Creem, "Face-Based Luminance Matching for Perceptual Color Map Generation," Proceedings of IEEE Visualization, pp 309–406, © 2002 IEEE.)

Figure 7.21. Example of varying the colors in face stimuli. Note the different cross-over points for the three colors, where the perception of a face flips from the left image to the right image. (Gordon Kindlmann, Erik Reinhard, and Sarah Creem, "Face-Based Luminance Matching for Perceptual Color Map Generation," Proceedings of IEEE Visualization, pp. 309–406, © 2002 IEEE.)

Figure 7.22. The two-tone pattern used in the minimally distinct border experiment. (Gordon Kindlmann, Erik Reinhard, and Sarah Creem, "Face-Based Luminance Matching for Perceptual Color Map Generation," Proceedings of IEEE Visualization, pp. 309–406, © 2002 IEEE.)

figure is easily recognized as a human face, whereas inverting this pattern breaks human face perception.

In this stimulus, it is possible to replace the black with a desired shade of gray. The white is replaced with a color, and the observer is given control over the intensity value of the color. By appropriately changing the intensity value of the color, the stimulus that was initially seen as a face will disappear, and the pattern on the other side will then be recognized as a face. The range of intensity

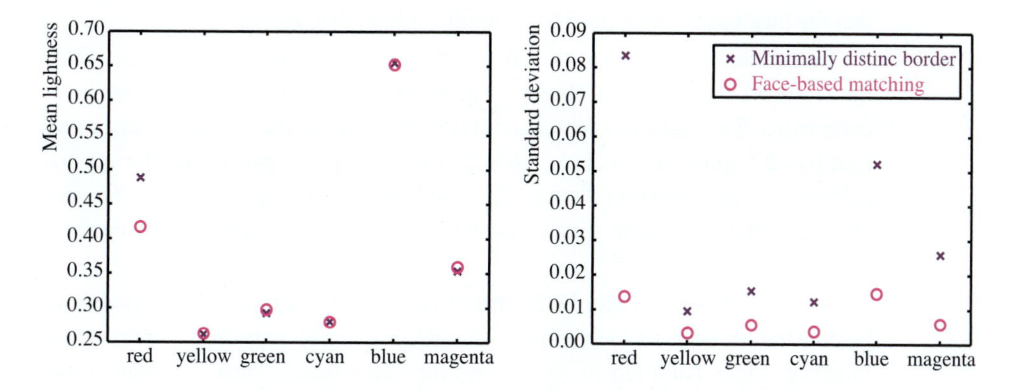

Figure 7.23. Mean and standard deviation for both the minimally distinct border and the face-based luminance matching techniques. (Gordon Kindlmann, Erik Reinhard, and Sarah Creem, "Face-Based Luminance Matching for Perceptual Color Map Generation," Proceedings of IEEE Visualization, pp. 309–406, © 2002 IEEE.)

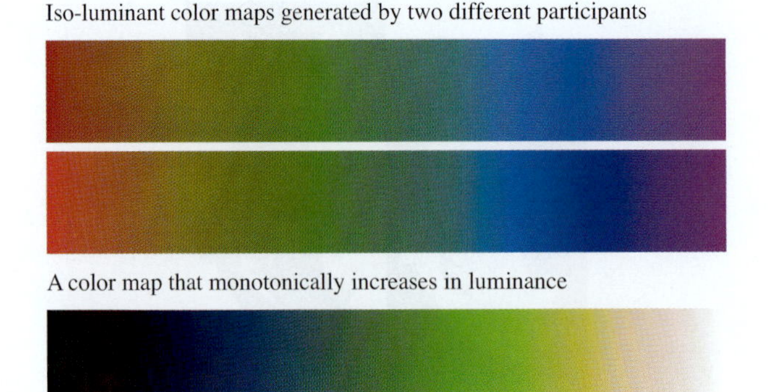

Iso-luminant color maps generated by two different participants
A color map that monotonically increases in luminance

Figure 7.24. Color maps generated using face-based luminance matching. (Gordon Kindlmann, Erik Reinhard, and Sarah Creem, "Face-Based Luminance Matching for Perceptual Color Map Generation," Proceedings of IEEE Visualization, pp. 309–406, © 2002 IEEE.)

values where both the left and right halves of the stimulus are perceived to be ambiguous tends to be small, suggesting a high level of accuracy. An example of a set of stimuli is shown in Figure 7.21.

In a user study, the face-based luminance-matching technique was compared against a conventional minimally distinct border technique, using a stimulus that can not be recognized as a human face, but has otherwise the same boundary length and a similar irregular shape (as shown in Figure 7.22). The results, shown in Figure 7.23, reveal that the two techniques produce the same mean values, but that the face-based technique has a significantly higher accuracy.

Color maps generated with this technique are shown in Figure 7.24. This figure shows that different observers generate somewhat different iso-luminant color maps. The variances between different observers have been measured using face-based luminance matching. Results for seven participants, each for normal color vision, are shown in Figure 7.25. Aside from showing differences between observers, this figure also shows that the standard rainbow color map is far from iso-luminant.

It is also shown in Figure 7.24 that monotonically increasing color maps can be constructed with this technique. This is achieved by choosing increasing luminance values from left to right along the color scale. Thus, different colors are matched to increasing luminance values, yielding the monotonically increasing luminance map shown in this figure. Note that face-based techniques have also been developed to test if a given color map is monotonically increasing or not [974].

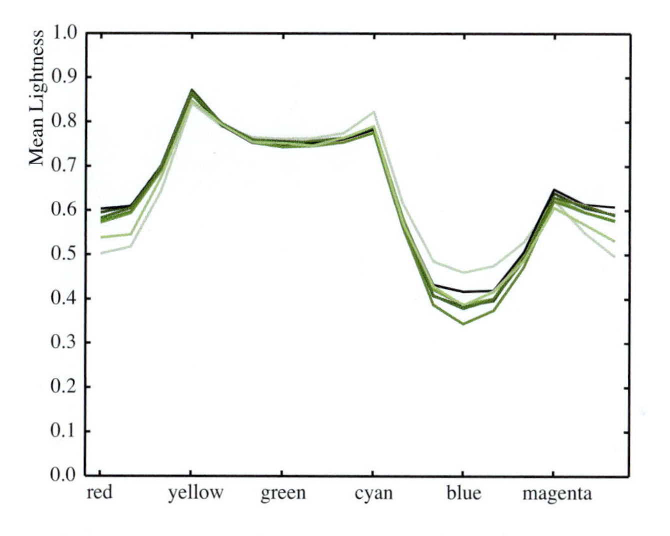

Figure 7.25. The perception of lightness varies between observers, as shown here for 18 different colors and seven participants. (Gordon Kindlmann, Erik Reinhard, and Sarah Creem, "Face-Based Luminance Matching for Perceptual Color Map Generation," Proceedings of IEEE Visualization, pp. 309–406, © 2002 IEEE.)

In summary, exerting luminance control over color maps used in visualization tasks is desirable. Taking the display, the viewing environment, as well as differences between observers into account, a simple perceptual test can be constructed to create iso-luminant color maps optimized for one specific viewing condition. This will help produce good visualizations and minimize errors in interpretation. The software for this technique is available online at http://www.cs.utah.edu/~gk/facelumin.html.

7.8 Further Reading

A recent book specifically on colorimetry is *Colorimetry: Fundamentals and Applications* by Ohta and Robertson [849]. More general texts are Hunt's book, *The Reproduction of Colour* [509] and, of course, Wyszecki and Stiles, *Color Science: Concepts and Methods, Quantitative Data and Formulae* [1262].

Chapter 8
Color Spaces

As shown in Chapter 7, due to the trichromatic nature of the human visual system, color may be represented with a color vector consisting of three numbers, or tristimulus values. The CIE XYZ color space is derived from spectral power distributions by means of integration with three *color-matching functions*. These color-matching functions can be considered approximate linear transformations of the human cone sensitivities. For the XYZ color space, the Y channel was derived to be identical with the photopic luminance response function.

Closely tied to all color-matching functions, and their corresponding tristimulus values, are the primaries or light sources that can be used to re-derive the functions in a color-matching experiment. One of the main goals in creating the the CIE XYZ color space was to have tristimulus values that are all positive. In order to do that, it was necessary to select an imaginary set of primaries. Essentially, this means that it is not possible to construct a set of only three positive light sources that, when combined, are able to reproduce all colors in the visible spectrum.

In general practice, the XYZ color space may not be suitable for performing image calculations or encoding data, as there are a large number of possible XYZ values that do not correspond to any physical color. This can lead to inefficient use of available storage bits and generally requires a higher processing bit-depth to preserve visual integrity. It is not possible to create a color display device that corresponds directly to CIE XYZ; it is possible to create a capture device that has sensor responsivities that are close to the CIE XYZ color-matching functions, though due to cost of both hardware and image processing, this is not very common. Therefore, since the corresponding primaries are imaginary, a transcoding into a specific device space is necessary. Depending on the desired task, the use of a different color space may be more appropriate. Some of the reasons for the development of different color spaces are:

Physical realizability. Camera and display manufacturers build sensors and displays that have specific responsivities (cameras) and primaries (displays). The output of any given camera is often implicitly encoded in a color space that corresponds to the camera's hardware, firmware, and software. For instance, most digital sensors capture data linearly with respect to amounts of light, but this data is *gamma corrected* by compressing the signal using a power function less than 1.0. This gamma-corrected data may or may not still correspond to the physical RGB sensors used to capture the image by the camera. The signal that is sent to a display device such as a cathode ray tube, which has its own *device-dependent* color space defined by its phosphors, may not be in the same color space as the cathode ray tube; it must first be transformed or color reproduction will be impaired. In a closed system, it is possible to directly capture images using a camera that is specifically tied to an output device, removing the need for transforming the colors between capture and display.

Efficient encoding. Some color spaces were developed specifically to enable efficient encoding and transmission of the data. This encoding and transmission may rely on certain behaviors of the human visual system, such as our decrease in contrast sensitivity for chromatic signals compared to achromatic signals. In particular, the color encodings used by the different television signals (NTSC, PAL, SECAM) exploit this behavior of the visual system in their underlying color models.

Perceptual uniformity. For many applications, we would like to measure the perceived differences between pairs of colors. In most linear color spaces, such as CIE XYZ, a simple Euclidean distance does not correspond to the perceptual difference. However, color spaces can be designed to be perceptually uniform, such that the Euclidean distance in those color spaces is a good measure of perceptual difference.

Intuitive color specification. Many color spaces are designed to be closely tied to output display primaries, such as the red, green, and blue primaries of a typical CRT display. However, the mapping from such values encoded in these spaces to the appearance of a color can be very non-intuitive. Some color spaces are designed to provide a more intuitive meaning to allow for ease of specifying the desired color. These spaces may encode the data in ways that more closely match our perception of color appearance, using terms such as hue, saturation, and lightness (HSL).

Color spaces are often characterized by their bounding volumes, or the range of colors that they are capable of representing. This is often referred to as the *gamut* of colors that a space can represent. This is only a valid assumption if the values that color vectors can store are limited, for example all positive with a range between 0 and 1. For instance, many (but not all) color spaces do not allow negative values—typically because the encoded values correspond to the light energy coming from a specific set of output primaries, and light cannot be negative. We generally assume that only output devices have color gamuts, while capture devices have a dynamic range or limitations in their spectral sensitivity. This is often an area of great philosophical and semantic debate and discussion in the color-imaging community.

The conversion of colors represented in one gamut to another, often smaller, gamut is a frequently occurring problem in color reproduction. This field of study is known as *gamut mapping*. There are many ways to remap colors from a larger to a smaller gamut, with some of the possible approaches outlined in Section 15.4. As display devices and color printers get more sophisticated, the opposite problem is also becoming an active area of research, often called gamut expansion. How we move colors that have been encoded in a limited range to a device capable of displaying a much larger range has both similar and unique problems to traditional gamut mapping.

The CIE XYZ color space is still actively used for converting between different color spaces, as many color spaces are actually defined by means of a transform to and from XYZ. We can think of CIE XYZ as a *device-independent* method for representing colors, and any given output device can be described by its relationship to XYZ. It is important to stress that when a color space that corresponds to physical primaries has a well-documented and specified relationship with CIE XYZ, then this device can also be thought of as device-independent. Although the space may be tied to physically realizable primaries, it is still possible to specify any color match using that space (allowing for both positive and negative encoded values), or also to derive a unique set of color-matching functions that correspond to those primaries.

For simple, linear and additive trichromatic display devices the transformation to and from CIE XYZ can usually be given by a 3×3 matrix. This transformation relies on Grassmann's Laws of additive color mixing. Often, color spaces are defined by the linear relationship to CIE XYZ, with additional non-linear processing. In most cases, such non-linear processing is designed to minimize perceptual errors when storing in a limited bit-depth without ample color precision, or to di-

	R	G	B	White
x	0.6400	0.3000	0.1500	0.3127
y	0.3300	0.6000	0.0600	0.3290

Table 8.1. The xy chromaticity coordinates for the primaries and white point specified by ITU-R BT.709. The sRGB standard also uses these primaries and white point.

rectly be displayed on devices that have a non-linear relationship between signal and energy emitted.

To construct such a linear 3×3 matrix transforming a specific additive display device to CIE XYZ, the three primaries as well as the white point of the device (or color space) need to be known. These are typically described as xy chromaticity coordinates, though they can also be specified as CIE XYZ tristimulus values.

The xy chromaticity coordinates of the primaries represent the colors obtained for the color vectors (1, 0, 0), (0, 1, 0), and (0, 0, 1). Basically, they represent chromaticity of the maximum value of any given channel when displayed alone. Recall from Chapter 7 that chromaticity coordinates are two-dimensional projections of a three-dimensional color space, and that they have been normalized to remove the luminance component from the specification. Without the luminance component, it is impossible to know the relative power or energy of the three primaries, and so we cannot know what color will be shown when all three primaries are displayed at their maximum. Thus, we need to include a fourth description of the device; the display or color space white point. The white point of a color space is the color obtained by the maximum of all three primaries, typically expressed as the color vector (1, 1, 1). As an example, the ITU-R BT.709 specifies a color space with primaries and white point as specified in Table 8.1. The white point is effectively D65.

Assuming that the primaries represent some form of red, green, and blue, the xy chromaticity coordinates can be extended to compute chromaticity triplets by setting $z = 1 - x - y$. This leads to chromaticity triplets (x_R, y_R, z_R), (x_G, y_G, z_G), and (x_B, y_B, z_B), as well as a triplet (x_W, y_W, z_W) for the white point. If we know the maximum luminance of the white point, we can compute its tristimulus value (X_W, Y_W, Z_W). Using these values, we solve the following set of equations for the luminance ratio scalers S_R, S_G, and S_B:

$$X_W = x_R S_R + x_G S_G + x_B S_B, \tag{8.1a}$$

$$Y_W = y_R S_R + y_G S_G + y_B S_B, \tag{8.1b}$$

$$Z_W = z_R S_R + z_G S_G + z_B S_B. \tag{8.1c}$$

The conversion between RGB and XYZ is then given by

$$\begin{bmatrix} X \\ Y \\ Z \end{bmatrix} = \begin{bmatrix} x_R S_R & x_G S_G & x_B S_B \\ y_R S_R & y_G S_G & y_B S_B \\ z_R S_R & z_G S_G & z_B S_B \end{bmatrix} \begin{bmatrix} R \\ G \\ B \end{bmatrix}. \tag{8.2}$$

More often than not, we do not have the maximum luminance of the white point provided to us, rather just the chromaticity coordinates. Recall from Chapter 7 that for a linear, additive display device we can determine the relationship between CIE XYZ and the device RGB using the following equation:

$$\begin{bmatrix} X \\ Y \\ Z \end{bmatrix} = \begin{bmatrix} X_{R_{\max}} & X_{G_{\max}} & X_{B_{\max}} \\ Y_{R_{\max}} & Y_{G_{\max}} & Y_{B_{\max}} \\ Z_{R_{\max}} & Z_{G_{\max}} & Z_{B_{\max}} \end{bmatrix} \begin{bmatrix} R \\ G \\ B \end{bmatrix}, \tag{8.3}$$

where $XYZ_{R_{\max}}$ represents the tristimulus values of the primary when displayed at its maximum power. When we do not have the full tristimulus values of the primaries, we can solve for the relative tristimulus ratios in order to convert between RGB and XYZ. We can expand Equation (8.3) to be a product of the chromaticity coordinates along with the luminance component:

$$\begin{bmatrix} \left(\frac{x_w}{y_w}\right) Y_w \\ Y_w \\ \left(\frac{z_w}{y_w}\right) Y_w \end{bmatrix} = \begin{bmatrix} \left(\frac{x_r}{y_r}\right) & \left(\frac{x_g}{y_g}\right) & \left(\frac{x_b}{y_b}\right) \\ 1 & 1 & 1 \\ \left(\frac{z_r}{y_r}\right) & \left(\frac{z_g}{y_g}\right) & \left(\frac{z_b}{y_b}\right) \end{bmatrix} \begin{bmatrix} Y_r & 0 & 0 \\ 0 & Y_g & 0 \\ 0 & 0 & Y_b \end{bmatrix} \begin{bmatrix} R \\ G \\ B \end{bmatrix}. \tag{8.4}$$

This expansion still requires the knowledge of both the luminance of the white, Y_w, as well as the luminance of the red, green, and blue channels Y_r, Y_g, Y_b. Since we do not have this information, we can calculate the luminance ratios Y^R that would be necessary to obtain the chromaticity of the given white point, with the chromaticities of the red, green, and blue channel we are given. First, we assume that the maximum luminance Y_w occurs when $R = G = B = 1$, and that it has a luminance ratio of 1.0. Equation (8.4) can then be reduced to

$$\begin{bmatrix} \left(\frac{x_w}{y_w}\right) \\ 1 \\ \left(\frac{z_w}{y_w}\right) \end{bmatrix} = \begin{bmatrix} \left(\frac{x_r}{y_r}\right) & \left(\frac{x_g}{y_g}\right) & \left(\frac{x_b}{y_b}\right) \\ 1 & 1 & 1 \\ \left(\frac{z_r}{y_r}\right) & \left(\frac{z_g}{y_g}\right) & \left(\frac{z_b}{y_b}\right) \end{bmatrix} \begin{bmatrix} Y_r^R \\ Y_g^R \\ Y_b^R \end{bmatrix}. \tag{8.5}$$

We can solve for the luminance ratios, represented by Y_{rgb}^R, by inverting the individual chromaticity matrix, as shown in Equation (8.6a). We then use these luminance ratios to solve for the 3×3 matrix from Equation (8.3) that will transform from device RGB into XYZ. This technique is also useful when you want to force a set of primaries with given chromaticities into having a specific white point chromaticity:

$$\begin{bmatrix} Y_r^R \\ Y_g^R \\ Y_b^R \end{bmatrix} = \begin{bmatrix} \left(\frac{x_r}{y_r}\right) & \left(\frac{x_g}{y_g}\right) & \left(\frac{x_b}{y_b}\right) \\ 1 & 1 & 1 \\ \left(\frac{z_r}{y_r}\right) & \left(\frac{z_g}{y_g}\right) & \left(\frac{z_b}{y_b}\right) \end{bmatrix}^{-1} \begin{bmatrix} \left(\frac{x_w}{y_w}\right) \\ 1 \\ \left(\frac{z_w}{y_w}\right) \end{bmatrix}; \tag{8.6a}$$

$$\begin{bmatrix} X_{R_{\max}} & X_{G_{\max}} & X_{B_{\max}} \\ Y_{R_{\max}} & Y_{G_{\max}} & Y_{B_{\max}} \\ Z_{R_{\max}} & Z_{G_{\max}} & Z_{B_{\max}} \end{bmatrix} = \begin{bmatrix} \left(\frac{x_r}{y_r}\right) & \left(\frac{x_g}{y_g}\right) & \left(\frac{x_b}{y_b}\right) \\ 1 & 1 & 1 \\ \left(\frac{z_r}{y_r}\right) & \left(\frac{z_g}{y_g}\right) & \left(\frac{z_b}{y_b}\right) \end{bmatrix} \begin{bmatrix} Y_r^R & 0 & 0 \\ 0 & Y_g^R & 0 \\ 0 & 0 & Y_b^R \end{bmatrix}. \tag{8.6b}$$

The conversion from XYZ to RGB can be computed by inverting the 3×3 matrix calculated above. For any matrix constructed in this manner, the luminance of any given color can by computed by evaluating

$$Y = y_R S_R + y_G S_G + y_B S_B. \tag{8.7}$$

Alternatively, the luminance can be calculated by adding up the luminance of each of the individual components, $Y = Y_R + Y_G + Y_B$. As discussed in previous chapters, the Y channel in XYZ color space is identical to the photopic luminance response function and since light is additive, we can calculate the luminance of any given display by simply adding the luminance of each of the color channels.

If an image is presented in an RGB color space for which the primaries or the white point are unknown, then it may be beneficial to assume that the image is encoded in a standard color space. Most standardized color spaces have well-defined linear matrix transformations to and from XYZ. Many standard color matrices are based upon actual primaries of output devices, or are designed to provide maximum precision to a desired range of colors. One common choice of encoding (and broadcasting) color spaces is that of HDTV, whose primaries are specified by ITU-R BT.709 (see Table 8.1). These primaries are also known as the Rec. 709 primaries; they were designed to be representative of the typical CRT display used for broadcast video and computer monitors [923]. It is also important to note that the primaries for ITU-R BT.709 are identical to those of the sRGB color space, which was also designed to be representative of the typical computer display connected to the Internet [524]. The transformation between the ITU-R

BT.709/sRGB primaries and the CIE XYZ tristimulus values is

$$\begin{bmatrix} X \\ Y \\ Z \end{bmatrix} = \begin{bmatrix} 0.4124 & 0.3576 & 0.1805 \\ 0.2126 & 0.7152 & 0.0722 \\ 0.0193 & 0.1192 & 0.9505 \end{bmatrix} \begin{bmatrix} R \\ G \\ B \end{bmatrix}; \tag{8.8a}$$

$$\begin{bmatrix} R \\ G \\ B \end{bmatrix} = \begin{bmatrix} 3.2405 & -1.5371 & -0.4985 \\ -0.9693 & 1.8706 & 0.0416 \\ 0.0556 & -0.2040 & 1.0572 \end{bmatrix} \begin{bmatrix} X \\ Y \\ Z \end{bmatrix}. \tag{8.8b}$$

Again, we can calculate the luminance of any given color in this color space by summing the values of the individual RGB channels:

$$Y = 0.2126R + 0.7152G + 0.0722B. \tag{8.9}$$

It is important to note that we can also calculate the white point of any given display by calculating the XYZ tristimulus values of the maximum RGB values:

$$\begin{bmatrix} X_W \\ Y_W \\ Z_W \end{bmatrix} = \begin{bmatrix} 0.4124 & 0.3576 & 0.1805 \\ 0.2126 & 0.7152 & 0.0722 \\ 0.0193 & 0.1192 & 0.9505 \end{bmatrix} \begin{bmatrix} R_{\max} \\ G_{\max} \\ B_{\max} \end{bmatrix}. \tag{8.10}$$

The ITU-R BT.709 primaries have a defined white point equivalent to that of CIE D65:

$$\begin{bmatrix} X_W^{709} \\ Y_W^{709} \\ Z_W^{709} \end{bmatrix} = \begin{bmatrix} 0.4124 & 0.3576 & 0.1805 \\ 0.2126 & 0.7152 & 0.0722 \\ 0.0193 & 0.1192 & 0.9505 \end{bmatrix} \begin{bmatrix} 100 \\ 100 \\ 100 \end{bmatrix} = \begin{bmatrix} 95.05 \\ 100.00 \\ 108.90 \end{bmatrix}. \tag{8.11}$$

This tristimulus value is equivalent to the chromaticity values of the white point listed in Table 8.1.

In this chapter, many of the more common color spaces used in various industries are discussed. These are generally grouped into device-specific and device-independent color spaces. While it is tempting to group the color spaces that are defined by a specific (and real) primary set as being device-dependent, if they have a well-defined transform to and from a device-independent space such as CIE XYZ, then these spaces are indeed device-independent. To facilitate implementation and ease of communication between the wide variety of color spaces, we include transformation matrices where possible.

8.1 RGB Color Spaces

The phosphors used in cathode ray tubes and plasma devices, as well as the illuminants and colored filters used in LCD and DLP displays (see Chapter 14), implicitly define the primaries and native white points of display devices. Typical additive display devices use a red, a green, and a blue primary. However, different display devices can have very disparate primaries, even if they are all described as being RGB. As a result, each display device has a different range of colors, or color gamut, that it is able to display. Nonetheless, these color spaces are all generically referred to as RGB color spaces.

It is important to remember that, as a result, there is no such thing as *the* RGB color space. Every device has its own color space, and these classes of RGB color spaces are therefore called *device-dependent*; if we send the same RGB values to each of the displays we will see different colors depending on their choice of primaries. This is in contrast to the XYZ color space, which is well defined, based on a set of imaginary primaries, and is an example of a *device-independent* color space. The practical implication is that an image can appear very different depending on the display, unless care is taken to process the data prior to output. The process of measuring how a display handles color is called device-characterization and is necessary to achieve color matches between different devices (see Section 14.18).

To enable some consistency in how images appear on a display, the images can be converted to a standardized set of primaries, with a standardized white point. This standardized set can be specified by the target display, e.g., a display requesting that all images be described using the ITU-R BT.709 primaries, which then performs an internal conversion from this space into the primaries of the display, or this conversion can be done using the model of modern color management. This model is achieved by converting the image into a device-independent space such as the XYZ or CIELAB color space and, subsequently, to the RGB color space of the target display. All too often, in general practice, such correction does not happen and images appear different on different displays. This may be because the color space of the image is unknown or the primaries and white point of the display are unknown.

Some standardization has occurred in recent years, for instance, through the use of the sRGB color space. Some cameras are able to output images in this color space, and some display devices are sRGB compliant. This means that in an imaging pipeline that employs the sRGB color space throughout, conversion between different device-dependent RGB color spaces is unnecessary. We can think of using the sRGB color space in this manner as a means for a default form of color management. If all devices work in sRGB, then all devices theoretically should match each other. This makes the sRGB color space suitable for such consumer applications as shooting a digital image and displaying this image on a CRT, or placing it on a website and having it appear somewhat consistent wherever it is viewed. If the output device is not assuming an sRGB input, however, then the colors may mismatch.

The sRGB color space, like many other RGB color spaces, is defined by a 3×3 matrix as well as a non-linear luminance encoding. A maximum luminance of 80 cd/m^2 is assumed, as this was representative of a typical display when the sRGB space was created. The matrix transforms between linearized sRGB and XYZ is the same as that specified by ITU-R BT.709, as given in (8.8). The non-linear encoding is given by

$$R_{\mathrm{sRGB}} = \begin{cases} 1.055\, R^{1/2.4} - 0.055 & R > 0.0031308, \\ 12.92\, R & R \leq 0.0031308; \end{cases} \tag{8.12a}$$

$$G_{\mathrm{sRGB}} = \begin{cases} 1.055\, G^{1/2.4} - 0.055 & G > 0.0031308, \\ 12.92\, G & G \leq 0.0031308; \end{cases} \tag{8.12b}$$

$$B_{\mathrm{sRGB}} = \begin{cases} 1.055\, B^{1/2.4} - 0.055 & B > 0.0031308, \\ 12.92\, B & B \leq 0.0031308. \end{cases} \tag{8.12c}$$

The non-linear encoding, which is also known as gamma encoding, and which is therefore frequently confused with gamma correction, is required to minimize quantization errors in digital applications. Since sRGB is an 8-bit quantized color space, it is beneficial to have a non-linear relationship between the color values and the luminance values that they represent. By compressing the data with a power function less than 1.0, we use more bits of precision for the darker colors where color differences are more perceptible. This helps keep quantization errors below the visible threshold where possible. It is important to note that although gamma correction and gamma encoding are different entities, the ultimate behavior is very similar. Gamma correction is necessary when an output display device responds non-linearly with regard to the input signal. Typical CRTs have a non-linear relationship between input voltage and output luminance, often described by a power function greater than 1.0 (generally around 2.4). Gamma correction is necessary to compress the signal prior to sending it to a display, in order to get a linear, or close to linear, response out. Since sRGB was defined to be representative of a typical CRT display, the gamma encoding also serves the purpose of gamma correction and the non-linearly encoded data can be sent directly to the output device.

For very small values in the sRGB space, the encoding is linear, as this produces better behavior for near-black values. When we combine the linear component with the exponent of $1/2.4$, we get a behavior that is very similar to having just an exponent of $1/2.2$, although it deviates for the dark colors. In practice, this short linear ramp is often abandoned for simplicity of implementation, and a single exponent of $1/2.2$ is used. Note that this simplified formulation is a deviation from the sRGB specification and is therefore not recommended.

The non-linear encoding given for the sRGB space can be parameterized as in Equation (8.13) [882]:

$$R_{\text{nonlinear}} = \begin{cases} (1+f)\,R^{\gamma} - f & t < R \le 1, \\ sR & 0 \le R \le t; \end{cases} \tag{8.13a}$$

$$G_{\text{nonlinear}} = \begin{cases} (1+f)\,G^{\gamma} - f & t < G \le 1, \\ sG & 0 \le G \le t; \end{cases} \tag{8.13b}$$

$$B_{\text{nonlinear}} = \begin{cases} (1+f)\,B^{\gamma} - f & t < B \le 1, \\ sB & 0 \le B \le t. \end{cases} \tag{8.13c}$$

Together with specifications for the primaries and the white point, the parameters f, s, and t specify a class of RGB color spaces that are used in various industries. The value of s determines the slope of the linear segment, and f is a small offset. The value of t determines where the linear slope changes into the non-linear encoding.

Table 8.2 lists the primaries and white points of a collection of commonly encountered device RGB spaces. The associated conversion matrices and non-linearities are given in Table 8.3. The two-dimensional gamuts spanned by each color space are shown in Figure 8.1, as projections in the CIE $u'v'$ chromaticity diagrams. The projected gamut for the HDTV color space is identical to the sRGB standard, and it is therefore not shown again. It is important to stress that device gamuts are limited by the range of colors that a real output device can create; however the encoding color space inherently has no gamut boundaries. Only when we impose a limitation on the range of values that can be encoded, such as [0, 1] or [0, 255] do color spaces themselves have gamuts. It is also important to point out that the two-dimensional gamut projection into a chromaticity diagram is inherently simplifying gamut descriptions. Since we need at least three dimensions to fully specify color appearance (see Chapter 11) color gamuts should also be at least three-dimensional. Since the non-linear encoding of HDTV and sRGB is different, it is possible for these spaces to have different gamuts when represented in a three-dimensional space, such as CIELAB. For additive display devices, however, the triangle formed by the primaries in a chromaticity diagram is often used in industrial applications to represent the ultimate gamut of the device.

The Adobe RGB color space was formerly known as SMPTE-240M, but was renamed after SMPTE's gamut was reduced. It has a larger gamut than sRGB, as shown in the chromaticity diagrams of Figure 8.1. This color space was developed with the printing industry in mind, as there were many colors that could be printed that could not be represented by a smaller gamut color space. It is important to note that when a limited precision is used, e.g. 8-bits for each color channel, then an extended gamut inherently means there will be less bits for all colors and quantization may occur. Thus, it is important to choose an encoding color space that is best representative of the most likely or most important colors, in order to not "waste bits" where they are not needed. Many digital cameras today provide an option to output images in either the Adobe RGB color, as well as sRGB.

High definition television (HDTV) and sRGB standards specify identical primaries, but they differ in their definition of viewing conditions. This difference is represented by the non-linear encoding. The sRGB space has a linear segment and a power function of $1/2.4$, while the ITU-R BT.709 has a linear segment and power function of $1/2.2$. These linear segments make the effective non-linearity approximately $1/2.2$ for sRGB and $1/2$ for HDTV. Thus, if improper assumptions are made between these color spaces, color mismatches can occur. The American National Television Systems Committee (NTSC) standard was created in 1953 and has been used as the color space for TV in North America. As phosphors

Color space		**Primaries**			**White point**	
		R	G	B	(Illuminant)	
sRGB/ITU-R BT.709	x	0.6400	0.3000	0.1500	D65	0.3127
	y	0.3300	0.6000	0.0600		0.3290
Adobe RGB (1998)	x	0.6400	0.2100	0.1500	D65	0.3127
	y	0.3300	0.7100	0.0600		0.3290
HDTV (HD-CIF)	x	0.6400	0.3000	0.1500	D65	0.3127
	y	0.3300	0.6000	0.0600		0.3290
NTSC (1953)	x	0.6700	0.2100	0.1400	C	0.3101
	y	0.3300	0.7100	0.0800		0.3161
PAL/SECAM	x	0.6400	0.2900	0.1500	D65	0.3127
	y	0.3300	0.6000	0.0600		0.3290
SMPTE-C	x	0.6300	0.3100	0.1550	D65	0.3127
	y	0.3400	0.5950	0.0700		0.3290
Wide Gamut	x	0.7347	0.1152	0.1566	D50	0.3457
	y	0.2653	0.8264	0.0177		0.3584

Table 8.2. Chromaticity coordinates for the primaries and white points of several commonly encountered RGB color spaces.

Color space	XYZ to RGB matrix	RGB to XYZ matrix	Non-linear transform
sRGB	$\begin{bmatrix} 3.2405 & -1.5371 & -0.4985 \\ -0.9693 & 1.8760 & 0.0416 \\ 0.0556 & -0.2040 & 1.0572 \end{bmatrix}$	$\begin{bmatrix} 0.4124 & 0.3576 & 0.1805 \\ 0.2126 & 0.7152 & 0.0722 \\ 0.0193 & 0.1192 & 0.9505 \end{bmatrix}$	$\gamma = 1/2.4 \approx 0.42$ $f = 0.055$ $s = 12.92$ $t = 0.0031308$
Adobe RGB (1998)	$\begin{bmatrix} 2.0414 & -0.5649 & -0.3447 \\ -0.9693 & 1.8760 & 0.0416 \\ 0.0134 & -0.1184 & 1.0154 \end{bmatrix}$	$\begin{bmatrix} 0.5767 & 0.1856 & 0.1882 \\ 0.2974 & 0.6273 & 0.0753 \\ 0.0270 & 0.0707 & 0.9911 \end{bmatrix}$	$\gamma = \frac{1}{2\frac{51}{256}} \approx \frac{1}{2.2}$ $f =$ N.A. $s =$ N.A. $t =$ N.A.
HDTV (HD-CIF)	$\begin{bmatrix} 3.2405 & -1.5371 & -0.4985 \\ -0.9693 & 1.8760 & 0.0416 \\ 0.0556 & -0.2040 & 1.0572 \end{bmatrix}$	$\begin{bmatrix} 0.4124 & 0.3576 & 0.1805 \\ 0.2126 & 0.7152 & 0.0722 \\ 0.0193 & 0.1192 & 0.9505 \end{bmatrix}$	$\gamma = 0.45$ $f = 0.099$ $s = 4.5$ $t = 0.018$
NTSC (1953)/ ITU-R BT.601-4	$\begin{bmatrix} 1.9100 & -0.5325 & -0.2882 \\ -0.9847 & 1.9992 & -0.0283 \\ 0.0583 & -0.1184 & 0.8976 \end{bmatrix}$	$\begin{bmatrix} 0.6069 & 0.1735 & 0.2003 \\ 0.2989 & 0.5866 & 0.1145 \\ 0.0000 & 0.0661 & 1.1162 \end{bmatrix}$	$\gamma = 0.45$ $f = 0.099$ $s = 4.5$ $t = 0.018$
PAL/SECAM	$\begin{bmatrix} 3.0629 & -1.3932 & -0.4758 \\ -0.9693 & 1.8760 & 0.0416 \\ 0.0679 & -0.2289 & 1.0694 \end{bmatrix}$	$\begin{bmatrix} 0.4306 & 0.3415 & 0.1783 \\ 0.2220 & 0.7066 & 0.0713 \\ 0.0202 & 0.1296 & 0.9391 \end{bmatrix}$	$\gamma = 0.45$ $f = 0.099$ $s = 4.5$ $t = 0.018$
SMPTE-C	$\begin{bmatrix} 3.5054 & -1.7395 & -0.5440 \\ -1.0691 & 1.9778 & 0.0352 \\ 0.0563 & -0.1970 & 1.0502 \end{bmatrix}$	$\begin{bmatrix} 0.3936 & 0.3652 & 0.1916 \\ 0.2124 & 0.7010 & 0.0865 \\ 0.0187 & 0.1119 & 0.9582 \end{bmatrix}$	$\gamma = 0.45$ $f = 0.099$ $s = 4.5$ $t = 0.018$
Wide Gamut	$\begin{bmatrix} 1.4625 & -0.1845 & -0.2734 \\ -0.5228 & 1.4479 & 0.0681 \\ 0.0346 & -0.0958 & 1.2875 \end{bmatrix}$	$\begin{bmatrix} 0.7164 & 0.1010 & 0.1468 \\ 0.2587 & 0.7247 & 0.0166 \\ 0.0000 & 0.0512 & 0.7740 \end{bmatrix}$	$\gamma =$ N.A. $f =$ N.A. $s =$ N.A. $t =$ N.A.

Table 8.3. Transformations for standard RGB color spaces (after [882]).

and other display technologies now allow much more saturated colors, it has been deprecated and replaced with SMPTE-C to match phosphors in current display devices, which are more efficient and brighter. Phase alternation line (PAL) and Séquentiel Couleur à Mémoire (SECAM) are the standards used for television in Europe. These spaces are discussed further in Section 8.4.

Finally, the Wide gamut color space is shown for comparison [882]. Its primaries are monochromatic light sources with wavelengths of 450, 525, and 700 nm. By moving the primaries closer to the spectrum locus, a larger triangle of representable colors is formed. In the limit, the primaries become monochromatic, as in the case of this color space.

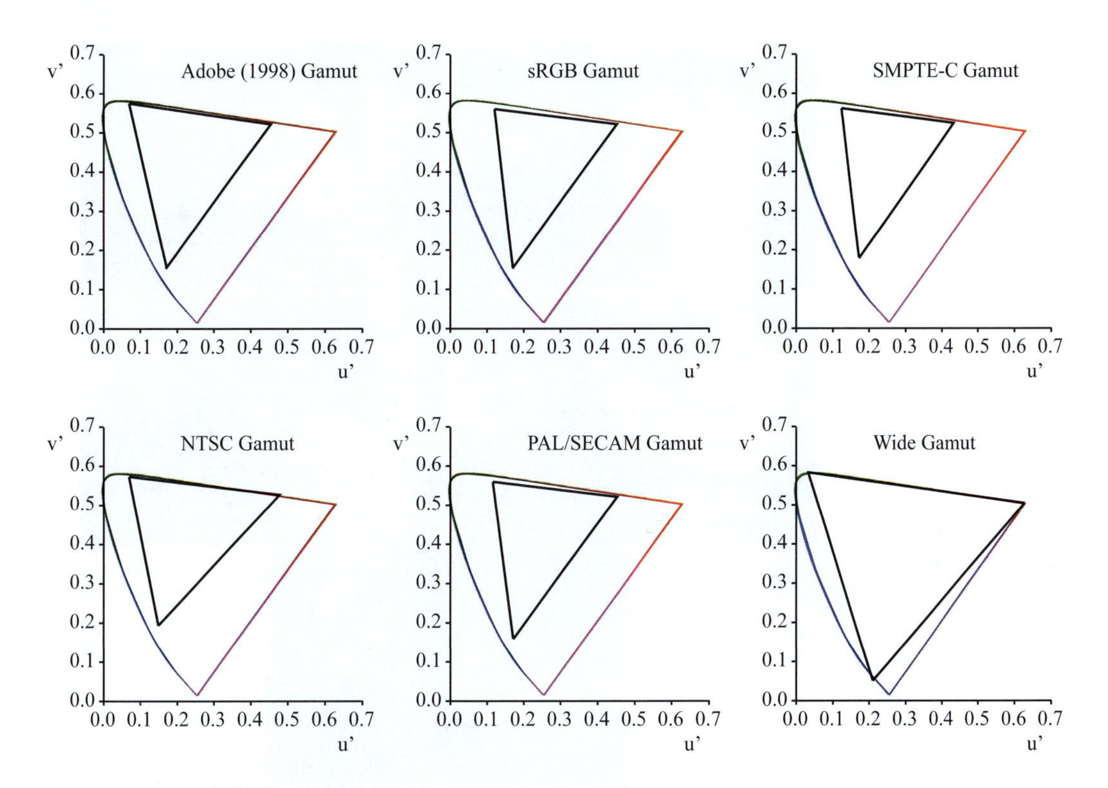

Figure 8.1. CIE (u′,v′) chromaticity diagrams showing the color gamuts for various color spaces.

The shift of primaries toward the spectrum locus yield primaries that, when plotted against wavelength, are more peaked. This is where the name sharpening stems from. Hence, color spaces formed by employing primaries with such peaked response functions are called sharpened color spaces. The associated widening of the gamut is beneficial in several applications, including white balancing, color appearance modeling (Chapter 11), and rendering (Section 8.12). Transforms for sharpened color spaces are introduced in Section 8.6.2.

Finally, to demonstrate that conversion between color spaces is necessary for correct color reproduction, Figure 8.2 shows what happens when an image's primaries are misinterpreted. In this figure, we have converted an image between different RGB color spaces. This has the same effect as displaying an image on different monitors without adjusting for the monitors' primaries. The error created with this procedure is what is commonly seen when color conversion is omitted. Note that even if care is taken to transform image data between color spaces, it is also important to assure that white points are correctly mapped to-

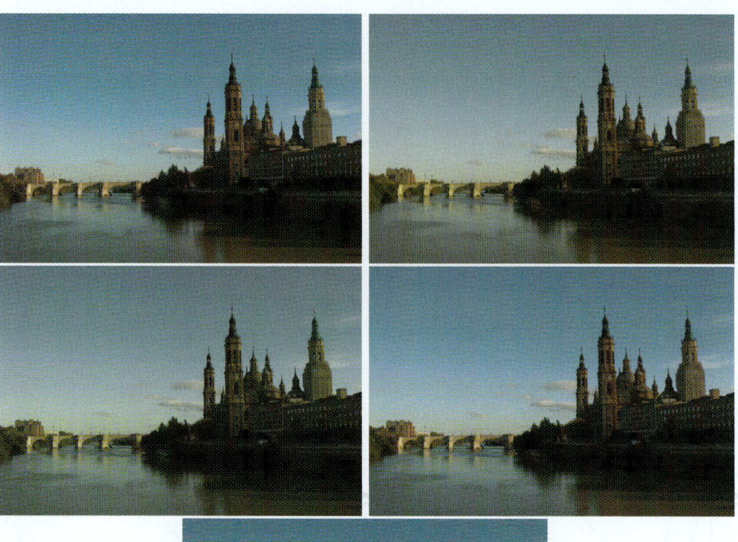

Figure 8.2. A set of images obtained by converting to different RGB color spaces. These images demonstrate the types of error that can be expected if the primaries of a display device are not taken into account. From left to right and top to bottom: the original image (assumed to be in sRGB space) and the image converted to Adobe RGB, NTSC, PAL, and SMPTE-C spaces; Zaragoza, Spain, November 2006.

gether. For instance, when moving from a space with a white point of D65 to one with a white point of D50, we cannot just move into CIE XYZ and then back out into the new RGB color space. We must first perform a chromatic-adaptation transform between the white points, as discussed in Chapter 10.

8.2 Printers

Printers are one technology that can be used to produce hardcopy images. We use the term hardcopy for printed images because once an image is created it cannot be changed, unlike the softcopy display device. Typically a printed image is formed by depositing specified amounts of colorants onto paper or some other substrate. If we assume that that three different inks are used to create a color, the amount of each colorant used to form a colored dot can be expressed as a triplet of percentages. This triplet forms another type of color cube, known as the *colorant space*. An example of colorant spaces are the well-known CMY and CMYK spaces, which are discussed in Section 8.2.1.

To encapsulate the range of colors that any given set of colorants can produce, it is generally necessary to first print those colorants onto paper and then measure the printed pages. These measurements, when expressed in a device-independent space such as CIE XYZ or CIELAB, form a three-dimensional volume essentially bounded by a deformed color cube. This irregular shape is then the gamut of the printing device. Unlike softcopy additive display devices, subtractive color mixing of printed colorants does not behave in a simple, linearly additive manner. The gamut boundary reflects this non-linearity, as the size and shape depends on the colorants used, the size of each dot, the opacity of the inks, the amount each dot overlaps with previously deposited dots, the order in which different colors are applied, and the paper used for printing.

It is often desirable to define for each printer a *printer model* which is a predictive model of the printer's gamut as well as the necessary amount of colorants necessary to generate a desired output color, given the above printer-dependent features. The Kubelka-Munk equations, given in Section 3.4.6, can for instance be used to map a printer's colorant space to a linear set of equations that can be used for predictive purposes. An alternative additive predictive model utilizes the Demichel and Neugebauer equations, which are discussed in Section 8.2.2.

8.2.1 CMY and CMYK Colorant Spaces

The CMY and CMYK colorant spaces are two common spaces utilized by printers. The ink deposited on paper reduces the amount of light that is reflected from the surface, and so this type of color creation is often called a subtractive color system. For example, the CMY color space uses cyan, magenta, and yellow primaries to generate colors. We can think of, as a loose approximation, cyan colorants as "subtracting" red light, magenta colorants as subtracting green light, and yellow colorants as subtracting blue light. Color values encoded in these type of spaces specify how much ink, toner, or dye needs be deposited for each point in the image to generate the desired output color. Often, a fourth black ink is used, which relaxes the constraints posed on the other three inks to produce neutral black. The extra black ink should be considered a practical addition for color reproduction using subtractive colorants. If black is used, colors can be specified in the CMYK color space, where the K indicates black. Many modern printers now use a large number of colorant primaries, in order to achieve accurate color reproduction, as well as an expanded gamut. It is not uncommon to see printers with six, ten, or even 12 different colorants or primaries. With so many primaries to choose from, it is generally up to the device manufacturers to determine the combinations of primaries, halftone patterns, and ink coverage necessary to generate any desired output color. For simplicity sake, we will focus on the more traditional CMY and CMYK printers as the methods for handling increasing numbers of primaries are similar.

The CMY and CMYK spaces are inherently device-dependent, as the actual tristimulus values of the output depend strongly on the types of colorants used, the technology of the printer itself, the properties of the halftone pattern used, the surface characteristics of the paper, and in many cases on the age of the print as well. In most cases, printing itself is inherently a binary process, as there is either colorant on a page or not. A wide range of colors can be produced by spatially varying the amount and location of colorants on the printed page. Furthermore, there is no reason to suspect that the transfer function between specified CMYK values and the amount of ink produced on paper is linear or uniform. In general, all of the variables that can affect the color produced by printers are not standardized, and thus differ on a per-printer type. Color management using a device-independent color space is one technique that has been used with success to achieve accurate and reliable colors across a wide variety of printing devices.

That said, a common conversion that needs to be carried out is between a given RGB color space and CMY(K). As both classes of color space can be highly device dependent, it is not possible to define a single transform that will provide acceptable results for all applications. A more appropriate way to perform such a task would be to first convert RGB values into a device-independent space, such as XYZ tristimulus values, and from there construct a conversion to CMY(K) through a printer model or an exhaustive series of measurements. This conversion is known as device characterization and forms the basis of most modern color-management systems.

For non-critical color reproduction, computationally cheap approximations are sometimes used. Common use of such conversions is one of the possible reasons that printed output rarely matches the softcopy output. The simple (and generally incorrect) conversion between RGB and CMY, recalling that a subtractive system removes light from a scene, is as follows:

$$C = 1 - R, \tag{8.14a}$$

$$M = 1 - G, \tag{8.14b}$$

$$Y = 1 - B; \tag{8.14c}$$

and

$$R = 1 - C, \tag{8.15a}$$
$$G = 1 - M, \tag{8.15b}$$
$$B = 1 - Y. \tag{8.15c}$$

If a black ink is used, we can calculate a simple approximation using what is often called gray component replacement (GCR) or undercolor removal (UCR). Traditionally, we would say that UCR reduces the total amount of colorant coverage necessary to achieve a neutral color, whereas GCR replaces the CMY primaries with the black primary. This may be done to remove any color cast from the CMY primaries and achieve a more neutral gray, as a monetary savings if black colorant is cheaper, or to minimize the color inconstancy of the printed page. Kang argues that GCR is a more sophisticated type of algorithm, and that UCR can actually be considered a subset of GCR [575]. Again, using a very simplified algorithm, one such conversion between CMY and CMYK using a type of GCR is

$$K = \min(C, M, Y), \tag{8.16a}$$
$$C' = C - K, \tag{8.16b}$$
$$M' = M - K, \tag{8.16c}$$
$$Y' = Y - K. \tag{8.16d}$$

The process therefore removes some of the cyan, magenta, and yellow ink, and replaces it with black. The most that can be removed is the minimum of the three C, M, and Y inks, so that neutral colors are preserved. The new values for C, M, and Y are as they were before, but with the value of K subtracted. Note that it would be possible to chose any value for K as long as

$$0 \le K \le \min(C, M, Y). \tag{8.17}$$

The reverse conversion is given by

$$C = C' + K, \tag{8.18a}$$
$$M = M' + K, \tag{8.18b}$$
$$Y = Y' + K. \tag{8.18c}$$

We must again stress that the CMY(K) values obtained in this manner are directly related to the primaries and white point of the original (perhaps unknown) RGB color space. These primaries will no doubt be very different from the CMY(K) primaries of whichever printing device will ultimately be used, and so the results of this type of RGB to CMY(K) conversion can only be considered approximate at best.

The color gamut of the CMY(K) spaces are typically said to be much smaller than the gamuts of RGB devices. This means that there are ranges of colors that can be displayed on a monitor that can not be reproduced in print. Section 15.4 discusses gamut mapping, a process that maps a larger gamut to a smaller one. It is important to stress that not all printers have a smaller gamut than all additive display devices, especially for colors closely associated with the printer primaries, such as bright yellows or dark cyan colors. The ultimate gamut of a printer device is heavily tied to the materials used to create the printed colors. For example, the same printer may produce a gamut much smaller than a typical LCD display when printing on plain or matte paper, but may have a gamut much larger than the LCD when printing on a high-quality glossy paper. In some instances, gamut-mapping algorithms may be best served by both reducing a larger gamut to a smaller one in some areas, and mapping a smaller to a larger (gamut expansion) in others. This just suggests that overarching statements such as "displays have larger gamuts than printers" should be met with a high degree of skepticism.

More recently, printers are using more than four inks to achieve a higher degree of color fidelity. For instance, some desktop printers have added two extra inks for light cyan and light magenta. This resulting device-dependent colorant space is known as CcMmYK, in which the c and m indicate light cyan and light magenta, respectively. This may be done to reduce the visibility of the half-toning process, and to create a more uniform printed surface. Other multiple ink configurations, such as the addition of green, red, orange, blue, or even clear inks, are certainly possible as well. These inks can be added to increase the overall gamut of the printer, increase the color constancy of the printer, or to increase overall image quality (perhaps by reducing the size of the halftone dots).

8.2.2 The Demichel-Neugebauer Printer Model

The Demichel-Neugebauer model can be used to infer the gamut boundaries of printed colorants on paper. It is therefore an empirical mapping between the device-colorant space and the achievable colors on paper. Typically, we consider printers and colorants to be a form of subtractive color mixing. Subtractive color mixing is generally a non-linear process, and more advanced mathematics are necessary to model and predict these non-linearities. If we consider that the binary process (either colorant or no colorant) of printers produces a series of dots that are then integrated by our visual system, then we can consider this integration to be an additive process. While the mixing of the colorants themselves are most likely non-linear, the mixing of the colors that the eye perceives can be modeled as an additive mixture of the colorant's spectral reflectance distributions. Essentially this is what the Demichel-Neugebauer equations, often referred to as just the Neugebauer equations, do. On a simplified level, the Demichel equations form a probabilistic model predicting how much of any given primary forms the basis of a color, and the Neugebauer equations model the additive color mixing of the primaries. The Neugebauer equations are often used for three-ink and four-ink processes, but can be extended for more inks. These equations are most appropriate to predict colors resulting from halftone printing.

In halftone printing, each of the colored inks is applied sequentially, with dots of ink either placed next to one another or partially overlapping. If three inks are utilized, then the paper may be covered by one ink, two inks, three inks, or may be left blank. There are eight combinations of overlap, assuming that the order in which inks are applied, is fixed. The eight colors associated with these combinations of overlap are called the *Neugebauer primaries*. We indicate these primaries with $X_W Y_W Z_W$, $X_1 Y_1 Z_1$, $X_2 Y_2 Z_2$, $X_3 Y_3 Z_3$, $X_{12} Y_{12} Z_{12}$, $X_{13} Y_{13} Z_{13}$, $X_{23} Y_{23} Z_{23}$, and $X_{123} Y_{123} Z_{123}$.

The tristimulus value of a color on paper is then a weighted combination of these eight primaries, given by the Neugebauer equations [828]:

$$X = a_W X_W + a_1 X_1 + a_2 X_2 + a_3 X_3 + a_{12} X_{12} + a_{13} X_{13} + a_{23} X_{23} + a_{123} X_{123}, \tag{8.19a}$$

$$Y = a_W Y_W + a_1 Y_1 + a_2 Y_2 + a_3 Y_3 + a_{12} Y_{12} + a_{13} Y_{13} + a_{23} Y_{23} + a_{123} Y_{123}, \tag{8.19b}$$

$$Z = a_W Z_W + a_1 Z_1 + a_2 Z_2 + a_3 Z_3 + a_{12} Z_{12} + a_{13} Z_{13} + a_{23} Z_{23} + a_{123} Z_{123}. \tag{8.19c}$$

Note that the Neugebauer equations are traditionally calculated as a weighted sum of spectral reflectances rather than tristimulus values, though due to the additivity of light and Grassmann's laws we can use either. The Neugebauer primaries consist of the printer primaries themselves, as well as the combinations of these primaries, which form *secondaries*. For a CMY printer, the secondaries are red $(M+Y)$, green $(C+Y)$, and blue $(M+C)$, while black is formed by adding all three primaries together and white is formed by just the paper itself. The relative weights for each of the Neugebauer primaries are calculated using the percentage coverage for each printer primary as a halftone dot percentage, given by c_1, c_2, and c_3. The relationship between the dot-area percentages and the weight factors

$a_W, a_1, \ldots$ are given by the Demichel equations [246, 247]:

$$a_W = (1-c_1)(1-c_2)(1-c_3), \tag{8.20a}$$

$$a_1 = c_1(1-c_2)(1-c_3), \tag{8.20b}$$

$$a_2 = (1-c_1)c_2(1-c_3), \tag{8.20c}$$

$$a_3 = (1-c_1)(1-c_2)c_3, \tag{8.20d}$$

$$a_{12} = c_1 c_2 (1-c_3), \tag{8.20e}$$

$$a_{13} = c_1(1-c_2)c_3, \tag{8.20f}$$

$$a_{23} = (1-c_1)c_2 c_3, \tag{8.20g}$$

$$a_{123} = c_1 c_2 c_3. \tag{8.20h}$$

We should consider the weights calculated by the Demichel equations to be probabilities of any given primary, based upon the percentage of dot-area coverage. For instance if any given area has a high percentage of cyan and yellow colorant, then the probability of that area being blue increases. The combined Demichel-Neugebauer model can thus be used to predict the XYZ tristimulus value for each of the eight Neugebauer primaries that are generated in a three-ink process. The equations can be extended for more than three inks, though the number of Neugebauer primaries increases exponentially with the number of printer inks used. In order to properly utilize the Demichel-Neugebauer equations, it is necessary to know the tristimulus values, or spectral reflectance of each of the Neugebauer primaries. Typically this is done through careful printing and measurement. The Neugebauer equations also have a fundamental assumption of additivity of the light, though in practice absorption and scattering of the inks, as well as optical diffusion of light through the inks and paper affect the additivity.

To test the additivity of any given printer a set of uniform color ramps can be generated and measured. If the system behaves in a relatively linear manner, and we know how much of each of the colorant will be applied, we can use the Demichel-Neugebauer equations to predict the output color on paper. Further information on printer color modeling can be found in the texts of Yule and Kang [575, 1287]. Wyble gives an excellent overview of more recent models of modeling color printers using the Demichel-Neugebauer equations along with other techniques [1258].

The opposite problem, namely that of color separation, is concerned with finding how much colorant to apply to reproduce a given color specified in XYZ color space. The nature of this problem is revealed by substituting the Demichel equations into the Neugebauer equations, yielding a polynomial in the dot percentages. Solving these equations for c_1, c_2, and c_3 leads to an expression mapping an XYZ color to dot percentages for each of the three colorants in terms of the eight Neugebauer primaries (which need to be measured for these equations to be useful).

As the solution is non-linear, an analytical inversion of the printer model can only be achieved if simplifying assumptions are made. Otherwise, regression techniques can be used to express each of the three dot percentages in terms of X, Y, and Z. First, we will assume that the color of paper results in a tristimulus value of $(X_W, Y_W, Z_W) = (1, 1, 1)$. Second, we assume that the densities of inks are additive, i.e., $X_{12} = X_1 X_2$ and so on for all other ink combinations [912–914]. This allows the Neugebauer equations to be factored as follows [1287]:

$$X = (1 - c_1 + c_1 X_1)(1 - c_2 + c_2 X_2)(1 - c_3 + c_3 X_3), \tag{8.21a}$$

$$Y = (1 - c_1 + c_1 Y_1)(1 - c_2 + c_2 Y_2)(1 - c_3 + c_3 Y_3), \tag{8.21b}$$

$$Z = (1 - c_1 + c_1 Z_1)(1 - c_2 + c_2 Z_2)(1 - c_3 + c_3 Z_3). \tag{8.21c}$$

A simplification is obtained by assuming that $X_2 = X_3 = Y_3 = 1$, so that we have

$$X = (1 - c_1 + c_1 X_1), \tag{8.22a}$$

$$Y = (1 - c_1 + c_1 Y_1)(1 - c_2 + c_2 Y_2), \tag{8.22b}$$

$$Z = (1 - c_1 + c_1 Z_1)(1 - c_2 + c_2 Z_2)(1 - c_3 + c_3 Z_3). \tag{8.22c}$$

Although these equations can be inverted, a further simplifying assumption will make inversion more straightforward. In particular, it may be assumed that $X_1 = Y_2 = Z_3 = 0$, so that we get

$$1 - c_1 = X, \tag{8.23a}$$

$$1 - c_2 = \frac{Y}{Y_1 + (1 - Y_1) X}, \tag{8.23b}$$

$$1 - c_3 = \frac{Z}{(Z_1 + (1 - Z_1) X)(Z_2 + (1 - Z_2)(1 - c_2))}. \tag{8.23c}$$

The three-ink model may be extended to four colors, for use with CMYK printer models [426]. While the procedure is straightforward [577], the equations become much more difficult to invert so that numerical techniques are required.

Real printing processes often deviate from the Demichel-Neugebauer model. The above equations therefore have to be used with care. The Neugebauer printer model can suffer from accuracy problems when the general assumptions are simplified beyond physical reality.

Various extentions have been proposed [540, 907], including the Clapper-Yule model for multiple inks [196, 197, 1287] and the spectral Neugebauer equations [1182]. The Clapper-Yule model itself was extended to model printing with embedded metallic patterns [465]. Another class of improved models is known as the *cellular* or *localized Neugebauer equations* [725, 864], where color space is subdivided into disjunct regions that are treated independently.

An extension that accounts for *dot gain* may also be incorporated into the Neugebauer equations. When a droplet of ink is deposited onto paper, it spreads out. This is called the *physical dot gain*. Light scattering inside the paper will also cause dots to absorb more light than the area coverage would predict, a phenomenon called *optical dot gain*. The desired coverage c_i can be recovered using the following empirical relation:

$$1 - c_i = (1 - \bar{c}_i)^{\gamma_i}, \tag{8.24}$$

where $\bar{c}_i$ is the specified dot percentage, and γ_i is an ink-specific parameter that can be derived from experimental data [1093]. Again, it is important to stress that this non-linear exponent is only an empirical approximation to better predict printed results and should not be taken as a model of a real physical process.

A further extension takes into account *trapping*. This is a phenomenon whereby the amount of ink that sticks to the page is less than the amount of ink deposited. Trapping thus reduces the dot percentage of coverage by a fraction t_i^p, where i indicates ink i sticking directly to the paper. For overprinting, the fraction t_{ij}^p models the trapping of ink i printed on top of ink j. The Neugebauer equations (given only for X) then become [1093]:

$$\begin{aligned}
X = {} & (1 - t_1^p c_1)(1 - t_2^p c_2)(1 - t_3^p c_3) X_W && \text{(8.25a)} \\
& + t_1^p c_1 (1 - t_{12}^p c_2)(1 - t_{13}^p c_3) X_1 && \text{(8.25b)} \\
& + (1 - t_1^p c_1) t_{12}^p c_2 (1 - t_{13}^p c_3) X_2 && \text{(8.25c)} \\
& + (1 - t_1^p c_1)(1 - t_2^p c_2) t_3^p c_3 X_3 && \text{(8.25d)} \\
& + t_1^p c_1 t_{12}^p c_2 (1 - t_{123}^p c_3) X_{12} && \text{(8.25e)} \\
& + t_1^p c_1 (1 - t_{12}^p c_2) t_{13}^p c_3 X_{13} && \text{(8.25f)} \\
& + (1 - t_1^p c_1) t_2^p c_2 t_{23}^p c_3 X_{23} && \text{(8.25g)} \\
& + t_1^p c_1 t_{12}^p c_2 t_{123}^p c_3 X_{123}. && \text{(8.25h)}
\end{aligned}$$

The Neugebauer equations assume that the primaries of the printing inks are known, or that they can be measured. This may not always be practical to do. In that case, it may be possible to compute an approximation of the primaries using a more complicated non-linear model such as the Kubelka-Munk theory used to calculate the absorption and scattering of translucent and opaque pigments [1093] (see Section 3.4.6).

8.3 Luminance-Chrominance Color Spaces

RGB color spaces are generally characterized by three physical primaries, or three primaries that are well defined in a device-independent manner. For any three-channel color space, we can say that tristimulus values that determine the amount of each of the three primaries are necessary to generate the final color. Although these spaces are useful for specifying and encoding color for direct transmission and encoding to real devices, it is often convenient to encode color into a luminance channel and two chrominance channels. The luminance channel then represents how light or dark the color is, and should (by definition) relate directly to the Y channel of the CIE XYZ color space. The two chrominance channels together then determine the chromatic content of any given color.

One utility of such color representation is the use of the chrominance channels to generate chromaticity diagrams. Recall from Chapter 7 that a chromaticity diagram is a two-dimensional projection of a three-dimensional space. Chromaticity diagrams can be useful for plotting the general location of chromatic primaries in an additive display device, or for determining complementary colors of a given stimulus. Common example of chromaticity diagrams are formed by plotting the x against the y channel of the Yxy color space, or the more perceptually uniform chromaticity diagram created by the u′ and v′ channels of the CIE Yu′v′ space. Although it is tempting to use chromaticity diagrams to plot the gamut of additive display devices, such as CRTs and LCDs, it should be readily evident that the two-dimensional projection immediately eliminates a large amount of information. In most luminance-chrominance spaces, this information loss is the actual luminance information, which means we cannot determine the actual color appearance of any given point in a chromaticity diagram. Representing device gamuts, which inherently need to have at least three color dimensions in a chromaticity space, can be misleading, at best, and completely inaccurate, at worst.

Another reason for using luminance-chrominance color spaces is that the human visual system naturally utilizes such a representation for transmission of signals between the retina and the visual cortex. The visual system utilizes a type of color opponent space, with a white-black channel, a reddish-greenish channel, and a bluish-yellowish channel. Due to the nature of generating these channels from the LMS cone signals, the visual system's spatial sensitivity (acuity and contrast sensitivity) is generally higher for variations in luminance than for variations in chrominance. As all cone types contribute to the formation of a luminance signal in the human visual system, its spatial resolution is generally determined by a single cone receptor. However, color opponent signals require at least two cones (four if the random distribution of L and M cones is taken into account [785]). As

Figure 8.3. The images on the left have all three channels subsampled, whereas the images on the right have only their chrominance channels subsampled. From top to bottom the subsampling is by factors of 2, 4, 8, 16, and 32; Long Ashton, Bristol, UK.

a result, color opponent signals have a spatial resolution corresponding to about four cone diameters [505].

It is therefore possible to use a lower spatial resolution for encoding and transmitting the chrominance channels than for the luminance channel, without greatly affecting how such encoded images are perceived. As luminance and chrominance are not encoded separately in RGB color spaces, the same efficient encoding could not be achieved in RGB.

An example of the effect of lowering the spatial resolution of the chrominance channels is shown in Figure 8.3. The images on the left in this figure have the spatial resolution of all three channels progressively lowered. On the right, only the chrominance channels are subsampled by corresponding amounts. The effect of lowering the spatial resolution is far more notable when all channels are subsampled.

This feature is exploited by the transmission encodings used in television signals to save bandwidth, an idea attributed to A. V. Loughren [507]. Television encodings are further discussed in Section 8.4. The ability to spatially subsample chrominance channels with minimal perceptual artifacts is also taken advantage of in most modern image-compression algorithms, such as JPEG.

8.3.1 CIE XYZ and Yxy

The CIE XYZ color space was introduced in Chapter 7. Due to the device-independent nature of the CIE XYZ space, along with its basis with the human visual system, CIE XYZ is often used to define many other color spaces. One important consideration in the derivation of the CIE XYZ space was the ability to specify all visible colors with positive values for X, Y, and Z. Due to the high degree of overlap between the cone responses, this is not possible using only three physically realizable primaries. As a result, the CIE XYZ space is defined using imaginary primaries.

XYZ tristimulus values can be plotted on a chromaticity diagram by computing chromaticity coordinates, as shown in Section 7.5.1. The combination of the XYZ luminance channel Y and two chromaticity coordinates x and y can be regarded as a color space. As chromaticity coordinates are normalized, they carry no information about luminance. In the Yxy color space, luminance is therefore separated from chrominance, though we cannot say that the xy chromaticities represent red-green and yellow-blue opponent color representations.

In chromaticity diagrams, the spatial difference between pairs of points is a poor indicator of perceptual difference between pairs of colors. An alternative way of describing this is that the xy chromaticity diagram is not perceptually uniform. This has given rise to several other color spaces, which can be regarded as improvements over Yxy in this regard. Examples are the CIE Yuv, CIE u′v′ uniform chromaticity scale (UCS) diagrams, and also the CIE L*a*b* (CIELAB) and CIE L*u*v* (CIELUV) color spaces. It should be noted that although the CIELAB color space was developed as a perceptually uniform luminance-chrominance space, it does not have a defined chromaticity diagram that goes along with it.

8.3.2 CIE Yuv (1960) and CIE Yu′v′

In the CIE Yuv color space, the luminance channel is the same as in the CIE XYZ and CIE Yxy color spaces (see Section 7.5.1). However, the chromaticity coordinates are defined differently, to achieve a more uniform chromaticity plane:

$$u = \frac{2x}{6y - x + 1.5}, \tag{8.26a}$$

$$v = \frac{3y}{6y - x + 1.5}. \tag{8.26b}$$

A further refinement was introduced by the CIE Yu′v′ color space, achieving yet greater uniformity in its chromaticity coordinates:

$$u' = \frac{2x}{6y - x + 1.5}, \tag{8.27a}$$

$$v' = \frac{4.5y}{6y - x + 1.5}. \tag{8.27b}$$

Note that the only difference between Yuv and Yu′v′ is between the v and v' channels, whereby $v' = 1.5\,v$. Finally, as both these color spaces are derived from CIE XYZ, they can also be considered to be well-defined device-independent color spaces. It should also be noted that the Yu′v′ space generally has superseded the Yuv space, although for historical compatibility that space is still used for calculated correlated color temperatures.

8.4 Television and Video

Broadcasting involves several stages, starting with the capture of video or film, followed by various stages of processing on the captured and encoded signal. This processed signal is then transmitted either by air, by satellite, by cable, or by a combination of all of these methods. Once received, the signal is decoded and displayed on a television.

To ensure that this series of events leads to an image that is reasonably close to the image captured at the start of this sequence, standard definitions for each of the processing steps are required. Such standards pertain to the color spaces used at various stages, the encoding of the signal, sample rates, the number of pixels in the image, and the order in which they are scanned.

Unfortunately, there is no single standard, and different broadcasting systems are currently in use. For instance, for scanning North America and Japan use the 480*i*29.97 system, whereas most of the rest of the world uses the 576*i*25 system.[1] The notation refers to the number of active scanlines (480 or 576), the *i* refers to interlaced video, followed by the frame rate. Note that the North American system has in total 525 scanlines. However, only 480 of them appear on screen. Aside from interlaced video, progressive scan systems are also in use. For instance, high definition television (HDTV) includes standards for both; 720*p*60 and 1080*i*30.

The transmission of signals requires the conversion of a captured RGB signal into a space that is convenient for transmission. Typically, a non-linear transfer function is applied first, leading to an $R'G'B'$ signal. The general form of this non-linearity is given in (8.13). Following Charles Poynton, in this section primed quantities denote non-linear signals [923].

The luminance-chrominance color spaces are suitable, as they allow the chrominance channels to be compressed, saving transmission bandwidth by exploiting the human visual system's relatively poor color acuity. Converting from a linear RGB space to one of these spaces yields luminance and chrominance signals. However, if the input to the transform is deliberately non-linear, as in $R'G'B'$, the conversion does not result in luminance and chrominance, but in a non-linear luminance and chrominance representation. These are called *luma* and *chroma*. Note that the non-linear luma representation is a rough approximation to the color appearance description of *lightness*, though the chroma bears little to no resemblance to the color appearance attribute of chroma. These appearance terms are defined later in Chapter 11.

An image encoded with chroma subsampling is shown in Figure 8.3. The sampling is usually described by a three-fold ratio, e.g., x:y:z which represents the horizontal and vertical subsampling. The first number represents the horizontal sampling reference of luminance with respect to a sampling rate of 3.375 MHz; thus a 4:4:4 represents a luminance sampling rate of 13.5 MHz. The second number refers to the chrominance horizontal factor with respect to the first number. The third number is either the same as the second or it is zero in which case

[1] We use the notation introduced by Charles Poynton in *Digital Video and HDTV: Algorithms and Interfaces* [923].

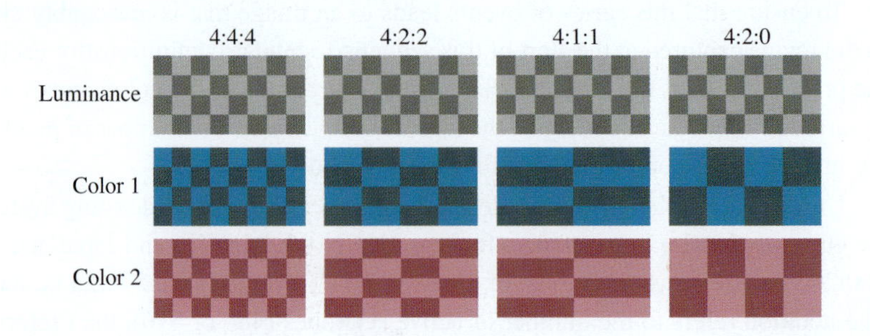

Figure 8.4. Four instances of chroma subsampling: 4:4:4, 4:2:2, 4:1:1, and 4:2:0.

the vertical color resolution is subsampled at a factor of 2. Examples of chroma subsampling are 4:4:4, 4:2:2, 4:1:1, and 4:2:0, as illustrated in Figure 8.4.

The encoding of signals is standardized by the NTSC standard for North America and PAL for Europe. These encodings turn RGB into a composite signal that is used in analog broadcast. On the other hand, newer video equipment uses component video, denoted $Y'P_BP_R$ for analog components or $Y'C_BC_R$ for digital components. Older equipment may use composite video. High definition television always uses either analog or digital component video.

The several systems in use today are identified by the number of active scanlines and characterized by the chromaticity coordinates of the primaries, the non-linear transfer function, and the coefficients used to compute luma and chroma. Each of these parts is governed by different standards, as summarized in Table 8.4.

With both standard definition television (SDTV; 480*i* and 576*i* systems) and HDTV (720*p*, 1080*p*, and 1080*i* systems) currently in use, the current situation is that there are two encoding systems for luma and chroma. This means, for instance, that there exist two different forms for YC_BC_R, dependent on which system is used. The same is true for its analog equivalent YP_BP_R. The accurate interchange of signals across different systems is therefore problematic.

System	Primaries	Transfer	Luma
480*i*	SMPTE RP 145	Rec. 709	Rec. 601
576*i*	EBU Tech. 3213	Rec. 709	Rec. 601
720*p*	Rec. 709	Rec. 709	Rec. 709
1080*p*	Rec. 709	Rec. 709	Rec. 709
1080*i*	Rec. 709	Rec. 709	Rec. 709

Table 8.4. Primaries, non-linear transfer function, and luma encoding for the different television systems currently in use (after [923]).

8.4.1 EBU Y′U′V′

Starting with linear RGB values, the EBU standard, which is used in European PAL and SECAM color encodings, assumes that the D65 white point is used, as well as the primaries listed under PAL/SECAM in Table 8.2. The associated transformation matrix, as well as the parameters for the non-linear transfer function are listed under the PAL/SECAM entry in Table 8.3. The R′G′B′ signal is then converted to Y′U′V′ transmission primaries. Here, the chromatic channels are formed by subtracting luma from blue and red, respectively [975]:

$$U' = 0.492111(B' - Y'), \tag{8.28a}$$

$$V' = 0.877283(R' - Y'). \tag{8.28b}$$

This leads to the following transformation matrix:

$$\begin{bmatrix} Y' \\ U' \\ V' \end{bmatrix} = \begin{bmatrix} 0.299 & 0.587 & 0.114 \\ -0.147141 & -0.288869 & 0.436010 \\ 0.614975 & -0.514965 & -0.100010 \end{bmatrix} \begin{bmatrix} R' \\ G' \\ B' \end{bmatrix}. \tag{8.29}$$

The inverse is given by

$$\begin{bmatrix} R' \\ G' \\ B' \end{bmatrix} = \begin{bmatrix} 1.000000 & 0.000000 & 1.139883 \\ 1.000000 & -0.394642 & -0.580622 \\ 1.000000 & 2.032062 & 0.000000 \end{bmatrix} \begin{bmatrix} Y' \\ U' \\ V' \end{bmatrix}. \tag{8.30}$$

8.4.2 NTSC (1953) Y′I′Q′

The color coding and transmission primaries for the American National Television Systems Committee (NTSC) standard are described in this section. The transmission primaries use a luminance-chrominance encoding, which was introduced in the early 1950s [698].

The color coding is no longer in wide use; it has been superseded by the SMPTE-C standard which is discussed in the following section. The RGB primaries for NTSC are listed in Table 8.2 under NTSC (1953). The corresponding conversion matrix and non-linear transfer function are listed in Table 8.3 as NTSC (1953). The NTSC standard uses Y′I′Q′ transmission primaries, where the letters I and Q stand for in-phase quadrature. The transmission primaries are constructed by differencing B′ and R′ from Y′ as follows:

$$I' = -0.27(B' - Y') + 0.74(R' - Y'), \tag{8.31a}$$

$$Q' = 0.41(B' - Y') + 0.48(R' - Y'). \tag{8.31b}$$

Figure 8.5. Visualization of the Y′I′Q′ color channels. The original image is shown in the top-left panel. The Y′ channel is shown in the top right and the bottom two images show the Y′+I′ channels (left) and Y′+Q′ channels (right); Zaragoza, Spain, November 2006.

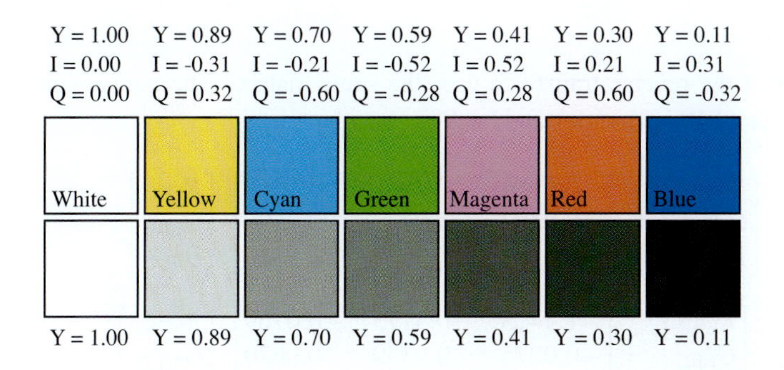

Figure 8.6. Standard NTSC color bars along with their corresponding luminance bars.

The I' is the orange-blue axis while Q' is the purple-green axis (see Figure 8.5 for a visualization). The reason for this choice is to ensure backward compatibility (the Y' channel alone could be used by black and white displays and receivers) and to utilize peculiarities in human perception to transmit the most visually pleasing signal at the lowest bandwidth cost. Human visual perception is much more sensitive to changes in luminance detail than to changes in color detail. Thus, the majority of bandwidth is allocated to encoding luminance information and the rest is used to encode both chrominance information and audio data. The color coding is given by

$$\begin{bmatrix} Y' \\ I' \\ Q' \end{bmatrix} = \begin{bmatrix} 0.299 & 0.587 & 0.114 \\ 0.596 & -0.275 & -0.321 \\ 0.212 & -0.523 & 0.311 \end{bmatrix} \begin{bmatrix} R' \\ G' \\ B' \end{bmatrix}, \tag{8.32}$$

and its inverse transform is

$$\begin{bmatrix} R' \\ G' \\ B' \end{bmatrix} = \begin{bmatrix} 1.000 & 0.956 & 0.621 \\ 1.000 & -0.272 & -0.647 \\ 1.000 & -1.107 & 1.704 \end{bmatrix} \begin{bmatrix} Y' \\ I' \\ Q' \end{bmatrix}. \tag{8.33}$$

The standard NTSC color bars are shown alongside their luminance bars in Figure 8.6.

For practical purposes, the transform between EBU and NTSC transmission primaries is possible because their RGB primaries are relatively similar. The transform is given by

$$\begin{bmatrix} I' \\ Q' \end{bmatrix} = \begin{bmatrix} -0.547 & 0.843 \\ 0.831 & 0.547 \end{bmatrix} \begin{bmatrix} U' \\ V' \end{bmatrix}; \tag{8.34a}$$

$$\begin{bmatrix} U' \\ V' \end{bmatrix} = \begin{bmatrix} -0.547 & 0.843 \\ 0.831 & 0.547 \end{bmatrix} \begin{bmatrix} I' \\ Q' \end{bmatrix}. \tag{8.34b}$$

Note that the reverse transform uses the same matrix. The conversion between NTSC and EBU primaries and back is given by

$$\begin{bmatrix} R_{\text{NTSC}} \\ G_{\text{NTSC}} \\ B_{\text{NTSC}} \end{bmatrix} = \begin{bmatrix} 0.6984 & 0.2388 & 0.0319 \\ 0.0193 & 1.0727 & 0.0596 \\ 0.0169 & 0.0525 & 0.8459 \end{bmatrix} \begin{bmatrix} R_{\text{EBU}} \\ G_{\text{EBU}} \\ B_{\text{EBU}} \end{bmatrix}; \tag{8.35a}$$

$$\begin{bmatrix} R_{\text{EBU}} \\ G_{\text{EBU}} \\ B_{\text{EBU}} \end{bmatrix} = \begin{bmatrix} 1.4425 & -0.3173 & -0.0769 \\ -0.0275 & 0.9350 & 0.0670 \\ -0.0272 & -0.0518 & 1.1081 \end{bmatrix} \begin{bmatrix} R_{\text{NTSC}} \\ G_{\text{NTSC}} \\ B_{\text{NTSC}} \end{bmatrix}. \tag{8.35b}$$

8.4.3 SMPTE-C RGB

Television in North America currently uses the SMPTE-C standard. Its RGB primaries, transformation matrices, and transfer functions are listed in Tables 8.2 and 8.3. The conversion matrix to $Y'I'Q'$ is the same as for the NTSC (1953) standard discussed in the preceding section. As a result, the conversion between EBU and SMPTE-C transmission components is given in (8.34). The conversion between SMPTE-C and EBU RGB signals is as follows:

$$\begin{bmatrix} R_{\text{SMPTE-C}} \\ G_{\text{SMPTE-C}} \\ B_{\text{SMPTE-C}} \end{bmatrix} = \begin{bmatrix} 1.1123 & -0.1024 & -0.0099 \\ -0.0205 & 1.0370 & -0.0165 \\ 0.0017 & 0.0161 & 0.9822 \end{bmatrix} \begin{bmatrix} R_{\text{EBU}} \\ G_{\text{EBU}} \\ B_{\text{EBU}} \end{bmatrix}; \tag{8.36a}$$

$$\begin{bmatrix} R_{\text{EBU}} \\ G_{\text{EBU}} \\ B_{\text{EBU}} \end{bmatrix} = \begin{bmatrix} 0.9007 & 0.0888 & 0.0105 \\ 0.0178 & 0.9658 & 0.0164 \\ -0.0019 & -0.0160 & 1.0178 \end{bmatrix} \begin{bmatrix} R_{\text{SMPTE-C}} \\ G_{\text{SMPTE-C}} \\ B_{\text{SMPTE-C}} \end{bmatrix}. \tag{8.36b}$$

8.4.4 ITU-R BT.601 $Y'P_BP_R$ (SDTV)

The ITU-R BT.601 standard defines the transform from $R'G'B'$ to transmission primaries. It does not define the primaries, white point, or transfer function of the $R'G'B'$ space. This standard is used for digital coding of SDTV signals, and the conversion is as follows:

$$\begin{bmatrix} Y' \\ P_B \\ P_R \end{bmatrix} = \begin{bmatrix} 0.299 & 0.587 & 0.114 \\ -0.169 & -0.331 & 0.500 \\ 0.500 & -0.419 & -0.089 \end{bmatrix} \begin{bmatrix} R' \\ G' \\ B' \end{bmatrix}; \tag{8.37a}$$

$$\begin{bmatrix} R' \\ G' \\ B' \end{bmatrix} = \begin{bmatrix} 1.000 & 0.000 & 1.403 \\ 1.000 & -0.344 & -0.714 \\ 1.000 & 1.772 & 0.000 \end{bmatrix} \begin{bmatrix} Y' \\ P_B \\ P_R \end{bmatrix}. \tag{8.37b}$$

8.4.5 ITU-R BT.709 $Y'C_BC_R$ (HDTV)

The ITU-R recommendation for HDTV signals uses the primaries listed under HDTV in Table 8.2. Its associated matrices and non-linear transfer function are given in Table 8.3. The transmission coefficients are computed by

$$\begin{bmatrix} Y' \\ C_B \\ C_R \end{bmatrix} = \begin{bmatrix} 0.2215 & 0.7154 & 0.0721 \\ -0.1145 & -0.3855 & 0.5000 \\ 0.5016 & -0.4556 & -0.0459 \end{bmatrix} \begin{bmatrix} R' \\ G' \\ B' \end{bmatrix}; \tag{8.38a}$$

$$\begin{bmatrix} R' \\ G' \\ B' \end{bmatrix} = \begin{bmatrix} 1.0000 & 0.0000 & 1.5701 \\ 1.0000 & -0.1870 & -0.4664 \\ 1.0000 & 1.8556 & 0.000 \end{bmatrix} \begin{bmatrix} Y' \\ C_B \\ C_R \end{bmatrix}. \tag{8.38b}$$

Further, the conversion between EBU and Rec.709 RGB signals is

$$\begin{bmatrix} R_{709} \\ G_{709} \\ B_{709} \end{bmatrix} = \begin{bmatrix} 1.0440 & -0.0440 & 0.0000 \\ 0.0000 & 1.0000 & 0.0000 \\ 0.0000 & -0.0119 & 1.0119 \end{bmatrix} \begin{bmatrix} R_{EBU} \\ G_{EBU} \\ B_{EBU} \end{bmatrix}; \tag{8.39a}$$

$$\begin{bmatrix} R_{EBU} \\ G_{EBU} \\ B_{EBU} \end{bmatrix} = \begin{bmatrix} 0.9578 & 0.0422 & 0.0000 \\ 0.0000 & 1.0000 & 0.0000 \\ 0.0000 & 0.0118 & 0.9882 \end{bmatrix} \begin{bmatrix} R_{709} \\ G_{709} \\ B_{709} \end{bmatrix}. \tag{8.39b}$$

The conversion between NTSC and Rec. 709 RGB is

$$\begin{bmatrix} R_{709} \\ G_{709} \\ B_{709} \end{bmatrix} = \begin{bmatrix} 1.5073 & -0.3725 & -0.0832 \\ -0.0275 & 0.9350 & 0.0670 \\ -0.0272 & -0.0401 & 1.1677 \end{bmatrix} \begin{bmatrix} R_{NTSC} \\ G_{NTSC} \\ B_{NTSC} \end{bmatrix}; \tag{8.40a}$$

$$\begin{bmatrix} R_{NTSC} \\ G_{NTSC} \\ B_{NTSC} \end{bmatrix} = \begin{bmatrix} 0.6698 & 0.2678 & 0.0323 \\ 0.0185 & 1.0742 & -0.0603 \\ 0.0162 & 0.0432 & 0.8551 \end{bmatrix} \begin{bmatrix} R_{709} \\ G_{709} \\ B_{709} \end{bmatrix}, \tag{8.40b}$$

and finally between SMPTE-C and Rec.709 RGB,

$$\begin{bmatrix} R_{709} \\ G_{709} \\ B_{709} \end{bmatrix} = \begin{bmatrix} 0.9395 & 0.0502 & 0.0103 \\ 0.0178 & 0.9658 & 0.0164 \\ -0.0016 & -0.0044 & 1.0060 \end{bmatrix} \begin{bmatrix} R_{SMPTE\text{-}C} \\ G_{SMPTE\text{-}C} \\ B_{SMPTE\text{-}C} \end{bmatrix}; \tag{8.41a}$$

$$\begin{bmatrix} R_{SMPTE\text{-}C} \\ G_{SMPTE\text{-}C} \\ B_{SMPTE\text{-}C} \end{bmatrix} = \begin{bmatrix} 1.0654 & -0.0554 & -0.0010 \\ -0.0196 & 1.0364 & -0.0167 \\ 0.0016 & 0.0044 & 0.9940 \end{bmatrix} \begin{bmatrix} R_{709} \\ G_{709} \\ B_{709} \end{bmatrix}. \tag{8.41b}$$

8.4.6 SECAM $Y'D_BD_R$

The transmission encoding of SECAM is similar to the YIQ color space. Its conversion from RGB is given by

$$\begin{bmatrix} Y' \\ D_B \\ D_R \end{bmatrix} = \begin{bmatrix} 0.299 & 0.587 & 0.114 \\ -0.450 & -0.883 & 1.333 \\ -1.333 & 1.116 & 0.217 \end{bmatrix} \begin{bmatrix} R' \\ G' \\ B' \end{bmatrix}; \tag{8.42a}$$

$$\begin{bmatrix} R' \\ G' \\ B' \end{bmatrix} = \begin{bmatrix} 1.0000 & 0.0001 & -0.5259 \\ 1.0000 & -0.1291 & 0.2679 \\ 1.0000 & 0.6647 & -0.0001 \end{bmatrix} \begin{bmatrix} Y' \\ D_B \\ D_R \end{bmatrix}. \tag{8.42b}$$

This color space is similar to the YUV color space, and the chromatic channels D_B and D_R may therefore also be directly derived as follows:

$$D_B = 3.059\,U, \tag{8.43a}$$

$$D_R = -1.169\,V. \tag{8.43b}$$

The luma channels in both spaces are identical.

8.4.7 Kodak's Photo Y′CC

Kodak's Photo Y′CC color space is used in the PhotoCD system. This system is intended to enable the display of imagery on both television, video, and computer monitors. It uses the RGB primaries and non-linear transfer function as ITU-R BT.709, which is listed under HDTV in Tables 8.2 and 8.3. Luma Y' and the two chroma channels are encoded by

$$\begin{bmatrix} Y' \\ C_1 \\ C_2 \end{bmatrix} = \begin{bmatrix} 0.299 & 0.587 & 0.114 \\ -0.299 & -0.587 & 0.886 \\ 0.701 & -0.587 & -0.114 \end{bmatrix} \begin{bmatrix} R' \\ G' \\ B' \end{bmatrix}. \tag{8.44}$$

The Kodak Y′CC color space includes an explicit quantization into an 8-bit format. The Y′CC channels are quantized using

$$\begin{bmatrix} {}^{8}Y' \\ {}^{8}C_1 \\ {}^{8}C_2 \end{bmatrix} = \begin{bmatrix} \dfrac{255\,Y'}{1.402} \\ 111.40\,C_1 + 156 \\ 135.64\,C_2 + 137 \end{bmatrix}. \tag{8.45}$$

To enable display of encoded imagery on various display devices, the inverse transform from Y′CC to RGB depends on the target display device. For display by computers on a cathode ray tube, the conversion is

$$\begin{bmatrix} Y' \\ C_1 \\ C_2 \end{bmatrix} = \begin{bmatrix} 1.3584\,{}^8Y' \\ 2.2179\,({}^8C_1 - 156) \\ 1.8215\,({}^8C_2 - 137) \end{bmatrix}. \tag{8.46a}$$

Non-linear R′G′B′ is then reconstructed:

$$\begin{bmatrix} R' \\ G' \\ B' \end{bmatrix} = \frac{1}{k} \begin{bmatrix} Y' + C_2 \\ Y' - 0.194C_1 - 0.509C_2 \\ Y' + C_2 \end{bmatrix}. \tag{8.47}$$

For computer monitors, the value of k equals 1. The range of values of R′G′B′ is between 0 and 346. To bring this range to the more conventional range of $[0, 255]$, a look-up table can be used. For display of YCC encoded images on televisions, we set $k = 353.2$. The resulting values are non-linear, which is fine if the display has a gamma of 2.2. However, for different displays, further gamma correction may have to be carried out.

8.5 Hue-Saturation-Lightness Spaces

As discussed in the introduction to this chapter, intuitive selection of colors requires the design and utilization of an appropriate color space. Ideally, for such a space, each axis should carry an easy to understand set of information. The hue, saturation, lightness (HSL) spaces exist in several variations, where the lightness channel is replaced with either intensity (HSI) or value channels (HSV). A variant with a chroma channel is also known (HCI).

In reality, these spaces should carry along an asterisk or quotation marks to differentiate them from spaces that are more closely tied to the actual color appearance attributes. They can be considered crude approximations of more visually precise color spaces. The definition of these color spaces are given in terms of an abstract RGB color space that is assumed to be normalized. The RGB color cube is rotated such that the diagonal between (0, 0, 0) and (1, 1, 1) forms the lightness axis. Note that this inherently suggests that "lightness" is a linear combination of R + G + B, without regard to the particular RGB color space used. Perpendicular to this axis is a circular plane defining color. The distance of a point in this plane to the lightness axis defines "saturation." The angle between a given defined point and the color of interest defines "hue," and thus ranges between 0° and 360°.

The color spaces in this section share all their disadvantages with using an assumed RGB. As the color spaces are defined with respect to RGB, HSL-type color spaces are generally device-dependent. If these spaces are defined using a specific set of RGB primaries, they could be considered to be device-independent, though in practice this is rarely done. Despite the suggestive names given to the axes, the hue, saturation, and lightness do not correlate well with the perceptual attributes of hue, saturation, and lightness as defined in Chapter 11. In particular, the lightness axis (or value, or brightness, or whatever other name is given to this channel) is not linear in any perceived quantity. To compute lightness, one should consider computing XYZ tristimulus values first and then converting to an appropriate perceptually uniform color space, such as CIE $L^*a^*b^*$.

Further, as the computation for hue is usually split into 60° segments, one may expect discontinuities in color space at these hue angles. Computations in these color spaces are complicated by the discontinuity that arises at the 0° / 360° hue angle [922].

In summary, the HSL-related color spaces discussed in the following sections could be used for selecting device-dependent colors, for instance in drawing programs. Nonetheless, it should be noted that the meaning of the axes does not correlate well with perceptual attributes of the same names, and care must be taken not to confuse these spaces with more advanced color appearance spaces.

8.5.1 HSI

Replacing the lightness axis with an intensity axis, which is computed somewhat differently, we arrive at the HSI space. The minimum of the three R, G, and B values is used as an intermediary variable:

$$v_{\min} = \min(R, G, B). \tag{8.48}$$

We also compute an intermediary parameter h that will be used to compute hue H:

$$h = \frac{1}{360} \cos^{-1} \left(\frac{\frac{R-G}{2} + (R-B)}{\sqrt{(R-G)^2 + (R-B)(G-B)}} \right). \tag{8.49}$$

The color space is then defined as

$$I = \frac{R+G+B}{3}, \tag{8.50a}$$

$$S = 1 - \frac{v_{\min}}{I}, \tag{8.50b}$$

(8.50c)

$$H = \begin{cases} 1-h & \text{if } \frac{B}{I} > \frac{G}{I} \wedge S > 0, \\ h & \text{if } \frac{B}{I} \leq \frac{G}{I} \wedge S > 0, \\ \text{undefined} & \text{if } S = 0. \end{cases} \tag{8.50d}$$

The hue angle H is undefined for colors with $S = 0$. This stems from the fact that in this case the color is monochromatic, and defining a value for hue would not be meaningful.

The conversion from HSI to RGB begins by converting H to degrees:

$$H' = 360H. \tag{8.51}$$

An RGB triplet is then formed as follows:

$$\begin{bmatrix} R \\ G \\ B \end{bmatrix} = \begin{cases} \begin{bmatrix} \frac{1}{3}\left(1 + \frac{S\cos(H')}{\cos(60-H')}\right) \\ 1-(B+R) \\ \frac{1}{3}(1-S) \end{bmatrix} & \text{if } 0 < H' \leq 120, \\ \begin{bmatrix} \frac{1}{3}(1-S) \\ \frac{1}{3}\left(1 + \frac{S\cos(H'-120)}{\cos(180-H')}\right) \\ 1-(R+G) \end{bmatrix} & \text{if } 120 < H' \leq 240, \\ \begin{bmatrix} 1-(G+B) \\ \frac{1}{3}(1-S) \\ \frac{1}{3}\left(1 + \frac{S\cos(H'-240)}{\cos(300-H')}\right) \end{bmatrix} & \text{if } 240 < H' \leq 360. \end{cases} \tag{8.52}$$

It should be noted that some of the R, G, and B values depend on the prior computation of other R, G, and B values in the same triplet. This implies that these values must be computed sequentially, as they are interdependent.

8.5.2 HSV

The conversion between RGB and HSV begins by finding the minimum and maximum values of the RGB triplet:

$$v_{\min} = \min(R, G, B), \tag{8.53a}$$

$$v_{\max} = \max(R, G, B). \tag{8.53b}$$

Saturation S and value V are defined in terms of these values:

$$S = \frac{v_{\max} - v_{\min}}{v_{\max}}, \tag{8.54a}$$

$$V = v_{\max}. \tag{8.54b}$$

Of course, when $R = G = B$ we have a gray value, and thereby a color for which the saturation is 0 since $v_{\max} = v_{\min}$. In that case, hue H is undefined. Otherwise, hue is computed as an angle ranging between 0° and 360°:

$$H = \begin{cases} 60 \dfrac{G - B}{v_{\max} - v_{\min}} & \text{if } R = v_{\max}, \\ 60 \left(2 + \dfrac{B - R}{v_{\max} - v_{\min}} \right) & \text{if } G = v_{\max}, \\ 60 \left(4 + \dfrac{R - G}{v_{\max} - v_{\min}} \right) & \text{if } B = v_{\max}. \end{cases} \tag{8.55}$$

If the hue angle is less than 0 after this computation, the angle should be brought within range by adding 360°. The inverse transformation, from HSV back to RGB, begins by dividing the hue angle by 60:

$$H' = \frac{H}{60}. \tag{8.56}$$

The fractional part of H' is given by

$$f = H' - \lfloor H' \rfloor. \tag{8.57}$$

Intermediate variables a, b, and c are then computed as follows:

$$a = V\,(1 - S), \tag{8.58a}$$

$$b = V\,(1 - S f), \tag{8.58b}$$

$$c = V\,(1 - S(1 - f)). \tag{8.58c}$$

Figure 8.7. Hue angles (right) computed for the image on the left; Zaragoza, Spain, November 2006.

Figure 8.8. Cardinal hue angles. From top to bottom, left to right: original image, followed by images with hue set to 0°, 60°, 120°, 180°, 240°, and 300°; Zaragoza, Spain, November 2006.

The integer part of H' determines which of the four quadrants the hue angle H lies in. It is used to select which values to assign to R, G, and B:

$$\begin{bmatrix} R \\ G \\ B \end{bmatrix} = \begin{cases} \begin{bmatrix} V & c & a \end{bmatrix}^T & \text{if } \lfloor H' \rfloor = 0, \\ \begin{bmatrix} b & V & a \end{bmatrix}^T & \text{if } \lfloor H' \rfloor = 1, \\ \begin{bmatrix} a & V & c \end{bmatrix}^T & \text{if } \lfloor H' \rfloor = 2, \\ \begin{bmatrix} a & b & V \end{bmatrix}^T & \text{if } \lfloor H' \rfloor = 3, \\ \begin{bmatrix} c & a & V \end{bmatrix}^T & \text{if } \lfloor H' \rfloor = 4, \\ \begin{bmatrix} V & a & b \end{bmatrix}^T & \text{if } \lfloor H' \rfloor = 5. \end{cases} \tag{8.59a}$$

An example of the hue angles present in an image is shown in Figure 8.7 (right), which was created by converting an RGB image to HSV, then setting the saturation and value components to 1 before converting back to RGB.

The hue angles are arranged in a hexcone, with cardinal hue angles located at every 60 degrees between 0° and 360°. The cardinal hue angles are visualized in Figure 8.8, where the saturation and value components are retained, but the hue of each pixel is set to the six cardinal angles.

8.6 HVS Derived Color Spaces

As discussed in Chapter 4, the early part of the human visual system carries out various color transforms, starting with the transduction of light at the photoreceptor level, and followed by the recombination into color opponent spaces at the ganglion level. In this section, we discuss various color spaces that are based on our knowledge of the human visual system. Color opponent spaces are presented in Section 8.7.

8.6.1 LMS Cone Excitation Space

The response of the cones in the retina are modeled by the LMS color space, with L, M and S standing for long, medium, and short wavelengths. Although the CIE XYZ color matching functions are closely related to a linear transform of the LMS cone signals, they are not exact. Schanda gives an excellent overview between the relationship of the cone fundamentals to CIE XYZ color-matching functions [1002]. Nevertheless, approximations can be used to convert between CIE XYZ and LMS cone signals. One such approximation is the Hunt-Pointer-Estevez cone fundamentals, used in many color appearance models. The matrices

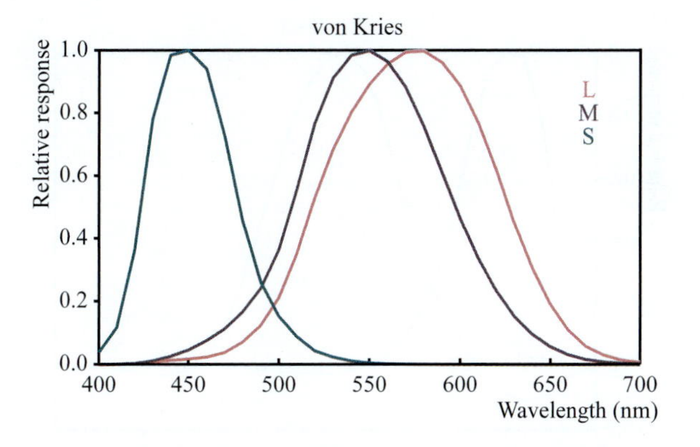

Figure 8.9. Relative response functions for the LMS cone-excitation space (after [318]).

that convert between XYZ and LMS are given by

$$\begin{bmatrix} L \\ M \\ S \end{bmatrix} = \begin{bmatrix} 0.3897 & 0.6890 & -0.0787 \\ -0.2298 & 1.1834 & 0.0464 \\ 0.0000 & 0.0000 & 1.0000 \end{bmatrix} \begin{bmatrix} X \\ Y \\ Z \end{bmatrix}; \tag{8.60a}$$

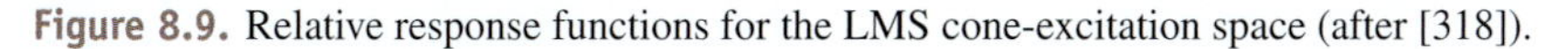

$$\begin{bmatrix} X \\ Y \\ Z \end{bmatrix} = \begin{bmatrix} 1.9102 & -1.1121 & 0.2019 \\ 0.3710 & 0.6291 & 0.0000 \\ 0.0000 & 0.0000 & 1.0000 \end{bmatrix} \begin{bmatrix} L \\ M \\ S \end{bmatrix}. \tag{8.60b}$$

As the LMS cone space represents the response of the cones in the human visual system, it is a useful starting place for computational models of human vision. It is also a component in the CIECAM02 and iCAM color appearance models (see Chapter 11). The relative response as a function of wavelength is plotted in Figure 8.9.

8.6.2 Sharpened Color Spaces

An alternative cone-response domain has been developed for use specifically in chromatic-adaptation transforms and color-constancy experiments, known as sharpened cone responses [581, 685] (see Figure 8.10 and 8.11). Two such examples are the cone signals used in the Bradford chromatic adaptation transform and, more recently, the CIECAT02 chromatic-adaptation transform. More details

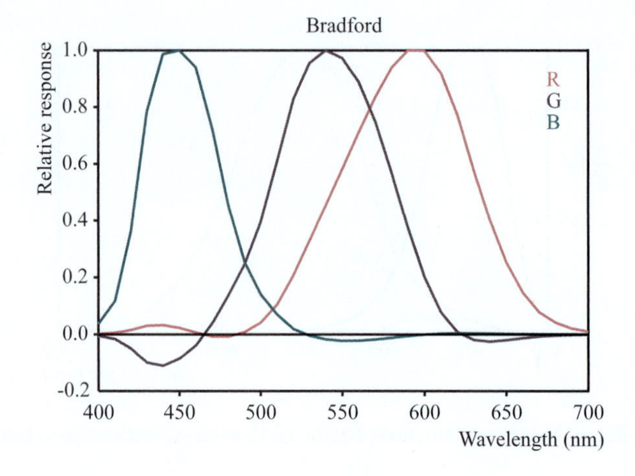

Figure 8.10. Relative response functions for the Bradford chromatic-adaptation transform (after [318]).

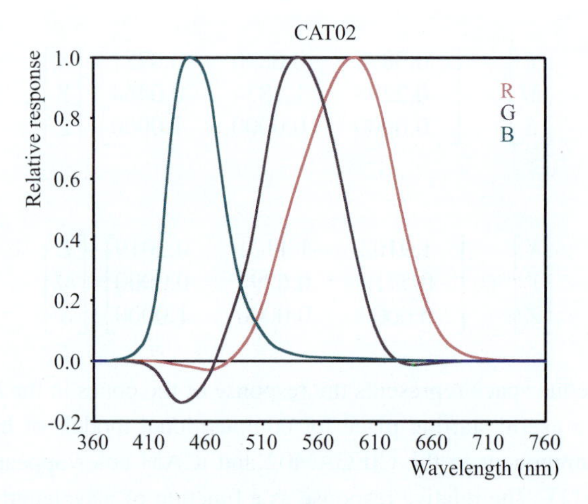

Figure 8.11. Relative response functions for the CAT02 chromatic-adaptation transform.

on these spaces can be found in Section 10.4.5:

$$M_{\text{Bradford}} = \begin{bmatrix} 0.8951 & 0.2664 & -0.1614 \\ -0.7502 & 1.7135 & 0.0367 \\ 0.0389 & -0.0685 & 1.0296 \end{bmatrix}; \tag{8.61a}$$

$$M_{\text{Bradford}}^{-1} = \begin{bmatrix} 0.9870 & -0.1471 & 0.1600 \\ 0.4323 & 0.5184 & 0.0493 \\ -0.0085 & 0.0400 & 0.9685 \end{bmatrix}; \tag{8.61b}$$

$$M_{\text{CAT02}} = \begin{bmatrix} 0.7328 & 0.4296 & -0.1624 \\ -0.7036 & 1.6975 & 0.0061 \\ 0.0030 & 0.0136 & 0.9834 \end{bmatrix}; \tag{8.62a}$$

$$M_{\text{CAT02}}^{-1} = \begin{bmatrix} 1.0961 & -0.2789 & 0.1827 \\ 0.4544 & 0.4735 & 0.0721 \\ -0.0096 & -0.0057 & 1.0153 \end{bmatrix}. \tag{8.62b}$$

8.6.3 MBDKL Color Space

The multi-stage color model proposed by de Valois et al. (see Section 4.6) is based on a color space developed to conveniently represent color signals as they are processed by early human vision. This color space was developed by MacLeod, Boynton, Derrington, Krauskopf, and Lennie [250, 722] and named accordingly the MBDKL space.

The three axes in this color space are chosen to coincide with human visual responses occurring in early human vision. In particular, the first axis corresponds to a luminance signal that is derived from the LMS color space (see Section 8.6.1) by summing the L and M channels. Here, it is assumed that the S cones do not contribute to the luminance signal.

The second and third channels are chromatic, with the second channel modulating color such that the excitation of S cones is unaffected. This may be termed the *constant S* axis. Along this dimension, the L and M components co-vary such that their sum remains constant. The third channel is chosen such that chromatic changes to the signal do not change the excitation of L cones or the M cones. Hence, this is the *constant* L + M axis. Changes along this channel will only affect the excitation of the S cones. The point where the three axes intersect is the *white point.*

This three-dimensional color space can be unfolded to lie in a plane for convenience of representation. This is achieved by converting to polar coordinates. The two chromatic channels are represented by azimuth $\phi \in [0,360)$, whereas luminance is represented by elevation $\theta \in [-90,90]$. A shift from white towards pinkish-red along the constant S axis is indicated with an azimuthal angle of 0° (see also Table 4.3 on page 237). The opposite direction, which realizes a shift from white towards cyan, is indicated with an angle of 180°. Similarly, a shift towards violet along the L + M axis is at an angle of 90°. The angle of 270° marks a shift toward greenish-yellow.

The elevation angle encodes luminance changes. An increase in luminance is given by positive elevation angles, whereas decreases are marked by negative elevation angles.

The azimuthal angle ϕ is computed from LMS signals as follows:

$$\phi = \tan^{-1}\left(\frac{L-M}{S-(L+M)}\right). \tag{8.63}$$

The luminance signal is computed by summing the contribution from L and M cones. However, the elevation angle is defined for changes in luminance, i.e., contrasts. For a pair of LMS signals, indicated with subscripts 1 and 2, the elevation angle is given by

$$\theta = \sin^{-1}(L_1 + M_1 - L_2 - M_2). \tag{8.64}$$

For any cell in the human visual system that linearly combines cone signals, there will be an azimuth and elevation pair (ϕ_0, θ_0) defining a *null plane*. The cell will not respond to any colors lying in this plane. Modulating colors along a direction perpendicular to this plane will maximally excite this cell. For a given color encoded by (ϕ, θ), the response R of a cell in spikes per second is

$$R = K\ (\sin(\theta)\cos(\theta_0) - \cos(\theta)\sin(\theta_0))\sin(\phi - \phi_0) \tag{8.65a}$$

$$= K\ \sin(\theta - \theta_0)\sin(\phi - \phi_0). \tag{8.65b}$$

Here, K is a suitably chosen scale factor. Plots showing R as function of azimuthal angle are shown in Section 4.6, where the MBDKL color space was used to analyze properties of cells in the early human visual system.

The unfolding of a color space into a 2D plane essentially returns us to plots that are in a sense equivalent to figures plotting variables directly against wavelength. However, the significance of the MBDKL color space is that the azimuthal angle does not directly correspond to wavelength, but instead is encoded such that signals eliciting maximal responses in LGN cells are at angles of 0°, 90°, 180°, and 270°, respectively.

8.7 Color Opponent Spaces

Color opponent spaces are characterized by a channel representing an achromatic signal, as well as two channels encoding color opponency. The two chromatic channels generally represent an approximate red-green opponency and yellow-blue opponency. In this section, we discuss the CIE 1976 L*a*b*, also called CIELAB, and the CIE 1976 L*u*v* color spaces CIELUV. These spaces also include a non-linear compression to achieve better perceptual uniformity throughout the color space. This perceptual uniformity, therefore, enables color differences between pairs of color to be assessed by simply computing their Euclidean distance. Improved color difference metrics are discussed in Section 8.8. We also discuss a color space IPT that was designed to minimize some of the perceptual errors of CIELAB, specifically with regards to lines of constant hue. The last color space discussed in this section is the L$\alpha\beta$ space, which was derived by applying principal components analysis (PCA) to a representative set of natural images, encoded first in LMS cone-response space. This space represents a generic orthogonal type of color representation, where each channel is decorrelated from the others.

Both the CIE 1976 L*a*b* and the CIE 1976 L*u*v* color spaces were standardized by the CIE when they recognized the need for a uniform color space in which color differences could be calculated. Two competing proposals were evaluated, and at the time both were deemed equally valid. As such, they were both formally recommended by the CIE [93, 1097]. Over time the CIE 1976 L*a*b* space has been recognized to be more accurate, especially in the printing, imaging, and materials industries. In part this is due to the more accurate chromatic-adaptation type model inherent in the calculation (CIELAB uses a normalization of XYZ, while CIELUV uses a subtraction of XYZ). Although it has fallen out of favor, the CIE 1976 L*u*v* space has found use in the display and television industries, mostly due to its accompanying chromaticity diagram [1097].

8.7.1 CIE 1976 L*a*b*

The input into the CIE 1976 L*a*b* model (CIELAB) are the stimulus (X,Y,Z) tristimulus values, as well as the tristimulus values of a diffuse white reflecting surface that is lit by a known illuminant, $(X_{\mathrm{n}}, Y_{\mathrm{n}}, Z_{\mathrm{n}})$. The CIELAB equations calculate values representing the lightness of a color L^*, as well as two opponent chromatic channels [577]:

$$\begin{bmatrix} L^* \\ a^* \\ b^* \end{bmatrix} = \begin{bmatrix} 0 & 116 & 0 & -16 \\ 500 & -500 & 0 & 0 \\ 0 & 200 & -200 & 0 \end{bmatrix} \begin{bmatrix} f(X/X_n) \\ f(Y/Y_n) \\ f(Z/Z_n) \\ 1 \end{bmatrix} \tag{8.66}$$

The function f is defined as

$$f(r) = \begin{cases} \sqrt[3]{r} & \text{for } r > 0.008856, \\ 7.787\,r + \dfrac{16}{116} & \text{for } r \le 0.008856. \end{cases} \tag{8.67}$$

The representation of lightness L^* is normalized between 0 and 100 for black and white. The non-linear compression function is defined by a linear component near black and a cube-root function above that. The lightness function is considered to be a perceptually uniform scale between black and white and is a function only of input luminance, Y and Y_n. The chromatic channels encode a value between reddish-magenta and green for a^* and a value between yellow and blue for b^*. These channels typically fall in the -128 to 128 range, though they are not explicitly constrained. All three of these channels can be considered to encode a measure of the human visual response to color stimuli. We can consider CIELAB to be a rudimentary color appearance space, as we will see in Chapter 11.

As can be seen from (8.66), the values of a^* and b^* depend on the relative luminance Y. This means that increasing or decreasing the luminance of the color will have an impact on the values of a^* and b^*. Although this is perceptually accurate, this does indicate that it is not possible to use the two chromatic channels as a chromaticity diagram.

The inverse transform between CIE L*a*b* and CIE XYZ can be computed directly[2]:

$$X = X_n \begin{cases} \left(\dfrac{L^*}{116} + \dfrac{a^*}{500} + \dfrac{16}{116} \right)^3 & \text{if } L^* > 7.9996, \\ \dfrac{1}{7.787} \left(\dfrac{L^*}{116} + \dfrac{a^*}{500} \right) & \text{if } L^* \leq 7.9996, \end{cases} \tag{8.68a}$$

$$Y = Y_n \begin{cases} \left(\dfrac{L^*}{116} + \dfrac{16}{116} \right)^3 & \text{if } L^* > 7.9996, \\ \dfrac{1}{7.787} \dfrac{L^*}{116} & \text{if } L^* \leq 7.9996, \end{cases} \tag{8.68b}$$

$$Z = Z_n \begin{cases} \left(\dfrac{L^*}{116} - \dfrac{b^*}{200} + \dfrac{16}{116} \right)^3 & \text{if } L^* > 7.9996, \\ \dfrac{1}{7.787} \left(\dfrac{L^*}{116} - \dfrac{b^*}{200} \right) & \text{if } L^* \leq 7.9996. \end{cases} \tag{8.68c}$$

Within the CIE L*a*b* color space, which is approximately perceptually linear, the difference between two stimuli may be quantified with the following color difference formula:

$$\Delta E^*_{\mathrm{ab}} = \left[(\Delta L^*)^2 + (\Delta a^*)^2 + (\Delta b^*)^2 \right]^{1/2}. \tag{8.69}$$

The letter E stands for difference in sensation (in German, Empfindung) [560].

[2]This inverse transform is approximate, but widely used in practice. It would only be inaccurate for colors that are both very dark and saturated. Such colors are rare in nature.

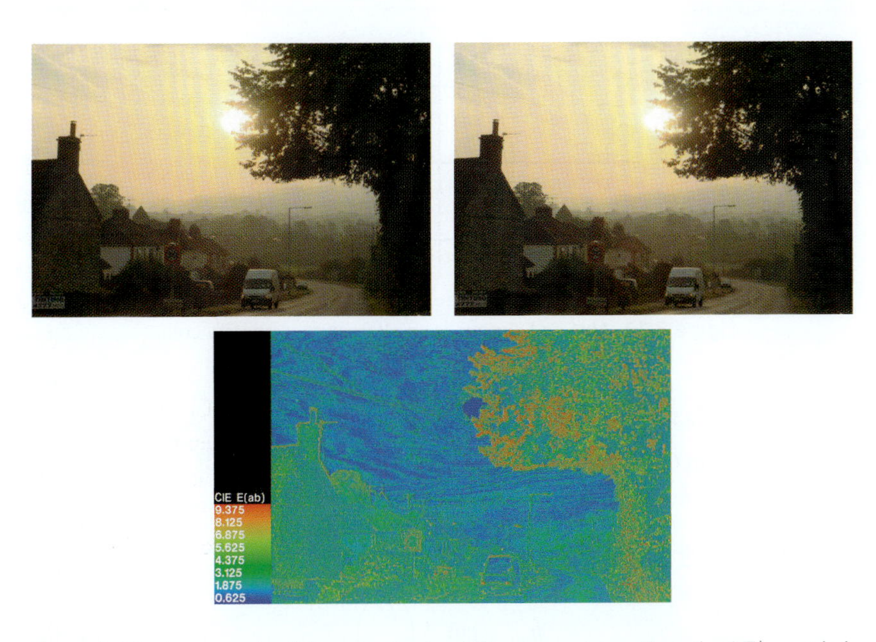

Figure 8.12. An example of the use of color-difference metrics. Here the ΔE^*_{ab} metric is applied to a pair of images with the top-left image encoded in a lossless file format (PPM), whereas the top-right image was stored as a jpeg file. The bottom image shows the per pixel ΔE^*_{ab} color difference; Wick, UK, October 2006.

An example of the use of color-difference metrics is given in Figure 8.12, where a pair of images is shown with one image subjected to lossy jpeg compression. The difference image encodes ΔE^*_{ab} as a set of gray levels, with lighter shades indicating larger errors. This image shows qualitatively where in the image the largest errors occur. For quantitative evaluation of errors, the actual ΔE^*_{ab} values should be examined.

It should be noted that this color-difference metric is only approximately linear with human visual perception. The reason is that the laboratory conditions under which this difference metric was tested, namely by observing flat reflection samples on a uniform background, has little bearing on typical imaging applications. Hence, this is a case where laboratory conditions do not extend to the real world. Also, it should be stressed that the CIELAB color-difference equations were not designed for predicting very large color differences. Further developments have led to improved color difference metrics, and these are outlined in Section 8.8.

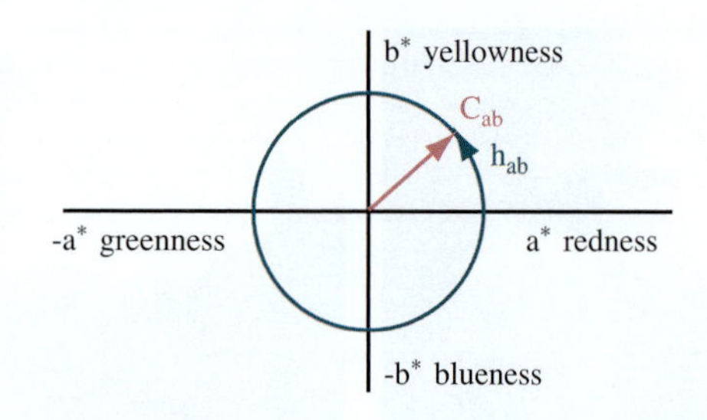

Figure 8.13. Hue and chroma is an alternative to specify color.

8.7.2 $L^*C^*h_{ab}$ Color Space

Conversion from CIE $L^*a^*b^*$ to $L^*C^*h_{ab}$ is given by:

$$L^*_{ab} = L^*, \tag{8.70a}$$

$$C^*_{ab} = \sqrt{a^{*2} + b^{*2}}, \tag{8.70b}$$

$$h_{ab} = \tan^{-1}\left(\frac{b^*}{a^*}\right). \tag{8.70c}$$

Essentially the transform from CIELAB to $L^*C^*h_{ab}$ space can be thought of as a rectangular to polar conversion and should be considered just an alternate representation. The C^* value defines the chroma of a stimulus, whereas h_{ab} denotes a hue angle. This transformation may be more intuitive to understand for users wishing to describe a color, and it is used in the CMC and CIE ΔE^* color difference metrics, which are presented in Sections 8.8.1 and 8.8.2. Note that the hue angle h_{ab} should be between 0° and 360°. If the result is not within this range, then 360 should be added or subtracted to the angle. For a fixed lightness, colors can be visualized along red-green and yellow-blue axes, or as hue-chroma pairs, as shown in Figure 8.13.

The CIELAB color difference can be specified in $L^*C^*h_{ab}$ color space by first calculating a rectangular hue difference, ΔH^*_{ab}:

$$\Delta H^*_{ab} = \left[(\Delta E^*)^2 - (\Delta L^*_{ab})^2 - (\Delta C^*_{ab})^2\right]^{1/2}, \tag{8.71a}$$

$$\Delta E^*_{ab} = \left[(\Delta L^*)^2 + (\Delta C^*_{ab})^2 + (\Delta H^*_{ab})^2\right]^{1/2}. \tag{8.71b}$$

This general form of CIELAB is used in color-difference metrics described in Section 8.8, where this basic formulation is expanded to include different weights for the three terms on the right-hand side. These different weights create ellipses of tolerances in $L^*C^*h_{ab}$ space. Rather than using a^* and b^* to approximate visual

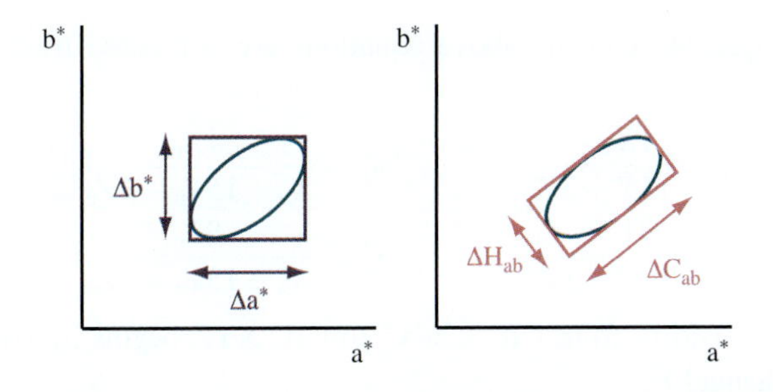

Figure 8.14. Tolerances in (a^*,b^*) space yield axis-aligned error metrics, whereas in CIELAB $L^*C^*h_{ab}$ space, the error metric is aligned with the principal axis of the ellipsoids.

tolerances, the use of weighted values for lightness, hue, and chroma yields closer approximations to these elliptical areas of equal tolerance [93]. This is visualized in Figure 8.14. Thus, color-difference metrics using well-chosen weights on lightness, hue, and chroma differences are considered to give more accurate results than the color-difference formulae associated with the CIELAB and CIELUV color spaces.

The inverse transform from $L^*C^*h_{ab}$ to CIE L^* a^* b^* is given by

$$L^* = L^*, \tag{8.72a}$$

$$a^* = C^* \cos(h_{ab}), \tag{8.72b}$$

$$b^* = C^* \sin(h_{ab}). \tag{8.72c}$$

A similar conversion is possible from CIE $L^*u^*v^*$ to $L^*C^*h_{uv}$. The equations are given in (8.70), except a^* is replaced with u^* and b^* is replaced with v^*.

8.7.3 CIE 1976 L*u*v*

Similar to the CIELAB model, the inputs to the CIELUV calculations are a stimulus (X,Y,Z) as well as the tristimulus values of white reflecting surface that is lit by a known illuminant, (X_n, Y_n, Z_n). The conversion from CIE 1931 tristimulus values to CIE 1976 $L^*u^*v^*$ (abbreviated as CIELUV) is then given by

$$L^* = \begin{cases} 116\left(\dfrac{Y}{Y_n}\right)^{1/3} - 16 & \dfrac{Y}{Y_n} > 0.008856, \\ 903.3\,\dfrac{Y}{Y_n} & \dfrac{Y}{Y_n} \leq 0.008856; \end{cases} \tag{8.73a}$$

$$u^* = 13L^*(u' - u'_n), \tag{8.73b}$$

$$v^* = 13L^*(v' - v'_n). \tag{8.73c}$$

The primed quantities in the above equations are computed from (X,Y,Z) as follows:

$$u' = \frac{4X}{X+15Y+3Z}, \qquad u'_{\mathrm{n}} = \frac{4X_{\mathrm{n}}}{X_{\mathrm{n}}+15Y_{\mathrm{n}}+3Z_{\mathrm{n}}}; \tag{8.74a}$$

$$v' = \frac{9Y}{X+15Y+3Z}, \qquad v'_{\mathrm{n}} = \frac{9Y_{\mathrm{n}}}{X_{\mathrm{n}}+15Y_{\mathrm{n}}+3Z_{\mathrm{n}}}. \tag{8.74b}$$

The inverse transform from CIE L*u*v* to CIE XYZ begins by computing the luminance channel Y,

$$Y = \frac{Y_n}{100}\left(\frac{L^*+16}{116}\right)^3 \tag{8.75}$$

and three intermediary parameters,

$$a = \frac{u^*}{13L^*} + u'_n, \tag{8.76a}$$

$$b = \frac{v^*}{13L^*} + v'_n, \tag{8.76b}$$

$$c = 3Y\,(5b-3). \tag{8.76c}$$

The Z coordinate is then

$$Z = \frac{(a-4)\,c - 15abY}{12b}, \tag{8.77}$$

and the X coordinate is

$$X = -\left(\frac{c}{b} + 3Z\right). \tag{8.78}$$

The transformation to CIE L*u*v* creates a more or less uniform color space, such that equal distances anywhere within this space encode equal perceived color differences. It is therefore possible to measure the difference between two stimuli (L_1^*, u_1^*, v_1^*) and (L_2^*, u_2^*, v_2^*) by encoding them in CIELUV space, and applying the following color difference formula:

$$\Delta E^*_{\mathrm{uv}} = \left[(\Delta L^*)^2 + (\Delta u^*)^2 + (\Delta v^*)^2\right]^{1/2}, \tag{8.79}$$

where $\Delta L^* = L_1^* - L_2^*$, etc.

In addition, u' and v' may be plotted on separate axes to form a chromaticity diagram. Equal distances in this diagram represent approximately equal perceptual differences. For this reason, where possible we show CIE (u′,v′) chromaticity diagrams rather than perceptually non-uniform CIE (x,y) chromaticity diagrams.

The reason for the existence of both CIELUV and CIELAB color spaces is largely historical. Both color spaces are in use today, with CIELUV more common in the television and display industries and CIELAB in the printing and materials industries [1097].

8.7.4 IPT Color Space

With more psychophysical data becoming available, including Hung and Berns [500] and Ebner and Fairchild [273], the limitations of CIELAB and CIELUV have become apparent. In particular, the uniformity of hue can be improved. A simple transform that better models the linearity of constant hue lines, while predicting other appearance attributes as well as or better than CIELAB and CIECAM97s (the latter is discussed in Chapter 11), is afforded by the IPT color space [274]. The abbreviation stands for image-processing transform, while the letters also carry meaning to indicate the three channels of this color space. The I channel represents intensity, whereas the P and T channels encode color opponency. The P channel, for *protan*, encodes an approximate red-green opponency, whereas the T channel, for *tritan*, encodes an approximate yellow-blue opponency.

The IPT color space is defined, as usual, as a transform from CIE XYZ. The conversion between XYZ and IPT begins by a matrix multiplication, taking the XYZ tristimulus values, specified according to the CIEXYZ 1931 2° observer model, to a variant of the LMS cone excitation space, which is close to the Hunt-Pointer-Estevez cone primaries (which are discussed in Section 10.4.1):

$$\begin{bmatrix} L \\ M \\ S \end{bmatrix} = \begin{bmatrix} 0.4002 & 0.7075 & -0.0807 \\ -0.2280 & 1.1500 & 0.0612 \\ 0.0000 & 0.0000 & 0.9184 \end{bmatrix} \begin{bmatrix} X \\ Y \\ Z \end{bmatrix}. \tag{8.80}$$

Note that the transformation into cone signals has assumed a normalization into CIE D65, rather than an equal energy stimulus. More details on this normalization can be found in Chapter 10. The LMS cone signals are then subjected to a nonlinearity:

$$L' = \begin{cases} L^{0.43} & \text{if } L \geq 0, \\ -(-L)^{0.43} & \text{if } L < 0; \end{cases} \tag{8.81a}$$

$$M' = \begin{cases} M^{0.43} & \text{if } M \geq 0, \\ -(-M)^{0.43} & \text{if } M < 0; \end{cases} \tag{8.81b}$$

$$S' = \begin{cases} S^{0.43} & \text{if } S \geq 0, \\ -(-S)^{0.43} & \text{if } S < 0. \end{cases} \tag{8.81c}$$

Although real cone signals would never have negative components, the linear matrix transform used in IPT can generate negative signals for real colors. This is why care must be taken to maintain the proper sign before applying the nonlinear compression. The IPT opponent space is then reached by a further matrix transform:

$$\begin{bmatrix} I \\ P \\ T \end{bmatrix} = \begin{bmatrix} 0.4000 & 0.4000 & 0.2000 \\ 4.4550 & -4.8510 & 0.3960 \\ 0.8056 & 0.3572 & -1.1628 \end{bmatrix} \begin{bmatrix} L' \\ M' \\ S' \end{bmatrix}, \tag{8.82}$$

where $I \in [0,1]$ and $P, T \in [-1,1]$ under the assumption that the input XYZ values were normalized. If the IPT axes are scaled by $(100, 150, 150)$, they become roughly equivalent to those found in CIELAB. The inverse transform begins by transforming IPT to non-linear cone excitation space:

$$\begin{bmatrix} L' \\ M' \\ S' \end{bmatrix} = \begin{bmatrix} 1.8502 & -1.1383 & 0.2384 \\ 0.3668 & 0.6439 & -0.0107 \\ 0.0000 & 0.0000 & 1.0889 \end{bmatrix} \begin{bmatrix} I \\ P \\ T \end{bmatrix}. \tag{8.83}$$

Linearization is then achieved by

$$L = \begin{cases} L'^{1/0.43} & \text{if } L' \geq 0, \\ -(-L')^{1/0.43} & \text{if } L' < 0; \end{cases} \tag{8.84a}$$

$$M = \begin{cases} M'^{1/0.43} & \text{if } M' \geq 0, \\ -(-M')^{1/0.43} & \text{if } M' < 0; \end{cases} \tag{8.84b}$$

$$S = \begin{cases} S'^{1/0.43} & \text{if } S' \geq 0, \\ -(-S')^{1/0.43} & \text{if } S' < 0, \end{cases} \tag{8.84c}$$

and the transform to CIE XYZ is given by

$$\begin{bmatrix} X \\ Y \\ Z \end{bmatrix} = \begin{bmatrix} 1.0000 & 0.0976 & 0.2052 \\ 1.0000 & -1.1139 & 0.1332 \\ 1.0000 & 0.0326 & -0.6769 \end{bmatrix} \begin{bmatrix} L \\ M \\ S \end{bmatrix}. \tag{8.85}$$

The IPT color space bears similarities with CIELAB, although with different coefficients. It was designed specifically to have the strengths of CIELAB, but to avoid the hue changes that could occur when compressing chroma along lines of constant hue. It is a suitable color space for many color-imaging applications, such as gamut mapping.

8.7.5 $L\alpha\beta$ Color Space

An interesting exercise is to determine what color space would emerge if an ensemble of spectral images were converted to an LMS cone space and, subsequently, subjected to principal components analysis (PCA) [984]. The general idea behind this approach is to determine if there is any link between the encoding of signals in the human visual system and natural image statistics. A set of natural images is used here, so that any averages computed are representative for a canonical natural scene. The photoreceptor output is simulated by converting the spectral images to the LMS color space.

The principal components analysis rotates the data so that the first principal component captures most of the variance. This can be understood by assuming that the data forms, on average, an elliptical point cloud. PCA rotates this ellipse so that its major axis coincides with the axis defined by the first principal component. Its second most important axis is aligned with the second principal component. The remaining axis is aligned with the third principal component. The principal components form an orthogonal coordinate system and, thus, form a new color space. In this color space, the data is maximally decorrelated.

By analyzing the resulting color space, it was found that this color space closely resembles a color opponent space. The 3×3 matrix that transforms between LMS and the PCA-derived color space consists of elements that, aside from a weight factor, recombine LMS signals in close-to-integer multiples. In addition, it was found that the orientation of the axes is such that they can be assigned meaning. The first principal component represents luminance, whereas the second and third principal components represent yellow-blue and red-green color opponent axes, respectively. For this reason, this color space is named L$\alpha\beta$.

Additionally, it was found that the point cloud obtained becomes more symmetrical and well behaved if the color space is derived from logarithmic LMS values.

The color opponency is shown in Figure 8.15, where the image is decomposed into its separate channels. The image representing the α channel has the β channel reset to 0 and vice versa. We have retained the luminance variation here for the purpose of visualization. The image showing the luminance channel only was created by setting both the α and β channels to zero. The transform between LMS and L$\alpha\beta$ is given by

$$\begin{bmatrix} L \\ \alpha \\ \beta \end{bmatrix} = \begin{bmatrix} \frac{1}{\sqrt{3}} & 0 & 0 \\ 0 & \frac{1}{\sqrt{6}} & 0 \\ 0 & 0 & \frac{1}{\sqrt{2}} \end{bmatrix} \begin{bmatrix} 1 & 1 & 1 \\ 1 & 1 & -2 \\ 1 & -1 & 0 \end{bmatrix} \begin{bmatrix} \log L \\ \log M \\ \log S \end{bmatrix}; \tag{8.86a}$$

Figure 8.15. The top-left image is decomposed into the L channel of the L$\alpha\beta$ color space, as well as $L+\alpha$ and $L+\beta$ channels in the bottom-left and bottom-right images; Rochester, NY, November 2004.

$$\begin{bmatrix} \log L \\ \log M \\ \log S \end{bmatrix} = \begin{bmatrix} 1 & 1 & 1 \\ 1 & 1 & -1 \\ 1 & -2 & 0 \end{bmatrix} \begin{bmatrix} \frac{\sqrt{3}}{3} & 0 & 0 \\ 0 & \frac{\sqrt{6}}{6} & 0 \\ 0 & 0 & \frac{\sqrt{2}}{2} \end{bmatrix} \begin{bmatrix} L \\ \alpha \\ \beta \end{bmatrix}. \tag{8.86b}$$

The diagonal matrix contains weight factors that scale the relative contribution of each of the three axes.

The theoretical importance of this color space lies in the fact that decorrelation of natural images, represented in LMS color space, yields a color opponent space which is very close to the one thought to be computed by the human visual system. In other words, for natural images, the human visual system appears to decorrelate the signal it transmits to the brain.

In practice, this color space can be used whenever natural images need to be processed. Although PCA only decorrelates the data, in practice, the axis tend to be close to independent. This means that complicated 3D color transforms on image data can be replaced by three simpler 1D transforms. An example application is discussed in Section 8.10.

8.8 Color Difference Metrics

Both the CIE $L^*a^*b^*$ and CIE $L^*u^*v^*$ are intended to be perceptually uniform color spaces, such that the Euclidean distance between two colors is a good estimation of their perceived difference. However, over time it has become clear that neither space is adequately uniform, and therefore the need for improved color difference metrics arose.

The Colour Measurement Committee of the Society of Dyers and Colourists developed the CMC color difference metric in 1984. The CIE also refined its color difference metrics, resulting in CIE 1994 $\Delta E*$ and later CIEDE2000.

8.8.1 CMC (l:c)

The CMC color difference metric [198, 765] is an elaboration of the CIE $L^*a^*b^*$ color difference formula. It uses a finer resolution for desaturated colors to account for the fact that small errors in the white balance of images are easily noticed [507]. The reason human vision is sensitive to small differences for near-achromatic signals is that a true achromatic stimulus silences the red-green and yellow-blue pathways.

CMC is derived from the CIE LCH color space, for which a transform was given in Section 8.7.2. Measurement of color differences is assumed to occur under D65 lighting, using the 10° standard observer function.

This color difference metric is parameterized by the ratio $l : c$, where l denotes a scale factor for lightness, and c is a scale factor for chroma. To determine the perceptability of the difference between two colors, the ratio is set to 1, i.e.,

8.8.3 CIEDE2000 Color Difference Metric

The CIEDE2000 color difference metric derives directly from the CIE L*a*b* color space. It was developed to improve the predictive power of the CIE94 metric in the saturated blue and near-neutral regions. The near-neutral color prediction is improved by scaling the a^* axis differently. The hue and chroma differences are weighted differently to aid in the prediction of blue color differences.

Given two colors specified in this space, we carry out the following calculations for each color. First, for both samples, we compute C^*_{ab}:

$$C^*_{\text{ab}} = \sqrt{(a^*)^2 + (b^*)^2}, \tag{8.93}$$

as well as the arithmetic mean of the two C^*_{ab} values, which we denote with $\bar{C}^*_{\text{ab}}$. A second intermediary value g is computed:

$$g = 0.5 \left(1 - \sqrt{\frac{\left(\bar{C}^*_{\text{ab}}\right)^7}{\left(\bar{C}^*_{\text{ab}}\right)^7 + 25^7}} \right). \tag{8.94}$$

With this value, we compute L', a', b', C', and h' for each of the two colors.

$$L' = L^*, \tag{8.95a}$$

$$a' = (1+g)\, a^*, \tag{8.95b}$$

$$b' = b^*, \tag{8.95c}$$

$$C' = \sqrt{(a')^2 + (b')^2}, \tag{8.95d}$$

$$h' = \frac{180}{\pi} \tan^{-1}\left(\frac{b'}{a'}\right). \tag{8.95e}$$

Note that h' and values derived from it are specified in degrees rather than in radians. The arithmetic means of the pairs of L', C', and h' values are also computed. These are denoted by $\bar{L}'$, $\bar{C}'$, and $\bar{h}'$. Checks should be made to ensure that hue angles remain positive. This also has implications for the computation of the arithmetic mean of the pair of h' values. Here, it is important to ensure that the hue difference remains below 180°. If a larger hue difference is found, then 360° is subtracted from the larger of the two hue angles, and the arithmetic mean is recomputed. Before computing S_L, S_C, and S_H, several intermediary values are calculated:

$$R_C = 2\sqrt{\frac{(\bar{C}')^7}{(\bar{C}')^7 + 25^7}}, \tag{8.96a}$$

$$R_T = -R_C \sin\left(60 \exp\left(-\left(\frac{\bar{h}' - 275}{25}\right)^2\right)\right), \tag{8.96b}$$

$$T = 1 - 0.17 \cos\left(\bar{h}' - 30\right) + 0.24 \cos\left(2\bar{h}'\right) \tag{8.96c}$$

$$+ 0.32 \cos\left(3\bar{h}' + 6\right) - 0.20 \cos\left(4\bar{h}' - 63\right), \tag{8.96d}$$

followed by

$$S_L = 1 + \frac{0.015\,(\bar{L}' - 50)^2}{\sqrt{20 + (\bar{L}' - 50)^2}}, \tag{8.97a}$$

$$S_C = 1 + 0.045\bar{C}', \tag{8.97b}$$

$$S_H = 1 + 0.015\bar{C}'\,T. \tag{8.97c}$$

The CIEDE2000 color difference metric, denoted with ΔE_{CIE00}, is computed by

$$\Delta L_{\text{CIE00}} = \frac{L_1^* - L_2^*}{k_L\, S_L}, \tag{8.98a}$$

$$\Delta C_{\text{CIE00}} = \frac{C_1^* - C_2^*}{k_C\, S_C}, \tag{8.98b}$$

$$\Delta H_{\text{CIE00}} = \frac{2 \sin\left(\frac{h_1' - h_2'}{2}\right)\sqrt{C_1' C_2'}}{k_H\, S_H}, \tag{8.98c}$$

$$\Delta E_{\text{CIE00}} = \sqrt{\Delta L^2_{\text{CIE00}} + \Delta C^2_{\text{CIE00}} + \Delta H^2_{\text{CIE00}} + R_T\, \Delta C_{\text{CIE00}}\, \Delta H_{\text{CIE00}}}. \tag{8.98d}$$

An informal comparison between the color difference metrics discussed in this section is given in Figure 8.16. The top-right image was created by applying a sharpening operator, which has increased contrast in the mid-range of values at the cost of a loss of detail in the dark and light regions. The color difference metrics typically show where these changes have occurred, although the CIE ΔE^*_{ab} and ΔE^*_{uv} predict a much larger visible difference than the other metrics.

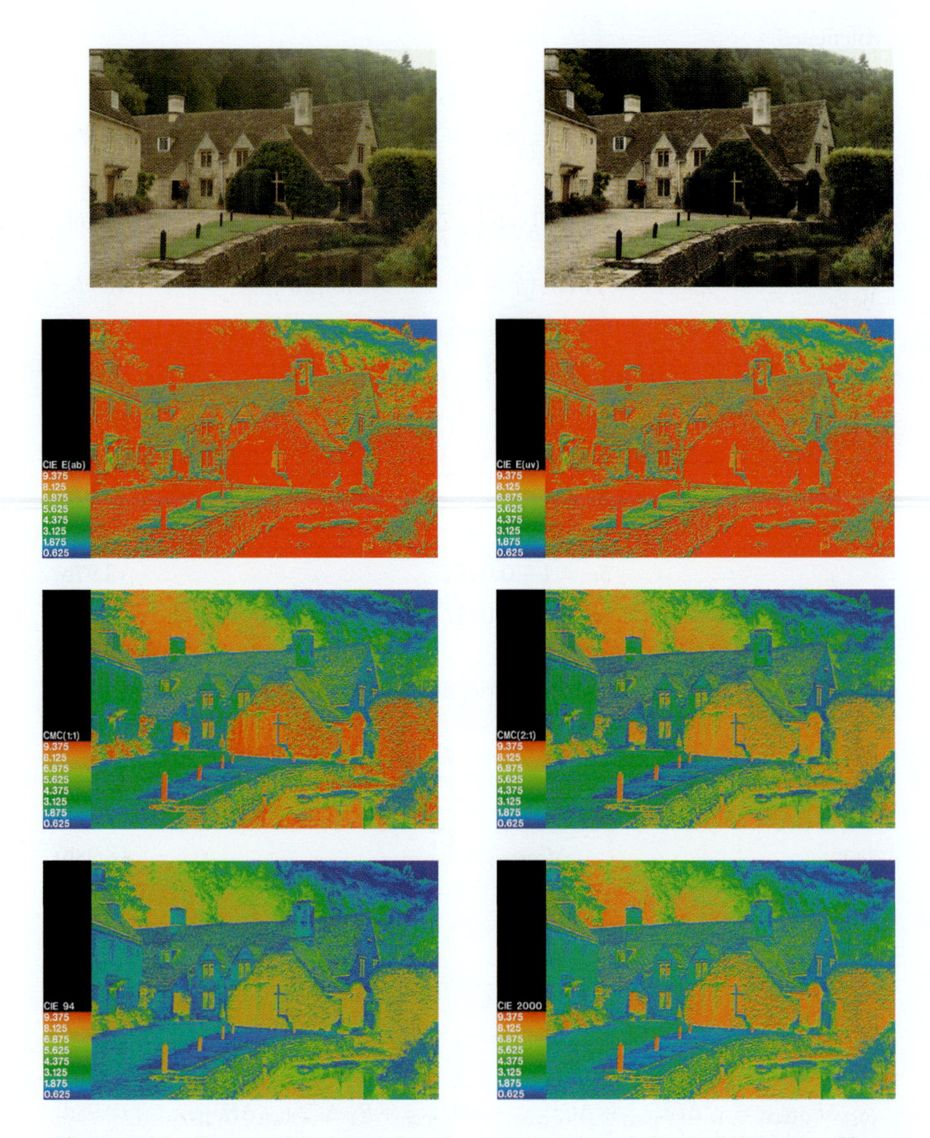

Figure 8.16. The top-right image is a sharpened version of the top-left image. They are compared in the remaining images using the color difference metrics discussed in this section; Castle Combe, UK, September 2006.

8.9 Color Order Systems

Color specification can be achieved in three fundamentally different ways. First, colors can be specified as an additive mixture. Given a set of primaries, we can then specify how much of each primary is added together to form the color. Examples are the RGB color spaces, as well as HLS-related spaces. Second, colors may be specified in systems derived from color-matching experiments. These include XYZ and its derived color spaces such as CIELAB and CIELUV. It should be noted that both the XYZ system, and its derivatives CIELAB and CIELUV also rely on additive mixtures of light, in the color-matching experiments themselves.

The third class of color specification methods is based directly on visual perception. Here, the aim is to produce color scales which are perceptually uniform. A *color order system* is defined as a conceptual system of organized color perception [93]. Such systems are separate from their physical incarnations, known as emphcolor order system exemplifications. For instance, the well-known Munsell color order system is published as an atlas of painted chips.

In color order systems, the objective is to subdivide color space into perceptually equal spacings. This can typically only be achieved under relatively controlled conditions. This includes the viewing environment, its geometry, as well as its illumination. The spectral distribution of the illuminant, as well as its strength, are defined. In a physical implementation of a color order system, the samples should remain constant over time, i.e., fading, yellowing, and changes with temperature and/or humidity should be avoided. If a color atlas is used to specify color, then in order to minimize the effects of metamerism, it is desirable that the sample to be specified is made of the same material as the paint chips in the atlas.

There are several color order systems with associated exemplifications available, including the Munsell Color System, the Natural Color System, the OSA Uniform Color Scales System, as well as several others. These are briefly introduced in the following sections.

8.9.1 Munsell Color System

Developed as a teaching aid for art students, the *Munsell color system* divides color space into equally spaced segments along the dimensions of value (lightness), hue, and chroma, which are then called *Munsell value* (V), *Munsell hue* (H), and *Munsell chroma* (C) [805, 808].

Rather than have the more conventional four primary hues of red, green, yellow, and blue, Munsell defined five principal hues by adding purple. This was because the general gap between the neighboring hues of red and blue was considered to be perceptually too large. The hue of any given stimulus can thus be designated by a single letter if it lies on one of the primaries, or a combination of two primaries: red (R), yellow-red (YR), yellow (Y), green-yellow (GY), green (G), blue-green (BG), blue (B), purple-blue (PB), purple (P), and red-purple (RP). These colors follow the same progression as a rainbow and can be placed on a circle. Each of the ten hues and hue combinations can be further subdivided into ten smaller steps, indicated with a number preceding the letter designation. For instance, purple ranges from 1P to 10P.

Munsell values range between black (0) and white (10). In addition, achromatic colors can be indicated with the prefix N, i.e., N0 ... N10. Hue and value together form a cylindrical shape if hue values are located on an axis orthogonal to the hue circle.

For each hue/value pair, the value of chroma represents how much the sample deviates from achromatic. The distance from the hue axis determines the chroma of a sample. The highest chroma value available is not uniformly distributed in color space, but instead depends on both hue and value. This system therefore leads to an asymmetrical shape in 3D space, known as the *Munsell color solid.*

In the Munsell notation system, a color is specified as H V/C. For instance, a bluish-green color can be written as 10BG 6/4. The Munsell color atlas [806] is now available as the Munsell book of color [807]

8.9.2 Natural Color System

Hering's hypotheses on color opponency and the eye's physiology leads to the general theory of six primary colors, which are black, white, red, green, yellow, and blue [461]. These ideas were later transformed into a color order system [554], as well as a color atlas [470]. After many years of improvements and refinements, this resulted in the *Natural Color System* (NCS) and the *Swedish Standard Color Atlas* [423–425, 1053, 1111].

In this system, the elementary colors are arranged in a hexagon, such that possible combinations are all allowable. Recall that colors on opposite ends of a color opponent channel cannot be perceived simultaneously (Section 5.3.2)

Beside the hexagon which encodes NCS hue, the two remaining perceptual quantities in NCS are NCS blackness (or whiteness), and NCS chromaticness.

8.9.3 OSA Uniform Color Scales System

Both the Munsell color system and NCS are cylindrical in the sense that hues are arranged in either a circle or a hexagon with the hue axis orthogonal, and Munsell chroma or NCS chromaticness defined as the distance to the hue axis. This means that the difference between two neighboring hues depends on the value for chroma or chromaticness. For a small change in hue, the distance between two colors of low chroma is much smaller than for colors with higher chroma. This just suggests that steps of hue in these systems cannot be considered perceptually uniform, though for any given value and chroma the steps should be close to uniform.

The *Optical Society of America (OSA) Uniform Color Scales System* (OSA-UCS) was designed to remedy this problem [718, 719, 836, 837]. It accomplished this by structuring the color space according to a crystal lattice, shaped as a cuboctahedron. The complexity of this system, however, requires a software implementation for it to be useful [93].

8.10 Application: Color Transfer between Images

In general, algorithms in computer graphics and computer vision may benefit from the appropriate choice of color space. In this chapter we have shown that the most common of all color spaces, namely RGB, is generally the least well defined. There is no one RGB color space, and many "standards" have emerged. Each of them have different primaries, a different white point, and a different non-linear encoding.

On the other hand, the CIE XYZ color space is well defined, and it is capable of encompassing the full range of colors visible by humans in an all positive manner. However, because the XYZ space also contains information that is well outside the visible range, it is not the most efficient means of encoding and transmitting information. If encoding and calculations are to be performed in the CIE XYZ space, then it is important to utilize a higher bit-depth representation in order to minimize quantization artifacts.

While many applications and algorithms continue to be defined in some RGB color space, we argue that in most cases an appropriate choice of color space can bring very significant benefits. As an example, in this section we outline an algorithm that crucially hinges upon the appropriate choice of color space. The task is the transfer of the color palette of one image to another. The question then is how to capture the mood of an image and apply it to a second image.

The most obvious approach is to measure certain statistical properties of one image and then modify a second image so that it obtains the same statistical properties. As color is represented with tristimulus values, and we have one such value per pixel, the problem is essentially a complicated three-dimensional problem. If the three color channels could be decorrelated first, the problem could be split into three one-dimensional sub-problems, which would arguably be easier to solve.

As we are predominantly interested in images of environments we would encounter in the real world, i.e., natural images, we require a color space for which the three color channels are decorrelated for natural images. Natural images have various statistical regularities that can be exploited in image processing. The most well-known is the $1/f$ shape of the amplitude spectrum to which natural images by and large adhere.[3]

A second, and very useful, statistic emerges when the correlation between color channels is considered. Converting an ensemble of spectral images of natural scenes to LMS color space, we essentially end up with an image represented in the same space as the photoreceptor's output.

A standard technique to decorrelate data is called *principal components analysis* (PCA) [579,696]. This technique rotates the data such that the data is aligned with the axes as well as possible. The first principle component is then the axis which accounts for the largest variance in the data, whereas each subsequent axis accounts for less and less variance. Applying PCA to an ensemble of images represented in LMS color space therefore rotates the data into a new color space in which the three color axes are maximally decorrelated.

The surprising result that emerges is that when PCA is applied in this manner to natural image ensembles, a new color space is found which strongly resembles a color opponent space similar to the opponent space computed by the retinal ganglion cells [984]. It is therefore apparent that the visual system decorrelates the visual signal before sending it to the brain. In practice, the three color channels are close to independent. This color space is the $L\alpha\beta$ color space discussed in Section 8.7.5.

For 2000 random samples drawn from each of the images shown in Figure 8.17, their distribution is plotted in Figure 8.18. The point clouds form more or less diagonal lines in RGB space when pairs of channels are plotted against each other, showing that the three color channels in RGB space are almost completely correlated for these images. This is not the case for the same pixels plotted in $L\alpha\beta$ space.

The practical implication for this is that by converting an image to the $L\alpha\beta$ color space, the three channels are decorrelated and, in practice, can often be treated as independent. The three-dimensional color transfer problem can therefore be recast into three independent one-dimensional problems [949].

[3]Natural image statistics are normally collected over large sets of images, or image ensembles. On average, the $1/f$ statistic holds to a high degree of accuracy. However, for individual images this statistic (and all other natural image statistics) may deviate from the ensemble average by an arbitrary amount.

Figure 8.17. Examples images used to demonstrate the correlation between channels. The first two images are reasonable examples of natural images, whereas the third image is an example of an image taken in a built-up area. Built environments tend to have somewhat different natural image statistics compared with natural scenes [1295, 1296]; Top left: Mainau, Germany, July 2005. Top right: Dagstuhl, Germany, May 2006. Bottom: A Coruña, Spain, August 2005.

To transfer the color palette of one image to another, both images are thus first converted to the $L\alpha\beta$ space. Within this space, along each of the axes a suitable statistic needs to be transferred. It turns out that a very simple matching of means and standard deviations is frequently sufficient to create a plausible result.

Naming the image from which its color palette is gleaned, the source image, and the image to which this palette is applied, the target image, we use the s and t subscripts to indicate values in source and target images. We first subtract the mean pixel value in each of the axes:

$$L'_s(x,y) = L_s(x,y) - \bar{L}_s, \tag{8.99a}$$

$$\alpha'_s(x,y) = \alpha_s(x,y) - \bar{\alpha}_s, \tag{8.99b}$$

$$\beta'_s(x,y) = \beta_s(x,y) - \bar{\beta}_s, \tag{8.99c}$$

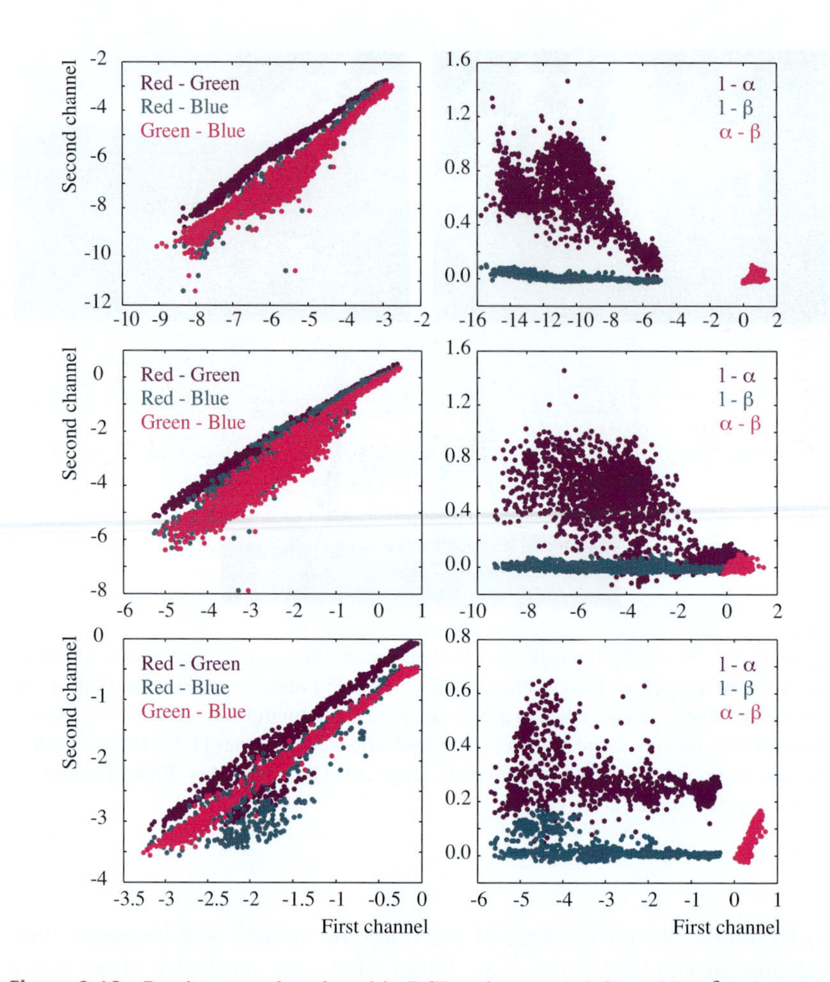

Figure 8.18. Random samples plotted in RGB color space (left) and $L\alpha\beta$ color space (right). The top to bottom order of the plots is the same as the order of the images in Figure 8.17.

$$L'_{\mathrm{t}}(x,y) = L_{\mathrm{t}}(x,y) - \bar{L}_{\mathrm{t}}, \tag{8.99d}$$

$$\alpha'_{\mathrm{t}}(x,y) = \alpha_{\mathrm{t}}(x,y) - \bar{\alpha}_{\mathrm{t}}, \tag{8.99e}$$

$$\beta'_{\mathrm{t}}(x,y) = \beta_{\mathrm{t}}(x,y) - \bar{\beta}_{\mathrm{t}}. \tag{8.99f}$$

Second, the six standard deviations σ_{s}^{L}, $\sigma_{\mathrm{s}}^{\alpha}$, $\sigma_{\mathrm{s}}^{\beta}$, σ_{t}^{L}, $\sigma_{\mathrm{t}}^{\alpha}$, and $\sigma_{\mathrm{t}}^{\beta}$ are computed. It is now possible to make the means and standard deviations in the three target

Figure 8.19. Color transfer between images. The left and middle images serve as input, and the right image was produced by matching means and standard deviations.

channels the same as those in the source channels:

$$L''_t(x,y) = \frac{\sigma_s^L}{\sigma_t^L} L'_t(x,y) + \bar{L}_s, \tag{8.100a}$$

$$\alpha''_t(x,y) = \frac{\sigma_s^\alpha}{\sigma_t^\alpha} \alpha'_t(x,y) + \bar{\alpha}_s, \tag{8.100b}$$

$$\beta''_t(x,y) = \frac{\sigma_s^\beta}{\sigma_t^\beta} \beta'_t(x,y) + \bar{\beta}_s. \tag{8.100c}$$

After conversion back to RGB, we obtain a displayable image. Examples of this approach are shown in Figure 8.19.

Matching means and standard deviations implies that the point cloud in each of the three dimensions is essentially Gaussian. This may be too simple a model. It may therefore be possible to match higher-order moments such as skew and kurtosis as well. Of course, the more high-order moments are taken into consideration, the better the shape of the point cloud in the source and target images are matched.

Alternatively, it would also be possible to match the histograms that can be computed along each axis. Histogram matching begins by computing the histograms of both the source and target channels. Next, the cumulative histograms are computed. Cumulative histograms are monotonically non-decreasing functions. The source histogram is then inverted. Pixels values in a particular channel in the target image are then converted by first looking up their corresponding fre-

Figure 8.20. The top two images served as input to the color transfer algorithm. The bottom-left image shows a result whereby means and standard deviations are matched, whereas the bottom-right image shows a result obtained with histogram matching. Top-left image: Turtle Mound, New Smyrna, FL, November 2004. Other images: Lake Konstanz, Germany, July 2005.

quency in the cumulative target histogram. This frequency is then mapped to a different pixel value by applying the inverted cumulative source histogram.

An example of this approach is shown in Figure 8.20. Although matching of means and standard deviations will provide plausible results for many image pairs, the example shown in this figure is not one of them. However, the histogram-matching technique produces a result that is much closer to what one might intuitively expect the result to look like.

There are several application areas for this simple technique. First, this method provides an aesthetically pleasing way to enhance images. Second, it has proven a useful method to create night-time images out of photographs taken by daylight.

Third, the technique is important in augmented reality applications where real environments are merged with computer-generated data. To make the computer-generated elements mesh with the real scene, a fast algorithm is required to account for mismatches in the choice of surface materials as well as differences in the lighting. While the color transfer algorithm cannot account for missing or wrongly placed shadows or other light interactions between the real and virtual components, the computer-generated elements can be adjusted to have the same

Figure 8.21. The University of Central Florida logo superimposed on a background [951]. The color transfer algorithm was applied to the image on the right and makes the logo blend in with the background better; UCF Campus, Orlando, FL, June 2004.

Figure 8.22. The color transfer algorithm benefits from the appropriate choice of color space. The top and left-most images are the input, whereas the bottom-middle image was computed in RGB space. The bottom-right image was computed in the decorrelated $L\alpha\beta$ space. Both transfers were computed using histogram matching; Nicosia, Cyprus, June 2006.

color statistics, which makes them fit in the scene better [951]. An example is shown in Figure 8.21.

It is interesting to compare the results obtained in $L\alpha\beta$ color space with images created in different color spaces. In particular, when the results are created in an RGB color space, it is expected that due to correlations between the axes, the results will not be as good. That this is indeed the case is shown in Figure 8.22.

As mentioned above, the color transfer algorithm implicitly assumes that the source and target images adhere to natural image statistics. Although image ensembles tend to show remarkable statistical regularities, individual images may have different statistical properties. It is therefore not guaranteed that this algorithm will work on any pair of images.

In particular, if the composition of the source and target images is very different, the algorithm may fail. In general, transformations whereby the source image has a fairly small color palette (such as night images, sunsets, and everything with a limited number of different colors) tend to work well.

If the source and/or the target image is not a natural image in the statistical sense, then the $L\alpha\beta$ color space is not an appropriate choice of color space. In that case, it may be possible to explicitly decorrelate the data in the source and target images separately, i.e., run PCA on both images [551]. It is then possible to construct a linear transform between the source and target image. This transform thus defines an image-specific color space, which may then be used to equalize the three means and standard deviations.

This section has shown that the appropriate choice of color space is sometimes crucial to achieve a particular goal. Most applications in computer vision and computer graphics will benefit from the appropriate choice of color space. Color should be an important consideration in the design of visual algorithms.

8.11 Application: Color-to-Gray Conversion

Conversion of color images to gray is often necessary, for instance in printing applications. Starting with an image defined in a given three-dimensional color space, grayscale conversion essentially constitutes a dimensionality reduction from three dimensions to one. Given the salience of luminance in human visual perception, a first approach would be to convert the color image to a color space that has a luminance channel. It is then possible to linearly remap the luminance values to the desired grayscale.

As an example, Figure 8.23 shows an image and its grayscale equivalent. The grayscale image was created by converting the color image to $L\alpha\beta$ space (see Section 8.7.5) and retaining only the luminance channel.

Figure 8.23. Conversion from color to grayscale using the luminance channel of the $L\alpha\beta$ color space; Castle Combe, UK, September 2006.

The use of the $L\alpha\beta$ color space is motivated by the fact that, for natural image ensembles, this color space is obtained by employing PCA, as outlined in the preceding section. As the luminance channel is the first principal component, this particular grayscale conversion will, on average, retain the most information.

Although an appropriate choice on average, for a specific image this conversion may yield a distinct loss of information. For instance, if the image contains large variations in chromatic content, without exhibiting large variations in its luminance channel, then the $L\alpha\beta$ color space will yield unconvincing results.

In such cases, we may apply PCA to each image individually, yielding the best possible color space achievable with linear transformations only. By taking only the first principal component and removing the second and third principal components, a grayscale image is obtained.

While this approach may produce somewhat better results for certain images, this is largely dependent on the distribution of points in color space. Principal components analysis is only able to rotate the point cloud such that the direction of greatest variance becomes axis-aligned. By projecting the data onto the first principal component, we are not guaranteed a plausible grayscale conversion. For instance, consider a dumbbell-shaped distribution of points such as in images of green foliage with red fruit. The greens will be clustered in one sub-volume of color space, and the reds will be clustered elsewhere. In the best case, PCA will rotate the data such that all the greens are mapped to very light colors, and all the reds are mapped to very dark colors. The mid-range of the grayscale will end up being under-utilized.

A better approach would be to preserve the color difference between pairs of pixels as much as possible. Here, it is important to start with a perceptually uniform color space, so that color differences have perceptual meaning. For instance, the CIE94 color difference metric could be used. Assuming that a tristimulus value $\mathbf{c}$ is mapped to grayscale by a function $T(\mathbf{c})$, and the color difference metric is written down as $\|\mathbf{c}_i - \mathbf{c}_j\|$, then the color difference between pairs of pixels should remain constant before and after the mapping [938]:

$$\frac{\|\mathbf{c}_i - \mathbf{c}_j\|}{C_{\text{range}}} = \frac{\|T(\mathbf{c}_i) - T(\mathbf{c}_j)\|}{T_{\text{range}}}. \tag{8.101}$$

A measure of the error introduced by the mapping T is given by

$$\varepsilon^2 = \sum_i \sum_{j=i+1} \left(\frac{\|\mathbf{c}_i - \mathbf{c}_j\|}{C_{\text{range}}} - \frac{\|T(\mathbf{c}_i) - T(\mathbf{c}_j)\|}{T_{\text{range}}} \right)^2. \tag{8.102}$$

Here, C_{range} and T_{range} are the maximum color differences in the color and grayscale images. Restricting the mapping T to linear transformations in the CIELAB color space, the general form of T is

$$T(\mathbf{c}) = (\mathbf{g} \cdot \mathbf{c}, 0, 0). \tag{8.103}$$

As two of the three dimensions are mapped to 0, the function T may be replaced by the scalar function $\mathbf{g} \cdot \mathbf{c}$, where $\mathbf{g}$ is the vector in color space that determines the axis to which all color values should be projected to obtain a grayscale image. Its direction will be determined shortly. The color difference metric for T is now the absolute value of the difference of two gray values. The error function thus simplifies to

$$\varepsilon^2 = \sum_i \sum_{j=i+1} \left(\frac{\|\mathbf{c}_i - \mathbf{c}_j\|}{C_{\text{range}}} - \frac{|\mathbf{g} \cdot (\mathbf{c}_i - \mathbf{c}_j)|}{T_{\text{range}}} \right)^2. \tag{8.104}$$

A solution to the grayscale conversion is then found by minimizing this error function, which is achieved with standard minimization techniques. For instance, the Fletcher-Reeves conjugate gradient method can be used. The optimization procedure can be seeded with an initial solution of $\mathbf{g} = (1, 0, 0)$. After convergence, the vector $\mathbf{g}$ for which the error is minimized then determines the axis onto which all colors are projected, leading to the desired grayscale image.

The power of this approach, however, is still limited by the fact that all pixel values are projected to a single dimension. Many different colors may project to the same gray value. A possible alternative approach can be derived by realizing that different colors should not be projected to the same gray value if they are neighboring. However, pixels separated from each other by differently colored image regions may successfully be mapped to the same gray value. Such an approach would help overcome the problem of having a limited number of different gray values to which all pixels must be mapped.

One such approach, called Color2Gray, begins by converting the image to the CIE L*a*b* space [378]. For a pair of pixels, indicated with subscripts i and j, the conventional color-difference metric computes the Euclidean distance between the two points. However, for the purpose of conversion to grayscale, it is better to split this color-difference metric into two metrics, one measuring the luminance difference and a second measuring the chromatic difference:

$$\Delta L^*_{ij} = L^*_i - L^*_j, \tag{8.105a}$$

$$\Delta \mathbf{C}^*{}_{ij} = (\Delta a^*_{ij}, \Delta b^*_{ij}) \tag{8.105b}$$

$$= (a^*_i - a^*_j, b^*_i - b^*_j). \tag{8.105c}$$

This separation allows iso-luminant colors with different chromaticities to be distinguished and, subsequently, to be mapped to different gray values. In addition, some user control is desirable, since, depending on the application, some colors should be preferentially mapped to lighter gray values than others. Such creative control would be difficult to automate.

For each pair of pixels, a desired difference in gray value δ_{ij} can now be computed. As each pair of pixels generates such a desired difference, there may not exist a grayscale image that satisfies each of these differences. Hence, these target differences are input to an optimization problem, with objective function:

$$\sum_{i,j} \left((g_i - g_j) - \delta_{ij}\right)^2. \tag{8.106}$$

Here, g_i is the gray value of pixel i which is found by minimizing this function. It is initialized by setting $g = L_i$. Once again, conjugate gradient methods are suitable for finding a minimum.

The target differences δ_{ij} are computed under user control. Just as the a^* and b^* channels span a chromatic plane, the color differences in these channels, Δa^* and Δb^*, span a chromatic color-difference plane. To determine which color differences are mapped to increases in gray value, the user may specify an angle Θ in this plane. The associate unit length vector is denoted with $\mathbf{v}_\Theta$.

A second user parameter α steers a non-linearity $c_\alpha(x)$ that is applied to the chromatic color difference $\Delta\mathbf{C}_{ij}$. This compressive function assigns relatively more importance to small color differences than large ones:

$$c_\alpha(x) = \alpha \tanh\left(\frac{x}{\alpha}\right). \tag{8.107}$$

The target color difference for a pair of pixels (i, j) is then

$$\delta_{ij} = \begin{cases} \Delta L_{ij} & \text{if } |\Delta L^*_{ij}| > c_\alpha(\|\Delta C^*_{ij}\|), \\ c_\alpha(\|\Delta C^*_{ij}\|) & \text{if } \Delta C^*_{ij} \cdot \mathbf{v}_\Theta \geq 0, \\ c_\alpha(\| - \Delta C^*_{ij}\|) & \text{otherwise.} \end{cases} \tag{8.108}$$

Figure 8.24. The red of this anturium stands out clearly in the color image. However, conversion to gray using the luminance channel of $L\alpha\beta$ space makes the flower nearly disappear against the background. The bottom-left image shows the result of applying PCA directly to the color image. The bottom-right image shows the result obtained by applying the spatial Color2Gray algorithm [378].

An example result is compared with a standard gray conversion in Figure 8.24. The image shown here has a dumbbell-shaped distribution along the red-green axis, which results in the red flower disappearing in the gray conversion for the standard luminance-channel approach. Applying PCA directly to this image does not improve the results notably, as shown in the same figure. However, the spatially variant technique described above, is able to assign a different range of gray levels to the red flower, making it distinctly visible in the gray-level image.

In summary, a good transformation to grayscale constitutes a dimensionality reduction from three to one dimensions, whereby some perceptually significant attribute is preserved. In particular, chromatic differences between pairs of isoluminant colors should be expressed in differences in gray level [939], especially if these colors are located in spatial proximity.

8.12 Application: Rendering

The rendering algorithms described in Section 2.10 quietly assumed that all lighting and materials are spectral, i.e., defined per wavelength. For some applications, a spectral renderer will be invaluable. As an example, it would be difficult to render a material with a reflectance that is sharply peaked at a small number of wavelengths when lit by, for example, a fluorescent light source, which itself has a sharply peaked emission spectrum.

However, most applications do not require a fully spectral renderer. As the human visual system under photopic conditions processes light in three different channels, it is often not required to render images using more than three channels. Thus, the emission spectra of lights and reflectance spectra of materials are represented by three values each. Each shading calculation is then carried out independently for the three components.

The choice of color space to use for rendering images is usually not considered. Lights and materials are frequently represented with an RGB triplet without concern for exactly which of the many RGB color spaces were intended. In the best case, one of a few standard color spaces is used, such as sRGB. Such a choice is motivated by the fact that the images produced by the renderer are already in the same color space that is used by conventional display devices.

However, the sRGB color space has a rather narrow color gamut (see Section 15.4). It is found that within a rendering application, the use of a wider gamut produces images that are closer to a full spectral reference image [1208]. Wider gamuts are available in the form of sharpened color spaces (see Section 8.6.2).

Given a spectral light source $L(\lambda)$ and a material $\rho(\lambda)$, a CIE XYZ tristimulus value can be computed by integrating using standard color matching functions $\bar{x}(\lambda)$, $\bar{y}(\lambda)$, and $\bar{z}(\lambda)$:

$$X = \int_\lambda L(\lambda)\,\rho(\lambda)\,\bar{x}(\lambda)\,d\lambda, \tag{8.109a}$$

$$Y = \int_\lambda L(\lambda)\,\rho(\lambda)\,\bar{y}(\lambda)\,d\lambda, \tag{8.109b}$$

$$Z = \int_\lambda L(\lambda)\,\rho(\lambda)\,\bar{z}(\lambda)\,d\lambda. \tag{8.109c}$$

The resulting XYZ values represent the material as seen under illuminant L. The dominant illuminant available within the environment should be chosen for L. This is possible, even if there are many light sources in an environment, as most of these light sources typically have the same spectral emission spectra.

Conversion to a sharpened color space then involves multiplication by a 3×3 matrix:

$$\begin{bmatrix} R' \\ G' \\ B' \end{bmatrix} = M_{\text{sharp}} \begin{bmatrix} X \\ Y \\ Z \end{bmatrix}. \tag{8.110}$$

A suitable conversion to a sharpened color space is given by the following matrix:

$$M_{\text{sharp}} = \begin{bmatrix} 1.2694 & -0.0988 & -0.1706 \\ -0.8364 & 1.8006 & 0.0357 \\ 0.0297 & -0.0315 & 1.0018 \end{bmatrix}; \tag{8.111a}$$

$$M_{\text{sharp}}^{-1} = \begin{bmatrix} 0.8156 & 0.0472 & 0.1372 \\ 0.3791 & 0.5769 & 0.0440 \\ -0.0123 & 0.0167 & 0.9955 \end{bmatrix}. \tag{8.111b}$$

After rendering in this color space, the resulting pixels need to be converted to the space in which the display device operates. This is achieved by converting each pixel first to XYZ and then to the desired display space. Assuming this is the sRGB color space, the conversion is then

$$\begin{bmatrix} R_d \\ G_d \\ B_d \end{bmatrix} = M_{\text{sRGB}}\, M_{\text{sharp}}^{-1} \begin{bmatrix} R' \\ G' \\ B' \end{bmatrix}. \tag{8.112}$$

Comparing renderings made in the sRGB and sharpened color spaces with a direct spectral rendering, it is found that the difference between the reference and the sharpened images remained at or below detectable levels. Measured using the CIE ΔE_{94}^* color difference metric (see Section 8.8.2), the average difference is 4.6 for the 98th percentile (with 5.0 visible in side-by-side image comparisons) [1208]. The sRGB color space does not perform as well, producing differences with the reference image that are more easily perceived at 25.4.

Aside from the choice of color space, color-accurate rendering will require white balancing, as spectral measurements of materials are typically made under different illumination than they will be rendered. Second, the dominant illuminant used in the rendering is usually different from the conditions under which the resulting image is viewed. Both are sources of error for which correction is needed. This topic is discussed in more detail in Section 10.7, where rendering is revisited.

8.13 Application: Rendering and Color-Matching Paints

Most image-synthesis applications render directly into an RGB space; usually the space spanned by the primaries of the targeted output device. However, this space is inadequate for some applications. One of them is simulating and rendering paint. As shown in Section 3.4.6, pigmented paints cause scattering and absorption, which can only be properly accounted for by employing a sufficiently accurate simulation. Using either RGB or CMY directly is not sufficient.

However, the Kubelka-Munk theory provides an alternative that may be employed in the computer simulation of paints. It is possible to derive the $K(\lambda)$ and $S(\lambda)$ functions by measuring the reflectance function $R(\lambda)$ for pigments in their masstone states (i.e., a pure pigment) as well as in tinted states (i.e., mixed with white in a known concentration). Using the techniques described in Section 3.4.6, reflectance functions for paints containing the same pigments but in different concentrations can be derived. Such spectral reflectance functions can be converted to XYZ by integrating with the appropriate color-matching functions. The XYZ triplets can in turn be converted to the color space used for rendering—for example, the sharpened RGB space discussed in Section 8.12.

Thus, by adding absorption and scattering functions for different pigments to derive a new spectral absorption and scattering function for the desired mixture, a spectral reflectance function can be computed, which is then converted to the preferred RGB rendering space. This approach leads to more accurate results than by adding RGB triplets directly to simulate a paint mixture. Hence, the use of Kubelka-Munk theory is recommended for simulating paint mixtures within rendering applications. It can also be used in drawing programs to simulate, for instance, airbrushing [412].

Finally, an important application lies in the conversion from RGB triplets to a recipe for mixing pigments in given concentrations to mix a paint that is as close as possible to the color defined by the RGB tristimulus value. This is the electronic equivalent of a procedure called *color matching* that should not be confused with the color-matching experiments discussed in Chapter 7. Color matching for paints refers to the matching of a paint mixture to a given sample of paint.

We assume that a database of spectral reflectance functions is available for a set of pigments. As Kubelka-Munk theory can predict the spectral reflectance of any mixture of these pigments, the conversion from a given spectral reflectance function to a mixture of paints can be posed as an optimization problem.

One approach to achieve this, called *spectrophotometric matching*, is to convert an RGB triplet to a spectral reflectance function, and then searching for a concentration for each pigment, such that when mixed, the resulting paint matches the target spectral reflectance function as well as possible [27, 81, 629, 874, 1198]. Algorithms to convert RGB triplets to spectral functions are discussed in Section 7.6.3.

Alternatively, color matching may proceed in tristimulus space and is then called tristimulus matching [27, 28, 345, 412, 629, 743]. The advantage of this approach is that only the tristimulus values of the sample need to be known for a paint to be matched. However, the number of different pigments that can be used in the mixture is limited to three [412].

The approach described here follows Haase and Meyer's technique and is based on minimizing the color difference between a trial match and the paint sample [412]. The CIE L*a*b* ΔE color difference, explained in Section 8.7.1, is used for this application.

As the aim is to find concentrations for each pigment, a mixture where one or more of the concentrations is negative, is not allowed. To avoid setting up an optimization problem where negative concentrations are a possible outcome, the problem is restated to find the square of concentrations for each pigment, labeled $q_i = c_i^2$ for the ith pigment.

Initially, a random XYZ tristimulus value is guessed, and the ΔE color difference with the target sample is computed. To enable a gradient descent method to be used, the gradient of ΔE^2 with respect to each of the squared concentrations q_i is computed:

$$\frac{\partial {\Delta E^*_{\text{ab}}}^2}{\partial q_i} = 2 q_i \frac{\partial {\Delta E^*_{\text{ab}}}^2}{\partial c_i}. \tag{8.113}$$

By the chain rule, we have

$$\frac{\partial {\Delta E^*_{\text{ab}}}^2}{\partial c_i} = \frac{\partial {\Delta E^*_{\text{ab}}}^2}{\partial X}\frac{\partial X}{\partial c_i} + \frac{\partial {\Delta E^*_{\text{ab}}}^2}{\partial Y}\frac{\partial Y}{\partial c_i} + \frac{\partial {\Delta E^*_{\text{ab}}}^2}{\partial Z}\frac{\partial Z}{\partial c_i}. \tag{8.114}$$

The right-hand side of this equation can be expanded by expressing each of the partials as follows:

$$\frac{\partial {\Delta E^*_{\text{ab}}}^2}{\partial X} = \frac{1000 \Delta a^*}{3 X^{2/3} X_0^{1/3}}, \tag{8.115a}$$

$$\frac{\partial {\Delta E^*_{\text{ab}}}^2}{\partial Y} = \frac{-1000 \Delta a^* + 232 \Delta L^* + 400 \Delta b^*}{3 Y^{2/3} X_0^{1/3}}, \tag{8.115b}$$

$$\frac{\partial {\Delta E^*_{\text{ab}}}^2}{\partial Z} = \frac{-400 \Delta b^*}{3 Z^{2/3} Z_0^{1/3}}. \tag{8.115c}$$

The remaining partials are expanded as follows:

$$\frac{\partial X}{\partial c_i} = k \sum_{\lambda} L(\lambda) \bar{x}(\lambda) \frac{\partial R(\lambda)}{\partial c_i}, \tag{8.116a}$$

$$\frac{\partial Y}{\partial c_i} = k \sum_{\lambda} L(\lambda) \bar{y}(\lambda) \frac{\partial R(\lambda)}{\partial c_i}, \tag{8.116b}$$

$$\frac{\partial Z}{\partial c_i} = k \sum_{\lambda} L(\lambda) \bar{z}(\lambda) \frac{\partial R(\lambda)}{\partial c_i}. \tag{8.116c}$$

Here, $\bar{x}(\lambda)$, $\bar{y}(\lambda)$, and $\bar{z}(\lambda)$ are standard color-matching functions, $L(\lambda)$ is the emission spectrum of the light source lighting the sample, and k is a normalization constant given by $100/\sum_\lambda L(\lambda)\bar{y}(\lambda)$. The partial $\partial R(\lambda)/\partial c_i$ is given by

$$\frac{\partial R(\lambda)}{\partial c_i} = \frac{S_M(\lambda)K_i(\lambda) - K_M(\lambda)S_i(\lambda)}{S_M^2(\lambda)} \left(1 - \frac{\dfrac{K_M(\lambda)}{S_M(\lambda)} + 1}{\sqrt{\left(\dfrac{K_M(\lambda)}{S_M(\lambda)}\right)^2 + 2\dfrac{K_M(\lambda)}{S_M(\lambda)}}} \right). \tag{8.117}$$

In this equation, K_M and S_M are the absorption and scattering functions for the current trial, and K_i and S_i are the absorption and scattering functions for the ith pigment.

With the current match trial a gradient vector is computed; this is then added to the current squares of the concentrations for each pigment q_i. This leads to new values for the concentrations c_i, which in turn are used to compute a new match trial. This process iterates until convergence is reached, i.e., all gradients are below a specified threshold. The concentrations found for each pigment can then be used to mix a paint that will look similar to the target XYZ tristimulus value.

The match is valid under the chosen illuminant L. However, it may be beneficial to minimize the effects of metamerism if the paint samples are to be viewed under different illuminants. For instance, if it is assumed that a match is required under CIE illuminant C, but metamerism needs to be minimized under CIE illuminant A with the initial match being four times as important, then the minimization function could be modified to

$$\Delta E^2 = 4\Delta E_C^2 + \Delta E_A^2. \tag{8.118}$$

The matching algorithm outlined here has important benefits over standard tristimulus matching approaches that produce a match in XYZ space. By using the uniform CIE L*a*b* color space, the match is found by minimizing the ΔE color-difference metric that corresponds to how humans perceive color differences. Second, an arbitrary number of pigments can be used in the match.

It would be possible to improve this method, though, as the steepest descent method is not guaranteed to find a global minimum. Second, the method does not take into account the fact that it may be both costly and inconvenient to mix as many pigments as there are reflectance functions in the database used to compute the match. As a result, it is likely that the color match involves small concentrations of many pigments. It would be beneficial to devise a method that would find the smallest number of pigments that would create a match.

8.14 Application: Classification of Edges

Image understanding refers to the process of obtaining high-level information about the scene that is captured in images and videos. Answers to questions such as What are the different objects in the scene? What actions are they performing? Are there humans in the scene? are all provided by this process. The main purpose of image understanding is to enable the user to make optimal decisions and take appropriate actions based on information from the scene. Examples of systems that use image understanding are surveillance systems and robots.

Image understanding often relies on lower-level information such as segmentation, motion estimation, and object tracking. Algorithms for these tasks often produce incorrect results due to the presence of shadows in the scene [217]. This is because pixel values of the same object across edges created by cast shadows are often very different and are, therefore, incorrectly recognized as belonging to different objects of the scene.

Clearly, removal of shadows from the input images and videos would improve the performance of these algorithms. Shadow removal may become easier if edges cast by shadows (henceforth called *shadow edges*) are somehow distinguishable from other edges, such as those cast by differences in reflectance properties of adjacent surfaces in the scene (also called *reflectance edges*).

Color is an important feature that may be exploited to distinguish between shadow and reflectance edges. Specifically, the choice of color space to represent data has been shown to be important [585]. The obvious difference across shadow edges is that regions under direct light appear brighter than the corresponding regions in shadows. The regions in shadow get light from their surroundings [337]. This light is referred to as *ambient* light. The difference across shadow edges is apparent only in the brightness of objects across shadow edges, if the spectral composition of ambient light is the same as that of the direct light. Such a phenomenon can occur in cases when the gray world assumption holds, that is, the average reflectance of objects in the scene is gray [513]. In this case, the reflectance of objects will not alter the spectral composition of light reflecting off of objects. Furthermore, there should be no participating media in the scene, as this could again change the spectral composition of ambient light.

When the above assumptions hold, and the only difference across shadow edges is in luminance, then an appropriate choice of color space can be employed to differentiate between shadow edges and reflectance edges [217, 357, 990]. Any color space that separately encodes luminance and chromatic information may be used. Edges that appear only in the luminance channel and not in color channels are more likely to be shadow edges, whereas edges appearing in both luminance

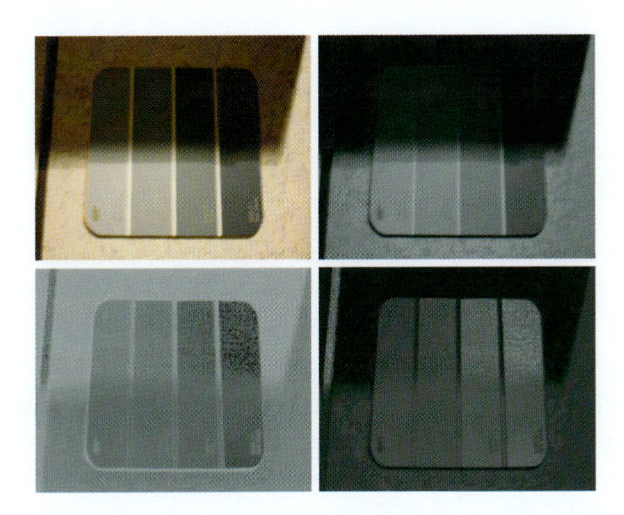

Figure 8.25. The image on the top left was captured in a scene where the gray world assumption was assumed to hold approximately. The image on the right shows the information captured by the luminance channel of the Yxy color space. The bottom two images show the information encoded by the color channels (x and y, respectively). Notice how most of the information across the shadow edge is contained in the luminance channel, whereas all channels encode the change across reflectance edges.

and color channels are probably reflectance edges. Figure 8.25 illustrates this difference between shadow and reflectance edges in scenes where the spectral component of ambient light is the same as that of direct light.

It is hypothesized that the difference across shadow edges in sunlit scenes will be in luminance as well as color, as the spectral component of ambient light is possibly different from that of direct light for such scenes. Sunlight, which is the direct light source, has a distinctly yellow spectrum, while the rest of the sky appears blue due to Rayleigh scattering (as discussed in Section 2.8.1). Ambient light therefore tends to have a bluish color. Figure 8.26 demonstrates this change across shadow edges in outdoor scenes. There appears to be a change in color as well as luminance across the edge. Choosing color spaces that separately encode luminance and color information is therefore not helpful in distinguishing between shadow and reflectance edges in outdoor scenes.

Given that the changes in color across shadow edges are possibly in blue and yellow (as the direct light is predominantly yellow and the ambient light is blue), color opponent spaces may be better choices for the purpose of edge classification than other color spaces, since these separately encode blue-yellow and red-green information. Shadow edges will appear in the luminance and blue-yellow

Figure 8.26. Colors appear to have a bluer shade in shadows and a yellower shade under direct light. The left image shows a color under direct sunlight, and the right image shows the same color in shadow. Images were derived from the photograph shown in Figure 8.27.

channels, but not in the red-green channel, while all three channels will tend to encode the difference across reflectance edges. The two edge types will therefore often be distinguishable in outdoor scenes. The invariance to shadows in the red-green channel was shown to have an evolutionary advantage in monkey foraging behavior [1144].

The performance of 11 color spaces were tested for the purpose of shadow-edge identification [585]. These include RGB, XYZ, Yxy, *normalized* rgb, LMS, Luv, Lab, $L\alpha\beta$, HSV, AC_1C_2, and the linear version of $L\alpha\beta$ (where the log of LMS values is not taken before conversion to $L\alpha\beta$). When an image is converted into a color space, one channel in the color space might encode the difference across reflectance edges only, while another channel might encode the difference across both shadow and reflectance edges. In that case, classification of edges in the image may be done by first converting the image to the color space and then using these two channels to distinguish between the two types of edges. The most suitable color space for distinguishing between the edge types will have one channel that does not distinguish between the edge types and one channel that best discriminates between the edge types when compared with all the channels in all the color spaces in the experiment.

Data for hundreds of edges of both types were collected and encoded by each of the 11 color spaces. Edge data was generated by photographing a variety of diffuse paint samples and then cropping these photographs to obtain uniform patches of each paint sample under direct light as well as in shadow. Figure 8.27 shows one of the photographs and the corresponding patches obtained from it. An edge may be represented by a pair of such patches. For a shadow edge, one patch will contain a color in direct light, and the second patch will contain the same color in shadow. For a reflectance edge, the patches will contain different colors in direct light; the patches might also contain different colors in shadow.

To compute the response of a color space to a shadow or reflectance edge, the two patches representing the edge are converted into that color space. The values of each patch are then averaged to obtain a single value representing color,

Figure 8.27. The image at the top shows a set of colored paint samples. The patches at the bottom were cropped from this photograph. The top row shows patches from the shadow region, and the bottom row shows corresponding patches from the lit region.

followed by an assessment of how different these values are for the two patches. If a channel shows large differences in the average values of the patches representing reflectance edges and small differences for patches representing shadow edges, then that channel is a good classifier.

The discrimination of each channel is computed using receiver operating characteristic (ROC) curves [773]. A ROC curve is a graphical representation of the accuracy of a binary classifier. They are created by thresholding each channel's response to reflectance and shadow edges. The true positive fraction is the fraction of correctly identified reflectance edges against the total number of reflectance edges, and the true negative fraction is the same fraction for shadow edges. The true positive fraction is plotted against the false positive fraction to give the ROC curve. The area under the ROC curve (AUC) is the most commonly used quantitative index describing the curve. This area has been empirically shown to be

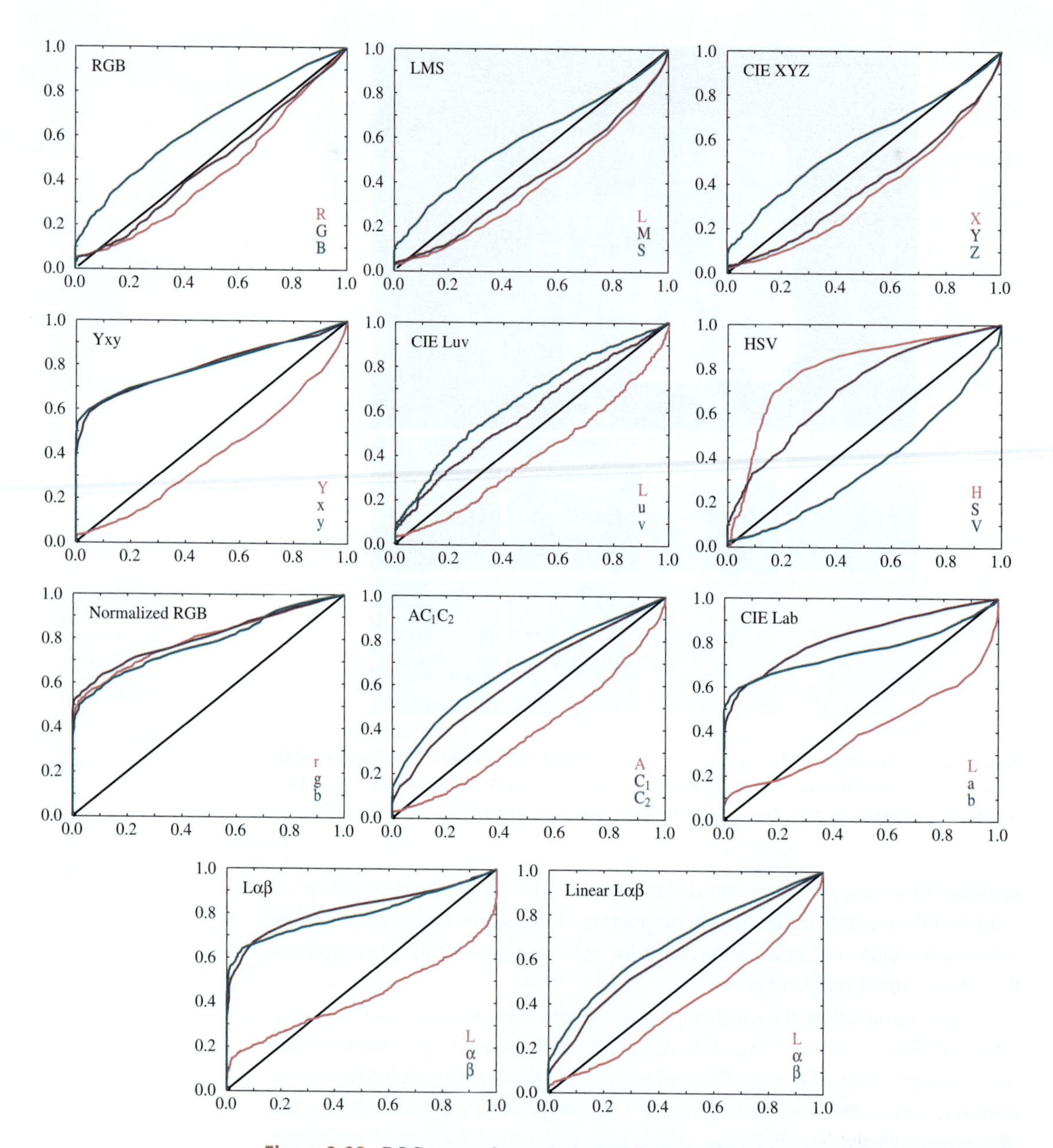

Figure 8.28. ROC curves for each channel in each of the 11 color spaces included in the experiment. (Erum Arif Khan and Erik Reinhard, "Evaluation of Color Spaces for Edge Classification in Outdoor Scenes," IEEE International Conference on Image Processing, Genova, Italy, pp 952–955, © 2005 IEEE.)

Channel	AUC (outdoors)	AUC (indoors)
rgb - r	0.435	0.598
rgb - g	0.477	0.573
rgb - b	0.649	0.567
XYZ - X	0.362	0.553
XYZ - Y	0.397	0.547
XYZ - Z	0.590	0.521
LMS - L	0.384	0.554
LMS - M	0.412	0.545
LMS - S	0.590	0.521
Yxy - Y	0.397	0.547
Yxy - x	0.795	0.902
Yxy - y	0.798	0.881
Luv - L	0.397	0.547
Luv - u	0.588	0.643
Luv - v	0.647	0.612
HSV - H	0.798	0.847
HSV - S	0.707	0.760
HSV - V	0.372	0.549
CIELAB - L	0.394	0.569
CIELAB - a	0.826	0.848
CIELAB - b	0.762	0.778
L$\alpha\beta$ - L	0.422	0.532
L$\alpha\beta$ - α	0.798	0.910
L$\alpha\beta$ - β	0.818	0.895
linear L$\alpha\beta$ - L	0.408	0.530
linear L$\alpha\beta$ - α	0.682	0.626
linear L$\alpha\beta$ - β	0.635	0.613
AC$_1$C$_2$ - A	0.388	0.509
AC$_1$C$_2$ - C_1	0.627	0.653
AC$_1$C$_2$ - C_2	0.679	0.601

Table 8.5. Channels and their AUCs for outdoor and indoor scenes.

equivalent to the Wilcoxon statistic and computes the probability of correct classification [421]. An area of 1 signifies perfect classification while an area of 0.5 means that the classifier performs at chance.

A suitable color space should have one channel that has an AUC close to 0.5, and another channel with an AUC close to 1. The most suitable color space is one for which the discriminating channel's AUC is closest to 1. The ROC curves computed for each channel for sunlit scenes are shown in Figure 8.28 and the AUCs for each channel (for both indoor and outdoor scenes) are shown in Table 8.5. Note that the ROC curve for *normalized* rgb shows the absence of any channel that does *not* discriminate between the two edge types. It is therefore removed from further consideration.

Four of the color spaces under consideration show color opponency. These color spaces are CIELAB, $L\alpha\beta$, linear $L\alpha\beta$ and AC_1C_2. While CIELAB is a perceptually uniform space, and $L\alpha\beta$ is perceptually uniform to a first approximation, the latter two are linear color spaces. The performance of the perceptually uniform color spaces is significantly better than their linear counterparts.

For indoor scenes however, where there is only a change in luminance across shadow edges, and all that is required of a color space is separate encoding of luminance and chromatic information, other color spaces tend to perform as well as color opponent spaces. For outdoor scenes, color opponent spaces perform well. The two perceptually uniform color opponent spaces prove to be the most suitable choices. CIELAB and $L\alpha\beta$ may be used to get the best discrimination between shadow and reflectance edges, followed by Yxy and HSV color spaces.

From this experiment, it is shown that color is an important cue in the classification of edges for sunlit scenes even when no assumptions are made about the gray world and the effects of participating media. Experimental results show that opponent color spaces are a good choice for the purpose of edge classification.

8.15 Further Reading

RGB color spaces are discussed in detail by Pascale [882], whereas transforms between different spaces are presented in [327]. A full account of standard and high definition television is presented by Charles Poynton [923]. The CIE colorimetric systems are discussed in various places, including Hunt's book on color reproduction [509] and Wyszecki and Stiles [1262]. Rolf Kuehni has written an entire text on the historical use of color spaces [631]. Bruce Lindbloom maintains a website with tools and equations dealing with different color spaces [685].

Chapter 9
Illuminants

The real world offers a wide range of illumination, which changes based on the atmospheric conditions of the moment, the time of the day, and the season of the year. Similarly, the range of illumination from man-made light sources is vast, ranging from the yellowish light of a tungsten filament to the bluish-white of a fluorescent light source.

The apparent color of an object is determined by the spectral power distribution of the light that illuminates the object and the spectral reflectance properties of the object itself. Therefore, the wide range of illumination under which an object may be viewed affects the perception of its color. This poses a significant problem, especially in industrial colorimetry, where the main concern is to match the color of a product to a given standard. The match achieved under one light source may no longer hold when the product is displayed under a different light source.

To alleviate this problem, the Commission Internationale de l'Éclairage (CIE) defines a set of specific spectral power distributions called *CIE Standard illuminants*. Using CIE illuminants in colorimetric computations ensures consistency across measurements. It also simplifies the color match problem and makes it independent of individual differences between light sources.

9.1 CIE Standard Illuminants and Sources

When we speak of an *illuminant*, we refer to a (relative) spectral power distribution, which may or may not be realizable by a light source. On the other hand, the term *source* refers to a physical emitter of light such as the sun and the sky, or a tungsten light bulb. As such, the definition of an illuminant precedes the construction of a light source that is aimed to represent that illuminant. The rationale is that new developments in light-source technology are likely to produce better light sources that more accurately represent any desired illuminant [194].

CIE illuminants represent general types of light sources commonly encountered in real life. For example, incandescent and fluorescent light sources and different phases of daylight are all represented by CIE illuminants.

9.1.1 CIE Standard Illuminant A

CIE Standard Illuminant A represents typical domestic tungsten-filament lighting. Its spectral radiant power distribution is determined by light emitted from a Planckian radiator (i.e., a blackbody or full radiator) operated at an absolute temperature of 2856 K according to the International Practical Temperature Scale [142].

The spectral radiant exitance $M_{e,\lambda}(T)$ of a Planckian radiator at temperature T (in Kelvin K) is approximated by

$$M_{e,\lambda}(T) = c_1 \lambda^{-5} \left[\exp\left(\frac{c_2}{\lambda T} - 1 \right) \right]^{-1} \; W m^{-3}, \tag{9.1}$$

where the radiation constants c_1 and c_2 are equal to

$$c_1 = 3.74150 \times 10^{-16} \, \mathrm{Wm}^2, \tag{9.2a}$$

$$c_2 = 1.4388 \times 10^{-2} \, \mathrm{mK}. \tag{9.2b}$$

The spectral power distribution of CIE illuminant A is then computed by inserting $T = 2856$ K into (9.1). However, in most colorimetric measurements, only relative spectral power distributions are necessary. As such, the spectral power distribution $S_A(\lambda)$ of CIE illuminant A (as is the case for other illuminants) is typically normalized to 100 at a wavelength of 560 nm:

$$S_A(\lambda) = 100 \frac{M_{e,\lambda}(2856)}{M_{e,560}(2856)}. \tag{9.3}$$

The spectral power distribution of CIE illuminant A is shown in Figure 9.1.

Correlated color temperature, tristimulus values, and chromaticity coordinates of illuminant A are listed in Table 9.1 along with other CIE illuminants. The other CIE illuminants listed in this table will be explained in the subsequent sections. Spectral data for all CIE illuminants are given in the accompanying DVD-ROM.

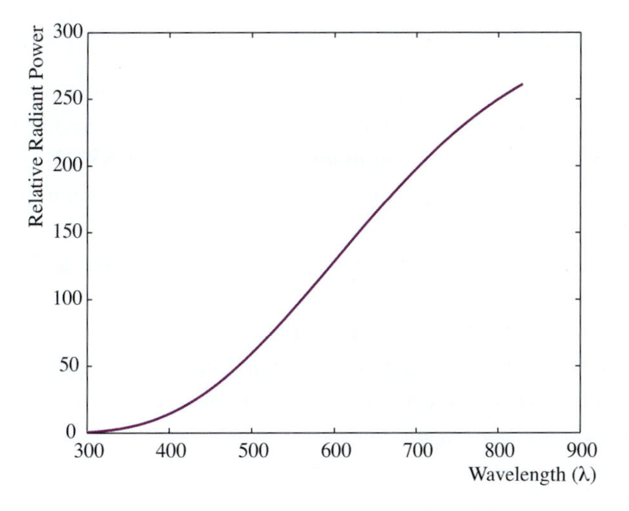

Figure 9.1. Relative spectral radiant power distribution of CIE standard illuminant A.

In practice, a light source approximating CIE illuminant A can be built by using a gas-filled coiled tungsten filament lamp operating at a correlated color temperature of $2856K$ [1262]. Correlated color temperature as well as other types of color temperature are explained in Section 9.2. For now, it suffices to say that the correlated color temperature of a light source equals the temperature of a Planckian radiator whose chromaticity coordinate is closest to that of the light source.

Illuminant	**CCT (K)**	**Tristimulus values**			**Chromaticity coordinates**			
		X	Y	Z	x	y	u'	v'
A	2856	109.85	100.00	35.58	0.45	0.41	0.26	0.52
B	4874	99.10	100.00	85.32	0.35	0.35	0.21	0.49
C	6774	98.07	100.00	118.22	0.31	0.32	0.20	0.46
D50	5003	96.42	100.00	82.51	0.35	0.36	0.21	0.49
D55	5503	95.68	100.00	92.14	0.33	0.35	0.20	0.48
D65	6504	95.04	100.00	108.88	0.31	0.33	0.20	0.47
D75	7504	94.97	100.00	122.61	0.30	0.32	0.19	0.46
E	5400	100.00	100.00	100.00	0.33	0.33	0.21	0.47
F2	4230	99.20	100.00	67.40	0.37	0.38	0.22	0.50
F7	6500	95.05	100.00	108.72	0.31	0.33	0.20	0.47
F11	4000	100.96	100.00	64.37	0.38	0.38	0.23	0.50

Table 9.1. Correlated color temperature (CCT), tristimulus values, CIE 1931 xy and CIE 1976 $u'v'$ chromaticity coordinates of commonly used CIE illuminants. Spectral data of all CIE illuminants are given in the accompanying DVD-ROM.

9.1.2 CIE Standard Illuminant D_{65}

CIE standard illuminant D_{65} represents average daylight with a correlated color temperature of approximately 6504 K. The reason to use a correlated color temperature of 6504 K, instead of the more rounded value of 6500 K, is a result of the slight adjustment of the radiation constant c_2 in Planck's formula (see (9.1)). Originally, $c_2 = 1.4380\,10^{-2}\,\text{mK}$ was accepted, but the value of c_2 is changed to $1.4388\,10^{-2}\,\text{mK}$ in the International Practical Temperature Scale 1968. The difference of 4 K follows from $6500\frac{1.4388}{1.4380} \approx 6504$.

Illuminant D_{65} is based on numerous spectro-radiometric measurements taken in the United States, Canada, and Great Britain [559]. Although the spectral distribution of daylight on the earth's surface is known to change as a function of season, time of day, geographic location, weather conditions, air pollution, etc., the CIE recommends the use of D_{65} pending the availability of additional information [193]. The relative spectral power distribution of CIE illuminant D_{65} is shown in Figure 9.2 together with other CIE D-series illuminants. These illuminants are explained in the next section.

As a result of the rather jagged spectral distribution of the CIE standard illuminant D_{65}, it has proven difficult to build a light source that faithfully represents this illuminant. Consequently, at present the CIE has not recommended any standard light source to realize illuminant D_{65}. Despite this, daylight simulators suitable for psychophysics, can be constructed by placing a suitable filter in front of a

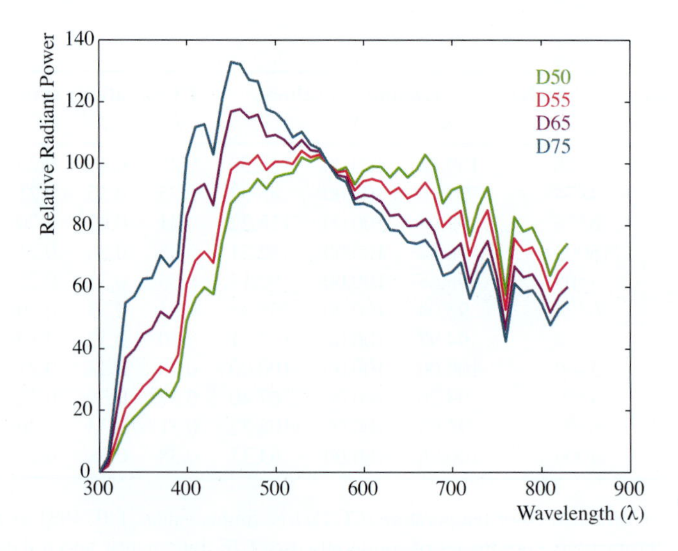

Figure 9.2. Relative spectral radiant power distributions of CIE D illuminants.

Figure 9.3. The Gretag Macbeth D65 daylight simulator consists of a halogen light source with a blue filter, shown here.

halogen light source. An example is the Gretag MacBeth D65 daylight simulator, shown in Figures 2.50 and 3.45. The filter is shown in Figure 9.3.

The CIE Colorimetry Committee has developed a method that can be used to assess the usefulness of existing light sources in representing CIE D illuminants [184, 192]. The method is too elaborate to reproduce here in full. Nonetheless, the main steps of the method are:

1. Measure the spectral power distribution of the light source under consideration by means of spectro-radiometry from $\lambda = 300$ to $\lambda = 700$ at 5-nm intervals.

2. Compute the CIE 1976 (u'_{10}, v'_{10}) chromaticity coordinates of the light source from the measured spectral data. These coordinates must fall within a circle of radius 0.015 centered at the chromaticity point of D_{65} on the same diagram. Otherwise the tested source does not qualify as a daylight simulator.

3. Even if the test in the previous step succeeds, the quality of the daylight simulator needs to be determined. For this purpose two tests are performed:

   (a) Compute the mean index of metamerism MI_{vis} (see Section 9.4) of five pairs of samples which are metamers under CIE illuminant D_{65}. The spectral radiance factors of these samples are provided by the CIE [1262]. These computations should use CIE 1964 (X_{10}, Y_{10}, Z_{10})

CIEL*a*b*	CIE L*u*v*	Category
< 0.25	< 0.32	A
0.25 to 0.50	0.32 to 0.65	B
0.50 to 1.00	0.65 to 1.30	C
1.00 to 2.00	1.30 to 2.60	D
> 2.00	> 2.60	E

Table 9.2. Categories of daylight simulators (from [1262]). Using the method explained in the text, MI_{vis} and MI_{uv} are computed for the test source. Depending on the color-difference formula, use either the CIE L*a*b* or CIE L*u*v* values to find the corresponding *category* for the MI_{vis} and MI_{uv} indices. The result is a two-letter grade, such as BD or AC, indicating the quality of the daylight simulator in the visible and ultra-violet regions of the spectrum.

tristimulus values. The resulting value indicates how good the metamerism of these five sample pairs are under the tested daylight simulator.

(b) Compute the mean index of metamerism MI_{uv} of three pairs of samples, where one member of each pair is non-fluorescent and the other member is fluorescent. The spectral values of these pairs are provided by the CIE [1262]. These pairs also yield identical tristimulus values under CIE illuminants D_{65}. All computations should once more use CIE 1964 (X_{10}, Y_{10}, Z_{10}) tristimulus values. The resulting value indicates how good the daylight simulator represents the standard illuminant D_{65} in the ultra-violet region of the spectrum.

4. Use MI_{vis} and MI_{uv} as an index to Table 9.2 to find the category under which the daylight simulator falls. A category of AA indicates that the daylight simulator has excellent performance both in the visible and ultraviolet region of the spectrum.

9.1.3 Other CIE Illuminants D

For standardization purposes, the CIE recommends use of illuminant D_{65} for all calculations that involve daylight. However, the CIE also defines a set of supplementary illuminants called D_{50}, D_{55}, and D_{75} which may be used according to the task at hand. These illuminants have correlated color temperatures of 5003 K, 5503 K, and 7504 K; therefore, their color ranges from yellowish-white to bluish-white.

Different disciplines have adopted different types of CIE daylight illuminants. For instance, D_{65} is commonly used by the paint, plastic, and textile industries, while the graphics art and photography communities favor D_{50} [93]. The relative spectral radiant power distributions of these illuminants, together with the CIE standard illuminant D_{65}, are shown in Figure 9.2.

9.1.4 CIE Method of Calculating D-Illuminants

The measurements by Judd et al. [559] revealed a simple relationship between the correlated color temperature of a CIE D-series illuminant and its relative spectral distribution. Denoting the correlated color temperature by T_c, the spectral radiant distribution $S(\lambda)$ of a daylight illuminant can be calculated as follows:

1. Use T_c to find the first chromaticity coordinate x_D. If T_c is between 4000 K and 7000 K, then

$$x_D = \frac{-4.6070 \times 10^9}{T_c^3} + \frac{2.9678 \times 10^6}{T_c^2} + \frac{0.09911 \times 10^3}{T_c} + 0.244063. \quad (9.4)$$

   If T_c is between 7000 K and 25000 K, then

$$x_D = \frac{-2.0064 \times 10^9}{T_c^3} + \frac{1.9018 \times 10^6}{T_c^2} + \frac{0.24748 \times 10^3}{T_c} + 0.237040. \quad (9.5)$$

2. Use x_D to compute the second chromaticity coordinate y_D:

$$y_D = -3.000 x_D^2 + 2.870 x_D - 0.275. \quad (9.6)$$

3. Use x_D and y_D to compute the factors M_1 and M_2:

$$M_1 = \frac{-1.3515 - 1.7703 x_D + 5.9114 y_D}{0.0241 + 0.2562 x_D - 0.7341 y_D}, \quad (9.7)$$

$$M_2 = \frac{0.0300 - 31.4424 x_D + 30.0717 y_D}{0.0241 + 0.2562 x_D - 0.7341 y_D}. \quad (9.8)$$

4. Using M_1, M_2, and the spectral distributions $S_0(\lambda)$, $S_1(\lambda)$, and $S_2(\lambda)$ (given in Table F.1), compute the relative spectral distribution $S'(\lambda)$:

$$S'(\lambda) = S_0(\lambda) + M_1 S_1(\lambda) + M_2 S_2(\lambda). \quad (9.9)$$

5. Normalize $S'(\lambda)$ to 100 at wavelength 560 nm:

$$S(\lambda) = 100 \frac{S'(\lambda)}{S'(560)}. \quad (9.10)$$

9.1.5 CIE F Illuminants

Fluorescent lighting is commonly used owing to its high electrical efficacy and its similarity to natural daylight. Therefore, the CIE originally defined 12 types of fluorescent illuminants denoted by F1 to F12 that represent typical fluorescent lamps used in practice [186]. F-series illuminants are categorized into three groups based on the emission spectra of the light sources they represent:

- F1–F6: standard fluorescent lamps;
- F7–F9: broadband fluorescent lamps;
- F10–F12: three-narrow-band fluorescent lamps.

From each of these groups, the CIE emphasizes one particular illuminant for use in colorimetric work. These are the F2 cool white single phosphor lamp; the F7, a daylight simulator; and the F11, which is a three-narrow-band fluorescent lamp [194]. Of these three illuminants, the cool white F2 is the most similar to typical fluorescent lights found in homes, offices, stores, and factories [735]. The spectral power distributions of all CIE F-series illuminants are given in the accompanying DVD-ROM.

Group	Illuminant	Commercial type	Chromaticity Coordinates		CCT (K)	CRI
			x	y		
Standard	F1	DAY	0.3131	0.3371	6430	76
	F2	CWF[a]	0.3721	0.3751	4230	64
	F3	WHITE	0.4091	0.3941	3450	57
	F4	WWF[b]	0.4402	0.4031	2940	51
	F5	DAY	0.3138	0.3452	6350	72
	F6	LWF[c]	0.3779	0.3882	4152	59
Broad-band	F7	D65	0.3129	0.3292	6500	90
	F8	D50	0.3458	0.3586	5000	95
	F9	CWX[d]	0.3741	0.3727	4150	90
Three-narrow-band	F10	TL85	0.3458	0.3588	5000	81
	F11	TL84	0.3805	0.3769	4000	83
	F12	TL83	0.4370	0.4042	3000	83

[a] Cool white fluorescent
[b] Warm white fluorescent
[c] Lite white fluorescent
[d] Cool white deluxe

Table 9.3. CIE F illuminants, their commercial equivalents, and several important properties. CCT: Correlated color temperature, CRI: Color-rendering index (see Section 9.3); (from [960]).

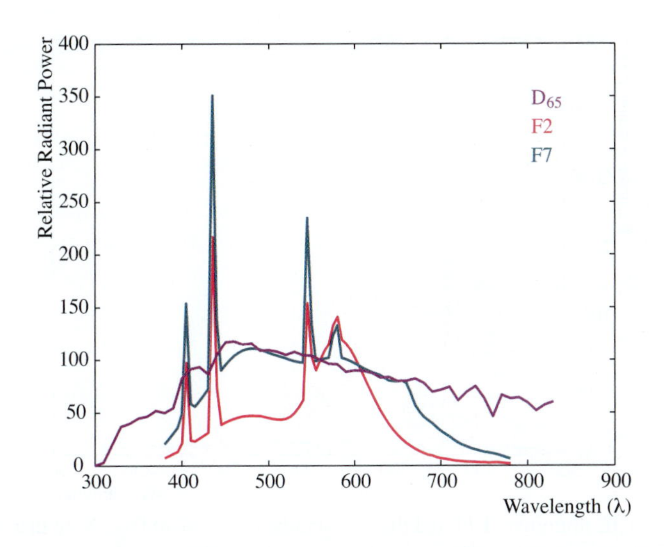

Figure 9.4. CIE illuminants F2 and F7 and the CIE standard illuminant D_{65}.

Although these illuminants were derived from actual fluorescent lamps, their commercial types were not disclosed. Recognizing that this may limit their usability, this issue has been addressed by work of other researchers [960]. Table 9.3 lists F-series illuminants and their commercial types given by Rich [960].

In this table, several abbreviations are used to denote the commercial type of each illuminant. For instance CWF, LWF, and WWF represent cool white, lite white, and warm white fluorescent lamps, respectively, based on their correlated color temperatures. CWX represents a cool white deluxe lamp which is constructed by supplementing a single phosphor lamp, such as CWF, with a second phosphor to enhance the under-represented parts of the emission spectrum [941].

Figure 9.4 shows the relative spectral radiant power of F2 and F7 along with the CIE standard illuminant D_{65}. F11 is shown in Figure 9.5. For this illuminant, normalization to 100 at 560 nm exaggerates its entire spectral distribution, since there is very little power at that wavelength. Therefore, for such illuminants, different normalization strategies, such as normalizing by the entire area under the spectral curve, are also employed [960].

Recognizing the need to catch up with the rapidly changing fluorescent industry the CIE released 15 new fluorescent illuminants in 2004 [194]. These new illuminants are denoted by symbols from FL3.1 to FL3.15 and are classified into several groups as follows:

- FL3.1 – 3.3: standard halo-phosphate lamps;
- FL3.4 – 3.6: deluxe type lamps;

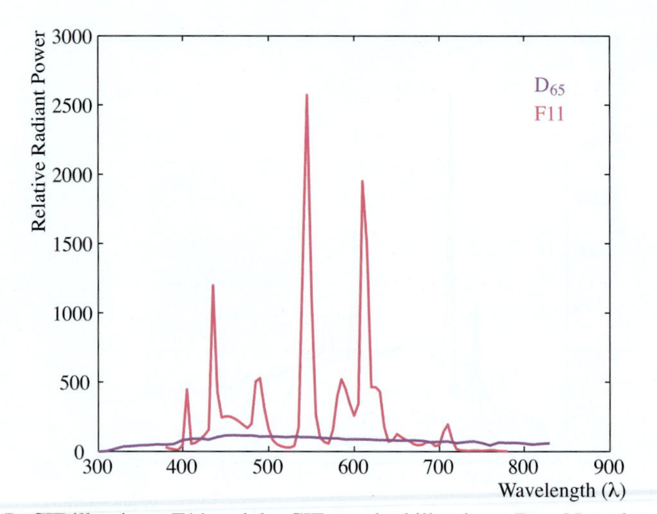

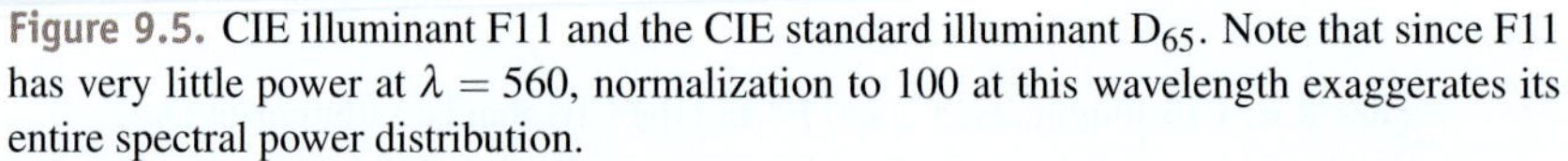

Figure 9.5. CIE illuminant F11 and the CIE standard illuminant D_{65}. Note that since F11 has very little power at $\lambda = 560$, normalization to 100 at this wavelength exaggerates its entire spectral power distribution.

- FL3.7 – 3.11: three band lamps;
- FL3.12 – 3.14: multi-band lamps;
- FL3.15: daylight simulator.

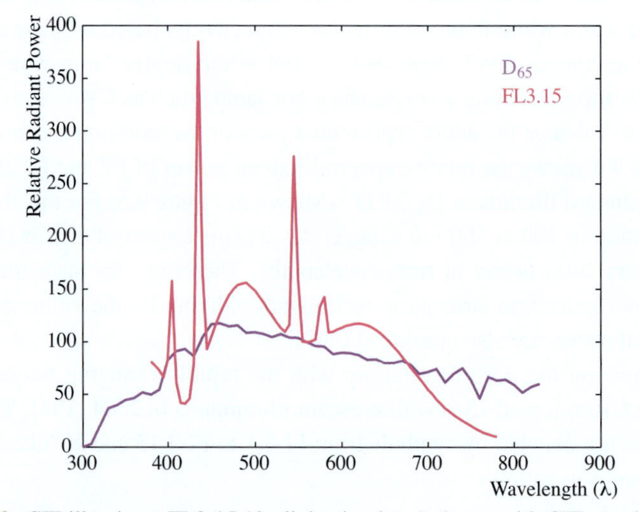

Figure 9.6. CIE illuminant FL3.15 (daylight simulator) shown with CIE standard illuminant D_{65}.

Spectral radiant distributions of these illuminants are given in the accompanying DVD-ROM, and a comparison of FL3.15 with the CIE standard illuminant D_{65} is shown in Figure 9.6.

9.1.6 Other Illuminants

An important theoretical illuminant is one whereby at each wavelength the same amount of light is emitted. Typically, this amount is assumed to be unity. This leads to a flat emission spectrum, which is called an *equal-energy stimulus*. It is also often referred to as an *equal energy spectrum on the wavelength basis* [1261]. Table 9.1 lists the correlated color temperature and chromaticity coordinates for the equal-energy stimulus under illuminant E.

The CIE defined two other illuminants, namely B and C, together with illuminant A, in 1931. Illuminant B is intended to represent direct sunlight with a correlated color temperature of 4874 K, while illuminant C is intended to model average daylight with a correlated color temperature of 6774 K.

Both of these illuminants can be realized by using special filters to steer the distribution of light emitted from a CIE standard source A [558]. Illuminant B can be realized by using a double layered colorless optical glass filter where each layer is 1 cm thick and contains one of the solutions B1 and B2 shown in Table 9.4.

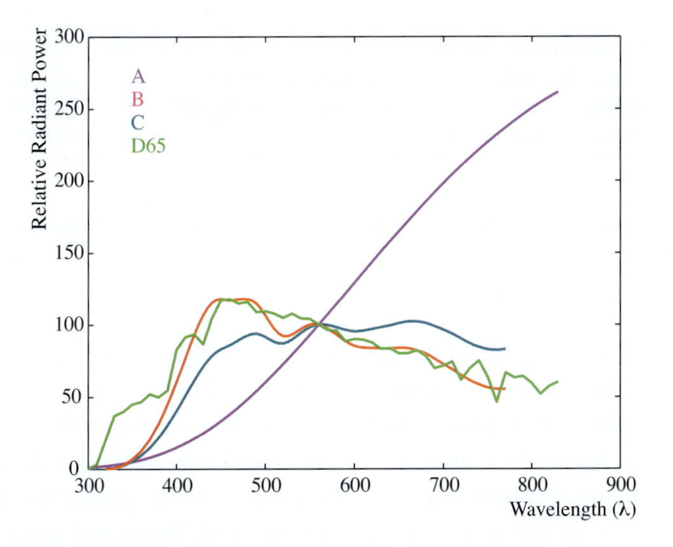

Figure 9.7. Relative spectral radiant power distributions of CIE standard illuminants A, D_{65}, and deprecated CIE standard illuminants B and C. Note the discrepancy of B and C from D_{65} especially in the ultra-violet region.

B1	Compound	Formula	Amount
	Copper Sulphate	$CuSO_4 . 5H_2O$	2.452 g
	Mannite	$C_6H_8(OH)_6$	2.452 g
	Pyridine	C_5H_5N	30.0 ml
	Distilled water to make		1000.0 ml
B2			
	Cobalt Ammonium Sulphate	$CoSO_4 . (NH_4)_2SO_4 . 6H_2O$	21.71 g
	Copper Sulphate	$CuSO_4 . 5H_2O$	16.11 g
	Sulphuric Acid	density 1.835 $g.ml^{-1}$	10.0 ml
	Distilled water to make		1000.0 ml

Table 9.4. Solutions B1 and B2 used for producing filters to realize illuminant B from illuminant A.

C1	Compound	Formula	Amount
	Copper Sulphate	$CuSO_4 . 5H_2O$	3.412 g
	Mannite	$C_6H_8 (OH)_6$	3.412 g
	Pyridine	C_5H_5N	30.0 ml
	Distilled water to make		1000.0 ml
C2			
	Cobalt Ammonium Sulphate	$CoSO_4 . (NH_4)_2SO_4 . 6H_2O$	30.58 g
	Copper Sulphate	$CuSO_4 . 5H_2O$	22.52 g
	Sulphuric Acid	density 1.835 $g.ml^{-1}$	10.0 ml
	Distilled water to make		1000.0 ml

Table 9.5. Solutions C1 and C2 used for producing filters to realize illuminant C from illuminant A.

Illuminant C can be realized similarly using the solutions C1 and C2 given in Table 9.5. These filters are commonly called *Davis-Gibson filters* [235]. The spectral distributions of illuminants B and C are shown in Figure 9.7.

However, both of these illuminants inadequately represent their intended spectral distributions. The chief deficiency is in the ultra-violet range of the spectrum, which plays an important role in the colorimetry of fluorescent materials. As such, they are dropped from the list of standards by the CIE, and illuminant D_{65} is recommended as a replacement. Currently illuminants B and C are not recommended for use in any application, but are included here for completeness.

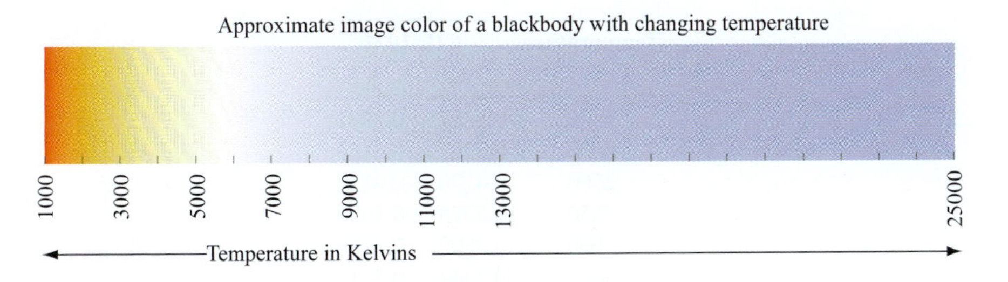

Figure 9.8. Various colors of a blackbody at different absolute temperatures. Note that these colors only approximate the true colors due to limitations in the printing process.

9.2 Color Temperature

When a blackbody is heated so that its temperature rises, its color changes from reddish to bluish (Section 3.5.1). This effect is shown in Figure 9.8. This figure shows that the color temperature of a blackbody is equal to its absolute temperature in Kelvin. Additionally, it is clear from this figure that the scale is non-linear with human visual perception, in that a very large part of the scale is taken up by blue colors. A related measure, known as the *reciprocal megakelvin* or *remek*,[1] is computed as the reciprocal color temperature. It's unit is $10^6 K^{-1}$, and is approximately perceptually linear [561, 927, 1262].

Most light sources and illuminants are not perfect blackbodies. Furthermore for non-incandescent sources, such as fluorescent lamps, the color they emit is not directly related to their temperature. Their color is determined by the composition of the gas mixture used in their tubes (see Section 3.5.4). Therefore, color temperature is used exclusively for blackbodies, and different terms are introduced for other types of sources. These are explained in the following sections.

9.2.1 Distribution Temperature

The distribution temperature is a property of a light source whose relative spectral distribution is similar to that of a blackbody radiator. The temperature of the blackbody radiator for which this match occurs is then called the *distribution temperature* of the light source. This implies that both the light source and the blackbody at that particular temperature emit light of the same chromaticity [1262].

Temperature T (in K)	CIE 1931 x	y
1500	0.5857	0.3931
2000	0.5267	0.4133
2500	0.4770	0.4137
2856	0.4475	0.4074
4000	0.3802	0.3767
5000	0.3444	0.3513
5500	0.3314	0.3405
6500	0.3118	0.3224
7500	0.2978	0.3083
10000	0.2762	0.2840

Table 9.6. Chromaticity coordinates of a blackbody radiator at various color temperatures.

9.2.2 Color Temperature of a Source

The *color temperature* of a source is the temperature (in Kelvin) of a blackbody radiator which emits light of the same (or almost the same) *chromaticity* as the light source. Note that the relative spectral power distributions of the source and the blackbody radiator need not be identical.

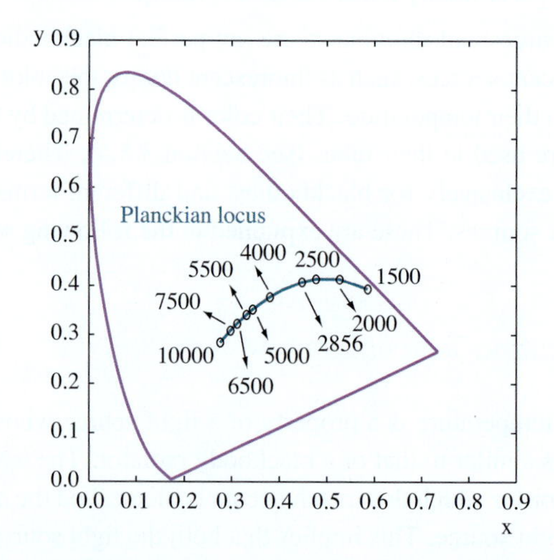

Figure 9.9. The Planckian locus sampled at various absolute temperatures.

[1]This quantity was formerly known as the *microreciprocal degree* or *mired* [1262].

To illustrate the concept of color temperature, let us consider the chromaticity coordinates of a blackbody radiator at various temperatures . These coordinates may be computed by integrating the spectral distribution of a blackbody radiator by the CIE 1931 color-matching functions. A set of such chromaticity coordinates is computed in the interval 1500 K–10000 K and tabulated in Table 9.6. By plotting these coordinates on the CIE 1931 chromaticity diagram, as shown in Figure 9.9, we find the *Planckian locus*, the line connecting subsequent color temperatures.

It is only appropriate to speak of a color temperature for a light source if its chromaticity coordinates lie on the Planckian locus. In general, only incandescent sources (e.g., a tungsten filament light bulb) satisfy this requirement. An example is the CIE standard source A which has a color temperature of 2856 K. For other light sources, the related attribute *correlated color temperature* may be determined, as discussed next.

9.2.3 Correlated Color Temperature

The term *correlated color temperature* is introduced when the chromaticity coordinates of a source are not identical to that of a full blackbody radiator (i.e., when the chromaticity coordinates of a light source do not lie on the Planckian locus). In that case, the correlated color temperature denotes the absolute temperature of a blackbody radiator whose perceived color is closest to that of the light source [1262].

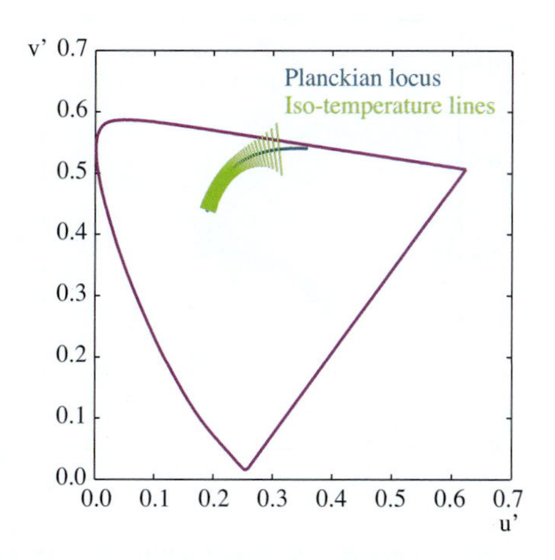

Figure 9.10. Iso-temperature lines shown on the CIE 1976 $u'v'$ chromaticity diagram.

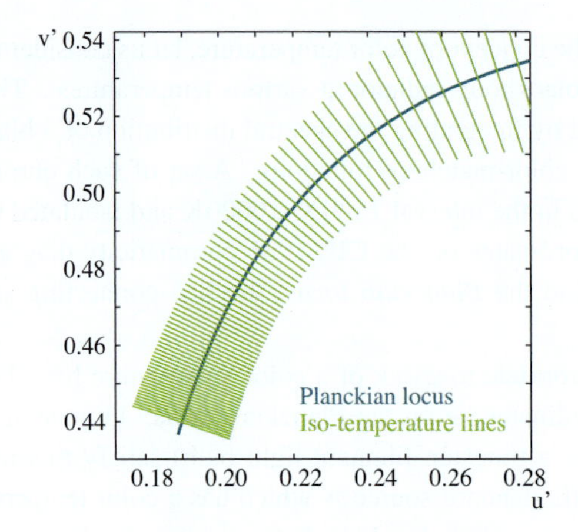

Figure 9.11. A close-up view of iso-temperature lines shown on the CIE 1976 $u'v'$ chromaticity diagram.

This temperature may be determined graphically using a chromaticity diagram. However, in this case, a uniform chromaticity diagram such as the CIE 1976 $u'v'$ Uniform Chromaticity Scale (UCS) needs to be employed. In UCS, there is a direct correlation between the distance between the chromaticity points and their perceived difference, a property missing in CIE 1931 xy chromaticity diagrams.

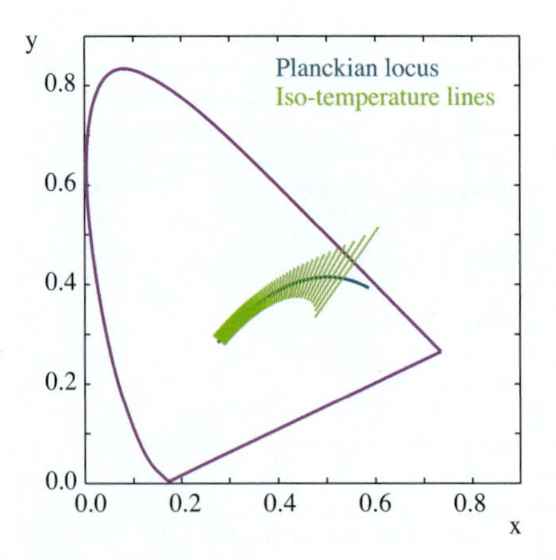

Figure 9.12. Iso-temperature lines transformed back to the CIE 1931 xy chromaticity diagram.

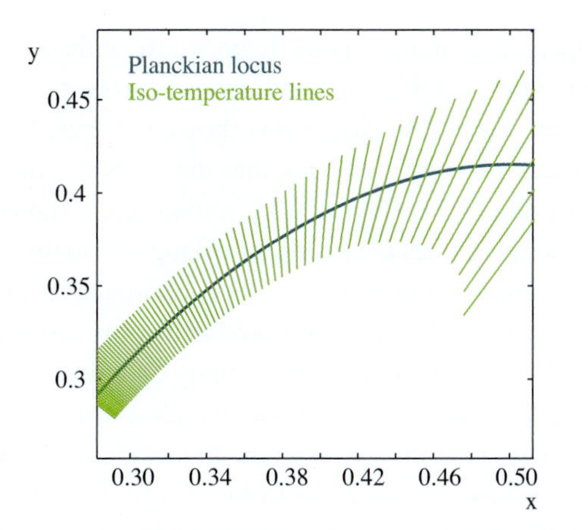

Figure 9.13. A close-up view of iso-temperature lines on the CIE 1931 chromaticity diagram. Note that intersections are not orthogonal as a result of the perceptual non-uniformity of this diagram.

The method of computing the correlated color temperature is illustrated in Figure 9.10. This figure shows the same data as Figure 9.9 except that it is transformed into the CIE $u'v'$ chromaticity diagram. The forward and backward transformations are given by Equations (7.13) and (7.14).

In Figure 9.10, the blue line shows the Planckian locus and the green lines show the so-called *iso-temperature lines*. Iso-temperature lines intersect the

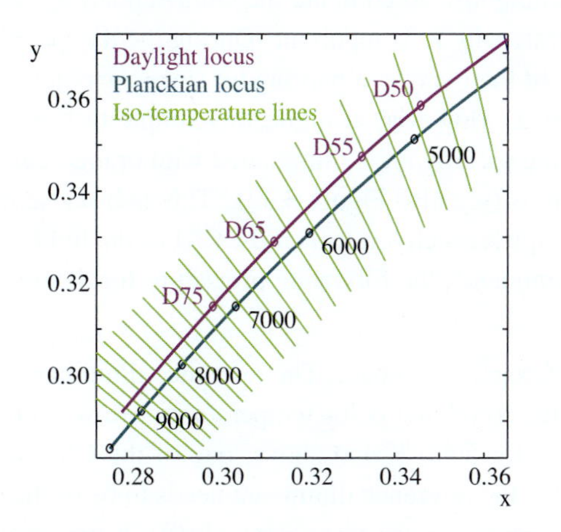

Figure 9.14. The daylight locus together with the Planckian locus sampled at various absolute temperatures. With the aid of iso-temperature lines, one may obtain the correlated color temperatures for daylight illuminants.

Planckian locus perpendicularly, and the temperature at the intersection gives the correlated color temperature for all stimuli whose chromaticity points lie on that iso-temperature line. A close-up diagram is shown in Figure 9.11.

This diagram can be converted back into the more familiar CIE 1931 chromaticity diagram using (7.14) (Figure 9.12; a close-up is shown in Figure 9.13). Note that iso-temperature lines are no longer orthogonal to the Planckian locus as a result of the non-uniform nature of the CIE xy chromaticity diagram.

As an example, consider finding the correlated color temperature of CIE daylight illuminants. In Figure 9.14, the chromaticity coordinates of CIE illuminants D_{50}, D_{55}, D_{65}, and D_{75} are shown on a line, which is also called the *daylight locus*. Since these coordinates do not lie on the Planckian locus, we can only speak of correlated color temperature for these illuminants. Correlated color temperatures may be found by tracing the iso-temperature lines to their intersection point with the Planckian locus.

9.3 Color-Rendering Index

Together with their reflectance properties, one of the key parameters that determines how objects appear to us is the color of the illuminating light source. The *color-rendering index* (CRI) of a light source thus is intended to quantify the effect the light source has on the appearance of objects in comparison with their appearance under a reference illuminant [1262].

Before explaining how to compute the color-rendering index, a practical example may illustrate why it is important. During the day people normally do not struggle to pick out their cars in a parking lot since the color of each car can easily be distinguished. However, at night, this simple task becomes significantly more difficult if the parking lot is illuminated with orange-colored lamps such as sodium vapor lamps (see also Figure 3.32). This is because the color-rendering index of these lamps is much lower than the CRI of daylight.

The CIE recommends the following procedure for determining the CRI of a test light source:

1. Select a reference illuminant. The reference illuminant appropriate for test sources with correlated color temperatures below 5000 K is a Planckian radiator and for 5000 K and above, one of the CIE D-series illuminants. Additionally, the reference illuminant needs to be of the same or nearly the same chromaticity as the test source [190]. A practical threshold for the chromaticity difference is $\Delta C_T = 5.4\,10^{-3}$, where ΔC is computed as

$$\Delta C = \left[(u_k - u_r)^2 + (v_k - v_r)^2 \right]^{0.5}. \tag{9.11}$$

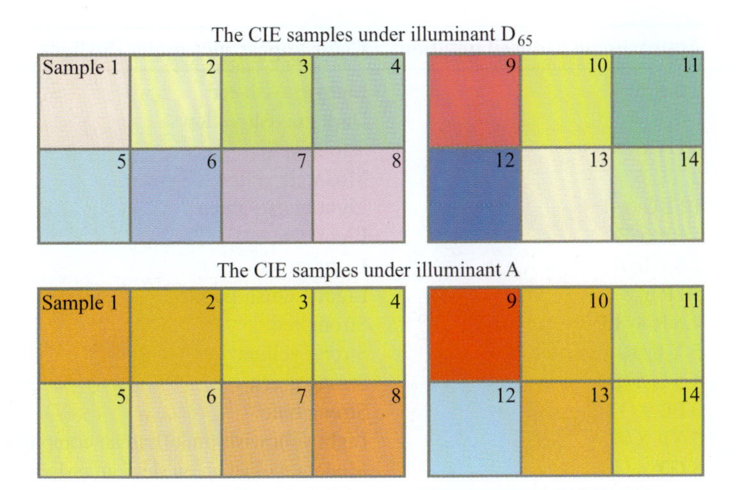

Figure 9.15. The appearance of the CIE samples under two different lighting conditions. These samples are specified by their radiance factors and are used to compute the color-rendering index of light sources.

Here (u_k, v_k) and (u_r, v_r) are the chromaticity coordinates of the test and the reference source, with respect to CIE 1960 UCS.

2. Obtain the spectral-radiance factors of standard samples provided by the CIE for computation of the color-rendering index. These samples are intended to span a large range of colors commonly encountered in practice. The CIE defines 14 samples; the first eight samples are moderately saturated, similar in lightness, and cover the hue circle [190]. In contrast, the last six samples vary widely in saturation and lightness. The appearance of these samples under daylight and incandescent light is illustrated in Figure 9.15. Their Munsell notations are given in Table 9.7. The full spectral data are provided on the accompanying DVD-ROM.

   As will be further detailed in the following discussion, the CRI obtained when any one of these samples is used, is called the CIE *special color-rendering index*. On the other hand, the mean CRI of the first eight samples gives the CIE *general color-rendering index.*

3. Determine the CIE 1931 XYZ tristimulus values and xy chromaticity coordinates of each sample under the test source and the reference illuminant, yielding $(X_{k,i}Y_{k,i}Z_{k,i})$ and $(X_{r,i}X_{r,i}X_{r,i})$ for the ith sample.

Sample No.	Approximate Munsell notation	Color appearance under daylight
1	7.5 R 6/4	Light grayish-red
2	5 Y 6/4	Dark grayish-yellow
3	5 GY 6/8	Strong yellow-green
4	2.5 G 6/6	Moderate yellowish-green
5	10 BG 6/4	Light bluish-green
6	5 PB 6/8	Light blue
7	2.5 P 6/8	Light violet
8	10 P 6/8	Light reddish-purple
9	4.5 R 4/13	Strong red
10	5 Y 8/10	Strong yellow
11	4.5 G 5/8	Strong green
12	3 PB 3/11	Strong blue
13	5 YR 8/4	Light yellowish-pink (human complexion)
14	5 GY 4/4	Moderate olive green (leaf green)

Table 9.7. Test color samples used for computation of the CIE color-rendering index of light sources.

4. Transform the tristimulus values into the CIE 1960 UCS *uv* chromaticity coordinates. This can be accomplished by either of the following equations:

$$u = \frac{4X}{X + 15Y + 3Z} \qquad v = \frac{6Y}{X + 15Y + 3Z}; \tag{9.12}$$

$$u = \frac{4x}{-2x + 12y + 3} \qquad v = \frac{6y}{-2x + 12y + 3}. \tag{9.13}$$

5. Account for the chromatic-adaptation difference (see Chapter 10) that would occur between the test source and the reference illuminant. This results in the following adapted-chromaticity coordinates:

$$u'_{k,i} = \frac{10.872 + 0.404 \frac{c_r}{c_k} c_{k,i} - 4 \frac{d_r}{d_k} d_{k,i}}{16.518 + 1.481 \frac{c_r}{c_k} c_{k,i} - \frac{d_r}{d_k} d_{k,i}}, \tag{9.14}$$

$$v'_{k,i} = \frac{5.520}{16.518 + 1.481 \frac{c_r}{c_k} c_{k,i} - \frac{d_r}{d_k} d_{k,i}}, \tag{9.15}$$

where subscripts k and r identify terms associated with the test source and the reference illuminant, respectively, and i denotes the sample number.

The terms c and d are computed by

$$c = \frac{1}{v}(4 - u - 10v), \tag{9.16}$$

$$d = \frac{1}{v}(1.708v + 0.404 - 1.481u). \tag{9.17}$$

6. Transform the adaptation-corrected $u'_{k,i}, v'_{k,i}$ values into the CIE 1964 uniform color space.[2] This is accomplished by

$$W^*_{r,i} = 25(Y_{r,i})^{1/3} - 17 \qquad W^*_{k,i} = 25(Y_{k,i})^{1/3} - 17, \tag{9.18}$$

$$U^*_{r,i} = 13W^*_{r,i}(u_{r,i} - u_r) \qquad U^*_{k,i} = 13W^*_{k,i}(u'_{k,i} - u'_k), \tag{9.19}$$

$$V^*_{r,i} = 13W^*_{r,i}(v_{r,i} - v_r) \qquad V^*_{k,i} = 13W^*_{k,i}(v'_{k,i} - v'_k). \tag{9.20}$$

Here $Y_{r,i}$ and $Y_{k,i}$ must be normalized such that $Y_r = Y_k = 100$.

7. Finally, the resultant color shift is computed by the CIE 1964 color difference formula:

$$\Delta E_i = \sqrt{(U^*_{r,i} - U^*_{k,i})^2 + (V^*_{r,i} - V^*_{k,i})^2 + (W^*_{r,i} - W^*_{k,i})^2} \tag{9.21}$$

$$= \sqrt{(\Delta U^*_i)^2 + (\Delta V^*_i)^2 + (\Delta W^*_i)^2}. \tag{9.22}$$

ΔE_i denotes the color shift that occurs for the ith sample when the test source is replaced by the reference illuminant. Once ΔE_i is obtained the CIE special color-rendering index, R_i is computed by

$$R_i = 100 - 4.6(\Delta E)_i. \tag{9.23}$$

This indicates that the maximum special CRI equals 100, and this condition only occurs if the color difference is zero for that particular sample when the test source is replaced by the reference illuminant. The constant 4.6 ensures that a standard warm white fluorescent lamp attains a CRI of 50 when compared against an incandescent reference [190].

8. The CIE general color-rendering index is computed as an average of the first eight special color-rendering indices:

$$R_a = \frac{1}{8}\sum_{i=1}^{8} R_i. \tag{9.24}$$

[2] Although the CIE 1964 uniform color space and color-difference formula has been replaced by the CIE 1976 uniform color space and color-difference formula, they have been retained for the time being for computing color-rendering indices [190].

The color-rendering index is an important metric especially in the lighting industry. Usually lamps with high color-rendering indices are preferred to those with low color-rendering indices.

The CIE procedure used to determine the color-rendering index has been subjected to various criticisms [1142]. One of the objections to the CIE method is that it uses an obsolete color space and an obsolete color-difference formula. Although improvements have been made in both of these areas, the recommendation has not been updated. Furthermore, the improvements in the study of chromatic-adaptation transforms are also not reflected in the recommended procedure.

In addition, it has been argued that even if its out-of-date components are replaced with the more improved versions, the procedure would still be inadequate [1142]. In this respect, the main criticism is directed at the possibility to choose more than one reference source and the insufficiency of the test samples [1142].

9.4 CIE Metamerism Index

If two object-color stimuli of different relative spectral radiant power distributions yield the same tristimulus values according to a given colorimetric observer and under a reference illuminant, these stimuli are said to be metameric to each other, or simply called *metamers*.

Denoting the spectral reflectance of objects as $\beta_1(\lambda)$ and $\beta_2(\lambda)$ and the spectral concentration of the illuminant as $S(\lambda)$, the necessary conditions for metamerism are

$$\int_\lambda \beta_1(\lambda)S(\lambda)\bar{r}(\lambda) = \int_\lambda \beta_2(\lambda)S(\lambda)\bar{r}(\lambda), \tag{9.25}$$

$$\int_\lambda \beta_1(\lambda)S(\lambda)\bar{g}(\lambda) = \int_\lambda \beta_2(\lambda)S(\lambda)\bar{g}(\lambda), \tag{9.26}$$

$$\int_\lambda \beta_1(\lambda)S(\lambda)\bar{b}(\lambda) = \int_\lambda \beta_2(\lambda)S(\lambda)\bar{b}(\lambda), \tag{9.27}$$

where $\bar{r}(\lambda)$, $\bar{g}(\lambda)$, and $\bar{b}(\lambda)$ are color-matching functions.

Metamers are perceived to be of the same color under a reference illuminant, $S(\lambda)$. However, these equations indicate that if either the illuminant or the color-matching functions are changed, these equalities may no longer hold; i.e., pairs of colors may no longer be metamers. Also, it is intuitively clear that the more different $\beta_1(\lambda)$ is from $\beta_2(\lambda)$, the more likely that this equality is distorted, should either the source or the observer change [1262]. The degree of this mismatch, as defined by the CIE when either the illuminant or the observer is changed, is called the *CIE metamerism index*.

Since the metamerism index indicates how much the perceived color of two samples will change when the lighting is changed, it is of particular importance for the paint and the textile industries. For instance, it is important to ensure that all parts of a car exterior look the same regardless of changes in illumination.

The CIE defines two types of metamerism index. The first index is used to quantify the degree of mismatch when the illuminant is changed for a constant observer. This is called the CIE *special metamerism index: change in illuminant* . The second index is used when the illuminant is fixed and the observer is allowed to change. This type of index is called the CIE *special metamerism index: change in observer* [188].

The CIE recommends the following procedure to compute the CIE special metamerism index of two samples:

1. Compute the tristimulus values $X_{m,i}, Y_{m,i}, Z_{m,i}$, $(i = 1, 2)$ and $(m = r)$, of the two samples under the reference illuminant (the preferred reference illuminant is D_{65}):

$$X_{m,i} = k \sum_{\lambda} \beta_i(\lambda) S_m(\lambda) \bar{x}(\lambda) \Delta\lambda, \tag{9.28}$$

$$Y_{m,i} = k \sum_{\lambda} \beta_i(\lambda) S_m(\lambda) \bar{y}(\lambda) \Delta\lambda, \tag{9.29}$$

$$Z_{m,i} = k \sum_{\lambda} \beta_i(\lambda) S_m(\lambda) \bar{z}(\lambda) \Delta\lambda, \tag{9.30}$$

   where $S_{m=r}(\lambda)$ is the spectral power distribution (SPD) of the reference illuminant, the β_i are the spectral reflectance of the samples, and $\bar{x}(\lambda), \bar{y}(\lambda)$, and $\bar{z}(\lambda)$ are the color-matching functions (CIE 1931 or 1964); $\Delta\lambda$ is the sampling resolution of the spectrum. The normalization factor k is calculated as

$$k = \frac{100}{\sum_{\lambda} S_m(\lambda) \bar{y}(\lambda) \Delta\lambda}. \tag{9.31}$$

   Normalization ensures that the Y-component of a perfect reflecting diffuser is equal to 100. Since the samples are metameric under the reference illuminant, their tristimulus values must be equal:

$$X_{m=r,1} = X_{m=r,2}, \tag{9.32}$$

$$Y_{m=r,1} = Y_{m=r,2}, \tag{9.33}$$

$$Z_{m=r,1} = Z_{m=r,2}. \tag{9.34}$$

2. Compute the tristimulus values $X_{m,i}, Y_{m,i}, Z_{m,i}$, $(i = 1,2)$ and $(m = t)$, of the two samples under the test illuminant using the procedure in the previous step. This yields two different sets of tristimulus values, namely one for each sample. Here, the choice of test illuminant depends on the target application. For instance, if the metamerism needs to be sustained under typical fluorescent lighting the CIE F_2 illuminant may be chosen [136, 137, 1262].

3. The CIE metamerism index M_{ilm} is equal to the CIELAB color difference ΔE^*_{ab} between the tristimulus values $X_{m=t,1}$, $Y_{m=t,1}$, $Z_{m=t,1}$ and $X_{m=t,2}$, $Y_{m=t,2}$, $Z_{m=t,2}$ (see Section 8.7.1):

$$M_{\text{ilm}} = \Delta E^*_{ab}. \tag{9.35}$$

   If a different color-difference formula, including the ones introduced in Section 8.8, is used this should be noted [194].

The CIE metamerism index for a change in observer can be computed similarly. The sole difference is that the illuminant is kept constant while the observer (i.e., its associated color-matching functions) is changed.

The CIE metamerism index is particularly important for the paint and textile industries where all products of a given type should match a standard. In this case, a lower degree of metamerism is desired [558].

9.5 Dominant Wavelength

The CIE tristimulus values of a color stimulus, or its related chromaticity coordinates, provide numerical information that can be used in colorimetric computations. However, they provide little information to the inexperienced with regard to what that color stimulus actually looks like. It is possible to infer that a stimulus with tristimulus values $X > Y > Z$ may look more reddish than bluish; it requires significant experience to determine what the actual color is.

Measures of *dominant wavelength* and *purity* are used in this context to describe colors in a more intuitive manner. To understand the appearance of a color stimulus from these two concepts, one should still know the progression of colors from the short to the long wavelengths in the light spectrum.

The *dominant wavelength* of a color stimulus is the monochromatic wavelength of the spectrum such that when light of this wavelength is mixed with an achromatic stimulus, the mixture matches the original stimulus in color. It is customary to use either CIE standard illuminant D_{65} or A as the achromatic stimulus.[3]

The dominant wavelength is denoted by λ_d and can be computed with the aid of a CIE chromaticity diagram. Recall, that in this diagram, all colors with chromaticity coordinates lying on a straight line can be produced by mixing, in appropriate proportions, the two colors with chromaticity coordinates defining the end points of this line. Therefore, to find the dominant wavelength of a color stimulus, we can draw a straight line starting from the chromaticity coordinate (x_w, y_w) of an achromatic stimulus (e.g, D_{65}) and passing through the chromaticity coordinate (x, y) of the stimulus in question. If we continue extending this line until it intersects with the spectrum locus, the wavelength corresponding to this intersection point gives the dominant wavelength of the color stimulus. This process is illustrated in Figure 9.16.

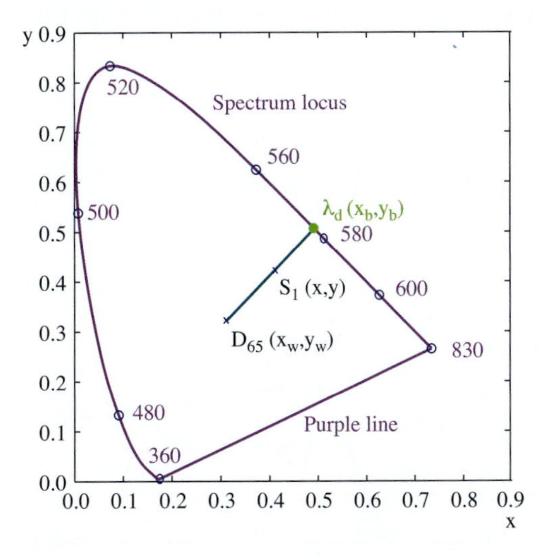

Figure 9.16. The dominant wavelength λ_d of stimulus S_1 is computed by finding the intersection with the spectrum locus of the straight line drawn from the chromaticity point of an achromatic stimulus (D_{65}) toward S_1.

[3]Although illuminant A has a yellowish hue, it still qualifies as achromatic under some circumstances due to the chromatic-adaptation mechanism of the human visual system [558].

Unfortunately, the spectrum locus is not represented with an analytical equation, so numerical or graphic solutions are required. One alternative is to first graphically determine the approximate position of the dominant wavelength by using the method described above. For instance, from Figure 9.16, we can see that the dominant wavelength of S_1 is between 560 nm and 580 nm. Then we can test the chromaticity coordinates of all monochromatic wavelengths in this interval such that the slope of the line drawn from S_1 to these points is closest to the slope of the line drawn from D_{65} to S_1. This procedure can be stated as an optimization problem, where the goal is to minimize the difference in these two slopes. The wavelength of the monochromatic stimulus that minimizes the slope then represents the dominant wavelength.

The dominant wavelength is useful to describe the approximate hue of a color stimulus. For instance, if one knows that the light of wavelengths between roughly 570 nm and 650 nm has an orange hue, a light source with its dominant wavelength falling in this interval will also tend to have an orange hue. The dominant wavelength is commonly used by LED manufacturers to specify the characteristics of light sources [1094].

Not all color stimuli have a dominant wavelength. For some stimuli, the straight line may intersect with the purple line rather than the spectrum locus;

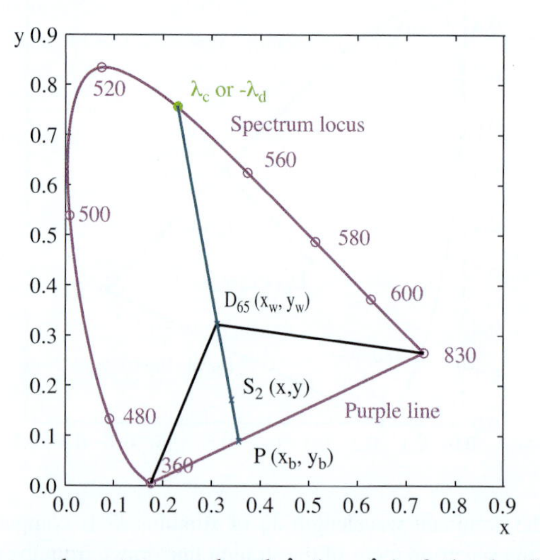

Figure 9.17. The complementary wavelength λ_c (or $-\lambda_d$) of stimulus S_2 is computed by finding the intersection with the spectrum locus of the straight line drawn from S_2 toward the achromatic stimulus (D_{65}). The intersection of this line with the purple line is shown by P and used for computing excitation or colorimetric purity. All data is shown on the CIE 1931 chromaticity diagram.

an example is shown in Figure 9.17. It is evident from the figure that this happens when the chromaticity coordinates of a stimulus lie in the triangle whose apex is the achromatic stimulus.

For such stimuli, we speak of *complementary wavelength* rather than dominant wavelength. The complementary wavelength λ_c (or $-\lambda_d$) indicates the monochromatic wavelength such that when the stimulus is mixed with a color of this wavelength, the mixture matches a reference achromatic stimulus in color. The value λ_c can be computed similarly to λ_d.

9.6 Excitation Purity

Although the dominant and complementary wavelengths give cues about the hue of a color stimulus, they do not convey information about how saturated a color stimulus is. That is, all color stimuli that lie on the same line as S_1 (see Figure 9.16) will have the same dominant wavelength, but they may differ significantly in saturation. For instance, we regard the stimuli close to the chromaticity coordinate of the achromatic stimulus as being desaturated. On the other hand, the stimuli close to the spectrum locus is considered be highly saturated.

Thus the saturation of a color stimulus is expressed by a different term called the *excitation purity*. It indicates how far a stimulus is from an achromatic stimulus on the CIE chromaticity diagram. The excitation purity is defined as a ratio of distances:

$$p_e = \begin{cases} \dfrac{x - x_w}{x_b - x_w} & \text{if } x - x_w \geq y - y_w, \\ \dfrac{y - y_w}{y_b - y_w} & \text{if } x - x_w < y - y_w, \end{cases} \tag{9.36}$$

Note that of the two forms on the right-hand side, the one with the largest numerator is chosen, as it results in the smallest precision error [194]. The excitation purity is only a loose measure of the *saturation* of a color stimulus [1262]. For more accurate estimations, more elaborate models must be used (see Chapter 11).

9.7 Colorimetric Purity

A related concept to excitation purity is the *colorimetric purity*, denoted by p_c. The colorimetric purity of a color stimulus expresses how much white light (i.e., achromatic light), with a wavelength equal to the dominant wavelength of the stimulus considered, must be added to the monochromatic stimulus so that their mixture matches the color of the stimulus in question. As such, the colorimetric purity p_c can be computed by

$$p_c = \frac{L_d}{L_d + L_n}, \tag{9.37}$$

where L_d and L_n represent the luminances of the monochromatic and achromatic stimuli, respectively. By using Grassmann's laws this equation can be rewritten as

$$p_c = \frac{L_d}{L_s}, \tag{9.38}$$

where L_s is the luminance of the stimulus for which the purity is computed, and clearly $L_s = L_d + L_n$. Colorimetric purity may also be derived from the excitation purity, and vice versa:

$$p_c = \begin{cases} \dfrac{y_b}{y} \dfrac{x - x_w}{x_b - x_w} & \text{if } x - x_w \geq y - y_w, \\ \dfrac{y_b}{y} \dfrac{y - y_w}{y_b - y_w} & \text{if } x - x_w < y - y_w, \end{cases} \tag{9.39}$$

or equivalently,

$$p_c = \frac{y_b}{y} p_e. \tag{9.40}$$

The coordinates (x,y), (x_b, y_b), and (x_w, y_w) have the same meanings as those used for excitation purity with one crucial exception. For stimuli that possess a complementary wavelength, (x_b, y_b) represents the chromaticity coordinates of the complementary wavelength λ_c, rather than the chromaticity coordinates on the purple line (see Figure 9.17).

In all, it is important to remember that the quantities discussed in this section are defined on a non-uniform CIE chromaticity diagram. This fact in general, may limit their use and should be kept in mind whenever a color is expressed by using these quantities [509].

9.8 Application: Modeling Light-Emitting Diodes

The technology for the construction of light-emitting diodes (LEDs) is rapidly advancing to the point that they have become useful in various novel applications, including car lights, illumination, and as backlights in LCD displays (see Section 14.2.1).

Important characteristics of LEDs are their luminous efficacy of radiation and their color-rendering index. In practical LED designs, there is a trade-off between these two characteristics: higher luminous efficacies can be achieved at the cost of

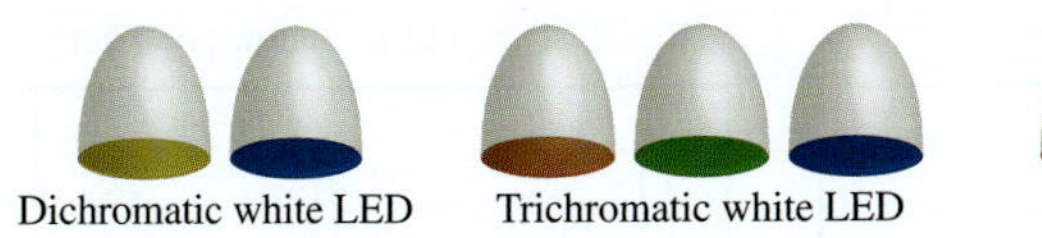

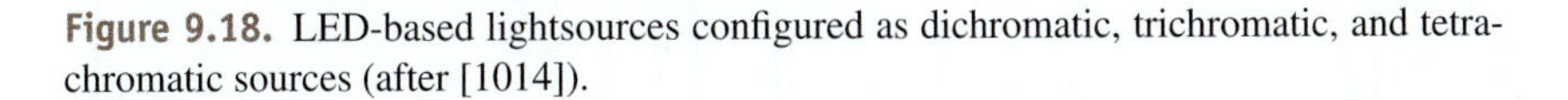

Figure 9.18. LED-based lightsources configured as dichromatic, trichromatic, and tetrachromatic sources (after [1014]).

a lower color-rendering index, and vice-versa [1014]. Nonetheless, the luminous efficacy follows Haitz's law, which states that the luminous output of LEDs is doubling every 18–24 months [1083][4].

White LEDs can be fabricated in several different ways. A main distinction between different types of LEDs is whether they contain phosphors or not, i.e., LEDs or phosphor-converted LEDS (pc-LEDs). Each of these types can be further classified into dichromatic, trichromatic, and tetrachromatic sources, as illustrated in Figures 9.18 and 9.19. The spectral power distribution functions for some of these types are shown in Figure 9.20.

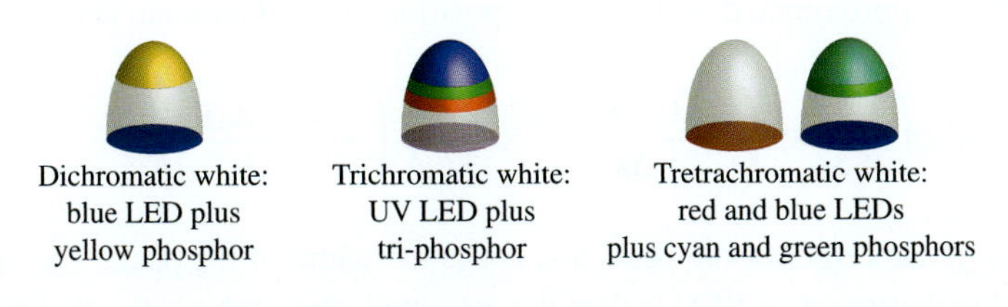

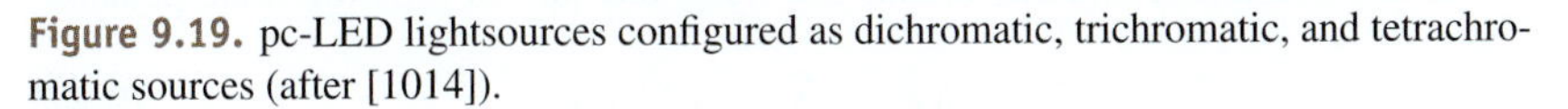

Figure 9.19. pc-LED lightsources configured as dichromatic, trichromatic, and tetrachromatic sources (after [1014]).

Dichromatic LEDs tend to have good luminous efficacy, whereas tetrachromatic LEDs have a better color-rendering index. Trichromatic LEDs are in between, with a good luminous efficacy as well as a good color-rendering index. In pc-LEDs, the conversion from short-wavelength photons to long-wavelength photons involves an energy loss, reducing their efficiency by between 10% and 30% [1014]. As tetrachromatic LEDs and pc-LEDs have four emission peaks, spread over the visible wavelength, their emission spectra are most uniform, leading to the highest color-rendering index of the types discussed here.

The spectral power distribution of an LED has two, three, or four peaks, associated with the specific semiconductors or phosphors used. The shape of each

[4]This law is analogous to Moore's law, which predicts that the number of transistors on chips doubles every 18–24 months.

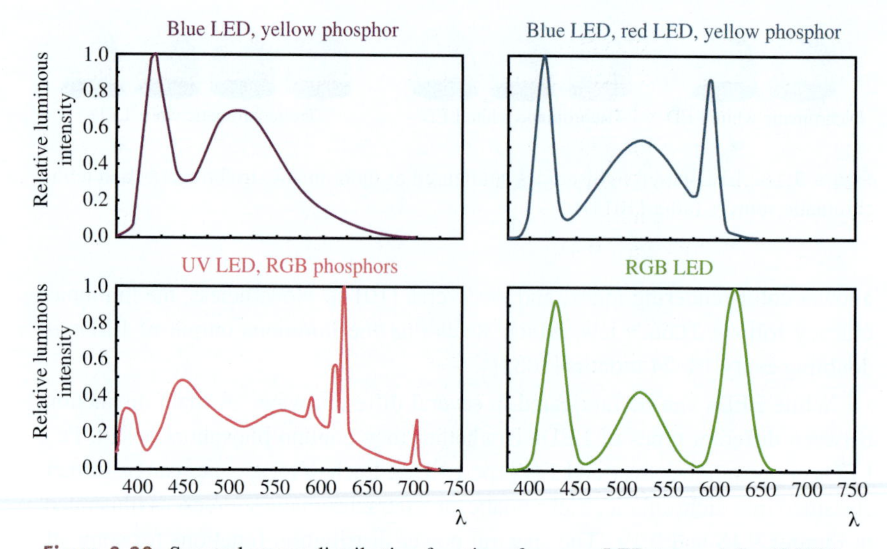

Figure 9.20. Spectral power distribution functions for some LED types. (after [1059]).

peak can be approximated with the superposition of two Gaussian curves [848]:

$$S_i(\lambda) = \frac{1}{3}\exp\left(-\frac{(\lambda-\lambda_0)^2}{\Delta\lambda}\right) + \frac{2}{3}\exp\left(-\frac{(\lambda-\lambda_0)^2}{\Delta\lambda}\right)^5, \tag{9.41}$$

where λ_0 is the center of the peak, and $\Delta\lambda$ is the width of the peak. The spectral power distribution of an LED is then the weighted sum of the individual peaks:

$$S(\lambda) = \sum_i w_i\, S_i(\lambda), \tag{9.42}$$

where the weights w_i adjust the relative magnitude of the peaks. The widths of the peaks, given by $\Delta\lambda$, are between approximately 20 and 30 nm. Typical peak wavelengths for a tetrachromatic LED are $\lambda_1 = 450$ nm, $\lambda_2 = 510$ nm, $\lambda_3 = 560$ nm, and $\lambda_4 = 620$ nm [1014].

9.9 Application: Estimating the Illuminant in an Image

Given an image, it is often desirable to know the spectral composition of the light source in the scene [1133]. For instance, object-segmentation algorithms can be based on color differences. If the nature of the illuminant is known, then such segmentation algorithms can be constructed in a more robust manner, as one can account for changes due to the illuminant as well as to highlights [56, 57, 132, 442, 595, 1236]. In addition, digital cameras typically employ corrective algorithms that make use of an estimate of the scene illuminant [153, 495, 1034, 1256]. A further application lies in image database retrieval, where knowledge of the illuminant will allow a better match between a source image and the collection of images in the database [266, 338, 1003, 1110].

To estimate the color of the illuminant on the basis of a single image, it is often convenient to make additional assumptions on the nature of the illuminants' spectra that are to be recovered. Given that the number of different illuminants encountered in average environments is limited to daylight spectra, as well as different types of incandescent and fluorescent artificial lights, this is not unreasonable. In fact, it allows the problem to be simplified from estimating the illuminant to classifying the illuminant to be one of a small set of spectra [1133, 1134].

Many different computational algorithms for separating the color of a surface and the color of the illuminant have been proposed, and they can be classified into different groups [1133], including those based on general linear models [141, 474, 662, 728, 1204], algorithms based on an analysis of diffuse inter-reflection and the appearance of highlights [339, 441, 596, 664], Bayesian methods [119, 994, 1301], multi-view methods [1149, 1298, 1299], and algorithms based on spectral images [1132, 1135, 1137].

The problem of illuminant estimation, or classification, emerges in human vision as well, where it is related to color constancy (Section 5.9). Humans are able to perceive the color of surfaces in a variety of everyday situations. The spectrum of the illuminant can vary without the perception of surface color being greatly affected [120, 123], which requires either implicit or explicit knowledge of the illuminant. Of course there are also viewing conditions where color constancy breaks down [454]. We will assume that under normal viewing conditions, color constancy holds sufficiently well that an estimation of the illuminant is possible.

Of the different computational techniques mentioned above, some are more plausible to be part of human vision than others. In particular, there is some evidence that the human visual system may use specular highlights to infer the color of the illuminant [1271]. As an example of how the color of the dominant illuminant may be estimated, we therefore focus our example on one of the algorithms that is are based on an analysis of highlights.

The nature of highlights visible on an object depends on its material. Many materials have a high oil or water content which causes specular reflections as a result of their constant index of refraction over the visible spectrum [578]. In addition, such materials may exhibit diffuse reflection. This includes, for instance, human skin. As a result, the specular highlight has the same color as the illuminant, whereas off-peak angles show the color of the surface. This behavior is captured in the *dichromatic reflection model*, which essentially models the bidirectional reflectance distribution function (BRDF) as an additive mixture of a specular and diffuse component [596, 1028, 1136].

It is now possible to analyze image areas that are known to contain highlights, for instance the nose in human faces [1098]. The pixels away from the highlight represent the diffuse reflection component; these are the darker pixels. They will lie on an approximately straight line in RGB color space. The lighter pixels will form a separate cluster, lying on a differently oriented straight line. Separating the two clusters can be achieved using a variation of principal components analysis [1098].

Alternatively, one may view each surface point as casting a vote for its illuminant [664]. Using the same dichromatic reflection model, whereby each point on the surface reflects partly its diffuse body color and partly the specular illuminant color, we can plot pairs of surface points in a CIE chromaticity diagram. Each pair forms a line segment which, when extended in the appropriate direction, will include the chromaticity point of the illuminant. In other words, colors from the same surface will have different purities, and they will all tend to lie on the same line. Of course, when a surface does not have a strong highlight, then the change in purity between different points on the surface can be small and, therefore, prone to error.

Repeating this procedure for a set of points drawn from a different surface, we find a second line in the chromaticity diagram. The intersection point with the first line will then yield an estimation of the chromaticity coordinate of the illuminant [664]. If there are more surfaces in a scene, then more lines can be drawn, which theoretically will all pass through the same intersection point. However, noise in the measurements, as well as diffuse inter-reflection between objects, may cause these lines to not pass exactly through the same point, thereby introducing inaccuracies. In that case, each surface in the scene creates a different line which can be seen as a vote for a particular illuminant color. The actual color of the illuminant is then estimated to be identical to the majority vote [664].

If it is known beforehand that the illuminant is a blackbody radiator, then the above algorithm can be simplified to the analysis of a single highlight-containing surface. Its line on the chromaticity diagram will intersect the Planckian locus (Figure 9.9).

An example image, shown in Figure 9.21, is informally analyzed by cropping highlight patches and plotting the pixels of these patches in a chromaticity diagram, shown in Figure 9.22. The analysis is only informal as the camera was not

Figure 9.21. The bull figurine is illuminated from behind by a white light and from the front by an incandescent source, leading to two spatially separate highlights that can be analyzed individually. (Bronze figurine by Loet Vanderveen.)

characterized, and therefore the conversion of pixels to chromaticity coordinates can only be approximate. These figures are only included to illustrate the method.

The highlight created by the white-light source forms a straight line that intersects the Planckian locus near a color temperature of 5500 K. The light source is, in fact, a D65 daylight simulator. The patch taken around the incandescent highlight yields a cluster in the chromaticity diagram on the right of Figure 9.22 that intersects the Planckian locus at a color temperature of under 2000 K.

In summary, the analysis of highlights in an image may be used to estimate the color of the illuminant, assuming that the image contains highlights that can be identified in some manner. There exists some evidence that the human visual system employs a similar strategy. However, several other computational techniques may be used instead of, or in conjunction with, highlight analysis. In any case, all techniques require all parts of the image, including the highlights, to be well exposed. Since the intensity of specular highlights can be arbitrarily high, high dynamic range imaging techniques (as discussed in Part III of this book) can be useful for illuminant estimation algorithms.

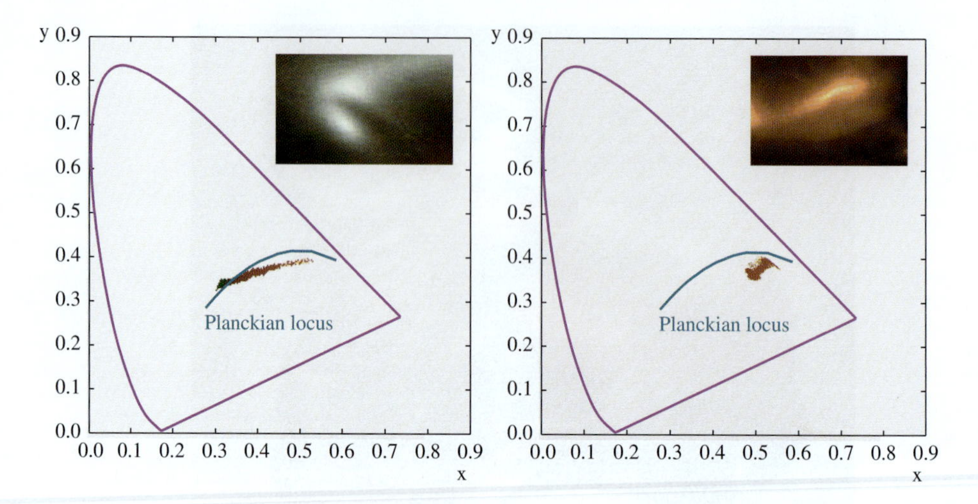

Figure 9.22. The two highlight patches in the insets are taken from Figure 9.21. The chromaticity coordinates of the white highlight form a line that intersects the Planckian locus (left), whereas the incandescent highlight does not (right).

9.10 Further Reading

The IESNA lighting handbook is an invaluable resource for the lighting engineer [941]. Wyszecki and Stiles have also collected a large amount of data [1262]. Many of the CIE publications are relevant to work involving illuminants.

Chapter 10
Chromatic Adaptation

In everyday life we encounter a wide range of viewing environments, and yet our visual experience remains relatively constant. This includes variations in illumination, which on any given day may range from starlight, to candlelight, incandescent and fluorescent lighting, and sunlight. Not only do these very different lighting conditions represent a huge change in overall amounts of light, but they also span a wide variety of colors. Objects viewed under these varying conditions generally maintain an overall constant appearance, e.g., a white piece of paper will generally appear white (or at least will be recognized as white).

The human visual system accommodates these changes in the environment through a process known as *adaptation*. As discussed briefly in Chapter 5, there are three common forms of adaptation: light adaptation, dark adaptation, and color or chromatic adaptation. Light adaptation refers to the changes that occur when we move from a very dark environment to a very bright environment, for instance going from a matinee in a darkened movie theater into the sunlight. When we first move outside, the experience is dazzling and may actually be painful to some. We quickly adapt to the bright environment, and the visual experience returns to normal. Dark adaptation is exactly the opposite, moving from a bright environment to a dark one. At first it is difficult to see anything in the dark room, but eventually we adapt and it is possible to see objects again.

In general the time required for dark adaptation is much longer than that required for light adaptation (see Section 3.4.2). Part of the light- and dark-adapting process can be explained physiologically by the dilation of the pupil. If we consider that the average pupil diameter can range from 2.5 mm in bright light, to 8 mm in darkness, this can account for about a 10:1 reduction in the amount of light striking the retina (considering a circular area of approximately 5 mm^2 to

50 mm^2) [1262] (see also Section 4.2.4). This reduction is not enough to account for the wide variety of dynamic ranges we typically encounter. The remainder of the adaptation occurs at the photoreceptor and retinal level, as well as at the cognitive level.

While changes in overall illumination levels are important for digital imaging systems, they are generally not considered the most important form of adaptation. Typically, we are more concerned with changes in the color of the lighting. As we learned in Section 2.11.2, the overall color of everyday illumination can vary widely, from orange sunlight at dawn or twilight, to the blue Northern sky. Despite these varieties of illumination, the colors of objects do not vary nearly as much. The human visual system is constantly evaluating the illumination environment and adapting its behavior accordingly. We can think of this process as the biological equivalent of the automatic white balancing available in today's digital cameras and camcorders (much as we can think of the automatic exposure adjustment akin to our light and dark adaptation).

This chapter details the process of chromatic adaptation, from examples and experimental data to computational models that predict human perception. A model for predicting corresponding colors, the stimuli necessary to generate a chromatic response under one lighting condition that is identical to that under a different lighting condition, is presented. These corresponding colors form an important component of today's color management systems. This chapter also outlines how we can use the computational models of chromatic adaptation in a digital-imaging system to form a basis for an automatic white-balancing algorithm.

10.1 Changes in Illumination

The wide range of viewing conditions that we may encounter in everyday life is staggering, and yet the diversity of conditions typically goes unnoticed. This is, in part, because our visual system is very adept at *normalizing* the viewing conditions and presenting an ultimate visual experience that is fairly consistent. Due to this normalization behavior, we often see the visual system described as being *color constant*, suggesting that object colors appear the same regardless of their illuminating conditions.

While the human visual system does provide some form of constancy to our everyday experiences, we are by no means color constant. Consider looking outside on a bright sunny day compared to looking outside on a cloudy overcast day. On the sunny day, all of the colors appear considerably more colorful, while they appear less chromatic on the cloudy day. Although the appearances have changed, we do not assume that all of the objects themselves have actually changed physical properties. Rather we recognize that the lighting has influenced the overall color appearance.

This change in color appearance has been carefully studied and quantified and is described in detail in Chapter 11. Artists also often have a keen eye to these overall changes in appearance caused by illumination. Claude Monet, in particular, was very fond of painting the same scene under a wide variety of illumination. His *Haystack* series and Rouen Cathedral paintings of the late 19th century are excellent examples. Although the actual contents of his paintings did not vary, he was able to capture an incredible variety of colors by painting at a variety of times of day and seasons. Interestingly enough, the actual appearance of these scenes may actually have looked more similar to him while he was painting through his own chromatic adaptation.

Although we are not fully color constant, the general concept of constancy does apply to chromatic content. A powerful example of chromatic adaptation is that most object colors do not significantly change appearance when viewed under different lighting conditions. This is perhaps best illustrated by thinking of reading a magazine or newspaper in different areas of your office or home. Under these conditions, you may have incandescent (tungsten) illumination, fluorescent lighting, and natural daylight through a window. Perhaps, you may also have candlelight or a fire providing some illumination. Throughout these different conditions, the appearance of the magazine paper will still be white. This is especially impressive considering the *color* of the illumination has gone from very red for the fire, toward yellow for the tungsten, and blue for the daylight. Our visual system is continuously adapting to these changes, allowing the paper to maintain its white appearance.

Another example of the vast changes in illumination can be seen in Figure 10.1, where we have the same scene photographed under very different conditions. The photographs in this figure were RAW digital camera images (i.e., minimally processed sensor data) with no automatic white balancing, so they provided an excellent approximation of the spectral radiances present in the scene. We can see that at sunset the color of the illumination was predominantly orange, while in the morning the color predominantly very blue.

Another interesting example is illustrated in Figure 10.2, where the wall of a building is illuminated by direct sunlight, with a bright blue sky behind it. In the photograph the wall looks very yellow, a perception that is enhanced by the simultaneous contrast of the blue sky behind it. In reality, the wall is in fact white, and the coloration is due entirely to the illumination. To an observer looking at

Figure 10.1. An example of the same scene viewed under two very different illuminating conditions.

Figure 10.2. A "white" building illuminated by direct sunlight.

the building it would be obvious that the building was indeed white, though the appearance of the wall would retain a slight yellow tint. In other words, if the person were asked to choose a paint to coat another wall such that it matched the building, they would easily select a white paint, but if they were asked to artistically reproduce the appearance of the building, they would select a yellow paint. Similarly to Monet and the Rouen Cathedral, they would recognize that although the building itself was white, the overall visual experience was actually yellow.

We have seen how the same scene can contain a huge variation in the color of lighting depending on the time of day it is viewed. Weather itself can also play

Figure 10.3. A building photographed under two very different weather conditions. (Photo courtesy of Ron Brinkmann.)

a large role in changing our everyday viewing conditions. Figure 10.3 shows the same building photographed on a relatively clear day, as well as on a foggy day. The building itself appears much more blue when illuminated on the clear day than it does on the foggy day. Despite these large changes in overall illumination, we do not think that the building itself is actually changing color. Chromatic adaptation accounts for much of the visual system's *renormalization*, allowing us to maintain a relatively large degree of color constancy.

Essentially, chromatic adaptation allows for a large number of lighting conditions to appear "white." Perhaps a better explanation is to say that a white paper is still recognized as being white when illuminated under lights of a wide range of chromaticities. For example, in Figure 10.4 we see a variety of real-world lights plotted in a CIE $u'v'$ chromaticity diagram. Also plotted in this diagram are the CIE Daylight coordinates. The colors of the points plotted in Figure 10.4 are the relative chromaticities when viewed in the sRGB color space with a *white point* of CIE D65. All of these light sources appear white to a person viewing them if they are viewed in isolation, due to chromatic adaptation. This is also why it is of critical importance to not use chromaticity coordinates as a method for describing color appearance! The color appearance of all of the light sources plotted in Figure 10.4 would most likely all be described the same, despite their markedly different chromaticities. Further, this is why it is important to not show

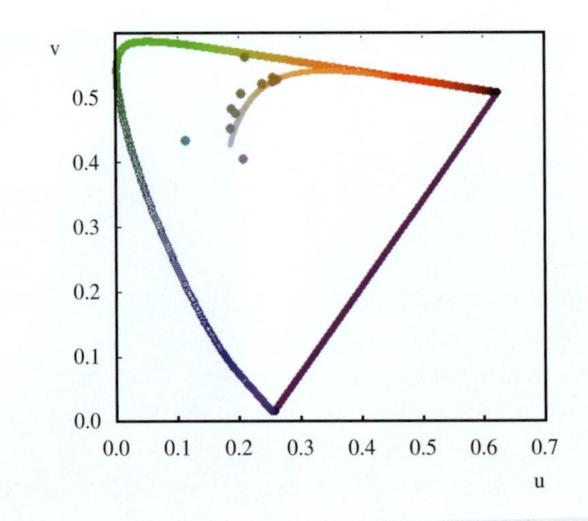

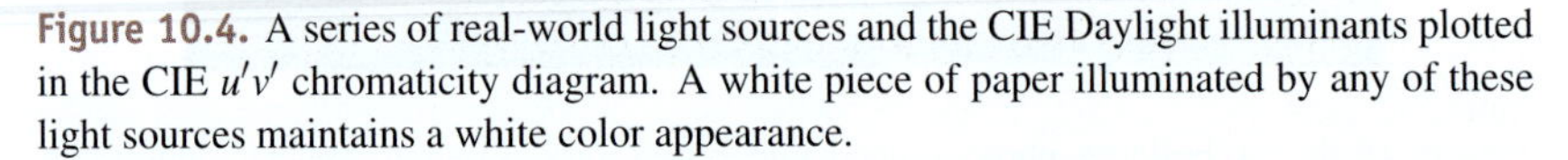

Figure 10.4. A series of real-world light sources and the CIE Daylight illuminants plotted in the CIE $u'v'$ chromaticity diagram. A white piece of paper illuminated by any of these light sources maintains a white color appearance.

chromaticity diagrams in color, since chromatic adaptation itself will change the appearance of any given point in a chromaticity space.

10.2 Measuring Chromatic Adaptation

Chromatic adaptation can be considered a form of normalization that occurs in the visual system, accounting for changes in the chromatic content of the viewing conditions. Ideally, we would like to create a computational model that can predict these changes for use in color-imaging systems. For example, we may want to know what colors to send to our printer in order to generate a match between an original viewed under one lighting condition and the reproduction that will be viewed under a different condition. Recall from Chapter 7 that colorimetry provides us with a method for determining if two colors match, with strict conditions that the viewing conditions of the two colors are identical. Models of chromatic adaptation can be used to extend colorimetry to predict visual matches when the viewing conditions are under different colored illumination.

In order to develop these models, it is important to measure the actual behavior of the chromatic adaptation of our visual system. Historically, several different types of psychophysical experiments have been used to test chromatic adaptation. Two types have been used with particular success: color matching and magnitude estimation. These experiments often generate what are known as corresponding colors, or stimuli that under one illumination "match" other stimuli viewed under different illumination.

Color-matching experiments are similar in nature to those used to generate the CIE standard observers. Essentially, stimuli are viewed under one condition and then a match is identified (or generated) under a different viewing condition. This can be done either through successive viewing or simultaneously using a haploscopic approach, as explained below.

In successive-viewing experiments, an observer looks at a stimulus under one illuminating condition and then determines a match under a different condition. Time must be taken to adapt to each of the viewing conditions. For example, an observer may view a color patch illuminated under daylight simulators for a period of time and then, in turn, determine the appearance match with a second stimulus illuminated with incandescent lights. Prior to making any match (or creating a match using the method of adjustment), the observer has to adapt to the second viewing condition. Fairchild and Reniff determined that this *time course* of chromatic adaptation should be about 60 seconds to be 90 percent complete [296]. It should be noted that this time course was determined for viewing conditions that only changed in chromatic content and not in absolute luminance levels, and more time may be necessary for changes in adapting luminance. In viewing situations at different luminances and colors, we would generally say that we are no longer determining corresponding colors based upon chromatic adaptation but rather full color-appearance matches. Details on models of color appearance will be found in Chapter 11.

The inherent need to delay the determination of a color match to allow for and to study chromatic adaptation leads some to coin this experimental technique *memory matching*. This need to rely on human memory to determine a color match adds inherent uncertainty to these measurements, when compared to direct color matching such as those found using a bipartite field. Wright suggested using a form of achromatic color matching to alleviate some of these uncertainties [1255]. An achromatic color matching experiment involves determining or generating a stimulus that appears "gray" under multiple viewing conditions. Determining when a stimulus contains no chromatic content should not involve color memory and should provide a direct measurement of the state of chromatic adaptation. Fairchild used this technique to measure chromatic adaptation for soft-copy (e.g., CRT) displays [302].

Another technique that attempts to overcome the uncertainties of memory matching is called *haploscopic* color matching. In these experiments the observer adapts to a different viewing condition in each eye and makes a color match while simultaneously viewing both conditions. These matches are made under the assumption that our state of chromatic adaptation is independent for each eye.

Fairchild provides an excellent overview of some of the historical experiments and devices used for these types of experiments [298].

Color matching is not the only technique that can be used to generate corresponding color data to study chromatic adaptation. Magnitude estimation has also been used with success to generate such data, most noticeable the LUTCHI data measured by Luo et al. [711, 712]. Magnitude estimation has observers generating descriptions of the overall color appearance of stimuli under multiple viewing conditions. Essentially, observers are asked to assign numerical values along color-appearance attributes such as lightness, chroma, and hue. In these types of experiments, it is important to adequately train the observers so that they understand and can assign ratings to each of the attributes. These color-appearance attributes themselves are described later in Chapter 11.

All of these chromatic-adaptation experiments are designed to generate sets of colors (often represented as tristimulus values or chromaticities) that appear to match under different viewing conditions. If there were no chromatic adaptation, we may expect that the tristimulus values would always be the same regardless of the illuminating conditions. Due to the renormalization caused by chromatic adaptation, what we generally see is that the color matches tend to "drift" toward the chromaticity of the adapting conditions themselves. This is another way of saying that our adapting *white point*, or the tristimulus values we consider to be achromatic, itself shifts as a function of the viewing conditions. Computational models that predict the degree of this shift are discussed later in this chapter.

10.3 Mechanisms of Chromatic Adaptation

In an ideal situation the models used to predict the behavior of chromatic adaptation would be based upon actual mechanisms of the human visual system. Alternatively, empirical models can be fit to the measured corresponding color data. These empirical models could be described as a type of *black-box* approach to chromatic adaptation. For a digital-imaging system, the black-box approach may be adequate, though it does not provide any insight into what is actually happening within the visual system with regard to chromatic adaptation. The black-box modeling is also very difficult to modify or fix if there is poor performance. In practice, the computational models we use should probably be considered a hybrid approach, as they can be considered *based* on our understanding of the visual system but also designed to fit the existing experimental data.

While the experimental techniques described in the previous section can provide insight into the ultimate behavior of the human visual system, they do not provide much information with regard to the underlying causes of the behavior. In fact, the full mechanisms of chromatic adaptation in the visual system are not yet understood. Typically the mechanisms are often broken into two categories: physiological and cognitive. Physiology itself may be too broad a description, as that could include the cognitive processes themselves. Perhaps a better description, as used by Fairchild, would be sensory and cognitive mechanisms [302].

Sensory mechanisms refer to those that occur automatically as a function of the physics of the input illumination. An example of this type of mechanism would be the automatic dilation of the pupils as a function of the overall amount of illumination. Other examples include the transition from rods to cones as a function of input luminance as well as gain-control mechanisms on the photoreceptors themselves. Gain control will be discussed in more detail below. While the sensory mechanisms may be automatic, they are not instantaneous. In fact, the 60-second adaptation time necessary for the memory-matching experiments described above is necessary to allow time for the sensory mechanisms [296]. Cognitive mechanisms can be considered higher-level mechanisms where the brain actually alters perception based upon prior experiences and understanding of the viewing conditions.

While pupil dilation and the transition from the rods to the cones plays a large role in our overall light and dark adaptation, these sensory mechanisms do not play a large role in chromatic adaptation. Although some imaging systems are designed for use at low light levels, for most common situations, the rods play a very small role in chromatic adaptation and color appearance. We will focus instead on the role of the cones, specifically, on gain control of the cone photoreceptors. One of the simplest descriptions of and explanation for chromatic adaptation can be thought of as an automatic and independent re-scaling of the cone signals based upon the physical energy of the viewing condition. This theory was first postulated by von Kries in 1902 and still forms the basis of most modern models of chromatic adaptation. In terms of equations, this *von Kries*-type adaptation is shown below:

$$L_a = \alpha L, \tag{10.1a}$$

$$M_a = \beta M, \tag{10.1b}$$

$$S_a = \gamma S, \tag{10.1c}$$

where L, M, and S are the cone signals, and α, β, and γ are the independent gain components that are determined by the viewing condition. While we know today that the hypothesis of von Kries is not entirely true, it is still remarkable how effective such a simple model can be for predicting chromatic adaptation.

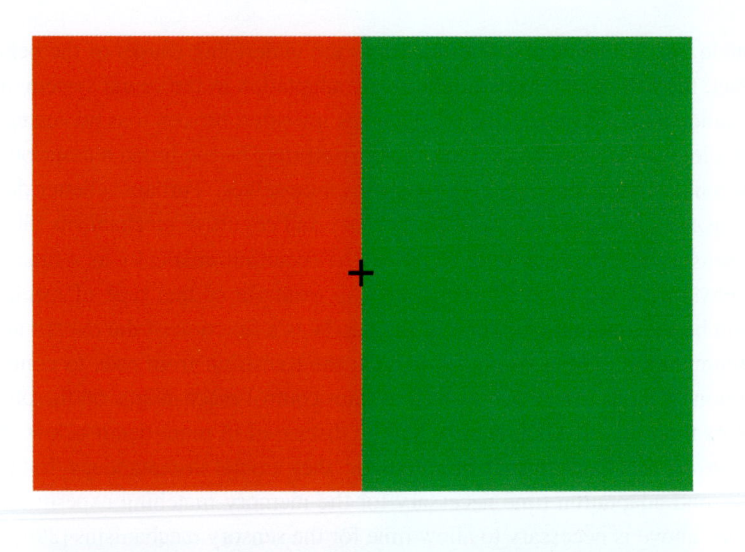

Figure 10.5. An example of sensory chromatic adaptation caused by photoreceptor gain control.

Evidence of the receptor gain control can be seen in the after-images that we encounter in everyday life. An example of this is shown in Figure 10.5. Focusing on the + symbol in the top frame of Figure 10.5 will cause half of the cones on your retina to adapt to the red stimulus while the other half will adapt to the green stimulus (interestingly enough, due to the optics of the eye, the left half of each retina will adapt to the green while the right half will adapt to the red). After focusing on the top frame for about 60 seconds, one should move one's gaze to the bottom frame. The image shown on the bottom should now look approximately uniform, despite the fact that the left side is tinted red and the right side is tinted green. This is because the photoreceptors in the retina have adapted to the red and green fields. Staring at a blank piece of paper should result in an after-image that is opposite to the adapting field (green/cyan on the left and reddish on the right). See also Figure 5.20 for another example.

Another way to think about the von Kries receptor gain control is to consider that the cones themselves become more or less sensitive based upon the physical amount of energy that is present in a scene. For instance, if a scene is predominantly illuminated by "blue" light, then we would expect the S cones to become less sensitive, while the L cones would become (relatively) more sensitive. Conversely, if there is more red energy in the illumination, such as with an incandescent light bulb, then we would expect the short cones to become relatively more sensitive than the long cones.

This example is illustrated in Figure 10.6. The spectral power distribution (SPD) of incandescent illumination (e.g., CIE Illuminant/Source A; see Section 9.1.1) is illustrated by the green curve in Figure 10.6, while the individual cone

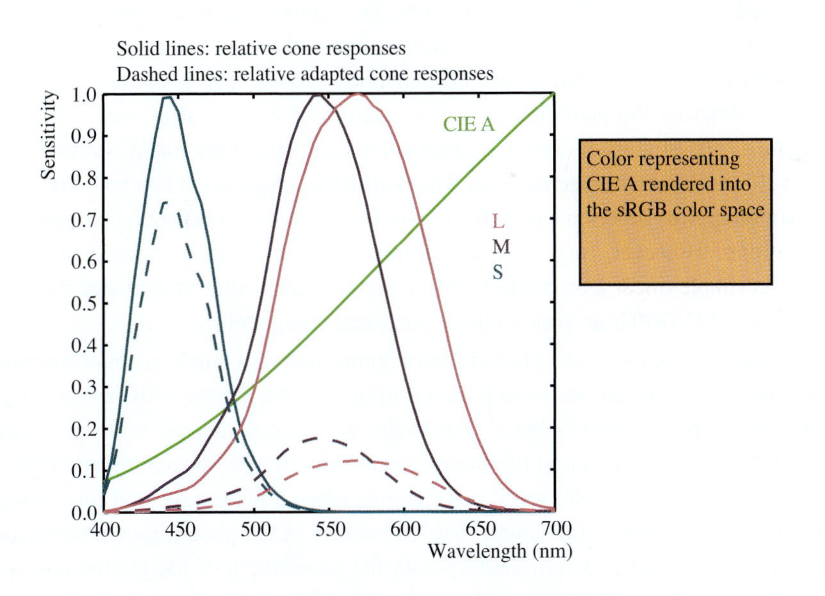

Figure 10.6. An example of von Kries-style independent photoreceptor gain control. The relative cone responses (solid line) and the relative adapted cone responses to CIE A (dashed) are shown. The background color represents CIE A rendered into the sRGB color space.

functions are shown with solid and dashed lines. The relative cone responses prior to chromatic adaptation are shown as solid lines, while the adapted cones are represented by the dashed lines. The patch of color in Figure 10.6 represents the spectral power distribution of CIE A approximately rendered into the sRGB color space with a white point of CIE D65; that is, the *tristimulus values* of the printed background when viewed under daylight illumination should be very similar to a blank page viewed under incandescent lighting.

Clearly, we can see that the patch of color in Figure 10.6 is a deep yellow-orange. The blank paper illuminated by incandescent light will not appear nearly as orange due to the effects of chromatic adaptation. Normalizing the photoreceptors independently as a function of the SPD of the light-source results in the dashed lines shown in Figure 10.6. If we were to then go ahead and integrate those new *adapted* cone responses with the background SPD, we would see that the relative integrated responses are about equal for all the cones, and thus the adapted perceptual response is approximately white. Alternative models based upon the von Kries-style of receptor gain control will be discussed in the next section.

Receptor gain control is not the only form of sensory adaptation for which we need to account. The visual system, as a generality, is a highly nonlinear imaging device. As a first-order assumption, many often say that the visual system behaves in a compressive logarithmic manner. While this is not entirely true, it does describe the general compressive nature of the visual system (compare Figure 17.9). Models of chromatic adaptation can include nonlinear compression as well as subtractive mechanisms. The nonlinear compression can take the form of simple power functions as well as more sophisticated hyperbolic or sigmoidal equations. However, for speed and simplicity, most current adaptation models rely on simple linear gain controls. More sophisticated color appearance models, such as CIECAM02, do contain these nonlinear components.

Although independent photoreceptor gain controls cannot account entirely for chromatic adaptation, models based upon this theory can still be very useful. From a purely physiological standpoint, we can consider photoreceptor gain control as a natural function of photopigment depletion, or bleaching, and regeneration. At a fundamental level, color begins when a photon of light hits a photopigment molecule in the cones. The probability of the photopigment molecule absorbing the light is determined by both the wavelength of the photon and the properties of the molecules, which are slightly different for the photopigments in each of the cone types.

When a photopigment molecule absorbs a photon, it undergoes a chemical transformation known as isomerization, which eventually leads to the transmis-

Figure 10.7. Example of a cognitive form of chromatic adaptation. In the middle frame, the yellow candy has a green filter placed over it causing its appearance to shift towards green. When the same green filter is placed over the entire image (bottom frame) the yellow candy appears yellow again (adapted from [509]).

and then transforming back to XYZ after adaptation. One commonly used linear transform from XYZ to LMS and its inverse are given by

$$\begin{bmatrix} L \\ M \\ S \end{bmatrix} = M_{\text{HPE}} \begin{bmatrix} X \\ Y \\ Z \end{bmatrix} \tag{10.4a}$$

$$= \begin{bmatrix} 0.38971 & 0.68898 & -0.07868 \\ -0.22981 & 1.18340 & 0.04641 \\ 0.00000 & 0.00000 & 1.00000 \end{bmatrix} \begin{bmatrix} X \\ Y \\ Z \end{bmatrix}; \tag{10.4b}$$

$$\begin{bmatrix} X \\ Y \\ Z \end{bmatrix} = M_{\text{HPE}}^{-1} \begin{bmatrix} L \\ M \\ S \end{bmatrix} \tag{10.4c}$$

$$= \begin{bmatrix} 1.91019 & -1.11214 & 0.20195 \\ 0.37095 & 0.62905 & 0.00000 \\ 0.00000 & 0.00000 & 1.00000 \end{bmatrix} \begin{bmatrix} L \\ M \\ S \end{bmatrix}. \tag{10.4d}$$

This transform is known as the *Hunt-Pointer-Estevez* transform [509]. Here, the cone sensitivities are normalized such that they produce an equal LMS value when presented with an equal-energy illuminant, or equal XYZ tristimulus values.

By combining (10.3) and (10.4), we can calculate the adapted tristimulus values for a given viewing illumination or white point:

$$\begin{bmatrix} L_w \\ M_w \\ S_w \end{bmatrix} = M_{\text{HPE}} \begin{bmatrix} X_{\max} \\ Y_{\max} \\ Z_{\max} \end{bmatrix}; \tag{10.5a}$$

$$\begin{bmatrix} X_{\text{adapted}} \\ Y_{\text{adapted}} \\ Z_{\text{adapted}} \end{bmatrix} = M_{\text{HPE}}^{-1} \begin{bmatrix} \frac{1}{L_w} & 0 & 0 \\ 0 & \frac{1}{M_w} & 0 \\ 0 & 0 & \frac{1}{S_w} \end{bmatrix} M_{\text{HPE}} \begin{bmatrix} X \\ Y \\ Z \end{bmatrix}. \tag{10.5b}$$

The adapting illumination can be either measured off of a white surface in the scene, ideally a perfect-reflecting diffuser, or alternatively for a digital image the illumination can often be approximated as the maximum tristimulus values of the scene, as shown in (10.5a). Since both the forward and inverse cone transforms are linear 3×3 matrix equations, as is the von Kries normalization, the entire chromatic adaptation transform can be cascaded into a single 3×3 matrix multiplication.

Now we know if the chromatically-adapted tristimulus values of two stimuli match, i.e., $XYZ1_{\text{adapt}} = XYZ2_{\text{adapt}}$, then the stimuli themselves should match. While this in itself can be useful, it is often more desirable to calculate what stimulus needs to be generated under one illumination to match a given color under a different illumination. For example, you may have a color patch illuminated by daylight, and you want to know what tristimulus values you need to generate to create a "matching" color patch that will be illuminated by an incandescent light.

This *corresponding-colors* calculation involves a round-trip or forward and inverse chromatic-adaptation calculation. The essence of von Kries chromatic adaptation is a normalization of the cones based upon the integrated energy (adapting white) of a given illumination condition. We calculate corresponding colors by first normalizing for one illumination and then setting these adapted signals to the second condition. This amounts to dividing out the color of one light and multiplying in the color of the second:

$$\begin{bmatrix} X \\ Y \\ Z \end{bmatrix}_{\text{adapted}} = M_{\text{HPE}}^{-1} \begin{bmatrix} L_{w,2} & 0 & 0 \\ 0 & M_{w,2} & 0 \\ 0 & 0 & S_{w,2} \end{bmatrix} \begin{bmatrix} \frac{1}{L_{w,1}} & 0 & 0 \\ 0 & \frac{1}{M_{w,1}} & 0 \\ 0 & 0 & \frac{1}{S_{w,1}} \end{bmatrix} M_{\text{HPE}} \begin{bmatrix} X \\ Y \\ Z \end{bmatrix}. \tag{10.6}$$

Again, as all of these calculations are simple linear 3×3 matrix transforms, they can all be cascaded into a single 3×3 transform for simple application to digital images.

We can use a von Kries-type adaptation to renormalize the Hunt-Pointer-Estevez transform itself such that it provides equal LMS values when presented with D65 illumination. We first calculate the "normalizing" white LMS values using Equation (10.5a) and then use those values in a von Kries adaptation on the cone signals:

$$\begin{bmatrix} L \\ M \\ S \end{bmatrix}_{\text{D65}} = M_{\text{HPE}} \begin{bmatrix} 0.9504 \\ 1.0000 \\ 1.0889 \end{bmatrix}_{\text{XYZD65}} ; \tag{10.7a}$$

$$M_{\text{HPED65}} = \begin{bmatrix} \frac{1}{L_{\text{D65}}} & 0 & 0 \\ 0 & \frac{1}{M_{\text{D65}}} & 0 \\ 0 & 0 & \frac{1}{S_{\text{D65}}} \end{bmatrix} M_{\text{HPE}} \tag{10.7b}$$

$$= \begin{bmatrix} 0.4002 & 0.7076 & -0.0808 \\ -0.2263 & 1.1653 & 0.0457 \\ 0.0000 & 0.0000 & 0.9184 \end{bmatrix}. \tag{10.7c}$$

The simple von Kries adaptation transform has been shown to provide a remarkably good fit to corresponding experimental color data, such as the data generated by Breneman in 1987 [126] and Fairchild in the early 1990s [302]. Although it is remarkable that such a simple model could predict the experimental data as well as the von Kries model does, there were some discrepancies that suggested that a better adaptation transform was possible.

10.4.2 CIELAB as a Chromatic-Adaptation Model

The CIELAB color space was introduced in Chapter 8, and provides a close-to-perceptually-uniform opponent color space that can also be used to derive color differences between two stimuli. While we do not generally describe the CIELAB model as a chromatic-adaptation model, there are elements of chromatic adaptation that can be considered. The CIELAB equations are presented again here, in a slightly different form, for clarity:

$$L^* = 116 f\left(\frac{Y}{Y_n}\right) - 16, \tag{10.8a}$$

$$a^* = 500\left[f\left(\frac{X}{X_n}\right) - f\left(\frac{Y}{Y_n}\right)\right], \tag{10.8b}$$

$$b^* = 200\left[f\left(\frac{Y}{Y_n}\right) - f\left(\frac{Z}{Z_n}\right)\right], \tag{10.8c}$$

$$f(r) = \begin{cases} r^{1/3} & r > 0.008856, \\ 7.787r + 16/116 & r \le 0.008856. \end{cases} \tag{10.8d}$$

Notice that for each of the CIELAB equations, the CIE XYZ tristimulus values are essentially normalized by the tristimulus values of *white*. This normalization is in essence a von Kries-type normalization, though for CIELAB this is performed on the actual tristimulus values rather than on the LMS cone signals. This type of transform was actually coined a *wrong von Kries* transform by Terstiege [1122].

Interestingly, one of the known weaknesses of the CIELAB color space, which was designed to be perceptually uniform, is changing hue when linearly increasing or decreasing chroma (these terms are explained in depth in Chapter 11). That is so say that lines of constant perceived hue are not straight in the CIELAB color space, most noticeably in the blue and red regions, as quantified by experiments such as those by Hung and Berns [500]. This often results in the *purple-sky* phenomenon when boosting chroma in the CIELAB color space. An example of this

Figure 10.8. The image on the left was modified by boosting its chroma in CIELAB color space by a factor of two, shown on the right. Note the undesired purple color which appears in parts of the water. The blockiness of the purple areas are JPEG compression artifacts; Grand Bahama, 2004.

is shown in Figure 10.8, which shows an image before and after boosting chroma by a factor of two.

In Figure 10.9 the Munsell renotation data is plotted in CIELAB for a given lightness value. Since the Munsell data was experimentally generated to be con-

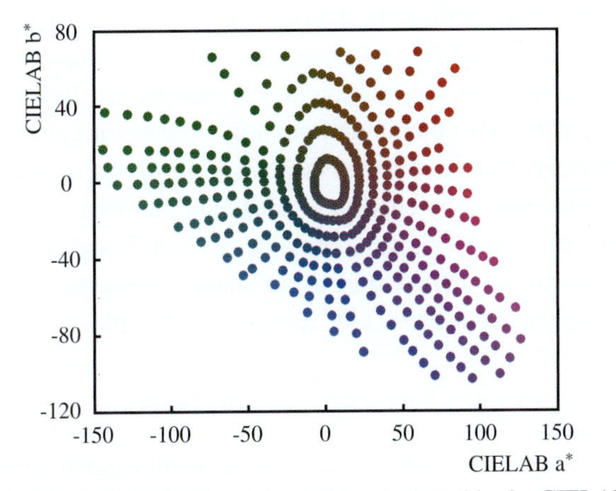

Figure 10.9. The Munsell Renotation data at Value 4 plotted in the CIELAB color space showing curvature for the lines of constant hue.

stant hue, the "spokes" corresponding to lines of constant hue should be straight [831]. Thus, the root cause of this type of color error was linked to the chromatic-adaptation transform using CIE XYZ instead of LMS in evaluations by Liu et al. in 1995 and Moroney in 2003 [690, 793].

10.4.3 Nayatani Chromatic-Adaptation Model

Further advances in modeling chromatic adaptation were developed by Nayatani et al. in the early 1980s [819, 820]. In addition to a von Kries-style photoreceptor gain control, they added a nonlinear exponential term as well as an additive noise term to handle low-light level adaptation for near threshold visual data. The generalized form of this chromatic adaptation model is given by:

$$L_a = \alpha_L \left(\frac{L + L_n}{L_w + L_n} \right)^{\beta_L}, \tag{10.9a}$$

$$M_a = \alpha_M \left(\frac{M + M_n}{M_w + M_n} \right)^{\beta_M}, \tag{10.9b}$$

$$S_a = \alpha_S \left(\frac{S + S_n}{S_w + S_n} \right)^{\beta_S}. \tag{10.9c}$$

Here, LMS_w are the cone excitations for the adapting field, or white point, LMS_n are the *noise* coefficients, typically a very small number, β_{LMS} are the nonlinear exponential values, and α_{LMS} are linear scaling factors. The von Kries-style photoreceptor gain control should be immediately obvious in the Nayatani equations, especially as the noise coefficients approach zero.

The exponential values β_{LMS} are functions of the adapting cone signals themselves, essentially meaning they monotonically increase as a function of the adapting luminance [820]. By combining both a photoreceptor gain control with the nonlinear exponential functions, the Nayatani et al. model is able to predict corresponding colors caused by chromatic adaptation, as well as color appearance changes caused by changes in overall luminance level. These include the increase in apparent colorfulness and contrast as luminance levels increase. More details on these color appearance changes are discussed in Chapter 11.

10.4.4 Fairchild Chromatic-Adaptation Model

The chromatic-adaptation model of Nayatani et al. illustrated the power of combining the simple von Kries-style linear photoreceptor gain control with additional components to create an extended model of chromatic adaptation capable of predicting additional color-appearance phenomena. Fairchild presented a similar extension designed for digital-imaging applications [299, 300]. The goal of this model is to have a linear chromatic-adaptation transform that is capable of predicting incomplete adaptation as well as the increase in colorfulness with luminance level, e.g., the Hunt effect as discussed in Chapter 11.

To this point, all of the models of chromatic adaptation we have discussed assume that we adapt completely to the reference viewing conditions. This implies that the adapting "white" will always appear white. The experiments of Breneman as well as further experiments by Fairchild showed that this is not always the case [126, 302].

Fairchild performed a series of experiments using a soft-copy CRT display to test the degree of chromatic adaptation and found that particularly for soft-copy displays, our adaptation is significantly less than complete. A simple example of this can be illustrated by setting a computer display to have a "white" of equal tristimulus values as CIE Illuminant A. The display itself will continue to appear very orange, even after significant adaptation time. Furthermore, if we place a piece of paper such that it is illuminated by the display, we perceive the paper to appear white, or more white, while the display itself, which is the only source of illumination for the paper, does not! This is illustrated in Figure 10.10. The

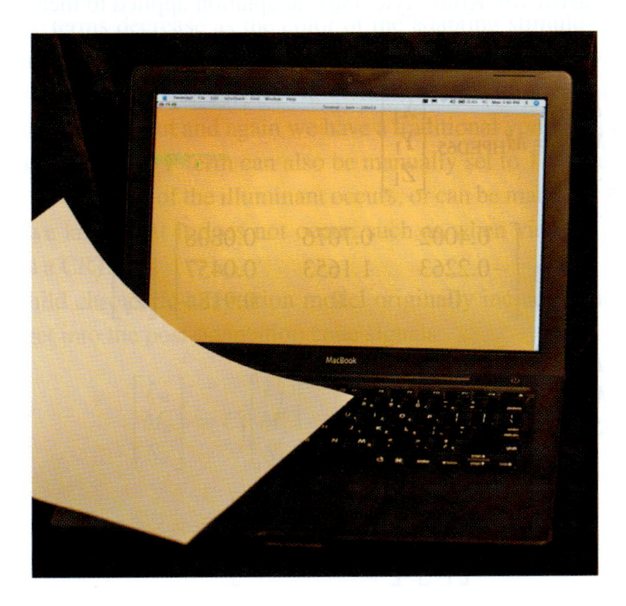

Figure 10.10. An example of discounting the illuminant: the laptop screen and the paper have identical chromaticities, yet the laptop appears more orange.

This transformation adds a luminance-dependent interaction between all of the cone signals to account for the boost in colorfulness when the adapting illumination is high. In practice, this C-matrix calculation introduced undesirable luminance-dependent artifacts in the predicted results and is generally eliminated [302].

Since the Fairchild chromatic-adaptation model was designed to be used on digital images, it was designed to collapse to a single 3×3 matrix transform much like the traditional von Kries model. Thus, for calculating corresponding colors for two different viewing conditions, the entire calculation can be expressed using

$$\begin{bmatrix} X \\ Y \\ Z \end{bmatrix}_2 = M_{\text{HPE}}^{-1} A_2^{-1} C_2^{-1} C_1 A_1 M_{\text{HPE}} \begin{bmatrix} X \\ Y \\ Z \end{bmatrix}_1 . \tag{10.14}$$

It should be noted that the A and C calculations only need to be performed once for each viewing condition, and then the entire transform can be applied to each pixel in a digital image. Pixel-specific chromatic-adaptation transforms that change as a function of the image content itself are an important element of *image-appearance models* and will be discussed in Chapter 11. To calculate corresponding colors using the Fairchild model, the forward A and C matrices must be determined for both of the viewing conditions and then inverted for the second condition to obtain A_2^{-1}, C_2^{-1}.

10.4.5 The Bradford and CIECAM97s Chromatic-Adaptation Models

The Bradford transform, first introduced to color-appearance modeling in the LLAB model, is a von Kries-style adaptation transform that includes a nonlinear component on the short-wavelength cone signals [641, 713]. This adaptation transform also includes a calculation for discounting the illuminant, modeling a form of incomplete chromatic adaptation. Perhaps the most important advance introduced by the Bradford transform is the use of *sharpened* cone signals instead of real physiologically plausible cone signals. The sharpened cones refers to the fact that the spectral sensitivity are more narrow than those of real cone sensitivities. The first stage of the Bradford model is a linear transform into these sharpened cone signals, designated RGB, from normalized CIE XYZ tristimulus values, as given by

$$\begin{bmatrix} R \\ G \\ B \end{bmatrix} = M_{\text{BFD}} \begin{bmatrix} \frac{X}{Y} \\ \frac{Y}{Y} \\ \frac{Z}{Y} \end{bmatrix} \tag{10.15a}$$

$$= \begin{bmatrix} 0.8951 & 0.2664 & -0.1614 \\ -0.7502 & 1.7135 & 0.0367 \\ 0.0389 & -0.0685 & 1.0296 \end{bmatrix} \begin{bmatrix} \frac{X}{Y} \\ \frac{Y}{Y} \\ \frac{Z}{Y} \end{bmatrix} \tag{10.15b}$$

Normalized tristimulus values must be used in this transform because of the non-linearity introduced to the short-wavelength B signal.

The sharpened "cone" signals are plotted along with the Hunt-Pointer-Estevez cone signals in Figure 10.11. The first thing to note is that these cone signals are not meant to be physiologically realistic, as evidenced by the negative lobes present. They can be thought to represent types of interactions (both inhibition and excitation) between the cone signals that may occur at the sensory level, though they were not designed specifically in that manner. There is experimental evidence that does suggest that the cone signals do interact during chromatic adaption and are not independent as von Kries suggested [244]. The sharpened cone signals effectively allow for a constancy of both hue and saturation across changes in chromatic adaptation.

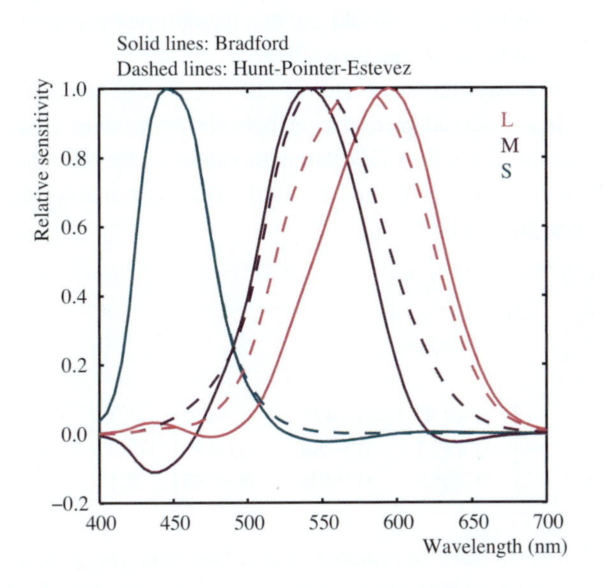

Figure 10.11. The sharpened cone signals of the Bradford chromatic-adaptation transform (solid) and the Hunt-Pointer-Estevez cone signals (dashed).

The adapted cone signals are then calculated using

$$R_a = \left[D \left(\frac{R_{wr}}{R_w} \right) + 1 - D \right] R, \tag{10.16a}$$

$$G_a = \left[D \left(\frac{G_{wr}}{G_w} \right) + 1 - D \right] G, \tag{10.16b}$$

$$B_a = \left[D \left(\frac{B_{wr}}{B_w} \right) + 1 - D \right] \left| B^{\beta} \right|, \tag{10.16c}$$

$$\beta = \left(\frac{B_w}{B_{wr}} \right)^{0.0834}. \tag{10.16d}$$

Note that the transform is based upon a reference illuminant of CIE D65 which is calculated by transforming the CIE XYZ of D65 using the Bradford matrix M_{BFD}. In (10.16), RGB_w are the sharpened cone signals of the adapting reference conditions (reference white) and RGB_{wr} are the cone signals of CIE D65. Note the absolute value in calculating the B_a signal, necessary because the sharpened cones can become negative.

The D term corresponds to the degree of discounting the illuminant. When D is set to one, the observer is completely adapted to the viewing environment, while when D is set to zero we would say that the observer is completely adapted to the *reference condition*, in this case CIE D65.

Since D65 is always the numerator of (10.16), we can say that if $D \in (0,1)$, the observer is in a mixed adaptation state, between the viewing condition and the reference condition. The CIE XYZ tristimulus values for the reference D65 can be calculated by multiplying by the inverse of the Bradford matrix and multiplying back in the luminance value:

$$\begin{bmatrix} X_a \\ Y_a \\ Z_a \end{bmatrix} = M_{\text{BFD}}^{-1} \begin{bmatrix} R_a Y \\ G_a Y \\ B_a Y \end{bmatrix} \tag{10.17a}$$

$$= \begin{bmatrix} 0.9870 & -0.1471 & 0.1600 \\ 0.4323 & 0.5184 & 0.0493 \\ -0.0085 & 0.0400 & 0.9685 \end{bmatrix} \begin{bmatrix} R_a Y \\ G_a Y \\ B_a Y \end{bmatrix}. \tag{10.17b}$$

The CIECAM97s color-appearance model borrowed the general form of the Bradford chromatic-adaptation transform (or more precisely the LLAB adaptation transform using the Bradford matrix), though they did not base the calculation on a CIE D65 reference which made it easier to calculate corresponding colors [191].

They also developed an empirical formula to calculate the D value from the luminance of the adapting conditions:

$$R_a = \left[D \left(\frac{1}{R_w} \right) + 1 - D \right] R, \tag{10.18a}$$

$$G_a = \left[D \left(\frac{1}{G_w} \right) + 1 - D \right] G, \tag{10.18b}$$

$$B_a = \left[D \left(\frac{1}{B_w^p} \right) + 1 - D \right] |B^p|, \tag{10.18c}$$

$$p = \left(\frac{B_w}{1.0} \right)^{0.0834}, \tag{10.18d}$$

$$D = F - \frac{F}{\left[1 + 2 \left(L_A^{1/4} \right) + \frac{L_A^2}{300} \right]}. \tag{10.18e}$$

Here, L_A is the adapting luminance, and F is a categorical *surround* function usually set to 1.0 for normal conditions and 0.9 for darkened conditions. More details on the effects of the luminance level and surround on color appearance will be discussed in Chapter 11.

The sharpened cone responses of the Bradford chromatic-adaptation transform proved to provide a better fit to the experimental correlated color data than models built using traditional cone signals [191]. The incomplete adaptation factors as well as the nonlinearity make the full CIECAM97s transform more difficult to use in a practical application. The *International Color Consortium* (ICC), the *de facto* standard body for color management, has recommended the use of a linearized version of the CIECAM97s/Bradford transform with complete chromatic adaptation [523]. Essentially, this is a simple von Kries transform, such as that shown in (10.6), to calculate corresponding colors from any given viewing condition into the ICC reference-viewing condition of CIE D50.

10.4.6 CIECAT02 Chromatic-Adaptation Transform

The ICC selection of a linearized version of the Bradford transform is one example that indicated the desire to have a practical chromatic-adaptation transform that was easily applied and inverted. The CIE recognized that and set out to optimize a linear transform that provided results that were similar to the performance of the nonlinear transform in CIECAM97s [195]. Several linear transforms were generated by various researchers to address this concern, many of which were summarized by Fairchild [301].

Calabria and Fairchild showed that all of the transforms that use *sharpened* cone responses behave similarly (and better than non-sharpened cones), and they are perceptually identical to each other for most practical applications [145]. The CIE then selected a sharpened transform that best fits the existing experimental data and has a high degree of backwards compatibility to the nonlinear transform in CIECAM97s [195]. This transform is known as CIECAT02, and it is essentially a von Kries-style normalization using the following sharpened cone transformation:

$$\begin{bmatrix} R \\ G \\ B \end{bmatrix} = M_{\text{CAT02}} \begin{bmatrix} X \\ Y \\ Z \end{bmatrix} \tag{10.19a}$$

$$\begin{bmatrix} X \\ Y \\ Z \end{bmatrix} = M_{\text{CAT02}}^{-1} \begin{bmatrix} R \\ G \\ B \end{bmatrix} \tag{10.19b}$$

$$M_{\text{CAT02}} = \begin{bmatrix} 0.7328 & 0.4296 & -0.1624 \\ -0.7036 & 1.6974 & 0.0061 \\ 0.0030 & -0.0136 & 0.9834 \end{bmatrix} \tag{10.19c}$$

$$M_{\text{CAT02}}^{-1} = \begin{bmatrix} 1.0961 & -0.2789 & 0.1827 \\ 0.4544 & 0.4739 & 0.0721 \\ -0.0096 & -0.0057 & 1.0153 \end{bmatrix} \tag{10.19d}$$

The CAT02 sharpened cone sensitivities are plotted along with the Bradford cone sensitivities in Figure 10.12, showing that the Bradford signals are very similar, though slightly sharper for the long wavelength cones.

To calculate corresponding colors using the CIECAT02 transform, assuming complete adaptation to both the viewing conditions, we can then just use the sharpened cone signals in a traditional von Kries adaptation matrix, as follows:

$$\begin{bmatrix} X_2 \\ Y_2 \\ Z_2 \end{bmatrix} = M_{\text{CAT02}}^{-1} \begin{bmatrix} \frac{R_{w,2}}{R_{w,1}} & 0 & 0 \\ 0 & \frac{G_{w,2}}{G_{w,1}} & 0 \\ 0 & 0 & \frac{B_{w,2}}{B_{w,1}} \end{bmatrix} M_{\text{CAT02}} \begin{bmatrix} X_1 \\ Y_1 \\ Z_1 \end{bmatrix}. \tag{10.20}$$

Here, $RGB_{w,1}$ and $RGB_{w,2}$ are the sharpened cone responses of the adapting conditions (white points).

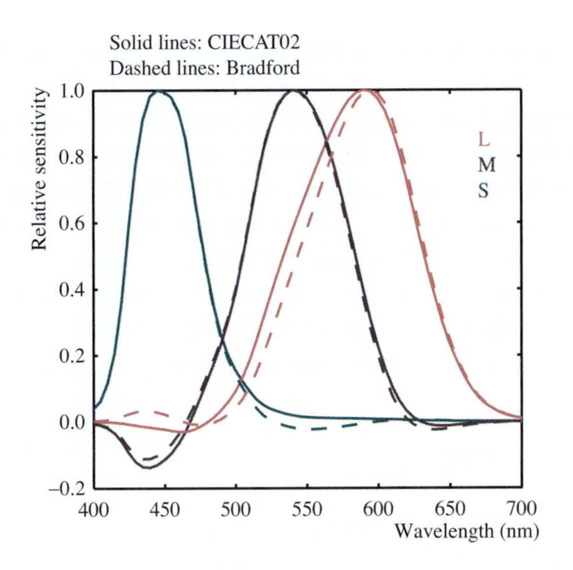

Figure 10.12. The sharpened cone signals of the CIECAT02 chromatic-adaptation transform (solid) and the Bradford cone signals (dashed).

10.5 Application: Transforming sRGB Colors to D50 for an ICC Workflow

Chromatic-adaptation transforms (CATs) are designed to essentially calculate tristimulus values that have been normalized by the conditions of the viewing environment. This can account for some degree of our perceived color constancy. By performing a forward and inverse transform, we can generate corresponding colors. These corresponding colors can be powerful tools in a digital-imaging library. For instance, if we have one characterized display device, we can determine what colors (tristimulus values) we need to generate on that device in order to match another device under a different viewing condition.

An example of this is to determine what colors need to be displayed on a CRT with a D65 white point in order to match a print viewed under D50 simulators. Of course, this cross-media comparison brings many more factors into play, such as the overall illumination levels and the color of the surround. Those factors are incorporated into more advanced color-appearance models, of which chromatic-adaptation transforms play an integral part. Traditional ICC color management uses chromatic-adaptation transforms to move colors from the native-device white into a standardized CIE-D50 white.

As we learned in Chapter 7, sRGB has become a "default" standard for color management on the Internet [524]. While sRGB was originally designed to approximate the typical viewing conditions and primaries of a CRT device, it has been adopted for a wide variety of other uses, including the encoding of digital-camera images for many consumer cameras. The sRGB transform assumes that the white point of most displays is equivalent to CIE D65, and so the encoded white value ($R = G = B$) will produce D65 tristimulus values.

The ICC, on the other hand, recommends using a white point of CIE D50 in the interim connection space, which is most often CIE XYZ. The sRGB transforms used to convert from sRGB (encoded as an 8-bit value) into CIE XYZ are given by

$$RGB_s' = \frac{RGB_s}{255.0}, \tag{10.21a}$$

$$RGB = f\left(RGB_s'\right), \tag{10.21b}$$

$$f(n) = \begin{cases} \left(\frac{n+0.055}{1.055}\right)^{2.4} & n > 0.04045, \\ 12.92n & n \leq 0.04045. \end{cases}, \tag{10.21c}$$

$$\begin{bmatrix} X \\ Y \\ Z \end{bmatrix}_{D65} = M_{\text{sRGB}} \begin{bmatrix} R \\ G \\ B \end{bmatrix}, \tag{10.21d}$$

where

$$M_{\text{sRGB}} = \begin{bmatrix} 0.4124 & 0.3576 & 0.1805 \\ 0.2126 & 0.7152 & 0.0722 \\ 0.0193 & 0.1192 & 0.9505 \end{bmatrix} \tag{10.21e}$$

It is important to note that the actual sRGB transform uses a piecewise-linear and nonlinear component rather than a standard *gamma* of 2.2 as is often expressed (see Section 8.1).

The forward transform from linearized RGB values into CIE XYZ tristimulus values will provide us with values normalized for D65. To use these values in an ICC color-managed workflow, we must apply a chromatic-adaptation matrix, transforming into D50. We will use the linear Bradford matrix, since that is recommended and used by the ICC [523]:

$$\begin{bmatrix} X \\ Y \\ Z \end{bmatrix}_{\text{D50}} = M_{\text{BFD}}^{-1} \begin{bmatrix} \frac{\rho_{\text{D50}}}{\rho_{\text{D65}}} & 0 & 0 \\ 0 & \frac{\gamma_{\text{D50}}}{\gamma_{\text{D65}}} & 0 \\ 0 & 0 & \frac{\beta_{\text{D50}}}{\beta_{\text{D65}}} \end{bmatrix} M_{\text{BFD}} \begin{bmatrix} X \\ Y \\ Z \end{bmatrix}_{\text{D65}}. \tag{10.22}$$

To avoid confusion between the traditional RGB used to describe sharpened cone signals, and the RGB of a digital device, we will use ρ, γ, and β in these equations. The sharpened cone signals for D50 and D65 are given by

$$\begin{bmatrix} \rho \\ \gamma \\ \beta \end{bmatrix}_{D50} = \begin{bmatrix} 0.9963 \\ 1.0204 \\ 0.8183 \end{bmatrix} = M_{BFD} \begin{bmatrix} 0.9642 \\ 1.0000 \\ 0.8249 \end{bmatrix}_{D50}; \tag{10.23a}$$

$$\begin{bmatrix} \rho \\ \gamma \\ \beta \end{bmatrix}_{D65} = \begin{bmatrix} 0.9414 \\ 1.0405 \\ 1.0896 \end{bmatrix} = M_{BFD} \begin{bmatrix} 0.9504 \\ 1.0000 \\ 1.0889 \end{bmatrix}_{D65}. \tag{10.23b}$$

By using the values calculated in (10.23) as well as the traditional sRGB color matrix in (10.22), we can generate a new matrix that will transform directly from linearized sRGB into CIE XYZ adapted for a D50 white:

$$M_{sRGBD50} = M_{BFD}^{-1} \begin{bmatrix} \frac{0.9963}{0.9414} & 0 & 0 \\ 0 & \frac{1.0204}{1.0405} & 0 \\ 0 & 0 & \frac{0.8183}{1.0896} \end{bmatrix} M_{BFD}\, M_{sRGB}, \tag{10.24a}$$

and the resulting matrix is

$$M_{sRGBD50} = \begin{bmatrix} 0.4361 & 0.3851 & 0.1431 \\ 0.2225 & 0.7169 & 0.0606 \\ 0.0139 & 0.0971 & 0.7140 \end{bmatrix}. \tag{10.24b}$$

Equation (10.24b) can now be used to directly take sRGB encoded values into the standard ICC working space. In fact, when you read any ICC profiles encapsulating sRGB, the "colorants" in those files will generally take the form of the pre-adapted transform from Equation (10.24b).

10.6 Application: White Balancing a Digital Camera

Another potential application for chromatic-adaptation transforms that can be very useful is *white balancing* digital camera data. Most current digital-imaging technologies use linear sensors, or sensors that transduce photons with a direct linear relationship to luminance. The data that are sensed must be processed in order to produce images that are pleasing to look at.

One necessary processing step is white balancing, which attempts to mimic the human behavior of chromatic adaptation. White balancing assures that when an image is captured under a very "red" illuminant, such as indoors under incandescent lighting, the resultant image appears more neutral. Since this image manipulation is akin to the human sensory process, it can actually be performed using chromatic-adaptation transforms to assure that the resultant images *appear* as the photographer remembers them based upon their own state of adaptation.

We will use a simulated digital camera to outline how this approach works. The input into the system is a full spectral image, a light source, and a digital camera with spectral sensitivities. A discussion of what to do when the nature of the illuminant is not known, is given in Section 9.9. Image formulation is calculated by multiplying these three signals together and then integrating (or summing) over all wavelengths. For simulation purposes, relative spectral sensitivities and spectral power distributions can be used. The image formation equation is given by

$$R = \frac{\sum_{\lambda} I(\lambda) R_s(\lambda) S(\lambda)}{\sum_{\lambda} S(\lambda)}; \tag{10.25a}$$

$$G = \frac{\sum_{\lambda} I(\lambda) G_s(\lambda) S(\lambda)}{\sum_{\lambda} S(\lambda)}; \tag{10.25b}$$

$$B = \frac{\sum_{\lambda} I(\lambda) B_s(\lambda) S(\lambda)}{\sum_{\lambda} S(\lambda)}. \tag{10.25c}$$

This equation is normalized by the relative energy of the light source, $S(\lambda)$. The full spectral image is represented by $I(\lambda)$ and $R_s(\lambda)$, $G_s(\lambda)$, and $B_s(\lambda)$ are the spectral sensitivities of the digital camera.

As example input into the system, we will use the METACOW image, which is a full-spectral synthetic image designed for imaging systems evaluation [293]. In addition to providing spectral information at each pixel, METACOW was designed to be maximally metameric to test the variation of an imaging system from the color-matching functions. Recall from Chapter 7 that a metameric pair of stimuli are stimuli that appear to match but that have different spectral reflectance or transmittance values. Due to the disparate spectral properties of the stimuli, the matches often break down when viewed under different lighting (illuminant metamerism) or when viewed by a different person or capture device (observer metamerism).

Figure 10.13. The METACOW image integrated with CIE 1931 Standard Observer and CIE D65, then transformed into the sRGB color space.

One half of each of the objects in the METACOW image was rendered to have the same spectral reflectance as the corresponding patch in the GretagMacbeth ColorChecker (shown in Figure 12.30), while the other half was designed to be a metameric match for the CIE 1931 Standard Observer when viewed under CIE D65 [757]. An example of METACOW integrated with CIE D65 and the 1931 Standard Observer and converted to sRGB is shown in Figure 10.13. The full spectral METACOW images are available on the DVD of this book.

For the sensor spectral sensitivities of our digital camera, we can choose a simple three-channel RGB camera with Gaussian spectral sensitivities centered at $c_r = 450$ nm, $c_g = 550$ nm, and $c_b = 600$ nm with a full-width half-max (FWHM) of 100 nm. The formulation for these sensitivities is given by

$$R_s(\lambda) = \exp\left(-\frac{(c_r - \lambda)^2}{2\sigma^2}\right); \tag{10.26a}$$

$$G_s(\lambda) = \exp\left(-\frac{(c_g - \lambda)^2}{2\sigma^2}\right); \tag{10.26b}$$

$$B_s(\lambda) = \exp\left(-\frac{(c_b - \lambda)^2}{2\sigma^2}\right). \tag{10.26c}$$

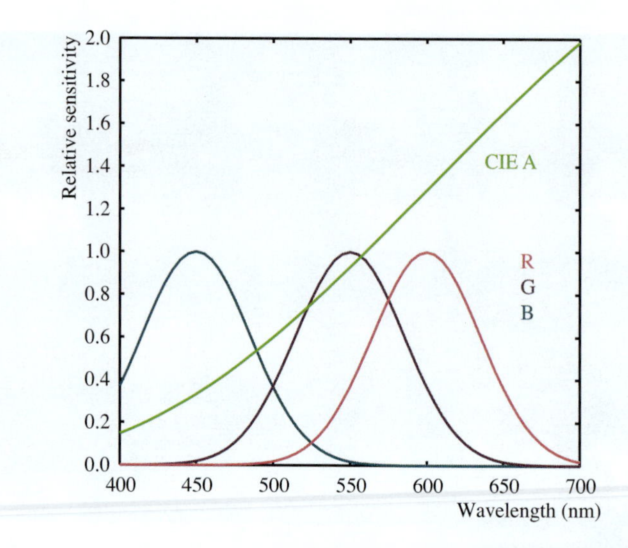

Figure 10.14. The simulated spectral sensitivities of our digital camera plotted along with CIE Illuminant A.

The full-width half-max value indicates the width of the Gaussian at half its maximum height. This value relates to σ according to [1232]:

$$FWHM = 2\sigma\sqrt{2\ln 2}. \tag{10.27}$$

This means that the parameter σ is approximately 42.47 nm to achieve a FWHM of 100 nm. The resulting sensitivities are plotted in Figure 10.14 along with CIE Illuminant A, which will be our capture illuminant.

Combining the spectral image with the CIE A light source and the digital-camera sensitivities, and integrating using (10.25) provides us with an RGB image. This image is shown, with a nonlinear gamma correction of 1/2.2 for display purposes, in Figure 10.15. There should be two things immediately evident from this image: the overall yellow tone caused by the CIE A illumination, and the fact that the two halves of the cows are now very different. The two halves no longer match because of both illuminant and observer metamerism, given that the spectral properties of the halves are very different.

Once an image has been captured the spectral information is lost, and there is no way to regain that information and make the two halves match again. We can, however, fix the overall yellow cast by applying a white-balancing based on a chromatic-adaptation transform. To do this, we must first derive a relationship between the digital camera sensors and CIE XYZ tristimulus values. This process is known as device characterization, and it is a crucial step for color management

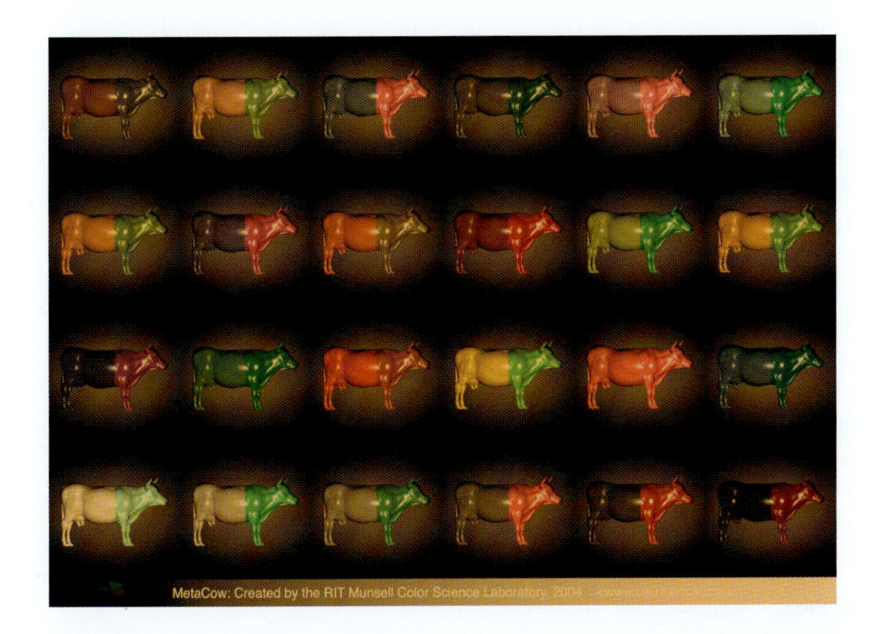

Figure 10.15. The scene values as captured by the digital camera simulation with no white balancing.

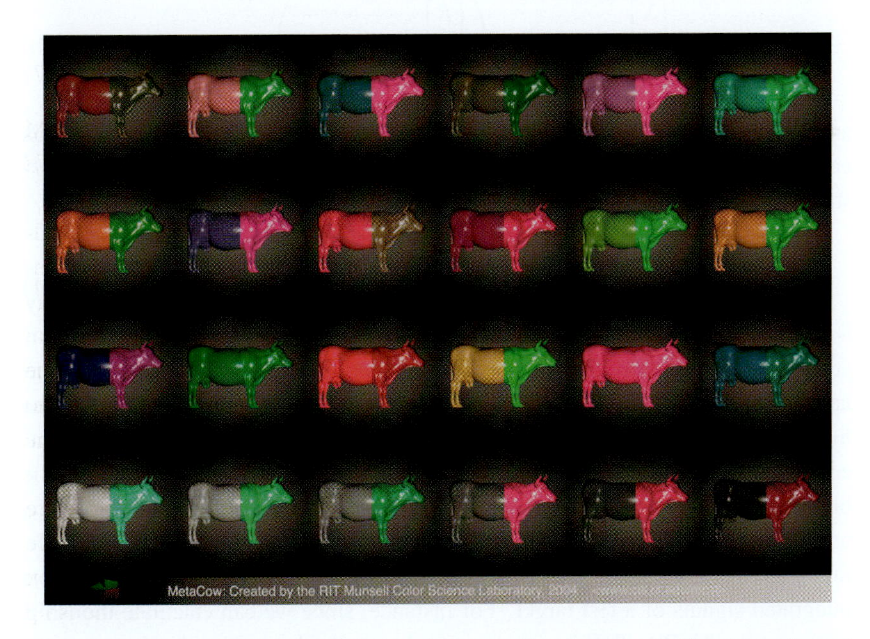

Figure 10.16. The digital-camera image chromatically adapted to D65 and transformed into the sRGB color space.

(see Section 12.12). Since we know both the spectral sensitivities of our digital camera as well as the spectral sensitivities of the CIE 1931 Standard observer, we can generate a transform to try to map one to the other. The simplest way to do this is to create a linear transform through least-squares linear regression. With $\hat{X}\hat{Y}\hat{Z}$ representing the predicted best estimate of CIE XYZ tristimulus values, the desired 3×3 transform is

$$\begin{bmatrix} \hat{X} \\ \hat{Y} \\ \hat{Z} \end{bmatrix}_\lambda = M \begin{bmatrix} R \\ G \\ B \end{bmatrix}_\lambda . \tag{10.28}$$

We are interested in determining the optimal matrix M that will minimize the error from predicted XYZ values to actual XYZ values. To do this, a linear regression can be performed using a *pseudo-inverse* transform. Since RGB is a $3 \times N$ matrix, we cannot take a direct inverse, and so a pseudo-inverse is required:

$$\mathbf{y} = M \cdot \mathbf{x}, \tag{10.29}$$

where

$$M = \mathbf{y} \cdot \mathbf{x}^T \cdot \left(\mathbf{x} \cdot \mathbf{x}^T\right)^{-1}, \tag{10.30a}$$

$$M = \begin{bmatrix} X \\ Y \\ Z \end{bmatrix} \begin{bmatrix} R & G & B \end{bmatrix} \left(\begin{bmatrix} R \\ G \\ B \end{bmatrix} \begin{bmatrix} R & G & B \end{bmatrix} \right)^{-1} , \tag{10.30b}$$

and $\mathbf{x}^T$ is the transpose of $\mathbf{x}$, $\mathbf{x} = (RGB)^T$, and $\mathbf{y} = (XYZ)^T$. Performing this calculation between the spectral sensitivities of the digital camera and the 1931 standard observer results in the following linear transform:

$$M_{\text{camera}} = \begin{bmatrix} 1.0198 & 0.0161 & 0.2456 \\ 0.2788 & 0.9160 & 0.0147 \\ 0.0582 & -0.1098 & 1.4368 \end{bmatrix} . \tag{10.31}$$

We can plot the differences between our optimized transformed camera signals and the actual 1931 Standard Observer, which is shown in Figure 10.17. The figure shows that the best linear fit provides a decent transform to get our camera into CIE XYZ, though there are some noticeable differences, especially in the blue channel. It is also possible to determine the optimized camera matrix M_{camera} using more terms such as squares or interactions in the linear regression (resulting in a larger than 3×3 transform matrix) or by doing a linear regression on the integrated signals of a test target. For instance, since we can calculate the XYZ values of the METACOW image, we could actually do a regression between the integrated RGB sensor values and the integrated XYZ values.

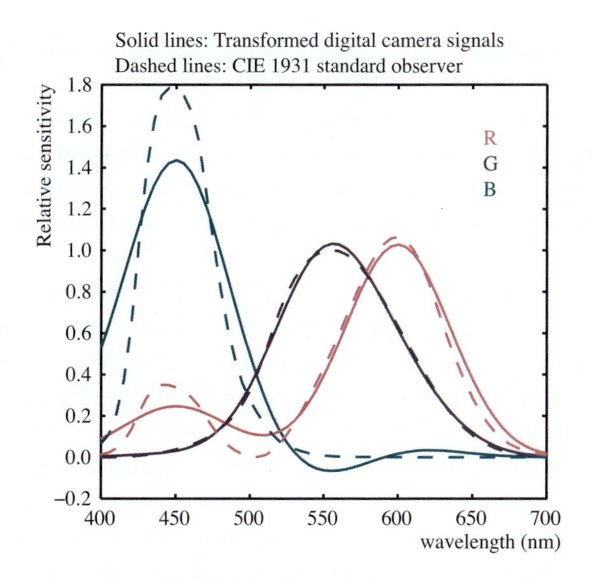

Figure 10.17. Our transformed digital-camera signals (solid) and the CIE 1931 Standard observer (dashed).

Once we have our M_{camera} transformation to CIE XYZ, we can then apply a standard chromatic-adaptation transform, such as CIECAT02. We can also move into a standard color space, such as sRGB, for ease of communication from our device-specific RGB values. In order to do that, we need to choose the "white" of our input image. We could use the integrated signals of just CIE A and our camera sensitivities, since we know that was the illuminant we imaged under, or we can use a white-point estimation technique. In this case, we will choose a simple white-point estimate, which is the maximum value of each of the RGB camera signals. The output white will be set to D65, the native white point of sRGB. The full transform to achieve a white-balanced sRGB signal from our unbalanced camera is then

$$\begin{bmatrix} R \\ G \\ B \end{bmatrix}_{\text{sRGB}} = M_{\text{sRGB}}^{-1} M_{\text{CAT02}}^{-1} \begin{bmatrix} \frac{\rho_{\text{D65}}}{\rho_{\max}} & 0 & 0 \\ 0 & \frac{\beta_{\text{D65}}}{\beta_{\max}} & 0 \\ 0 & 0 & \frac{\gamma_{\text{D65}}}{\gamma_{\max}} \end{bmatrix} M_{\text{CAT02}} M_{\text{camera}} \begin{bmatrix} R \\ G \\ B \end{bmatrix}. \tag{10.32}$$

The resulting image is shown in Figure 10.16.

What we should see now is that the chromatic-adaptation transform did a very good job of removing the overall yellow tint in the image, and the left halves of the cows look much like the traditional Color Checker. The metameric halves still look very strange and clearly do not match. This is because the sRGB color space has a much smaller *gamut* than the metameric samples in the METACOW image, and so the output image shows clipping in the color channels (below zero and greater than 255).

The gamut of an output device essentially describes the range of colors that the device is capable of producing. The process of moving colors from the gamut of one device to that of another is known as *gamut mapping*. In this case, we used a simple clipping function as our gamut-mapping algorithm. A more clever gamut-mapping algorithm may be required to prevent such artifacts or to make the image appear more realistic. Details of more sophisticated gamut-mapping algorithms can be found in [798], and an overview is given in Section 15.4.

This example is meant to illustrate the power that a chromatic-adaptation transform can have for white balancing digital cameras. Through device-characterization and the CIECAT02 transform we are able to transform our simulated digital-camera signals into the sRGB color space with successful results. For most cameras and spectral reflectances, these techniques should prove to be very useful.

10.7 Application: Color-Accurate Rendering

Image synthesis was discussed previously in Section 2.10 where its basic principles were outlined. The assumption was made that each wavelength of interest was rendered separately, and that therefore the renderer was fully spectral. In Section 8.12 the choice of sharpened RGB color space was motivated for use in a renderer. Such a color space allows images to be created that are perceptually close to their spectral counterparts, without the cost associated with a spectral renderer.

When creating images in this manner, the results will often have a strong color cast that may be unexpected. There are at least two sources for such color casts to occur, both relatively straightforward to correct, once their sources are known.

The first inaccuracy stems from a possible difference in measuring and rendering conditions for the materials in the scene. Especially when measured BRDFs are used, it should be noted that such BRDFs are measured under a specific illuminant. If this illuminant is different from the ones used in the geometric model, the appearance of the material will be inaccurate [1208]. As an example, one may read the Yxy chromaticity coordinates for colored patches of a Gretag-Macbeth color checker and convert these values to the chosen rendering color space (for instance the sharpened RGB color space discussed in Section 8.12). However, the Yxy coordinates listed on the Gretag-Macbeth ColorChecker chart are valid for illumination under CIE standard illuminant C. Unless the light sources in the scene to be rendered are also of type C, the chosen patches will appear different from the ones on a real Macbeth color checker.

Under the assumption that the scene to be rendered has a single dominant light source, i.e., there may be many light sources, but the ones providing the main illumination are of the same type, it is possible to transform the specification of all materials in the scene from the illuminant under which they were measured to the dominant illuminant under which they will be rendered. Such a transform involves a standard von Kries-style transformation as presented in this chapter, which can be carried out prior to rendering as a pre-filtering process.

The preferred approach is to replace the dominant light sources in the scene with neutral light sources of equal strength and change spectrally distinct light sources accordingly. If in the sharpened RGB color space, the white point of the dominant illuminant is given by (R_w, G_w, B_w), the remaining illuminants are transformed with

$$R'_s = \frac{R_s}{R_w}, \tag{10.33a}$$

$$G'_s = \frac{G_s}{G_w}, \tag{10.33b}$$

$$B'_s = \frac{B_s}{B_w}. \tag{10.33c}$$

The surface materials are also transformed to account for the change in illuminant. The transform from CIE XYZ to sharpened RGB is then given as

$$\begin{bmatrix} R'_m \\ G'_m \\ B'_m \end{bmatrix} = \begin{bmatrix} \frac{1}{R_w} & 0 & 0 \\ 0 & \frac{1}{G_w} & 0 \\ 0 & 0 & \frac{1}{B_w} \end{bmatrix} M_{\text{sharp}} \begin{bmatrix} X_m \\ Y_m \\ Z_m \end{bmatrix}. \tag{10.34}$$

Rendering then proceeds in the sharpened RGB color space as normal.

The second inaccuracy stems from a possible difference in dominant illuminant in the rendered scene and the viewing environment. This once more produces an image with an observable, and potentially distracting, color cast. The final image may therefore have to undergo a further white-balancing operation. Hence, after rendering, the transform from sharpened RGB to display values is [1208]

$$\begin{bmatrix} R_d \\ G_d \\ B_d \end{bmatrix} = M_d M_{\text{sharp}}^{-1} \begin{bmatrix} R'_w & 0 & 0 \\ 0 & G'_w & 0 \\ 0 & 0 & B'_w \end{bmatrix} \begin{bmatrix} R' \\ G' \\ B' \end{bmatrix}. \tag{10.35}$$

where M_d is the transform that takes XYZ tristimulus values to the display's color space, and (R'_w, G'_w, B'_w) represents the white balance associated with the viewing environment.

After rendering and white balancing for the display environment, we assume that further corrections are carried out to prepare the image for display. These will involve tone reproduction to bring the dynamic range of the image within the range reproducible by the display device, gamma correction to account for the nonlinear response of the display device to input voltages, and possibly correction for the loss of contrast due to ambient illumination in the viewing environment. Each of these issues is further discussed in Chapter 17.

While many computer-graphics practitioners do not follow this procedure, we hope to have shown that maintaining color accuracy throughout the rendering pipeline is not particularly difficult, although it requires some knowledge of the illuminants chosen at each stage. The combined techniques described in this section, as well as those in Section 8.12, enable a high level of color fidelity without inducing extra computational cost.

Finally, the white-balancing operation carried out to account for the illumination in the viewing environment can be seen as a rudimentary form of color-appearance modeling. Under certain circumstances, it may be possible to replace this step with the application of a full color-appearance model, such as iCAM, which is the topic of Chapter 11.

10.8 Further Reading

Further information on chromatic adaptation models can be found in Fairchild, *Color Appearance Models* [302].

Chapter 11
Color and Image Appearance Models

Colorimetry, as described in Chapter 7, is a standardized method of measuring colors and determining if color stimuli match. Using colorimetry, we can easily communicate information and specify colors in a *device-independent* manner. We know that if two stimuli have identical CIE XYZ tristimulus values and are viewed in identical conditions, then the two stimuli will match. We do not know, however, what the appearances of those stimuli are. Furthermore, we cannot guarantee if those stimuli will still match if any element of the viewing environment changes.

Chromatic adaptation, as discussed in Chapter 10, is one method that our visual system uses to alter its behavior based upon the color of the prevailing illumination. Chromatic-adaptation models extend basic colorimetry to predict color matches across changes in illumination color, essentially giving us *device- and illuminant-color-independent* representation. We still are not able to describe what the stimuli look like.

Color appearance extends upon chromatic adaptation and colorimetry to provide a description of how a stimulus actually appears to a human observer. In essence, we can say that color appearance allows us to take the numbers arriving from color measurement and translate these into words describing how a color looks.

In order to predict the actual perception of on object, it is necessary to know information about the viewing environment itself. As we saw with chromatic adaptation, our visual system is constantly adapting to the environment and, thus, the appearance of any given stimulus will be strongly influenced by these changes.

Color appearance models take into account measurements, not only of the stimulus itself, but also of the the viewing environment, and they generate, in essence, *viewing condition independent* color descriptions. Using color appearance models, we can determine whether or not two stimuli match even when viewed under markedly different conditions. Consider the problem of viewing an image on a LCD display and trying to create a print that matches the image, but that will be viewed in a very different context. Color appearance models can be used to predict the necessary changes that will be needed to generate this *cross-media color reproduction*.

This chapter will provide an overview of the terminology used in color science and color appearance, as well as discuss some well-known color appearance phenomena. The CIECAM02 color appearance model and its uses will be discussed, as well as two spatial or image appearance models: S-CIELAB and iCAM.

11.1 Vocabulary

As with all areas of study, it is important to have a consistent terminology or vocabulary to enable accurate and precise communication. When studying color, this is all the more important, as color is something we all grew up with and most likely acquired different ways of describing the same thing. Color and appearance vocabulary is often interchanged haphazardly, for example, easily swapping terms such as luminance, lightness, and brightness. This may be more prevalent when discussing a subject such as color, as it is easy to assume that since we have been exposed to color our entire life, we are all experts.

Even in our education, treatment of color is highly varied. To a child, color might be made up of their three primary paints or crayons: red, blue, and yellow. Many graphic artists or printers work under the premise of four primaries: cyan, magenta, yellow, and black. Computer-graphics researchers may assume that color is made up of red, green, and blue. Physicists might assume that color is simply the electromagnetic spectrum, while chemists may assume it is the absorption or transmission of this spectrum. All of these assumptions can be considered to be correct, and to each of the disciplines, all of the others may be considered quite wrong.

In the field of color appearance, the standard vocabulary comes from the *International Lighting Vocabulary*, published by the CIE, and the *Standard Terminology of Appearance*, published by the American Society for Testing and Materials (ASTM) [49, 187]. The definition of terms presented below comes directly from these works.

11.1.1 Descriptions of Color

The concept of color itself is well known to almost everyone, though when asked, few can give a precise definition of what exactly color is. Much of the confusion when studying color appearance stems from the very nature and definition of color itself. It is almost impossible to describe or define color without using an example or the word color itself, as evident from the CIE definition.

Color. The attribute of visual perception consisting of any combination of chromatic and achromatic content. This attribute can be described by chromatic color names such as yellow, orange, brown, red, pink, green, blue, purple, etc., or by achromatic color names such as white, gray, black, etc., and qualified by bright, dim, light, dark, etc., or by combinations of such names.

That the definition of color requires the use of the word it is trying to define is an uncomfortable circular reference that can lead to confusion. The original authors of this vocabulary were clearly concerned with this potential confusion, and they added a note that summarizes well the study of color appearance:

> Perceived color depends on the spectral distribution of the color stimulus, on the size, shape, structure, and surround of the stimulus area, on the state of adaptation of the observer's visual system, and on the observer's experience of prevailing and similar situations of observations.

Much of what is described in this note is addressed in further detail in this chapter.

The most important aspect in the definition of color is the first sentence: color is an attribute of visual perception. We typically say that color is a function of the lighting, the environment, and the observer, and without the visual system's involvement, color cannot exist. All the color terminology discussed in this section involves attributes of visual perception. That is to say, without the observer, there can be no discussion of color. The study of color appearance and color appearance models is an attempt to generate physically realizable measurements that correlate with these perceptual attributes.

The definition of color is extended to the notion of related and unrelated colors. The concept of these types of colors is simple, but also critical to gaining a further understanding of color appearance.

Related color. Color perceived to belong to an area of object seen in relation to other colors.

Unrelated color. Color perceived to belong to an area of object seen in isolation from other colors.

Related colors are viewed in the same environment or area as other stimuli, while unrelated colors are viewed in isolation. We generally think that color stimuli are not viewed in complete isolation (unless you are an astronomer), and as such most color appearance models have been designed to primarily predict related colors. The pedigree of these models, however, are the many color-vision experiments performed that used stimuli viewed in isolation. These experiments have been powerful tools for gaining an understanding of how the human visual system behaves. It is also important to understand the differences between these types of stimuli when utilizing models designed to predict one specific type of color appearance.

There are several important color perceptions that only exist when viewed as related or unrelated stimuli. Perhaps the most fascinating of these cases is the so-called "drab" colors such as brown, olive, khaki, and gray. These colors can only exist as related colors. It is impossible to find an isolated brown or gray stimulus, as evidenced by the lack of a brown or gray light bulb. Even if the surface of these lights looked brown or gray, they would appear either orange or white when viewed in isolation.

This is not to say that we cannot make a stimulus look brown or gray using only light, as it is quite easy to do so using a simple display device such as a CRT or LCD. We can only generate those perceptions when there are other colors being displayed as well, or if we can see other objects in the viewing environment. When we talk about relative attributes of color appearance, such as *lightness* or *chroma*, then by definition, these attributes only exist for related colors. The absolute color appearance terminology can be used for either related or unrelated colors.

Of all the color appearance attributes, hue is perhaps the easiest to understand. When most people are asked to describe the color of an object, they will generally describe the hue of the object. That said, it is almost impossible to define hue without using examples of what a hue is.

Hue. Attribute of a visual sensation according to which an area appears to be similar to one of the perceived colors: red, yellow, green, and blue, or to a combination of two of them.

Achromatic color. Perceived color devoid of hue.

Chromatic color. Perceived color possessing a hue.

Hue is most often described as a continuous *hue-circle* or *color-wheel*, as shown in Figure 11.1. The standard definition of hue given by the CIE contains a notion of unique or physiological hues. Specifically these hues are red, yellow,

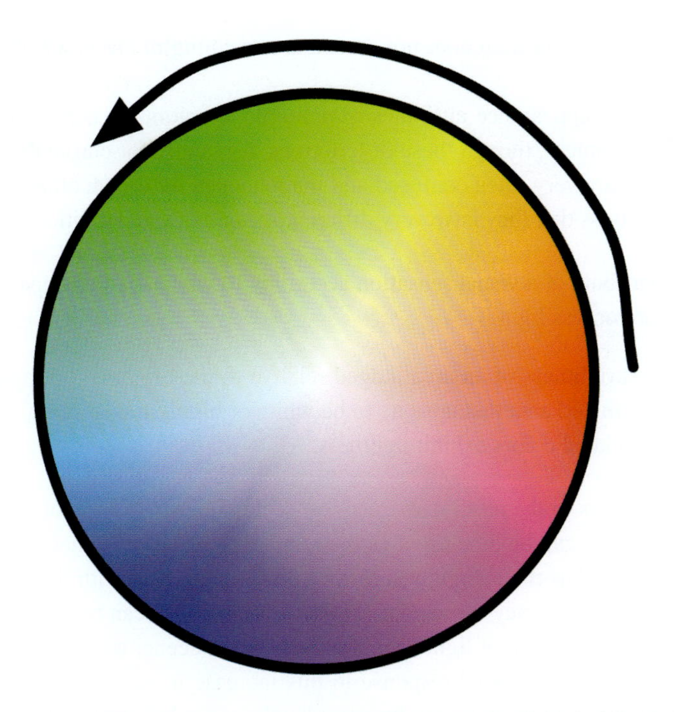

Figure 11.1. The concept of a hue-circle or wheel. The physiological hues of red, yellow, green, and blue form a closed ring with other colors falling between (a combination of) adjacent pairs [49].

green, and blue. These unique hues follow the opponent color theory of vision postulated by Hering in the late 19th century [462]. It has been noted that several of these particular hues are never perceived together, and other colors can be described as a combination of two of them. For example, there is no stimulus that is perceived to be reddish-green or a yellowish-blue, but purple can be described as a reddish-blue and cyan as a bluish-green. This opponent theory suggests the fundamental notion that human color vision is encoded into *roughly* red-green and blue-yellow channels. Thus, opponent color representation is a crucial element of all color appearance models.

The definitions for the achromatic and chromatic colors is also important, as the achromatic colors represent the third channel, the white-black channel theorized by Hering. Although hue, or more precisely hue-angle, is often described as a location on a circle, there is no natural meaning for a color with a hue of zero, and so that value can be considered an arbitrary assignment (as is a hue angle of 360 degrees). Achromatic colors describe colors that do not contain any hue information, but this definition does not extend to a meaningful placement on the hue circle.

The only color appearance attributes of achromatic colors are defined to be brightness and lightness, though these descriptions also apply for chromatic colors. These terms are very often confused and interchanged with each other, but it is important to stress that they have very different meanings and definitions.

Brightness. Attribute of a visual sensation according to which an area appears to emit more or less light.

Lightness. The brightness of an area judged relative to the brightness of a similarly illuminated area that appears to be white or highly transmitting. Note that only related colors exhibit lightness.

Brightness refers to the absolute perception of the amount of light of a stimulus, while lightness can be thought of as the relative brightness. Remember that *luminance* is defined as the energy of light in a scene integrated with the human luminous photopic-efficiency function, CIE $V(\lambda)$. Luminance can be considered to be linear with the amount of light energy present in the scene, while brightness and lightness are our perceptual response to this increase in energy (typically a nonlinear response). In practice, the human visual system behaves as a relative lightness detector.

We can understand the relative lightness behavior of our visual system by imagining the viewing of a black-and-white photograph, both indoors and outdoors. When viewed indoors, the black portions of the image look black, and the white portions look white. This same relationship holds when we look at the picture outdoors on a sunny day. If we were to measure the amount of light coming off the photograph, we may find that the amount of light coming off the black areas of the photograph is hundreds of times larger than the amount of light coming off the white areas of the photograph when viewed indoors. Yet our relative perception of the *lightness* of the black and white areas remains constant. In practice, we find that the black areas of the photograph actually appear darker when viewed in bright sunlight while the white areas appear lighter, a phenomena known as the *Stevens effect*. This and other color appearance phenomena are discussed in more detail in Section 11.2.

Like the definition of color itself, the definitions of lightness and brightness include a note, stating that only relative colors can exhibit lightness. This is because lightness is brightness judged in comparison to a similarly illuminated white object. This is the reason that there cannot be a gray light source, because if it were viewed in isolation it would be the brightest stimulus in the field of view and would thus appear white. As a simple mathematical relation, we can describe the general concept of lightness by

$$\text{Lightness} = \frac{\text{Brightness}}{\text{Brightness of White}}. \tag{11.1}$$

Lightness and brightness are used to describe the relative amount of light a colored stimulus appears to reflect or emit. Hue describes the general "color" of a given stimulus. The terms used to describe the amount of hue or chromatic content a stimulus has are also often confused and interchanged. These terms, *colorfulness*, *chroma*, and *saturation* have very distinct meanings, and care should be taken when choosing which attribute to describe the color of a stimulus.

Colorfulness. Attribute of a visual sensation according to which the perceived color of an area appears to be more or less chromatic. For a color stimulus of a given chromaticity and, in the case of related colors, of a given luminance factor, this attribute usually increases as the luminance is raised, except when the brightness is very high.

Chroma. Colorfulness of an area judged as a proportion of the brightness of a similarly illuminated area that appears white or highly transmitting. For given viewing conditions and at luminance levels within the range of photopic vision, a color stimulus perceived as a related color, of a given chromaticity, and from a surface having a given luminance factor, exhibits approximately constant chroma for all levels of luminance except when the brightness is very high. In the same circumstances, at a given level of illuminance, if the luminance factor increases, the chroma usually increases.

Saturation. Colorfulness of an area judged in proportion to its brightness. For given viewing conditions and at luminance levels within the range of photopic vision, a color stimulus of a given chromaticity exhibits approximately constant saturation for all luminance levels, except when brightness is very high.

Colorfulness describes the absolute amount or intensity of the hue of a given stimulus. Similarly, chroma is to colorfulness as lightness is to brightness, that is, a relative amount of colorfulness for a given brightness. Whereas chroma is defined to be colorfulness of an area relative to the brightness of a similarly illuminated white stimulus, saturation is colorfulness relative to the brightness of the stimulus itself. So while only a related color can exhibit chroma, both related and unrelated colors can exhibit saturation. The differences between chroma and saturation are subtle, but distinct.

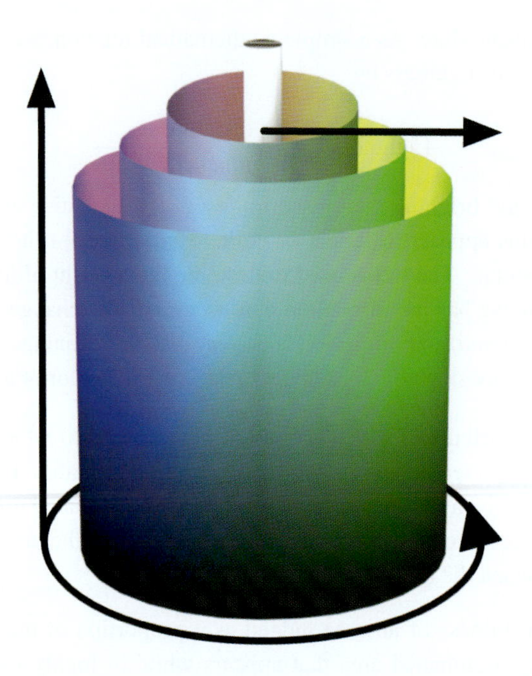

Figure 11.2. The appearance attributes of lightness, chroma, and hue can be represented as coordinates in a cylindrical color space. Brightness, colorfulness, and hue can also be represented as such a space.

We can generalize colorfulness or chroma along with hue and lightness or brightness as attributes that form a cylindrical color space, as seen in Figure 11.2. That is, for any given lightness or brightness, increasing chroma or colorfulness is represented by a perpendicular line away from the lightness axis. Colors at a constant chroma or colorfulness are then represented by cylinders of varying radii in this space.

Saturation can be best described as a *shadow series* of colors. This is perhaps best explained with an example. If we have rounded colored objects that are illuminated at a glancing angle, these objects will have high lightness and chroma on the side that is facing the light, and the brightness and colorfulness will decrease as we move to the side away from the light, eventually becoming black where the object sees no light. This is illustrated by a series of rendered globes in Figure 11.3. Although the brightness and colorfulness have both decreased, as have the lightness and chroma, the saturation has remained constant. Since saturation is the relationship between brightness and colorfulness of the object itself, we can say that in a shadow series, because brightness and colorfulness co-vary, the

Figure 11.3. A rendered shadow series illustrating constant saturation. The glancing illumination causes a decrease in brightness and colorfulness (as well as lightness and chroma) across the surface of each globe, but saturation remains constant.

saturation is constant. Thus, we can think of the appearance attributes of lightness, hue, and saturation as forming coordinates in a conical color space, shown in Figure 11.4. In this case lines of constant saturation form the surface of cones of varying radii originating from a single point at black.

We can once again generalize the formulas for chroma and saturation, which will be useful when we move forward and discuss color appearance models. Recall that chroma is considered relative colorfulness, while lightness is considered relative brightness. Thus, by considering the brightness of a scene as white, we can also define relative saturation as the relationship between a stimulus' chroma to its lightness. These relationships are shown in (11.2).

$$\text{Chroma} = \frac{\text{Colorfulness}}{\text{Brightness (white)}}, \tag{11.2a}$$

$$\text{Saturation} = \frac{\text{Colorfulness}}{\text{Brightness}} = \frac{\dfrac{\text{Colorfulness}}{\text{Brightness(white)}}}{\dfrac{\text{Brightness}}{\text{Brightness (white)}}}, \tag{11.2b}$$

$$\text{Saturation} = \frac{\text{Chroma}}{\text{Lightness}}. \tag{11.2c}$$

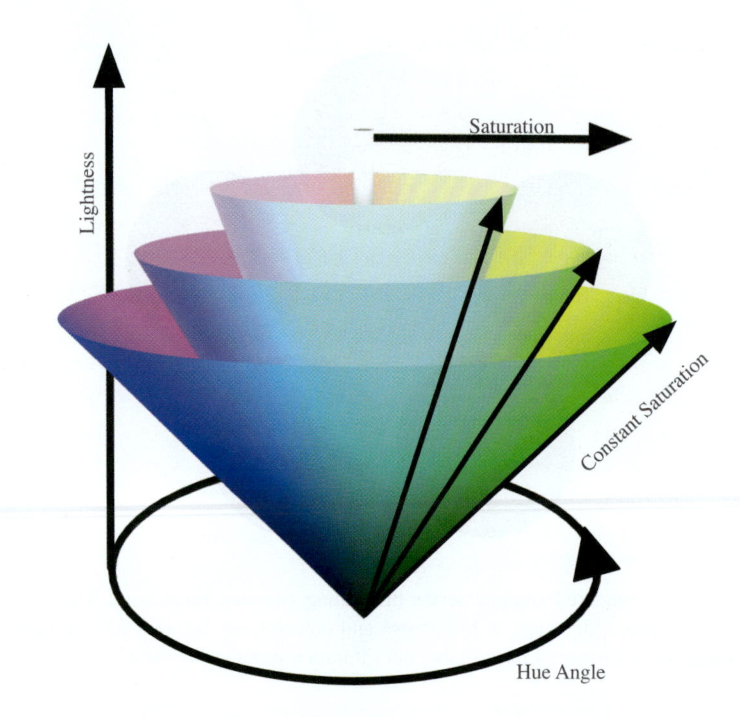

Figure 11.4. The appearance attributes of lightness, saturation, and hue can be represented as coordinates in a conical color space originating at a single point at black.

Similarly to our behavior with lightness and brightness, the human visual system generally behaves more like a relative chroma detector. As the luminance of the viewing conditions increases, the perceptions of lightness and chroma tend to remain constant while the brightness of a white stimulus increases. In this same situation the colorfulness also increases as a function of luminance. This can be visualized by thinking of an typical outdoor scene. On a sunny day, everything lo¡oks very colorful, while the same scene viewed on cloudy day appears much less colorful. This appearance phenomenon is known as the *Hunt effect* and will be discussed in more detail Section 11.2.2.

11.1.2 Relative or Absolute Color Reproduction?

When our goal is digital color imaging, reproduction, or synthesis, it is often easiest and most desirable to represent color as coordinates in a trichromatic space. The success of CIE-based colorimetry for many applications illustrates the potential strengths of this approach. Colorimetry, as we recall, is really only valid when dealing with predicting or generating color matches in identical viewing conditions. If the viewing conditions change, such as when going from a CRT monitor to a print medium or moving a print medium from indoors to outdoors, colorimetry becomes insufficient.

Thus, for accurate color-imaging applications, it becomes necessary to specify the stimuli in terms of color appearance. Complete specification requires at least five of the perceptual attributes described above: brightness, lightness, colorfulness, chroma, and hue. It is generally suggested that the specification of saturation is not necessary, as that can be inferred from the other attributes. Hunt has recently made a compelling argument that saturation should not be ignored, and that it may actually provide a wealth of useful information as we start to focus on the color reproduction of three-dimensional objects [508]. His argument stems from the fact that for a shaded object the chroma or colorfulness changes while saturation remains constant, and humans recognize that the object's color also remains constant.

When designing color reproduction or rendering systems it might appear that specifying more than the three relative color appearance attributes is also redundant. This is not always the case, however, as described in detail by Nayatani et al. [821]. They discuss the distinction between choosing an absolute brightness-colorfulness match or a relative lightness-chroma match. In practice, for most color-reproduction applications, it is often sufficient and desirable to generate lightness-chroma appearance matches, rather than the absolute brightness-colorfulness match.

We can return to our example of a photograph captured from an outdoor scene in very bright sunlight. The absolute luminance levels and brightness values of the outdoor scene are incredibly high. The photograph is then reproduced and viewed in an indoor environment, at much lower luminance levels. Regardless whether the reproduction is a hard-copy print, or just an image on a display device, it is probably physically impossible to achieve the luminance levels necessary to generate an absolute brightness-colorfulness match such that the measured energy reflected off the print or emitted from the display is the same as the original outdoor environment.

Likewise, if we were to generate an image on our display device that has a limited luminance range, we may not want to create an absolute appearance match when we display that image on a high-dynamic range display viewed in a very bright environment. While it may now be physically possible to reproduce the absolute attributes, it would be undesirable since the image would probably appear dark and of very low contrast. Thus, for most general color-reproduction applications, it is often more desirable to create a relative lightness-chroma appearance match, where the relationship between objects in the scene is held constant. We can think of this as analogous to dynamic range reduction for image capture and rendering (discussed in Chapter 17), since we want to generate a relative color appearance match that closely matches in appearance to the original high dynamic range scene.

11.1.3 Descriptions of Viewing Environments

When discussing color appearance and color appearance modeling, it is important to have a consistent vocabulary to describe the viewing environment, or field, just as it is important for the color attributes themselves. As we learned from the definition of color itself, the perception of color is heavily dependent on the environment that the stimulus is viewed in. In color appearance terminology, the viewing environment is also generalized as the *viewing conditions*. Typically, in color appearance, we divide the viewing field into four generalized components: stimulus, proximal field, background, and surround. For a simple color stimulus, the viewing field is specified as in Figure 11.5.

Stimulus. The stimulus is the color element of interest. As with standard colorimetry, the stimulus used for most color appearance research is a small uniform color patch, which subtends 2° of visual angle.

As we learned in Chapter 7, the CIE has separate systems of colorimetry designed to handle small or large color patches, the CIE 1931 Standard Observer and the CIE 1964 supplemental Standard Observer. Most of the experimental data used in color appearance modeling and research has been performed using small uniform color patches, though this is a very active research field and is constantly

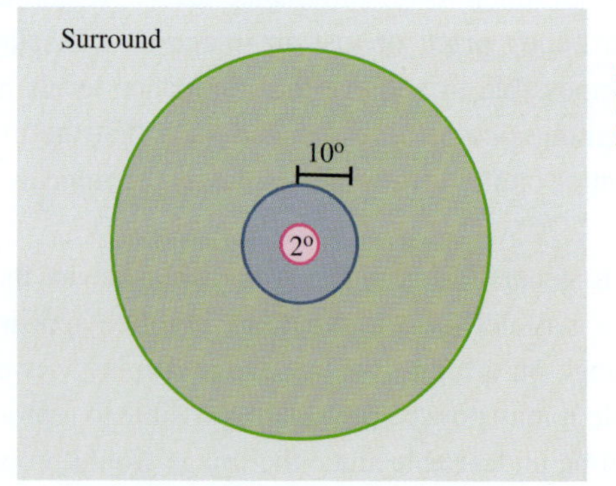

Figure 11.5. The components of the viewing field used for color appearance specification.

being extended. Ideally, the stimulus is described by the full spectral representation in absolute radiance terms, though in practice this is difficult to achieve. When the spectral power distribution is unavailable, the stimulus should be described using a standard device-independent space, such as CIE XYZ tristimulus values, or LMS cone responsivities.

As we progress and apply color appearance modeling to color imaging and synthesis, the definition of the stimulus can get somewhat blurred. Is the stimulus a single pixel, a region of pixels, or the entire image? While it may be most convenient to simply assume that the entire image is the stimulus, that is surely an oversimplification. Currently there is no universally correct definition of the stimulus for spatially complex scenes such as images. Therefore when using images with color appearance models, care should be taken to fully understand the manner in which the models are applied. CIECAM02, the current CIE-recommended color appearance model, is designed to be used for color management and imaging; most applications assume that each pixel can be treated as a separate stimulus. Spatial models of color appearance, also known as *image appearance models*, are designed for use with complex images and will be discussed in more detail in Section 11.4.

Proximal field. The proximal field is considered to be the immediate environment extending from the stimulus for about 2° in all directions.

The proximal field can be useful for measuring local contrast phenomena such as spreading and crispening, which are introduced in Section 11.2.1. Again the question of defining the locally spatial proximal field becomes very difficult when considering digital color images. Should the proximal field for any given pixel be considered to be all of the neighboring pixels, or perhaps just the edge border of the image itself? While it is an important factor in determining the color appearance of a stimulus, in most real-world applications, the proximal field is assumed to be the same as the background. Again, the ideal description of the proximal field would be a measure of both spatial and spectral properties, though these are often unavailable.

Background. The background is defined to be the environment extending from the proximal field, for approximately 10° in all directions. If there is no proximal field defined, then the background extends from the stimulus itself.

Specification of the background is very important in color appearance, as it strongly influences the appearance of any given stimulus and is necessary to model phenomena such as simultaneous contrast. Specifying the background when dealing with simple color patches is relatively straightforward. At the minimum, the background should be specified with absolute CIE XYZ tristimulus values, since the luminance of the background will also have an effect on color appearance. Although the background is defined to span at least 10 degrees of visual angle, typically it is specified using the same CIE standard observer as the stimulus itself.

Again, specifying the background when calculating the appearance of color images can be as confusing as specifying the stimulus and the proximal field. For any given image pixel, the background actually consists of many of the neighboring pixels as well as the area outside the image. Generally, there are two different assumptions that people use for specifying the background in color-imaging applications. The first and easiest is to assume that the entire image is the stimulus, so that the background is the area extending 10° from the image edge. Another assumption is that the background is constant and of some medium chromaticity and luminance, e.g., a neutral gray. This can be chosen as a non-selective gray viewed in the same lighting conditions or can be taken to be the mean color of the image itself.

Many imaging applications strive to reproduce images of constant spatial structure and size, and thus these concerns are mitigated or disappear entirely. Care must be taken when calculating color appearances across changes in image sizes, though. Braun and Fairchild describe the impact on some of these background decisions [124]. More recently, Nezamabadi et al. and Xiao et al. have investigated the effect of image size on color appearance [833, 1263].

Surround. The surround is considered anything outside of the background. For most practical applications, the surround is considered to be the entire room that the observer is in.

For practical applications, color appearance models tend to simplify and express the surround in distinct categories: dark, dim, and average. For instance, movie theaters are usually considered to have a dark surround, while most people tend to view televisions in a dim surround. A typical office or computing environment is often said to be an average surround. A more detailed discussion on the effect of the surround on color appearance is available in Section 11.2.4.

11.1.4 Modes of Viewing

Any changes to the viewing fields discussed above may quickly result in a change in the color appearance of a stimulus. The discussion of color appearance phenomena in the following section explains some of these changes in detail. Other factors that cannot be readily explained using our model of a simplified viewing field can also have a large effect on the perceived appearance of a stimulus. As we have established by now, the perception of color cannot be adequately explained by the physics of light alone. The human observer is the critical component that is ultimately responsible for any sensation or perception. The human visual system relies both upon sensory mechanisms and automatic changes based on biological and physiological factors, as well as on cognitive interpretations.

The full scope of cognitive mechanisms is not yet understood, though we are able to recognize and model some of the ultimate behavior and responses. One of the most important cognitive behaviors affecting color appearance is the mode of appearance. Like much of color itself, the mode of color appearance is a difficult concept to grasp at first and might be best described with an example.

Often times, we catch a glimpse of colors that at first appear vibrant and highly chromatic and then suddenly become mundane. When walking past a semi-darkened room recently, we were struck by what appeared to be a wall that appeared painted with a mural of abstract colored rectangles of various shades of brown and orange. The appearance of the wall was quite striking and highly chromatic. Upon entering the room, it became obvious that the colors were the result of the street-lights, light from very orange low-pressure sodium lamps, entering through an open window. Almost immediately, the appearance of the various highly chromatic rectangles changed and appeared to be the same white wall illuminated by the orange light.

This scene was captured with a RAW (non-white balanced) camera, and is shown at the top of Figure 11.6. The approximation of the immediate change in appearance upon realizing it was the light from the window is shown in the bottom frame of Figure 11.6. This immediate change in appearance was entirely cognitive, and it can be attributed to a switch from a surface mode of viewing to an illumination mode.

There are five modes of viewing that affect color appearance: illuminant, illumination, surface, volume, and film. A brief overview of these modes is presented here, though a more complete description can be found in the texts by the Optical Society of America [860], or more recently by Fairchild [302].

The *illuminant mode of appearance* is color appearance based on the perception of a self-luminous source of light. Color appearance perceptions in the illuminant mode typically involve actual light sources, and more often then not, these are the brightest colors in the viewing field, for example, looking at a traffic light at night or at a light fixture indoors.

The cognitive assumption that the brightest objects in a scene are light sources can lead to some interesting appearance phenomena when non-illuminant objects

Figure 11.6. The wall of a room that appeared to be a mural of highly chromatic different colored squares (top) and the change in appearance once it was obvious that it was orange light entering from an open window (bottom).

in a scene appear much brighter than the surrounding scene. These objects can sometimes be perceived in an illuminant mode and may be described as glowing. Evans coined this perception *brilliance*, though Nayatani and Heckaman have more recently touched upon on this topic [288, 448, 822].

Another example of an object appearing to glow may occur when there are fluorescent objects in a scene. Fluorescence is found in materials that absorb energy at one wavelength and emit that energy as light at much longer wavelengths. Fluorescent objects can appear to glow, in part because they absorb energy from non-visible UV portions of the spectrum and emit this as light in the visible portions. Thus, these objects can appear to be much brighter than the objects in the surrounding scene, and they are often said to take on the appearance of a light source.

The *illumination mode of appearance* is similar to the illuminant mode, except that perceived color appearance is thought to occur as a result of illumination rather than the properties of the objects themselves. Consider the example given in Figure 11.6. Upon entering the room, it became evident that it was a single color wall that was being illuminated through a series of windows. The orange color was instantly recognized as coming from an illumination source. There are many clues available to a typical observer of a scene when determining if color is a result of illumination. These clues include the color of the shadows, the color of the entire scene, complex color interactions at edges and corners, as well as the color of the observer themselves.

The perceived color of the wall, as described above, is an example of the *surface mode of appearance*. At first it appeared that the surface of the wall was painted with a colored mural. In this mode of appearance, the color of a surface is perceived as belonging to the object itself. For example, an observer may know that the skin of an apple is red, and the apple will maintain a certain degree of "redness" despite large changes in illumination. This is an example of *discounting-the-illuminant*, as we discussed in Chapter 10. Any recognizable object provides an example of the surface mode of appearance. For this mode of appearance, we need both a physical surface and an illuminating light source.

The *volume mode of appearance* is similar in concept to the surface mode, except that the color is perceived to be a part of a bulk or volume of a transparent substance. An example of the volume mode of appearance can be found in the perceived color of liquids, such as a cup of coffee. Coffee is often classified into categories such as light, medium, and dark, based upon the color of the brewed coffee itself. The color of the coffee is not thought to be just a characteristic of the surface, but rather it exists throughout the entire volume. Adding milk to the coffee increases the scattering and the beverage becomes opaque. This is an example of a volume color changing into a surface color. Volume color requires transparency as well as a three-dimensional shape and structure.

The final mode of appearance, called either the *aperture or film mode*, encompasses all remaining modes of appearance. In the film mode, color is perceived as an aperture that has no connection with any object. One example is a photographic transparency on a light-box. Any object can switch from surface mode to aperture mode, if there is a switch in focus from the surface itself. This can be accomplished purposely, by using a pinhole or a camera lens system. The differences in our perception between a soft-copy display device and a hard-copy print may be associated with the different modes of viewing. The print is generally considered to be viewed as a surface mode, while the self-luminous display may be viewed in film mode or perhaps in an illuminant mode.

11.2 Color Appearance Phenomena

This section deals with common examples of colored stimuli that do not follow the predictions of basic colorimetry. Certain changes in the viewing field cause changes in appearance that cannot be predicted with XYZ tristimulus matching. The CIE system of colorimetry was developed using a color-matching experiment, as discussed in Chapter 7. To recap, colorimetry states that if two stimuli have identical tristimulus values then those two stimuli will match each other for a single given viewing condition. Colorimetry does not attempt to predict if the colors will match if any aspect of the viewing condition changes.

This section will illustrate several examples of where the color matches will indeed breakdown, as various elements of the viewing field are changed. Among the changes in viewing condition are changes in illumination level, illumination color, surround, background, and viewing mode. The examples shown here illustrate the limitations of basic colorimetry and the need for advanced colorimetry, also called color appearance modeling. The foundations of most color appearance models stem from the study of these individual phenomena, so it is important to review and understand our behavior. The recognition and understanding of these color appearance phenomena are also important for a color imaging-system designer or computer-graphics researcher, as many of these examples are very common in their occurrence. This section will describe several common color appearance phenomena, including those caused by local spatial structure, luminance or illuminant color changes, and changes in the viewing surround.

11.2.1 Changes in Appearance Caused by Spatial Structure

Perhaps the most easily recognized color appearance phenomenon is that of simultaneous contrast. This was introduced briefly in Section 5.4.2. Figure 11.7 illustrates an example of simultaneous lightness contrast. Each of the rows of the

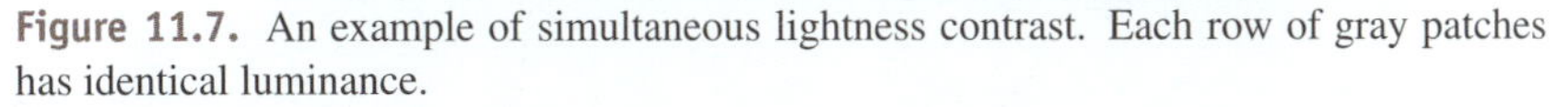

Figure 11.7. An example of simultaneous lightness contrast. Each row of gray patches has identical luminance.

figure are gray patches of identical luminance, with each subsequent row decreasing in luminance. When placed upon a background with a white-to-black gray gradient, we see that the patches in each row actually appear of a different lightness (or perception of luminance) creating the illusion of a dark to light contrast or tone scale.

Similarly, the relative *contrast* of the tone scale in each of the columns also appears different. These observed differences are a result of the background in the viewing field. It is for this reason that we should not consider the perception of contrast to be a global property of an image measured by the ratio of maximum white to black, also known as *Michelson contrast*. Rather, we should consider contrast to be both a global and local perception. Peli gives an excellent overview of contrast in complex images, and more recently Calabria and Fairchild measured contrast perception for color imaging [146, 147, 888].

Simultaneous contrast is not limited to changes in lightness perception, as it can also cause the color of a stimulus to shift in appearance when the color of the background changes. This is illustrated in Figure 11.8, where each of the small patches has an identical XYZ tristimulus value. The change in color of a stimulus tends to follow the opponent color theory of vision. The patches on the yellow gradient should appear more blue, while the patches on the blue gradient should appear more yellowish-green. Interesting to note in Figure 11.8 is that the colors of any neighboring patches do not appear very different, though the appearance change across the entire field is readily apparent. This again suggests a very localized perception of appearance. Shevell et al. present some very striking examples of this spatially localized simultaneous contrast; they also developed visual models to predict this spatial relationship [67, 1039]. In general, as the spatial frequency of a stimulus increases, the contrast effect actually ceases and in some cases reverses and may assimilate the background color or spread its own color to the background.

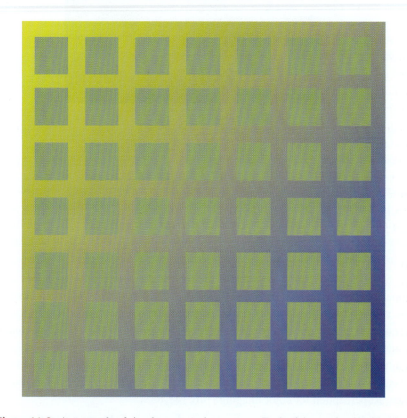

Figure 11.8. An example of simultaneous color contrast. Each of the patches has identical CIE XYZ tristimulus values.

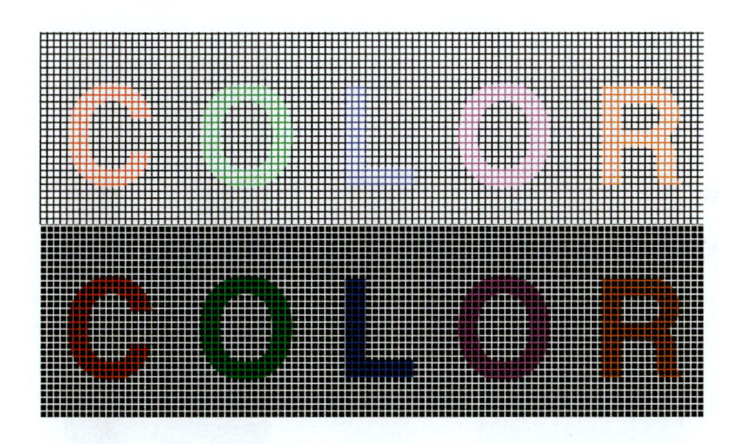

Figure 11.9. An example of spreading. Only the lines contain color, though that color appears in the background as well (adapted from [630]).

At a high-enough spatial frequency, simultaneous contrast is replaced with *spreading*. With spreading, the color of a stimulus mixes with the color of the background. Recall that in the case of simultaneous contrast, the color of a stimulus takes on the opposite color of the background. One hypothesis is that spreading is caused by optical blurring of the light coming from the stimulus and mixing with the light coming from the background. While this might be true for very high-frequency stimuli, such as half-tone dots, it does not fully explain the spreading phenomenon. Spreading often occurs when the stimuli are very distinct from the background. An example is shown in Figure 11.9. The only colors present in this figure are the black or white backgrounds, and the colored lines, though it should appear as if the color portions of the lines have spread into the background.

Research is ongoing to understand the transition point between simultaneous contrast and spreading and the overall effects of spatial frequency on color appearance [67, 1039]. Related appearance phenomena, though arguably more complex, include neon spreading and the watercolor effect. Neon spreading combines spreading with the perceptual attribute of transparency, while the watercolor effect generates the appearance of a surface color based entirely on edge boundaries. Bressan provides an excellent review of neon spreading [127], and Pinna does the same for the watercolor effect [898].

Simultaneous contrast can also give rise to an increase in perceived color difference between color stimuli. This effect is known as crispening and can be observed in Figure 11.10. Crispening causes an increase in perceived color difference between two stimuli when the viewing background is similar to the color of

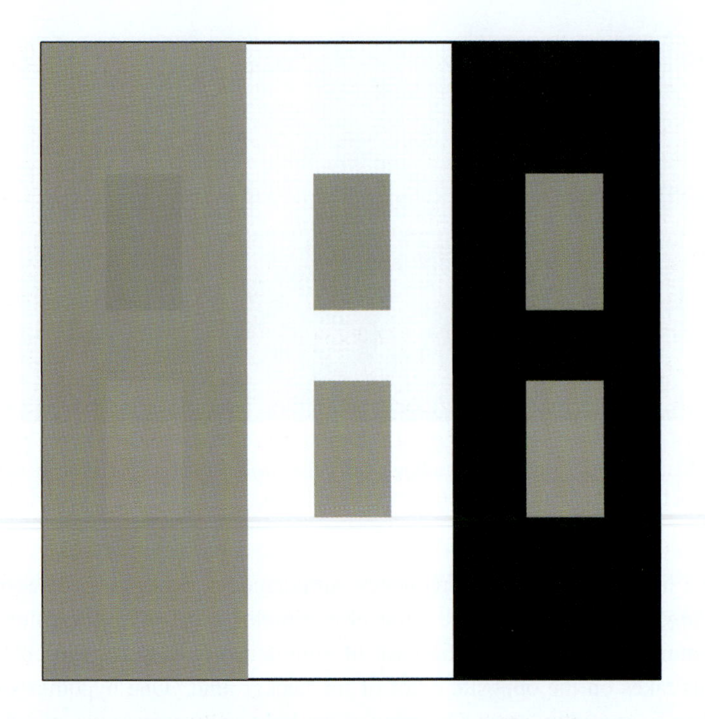

Figure 11.10. An example of lightness crispening. The color difference between the two small patches is the same for all backgrounds, but should appear greatest when the background is close in color to the stimuli.

the stimuli. In the figure, the differences between the small gray patches are the same for all three backgrounds, but the difference looks the largest on the gray background. Similar effects can be seen for color patches, known as chromatic or chroma crispening. More details on lightness and chroma crispening can be found in Semmelroth [1025] and Moroney [792].

11.2.2 Luminance-Induced Appearance Phenomena

The above phenomena involve appearance changes as a function of spatial structure or color-background viewing field. Large color shifts can also occur when the viewing illumination changes. We have already introduced the color changes caused by chromatic adaptation in Chapter 10. Color appearance is also dictated by the overall amount of energy or illumination. Luminance changes are very common in everyday life, and we are probably all familiar with the effect on color appearance, even if we do not realize it at first. The classic example is the difference between looking outside on a bright sunny day or on a dark overcast day. Objects tend to appear very bright and colorful on a sunny day and somewhat subdued on an overcast day. These occurrences can be well described by both the Hunt effect and the Stevens effect.

The Hunt effect states that as the luminance of a given color increases, its perceived colorfulness also increases. This effect was first identified in a seminal study by Hunt on the effects of light and dark adaptation on the perception of color [504]. A haploscopic matching experiment was performed, similar to that used to generate the CIE Standard Observers. However, in haploscopic matching, observers are presented with one viewing condition in their left eye and another in their right eye.

In the Hunt experiment, observers used the method of adjustment to create matches on stimuli viewed at different luminance levels for each eye. When the adjusting stimulus was present at low luminance levels, the observers required a significant increase in the colorimetric purity to match a stimulus viewed at a very high luminance level. This indicated that the perception of colorfulness is not independent of luminance level. If we think about this in terms of a chromaticity diagram, the *perceived* chromaticity shifts toward the spectrum locus as the luminance levels increase, or shifts toward the adapted white as luminance levels decrease.

Going back to the sunny day example, the Hunt effect partially explains why objects appear much more vivid or colorful when viewed in a bright sunny environment. To summarize, the Hunt effect states that as absolute luminance levels increase the colorfulness of a stimulus also increases.

Similar to the Hunt effect, we also see that brightness is a function of chromatic content. The perception of brightness or lightness is often erroneously assumed to be a function of luminance level alone. This is not the case and is well illustrated by the Helmholtz-Kohlrausch effect. The Helmholtz-Kohlrausch effect shows that brightness changes as a function of the saturation of a stimulus; i.e., as a stimulus becomes more saturated at constant luminance, its perceived brightness also increases.

Another way to describe this effect is say that a highly chromatic stimulus will appear brighter than an achromatic stimulus at the same luminance level. If brightness were truly independent of chromaticity, then this effect would not exist. It is important to note that the Helmholtz-Kohlrausch effect is a function of hue angle as well. It is less noticeable for yellows than purples, for instance. Essentially this means that our perception of brightness and lightness is a function of saturation and hue, and not just luminance. Fairchild and Pirrotta published a general review of the Helmholtz-Kohlrausch effect as well as some models for predicting the effect [295].

Another interesting relationship between luminance level and chromatic colors is the Bezold-Brücke hue shift. This phenomenon relates the perceived hue of monochromatic light sources with luminance level. It is often assumed that the hue of monochromatic light can be described completely by its wavelength. This is not the case, as the hue of a monochromatic light will shift as the luminance of the light changes. The amount of hue shift also changes both in direction and magnitude as a function of the wavelength. Experimental results regarding the Bezold-Brücke hue shift can be found in work published by Purdy [929].

One very important consideration for these hue shifts is that all of the experimental data were obtained using unrelated colors. Recall that unrelated colors are stimuli viewed in complete isolation and that they occur rarely in everyday life. Perhaps luckily for digital-imaging applications, Hunt published a report indicating that the Bezold-Brücke hue shift disappears for related colors [506]. As such, most modern color appearance models designed for imaging applications do not take this phenomenon into account.

A further very important color appearance phenomenon that must be accounted for is the overall change in contrast as a function of luminance level. This increase in contrast was examined closely in a classic study by Stevens and Stevens [1085]. Their study indicated that as luminance level increases so too does the brightness contrast (see Section 6.4); this effect has come to be referred to as the Stevens effect.

In their original study, observers performed magnitude-estimation experiments on the brightness of stimuli across many different luminance-adapting conditions. The results showed that brightness tended to follow a power-law relationship with luminance, thus forming the basis for Stevens' power law. This nonlinear compressive relationship between luminance and brightness, as well as lightness, is a very important behavior of our visual system. However, this study also shows that the exponent of the power function changes as a function of adapting-luminance level. Essentially, as the adapting-luminance level increases, bright colors tend to look brighter, and darker colors tend to look darker. So, as the adapting-luminance level increases, the rate of change between the brightness of the dark and light colors increases. This rate of change is often considered the contrast of the scene; we can therefore say that contrast increases with absolute luminance levels.

It is interesting to note that most of these color appearance phenomena can influence our perception of the contrast in images. The Stevens effect is the most obvious of these, but both the Hunt effect and the Helmholtz-Kohlrausch effect can also increase perceived contrast. This suggests that there are strong interactions between color appearance attributes, and they should not be taken to be orthogonal or independent perceptions. Any change in one of these attributes may actually result in undesirable (or desirable) changes in the others.

11.2.3 Hue Phenomena

Clearly luminance changes can cause large shifts in the appearance of colored stimuli, indicating the importance of absolute luminance measurements for predicting appearance rather than just relative colorimetric data. This section examines several phenomena that result from changing the hue of the viewing conditions, unrelated to the overall changes caused by chromatic adaptation. These hue changes are much less common than luminance changes and often not very perceptible or negligible for most color-imaging applications. They are included here as a historical note, and because many models of color appearance have been developed that are capable of compensating for these effects.

As shown above, the Bezold-Brücke hue shift illustrated that we cannot simply use the wavelength of monochromatic light sources as a predictor of perceived hue. As luminance levels change, the perceived hue can also change. A similar effect is known as the Abney effect, which states that adding a broad spectrum "white" light to a monochromatic light does not preserve constant hue. Another way of expressing this is to say that straight lines in a chromaticity diagram, radiating from the chromaticity of the white point to the spectral locus, are not necessarily lines of constant hue. Unlike the Bezold-Brücke hue shift, this effect is valid for related colors as well as unrelated colors. The Abney effect indicates another area of the visual system that behaves as a nonlinear system. More details of both the Bezold-Brücke and the Abney effect can be in found in [929].

Another interesting, though difficult to reproduce, phenomenon involving monochromatic illumination is the Helson-Judd effect [455]. In essence, this effect says that non-selective (gray) stimuli viewed under highly chromatic illumination take on the hue of the light source if they are lighter than the background, and take on the complementary hue if they are darker than the background. So a dark-gray sample viewed on a medium-gray background under red illumination will look somewhat green, while a light-gray sample would appear pinkish. This effect almost never occurs in common practice, and it is very difficult to reproduce even in a controlled experiment. Some color appearance models do take this effect into account in their predictions, and so it is worth mentioning. More recently, Fairchild has shown that this effect disappears when dealing with complex stimuli, such as images [302].

11.2.4 Changes in Appearance Caused by the Surround

The Stevens effect demonstrated that contrast for simple patches increases as a function of the adapting luminance and the luminance of the stimulus itself. Bartleson and Breneman studied the effects of the luminance level of the surround on complex stimuli, specifically images [72]. Recall that in color appearance terminology, the surround of the viewing field is considered to be the entire field outside of the background and can generally be considered the entire viewing room. Bartleson and Breneman found results that were very similar to Stevens with regard to changes in overall luminance level.

More interestingly, they noticed interesting results with respect to the change in the relative luminance of the image surround when compared to the luminance levels of the stimulus itself. It was found that perceived contrast in images increases as the luminance of the surround increases but the luminance of the stimulus remains constant. Another way of saying this is that when an image is viewed in a dark surround, the black colors look lighter while the light colors remain relatively constant. As the surround luminance increases, the blacks begin to look darker, causing overall image contrast to increase. More recently, Liu found similar results using a different approach of allowing observers to adjust the luminance and color of the surround for optimal image quality [689].

This behavior closely matches a practice that historically has taken place in the photographic world. For optimum tone reproduction, photographic-film transparencies designed for viewing in a darkened room have a much higher contrast than those designed for viewing as a print in a bright room. Hunt and Fairchild provide more in-depth analysis of the history and prediction of optimal tone reproduction for complex images, an important consideration for overall image quality [302, 509]. In their original publication, Bartleson and Breneman published equations that predicted their results well, though these equations were later simplified to create equations for calculating optimal tone reproduction. Such equations have been adapted and are included in the predictions of many models of color appearance, including CIECAM02 [302].

Surround compensation can play a key part in the design and implementation of a color-reproduction system, for example, producing a computer-rendered feature film that will be viewed originally in a darkened movie theater and then later at home on a television in a dim or average surround. If care is not taken, the movie may appear to have too much contrast when viewed at home. This phenomenon actually seems counter-intuitive to most people, as many of us actually turn off the surrounding lights when viewing movies at home to obtain a higher contrast. While it is true that turning off the surround lights will actually reduce the perception of contrast, what most people are actually doing is removing flare from the surface of the screen. Flare can be considered stray light that is not part of the desired stimulus; it has the effect of reducing overall contrast by essentially increasing the luminance of the black. We should not confuse the removal of flare with the change in perception caused by the luminance of the surround.

11.3 Color Appearance Modeling

CIE tristimulus colorimetry was originally created for a single, relatively simple, purpose, namely predicting when two stimuli will match for the average observer, under a single viewing condition. For this task, colorimetry has been remarkably successful for science and engineering purposes. We have already seen the limitations of basic colorimetry with some of the color appearance phenomena described above, since altering any part of the viewing field can drastically change the appearance of a stimulus. Chromatic-adaptation transforms, as described in Chapter 10, extend basic colorimetry so that it is possible to predict matches across changes in the color of the illumination across viewing conditions. Chromatic-adaptation transforms do not yet provide enough information to allow us to describe the actual color appearance of a stimulus.

To accurately describe the color appearance of a stimulus, we can use the color terminology described above, including the relative terms of lightness, saturation, and chroma, as well as the absolute terms of brightness and colorfulness. The attribute of hue can be used to describe either absolute or relative appearance. Even with the addition of a chromatic-adaptation transform, CIE tristimulus colorimetry is not able to predict any of these appearance terms. In order to generate measurements corresponding to these attributes, we must use a *color appearance model*.

So what is a color appearance model? In its simplest form, we can think of a color appearance model as a tool to translate numbers and measurements into words describing a color. The CIE Technical Committee TC1-34, *Testing Colour Appearance Models*, defined exactly what is necessary to be considered a color appearance model [191]. The definition agreed upon is as follows: *A color appearance model is any model that includes predictors of at least the relative color appearance attributes of lightness, chroma, and hue* [191].

This is a very lenient definition of what constitutes a color appearance model, though it does capture the essence. At the very least, we can say that color appearance models minimally require some form of a chromatic-adaptation transform, as well as some form of a *uniform color space*. Color appearance models that are more complicated are capable of predicting absolute attributes, such as brightness and colorfulness, as well as luminance-dependent effects such as the Hunt and Stevens effects. Spatially structured phenomena, such as crispening and simultaneous contrast, require models of spatial vision as well as color appearance.

There are many color appearance models that have been developed, each designed with specific, and perhaps different, goals in mind. Some of these models that one may come across include CIELAB, Hunt, Nayatani, ATD, RLAB, LLAB, ZLAB, CIECAM97s, and CIECAM02. The interested reader is encouraged to examine Fairchild's thorough examination on the history and development of many of these models [302]. Most of these models are beyond the scope of this text, though we will examine CIELAB as well as CIECAM02 as simple color appearance predictors. The latter is the current recommended color appearance model from the CIE; it is based upon the strengths of many of the other models mentioned above.

11.3.1 CIELAB as a Color Appearance Model

As we first introduced in Chapter 8, and revisited in Chapter 10, CIELAB is a color space designed for predicting small color differences. It was designed to be perceptually uniform for the ease of calculating these color differences. Despite this design intention, CIELAB has been used with varying degrees of success as a color appearance model. This is because CIELAB does include predictors of lightness, chroma, and hue. These attributes are generated by performing a rectangular-to-cylindrical transformation on the CIE a^* and b^* coordinates. These predictors allow CIELAB to be labeled as a color appearance model. We will use it here as a simple model to illustrate the design of more complicated color appearance models.

CIELAB calculations require a pair of CIE XYZ tristimulus values: those of the stimulus itself, as well as those of the reference white point. Already we see that CIELAB requires more information about the viewing conditions than standard tristimulus color matching, as do all color appearance models. The reference white-point values are used in a von Kries-type chromatic-adaptation transform. Recall from Chapter 10 that this chromatic-adaptation transform is less than optimal and may be responsible for some of the hue nonlinearity associated with the CIELAB space.

The adaptation transform is followed by a compressive cube-root nonlinearity and an opponent color transformation. The cube-root power functions attempt to model the compressive relationship between physical measurements of luminance and psychological perceptions. These compressive results were discussed above with regard to the Stevens effect. The cube-root function is replaced by a linear function for very dark stimuli. The CIELAB L^* coordinate, as a correlate to perceived lightness, is a function purely of luminance. It ranges from 0.0 for absolute black stimuli to 100.0 for diffuse white stimuli. It is possible to calculate L^* values greater than 100 for specular highlights or fluorescent materials that have a higher luminance than the input white point, though care should be taken to fully comprehend the meaning of these values.

The a^* and b^* coordinates approximate, respectively, the red-green and yellow-blue of an opponent color space. A positive a^* value approximates red while a negative approximates green. Similarly, a positive b^* correlates with yellow while negative values correlate to blue. It should be noted that only the physiological unique hue of yellow lines up with the b^* axis while the other unique hues are a combination of both the chromatic components. Purely achromatic stimuli such as whites, grays, and blacks should have values of 0.0 for both a^* and b^*.

The definition of a color appearance model requires a minimum of predictions for lightness, chroma, and hue. The CIELAB L^* provides us with a lightness prediction, but a^* and b^* do not provide us with correlates of chroma and hue. These correlates can be calculated by transforming the Cartesian coordinates of a^* and b^* into cylindrical coordinates of $C_{ab}{}^*$ and h_{ab}, where $C_{ab}{}^*$ represents chroma and h_{ab} represents hue angle (see Section 8.7.2).

With the cylindrical coordinates of chroma and hue angle, we now have enough information to predict or describe the general color appearance of a stimulus. There are several important caveats, however, that should be considered. The wrong von Kries transform is clearly a source of color appearance errors. CIELAB is also incapable of predicting many of the color appearance phenomena described above. These include all luminance, surround, background, and discounting-the-illuminant effects. CIELAB also assumes 100 percent complete adaptation to the white point.

Since CIELAB was really only designed to predict small color differences between similar objects under a single viewing condition, it is impressive that it can be used as a color appearance model at all. Many color-management utilities actually use CIELAB as a method of specifying device-independent colors as well as a method for transferring between different white points. For such applications, a user may be better off using a more advanced chromatic-adaptation transform such as those described in Chapter 10 along with CIELAB, rather than relying on the chromatic-adaptation transform in CIELAB.

11.3.2 The CIECAM02 Color Appearance Model

Color appearance research has been ongoing for many decades, and the fruition of this research resulted in the formulation of many different color appearance models. Many of these models were designed with different goals using a variety of experimental data. Until the late 1990s, it was often difficult to choose which model to use for any given task, resulting in industry confusion. This confusion was alleviated in 1997 with the formulation of the CIE recommended CIECAM97s color appearance model [191]. The CIECAM97s model was designed to work at least as well as all of the existing models for the color appearance phenomena that it predicts. It can be viewed as essentially a hybrid of the best parts of many different models. The general pedigree of CIECAM97s arose from components in the Hunt, Nayatani, RLAB, and LLAB models [191].

The *s* in CIECAM97s represented the initial attempt to create a simplified color appearance model, for general-purpose imaging applications. The initial goal had been to create a comprehensive (*CIECAM97c*) version of the model that included predictions for a wider array of appearance phenomena, as well as including rod contributions so it could be used for very low-level scotopic and mesopic applications. The industry demands for the more comprehensive model were not there, and as such this model was never developed [302].

That the year 1997 was included in CIECAM97s was an indicator that the CIE recognized that this was an interim recommendation designed to focus color appearance research on a single unified model and provide a benchmark for further appearance-model development. This approach was very successful and has led directly to the creation of the CIECAM02 color appearance model. CIECAM02 is the currently recommended color appearance model, specifically designed for color-management systems and digital color imaging [195]. CIECAM02 was designed specifically to account for some weaknesses that were discovered in the CIECAM97s model and was intended to be simpler and more easily inverted [195]. That the year is still included in the name of CIECAM02 indicates that the CIE still considers this an interim model in this very active field. Though it is simpler than CIECAM97s, it is a more accurate model based upon existing visual data and is recommended to replace CIECAM97s in most applications.

The forward and inverse calculations of CIECAM02 are presented below. As a general workflow, the input to CIECAM02 consists of XYZ tristimulus values. These are transformed into a sharpened cone space for the chromatic-adaptation transform. The adapted signals are then transformed into cone signals and processed through a nonlinear hyperbolic-type compressive transform. The compressed signals are transformed into the appearance attributes discussed above. CIECAM02 is capable of predicting all six appearance attributes: light-

Surround condition	c	N_C	F
Average	0.69	1.0	1.0
Dim	0.59	0.9	0.9
Dark	0.525	0.8	0.8

Table 11.1. Input parameters for the surround conditions in CIECAM02.

ness, brightness, chroma, colorfulness, saturation, and hue. CIECAM02 is also designed to predict the changes in color appearance as a function of luminance, including the Hunt and Stevens effects, as well as the changes in appearance as a function of the surround. Additionally CIECAM02 is designed to predict the changes in lightness and brightness perception as a function of chromatic content.

The inputs to CIECAM02 include the CIE XYZ tristimulus values of the stimulus (X, Y, and Z), and like CIELAB, the CIE XYZ tristimulus values of the adapting white (X_w, Y_w, and Z_w). In addition, the absolute luminance in cd/m^2 of the adapting white, L_A, is input. The luminance of the adapting white is typically taken to be 20 percent of the luminance of a white object in the scene, or 20 percent of the maximum output luminance of a self-emitting display. The relative background-luminance factor, Y_b, is also input into CIECAM02. The surround conditions are generally defined to be one of three categories: average, dim, or dark. If the user is unsure of which category to select, they can calculate the surround ratio, S_R, between the luminance of a white in the surround L_{SW} and the luminance of the scene or device white L_{DW} as follows:

$$S_R = \frac{L_{\text{SW}}}{L_{\text{DW}}}. \tag{11.3}$$

If the surround ratio is zero, then a dark surround should be chosen; if it is less than 0.2, a dim surround should be selected. An average surround is selected for all other values. The values c, N_c, and F are then input from Table 11.1. It is possible to linearly interpolate the values from two surround conditions, though it is generally recommended to use the three distinct categories.

11.3.3 Chromatic Adaptation

The first stage of the CIECAM02 calculations is the chromatic-adaptation transform. This process is identical to that discussed in Section 10.4.6; it is repeated here for completeness. The XYZ tristimulus values of the stimulus and the white point are first transformed into sharpened cone signals, represented by RGB in

CIECAM02:

$$\begin{bmatrix} R \\ G \\ B \end{bmatrix} = \mathbf{M}_{\mathrm{CAT02}} \begin{bmatrix} X \\ Y \\ Z \end{bmatrix}, \tag{11.4}$$

with M_{CAT02} defined as

$$M_{\mathrm{CAT02}} = \begin{bmatrix} 0.7328 & 0.4296 & -0.1624 \\ -0.7036 & 1.6975 & 0.0061 \\ 0.0030 & 0.0136 & 0.9834 \end{bmatrix}. \tag{11.5}$$

The next stage is to calculate the degree of chromatic adaptation D, which is taken to be a linear scale between the adaptation to the current viewing condition and the reference illuminant, which is CIE Illuminant E (an illuminant with an equal-energy spectrum) for CIECAM02. A D value of 1.0 implies we are completely adapted to the current viewing condition, while a factor of 0.0 implies we are fully adapted to the reference condition. The degree of adaptation D can be explicitly set to a value based upon an understanding of the viewing conditions, or can be implicitly calculated. If it is implicitly calculated, it is taken to be a function of only the adapting luminance L_A and surround factor F.

$$D = F \left[1 - \left(\frac{1}{3.6} \right) \exp \left(\frac{-(L_A + 42)}{92} \right) \right]. \tag{11.6}$$

This equation was designed to produce a maximum value of 1.0 for a stimulus viewed at 1000 lux.

The adapted cone signals are then calculated using a simple linear von Kries-style adaptation transform:

$$R_c = \left[\frac{Y_w D}{R_w} + 1 - D \right] R, \tag{11.7a}$$

$$G_c = \left[\frac{Y_w D}{G_w} + 1 - D \right] G, \tag{11.7b}$$

$$B_c = \left[\frac{Y_w D}{B_w} + 1 - D \right] B. \tag{11.7c}$$

Note that these equations multiply back in the relative luminance of the adapting white Y_w. While typically this is taken to be 100, there are some applications where this is not the case. Fairchild presents a compelling argument that Y_w should always be replaced with 100.0 in these calculations, especially when using the chromatic-adaptation transform outside of CIECAM02 [302].

There are several other factors that can be considered part of the greater chromatic-adaptation transform. These terms, although not explicitly part of the von Kries adaptation, account for changes in chromatic adaptation as a function of overall luminance levels, as well as some background luminance and induction factors that can account for some color appearance changes as a function of the background luminance. These factors are calculated using

$$k = \frac{1}{(5L_A + 1)}, \tag{11.8a}$$

$$F_L = 0.2k^4 (5L_A) + 0.1 \left(1 - k^4\right)^2 (5L_A)^{1/3}, \tag{11.8b}$$

$$n = \frac{Y_b}{Y_w}, \tag{11.8c}$$

$$N_{bb} = N_{cb} = 0.725 \frac{1}{n^{0.2}}, \tag{11.8d}$$

$$z = 1.48 + \sqrt{n}. \tag{11.8e}$$

The adapted RGB values are then converted into cone signals prior to the nonlinear compression. This is done by inverting the M_{CAT02} matrix to obtain adapted XYZ tristimulus values and then transforming those into cone signals using the Hunt-Pointer-Estevez transform, normalized for the reference equal-energy condition. Note that the RGB' terminology is used in CIECAM02 rather than LMS to maintain consistency with older color appearance models. This transform is given by:

$$\begin{bmatrix} R' \\ G' \\ B' \end{bmatrix} = M_{\text{HPE}}\, M_{\text{CAT02}}^{-1} \begin{bmatrix} R_c \\ G_c \\ B_c \end{bmatrix}; \tag{11.9a}$$

$$M_{\text{HPE}} = \begin{bmatrix} 0.38971 & 0.68898 & -0.07868 \\ -0.22981 & 1.18340 & 0.04641 \\ 0.00000 & 0.00000 & 1.00000 \end{bmatrix}; \tag{11.9b}$$

$$M_{\text{CAT02}}^{-1} = \begin{bmatrix} 1.096124 & -0.278869 & 0.182745 \\ 0.454369 & 0.473533 & 0.072098 \\ -0.009628 & -0.005698 & 1.015326 \end{bmatrix}. \tag{11.9c}$$

The cone signals are then processed through a hyperbolic nonlinear compression transform. This transform was designed to be effectively a power function of a wide range of luminance levels [302]. These compressed values RGB'_a are then used to calculate the appearance correlates:

$$R'_a = \frac{400\left(F_L \frac{R'}{100}\right)^{0.42}}{27.13 + \left(\frac{F_L R'}{100}\right)^{0.42}} + 0.1, \tag{11.10a}$$

$$G'_a = \frac{400\left(F_L \frac{G'}{100}\right)^{0.42}}{27.13 + \left(\frac{F_L G'}{100}\right)^{0.42}} + 0.1, \tag{11.10b}$$

$$B'_a = \frac{400\left(F_L \frac{B'}{100}\right)^{0.42}}{27.13 + \left(\frac{F_L B'}{100}\right)^{0.42}} + 0.1. \tag{11.10c}$$

11.3.4 Calculations of Hue

The adapted cone signals can now be used to calculate the appearance correlates. The first stage is to move into a temporary opponent color space, similar to a^* and b^* in the CIELAB calculations. A more accurate rectangular opponent color space will be calculated at a later stage:

$$a = R'_a - \frac{12\,G'_a}{11} + \frac{B'_a}{11}, \tag{11.11a}$$

$$b = \left(\frac{1}{9}\right)\left(R'_a + G'_a - 2B'_a\right). \tag{11.11b}$$

Hue angle is then calculated from these opponent changes in a manner identical to the calculation in CIELAB:

$$h = \tan^{-1}\left(\frac{b}{a}\right). \tag{11.12}$$

It is important to note that the CIECAM02 calculations expect hue angle to be represented in degrees rather than radians. Also, in practical implementations, an atan2() function that ranges from 0–360 degrees should be used rather than -180–180. If this function is not available, the hue angle can be calculated by adding 180 if a is a negative number and adding 360 if a is positive and b is negative.

	Red	Yellow	Green	Blue	Red
i	1	2	3	4	5
h_i	20.14	90.00	164.25	236.53	380.14
e_i	0.8	0.7	1.0	1.2	0.8
H_i	0.0	100.0	200.0	300.0	400.0

Table 11.2. Values used to calculate hue quadrature.

Next, an eccentricity factor e_t is computed:

$$e_t = \frac{1}{4}\left[\cos\left(h\frac{\pi}{180}+2\right)+3.8\right]. \tag{11.13}$$

This eccentricity factor is necessary, because the four physiological hues, red, green, yellow, and blue, are not perceptibly equidistant from each other. This is especially true between the blue and red hues, as there is a large range of perceptible purple colors in that range. Note, that for this calculation, the cosine function is expecting radians while the hue angle is represented in degrees; hence, the $\frac{\pi}{180}$ term.

The data in Table 11.2 are then used to determine hue quadrature through a linear interpolation between the calculated hue value and the nearest unique hue:

$$H = H_i + \frac{100\,(h-h_i)\,/e_i}{(h-h_i)\,/e_i + (h_{i+1}-h)\,/e_{i+1}}. \tag{11.14}$$

11.3.5 Brightness and Lightness Calculations

The appearance correlates of lightness and brightness are calculated by first determining the achromatic signal, which is essentially a linear combination of the "cone" signals, along with the background-induction factor N_{bb}:

$$A = \left[2R'_a + G'_a + (1/20)B'_a - 0.305\right] N_{bb}. \tag{11.15}$$

The relative attribute of lightness J is then calculated with this achromatic signal, processed through another nonlinear function that is determined by the surround input factors c along with z, the base nonlinear component calculated earlier:

$$J = 100\left(\frac{A}{A_w}\right)^{cz}. \tag{11.16}$$

The absolute appearance attribute of brightness Q is then calculated from the lightness signal. This calculation is a function of the viewing surround c along with the absolute luminance level function F_L. This accounts for both the Stevens effect as well as the changes in perceived contrast as a function of the surround:

$$Q = \left(\frac{4}{c}\right)\sqrt{\frac{J}{100}}\,(A_w + 4)\;F_L^{0.25}. \tag{11.17}$$

11.3.6 Chroma, Colorfulness, and Saturation

The chromatic attributes of chroma, colorfulness and saturation can now be calculated. They are functions of the lightness and brightness to account for the Hunt effect. First, a temporary magnitude value t is calculated:

$$t = \frac{(50000/13)\,N_c N_{cb} e_t \left(a^2 + b^2\right)^{1/2}}{R'_a + G'_a + (21/20)B'_a}. \tag{11.18}$$

This value has components that are very similar to the CIELAB calculation of chroma, but it also includes factors accounting for the background- and surround-induction values, as well as for the hue-eccentricity value.

The temporary magnitude value t is then used along with the lightness calculation and the background exponent of n to ultimately calculate the relative attribute of chroma C:

$$C = t^{0.9}\sqrt{\frac{J}{100}}\,(1.64 - 0.29^n)^{0.73}. \tag{11.19}$$

From the calculation of chroma, along with the luminance-adaptation factor F_L, we can calculate the absolute correlate of colorfulness M. We can also use the colorfulness and brightness values to calculate a value for saturation s:

$$M = C F_L^{0.25}, \tag{11.20a}$$

$$s = 100\sqrt{\frac{M}{Q}}. \tag{11.20b}$$

11.3.7 Opponent Color Representation

Sometimes it may be desirable to have the rectangular, or Cartesian, representation of the appearance coordinates. These can be calculated from the hue angle, along with the chroma, colorfulness, or saturation attributes:

$$a_C = C\cos(h), \tag{11.21a}$$

$$b_C = C\sin(h), \tag{11.21b}$$

$$a_M = M\cos(h), \tag{11.21c}$$

$$b_M = M\sin(h), \tag{11.21d}$$

$$a_s = s\cos(h), \tag{11.21e}$$

$$b_s = s\sin(h). \tag{11.21f}$$

11.3.8 Inverting CIECAM02

Many times all we are interested in is determining the appearance attributes of a given stimulus. We can consider these attributes as representing not only a device-independent color, but also a viewing-condition-independent color. If all we want is to explain what a stimulus looks like, then the forward model of CIECAM02 is sufficient.

However, one of the most powerful applications of a color appearance model involves the prediction of corresponding colors through both a forward and inverse transform. For example, suppose we have an image displayed on a computer screen that we are viewing in our office, and we want to print that image. We can use a color appearance model to account for the changes in these disparate viewing conditions and calculate the tristimulus values necessary to get a print that matches the screen. In most typical color-management situations, we desire an appearance match rather than a direct colorimetric match.

The surround parameters of the output viewing condition are used for the inversion, as determined from Table 11.1. The first stage is to calculate the achromatic signal for the output white A_w using (11.15). There may be times when we start out with only a subset of the appearance attributes. In these cases, we must first calculate the relative appearance attributes before inverting. As such, if the input to the inverse model is absolute brightness Q, we must first calculate the relative lightness response J:

$$J = 6.25 \left(\frac{cQ}{(A_w + 4)\, F_L^{0.25}} \right)^2. \tag{11.22}$$

If we start with colorfulness M, then we first calculate chroma:

$$C = \frac{M}{F_L^{0.25}}. \tag{11.23}$$

If we start with lightness J and saturation s, then we first must calculate brightness in order to calculate chroma:

$$Q = \frac{4}{c} \sqrt{\frac{J}{100}}\, (A_w + 4)\, F_L^{0.25}, \tag{11.24a}$$

$$C = \left(\frac{s}{100} \right)^2 \frac{Q}{F_L^{0.25}}. \tag{11.24b}$$

If we have the hue angle h, then we can proceed, or we can calculate it from hue quadrature H and Table 11.2:

$$h' = \frac{(H - H_i)\,(e_{i+1} h_i - e_i h_{i+1}) - 100\, h_i e_{i+1}}{(H - H_i)\,(e_{i+1} - e_i) - 100\, e_{i+1}}, \tag{11.25a}$$

$$h = \begin{cases} (h' - 360) & \text{if } h' > 360, \\ h' & \text{otherwise.} \end{cases} \tag{11.25b}$$

From there, we calculate the temporary magnitude value t and the eccentricity value e_t. It is important to note that for the remainder of these equations we use the parameters from the output viewing conditions:

$$t = \left(\frac{C}{\sqrt{\frac{J}{100}}\,(1.64 - 0.29^n)^{0.73}} \right)^{1/0.9}, \tag{11.26a}$$

$$e_t = \frac{1}{4}\left(\cos\left(h\frac{\pi}{180} + 2\right) + 3.8\right). \tag{11.26b}$$

Next we calculate the achromatic signal A, along with the viewing-condition-dependent components p_1, p_2, and p_3:

$$A = A_w \left(\frac{J}{100}\right)^{1/cz}, \tag{11.27a}$$

$$p_1 = \frac{(50000/13)\, N_c N_{cb}\, e_t}{t}, \tag{11.27b}$$

$$p_2 = \left(\frac{A}{N_{bb}}\right) + 0.305, \tag{11.27c}$$

$$p_3 = \frac{21}{20}. \tag{11.27d}$$

We then convert hue angle into radians:

$$h_r = h\frac{\pi}{180}. \tag{11.28}$$

If $|\sin(h_r)| \geq |\cos(h_r)|$,

$$p_4 = \frac{p_1}{\sin(h_r)}, \tag{11.29a}$$

$$b = \frac{p_2 (2+p_3) \left(\frac{460}{1403}\right)}{p_4 + (2+p_3)\left(\frac{220}{1403}\right)\left[\frac{\cos(h_r)}{\sin(h_r)}\right] - \left(\frac{27}{1403}\right) + p_3 \left(\frac{6300}{1403}\right)}, \tag{11.29b}$$

$$a = b \left[\frac{\cos(h_r)}{\sin(h_r)}\right]. \tag{11.29c}$$

Otherwise, if $|\sin(h_r)| < |\cos(h_r)|$,

$$p_5 = \frac{p_1}{\cos(h_r)}, \tag{11.30a}$$

$$a = \frac{p_2 (2+p_3) \left(\frac{460}{1403}\right)}{p_5 + (2+p_3)\left(\frac{220}{1403}\right) - \left[\left(\frac{27}{1403}\right) - p_3 \left(\frac{6300}{1403}\right)\right]\left[\frac{\sin(h_r)}{\cos(h_r)}\right]}, \tag{11.30b}$$

$$b = a \left[\frac{\sin(h_r)}{\cos(h_r)}\right]. \tag{11.30c}$$

We now calculate the RGB'_a signals:

$$R'_a = \frac{460}{1403} p_2 + \frac{451}{1403} a + \frac{288}{1403} b, \tag{11.31a}$$

$$G'_a = \frac{460}{1403} p_2 - \frac{891}{1403} a - \frac{261}{1403} b, \tag{11.31b}$$

$$B'_a = \frac{460}{1403} p_2 - \frac{220}{1403} a - \frac{6300}{1403} b. \tag{11.31c}$$

Then we undo the nonlinearities, again using the F_L calculation of the output viewing condition:

$$R' = \frac{100}{F_L} \left(\frac{27.13\,|R'_a - 0.1|}{400 - |R'_a - 0.1|}\right)^{1/0.42}, \tag{11.32a}$$

$$G' = \frac{100}{F_L} \left(\frac{27.13\,|G'_a - 0.1|}{400 - |G'_a - 0.1|}\right)^{1/0.42}, \tag{11.32b}$$

$$B' = \frac{100}{F_L} \left(\frac{27.13\,|B'_a - 0.1|}{400 - |B'_a - 0.1|}\right)^{1/0.42}. \tag{11.32c}$$

Note that if any of the values of $R'_a - 0.1$, $G'_a - 0.1$, $B'_a - 0.1$ are negative, then the corresponding RGB' must be made negative.

At this point, we leave the Hunt-Pointer-Estevez cone space and move into the sharpened cone space for the inverse chromatic-adaptation transform:

$$\begin{bmatrix} R_c \\ G_c \\ B_c \end{bmatrix} = M_{\text{CAT02}} M_{\text{HPE}}^{-1} \begin{bmatrix} R' \\ G' \\ B' \end{bmatrix}; \tag{11.33a}$$

$$M_{\text{HPE}}^{-1} = \begin{bmatrix} 1.910197 & -1.112124 & 0.201908 \\ 0.370950 & 0.629054 & -0.000008 \\ 0.000000 & 0.000000 & 1.000000 \end{bmatrix}. \tag{11.33b}$$

We calculate the adapted sharpened signals, again using a degree of adaptation factor D that is explicitly determined for the output viewing conditions, or implicitly calculated using Equation 11.6:

$$R = \frac{R_C}{\left(\dfrac{Y_w D}{R_w} + 1 - D\right)}, \tag{11.34a}$$

$$G = \frac{G_C}{\left(\dfrac{Y_w D}{G_w} + 1 - D\right)}, \tag{11.34b}$$

$$B = \frac{B_C}{\left(\dfrac{Y_w D}{B_w} + 1 - D\right)}. \tag{11.34c}$$

Finally we can transform the adapted signals back into CIE XYZ tristimulus values. These values can then be further transformed into a device-specific space for accurate color appearance reproduction between devices:

$$\begin{bmatrix} X \\ Y \\ Z \end{bmatrix} = M_{\text{CAT02}}^{-1} \begin{bmatrix} R \\ G \\ B \end{bmatrix}. \tag{11.35a}$$

We have included a C++ implementation of both forward and reverse versions of CIECAM02 on the DVD included with this book.

11.3.9 Final Thoughts on CIECAM02

As we can see from this section, the CIECAM02 model is a fairly complicated color appearance model that is capable of predicting a variety of changes in appearance caused by changes in viewing conditions. The CIECAM02 model was designed specifically for color-management applications, such as reproducing prints that accurately match a monitor. When the desired outcome for device color management is appearance match, then CIECAM02 should be considered a valuable tool in any color-engineer's toolkit. Although CIECAM02 was designed specifically for image-processing use, it should not be considered a spatial model of color appearance. CIECAM02 treats all pixels in an image as individual color stimuli and does not allow for complex spatial interactions, or calculations of local contrast phenomena. Several spatial extensions of color appearance models are mentioned in the next section.

11.4 Image Appearance Modeling

Color appearance modeling has been successful in facilitating device-independent color representation across a variety of viewing conditions and is incorporated into many modern color-management systems. While CIECAM02 is a very powerful color appearance model, there remains significant room for improvement and extension of its capabilities. Image appearance modeling is one such technique designed to address issues with respect to spatial properties of vision, image perception, and image quality. Just as color appearance models can be considered extensions of basic CIE colorimetry, image appearance models can be considered further extensions. These models evolved to combine attributes of color appearance along with attributes of spatial vision models. Most have also evolved specifically with attributes of color difference equations in mind, for the purpose of better predicting perceived differences between spatially complex stimuli.

Spatial vision models that have been historically used for image-quality metrics generally focus on black-and-white quality. As mentioned in the previous section, color appearance models have largely ignored spatial vision while spatial vision models have largely ignored color. Examples of these types of spatial vision models include the Daly Visible Differences Predictor (VDP) and, more recently, Watson's Spatial Standard Observer [229, 1219, 1221]. These models predict image differences based only upon luminance information. Some spatial models of vision that contain both color and spatial properties of stimuli include the Retinex model as described in Section 5.8.4 by Land and McCann and its various derivatives and the S-CIELAB model [1292]. The CVDM metric is a further refinement of the S-CIELAB model combined with attributes of visual masking from VDP [550]. The Retinex model was not originally designed as a complete model of image appearance and quality, though its spatially variable mechanisms of chromatic adaptation and color constancy serve some of the same purposes in image rendering and certainly provide some of the critical groundwork for recent research in image appearance modeling.

This section will focus on two models that have their pedigree in CIE colorimetry and color appearance modeling: S-CIELAB and iCAM. S-CIELAB is a simple spatial extension to the CIELAB color space, and was designed to measure color differences of complex images. Fairchild and Johnson based iCAM on S-CIELAB, combined with a spatially variant color appearance model [291,292].

11.4.1 S-CIELAB

The S-CIELAB model was designed as a spatial pre-processor to the standard CIE color difference equations [1292]. The intention of this model was to account for color differences in complex color stimuli, such as those between a continuous tone image and its half-tone representation.

The spatial pre-processing uses separable convolution kernels to approximate the contrast-sensitivity functions (CSF) of the human visual system. These kernels behave as band-pass functions on the luminance channels and low-pass functions for the chromatic opponent channels. It is important to emphasize that the CSF kernels chosen are tied heavily into the color space in which they were designed to be applied. The opponent color space chosen in S-CIELAB is based upon color appearance data from a series of visual experiments on pattern color performed by Poirson and Wandell [908,909]. This space, and the corresponding measured contrast-sensitivity functions, were both fit to the visual data and should be used together.

For S-CIELAB, the CSF serves to remove information that is imperceptible to the human visual system and to normalize color differences at spatial frequencies that are perceptible. This is especially useful when comparing images that have been half-toned for printing. For most common viewing distances, these dots are not resolvable and tend to blur into the appearance of continuous colors. A pixel-by-pixel color-difference calculation between a continuous image and a half-tone image results in extremely large errors, while the perceived difference can actually be small. The spatial pre-processing stage in S-CIELAB blurs the half-tone image so that it more closely resembles the continuous tone image.

The original S-CIELAB implementation uses separable convolution kernels to perform the spatial filtering, along with the traditional CIELAB color difference equations. Johnson and Fairchild introduced a slight refinement to this model based upon spatial filtering in the frequency domain and the use of the CIEDE2000 color-difference metrics [555]. This implementation has been folded

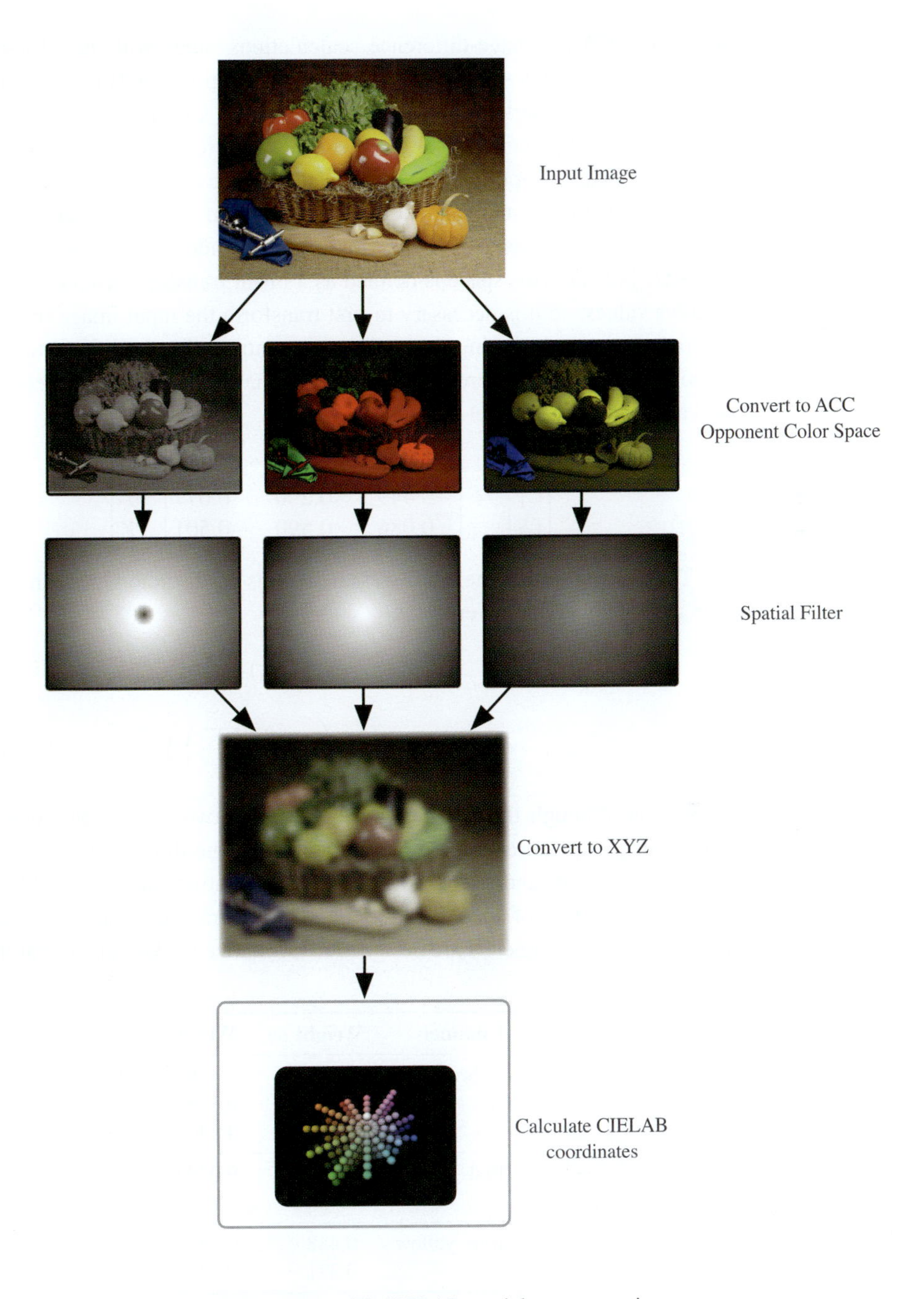

Figure 11.11. The general flow chart of S-CIELAB spatial pre-processing.

into the iCAM image-difference calculations and will be discussed in Section 11.4.2. This section will focus on the original S-CIELAB calculations. The general flowchart of the S-CIELAB model is shown in Figure 11.11 for a single image. In practice, this process is usually performed for an image pair, and a pixel-by-pixel color difference is calculated on the output image pair.

The first stage in the calculation is to transform the input image into an opponent color space based upon the Poirson and Wandell data, known as AC_1C_2 or $O_1O_2O_3$ [1292]. This space is defined as a linear transform from CIE XYZ tristimulus values, so it is necessary to first transform the input image into a colorimetric representation using device characterization. The transformation into the opponent color space approximating an achromatic channel, a red-green channel, and a blue-yellow channel is given by

$$\begin{bmatrix} A \\ C_1 \\ C_2 \end{bmatrix} = \begin{bmatrix} 0.279 & 0.720 & -0.107 \\ -0.449 & 0.290 & 0.077 \\ 0.086 & 0.590 & -0.501 \end{bmatrix} \begin{bmatrix} X \\ Y \\ Z \end{bmatrix}. \tag{11.36}$$

The convolution kernels are weighted sums of discrete Gaussian functions. The general form of the kernels is

$$f(x,y) = k \sum_i \omega_i E_i(x,y), \tag{11.37a}$$

$$E_i(x,y) = k_i \exp\left(-\left(\frac{x^2+y^2}{\sigma_i^2}\right)\right). \tag{11.37b}$$

Note that although these equations are given in two dimensions, due to their separability they are computed in practice using one-dimensional convolutions. The parameters ω_i and σ_i for (11.37) are the weighting functions and half width expressed in terms of spatial frequency. These parameters can be found in Table 11.3. The k parameters are designed to normalize the kernels so that they sum

Channel	Weight ω_i	Width σ_i
Achromatic	0.921	0.0283
	0.105	0.133
	-0.108	4.336
Red-green	0.531	0.0392
	0.330	0.494
Blue-yellow	0.488	0.0536
	0.371	0.386

Table 11.3. Convolution kernel parameters for the S-CIELAB model.

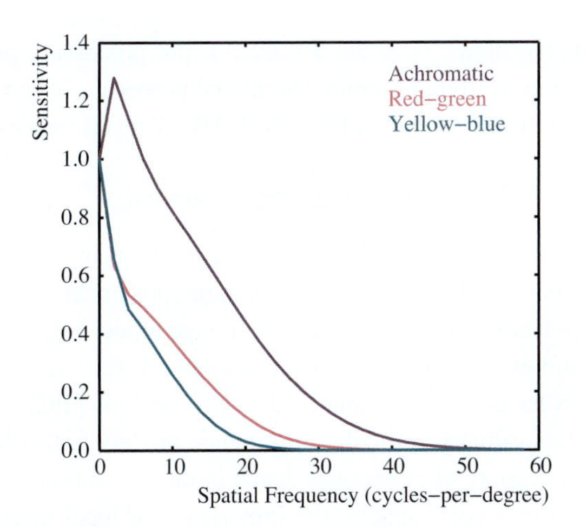

Figure 11.12. The convolution kernels of S-CIELAB represented in the frequency domain. Notice the general approximation of the human *contrast-sensitivity functions*.

to 1.0. This assures that when solid patches are filtered and used as stimuli, the S-CIELAB model reduces to a standard color difference equation.

The final kernel size is determined to be a function of the viewing conditions, primarily the resolution of the display, the number of pixels in the image, and the viewing distance. Equation (11.38) shows this relationship for an image size expressed in pixels, a display resolution expressed in dots-per-inch, and a viewing distance expressed in inches. The spatial frequency of the kernel size is expressed as a function of samples/pixels per degree of visual angle. In practice, the spread of the Gaussian σ_i is multiplied by samples per degree to generate the convolution kernel:

$$\text{samples per degree} = \tan^{-1}\left(\frac{\frac{\text{size}}{\text{dpi}}}{\text{distance}}\right)\frac{180}{\pi}. \tag{11.38}$$

We can get a general idea of the type of spatial filtering performed by these convolution kernels by transforming them into the frequency domain using a 1D Fourier transform. If we do this for a large enough kernel, greater than 120 samples per degree, we can see the full extent of the filtering. This is shown in Figure 11.12. We can see that the achromatic filter behaves in a band-pass nature, while the two chromatic filters are low-pass in nature. The achromatic filter essentially blurs less than the red-green, which in turn blurs less than the blue-yellow. This follows the generally known behavior of the contrast-sensitivity functions of the human visual system. Thus the S-CIELAB pre-processing provides a relatively simple way to take into account the spatial properties of the visual system, while also relying on the strength of the CIELAB color difference equations.

11.4.2 The iCAM Image Appearance Framework

Research in image appearance modeling is essentially trying to tie in our knowledge of spatial vision with our knowledge of color appearance, in order to create a single model applicable to color appearance, image rendering, and image-quality specifications and evaluations. One initial framework for such a model for still images, referred to as iCAM, is outlined in this section [292]. In addition to traditional CIE colorimetry and color appearance modeling, this framework was built upon a wide variety of previous research in the perception of images. These areas include uniform color spaces, the importance of local image contrast and surround, algorithms for image-difference and image-quality measurement, and color and spatial adaptation to natural scenes. The iCAM framework is based upon earlier models of spatial and color vision applied to color appearance problems, as well as on models for high dynamic range (HDR) imaging [292].

As we learned earlier in this chapter, a color appearance model is necessary to extend CIE colorimetry to fully describe the appearance of color stimuli across a variety of viewing conditions. Likewise, an image appearance model is necessary to describe the more spatially complex color stimuli. For simplicity, we are generically using the term *image* to describe any complex scene that has been discreetly sampled or rendered. Image appearance models attempt to extend upon these traditional color appearance descriptors to also predict such perceptual attributes as sharpness, graininess, and contrast. Since image appearance also generates the standard attributes such as lightness, chroma, and hue, these models should inherently be able to predict the overall perceived color difference between complex image stimuli.

At the heart of all CIE methods for describing color match, difference, and appearance lies a uniform color space. For example, S-CIELAB and traditional CIE color difference equations rely on CIELAB as their color space. A uniform color space also lies at the heart of an image appearance model. This space, along with per-pixel chromatic adaptation and local contrast detection are the essence of image appearance modeling. It should be emphasized that iCAM was never designed to be a complete model of image appearance, but rather a generic framework or foundation to aid in the research and development of more complete models.

The general process for using the iCAM model for predicting the overall appearance of still images is shown in Figure 11.13. Ideally, the model takes colori-

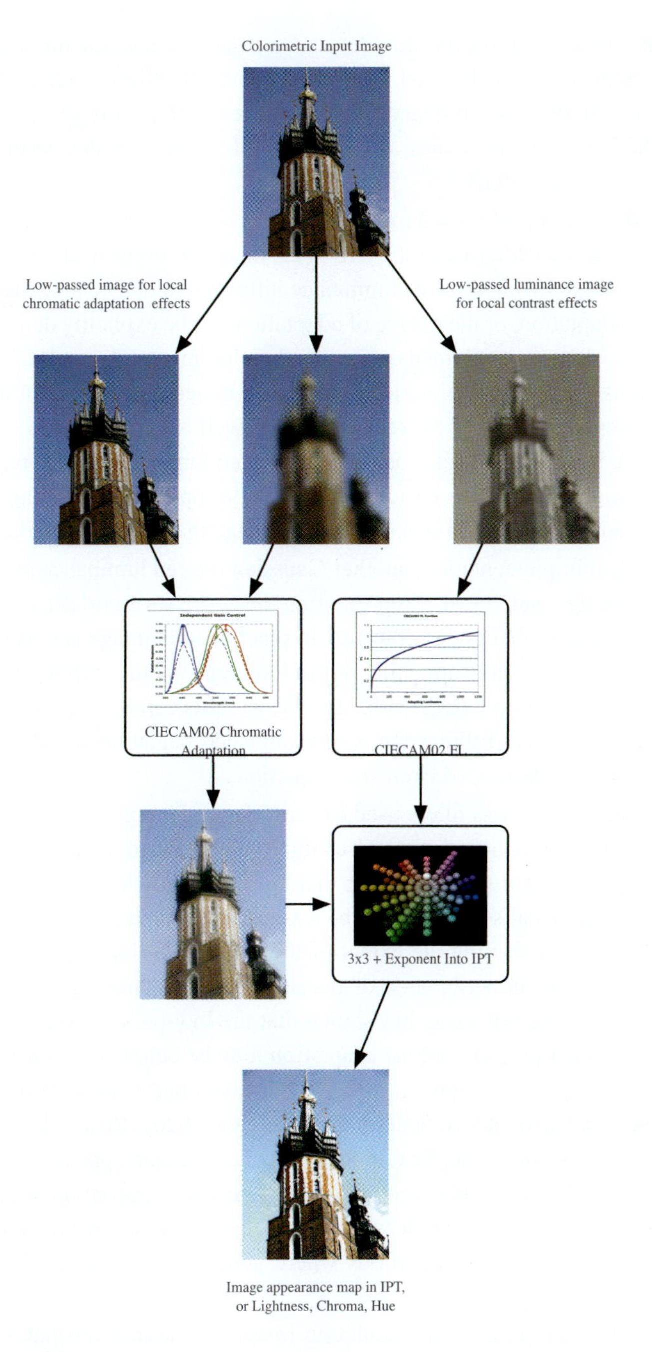

Figure 11.13. An iconic flow chart of the iCAM framework.

metric images as input for the stimulus and surround in absolute luminance units. Relative colorimetry can be used, along with an absolute luminance scaling or approximation of absolute luminance levels. Images are typically specified in terms of CIE XYZ tristimulus values, or another well-understood device-independent RGB space such as sRGB.

The adapting stimulus used for the chromatic-adaptation transform is then calculated by low-pass filtering of the CIE XYZ image stimulus itself. This stimulus can also be tagged with absolute luminance information to calculate the degree of chromatic adaptation, or the degree of adaptation can be explicitly defined. A second low-pass filter is performed on the absolute luminance channel, CIE Y, of the image and is used to control various luminance-dependent aspects of the model. This includes the Hunt and Stevens effects, as well as simultaneous contrast effects caused by the local background. This blurred luminance image representing the local neighborhood can be identical to the low-pass filtered image used for chromatic adaptation, but in practice it is an image that is less blurred.

In an ideal implementation, another Gaussian filtered luminance image of significantly greater spatial extent representing the entire surround is used to control the prediction of global image contrast. In practice, this image is generally taken to be a single number indicating the overall luminance of the surrounding viewing conditions, similarly to CIECAM02. In essence, this can be considered a global contrast exponent that follows the well-established behavior of CIECAM02 for predicting the Bartleson and Breneman equations.

The specific low-pass filters used for the adapting images depend on viewing distance and application. A typical example might be to specify a chromatic-adaptation image as a Gaussian blur that spans 20 cycles per degree of visual angle, while the local surround may be a Gaussian blur that spans 10 cycles per degree [1267]. Controlling the extent of the spatial filtering is not yet fully understood, and it is an active area of research. For instance, recent research in high dynamic range rendering has shown that the low-pass filtering for the local chromatic adaptation and contrast adaptation may be better served with an edge-preserving low-pass function, such as the bilateral filter as described by Durand and Dorsey, and also later in Section 17.5 [270, 625, 626]. Research in HDR rendering is one example of application-dependence in image appearance modeling. A strong local chromatic adaptation might be generally appropriate for predicting actual perceived image-differences or image-quality measurements, but inappropriate for image-rendering situations where generating a pleasing image is the desired outcome.

A step-by-step process of calculating image appearance attributes using the iCAM framework as described above follows. The first stage of processing in iCAM is calculating chromatic adaptation, as is the first stage for most general color appearance models. The chromatic-adaptation transform embedded in CIECAM02 has been adopted in iCAM, since it has been found to have excellent performance with all available visual data. This transform is a relatively simple, and easily invertible, linear chromatic-adaptation model amenable to image-processing applications.

The general CIECAM02 chromatic-adaptation model is a linear *von Kries* normalization of sharpened RGB image signals from a sharpened RGB adaptation *white*. In traditional chromatic-adaptation and color appearance modeling that we have discussed up to this point, the adapting *white* is taken to be a single set of tristimulus values, often assumed to be the brightest signal in the image or the XYZ tristimulus values of the scene measured off a perfect reflecting diffuser.

It is at this stage that image appearance deviates from traditional color appearance models. Instead of a single set of tristimulus values, the adapting signal is taken to be the low-pass filtered adaptation image at each pixel location. This means that the chromatic adaptation is actually a function of the image itself. These adapting *white* signals in the blurred image can also be modulated by the global white of the scene, if that is known.

The actual calculations of the adapted signals are identical to those in CIECAM02, but will be presented again here for clarity. The von Kries normalization is modulated with a degree-of-adaptation factor D that can vary from 0.0 for no adaptation to 1.0 for complete chromatic adaptation. The calculation of D is also identical to the CIECAM02 formulation and can be used in iCAM as a function of adapting luminance L_A for various viewing conditions. Alternatively, as in CIECAM02, the D factor can be established explicitly.

The chromatic-adaptation model is used to compute pixel-wise corresponding colors for CIE Illuminant D65 that are then used in the later stages of the iCAM model. It should be noted that, while the chromatic-adaptation transformation is identical to that in CIECAM02, the iCAM model is already significantly different since it uses the blurred image data itself to spatially-modulate the adaptation white point. Essentially, any given pixel is adapted to the identical pixel location in the blurred image.

$$\begin{bmatrix} R \\ G \\ B \end{bmatrix} = \mathbf{M}_{\text{CAT02}} \begin{bmatrix} X \\ Y \\ Z \end{bmatrix}; \tag{11.39a}$$

$$D = F\left[1 - \left(\frac{1}{3.6}\right) \exp\left(\frac{-L_A - 42}{92}\right)\right]; \tag{11.39b}$$

$$R_c(x,y) = \left[D\left(\frac{R_{\text{D65}}}{R_W(x,y)}\right) + (1-D)\right] R(x,y); \tag{11.39c}$$

$$G_c(x,y) = \left[D\left(\frac{G_{\text{D65}}}{G_W(x,y)}\right) + (1-D)\right] G(x,y); \tag{11.39d}$$

$$B_c(x,y) = \left[D\left(\frac{B_{\text{D65}}}{B_W(x,y)}\right) + (1-D)\right] B(x,y). \tag{11.39e}$$

The use of the blurred XYZ image as a spatially modulated adapting white point implies that the content of an image itself, as well as the color of the overall illumination, controls our state of chromatic adaptation. In this manner, iCAM behaves similar, with regards to color constancy, to the spatial modulations of the Retinex approach to color vision. This behavior can result in a decrease in overall colorfulness or chroma for large uniform areas, such as a blue sky. While this may be the correct prediction for the overall image appearance, it may produce undesirable results when using this type of model for image-rendering applications.

Another example of the localized spatial behavior inherent in an image appearance model is the modulation of local and global contrast using the absolute luminance image and surround luminance image or value. This modulation is accomplished using the luminance level control function F_L from CIECAM02. This function is essentially a compressive function that slowly varies with absolute luminance; it has been shown to predict a variety of luminance-dependent appearance effects in CIECAM02 and earlier models [302]. As such, it was also adopted for the first incarnation of iCAM. However, the global manner in which the F_L factor is used in CIECAM02 and the spatially localized manner used in iCAM is quite different; more research is necessary to establish its overall applicability for image appearance.

The adapted RGB signals, which have been converted to corresponding colors for CIE Illuminant D65, are transformed into LMS cone responses. These LMS cone signals are then compressed using a simple nonlinear power function that is modulated by the per-pixel F_L signal. Then, the cone signals are transformed into an opponent color space, approximating the achromatic, red-green, and yellow-blue space analogous to higher-level encoding in the human visual system. These opponent channels are necessary for constructing a uniform perceptual color space and correlates of various appearance attributes. In choosing this transformation for the iCAM framework, simplicity, accuracy, and applicability to image processing were the main considerations.

The uniform color space chosen was the IPT space previously published by Ebner and Fairchild and introduced in Section 8.7.4 [274]. The IPT space was derived specifically for image-processing applications to have a relatively simple formulation and to have a hue-angle component with good prediction of constant perceived hue. Predicting lines of constant hue has traditionally been very important in gamut-mapping applications and will be increasingly important with any gamut-expansion algorithms that are desired for new high dynamic range and wide-gamut displays. The mathematical transformation into the IPT opponent space is far simpler than the transformations used in CIECAM02:

$$\begin{bmatrix} L \\ M \\ S \end{bmatrix} = \begin{bmatrix} 0.4002 & 0.7075 & -0.0807 \\ -0.2280 & 1.1500 & 0.0612 \\ 0.0000 & 0.0000 & 0.9184 \end{bmatrix} \begin{bmatrix} X_{\mathrm{D65}} \\ Y_{\mathrm{D65}} \\ Z_{\mathrm{D65}} \end{bmatrix}; \tag{11.40a}$$

$$L'(x,y) = \begin{cases} L(x,y)^{0.43F_L(x,y)} & \text{if } L(x,y) \geq 0, \\ -\left|L(x,y)\right|^{0.43F_L(x,y)} & \text{if } L(x,y) < 0;, \end{cases} \tag{11.40b}$$

$$M'(x,y) = \begin{cases} M(x,y)^{0.43F_L(x,y)} & \text{if } M(x,y) \geq 0, \\ -\left|M(x,y)\right|^{0.43F_L(x,y)} & \text{if } M(x,y) < 0; \end{cases} \tag{11.40c}$$

$$S'(x,y) = \begin{cases} S(x,y)^{0.43F_L(x,y)} & \text{if } S(x,y) \geq 0, \\ -\left|S(x,y)\right|^{0.43F_L(x,y)} & \text{if } S(x,y) < 0; \end{cases} \tag{11.40d}$$

$$\begin{bmatrix} I \\ P \\ T \end{bmatrix} = \begin{bmatrix} 0.4000 & 0.4000 & 0.2000 \\ 4.4550 & -4.8510 & 0.3960 \\ 0.8056 & 0.3572 & -1.1628 \end{bmatrix} \begin{bmatrix} L' \\ M' \\ S' \end{bmatrix}. \tag{11.40e}$$

Although it seems counter-intuitive to have negative LMS cone responses, it is possible to have negative LMS values because of the chromatic adaptation on sharpened RGB values as well as the linear transformation from CIE XYZ values; care must be taken when applying the nonlinear power function.

Once the IPT values are computed for the image data, a simple coordinate transformation from the rectangular opponent space to cylindrical coordinates is applied to obtain image-wise predictors of the traditional color appearance attributes such as lightness J, chroma C, and hue angle h:

$$J = I, \tag{11.41a}$$

$$C = \sqrt{P^2 + T^2}, \tag{11.41b}$$

$$h = \tan^{-1}\left(\frac{P}{T}\right), \tag{11.41c}$$

$$Q = {F_L}^{1/4} J, \tag{11.41d}$$

$$M = {F_L}^{1/4} C. \tag{11.41e}$$

Differences in these dimensions can be used to compute differences between two images. In some instances, correlates of the absolute appearance attributes of brightness Q and colorfulness M are required. These are obtained by scaling the relative attributes of lightness and chroma with the appropriate function of F_L, again derived from the image-wise luminance map as shown in (11.41).

11.4.3 Image-Difference Calculations in iCAM

For image-difference and image-quality predictions within the iCAM framework, it is also necessary to apply spatial filtering to the image data to eliminate any image variations at spatial frequencies too high to be perceived. This is performed, as in S-CIELAB, as a pre-processing stage. The spatial pre-processing serves to eliminate information that is imperceptible to the visual system and to normalize color differences at spatial frequencies that are visible.

Like the filtering performed in S-CIELAB, this computation is highly dependent on viewing distance and is based on filters derived from human contrast-sensitivity functions. Since the human contrast-sensitivity functions vary with respect to luminance, band-pass for the low and high frequencies, and low-pass for the chromatic channels, it is necessary to apply these filters in an opponent color space.

The choice of opponent color space is crucial for this step as it is highly desirable that the luminance and chromatic channels be mathematically orthogonal. Problems such as chromatic fringing can arise when this is not the case. As such, the iCAM framework performs spatial filtering in a specially designed orthogonal color space called $\mathrm{Y'C_1C_2}$. More details on the development of this opponent space can be found in [556, 1070]. The linear transform from CIE XYZ into $Y'C_1C_2$ is given by

$$\begin{bmatrix} Y' \\ C_1 \\ C_2 \end{bmatrix} = \begin{bmatrix} 0.0556 & 0.9981 & -0.0254 \\ 0.9510 & -0.9038 & 0.0000 \\ 0.0386 & 1.0822 & -1.0276 \end{bmatrix} \begin{bmatrix} X \\ Y \\ Z \end{bmatrix}_{\mathrm{D65}}. \tag{11.42}$$

It is important to note that this space is designed to approximate an iso-luminant and opponent space as best as possible with a linear and orthogonal space, and requires CIE tristimulus values specified for CIE Illuminant D65.

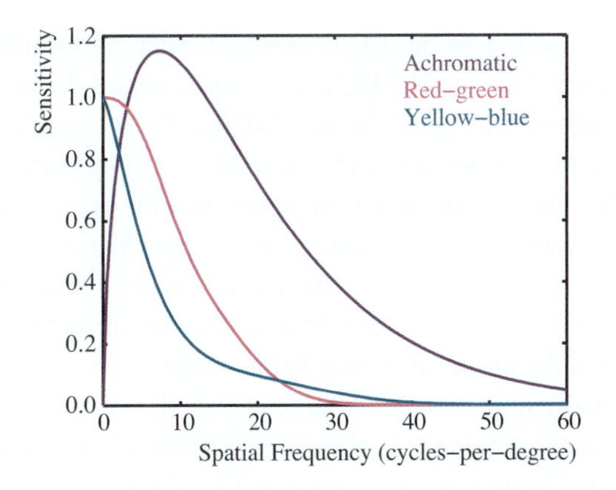

Figure 11.14. The spatial filters used to calculate image differences in the iCAM framework.

The spatial filters are designed to be used in the frequency domain, rather than as convolution kernels in the spatial domain. The general form for these filters is

$$\mathrm{CSF}_{\mathrm{lum}}(f) = a \cdot f^{c} \cdot e^{-b \cdot f}, \tag{11.43a}$$

$$\mathrm{CSF}_{\mathrm{chrom}}(f) = a_1 \cdot e^{-b_1 \cdot f^{c_1}} + a_2 \cdot e^{-b_2 \cdot f^{c_2}}. \tag{11.43b}$$

The parameters, a, b, and c, in (11.43) are set to 0.63, 0.085, and 0.616, respectively, for the luminance CSF for application on the Y' channel. In (11.43), f is the two-dimensional spatial frequency defined in terms of cycles per degree (cpd) of visual angle. To apply these functions as image-processing filters, f is described as a two-dimensional map of spatial frequencies of identical size to the image itself. For the red-green chromatic CSF, applied to the C_1 dimension, the parameters $(a_1, b_1, c_1, a_2, b_2, c_2)$ are set to (91.228, 0.0003, 2.803, 74.907, 0.0038, 2.601). For the blue-yellow chromatic CSF, applied to the C_2 dimension, the parameters are set to (5.623, 0.00001, 3.4066, 41.9363, 0.083, 1.3684). The resulting spatial filters for a single dimension are shown in Figure 11.14.

The band-pass nature of the luminance contrast-sensitivity function, as well as the low-pass nature of the two chromatic channels can be seen in Figure 11.14. These filters should look similar to those used in S-CIELAB (see Figure 11.12), though slightly smoother. The smoothness is a result of defining the filters in the frequency domain rather than as discrete convolution kernels of limited extent. Two other important features can be seen with regard to the luminance CSF: its behavior at 0 cycles per degree (the DC component) and that the response goes above 1.0.

Care must be taken with the DC component when performing spatial filtering in the frequency domain. The DC component contains the mean value of the image for that particular channel. As with S-CIELAB, we would like the image-difference metric to collapse down into a traditional color-difference metric for solid patches. For this to happen, it is important that the mean value of any given channel does not change. The luminance spatial filter described by (11.43) and shown in Figure 11.14 goes to zero at the DC component. Therefore, it is necessary to first subtract the mean value of the luminance channel, apply the spatial filter, and then add the mean value back to the image.

The other important feature of the luminance CSF is that it goes above 1.0 for a band of frequencies ranging roughly between 3 and 15 cycles per degree. This is where the visual system is most sensitive to color differences, and, as such, these regions are more heavily weighted. Care must be taken when applying a frequency-domain filter that goes above 1.0, as this can often lead to severe ringing. However, when the filter is sufficiently broad, this is often not a problem. When the filter itself becomes very narrow, such as when applied to a large high-resolution image, it may be necessary to normalize the luminance CSF such that the maximum is at 1.0. Even then, care must be taken to minimize ringing artifacts when using a narrow image-processing filter.

The images are filtered by first transforming into the frequency domain by use of a Fast Fourier Transform, FFT, multiplying with the CSF filter, and then inverting the FFT to return to the spatial domain:

$$\begin{aligned} \text{FiltImage}_{\text{lum}} &= \text{FFT}^{-1}\left\{\left(\text{FFT}\left\{\text{Image} - \text{mean(Image)}\right\}\right) \cdot \text{CSF}_{\text{lum}}\right\} \\ &\quad + \text{mean(Image)}, \qquad (11.44a) \\ \text{FiltImage}_{\text{chrom}} &= \text{FFT}^{-1}\left\{\left(\text{FFT}\left\{\text{Image}\right\}\right) \cdot \text{CSF}_{\text{chrom}}\right\}. \qquad (11.44b) \end{aligned}$$

The contrast-sensitivity functions described in this framework serve to modulate spatial frequencies that are not perceptible and to enhance certain frequencies that are most perceptible. Generally CSFs are measured using simple grating stimuli with care taken to avoid spatial frequency adaptation. Spatial frequency adaptation is very similar to chromatic adaptation and essentially decreases sensitivity to certain frequencies based upon information present in the visual field. This decrease in sensitivity to some frequencies can serve to enhance the sensitivity to other frequencies, through a form of CSF normalization. An excellent description of the mechanisms of spatial frequency adaptation can be found in Blakemore and Campbell, and more recently in Webster [101, 1228, 1229].

It should be noted that a multi-scale, or multi-channel, spatial vision model is not required to predict spatial frequency adaptation. Instead, all that is required is that the CSF functions be allowed to change shape as a function of the adapting field, though this is clearly an indicator of the existence of multi-scale mechanisms in the visual system.

Since spatial frequency adaptation cannot be avoided in real-world viewing conditions, several models of spatial frequency adaptation or contrast-gain control have been described for practical applications [229, 290, 310, 311, 884, 1220]. These models generally alter the nature or behavior of the CSF based upon either assumptions on the viewing conditions, or based upon the information contained in the images themselves. A simplified image-dependent mechanism for spatial frequency adaptation is given by

$$\mathrm{CSF}_{\mathrm{adapt}} = \frac{\mathrm{CSF}}{\alpha \cdot \mathrm{FFT}(\mathrm{Image}) + 1}, \tag{11.45a}$$

$$\alpha = \frac{1}{D \cdot X_{\mathrm{size}} \cdot Y_{\mathrm{size}}}. \tag{11.45b}$$

This model essentially normalizes the contrast-sensitivity function based upon the amount of information present in the image itself. It can almost be thought of as a von Kries-style frequency adaptation. More details on the development of this spatial frequency adaptation model can be found in [294].

When using this type of model, the frequency representation of the image itself is typically blurred to represent spatial frequency channels. This blurring can be done by performing a Gaussian convolution on the frequency image, which can be thought of as multiplying the spatial image by a Gaussian envelope. The scaling function α converts the frequency representation into absolute units of contrast at each spatial frequency.

The D factor is similar to the degree of chromatic-adaptation factor found in CIECAM02 and is set to 1.0 for complete spatial frequency adaptation. Spatial frequency adaptation is important when calculating image differences between images that may have regular periodic patterns, such as stochastic half-tone pattern jpeg-compressed images with an 8-pixel blocking pattern. The regular period of these patterns reduces the visual sensitivity to the spatial frequency of the pattern itself, thus making the pattern less visible.

One potential benefit of spatial frequency adaptation is the ability to predict visual masking without the need for multi-scale approaches. If a masking frequency is present in an image, the contrast-sensitivity function for that particular frequency region will become less sensitive.

11.4.4 Summary of Image Appearance

Just as color appearance models were created to extend CIE colorimetry for predicting changes in appearance across disparate viewing conditions, image appearance models aim to further extend this ability across spatially complex images. The goal of image appearance modeling is to augment the traditional color appearance attributes described in Section 11.1.1 with other perceptual attributes of images such as sharpness, graininess, and contrast. S-CIELAB and iCAM were both designed as simple extensions to traditional color appearance models. Neither of these models are yet ready to fully account for all of the unique perceptions the human visual system has for complex stimuli. We can consider the field of image appearance modeling to be in its infancy, and we hope for great strides in the future. Applications for these types of models can include high dynamic range image rendering, gamut mapping, digital watermarking, and image-quality measurements. These models are not limited to static imagery and can easily be extended into the temporal domain by adding spatio-temporal filtering and models for the time-course of chromatic adaptation.

11.5 Applications of Color and Image Appearance Models

Color appearance modeling is designed to predict the change in appearance of a stimulus when the viewing conditions change. This makes color appearance models useful for various applications. CIECAM02 was designed with color management applications in mind [195]. Desktop imaging allows us to view, synthesize, and capture images from a variety of sources. These images can then be printed and viewed in a wide array of conditions, or even posted online or emailed around the globe. Color appearance models allow us to essentially represent colors in a viewing-condition independent manner and allow us to calculate the changes necessary to get accurate color matches in many different environments.

This section outlines one such practical application: calculating the colors necessary to send to a printer viewed in a bright environment to match a monitor viewed in a dark environment. This section also touches on a practical application of a per-pixel chromatic-adaptation transform for white balancing a digital photograph, as well as calculating image differences using the iCAM framework described above.

11.5.1 Application: Creating an Appearance Match between a Monitor and a Printer

Perhaps the most common use for a color appearance model is for cross-media color reproduction, e.g., color management. If we have a stimulus presented on a computer monitor, how do we reproduce the same appearance when we send the stimulus to a printer? In order to do that, we must take into account the viewing conditions that the original was viewed in, as well as the viewing conditions of the reproduction. For this example, we will start with an original image that is viewed on a D65 sRGB display, with a maximum luminance of 100 cd/m^2. We will assume that the screen is being viewed in a *dark* surround, and that the relative luminance of the background Y_b is 10 percent. The adapting luminance L_A will be taken to be 20 percent of the maximum luminance of the display. The original image is shown in Figure 11.15.

To fully illustrate the changes in perception predicted by CIECAM02, we will choose an output viewing condition at a very high luminance level of 1000 cd/m^2, with a 20 percent relative luminance of the background. Thus, the adapting luminance L_A is 200 cd/m^2 and the background luminance Y_b is 20. Since this is a print viewed in a typical environment, we will choose an *average* surround with a D50 white point. If we calculate the appearance correlates for the monitor environment using the forward CIECAM02 calculations, and then invert the model using the print-viewing conditions, we achieve the image shown in Figure 11.16. It should be noted that for this calculation we allowed CIECAM02 to calculate the degree of adaptation factor D, based upon the adapting luminance.

There should be a few things immediately evident. The first, and most obvious, change is that the entire image appears more yellow. This is because the D50 illumination of the print-viewing environment is significantly more yellow than the D65 of the monitor. If we were to stand in a room that is illuminated with D50 simulators, especially at the light levels used in this example, we would quickly adapt to this more yellow environment. As such, our sensitivity to yellow would decrease.

In order to achieve an appearance match back to our monitor, it is necessary to increase the yellow content of the image to account for this change in sensitivity. Note that for illustrative purposes in this example we actually transformed the image back into sRGB, which does have a native white of D65. In practice we send the adapted XYZ tristimulus values to a printer, which then reproduces those desired tristimulus values through a printer characterization, specifically measured for a D50 viewing environment.

Figure 11.15. Original image encoded in sRGB. (Photo courtesy of Ron Brinkmann.)

The next thing to notice is that both the contrast and the overall colorfulness of the reproduced image are markedly decreased. This is because our original was viewed at a very low luminance level in a darkened environment. When we view the printed reproduction in an environment with a very high luminance level and surround, the Hunt and Stevens effects, as well as the surround effects, will increase the perceived contrast and colorfulness. It is necessary to decrease the actual image contrast to account for these changes in the perceived contrast.

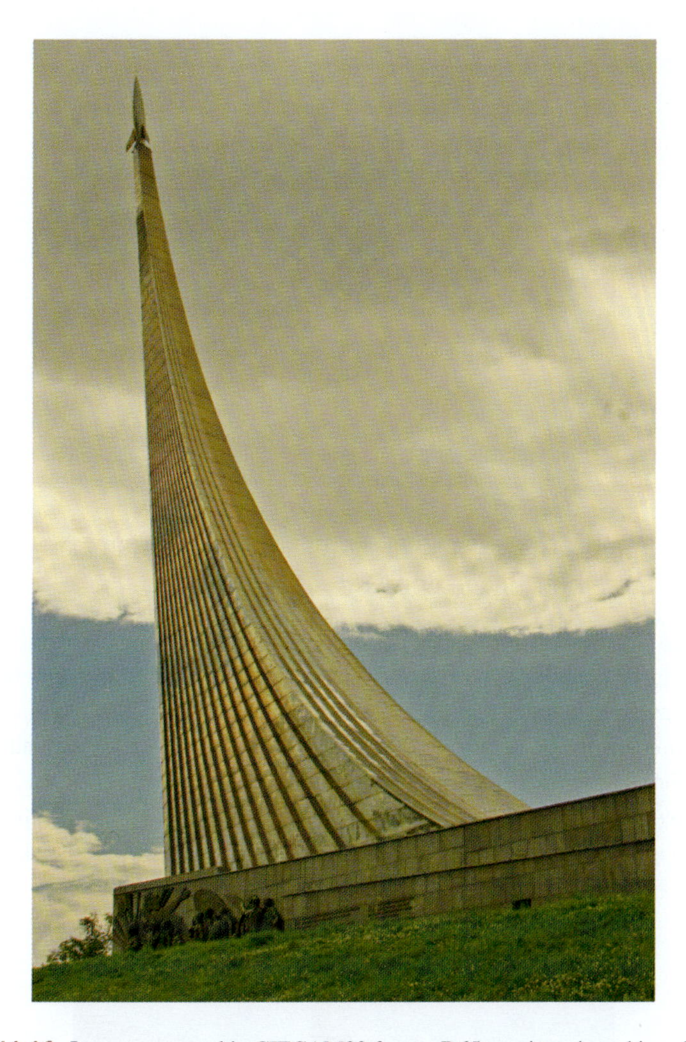

Figure 11.16. Image processed in CIECAM02 from a D65 monitor viewed in a dark surround to a D50 print viewed in an average surround. (Photo courtesy of Ron Brinkmann.)

This example is meant to illustrate the predictive power of using a color appearance model for cross-media color reproduction. In general, CIE colorimetric matches will not be adequate for reproducing accurate colors across a wide variety of viewing conditions. In these situations, color appearance matches are a much more desirable outcome. CIECAM02 was designed specifically for color management applications, such as the example described here, and is an excellent choice for many color reproduction or rendering systems.

Figure 11.17. An image of a white building illuminated by the setting sun.

11.5.2 Application: Per-Pixel Chromatic Adaptation Based on Image Content

There are many occasions when we would like to automatically white balance an image using a color appearance or chromatic-adaptation transform. This can be very useful when there is a global color shift caused by the illumination, as we saw in Chapter 10. Other times the color shift is more localized, and using a global chromatic-adaptation transform can result in undesirable effects.

An example is shown in Figure 11.17 and Figure 11.18. The original white building is illuminated with a very orange light at sunset, while the sky behind

Figure 11.18. Processing the image through the CIECAT02 global chromatic-adaptation transform using the tristimulus values of the building itself as the adapting white.

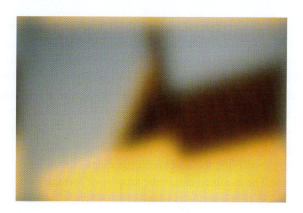

Figure 11.19. The low-pass filtered image used in a per-pixel chromatic-adaptation transform.

the image is still very blue. Since we know that the building is white, we can take the CIE XYZ tristimulus values averaged across a number of pixels in the building and use those as our adapting white in a standard CIECAT02 chromatic-adaptation transform. The results of this calculation are shown in Figure 11.18. We can see immediately that the building does indeed look white, but at the same time the sky looks far too blue. This is because the sky is also being adapted by the orange illumination, when in reality an observer would adapt locally when viewing this scene.

In a situation like this, we would like to use a local chromatic-adaptation transform, similar to that used in the iCAM image appearance framework. This can be

Figure 11.20. The adapted image using a per-pixel chromatic-adaptation transform with complete adaptation to the low-passed image.

accomplished by using (11.39), along with the blurred adapting image shown in Figure 11.19. The results of this calculation are shown in Figure 11.20.

We can see immediately that the resulting image shows some of the desired localized adaptation results, as the building looks mostly white while the sky looks less blue. Unfortunately, there are also severe haloing artifacts around the edges of the image, as well as undesirable color de-saturation in areas such as the roof.

It would be better to combine some aspects of the global chromatic-adaptation transform with some aspects of the per-pixel chromatic adaptation. This can be accomplished by modifying the adaptation equations slightly to include a global adaptation factor. The modified chromatic adaptation formulae are then

$$R_c(x,y) = \left[D\left(\frac{R_{D65}}{R_W(x,y)+R_g}\right)+(1-D)\right]R(x,y), \tag{11.46a}$$

$$G_c(x,y) = \left[D\left(\frac{G_{D65}}{G_W(x,y)+G_g}\right)+(1-D)\right]G(x,y), \tag{11.46b}$$

$$B_c(x,y) = \left[D\left(\frac{B_{D65}}{B_W(x,y)+B_g}\right)+(1-D)\right]B(x,y). \tag{11.46c}$$

In these modified adaptation values R_g, G_g, and B_g represent a single global adaptation white point. We can now calculate an image that has been adapted with a mix of local and global illumination. This is shown in Figures 11.21 and 11.22. For the calculations of Figure 11.21, we essentially assumed that the

Figure 11.21. The adapted image using a combination of the localized per-pixel chromatic-adaptation transform as well as a global RGB_g equal energy white.

Figure 11.22. The adapted image using a combination of the localized per-pixel chromatic-adaptation transform as well as a global RGB_g derived from the tristimulus values of the building itself. In essence, a combination of Figure 11.18 and Figure 11.20.

image was locally adapted to the low-passed image shown in Figure 11.19 and globally adapted to an equal energy illumination, where RGB_g equaled 1.0. Figure 11.22 was calculated with the same localized adaptation, but with the global white point set to the tristimulus values of the building itself, the same tristimulus values used to calculate Figure 11.18.

We can see from these results that combining the per-pixel chromatic-adaptation transform with the global adaptation based on (11.46) allows for a wide variety of white balancing effects.

11.5.3 Application: Calculating Image Differences

The general purpose of S-CIELAB and one of the initial goals of the iCAM framework is to predict overall image differences between two spatially complex stimuli. A classic example of why this is necessary is described in this section. An original image is shown in Figure 11.23. Two reproductions are shown in Figure 11.24 and Figure 11.25.

One of the reproductions has a very noticeable red arch, while the other has noise added to it. The noise was added in such a way that the mean difference in CIE XYZ tristimulus values was exactly the same for the two reproductions. The image with the arch is clearly different, while the image with the noise is imperceptible. If we were to calculate a CIELAB color difference, using the CIE ΔE^*_{94} color-difference equation, we would find that the noise image has an average color difference of about 1.05 units from the original, while the red arch image has

Figure 11.23. An original image used to study image-difference calculations. (Photo courtesy of Ron Brinkmann.)

Figure 11.24. The same image as Figure 11.23 altered by changing the color of the arch. (Photo courtesy of Ron Brinkmann.)

an average color difference of about 0.25 units. This means that the noise image, though barely perceptible, has on average a four times larger color difference.

We can process these images through the iCAM image difference framework, using 11.43–11.44, for a viewing distance corresponding to 80 cycles per degree.

Figure 11.25. The same image as Figure 11.23 altered by adding noise such that the colorimetric error is the same as Figure 11.24. (Photo courtesy of Ron Brinkmann.)

When we do that, the mean color difference in IPT space for the red arch image becomes approximately 0.75, while the mean color difference for the noise image becomes 0.25. These units have been scaled by 100 on the I dimension, and 150 on the P and T dimensions, to match the units of CIELAB [274]. We see that by applying the spatial filters as a pre-processing stage, the image difference calculations match up with what we would intuitively expect, while a color difference calculation without the filters does not.

11.6 Further Reading

A full account of the history and development of color appearance models is given by Mark Fairchild in *Color Appearance Modeling* [302]. Robert Hunt's book, *The Reproduction of Colour*, provides in-depth information on issues pertaining to color reproduction.

Part III
Digital Color Imaging

Chapter 12
Image Capture

Previous chapters presented the theory of light propagation and the perception of color, as well as its colorimetric representation. The final part of this book details how man-made devices capture and display color information in the form of images.

When devices have sensors that individually respond to the same range of wavelengths of the electromagnetic spectrum to which human sensors are responsive, they generate images that are similar to what humans perceive when they view their environment. Such devices enable the preservation of scenes of value in people's lives, in the form of images and movies. Images and videos are also used for art, for entertainment, and as information sources. This chapter presents image-capture techniques for electromagnetic radiation in the visible part of the spectrum. Although we focus predominantly on digital cameras and the radiometry involved with image capture, we also briefly describe photographic film, holography, and light field techniques. We begin with a description of the optical imaging process which enables images to be projected onto sensors.

An image starts with a measurement of optical power incident upon the camera's sensor. The camera's response is then correlated to the amount of incident energy for each position across the two-dimensional surface of its sensor. This is followed by varying levels of image processing, dependent on the type of camera as well as its settings, before the final image is written to file. A typical diagram of the optical path of a digital single-reflex camera (DSLR) is shown in Figure 12.1. The main optical components of a DSLR camera are:

Lens. The lens focuses light onto an image plane, where it can be recorded.

Mirror. The mirror reflects light into the optical system for the view finder. This ensures that the image seen through the view-finder is identical to the im-

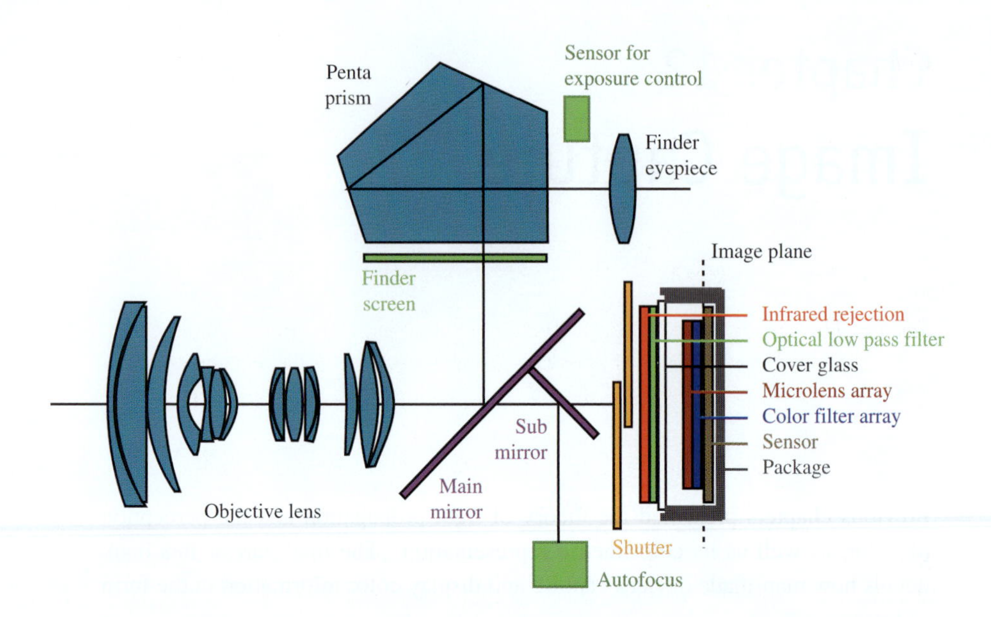

Figure 12.1. The components of a typical digital single lens reflex camera (after [1141]).

age that will be captured by the sensor. During image capture, the lens is temporarily lifted to allow light to reach the sensor; in that position, it also blocks stray light from entering the camera through the view-finder.

View-finder. Light reflected off the mirror is passed through a penta-prism to direct it toward the finder eyepiece.

Shutter. The shutter determines the duration of the exposure.

Infrared rejection filter. Since humans are not sensitive to infrared light, but camera sensors are, an infrared rejection filter is typically present to ensure that the camera only responds to light that can be detected by the human visual system.

Optical low-pass filter. This filter ensures that high frequencies in the scene are removed before spatial sampling takes place, in cases where these frequencies would cause aliasing.

Micro-lens array. Since for each picture element (pixel) the light-sensitive area is smaller than the total area of the pixel, a micro-lens can help ensure that all light incident upon the pixel surface is directed towards the light-sensitive part. A micro-lens therefore improves the efficiency of the camera.

Color filter array. To enable a single sensor to capture a color image, the sensor is fitted with a color filter array, so that each pixel is sensitive to its own set of wavelengths. The pattern repeats itself over small clusters of pixels, of for instance 2×2 or 4×4 pixels. The camera's firmware will perform interpolation (demosaicing) to reconstruct a color image. The nature of the filter, as well as the signal processing, determine the primaries and the white point of the camera.

Sensor. The sensor is the light sensitive device that is placed in the image plane. It records the incident number of photons for each of its pixels and converts it to charge. After an exposure is complete, the charge is converted, pixel by pixel, to a voltage and is then quantized. The camera's firmware then applies a variety of image processing algorithms, such as demosaicing, noise reduction, and white balancing.

Some camera types, such as point-and-shoot cameras, have no mirrors, and instead have either an electronic view-finder, or have separate optics for the view-finder. Cameras for motion imaging largely consist of the same components, bar the mirrors, and may rely on an electronic shutter to determine the length of light measurement for each frame. Differences can also be found in the in-camera image processing algorithms. Digital video cameras, for instance, may use noise reduction techniques that work over a sequence of frames, rather than by necessity having to rely on the information present in a single frame [1000]. Digital cinema cameras do not include elaborate signal processing circuitry and rely on off-camera post-processing to shape and optimize the signal.

12.1 Optical Image Formation

Images are formed by optical imaging systems that focus light onto a sensor. For any real system, not all light available in a scene can be collected and imaged. Since camera systems collect only a segment of any wavefront, some diffraction will occur, and this limits the achievable quality. Thus, optical imaging systems are considered *diffraction-limited.* For sensor elements that are large relative to the wavelengths being imaged, diffraction becomes relatively unimportant, and geometrical optics can be used to model such system behavior (see Section 2.9). In practice, the behavior of such systems can therefore be modeled using geometrical optics, without the need to resort to physical optics. The basis of geometrical optics is formed by Fermat's principle, Snell's law, the eikonal equation, and the ray equation [666]; each is described in Chapter 2.

Without reference to any specific imaging system, consider a point light source, emitting light equally in all directions. In terms of ray tracing (see Section 2.10.1), rays emanating from this light source diverge. The point source itself is called the *focus* of a bundle of rays. If, by placing an optical system near a point light source, a ray bundle can be made to converge into an arbitrary point, then this point is also called a *focus point* [447].

If a ray bundle emanating from a point $\mathbf{P}_0$ converges in its entirety upon a second point $\mathbf{P}_1$, then the optical system that caused this convergence is called *stigmatic* (sharp) for these two points. The points $\mathbf{P}_0$ and $\mathbf{P}_1$ are *conjugate points*; reversing their roles would allow a perfect image to be created at $\mathbf{P}_0$ for rays emanated at $\mathbf{P}_1$. If the bundle of rays converges upon a small area rather than a point, a blur spot is created, and the image is no longer perfect.

Of course, the system may allow points nearby $\mathbf{P}_0$ to be stigmatically imaged to points that are nearby $\mathbf{P}_1$. In an *ideal optical system*, the region of points that are stigmatically imaged is called *object space*, whereas the region of points into which object space is stigmatically imaged is called *image space*. Note that both spaces are three-dimensional.

If we assume that a given set of points in object space spans a curve, then this curve will be mapped to another curve in image space. If both curves are identical, then the image is called *perfect*. An optical imaging system that is stigmatic as well as perfect is called an *absolute instrument*. Maxwell's theorem for absolute instruments as well as Carathéodory's theorem applies to such instruments [112]. The former states that the optical length of any curve in object space equals the optical length of its image. The latter theorem says that the mapping between object and image space of an absolute instrument is either a projective transformation, an inversion, or a combination of both.

The constraints placed upon optical systems before they can be called absolute instruments are rather limiting. In most real imaging systems, the image space is a region on a plane or curved surface, named the *image plane* (although only a small region of the plane is used).

12.1.1 Imaging Geometry

In the following, we make the assumption that the optical imaging system is constructed such that all rays only make a small angle ϕ with respect to a reference axis. Such rays are called *paraxial*. Systems for which this assumption holds, i.e., ϕ is small, enable sine and cosine terms to be simplified as $\sin(\phi) \approx \phi$ and $\cos(\phi) \approx 1$. This simplification to geometric optics is called *linear optics*.

If, in addition, all optical elements involved in the system are arranged along the reference axis, which is then called the *optical axis*, and all elements are rotationally symmetric with respect to this central axis, then further simplifications are possible. The study of such systems is called *Gaussian*, *first-order*, or *paraxial optics*.

According to Maxwell's and Carathéodory's theorems, in Gaussian optics imaging may be approximated as a projective transformation. An object point $\mathbf{P} = (p_x\ p_y\ p_z)^T$ generally projects to $\mathbf{P}' = (p'_x\ p'_y\ p'_z)^T$ according to [666]

$$p'_x = \frac{m_{11}p_x + m_{12}p_y + m_{13}p_z + m_{14}}{m_{41}p_x + m_{42}p_y + m_{43}p_z + m_{44}}, \tag{12.1a}$$

$$p'_y = \frac{m_{21}p_x + m_{22}p_y + m_{23}p_z + m_{24}}{m_{41}p_x + m_{42}p_y + m_{43}p_z + m_{44}}, \tag{12.1b}$$

$$p'_z = \frac{m_{31}p_x + m_{32}p_y + m_{33}p_z + m_{34}}{m_{41}p_x + m_{42}p_y + m_{43}p_z + m_{44}}. \tag{12.1c}$$

Using homogeneous coordinates (see Section A.10) and applying symmetry arguments, this projection can be represented with the following transformation:

$$\begin{bmatrix} p'_x \\ p'_y \\ p'_z \\ p'_w \end{bmatrix} = \begin{bmatrix} f & 0 & 0 & 0 \\ 0 & f & 0 & 0 \\ 0 & 0 & z'_0 & ff' - z_0 z'_0 \\ 0 & 0 & 1 & -z_0 \end{bmatrix} \begin{bmatrix} p_x \\ p_y \\ p_z \\ 1 \end{bmatrix}, \tag{12.2}$$

where we have chosen two focal points z_0 and z'_0 and two focal lengths f and f'. The 3D position of the transformed point is found by applying the homogeneous divide, yielding $\mathbf{P}' = (p'_x/p'_w\ \ p'_y/p'_w\ \ p'_z/p'_w)^T$.

As we are still discussing a general transformation effected by an arbitrary rotationally symmetric optical system, we will only assume that this system sits somewhere between the object point $\mathbf{P}$ and the image point $\mathbf{P}'$ and is centered around the z-axis. This geometric configuration is shown in Figure 12.2. Note that we use a right-handed coordinate system with the positive z-axis, i.e., the optical axis, pointing toward the right. The positive y-axis points up. The $x = 0$-plane is called the *meridional plane*, and rays lying in this plane are called *meridional rays*. All other rays are called *skew rays*. Meridional rays passing through an optical system will stay in the meridional plane.

Due to the fact that the optical system is assumed to be circularly symmetric around the z-axis, we can drop the x-coordinate and focus on the y- and z-coordinates. The former is given by

$$p'_y = \frac{f p_y}{z - z_0}, \tag{12.3}$$

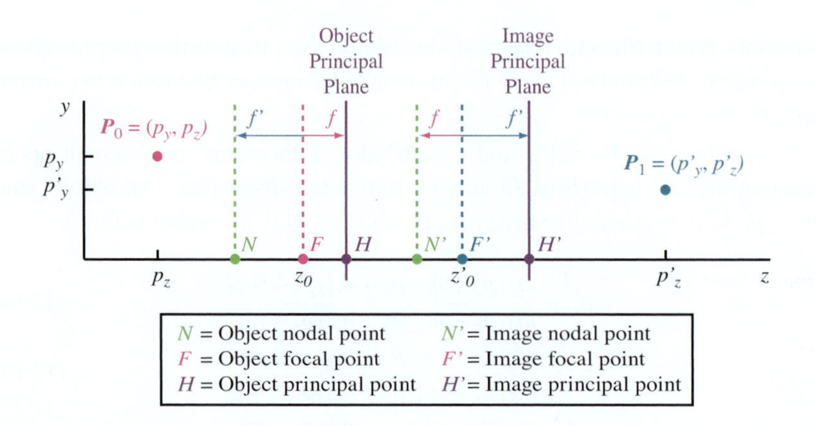

Figure 12.2. Projective transformation from $\mathbf{P} = (p_y \;\; p_z)^T$ to $\mathbf{P}' = (p'_y \;\; p'_z)^T$. The transformation is characterized by the object focal point at z_0, the image focal point at z'_0, the object focal length f, and the image focal length f'.

which is known as *Newton's equation*. The z-coordinate is given by

$$p'_z - z'_0 = \frac{f f'}{z - z_0}, \tag{12.4}$$

yielding the formula for the perspective transformation of a pinhole camera. A pinhole camera is an optical system consisting of a small hole in a surface that separates image space from object space. It is a camera model that is frequently used in computer graphics. In addition, a real pinhole camera can be built cheaply [956].

Associated with an optical imaging system that effects this transformation are several important points located along the z-axis. These points are collectively referred to as *cardinal points* [666]:

Object focal point. Located at $\mathbf{F} = (0 \;\; 0 \;\; z_0)^T$, it is also known as the first focal point or the front focal point. An object located at this point will be imaged infinitely far away. This result is obtained by substituting $z = z_0$ in (12.4).

Image focal point. Located at $\mathbf{F}' = (0 \;\; 0 \;\; z'_0)^T$, it ia also known as the *back focal point* or the *second focal point*. From (12.4), we have that for an image located at $\mathbf{P}_1 = \mathbf{F}'$, that its conjugate object point is located at infinity.

Object principal point. Located at position $\mathbf{H} = (0 \;\; 0 \;\; z_0 + f)^T$, it is also called the *first principal point* or the *front principal point*. The plane parallel to the x-y axis through this point is called the object principal plane. An object located in this plane will be imaged with a lateral magnification of 1, i.e., its size after mapping into the corresponding image principal plane will remain unaltered. Note that, for now, we will assume that the principal plane is, as the name suggests, a planar surface. In a real optical system, this is not necessarily true, leading to optical errors which are discussed in Section 12.3.

Image principal point. This point is located at position $\mathbf{H}' = (0 \;\; 0 \;\; z'_0 + f')^T$.

Object nodal point. This point is located at position $\mathbf{N} = (0 \;\; 0 \;\; z_0 - f')^T$. A ray passing through $\mathbf{N}$ at an angle θ with the optical axis will pass through the image nodal point $\mathbf{N}'$ at the same angle. For such rays, the angular magnification is 1.

Image nodal point This point is located at position $\mathbf{N}' = (0 \;\; 0 \;\; z'_0 - f)^T$.

In a physical system, the radii of the lens elements are limited in spatial extent. This means that only a fraction of the amount of light emitted by a point source can ever be imaged. Further, the smallest diameter through which light passes is either determined by fixed lens elements, or by an adjustable diaphragm. The limiting element in this respect is called the *aperture stop*. Correspondingly, the element limiting the angular extent of the objects that can be imaged is called the *field stop*. This element determines the field of view of the camera.

Further important parameters characterizing an optical imaging system are the *entrance pupil* and *exit pupil*. The entrance pupil is the aperture stop as seen from a point both on the optical axis and on the object. An example with a single lens and an aperture stop is shown in Figure 12.3. This means that the size of the entrance pupil is determined both by the size of the aperture stop as well as the lenses that may be placed between the object and the aperture stop. The axial object point forms the apex of a cone whose base is determined by the entrance pupil. The entrance pupil therefore determines how much light emanated from this point is imaged by the system.

The exit pupil is the aperture stop as seen from the image plane through any lenses that may be located between the aperture stop and the image plane (see Figure 12.3). The exit pupil forms the base of a second cone, with the axial point in the image plane as the apex. The entrance and exit pupils are generally not the same, given that the presence of lenses may magnify or minify the apparent size of the aperture stop. The ratio of the entrance and exit pupil diameters is called the *pupil magnification* or *pupil factor*, a parameter used in determining the radiometry associated with the optical system.

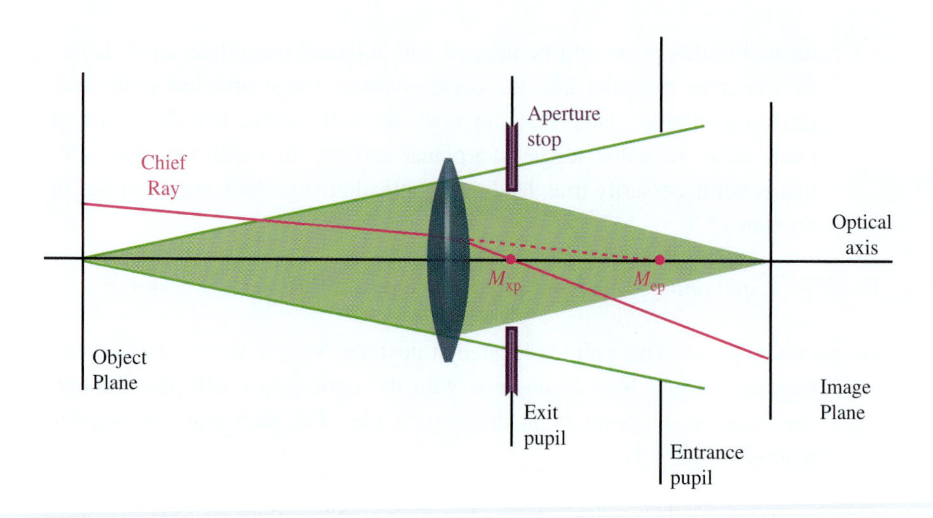

Figure 12.3. The entrance and exit pupils of an optical system with a single lens and an aperture stop. Also shown is a chief ray, which passes through the center of the aperture stop.

Finally, rays starting from any off-axis point on the object that passes through the center of the aperture stop are called *chief rays* [447]. An example is shown in Figure 12.3. The chief ray is aimed at the center of the entrance pupil M_{ep} and passes through the center of the exit pupil M_{xp}. The *marginal ray*, on the other hand, starts from the on-axis object point and passes through the rim of the aperture stop (or entrance pupil).

12.1.2 Imaging Radiometry

A camera consists of an optical system and a sensor. The sensor measures image irradiance E_e as a result of scene radiance L_e incident upon the optical system (see Section 6.2). The relationship between these two quantities is very important in imaging.

We assume that the object distance is large relative to the object focal length. The image irradiance is then proportional to the scene radiance. Further, E_e is proportional to the size of the entrance pupil, meaning that the larger the entrance pupil, the more light the system collects. Finally, E_e is inversely proportional to f^2, the square of the object's focal length. This comes from the fact that the lateral magnification is proportional to the focal length. Hence, the longer the focal length, the larger the area in the image plane over which light is distributed.

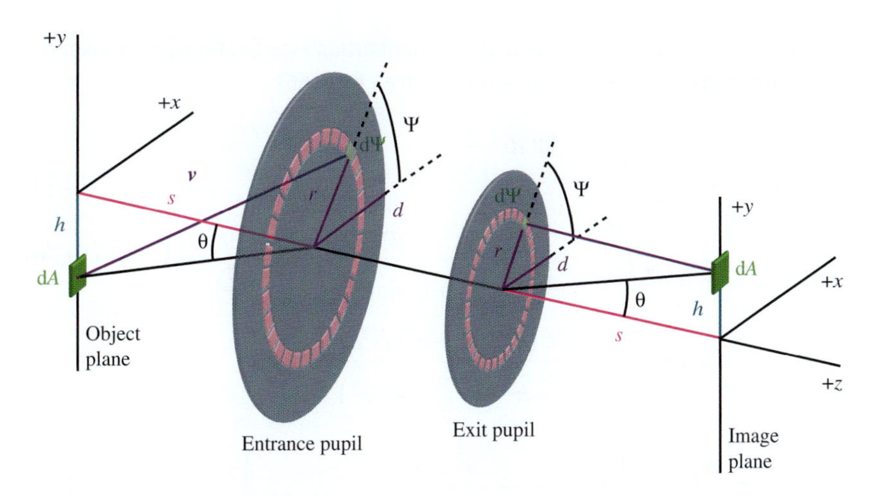

Figure 12.4. A differential area dA is imaged onto the area dA' on the image plane by a system characterized by its entrance and exit pupils (after [666]).

Following Lee [666], we analyze a differential area dA, located off-axis in the object plane, which projects to a corresponding differential area dA' on the image plane. In between these two areas is the optical system, which is characterized by the entrance and exit pupils. The geometric configuration is shown in Figure 12.4. The chief ray starting at dA makes an angle θ with the optical axis. The distance between dA and the entrance pupil along the optical axis is indicated by s; this patch is located a distance h from the optical axis.

Further, d is the radius of the entrance pupil, and the differential area $d\Psi$ on the entrance pupil is located a distance r from the optical axis. We are interested in integrating over the entrance pupil, which we do by summing over all differential areas $d\Psi$. The vector $\mathbf{v}$ from dA to $d\Psi$ is given by

$$\mathbf{v} = \begin{bmatrix} r\cos(\Psi) \\ r\sin(\Psi) - h \\ s \end{bmatrix}. \tag{12.5}$$

This vector makes an angle α with the optical axis, which can be computed from

$$\cos(\alpha) = \frac{s}{\|\mathbf{v}\|}. \tag{12.6}$$

Assuming that the differential area dA is Lambertian (see Section 2.9.2), then the flux incident upon differential area dA' is given by [666]

$$d\Phi_0 = L_e \int_{r=0}^{d} \int_{\Psi=0}^{2\pi} \frac{r\, d\Psi\, dr \,\dfrac{s}{\|\mathbf{v}\|}}{\|\mathbf{v}\|^2}\, dA\, \frac{s}{\|\mathbf{v}\|} \tag{12.7a}$$

$$= L_e \int_{r=0}^{d} \int_{\Psi=0}^{2\pi} \frac{r\, s^2\, d\Psi\, dr}{\left(r^2 \cos^2(\Psi) + (r \sin(\Psi) - h)^2 + s^2\right)^2}\, dA \tag{12.7b}$$

$$= L_e\, dA \int_{r=0}^{d} \frac{2\pi\left(s^2 + h^2 + r^2\right) r s^2 dr}{\left((s^2 + h^2 + r^2)^2 - 4h^2 r^2\right)^{3/2}} \tag{12.7c}$$

$$= \frac{\pi}{2} L_e\, dA \left(1 - \frac{s^2 + h^2 - d^2}{\left((s^2 + h^2 + d^2)^2 - 4h^2 d^2\right)^{1/2}}\right). \tag{12.7d}$$

As shown in Figure 12.4, all quantities used to compute the flux over the entrance pupil have equivalent quantities defined for the exit pupil; these are all indicated with a prime. Thus, the flux over the exit pupil can be computed analogously:

$$d\Phi_1 = \frac{\pi}{2} L'_e\, dA' \left(1 - \frac{s'^2 + h'^2 - d'^2}{\left((s'^2 + h'^2 + d'^2)^2 - 4h'^2 d'^2\right)^{1/2}}\right). \tag{12.8}$$

Under the assumption that the optical system does not suffer light losses, the flux at the entrance and exit pupils will be the same. The irradiance at the exit pupil is then given by

$$E'_e = \frac{d\Phi_0}{dA'} \tag{12.9a}$$

$$= \frac{\pi}{2} L_e \frac{dA}{dA'} \left(1 - \frac{s^2 + h^2 - d^2}{\left((s^2 + h^2 + d^2)^2 - 4h^2 d^2\right)^{1/2}}\right), \tag{12.9b}$$

which is equivalent to

$$= \frac{\pi}{2} L'_e \left(1 - \frac{s'^2 + h'^2 - d'^2}{\left((s'^2 + h'^2 + d'^2)^2 - 4h'^2 d'^2\right)^{1/2}}\right). \tag{12.9c}$$

If the index of refraction at the object plane is n, and the index of refraction at the image plane is n', then it can be shown that the irradiance at the exit pupil is given by

$$E'_e = \frac{\pi}{2} L'_e \left(\frac{n'}{n}\right)^2 \left(1 - \frac{s'^2 + h'^2 - d'^2}{\left((s'^2 + h'^2 + d'^2)^2 - 4h'^2 d'^2\right)^{1/2}}\right) \tag{12.10a}$$

$$= \frac{\pi}{2} L'_e \left(\frac{n'}{n}\right)^2 G, \tag{12.10b}$$

where

$$G = 1 - \frac{s'^2 + h'^2 - d'^2}{\left((s'^2 + h'^2 + d'^2)^2 - 4h'^2 d'^2\right)^{1/2}}. \tag{12.10c}$$

Equation (12.10a) is called the *image irradiance equation*. While this equations is general, the quantities involved are difficult to measure in real systems. However, this equation can be simplified for certain special cases. In particular on-axis imaging, as well as off-axis imaging, whereby the object distance is much larger than the entrance pupil diameter form simplified special cases. These are discussed next.

12.1.3 On-Axis Image Irradiance

When the object of interest is located on the optical axis, then h as well as h' are equal to 0. Equation (12.10a) then simplifies to

$$E'_e = \pi L_e \left(\frac{n'}{n}\right)^2 \left(\frac{d'^2}{s'^2 + d'^2}\right). \tag{12.11}$$

If we consider the cone spanned by the exit pupil as the base and the on-axis point on the image plane as the apex, then the sine of the half-angle β of this cone is given by

$$\sin(\beta) = \frac{d'^2}{\sqrt{s'^2 + d'^2}}. \tag{12.12}$$

We can therefore rewrite (12.11) as follows:

$$E'_e = \frac{\pi L_e}{n^2} \left(n' \sin(\beta)\right)^2. \tag{12.13}$$

The quantity $n' \sin(\beta)$ is called the *numerical aperture*. Since E'_e is proportional to the numerical aperture, we see that the larger this numerical aperture, the lighter the image will be. Thus, it is a value that can be used to indicate the speed of the optical system.

A related measure is the relative aperture F, which is given by

$$F = \frac{1}{2n' \sin(\beta)}. \tag{12.14}$$

The relative aperture is also known as the *f-number*. In the special case that the image point is at infinity, we may assume that the distance between image plane and exit pupil s' equals f', i.e., the image focal length. The half-angle β can then be approximated by $\tan^{-1}(d'/f')$, so that the relative aperture becomes

$$F_\infty \approx \frac{1}{2n' \sin(\tan^{-1}(d'/f'))} \tag{12.15a}$$

$$\approx \frac{1}{n'} \frac{f'}{2d'}. \tag{12.15b}$$

By using the pupil magnification $m_p = d/d'$, this equation can be rewritten as

$$F_\infty \approx \frac{1}{m_p n} \frac{f}{2d}. \tag{12.15c}$$

If the object and image planes are both in air, then the indices of refraction at both sites are approximately equal to 1. If the magnification factor is assumed to be close to 1 as well, then the relative aperture for an object at infinity may be approximated by

$$F_\infty = \frac{f}{D}, \tag{12.16}$$

where D is the diameter of the entrance pupil. An alternative notation for the f-number is f/N, where N is replaced by the ratio f/D. Thus, for a lens with a focal length of 50 mm and an aperture of 8.9 mm, the f-number is written as f/5.6. The image irradiance E'_e can now be written as

$$E'_e = \frac{\pi D^2 L_e}{4} \left(\frac{m_p}{f}\right)^2. \tag{12.17}$$

We note that the quantity $\pi D^2/4$ is the area of the entrance pupil.

12.1.4 Off-Axis Image Irradiance

For object points that are not on the optical axis, we can simplify (12.10a) if the distance of the object to the entrance pupil is much larger than the radius of the entrance pupil, i.e., $s \gg d$. The image irradiance then becomes

$$E'_e \approx \pi L_e \frac{s^2 d^2}{\left(s^2 + d^2 + h^2\right)^2} \frac{dA}{dA'} \tag{12.18a}$$

$$\approx \pi L_e \frac{s^2 d^2}{\left(s^2 + h^2\right)^2} \frac{dA}{dA'}. \tag{12.18b}$$

This equation can be rewritten by noting that the cosine of the off-axis angle θ (see Figure 12.4) is given by

$$\cos(\theta) = \frac{s}{\sqrt{s^2 + h^2}}. \tag{12.19}$$

The image irradiance is therefore approximated with

$$E'_e \approx \pi L_e \cos^4(\theta) \left(\frac{d}{s}\right)^2 \frac{dA}{dA'}. \tag{12.20}$$

The ratio dA/dA' is related to the lateral magnification m of the lens through

$$m = \sqrt{\frac{dA}{dA'}}, \tag{12.21}$$

so that we have

$$E'_e \approx \pi L_e \cos^4(\theta) \left(\frac{d}{s}\right)^2 m^2. \tag{12.22}$$

The lateral magnification satisfies the following relation:

$$\frac{m}{m-1} = \frac{f'}{s}. \tag{12.23}$$

The image irradiance can therefore be approximated with

$$E'_e \approx \pi L_e \cos^4(\theta) \left(\frac{d}{(m-1)\, f'}\right)^2, \tag{12.24}$$

or, in terms of the f-number,

$$E'_e \approx \frac{\pi L_e}{4 F^2\, n'^2} \frac{1}{(m-1)^2\, m_p^2} \cos^4(\Theta). \tag{12.25}$$

If we assume that the lateral magnification m is 2, the pupil magnification m_p is 1, and the index of refraction at the image plane is $n' = 1$, then we can write

$$E'_e \approx \frac{\pi L_e}{4 F^2} \cos^4(\Theta) \approx \frac{\pi L_e}{4} \frac{d^2}{f^2} \cos^4(\Theta). \tag{12.26}$$

Note that the fall-off related to Θ behaves as a cosine to the fourth power. This fall-off is a consequence of the geometric projection. Even in systems without vignetting, this fall-off can be quite noticeable, the edges of the image appearing significantly darker than the center [666]. However, lens assemblies may be designed to counteract these effects, so that in a typical camera system the fall-off with eccentricity does not follow the cosine to the fourth power fall-off.

12.1.5 Vignetting

Consider a simple optical configuration as shown in Figure 12.5. Although no aperture stop is present, the finite diameters of the lenses impose an effective aperture. The cross-section of this aperture depends on which point on the object plane is considered. The further off-axis this point is taken, the smaller the cross-section. This means that less light reaches the image plane dependent on the distance between the point and the optical axis. This fall-off is independent of and in addition to the $\cos^4$ fall-off discussed in the preceding section. It is known as *vignetting* [15].

As the amount of vignetting depends on the distance to the optical axis, we introduce a spatial dependency on points in the object plane (x, y) and corresponding points on the image plane (x', y'), as well as an attenuation factor $V(x', y')$ that takes vignetting into consideration. The general image irradiance equation then

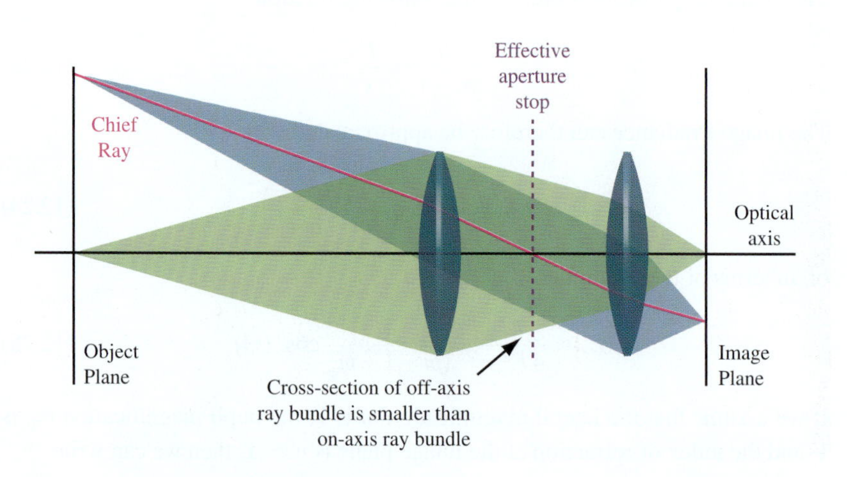

Figure 12.5. In this optical system that consists of two lenses, the effective aperture of the off-axis ray bundle is smaller than the effective aperture of the on-axis ray bundle.

becomes

$$E'_e(x',y') = \frac{\pi}{2} L'_e(x',y')\; T\; V(x',y') \left(\frac{n'}{n}\right)^2 G, \tag{12.27}$$

where G is defined in (12.10c). We have also introduced the transmittance T of the lens assembly to account for the fact that lenses absorb some light.

12.1.6 Veiling Glare

Optical systems in practice suffer from several limitations and imperfections that are so far not accounted for in the image irradiance equation.

The lens assembly, as well as the inside of the lens barrel and the blades of the aperture, may scatter and reflect light to some extent. Theoretically, a point light source should be imaged onto a single pixel of the sensor, but due to limitations of the camera design, some light will be imaged onto all other pixels according to some distribution. This distribution is known as the *glare spread function* (or GSF).

The phenomenon itself is known as *veiling glare* or *lens flare* [1246], dependent on the effect it has on the resulting image. In particular, strong light sources

Figure 12.6. Example of an image showing veiling glare; Bristol Balloon Fiesta, Ashton Court Estate, August 2007.

that may even be outside the field of view may result in a spatially uniform increase in image irradiance and thereby in a corresponding reduction of contrast. This effect is called veiling glare.

Specific undesirable patterns may be imaged as a result of reflections inside the lens assembly. These patterns are collectively called flare sources. For instance, light reflected off the blades of the diaphragm may produce an image of the diaphragm, resulting in *iris flare*.

An example of an image suffering from flare is shown in Figure 12.6. Under normal exposure settings, and with good quality hardware, flare should remain invisible. In this particular example, the sun was directly visible, causing reflections inside the camera, which are seen here as bright patches that are not part of the environment.

Accounting for flare sources with a spatially dependent function $g(x',y')$, the image irradiance is therefore more generally modeled with

$$E'_e(x',y') = \frac{\pi}{2} L'_e(x',y')\; T\; V(x',y') \left(\frac{n'}{n}\right)^2 G + g(x',y'). \tag{12.28}$$

The more components a lens has, the more surfaces there are that could reflect and refract light. Zoom lenses, which can focus light onto the image plane at a variable range of magnifications [545, 680, 873], as shown in Figure 12.7, tend to have more internal surfaces than other lens types, so that they are also more liable to produce flare. Within a camera, an extra source of flare is the sensor itself, which may reflect as much as 30% of all light incident upon it. This is due to the different indices of refraction of the Si ($n_{\mathrm{Si}} \approx 3$) and $\mathrm{SiO_2}$ ($n_{\mathrm{SiO_2}} = 1.45$) layers in the sensor itself [812]. Substituting these values into (2.138a) yields a reflectance of $R = 0.35$. Most of this reflected light will re-enter the lens, possibly to be reflected and refracted again.

To minimize the occurrence of flare, lenses can be coated with an anti-reflective coating, as discussed further in Section 12.6.1. Similarly, the anti-

Figure 12.7. From left to right: the same scene was captured with different magnification settings; Long Ashton, Bristol, May 2007.

reflective coating can be sandwiched between the Si/SiO_2 interface to reduce reflections off of the sensor. Photographers may further reduce the effect of lens flare by employing a lens hood, so that light entering the optical system from oblique angles is blocked.

Finally, a related phenomenon is known as *glare*. Glare is caused by stray reflections entering the lens directly due to strongly reflective surfaces being imaged [335]. If these reflections are off dielectric materials, then they will be polarized and can therefore be subdued by using polarization filters.

12.2 Lenses

So far we have examined optical systems more or less as a black box and under several assumptions, such as Gaussian optics, derived the image irradiance that is the basis of optical image formation. The components of optical systems typically consist of refractive elements as well as an aperture. In this section we discuss lenses as well as some of their attributes.

If a lens consists of a single element, it is called a *simple lens*. If it is made up of multiple components, it is called a *compound lens*. Simple lenses are characterized by their index of refraction as well as the shape of their front and back surfaces. They may also be coated to improve their optical properties. For now we concentrate on their shape.

The front and back surfaces of a lens may be either planar, spherical, or curved in some different way. In the latter case, such lens elements are called *aspherics*. Lenses with spherical surfaces are important because they are relatively straightforward to produce and polish, while still allowing high accuracy and minimal artifacts [447]. Indicating the radius of the surface on the side of the object plane with d_0 and the radius of the surface facing the image plane with d_1, then dependent on the sign of the radius different lens types may be distinguished, as shown in Figure 12.8.

Note that an infinite radius means that the surface is planar. The sign of the radius indicates the direction of curvature in relation to the vector along the optical axis that points from the object plane to the image plane. A positive radius is for a spherical surface that has its center on the side of the object plane, whereas a negative radius is for a surface with its center on the side of the image plane.

Lenses that are thicker at the center, such as the three convex lenses in Figure 12.8, are variously called convex, converging, or positive. They tend to direct light towards the optical axis. If the opposite holds, i.e., they are thinner at the center, they are called concave, diverging, or negative [447].

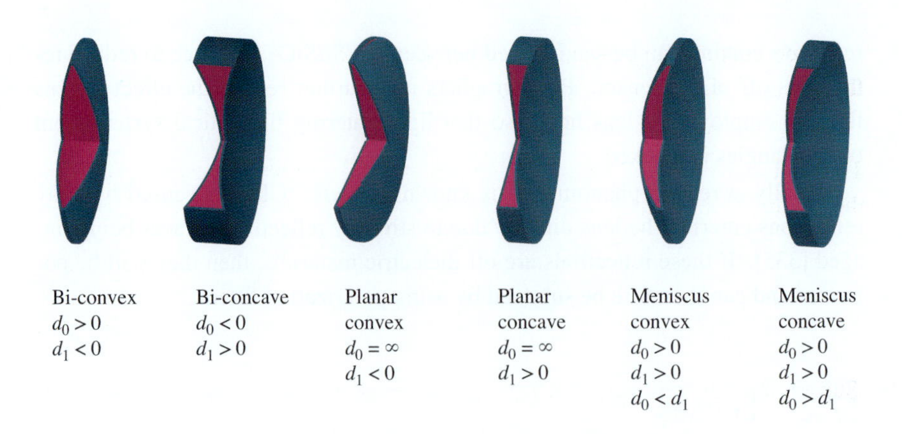

Figure 12.8. Different single lens types.

12.2.1 A Single Spherical Surface

Consider a spherical surface with radius $d = \|\mathbf{P}_2 - \mathbf{C}\|$ as shown in Figure 12.9. The axial point $\mathbf{P}_0$ is assumed to be on the object plane. A ray with length l_o intersects the spherical surface at point $\mathbf{P}_2$, and is refracted toward the surface normal $\mathbf{n}$ (under the assumption that the index of refraction of the lens n_1 is larger than the index of refraction n_0 of the surrounding medium) so that it intersects the optical axis at $\mathbf{P}_1$.

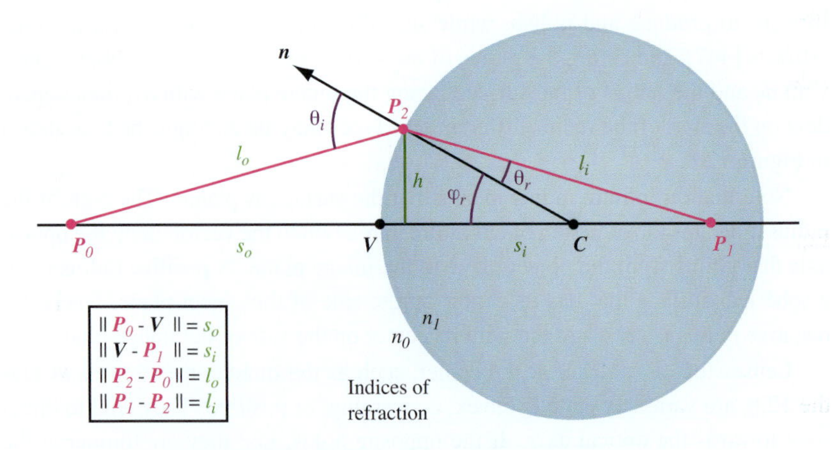

Figure 12.9. Geometry associated with refraction through a spherical surface.

According to Fermat's principle (see Section 2.9.1), the optical path length of this ray is given by

$$n_0\, l_o + n_1\, l_i. \tag{12.29}$$

The path lengths of the two ray segments l_o and l_i can be expressed in terms of the radius of the sphere d, the object distance s_o, and the image distance s_i:

$$l_o = \sqrt{d^2 + (s_o + d)^2 - 2d\,(s_o + d)\cos(\phi)}, \tag{12.30a}$$

$$l_i = \sqrt{d^2 + (s_i - d)^2 + 2d\,(s_i - d)\cos(\phi)}. \tag{12.30b}$$

Substituting this result into the optical path length (12.29) yields

$$\begin{aligned} &n_0\sqrt{d^2 + (s_o + d)^2 - 2d\,(s_o + d)\cos(\phi)} \\ &+ n_1\sqrt{d^2 + (s_i - d)^2 + 2d\,(s_i - d)\cos(\phi)}. \end{aligned} \tag{12.31}$$

Applying Fermat's principle by using the angle ϕ as the position variable results in [447]

$$\frac{n_0\, d\,(s_o + d)\sin(\phi)}{2l_o} - \frac{n_1\, d\,(s_i - d)\sin(\phi)}{2l_i} = 0, \tag{12.32}$$

which can be rewritten to read

$$\frac{n_0}{l_o} + \frac{n_1}{l_i} = \frac{1}{d}\left(\frac{n_1\, s_i}{l_i} - \frac{n_0\, s_o}{l_o}\right). \tag{12.33}$$

Any ray that travels from $\mathbf{P}_0$ to $\mathbf{P}_1$ by means of a single refraction through the spherical surface obeys this relationship. Of course, there exist angles ϕ (and therefore intersection points $\mathbf{P}_2$) for which the point $\mathbf{P}_1$ is never reached.

If we apply the principle of Gaussian optics, i.e., $\cos(\phi) \approx 1$, and only consider paraxial rays, then we can simplify (12.30) to approximate

$$l_o \approx s_o, \tag{12.34a}$$

$$l_i \approx s_i. \tag{12.34b}$$

Equation (12.33) then becomes

$$\frac{n_0}{s_o} + \frac{n_1}{s_i} = \frac{n_1 - n_0}{d}. \tag{12.35}$$

In the special case that the image point is at infinity, i.e., $s_i = \infty$, then we find the object focal length f from (12.35):

$$f = s_o = \frac{n_0}{d\,(n_1 - n_0)}. \tag{12.36}$$

Similarly, the image focal length is obtained by setting $s_o = \infty$, giving

$$f' = s_i = \frac{d\,n_1}{n_1 - n_0}. \tag{12.37}$$

12.2.2 Thin Lenses

In the preceding discussion we have ignored the fact that a lens has both a front and a back surface. We have only analyzed the front surface and derived equations relating the object and image path lengths to the indices of refraction and the radius of the spherical surface.

We now consider the combined effect of both the front and back surface. Assume that the lens is being constructed of two spherical surfaces, with radii d_0 and d_1, as shown in Figure 12.10. The analysis of refraction at both the front and back surfaces yields the following relation:

$$\frac{n_0}{s_0} + \frac{n_0}{s_i} = (n_1 - n_0)\left(\frac{1}{d_0} - \frac{1}{d_1}\right) + \frac{n_1\,t}{(s_n - t)\,s_n}. \tag{12.38}$$

If the lens is thin, we may approximate its thickness as $t \approx 0$, so that the second term on the right-hand side vanishes:

$$\frac{n_0}{s_0} + \frac{n_0}{s_i} = (n_1 - n_0)\left(\frac{1}{d_0} - \frac{1}{d_1}\right). \tag{12.39}$$

Further, if the surrounding medium is air, so that $n_0 \approx 1$, we arrive at the *thin-lens equation*, which is also known as the *lens-maker's formula*:

$$\frac{1}{s_o} + \frac{1}{s_i} = (n_1 - 1)\left(\frac{1}{d_0} - \frac{1}{d_1}\right). \tag{12.40}$$

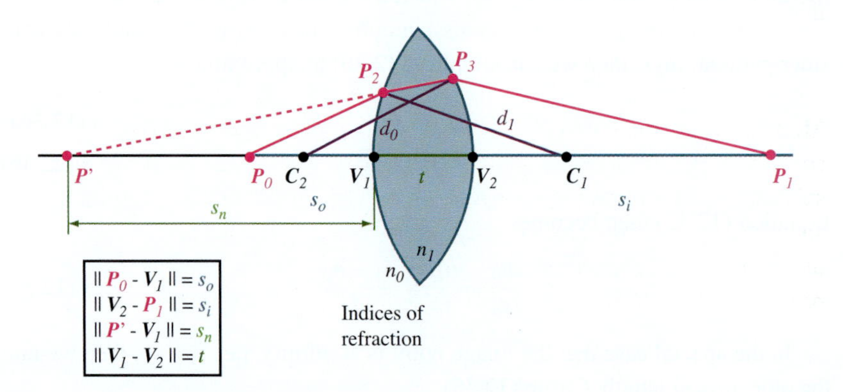

Figure 12.10. Two facing spherical surfaces form a thin lens, which projects light emanating from on-axis point $\mathbf{P}_0$ to point $\mathbf{P}_1$.

In the limit that the object distance s_o becomes infinite, the image distance becomes the image focal length $s_i = f_i$. Conversely, a similar argument can be made for the object focal length for points projected on an infinitely far-away image plane: $s_o = f_o$. But since we are dealing with a thin lens, we may set $f_i = f_o$. Replacing either with a generic focal length f, we have

$$\frac{1}{f} = (n_1 - 1)\left(\frac{1}{d_0} - \frac{1}{d_1}\right). \tag{12.41}$$

Note that all light rays passing through the focal point, located a distance f in front of the lens, will travel through space in parallel after passing through the lens. Such light is called *collimated*. Combining (12.40) with (12.41), we find the well-known *Gaussian lens formula*:

$$\frac{1}{f} = \frac{1}{s_o} + \frac{1}{s_i}. \tag{12.42}$$

Related to this formula is the concept of lateral magnification m_t, which is also known as *transverse magnification* and is given by

$$m_t = -\frac{s_i}{s_o}. \tag{12.43}$$

This measure denotes the ratio of the size of the image to the size of the object. The sign indicates whether the image is inverted (upside-down) or not: a positive lateral magnification means the image is right side up, whereas a negative lateral magnification indicates an inverted image. The object and image distance s_o and s_i are positive for real objects and images[1], so that all real images created by a thin lens are inverted ($m_p < 0$).

12.2.3 Thick Lenses

Most lenses do not have a thickness that can be safely ignored. In such cases, we speak of *thick lenses*. For simplicity, we think of thick lenses as consisting of two spherical surfaces, spaced a distance t apart. Such lenses behave essentially as an optical system and are characterized by their six cardinal points, as described in Section 12.1.1. If the effective focal length is measured with respect to the principal planes, then we have [447]

$$\frac{1}{f} = (n_1 - 1)\left(\frac{1}{d_0} - \frac{1}{d_1} + \frac{(n_1 - 1)\,t}{n_1\, d_0\, d_1}\right). \tag{12.44}$$

[1]If rays leaving an optical system are converging, then the resulting image is termed *real*, whereas the image is called *virtual* if the rays leaving the optical system are divergent.

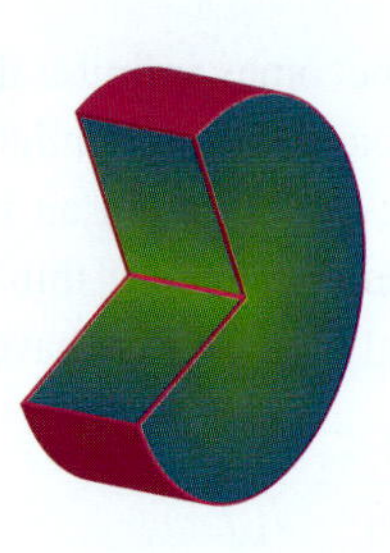

The index of refraction of a GRIN lens varies radially.

Figure 12.11. A cylindrical gradient index lens.

12.2.4 Gradient Index Lenses

For the lenses described so far, the design parameters available are the index of refraction, achieved through the appropriate choice of dielectric material, and the curvature of the front and back faces of simple lenses. The placement of multiple simple lenses and other optical elements along the path taken by the light from the object plane to the image plane gives further freedom in lens design.

However, it is also possible to construct a lens by fabricating materials that have a spatially varying index of refraction. Such lenses are known as *gradient index lenses* [734]. Fabrication typically proceeds by immersing a rod of homogeneous glass in a bath of molten salts. Over a period of time, ionic diffusion then causes the index of refraction to change. A consequence is that the lens can be a circular disk, with planar front and back surfaces, as shown in Figure 12.11. Such lenses are known as *GRIN rods*.

The analysis of gradient index lenses is more complicated than for other lenses, as light incident at oblique angles will follow a curved ray path. Ray tracing techniques may be used to analyze the behavior of such lenses by evaluating the ray equation (2.180) as rays traverse the lens (discussed in Section 2.9.1).

12.3 Aberrations

Much of the analysis of lens systems presented so far uses approximations that are only accurate to a first order. A full lens design requires more complicated analysis that is beyond the scope of this book [1061, 1062]. Lens design programs have taken much of the burden of analysis out of the hands of the lens designer. In addition, ray tracing, discussed in the context of computer graphics in Section 2.10.1, can now be effectively used to evaluate lens design.

Even so, deviations from idealized conditions do occur in practice. They are collectively known as *aberrations* and can be classified into two main groups: chromatic aberrations and monochromatic aberrations.

Chromatic aberrations occur when the material used to manufacture the lens has a wavelength-dependent index of refraction. Such materials show dispersion of light (see Section 3.6.2), which is clearly unwanted in lens design. *Monochromatic aberrations* form a group of distortions that make the image either unclear (spherical aberration, astigmatism, and coma), or warped (distortion, Petzval field curvature). Both classes of aberrations are discussed next.

The assumption that ray paths make only small angles with the optical axis, i.e., the limitation to paraxial rays, enabled sine and cosine functions to be approximated as linear and constant, respectively. Thus, we used only the first terms of their respective Taylor expansions:

$$\sin(x) = \sum_{n=0}^{\infty} \frac{(-1)^n}{(2n+1)!}\, x^{2n+1}; \tag{12.45a}$$

$$\cos(x) = \sum_{n=0}^{\infty} \frac{(-1)^n}{(2n)!}\, x^{2n}. \tag{12.45b}$$

If we were to use the first two terms of these expansions, we would get the so-called *third-order theory*, which is more accurate, but more complicated. The aberrations discussed here, however, are those that result from a deviation from first-order (Gaussian) optics.

12.3.1 Spherical Aberrations

For a spherical refracting surface, (12.35) gave the relation between its radius, indices of refraction, and the image- and object-distances. It made the assumption that the ray path has the same length as the path from object to image plane along the optical axis. If the ray path is recalculated using third-order theory, then the difference with (12.35) is a partial measure of the error introduced by assuming Gaussian optics. If the second term of the Taylor expansion is retained, then this equation becomes [447]

$$\frac{n_0}{s_o} + \frac{n_1}{s_i} = \frac{n_1 - n_0}{d} + h^2 \left(\frac{n_0}{2s_o} \left(\frac{1}{s_0} + \frac{1}{d} \right)^2 + \frac{n_1}{2s_i} \left(\frac{1}{d} - \frac{1}{s_i} \right)^2 \right). \tag{12.46}$$

With respect to (12.35), the extra term varies with h^2, where h is the distance from the point on the lens where the ray of interest strikes to the optical axis. This term causes light passing through the periphery of the lens to be focused nearer the

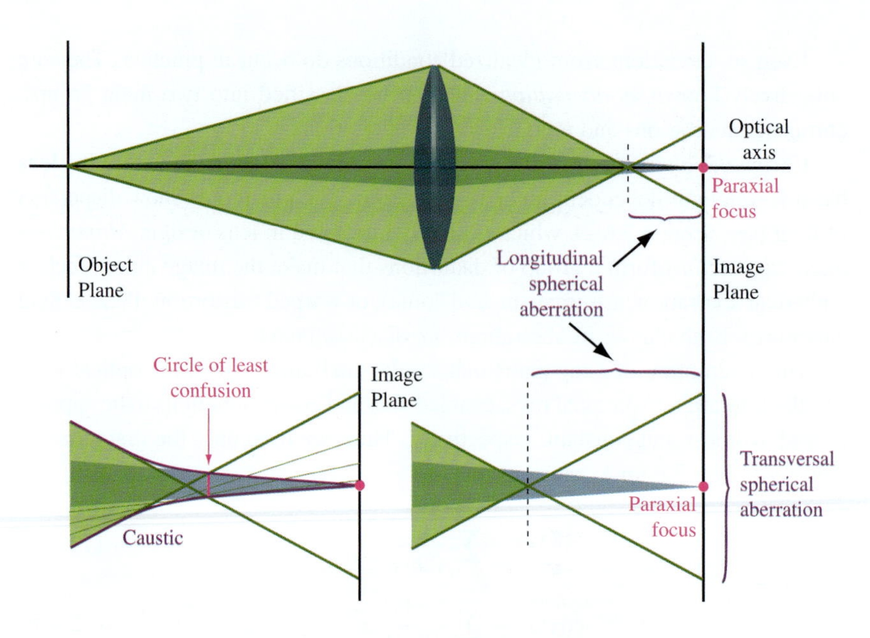

Figure 12.12. Diagram of spherical aberration.

lens than light that passes through the lens nearer the optical axis. This process is called *spherical aberration* and is illustrated in Figure 12.12. Here, the rays intersect the optical axis over a given length, and this phenomenon is therefore called *longitudinal spherical aberration*. These same rays intersect the image plane over a region, and this is called *transversal spherical aberration*.

Finally, in the presence of spherical aberration, the rays form a convex hull which is curved. This shape is called a *caustic*. The diameter of the projected spot is smallest when the image plane is moved to the location associated with the so-called *circle of least confusion*, which in Figure 12.12 is in front of the focus point that would be computed if only paraxial rays were taken into account.

12.3.2 Comatic Aberrations

A second type of aberration stems from the fact that the principal planes discussed in Section 12.1.1 are only well approximated with planar surfaces very close to the optical axis. Further away from the optical axis, the principal planes are in effect principal curved surfaces. Even a small distance away from the optical axis, these surfaces curve away from the idealized planar shape. The effect is called *coma* or *comatic aberration* and is shown in Figure 12.13. If the marginal rays focus farther from the optical axis than the principal ray (the off-axis ray that passes

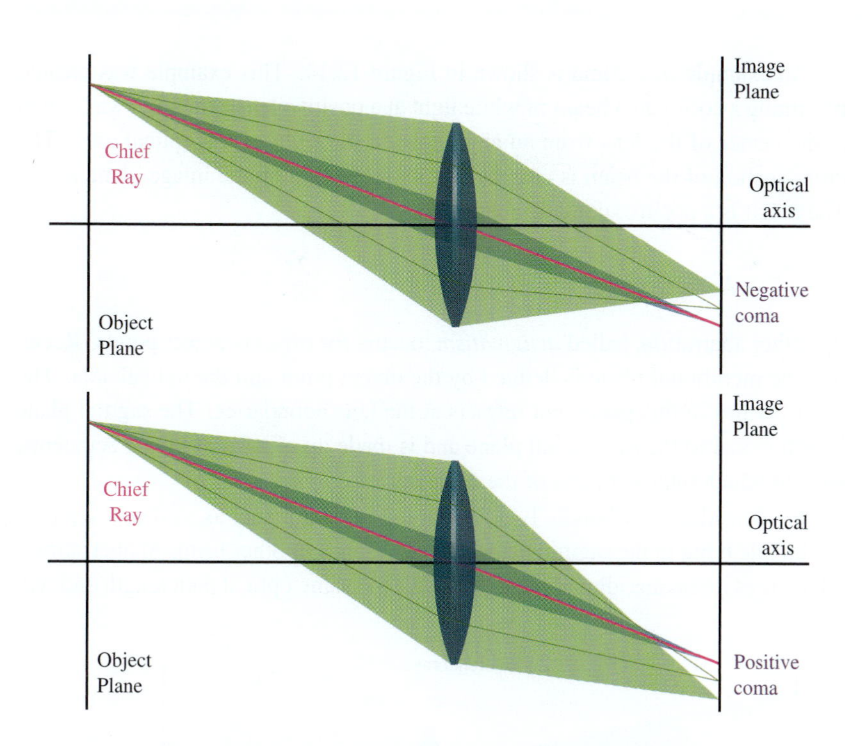

Figure 12.13. Diagram of a positive and a negative coma.

through the principal points), then the result is a positive coma. If the marginal rays hit the image plane closer to the optical axis than the principal ray, the coma is called negative.

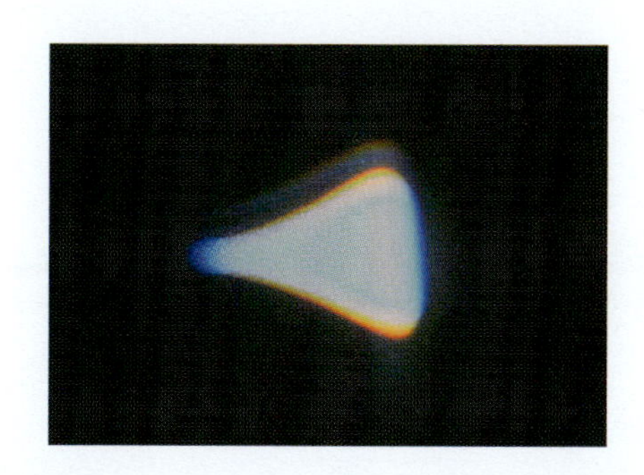

Figure 12.14. An example of a positive coma created by aiming an off-axis collimated beam of light at the center of a positive lens. The colored fringes suggest the presence of chromatic aberrations as well.

An example of a coma is shown in Figure 12.14. This example was created by aiming a collimated beam of white light at a positive lens. The light was aimed at the center of the lens from an angle toward the right of the optical axis. The cross-section of the beam is circular, yet its projection on the image plane is not. The result is a positive comatic aberration.

12.3.3 Astigmatism

A further aberration, called *astigmatism*, occurs for off-axis object points. Recall that the meridional plane is defined by the object point and the optical axis. The chief ray lies in this plane, but refracts at the lens boundaries. The sagittal plane is orthogonal to the meridional plane and is made up of a set of planar segments, each of which intersect parts of the chief ray.

We consider a ray bundle lying in the meridional plane, as well as a separate ray bundle lying in the sagittal plane. In the absence of other forms of aberrations, all the rays in the meridional plane will have the same optical path length and will

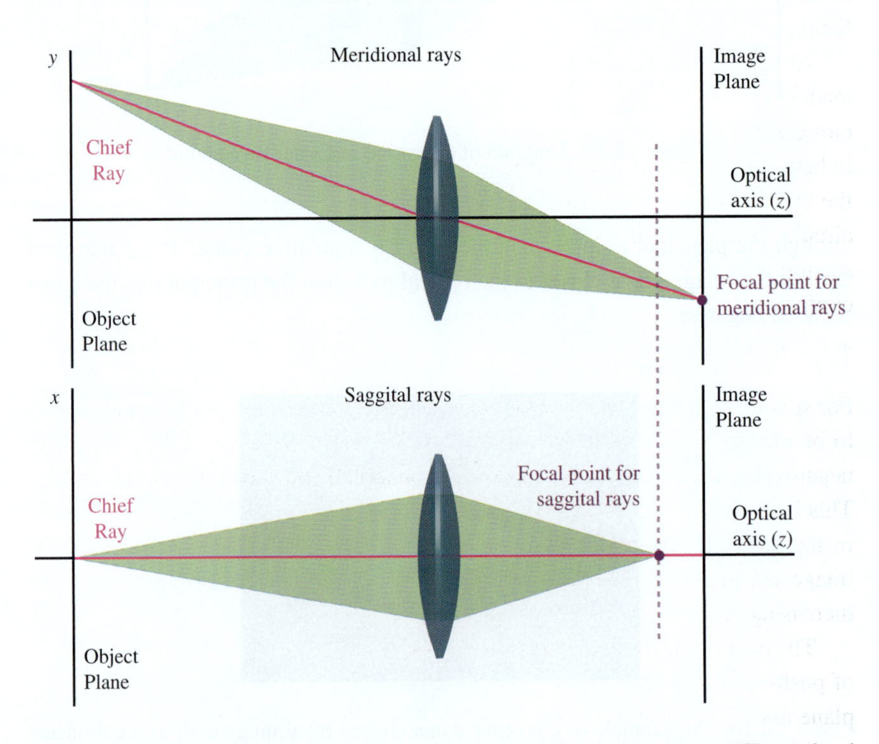

Figure 12.15. A flat bundle of rays located in the meridional plane has a different focal point than a flat bundle of rays lying in the sagittal plane, leading to astigmatism.

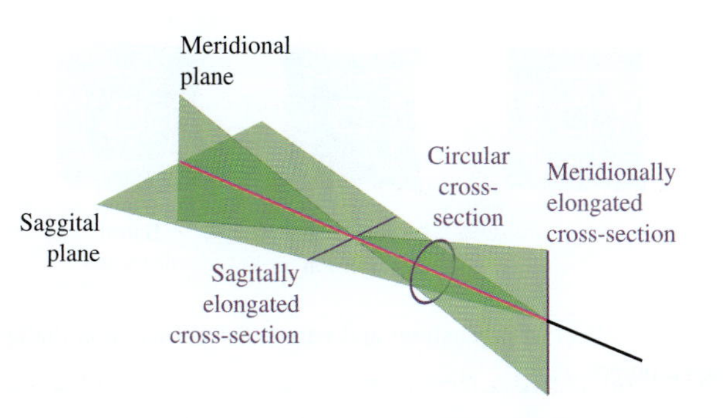

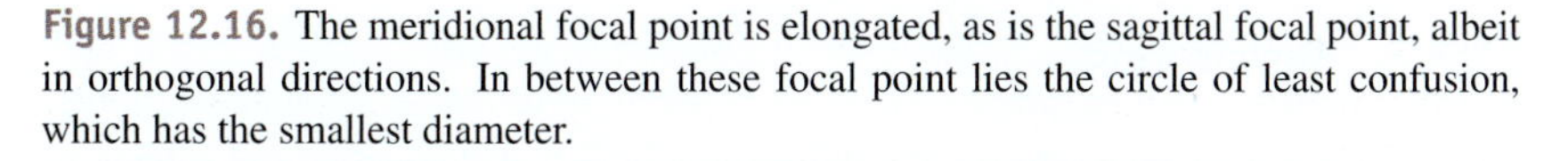

Figure 12.16. The meridional focal point is elongated, as is the sagittal focal point, albeit in orthogonal directions. In between these focal point lies the circle of least confusion, which has the smallest diameter.

therefore converge at the same focal point. However, this optical path length can be different from the optical path length of the rays lying in the sagittal plane. In turn, this causes the focal point of sagittal rays to lie either before or after the focal point associated with meridional rays, as shown in Figure 12.15.

At the sagittal focal point, the meridional rays will still not have converged, leading to an elongated focal point. Similarly, the meridional focal point will be elongated. For rays that are neither sagittal nor meridional, the focal point will be in between the sagittal and meridional focal points. Somewhere midway between the two focal points, the cross-section of all rays passing through the lens will be circular (see Figure 12.16). In the presence of astigmatism, this circular cross-section will be the sharpest achievable focal point; it is called the circle of least confusion.

12.3.4 Petval Field Curvature

For spherical lenses, the object and image planes, which were hitherto considered to be planar, are in reality curving inwards for positive lenses and outwards for negative lenses. In fact, both "planes" are spherical and centered around the lens. This is known as *Petzval field curvature*. The planar approximation is only valid in the paraxial region. If a flat image plane is used, for instance because the image sensor is flat, then the image will only be sharp near the optical axis, with increasing blur evident in the periphery.

The occurrence of field curvature can be counteracted by using a combination of positive and negative lenses. For instance, the inward curvature of the image plane associated with a positive lens can be corrected by placing a negative lens near the focal point of the positive lens. The negative lens is then known as a *field flattener*.

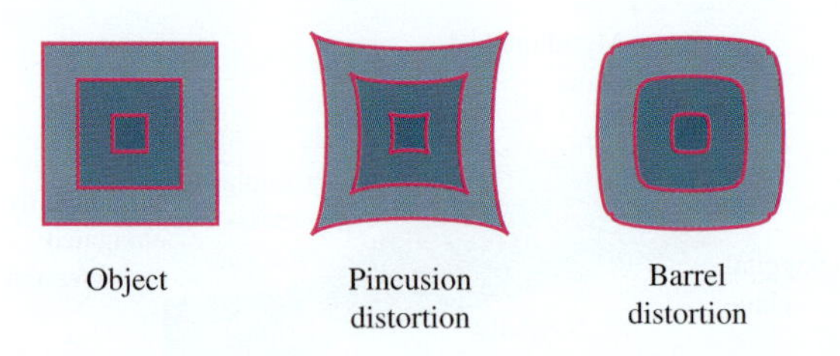

Figure 12.17. The effect of pincushion and barrel distortions upon the appearance of a regularly shaped object.

12.3.5 Distortions

Another source of monochromatic aberrations is called *distortion* and relates to the lateral magnification of the lens (defined for thin lenses in Section 12.2.2). While this measure was defined as a constant for the whole lens, it is only valid under paraxial conditions, i.e., near the optical axis. In real spherical lenses, the lateral magnification is a function of distance to the optical axis.

If the lateral magnification increases with distance to the optical axis, then a *positive distortion* or *pincushion distortion* results. For lenses with a decreasing lateral magnification, the result is a *negative distortion* or *barrel distortion*. Both distortions affect the appearance of images. In the case of a pincushion distortion, straight lines on the object appear to be pulled toward the edge of the image.

Figure 12.18. An example of a lens with barrel distortion.

A barrel distortion causes straight lines to be pushed toward the center of the image as shown in the diagram of Figure 12.17 and the photographic example of Figure 12.18 [1235].

Single thin lenses typically show very little distortion, with distortion being more apparent in thick lenses. Pincushion distortion is associated with positive lenses, whereas barrel distortion occurs in negative lenses. The addition of an aperture stop into the system creates distortion as well. The placement of the aperture stop is important; in a system with a single lens and an aperture stop, the least amount of distortion occurs when the aperture stop is placed at the position of the principal plane. The chief ray is then essentially the same as the principal ray.

Placing the aperture stop after a positive lens will create pincushion distortion, whereas placing the aperture stop in front of the lens at some distance will create barrel distortion. For a negative lens, the position of the aperture stop has the opposite effect. The effects of distortion in a compound lens can be minimized by using multiple single lenses which negate each other's effects and by placing the aperture half way between the two lenses.

Other types of geometric distortion are *decentering* distortion and *thin prism* distortion; the former is caused by optical elements that are not perfectly centered with respect to each other, and the latter is due to inadequate lens design, manufacturing, and assembly and is characterized by a tilt of optical elements with respect to each other.

12.3.6 Chromatic Aberrations

As discussed in Section 3.6.2, we may expect dielectric materials, and therefore lenses, to have a wavelength-dependent index of refraction. The focal length of lenses in turn is dependent on the index of refraction of the lens, as shown in (12.41) for thin lenses and (12.44) for thick lenses. Assuming a collimated beam of white light aimed at a positive lens along its optical axis, the focal point for wavelengths in the blue range lies closer to the lens than for wavelengths near the red range. The distance between these two focal points along the optical axis is called the *longitudinal* or *axial chromatic aberration*.

As light is refracted by greater or lesser angles dependent on wavelength, rays exiting the lens from the same off-axis position will hit the image plane at different positions. The wavelength-dependent lateral displacement of light creates a region along the image plane known as the *lateral chromatic aberration*. The colored fringes are shown in Figure 12.14. The distinction between longitudinal and lateral chromatic aberration is analogous to the longitudinal and transversal effects shown for spherical aberrations in Figure 12.12.

One can correct for chromatic aberration by applying a pair of thin lenses having different refractive indices, a configuration termed *thin achromatic doublets*. The wavelength-dependent focal length $f(\lambda)$ of the pair of lenses is then given by

$$\frac{1}{f(\lambda)} = \frac{1}{f_1(\lambda)} + \frac{1}{f_2(\lambda)} - \frac{d}{f_1(\lambda)\, f_2(\lambda)}, \tag{12.47}$$

assuming that the lenses are a distance d apart, their respective focal lengths are $f_1(\lambda)$ and $f_2(\lambda)$, and their indices of refraction are $n_1(\lambda)$ and $n_2(\lambda)$. If the index of refraction of the surrounding medium is taken to be 1, then the focal lengths of the two lenses are given by

$$\frac{1}{f_1(\lambda)} = k_1\,(n_1(\lambda) - 1), \tag{12.48a}$$

$$\frac{1}{f_2(\lambda)} = k_2\,(n_2(\lambda) - 1). \tag{12.48b}$$

In these equations, we have replaced the factor that depends on the radius of the front and back surfaces of the two lenses by the constants k_1 and k_2, which do not depend on wavelength. Substituting (12.48) into (12.47) gives

$$\frac{1}{f(\lambda)} = k_1\,(n_1(\lambda) - 1) + k_2\,(n_2(\lambda) - 1) - \frac{d}{\dfrac{1}{k_1\,(n_1(\lambda) - 1)}\;\dfrac{1}{k_2\,(n_2(\lambda) - 1)}}. \tag{12.49}$$

For the focal lengths $f(\lambda_R)$ and $f(\lambda_B)$ to be identical, the two thin lenses should be spaced a distance d apart. This distance can be found by solving (12.49) for d, giving

$$d = \frac{1}{k_1\, k_2}\;\frac{k_1\,(n_1(\lambda_B) - n_1(\lambda_R)) + k_2\,(n_2(\lambda_B) - n_2(\lambda_R))}{(n_1(\lambda_B) - 1)\,(n_2(\lambda_B) - 1) - (n_1(\lambda_R) - 1)\,(n_2(\lambda_R) - 1)}. \tag{12.50}$$

In the special case that the two lenses are touching, we have that $d = 0$, giving

$$\frac{k_1}{k_2} = -\frac{n_2(\lambda_B) - n_2(\lambda_R)}{n_1(\lambda_B) - n_1(\lambda_R)}. \tag{12.51}$$

It is now possible to specify the focal length of the compound lens as that of yellow light, which has a wavelength nearly half way between the wavelengths for blue and red light. The focal lengths for the two thin lenses are given by

$$\frac{1}{f_1(\lambda_Y)} = k_1\,(n_1(\lambda_Y) - 1), \tag{12.52a}$$

$$\frac{1}{f_2(\lambda_Y)} = k_2\,(n_2(\lambda_Y) - 1). \tag{12.52b}$$

Dividing and rearranging these two equations leads to

$$\frac{k_1}{k_2} = \frac{n_2(\lambda_Y) - 1}{n_1(\lambda_Y) - 1} \frac{f_2(\lambda_Y)}{f_1(\lambda_Y)}. \tag{12.53}$$

From Equations (12.51) and (12.53), we find that

$$\frac{f_2(\lambda_Y)}{f_1(\lambda_Y)} = \frac{(n_2(\lambda_B) - n_2(\lambda_R)) / (n_2(\lambda_Y) - 1)}{(n_1(\lambda_B) - n_1(\lambda_R)) / (n_1(\lambda_Y) - 1)} = \frac{w_2}{w_1}. \tag{12.54}$$

The quantities w_1 and w_2 are the *dispersive powers* associated with the indices of refraction n_1 and n_2. If we take the following wavelengths,

$$\lambda_F = 486.1 \text{ nm}, \tag{12.55a}$$

$$\lambda_D = 589.2 \text{ nm}, \tag{12.55b}$$

$$\lambda_C = 656.3 \text{ nm}, \tag{12.55c}$$

that coincide with the Fraunhofer F, D, and C spectral lines (see Table 3.8 on page 169), then we can define the dispersive power of an optical material more precisely. Rather than dispersive power, its reciprocal $V = 1/w$ is often used, which is variously known as the *V-number*, the *Abbe number*, the *dispersive index*, or the *constringence* of the material. The Abbe number V of a material is thus defined as

$$V = \frac{n_D - 1}{n_F - n_C}, \tag{12.56}$$

where $n_D = n(\lambda_D)$, $n_F = n(\lambda_F)$, and $n_C = n(\lambda_C)$. It is desirable in lens design to use materials that have low dispersion, and therefore high Abbe numbers, as well as high indices of refraction.

12.3.7 Blur Circle

An ideal optical system would image a point source onto a single point on the image plane. However, the various types of aberrations discussed above lead to a blurred shape on the image plane, which can be approximated as a circle. This *blur circle* then has a radius b that can be determined as follows.

The configuration and distances (all assumed positive) are as indicated in Figure 12.19. Note that the object is placed a height h above the optical axis, which coincides with the radius of the aperture d of the lens. The radius b of the blur

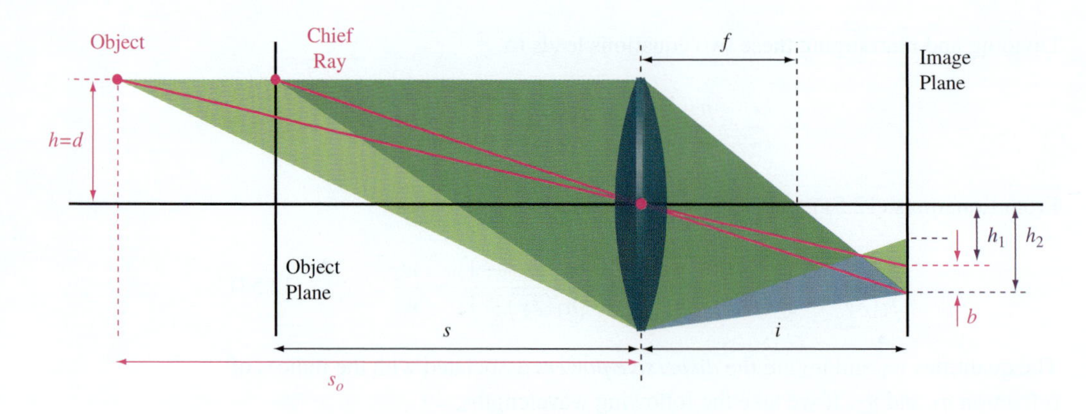

Figure 12.19. The geometric configuration assumed for computing the blur circle.

circle is then

$$b = h_2 - h_1 \tag{12.57a}$$

$$= di\left(\frac{1}{s} - \frac{1}{s_o}\right) \tag{12.57b}$$

$$= di\frac{s_o - s}{s\,s_o}. \tag{12.57c}$$

The Gaussian lens formula (12.42) for this configuration is

$$\frac{1}{f} = \frac{1}{s} + \frac{1}{i}, \tag{12.58}$$

and therefore we have

$$i = \frac{sf}{s-f}. \tag{12.59}$$

Substitution into (12.57c) yields

$$b = \frac{sfd}{s-f}\,\frac{s_o - s}{s\,s_o} = \frac{pfd}{s-f}. \tag{12.60}$$

In this equation, the quantity $p = (s_o - s)/s\,s_o$ can be seen as the percentage focus error. We use the blur circle next in the calculation of depth of field.

12.3.8 Depth of Field

Only points in the object plane are maximally sharply focused on the image plane. However, objects in a scene are more often than not located at various distances before and after the object plane. While all these points are imaged less sharply,

Figure 12.20. The limited depth of field causes a narrow range of depths to appear in focus. Here, the background is out of focus as a result. Macro lenses, in particular, have a very narrow depth of field, sometimes extending over a range of only a few millimeters; Orlando, FL, May 2004.

there is a region of distances between the lens and object points that are all focused reasonably sharply. This range of depths is called *depth of field*. As an example, an image with a limited depth of field is shown in Figure 12.20.

What constitutes an acceptably sharp focus depends on the size of the image plane, the resolution of the sensor, the size at which the image is reproduced, and all of this in relation to the angular resolution of the human visual system. As such, there is not a simple calculation as to range of values that are considered to be in focus.

However, if for some system it is determined that a circle with a radius b on the image plane leads to a reproduction that displays this circle as a single point. We then assume that the camera is focused on an object at distance s_o and that the blur circle is much smaller than the aperture, i.e., $d \gg b$. The distance from the lens to the nearest point s_{near} that has acceptable focus is then

$$s_{\mathrm{near}} = \frac{s_o f}{f + \frac{b}{d}\,(s_0 - f)}. \tag{12.61a}$$

Similarly, the furthest distance s_{far} still in acceptable focus is

$$s_{\mathrm{far}} = \frac{s_o f}{f - \dfrac{b}{d}\,(s_o - f)}. \tag{12.61b}$$

With $s_{\mathrm{near}} < s_o < s_{\mathrm{far}}$, the depth of field is now simply the distance between the near and far planes that produce an acceptable focus, i.e., $s_{\mathrm{far}} - s_{\mathrm{near}}$.

We see from (12.61b) that the far plane will become infinite if the denominator goes to zero, which happens if

$$f = \frac{b}{d}\,(s_o - f). \tag{12.62}$$

Solving for s_o gives the *hyperfocal distance*:

$$s_o = f\left(\frac{d}{b} + 1\right). \tag{12.63}$$

The corresponding near plane is given by

$$s_{\mathrm{near}} = \frac{f}{2}\left(\frac{d}{b} + 1\right). \tag{12.64}$$

Thus, if the camera is focused upon the hyperfocal plane, all objects located between this value for s_{near} and beyond will be in sharp focus.

Finally, we see that the size of the aperture d affects the depth of field. An example of a pair of images photographed with different aperture sizes is shown in Figure 12.21. In particular in macro photography the depth of field can become very limited—as small as a few millimeters in depth.

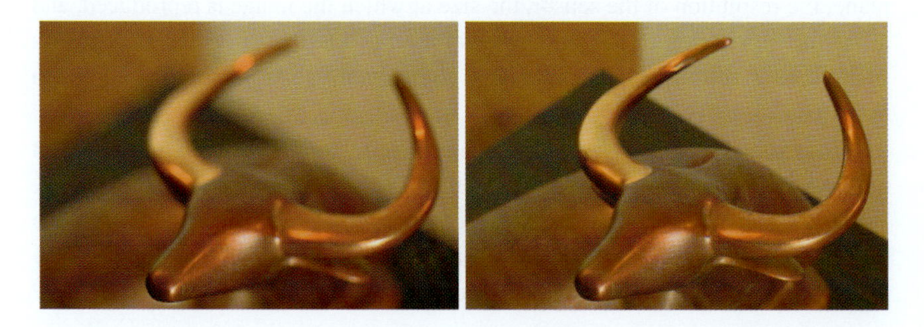

Figure 12.21. Images photographed with a large aperture (f/3.2, left) and a small aperture (f/16, right). The depth of field is reduced to less than a centimeter in the left image, due to the use of a 60 mm macro lens. The figurine is a Loet Vanderveen bronze.

While depth of field is a concept related to the object side of the lens, a similar concept exists for the image side of the lens. Here, it is called *depth of focus* or *lens-to-film tolerance* and measures the range of distances behind the lens where the image plane would be in sharp focus. The depth of focus is typically measured in millimeters, or fractions thereof.

12.4 The Diaphragm

The aperture stop discussed in the preceding sections determines how much light reaches the image plane. Most camera designs feature a variable aperture stop, called the *diaphragm*, giving the photographer control over how light or dark the image will be. An example is shown in Figure 12.22.

In addition to regulating the amount of light incident upon the image plane, the diaphragm also has an impact on the severity of different types of aberrations. An optical system with a very small aperture stop operates largely in the paraxial region, limiting monochromatic aberrations. Thus, although the total amount of light collected at the image sensor depends on both the size of the diaphragm and the exposure time, for every camera there exists an optimal aperture size that minimizes the occurrence of artifacts, including vignetting. Whether practical conditions allow that setting to be used, depends on how much light and how much movement is available in the scene.

The maximum aperture diameter afforded by a lens, and therefore the minimum f-number, determines the flux reaching the image plane. The larger this aperture, the shorter the exposure time required to collect the desired image irradiance. Thus, the maximum aperture diameter determines the *lens speed*.

Figure 12.22. A diaphragm with three size settings. This diaphragm does not allow more than three settings and is suitable for budget cameras. More typical diaphragms are constructed using a set of blades which together form a variable-sized opening.

Figure 12.23. A partly opened two-blade shutter.

12.5 The Shutter

The shutter is a mechanism which can be opened and closed to allow light to reach the sensor for a pre-determined time. It can be either a physical device, or it can be implemented electronically. In the latter case, the firmware of the camera simply instructs the sensor to start and stop recording at the appropriate times.

When the shutter is a mechanical device, there is choice of where to place it in the optical path. In principle, the shutter can be placed anywhere: before the lens, in between the lens elements, or in front of the sensor. When placed in front of the sensor, it is called the *focal plane shutter*; when placed in between the lens elements, it is called an *inter-lens shutter*. Both types of shutter have an effect on the type of artifacts that may occur. As such, shutters may be designed to have a beneficial effect on the quality of the imagery [906].

As an example, an ideal shutter opens in zero time, remains fully open for the duration of the exposure, and then shuts again in zero time. This ideal is clearly not achievable with a mechanical shutter. However, a shutter may be designed to open over some amount of time and then close again. If this results in the center of the aperture being exposed longer than its margin, the glare associated with the aperture will be softer than with an ideal shutter. This can be visually more pleasing.

Inter-lens shutters are usually center-opening, meaning that a set of blades parts to let light through. An example of a two-blade shutter is shown in Figure 12.23. They are normally placed as near as possible to the diaphragm, as this is where the aperture of the system is smallest. During opening and closing, which may take a substantial fraction of the total exposure time, both the size and the shape of the shutter opening may vary over time. Thus, the effective aperture of the camera becomes a function of time.

Focal plane shutters are normally made of a pair of curtains that slide across the sensor at some distance from each other, forming a moving slit. The closer this type of shutter is mounted to the sensor, the less diffraction off the edge of the shutter will have an effect on the image quality. Focal plane shutters are often used in single-lens reflex (SLR) cameras.

12.6 Filters and Coatings

In addition to lenses to focus the light, other elements may be placed in the optical path. These elements consist of coatings applied to the surface of the lens elements, as well as filters. In particular, coatings are applied to reduce unwanted reflections of the lenses, as well as the sensors which tend to reflect a significant portion of the incident light. Filters may be added to the optical system to reduce aliasing, and to split the light into color bands. The latter is called the color filter array and requires subsequent signal processing to reconstruct a color image. We therefore defer a discussion of this filter to Section 12.8.1.

12.6.1 Anti-Reflective Coatings

The surface reflectance of a lens is typically reduced when a thin layer of coating is applied. Under the assumption that the index of refraction for the surrounding medium is 1, and that the angle of incidence is perpendicular to the lens surface, the reflectance of the lens is given by (2.139a). If we introduce a single layer of coating with an index of refraction given by n_c, the reflectance of the lens becomes

$$R = \left(\frac{n - n_c^2}{n + n_c^2} \right)^2, \tag{12.65}$$

where n is the refractive index of the glass used in the fabrication of the lens. We also assume that the thickness of the coating is a quarter of a wavelength. To reduce the reflectance to 0, we would have to choose a coating with a refractive index of $n_c = \sqrt{n}$. For glass with a refractive index of around 1.5, this means that a coating with an index of refraction of around 1.22 should be sought. Unfortunately, such coatings do not exist. The most suitable coatings have an index of refraction of around 1.33 [613].

For a two-layer coating, the reflectance is given by

$$R = \left(\frac{n_{c,1}^2\, n - n_{c,2}^2}{n_{c,1}^2\, n + n_{c,2}^2} \right)^2, \tag{12.66}$$

where the coating applied to the lens has a refractive index of $n_{c,2}$ and the second, outer layer has an index of refraction of $n_{c,1}$. Once more, the coatings are assumed to have a thickness of a quarter of a well-chosen wavelength. To guarantee no reflections, this equation shows that we should seek a combination of coatings for which the following relation holds:

$$n = \frac{n_{c,2}}{n_{c,1}}. \tag{12.67}$$

In practice, solutions of this form are feasible, and lenses employed in digital cameras are routinely coated with multiple coats to suppress unwanted reflections.

12.6.2 Optical Anti-Aliasing

The process of demosaicing[2] makes implicit assumptions on the scene being imaged. In particular, the spatial frequencies incident upon the sensor must be low enough to enable the spatial averaging algorithms to produce sensible results. If the spatial frequency reaches beyond the Nyquist limit, which is half the spatial frequency of the sensor, the result is that these high frequencies are mapped to frequencies below the Nyquist limit. This is called *aliasing* and results in visible artifacts such as jaggies along edges and Moiré patterns in places where textures have frequencies that are too high [106, 107].

A good camera design therefore includes counter-measures against spatial frequencies that cannot be represented in the digital read-out of the sensor. One of the ways to achieve this is to apply a low-pass filter. In cameras, this is usually achieved with an optical filter, placed in front of the sensor. Such an anti-aliasing filter can be constructed for instance by using polarization [390]. By placing two birefringent elements in succession, for instance quartz, calcite (see Figure 2.26), liquid crystal, or lithium niobate [613], the optical signal can be low-pass filtered in horizontal and vertical directions before it is projected onto the sensor.

The displacements Δx and Δy achieved by the two birefringent filters can be computed from the thickness Δd of the filter and the two indices of refraction n_e and n_o of the birefringent material as follows:

$$\Delta x = \Delta y = \Delta d \, \frac{n_e^2 - n_o^2}{2 n_e n_o}. \tag{12.68}$$

The displacement Δy is achieved by the second filter, which is rotated 90° with respect to the first filter.

Unfortunately, high-quality birefringent elements are expensive. An alternative is to base the anti-aliasing filter on phase delay. Remember that the speed

[2]Note that the “c” is pronounced as “k”.

of light propagation depends on the type of dielectric material the light passes through. By etching a shallow pattern into a dielectric element, some of the light passes through a thicker layer of material, emerging with some phase delay. By varying the depth of the etched pattern, the amount of phase delay can be controlled. The different amounts of phase-delayed light will interfere, thereby canceling high-frequency content more than low-frequency content. The type of pattern edged into the dielectric element can be a set of discs which are randomly placed and have randomly chosen diameters. The advantage of this approach is cost of manufacture, whereas the disadvantage is that unless the filter is carefully designed, this type of phase-delay filter can act as a diffraction grating, causing significant artifacts [5].

12.6.3 Infrared Rejection

Filters and other elements may be placed in the optical path to alter attributes of the light before it hits the sensor. One reason for altering the light is to remove the wavelengths to which the human visual system is not sensitive but to which the camera's sensor is sensitive. Failing to remove these wavelengths will create incorrect colors. For instance, some flowers as well as some types of black clothing reflect substantially in the infrared [5]. The sensing of the infrared component is liable to turn blue in the scene into purple in the image.

Most infrared rejection filters are absorption-type, i.e., infrared wavelengths are absorbed and converted to heat. Some filters reflect the infrared component using a dichroic element, or they use a combination of the two techniques. The

Figure 12.24. A CCD sensor with the infrared filter fitted (left) and removed (right).

which is the reciprocal of the input required to generate a response that is just above the sensor's noise level.

Linearity. The collection of charge in response to incident photons, and the subsequent conversion to a digital signal should maintain a linear relationship between the number of photons and the value of the resulting signal.[3] For most sensors, this relationship holds to a high degree of accuracy. However, the linearity of transistors may be compromised in sensor designs that utilize ultra-thin electrical paths (0.35 μm and below) [688].

Pixel uniformity. Due to variations in the fabrication process, a uniformly illuminated sensor may produce a signal that is spatially variant. Such undesired non-uniformities may sometimes be corrected for by the firmware. In particular, two forms of non-uniformity are distinguished—dark signal non-uniformity (DSNU), which can be measured in the absence of illumination, and photo response non-uniformity (PRNU), measured when the sensor is illuminated. DSNU is discussed further in Section 12.10.2, while PRNU is discussed in Section 12.11.4.

In the following, we briefly discuss the operating principles of CCD and CMOS sensors.

12.7.1 CCD Devices

In a charge-coupled device (CCD), the photosensitive elements are photodiodes or photogates that are arranged in an array. A photodiode can absorb photons, attracting electrons which reduce the voltage across the diode proportionally to the amount of incident power.

When the exposure starts, these photosites collect charge as the result of incident photons. Each photosite has a maximum amount of charge that can be collected, after which the photosite saturates. This is called the *full-well capacity*. After the exposure has finished, the collected charges are moved towards the analog to digital (A/D) converter. The image is thus read one pixel at a time, which is a relatively slow process.

There are three types of CCD architecture, called full-frame, frame-transfer, and inter-line sensors. In a full-frame sensor, the full surface area of the sensor is devoted to collecting light. A full-frame CCD does not contain an electronic shutter, so that a camera with a full-frame sensor is typically fitted with a mechanical shutter, as discussed in Section 12.5.

[3] For logarithmic sensors, the relationship between output signal and number of incident photons should be as close as possible to a logarithmic function.

A frame-transfer architecture is composed of a light-sensitive area for charge collection and a separate storage area that is masked by an opaque material, such as aluminium. After the exposure finishes, the collected charge is transferred to the masked area before read-out commences. The transfer to the opaque storage area effectively implements an electronic shutter, despite the slow read-out process. A disadvantage of this approach is that the size of the light-sensitive area is only half that of the sensor, so that compared to a full-frame design, an equivalent sensor would be twice as large and, therefore, more expensive to produce. A further disadvantage is that the process of transferring charge from one area of the chip to another may cause some smear. The fill factor of the light-sensitive area can be as high as that of a full-frame sensor.

An inter-line design redistributes the storage area of a frame-transfer sensor into strips, so that columns of pixels are alternated with columns of storage areas. This reduces the fill-factor of each pixel to about 0.5, but this can be improved to 0.9 or higher by placing a micro-lens that which focuses light onto the sensor in front of each pixel, and away from the aluminium strips. As each pixel has its own adjacent storage area, the transfer of charge requires only one step, thereby vastly reducing the occurrence of smear.

In addition to moving charges underneath an opaque strip, it is possible to implement a fast electronic shutter by fitting each pixel with a shutter drain. This is an extra line which can be biased to a high voltage, thus draining away photon-generated electrons. Lowering the voltage on this shutter drain and increasing the voltage on the imaging-array gate then opens the shutter [940].

The read-out procedure of a CCD sensor requires the collected charges to be moved to a single on-chip site, where the charge is amplified and presented to the external interface pin. The basic principle used to achieve this transfer is unique to CCDs and is shown in Figure 12.25. Each pixel contains a potential well that can collect and hold the charge as long as a bias voltage is applied. To move the charge from one potential well to the next, first the bias voltage for the neighboring pixel is raised, so that the pair of pixels form a single larger potential well. Then, the bias voltage is removed from the first pixel, so that the charge concentrates in the potential well associated with the second pixel.

This process of charge transfer can be used to move the charges step-by-step to one or more amplifiers, typically consisting of a few transistors, which convert the charge to a voltage. Analog-to-digital conversion takes place off-chip, and several further steps of image processing may then follow in the digital domain. These are typically implemented in a separate integrated circuit (IC).

CCD sensors typically require different voltages for different parts comprising the chip. The clocking voltage in particular needs to be relatively high, leading

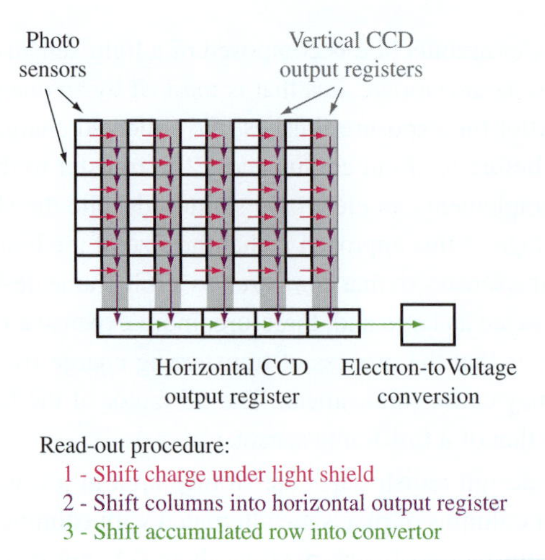
Photo sensors
Vertical CCD output registers
Horizontal CCD output register
Electron-toVoltage conversion
Read-out procedure:
1 - Shift charge under light shield
2 - Shift columns into horizontal output register
3 - Shift accumulated row into convertor

Figure 12.25. The read-out procedure on an inter-line CCD sensor is that first all charges are moved underneath the opaque strip. In parallel, all columns are shifted down by one position, moving the bottom scan-line into the horizontal read-out register. This register is then read step-by-step. This process is repeated until all scan-lines are read.

to both high system complexity and high power consumption [103], although the trend in CCD design is towards lower voltages and less power consumption. Additionally, it is possible to design CCD sensors with a single power supply voltage, which is used to internally derive other required voltages using an integrated charge pump [154].

Finally, when the light incident upon a pixel is very bright, the potential well may fill up and charge may leak away into neighboring pixels. When this happens, the result is a streak of light called *bloom*. Moderns CCDs have anti-blooming mechanisms at each charge detection node; these allow extraneous charge to leak away in a controlled manner. Such gates typically reduce the fill factor by about 30%, reducing sensor sensitivity.

Bloom can also be avoided manually, by avoiding long exposures. To obtain the desired image, several short exposures, each without bloom, may be stacked in software. The noise characteristics of the resulting image are not affected by this method.

12.7.2 CMOS Devices

Complementary metal-oxide semiconductor (CMOS) processing is used to fabricate standard integrated circuits for computing and memory applications, but can be adapted for use as sensors. With CMOS technology, it is possible to integrate the sensor array, the control logic, and potentially analog-to-digital conversion on the same chip. CMOS sensors are mixed-signal circuits, containing both analog circuitry, possibly A/D converters, and digital control logic. The main building blocks are a pixel array, analog signal processors, row and column selectors, and a timing and control block.

The pixel array itself can be constructed in many different ways. In contemporary CMOS image-sensor devices, the photosensor at each pixel is augmented with additional circuitry, such as a buffer/amplifier, yielding an *active pixel sensor* (APS). With such designs, high frame-rates are possible [341]. As each amplifier is only activated during readout, the power dissipation can be kept low. Early active pixel designs suffered from fixed pattern noise [95], which is discussed in Section 12.10.2. Many variations of active pixel sensors are currently in existence, based upon either photodiodes or photogates. Examples of technologies that have been demonstrated include *thin film on ASIC* (TFA) pixel designs, which yield high sensitivity and dynamic range [707, 708], and complementary active pixel sensors (CAPS) [1264]. There is an ongoing effort to develop sensors with higher fill factors, lower power consumption, better noise characteristics, and better sensitivity.

Analog-to-digital (A/D) conversion circuitry may be included for each pixel. The design is then called a *digital pixel sensor* (DPS) and benefits high-speed imaging applications, as some of the bottlenecks associated with A/D conversion and readout are reduced. Digital pixel sensors also potentially eliminate some of the problems that occur with the analog circuitry associated with APS systems, including a reduction of column fixed-pattern noise and column readout-noise (see Section 12.10.2). The drawback of the DPS approach is that the extra circuitry causes a reduction of the fill factor, although this problem reduces as the technology scales down.

CMOS sensors sometimes include circuitry that allows each pixel to be read in an arbitrary order. Electronic scan circuits can be added that determine the pixel read order. As a result, when the exposure is finished, either the whole image can be read, or specific regions may be selected. This can, for instance, be used to implement region-of-interest (ROI), electronic zoom, pan, and tilt.

The sensor design may incorporate additional analog systems, for instance to improve noise, increase the signal-to-noise ratio, or perform A/D conversion. If an A/D converter is included, then it will be subject to several design considerations [330]. The throughput should be high enough to yield a frame-rate that supports the intended application. The converter should not introduce non-linearities or other artifacts and should dissipate only a small amount of power. If too much power is used, then the heat generated by the A/D converter may create temperature variations across the chip's surface, which in turn has a negative effect on the sensor's noise characteristics. Finally, the A/D converter should not introduce noise into the system through cross-talk.

To improve the dynamic range of CMOS systems, a variety of techniques may be incorporated. For instance, the sensor may be designed to produce a logarithmic response [177, 330], employ conditional resets [1273], switch between pixel circuit configurations [850], change saturation characteristics [241], use negative feedback resets [525], employ an overflow capacitor per pixel [22], or use multiply exposed pixels [2, 748, 809, 995, 1266, 1272]. While the dynamic range of image sensors continues to improve, photographers can also create high dynamic range images with current digital cameras by employing the multiple exposures techniques discussed in Section 13.1.

A range of other image processing functionalities may be included on the same chip, including motion detection, amplification, multi-resolution imaging, video compression, dynamic range enhancement, discrete cosine transforms, and intensity sorting [95, 330], making CMOS technology suitable for a wide variety of applications.

12.8 In-Camera Signal Processing

The sensor produces a signal that will require further processing to form a presentable color image. First, a color image must be reconstructed. Further processing may include white balancing and sharpening amongst other. In the following, we discuss demosaicing, white balancing, and sensitivity settings in more detail.

12.8.1 Color Filter Arrays

Digital color cameras can be constructed by placing spatially varying colored filters in front of the image sensor. As a result, the light will be filtered such that each pixel will be sensitive to a particular distribution of wavelengths. It is therefore possible to create a color image with a single sensor.

Each cluster of pixels on the sensor is illuminated with a pattern of colored light, with many patterns being available to the sensor designer [5, 880]. The most often used pattern is the *Bayer mask*, which contains 2×2 clusters of pixels with one red, two green, and one blue filter [76]. The two green pixels are placed diagonally with respect to each other. The double weight for the green range of wavelengths is chosen because the human visual system has highest visual acuity

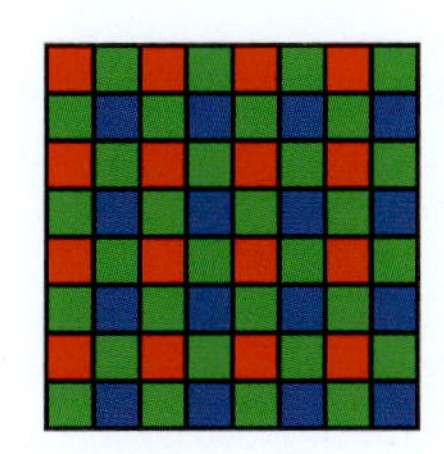

Figure 12.26. The Bayer color filter array.

at medium-wavelength light (see Section 6.1). An example of a Bayer color filter array (CFA) is shown in Figure 12.26.

The output of the sensor can be considered to be a composite, monochromatic image, with each pixel filtered by the CFA. To reconstruct a color image, it would be possible to take each cluster of four pixels and recombine these into a single colored output pixel. However, such a naïve approach would exhibit poor spatial resolution.

A better alternative is to reconstruct an image at the same resolution as the sensor. Under the assumption that object features are projected onto large sets of pixels, interpolation schemes may be employed to reconstruct the image. This process is known as *CFA interpolation* or *demosaicing* [5].

There are several trade-offs that a camera designer faces when reconstructing a color image [589, 590, 633, 1104]. First, the processing taking place by the firmware must be as fast and uncomplicated as possible. Second, the processing should introduce as few artifacts as possible.

For the demosaicing process, this typically means that the luminance values are taken to be estimated by the values sensed for the green channel. To obtain luminance values for the pixels that are not green, linear interpolation may be used. As this creates a blurry image, a better strategy is to use non-linear adaptive averaging [7], for instance by employing a simple edge detector and ensuring that averaging does not take place across edges [512]. The kernel size of the edge detector is an important trade-off between processing time and quality of the results. The edge-detection and averaging process can be further refined by making the assumption that red, green, and blue color values are correlated within a local image region. This means that, for instance, blue and red are related as follows:

$$B(x,y) = R(x+1,y) + k, \tag{12.70}$$

where k is a constant. This assumption is not unreasonable given that strong correlations between red, green, and blue channels exist in natural images. Statistical regularities in natural images are further discussed in Section 15.1. By using

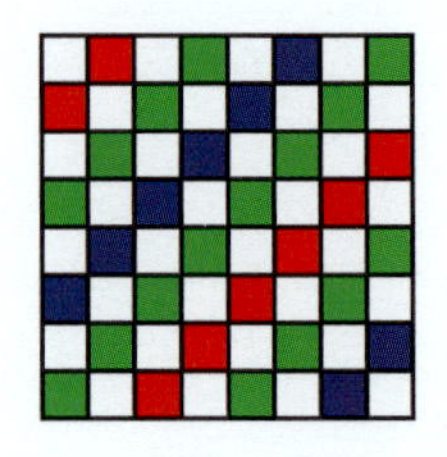

Figure 12.27. Kodak's proposal for a new color filter array, including unfiltered panchromatic pixels to efficiently derive an estimated luminance signal.

this particular correlation, the edge-detection process will have more values in a small neighborhood of pixels to consider, so that a more accurate result can be obtained [5].

In a camera design, the luminance channel (often approximated by just using the green channel) may be augmented by two chrominance channels, which are usually taken to be red minus green (C_R) and blue minus green (C_B). Note that the green value was computed using the CFA interpolation scheme. This very simple solution is intended to minimize firmware processing times. To compute the missing chrominance values, linear interpolation is employed between neighboring chrominance values (the nearest neighbors may be located along the diagonals).

Luminance and chrominance values are then converted to RGB. This conversion, as discussed in Section 8.1, is device-dependent as a result of the dyes chosen in the manufacture of the CFA, as well as the firmware processing applied to the image.

The process of demosaicing is intimately related to the choice of color filter array. While the Bayer pattern is currently favored in most camera designs, it is not optimal in the sense that the filters absorb much of the incident light, leading to a reduced quantum efficiency of the camera system. In the summer of 2007, Kodak announced an alternative color filter array which consists of four types of filter, namely red, green, blue, and panchromatic. The latter is effectively an unfiltered pixel. The advantage of this approach is that a luminance channel can be derived from the CFA with a much higher efficiency. In addition, the pattern is constructed such that when the panchromatic pixels are ignored, the result is a Bayer pattern that allows conventional color interpolation algorithms to be employed. The Kodak CFA is shown in Figure 12.27.

The placement of a CFA in the optical path significantly reduces the amount of light that reaches the sensor, requiring an increase in aperture size, exposure time, or sensor sensitivity to compensate. An alternative to the use of CFAs is to use multiple sensors. A beam splitter is then placed in the optical path to split the light beam into three directions so that light can be focused onto three sensors, one for each of the red, green, and blue ranges of wavelengths. The result is a higher quantum efficiency and better color separation. Due to cost factors associated with the perfect registration of the three light beams, this approach is used only in high-end equipment such as studio video cameras.

12.8.2 White Balancing

Further in-camera processing may be required, dependent on user preferences and camera settings. This includes, for instance, white balancing, which can be either automatic or user-specified [275]. In the case that the white balance factor is set to a specific color temperature, the camera will multiply the pixel values by a 3×3 matrix, often derived from a pre-calculated look-up table.

When the camera is set to "auto," the camera will have to infer the illuminant from the image it has just captured. While algorithms to achieve this were discussed in Section 9.9, in-camera processing must be very fast, and that condition limits the range of algorithms that can be sensibly employed. A very simple and fast approach is to make the gray-world assumption, which states that averaging all the colors found in an environment yields gray. Specifically, the average image color is typically assumed to be 18% gray. Thus, the color found by computing this average can be used as the correction factor. In motion imaging cameras, the average color can be computed over many frames, rather than for each frame individually. There are, of course, many situations where the gray-world assumption fails, and in those cases the white balancing will also fail. This can be remedied by employing a more advanced algorithm, for example by excluding very saturated pixels during the averaging [5].

A second approach is to assume that the pixel with the highest value represents a white surface. The color of this pixel is then assumed to be that of the light source. This heuristic fails if self-luminous surfaces are imaged. For instance, colored light sources such as traffic lights would cause this approach to fail.

The third approach to automatic white balancing using in cameras is to analyze the color gamut of the captured image. Based on statistical assumptions on the average surface reflectance and the emission spectra of typical light sources, the color gamut of the captured image can be compared to a database of color gamuts [503].

Practical in-camera white balancing algorithms employ a combination of these three techniques. Many professional photographers prefer to set their cameras to record "raw" (unprocessed) image data and to perform processing (including demosaicing, color correction, white balance, and picture rendering) off-line in application software such as Adobe Lightrom, Photoshop, or Apple Aperture. This opens up the possibility of using the advanced algorithms that were described in Section 10.6. Note that off-line white balancing may suffer from gamut limitations, a problem that can be largely overcome by using high dynamic range imaging [952].

12.8.3 ISO Sensitivity

Digital sensors usually have a parameter called *ISO sensitivity* that allows the analog gain to be adjusted prior to analog-to-digital conversion [534]. For high ISO values, more amplification is applied so that less incident light is necessary to generate output values that lie within the camera's operating range. However, noise in the signal is also amplified more, and the resulting image is liable to appear grainy. Figure 12.28 shows the impact of changing the ISO values on the final image.

Figure 12.28. Both images were captured with a Nikon D2H; the left image with ISO 200 (exposure 1/20 s at f/20) and the right image using the Hi-2 setting, which is equivalent to ISO 6400 (1/1000 s at f/20). The increase in sensor sensitivity is paired with a marked increase in noise; Long Ashton, Bristol, May 2007.

12.9 A Camera Model

Mathematical models of a camera are useful for understanding the sources of error that may accumulate during image acquisition. In the following description of a camera model, the analysis of noise sources, and calibration scheme, we follow the work of Healey and Kondepudy [439]. It may be assumed that the number of electrons collected at a photosite (x, y) is obtained by integrating over the surface area of the pixel (u, v) and over all wavelengths λ to which the sensor is sensitive. As each electron represents a unit of charge, the total charge Q collected at a photosite is then

$$Q(x,y) = \Delta t \int_{\lambda} \int_{x}^{x+u} \int_{y}^{y+v} E_e'(x,y,\lambda)\, S_r(x,y)\, c_k(\lambda)\, dx\, dy\, d\lambda. \tag{12.71}$$

Here, Δt is the integration time in seconds, E_e' is the spectral irradiance incident upon the sensor (W/m^2), S_r is the spatial response of the pixel, and $c_k(\lambda)$ is the ratio of charge collected to the light energy incident during the integration (C/J).

In an ideal system, there is no noise and there are no losses, so that the charge collected at a photosite is converted to a voltage, which may be amplified with a gain of g, leading to a voltage V:

$$V(x,y) = Q(x,y)\, g. \tag{12.72}$$

The A/D converter then converts this voltage into a digital count n. Assuming that there are b bits, n will be in the range

$$0 \leq n \leq 2^b - 1. \tag{12.73}$$

If the quantization step is given by s, this means that the voltage is rounded to $D = n\,s$, relating n, s, and V as follows:

$$(n(x,y) - 0.5)\, s < V(x,y) \leq (n(x,y) + 0.5)\, s. \tag{12.74}$$

The digital image $n(x,y)$ can then be further processed in the camera firmware.

Of course, in practice, sensors do not follow this ideal model, and, in particular, noise is introduced at every stage of processing. In the following sections, we discuss the different sources of noise, followed by an outline of a procedure to characterize a camera's noise response.

12.10 Sensor Noise Characteristics

One of the greatest design challenges for any type of image sensor is the suppression of noise, which can be defined as any component of the output signal that is unwanted and does not represent the image irradiance incident upon the sensor. The first of two classes of noise is temporal noise, which can be induced by the various components making up the sensor. This type of noise will be different every time an image is captured. On the other hand, fixed pattern noise arises due to irregularities during the fabrication of the sensor and will induce an unwanted spatially varying signal that is the same every time an image is captured.

12.10.1 Reset Noise

Before image sensing can begin, the potential wells have to be reset. The procedure to accomplish this requires a certain amount of time, which is known as the reset-time constant. In high-speed applications, clock speeds may be chosen such that the reset-time interval, the time allocated to resetting the charge-detection nodes, is longer than the reset-time constant. This means that not all potential wells are fully empty when the capture of the image begins. The residual charge forms a spatially varying signal known as *reset noise* [521].

This type of noise can be canceled using *correlated double sampling* (CDS) circuitry. First, each pixel is reset. The remaining signal is then read, yielding a *bias frame* containing a measure of the reset noise. After image capture, the signal is read again. Subtraction of the two signals yields a reset-noise-free image.

12.10.2 Fixed Pattern Noise

There are several sources of noise in a camera that show up as inaccuracies in the camera's output. First, when the sensor is uniformly lit, the charge collected at every pixel will vary due to fabrication errors. This type of variation is a form of *fixed-pattern noise* known as *dark signal non-uniformity* (DSNU). It can be corrected for by taking a photograph with the lens cap fixed. The resulting image is a measure of fixed-pattern noise, and it can be subtracted from every image taken. Some cameras have a built-in function to remove this type of noise, which is achieved by taking a second exposure with the shutter closed and subtracting it from the real exposure.

In our camera model, the fixed pattern noise shows up as a variation in sensor response S_r, as well as a variation in the quantum efficiency $c_k(\lambda)$. These two sensor characteristics are each scaled by a constant which is fixed (but different) for each pixel. Naming these constants $k_1(x,y)$ and $k_2(x,y)$, we assume that they are properties of each pixel and can therefore be taken out of the integration of (12.71). Taking $k(x,y) = k_1(x,y)\, k_2(x,y)$, our camera model now produces a charge Q_n for charge collection site (x,y):

$$Q_\text{n}(x,y) = Q(x,y)\, k(x,y), \tag{12.75}$$

where the subscript n is introduced to indicate that this electron/charge count includes a noise factor. The fixed pattern noise $k(x,y)$ is taken to have a mean of 1 and a variance of σ_k^2 which depends on the quality of the camera design. This model is valid in the absence of bloom; i.e., we assume that charge collection at each photosite is independent from collection at neighboring sites.

12.10.3 Dark Current and Dark Current Shot Noise

In the absence of any light, electrons may still make their way into the potential wells that collect charge, for instance due to thermal vibrations. The uniform distribution of electrons over the sensor collected in the dark is known as the *dark current*. Post-processing may correct for the average level of dark signal, but a certain amount of noise will be superimposed on the dark signal. In addition, there may be a non-uniformly distributed source of noise which accumulates in the dark [203], known as *dark current shot noise*, which cannot be corrected for by post-processing, but should be minimized through appropriate sensor design and operation. In critical applications, dark current may be reduced by cooling the sensor [546], which is especially important in low-light photography which for instance occurs in astronomy.

Dark current is independent of the number of photoelectrons generated. The charge associated with the number of dark electrons Q_{dc} produced as a result of dark current is therefore added to the signal, giving a total charge Q_{n}:

$$Q_{\mathrm{n}} = Q(x,y)\, k(x,y) + Q_{\mathrm{dc}}(x,y). \tag{12.76}$$

12.10.4 Photon Shot Noise

The arrival times of photons at the image sensor is random and follows Poisson statistics. This means that if the mean number of photons increases, the variance in the signal increases by the same amount. As such, above a given threshold, photon shot noise will dominate the other types of noise present in the output signal. However, as all other types of noise are independent of the input signal, below this threshold other types of noise will be larger.

The impact of photon shot noise can be reduced by collecting a higher number of photo-electrons, for example by widening the aperture, or increasing exposure time—up to the limit imposed by the full-well capacity of the sensor.

Under the assumption that photons adhere to a Poisson distribution, and that the number of photoelectrons collected at each charge collection site is a linear function of the number of incident photons, the distribution of photoelectrons also follows Poisson statistics. As such, the uncertainty in the number of collected electrons can be modeled with a zero mean random variable Q_{s} which has a variance that depends on the number of photoelectrons, as well as the number of dark electrons. The total number of electrons then becomes

$$Q_{\mathrm{n}} = Q(x,y)\, k(x,y) + Q_{\mathrm{dc}}(x,y) + Q_{\mathrm{s}}(x,y). \tag{12.77}$$

12.10.5 Transfer Noise

The process of transferring the accumulated charges to the output amplifier may cause errors, leading to transfer noise [155,156]. However, the read-out efficiency of modern CCD devices can be greater than 0.99999 [439]. For practical purposes, transfer noise can therefore be neglected. The amplifier of a CCD device generates additional noise with zero mean. This *amplifier noise* is independent of the amount of charge collected and therefore determines the noise floor of the camera.

The amplifier applies a gain g to the signal, but also introduces noise and applies low-pass filtering to minimize the effects of aliasing. The amplifier noise (also known as read noise) is indicated with Q_r^-, so that the output voltage of the amplifier is given by

$$V(x,y) = (Q(x,y)\, k(x,y) + Q_{dc}(x,y) + Q_s(x,y) + Q_r(x,y))\; g. \tag{12.78}$$

If compared with (12.72), we see that the fixed pattern noise is multiplied with the desired signal, whereas the remaining noise sources are added to the signal. The combined signal and noise is then amplified by a factor of g.

12.10.6 Quantization Noise

The voltage V produced by the amplifier is subsequently sampled and digitized by the A/D converter. This leads to a further additive noise source Q_q, which is independent of V, giving

$$D(x,y) = (Q(x,y)\, k(x,y) + Q_{dc}(x,y) + Q_s(x,y) + Q_r(x,y))\; g + Q_q(x,y). \tag{12.79}$$

The quantization noise Q_q is a zero-mean random variable which is uniformly distributed over the range $[-0.5q, 0.5q]$ and has a variance $q^2/12$ [857, 858].

12.10.7 Implications of Noise

Most of the noise sources discussed here are independent of the image irradiance, with the exception of photon shot noise. This means that all noise sources become increasingly important when photographing dark environments, requiring longer exposure times. Figure 12.29 shows a plot on log-log axes where the number of photon electrons increases linearly with image irradiance (the slope of the curve is 1). This trend continues until the full-well capacity is reached. At this stage, the photosite cannot convert more photon electrons, and the value of Q does not rise any further. All noise sources, except photon shot noise, lie on a horizontal line in this plot, indicating independence from the image irradiance.

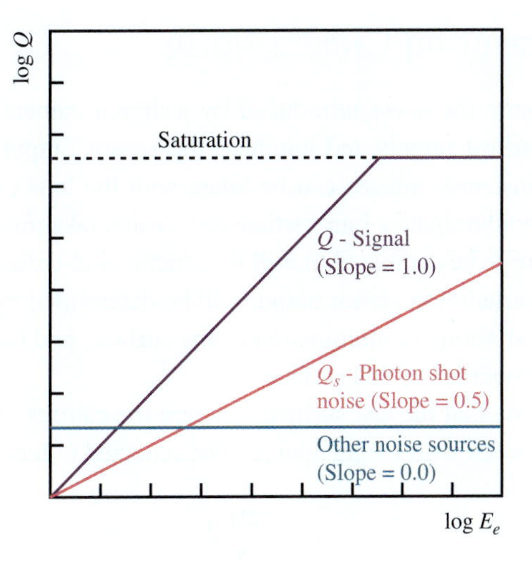

Figure 12.29. Signal and noise as function of image irradiance. The axis are on a log-scale, with, for this example, arbitrary units.

The standard deviation of photon shot noise, however, increases as the square root of the signal level, i.e., in Figure 12.29 the amount of photon shot noise is indicated with a straight line with a slope of 0.5. To limit the effect of photon shot noise, it is therefore desirable to design sensors with a large full-well capacity, since at the saturation point, the impact of photon shot noise is smallest. On the other hand, the full-well capacity of a photosite depends directly on its surface area.

With the exception of high-end digital cameras, where full-frame sensors are used, the trend in sensor design is generally toward smaller sensors, as this enables the use of smaller optical components which are cheaper. This is useful for applications such as cell phones and web cams. In addition, the trend is toward higher pixel counts per sensor. This means that the surface area of each photosite is reduced, leading to a lower full-well capacity, and therefore saturation occurs at lower image irradiances.

In addition, the dynamic range of a sensor is related to both the full-well capacity and the noise floor. For the design of high dynamic range cameras, it would be desirable to have a high full-well capacity, because, even in the case that all other noise sources are designed out of the system as much as possible, photon shot noise will limit the achievable dynamic range.

12.11 Measuring Camera Noise

In order to measure the noise introduced by a digital camera, we can image a variety of known test targets and compare the sensor output with known values [439]. For instance, images can be taken with the lens cap fitted. Second, we can uniformly illuminate a Lambertian surface and take images of it. If sensor characteristics are to be estimated, it will be beneficial to defocus such an image. In practice, the variation in sensor output will be determined by camera noise, as well as (small) variations in illumination of the surface, and (small) variations in the reflectance properties of this surface.

We will assume that the test surface is facing the camera and that it is nearly uniformly illuminated with a light source. The reflected radiance L_e is then given by

$$L_e = \frac{\rho_d}{\pi} L, \tag{12.80}$$

where the light source has strength L and the Lambertian surface reflects a fraction of ρ_d / π into the direction of the camera. This relation was obtained by applying Equation (2.209) to (2.206) and carrying out the integration assuming that the light source is a point light source positioned infinitely far away.

The lens in front of the sensor then focuses the light onto the sensor so that the irradiance incident upon a pixel is given by plugging (12.80) into (12.10c):

$$E'_e = \frac{\rho_d}{2} L \left(\frac{n'}{n}\right)^2 G, \tag{12.81}$$

which we will rewrite as

$$E'_e = k_s \, \rho_d \, L. \tag{12.82}$$

To model the (small) variation of the illumination and reflectance as a function of position on the image plane, the factors in this equation are parameterized as a function of both wavelength λ and position (x,y) on the sensor:

$$E'_e(x,y,\lambda) = k_s \, \rho_d(x,y,\lambda) \, L(x,y,\lambda). \tag{12.83}$$

The average illumination on the test target is $\bar{L}(\lambda)$ and the average reflectance is given by $\bar{\rho}_d(\lambda)$. The illumination onto the test card surface that is ultimately projected onto pixel (x,y) is then

$$L(x,y,\lambda) = \bar{L}(\lambda) + L_r(x,y,\lambda), \tag{12.84}$$

where L_r is the deviation from the average $\bar{L}(\lambda)$ for this pixel. The expected value of $L(x,y,\lambda)$ is then

$$\mathrm{E}\left(L(x,y,\lambda)\right) = \bar{L}(\lambda), \tag{12.85}$$

and the expected value of the residual $L_r(x,y,\lambda)$ is zero:

$$\mathrm{E}\left(L_r(x,y,\lambda)\right) = 0. \tag{12.86}$$

Similarly, the reflectance of the test card at the position that is projected onto sensor location (x,y) can be split into an average $\bar{\rho}_d$ and a zero mean deviation $\rho_r(x,y,\lambda)$:

$$\rho_d(x,y,\lambda) = \bar{\rho}_d(\lambda) + \rho_r(x,y,\lambda), \tag{12.87a}$$

where

$$\mathrm{E}\left(\rho_d(x,y,\lambda)\right) = \bar{\rho}_d(\lambda), \tag{12.87b}$$

$$\mathrm{E}\left(\rho_r(x,y,\lambda)\right) = 0. \tag{12.87c}$$

It is reasonable to assume that there is no correlation between the illumination on the test card surface and its reflectance. This means that the image irradiance can be written as

$$E'_e(x,y,\lambda) = k_s\left(\bar{L}(\lambda)\bar{\rho}_d(\lambda) + \varepsilon(x,y,\lambda)\right), \tag{12.88a}$$

with

$$\begin{aligned}\varepsilon(x,y,\lambda) = & \rho_r(x,y,\lambda)\,\bar{L}(\lambda)\\ & + \bar{\rho}_d(\lambda)\,L_r(x,y,\lambda)\\ & + \rho_r(x,y,\lambda)\,L_r(x,y,\lambda),\end{aligned} \tag{12.88b}$$

The expected value of ε is then 0. The charge collected by the sensor is given by (12.71) and can now be split into a constant component Q_c, and a spatially varying component $Q_v(x,y)$:

$$Q(x,y) = Q_c + Q_v(x,y), \tag{12.89}$$

where

$$Q_c = k_s \Delta t \int_\lambda \int_x^{x+u} \int_y^{y+v} \bar{L}(\lambda)\bar{\rho}_d(\lambda)\,S_r(x,y)\,c_k(\lambda)\,dx\,dy\,d\lambda; \tag{12.90a}$$

$$Q_v(x,y) = k_s \Delta t \int_\lambda \int_x^{x+u} \int_y^{y+v} \varepsilon(x,y,\lambda)\,S_r(x,y)\,c_k(\lambda)\,dx\,dy\,d\lambda. \tag{12.90b}$$

As $\mathrm{E}(\varepsilon(x,y,\lambda)) = 0$, we have that $Q_v(x,y)$ has a zero mean and a variance that depends on the variance in $\bar{L}(\lambda)$ and $\bar{\rho}_d(\lambda)$. During a calibration procedure, it is important to control the illumination of the test card to achieve illumination that is as uniform as possible. Similarly, the test card should have as uniform a reflectance as possible to maximally reduce the variance of $Q_v(x,y)$. Most of the variance in the resulting signal will then be due to the sensor noise, rather than to non-uniformities in the test set-up.

12.11.1 Noise Variance

The variance of the system modeled by (12.79) consists of several components. We will first discuss these components, leading to an expression of the total variance of the sensor. We can then estimate its value by photographing a test card multiple times [439].

The quantized values $D(x,y)$ can be modeled as random variables as follows:

$$D(x,y) = \mu(x,y) + N(x,y). \tag{12.91}$$

The expected value $\mathrm{E}(D(x,y))$ is given by $\mu(x,y)$:

$$\mu(x,y) = Q(x,y)\,k(x,y)\,g + \mathrm{E}\left(Q_{\mathrm{dc}}(x,y)\,g\right). \tag{12.92}$$

The zero-mean noise is modeled by $N(x,y)$:

$$N(x,y) = Q_{\mathrm{s}}(x,y)\,g + Q_{\mathrm{r}}(x,y)\,g + Q_{\mathrm{q}}(x,y). \tag{12.93}$$

The noise sources can be split into a component that does not depend on the level of image irradiance and a component that does. In particular, the photon shot noise, modeled as a Poisson process, increases with irradiance (see Section 12.10.4):

$$Q_{\mathrm{s}}(x,y)\,g. \tag{12.94}$$

Accounting for the gain factor g, the variance associated with this Poisson process is given by

$$g^2\left(Q(x,y)\,k(x,y) + \mathrm{E}(Q_{\mathrm{dc}}(x,y))\right). \tag{12.95}$$

The signal-independent noise sources, amplifier noise and quantization noise, are given by

$$Q_{\mathrm{r}}(x,y)\,g + Q_{\mathrm{q}}(x,y) \tag{12.96}$$

and have a combined variance of

$$g^2\sigma_r^2 + \frac{q^2}{12}, \tag{12.97}$$

where σ_r^2 is the variance of the amplifier noise. The total variance σ^2 in the noise introduced by the sensor is then the sum of these two variances:

$$\sigma^2 = g^2 \left(Q(x,y)\, k(x,y) + \mathrm{E}(Q_{\mathrm{dc}}(x,y))\right) + g^2 \sigma_r^2 + \frac{q^2}{12}. \tag{12.98}$$

The expected value of the dark current for a given pixel can be replaced by the sum of the average expected value over the whole sensor and the deviation from this expected value for a given pixel. Introducing the notation $Q_{\mathrm{E(dc)}}(x,y) = \mathrm{E}(Q_{\mathrm{dc}}(x,y)$, we can write

$$Q_{\mathrm{E(dc)}}(x,y) = \bar{Q}_{\mathrm{E(dc)}} + Q_{d\mathrm{E(dc)}}(x,y), \tag{12.99}$$

where $\bar{Q}_{\mathrm{E(dc)}}$ is the average expected value of the dark current and $Q_{d\mathrm{E(dc)}}(x,y)$ is the deviation from the expected value. Equation (12.98) can now be rewritten as follows:

$$\begin{aligned}\sigma^2 =& \, g^2 \left(Q(x,y) + \bar{Q}_{\mathrm{E(dc)}}\right) + g^2 \left(k(x,y) - 1\right) Q(x,y) \\ &+ g^2 \left(Q_{d\mathrm{E(dc)}}(x,y) - \bar{Q}_{\mathrm{E(dc)}}\right) + g^2 \sigma_r^2 + \frac{q^2}{12}.\end{aligned} \tag{12.100}$$

Under the assumptions that $|k(x,y) - 1| \ll 1$ and $|Q_{d\mathrm{E(dc)}}(x,y) - \bar{Q}_{\mathrm{E(dc)}}| \ll \bar{Q}_{\mathrm{E(dc)}}$, the variance of the sensor noise can be approximated with

$$\sigma^2 \approx g^2 \left(Q(x,y) + \bar{Q}_{\mathrm{E(dc)}}\right) + g^2 \sigma_r^2 + \frac{q^2}{12}. \tag{12.101}$$

12.11.2 Estimating the Total Variance

By photographing a uniformly illuminated test card twice, we can expect to obtain two quantized images, both with the same expected value $\mu(x,y)$ for each pixel, but with a differing random noise $N(x,y)$ [439]:

$$D_1(x,y) = \mu(x,y) + N_1(x,y); \tag{12.102a}$$

$$D_2(x,y) = \mu(x,y) + N_2(x,y). \tag{12.102b}$$

By subtracting the two images, we get a difference image with zero mean and a variance of $2\sigma^2$. The expected value of either image D_1 and D_2 is given by

$$\mu = Q(x,y)\, g + \bar{Q}_{\mathrm{E(dc)}}\, g. \tag{12.103}$$

If the spatial variance of $Q(x,y)$ is sufficiently minimized, then we may substitute the spatial mean $\bar{Q}$:

$$\mu = \bar{Q}\, g + \bar{Q}_{\mathrm{E(dc)}}\, g. \tag{12.104}$$

Using (12.101), the variance in terms of μ is then

$$\sigma^2 = g\mu + g^2\sigma_r^2 + \frac{q^2}{12}. \tag{12.105}$$

Given two captured images D_1 and D_2, an estimate of μ can be computed by

$$\hat{\mu} = \frac{1}{XY}\sum_{x=0}^{X}\sum_{y=0}^{Y} D_1(x,y) + \frac{1}{XY}\sum_{x=0}^{X}\sum_{y=0}^{Y} D_2(x,y). \tag{12.106}$$

An estimate of the variance σ^2 is then computed with

$$\hat{\sigma}^2 = \frac{1}{XY-1}\sum_{x=0}^{X}\sum_{y=0}^{Y}\left(D_1(x,y) - D_2(x,y) - \hat{\mu}_d(x,y)\right)^2, \tag{12.107}$$

where $\hat{\mu}_d$ is the mean of the difference image.

The illumination used for the pair of images was so far left unspecified. In order to compute estimates of the amplifier gain g and the variance of the signal-independent noise sources $\sigma_c^2 = g^2\sigma_r^2 + q^2/12$, we can now vary the illuminance on the test card and collect a pair of images for each illumination level. Using (12.105) the parameters $\hat{g}$ and $\hat{\sigma}_c^2$ can be found by minimizing the following expression using a line fitting technique:

$$\sum_i \frac{\left(\hat{\sigma}_i^2 - (\hat{g}\hat{\mu}_i + \hat{\sigma}_c^2)\right)^2}{\operatorname{var}\left(\hat{\sigma}_i^2\right)}, \tag{12.108}$$

where the sum is over the set of image pairs which are indexed by i. The variance in $\hat{\sigma}_i$ can be estimated by

$$\operatorname{var}\left(\hat{\sigma}_i^2\right) \approx \frac{2\left(\hat{\sigma}_i^2\right)^2}{XY-1} \tag{12.109}$$

12.11.3 Estimating Dark Current

In the absence of light, a camera will still output a signal as a result of dark current. By fitting the lens cap, the number of photo-generated electrons will be zero, so that the quantized value output by the sensor is given by [439]

$$D(x,y) = \left(Q_{\mathrm{dc}}(x,y) + Q_{\mathrm{s}}(x,y) + Q_{\mathrm{r}}(x,y)\right) g + Q_{\mathrm{q}}(x,y). \tag{12.110}$$

The mean value of this signal is $Q_{\mathrm{E(dc)}}(x,y)\, g$, as all other noise sources have zero mean, and its variance is $\sigma^2(x,y)$. This variance can be estimated with the technique discussed in the preceding section. By averaging a set of n images obtained in this manner, the variance will be reduced to $\sigma^2(x,y)/n$. For large enough values of n, the pixel values will then converge to an accurate estimate of the dark current $Q_{\mathrm{E(dc)}}(x,y)\, g$.

12.11.4 Estimating Fixed Pattern Noise

We now have estimates of the dark current, the amplifier gain, and the total variance in the noise. To estimate the fixed pattern noise, it is not sufficient to vary the illumination level uniformly across the test card, as this would not enable us to disentangle the spatial variation due to fixed pattern noise and the spatial variation as a result of non-uniform illumination and/or reflectance [439].

However, by taking a sequence of images by varying the illumination non-uniformly, as well as varying the orientation of the test card with respect to the camera, we can average out the variation due to illumination and retain the fixed pattern noise $k(x,y)$, here called *photo response non-uniformity* (PRNU). Suppose that we create n_1 illumination conditions and capture n_2 images for each of these conditions. As the total noise $N(x,y)$ has zero mean, averaging the frames for imaging condition i will lead to the following estimate:

$$k(x,y)\,Q_i(x,y)\,g + Q_{\mathrm{E(dc)}}(x,y)\,g, \tag{12.111}$$

with variance $\sigma_i^2(x,y)/n^2$. From Section 12.11.3 we already have an estimate of the dark current, which we can subtract. This yields the corrected estimate

$$d(x,y) \approx k(x,y)\,Q_i(x,y)\,g. \tag{12.112}$$

This estimate varies over the sensor surface as a result of fixed pattern noise as well as variations in reflectance and illumination. However, for relatively small pixel neighborhoods, the variation due to reflectance and illumination of the test card may be assumed to be small. It is therefore possible to compute the mean $\bar{d}(x,y)$ over small windows for each pixel. Regions of 9×9 pixels tend to work well [439], leading to estimates of

$$\bar{d}(x,y) \approx Q_i(x,y)\,g. \tag{12.113}$$

The ratio between a single pixel estimate and the windowed estimate is then a rough approximation of the fixed pattern noise $k_e(x,y)$:

$$\frac{d(x,y)}{\bar{d}(x,y)} \approx \frac{k(x,y)\,Q_i(x,y)\,g}{Q_i(x,y)\,g} \approx k_e(x,y). \tag{12.114}$$

To refine this approximation, the average ratio over the n_1 imaging conditions is computed:

$$k_e(x,y) \approx \frac{1}{n_1} \sum_{i=0}^{n_1} \frac{d_i(x,y)}{\bar{d}_i(x,y)}. \tag{12.115}$$

A further refinement can be obtained by removing outliers from this summation.

12.12 Radiometric Camera Calibration

Images produced with digital cameras are represented with device-dependent RGB pixel values. Although camera designers know the correspondence between the RGB values produced by the camera and the image irradiances that gave rise to these values, this information is not generally available to users of digital cameras. As shown in the preceding sections, the optics, the CFA, the sensor as well as the processing occurring in the firmware, all contribute to distortions in the signal.

Camera characterization involves recovering either the camera response curves f_k or the sensor's responsivities. In this section, we discuss methods to retrieve these functions on the basis of calibrated input. In the context of high dynamic range imaging, uncalibrated input in the form of a set of multiply exposed images may also be used to recover the camera response curve (as discussed in Section 13.2).

For most applications, it is desirable to know the relation between pixel values and scene irradiances accurately, and for many applications this information is vital [804]. For instance, accurate color reproduction is not possible without beginning with accurate capture of irradiance values. Many color-based computer vision algorithms lose accuracy without accurate capture.

To recover a linear relationship between the irradiance values and the pixel encoding produced by the camera, we need to model the non-linearities introduced by in-camera processing. The process of recovering this relationship is known as *camera characterization* or *color characterization* [1187].

There are two general approaches to color characterization; they are based on spectral sensitivities and color targets [480]. Methods based on spectral sensitivities require specialized equipment such as a monochromator (a device that emits light with a single user-specified wavelength) and a radiance meter [89]. With such equipment, the response of the camera to each wavelength can be measured, and from it the relationship between the spectral responsivities of the camera and the CIE color matching functions can be estimated.

Target-based techniques make use of a set of differently colored samples that can be measured with a spectrophotometer. A digital image is then captured of the target. For each patch, we then have measured CIE XYZ values and captured RGB values. The most precise calibration for any given camera requires recording its output for all possible stimuli and comparing it with separately measured values for the same stimuli [1201]. However, storage of such a quantity of data is impractical, and, therefore, the response of the device is captured for only a limited set of stimuli—perhaps between a few dozen to a few hundred patches. The responses to these representative stimuli can then be used to calibrate the device for input stimuli that were not measured.

To find the transformation between RGB and XYZ values, several techniques have been developed [480], including look-up tables which can be used to interpolate or extrapolate new values [501, 502], least squares polynomial modeling [480, 574], neural networks [572], and fitting data to models of cameras and image formation [65, 162, 439].

12.12.1 Look-Up Tables

If the stimuli densely sample the device gamut, then the transformation for device output to the ideal response may be estimated by three-dimensional interpolation of measured responses [501, 575]. The dense sampling of stimuli responses is stored in *look-up tables* (LUTs). The advantage of this technique over others is that it avoids estimation of the mapping with a function, a process that is complex and computationally expensive.

Look-up tables require a few hundred measured responses. It may not always be possible to densely sample the entire device gamut, and only a sparse sampling of responses to stimuli may be available. In such cases, a function has to be estimated from known responses, providing a mapping from the output of the device being calibrated to the "ground truth."

An example of a sparse set of stimuli is given in Figure 12.30, which shows a chart of different colors that can be used as input to calibrate capturing devices.

Figure 12.30. The Gretag-Macbeth ColorChecker can be used to characterize a digital camera.

Number	Name	CIE 1931 Yxy Y	x	y	Munsell notation Hue		Value / Chroma
1	Dark skin	10.1	0.400	0.350	3	YR	3.7 / 3.2
2	Light skin	35.8	0.377	0.345	2.2	YR	6.47 / 4.1
3	Blue sky	19.3	0.247	0.251	4.3	PB	4.95 / 5.5
4	Foliage	13.3	0.337	0.422	6.7	GY	4.2 / 4.1
5	Blue flower	24.3	0.265	0.249	9.7	PB	5.47 / 6.7
6	Bluish green	43.1	0.261	0.343	2.5	BG	7 / 6
7	Orange	30.1	0.506	0.407	5	YR	6 / 11
8	Purplish blue	12.0	0.211	0.175	7.5	PB	4 / 10.7
9	Moderate red	19.8	0.453	0.306	2.5	R	5 / 10
10	Purple	6.6	0.285	0.202	5	P	3 / 7
11	Yellow green	44.3	0.380	0.489	5	GY	7.1 / 9.1
12	Orange yellow	43.1	0.473	0.438	10	YR	7 / 10.5
13	Blue	6.1	0.187	0.129	7.5	PB	2.9 / 12.7
14	Green	23.4	0.305	0.478	0.25	G	5.4 / 8.65
15	Red	12.0	0.539	0.313	5	R	4 / 12
16	Yellow	59.1	0.448	0.470	5	Y	8 / 11.1
17	Magenta	19.8	0.364	0.233	2.5	RP	5 / 12
18	Cyan	19.8	0.196	0.252	5	B	5 / 8
19	White	90.0	0.310	0.316		N	9.5 /
20	Neutral 8	59.1	0.310	0.316		N	8 /
21	Neutral 6.5	36.2	0.310	0.316		N	6.5 /
22	Neutral 5	19.8	0.310	0.316		N	5 /
23	Neutral 3.5	9.0	0.310	0.316		N	2.5 /
24	Black	3.1	0.310	0.316		N	2 /

Table 12.1. CIE 1931 chromaticity coordinates (xy) and luminous reflectance factor Y for each of the 24 patches of the Gretag-Macbeth ColorChecker, which is shown in Figure 12.30. Also given is the equivalent Munsell notation. All values are for patches illuminated with CIE illuminant C.

This particular example is the Gretag-Macbeth ColorChecker, which is formulated to minimize metamerism under different illumination conditions [757]. If illuminated under standard illuminant C, the Yxy chromaticity coordinates for each of the 24 patches are given in Table 12.1.[4]

[4]Note that the photograph of Figure 12.30 was taken under D65 illumination, which was not evenly distributed over the chart. This particular photograph is only intended as an example of a chart that can be used for calibration—not to allow accurate color reproduction on the basis of this figure. To achieve good color reproduction, a color rendition chart must be placed in the scene that is to be photographed.

12.12.2 Error Metric

When relatively few responses have been obtained, the measured data may be viewed in terms of the color difference (or error) in the device and ideal responses for each stimulus. A transformation can be computed such that the difference between the transformed device output and the ideal response is minimized. Thus, the following function has to be minimized:

$$\underset{f_k}{\operatorname{argmin}} \sum_{n=1}^{N} ||f_k(p_n) - P_n||^2, \tag{12.116}$$

where p_n is the pixel value recorded by the camera for the nth stimulus, P_n is the corresponding measured response, N is the total number of samples, and f_k is the transformation being estimated for the kth color channel. The notation $||.||^2$ is introduced to refer to the ΔE^*_{ab} color-difference metric defined in the CIELAB color space, as discussed in Chapter 8.7.1. The objective function above could be augmented with additional constraints, for instance to enforce smoothness of the function f_k [65].

Typical techniques for finding the mapping from known data include linear and polynomial regression, as well as neural networks [575].

Once the camera is calibrated, the recovered function f_k can be applied to all images captured with this camera. Color calibration is necessary for users of digital cameras to generate high quality images that are faithful representations of actual input data and are free of any unwanted effects of various processes inside the capturing device. The data after the mapping is termed *device independent*. Data captured from two different devices can be compared after their appropriate mappings, which may be different for the two devices.

12.13 Light Field Data

The radiance at a point in a scene may be represented by the *plenoptic function* [11]. This is a seven-dimensional function $L(x,y,z,\theta,\phi,\lambda,t)$, where x, y, and z denote the location of a point in space, θ and ϕ represent the direction of the ray, λ represents the wavelength of the ray, and t represents time. The plenoptic function therefore models any ray of any wavelength and direction, originating at any point in space, at any time. When we consider a monochromatic version of this function, at a fixed time t, then this function becomes five-dimensional: $L(x,y,z,\theta,\phi)$.

The plenoptic function may be further simplified if the environment is assumed to have no occlusions. The environment may be represented by a cube

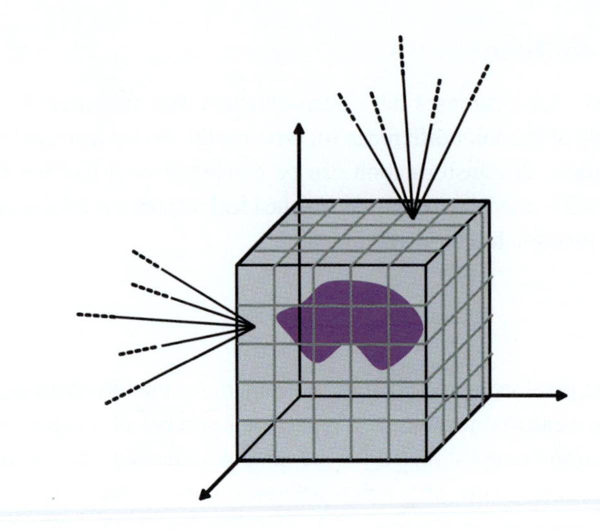

Figure 12.31. Rays emanating from the surface of the cube can represent the radiance at any point within the scene (after [11]).

with rays emanating from it in all directions, as shown in Figure 12.31. Now for any point in the cube, the value of rays may be found by intersecting them with the two-dimensional faces of the environment cube. This simplified plenoptic function is only four-dimensional. Rays originating from each face of the cube

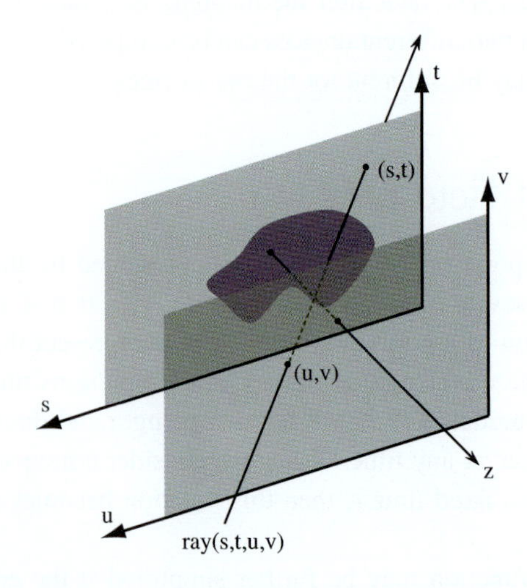

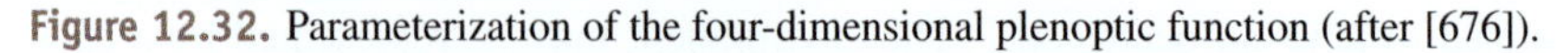

Figure 12.32. Parameterization of the four-dimensional plenoptic function (after [676]).

can be represented by $L(u, v, s, t)$, where s and t specify the location on the cube face, and u and v parameterize the direction of the ray, as shown in Figure 12.32.

This four-dimensional function is variously called a *lumigraph*, or a *light field* [381, 676]. If this function is captured, then new images of the scene from different viewpoints may be generated from a single capture. To obtain the intensity of a ray emanating from a point in the scene, the intersection of the ray is found with the (u, v) and (s, t) planes, and the value of the function $L(u, v, s, t)$ is the intensity of that ray.

Standard cameras are designed so that rays from several different directions are incident on each sensor location, dependent on the focusing achieved with the lens assembly. The sensor integrates light over a small solid angle. By modifying standard camera designs, it is possible to reduce the solid angle over which a sensor records, thereby separately recording each ray of light passing through the aperture. Devices that do this are able to capture light field data, i.e., the four-dimensional plenoptic function. Such devices are briefly introduced in the following sections.

12.13.1 The Plenoptic Camera

The *plenoptic camera* was introduced by Adelson and Wang [12]. The main difference between a standard camera and the plenoptic camera is that it separates rays before they reach the sensing array. This separation is typically achieved by placing a micro-lens array in front of the sensing array, as illustrated in Figure 12.33. The captured light field data is sufficient to generate images of the scene from a variety of viewpoints and can also be used to infer the depths of objects in the scene. The latter is called *single lens stereo*.

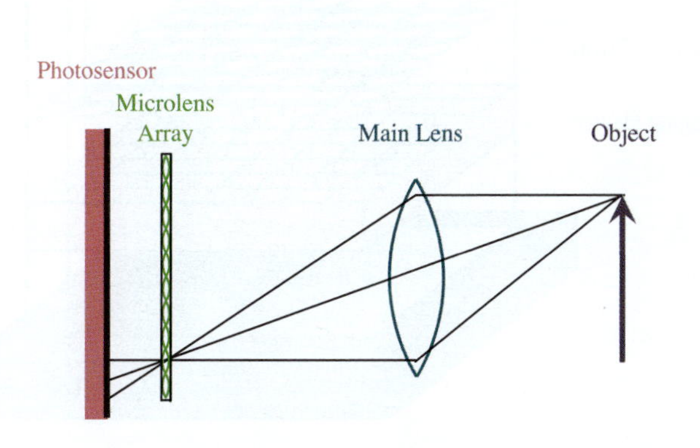

Figure 12.33. Structure of the plenoptic camera (after [12]).

A portable version of the plenoptic camera, producing better quality images, was introduced by Ng et al. [834]. From a single capture of light field data, they generate sharp photographs, focused at different depths.

12.13.2 Artificial Compound Eye Camera

Another imaging system that captures light field data mimics an insect's compound eye structure [843, 1118, 1119]. The design of this system is similar to that of the plenoptic camera, except that this system does not have a main lens (Figure 12.34). Such a design results in the formation of many small images of the scene (each corresponding to a different micro-lens in the micro-lens array). These images together form a *compound* image. As shown in the figure, a unit constitutes a single micro-lens and its corresponding sensing array. The signal separator prevents cross-talk between adjacent units.

Once the compound image has been captured, it can be used to generate a higher-resolution image of the scene. The specific design of such systems allows extremely compact systems to be made; the minimum thickness achieved is no more than a piece of paper. However, as noted by Barlow while comparing human and insect eyes, the spatial resolution that can be resolved by small lens arrays is severely limited by diffraction [61].

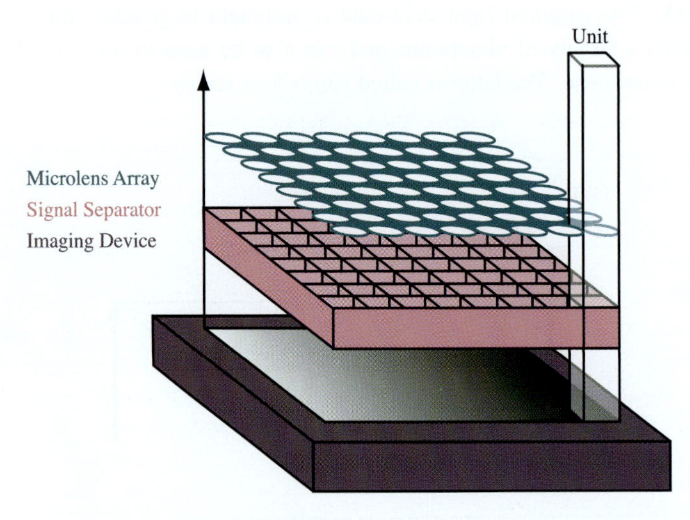

Figure 12.34. Structure of the compound eye camera (after [1118]).

12.14 Holography

Conventional image-recording techniques, for instance photography, record light for a given exposure time. During that time, light waves impinge on the sensor, which thereby records intensity information. The recording medium is only sensitive to the power of the light wave. In the particle model of light, the number of photons is counted. In the wave model of light, amplitude information is retained.

An image sensor does not record where the light came from, other than the focusing effected by the lens. The image is therefore a flat representation of the 3D world it captures. Holographic imaging is a technique that attempts to photographically capture 3D information by recording both amplitude and phase information.

Recall that most light sources emit incoherent light, i.e., the phase of each of the superposed waves is uncorrelated (see Section 2.7.1). Thus, during an exposure, light waves arrive at the photographic plate with random phases. The exception to this are lasers, which hold the key to holographic imaging. Shining a laser beam at a photographic plate will cause all the light to arrive in phase. Such a beam is called the *reference beam.* As the cross-section of a laser beam is non-diverging, it has the capability to illuminate only one point of a photographic plate. However, by inserting a lens into its optical path, the beam can be made to diverge. This is a process known as *decollimation.* It then illuminates the entire photographic plate with coherent light.

If the same laser beam is split, with some part directly aimed at the photographic plate and the other part reflected by an object, then the path length taken by the light that is reflected will be longer than the path length of the direct beam. This second beam is called the *object beam.* As its path length is longer, the light will take slightly longer to arrive at the plate and will, therefore, be out of phase with the reference beam. Of course, each point on the object will create a slightly different path length.

The phase differences will cause an interference pattern in space, including on the surface of the photographic plate. Section 2.7.1 discusses interference, which in summary, are changes in amplitude due to light waves summing or canceling out. As a result, the phase difference between object and reference beams causes local changes in amplitude to which photographic materials are sensitive. Hence, phase information is encoded as an interference pattern. Such an encoding, invented by Dennis Gabor in 1948 [342], is called a hologram; an example is shown in Figure 12.35.

To create a hologram, a reference beam of laser light must be aimed at the photographic plate directly, as well as at the object to be imaged. To reach the

Figure 12.35. A photograph of a hologram, which was lit with a D65 daylight simulator at an angle of 45° from above.

entire plate, as well as the entire object, lenses are typically used to make the laser light divergent. A beam splitter, typically a half-silvered mirror, can be used to split the light into object and reference beams. It is not possible to use a pair of lasers for this task, as this would create phase differences unrelated to the object. An example configuration is shown in Figure 12.36.

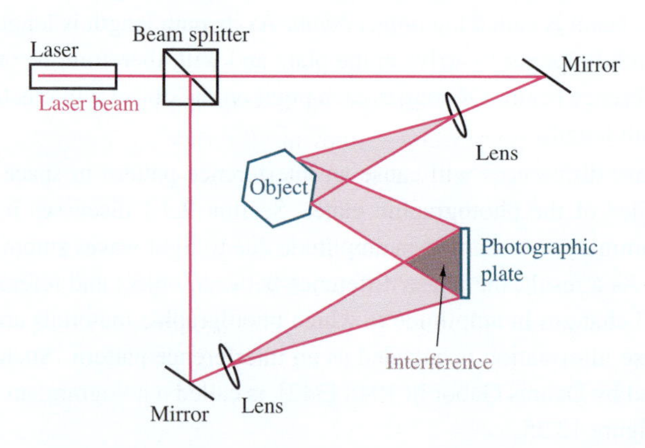

Figure 12.36. Configuration for creating a transmission hologram. The object reflects in many more directions than indicated, but only the directions toward the photographic plate are indicated.

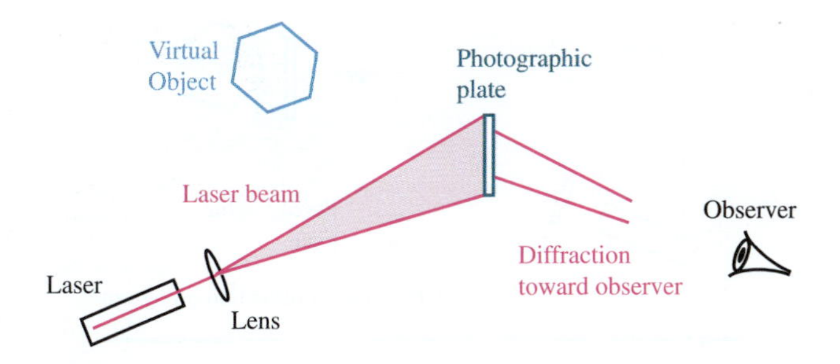

Figure 12.37. Configuration for viewing a transmission hologram.

To view the hologram, a laser is once more used to illuminate the photographic plate, as shown in Figure 12.37. In this case, the observer and the illuminant are located on opposite sides of the photographic plate; thus, this is called a *transmission hologram* [267, 427].

Alternatively, it is possible to create a hologram by using the light that is transmitted through the photographic plate to illuminate the object. The light reflected off the object will then create an interference pattern with the light directly incident upon the photographic plate. Such a configuration, shown in Figure 12.38, requires the illumination and the observer to be located on the same side of the holographic plate. Holograms created in this manner are called *reflection holograms*. Instead of using a laser to illuminate the hologram, reflection holograms can also be viewed under white light, because white light incorporates the wavelength of the laser used to create the hologram [267, 427].

In all cases, the recording environment needs to be carefully controlled for vibrations, because even very small movements during the recording process will create large problems due to the nanometer scale at which interference fringes occur. Usually, optical workbenches are used to combat vibration. In addition, it is possible to use pulsed lasers, allowing for very short exposure times. To enable recording of interference patterns in the first place, the spatial resolving power of the photographic material needs to be very high. The grain of ordinary photographic materials is too large to record holograms.

For illustrative purposes, we will assume that the object being imaged is the size of a single point, and that both the light source and the object are far away from the photographic plate. Then, both the object and reference beams can be considered plane waves. As discussed in Section 2.7.1, the interference pattern created by a pair of coherent monochromatic time-harmonic plane waves $\mathbf{E}_o$ and $\mathbf{E}_r$, where the subscripts denote the object or reference beam, is expressed by the irradiance at any point in space. Related to this quantity is the optical intensity,

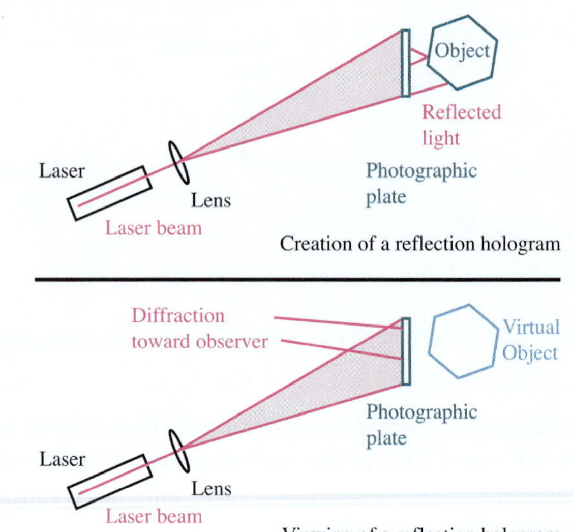

Figure 12.38. Configuration for creating (top) and viewing (bottom) a reflection hologram.

which is defined as

$$I_o = \underline{E}_o \underline{E}_o^* = |\underline{E}_o|^2, \tag{12.117a}$$

$$I_r = \underline{E}_r \underline{E}_r^* = |\underline{E}_r|^2, \tag{12.117b}$$

where $\underline{E}_o$ and $\underline{E}_r$ are the complex amplitudes of their corresponding electric vectors. Adding two waves of the same frequency causes their amplitudes to sum, yielding the following optical intensity [427]:

$$\begin{aligned} I &= |\underline{E}_o + \underline{E}_r|^2 && (12.118a) \\ &= |\underline{E}_o|^2 + |\underline{E}_r|^2 + \underline{E}_o \underline{E}_r^* + \underline{E}_o^* \underline{E}_r && (12.118b) \\ &= |\underline{E}_o|^2 + |\underline{E}_r|^2 + 2\sqrt{I_o I_r} \cos(\varphi_o - \varphi_r). && (12.118c) \end{aligned}$$

The third term in this equation is called the interference term and corresponds to the interference term discussed in Section 2.7.1. In other words, the optical intensity at each point depends on the magnitude of each of the two waves, plus a quantity that depends on the phase difference which is only the result of the difference in path length between the reference and object beams.

Dependent on the quality of the laser and the total path lengths involved in the holographic imaging process, some of the laser's coherence may be lost along the way. The factor $V \in [0,1]$ accounts for this loss of coherence. It is known as the *van Cittert-Zernike coherence factor* and is computed as follows [343]:

$$V = \frac{I_{\text{max}} - I_{\text{min}}}{I_{\text{max}} + I_{\text{min}}}. \quad (12.119)$$

The values I_{min} and I_{max} are the minimum and maximum optical intensity in an interference pattern when the complex amplitudes of the two plane waves are equal. The value of V is 1 in the ideal case when coherence is maximal and 0 in the absence of coherence.

As reflection off a surface may depolarize the light, the optical intensity at a given point in space may be reduced by an amount linear in the angle between the two electrical vectors. This is accounted for by the $\cos(\theta)$ factor in (12.120). If no depolarization occurs, this angle is zero and no attenuation occurs. At right angles, the irradiance becomes zero. Accounting for the loss of coherence, as well as depolarization, the optical intensity becomes

$$I = |\underline{E}_o|^2 + |\underline{E}_r|^2 + 2V\sqrt{I_o I_r}\cos(\varphi_o - \varphi_r)\cos(\theta) \quad (12.120a)$$

$$= |\underline{E}_o|^2 + |\underline{E}_r|^2 + V(\underline{E}_o\underline{E}_r^* + \underline{E}_o^*\underline{E}_r)\cos(\theta). \quad (12.120b)$$

After processing of the plate, the transmittance $t(x,y)$ of each point of the plate is given by a function linear in the optical intensity [427]:

$$t(x,y) = t_0 + \beta\,\Delta t\,I, \quad (12.121)$$

where t_0 is a constant transmittance term, β is a constant related to the photographic material used, and Δt is the exposure time.

The complex amplitude transmitted at each point of the hologram, when it is illuminated by a laser of the same wavelength as the laser used to create the hologram, is then

$$u(x,y) = \underline{E}_r t(x,y) \quad (12.122a)$$

$$= \underline{E}_r t_0 + \beta\,\Delta t\,\underline{E}_r\left(|\underline{E}_o|^2 + |\underline{E}_r|^2 + 2V\sqrt{I_o I_r}\cos(\varphi_o - \varphi_r)\cos(\theta)\right) \quad (12.122b)$$

$$= \underline{E}_r t_0 \quad (12.122c)$$

$$+\beta\,\Delta t\,\underline{E}_r\left(|\underline{E}_o|^2 + |\underline{E}_r|^2\right)$$

$$+\beta\,\Delta t\,V\underline{E}_o\,|\underline{E}_r|^2\cos(\theta)$$

$$+\beta\,\Delta t\,V\underline{E}_o^*\underline{E}_r^2\cos(\theta). \quad (12.122d)$$

The transmitted amplitude therefore consists of several terms. The first term is the constant background transmittance, the second term constitutes the unmodified illuminating beam, the third term is the reconstructed image, and the fourth term is a twin image [343].

The twin image is undesirable if it overlaps with the reconstructed image. However, such overlap can be avoided by aiming the reference beam at the photographic plate at an angle ϑ. The amplitude E_r is then given by

$$E_r = |\underline{E}_r| \exp\left(\frac{2\pi i x \sin(\vartheta)}{\lambda}\right). \tag{12.123}$$

The implications of this are that the magnitude of the reference beam's amplitude remains constant, i.e., $|\underline{E}_r|^2 = \text{constant}$, but the square of E_r becomes

$$E_r^2 = |\underline{E}_r|^2 \exp\left(\frac{4\pi i x \sin(\vartheta)}{\lambda}\right). \tag{12.124}$$

This means that the diffracted image is angled away from its twin image by an angle of 2ϑ. As long as the object being imaged subtends a solid angle smaller than this, the twin image will not overlap with the desired reconstruction [343].

The interference fringes recorded by the photographic plate cause diffraction when illuminated with a (coherent) light from the same direction as the reference beam. Many different types of hologram have been developed; they can be classified according to whether the photographic plate is a thin layer or volumetric and whether the hologram is exposed once or multiple times. Although computationally very expensive, interference patterns can be computed with modified rendering algorithms and followed by transfer to photographic plates.

12.15 Further Reading

There are many general books on image processing [116, 376, 543, 893, 1071]. Charles Poynton's book *Digital Video and HDTV: Algorithms and Interfaces* contains indispensable information for the broadcast, television, and video industries [923].

Camera sensors and the signal processing in digital cameras is described by Nakamura's book, *Image Sensors and Signal Processing for Digital Still Cameras* [813]. The inner workings of film cameras are described in Goldberg's *Camera Technology: The Dark Side of the Lens* [374]. Ansel Adams's books *The Camera, The Negative*, and *The Print* are indispensable for understanding photography [8–10].

Further reading on holography is available in the following references [149, 427, 634, 1060]. Electron holography is discussed in [1184]. Non-technical (i.e., non-mathematical) introductions to holography are available in [267, 580, 600].

Chapter 13
High Dynamic Range Image Capture

The dynamic range that can be acquired using conventional cameras is limited by their design. Images are typically gamma encoded to eight bits per color channel. This means that a luminance range of around two orders of magnitude can be captured and stored. Many camera manufacturers allow Raw data to be exported from their cameras as well. These are images that have undergone minimal in-camera processing and are typically linearly encoded to a bit-depth of around 10 to 14. Note, here, that a linear encoding of 10 bits does not present a larger range of luminance values than an 8-bit non-linear encoding. Even if Raw data gives a higher dynamic range than conventional 8-bit images, at this point in time, we cannot speak of a truly high dynamic range. In other words, limitation in current camera designs do not allow us to capture the full range of intensities available in typical real-world scenes.

Some of the trade-offs in camera design are outlined in the preceding chapter, and include noise characteristics, the capacity to hold a limited amount of charge before a pixel saturates, veiling glare, and the fact that as new sensor designs incorporate smaller pixel sizes, quantum efficiency is reduced. All these factors contribute to a limit on the range of intensities that can be captured in any given exposure.

To capture the vast range of light present in many scenes, we must therefore resort to either using dedicated, unconventional hardware, or to reconstructing a high dynamic range (HDR) image from a set of differently exposed conventional images. The techniques used for the acquisition of high dynamic range images are discussed in this chapter.

13.1 Multi-Exposure Techniques

The output generated by standard cameras has several limitations. First, sensors in standard cameras are unable to capture the wide range of radiance values often found in our environment. For instance, the response for very light scene elements, such as specular highlights and light sources, is saturated for radiance values that are larger than the camera is capable of capturing. The second limitation is that the output is quantized, so that all displayable radiance values are represented by a total of 256 values in the output. Third, the camera's firmware may introduce non-linearities before the photographer is presented with an image. Much of the image processing that occurs in the firmware can now be avoided by using the camera's Raw format.

A variety of techniques and cameras have been developed to obtain high dynamic range data. Currently, the cheapest and therefore dominant technique, uses a conventional camera to take multiple, differently exposed images of the same scene. These exposures are later recombined in software into a single HDR image. The most accurate approach is to vary the exposure time, rather than change the aperture or apply neutral density filters [950]. Many cameras can be config-

Figure 13.1. A sequence of exposures taken with different exposure times; St. Barnabas Monastery, North Cyprus, July 2006.

Figure 13.2. The final result, obtained after merging the exposures of Figure 13.1 into a single high dynamic range image, followed by applying dynamic range reduction to enable the image to be printed (discussed in Chapter 17); St. Barnabas Monastery, North Cyprus, July 2006.

ured to take several exposures using auto-bracketing; in some cases up to nine shots with a single press of the button, each spaced up to one f-stop apart. This enables the convenient capture of a high dynamic range scene, albeit that the full range is represented in a sequence of exposures that then need to be recombined. An example of such a set of exposures is shown in Figure 13.1, and the combined result is displayed in Figure 13.2. Note that the shorter exposures are necessary to adequately capture the sky, whereas the longer exposures capture the details of the tree. The middle exposure of the nine appears perhaps reasonable at first sight, but note that in this image the sky is already over-exposed. This is a problem that is resolved by the techniques discussed in this and the following chapters.

Multiple exposure techniques allow a much larger dynamic range to be captured than can be achieved with single exposures. However, the achievable range is not infinite due to the occurrence of veiling glare (see Section 12.1.6). In a sequence of exposures, some of the images are deliberately under- or over-exposed. In essence, the camera is used outside its normal operating range, and especially for the over-exposed images, which are captured to recover dark areas of the environment, veiling glare may occur. Thus, veiling glare places a limit on the achievable dynamic range [758,1115], which may either be mitigated by employing deconvolution techniques [1080], or by placing a mesh between the environment and the camera, and recombining multiple high dynamic range images shot through the mesh with small lateral offsets [1115].

One of the disadvantages of multiple exposure techniques is that the set of exposures should vary only in exposure time. As such, the scene is assumed to be static, and the camera position is assumed to be fixed. In practice, these requirements may be difficult to maintain. Ensuring the camera is steady during the sequence of exposures is relatively straightforward. The use of a tripod mitigates the worst artifacts associated with moving the camera, while image alignment techniques can be used as a post-process (see Section 13.5). In the case that handheld exposures are taken, aside from image alignment, the effects of camera shake should be taken into account [309].

A more complicated problem is that the scene needs to remain static. This requirement can be met in the studio, but especially outdoor scenes can be difficult to control. In particular, movement of clouds, vegetation, as well as humans and animals, causes each of the exposures to depict a different scene. Recombining multiple exposures may then lead to ghosting artifacts. These problems can be reduced by employing a ghost removal algorithm (see Section 13.4), even though given the current state-of-the-art, it is still preferrable to ensure the environment being captured is, in fact, static.

13.1.1 Merging Exposures

To allow multiple exposures to be merged into a single image, each of the exposures needs to be linearized, which is a straightforward task if the camera's response curve is known. When this curve is not known, the camera response function may be computed using the techniques discussed in this section. The reason that recovery of the camera response function is an important step in high dynamic range imaging (HDRI) can be seen in Figure 13.3. The response curve of the green channel for two different cameras is compared with a linear ramp, as well as with the encoding of the green channel in the sRGB color space. The four curves are all different from each other. It is therefore not appropriate to assume that the output generated by a camera is either linear or adheres to any other color standard, such as the encoding used in sRGB.

Once the response curve is obtained, each pixel's image irradiance value E'_e can be computed. Considering a single pixel, its value p is due to some incident irradiance E'_e, recorded for an exposure time of Δt. The relationship between the variables E'_e and p can be expressed as

$$p = g(E'_e \Delta t). \tag{13.1}$$

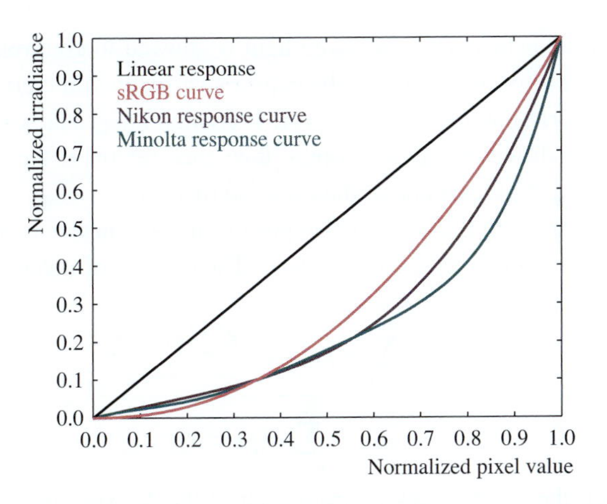

Figure 13.3. Camera response curves (green channel only) compared with sRGB and linear response functions. The Nikon response curve is for a Nikon D2H, whereas the Minolta curve is for a DiMAGE A1 camera.

The product $E_e' \Delta t$ is also referred to as the *exposure*. The camera response function g represents the processing of the irradiance values within the camera. As camera manufacturers typically do not make the nature of this processing available to their customers, the camera response curve g needs to be reverse-engineered from the captured images.

To find E_e' from the corresponding pixel values p, the function g needs to be inverted to give

$$E_e' = \frac{g^{-1}(p)}{\Delta t}. \tag{13.2}$$

The following section discusses how high dynamic range imagery may be obtained using multiple linearized images of the same scene; the remaining sections discuss techniques to compute the camera response function.

13.1.2 Multiple Linearized Exposures

When an image has been linearized, its dynamic range is limited by the range that the camera sensor is able to capture at one time. One method to obtain a higher dynamic range is to use multiple linearized images that were captured using different exposures. By altering the exposure settings of the camera, it is possible to capture a different range of radiances of the scene. Exposure settings that allow less light to hit the sensor will cause lighter regions to be exposed correctly in the resulting image, while the darker regions will appear too dark. Similarly, when the exposure time is increased and more light is allowed to go through the aperture, darker regions of the scene will be exposed correctly while brighter regions will appear as burnt-out areas in the image. If enough images are captured of the same scene with different exposure times, then each region of the scene will be correctly exposed in at least one of the captured images.

Information from all the individual exposures in the image is used to compute the high dynamic range image for the scene. The process of image generation is as follows:

$$E'_e(x,y) = \frac{\sum_{n=1}^{N} w(p_n(x,y)) \frac{g^{-1}(p_n(x,y))}{\Delta t}}{\sum_{n=1}^{N} w(p_n(x,y))} \tag{13.3}$$

For each pixel, the pixel values p_n from each of the N exposures are linearized by applying the inverse of the camera response g. The resulting values are then normalized by dividing by the exposure time Δt. The irradiance value E'_e of the high dynamic range image is a weighted average of these normalized values.

Both the weight function and the camera response need to be determined before this equation can be used effectively. In the following section, we discuss possible choices for the weight function. Different algorithms to recover the camera response g are discussed afterward.

13.1.3 Weight Functions

The weights in (13.3) are used to decrease the impact of pixel values that are incorrectly exposed. For instance, low weights may be given to pixel values that are close to 0 or 255, as these values are possibly the result of under- or overexposure. Following this argument, Debevec et al. [238] use a hat function for their weights:

$$w(p_n) = \begin{cases} p_n - p_{\min} & \text{if } p_n \leq \dfrac{p_{\max} + p_{\min}}{2}, \\ p_{\max} - p_n & \text{if } p_n > \dfrac{p_{\max} + p_{\min}}{2}. \end{cases} \tag{13.4}$$

A possible function for the weights may be derived from the signal to noise ratio θ, which for an irradiance value E'_e is

$$\theta = E'_e \frac{dp_n}{dE'_e} \frac{1}{\sigma_N(p_n)} \tag{13.5a}$$

$$= \frac{g^{-1}(p_n)}{\sigma_N(p_n) \dfrac{dg^{-1}(p_n)}{dp_n}}, \tag{13.5b}$$

where $\sigma_N(p_n)$ is the standard deviation of the measurement noise. If the assumption is made that the noise is independent of pixel value p_n (a valid assumption for all noise sources bar photon shot noise, see Section 12.10.4), then the weight function can be defined as [784]:

$$w(p_n) = \frac{g^{-1}(p_n)}{\frac{dg^{-1}(p_n)}{dp_n}}. \tag{13.6}$$

13.2 Response Curve Recovery

The following sections discuss four different techniques to recover the camera response function g and recombine images taken with multiple exposures into high dynamic range images. The recovery of the camera response curve is necessary to enable the construction of high dynamic range images. In addition, it is a measure of the non-linearity in the output of a camera and can, therefore, be seen as a partial calibration of the camera.

13.2.1 Method of Mann and Picard

Mann and Picard propose to estimate the camera's response function using as few as two images of the same scene, captured with different exposure times [731]. Starting with a dark pixel value in the image with the shorter exposure time, this pixel value is related to the image irradiance as specified in (13.1). The corresponding pixel in the second image is related to the image irradiance as follows:

$$p_{n+1} = g(k E'_e \Delta t), \tag{13.7}$$

where the exposure $E'_e \Delta t$ is k times larger than the first exposure. We then look for the pixel in the darker image that has the same value p_{n+1}, and we also record the value of the corresponding pixel in the second image. This process is repeated until the maximum output value of the function g is reached. We now have a set of pixels that are all related to each other: $(g(E'_e \Delta t)$ relates to $g(k E'_e \Delta t))$, and $(g(k E'_e \Delta t)$ in turn relates to $g(k^2 E'_e \Delta t))$, while $(g(k^2 E'_e \Delta t)$ relates to $g(k^3 E'_e \Delta t))$, etc.

These pixel pairs are then used to generate a *range-range* plot, as shown in Figure 13.4. To obtain the camera response function from these points, some assumptions about the nature of this function are made. The function is assumed to be a semi-monotonic function, and $g(0) = 0$. The function is also assumed to follow a power curve [731]:

$$g(E'_e \Delta t) = \alpha + \beta \left(E'_e \Delta t\right)^{\gamma}. \tag{13.8}$$

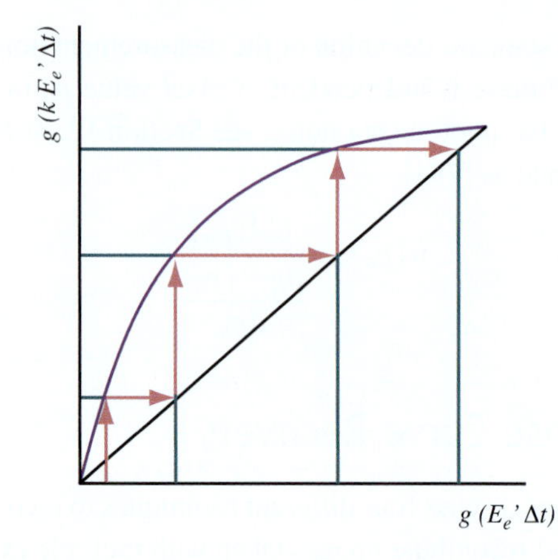

Figure 13.4. An illustration of the range-range plot defined by Mann and Picard (after [731]).

Here, α is 0 (from the assumption that $g(0) = 0$), and β and γ are scalar values. The pixel values in each pair can be related to each other as follows:

$$p_{n+1} = \beta \left(kE'_e \Delta t\right)^{\gamma} \tag{13.9a}$$

$$= k^{\gamma} p_n. \tag{13.9b}$$

The value of γ may be computed by applying regression to the known points in the range-range plot. There is now sufficient information to compute the camera's response function. Note that this solution will give results accurate up to a scale factor.

13.2.2 Method of Debevec and Malik

Debevec and Malik propose to compute the function g by solving a system of linear equations [238]. The input into their algorithm is also a set of N images of the same scene, with varying exposures. The scene is assumed to be static, so that the irradiance at each point on the sensor remains the same across the image set.

To compute the inverse camera response function g^{-1}, the reciprocity assumption needs to hold, i.e., halving E'_e in (13.1) and simultaneously doubling Δt will result in the same exposure value. Under extreme circumstances, such as a very long or short exposure time, this reciprocity assumption may not hold. For typical analog sensors, reciprocity holds to within one-third f-stop for exposure times between $1/10,000$ to 10 seconds. The reciprocity assumption holds for digital cameras under the assumption that the sensor measures the total number of incident photons in the exposure time [238]. If we make the assumption that g is a monotonically increasing function, its inverse becomes well defined and may be computed given the specified input.

The linear system of equations is derived as follows. Taking the log on both sides of (13.1), the following equation is obtained:

$$\ln(g^{-1}(p_n(x,y))) = \ln(E'_e(x,y)) + \ln(\Delta t_n). \tag{13.10}$$

In this equation, the unknowns are $\ln(g^{-1})$ and $\ln(E'_e(x,y))$. As the function g^{-1} is only applied to 256 values, essentially it represents 256 unknowns. The other set of unknowns, $\ln(E'_e(x,y))$, has as many unknowns as the number of pixels in a single image from the input. All these unknowns may be estimated by minimizing the following objective function:

$$\begin{aligned} O = & \sum_x \sum_y \sum_{n=1}^{N} \left[\ln(g^{-1}(p_n(x,y))) - \ln(E'_e(x,y)) - \ln(\Delta t_n)\right]^2 \\ & + \alpha \sum_{p=0}^{255} (\ln(g^{-1}))''(p)^2. \end{aligned} \tag{13.11}$$

Here, α is a user-specified weight that determines the relative importance of each part of the objective function. The first term of (13.11) is derived from (13.10), while the second term is the sum of the squared values of the second derivative of $\ln(g^{-1})$, and it is used to ensure smoothness in the function. It is computed by

$$(\ln(g^{-1}))''(p) = \ln(g^{-1}(p-1)) - 2\ln(g^{-1}(p)) + \ln(g^{-1}(p+1)). \tag{13.12}$$

Minimizing the objective function O is a linear least-squares problem and can be solved by singular value decomposition [163]. To ensure that the system is over-constrained, the number of knowns needs to be larger than the number of unknowns, which for images of dimensions h by w, is equal to $256 + hw$. This means that not all the pixels in an image need to be used as known values in this system of equations, as long as enough pixels are used to over-constrain the system. Pixels should be selected for inclusion in the system of equations on the basis of how well they represent the range of displayable values. Good candidates are those that belong to uniform regions in the image and those that cover as wide a range of luminance as possible.

Finally, the equations in the system may be weighted, based on their probability of being correct. Debevec and Malik use the weight function described in (13.4). After incorporating these weights, the objective function becomes

$$O = \sum_x \sum_y \sum_{n=1}^{N} \left\{ w(p_n(x,y)) \left[\ln(g^{-1}(p_n(x,y))) - \ln(E'_e(x,y)) - \ln(\Delta t_n) \right] \right\}^2 + \alpha \sum_{p=0}^{255} w(p) \left[(\ln(g^{-1}))''(p) \right]^2. \tag{13.13}$$

This algorithm can be used to obtain the function g^{-1} for a single channel. The above process, however, needs to be repeated for each of the three color channels.

13.2.3 Method of Mitsunaga and Nayar

Another technique to compute the camera response was proposed by Mitsunaga and Nayar [784]. This technique models the camera response with a flexible high-order polynomial and recovers the parameters for this model using a set of images of the same scene captured with different exposures. The advantages of this technique are that it is able to compute an accurate camera response despite noise in the input images. It also does not require exact values of the exposures of images. Given rough estimates, it computes accurate exposure information along with the camera response function. The model of the inverse of camera response g^{-1} is as follows:

$$g^{-1}(p) = \sum_{k=0}^{K} c_k p^k. \tag{13.14}$$

The response function is represented by a polynomial of order K applied to image values. The minimum order K required to represent a function will depend on the complexity of the function, so finding the polynomial representation of the function will include finding the value of K as well as the values of all coefficients c_k.

Whereas in most approaches the image irradiance E'_e is assumed to equal the scene radiance L_e, Mitsunaga and Nayar's technique takes into account the size of the camera's aperture to relate E'_e to L_e. The pixel value p is a function of the image irradiance:

$$E'_e = g^{-1}(p). \tag{13.15}$$

The relation between E'_e and L_e is given by (12.26) [488]. The image values p can now be expressed as a function of scene radiance and shutter speed as well as the aperture size:

$$p = g\left(L_e \frac{\pi}{4} \left(\frac{d}{f} \right)^2 \cos^4(\Theta) \Delta t \right) \tag{13.16a}$$

$$= g(L_e \rho e), \tag{13.16b}$$

where ρ is a constant that is determined by properties of the camera, and e is the expression that encapsulates the variation in image irradiance due to both the shutter speed and the aperture size:

$$\rho = \frac{cos^4(\Theta)}{f^2}, \tag{13.17}$$

$$e = \left(\frac{\pi d^2}{4}\right) \Delta t. \tag{13.18}$$

The ratio R of the linearized values is used to generate the objective function. For a pixel at location (x,y) in the nth exposure, this ratio can be written as

$$\frac{E'_{e,n}(x,y)\Delta t_n}{E'_{e,n+1}(x,y)\Delta t_{n+1}} = \frac{L_e(x,y)\,\rho(x,y)\,e_n}{L_e(x,y)\,\rho(x,y)\,e_{n+1}} \tag{13.19}$$

$$= \frac{g^{-1}(p_n(x,y))}{g^{-1}(p_{n+1}(x,y))} \tag{13.20}$$

$$= R_{n,n+1}. \tag{13.21}$$

Substituting the inverse camera response with the polynomial model (13.14) into (13.21), the following relationship is obtained:

$$R_{n,n+1} = \frac{\sum_{k=0}^{K} c_k\, p_n^k(x,y)}{\sum_{k=0}^{K} c_k\, p_{n+1}^k(x,y)}. \tag{13.22}$$

Using Equation 13.22, the following minimization function is obtained:

$$O = \sum_{n=1}^{N-1} \sum_{x} \sum_{y} \left[\sum_{k=0}^{K} c_k\, p_n^k(x,y) - R_{n,n+1} \sum_{k=0}^{K} c_k\, p_{n+1}^k(x,y) \right]. \tag{13.23}$$

The minimum of the function occurs when the derivative of the function O with respect to c_k equals zero:

$$\frac{\delta O}{\delta c_k} = 0. \tag{13.24}$$

As with previous techniques, this gives a solution up to a scale factor. However, when a single constraint is added, the system has a unique solution.

Thus, given approximate exposure ratios $R_{n,n+1}$, the unknown parameters of the inverse response function may be computed. After this computation, the actual values of the exposure ratios may be found by searching in the neighborhood of the initial ratio estimates. When more than two images are provided, searching for the ratios can be time consuming, and, therefore, an iterative version of the above process is used. First, user-specified approximate ratio estimates are used to compute function parameters as discussed above. Next, these parameters are used to compute an updated value of the ratio, using the following equation:

$$R_{n,n+1} = \sum_x \sum_y \frac{\sum_{k=0}^{K} c_k \, p_n^k(x,y)}{\sum_{k=0}^{K} c_k \, p_{n+1}^k(x,y)}. \tag{13.25}$$

The updated ratio is again used to compute function parameters, and this cycle continues until the change in linearized values (caused by the change in function parameter estimates of successive iterations) for all pixels in all the different images is smaller than a small threshold:

$$\left| g_{(i)}^{-1}(p_n(x,y)) - g_{(i-1)}^{-1}(p_n(x,y)) \right| < \varepsilon, \tag{13.26}$$

where ε is the threshold on the change across iterations, and i is the iteration number.

The order K of the polynomial representation of the camera response function can be computed by repeating the above process several times, each time with a different test value of K. For instance, the function parameters may be estimated for all K values less than 10, and the solution with the smallest error is chosen. For color images, the algorithm may be applied separately for each channel. Per-

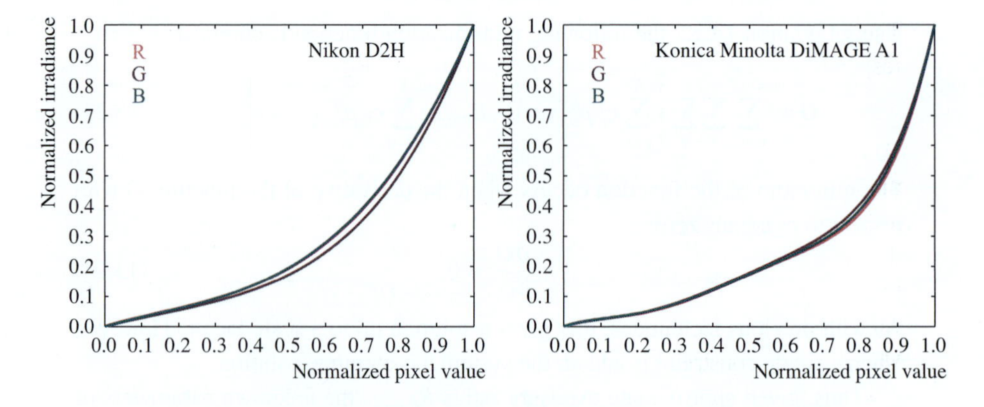

Figure 13.5. The response curves recovered for each of the red, green, and blue channels of a Nikon D2H and a Konica Minolta DiIMAGE A1 camera.

channel response curves for the Nikon D2H and Konica Minolta DiMAGE A1 cameras generated with this technique are shown in Figure 13.5.

13.2.4 Method of Robertson et al.

An alternative algorithm was proposed by Robertson et al. [967]. Their HDR assembly algorithm weights each pixel by a weight factor as well as the square of the exposure time:

$$E'_e(x,y) = \frac{\sum_{n=1}^{N} g^{-1}(p_n(x,y))\, w(p_n(x,y))\, t_n}{\sum_{n=1}^{N} w(p_n(x,y))\, t_n^2}. \tag{13.27}$$

Here, the summation is over each of the N exposures; t_n is the exposure time for the nth exposure, and g is the camera's response curve. This approach is motivated by the fact that longer exposures tend to produce a better signal-to-noise ratio. This weighting therefore tends to produce less noise in the resulting HDR image.

The camera response curve can be recovered by solving the following objective function O, for instance using Gauss-Seidel:

$$O = \sum_{n=1}^{N} \sum_{x} \sum_{y} \bar{w}(p_n(x,y)) \left(g^{-1}(p_n(x,y)) - t_n E'_e(x,y)\right)^2, \tag{13.28}$$

$$\bar{w}(p_n(x,y)) = \exp\left(-4 \frac{(p_n(x,y) - 127.5)^2}{127.5}\right). \tag{13.29}$$

The weight function $w(p_n(x,y))$ can then be set to the derivative of the camera response function:

$$w(p_n(x,y)) = g'(p_n(x,y)). \tag{13.30}$$

To facilitate differentiating the camera response curve, a cubic spline is fit to the camera response function. The appropriate selection of the spline's knot locations is crucial to obtain sensible results. As under- and over-exposed pixels do not carry information, it is desirable to set the weight function for these pixels to zero. In Robertson's method, this means that the derivative of the camera response needs to be close to zero for both small and large pixel values.

To achieve this, the spline that is fit to the camera response curve must have a slope of zero near 0 and 255. This makes the shape of the camera response curve recovered with Robertson's method markedly different from the curve obtained with the techniques presented in the preceding sections. Each camera can only have one response curve, and it is therefore reasonable to expect that different response curve recovery algorithms produce qualitatively similar results. As a result, HDR assembly may proceed according to (13.27), provided that the weights $w(p_n(x,y))$ are not computed as the derivative of the camera response curve.

13.3 Noise Removal

Night scenes often have a high dynamic range, for instance, due to the presence of street lights. Due to the overall low light levels, the exposure times tend to be relatively long. However, it is desirable to keep the exposure times relatively short, so that object movement is kept to a minimum. One way to achieve this is to increase the sensitivity of the sensor. Increasing the ISO setting in digital cameras is equivalent to amplifying the analog signal before it is passed through the A/D converter. This procedure invariably increases the noise present in the individual exposures.

In particular, the short exposures in the sequence will show relatively high noise levels. Due to the division by the exposure time in the HDR assembly process (see Equation (13.3)), noise in short exposures is amplified relative to longer exposures.

Assuming that each of the images in the sequence captures the same scene, and that the only difference is the exposure time, a modified form of image averaging may be used. In standard image averaging, a set of identical photographs of the same scene is "averaged" into a single, low-noise image. Accounting for the different exposure times, HDR image assembly and noise removal may be combined [24]:

$$c_a = \frac{\displaystyle\sum_{n=a}^{a+s} \frac{g^{-1}(p_n(x,y))\, h(p_n(x,y),a)}{t_n}}{\displaystyle\sum_{n=a}^{a+s} h(p_n(x,y),a)}, \tag{13.31}$$

$$E_e'(x,y) = \frac{\displaystyle\sum_{n=1}^{N} c_n\, w(g(t_n\, c_n), Y_{c_n})}{\displaystyle\sum_{n=1}^{N} w(g(t_n\, c_n), I_{c_n})}. \tag{13.32}$$

Here, the noise averaging is carried out for each pixel in each exposure over a set of corresponding pixels in s subsequent exposures. In general, using a larger value of s will give better noise reduction without introducing artifacts. However,

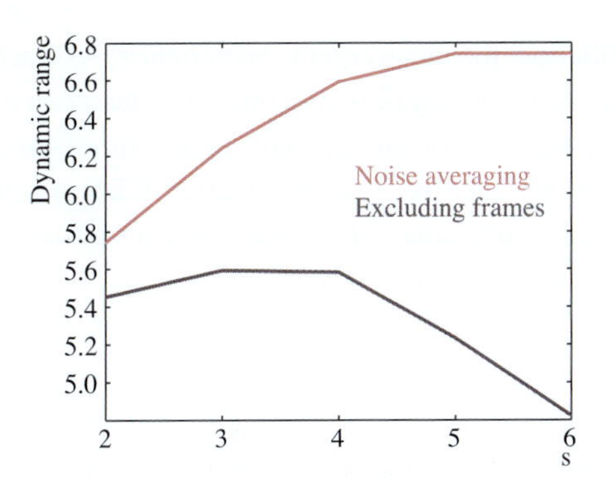

Figure 13.6. The dynamic range obtained for different values of s, the parameter that determines how many exposures are averaged in the noise-reduction step.

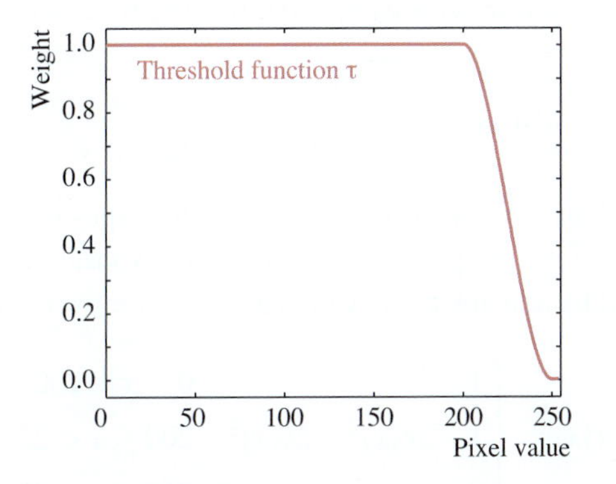

Figure 13.7. The multiplier reducing the weight for over-exposed pixels.

there is a maximum value for s beyond which no further image enhancement is obtained, as shown in Figure 13.6. Limiting the number of exposures to s, rather than averaging over all exposures longer than the one being filtered, will therefore save some computation time.

Figure 13.6 also shows the resulting dynamic range obtained by excluding the s shortest exposures in the sequence without noise averaging. Here, the dynamic range is reduced as more exposures are removed, caused by a lack of range in the light areas of the image.

The weight function $h(p_n, a)$ is set so that the noise is minimized. We assume that the main source of noise in dark environments is *photon shot noise*, which is noise arising from the fact that photons arrive at the sensor at random intervals. The Poisson distribution is a good model for photon shot noise [116].

In such a distribution, the mean equals the variance. Assuming that a 1-second exposure results in the recording of N photons, then the mean μ equals N, and the variance σ also equals N. For an exposure time t times longer, the mean will be $\mu = tN$, and the variance is $\sigma = tN$. During HDR assembly, pixel values are divided by the exposure time. This means that for a t-second exposure, after division the mean and variance are [116]

$$\mu = \frac{tN}{t} = N. \tag{13.33}$$

$$\sigma = \frac{tN}{t^2} = \frac{N}{t}. \tag{13.34}$$

As a result, increasing the exposure time by a factor of t reduces the variance, and, therefore, the noise, by a factor of t. This means that the weight factor applied to each pixel should be linear in the exposure time. The function $h(p_n, a)$ is therefore given by

$$h(p_n(x,y), a) = \begin{cases} t_n & n = a, \\ \tau(p_n(x,y))\, t_n & n \neq a. \end{cases} \tag{13.35}$$

Thus, if the pixel $p_n(x,y)$ is from exposure a, the exposure time itself is taken as the weight function. For pixels from a different exposure than a, the exposure time is multiplied by a factor $\tau(p_n)$ which accounts for over-exposed pixels:

$$\tau(x) = \begin{cases} 1 & 0 \le x < 200, \\ 1 - 3k(x)^2 + 2k(x)^3 & 200 \le x < 250, \\ 0 & 250 \le x \le 255, \end{cases} \tag{13.36}$$

$$k(x) = 1 - \frac{250 - x}{50} \tag{13.37}$$

The multiplier τ is plotted in Figure 13.7.

Equation (13.32) uses a weight function w that depends on both the pixel value $p_n(x,y)$, which is taken to be either a red, green, or blue value. In addition, a representation of the pixel's luminance Y_{c_n}, derived from the recorded pixel values using Equation (8.9), is part of this weight function to avoid saturated pixels that might give an unnatural color cast. The weight function itself is given by:

$$w(p_n(x,y), Y_{c_n}) = g^{-1}(p_n(x,y))\, g'(p_n(x,y)) \left[1 - \left(\frac{Y_{c_n} - 127.5}{127.5}\right)^{12}\right]. \tag{13.38}$$

A result with and without application of the noise-removal scheme is shown in Figure 13.8.

Figure 13.8. The image on the left was created without noise averaging, whereas the frames were first noise averaged for the image on the right.

Figure 13.9. Movement of scenery between the capture of multiple exposures (shown at top) lead to ghosting artifacts in the assembled high dynamic range image (bottom). This is apparent in the lack of definition of the sheep's head, but also note that the movement in the long grass has resulted in an impression of blurriness; Carnac, France, July 2007.

short exposures; we denote these by S_{k-1} and S_{k+1}. These two frames are used to generate an approximation of S_k that is the kth frame with a short exposure. The process by which this information is estimated is illustrated in Figure 13.11 and described below.

To estimate the information contained in the missing frame S_k, first bidirectional flow fields are computed using a gradient-based approach. Bidirectional flow fields consist of the forward warp $f_{k,F}$ that is applied to S_{k-1} to give S_{k*}^{F0} and the backward warp $f_{k,B}$ that is applied to S_{k+1} to give S_{k*}^{B0}. To compute these flow fields, the exposures of S_{k-1} and S_{k+1} are boosted to match the long exposure. Then, motion estimation between these boosted frames and the kth frame is performed to generate the flow fields. Kang et al. use a variant of the Lucas and Kanade technique [703] to estimate motion between frames.

The two new frames S_{k*}^{F0} and S_{k*}^{B0} are interpolated to give S_{k*}. The exposure of this frame is adjusted to match the exposure of L_k, to give L_{k*}. The flow field f_{k*} between L_{k*} and L_k is computed using hierarchical global registration. After computation, it is applied to S_{k*} to give S'_{k*}. To compute the final HDR image at time t, the following frames are used: L_k, S_{k*}^{F0}, S_{k*}^{B0}, and S'_{k*}. Note, that if the frame at time t has a short exposure, the same process may be used to estimate the long exposure.

To compute an HDR image from captured and computed frames, the following procedure may be used:

- The frames L_k, S_{k*}^{F0}, S_{k*}^{B0}, and S'_{k*} are converted to radiance values by using the camera response function and known exposure times. The radiance values of the above frames may be denoted by $\hat{L}_k$, $\hat{S}_{k*}^{F0}$, $\hat{S}_{k*}^{B0}$, and $\hat{S}'_{k*}$, respectively. The response function may be computed using any of the techniques discussed earlier in this section.

- For all pixel values in $\hat{L}_k$ that are higher than a certain threshold, the corresponding values in the HDR image are taken directly from $\hat{S}'_{k*}$, since high values in $\hat{L}_k$ are considered to be saturated and not good enough to provide reliable registration with adjacent frames. Note that in the case where the frame at time t has a short exposure, all values in $\hat{S}_k$ *below* a specified threshold are assumed to provide unreliable registration, and corresponding values in the HDR image are simply taken from $\hat{L}'_{k*}$.

- In the remaining regions, the radiance map values are computed by

$$E_{(x,y)} = \frac{f_{MW}(\hat{S}_{*(x,y)}^{F0}, \hat{L}_{(x,y)})\hat{S}_{*(x,y)}^{F0} + f_{MW}(\hat{S}_{*(x,y)}^{B0}, \hat{L}_{(x,y)})\hat{S}_{*(x,y)}^{B0} + f_W(\hat{L}_{(x,y)})}{f_{MW}(\hat{S}_{*(x,y)}^{F0}, \hat{L}_{(x,y)}) + f_{MW}(\hat{S}_{*(x,y)}^{B0}, \hat{L}_{(x,y)}) + f_W(\hat{L}_{(x,y)})} \tag{13.39}$$

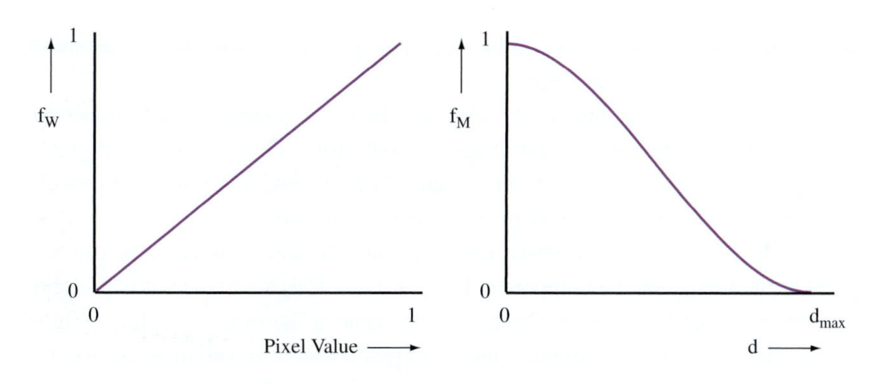

Figure 13.12. The flow fields f_W (left) and f_M (right).

where the subscript k has been dropped as it is common to all frames, (x,y) denotes pixel location, f_W denotes a weight function which is illustrated in Figure 13.12, and $f_{MW}(a,b) = f_M(|a-b|) \times f_W(a)$, where f_M is given by

$$f_M(\delta) = \begin{cases} 2\left(\frac{\delta}{\delta_{max}}\right)^3 - 3\left(\frac{\delta}{\delta_{max}}\right)^2 + 1 & \text{if } \delta < \delta_{\max}, \\ 0 & \text{otherwise.} \end{cases} \tag{13.40}$$

This function is also illustrated in Figure 13.12. Its main purpose is to ensure that if the warped radiance values are very different from the captured radiance values, then they should be given a relatively small weight in computation of the final radiance values.

13.4.2 Variance-Based Ghost Removal

Techniques that rely on accurate motion estimation to remove ghosts from HDR images may not work well whenever motion estimation fails. Scenes in which motion estimation cannot as yet be performed accurately, include those where an object is visible in some frames but not in others [1206]. Motion estimation is also prone to errors when the scene includes phenomena like inter-reflections, specularity, and translucency [573].

Ward introduced an alternate technique that does not rely on motion estimation [953]. The underlying assumption is that for each pixel location, a single correctly exposed pixel value is adequate to represent the scene radiance at that location. Therefore, in regions where there is movement in the scene, a single exposure may be used to provide values for the HDR image, as opposed to a weighted average of several exposures. If this approach is followed, ghosts will not appear in the final HDR image.

Ghost removal may thus be divided into two sub-problems. First, ghost regions need to be identified in the image. Second, for each pixel in these regions, an appropriate exposure needs to be selected from the image set that will provide the scene radiance value for that pixel in the HDR image.

To identify the regions where ghosts occur, the weighted variance of each pixel in the image can be computed. If a pixel has a high variance, it is possible that this variance is due to movement in the scene at the time of capture. Ghost regions are, therefore, identified as the set of pixels that have variance above 0.18.

Values from a single exposure may be used in the HDR image for ghost regions. However, replacing regions of high dynamic range with values from a single exposure will reduce the dynamic range of the HDR image. Another way to deal with ghost regions is to find the most appropriate exposure for each pixel in the ghost region and to use its value in the HDR image. A possible problem with this approach is that a moving object may be broken up, since the most appropriate pixels that represent the object may come from different exposures.

To avoid these problems, the ghost regions may first be segmented into separable patches. The most appropriate exposure that will provide radiance values for each of these patches in the HDR image is then determined. For each patch, a histogram is generated using the corresponding values in the original HDR image. The top 2% are discarded as outliers, and the maximum in the remaining values is considered. The longest exposure that includes this value in its valid range is selected to represent that patch.

Finally, interpolation is performed between these patches and their corresponding patches in the original HDR image. Each pixel's variance is used as the interpolation coefficient. This ensures that pixels that have low variances, but lie within a patch identified as a ghost region, are left unaltered.

13.4.3 Ghost Removal in Panoramic Images

A similar approach to removing ghosts from images has been proposed by Uyttendaele et al. [1167]. This approach is not designed specifically to remove ghosts from high dynamic range images. However, this approach for ghost removal is easily adapted for HDR images. Similar to the preceding algorithm, this method is not able to handle situations in which there is high contrast *within* a single ghost region.

Given a set of input images of a scene, Uyttendaele et al.'s algorithm creates panoramas of the scene in which ghosting artifacts have been removed. Once the images are registered, the first task is to identify regions in each input image that belong to moving objects. Once these are found, a single instance of each moving object is selected and used in the panorama.

Regions in each input image that belong to moving objects are found by comparing each pixel in an image with corresponding pixels in the remaining images. If a pixel differs from corresponding pixels by more than a threshold amount, it is identified as representing a moving object. Once all pixels in all the images have been evaluated in this manner, a region-extraction algorithm is applied to each image to identify and label contiguous regions. These are known as regions of difference (ROD).

RODs in different images are assumed to represent the same moving object if they overlap. Thus, to have a single instance of a moving object in the panorama, only one of the overlapping RODs is used in the final image. To find a single instance of each object, RODs are represented by a graph. Each ROD is a vertex in this graph, and if two RODs overlap, their corresponding vertices in the graph are connected with an edge. To obtain a single instance of a moving object, the vertex cover can be found and then removed from this graph. This leaves behind a set of disconnected vertices, representing a single instance of all moving objects in the scene.

As there are multiple instances of each moving object, any one of these may be used in the panorama. To ensure that the best possible instance is used, each ROD or vertex is given a weight. The closer the ROD is to the center of the image and the larger its size, the higher its corresponding vertex weight. Thus finding and removing the vertex cover with the minimum weight will leave behind the most appropriate vertex or instance of object in the panorama.

13.4.4 Ghost Removal using Kernel Density Estimation

Khan et al. [586] have proposed a way to remove ghosts in high dynamic range images even if the ghosting appears in a high contrast region. For example, this method makes it easier to capture a candle flame. In addition, this approach is not dependent on explicit motion detection or detection of regions within the image where ghosting is likely to occur. Ghost removal is achieved by making a straightforward change to the high dynamic range image-generation process as given by (13.3); a change in the way that the weights are computed.

As discussed previously, a weight is assigned to a pixel based on the probability that the pixel represents a correctly exposed value. Here, weights are computed that are also based on the probability that pixels belong to the static part of the scene. If pixels representing moving objects in the scene are given small weights, then the corresponding high dynamic range image will show less ghosting.

The probability that a pixel belongs to the static part of the scene may be computed statistically by representing the static scene with a distribution and then finding the probability that the candidate pixel belongs to this distribution. If such a probability is very small, it will indicate that the pixel does not belong to the static scene and, instead, represents a moving object. Such a pixel is given a small weight. On the other hand, a pixel with a large probability is given a large weight. As the probability is correlated with a suitable weight for each pixel, the weight of a pixel may be set equal to this probability.

If it is assumed that most of the pixels in any local region in the image set represent the static part of the scene, then the distribution of the static part may be approximated by the pixels in the immediate neighborhood of the pixel for which the weight is being computed. Using weighted kernel density estimation [881, 980], this probability is given by

$$P(\mathbf{x}|F) = \frac{\sum\limits_{\mathbf{y} \in F} w(\mathbf{y}) K_{\mathbf{H}}(\mathbf{x} - \mathbf{y})}{\sum\limits_{\mathbf{y} \in F} w(\mathbf{y})}, \tag{13.41}$$

where F is the distribution representing the static scene, $\mathbf{y}$ is an element of F that represents a pixel, and $K_{\mathbf{H}}$ is the kernel function with a bandwidth specified by $\mathbf{H}$. Each element $\mathbf{y}$ is a five-dimensional vector, where three dimensions denote the color and two dimensions denote the distance of the pixel from the candidate pixel $\mathbf{x}$. Khan et al. use the identity matrix to specify the bandwidth. The initial weights $w(\mathbf{y})$ are based on each pixel's probability of representing a correctly exposed value as discussed in Section 13.1. The neighborhood of pixel $\mathbf{x}$ is represented by all the pixels in its immediate vicinity, in all the differently exposed images.

The final weight of a pixel can now be computed by multiplying its initial weight with the probability computed above, to give

$$w_{t+1}(\mathbf{x}) = w_0(\mathbf{x}) P(\mathbf{x}|F). \tag{13.42}$$

While this procedure reduces ghosting artifacts in the image, they are still apparent in the final HDR image, and further processing is typically required. The algorithm can improve weights iteratively by using the updated weights from (13.42) as the initial weights of pixels $\mathbf{y}$ belonging to the distribution in (13.41). The updated probability can be used once more to compute the weights for each pixel, and the process can be repeated in this manner until the weights converge. The final weights will show little if any ghosting, as illustrated by Figure 13.13.

Figure 13.13. Ghost removal using (in reading order): 1, 4, 7, and 10 iterations. (Khan et al., "Ghost Removal in High Dynamic Range Imaging," IEEE International Conference on Image Processing, 2005–2008, © 2006 IEEE [586].)

13.5 Image Alignment

If the image set is captured with a hand-held camera, the images are often misaligned due to hand motion at the time of capture. This causes objects to appear blurred at the edges, and in cases of severe camera movement, can mean significant ghosting in the final high dynamic range image. Thus, a multiple-exposure technique to combine several images into a single high dynamic range image should consider alignment of the separate exposures.

While an obvious solution is to consider standard image alignment techniques, this is often not possible, because the exposures in the set are all different. This causes different parts of the scene to be properly exposed in different images, making alignment a more difficult problem.

One method of aligning images is to use optic flow, as discussed in Section 13.4.1. Motion is estimated between scenes captured by two consecutive images. For moving objects, motion vectors may exist only for certain regions in the image (those representing the object). For misaligned images, the motion vectors are likely to exist for larger portions of the image and, potentially, for the whole image.

A second technique was proposed by Ward et al. [1214]. This method is able to align images in the case of translation of the camera at the time of image capture. It does not rely on motion estimation or a known camera response function. First, binary maps of the images are computed by setting all of the pixels above a threshold value to one and the remaining pixels to zero. The threshold for an im-

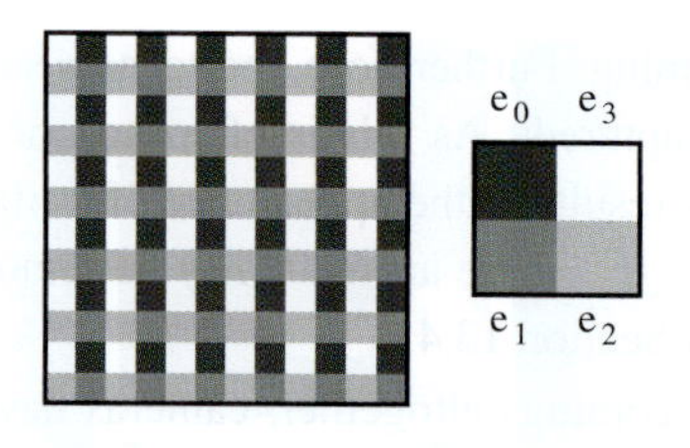

Figure 13.15. Multiple exposures may be captured by placing this filter in front of the sensing array (after [817]).

13.6.2 Spatially Varying Pixel Exposures

Devices with spatially varying pixel exposures avoid some of the problems encountered with split aperture imaging. These devices capture multiple exposures of the same scene simultaneously on a single sensor. This is achieved by sampling across both the spatial and exposure dimensions of image irradiance at the same time [817]. This can be achieved by placing a spatially varying filter in front of the sensor. Such a filter is shown in Figure 13.15. Here, a 2×2 window in this filter will have four different transmission coefficients, where $e_0 < e_1 < e_2 < e_3$. This pattern is repeated across the filter.

If such a filter is placed in front of the camera sensor, it will affect the irradiance at each sensor location. Four different exposures, with reduced spatial resolution, will be captured on the sensor.

The post-processing of images captured with this method begins by discarding over- and under-exposed values in the four images and then finding the surface that best fits the undiscarded values [817]. Such a surface representation that treats the image as a height field is then resampled to construct four complete images with appropriate resolution. These images are then recombined into a single HDR image.

13.6.3 Adaptive Dynamic Range Imaging

An adaptive dynamic range (ADR) device adapts to the different radiances from different parts of the scene. This adaptation is done separately for each pixel, and, therefore, each pixel sensor is optimally exposed for the radiance that it captures [816]. The advantage of such devices over those discussed in the previous section is that the spatially varying filter is not static, and they can therefore capture a larger class of scenes well.

An optical attenuator is placed in front of the sensor array, and its transmittance depends on two factors: its transmittance in the previous iteration and the

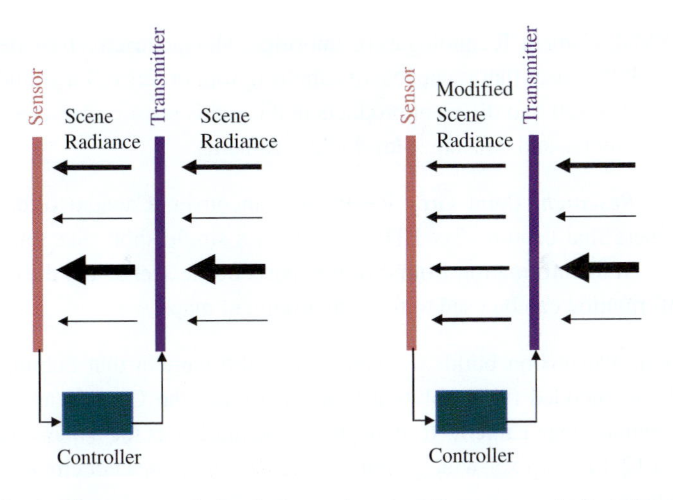

Figure 13.16. The concept of adaptively varying pixels' exposures. At first, the attenuator uniformly transmits the light (left). Once the sensors have accumulated some irradiance, their response is used by the controller to adjust the optical attenuator (right) (after [816].

corresponding sensor's value. If the pixel value is saturated, then transmittance is decreased. Therefore, in the next iteration, the pixel's value will have a smaller probability of being saturated. The transmittance is continually adjusted in this manner until pixel values are no longer saturated. A similar process is applied for under-exposed pixels. Transmittance is increased in such cases to allow higher irradiance for such pixels. Figure 13.16 illustrates this concept.

13.7 Direct High Dynamic Range Capture

The techniques to capture high dynamic range images introduced so far all depend on multiple exposures, even if they are generated and recombined in-camera. A more durable approach would be to develop cameras that can directly capture the full dynamic range available in real-world environments. Such cameras are both rare and expensive and have only recently started to appear on the market. Several such devices are being developed simultaneously by different companies:

Grass Valley. One of these devices is the Viper FilmStream camera, which is produced by Grass Valley (a division of Thomson). This film camera is capable of capturing about three orders of magnitude, which is well matched to the dynamic range reproducible in movie theaters, but is otherwise known as medium dynamic range.

SMaL. SMaL Camera Technologies (Cambridge, Massachusetts) have developed a CMOS sensor that is capable of capturing four orders of magnitude. They currently sell two different products that use this sensor, marketed as security cameras and consumer-level photo cameras.

Point Grey Research. Point Grey Research (Vancouver, Canada) offer a video camera that captures six HDR images in a single shot. Six sensors capture HDR information from different parts of the scene, and the combined information can be used to form environment maps.

Panavision. Panavision builds the Genesis, a film camera that outputs 10 bits of log-encoded full resolution film, qualifying the Genesis as a medium dynamic range camera. It serves the same market as the Viper. The sensor is a 12.4 megapixel-wide gamut CCD device that does not employ a Bayer pattern.

Pixim. Pixim (Mountain View, California) offer two CMOS sensors that can each capture a dynamic range of roughly four orders of magnitude.

SpheronVR. SpheroCam HDR, a camera offered by SpheronVR (Kaiserslautern, Germany) is the highest performance HDR camera currently on the market. It captures nearly eight orders of magnitude of dynamic range, at a very high spatial resolution. It uses a line scan CCD for capture and, therefore, needs to physically rotate to capture a scene. This means that full spherical

Figure 13.17. A full 360° scan taken with the SpheroCam HDR camera. The result was tone mapped with the photographic operator (see Section 17.3). Sunrise in Tübingen, Germany, taken from the rooftop of the Max Planck Institute (MPI), November 2006. (Photograph taken by Narantuja Bujantogtoch as part of the MPI for Biological Cybernetics Database of Natural Illuminations.)

captures can be taken, but a disadvantage is that it takes from fifteen to thirty minutes to complete a full 360° scan. An example image taken with this camera is shown in Figure 13.17.

It is anticipated that when these devices become more affordable, they will replace other forms of HDR generation. Their use will not require HDR assembly as a post-processing step, thereby removing a limitation that prevents multi-exposure techniques to reach the mass market. Further HDR sensor and camera designs are discussed in Hoefflinger's book [475].

13.8 Application: Drawing Programs

Many of the limitations associated with conventional photography are addressed and removed in high dynamic range imaging (HDRI). High dynamic range (HDR) images are often called *radiance maps*. Being able to capture real-world luminance has many advantages in photography and film [952], as well as for novel applications such as image-based lighting [239, 240] and image-based material editing (discussed in Section 13.9) [587].

Figure 13.18. An interactive brush showing a histogram of the pixels under the circle. This user interface allows the user to accurately select absolute luminance values, even if they are outside the display range. The area under the brush is displayed directly, after appropriate exposure control is applied, whereas the remainder of the image is tone-mapped prior to display. (Background image courtesy of Paul Debevec. This image was published in Mark Colbert, Erik Reinhard, and Charles E Hughes, "Painting in High Dynamic Range," *Journal of Visual Communication and Image Representation,* © Elsevier 2007.)

On the other hand, new challenges emerge. For instance, how can one manipulate high dynamic range images when they cannot be displayed directly on conventional display devices? This is a new issue that arises in drawing programs that offer support for high dynamic range images. With a luminance range that exceeds the display range, drawing directly onto an HDR canvas can be likened to key-hole surgery: only part of the display range can be visualized at any given time. Typically, one may solve such an issue by drawing on tone-mapped representations of images (see Chapter 17 on dynamic range reduction). Unfortunately, it is not clear that this approach is either intuitive or accurate.

In such cases, an alternative visualization may help, for instance by attaching a histogram visualization for the pixels under a region of interest (indicated by the size and position of a brush), allowing accurate selection of absolute luminance values. An example is shown in Figure 13.18. A qualitative approach may exploit visual phenomena familiar to artists, such as glare to allow the user to intuitively map a visual representation of glare back to the luminance value with which the brush paints [202].

13.9 Application: Image-Based Material Editing

There are several advantages to using high dynamic range images that are only just beginning to be explored. Although veiling glare may limit the range of values that can be captured, there is no limit on the range of values that can be represented in a suitably chosen file format. As a result, image histograms tend to have a long tail, which is a significantly different shape than histograms of conventional images tend to have. This makes it easier to determine, for instance, which pixels potentially represent highlights.

Further, the lack of quantization means that a much richer amount of information is available in images, which may simplify image processing tasks. For instance, gradient fields computed in images are smoother. As an example, the use of high dynamic range images for the purpose of estimating the shape of objects (shape from shading [1293]) does not have to overcome the limitations imposed by quantization.

These advantages, along with the fact that some distortions in images are not easily perceived, have led to the development of an image-editing technique that enables substantial changes to the apparent material of objects in images [587]. The input to the algorithm is a single high dynamic range image, along with an alpha matte which defines the set of pixels belonging to the object of interest.

The algorithm then extracts an approximate 2.5D shape of the object, i.e., a depth map, and separately computes a representation of the environment, based on

Figure 13.19. The pixels representing the bronze statue (left) were processed to produce the impression of transparency (right); Loet Vanderveen bronze, Native Visions Art Gallery, Winter Park, FL. (© 2006 ACM, Inc. Included here by permission [587].)

the pixels representing the background. The object shape as well the environment map are then plugged into the rendering equation (2.207). The user is then free to choose an arbitrary BRDF f_r, thereby changing the apparent material of the object.

In many cases, it is beneficial to simplify this rendering procedure by filtering the environment directly in several different ways before texture mapping the result onto the object. This has the added benefit of introducing desirable high frequency content which helps mask inaccuracies in the recovered shape, as well as inaccuracies in the environment. It allows objects to be converted to either transparent or translucent objects. The result is that plausible material transformations can be achieved (shown in Figure 13.19), without requiring 3D scanning equipment or multi-view techniques to derive more physically accurate 3D geometry. However, it is noted that without high dynamic range images as input, this approach would fail.

13.10 Further Reading

Hardware for high dynamic range image and video capture is discussed in detail in the book edited by Hoefflinger [475]. A more general discussion on high dynamic range imaging is available in *High Dynamic Range Imaging* [952].

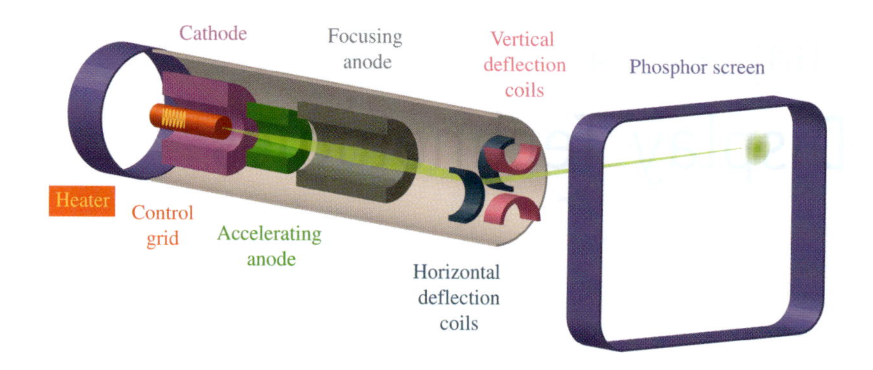

Figure 14.1. The components of a CRT display device.

A CRT display is made up of four key components encapsulated in an evacuated glass envelope: an electron gun, a deflection system, a shadow mask (or aperture grille), and a phosphor-coated screen. Visible light is produced when the electrons produced by the electron gun hit the phosphor coated screen at a position determined by the deflection system. This operation is illustrated in Figure 14.1.

The electron gun is responsible for producing and focusing the electron beam. When the cathode is heated by means of a heater, electrons escape from its surface by a phenomenon called *thermionic emission*. The control grid has a negative potential with respect to the cathode, pushing electrons towards its center. A positive potential (around 20 kV) is applied to the accelerating anode which causes the electrons to exit the control grid at high velocities. These electrons are focused to form an electron beam using the focusing anode [1120].

In the absence of vertical and horizontal deflection, such an electron beam would hit the center of the screen causing the center to glow. However, usually the aim is to produce visible light over the entire screen surface. Steering the beam toward any desired point on the screen surface may be accomplished by deflecting the beam using magnetic or electrostatic forces [1120]. *Magnetic deflection* employs two pairs of coils, one pair in the horizontal and the other pair along the vertical axis of the electron gun. The magnetic field produced by each pair is perpendicular, and, by adjusting the current passing through each coil, the electron beam can be deflected towards any direction on the phosphor screen.

The deflected electron beam travels in a vacuum until it collides with the phosphors. Upon collision, some of the electrons in the phosphor material jump to a higher energy state and then return back to their original states. Typical phosphor decay times are between 20 and 50 μs. Depending on the type of the phosphor used, this causes the emission of a photon with a wavelength determined by the

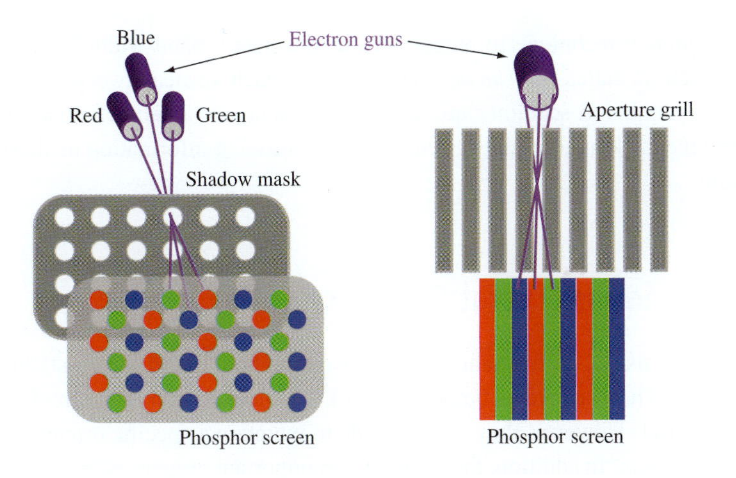

Figure 14.2. Shadow mask and aperture grill used in color CRTs ensure that each of the three types of phosphor receives light from the correct beam.

change in energy levels. This phenomenon is called *cathodoluminescence* which is a type of luminescence (see Section 3.5.6).

The color from a CRT is obtained by using three types of phosphors that emit light predominantly in the long-, medium, and short-wave regions of the visible spectrum (see Figure 7.17). Each type of phosphor is excited by separate beams produced by three electron guns. To ensure that each gun excites the correct type of phosphor, a masking mechanism is used [464]. The most common types of masks, the shadow mask and the aperture grill, are shown in Figure 14.2. A close-up photograph of a CRT display is shown in Figure 3.38.

To obtain an image on the CRT screen that appears to be steady, the electron beam must sweep through the screen multiple times every second. Conventionally, a sweep cycle starts from the top-left corner of the screen and scans from left to right and top to bottom. When all of the scanlines are swept, the beam is deflected back to the top-left corner for the next cycle. Such a systematic sweeping pattern is called a *raster*. For television, scanning is normally performed with very nearly the same frame rate as the alternating current power that supplies the display device (60 Hz in the US and 50 Hz in Europe).

Sensitivity to flicker increases with ambient light level. "Flicker-free" televisions use signal processing to create interpolated fields or frames and drive the display at double the prevailing main frequency. Because they are used in relatively bright environments, computer monitors have a higher *refresh rate*, achieving for instance 85 Hz.

A common technique to reduce information rate ("band-width") for a given refresh rate is called *interlacing*. Here, during each vertical sweep the odd- or the even-numbered scanlines are alternately displayed. This technique exploits the fact the the human visual system fills in the missing information in alternating scanlines.

14.2 Liquid Crystal Displays (LCDs)

Liquid crystals (LC) are organic fluids that exhibit both liquid and crystalline-like properties. They behave as liquids in that LC molecules can move freely within their material. However, they can be made to maintain a specific orientation, unlike true liquids. In addition, they exhibit two important properties that make them suitable for use in display devices. First, liquid crystals are transparent. Second, their orientation, and hence their effects on light polarization, can be changed by applying an external electrical field [810] (see Section 2.3 for polarization).

These physical properties form the basis of liquid crystal display technologies. In this section, we focus on displays that are operated in transmissive mode, i.e., a back-light (uniformly) illuminates an LCD screen, which modulates per pixel how much light is transmitted. Each pixel contains either a red, a green, or a blue filter, so that color images are formed by appropriately addressing triplets of pixels[1]. Each of the components of LCD displays are discussed in the following.

14.2.1 Back-Lights

The light sources used in LCD displays are most typically cold-cathode fluorescent lamps (CCFLs), although LED-based back-lights are just entering the market. Both types are discussed here.

A CCFL consists of a glass-tube with electrodes at both ends. The tube contains an inert gas and a small amount of mercury which emits ultraviolet radiation when stimulated by an applied field. The inside of the tube is coated with a blend of phosphors which luminesce when excited by the ultraviolet radiation, such that the spectrum produced by the light source is seen as white, although the spectrum may comprise of bands that are spectrally distinct (see Figure 14.3) [170]. A diffuser is placed between the light source and the LCD screen to ensure uniform illumination.

[1] Note that in our definition, a pixel is a single unit which emits light according to a single spectral power distribution. Another definition is possible, whereby a pixel consists of three units, peaked at long, medium, and short wavelengths.

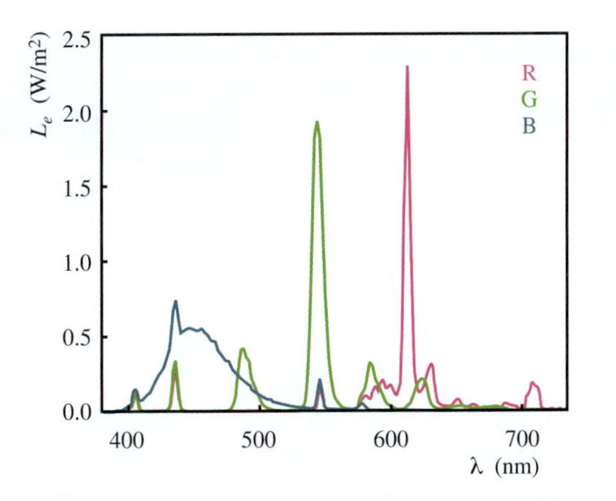

Figure 14.3. The spectral power distribution for typical phosphors used in CCFL back-lighting (adapted from a plot provided by Louis Silverstein, VCD Sciences, Inc., Scottsdale AZ).

In a typical LCD display, the power consumption is dominated by the CCFL. Dynamic dimming techniques are introduced to minimize the energy use [170, 175, 349]. Based on the imagery displayed, it may be possible to dim the back-light and increase the transmittance of the LCD panel to compensate.

Another approach which can effectively reduce power requirements is the use of LED technology for back-lighting. This technology offers further advantages in the form of enhanced ruggedness and stability under a wide range of environment temperatures. LEDs are mercury and lead-free.

Further, with LEDs it is possible to create much wider color gamuts, due to the use of saturated primaries [569], and to also increase the light output [1237]. While a color gamut of around the same size as the NTSC color gamut is achievable with pc-LEDs, mixed RGB LEDs can produce a wider color gamut. Recall from Section 9.8 that LEDs can be constructed with sharp spectral peaks having a width of around 20–30 nm. As a result, their primaries will lie close to the spectral locus of a chromaticity diagram. In addition, it is possible to construct back-lights with more than three primaries (for instance five or six primaries) to further extend the available gamut [1059].

The luminance requirements of the back-light have an impact on its design. For standard displays emitting less than 750 cd/m^2, the back-light can be mounted along the edge of the display, whereas for brighter displays a direct-view back-light will be required. Edge-lit and direct-view displays are diagrammed in Figure 14.4.

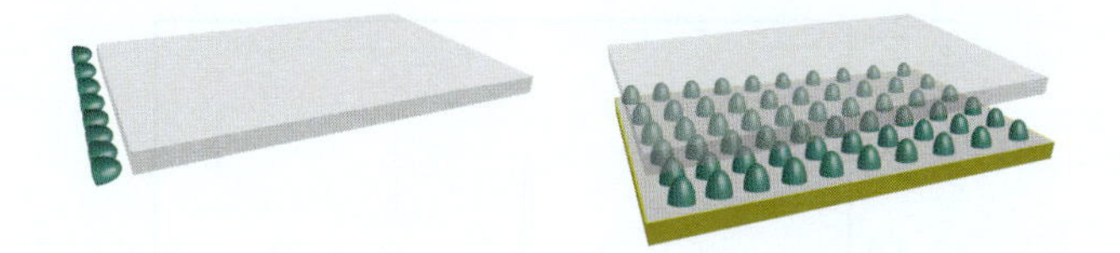

Figure 14.4. Example diagrams of edge-lit (left) and direct-view back-lights (right).

One of the problems with LED back-light design is temperature dissipation. The printed circuit board acts as a heat sink. The thermal resistance of recent high-flux LEDs is in the order of 13–15 °C/W for the Luxeon emitter, 9 °C/W for the Osram Golden Dragon, and 2.6 °C/W for the Lamina BL2000. This means for example that for a 3 W Luxeon emitter, the LED die temperature would rise to around 39 °C above ambient. Thus, thermal resistance puts an upper bound on luminous output, as the associated temperature rise will limit device reliability [1059].

14.2.2 Liquid Crystal Phases

Liquid crystals may be found in several phases such as *nematic*, *cholesteric*, and *smectic* [1120]. For a given material, the phase it assumes depends on temperature, as shown in Figure 14.5 [706]. Below temperature T_m, the material is anisotropic and crystalline. Above temperature T_c, the material is isotropic and liquid. In between, three phases can be distinguished, which increase from more ordered to less ordered structures and are collectively known as the *mesophase*. These phases are named smectic C, smectic A, and nematic. The molecules in

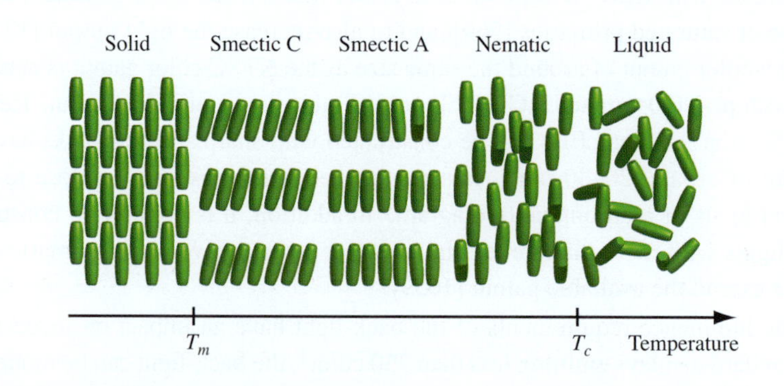

Figure 14.5. The different phases that liquid crystals can assume depend on temperature. Inbetween temperatures T_m and T_c the material behaves as a liquid, albeit with ordered structure, and has a milky appearance.

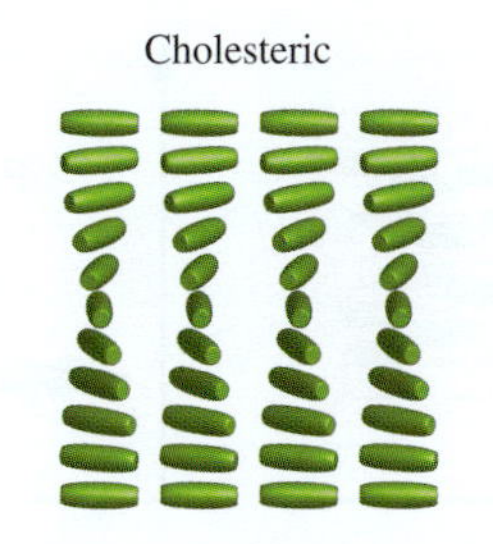

Figure 14.6. In the cholesteric phase, the liquid crystal molecules are aligned parallel to the front and back substrates (i.e., the tilt angle is 0), and form a helical structure.

the smectic phases are oriented alike, and additionally form layers. The smectic C phases is characterized by molecules showing small random deviations from a preferred direction which is angled with respect to the orientation of the layers. In the smectic A phase, this *tilt angle* becomes perpendicular to the surfaces of the layers. In the nematic phase, the molecules are oriented alike, but do not form layers. All phases are anisotropic, except for the liquid phase, which due to the random orientation of the molecules is isotropic.

It is possible to add chiral compounds, which change the nematic phase to the cholesteric phase. In this phase, the directors of calamitic (i.e., rod-shaped) molecules in subsequent layers form a helical structure, as shown in Figure 14.6. The *director* is the vector describing the long axis of a liquid crystal molecule. Additional phases may be distinguished, but these are less important for the manufacture of liquid crystal displays.

The director of each molecule makes an angle θ with the average direction. The wider the spread of angles around the average, the more chaotic the ordering of the molecules, and the more isotropic the appearance of the material. The order parameter S of a phase is a measure of the optical properties that may be expected, and is given by [706]

$$S = 0.5 \left\langle 3 \cos^2(\theta) - 1 \right\rangle, \tag{14.1}$$

where θ is the angular deviation of individual liquid crystal molecules from the average director, and $\langle . \rangle$ indicates an average over a large number of molecules. In a fully ordered state, the molecules all perfectly align, yielding $S = 1$. In a fully unordered state, we obtain $S = 0$. In a typical nematic phase, the order parameter S is between 0.4 and 0.7 [706]. The ordering of calamitic molecules in smectic and nematic phases gives rise to birefringence, as discussed in Section 2.6. This optical property is exploited in liquid crystal displays.

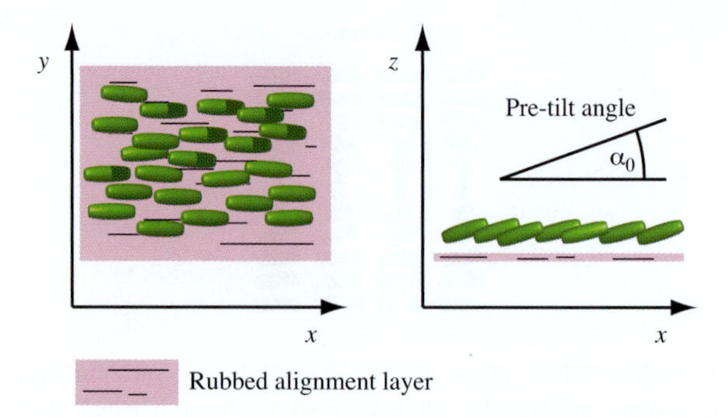

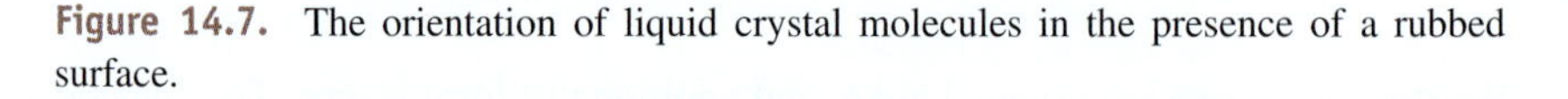

Figure 14.7. The orientation of liquid crystal molecules in the presence of a rubbed surface.

14.2.3 Rubbing

The orientation of liquid crystal molecules is non-random in the mesophases. In the construction of a liquid crystal display, it is important to force the alignment of the molecules with respect to each other, but also with respect to a substrate. Liquid crystals are normally sandwiched between glass plates. Before any electric field is applied, the molecules nearest to the glass plates can be oriented in a preferred direction by creating microscopic grooves. This is achieved by coating the glass plates with a layer of polyimide material, which is then rubbed with a fine cloth with short fibers [1112]. The coated and rubbed polyimide layer is called the *alignment layer*.

This rubbing procedure creates microscopic grooves to which liquid crystal molecules will align. The orientation is in the plane of the glass substrate, but also causes the molecules to orient slightly away from the substrate, known as the *pre-tilt angle* α_0. The orientation of liquid crystal molecules in the rest-state is shown in Figure 14.7.

14.2.4 Twisted Nematic Operation in the Field-Off State

In a *twisted nematic* display, both glass substrates are coated and rubbed. However, the rubbing direction of the second substrate is orthogonal to the direction of the first substrate (see Figure 14.8). The molecules near the surface of the first substrate are therefore aligned with an angle of 90° to the molecules near the surface of the second substrate. The molecules in between are twisted in a helix structure that rotates over a *twist angle* of $\beta = 90°$ between the first and second substrate. This alignment is visualized in Figure 14.9.

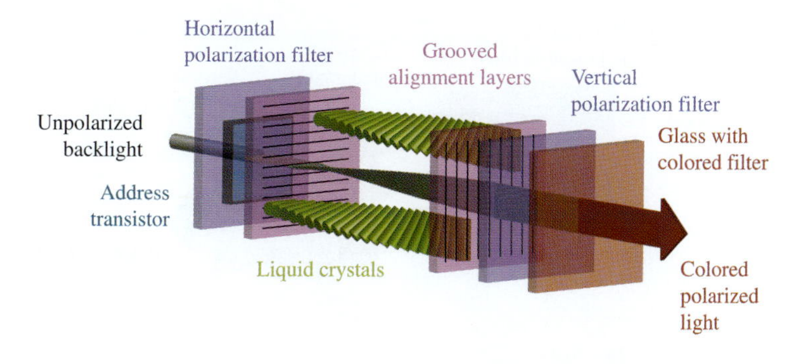

Figure 14.8. A pixel in a twisted nematic active matrix LCD display.

The next step towards understanding the operation of a twisted nematic liquid crystal display is to consider how polarized light propagates through a twisted nematic liquid crystal layer. We assume that the light is incident at a right angle, i.e., along the z-axis in Figure 14.9. The birefringence of the aligned molecules polarizes light into two orientations. By placing linear polarizing filters before the first and after the second alignment layer, light entering and exiting the liquid crystal layer can be linearly polarized.

The first polarizing filter is known as the *polarizer*, whereas the second polarizing filter is known as the *analyzer*. The polarizer is required, as without it, the liquid crystal cell by itself would not show the desired optical effect. The amount of light transmitted through the analyzer depends on the angle of polarization produced by the liquid crystal layer with respect to this filter (as given by Malus' law in (2.86)).

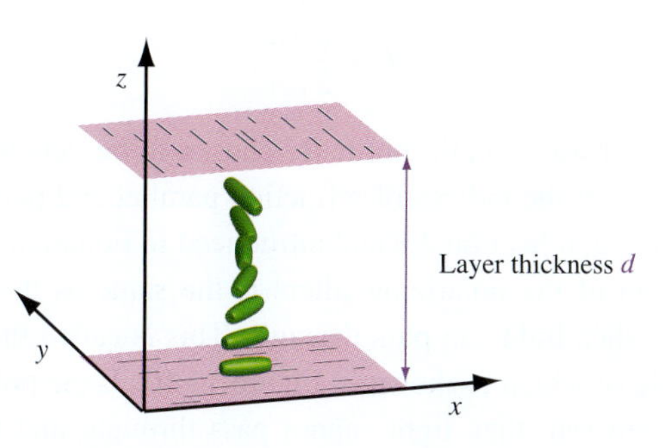

Figure 14.9. The orientation of liquid crystal molecules in the presence of a pair of orthogonally oriented rubbed surfaces.

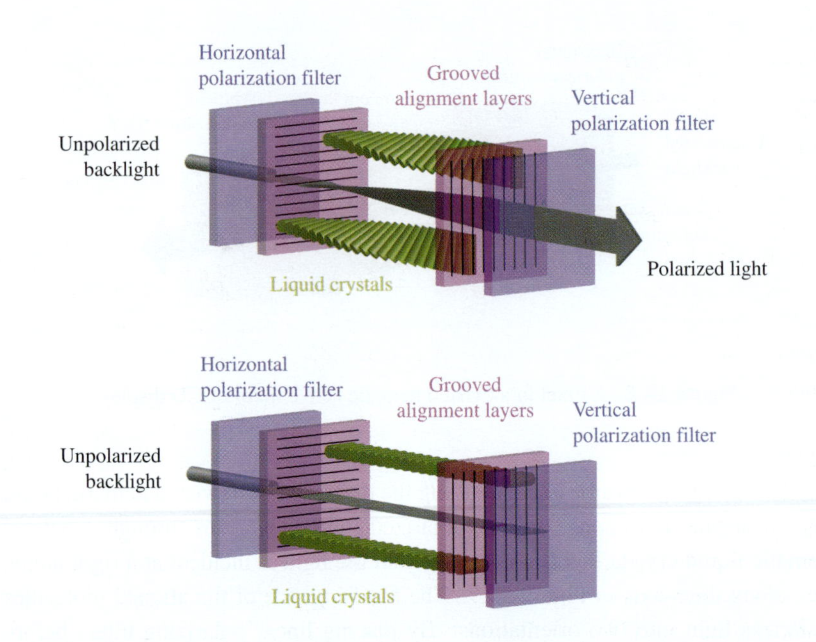

Figure 14.10. The stack of LCs takes a twisted shape when sandwiched between two glass plates with orthogonal surface etching (top). They rotate the polarization of light passing through them. Under an electric field, liquid crystal molecules can be realigned, and light transmission can be blocked as a result of the light having polarization orthogonal to the second polarizer (bottom). This example shows the normally white mode.

When propagating through the liquid crystal layer, which is assumed to have a thickness of d (typically in the range of 3.5 to 4.5 μm), the birefringence of the material changes the light's polarization from linear to circular to linear again, provided that the layer thickness is chosen to be

$$d = \frac{\sqrt{3}\,\lambda}{2\,\Delta n}, \tag{14.2}$$

where λ is a typical wavelength, normally chosen to be yellow at 505 nm, and Δn is the difference of the indices of refraction parallel and perpendicular to the director, as discussed in Section 2.6 and introduced in Equation (2.141).

If the direction of the polarizing filters is the same as their corresponding alignment layers, then light can pass through. This is called the *normally white mode*, an example of which is shown in Figure 14.10. If the polarizer and alignment layers are crossed, then light cannot pass through, and this is called the *normally black mode*.

14.2.5 Realigning Liquid Crystals

The helically twisted shape of the liquid crystals, as determined by the two crossed alignment layers, can be modified by applying a voltage. To this end, electrodes can be added to the sandwich. The configuration of the electrodes across the display is intimately linked with the addressing scheme used to switch individual pixels. This is discussed in Section 14.2.9.

For now, we assume that a single pixel is formed by augmenting the layered structure with two electrodes, across which a voltage can be applied. Such electrodes must be transparent to enable light to pass through. They are typically manufactured using a thin layer of indium tin oxide (ITO), typically 100 nm thick.

Due to the anisotropic dielectric constant associated with the liquid crystal molecules ($\Delta\varepsilon$, see Section 2.6), the electric field created by applying a voltage across the electrodes exerts a small force, which realigns the molecules to the electric field. This alignment occurs throughout the liquid crystal layer, with the exception of those molecules that are in close proximity to the alignment layers.

Inter-molecular forces resist this realignment, so that rotation of the molecules begins at a threshold voltage V_{th}. Full alignment is reached at a voltage V_{max}, where all molecules, except those in contact with the alignment layers, are rotated parallel to the field (assuming that $\Delta\varepsilon > 0$). Note that the electric field changes both the tilt angle α (perpendicular to the x-y plane of Figure 14.9) and the twist angle β (in the x-y plane).

As the realignment causes light to reach the second polarizer at different angles dependent on the applied voltage, the amount of light transmitted through the

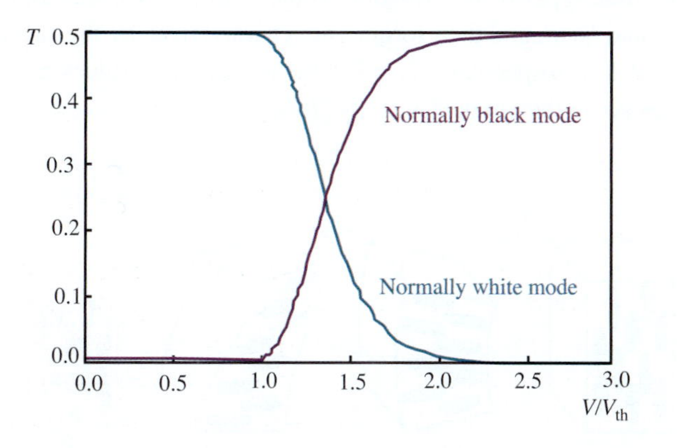

Figure 14.11. The transmittance T of a twisted nematic cell as a function of relative driving voltage V (after [706]). As discussed further in Section 14.2.7, the transmittance has a maximum of 0.5.

display depends on this voltage. The transmittance T as a function of voltage V for both the normally white and normally black modes is shown in Figure 14.11.

When a voltage is applied, the liquid crystal molecules realign themselves as function of inter-molecular forces and the externally applied electric field. As these molecules are relatively large, their physical movement takes a certain amount of time. The dynamic viscosity η_d, measured in Ns/m^2, of the liquid crystal material depends on the force F required to move a molecule with area A over a distance d with velocity v, as follows:

$$\eta_d = \frac{F\,d}{A\,v}. \tag{14.3}$$

The kinematic viscosity η, measured in mm^2/s, equals the dynamic viscosity η_d divided by the density δ of the material:

$$\eta = \frac{\eta_d}{\delta}. \tag{14.4}$$

However, the density of liquid crystal materials hovers around 1 Ns^2/mm^4, so that the value of both types of viscosity in liquid crystal displays are close to identical. The viscosity of liquid crystal materials is typically in the 20 to 45 mm^2/s range [706]. The effect of the viscous properties of liquid crystal materials is that the time to realign the molecules for each pixel is not instant, but takes a little while, typically at least a few milliseconds and perhaps tens of milliseconds.

In the presence of electric and magnetic fields, the equilibrium that is reached depends on three types of deformation to the liquid crystal material. These are *splay*, *twist*, and *bend*, and their torques are characterized by the elastic constants K_{11}, K_{22}, and K_{33}, respectively (see Figure 14.12). The values associated with these torques are in the order of 10^{-3} N [706].

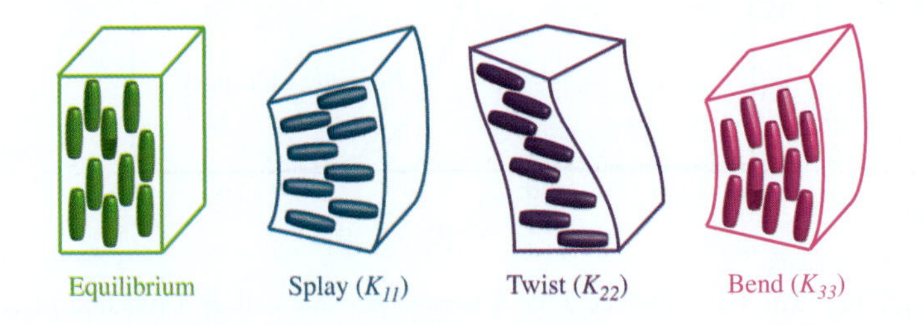

Figure 14.12. Splay, twist, and bend in liquid crystal materials (after [706]).

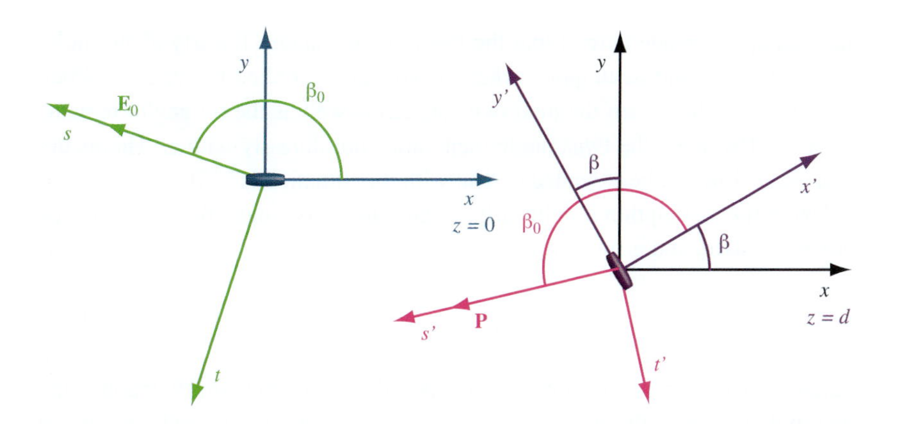

Figure 14.13. Coordinate system at $z = 0$ (left) and $z = d$, where d is the thickness of the cell.

14.2.6 Light Propagation

We consider a pixel configured as a twisted nematic cell. The coordinate system assumed puts the surface of the cell in the x-y-plane. The light incident on the cell is then in the z-direction. The twist angles β are also in the x-y-plane, whereas the tilt angles are perpendicular to this plane. The liquid crystal molecules form a helical shape with its axis in the z-direction.

Further, we assume that the light incident upon the cell is linearly polarized with the polarization oriented with an angle β_0 with respect to the x-axis. The coordinate system assumed in the following computations is sketched in Figure 14.13. Thus, we no longer require the polarizing filter at the first substrate to be aligned with the grooves in the first alignment layer. The electric field vector $\mathbf{E}_0$ therefore gives rise to the Jones vector $\mathbf{J}$, given by

$$\mathbf{J} = \begin{bmatrix} J_x \\ J_y \end{bmatrix} = R(-\beta_0) \begin{bmatrix} \mathbf{E}_0 \\ \mathbf{0} \end{bmatrix}, \tag{14.5}$$

where $R(\theta)$ is a rotation matrix,

$$R(\theta) = \begin{bmatrix} \cos(\theta) & \sin(\theta) \\ -\sin(\theta) & \cos(\theta) \end{bmatrix}. \tag{14.6}$$

The Jones vector is defined in Section 2.6. The twist angle β_z at position z along the z-axis is equal to

$$\beta_z = \beta_0 z, \tag{14.7}$$

so that the final twist angle after passing through the cell with thickness d is

$$\beta_d = \beta_0 d. \tag{14.8}$$

The assumption made here is that the twist angle changes linearly along the z-axis. This is a valid assumption when no voltage is applied to the cell. When a voltage is applied, then the total twist angle (as well as the tilt angle) remains the same. However, the twist angle then varies non-linearly with z. The model discussed below can be extended to deal with this situation [1270].

Under the assumption of a linearly progressing twist angle, the pitch p of the helix is constant, and is given by

$$p = \frac{2\pi}{\beta_0}. \tag{14.9}$$

Although we omit the derivation, it is now possible to compute the direction of the linear polarization with respect to the analyzer after light has passed through the cell, which has thickness d. Assume that the analyzer is oriented in direction $\mathbf{P}$, as shown in Figure 14.13. The Jones vector, given in the s'-t' coordinate system, which is the coordinate system associated with the analyzer, is [706, 1270]

$$\begin{bmatrix} J_{s'} \\ J_{t'} \end{bmatrix} = R(-\Psi)\, \exp\left(\frac{-2\pi i \bar{n} d}{\lambda}\right)\, M\, R(-\beta_0) \begin{bmatrix} \mathbf{E}_0 \\ \mathbf{0} \end{bmatrix}, \tag{14.10}$$

where the *key matrix* M is given by

$$M = \begin{bmatrix} \cos(\gamma) + i a \beta \sin(\gamma) & \beta \sin(\gamma) \\ -\beta \sin(\gamma) & \cos(\gamma) - i a \beta \sin(\gamma) \end{bmatrix}. \tag{14.11}$$

The unknowns in these equations are

$$\bar{n} = \frac{n_{\parallel} + n_{\perp}}{2}, \tag{14.12a}$$

$$a = \frac{p\,\Delta n}{2\lambda}, \tag{14.12b}$$

$$\beta = \frac{2\pi d}{p}, \tag{14.12c}$$

$$\gamma = \beta\, \sqrt{1 + a^2}, \tag{14.12d}$$

$$s = \frac{2\pi d}{p\varepsilon}. \tag{14.12e}$$

Equation (14.10) is a general expression that is valid for several types of nematic display. In particular, if the tilt angle α is chosen to be 0, and the twist angle is $\beta = \pi/2$, then we have a classic twisted nematic display. For $\alpha = 0$ and $\beta > \pi/2$, the display is called a super-twist nematic display. Finally, $\alpha \neq 0$ and $\beta \neq \pi/2$ yields a so-called mixed-mode nematic display.

14.2.7 Transmittance of a Twisted-Nematic Cell

As mentioned above, a single pixel (cell) of a twisted nematic display can be described with (14.10) where $\alpha = 0$ and $\beta = \pi/2$. For a cell with thickness d, the pitch p can then be derived from (14.9):

$$p = 4d. \tag{14.13}$$

This leads to the two parameters a and γ with values

$$a = \frac{2d\Delta n}{\lambda}, \tag{14.14a}$$

$$\gamma = \frac{\pi}{2}\sqrt{1+\left(\frac{2d\Delta n}{\lambda}\right)^2}. \tag{14.14b}$$

Further, the s' and t' axes are rotated over 90° with respect to the x- and y-axes. By substitution of (14.14) into (14.11) and subsequently into (14.10), we obtain [706]

$$\begin{bmatrix} J_{s'} \\ J_{t'} \end{bmatrix} = \begin{bmatrix} e\left\{\cos\left(\frac{\pi}{2}\sqrt{1+\left(\frac{2d\Delta n}{\lambda}\right)^2}\right) + \frac{\pi i d\Delta n}{\lambda}\sin\left(\frac{\pi}{2}\sqrt{1+\left(\frac{2d\Delta n}{\lambda}\right)^2}\right)\right\} E_0 \\ -e\frac{\pi}{2}\sin\left(\frac{\pi}{2}\sqrt{1+\left(\frac{2d\Delta n}{\lambda}\right)^2}\right) E_0 \end{bmatrix}. \tag{14.15}$$

In this equation, we have

$$e = \exp\left(\frac{-2\pi i \bar{n} d}{\lambda}\right), \tag{14.16}$$

where Δn is the difference between the ordinary and extraordinary indices of refraction of the birefringent material and is taken from (2.141). The value $\bar{n}$ is the average of these indices of refraction:

$$\bar{n} = \frac{n_{\parallel} + n_{\perp}}{2}. \tag{14.17}$$

A parallel analyzer would be oriented along the t'-axis, leading to a normally black mode. The irradiance E_e of the light passing through the analyzer is derived from the time averaged value of $J_{t'}$, analogously to the expression for irradiance

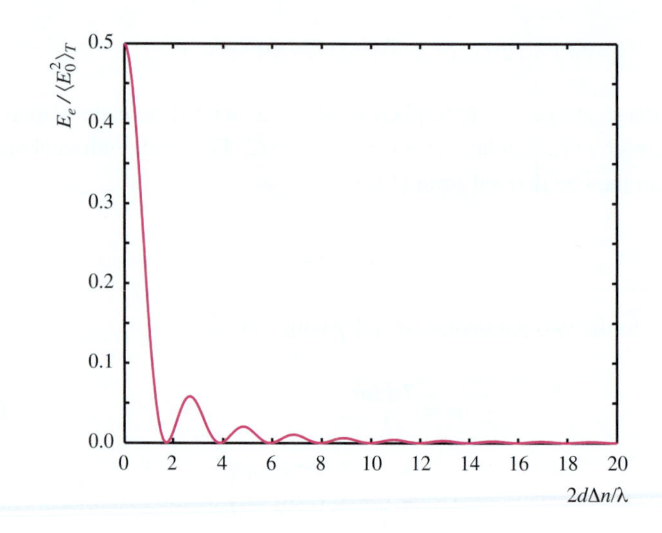

Figure 14.14. The reduced irradiance $E_e / \left\langle E_0^2 \right\rangle_T$ plotted against $a = (2d\,\Delta n)/\lambda$ for the normally black mode.

discussed in Section 2.4 [706]:

$$E_e = \frac{1}{2} |J_{t'}|^2, \tag{14.18a}$$

$$= \frac{1}{2} \frac{\sin^2\left(\frac{\pi}{2}\sqrt{1+\left(\frac{2d\,\Delta n}{\lambda}\right)^2}\right) \langle E_0^2 \rangle_T}{1+\left(\frac{2d\,\Delta n}{\lambda}\right)^2}, \tag{14.18b}$$

where the time average of E_0 is taken. As can be seen from this equation, as well as from Figure 14.14 which shows the reduced irradiance (effectively the transmittance), the amount of light transmitted reaches a minima when the square root becomes a multiple of 2. The first minimum is obtained for

$$\sqrt{1+\left(\frac{2d\,\Delta n}{\lambda}\right)^2} = 2, \tag{14.19}$$

so that we find the relation

$$d = \frac{\sqrt{3}\,\lambda}{2\,\Delta n}, \tag{14.20}$$

a result which was already anticipated in (14.2). Thus, a given wavelength, together with the material-dependent parameter Δn, determines the optimal wavelength to block all light. As every wavelength in the visible range would require

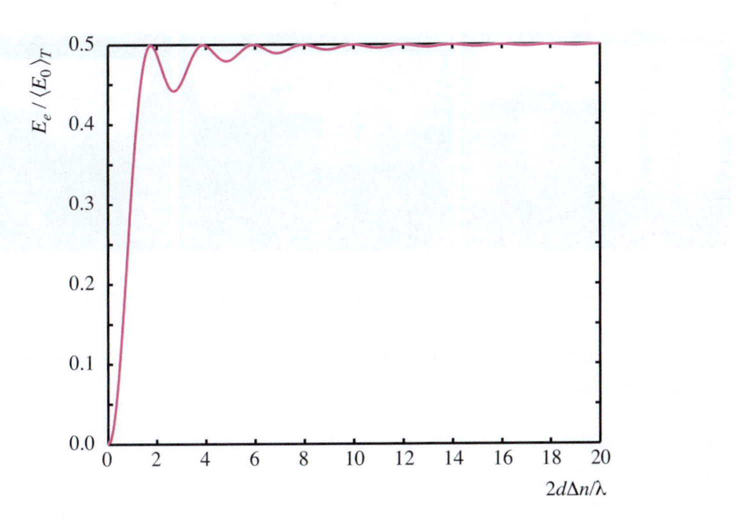

Figure 14.15. The reduced irradiance $E_e / \langle E_0^2 \rangle_T$ plotted against $a = (2d\,\Delta n)/\lambda$ for the normally white mode.

a somewhat different thickness of the liquid crystal cell, the normally black mode can only be fully realized for one wavelength. As a result, this mode leaks light which is seen as a bluish-yellowish color.

For the normally white mode, the analyzer is aligned with the s'-axis, so that the irradiance becomes

$$E_e = \frac{1}{2} \left| J_{s'} \right|^2, \tag{14.21a}$$

$$= \frac{1}{2} \frac{\cos^2\left(\frac{\pi}{2}\sqrt{1+\left(\frac{2d\,\Delta n}{\lambda}\right)^2}\right) + \left(\frac{2d\,\Delta n}{\lambda}\right)^2}{1+\left(\frac{2d\,\Delta n}{\lambda}\right)^2} \langle E_0^2 \rangle_T. \tag{14.21b}$$

Here, maxima occur for the same values of d for which minima occur in the normally black mode. As a result, the thinnest cells for which maximal transmittance happens have a thickness of $d = \sqrt{3}\,\lambda / 2\,\Delta n$. A plot of the reduced irradiance (i.e., $E_e / \langle E_0^2 \rangle_T$) in the normally white mode is shown in Figure 14.15.

In both the normally black and the normally white mode, the irradiance transmitted through the cell depends on wavelength, as can be determined from (14.21). This results in color shifts that depend on the selected gray level. However, when the maximum voltage is applied in the normally white mode, the cell becomes

Figure 14.16. Shown here is a Toshiba 32-inch LCD television and a 12-inch Apple iBook with LCD screen, photographed at different angles. The iBook is several years older than the television and has a pronounced angular dependence. The television is virtually free of angular artifacts.

black independent of wavelength. As the black state of the normally white mode does not have a colored tinge, it is the preferred mode [706].

The equations shown in this section are valid for light incident at right angles. For obliquely incident light, the Jones matrix method can be refined and extended [168, 398, 684, 706, 1274, 1284], showing that dependent on the type of cell, the transmittance shows a more or less pronounced fall-off with viewing angle, as shown in Figure 14.16.

Moreover, these equations assume that the cell is in the off-state, i.e., no voltage is applied. Under this assumption, the twist angle varies linearly along the z-axis, and the tilt angle can be assumed to be constant (see Figure 14.17). When a voltage is applied, the total twist angle remains the same, due to the anchoring effect imposed by the rubbed substrates. However, the distribution of twist and

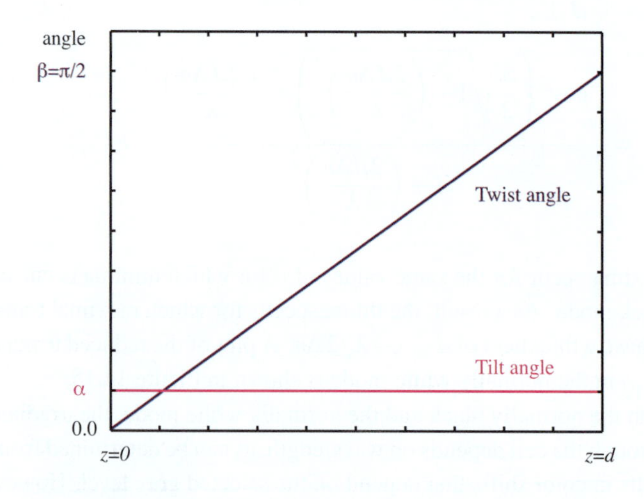

Figure 14.17. Twist and tilt angles in the off-state of a twisted nematic cell (after [1270]).

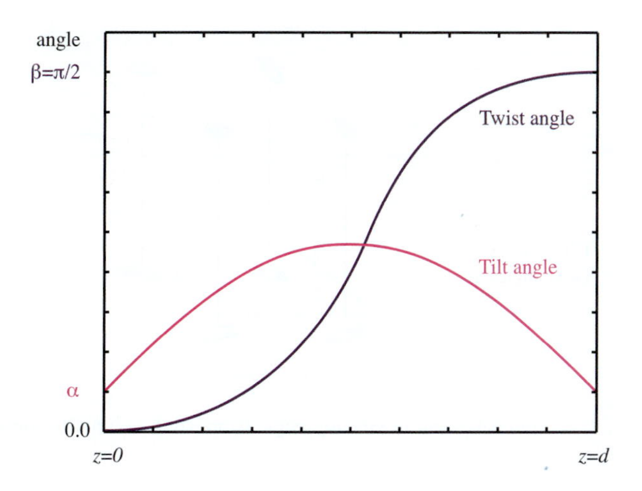

Figure 14.18. Twist and tilt angles in the on-state of a twisted nematic cell (after [1270]).

tilt angles along the z-axis will vary non-linearly, as shown in Figure 14.18. A computational model of these phenomena is discussed by Yamauchi [1270].

14.2.8 Multi-Domain and In-Plane Switching Cells

Improvements in viewing angle for different LCD types can be achieved with various approaches, including the use of lenslet arrays [547], in-plane switching (which keeps the tilt angle constant throughout the liquid crystal medium) [562], vertical alignment [481], retardatio-film-compensated twisted nematic LCDs, optically compensated bending (OCB), and multi-domain twisted nematic LCDs [375]. In this section, we briefly discuss multi-domain and in-plane switching solutions.

In the preceding sections we have omitted the analysis for off-axis viewing of twisted nematic cells. However, it can be shown that under such circumstances a significant fall-off in transmittance can be observed. Further, contrast reversals may occur. One of the reasons is that the tilt angle is non-zero, meaning that certain viewing angles coincide with the direction of the director, so that transmittance is governed by a different index of refraction than other viewing angles. Twisted nematic displays are normally oriented such that viewing the display from below yields the most artifacts, including contrast inversion.

To minimize such viewing angle-dependent fall-off, it is therefore desirable to construct displays such that the viewing angle and the director are either never aligned, or such that the angle between view vector and director has a uniform distribution over the surface area of the pixel.

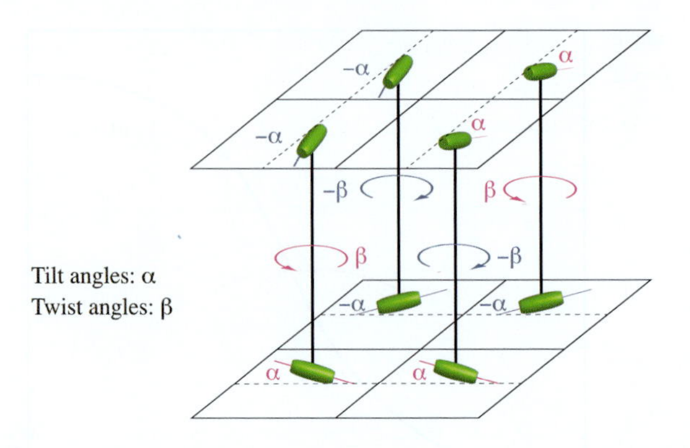

Figure 14.19. An improvement in viewing angle is realized with multi-domain cells.

The latter approach is taken in multi-domain techniques, where each pixel is subdivided into preferably four domains. A pair of sub-domains has identical tilt angles, whereas the other pair has opposite tilt angles. Further, the tilt angle progresses pairwise through the cell in opposite directions. This scheme is visualized in Figure 14.19. It is also possible to construct multi-domain pixels with only two domains, or with more than four domains. However, fewer than four domains results in lower efficiency, whereas more than four domains does not lead to a noticeable improvement in viewing angle [706].

Finally, in-plane switching techniques are constructed with both electrodes on one side of the cell, rather than on opposite sides, as shown in Figure 14.20. The

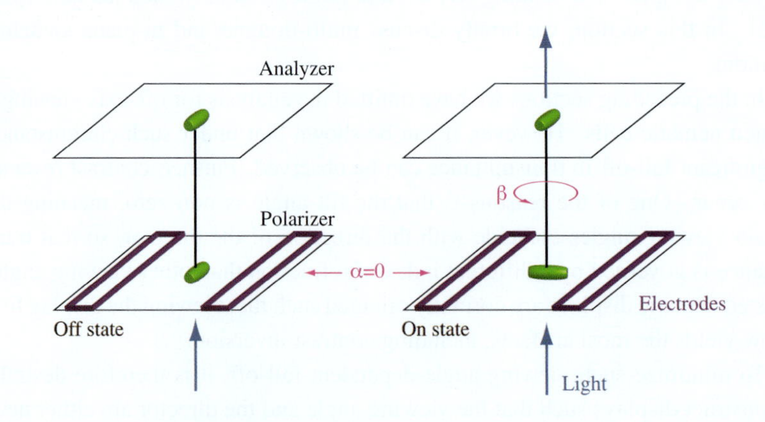

Figure 14.20. In-plane switching cells are constructed with both electrodes on one side of the cell, so that the tilt angle remains constant, irrespective of the applied voltage.

advantage of this configuration is that the tilt angle is constant and close to zero, independent of the voltage applied to the electrodes. As a result, the transmittance T varies little with viewing angle. It is given by

$$T = \frac{\sin(2\beta)\sin^2(\gamma)}{2}, \tag{14.22}$$

where β is the twist angle, and γ is given by

$$\gamma = \frac{2\pi \Delta n d}{\lambda}. \tag{14.23}$$

The maximum transmittance of 0.5 is reached when $\beta = \pi/4$ and $\gamma = \pi/2$, whereas for $\beta = 0$ the transmittance is 0.

While in-plane switching solves the attenuation of transmittance with viewing angle, there are two problems associated with this technique. First, the torque required to rotate the liquid crystal molecules is larger than with other techniques, and this leads to a longer response time. Further, the placement of the electrodes reduces the aperture of the system, resulting in less light being transmitted.

14.2.9 Active Matrix Addressing

To modulate the transmittance of a single cell, we have seen in Figure 14.11 that a voltage must be applied across the liquid crystal layer. In a twisted nematic LCD display, a large array of pixels should be individually addressed to form an image. Although different addressing schemes exist, we focus on active matrix addressing.

A pair of pixels with associated addressing lines is shown in Figure 14.21. The addressing circuitry is fabricated on the glass substrate nearest the back-light

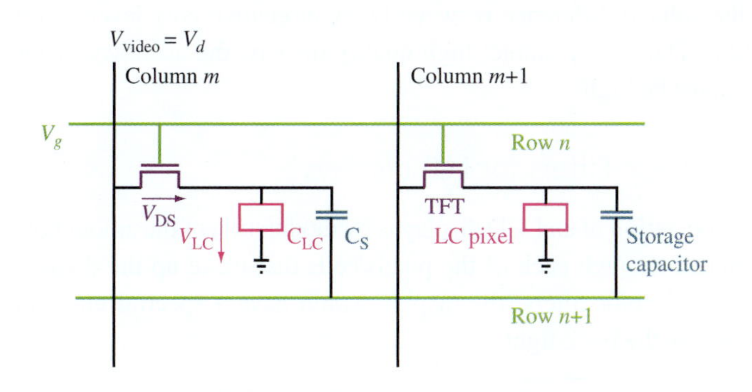

Figure 14.21. Two pixels of an active matrix display.

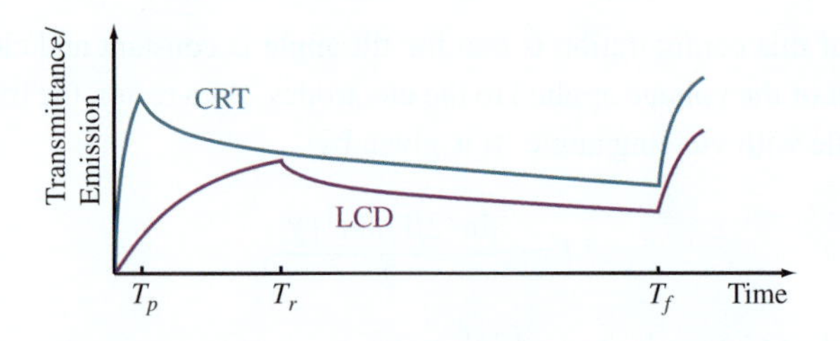

Figure 14.22. A comparison of LCD and CRT addressing times (after [84]). Note that the emission time for a CRT dot is $T_p \approx 50\ \mu$s, whereas an LCD pixel transmits light for $T_f \approx 16$ ms.

using thin film transistor (TFT) technology. Pixels are addressed one row at a time by raising the control voltage V_g for the row (n in Figure 14.21). The transistors for the pixels in this row are now conductive.

The image signal voltage V_d applied to each of the columns then charges the liquid crystal cells C_{LC} for the pixels in row n only. The auxiliary capacitor C_S is also charged, which helps keep the pixel in its state during the time when all other rows are addressed. As a consequence of charge storage in the capacitance of each LCD pixel, the pixel modulates light until it is next addressed (unlike a CRT, which emits light at each phosphor dot only for a brief instant).

The time to address a full frame, i.e. N rows, is T_f. As a result, the time available for addressing one row of pixels is $T_r = T_f/N$. A TFT-addressed twisted nematic display reaches the desired voltage over the liquid crystal cell, and therefore the desired transmittance, more slowly than in the per-pixel addressing of a CRT display. The relative timing of the two technologies is shown in Figure 14.22.

In Figure 14.11 the transmittance of the liquid crystal cell against driving voltage is shown. The relationship between the two is more or less linear over a voltage swing of around 2 V. To construct a display that can handle 256 gray levels, the voltage difference between two consecutive gray levels is around 7.8 mV [706]. Thus, for a stable, high quality display, the accuracy of the driving voltages must be high.

14.2.10 Color Filters for LCD displays

The basic operation of an LCD display is by modulation of the amount of light that is transmitted through each of the pixels/cells that make up the display[2]. Without the use of color filters, the display would have a spectral emission mostly determined by the back-light.

[2]We limit our discussion to transmissive displays. However, it should be noted that LCD displays can also be constructed reflective or a combination of transmissive and reflective.

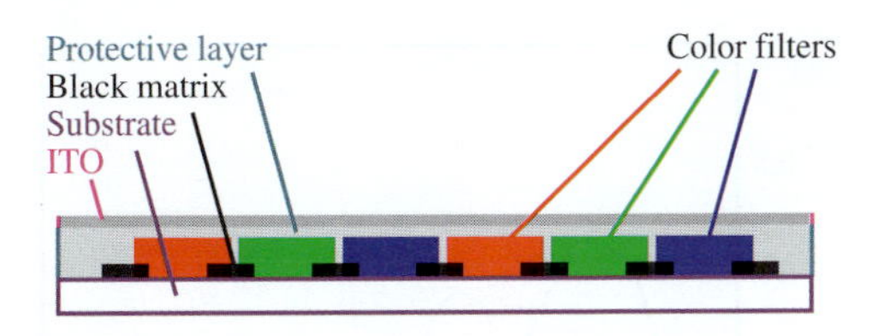

Figure 14.23. Cross-section of a color filter layer for LCD displays (after [706]).

To construct a color display, a layer of additive color filters is added to the substrate. A black matrix may be added as well, to help minimize light leakage between pixels, thereby improving the reproduction of dark tones. A cross-section of the color filter layer is shown in Figure 14.23.

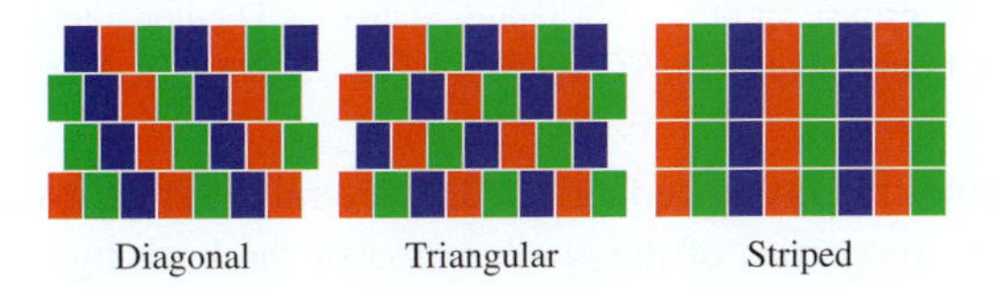

Figure 14.24. Spatial layout schemes for color filters used in LCD display devices (after [706]).

The spatial layout of pixels depends on the application. For moving images, a triangular or diagonal arrangement is preferred, whereas for still images (i.e., computer displays), a striped configuration is better. These layouts are shown in Figure 14.24. A photograph of an LCD display is presented in Figure 14.25 and shows a striped pattern.

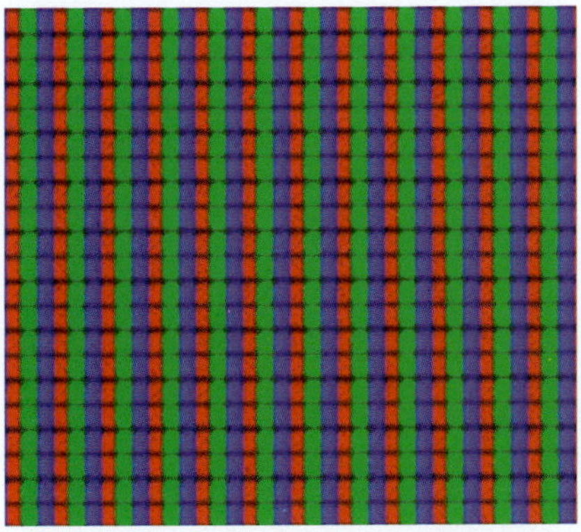

Figure 14.25. A close-up photograph of a Toshiba LCD television.

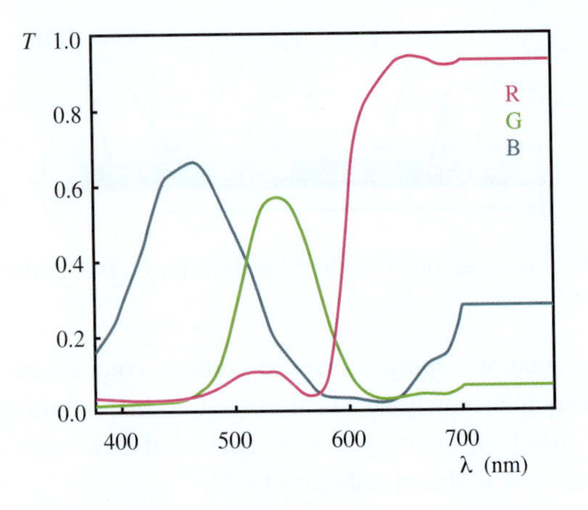

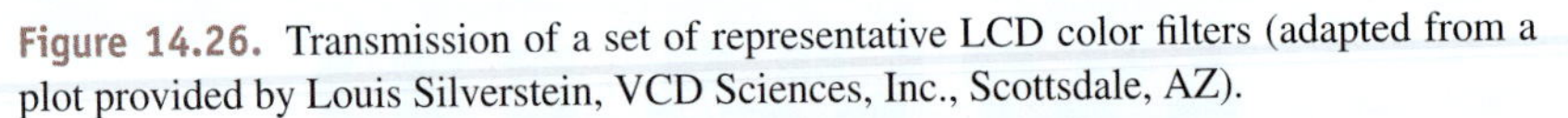

Figure 14.26. Transmission of a set of representative LCD color filters (adapted from a plot provided by Louis Silverstein, VCD Sciences, Inc., Scottsdale, AZ).

Additive color filters can be fabricated in several different ways. Regardless, the colorimetric properties of the combination of the back-light and the filters are chosen to conform to the display standard (e.g., sRGB). An example of a set of spectral transmissions are shown in Figure 14.26. Further, the loss of light through the filter should be minimized. The transmittance should be uniform across the surface of the filter. Further requirements for color filters are related to the manufacturing process, and include for instance thermal stability, chemical resistance, reliability and light stability [855].

Color filters can be manufactured by dyeing, pigment dispersion and printing. Pigment dispersion is now most commonly used [706]. The resulting color gamuts for typical instances of these techniques are shown in Figure 14.27.

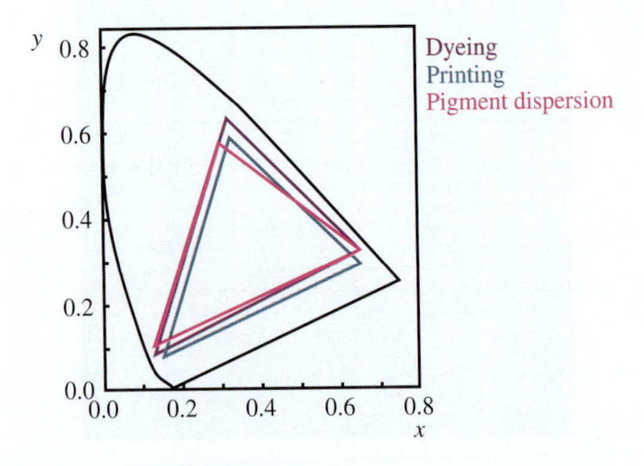

Figure 14.27. Color gamuts for different color filter technologies (after [855]).

14.2.11 Electronically Controlled Birefringence

A different approach to generating color requires a mixed-mode twisted nematic display. In such a display, there is control over both the tilt angle α and the twist angle β, whereas in a conventional twisted nematic display, applying a voltage only changes the twist angle.

If the tilt angle is changed, then light incident upon the cell hits the liquid crystal molecules under increasing angles. The apparent index of refraction is then no longer $n_{\parallel}$, but is dependent on the tilt angle and, indirectly, on the voltage V_{LC} applied to the cell. At the same time, this means that $\Delta n = n_{\parallel} - n_{\perp}$ depends on the applied voltage. In a mixed mode twisted nematic display the reduced irradiance is given by [706]

$$\frac{E_e}{\left\langle E_0^2 \right\rangle_T} = \cos^2\left(\beta\sqrt{1 + \frac{\pi\, d\, \Delta n}{\lambda}}\right). \tag{14.24}$$

The first maximum of this function occurs for $\pi\, d\, \Delta n / \lambda = \sqrt{3}$. As a result, the maximum energy transmitted by the cell occurs for the wavelength λ,

$$\lambda = \frac{\pi\, d\, \Delta n}{\sqrt{3}}. \tag{14.25}$$

By controlling the tilt angle, and thereby Δn, the wavelength of maximum transmittance can be controlled. This phenomenon is called *electronically controlled birefringence* (ECB) and enables the formation of color by shifting the peak wavelength of light transmitted through the cell.

14.3 Transflective Liquid Crystal Displays

A variation of LCD displays can be constructed such that they can be viewed by applying a back-light as well as in reflective mode, i.e., lit from the front by whatever ambient light is available. This means that the back-light is only used in dark environments, which leads to a very energy efficient display that finds application for instance in mobile devices.

In reflective mode, light incident upon the front of the display passes through the liquid crystal layer, is reflected by a diffuse reflector, and then passes through the liquid crystal layer once more before exiting the display. In transmissive mode, light from the back-light passes through the display as discussed in the preceding section. Each pixel is subdivided into two areas, with one serving transmissive mode and the other reflective mode.

In dual cell gap types, the gap between the front and back substrates is halved for the pixel fragment serving reflective mode to account for the fact that light passes through this section twice. The areas of the pixel devoted to the transmissive and reflective modes are given as a ratio, which can be varied dependent on the application. It typically ranges from 1:4 to 4:1 [1294]. The response time in reflective mode will be much faster than for transmissive mode for a dual cell gap type transflective display, as the time to set a pixel depends on the square of the thickness of the liquid crystal layer (i.e., the cell gap). Typically, the response time is four times slower for the transmissive mode of operation.

Single cell gap transflective liquid crystal displays have liquid crystal layers with the same thickness for both transmissive and reflective layers. This makes manufacturing more straightforward and creates a uniform response time for both modes of operation. However, the light efficiency is now lower.

14.4 Plasma Display Panels (PDPs)

Plasma display panels [99, 1280] are based upon *gas discharge* (see also Section 3.5.3). Gas discharge is a sequence of reactions that occur when a voltage is applied through a tube or a cell that contains certain chemically inert gases. As a result of this potential difference, electrons accelerate, and collide with neutral gas atoms. These collisions may ionize the gas, i.e., remove electrons from it, if electrons impart sufficient energy to the gas atoms.

The excited atom remains excited for a short duration (around 10^{-8} s), after which it immediately returns to the ground state. This causes the emission of a photon with a wavelength depending on the energy difference between the two orbitals, which, in turn, depends on the mixture of the inert gas used in the reaction (see Section 3.5.3).

Early monochrome plasma displays used a mixture composed of neon to which a small percentage of argon (about 0.1%) was added [1120]. This mixture caused the display to emit an orange color. One alternative to obtain other colors is to change the composition of the gas mixture, although other inert gas mixtures tend to be luminously less efficient than those using neon as the main component [17].

Modern plasma display panels use UV-emitting plasma and introduce phosphors into the display. In this case, a gas mixture composed of helium and xenon ($\approx$ 5%–10%) is used [1066]. When excited by electrons, this mixture emits ultraviolet light with a sharp peak at 147 nm and a broader emission spectrum centered around 172 nm. This light, in turn, excites the phosphors causing glow in the vis-

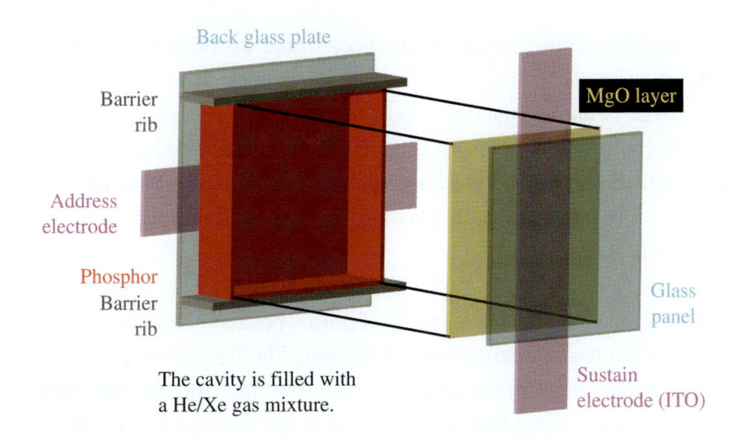

Figure 14.28. An exploded view of a red pixel of a plasma display panel.

ible region of the spectrum. If three phosphors with peaks around the red, blue, and green parts of the spectrum are used, their combination yields a sufficiently large color gamut.

Figure 14.28 shows the structure of a pixel in a plasma display panel. Here, the phosphor-coated cavities contain the gas mixture. The barrier ribs minimize light leakage between pixels; differently shaped barriers have been developed for improved efficiency [999, 1276]. The sustain electrodes are made from transparent indium tin oxide (ITO) to enable as much light to escape the cavity as possible [276]. The phosphors are at the back side of the display. The MgO layer protects the sustain electrodes and the dielectric layer into which these electrodes are embedded [1160]. Due to the path taken by the light, this configuration is known as the *reflected phosphor geometry* [1159].

The screen of a plasma display is composed of a large array of pixels laid out in a matrix fashion. Each pixel is sandwiched between row and column electrodes and can be turned on by applying a voltage to the corresponding row and column. The display can be driven by using alternating or direct current [1120]. Pixel pitch is typically around 800–1000 μm^2. Plasma displays are predominantly built in larger frames (up to a 102-inch diagonal with 1920 $\times$ 1080 pixels. PDPs typically have peak luminance values of 300 cd/m^2 with contrast ratios of 300:1, although displays having a maximum luminance of 1000 cd/m^2 and a contrast ratio of 2000:1 have been demonstrated [276]).

Plasma displays are typically power-limited: not all cells can produce maximum luminance simultaneously. Typical video material has an average luminance of 20%; this average luminance can be distributed arbitrarily across the panel.

However, beyond some average luminance, signal processing circuits attenuate the RGB drive signals so as to limit average luminance[3].

If discharge is taking place in a cell, dropping the sustain voltage will extinguish the discharge. The cell will remain off until addressed and ignited. Once ignited, it will remain in the discharge state until the sustain voltage is dropped again. Each pixel in a plasma display is thereby electrically bi-stable, having in effect one bit of memory.

The intensity modulation of pixels in plasma displays is achieved by pulse-width modulation (PWM). The number of pulses per second is far in excess of the temporal resolution of the human visual system. The fraction of time that the pixels are on therefore determines which average light intensity is reproduced [258]. The phosphors used in a plasma display panel are comparable to those used in CRTs, so the reproduction of color in PDPs approximates that of CRTs.

The main disadvantage of plasma displays is their low energy efficiency, which is less than 1% [276]. The reasons for this low efficiency are the low conversion from power to ultra-violet radiation (around 10% efficiency), as well as an inefficient conversion from ultra-violet radiation to visible light by means of phosphorescence (around 20%–40% efficiency). As a result, the luminous efficacy is less than 2 lm/W. The display industry is currently working to improve this number to around 5 lm/W which is what CRTs achieve [276].

One of the disadvantages of PDP technology, as with all phosphor-based displays, is that the phosphors themselves degrade with use. If, for instance, a logo or a menu-bar is displayed for prolonged periods of time, then this results in an after-image that can not be removed. This phenomenon is called *burn-in*.

14.5 Light-Emitting Diode (LED) Displays

A light-emitting diode (LED) is based upon a semiconductor containing a p-n junction as shown in Figure 14.29 (see also Section 3.5.7). The p-type region has an excess amount of holes and the n-type has an excess amount of electrons. Excess electrons and holes are introduced into the semiconductor by doping it with other atoms that readily give out or accept electrons. For instance, if the semiconductor is made of silicon (Si), which is from the IVA group of the periodic table, doping it with a IIIA atom makes it a p-type (acceptor) and a VA atom makes it an n-type (donor).

Under forward bias, electrons and holes diffuse toward the junction where they may recombine. The recombination may be radiative (light is emitted), or

[3]When operated in the luminance-limiting regime, PDPs behave similarly to RGB+W DLP® projectors.

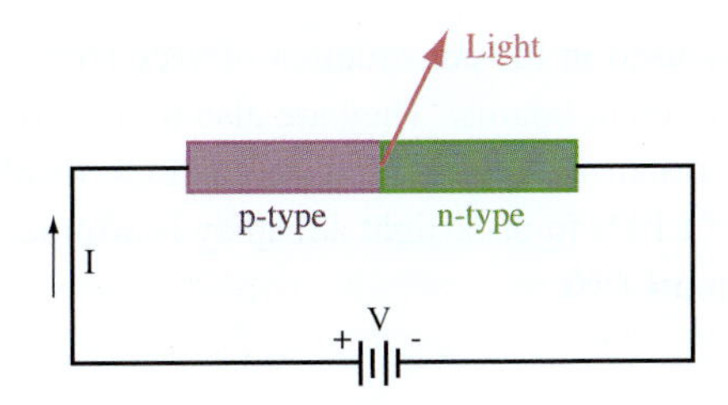

Figure 14.29. An LED is a simple semiconductor containing a p-n junction that is driven with a forward bias. Light is emitted from the junction through electron-hole recombination.[4]

non-radiative (heat is produced) (see Figure 14.30). If radiative recombination occurs, the wavelength of the emitted light depends on the band-gap energy between the conductance and valence bands of the semiconductor material and the impurities used. Therefore, by using different materials, different colors of light can be obtained.

The operation of an LED involves *electroluminescence*, whereby a material emits light in response to an electrical current. In this sense, an LED is the opposite of a solar cell which produces electrical current in response to light.

The most common types of semiconductors used in LEDs are gallium arsenide (GaAs) or its related ternary compounds such as GaAsP and GaAlAs. Blue LEDs are typically based upon gallium nitride (GaN) [918, 1120].

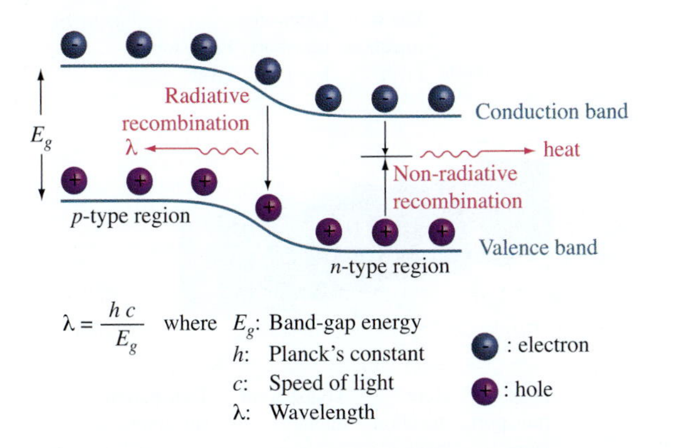

Figure 14.30. An electron and a hole may recombine radiatively or non-radiatively. Radiative recombination results in light whose wavelength depends on the band-gap energy of the semiconductor. Non-radiative recombination produces heat.

[4]In this diagram, the electrons flow form the negative post of the power supply to the positive post, which is opposite to the direction of conventional current, which is drawn here.

LEDs are usually used in the construction of very large scale displays, such as interactive advertisement boards. They are also used as back-lights for liquid crystal displays. For instance, Apple's iPod uses an LCD display with LED back-light. Another use of LEDs to back-light a display is afforded by HDR displays, as discussed in Section 14.16.

14.6 Organic Light-Emitting Diode Displays

An *organic light-emitting diode* (OLED) display is a relatively recent but highly promising display technology, which produces visible light when an electrical current is passed through it. Light emission from an OLED is similar to inorganic LEDs in that photons are produced when a hole-electron pair radiatively recombines. However, in an OLED radiative recombination takes places in an organic layer rather than in an inorganic p-n junction.[5]

The interest in OLEDs dates back to the mid-1960s [451, 452], although the breakthrough in OLED technology came after Tang and van Slyke [1117], recognized that the efficiency of an OLED can be greatly increased by using a bi-layered structure where each layer is responsible for either electron or hole transport and injection (Figure 14.31). A multi-layer structure allows each layer to be optimized separately for different tasks.

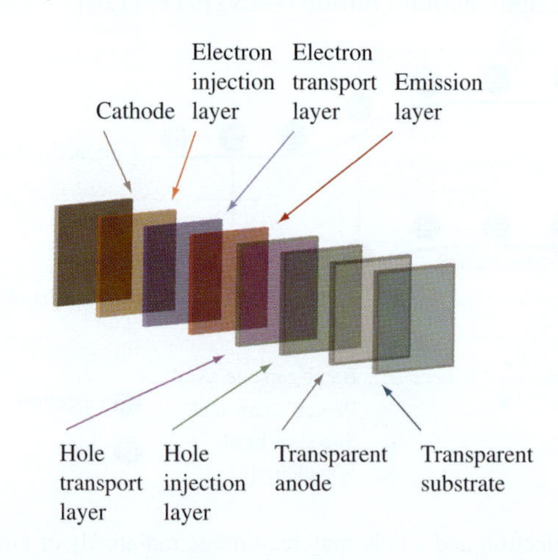

Figure 14.31. An OLED is composed of a stack of layers sandwiched between a cathode and an anode. Each layer can be optimized separately.

[5]Organic materials are those primarily made of compounds of carbon, hydrogen, oxygen, and nitrogen.

In an OLED, electrons and holes injected from the cathode and anode diffuse toward the emissive layer under an electrical field, where they form electron-hole pairs called *excitons* [1275]. An exciton may radiatively recombine resulting in emission of a photon whose wavelength depends on the band energy of the organic material in the emissive layer [1168]. To increase the emission efficiency, the emissive layer may be doped with other organic materials [1040].

OLEDs can be made of small organic molecules or larger conjugated polymers [143]. In the former type, organic layers are prepared by depositing each layer on a transparent substrate by thermal evaporation. As polymers are too large to evaporate and may disintegrate upon excessive heat, *polymer-based OLEDs* (PLEDs) are usually created by spin-coating the polymer liquid onto indium tin oxide coated substrates which are solidified by heating [328]. The fabrication process of PLEDs is therefore simpler than that for OLEDs, although the manufacturing of the top electrodes for PLEDs is more complicated. In addition, the production of PLEDs is more wasteful of materials [924].

OLED displays are usually fabricated on a glass substrate, but they can also be produced on flexible plastic [400, 1103]. In addition, OLED displays can be transparent when in the off-state, making them useful for augmented reality applications and heads-up displays [399]. In addition, transparency is necessary in

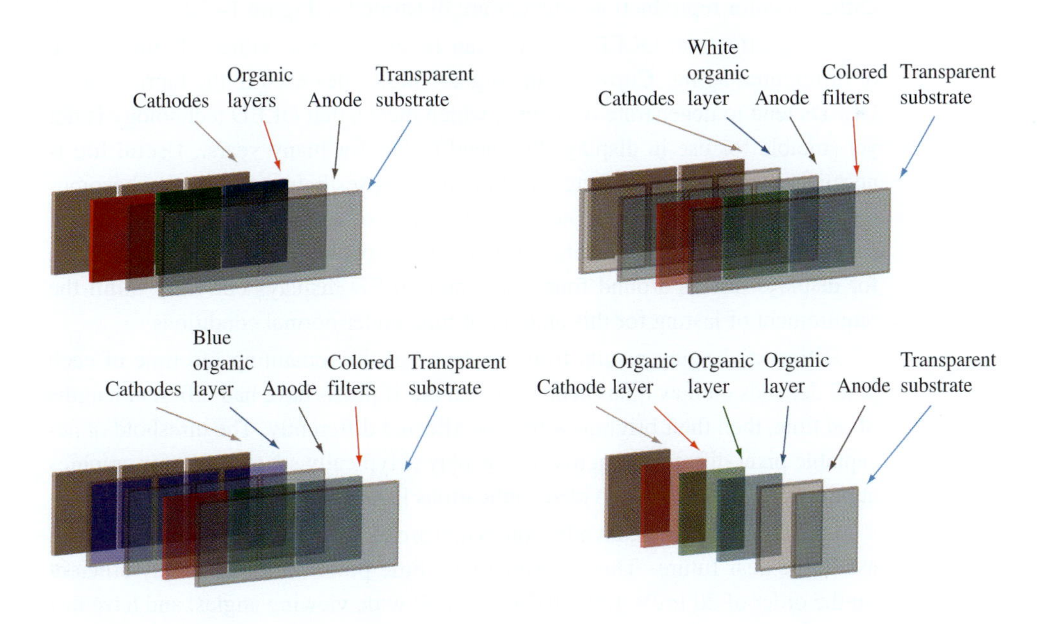

Figure 14.32. Various methods to obtain a range of colors using OLEDs.

as shown at the bottom of Figure 14.32 [401]. In either case, the anodes will also have to be transparent [740].

A full-color OLED display device can be realized using a matrix layout similar to that of an LCD display. Different from an LCD display, OLEDs themselves emit light and, therefore, they do not require a back-light. An OLED display can use passive or active matrix addressing. In a passive configuration, the intensity emitted for each pixel needs to be very high and depends on the number of scanlines in the display—more scanlines means that each flash will have to be correspondingly brighter due to the fact that only one line is active at any given time. If there are N scanlines, then the intensity of each pixel will be N times higher than the average perceived intensity [1026]. This places an upper limit on the size of the display that can be addressed in passive matrix addressing.

Full-color can be produced by subdividing each pixel into three OLEDs emitting red, green, and blue colors. Alternatively, an OLED emitting white light can be filtered by three color filters, or an OLED emitting blue or ultraviolet light may be used to illuminate three fluorescent materials which re-emit light with the desired spectral composition [328]. One final method is to stack up OLEDs on top of each other by separating them with transparent films. This allows for a higher spatial resolution in the same display area and is, therefore, a favorable technique. Different color reproduction schemes are illustrated in Figure 14.32.

Aging effects in OLED displays can be classified as either life-time aging or differential aging. Currently, the organic molecules used in the fabrication of OLEDs tend to degenerate over time, which means that OLED technology is not yet suitable for use in displays that need to last for many years. Useful life is normally measured as the length of time that elapses before the display becomes half as bright as it was when new. The life span of an OLED micro-display also depends on the luminance value at which it is operated. The product life cycle for displays hovers around four years, and OLED displays currently fulfill the requirement of lasting for this amount of time under normal conditions.

Differential aging results from the fact that the remaining life time of each pixel depends on how it has been switched on. If pixels have had different lengths of on time, then their brightness may be affected differently. The threshold of acceptable pixel differences across the display is typically set to 2–3% for graphics applications and 5–8% for video applications [924].

OLEDs offer significant advantages that are likely to result in their widespread use in the near future. They consume very little power, are luminously efficient (in the order of 20 lm/W [894, 1029]), exhibit wide viewing angles, and have fast response times (in the order of a few microseconds). OLEDs are also very thin and lightweight, and can be deposited on flexible surfaces. This opens up the eventual possibility for display devices that can be rolled-up or stuck onto curved surfaces.

14.7 Field Emission Displays

Field emission displays (FEDs) work on the same principle as cathode ray tubes in that a cathode generates electrons in a vacuum [1114]. The electrons are accelerated until they hit a phosphorescent layer which, as a result, produces visible light. Unlike a CRT, however, a field emission display incorporates one or more cathodes per pixel which are constructed of a metal tip or more recently of carbon nanotubes [346, 782] or diamond-like carbon [535, 682]. Different from a CRT, electrons are generated by field emission, for which very thin, sharply pointed electrodes are required.

Field emission displays can be made with an extremely shallow depth (around 2 mm). The power consumption is projected to be lower than that of LCDs and CRTs [535]. The behavior of a FED is otherwise similar to a CRT, including its advantage of a wide viewing angle (around 170°).

The cathodes of developmental FED devices were made of molybdenum, which had several disadvantages. The most important of these is that the energy required to produce light may create enough heat to melt the cathodes. As a result, the pointed tips would be deformed, reducing the field effect and thereby the device's efficiency. In addition the cathodes may react with the residual gases within the vacuum, which would further reduce the efficiency.

As a result, the traditional FED devices have never become a commercial success. However, recent developments in nanotechnology has enabled the de-

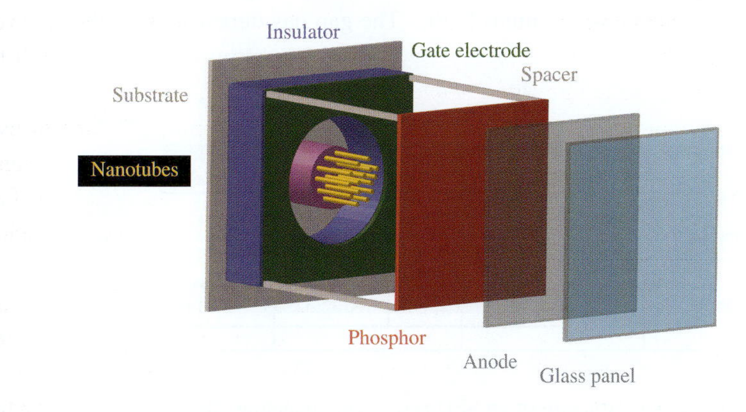

Figure 14.33. The components of a FED pixel.

velopment of nanotubes, which are made of carbon atoms configured in covalent bonds (see Section 3.3).

In a FED no beam deflectors are required to steer the electron beam in scan-line order across the display. Each pixel produces its own electron beam. A diagram showing the components of a single pixel is shown in Figure 14.33. The chemically stable nature of these materials has solved the problem of overheating, melting, and reacting with residual gases. As a result, there exists renewed interest in FED devices as a potential technology that may be incorporated in display devices.

14.8 Surface-Conduction Electron-Emitter Displays

Surface-conduction electron-emitter display (SED) technology, in development by Canon, is based on the same principle as the CRT and the FED: photons are emitted when phosphors are bombarded by electrons. However, different from a typical CRT and similar to FEDs, each red, green, and blue component of each SED pixel contains its own miniature electron gun.

Reserving a separate electron gun per pixel component obviates the need to have a beam deflection system typical of standard CRTs. SEDs can therefore be constructed as flat panels. However, the construction of each pixel in a SED is different from a FED as electrons are produced by a tunneling mechanism. As a consequence, device depths of a few centimeters can be achieved [276].

Light emission from an SED pixel is illustrated in Figure 14.34. When an electrical voltage of 16–18 volts is applied at V_f, electrons jump between electron emitters owing to the tunneling effect (i.e., passage of electrons through an insulator—in this case vacuum) [770]. The gap has dimensions on the order of

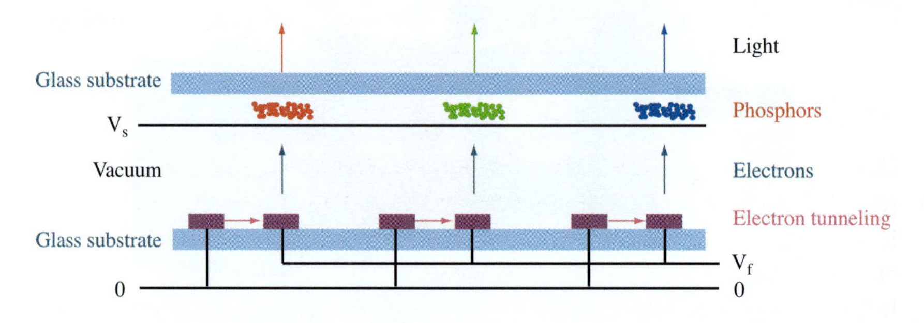

Figure 14.34. The structure of an SED pixel. The tunneling effect and the attraction of electrons toward the phosphors are depicted.

several nanometers. Some of the freed electrons are accelerated toward the phosphors as a result of the large potential difference ($\approx$10 kV) applied at V_s. When these electrons strike the phosphor molecules, photons are emitted due to cathodoluminescence. The efficacy of current prototypes reaches that of conventional CRT displays and is of the order of 5 lm/W.

Since phosphors produce light into a wide angle, SED image quality does not degrade at wide viewing angles. Since the phosphor technology has been successfully used for accurate color reproduction in standard CRTs, the SED technology inherits this desired property as well. Finally, owing to its slim and ergonomic design, the SED technology may become an interesting competitor in the flat panel display market in the near future. Displays with a diagonal of 36 inches have been announced [845]. However, there remain significant challenges to high-volume manufacturing.

14.9 Microcavity Plasma Devices

Plasma devices are efficient for large displays, but not so for smaller displays. This is largely due to limitations related to the minimum size of each pixel. However, a recent development borrows manufacturing techniques from VLSI and microelectromechanical systems to produce microplasma devices. These devices consist of regularly spaced microcavities filled with a helium/xenon mixture and surrounded by electrodes. They therefore operate on the same principle as conventional plasma display devices, albeit at a much smaller scale due to the new fabrication techniques employed.

Microcavity plasma devices (or microplasma devices for short) are still at an early stage of development, but they promise very high packing densities (of more than 10^4 pixels per square centimeter), peak luminances of over 3000 $\mathrm{cd/m^2}$, and luminous efficacies of more than 6 $\mathrm{lm/W}$ [276].

14.10 Interferometric Modulator (IMOD) Displays

IMOD is an emerging reflective display technology that utilizes interference to reproduce color. As discussed in Section 2.7.1, interference causes certain wavelengths to cancel out if they are out of phase (i.e., destructive interference) and other wavelengths to add if they are in phase (i.e., constructive interference). An IMOD display utilizes both effects to reproduce either a black or a colored pixel. Color generation from an IMOD display resembles the iridescence effect on some butterflies' wings—interference at the scale of visible wavelengths.

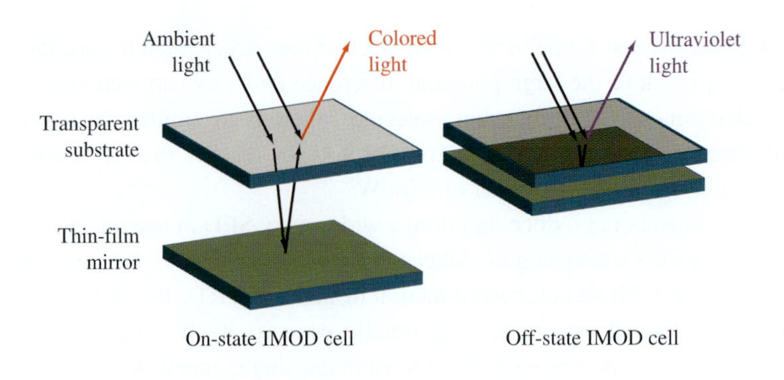

Figure 14.35. On and off states of an IMOD cell.

An IMOD pixel consists of a transparent substrate on top of a mechanically deformable thin-film mirror; these are separated by a small cavity [324, 1195]. The height of this cavity is in the order of a few hundred nanometers to allow interference of light in the visible range. Ambient light that hits the transparent surface is both reflected off the surface and transmitted toward the thin film mirror. The portion of the light that is transmitted toward the mirror goes through several inter-reflections between the mirror and the substrate. At each inter-reflection some portion of light exits the cavity and interferes with the light that was initially reflected off the surface. The height of the cavity determines which wavelengths interfere destructively, hence cancel each other out, and which wavelengths interfere constructively, according to the principle shown in Figure 2.31. When the cavity is collapsed by applying a voltage, constructive interference occurs only in the ultraviolet range, which is outside the visible spectrum, and therefore results in a black cell (Figure 14.35). By using cavities of different heights various primary colors can be obtained, as illustrated in Figure 14.36.

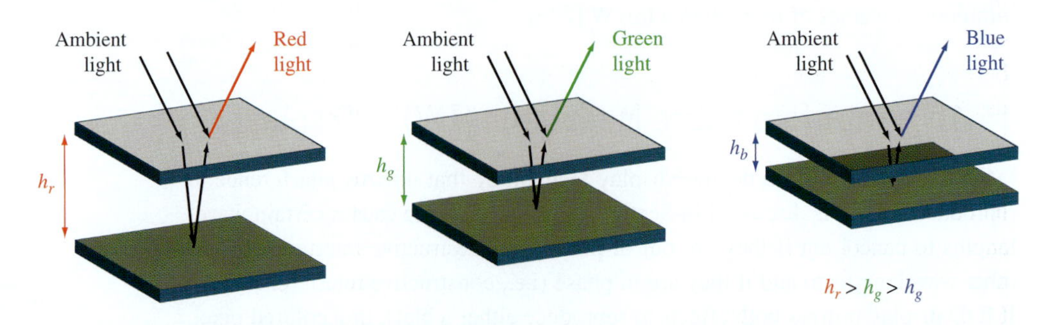

Figure 14.36. Three primary colors can be obtained by constructing cavities of different heights.

To construct a full color IMOD pixel, three types of subpixels are used with each type having a different gap size. To reproduce various shades of color, each subpixel type is further divided into sub-subpixels, or subpixel elements. The desired intensity for each primary color can then be obtained by spatial or temporal dithering.

The chief advantages of IMOD displays over other flat panels include a significantly reduced power consumption and the ability to reproduce bright colors under direct sunlight. These factors may render IMOD technology an attractive alternative particularly for hand-held applications. In dark environments, supplemental illumination is provided by a low power front light. At the time of writing, IMOD displays have a reflectivity of around 50% and a contrast ratio of 8 : 1. This compares favorably with typical newspaper which has approximately 60% reflectivity and a 4 : 1 contrast ratio [527].

14.11 Projection Displays

In projection displays, the final image is not displayed on the light modulator itself but is projected on a screen internal or external to the end-user display device. This enables a large image to be produced by a relatively small light modulating device, often leading to low manufacturing costs.

Projection displays may be distinguished as *rear* and *front* projection displays. A rear projector uses a transmissive screen that must have dispersive characteristics in order to scatter the incident light into a wide viewing angle. A rear projection display system is typically a single unit comprised of the projector, the screen, and an optical system in between that ensures a sharp image on the screen.

In a front projection system, also called *beamer*, the projector is placed in front of a diffusely reflective screen which scatters light in a wide viewing angle. The distance between the projector and the screen is usually much larger in comparison to a rear projection system. In a front projection system, the coupling between the screen and the projector is minimal, that is, the same projector can be used with different screens placed at different distances within the effective focal length of the projector.

Several possible types of lamp are available for use in projection displays, including metal halide and ultra-high performance (UHP) varieties. Metal-halide additives improve the color properties of gas-discharge lamps, but at the same time reduce the luminance output, making them less ideal for projection displays. Ultra-high performance (UHP) lamps are nowadays the norm, due to their high light output combined with a small physical size [249]. This has enabled the size of projection systems to become significantly smaller.

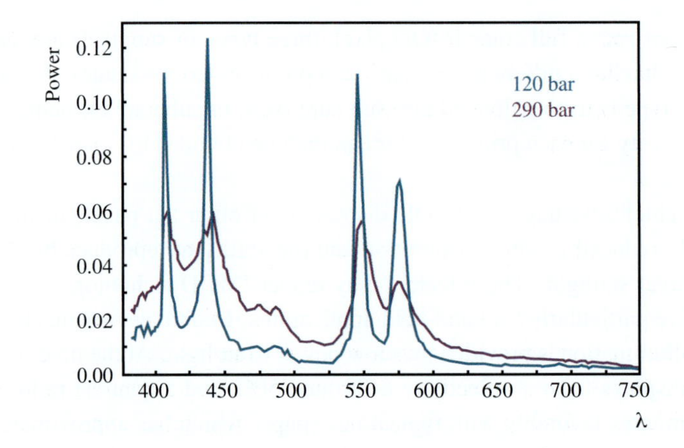

Figure 14.37. The emission spectra of an UHP lamp at different pressures (after [249]).

An UHP lamp relies on mercury discharge only: no rare earth gases are used. This gives high brightness and high luminous efficiency [249]. The emission spectra of UHP lamps depends on the pressure realized within the bulb. At higher pressures, the spectrum becomes flatter, giving a better color-balanced output with a higher color rendering index. The emission spectra of an UHP lamp at different pressures is plotted in Figure 14.37.

14.11.1 Pulse-Width Modulation

Some of the projection display types discussed below can switch each pixel into either an on-state or an off-state. This includes DLP projectors and LCoS devices (Section 14.13 and 14.14). To create intermediate levels, the switching between on- and off-states is achieved at rates much higher than the frame rate. Thus, to achieve gradation, the bit-planes of a pixel value are considered in turn. The least significant bit is loaded first and illuminated for a brief period of time, namely a fraction of $1/(2^n - 1)$ of the frame time, assuming that n is the number of bits per color. Then the second bit is loaded and illuminated for twice the amount of time. This is repeated until all bit-planes have been processed. The fraction of a frame t_i that the ith bit is processed is therefore

$$t_i = \frac{2^i}{2^n - 1}, \qquad 0 \leq i < n, \tag{14.26}$$

whereby each t_i constitutes the duration of one of the n sub-frames. This approach is feasible due to the very short switching times that are possible, combined with the comparatively slow response of the human visual system which effectively integrates the flashes of light over periods exceeding the frame time.

In the following sections, an overview of the most commonly used projector technologies is given.

14.12 Liquid Crystal Display (LCD) Projectors

An LCD projector contains an ultra-bright lamp which emits white light. The wide spectrum is separated into three relatively narrow bands with the peak of each band in the red, green, and blue regions of the spectrum. Separation is accomplished by passing the light through multiple dichroic mirrors (see Section 3.7), each of which transmits a certain band of wavelengths and reflects the remainder (Figure 14.38).

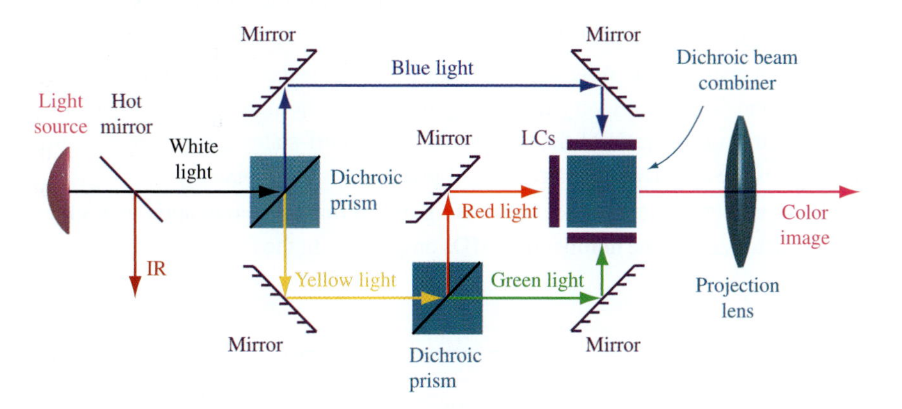

Figure 14.38. The main internal elements of an LCD projector.

The red, green, and blue beams are then directed toward individual liquid crystal panels, which modulate their intensity according to the pixel values of the frame that is to be displayed. The modulated beams are then combined by a prism and reflected toward a lens that focuses the image at a certain distance. The operating principle of an LCD projector is shown in Figure 14.38.

14.13 Digital Light Processing (DLP®) Projectors

Digital light processing® technology is based on light reflection from individually controllable microscopic mirrors. Fabricated on a digital micromirror device

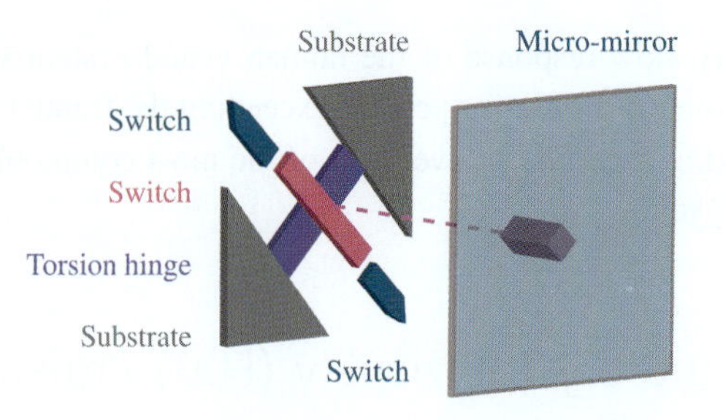

Figure 14.39. A simplified diagram of the components of a single micromirror.

(DMD), each mirror is hinged and can be tilted between $\pm 10^\circ$ to change the direction of the reflected light (see Figure 14.39). When tilted to $+10$ degrees the incident light is reflected toward a lens which projects the light onto a screen. This is called the *on-state*. If tilted by -10 degrees, the incident light is reflected toward a light absorber (the *off-state*).

The intensity of the reflected light can be controlled by alternating between the "on" and "off" states very rapidly, up to $50,000$ alternations per second. For each tiny mirror, the ratio of the number of on- and off-states across a frame time determines the effective intensity of light produced for that pixel.

DLP projectors can produce color using one of three different approaches that are characterized by the number of DMDs employed. In the first approach, using a

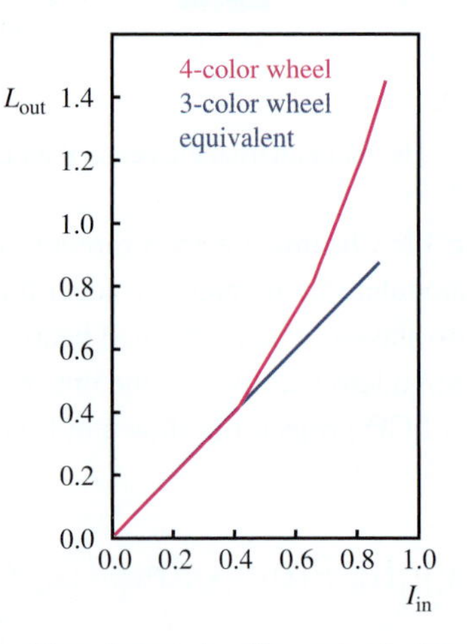

Figure 14.40. Compared with a three-color filter wheel, the extra white improves the luminance output (after [1096]).

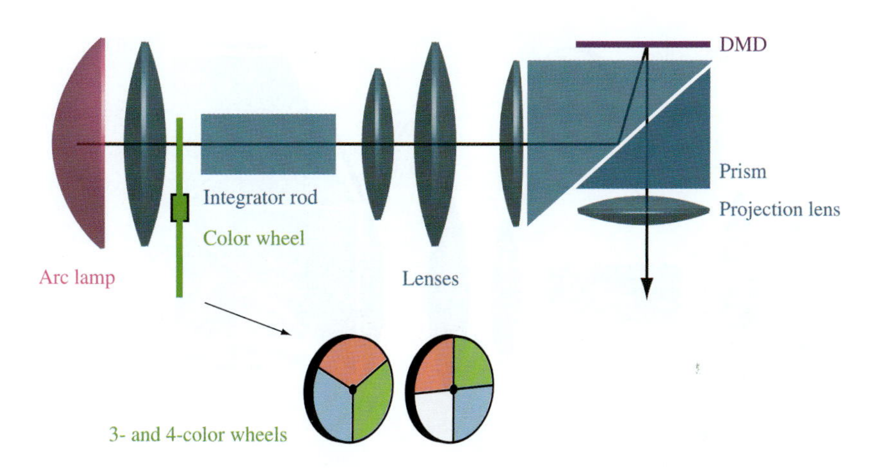

Figure 14.41. The optical components of a 1-chip DLP projector (adapted from [489]).

single DMD, a color filter wheel rotates rapidly in front of a broadband bright light source. When the blue region of the wheel is in alignment with the light source, it only allows blue light to reach the DMD. Synchronously, the micromirrors align themselves according to the blue channel of the image to be displayed. These steps are repeated rapidly for the red and green channels, resulting in a full color output. This RGB sequential 1-DMD design is generally not able to reproduce very high luminances, due to the absorption of energy by the red, green, and blue color filters.

To achieve higher luminance, the color wheel is frequently fitted with an additional white channel (see Figure 14.40), and thus this approach uses a total of four channels (RGBW) [92]. Since one DMD is used in both RGB and RGBW designs, these set-ups are also called 1-chip DLP projection systems (Figure 14.41). Their advantages include portability and low cost.

The disadvantage of using a white channel is that the color gamut has an unusual shape, in that the volume spanned by the red, green, and blue primaries, is augmented with a narrow peak along the luminance axis. This means that colors reproduced at high luminances will be desaturated. At low luminances, saturated colors can be reproduced. The lack of accuracy of RGBW systems at high luminance values makes this approach useful for business graphics, but less so for color-critical applications.

A second alternative, which has a higher light efficiency and optimizes the lifetime of the light bulb, uses a 2-chip design. Bulbs with a long lifetime may be deficient in the red end of the spectrum. To obtain a high light efficiency, red light is continuously directed towards one DMD, whereas the second DMD al-

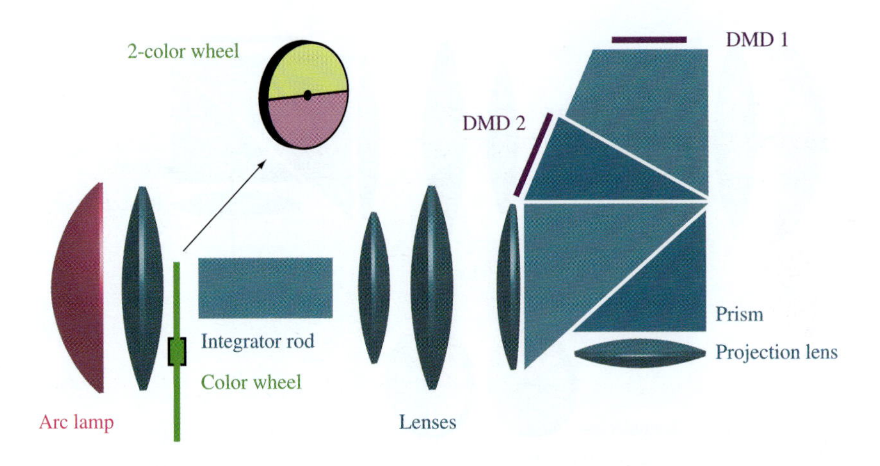

Figure 14.42. The optical components of a 2-chip DLP projector (adapted from [489]).

ternates between reflecting green and blue light. This is achieved by designing a color wheel with yellow (red + green) and magenta (red + blue) segments. The dichroic prism spatially separates the red component from alternatively the green and blue components. A 2-chip projection system is outlined in Figure 14.42

In general, 1- and 2-chip DLP devices suffer from artifacts which stem from the rotation of the color wheel and the corresponding sequential reproduction of the three color channels. If this speed is too low, then the time for a boundary between two neighboring colors of the filter to pass through the optical path is relatively long. When a projection is viewed, the viewer's eye-tracking (see Section 4.2.1) may coincide with the movement of the chromatic boundary that is passing across the image [252]. Dependent on the relative speeds of the rotation of the color wheel and the eye movement itself, this may be seen as a colored fringe. The leading and trailing edges will exhibit color fringes. These color separation artifacts are known as the *rainbow effect*.

The third alternative involves the use of three chips, with each chip controlling a different color channel (Figure 14.43). Similar to an LCD projector, a broadband light can be separated into three components using a prism, or, alternatively, red, green, and blue LEDs can be used to illuminate each chip. As this approach requires three DMDs, this alternative is usually more costly. However, 3-DMD systems do not exhibit color fringing. They are used in DLP cinema projectors.

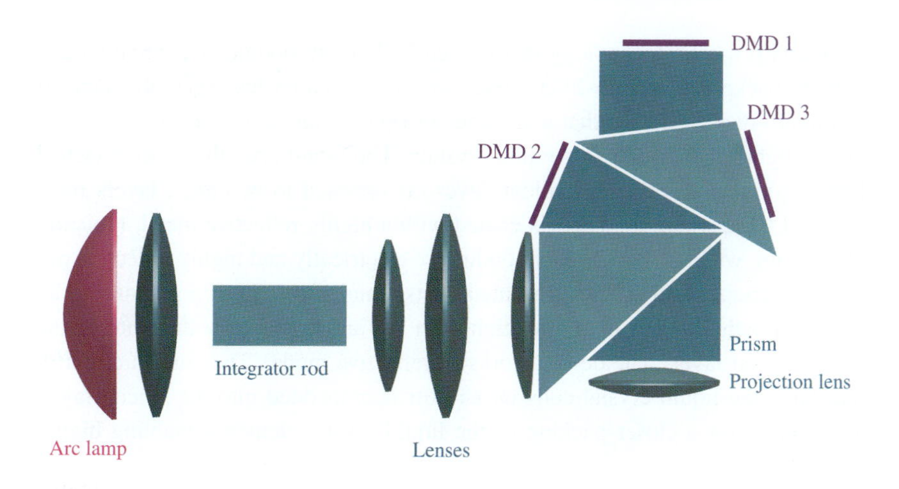

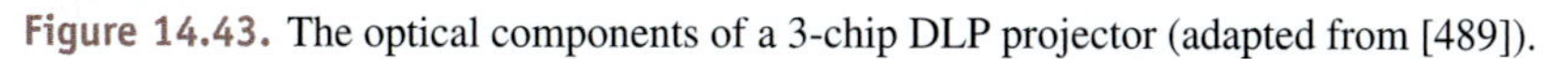
Figure 14.43. The optical components of a 3-chip DLP projector (adapted from [489]).

14.14 Liquid Crystal on Silicon (LCoS) Projectors

An LCoS projector is very similar to that of an LCD projector (Figure 14.44). However, instead of using transmissive LCD panels, an LCoS projector utilizes reflective LCoS modulators. LCoS projectors can also be configured to form color by use of a color filter wheel [1163].

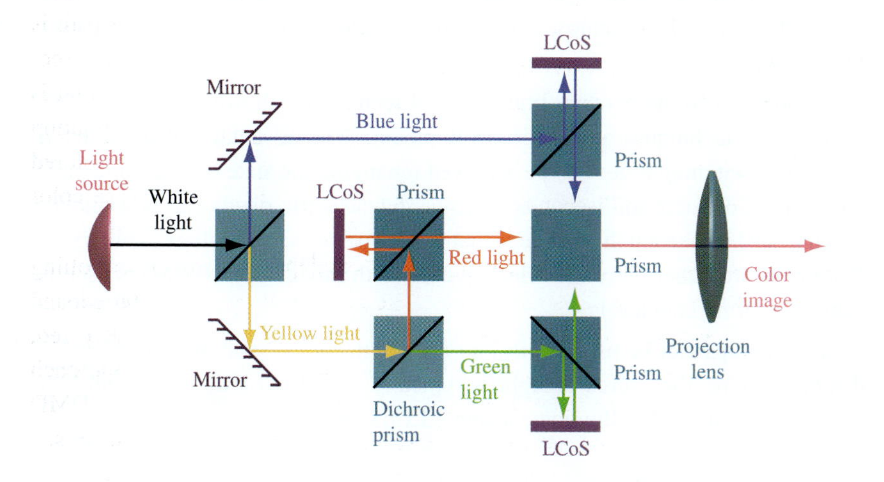

Figure 14.44. The components of an LCoS projector.

LCoS projector technology involves an LCD light modulator fabricated on a silicon "backplane" upon which active electronic circuitry has been fabricated. It is a micro-display device that is used under optical magnification. The modulation of light is achieved using liquid crystals. The liquid crystals are sandwiched between a glass layer and a silicon layer (as opposed to two glass layers for a typical LCD), where the latter is coated with a highly reflective metal, typically aluminium, which is both highly conductive electrically and highly reflective optically (Figure 14.45). Thus, the metal acts simultaneously as a mirror and an electrical cathode, and in conjunction with the anode, controls the state of the liquid crystal layer. The device works in reflective mode. The control circuitry that turns the liquid crystal cells on and off is embedded into the silicon layer. This allows for a closer packing of the liquid crystal elements enabling higher resolutions.

An LCoS chip typically does not have colored filters to achieve color reproduction, but, if one chip is used, it is sequentially flashed with red, green, and blue light. This is synchronized with altering the states of the chip, so that when illuminated with red light, the image corresponding to the red channel is loaded, and similarly for green and blue. This is called *sequential color*. The advantages of this approach are that only one third the number of pixel elements are required, and that even at high magnification, there will be no spatial separation of the red, green, and blue channels. It should be noted that spatial accuracy is traded for the potential for temporal artifacts, including eye-tracking induced color fringing.

Three-chip LCoS projectors do not use field sequential color, but use dichroic mirrors to separate a white light beam into red, green, and blue beams, which are aimed at separate LCoS chips, before being recombined into a color signal (see Figure 14.44).

In addition to the nematic liquid crystal technology shown in Figure 14.45, ferroelectric techniques are under development. The advantage of the latter is that faster switching times can be achieved (changing the state of a nematic pixel takes in the order of milliseconds, whereas ferroelectric pixels can be switched much faster). A second difference is that the ferroelectric technology is binary, so that pixels are either on or off. Here, pulse-width modulation (PWM) is used to achieve intermediate levels.

LCoS chips can be used in *near-to-eye* applications, such as head-mounted displays. Here, the source of light is typically a triple of red, green, and blue LEDs that sequentially illuminate a single LCoS chip.

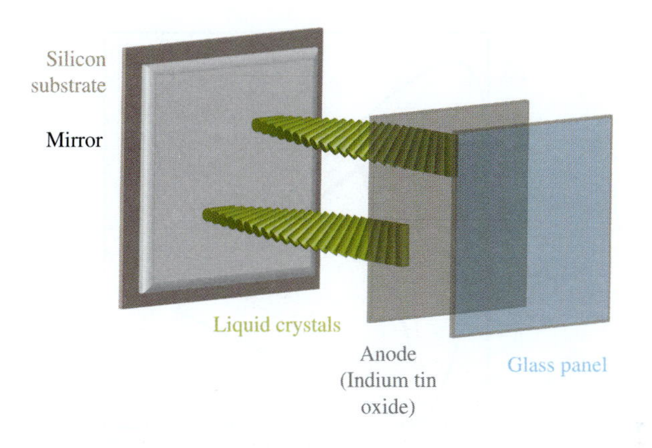

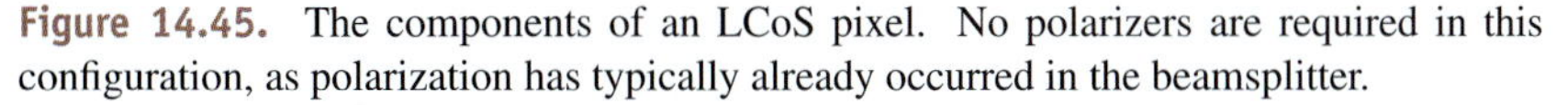

Figure 14.45. The components of an LCoS pixel. No polarizers are required in this configuration, as polarization has typically already occurred in the beamsplitter.

14.15 Multi-Primary Display Devices

The range of colors reproducible by a display device is called its color gamut (see Chapter 8), and this usually corresponds to only a fraction of colors that the human eye can distinguish. Figure 14.46 shows the gamuts of several display standards in a CIE 1931 chromaticity diagram. As this figure shows, the standard gamuts cover only some part of the total range of visible colors.[6]

In many digital imaging applications it would be desirable to display a wider array of colors than the Rec. 709/sRGB gamut typical of most of today's displays. Colors outside the display gamut can be replaced with well-chosen stand-in colors by using gamut mapping techniques (Section 15.4), or, in professional imaging, by skilled craftspeople who can manually execute gamut mapping. However, it is impossible to convey the original color information. A better solution would be to increase the gamut of the display device and avoid the gamut mapping problem altogether.

Inspection of Figure 14.46 shows that the gamut of a display device can be increased if it uses more highly saturated primaries, which amounts to stretching the corners of the triangle towards the spectrum locus [497,588]. This requires the use of almost monochromatic primaries such as lasers [478] or LEDs. Another solution is to use color filters to narrow the emission bandwidth of an excited phosphor. However, these filters block a significant portion of light causing a

[6]Limited coverage is also observed in the dimension of luminance, which is orthogonal to the chromaticity plane drawn in Figure 14.46. However, the aim of multi-primary display devices is to increase the coverage of reproducible colors in the chromaticity plane, rather than extend the dynamic range of reproducible luminance values.

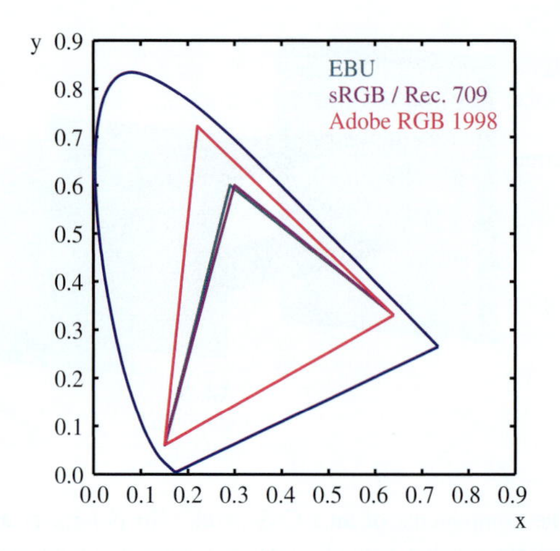

Figure 14.46. Several standard gamuts are shown in a CIE 1931 chromaticity diagram.

reduction in the maximum display brightness [588, 847]. Alternatively, a diffraction grating could be inserted into the optical path to separate white light into its constituent components [20].

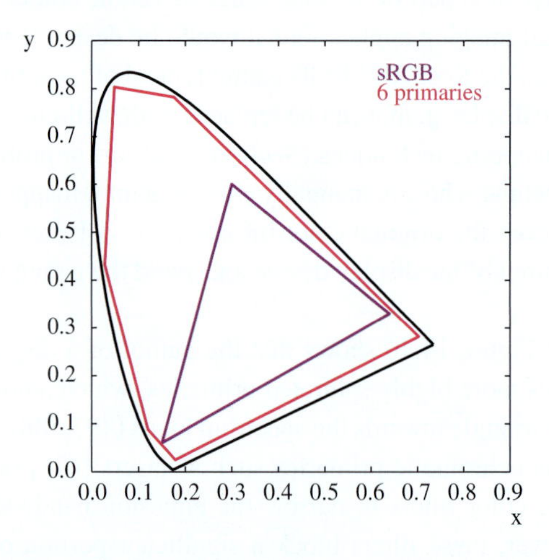

Figure 14.47. The concept of extending gamut with multiple primaries. Each vertex of the red polygon represents the chromaticity coordinates of one of the six primaries of an experimental six-primary display device. This approach alleviates the gamut limitations imposed by a three-primary display. The sRGB gamut is shown for reference.

A different approach is to use display devices that use more than three primaries. Such devices are called *multi-primary display devices*, and they have been actively studied in recent years [131, 981, 1269]. The advantage of using more than three primaries is that the color gamut ceases to be restricted to a triangle, but may take arbitrary polygonal shapes. This affords the flexibility necessary to better cover the horseshoe shape of the visible color gamut (see Figure 14.47).

A multi-primary display device can be constructed by superposing the images produced by two conventional LCD projectors[7]. Each projector is fitted with three additional interference filters placed in the optical paths of the red, green, and blue

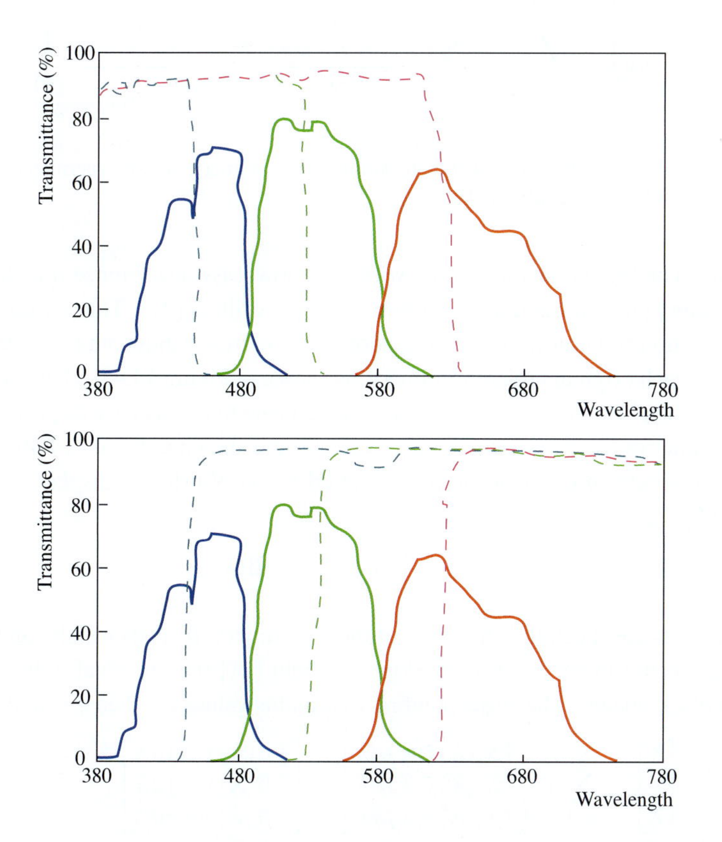

Figure 14.48. Low-pass filters (shown with dashed lines) are used to select the lower bandwidth regions of the original primaries of the first LCD projector (top). High-pass filters (bottom) are used to select the higher bandwidth regions of the primaries of the second projector (after [21]).

[7]The technique described here is currently of an experimental nature.

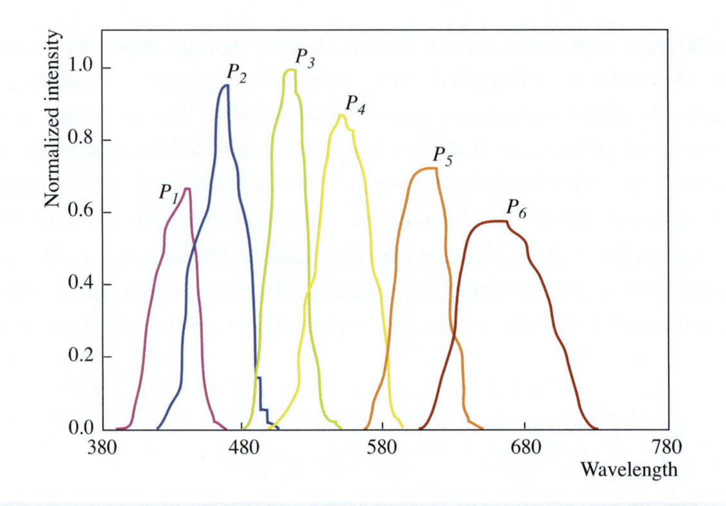

Figure 14.49. The spectral power distributions of six primaries of an experimental multi-primary display device (after [21]).

beams. The first projector is fitted with three low-pass interference filters, while the second projector is fitted with three high-pass filters [21]. This set-up allows the selection and transmittance of light of only a narrow bandwidth from the spectrum of each original projector primary, as shown in Figure 14.48. These selected regions then become the primaries of the composite multi-primary display device. The primaries for a six-primary display device are shown in Figure 14.49.

The spectrum $C(\lambda)$ of the light emitted by an N-primary display device is given by

$$C(\lambda) = \sum_{j=1}^{N} \alpha_j S_j(\lambda) + \beta(\lambda), \tag{14.27}$$

where S_j is the spectral distribution of the jth primary, α_j is its weight, and $\beta(\lambda)$ is the spectral intensity of the background light [21] (i.e., residual light emitted for a black image). The corresponding tristimulus values are then given by

$$\begin{bmatrix} X \\ Y \\ Z \end{bmatrix} = \sum_{j=1}^{N} \alpha_j \begin{bmatrix} \int S_j(\lambda)\bar{x}(\lambda)\,d\lambda \\ \int S_j(\lambda)\bar{y}(\lambda)\,d\lambda \\ \int S_j(\lambda)\bar{z}(\lambda)\,d\lambda \end{bmatrix} + \begin{bmatrix} \int \beta(\lambda)\bar{x}(\lambda)\,d\lambda \\ \int \beta(\lambda)\bar{y}(\lambda)\,d\lambda \\ \int \beta(\lambda)\bar{z}(\lambda)\,d\lambda \end{bmatrix}, \tag{14.28a}$$

$$= \sum_{j=1}^{N} \alpha_j \begin{bmatrix} P_{X_j} \\ P_{Y_j} \\ P_{Z_j} \end{bmatrix} + \begin{bmatrix} X_\beta \\ Y_\beta \\ Z_\beta \end{bmatrix}, \tag{14.28b}$$

where P_{X_j}, P_{Y_j}, and P_{Z_j} are the tristimulus values of each primary and X_β, Y_β, Z_β are the tristimulus values of the background light.

To exploit the full potential of multi-primary display devices in the absence of a wide-gamut interchange color space, they should be driven by multi-spectral or hyper-spectral image data. For multi-spectral images, the (visible) spectrum is divided into a maximum of typically 16 bands. Hyper-spectral images are represented with more than 16 bands[8].

Multi- and hyper-spectral images can be produced by a spectral renderer. Spectral images can also be created with the aid of multi-spectral cameras and capture techniques [1269]. However, in many cases one only has access to three-channel image data (i.e., RGB pixel values) [851]. Therefore, solutions are sought for finding the weights α for each primary from a three-dimensional input signal.

This problem can be solved by inverting (14.28b), although several degrees of freedom exist since an N-dimensional vector $[\alpha_1, \ldots, \alpha_N]^T$ is computed from a three-dimensional vector [1268]. Another method is to use an $N \times 3$ look-up table.

14.16 High Dynamic Range Display Devices

The ratio of the maximum and the minimum luminance values a display device can produce is usually termed its dynamic range. Most typical display devices (CRTs, LCDs, etc.) have a dynamic range between two and three orders of magnitude.[9] In contrast, most real-world environments contain a significantly higher dynamic range, approximately 10 orders of magnitude from starlight to sunlight. Figure 14.50 shows typical luminance values found in natural environments together with what typical monitors can reproduce. This discrepancy is further emphasized by the capability of the eye to adapt to 10 orders of dynamic range at different times and more than 5 orders of magnitude at a single state of adaptation [312].

While the range of luminances that can be produced using phosphor-based display devices is high, some of this range is at low levels where the human visual system is unable to detect contrasts [923]. Thus, the effective dynamic range is limited. As for LCDs, the liquid crystal cells still transmit light even in their off state, increasing the black level. This limits their effective dynamic range.

High dynamic range imaging is about to become main stream, as it is adopted by key industries, including film and games [953]. Many algorithms have been developed to compress the luminance of HDR images, so that they may be displayed on typical monitors (see Chapter 17). Although these algorithms enable

[8]Placing the separation between multi-spectral and hyper-spectral imagery at 16 bands is somewhat arbitrary, and other definitions may be encountered in the literature.

[9]An order of magnitude represents a factor of 10.

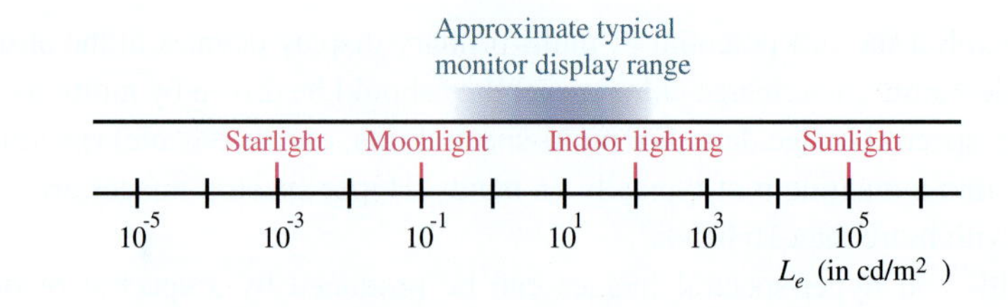

Figure 14.50. Typical luminance levels found in natural environments and the range of luminances a typical monitor can display.

us to view HDR images on standard monitors, the full potential of an HDR image is not realized in this manner.

This problem has led to the development of experimental high dynamic range display devices [1021]. Such devices can be constructed with a two-layered architecture, where the first layer is emissive and the second layer is transmissive. Both layers can be spatially modulated, resulting in a good trade-off between spatial resolution and dynamic range.

Using this principle, Seetzen et al. produced two types of HDR displays [1021]. In the first type, a DLP device is used as a back projector to provide the required light. The front layer consists of an LCD panel used to modulate the light from the projector. Between the layers, a Fresnel lens is installed to couple the two layers. The projector-based HDR display provides a proof of concept for the system. In addition, the use of a projector means that high contrasts can be produced at high frequencies, limited only by the resolution of the projector and the LCD panel, and the accuracy of the alignment between them. This is a distinct advantage in applications such as psychophysical experiments. However, its large depth makes the device impractical in more mainstream applications.

To reduce the depth, a second model was constructed by replacing the projector with an array of ultra-bright LEDs. In the first prototypes, a hexagonal grid of 760 white LEDs is used, although the currently commercialized 37-inch display contains 1380 white LEDs. Each LED can be individually controlled. The LCD front panel has a resolution of 1920 by 1080 pixels (Figure 14.51).

In a darkened room the minimum and maximum display luminance were measured to be 0.015 cd/m^2 and 3450 cd/m^2, respectively. In theory, the minimum luminance can be zero when all LEDs are turned off. However, in practical situations, some stray light will always reflect off the screen, thereby increasing the black level. The minimum value corresponds to the smallest driving signal of LEDs larger than zero. The dynamic range of the device is therefore $230,000:1$, i.e., more than five orders of magnitude.

Figure 14.51. The BrightSide HDR display at the Max Planck Institute for Biological Cybernetics; Tübingen, Germany, November 2006.

The HDR display is addressed by deriving two sets of signals from an input linear-light HDR image. The first set of signals drives the LEDs while the second set controls the LCD. In principle, there are many combinations of LED and LCD values that lead to the same net emitted luminance. It is desirable to choose the lowest LED level possible and adjust the LCD level accordingly. This minimizes power consumption of the display as well as aging of the LEDs.

The fact that a two-layer system allows for different combinations of LED and LCD levels to yield the same output can also be leveraged to calibrate the display [1023]. By adding a lightguide, some of the emitted light can be guided towards the side of the screen, where image sensors can be fitted. A calibration procedure can then measure the light output of each individual LED and correct for variations due to the manufacturing process, including LED chromaticity and non-uniformities in the LCD screen, as well as operating variations, including the differential aging of LEDs, and thermal effects.

The low spatial resolution of the back-light has two important implications. First, the full dynamic range can only be achieved at low spatial frequencies. In other words, in every local neighborhood covered by a single LED, the dynamic range will be limited to that of the LCD panel. However, this problem is significantly alleviated by the fact that the human eye is not sensitive to extreme dynamic ranges over small visual angles [1021, 1186].

Second, the image data presented to the LCD panel needs to be adjusted for the low resolution produced by the LED array. This is performed by artificially enhancing the edges of the input sent to the LCD layer, where the amount of enhancement depends on the point spread function of the LED elements.

14.17 Electronic Ink

A reflective display can be constructed from pigments that are electrically charged [169,205,348]. By making white pigments positively charged and black pigments negatively charged while suspending them in a clear liquid capsule that is sandwiched between a pair of electrodes, a display can be constructed, as diagrammed in Figure 14.52. The capsule can be as small as 10 μm^2. The white ink is made of titanium oxide.

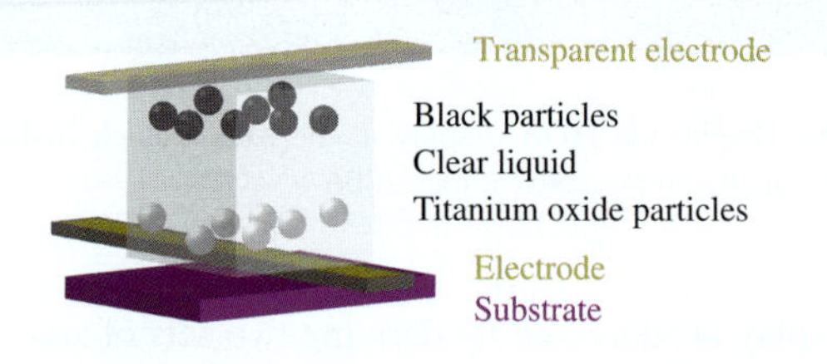

Figure 14.52. A diagram of an electronic ink pixel.

Fabrication of electronic ink employs standard printing processes and can therefore be constructed onto several different materials, including plastic, fabric, paper, and glass. The power consumption of electronic ink devices is very low due to the fact that energy is only required to change the state of a pixel, but not to maintain it. Therefore, the image being displayed remains after the power is switched off.

14.18 Display Characterization

To produce colorimetrically accurate images or stimuli, the properties of the display device must be taken into account. *Display characterization* refers to the process of measuring display properties. *Display calibration* involves adjusting the display device (or an associated subsystem, such as a graphics adapter), to bring the display subsystem into a desired configuration [122].

In the most general sense, a display is said to be characterized if for every pixel value, the corresponding CIE tristimulus values emitted by the display are known.

The most direct, albeit impractical, method to achieve this would be to measure the distribution of emitted light by a spectroradiometer for all pixel values. The CIE XYZ tristimulus values can then be found using

$$X_{\text{RGB}} = \int_\lambda \bar{x}(\lambda)\, f(\lambda; R, G, B)\, d\lambda, \tag{14.29a}$$

$$Y_{\text{RGB}} = \int_\lambda \bar{y}(\lambda)\, f(\lambda; R, G, B)\, d\lambda, \tag{14.29b}$$

$$Z_{\text{RGB}} = \int_\lambda \bar{z}(\lambda)\, f(\lambda; R, G, B)\, d\lambda, \tag{14.29c}$$

where $f(\lambda; R, G, B)$ is the spectral power distribution given the pixel triplet (R, G, B). The CIE color matching functions $\bar{x}$, $\bar{y}$, $\bar{z}$ are used here. This method is impractical because would require approximately 16 million measurements for a display having 8 bits per channel. Even if attempted, the characteristics of the display device are liable to change significantly during the time taken to measure the device [211].

The number of measurements can be significantly reduced if the display device satisfies certain assumptions. The first of these assumptions is called *channel independence*. This assumption of channel independence states that driving a color channel with different signal values does not affect the distribution and amount of the light emitted from other channels. Under this assumption, the following relationship holds:

$$f(\lambda, R, G, B) = f_r(\lambda; R) + f_g(\lambda; G) + f_b(\lambda; B) + f_0. \tag{14.30}$$

Here, f_0 denotes the sum of the light reflected by the display surface due to unwanted external light and the light emitted when all pixels are zero. This assumption reduces a three-dimensional problem to three simpler one-dimensional problems. Equation (14.30) can be rewritten using tristimulus values instead of spectral distributions:

$$\begin{bmatrix} X_{\text{RGB}} \\ Y_{\text{RGB}} \\ Z_{\text{RGB}} \end{bmatrix} = \begin{bmatrix} X_R \\ Y_R \\ Z_R \end{bmatrix} + \begin{bmatrix} X_G \\ Y_G \\ Z_G \end{bmatrix} + \begin{bmatrix} X_B \\ Y_B \\ Z_B \end{bmatrix} + \begin{bmatrix} X_0 \\ Y_0 \\ Z_0 \end{bmatrix}. \tag{14.31}$$

Display characterization can be further simplified if a second assumption known as *channel chromaticity constancy* also holds. This assumption states that the basic spectral shape of the emitted light remains the same regardless of the value of a color channel. Thus, the spectral distribution only undergoes an amplitude scaling:

$$f_r(\lambda;R) = v_r(R)\, f_r(\lambda;R_{\max}), \tag{14.32a}$$

$$f_g(\lambda;G) = v_g(G)\, f_g(\lambda;G_{\max}), \tag{14.32b}$$

$$f_b(\lambda;B) = v_b(B)\, f_b(\lambda;B_{\max}). \tag{14.32c}$$

Here, $f_r(\lambda;R_{\max})$ represents the emission spectra when the red channel is full on, and $v_r(R)$ is the tone response curve when the channel is driven with R instead of $R_{\max}$. With this formulation, all tone response curves must lie in the interval $[0,1]$. It is possible to rewrite (14.32) using the tristimulus values:

$$\begin{bmatrix} X_R \\ Y_R \\ Z_R \end{bmatrix} = v_r(R) \begin{bmatrix} X_{R_{\max}} \\ Y_{R_{\max}} \\ Z_{R_{\max}} \end{bmatrix}, \tag{14.33a}$$

$$\begin{bmatrix} X_G \\ Y_G \\ Z_G \end{bmatrix} = v_g(G) \begin{bmatrix} X_{G_{\max}} \\ Y_{G_{\max}} \\ Z_{G_{\max}} \end{bmatrix}, \tag{14.33b}$$

$$\begin{bmatrix} X_B \\ Y_B \\ Z_B \end{bmatrix} = v_b(B) \begin{bmatrix} X_{B_{\max}} \\ Y_{B_{\max}} \\ Z_{B_{\max}} \end{bmatrix}, \tag{14.33c}$$

where the subscripts $R_{\max}$, $G_{\max}$, and $B_{\max}$ are used to distinguish the tristimulus values produced when each channel is driven by the maximum signal value. If a monitor satisfies both of these assumptions, a *forward model* can be written to simulate its operation. For this, we can combine (14.31) and (14.33):

$$\begin{bmatrix} X_{\mathrm{RGB}} \\ Y_{\mathrm{RGB}} \\ Z_{\mathrm{RGB}} \end{bmatrix} = \begin{bmatrix} X_{R_{\max}} & X_{G_{\max}} & X_{B_{\max}} \\ Y_{R_{\max}} & Y_{G_{\max}} & Y_{B_{\max}} \\ Z_{R_{\max}} & Z_{G_{\max}} & Z_{B_{\max}} \end{bmatrix} \begin{bmatrix} v_r(R) \\ v_g(G) \\ v_b(B) \end{bmatrix} + \begin{bmatrix} X_0 \\ Y_0 \\ Z_0 \end{bmatrix}, \tag{14.34a}$$

$$\begin{bmatrix} X_{\mathrm{RGB}} \\ Y_{\mathrm{RGB}} \\ Z_{\mathrm{RGB}} \end{bmatrix} = M \begin{bmatrix} v_r(R) \\ v_g(G) \\ v_b(B) \end{bmatrix} + \begin{bmatrix} X_0 \\ Y_0 \\ Z_0 \end{bmatrix}, \tag{14.34b}$$

where the matrix M is the 3×3 tristimulus matrix. The forward model is useful to find the tristimulus value corresponding to an (R,G,B) triplet used to drive the display. In most situations though, an inverse model is required to find the (R',G',B') values that yield the desired tristimulus values [1035]. This model can be obtained by inverting the forward model:

$$\begin{bmatrix} R_{\mathrm{XYZ}} \\ G_{\mathrm{XYZ}} \\ B_{\mathrm{XYZ}} \end{bmatrix} = M^{-1} \begin{bmatrix} X - X_0 \\ Y - Y_0 \\ Z - Z_0, \end{bmatrix}, \tag{14.35a}$$

$$\begin{bmatrix} R' \\ G' \\ B' \end{bmatrix} = \begin{bmatrix} v_r^{-1}(R_{\mathrm{XYZ}}) \\ v_g^{-1}(G_{\mathrm{XYZ}}) \\ v_b^{-1}(B_{\mathrm{XYZ}}) \end{bmatrix}. \tag{14.35b}$$

Note that the inverse model is valid only if the matrix M as well as the tone response curves are invertible. For M to be invertible, the individual color channels of the monitor must be independent. This condition is satisfied by most typical desktop computer monitors [1034, 1035], but it is not satisfied by LCD monitors to be used in demanding graphics arts, studio video, or digital cinema production work. Thus, 3×3 linear models have been found to work well with both CRT and LCD displays [74, 92]. However, they are not good predictors for four-channel DLP devices [1096, 1259, 1260]. In addition, tone response curves are usually monotonically increasing functions and are therefore invertible [1035].

As (14.35b) shows, the three key components necessary for display characterization are

1. the color conversion matrix M^{-1};

2. the tone response curves v_r, v_g, v_b;

3. The sum of the emitted light when all pixels are zero and the light reflected off the surface of the monitor.

In the following we focus on the details of how each component can be obtained.

Before carrying out any measurements, the first step must be to set the brightness and contrast controls of the monitor since these controls affect the tone response curves. For a CRT monitor, the brightness control affects the black level of the monitor, which is the amount of light emitted when displaying a black image. It would therefore be better to refer to this control as the black level setting. It operates on non-linear RGB values and, therefore, slides the range of values along the electro-optic conversion curve (EOCF).

This control should first be set to minimum and then slowly increased until the display just starts to show a hint of gray for a black input [923]. The setting is then reduced one step to full black. Setting the brightness control to a lower value causes information to be lost in the dark regions of the displayed material. Setting

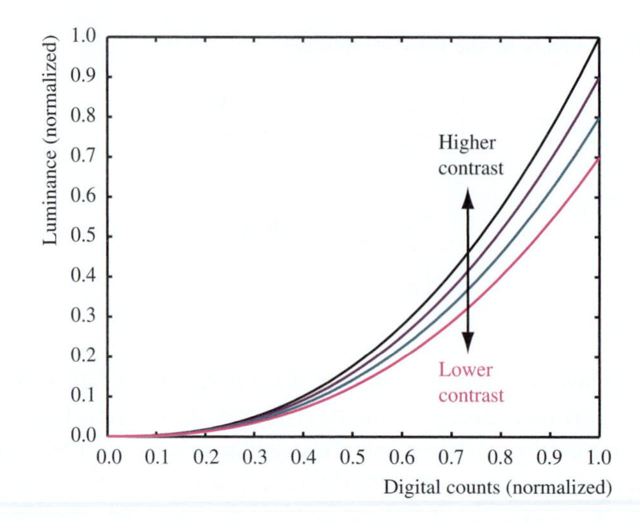

Figure 14.53. The contrast control of CRTs and the brightness control of LCDs are used to maximize the luminance range output by the display device (after [923]).

it to a higher value may significantly reduce the contrast of displayed imagery. If the black level is set appropriately, it may be assumed for the purpose of further calculations that the black level is zero. This simplifies the calibration procedure.

The meaning of the brightness control in an LCD display device is different than for a CRT, as here the control affects the intensity of the back-light [923].

After setting the correct brightness value, the contrast knob can be adjusted to maximize the available display luminance. Increasing the contrast amounts to scaling the displayed luminance, as shown in Figure 14.53.

Once brightness and contrast settings are made, the next step involves measuring the emitted light when all pixels are set to zero. Ideally, this measurement should be made under the lighting conditions under which the display will subsequently be used. The measured values are then converted to tristimulus values, which represent the monitor's black level.

The next step is to measure the tristimulus values corresponding to the maximum drive value for each channel. For this, each channel is individually driven with the maximum pixel value and its corresponding tristimulus values are computed. These are then adjusted by subtracting the tristimulus values of the black level. The adjusted tristimulus values constitute the matrix M in (14.34b).

The last step involves the measurement of the tone response curves. This is typically achieved by ramping each channel from zero to its maximum signal value and taking measurements at several sample points:

$$v_r(R_i) = \frac{\int_\lambda \bar{x}(\lambda) f_r(\lambda; R_i) d\lambda}{\int_\lambda \bar{x}(\lambda) f_r(\lambda; R_{max}) d\lambda} \tag{14.36a}$$

$$= \frac{X_{R_i}}{X_{R_{\max}}}, \tag{14.36b}$$

$$v_g(G_i) = \frac{Y_{G_i}}{Y_{G_{\max}}}, \tag{14.36c}$$

$$v_b(B_i) = \frac{Z_{B_i}}{Z_{B_{\max}}}. \tag{14.36d}$$

Ordinarily, a power-law corresponding to the display's intended standard will result, e.g., a 2.2 power law for an sRGB-compliant display. In principle, we could have used the Y tristimulus value to compute all three response curves. However, using X for the red channel, Y for the green channel, and Z for the blue channel ensures that we collect data where the emissivity of each channel is the highest. This makes the method more robust against measurement noise [92]. The full curve can then be constructed by either linear or non-linear interpolation [92]. For an 8-bit/channel display, 17 measurements are typically sufficient (i.e., $1 \leq i \leq 17$) [92, 704, 920, 1243], that is driving with pixel component values 0, 15, 30, ..., 255.

With the computation of the tone response curves, all the necessary information for monitor characterization is established, assuming that the channel independence and channel chromaticity constancy properties hold. However, it is essential to remember that this information is valid as long as the monitor is used with comparable settings that were used during characterization. Also if the monitor will be in an illuminated room, the light reflected off the screen must be measured and incorporated into the offset term $[X_0, Y_0, Z_0]^T$ of (14.35a).

14.18.1 CRT Monitors

In the previous section, we explained the most generic way of monitor characterization without discussing the specifics of different display devices. The accuracy of this method depends on how well the display device meets the channel independence and chromaticity constancy requirements. In the domain of CRT monitors, these measurements are also called *gun independence* and *phosphor constancy* [1035]. Fortunately, most CRT display devices satisfy these assumptions sufficiently well [91, 210].

The tone response curves of CRT monitors are often modeled by a *gain-offset gamma* (GOG) model . For the red channel, this is expressed as

$$v_r(R) = \left(k_{g,r} \frac{R}{R_{\max}} + k_{o,r} \right)^{\gamma_r}, \tag{14.37}$$

where $k_{g,r}$, $k_{o,r}$, and γ_r denote the gain, offset, and gamma for the red channel [92]. In this equation, the gamma term γ models the inherent non-linearity in the cathode ray tube. It is caused by the space charge effect due to the accumulation of electrons between the cathode and the accelerating anode (see Section 14.1) [173, 652]. This parameter, which is typically between 2.2 and 2.4, cannot be altered by the user.

However, the other two parameters, gain and offset, can usually be adjusted by the user by altering the contrast and brightness knobs, respectively. Under proper settings (14.37) simplifies to

$$v_r(R) = \left(\frac{R}{R_{\max}} \right)^{\gamma_r}. \tag{14.38}$$

Similar equations can be written for the green and blue channels. With this simplification, we can rewrite (14.35b) for a CRT monitor as follows:

$$\begin{bmatrix} R \\ G \\ B \end{bmatrix} = \begin{bmatrix} (R'_{\mathrm{XYZ}})^{1/\gamma_r} \\ (G'_{\mathrm{XYZ}})^{1/\gamma_g} \\ (B'_{\mathrm{XYZ}})^{1/\gamma_b} \end{bmatrix}. \tag{14.39}$$

In computer graphics, this operation is commonly called *gamma correction*. In video and HDTV, gamma correction takes place at the camera and incorporates a power function exponent that effects not only pre-compensation for the non-linearity of the eventual display device, but also includes compensation for appearance effects at the display.

14.18.2 LCD Monitors

As discussed in Section 14.2, LCD monitors work on a different principle than that of CRT monitors. The assumptions for CRT monitors are not necessarily valid for LCD displays. However, several characterization studies have shown that the channel independence and chromaticity constancy assumptions hold to a large degree also for LCD displays [297, 1035]. Nevertheless, displays produced by different manufacturers may differ significantly in the degree they satisfy these assumptions, and therefore some displays may not be characterizable with the standard technique explained in the previous section [359]. For such monitors, more complex methods can be employed [237, 1116, 1277].

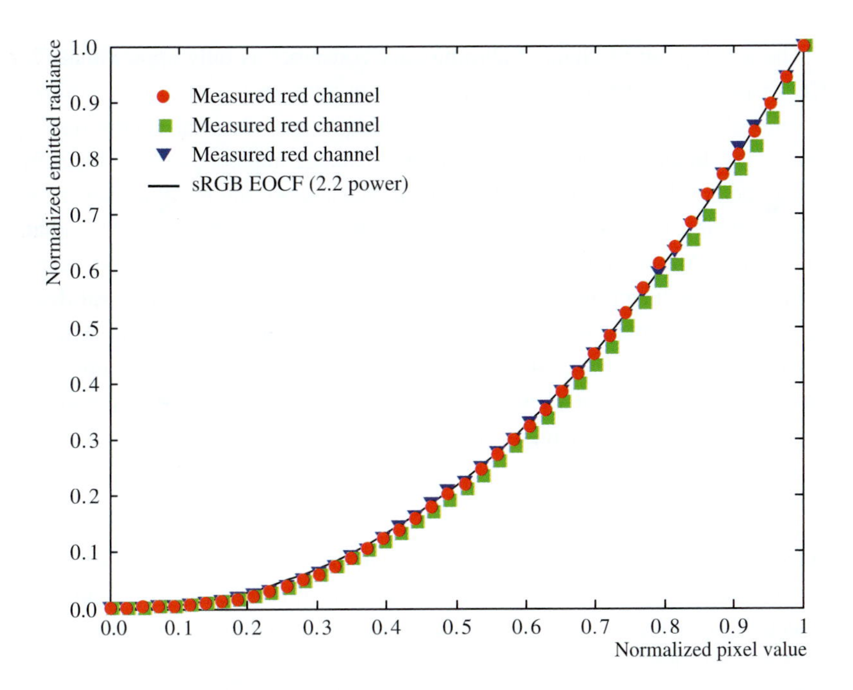

Figure 14.54. Measured electro-optic conversion curve (EOCF) of the red, green, and blue channels of a Dell UltraSharp 2007FP 20.1-inch LCD monitor. A typical gamma of 2.2 is shown for comparison.

One of the important differences between LCDs and CRTs lies in their tone response curves. These functions are usually called *electro-optic conversion curves* (EOCFs). The native response of LCDs is shown in Figure 14.11, indicating that the intrinsic, physical electro-optic conversion curves are essentially sigmoidal [368, 1035].

However, to make LCD monitors compliant with the existing image, video, and PC content, and to make them compatible with CRT display devices, they are typically equipped with look-up tables to transform their response curves to follow a 2.2 power law [297, 1035], as standardized by the sRGB standard (IEC 61966-2-1). In Figure 14.54, the measured curves of the individual color channels of a current LCD monitor are shown. Note that both axes in this graph are normalized, so that the black level of this display cannot be inferred from this figure. This specific monitor closely approximates the response curve of a typical CRT with an EOCF of 2.2. For some monitors though, critical deviations from the gain-offset gamma model are sometimes observed, and therefore the use of external look-up tables is recommended for color critical applications [237, 297].

The assumption of channel chromaticity constancy is only approximate for LCD displays. With two color channels set to zero, in the ideal case, varying the value of the remaining color channel should produce identical xy chromaticity coordinates. In an LCD panel, however, light leakage impairs this ideal behavior. By subtracting the tristimulus values of the black level, the resulting values are likely to exhibit much improved channel chromaticity constancy, as for instance shown for an Apple Studio Display [297].

Finally, dependent on the specific technology employed, the spectral distribution of light emitted from an LCD panel may change significantly with the viewing angle. Therefore, characterization performed from one viewing angle will only be valid for that viewing angle. Significant colorimetric variations are observed for LCDs with changes in viewing angle [1035].

14.18.3 LCD Projectors

An LCD projector can be characterized in much the same way as an LCD monitor. However, here the contribution of the black level is more important and cannot be neglected. Thus, for an LCD projector display, the procedure for characterization is as follows [1096].

First, the pixel values are linearized. It should be noted that most projectors are built for displaying video or PC content and, therefore, follow a power-law curve. Thus, linearization can be achieved by applying (14.39), leading to the linear triplet $(R, G, B)^T$.

Next, the triplets for the primaries $(X_R, Y_R, Z_R)^T$, $(X_G, Y_G, Z_G)^T$, and $(X_B, Y_B, Z_B)^T$, as well as the black point $(X_K, Y_K, Z_K)^T$ are measured. Using homogeneous coordinates (see Appendix A.10), the transformation is given by

$$\begin{bmatrix} X \\ Y \\ Z \\ 1 \end{bmatrix} = \begin{bmatrix} X_R - X_K & X_G - X_K & X_B - X_K & X_K \\ Y_R - Y_K & Y_G - Y_K & Y_B - Y_K & Y_K \\ Z_R - Z_K & Z_G - Z_K & Z_B - Z_K & Z_K \\ 0 & 0 & 0 & 1 \end{bmatrix} \begin{bmatrix} R \\ G \\ B \\ 1 \end{bmatrix}, \tag{14.40}$$

yielding the desired transformation.

14.18.4 Calibration in Practice

Calibration of display devices is an involved procedure that typically requires dedicated measuring equipment. This task can be simplified by using software and/or hardware solutions, either measuring equipment provided with the display subsystem, or third-party equipment, such as that depicted in Figure 14.55. A sensor is placed in front of the display, measuring its output. By correlating the

Figure 14.55. A display calibration device.

values sent to the display with the measured values, the display is characterized. The output of the calibration procedure is typically an ICC profile, which is used either by the operating system, or by application software to adjust imagery for display on the characterized display. Such profiles are commonly in use to ensure appropriate color reproduction (see Chapter 16).

14.19 Further Reading

Display devices are discussed in many textbooks. Particularly relevant to color is Widdel et al., *Color in Electronic Displays* [1243]. Poynton's *Digital Video and HDTV: Algorithms and Interfaces* contains a wealth of color-related display information [923]. Specific display technologies are presented in *Liquid Crystal TV Displays: Principles and Applications of Liquid Crystal Displays* by Kaneko [571], *Organic Light-Emitting Devices: A Survey* by Shinar [1040], and *Flat-Panel Displays and CRTs* by Tannas [1120]. The electro-optical effects in different types of liquid crystal displays, as well as the various addressing schemes in use for such displays, are discussed by Ernst Lueder in *Liquid Crystal Displays: Addressing Schemes and Electro-Optical Effects* [706]. The desire to manufacture flexible flat-panel displays has led to a considerable amount of research, which is discussed in Gregory P Crawford (ed.), *Flexible Flat Panel Displays* [213].

Chapter 15

Image Properties and Image Display

In this book, the focus is on light and color, and its implications for various applications dealing with images. In previous chapters, some of the fundamentals pertaining to image capture and image display were discussed. This leaves us with several attributes of images that are worth knowing about. In particular, images of existing environments tend to have statistical regularities that can be exploited in applications such as image processing, computer vision, and computer graphics. These properties are discussed in the following section.

Further, we discuss issues related to the measurement of dynamic range. Such measurements are becoming increasingly important, especially in the field of high dynamic range imaging, where knowledge about the dynamic range of images can be used to assess how difficult it may be to display such images on a given display device.

The third topic in this chapter relates to cross-media display. As every display has its own specification and properties and is located in a differently illuminated environment, the machinery required to prepare an image for accurate display on a specific display device, should involve some measurement of both the display device and its environment. One of the tasks involved in this process, is matching the gamut of an image to the gamut of the display device. Finally, we discuss gamma correction and possible adjustments for ambient light which is reflected off the display device.

15.1 Natural Image Statistics

Images of natural scenes tend to exhibit statistical regularities that are important for the understanding of the human visual system which has evolved to interpret such natural images. In addition, these properties may prove valuable in image-processing applications. An example is the color transfer algorithm outlined in Section 8.10, which relies on natural image statistics to transfer the color mood of an image to a second image.

If all greyscale images that consist of 1000×1000 pixels are considered, then the space that these images occupy would be a million-dimensional space. Most of these images never appear in our world: natural images form a sparse subset of all possible images. In recent years, the statistics of natural images have been studied to understand how properties of natural images influence the human visual system. In order to assess invariance in natural images, usually a large set of calibrated images are collected into an *ensemble* and statistics are computed on these ensembles, rather than on individual images.

Natural image statistics can be characterized by their order. In particular, first-order, second-order, and higher-order statistics are distinguished [1001]:

First-order statistics treat each pixel independently, so that, for example, the distribution of intensities encountered in natural images can be estimated.

Second-order statistics. measure the dependencies between pairs of pixels, which are usually expressed in terms of the power spectrum (see Section 15.1.2).

Higher-order statistics are used to extract properties of natural scenes which can not be modeled using first- and second-order statistics. These include, for example, lines and edges.

Each of these categories is discussed in some detail in the following sections.

15.1.1 First-Order Statistics

Several simple regularities have been established regarding the statistical distribution of pixel values; the gray-world assumption is often quoted in this respect. Averaging all pixel values in a photograph tends to yield a color that is closely related to the blackbody temperature of the dominant illuminant [902]. The integrated color tends to cluster around a blackbody temperature of 4800 K.

For outdoor scenes, the average image color remains relatively constant from 30 minutes after dawn until within 30 minutes of dusk. The correlated color temperature then increases, giving an overall bluer color.[1]

The average image color tends to be distributed around the color of the dominant scene illuminant according to a Gaussian distribution, if measured in log space [665, 666]. The eigenvectors of the covariance matrix of this distribution follow the spectral responsivities of the capture device, which for a typical color negative film are [666]

$$L = \frac{1}{\sqrt{3}} \left(\log(R) + \log(G) + \log(B)\right), \tag{15.1a}$$

$$s = \frac{1}{\sqrt{2}} \left(\log(R) - \log(B)\right), \tag{15.1b}$$

$$t = \frac{1}{\sqrt{6}} \left(\log(R) - 2\log(G) + \log(B)\right). \tag{15.1c}$$

For a large image database, the standard deviations in these three channels are found to be 0.273, 0.065, and 0.030. Note that this space is similar to the $L\alpha\beta$ color space discussed in Section 8.7.5. Although that space is derived from LMS rather than RGB, both operate in log space, and both have the same channel weights. The coefficients applied to each of the channels are identical as well, with the only difference being that the role of the green and blue channels has been swapped.

15.1.2 Second-Order Statistics

The most remarkable and salient natural image statistic that has been discovered so far is that the slope of the power spectrum tends to be close to 2. The power spectrum of an M $\times$ M image is computed as

$$S(u,v) = \frac{|F(u,v)|^2}{M^2}, \tag{15.2}$$

where F is the Fourier transform of the image. By representing the two-dimensional frequencies u and v in polar coordinates ($u = f\cos\phi$ and $v = f\sin\phi$) and averaging over all directions ϕ and all images in the image ensemble, it is found that on log-log scale amplitude as a function of frequency, f lies approximately on a straight line [144, 314, 983, 986, 1001]. This means that spectral power as a function of spatial frequency behaves according to a power law function. Moreover, fitting a line through the data points yields a slope α of approximately 2 for

[1]Despite the red colors observed at sunrise and sunset, most of the environment remains illuminated by a blue skylight, which increases in importance as the sun loses its strength.

natural images. Although this spectral slope varies subtly between different studies [144, 259, 315, 983, 1130], and with the exception of low frequencies [651], it appears to be extremely robust against distortions and transformations. It is therefore concluded that this spectral behavior is a consequence of the images themselves, rather than of particular methods of camera calibration or exact computation of the spectral slope. We denote this behavior by

$$S(f) \propto \frac{A}{f^{\alpha}} = \frac{A}{f^{2-\eta}}, \tag{15.3}$$

where A is a constant determining overall image contrast, α is the spectral slope, and η is its deviation from 2. However, the exact value of the spectral slope depends somewhat on the type of scenes that make up the ensembles. Most interest of the natural image statistics community is in scenes containing mostly trees and shrubs. Some studies show that the spectral slope for scenes containing man-made objects is slightly different [1295, 1296]. Even in natural scenes, the statistics vary, dependent on what is predominantly in the images. The second-order statistics for sky are, for example, very different from those of trees.

One of the ways in which this becomes apparent is when the power spectra are not circularly averaged, but when the log average power is plotted against angle. For natural image ensembles, all angles show more or less straight power spectra, although most power is concentrated along horizontal and vertical directions [986, 1001] (see also Figure 15.3). The horizon and the presence of tree trunks are said to be factors, although this behavior also occurs in man-made environments.

The power spectrum is related to the auto-correlation function through the Wiener-Khintchine theorem, which states that the auto-correlation function and the power spectrum form a Fourier transform pair [840]. Hence, power spectral behavior can be equivalently understood in terms of correlations between pairs of pixel intensities.

A related image statistic is contrast, normally defined as the standard deviation of all pixel intensities divided by the mean intensity (σ/μ). This measure can either be computed directly from the image data, or it can be derived from the power spectrum through Parceval's theorem [1001]:

$$\frac{\sigma^2}{\mu^2} = \sum_{(u,v)} S(u,v). \tag{15.4}$$

This particular contrast computation can be modified to compute contrast in different frequency bands. Frequency-conscious variance can then be thresholded, yielding a measure which can detect blur [313]. This is useful as lack of contrast can also be caused by the absence of sufficient detail in a sharp image.

The above second-order statistics are usually collected for luminance images only, as luminance is believed to carry the greatest amount of information. However, chrominance channels are shown to exhibit similar spectral behavior [878], and therefore, all subsequent qualitative arguments are expected to be true for color as well.

The fact that the power spectral behavior of natural images, when plotted in log-log space, yields a straight line with a slope of around -2 is particularly important, since recent unrelated studies have found that in image interpretation tasks, the human visual system performs best when the images conform to the above second-order statistic. In one such study, images of a car and a bull were morphed into each other, with varying distances between the images in the sequence [879]. Different sequences were generated with modified spectral slopes. The minimum distance at which participants could still distinguish consecutive images in each morph sequence was then measured. This distance was found to be smallest when the spectral slope of the images in the morph sequence was close to 2. Deviation of the spectral slope in either direction resulted in a deteriorated performance to distinguish between morphed images.

In a different study, the effect of spectral slope on the detection of mirror symmetry in images was assessed [933]. Here, white noise patterns with varying degrees of vertical symmetry were created and subsequently filtered to alter the spectral slope. An experiment, in which participants had to detect if symmetry was present, revealed that performance was optimal for images with a spectral slope of 2.

These studies confirm the hypothesis that the HVS is tuned to natural images. In fact, the process of whitening (i.e., flattening the power spectrum to produce a slope α of 0) is consistent with psychophysical measurements [52], which indicates that the HVS expects to see images with a $1/f^2$ power spectrum.

15.1.3 Power Spectra

In this section, we show how the power spectrum of an ensemble of images can be established. We used 133 images from the van Hateren database of natural images [433]. The images in this ensemble are available as luminance images. Example images are shown in Figure 15.1.

For this image ensemble, the power spectrum was computed and the spectral slope was estimated. The power spectrum computation proceeds as follows.[2]

A window of 512×512 pixels was cut out of the middle of the image upon which further processing was applied. Then, the weighted mean intensity μ was

[2]This calculation closely follows the method described in [1001].

Figure 15.1. Example images drawn from the van Hateren database [433]. (Images courtesy of Hans van Hateren.)

subtracted to avoid leakage from the DC-component of the image, where μ is defined as

$$\mu = \frac{\sum_{(x,y)} L(x,y)w(x,y)}{\sum_{(x,y)} w(x,y)}. \tag{15.5}$$

Next, the images were prefiltered to avoid boundary effects. This is accomplished by applying a circular Kaiser-Bessel window function (with parameter $\alpha = 2$) to the image [429]:

$$w(x,y) = \frac{I_0\left(\pi\alpha\sqrt{1.0 - \left(\frac{x^2+y^2}{(N/2)^2}\right)}\right)}{I_0(\pi\alpha)}, \qquad 0 \le \sqrt{x^2+y^2} \le \frac{N}{2}. \tag{15.6}$$

Here, I_0 is the modified zero-order Bessel function of the first kind and N is the window size (512 pixels). In addition, this weight function was normalized by letting

$$\sum_{(x,y)} w(x,y)^2 = 1. \tag{15.7}$$

This windowing function was chosen for its near-optimal trade-off between side-lobe level, main-lobe width, and computability [429]. The resulting images were then compressed using the Fourier transform,

$$F(u,v) = \sum_{(x,y)} \frac{L(x,y) - \mu}{\mu} e^{2\pi i(ux+vy)}. \tag{15.8}$$

Finally, the power spectrum was computed as per (15.2) and the resulting data points plotted. Although frequencies up to 256 cycles per image are computed, only the 127 lowest frequencies were used to estimate the spectral slope. Higher frequencies may suffer from aliasing, noise, and low modulation transfer [1001]. The estimation of the spectral slope was performed by fitting a straight line through the logarithm of these data points as a function of the logarithm of $1/f$. This method was chosen over other slope estimation techniques, such as the Hill estimator [472] and the scaling method [215], to maintain compatibility with [1001]. In addition, the number of data points (127 frequencies) is insufficient for the scaling method, which requires at least 1,000 data points to yield reasonably reliable estimates.

With this method, second-order statistics were extracted. The 1.87 spectral slope reported for the van Hateren database was confirmed (we found 1.88 for our subset of 133 images). The deviations from this value for the artificial image ensemble are depicted in Figure 15.2. The angular distribution of power tends to show peaks near horizontal and vertical angles (Figure 15.3). Finally, the distribution of spectral slopes for the 133 images in this ensemble is shown in Figure 15.4.

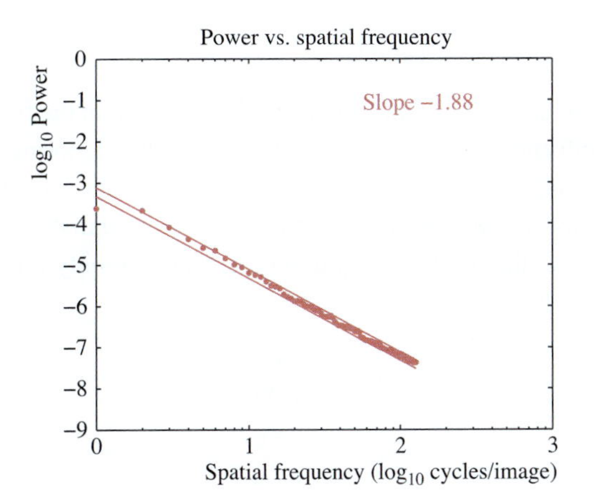

Figure 15.2. Spectral slope for the image ensemble. The double lines indicate ±2 standard deviations for each ensemble.

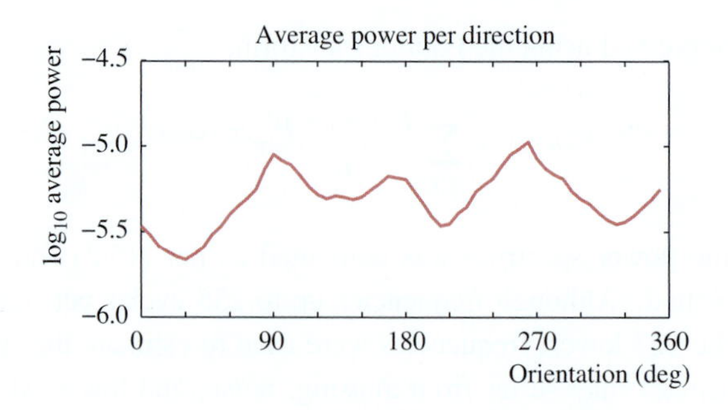

Figure 15.3. Log average power as function of angle.

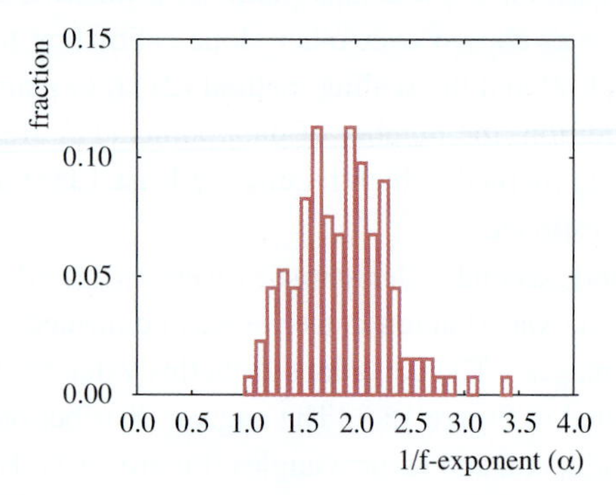

Figure 15.4. A histogram of spectral slopes, derived from an image ensemble of 133 images taken from the van Hateren database.

15.1.4 Higher-Order Statistics

One of the disadvantages of using amplitude information in the frequency domain is that phase information is completely discarded, thus ignoring the position of edges and objects. For this reason, higher-order statistics have been applied to natural image ensembles. The simplest global nth-order statistics that may capture phase structure are the third and fourth moments, commonly referred to as skew and kurtosis [840]:

$$s = \frac{E\{x^3\}}{E\{x^2\}^{3/2}}, \tag{15.9a}$$

$$k = \frac{E\{x^4\}}{E\{x^2\}^2} - 3. \tag{15.9b}$$

These dimensionless measures are, by definition, zero for pure Gaussian distributions. Skew can be interpreted as an indicator of the difference between the mean and the median of a data set. Kurtosis is based on the size of a distribution's tails relative to a Gaussian distribution. A positive value, associated with long tails in the distribution of intensities, is usually associated with natural image ensembles. This is for example evident when plotting log-contrast histograms, which plot the probability of a particular contrast value appearing. These plots are typically non-Gaussian with positive kurtosis [986].

Thomson [1126] has pointed out that for kurtosis to be meaningful for natural image analysis, the data should be decorrelated, or whitened, prior to any computation of higher-order statistics. This can be accomplished by flattening the power spectrum, which ensures that second-order statistics do not also capture higher-order statistics. Thus, skew and kurtosis become measures of variations in the phase spectra. Regularities are found when these measures are applied to the pixel histogram of an image ensemble [1125, 1126]. This appears to indicate that the HVS may exploit higher-order phase structure after whitening the image. Understanding the phase structure is therefore an important avenue of research in the field of natural image statistics. However, it appears that so far it is not yet possible to attach meaning to the exact appearance of higher-order statistics.

15.1.5 Gradient Field Statistics

Rather than study the statistical properties of pixels, we can also look at derived quantities, such as gradients. The gradient field of an image is a vector-valued quantity ∇L, defined as

$$\nabla L = (L(x+1,y) - L(x,y), L(x,y+1) - L(x,y)). \tag{15.10}$$

Natural images typically have strong discontinuities in otherwise relatively smooth regions. This means that large image areas have only small gradients, with larger gradients happening more sparsely. Thus, the components of the gradient field will show a characteristic distribution, which can be observed by computing its histogram. The gradient distribution of a set of four example images, shown in Figure 15.5, is plotted in Figure 15.6. The images used here were captured in high dynamic range (see Chapter 17), and therefore contain much larger gradients than encountered in conventional 8-bit images. In addition, these images are linear by construction, having a gamma of 1.0. Conventional images are likely to have gamma correction applied to them, and this may affect the distribution of gradients in images.

Figure 15.5. Example images used for computing gradient distributions, which are shown in Figure 15.6. From left to right, top to bottom: Abbaye de Beauport, Paimpol, France, July, 2007; Combourg, France, July, 2007; Fougères, France, June, 2007); Kermario, Carnac, France, July, 2007.

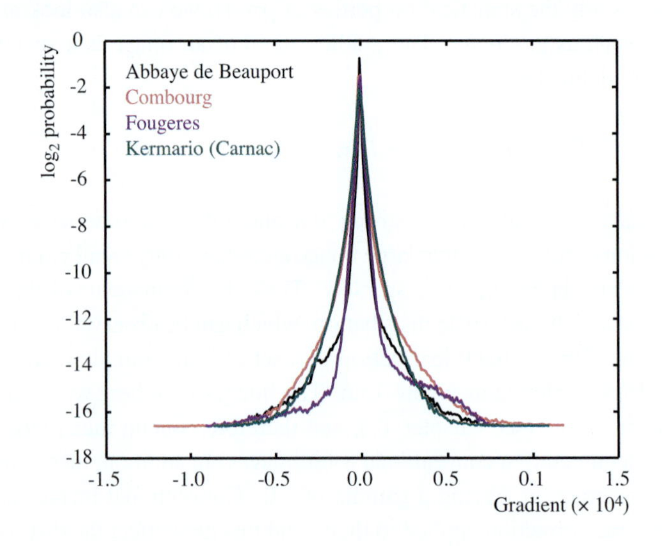

Figure 15.6. The probability distribution of gradients for the four images depicted in Figure 15.5.

15.1.6 Response Properties of Cortical Cells

One of the reasons to study the statistics of natural images is to understand how the HVS codes these images. Because natural images are not completely random, they incorporate a significant amount of redundancy. The HVS may represent such scenes using a sparse set of active elements. These elements will then be largely independent.

Here, it is assumed that an image can be represented as a linear superposition of basis functions. Efficiently encoding images now involves finding a set of basis functions that spans image space and ensures that the coefficients across an image ensemble are statistically as independent as possible [853]. The resulting basis functions (or filters) can then be compared to the response properties of cortical cells in order to explain the early stages of human visual processing. This has resulted in a number of different representations for basis functions, including principal components analysis (PCA) [55,420], independent components analysis (ICA) [82,83,433], Gabor basis functions [314,315], and wavelets [315,514,853].

The PCA algorithm finds a set of basis functions that maximally decorrelates pairs of coefficients; this is achieved by computing the eigenvalues of the covariance matrix (between pairs of pixel intensities). The corresponding eigenvectors represent a set of orthogonal coefficients, whereby the eigenvector with the largest associated eigenvalue accounts for the largest part of the covariance. By using only the first few coefficients, it is possible to encode an image with a large reduction in free parameters [314]. If stationarity is assumed across all images in an ensemble and between all regions in the images, i.e., the image statistics are the same everywhere, then PCA produces coefficients that are similar to the Fourier transform [113]. Indeed, PCA is strictly a second-order method assuming Gaussian signals.

Unfortunately, decorrelation does not guarantee independence. Also, intensities in natural images do not necessarily have a Gaussian distribution, and therefore PCA yields basis functions that do not capture higher-order structure very well [853]. In particular, the basis functions tend to be orientation and frequency sensitive, but with global extent. This is in contrast to cells in the human primary cortex which are spatially localized (as well as being localized in frequency and orientation).

In contrast to PCA, independent components analysis constitutes a transformation resulting in basis functions that are non-orthogonal, localized in space, and selective for frequency and orientation. They aim to extract higher-order information from images [82,83]. ICA finds basis functions that are as independent as possible [522]. To avoid second-order statistics from influencing the result of ICA, the data is usually first decorrelated (also called whitened), for example using a PCA algorithm. Filters can then be found that produce extrema of the kurtosis [1001]. A kurtotic amplitude distribution is produced by cortical simple cells, leading to sparse coding. Hence, ICA is believed to be a better model than PCA for the output of simple cortical cells.

The receptive fields of simple cells in the mammalian striate cortex are localized in space, oriented, and bandpass. They are therefore similar to the basis functions of wavelet transforms [315, 853]. For natural images, strong correlations between wavelet coefficients at neighboring spatial locations, orientations, and scales, have been shown using conditional histograms of the coefficients' log magnitudes [1049]. These results were successfully used to synthesize textures [919, 1047] and to denoise images [1048].

15.1.7 Dimensionality of Reflectance and Daylight Spectra

The reflectance spectra of large numbers of objects have been analyzed for the purpose of determining their dimensionality. We have so far assumed that a spectral distribution is sampled at regular intervals, for instance from 380 to 780 with a 5 or 10 nm spacing. This leads to a representation of each spectrum by 43 or so numbers.

Values in between sample points can be approximated, for instance, by linear interpolation. However, it is also possible to use higher-order basis functions and, thereby, perhaps reduce the number of samples. The question then becomes which basis functions should be used, and how many are required.

It has been found that the eigenvalues of the covariance matrix decay very fast, leading to only a few principal components that account for most of the variance [201]. Similar results have been found for daylight spectra [559], suggesting that spectral distributions can be represented in a low-dimensional space, for instance having no more than six or seven dimensions [728, 729].

15.2 Dynamic Range

The dynamic range afforded by an image encoding depends on the ratio between smallest and largest representable value, as well as the step size between successive values. Conventional images are stored with a byte per pixel for each of the red, green, and blue components. Thus, for low dynamic range images, there are only 256 different values per color channel.

High dynamic range images encode a significantly larger set of possible values. The maximum representable value may be much larger, and the step size between successive values may be much smaller. The file size of high dynamic range images is therefore generally larger as well, although at least one standard (the OpenEXR high dynamic range file format [565]) includes a very capable compression scheme, and JPEG-HDR essentially occupies the same amount of space as an equivalent low dynamic range image would [1209, 1210]. To construct an image in this file format, the HDR image is tone-mapped and encoded as a standard JPEG. The ratio between the original and the JPEG image is subsampled and stored in a meta-tag. HDR-aware viewers can reconstruct the HDR from this data, whereas other image viewers simply ignore the meta-tag and display the JPEG directly. Encoding of a residual image, such as the ratio image, can also be used to store an extended gamut [1073].

Recently, the problem of HDR video encoding has received some attention. This has led to MPEG-HDR, a high dynamic range video encoder/decoder promising high compression ratios [733], as well as a second video-encoding scheme that is based on a model of human cones [435], which was briefly discussed in Section 4.8.

In general, the combination of smallest step size and ratio of the smallest and largest representable values determines the dynamic range that an image encoding scheme affords. Images have a dynamic range that is bounded by the encoding scheme used. The dynamic range of environments is generally determined by the smallest and largest luminance found in the scene. Assuming this range is smaller than the range afforded by the encoding scheme, such an image can be captured and stored. The question then remains: What is the actual dynamic range of such a scene?

A simplistic approach to measure the dynamic range of an image may, therefore, compute the ratio between the largest and smallest pixel value of an image. Sensitivity to outliers may be reduced by ignoring a small percentage of the darkest and brightest pixels. Alternatively, the same ratio may be expressed as a difference in the logarithmic domain.

However, the recording device or rendering algorithm may introduce noise that will lower the useful dynamic range. Thus, a measurement of the dynamic range of an image should factor in noise. A better measure of dynamic range is, therefore, a signal-to-noise ratio, expressed in decibels (dB), as used in signal processing. This concept is discussed further in the following section.

Assuming we have a reasonable measure of dynamic range, it then becomes possible to assess the extent of the mismatch between the dynamic range of an image and the dynamic range of a given display device. A good tone-reproduction operator will be able to reduce the dynamic range of an image to match that of the display device without producing visible artifacts.

15.2.1 Signal-to-Noise Ratio

Assuming that the signal and the noise are interdependent, a measure of the signal-to-noise ratio (SNR) may be computed as follows:

$$\mathrm{SNR}^{\mathrm{dep}} = 20 \log_{10}\left(\frac{\mu_s}{\sigma_n}\right), \tag{15.11}$$

where μ_s is the mean average of the signal, and σ_n is the standard deviation of the noise. In the case that signal and noise are independent, the SNR may be computed by

$$\mathrm{SNR}^{\mathrm{indep}} = 20 \log_{10}\left(\frac{\sigma_s}{\sigma_n}\right). \tag{15.12}$$

For high dynamic range images, which typically have long-tailed histograms, using the mean signal to compute dynamic range may not be the most sensible approach, since small but very bright highlights or light sources tend to have little impact on this measure. Thus, rather than use σ_s or μ_s to compute one of the above SNR measures, it may be a better approach to compute the peak signal-to-noise ratio, which depends on the maximum luminance $L^{\max}$ of the image:

$$\mathrm{SNR}^{\mathrm{peak}} = 20 \log_{10}\left(\frac{(L^{\max})^2}{\sigma_n}\right). \tag{15.13}$$

While μ_s, σ_s, and $L^{\max}$ are each readily computed for any given image, the estimation of the noise floor σ_n in an image is a notoriously difficult task, for which many solutions exist. In particular, assumptions will have to be made about the nature of the noise.

In some specific cases, the noise distribution may be known. For instance, photon shot noise, which is characterized by the random arrival times of photons at image sensors, has a Poisson distribution. For a sensor exposed for T seconds, and a rate parameter of ρ photons per second, the probability P of detecting p photons is given by

$$P(p) = \frac{(\rho\, T)^p\, e^{-\rho T}}{p!}. \tag{15.14}$$

The mean μ_n and standard deviation σ_n are then

$$\mu_n = \rho\, T, \tag{15.15a}$$

$$\sigma_n = \sqrt{\rho\, T}. \tag{15.15b}$$

Thus, for photon shot noise, the signal-to-noise ratio can be computed by [1282]

$$\mathrm{SNR}^{\mathrm{Poisson}} = 10 \log_{10}(\rho\, T), \tag{15.16}$$

which is again measured in decibels.

When the source of the noise in an image is unknown, further assumptions about the noise will have to be made before a signal-to-noise ratio can be computed. In particular, a model of the distribution of the noise must be assumed. In the absence of further information, one may, for instance, assume that the noise has zero mean and has a Gaussian distribution with a standard deviation of σ_n.

Under this assumption, one could apply a Gaussian filter to the image L with a kernel width of σ_n, producing a second, blurred image L^{blur}. Alternatively, rather than blurring the image, a median filter may be applied that minimizes excessive blur and, thereby, potentially yields a more accurate measure of dynamic range. It would also be possible to apply more advanced noise reduction algorithms, such as level set methods [727].

An indication of the noise floor is then given by the pixel-wise mean squared difference between the blurred and original images:

$$\sigma_n = \sqrt{\frac{1}{N} \sum_{i=0}^{N} \left(L_i - L_i^{\mathrm{blur}}\right)^2}. \tag{15.17}$$

The accuracy of the noise floor and, therefore, the measure of dynamic range, is dependent on the noise reduction technique, as well as on the filter parameter settings chosen. This makes these techniques limited to comparisons of the dynamic range between images that are computed with identical techniques and parameter settings.

On the other hand, the signal-to-noise ratio may be computed by relying on the minimum and maximum luminance ($L^{\min}$ and $L^{\max}$):

$$\mathrm{SNR}^{\max} = 20 \log_{10} \left(\frac{L^{\max}}{L^{\min}} \right). \tag{15.18}$$

In this case, there is no dependency on the choice of noise-reduction technique. Instead, there is a dependency on $L^{\min}$ and $L^{\max}$, which are then taken to be representative of the signal and noise levels. However, by definition, these values are outliers. As this is a logarithmic measure, its sensitivity to outliers is somewhat ameliorated compared with the common approach of simply listing the dynamic range as $L^{\max}/L^{\min}$.

Finally, whether an image's dynamic range is measured in terms of signal-to-noise ratios or simply by comparing the maximum and minimum luminance, there is always a trade-off between accuracy and sensitivity to outliers or parameter settings. In the following section we show measurements performed on high dynamic range images using the metrics outlined in this section.

15.2.2 Gaussian Filter

In this section we show, for both the Gaussian and median filters, how their filter-parameter settings affect the dynamic range measurements. The example images used are shown in Figure 15.7.

For Gaussian filters, the kernel is specified with two parameters that both determine the result. First, the width of the Gaussian filter is determined by σ, such that for an image region centered around the origin, the kernel is given by

$$R_\sigma(x,y) = \frac{1}{2\pi\sigma^2} e^{-\frac{x^2+y^2}{2\sigma^2}}. \tag{15.19}$$

Figure 15.7. Example images used for assessing the impact of filter-kernel parameters on SNR measures; Tower of Hercules Roman Lighthouse, A Coruña, Spain, August 2005; Rennes, France, July 2005; Insel Mainau, Germany, June 2005.

As any Gaussian filter has infinite non-zero extent, a second parameter determines into how many pixels the kernel is discretized. This number, k, is typically odd. While the continuous Gaussian filter has unit area, after discretization, the filter no longer sums to 1. We therefore normalize the discretized filter kernel before starting the computations.

As an example, for k = 3, 5, 7, the filter kernels are

$$\begin{bmatrix} 0.075 & 0.124 & 0.075 \\ 0.124 & 0.204 & 0.124 \\ 0.075 & 0.124 & 0.075 \end{bmatrix}; \quad \begin{bmatrix} 0.003 & 0.013 & 0.022 & 0.013 & 0.003 \\ 0.013 & 0.060 & 0.098 & 0.060 & 0.013 \\ 0.022 & 0.098 & 0.162 & 0.098 & 0.022 \\ 0.013 & 0.060 & 0.098 & 0.060 & 0.013 \\ 0.003 & 0.013 & 0.022 & 0.013 & 0.003 \end{bmatrix};$$

$$\begin{bmatrix} 0.000 & 0.000 & 0.001 & 0.002 & 0.001 & 0.000 & 0.000 \\ 0.000 & 0.003 & 0.013 & 0.022 & 0.013 & 0.003 & 0.000 \\ 0.001 & 0.013 & 0.059 & 0.097 & 0.059 & 0.013 & 0.001 \\ 0.002 & 0.022 & 0.097 & 0.159 & 0.097 & 0.022 & 0.002 \\ 0.001 & 0.013 & 0.059 & 0.097 & 0.059 & 0.013 & 0.001 \\ 0.000 & 0.003 & 0.013 & 0.022 & 0.013 & 0.003 & 0.000 \\ 0.000 & 0.000 & 0.001 & 0.002 & 0.001 & 0.000 & 0.000 \end{bmatrix}.$$

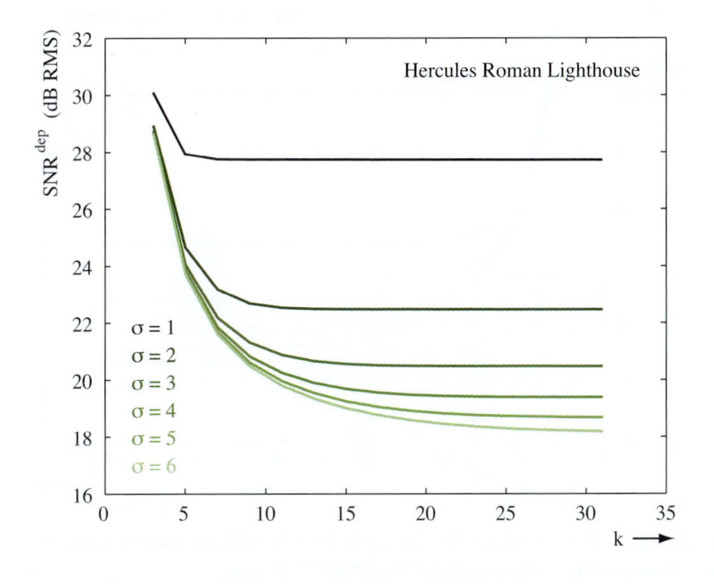

Figure 15.8. Different values for Gaussian filter-kernel parameters k and σ computed using the Tower of Hercules image for the SNR^{dep} metric.

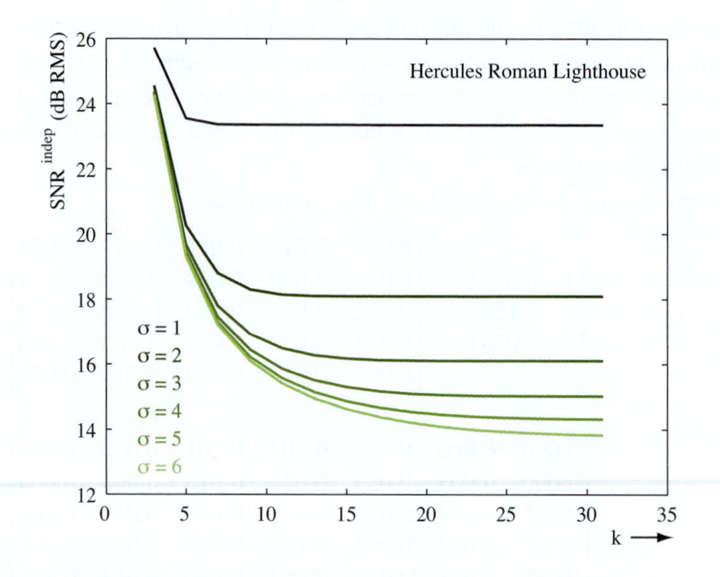

Figure 15.9. Different values for Gaussian filter-kernel parameters k and σ computed using the Tower of Hercules image for the $\mathrm{SNR}^{\mathrm{indep}}$ metric.

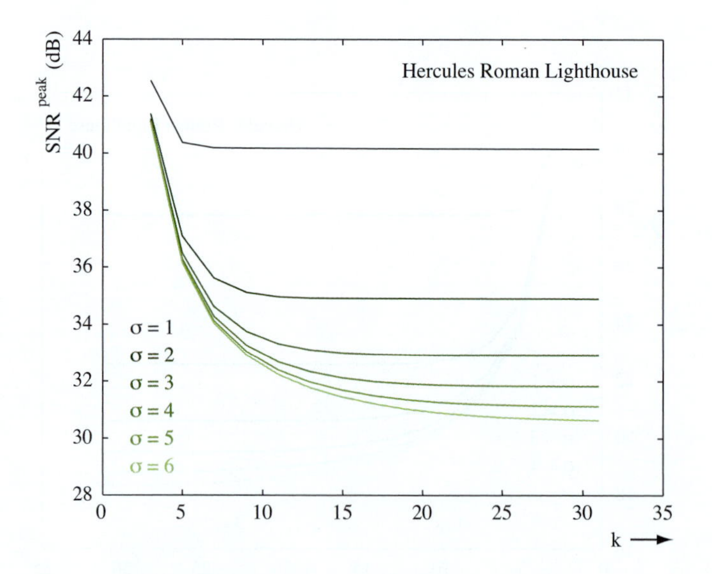

Figure 15.10. Different values for Gaussian filter-kernel parameters k and σ computed using the Tower of Hercules image for the $\mathrm{SNR}^{\mathrm{peak}}$ metric.

To show the impact of discretization, the k parameter is increased from 3 in steps of 2 for a Gaussian filter with widths $\sigma = 1 \ldots 6$, producing a sequence of blurred images $L_{k,\sigma}^{\text{blur}}$. These images are then used to compute the various SNR metrics outlined in the previous section. The results for the Tower of Hercules image of Figure 15.7 are shown in Figures 15.8, 15.9, and 15.10.

The trends that can be observed in these figures are typical, also for the other two example images. It is expected that if the image is blurred more, by using wider filter kernels, the difference between the blurred and unblurred versions becomes larger, and this increases the measured noise floor. As a result, the measured dynamic range will be lower. A second effect that is to be expected with this approach is that when a larger filter kernel is used, a larger number of samples is required before the results become stable. These figures thus highlight the fact that all three measures depend on the choice of parameter settings.

15.2.3 Median Filter

In computer vision, median filters are often used to remove noise in images. The median filter may therefore also be employed to aid in the computation of the dynamic range of an image.

The chief advantage of a median filter over a Gaussian filter is that it does not tend to blur over edges. It therefore tends to leave the signal intact, while only removing noise. This is achieved by considering a neighborhood of k by k pixels, sorting the values in ascending order, and selecting the median value. Each pixel is thus replaced by the median value of a neighborhood of pixels.

This leaves one user parameter to specify the size of the local neighborhood. We have measured the signal-to-noise ratios as before, but now on the basis of the median filter. For the three example images, the SNR measures are given in Figures 15.11 through 15.13.

These figures show that wider filter kernels reduce each of the SNR measures before settling into a stable value. However, when a stable value is reached, the image obtained by applying a median filter with a large kernel will have deteriorated significantly. An example is shown in Figure 15.14. As a result, the SNR measures that are based on a median filter affect both the noise as well as the signal. We therefore do not recommend using large filter kernels to measure noise. In the following, we will therefore use a value of $k = 3$ when a median filter is used for SNR measures.

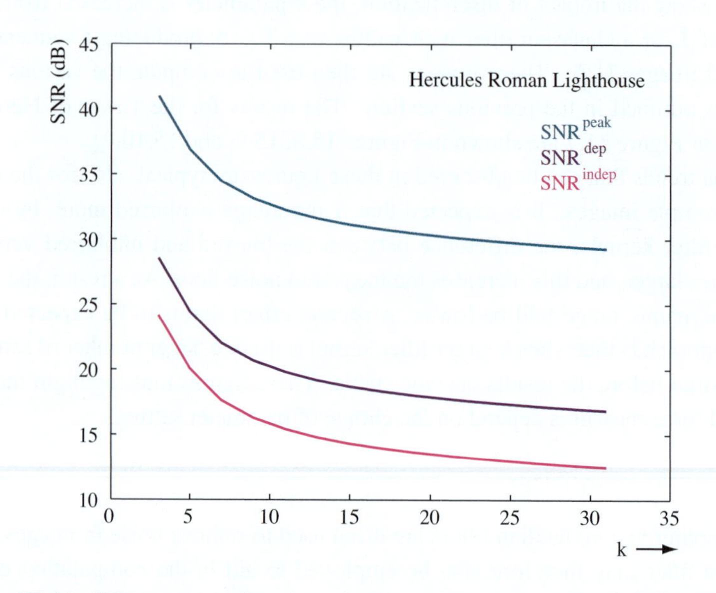

Figure 15.11. SNR measures plotted as a function of the median-filter parameter k for the Tower of Hercules image.

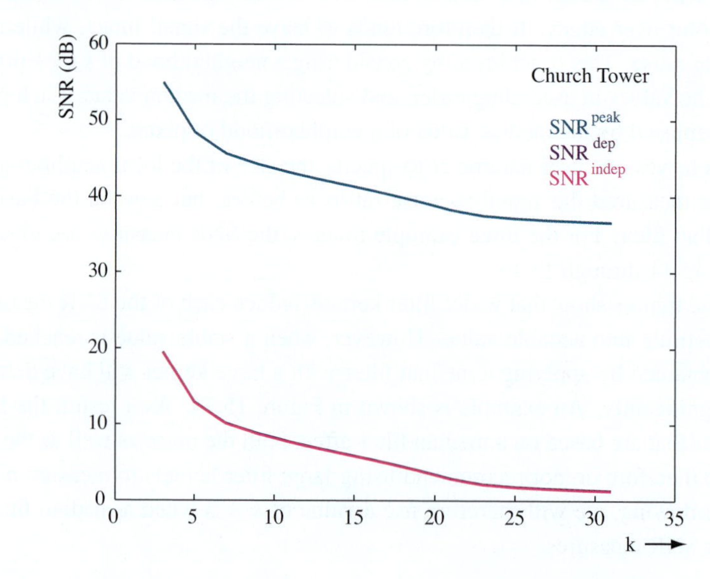

Figure 15.12. SNR measures plotted as a function of the median-filter parameter k for the Church Tower image.

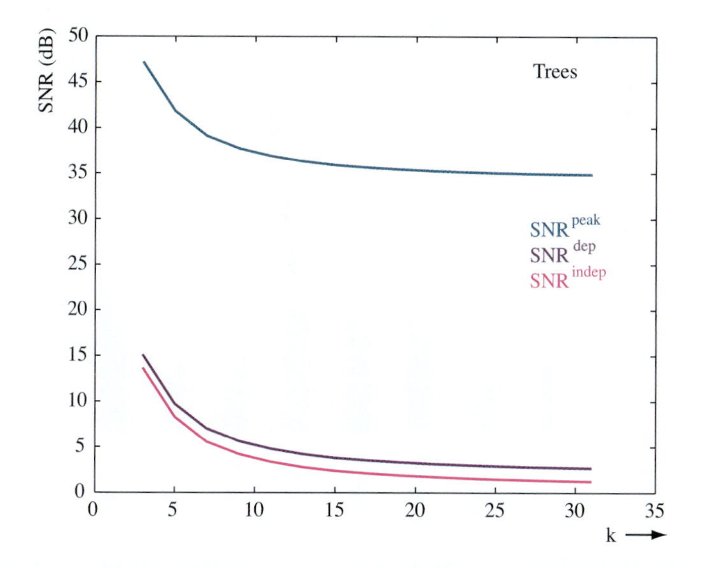

Figure 15.13. SNR measures plotted as a function of the median-filter parameter k for the Tree image.

Figure 15.14. Luminance values after filtering with a median filter with kernel size $k = 3$ (left) and $k = 15$ (right).

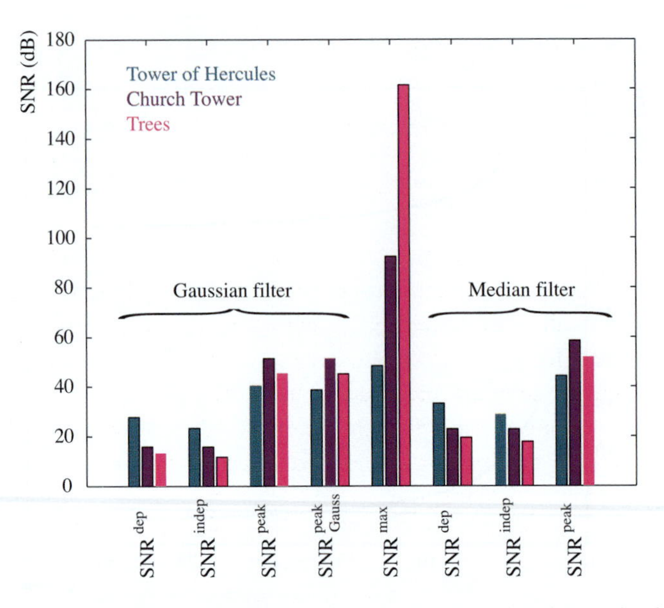

Figure 15.15. Different signal-to-noise metrics compared against each other.

15.2.4 Dynamic Range Measurements

To enable further comparison, we select a relatively small kernel size ($\sigma = 1$) for Gaussian filters, because the amount of noise present in these images is small. For this filter, the value of k needs to be at least 7 to obtain stable results. In the following, the median filter is applied with a kernel size of $k = 3$ to minimize affecting the signal, and thereby the signal-to-noise ratio.

We also compare two different version of SNR^{peak}. The first is based on $L^{\max}$, whereas the second is based on a filtered version of $L^{\max}$. The filter used here is a Gaussian filter with $\sigma = 1$ and $k = 7$. Such a small filter kernel should remove single-pixel outliers, and is therefore likely to be a more stable measure for the peak of the signal than the unfiltered maximum luminance.

Given these parameter settings, the various SNR measures are compared for all three example images in Figure 15.15. There are several observations to be made on the basis of this comparison. First, it can be seen that the SNR measure based on the minimum and maximum luminance overestimates the dynamic range in the images. In addition, $\text{SNR}^{\max}$ shows the largest range for the tree image, which can be attributed to a spurious pixel. It is therefore not a good measure to use in general. This also means that reporting the dynamic range as the ratio of $L^{\max}/L^{\min}$ is not a sensible approach (despite the common practice of using exactly this measure).

On the other hand, the RMS-based measures rely on average values in the image. This results in a tendency to underestimate the dynamic range, as small but very bright features do not contribute much to this measure. However, it is precisely these features that have a significant impact on the dynamic range.

The best approaches therefore appear to be afforded by the peak signal-to-noise measures. Note that each of these measures, whether guided by a Gaussian or a median filter, rate the night scene to have the highest dynamic range. This is commensurate with visual inspection, which would rank this scene first, followed by the Tower of Hercules, and then the tree scene (which is expected to have the lowest dynamic range of the three example images).

The Gaussian-based peak-SNR measures provide very similar results, independent of whether the image is first blurred by a small filter kernel prior to determining $L^{\max}$. The median-based peak-SNR metric shows a somewhat higher dynamic range, presumably because this filter is better able to leave edges intact.

As such, of the filters and measures included in this comparison, we would advocate the use of a median filter-based measure of dynamic range, employing a peak-SNR measure, rather than a RMS measure.

15.3 Cross-Media Display

In many cases, the origins of an image are unknown. This means that even if the display device and the viewing environment have been appropriately characterized and corrected for, the image may not appear as intended. For color-accurate reproduction of images, we need several pieces of information about the image, as well as about the display device and its environment. These include display primaries, white point, gamma encoding, as well as a characterization of the display environment.

In the following sections, we assume that the original image is defined in a display-defined space, rather than a scene-referred space. As a result, in the preparation of an image for a particular display, the (non-linear) processing applied to the image must first be undone before the processing specific to the target display can be applied.

15.3.1 LDR to LDR Display Pipeline

A low dynamic range (LDR) image will be in a given color space, including its associated gamma encoding. The first step toward display of such an image is to

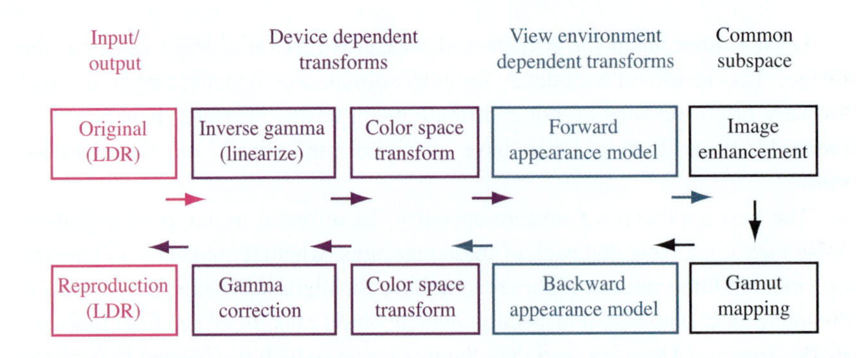

Figure 15.16. The processing steps necessary to prepare a low dynamic range image for display on a specific display device.

undo all the device-dependent encoding steps and bring the image into a common subspace where further processing may occur to account for a specific rendering intent. In addition, gamut mapping may occur in this subspace. The image is then ready to be transformed into the device-dependent space associated with the display device for which the image is being prepared. An overview of the steps involved is given in Figure 15.16 [795].

The first step in the transform from original low dynamic range image to the device-independent subspace is to linearize the data by undoing the gamma encoding. With known primaries and white point, the image can then be transformed into a device-independent color space (see Chapter 8). A color appearance model may then be applied to account for the viewing environment for which the image was originally intended (see Chapter 11).

The image is now in a common subspace, where further image processing may be applied, such as enhancing for accurateness, pleasantness, etc., depending on the reason for displaying a particular image; this step may be omitted depending on rendering intent.

The next step in the common subspace is to account for the difference in color gamut between the display device and the gamut defined by the image's primaries. In particular, if the color gamut of the display device is smaller than the image's, as for example in printing devices, the range of colors in the image must be shoe-horned into a smaller gamut. It should be noted here that source and target gamuts may overlap, so that in practice gamut compression may be applied for some colors and gamut expansion may be applied for other colors, even within the same image. Algorithms to accomplish this are called *gamut mapping algorithms* and are discussed in Section 15.4.

After gamut mapping, the linear, enhanced, and gamut-mapped image must be encoded for the display device on which the image is to be displayed. The first step is to apply a color appearance model in backward direction, using appearance parameters describing the viewing environment of the display. The image is then transformed into the display's device-dependent color space, followed by gamma correction to account for the non-linear relationship between input values and light intensities reproduced on the display..

The human visual system's adaptability is such that if one or more of the above steps are omitted, chances are that inaccuracies go largely unnoticed. Many applications are not color critical, and this helps ameliorate the effects of inaccurate color reproduction. However, if two color reproductions are shown side-by-side, for instance when a print is held next to a computer monitor, large to very large differences may become immediately obvious. For accurate cross-media color reproduction, all the steps summarized in Figure 15.16 need to be executed.

15.3.2 HDR to LDR Display Pipeline

In the case of high dynamic range images, the steps required to prepare for display are different. To begin, HDR images are linear by construction, so that linearization is not required. The sequence of events leading to color-accurate reproduction of high dynamic range images is currently not fully understood, but is an active area of research. However, it is believed that the sequence of steps outlined in Figure 15.17 will lead to the acceptable reproduction of HDR images.

We will assume the HDR data to be in a known device-independent color space. However, this is by no means guaranteed, and color space transforms may have to be applied before executing further device-independent processing. Image processing for rendering intent of HDR images is currently not an active research area. The same is true for gamut mapping of HDR images. As a result, the application of standard image processing and gamut mapping to HDR images is currently unexplored territory. The issue of HDR color is further discussed in Section 17.10.1.

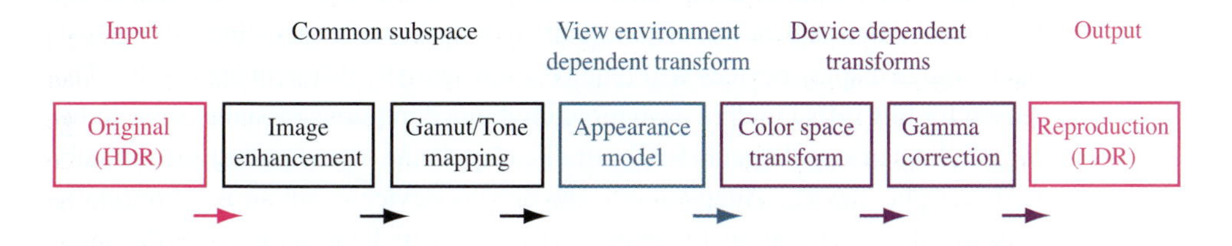

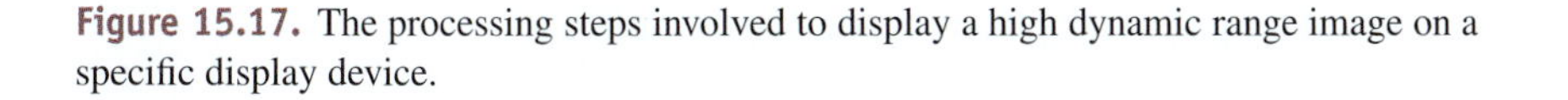

Figure 15.17. The processing steps involved to display a high dynamic range image on a specific display device.

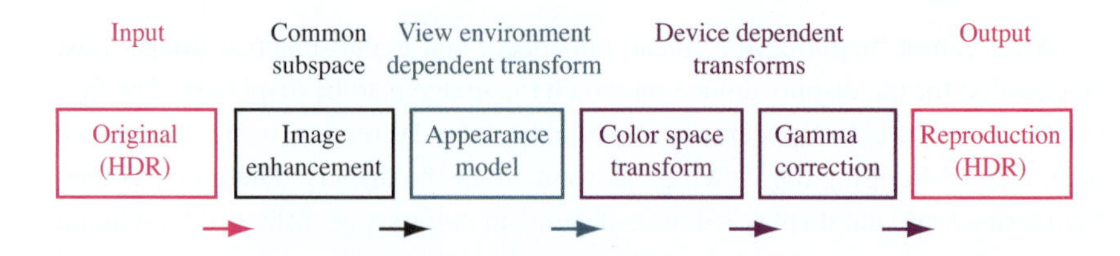

Figure 15.18. The processing steps involved to display a high dynamic range image on a high dynamic range display device.

The range of luminance values of an HDR image tends to be much larger than the display range of current common display devices. Thus, recently much effort has been expended on developing techniques to reduce the dynamic range of HDR images for the purpose of displaying them on low dynamic range display devices [952]. Dynamic range reduction algorithms are known as *tone mapping* or *tone reproduction operators*. They are the topic of Chapter 17.

A color appearance model should then be applied to account for the display environment. Here, a further interesting complication occurs, as most color appearance models were developed to account for changes in viewing environment over a relatively small range of luminance values. Whether color appearance models can be extended to operate over much larger luminance ranges is currently unknown.

The final two steps, a device-dependent color space transform and gamma correction, are identical for both low dynamic range and high dynamic range content.

15.3.3 HDR to HDR Display Pipeline

The emergence of high dynamic range display devices is further complicating the issue of color-accurate rendering. In theory, high dynamic range image content can be displayed directly without the need for tone reproduction, as shown in Figure 15.18. However, if the dynamic range of the input image exceeds the display capability, then some dynamic range reduction may still be needed. A much larger unknown in this sequence of steps is that color appearance modeling for high dynamic range display devices will require a substantially different approach. The reason for this is that under conventional lighting conditions (assume office lighting) the human visual system will partially adapt to the room lighting and partially to the display device. Adaptation to the display device is not an issue normally considered for conventional display devices, as their light output is of the same order as the lighting in a normal office environment. It is currently unknown how to account for partial adaptation to the display device.

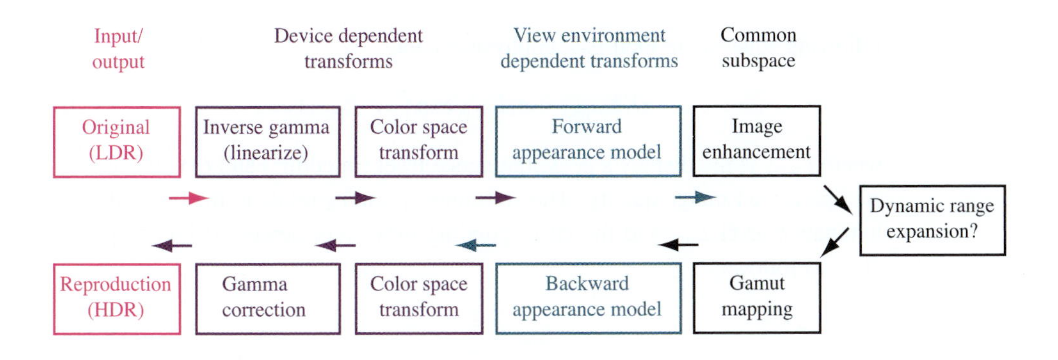

Figure 15.19. The processing steps involved to display a low dynamic range image on a high dynamic range display device.

15.3.4 LDR to HDR Display Pipeline

Finally, the display of low dynamic range images on a high dynamic range display requires careful consideration. As the input image now has a much lower dynamic range, an extra step involving range expansion may be required. The reason that such a step may be necessary is that the low dynamic range image may be a representation of a high dynamic range environment. The image may have originally been captured in high dynamic range and subsequently stored in a low dynamic range file format. Alternatively, the image may be created with conventional techniques, possibly exhibiting under- and/or over-exposed areas. In each of these cases, though, a more accurate reproduction may be obtained by expanding the range in a suitable way. However, dynamic range expansion for the display of conventional images, i.e., legacy content, on high dynamic range display devices is only beginning to be explored.

Nonetheless, a relatively consistent picture is starting to emerge. User studies have shown that the perceived image quality is positively affected by increasing both the contrast and the absolute luminance levels [1022, 1279]. Increasing both simultaneously yields the largest increase in perceived quality, although participants tend to prefer an increase in luminance level first, and only if the mean luminances of two images are the same, then the image with the higher contrast is preferred [26].

Further, Seetzen et al. have found a relationship between the perceived quality of images and the peak luminance and contrast ratio afforded by the display [1022]. For a given contrast ratio, the perceived image quality rises with peak luminance level up to a maximum, after which it falls again. They fit the following equation to their psychophysical data:

$$Q(L_{\text{peak}}) = c_1 L_{\text{peak}} \exp\left(-c_2 L_{\text{peak}}\right), \tag{15.20}$$

where L_{peak} is the display's peak luminance level (in cd/m^2) and Q is the associated perceived image quality. The constants c_1 and c_2 relate to the optimal peak luminance level L_{opt} and the corresponding maximum perceived image quality Q_{max} as follows:

$$L_{\text{opt}} = \frac{1}{b}, \tag{15.21a}$$

$$Q_{\text{max}} = \exp(-1)\frac{a}{b}. \tag{15.21b}$$

Values of L_{opt} and Q_{max} for various contrast ratios are given in Table 15.1. Further, it was found that the optimal contrast ratio ΔL_{opt} as a function of peak luminance L_{opt} can be broadly modeled with the following relation:

$$\Delta L_{\text{opt}} \approx 2862 \ln\left(L_{\text{opt}}\right) - 16283. \tag{15.22}$$

All studies suggest that simply emulating LDR display hardware when displaying legacy content on HDR displays would be a sub-optimal solution. Increasing luminance and contrast would be preferred. Algorithms to achieve appropriate up-scaling are termed *inverse tone reproduction operators*. Fortunately, provided that the LDR image is properly exposed and does not contain compression artifacts, the procedure for up-scaling can be straightforward. In fact, user studies indicate that linear up-scaling is adequate and is preferred over non-linear scaling [26].

In line with these findings, Rempel et al. propose a linear scaling for pixels representing reflective surfaces and a non-linear scaling for pixels representing light sources and other scene elements that are perceived as bright [955]. A similar technique was proposed specifically for the treatment of light sources [776, 778].

Contrast	L_{opt}	Q_{max}
2500	856	0.4756
5000	1881	0.5621
7500	3256	0.6602
10000	12419	1.0000

Table 15.1. Optimal display peak luminance values and corresponding maximum perceived image quality for various contrast ratios (after [1022]).

In video encodings, values over 235 (the white level) are typically assumed to represent light sources or specular highlights, and this value is therefore used to determine which type of up-scaling is used. Given that the conversion from low dynamic range to high dynamic range values contains two segments, there is a risk of introducing spatial artifacts when neighboring pixels are scaled by different methods. These effects can be mitigated by applying spatial filtering [955]. If the input image was not linear to begin with, it is first linearized by undoing the gamma encoding.

Input images are not always well-behaved; for instance they may contain under- or over-exposed areas. In certain cases, it may be possible to apply texture synthesis techniques to infer and re-synthesize the missing image areas [1207]. In addition, images may have been encoded using lossy compression. In this case, the up-scaling to a higher display range may amplify the artifacts that were initially below visible threshold to values that are easily perceived. Additional processing is therefore likely to be necessary to deal with problem images.

15.4 Gamut Mapping

One of the attributes of a color space is its gamut, the volume of colors that can be reproduced. The boundary of a color space, the gamut boundary, is a three-dimensional shape which is different for different color spaces. Given that each device has its own color space, it is evident that while we can expect considerable overlap between the gamut of different devices, there are ranges of colors in one device that cannot be reproduced by another device [795, 798, 1095]. As an example, Figure 8.1 shows that even for a single luminance value, the triangular boundaries imposed by several RGB color spaces are different.

In practice, this means that the conversion between two RGB color spaces may yield colors that cannot be represented in the target color space. Starting, for example, with the Adobe RGB color space which has a relatively large gamut, conversion to the smaller gamut of the sRGB color space tends to yield problem colors, as shown in Figure 15.20. Even larger differences in color gamut exist between display devices and printing devices. In particular, the CMYK gamut used by softcopy display devices is much smaller than typical RGB spaces.

However, it should be noted that even if the 3D volume spanned by the gamut boundaries leads to a smaller volume, the print gamut may be displaced with respect to the display gamut yielding non-overlapping regions. The print gamut may therefore include colors that the display gamut cannot reproduce (and vice versa). The implication is that gamut mapping algorithms will have to compress

Figure 15.20. The images on the left are assumed to be specified in the Adobe RGB color space. Conversion to the sRGB color space creates some pixels that are out-of-gamut. These are encoded as red pixels on the right. Top: Castle Coombe, UK, September 2006. Bottom: Stained glass window in Merchant Venturers Building, Bristol, UK, March 2006.

some colors and expand others. The discussion in this section outlines the issues related to gamut mapping, and follows the work by Ján Morovic [795, 797, 798].

To achieve a consistent output across media it is therefore necessary to map an image from one color space into a different one. Simple color space conversion is appropriate for colors that are within the overlapping region in both source and target color spaces. Colors that are outside the target color gamut need to be mapped to colors that are within the target device's gamut. The gray region in Figure 15.21 indicates the set of colors that cannot be reproduced given that the two example gamuts only partially overlap.

Remapping out-of-gamut colors to colors that are exactly at the boundary of the target device's gamut is called *gamut clipping*. This approach has the advantage that most colors remain unaffected and will therefore be accurately reproduced. However, if only out-of-gamut colors are remapped, then undesirable visible discontinuities may be introduced at the boundaries.

An alternative approach is to warp the entire gamut, thereby distorting all colors. This is called gamut mapping. The aim is then to keep the distortion of all colors to a minimum and, hopefully, below the visible threshold. Such an approach avoids jarring discontinuities, however, the trade-off lies in the fact that all colors will be distorted by some amount.

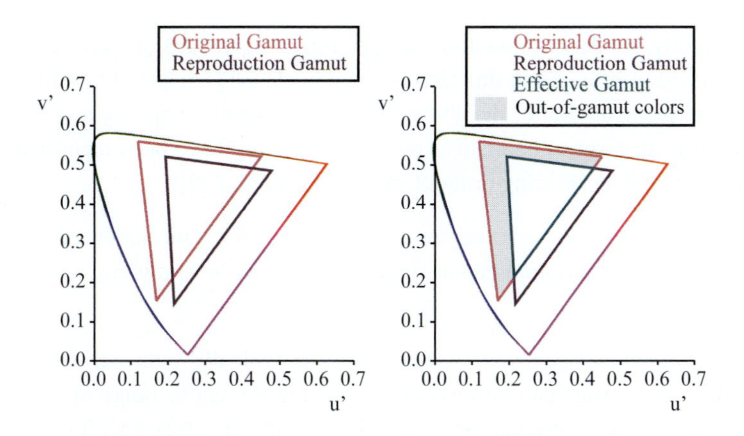

Figure 15.21. Partially overlapping gamuts (left) cause the colors located in the gray region to be out-of-gamut.

Various alternatives to these two extremes exist. For instance, in CIELAB space, it is possible to keep the luminance values unaffected and only warp a^* and b^* values. It is also possible to clip some values, leave others unaffected, and have a non-linear mapping for intermediate values.

Gamut mapping and gamut clipping algorithms require several stages of computation. First, the boundary of the reproduction gamut needs to be established. Second, a color space is to be selected in which the mapping is carried out (such as CIELAB). The type of mapping spans a further range of algorithms. While gamut clipping is relatively straightforward, there are many approaches to gamut mapping.

The image to be reproduced is given in a particular color space. However, the image's pixels form a color gamut that may be a subset of this color space's gamut. Hence, it may be beneficial to compute the image's gamut and map that to the reproduction device's gamut.

15.4.1 Aims of Gamut Mapping

When converting an image between different media, for instance by preparing an sRGB image for printing on a CMYK printer, the most appropriate gamut mapping algorithm may well depend on the application for which the image is printed. In general, the aim of gamut mapping is to ensure a good correspondence of overall color appearance between the original and the reproduction by compensating for the mismatch in the size, shape and location between the original and reproduction gamuts [795, 797].

Currently, models to quantify the appearance of images are still in their infancy, although progress in this area is being made (see Chapter 11). This makes a direct implementation of the above definition of gamut mapping more complicated. Instead, a set of heuristics for gamut mapping have been identified; they are listed here in decreasing order of importance [721, 795]:

Ensure that gray values are mapped to gray values. This means that achromatic values in the original gamut should remain achromatic in the reproduction gamut.

Maximize luminance contrast. As the gamut is typically reduced, some of the loss in contrast can be recovered by using the entire range of luminance values in the reproduction gamut. Even if this involves a shift in mean luminance, it is desirable to map the image's black and white points to the minimum and maximum reproducible luminance.

Reduce the number of out-of-gamut colors. Although ideally all the colors are brought within range, the overall appearance of an image can be improved by clipping some values against the boundaries of the reproduction gamut. In the extreme case, all the out-of-gamut colors are clipped, although this is unlikely to be the optimal case.

Minimize hue shifts. It is desirable to leave the hue of colors unmodified.

Increase saturation. Given that the reproduction gamut is smaller than the original gamut, it is often preferable to maximize the use of this smaller gamut, even if this means increasing saturation. This is similar to maximizing luminance contrast, albeit along a different dimension.

Maintain the relationship between colors [1095]. If the mapping of colors requires shifts in luminance, saturation, and/or hue, then it is important to shift related colors in similar ways.

It should be noted that this list of requirements, and their ordering, is largely based on practice and experience in color reproduction.

15.4.2 Gamut Boundaries

Before any gamut mapping can take place, it is necessary to compute the boundaries of the reproduction gamut. Dependent on the device, this can be achieved by means of the Kubelka-Munk equations [283] (see Section 3.4.6) or the Neugebauer model [725]. The most common approach, however, is to use sampling algorithms to derive gamut boundaries. Further methods for computing either media gamuts or image gamuts have also been discussed in the literature [176, 528, 621, 796, 989].

To achieve effective gamut clipping, as well as gamut mapping, an algorithm to intersect a line along which the mapping is to be carried out with the calculated gamut must be applied. Such problems can be solved with techniques akin to ray tracing [1042], although specific gamut intersection algorithms have been proposed as well [124, 467, 468, 794].

15.4.3 Color Spaces

Gamut mapping algorithms are commonly designed to preserve some perceptual attributes, while changing others, with the intended end result that all colors fall within the reproduction gamut. In addition, the visible difference between the original image and the reproduced image should be minimal.

Perceptual attributes, including hue, chroma, saturation, brightness, lightness, and colorfulness, are usually computed on the basis of a device-independent color space such as CIE $L^*a^*b^*$ or CIE $L^*u^*v^*$, or on the basis of a color appearance model such as CIECAM97s.

The quality of gamut mapping then depends, in part, on how accurate the predictors for these perceptual attributes are. For instance, CIE $L^*a^*b^*$ has deficiencies in the uniformity of its hue angle predictor in the blue region, i.e., around hue angles of 290°. Under ideal circumstances, the predictors for the different perceptual attributes should be independent. However, in this case, sets of colors that differ only in luminance or chroma may also cause differences in their hue prediction. Problems in this regard are smaller with the CIECAM97s color appearance model [795].

15.4.4 Gamut Clipping

Gamut clipping algorithms only affect colors that are within the boundaries of the original gamut, but outside the reproduction gamut. These colors are moved to lie on the reproduction gamut boundary. Colors that already lie within the reproduction gamut are unaffected.

Two variations of gamut clipping are shown in Figures 15.22 and 15.23. Note that although we show gamut clipping in a two-dimensional chromaticity diagram, in practice gamut clipping (and gamut mapping in general) is a three-dimensional problem. We only show two-dimensional examples because these are simpler to illustrate.

Figure 15.22 shows an example where out-of-gamut values are moved along the color axes until they reach the boundary. If the gamut boundary is aligned

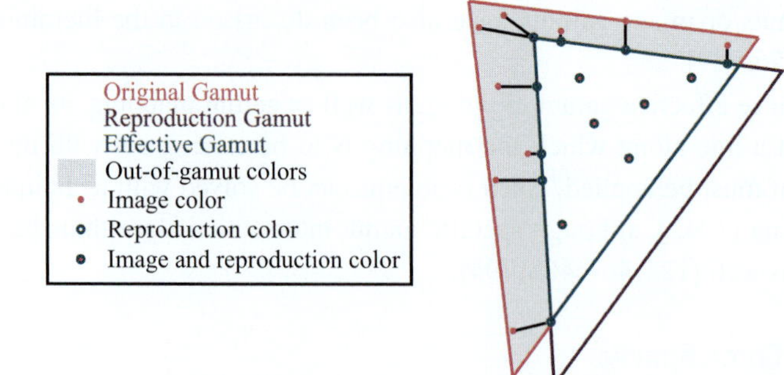

Figure 15.22. A simple gamut clipping scheme, whereby out-of-gamut colors are moved toward the reproduction gamut boundary along the axes of the color space in which gamut clipping is executed.

with the color space's axes, then this constitutes the smallest possible shift and, therefore, the smallest possible distortion in the final result.

Gamut boundaries, however, rarely ever align with the axes of any color space. A more accurate gamut clipping approach therefore shifts out-of-gamut colors along a direction orthogonal to the nearest gamut boundary, as shown in Figure 15.23.

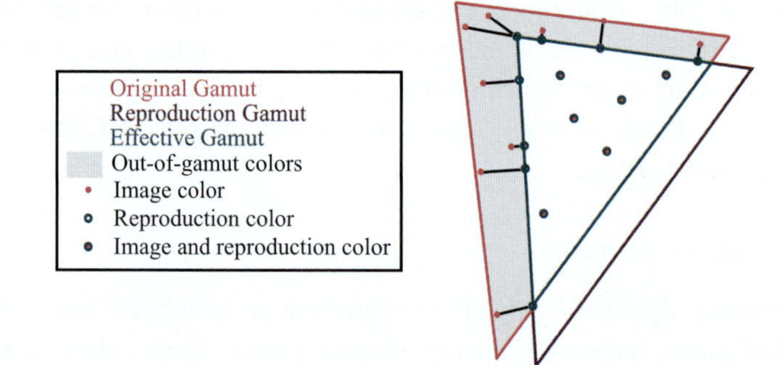

Figure 15.23. Gamut clipping where out-of-gamut colors are moved toward the nearest point on the reproduction gamut boundary.

15.4.5 Gamut Compression

Gamut mapping strategies, other than gamut clipping, will affect colors that are both inside and outside the reproduction gamut. It may be necessary to remap col-

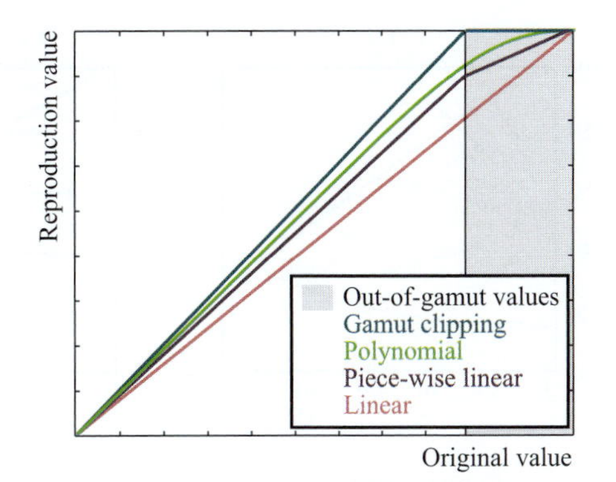

Figure 15.24. Gamut compression types (adapted from [795]).

ors inside the reproduction gamut to maintain their relationship with neighboring colors which may be outside the reproduction gamut. This means that all colors are remapped by some amount. Typically, colors that are outside the reproduction gamut are remapped over larger distances than colors inside the reproduction gamut. Mappings can be classified either as linear, piece-wise linear, polynomial, or a combination of these methods. Example remapping functions are shown in Figure 15.24.

15.4.6 Mapping Directions

Once a particular type of mapping is chosen, the gamut boundaries of the original and reproduction gamuts are known, and the direction of the mapping needs to be chosen. This direction can be chosen to be towards a specific point, such as the lightness value for which a cusp occurs, as shown in the top left plot of Figure 15.25. Alternatively the middle value on the lightness axis (in L*C*h* space for instance) can be chosen as the point toward which the mapping is directed (middle figure on the top row of Figure 15.25).

A different strategy remaps colors such that either lightness and hue, or saturation and hue remains constant (top right and bottom left of Figure 15.25). Finally, the direction of the mapping can be toward the nearest point on the boundary of the reproduction gamut (Figure 15.25, bottom right).

15.4.7 Comparison of RGB Spaces

The range of colors that can be reproduced in each RGB color space depends on the chosen primaries. Converting colors from one space to another is frequently required, for instance to prepare an image for display on a particular device. The

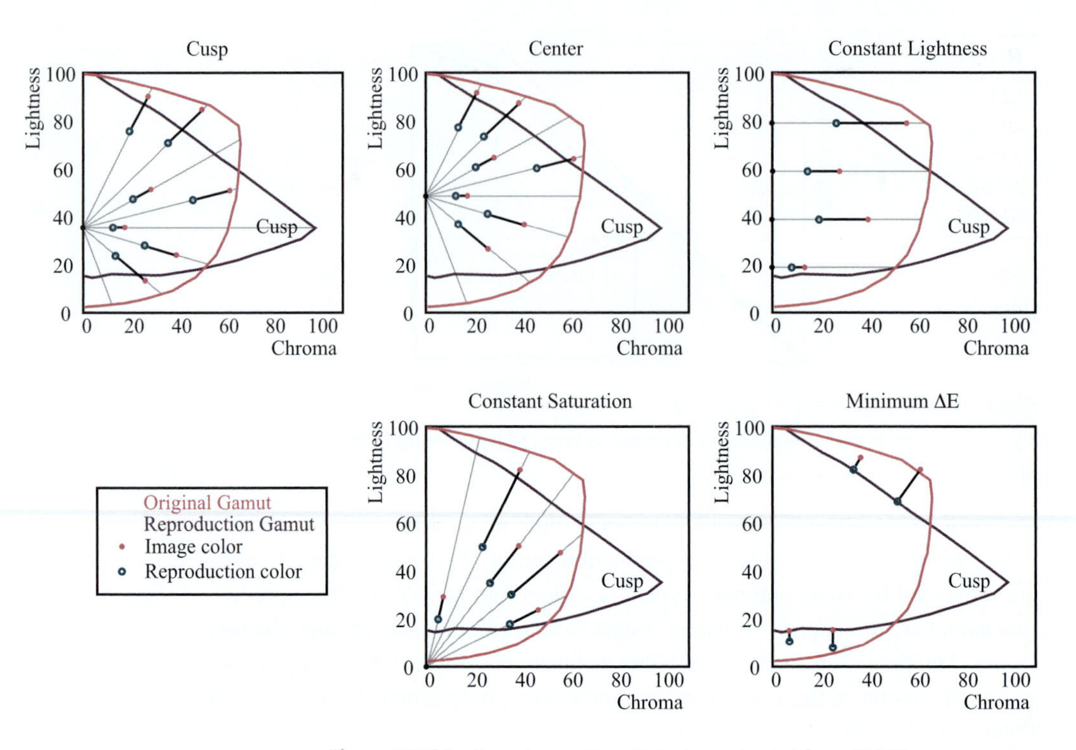

Figure 15.25. Gamut mapping strategies (adapted from [795]).

performance of different RGB color spaces has been measured in an experiment by Henry Kang [576, 577]. Prints are made of 150 patches of color using an inkjet printer. The prints are then measured under D65 illumination to obtain CIE XYZ tristimulus values, which are converted to CIE LAB. This ensures that the test set covers a range of colors that are commonly encountered, as they can all be reproduced on standard printers.

The measured XYZ values are then converted to one of the RGB color spaces with their associated non-linear encodings. The RGB data thus obtained is subsequently converted back to CIE LAB, for comparison with the measured CIE LAB values. Pairs of CIE LAB values are then subjected to the ΔE_{ab} color-difference metric (see Section 8.7.1) to assess the magnitude of the error for each RGB space/encoding. The percentage of colors that ended up being out-of-range after the RGB encoding process, as well as the average and maximum ΔE_{ab} errors are listed in Table 15.2.

If monochromatic lasers are used as primaries, then the vertices of the triangle spanned in a chromaticity diagram lie on the spectrum locus, thus giving a much

RGB space	Percentage out-of-range	Average ΔE_{ab}	Maximum ΔE_{ab}
SMPTE-C	30.0%	2.92	27.42
sRGB	25.3%	2.22	28.26
Adobe RGB (1998)	10.0%	1.17	19.83
NTSC	6.0%	0.87	12.19
SMPTE-240M	5.3%	0.74	9.60

Table 15.2. Percentage of out-of-gamut errors accumulated for different color spaces (after [577]).

wider gamut. This would result in fewer out-of-gamut problems. For instance, using monochromatic primaries with xy chromaticity coordinates (0.7117, 0.2882), (0.0328, 0.8029), (0.1632, 0.0119) for red, green and blue, respectively, yields 5.3% of out-of-range colors with average and maximum errors of 1.05 and 28.26, respectively. Although the average error is now much smaller, the maximum error remains large. This is due to the fact that the choice of primaries is important. In this case, the primaries are such that the line between the green and red primaries excludes a significant portion of yellow colors [577].

Finally, it is possible to select primaries that are outside the spectral locus. An example is the RIMM-ROMM space [1072]. In this color space, the red primary is at 700 nm, but the green and blue primaries are outside the spectral locus, having xy chromaticity coordinates of (0.7347, 0.2653), (0.1596, 0.8404), and (0.0366, 0.0001). In the above experiment, no patches were out-of-gamut, and the average and maximum ΔE_{ab} errors are 0.71 and 4.29, respectively [577]. Although it is clear that such a color space cannot be realized with any physical device, it is a useful intermediate color space, as it largely avoids gamut conversion problems.

15.5 Gamma Correction

Normally, final step in the preparation of an image for display is gamma correction. This step attempts to correct for the non-linear relationship between input voltage and strength of the light emitted by the phosphors of a cathode ray tube (CRT), known as the electro-optic conversion curve (EOCF).

This non-linearity is frequently well modeled by a power law whereby the exponent is given by the constant γ. This is especially true for CRTs. On the other hand, LCDs can have very different EOCFs, and may not necessarily be modeled by a power law, unless corrective look-up tables are included in the design.

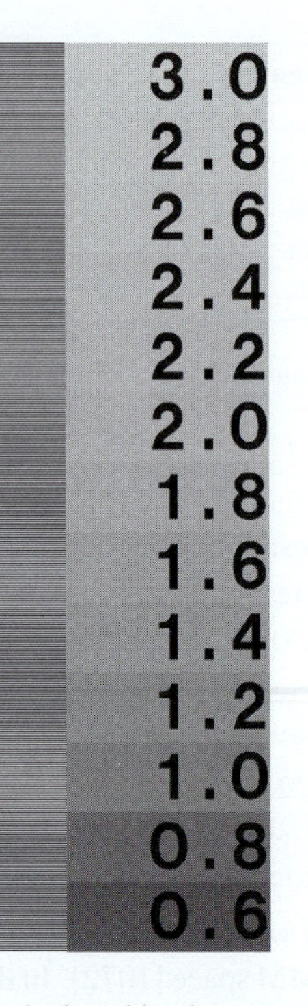

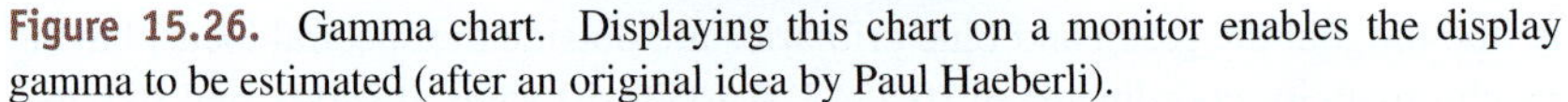

Figure 15.26. Gamma chart. Displaying this chart on a monitor enables the display gamma to be estimated (after an original idea by Paul Haeberli).

To correct for the non-linear response of power-law displays, gamma correction is applied using:

$$L'_d = L_d^{1/\gamma}, \tag{15.23}$$

where γ is the gamma value associated with the display device [923].

The EOCF of a monitor can be measured with a photometer. However, under the assumption of power law behavior, it is also possible to estimate the gamma exponent by displaying the image shown in Figure 15.26. The alternating black and white scanlines on one side should produce a middle gray as these lines are fused by the human visual system. This middle gray can be matched to an entry in the chart on the other side, which consists of patches of gray of varying intensities. The best match in the chart is the gray value that the monitor maps to middle gray. From the actual value in that patch $p \in [0,1]$ the gamma exponent of the monitor can be derived:

$$\gamma = \frac{\log(0.5)}{\log(p)}. \tag{15.24}$$

15.6 Ambient Light

After gamma correction, there are two further complications that should be addressed in critical applications. The appearance of an image is affected by both the state of adaptation of the observer—a problem addressed by color appearance modeling—and also by the fact that display devices are made of reflective materials. The former is discussed in detail in Chapter 11, while the effect of ambient reflections is discussed in this section.

Since a typical display device is located in a lit environment, such as a home or an office, some of that light falls on the monitor is reflected towards the viewer. This ambient light could be crudely modeled by adding a constant value to the light that is emitted from the screen. If L'_d is the light emitted by the display device, then L''_d is the light that reaches the observer:

$$L''_d = L'_d + L_R, \tag{15.25}$$

where L_R is the amount of reflected ambient light. While this component can once more be measured with a photometer with the display device switched off, there also exists a technique to estimate the reflections off a screen, similar to reading a gamma chart, albeit slightly more complicated [251].

This technique is based on Weber's law (see Section 5.5.2). It requires a method to determine a just noticeable difference (JND) ΔL_d for a particular display value L'_d, obtained with all the light sources in the room switched off. We can repeat this experiment with the room lights on, yielding a second JND ΔL_b.

To determine the just noticeable contrast, a test pattern with a set of patches increasing in contrast can be displayed. By choosing $1/f$ noise as the test pattern (as shown in Figure 15.27), the pattern matches the second-order image statistics of natural scenes. It is therefore assumed that the calibration is for tasks in which natural images are involved.

By reading this chart, the viewer is able to select which patch holds the pattern that is just visible. It was found that this task is simple enough to be fast, yet accurate enough to be useful in the following computations [251].

This pattern is read twice, once with the lights on, and once with the lights off. With the two JNDs obtained using the above method, Weber's law then states that

$$\frac{\Delta L_d}{L'_d} = \frac{\Delta L_b}{L'_d + L_R}. \tag{15.26}$$

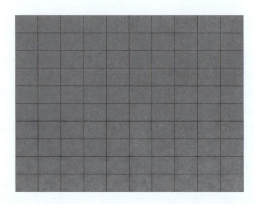

Figure 15.27. Test pattern used to determine the smallest amount of contrast visible on a particular display device. (Image from [251], © 2006 ACM, Inc. Reprinted with permission.)

From this equation, we can compute how much light L_R is reflected off the screen:

$$L_R = L'_d \left(\frac{\Delta L_b}{\Delta L_d} - 1 \right), \tag{15.27}$$

where all the variables on the right-hand side are known.

Knowing how much light is reflected from a particular display device when placed in a specific environment is sufficient for the development of an algorithm to correct for ambient light. As the screen reflections add light to the signal that is propagated from the screen to the viewer, it should be possible to subtract the same amount, i.e., L_R, from all pixels prior to display. However, this leads to two problems.

First, dark pixels may be mapped to negative values, which would then have to be clamped to zero. This is unacceptable as it will lead to uncontrolled loss of detail. Second, by subtracting a quantity from each pixel, the dynamic range of the display device is under-utilized as the range between $[L_{\max} - L_R, L_{\max}]$ is no longer used.

Alternative solutions to subtracting L_R from each pixel include modifying the gamma value used in gamma correction to account for ambient light [1216], using hyperbolic functions [700], or using histogram equalization [1230]. However, none of these techniques take into account the value of L_R we have just computed.

With L_R valid for a display value of L'_d, we seek a monotonically non-decreasing function f that maps L'_d to $L'_d - L_R$, has a slope of 1 at L'_d to maintain contrast at this display value, and maps 0 to 0 and m to m (where m is the maximum value). The constraints on this function $f : [0, m] \rightarrow [0, m]$ can be summarized as follows:

$$f(0) = 0, \tag{15.28a}$$
$$f(m) = m, \tag{15.28b}$$
$$f(L) = L - L_R, \qquad L_R \leq L \leq m \tag{15.28c}$$
$$f'(L) = 1, \tag{15.28d}$$
$$f'(x) \geq 0, \tag{15.28e}$$

By splitting this function into two domains, the first spanning $[0, L]$ and the second spanning $[L, m]$, these constraints can be met by setting [1006]

$$f(x) = \frac{px}{(p-1)x + 1}. \tag{15.29}$$

We first make the appropriate substitutes to yield the following pair of functions:

$$f_{[0,L]}(x) = \frac{(L - L_R)\, px}{x(p-1) + L}, \tag{15.30a}$$

$$f_{[L,m]}(x) = \frac{p \dfrac{x - L}{m - L} (m - L + L_R)}{(p-1) \dfrac{x - L}{m - L} + 1} + L - L_R. \tag{15.30b}$$

We can now solve for the free parameter p by using the restrictions on the derivative of f. For the $[0, L]$ range we have

$$f'_{[0,L]}(x) = \frac{p(L - L_R)}{L(p-1)x/L + 1} - \frac{(p-1)px(L - L_R)}{L_2\left((p-1)x/L + 1\right)^2} \tag{15.31a}$$

$$= \frac{p(L - L_R)L}{(xp - x + L)^2} \tag{15.31b}$$

$$= 1. \tag{15.31c}$$

By substituting $x = L$, we find $p_{[0,L]} = (L - L_R)/L$. Similarly, for the range $[L,m]$ the derivative $f_{[L,m]'}$ is given by

$$f_{[m,L]}(x) = \frac{p\,(m-L+L_R)}{(m-L)\left(\dfrac{(p-1)\,(x-L)}{m-L}+1\right)} \tag{15.32a}$$

$$= \frac{(p-1)\,p\,(x-L)\,(m-L+L_R)}{(m-L)^2\left(\dfrac{(p-1)\,(x-L)}{m-L}+1\right)^2} \tag{15.32b}$$

$$= 1. \tag{15.32c}$$

Solving for p yields

$$p_{[L,m]} = \frac{m-L}{m-L+L_R}. \tag{15.33}$$

Substitution into $f(x)$ then yields the final form of the remapping function:

$$f(x) = \begin{cases} \dfrac{(L-L_R)^2\,x}{L^2 - L_R\,x} & 0 \le x \le L, \\ \dfrac{x-L}{1 - \dfrac{L_R\,(x-L)}{(m-L+L_R)\,(m-L)}} + L - L_R & L \le x \le m. \end{cases} \tag{15.34}$$

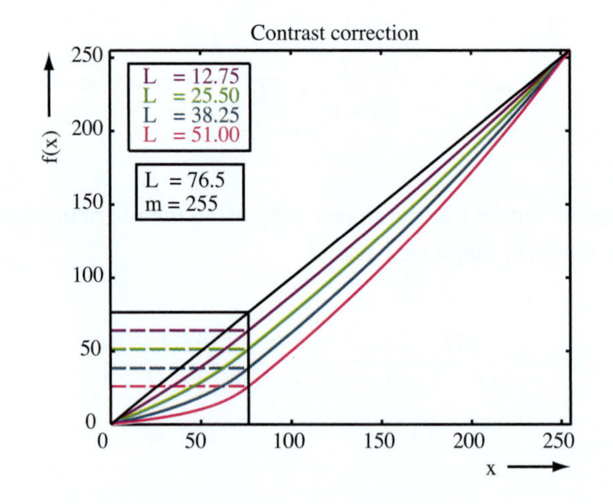

Figure 15.28. Example remapping functions for JNDs of 5%, 10%, 15%, and 20% of the maximum display value m. (Image from [251], © 2006 ACM, Inc. Reprinted with permission.)

Choosing the maximum value $m = 255$ and the pivot point L at one third the maximum, example mappings are plotted in Figure 15.28. It will be assumed that the pivot point L is the same as the luminance value for which the JNDs were computed, i.e., $L = L'_d$. As the correction algorithm is optimal for this particular value, and necessarily approximate for all others, the choice of pivot point, and therefore the display value L'_d for which the JNDs are computed, should be chosen such that the displayed material is optimally corrected. For very dark images, the JNDs should be measured for lower values of L'_d, whereas for very light images, higher values of L'_d could be chosen. In the absence of knowledge of the type of images that will be displayed after correction, the JNDs can be computed for a middle gray, i.e., $L = L'_d \approx 0.2\,m$.

In summary, by applying this function to display values, we get new display values which are corrections for ambient reflections off of the screen. The light that reaches the retina is now given by

$$L''_d = f(L'_d) + L_R. \tag{15.35}$$

For the pivot point $L = L'_d$, we have $f(L'_d) = L'_d - L_R$. The light that reaches the eye is therefore exactly $L''_d = L'_d$, which is the desired result. For values larger or smaller than L'_d, the correction is only partial.

Chapter 16
Color Management

Color management is arguably the most important practical application of many of the concepts we have discussed so far. As an abstract concept, color management describes how to manipulate color to achieve appearance matches from one physical device to another color imaging device. In order to achieve these matches, it is necessary to combine elements of color measurement, colorimetry, chromatic adaptation, and color appearance. This chapter will describe the general concept of a color management system and go into some detail about the de-facto industry standard *ICC Profile Format* for color management. A brief overview of color management using color look-up tables (LUTs) is also described. This chapter should be considered an introduction to the general concepts of color management and should not be considered a blueprint for creating or manipulating device color profiles. Rather, we hope this chapter brings a greater understanding of what a color profile is and what types of data it may contain.

16.1 A Generic Color Management System

Color imaging devices all inherently have a device-native color space based upon the characteristics of the device itself. These characteristics may be the light-producing primaries of a display device such as a CRT or LCD, the spectral sensitivities of a camera or scanner, or the inks, dyes, or toners of a color printer. Even if these devices have the same type of native color space such as an RGB space of a camera and a computer display, there is no reason to believe that matching device units in those RGB spaces will result in identical colors. The general concept of color management involves moving those device units from all the individual color spaces into a unified *device-independent* space. We have already been in-

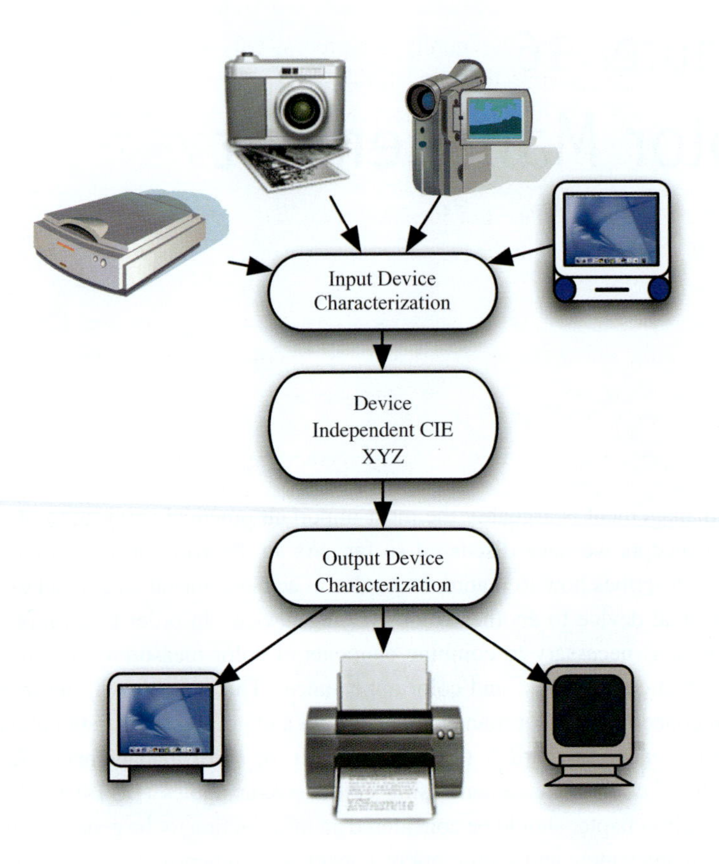

Figure 16.1. An example of device-independent color management using CIE colorimetry.

troduced to these types of color spaces, based upon CIE XYZ colorimetry and extended to include color appearance spaces such as CIELAB and CIECAM02. Recall our introduction to device-independent color from Chapter 7 using the CIE XYZ system of colorimetry, redrawn here in Figure 16.1.

For this type of color management system, we bring the individual device units into a common reference color space. The common reference color space, or connection space, must be well understood and unambiguously defined in order to be considered device-independent. We bring the colors into this space through a process known as device characterization, which involves careful measurement of the individual device properties. Conceptually, we can think of converting the device units into the reference color space as a forward characterization. From the reference color space we can then move to any output device using an inverse device-characterization. Essentially this process tells us what exact device units are necessary to generate the same XYZ tristimulus values. Again, the inverse characterization requires careful measurement of the properties of the output device. In reality, we may not have a specific one-directional nature for a color management system. We can think of many devices that could be considered on both the input and output side. For instance, a digital camera may capture an image and move that image into the reference space. From there, we may want to move that image to a CRT or LCD monitor for digital preview and manipulation. The final stage may involve printing the image, such that it matches our input device or matches the image that was displayed on the computer screen. In this situation it is necessary to have both the forward and inverse device characterization for moving into and out of the reference color space. There are many different techniques for characterizing color imaging devices, including modeling the physical properties of the device or through the use of color look-up tables (LUTs). An introduction to these techniques is given in this chapter; for more information the interested reader should see [93, 577, 1035].

Recall that CIE XYZ colorimetry only guarantees a color match under a single viewing condition; it tells us nothing about what that color actually looks like. Almost by definition, color management requires the generation of color matches with very disparate devices and viewing conditions, such as a computer display viewed in a dim surround and a printer viewed in an average or bright surround. In these situations, it may be desirable to have your reference color space be a color appearance space, or at least use certain aspects of color appearance models such as chromatic adaptation transforms or lightness rescaling. We may not actually generate colorimetric matches between the input and output devices, but are rather more interested in generating color appearance matches.

16.2 ICC Color Management

One of the de-facto industry standards for device color management is the ICC color management system. The International Color Consortium®, or ICC, was founded in 1993 to create an open color management system. The founding members of the ICC were Adobe Systems Inc., Agfa-Gevaert N.V., Apple Computer, Inc., Eastman Kodak Company, FOGRA (Honorary), Microsoft Corporation, Silicon Graphics, Inc., Sun Microsystems, Inc. and Taligent, Inc. [523]. Today the ICC has expanded to over 60 members in industry and academia. Specifically, the ICC is focused on developing and maintaining a color profile format, as well as the registration of specific tags and descriptions of elements for these profiles [523]. As of this printing, the current ICC profile format stands at Version 4.2.0.0, which is defined in the ICC.1:2004-10 profile specification and also at their website (http://www.color.org).

Much like our generic color management example given above, the profile-based system recommended by the ICC is grounded in the use of a reference color space which is "unambiguously defined" [523]. The reference color space chosen by the ICC is the CIE XYZ system of colorimetry, based upon the 1931 Standard Observer. Since we know the CIE XYZ system still contains some ambiguity with regard to the appearance of colors, further assumptions based upon a reference viewing condition and standard color measurement techniques were also defined. The reference viewing condition follows the recommendations of ISO 13655:1996, *Graphic Technology—Spectral Measurement and Colorimetric Computation for Graphic Arts Images* [533]. The reference medium is defined to be a reflection print, viewed under CIE D50 illumination at 500 lux, with a 20% surround reflectance. Additionally, the reflectance media should be measured using a standard 45/0 or 0/45 measurement geometry with a black backing placed behind the media substrate. This measurement geometry places the light source and color measurement sensor 45 degrees apart to minimize specular reflectance contributions. In addition, the ICC and ISO standards define the relative reflectance of the reflectance substrate to be 89% relative to a perfect reflecting diffuser, with the darkest value having a reflectance of 0.30911% [523]. These values correspond to a linear dynamic range of approximately 290–1. Additionally there are specifications for viewing flare, or stray light hitting the reference medium, of about 1.06 candelas per square meter. These standardized viewing conditions, along with the reference color space of the 1931 CIE XYZ tristimulus values, are called the Profile Connection Space (PCS). The PCS is the basis for all the color profile based transformations used in ICC color management, as shown in Figure 16.2. Additionally, the ICC allows the use of CIELAB as a PCS, calculated based upon the standard CIE XYZ tristimulus values as well as on the reference viewing conditions.

The color transforms represented by the circular T in Figure 16.2 represent a road map for converting between specific device units and the reference CIE XYZ color space. These transforms form the basis of the ICC color profiles. These transforms could be based entirely on color measurements of the particular imaging devices and conform to generating CIE XYZ colorimetric matches. Recall once again, from Chapter 7, that colorimetric matches are only guaranteed to hold up for a single viewing condition. When performing cross-media color reproduction, we are almost certain to attempt to generate color matches across vastly disparate viewing conditions, using a wide variety of devices. By providing

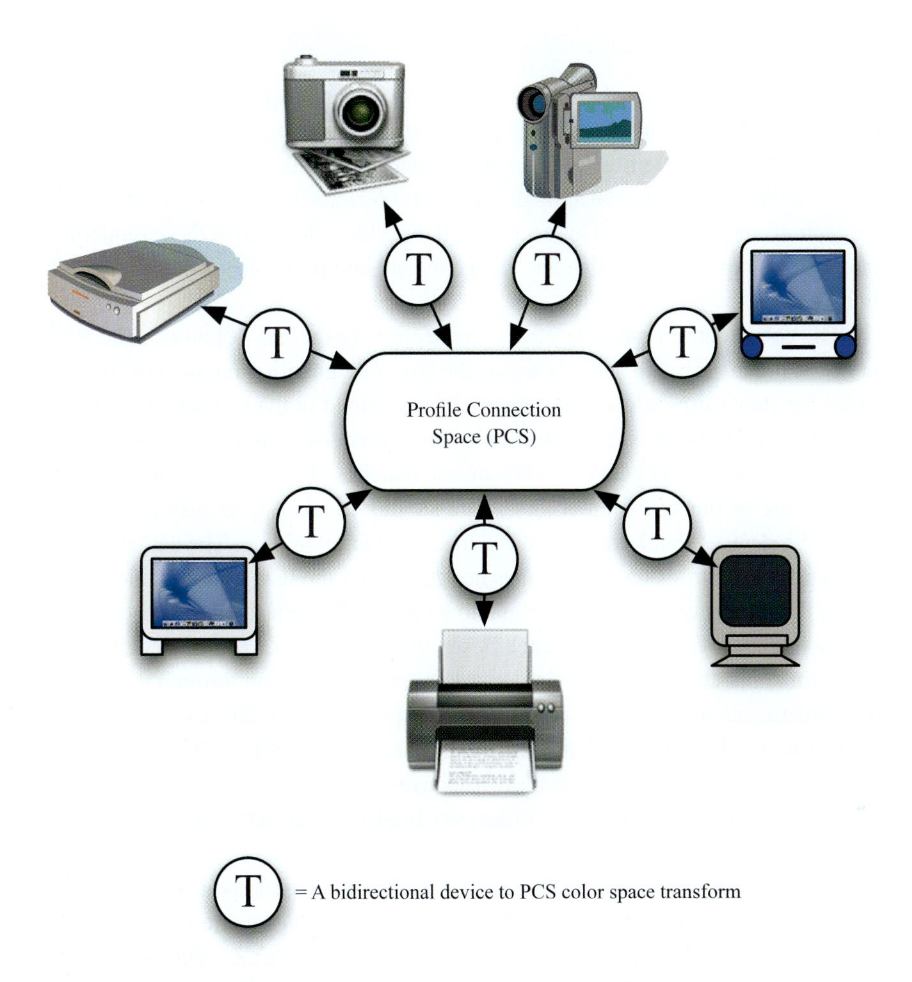

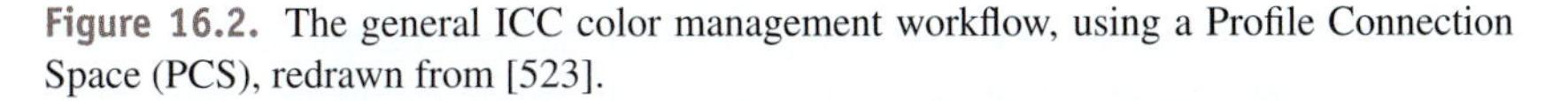

Figure 16.2. The general ICC color management workflow, using a Profile Connection Space (PCS), redrawn from [523].

additional information for the PCS, namely by providing a reference viewing condition and medium as well, the ICC has created a system for color management that can account for these changes in color appearance. This includes changes in the color of the viewing environment, using chromatic adaptation transforms, as well as changes in the surround, media-white and black points, and overall tone and contrast. How these color transforms are chosen and applied is specified by the *rendering intent* of the end-user.

16.2.1 Rendering Intents

One of the biggest areas of confusion that many users experience when they first encounter ICC color management is the choice of rendering intent offered by the system. The ICC specifies four distinct rendering intents, also called rendering styles, in their profile standard: media-relative colorimetric, absolute colorimetric, perceptual, and saturation. This section will describe what each of these rendering intents is trying to accomplish and the goals and compromises of these different intents.

The colorimetric rendering intents are seemingly the most obvious. The ICC profile standard does not actually define the absolute rendering intent in the profile itself, but rather uses additional information contained in the profiles to calculate ICC absolute transforms. The two colorimetric rendering intents are designed to faithfully preserve and generate colorimetric matches of *in-gamut* colors, often at the expense of out-of-gamut colors. Care should be taken when using these intents with devices that have large differences in gamut volume. The actual gamut mapping of the out-of-gamut colors is not defined by the ICC or the profile and is left to the discretion of the device manufacturer. Although the colorimetric rendering intents suggest a strict match based upon CIE XYZ colorimetry, they do allow for some changes in viewing or measurement condition through the use of a chromatic adaptation transform. When the input or output data are measured or viewed under illumination that differs from the reference CIE D50, a chromatic adaptation transform should be used to convert those data into CIE D50. The ICC recommends the use of the linearized Bradford transform, as described in Chapter 10.

The *media-relative colorimetric rendering intent* is based upon color data that is normalized relative to the media white point for reflecting and transmitting data. This implies that the relative media white will have CIE XYZ values of D50 in the PCS, or have CIELAB values of 100 for L^*, and 0, 0 for a^* and b^*. This transform rescales the colors, accounting for chromatic adaptation into the reference viewing conditions, such that the white matches the reference white. It should be noted that this rescaling is technically only appropriate for the in-gamut colors. The goal of this transform is to assure that the highlight detail in color images is maintained, by scaling all colors into the range of the reference medium, though this rescaling may introduce chromatic shifts for other colors [523]. To prevent these shifts from happening, it may be desirable to use the absolute colorimetric intent.

The *absolute colorimetric intent* does not necessarily rescale the data such that the whites match for a specific device and the reference medium. One of the requirements of an ICC profile is that it contains the PCS values of the media white, relative to a perfect reflecting diffuser. Whereas the reference viewing condition is said to have a relative D50 white of [0.9642, 1.0, 0.8249], the actual media white may be considerably different. In this situation the ICC absolute CIE XYZ tristimulus values are rescaled using

$$\mathbf{X}_{\text{abs}} = \left(\frac{\mathbf{X}_{\text{mw}}}{\mathbf{X}_{\text{D50}}}\right)\mathbf{X}_{\text{rel}}, \tag{16.1a}$$

$$\mathbf{Y}_{\text{abs}} = \left(\frac{\mathbf{Y}_{\text{mw}}}{\mathbf{Y}_{\text{D50}}}\right)\mathbf{Y}_{\text{rel}}, \tag{16.1b}$$

$$\mathbf{Z}_{\text{abs}} = \left(\frac{\mathbf{Z}_{\text{mw}}}{\mathbf{Z}_{\text{D50}}}\right)\mathbf{Z}_{\text{rel}}, \tag{16.1c}$$

where $\mathbf{X}_{\text{mw}}$ is the CIE tristimulus value of the media white, as specified by the *mediaWhitePointTag* and $\mathbf{X}_{\text{rel}}$ is the relative CIE tristimulus value being normalized. It is important to note that while this is referred to as ICC absolute colorimetric intent, it is actually closer to relative colorimetric scaling as defined by the CIE, since we are dealing with normalized units relative to a perfect reflecting diffuser. This alone has probably been a point of confusion for many users. The ICC absolute colorimetric intent may be most useful for simulating different output devices or mediums on a single device such as a computer LCD display, a process generally referred to as color proofing. A similar strategy could be utilized to simulate one type of printing press or technology on another printer. For practical applications, the ICC states that the media-relative colorimetric intent can be used to move in and out of the PCS, and absolute colorimetric transforms can be calculated by scaling these media-relative PCS values by the ratio of the destination media white to the source media white [523].

The ICC colorimetric intents should be based upon actual color measurements of the imaging devices, allowing for chromatic adaptation to the reference viewing conditions. It is possible that these color measurements do not represent actual color measurements but rather measurements that may have been altered in order to minimize color error for important colors, such as skin tones, or even to create preferred color reproduction of the in-gamut colors. Although these types of color measurements may not represent true CIE XYZ colorimetry, they will be interpreted as colorimetric transformations in an ICC workflow.

The *saturation* and *perceptual intents* are much less stringently defined than the two colorimetric rendering intents. These are purposely left to be vendor-specific, to allow for proprietary color rendering strategies. Of these two, the saturation intent is more specific. This intent was designed to allow for the reproduction of simple images where color saturation, vividness, or chroma is most important, such as business graphics and charts. In these situations, it may be desirable to perform non-colorimetric rendering in order to maintain the overall color saturation. An example of this may be actually changing the hue of a graphic in order to get the most pleasing image, e.g., using the full"red" primary of one computer display to reproduce a graphic created while viewing a second display whose red primary is different.

The *perceptual rendering intent* was designed to produce color reproductions that best match the *appearance* of images when viewed on different display devices. It is specifically for this rendering intent that it was necessary to accurately define the reference viewing conditions and medium, and those conditions are used only for perceptual re-rendering. This rendering intent allows for the rescaling of both the out-of-gamut and in-gamut colors, as necessary, in order to generate appearance matches. The ICC suggests that for perceptual intent the PCS should represent the appearance of an image as viewed by a human adapted to the reference conditions [523]. In addition, this rendering intent allows for appearance changes based upon the viewing conditions themselves, such as changes in the surround or illumination amounts, as discussed in Chapter 11. Since the relative white and black levels, and as such the linear dynamic range, are well defined in the reference viewing conditions this allows the perceptual intent to adjust the tone scale based upon these levels. For example, an input profile from a device with a larger dynamic range than the reference viewing medium could use something like a smooth sigmoid function to maintain details in the shadows and highlights. Sophisticated gamut-mapping algorithms or gamut-expanding algorithms can be used to shrink or expand the gamut in and out of the PCS.

As we learned in Chapter 11, viewing conditions can greatly alter the perception of color appearance. As illumination levels increase, we know that both the colorfulness and the contrast increase via the Hunt and Stevens effects. We know that, additionally, the surround levels can also change the apparent contrast, as demonstrated by Bartleson and Breneman. We know that our chromatic adaptation changes as a function of the mode of viewing, as well as illumination levels and other factors. It is possible to use a color appearance model to aid in the generation of the transforms used in the perceptual rendering intent. Many times we do not actually want our color reproductions to generate actual color appearance matches, but perhaps would rather they produce a pleasing color image that best utilizes the input or output device space. A conceptual flowchart of using the perceptual rendering intent for a single input and output imaging device is shown in Figure 16.3.

Whereas the colorimetric rendering intents should build the device-to-PCS and PCS-to-device transforms based upon measurements and device characterization, this is not absolutely necessary in perceptual intent transforms. To achieve

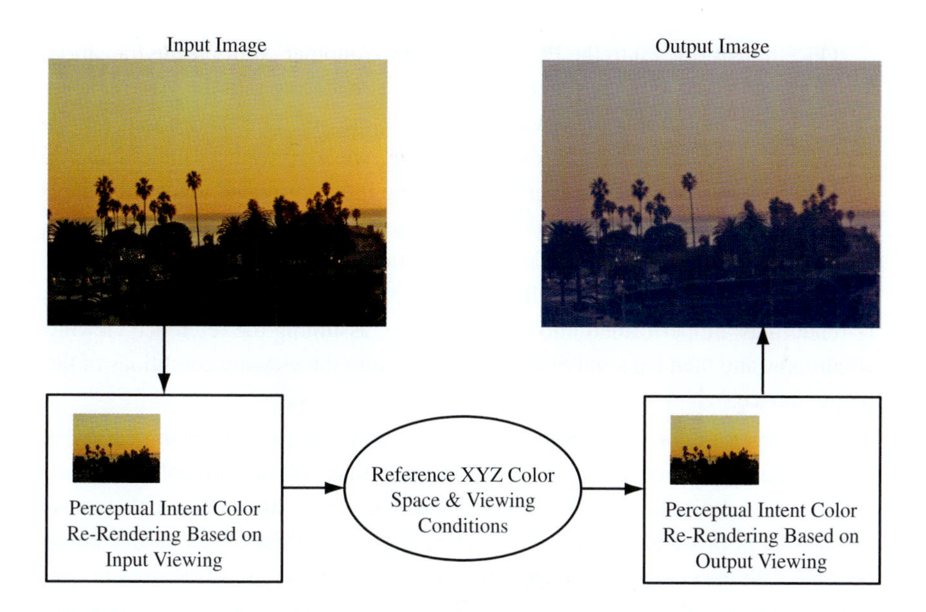

Figure 16.3. An ICC perceptual rendering intent workflow with a single input and output device.

any degree of color reproduction accuracy, however, many transforms do start with colorimetric measurements as a baseline. In the example shown in Figure 16.3, an input image is re-rendered into the profile connection space, based upon the defined reference viewing conditions. This re-rendering should take into account the properties of the scene, as well as properties of the device itself. For instance if the image taken is a brightly illuminated outdoor scene using a camera capable of capturing a wide dynamic range, then the perceptual intent re-rendering would need to rescale, or tone-map the image down into the reference color space. Chapter 17 gives more details on how this type of dynamic range reduction transform may be accomplished using an automatic algorithm.

A typical re-rendering would have to at least map the tone-scale into the lower range of the reference media, taking care to preserve highlight and shadow detail. Careful consideration should be given to areas of the image that cannot be faithfully reproduced by the reference medium. For example, since the reference medium is defined to be a typical reflection print, there is a maximum reflectance that is bounded by the printing substrate or paper—defined to be 89% of a perfect reflecting diffuse surface. In many real images, the specular highlights may be well above even a 100% diffuse surface. We can either scale these highlights down into the range of the PCS or allow the PCS to have values greater than 1.0 for CIE Y, or 100 for CIELAB L^*.

Once transformed into the PCS, we then have another color profile transform to an output device. This is generally another re-rendering into the viewing conditions of the output display device. If we are moving to a device that also is capable of reproducing a large dynamic range, we may wish to undo some of the tone compression that was performed on the input, provided this can be done in a visually lossless manner. Additionally, the perceptual re-rendering can take into account properties of the viewing conditions, adjusting the overall brightness or contrast as a function of illumination level and the surround. Again, these re-renderings are performed into the PCS space, assuming the reference viewing conditions, and then back out of that PCS space into the viewing conditions of the output device.

Color measurement and color appearance models may be used to generate a baseline perceptual rendering transform, but they may still not generate an overall desired color reproduction. As such, it is possible to manipulate the color transformations in order to maximize overall image pleasantness. Just as this could be done with the colorimetric transforms, for example, to minimize errors in specific regions of color space, the PCS will assume that color appearance matches are being generated. As such, the creation of perceptual transforms used in the ICC color management system can almost be classified as part science and technology and part creative or artistic choice. Commercial software is available to generate colorimetric profiles and transforms for most common color imaging devices, though there are far fewer resources available for generating these perceptual transforms.

To reemphasize this point, remember that the perceptual rendering intent is used to generate *desired or pleasing* color appearance matches, and the specific definition of these transforms is left to the actual manufacturers, vendors, and implementers of the ICC profiles. It is quite likely that the color-managed image reproduction between multiple devices using the perceptual intent will be different based upon the different manufacturers' choices of implementation. Although this does sound slightly ominous with regard to accurate color reproduction across different device manufacturers, by forcing all the perceptual rendering transforms to first go into the viewing conditions of the reference PCS and then using these same reference conditions as well on output, the ICC system does mitigate in practice manufacturer differences. It should also be noted that although the reference viewing conditions are designed to emphasize the graphic arts and printing industry, similar concepts can be utilized or extended for pure soft-copy reproduction, such as digital camera and computer display, or a film and video workflow.

All of the color transforms for these different rendering intents are actually encoded into the color profiles themselves. Many, but not all, profiles supply all of these transforms internally and allow the color management engine to choose the appropriate one based upon user input. This potentially makes the generation of color profiles a complicated, and even artistic, procedure. Many end users of color management systems are also often confused about which rendering intent to choose. It is our goal that this chapter will help aid them in making their decisions. The ICC also provides some guidance, generally recommending the perceptual rendering intent when strict adherence to colorimetric matching is not necessary, such as when creating pleasing natural image reproductions [523].

They also recommend using the perceptual transform when the input and output devices have largely disparate gamuts or properties, as the perceptual intent should do a better job of gamut-mapping and taking into account changes in viewing conditions. When working in a proofing environment, e.g., when simulating one type of device on another, the ICC recommends using the colorimetric intents. Remember that the colorimetric intents are only attempting to generate colorimetric matches for colors that are within the gamut of the devices; they do not have defined gamut-mapping algorithms associated with them.

16.2.2 The Color Management Module

In the ICC framework of color management, the profiles themselves provide all of the information necessary to generate the color transformations for all of the different rendering intents. These profiles are then linked together by a color management engine, or color management module (CMM). The general architecture of the ICC color management system is shown in Figure 16.4. The job of the CMM is to parse the profile data structure, interpret the correct set of profile tags necessary to generate a set of color transforms based upon the rendering intent, and link profiles together to create an end-to-end color reproduction system. The ICC does allow for the use of third-party color management modules, as well as private data tags in the color profiles themselves, in order to allow for proprietary discrimination or value.

The example architecture shown in Figure 16.4 is designed to operate at the computer operating system level, where an application sits on top of, or can refer to, the color management interface. Additionally the application can call certain graphics or imaging libraries that also sit on top of the color management interface. The interface also handles the file reading, and perhaps parsing, of the ICC profiles and passes that information to the color management module. In order to function properly, there must always be a default color management module that handles the generation and application of the appropriate color transform, which is then passed back through the color management interface. By allowing third-party developers to plug into the color management interface rather than to

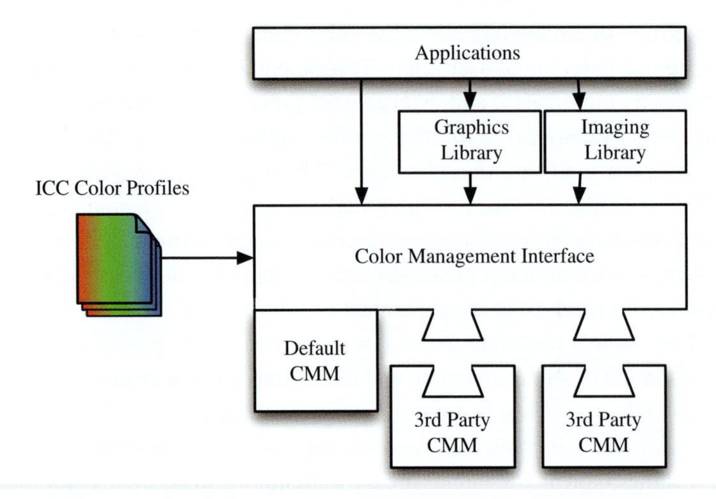

Figure 16.4. A general overview architecture of the ICC color management framework, redrawn from [523].

applications and libraries directly, the system allows dynamic swapping of the actual CMM used. When used at the operating system level, this type of architecture allows all computer applications to receive identical color management and not worry about applying the color transforms themselves.

16.2.3 ICC Profiles

As we have already discussed, the most important pieces of the ICC color management system are the color profiles themselves. These are essentially binary files that contain the information necessary for the CMM to generate the color transforms. These are required to contain a 128-byte header, a profile *tag table*, and the profile tagged element data. These tags and the data they point to represent the essence of the profile, and a registry of them is maintained by the ICC. The ICC specification does allow for the use of private data tags for use with third-party color management modules. It is important to note that Version 4 of the ICC specification allows for the use of 8-bit and 16-bit integers for the data elements in the profile, as well as for the profile connection space. Care must be taken when performing color transformations with limited bit-depth, as it is very easy to introduce quantization errors. There are far too many defined tags for the scope of this text, so this section will outline several important tags as an introduction to the ICC profile format. The interested reader is encouraged to examine the full ICC specification for more details [523].

The profile header contains useful information about the profile itself, as well as the type of transforms defined. Among other elements, it contains information about the canonical input space, e.g., RGB or CMY space and the desired profile connection space or canonical output space, e.g., XYZ or CIELAB. It can also include information about the preferred CMM, the manufacturer and model of the device being profiled, the profile creator, etc. Additionally the profile header contains the *class* of the profile. There are seven distinct classes of profiles, and these are defined to make it easier for the CMM to implement certain types of color transforms. The seven classes are:

- input device profile;
- display device profile;
- output device profile;
- DeviceLink profile;
- ColorSpace conversion profile;
- abstract profile;
- named color profile.

Of these classes, the input, output, and display profiles are by far the most common among standard color imaging systems. The device link profile is more exotic, and is used to link two devices together without using the PCS. This may be beneficial when performing color reproduction in a closed system, such as tying a scanner or camera directly to a printer. Additionally these classes of profiles may be used for color proofing of multiple printers, while avoiding any cross-contamination, for example, between the CMYK primaries of the two printers. The color space conversion profile is designed for transforming between a given color space that does not describe a specific device and the PCS. For example, it would be possible to use this type of profile to convert from images encoded in an LMS cone space into the PCS XYZ space. The abstract profile and named profile are even more rare. The abstract profile can contain color transformations from the PCS to the PCS and could be used as a form of color re-rendering. The named color profile is used to transform from device units into a specific categorical color space, corresponding to color names. This could be used for a specific device that has a set of named colorants or primaries.

Following the profile header is the tag table, which serves as an index into the actual transform information contained in the profile. The tag table includes

Input value	LUT array index	Output value
0.0000	0	0.0000
0.1000	1	0.0063
0.2000	2	0.0290
0.3000	3	0.0707
0.4000	4	0.1332
0.5000	5	0.2176
0.6000	6	0.3250
0.7000	7	0.4563
0.8000	8	0.6121
0.9000	9	0.7931
1.0000	10	1.0000

Table 16.1. Example one-dimensional look-up table (1D-LUT) approximating a gamma of 2.2.

elements form the first, second, and third columns of this matrix, respectively:

$$\begin{bmatrix} X \\ Y \\ Z \end{bmatrix} = \begin{bmatrix} X_{\text{red}} & X_{\text{green}} & X_{\text{blue}} \\ Y_{\text{red}} & Y_{\text{green}} & Y_{\text{blue}} \\ Z_{\text{red}} & Z_{\text{green}} & Z_{\text{blue}} \end{bmatrix} \cdot \begin{bmatrix} R \\ G \\ B \end{bmatrix}_{\text{device}} . \tag{16.5}$$

The red, green, and blue *TRCTags* represent transforms for undoing any nonlinear encoding of the device color space, and as seen in Figure 16.5, they are applied before the linear transform to CIE XYZ. The curve types allowed by the *TRCTags* include a simple power or gamma function of the form $Y = X^{\gamma}$. These values are stored as 16-bit fixed numbers, with eight fractional bits. In practice, this means that the numbers encoded would be divided by 2^8 or 256 to get the appropriate gamma value.

Alternatively, the *TRCTags* can encode sampled one-dimensional functions of up to 16-bit unsigned integer values. These sampled functions can be thought of as one-dimensional look-up tables, or 1D-LUTs. Perhaps the easiest way to think about 1D-LUTs is to represent the table as an array of numbers. The input color, in device space, is then the index into this array, and the output is the value contained at that index. A simple example is given in Table 16.1, using an 11-element LUT to approximate a gamma function of 2.2. The ICC specification states that function values are between entries; linear interpolation should be used to calculate the intermediate entries. We can plot the 11-element 1D-LUT, linearly interpolated to 256 entries against an actual gamma function, as shown in Figure 16.6.

This figure suggests that there is a reasonable relationship between the small 1D-LUT approximating the 2.2 gamma function and the actual function. This re-

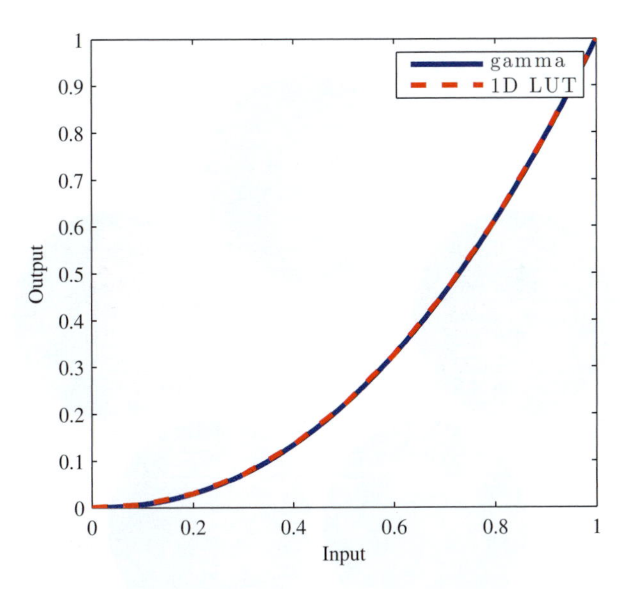

Figure 16.6. The differences between an 11-element 1D-LUT interpolated to 256 elements and an actual gamma curve. Although the functions appear similar, the differences between the two can lead to visual errors.

sult is actually slightly misleading, as we can find out by calculating the color differences with an 8-bit image encoded with a 1/2.2 gamma function. An example image is shown in Figure 16.7. We can convert this image into XYZ by applying the two different TRC functions, followed by a matrix transformation. From there we can calculate CIE $\delta E^{*}{}_{a}b$ color differences to get an idea of how large the error may be. Assuming Figure 16.7 is encoded using the sRGB primaries, with a gamma correction of 1/2.2, we find that the maximum color difference between using a real gamma function and an 11-element 1D-LUT is about 3.7. The complete range of errors is shown in Figure 16.8. We can see that there are some quantization artifacts that would be clearly visible. This suggests that when using 1D-LUTs, care should be taken to use as large a LUT as possible to avoid errors introduced by the interpolation. This idea will be revisited later in the chapter with regards to 3D-LUTs.

Again, these *MatTRC* transformations are only really valid for certain types of devices and are only defined for input and display classes of profiles. They are also only used for colorimetric transformations into and out of the CIE XYZ version of the PCS. For the other rendering intents, or for other color imaging devices, the transforms used can be generically referred to as the *AToB* and *BToA* transform, for those going into and out of the PCS, respectively. The specific tags

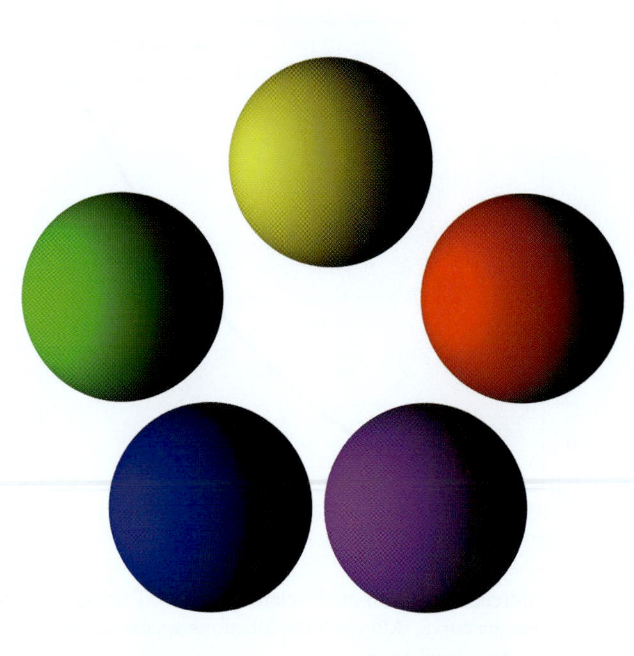

Figure 16.7. An example image encoded with sRGB primaries and a 1/2.2 gamma compression.

Figure 16.8. The CIE δE^*_{ab} color differences between using a small interpolated 1D-LUT and an actual gamma function to move into the ICC PCS.

Profile class	MatTRC	AToB0Tag	AToB1Tag	AToB2Tag
Input	colorimetric	device to PCS perceptual	device to PCS colorimetric	device to PCS saturation
Display	colorimetric	device to PCS perceptual	device to PCS colorimetric	device to PCS saturation
Output	undefined	device to PCS perceptual	device to PCS colorimetric	device to PCS saturation
ColorSpace	undefined	device to PCS perceptual	device to PCS colorimetric	device to PCS saturation

Profile class	BToA0Tag	BToA1Tag	BToA2Tag
Input	PCS to device perceptual	PCS to device colorimetric	PCS to device saturation
Display	PCS to device perceptual	PCS to device colorimetric	PCS to device saturation
Output	PCS to device perceptual	PCS to device colorimetric	PCS to device saturation
ColorSpace	PCS to device perceptual	PCS to device colorimetric	PCS to device saturation

Table 16.2. The different color transformation tags and rendering intents available in ICC profiles.

used by the ICC are summarized in Table 16.2, for the input, display, output, and color space classes of transform. The three other classes have specific rules and tags and are beyond the scope of this text.

Conceptually, we can refer to all of the *AToB* and *BToA* transforms as those based upon color look-up tables (cLUTs). This is perhaps an oversimplification, as we can see in Figure 16.9, as these transforms can contain 1D-LUTs, matrix transforms, multi-dimensional cLUTs, or combinations of these types of transforms. Specifically, the AToB transforms contain up to five processing elements that are applied in a specific order: 1D-LUTs "B" Curves, 3×3 matrix with an offset (represented by a 3×4 matrix), 1D-LUTs "M" curves, n-dimensional color LUTs, and 1D-LUTs "A" curves. The four combinations of these, shown in Figure 16.9, can be made by setting specific individual transforms to identity.

We have already discussed how the one-dimensional look-up tables can be applied in the *MatTRC* ICC transforms. Conceptually, the "A", "M", and "B" curves work the same way. It is important to note that the "A" curves, of which there are as many as there are color channels in our imaging device, can only be used in conjunction with the multi-dimensional cLUT. The matrix transform is very similar to that used in the *MatTRC*, though there is an additional offset element. This additive offset could be useful to account for differences in the black of the device space and that of the PCS, or it could be used to account for flare in the viewing conditions. The multi-dimensional look-up table (cLUT) is perhaps the most important element of the AToB transforms. As such, a quick overview of how multi-dimensional look-up tables work is in order. We should also note that when our device-space has more than 3 primaries, it is the cLUT that reduces the dimensionality of those channels down to the three dimensions of the CIE XYZ or more often CIELAB PCS.

We can think of a cLUT as a dictionary that translates specific device primaries from the device space into another color space. Consider a simple color

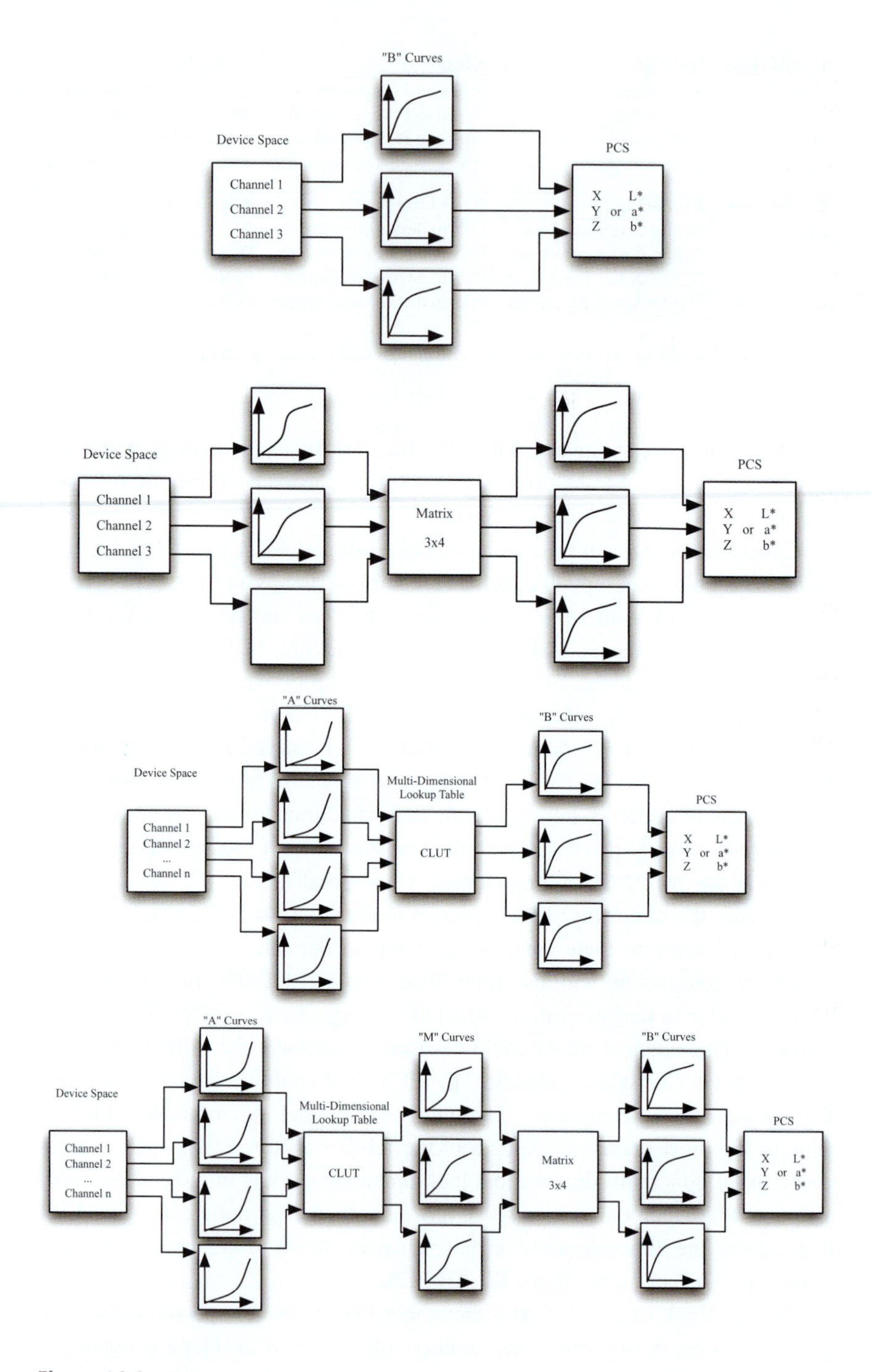

Figure 16.9. Example AToB LUT transforms from device coordinates to the ICC PCS, redrawn from [523].

printer, that has cyan, magenta, and yellow primaries (CMY). As we know, the goal of the ICC profile is to provide a transform from those specific device primaries into the PCS. We could accomplish this using mathematical and empirical models, such as the Kubelka-Munk model discussed in Chapter 8. We could also accomplish this using a brute-force color measurement method along with a color look-up table. In an extreme case, if we state that our printer is capable of generating 256 levels (8 bits) of each of the cyan, magenta, and yellow colors, then we need to print each of those colors separately, and then all of the combinations of colors. We could then take a color measuring device such as a spectrophotometer and measure each one of the printed combinations. With three channels that would equal $256 \times 256 \times 256$, or over 16 million samples to measure! Assuming we have the time and energy necessary to accomplish this task, however, we would then have a direct map that could tell us for any given combination of CMY what the associated XYZ or CIELAB values are.

Obviously, this is not the most practical approach as the amount of time, let alone printer media, would not be conducive to such an exercise. If we consider devices with more than three channels, we can see that the number of measurements necessary to completely fill our LUT would grow exponentially. For practical implementations we can use a lower number of measurements in each of the channels and perform some type of interpolation for values in between. As we saw with the simple 1D-LUTs, if we use too few samples the color error can be very large. Therefore, it is desirable to achieve a balance between the number of samples, type of interpolation, and acceptable color error.

Figure 16.10 shows an example of building a cLUT using five sample points for each of the CMY levels. This creates what is known as a $5 \times 5 \times 5$ LUT, with a total of 125 samples. We could print each of these samples and measure the corresponding CIELAB or CIE XYZ values to populate the table. Alternatively we can use a printer model to estimate what the sample values would be. Essentially at each of the colored bulbs, or nodes, in Figure 16.10, we can store the

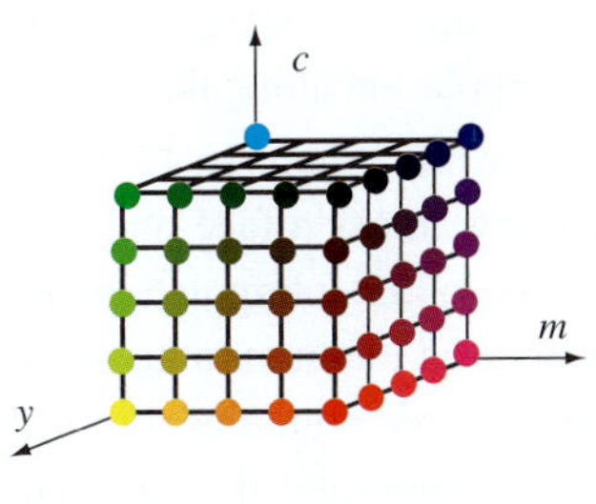

Figure 16.10. Mapping a device CMY space using a $5 \times 5 \times 5$ color look-up table.

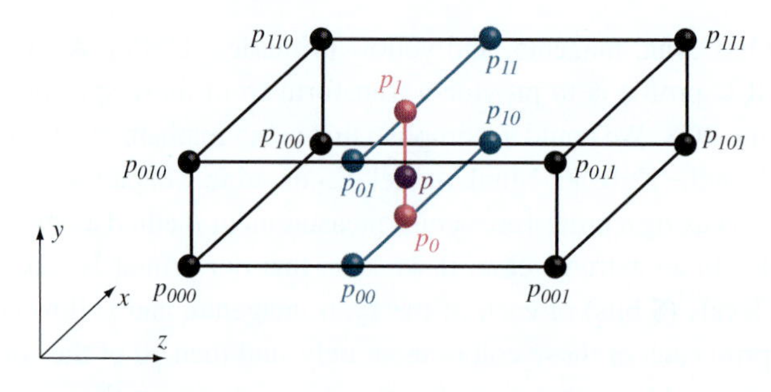

Figure 16.11. An example of trilinear interpolation.

corresponding color measurement information. To determine what the approximate color measurement data would have been in between the nodes, we perform a three-dimensional interpolation.

Common forms of 3D interpolation include trilinear, prism, pyramid, and tetrahedral. Essentially, we can think of the different types of interpolation as having trade-offs between accuracy, speed, robustness to noise in the imaging system, and complexity. They also have different properties with regard to the number of individual nodes or vertices they use and their ability to maintain local curvature in the device color space. In general, the fewer vertices used, the better they are able to handle local changes in the color space, though this may come at the expense of color accuracy when noise is present.

Trilinear interpolation is the three-dimensional equivalent to linear interpolation; it uses all eight nodes of the three-dimensional color space. An example of trilinear interpolation is shown in Figure 16.11. If we are interested in finding the values of point p in Figure 16.11, we must first calculate the linear distance along each of the three dimensions. Essentially we can think of trilinear interpolation as performing three separate interpolations along each of the color channels. First, we calculate the points $p_{00} - p_{11}$ via linear interpolation along the z-axis, then we calculate the points $p_0 - p_1$ through interpolation along the x-axis, and finally calculate point p through interpolation along the y-axis.

Prism interpolation, as shown in Figure 16.12, first subdivides the color space cube into two pieces by cutting across a face diagonal. If possible, this cutting should contain the neutral axis so that interpolation along that line is maintained, and color errors are not introduced into neutral tones. A test is then performed to determine in which prism the desired point lies. From there, the point is projected onto the two triangle faces and interpolated using triangle interpolation. Simple linear interpolation is then done between two projected points. As such, prism

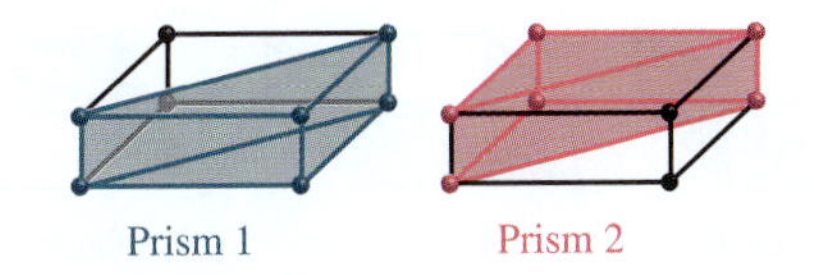

Figure 16.12. An example of sub-dividing the color space cube for prism interpolation.

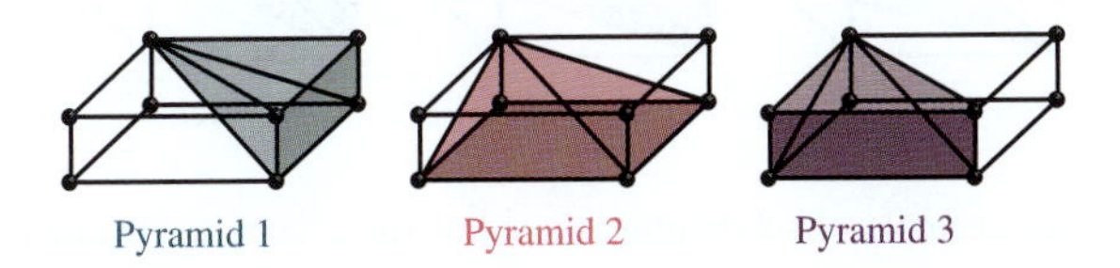

Figure 16.13. An example of subdividing the color space cube for pyramid interpolation.

interpolation uses only six nodes or vertices of our sampled color space and is more capable of following local curvature in the color space.

Similarly, pyramid interpolation splits the color cube into three pyramids, each containing five vertices, as shown in Figure 16.13. These pyramids all come together at a single point, and that point should be chosen to correspond to neutral colors. Again tests are performed to determine in which of the pyramids the point lies, and from there the colors are interpolated.

Tetrahedral interpolation subdivides the space into six tetrahedra, each using only four vertices. All of the tetrahedra share a common edge, which ideally should be pointed along the neutral axis. An example of this sub-division is shown in Figure 16.14. Due to its computational simplicity, as well as ability to maintain the neutral axis, this is a common form of interpolation used and will be discussed in further detail later in this chapter. It should be noted that tetrahedral interpolation for color space transformation has been covered by certain patents since the 1970s and 1980s, as detailed by Kang [575]. More details on specific interpolations can also be found in Kang and Sharma et al. [575, 577, 1035].

Color look-up tables can be powerful tools for transforming colors from device specific spaces into the ICC reference color space. These cLUTs can be constructed in such a way as to provide color re-rendering for perceptual rendering intents, or they can also be used to guide certain regions of color space to other regions. For instance, it would be possible to use cLUTs to shift most "green" colors to a more saturated green, to create pleasing grass and tree reproduction. Likewise, we could shift the "blue" colors away from purple to create pleasing sky colors. These types of manipulations may be valid for the perceptual intent. Color look-up tables are not limited to just three dimensions, as we have illustrated in these examples. The same concepts and interpolation techniques can

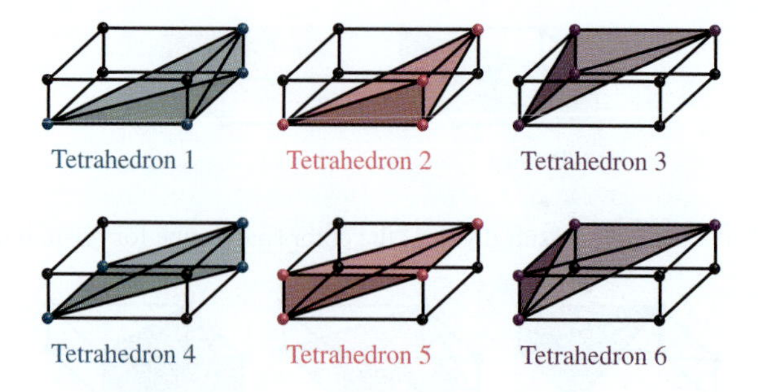

Figure 16.14. An example of subdividing the color space cube for tetrahedral interpolation.

be easily extended to handle n-dimensional color imaging devices. The cLUT is then used to reduce this dimensionality to the three-dimensional nature of the ICC PCS.

What we have described so far has been limited to creating a regular sampling grid in the device color space. Sometimes this is not ideal, due to the nonlinear relationship between the device colorants and the PCS. Often times we would like to have the interpolation and sampling performed in a perceptually uniform color space to minimize visual artifacts. This is one of the reasons CIELAB is offered as the ICC profile connection space and why many output ICC profiles choose to use CIELAB rather than CIE XYZ as the PCS. We could use a non-uniform sampling grid in the cLUT, by essentially changing the distance between each of the nodes as a function of the device color space. For example, we could place more nodes toward the dark colors, knowing that the human visual system is more sensitive to errors in dark regions. This non-uniform sampling can make it difficult to use some of the geometric interpolation methods as described above, since the tests to determine which bounding geometry would be different for each sample cube. Alternatively, we can use 1D-LUTs on the device channels to provide an expansion or compression of the channels. This allows us to use regularly sampled grids and standard interpolation techniques, which is why the ICC allows the specification and application of 1D-LUTS, shown as the "A" curves in Figure 16.9, prior to application of the cLUTs.

It should be quite clear how the AToB and BToA transforms offered by the ICC specification can provide a powerful set of tools to generate color reproduction matches from device spaces into the reference color space. It should also be quite clear that the shear number of different types of transforms that must be generated for specific devices can be overwhelming, and therefore beyond the scope of most casual users. As such, third party profile-creation software tools have been developed, often in association with color measuring devices. These tools handle all the appropriate look-up table generation, color measurement, and binary encoding of the profiles.

16.2.4 Black Point Compensation

There are many more tags than those we have described available in ICC profiles, most of which are outside the scope of this text. One last tag that should be mentioned, however, is the *mediaBlackPointTag*. Much like the *mediaWhitePointTag*, which is used to transform the data from the relative colorimetric intent to the absolute colorimetric intent, the black point contains the measured CIE XYZ tristimulus values of the darkest area of the actual medium. It is recommended that the chromaticities of the black point match those of the white, although this is not strictly enforced. If the black has been measured under different illumination than the CIE D50 of the reference viewing conditions, then these tristimulus values must be converted into D50 using a chromatic adaptation transform.

Ideal imaging devices would all have identical black values with a luminance of zero. Unfortunately, real imaging devices have a substantially higher, and often different, black value. This can come from light leakage, internal and external flare, or the inability to absorb all the incident light. For reflection media, there is a limit to how dark a colorant can actually be, and different colorants may have very different limitations. As such, it is quite common to have black point differences between two real imaging devices, especially across different types of media. This is illustrated in Figure 16.15. Essentially this can be thought of as a gamut-mapping issue, as many dark colors from the input are mapped or clipped into a single black in the output. This mapping would result in a loss of information in the shadow regions, which may be undesirable.

In many situations, we could simply request using a perceptual rendering intent that should take these differences into account. Many times, however, we are interested in maintaining relative colorimetric matches for most of the data, but would still like to maintain details in the shadows. This correction is known as *black point compensation*. Although the ICC does not specify how this transform is performed, they do allow it to happen for relative colorimetric and saturation rendering intents. It is not allowed for absolute colorimetric and should be already built into the perceptual rendering intent. What black point compensation should do is compress the shadow data in the input space, or PCS, using a simple tone-mapping algorithm. This mapping should attempt to maintain as much linearity for the remaining colors as possible.

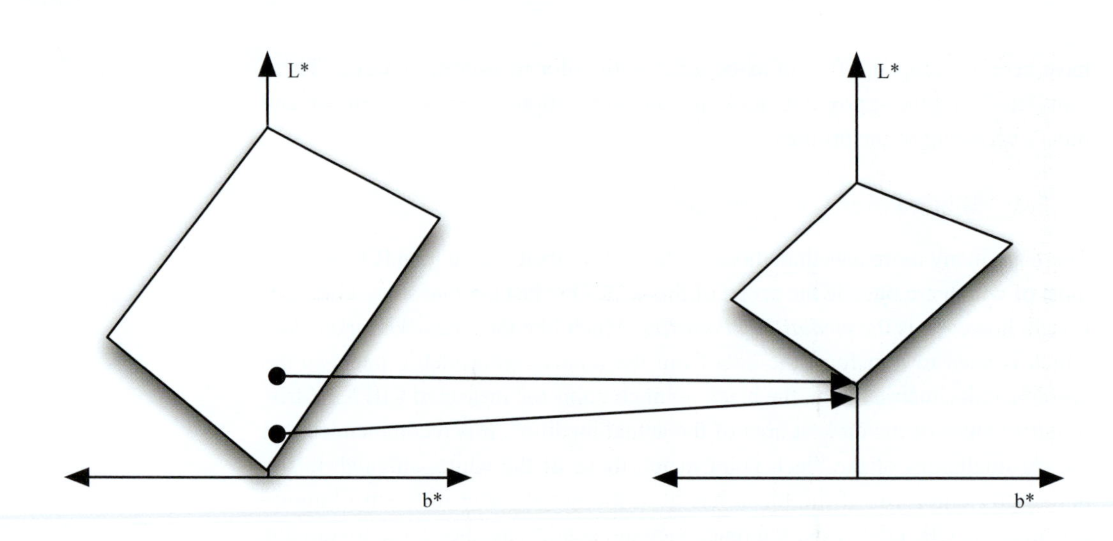

Figure 16.15. Two iconic projections of device-gamuts with different blacks in the CIELAB space.

16.3 Practical Applications

Color management, and specifically the ICC color management framework, is in essence a practical application of many of the concepts we have discussed in this text. It combines elements of color measurement and colorimetry, chromatic adaptation, and color appearance to generate reproduction matches across a variety of color imaging devices. In this section, a brief overview of color management from a digital camera to a computer display device is given. Additionally we will discuss an example of trilinear and tetrahedral interpolation.

16.3.1 Color Management Between a Digital Camera and a Computer Display

Today, ICC profiles can be embedded in almost all major image file formats. Most commercial digital cameras have the option of, or default to, saving images in the sRGB format and embedding this profile in the image file. An example of such an image is shown in Figure 16.16. To display this image such that it is appropriately reproduced, we can use the standard ICC transforms. Although we are using sRGB as an image encoding and input into our color management system, most sRGB profiles are actually of the display class and contain the *MatTRCTag* indicating that the transform into the CIE XYZ of the PCS is handled by three tone-reproduction curves. Parsing the sRGB profile, which ships with most major

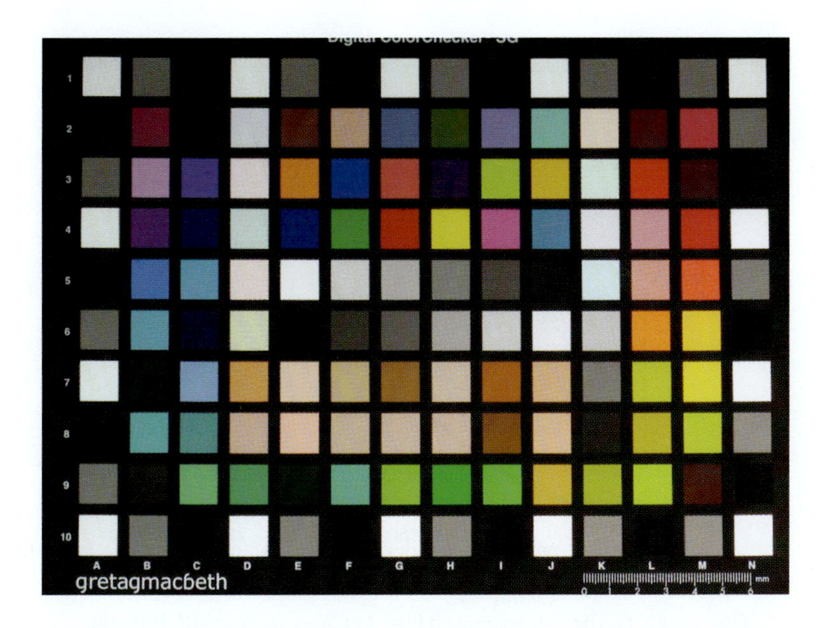

Figure 16.16. An image captured with a digital camera and tagged with an sRGB profile.

computer operating systems, we can examine the transforms used to move into the PCS.

First we consider the nonlinear red, green, and blue *TRCTags*. Recalling from Chapter 8, we know that the transfer function defined for sRGB is broken into two distinct parts: a linear segment for dark colors and a gamma-offset-gain function for the remaining parts. This function is encoded as 16-bit values in a 1024 element 1D-LUT in the ICC profile as shown in Figure 16.17. We first apply this LUT to our input sRGB data to get linear RGB units. Next we can examine the red, green, and blue *MatrixColumnTags* and combine these into our 3×3 matrix to transform into CIE XYZ:

$$\begin{bmatrix} X \\ Y \\ Z \end{bmatrix}_{\text{D50}} = \begin{bmatrix} 0.4361 & 0.3851 & 0.1431 \\ 0.2225 & 0.7169 & 0.0606 \\ 0.0139 & 0.0971 & 0.7141 \end{bmatrix} \cdot \begin{bmatrix} R \\ G \\ B \end{bmatrix}_{\text{sRGB\_linear}} . \tag{16.6}$$

Note that this transform appears to be significantly different than the standard sRGB transform. This is because it has been chromatically adapted from the standard white of CIE D65 to the reference D50 transform using a linear Bradford transform as shown in Equation (16.4). We apply this matrix to our linearized

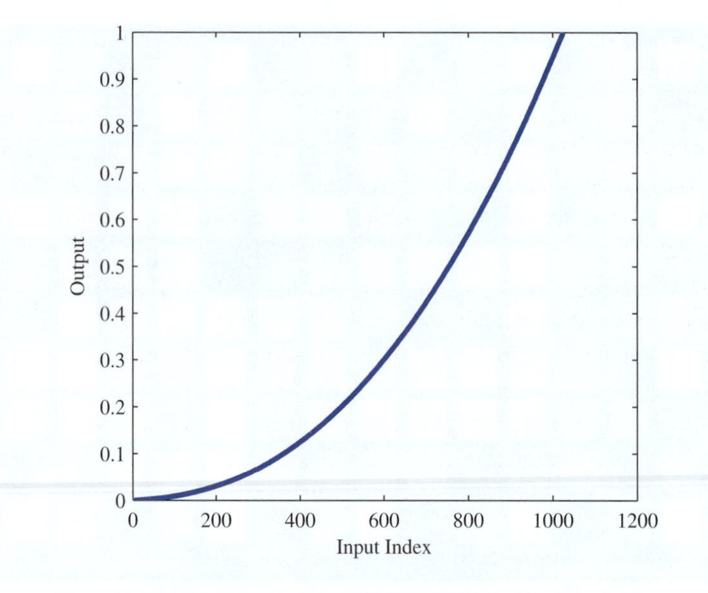

Figure 16.17. The red, green, and blue *TRCTags* from an sRGB profile.

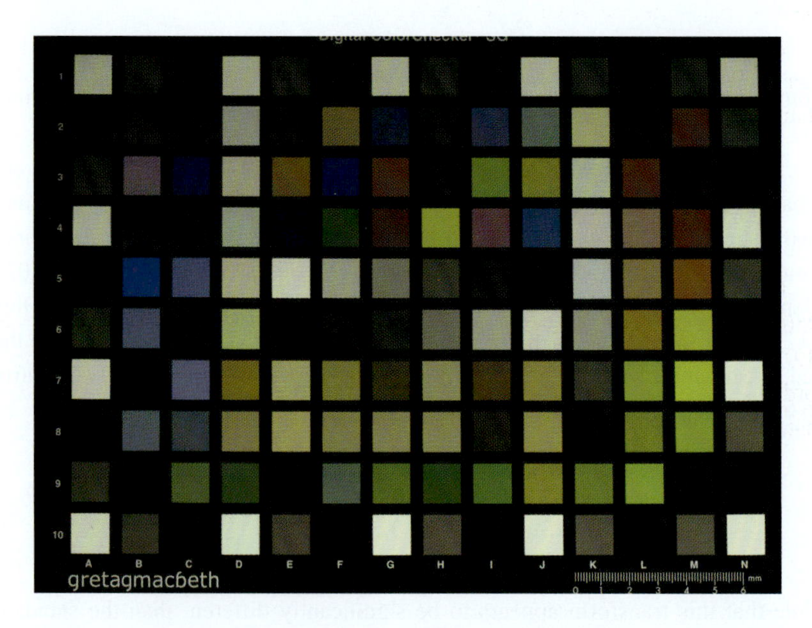

Figure 16.18. An image transformed into the ICC XYZ reference color space.

RGB data, and our image is now in the reference CIE XYZ PCS as shown in Figure 16.18.

Once we are in the reference color space, we need to move back out onto a computer display. Essentially, we want to figure out what digital counts should be displayed on our monitor to achieve a colorimetric match to our input image. In order to do this, we need to characterize our monitor. We can do this using commercial software and a spectrophotometer. We first measure the red, green, and blue primaries at their full intensity. This will help generate the RGB to XYZ color matrix as shown in Equation (16.6). We also measure the combination of the three primaries at maximum value to get a measurement of the white of our computer display. Typically, the native white will be around CIE D65. It is a good idea to perform a check of your monitor, adding up the individual tristimulus values of each of the three channels (equivalent to passing R = G = B = 1 into the 3×3 color matrix) to see if they match the measured tristimulus values of your white. This will determine if your display is behaving in an additive way. From there, we need to measure the individual tone-reproduction curves of each of the primaries. This is done by presenting individual ramps on each of the channels. We can fit a simple gamma function to these measurements, or we can use the measurements themselves as a 1D-LUT.

For this particular display, the profile is included on the DVD; it was found that a simple gamma of approximately 1.8 was adequate to describe all three color channels. The measured color conversion matrix is

$$\begin{bmatrix} X \\ Y \\ Z \end{bmatrix}_{\mathrm{D50}} = \begin{bmatrix} 0.4735 & 0.3447 & 0.1460 \\ 0.2517 & 0.6591 & 0.0891 \\ 0.0058 & 0.0677 & 0.7515 \end{bmatrix} \cdot \begin{bmatrix} R \\ G \\ B \end{bmatrix}_{\mathrm{display}}. \tag{16.7}$$

Note that this matrix also has a native white (R=G=B=1) equivalent to CIE D50, despite the fact that the monitor was measured to have a white of CIE XYZ (1.0104, 1.0300, 1.0651). This is because, once again, we applied a linear Bradford chromatic adaptation transform to get into the PCS color space. This chromatic adaptation transform is given by

$$\begin{bmatrix} X \\ Y \\ Z \end{bmatrix}_{\mathrm{PCS}} = \begin{bmatrix} 1.042633 & 0.030930 & -0.052887 \\ 0.022217 & 1.001877 & -0.021103 \\ -0.001160 & -0.003433 & 0.761444 \end{bmatrix} \cdot \begin{bmatrix} X \\ Y \\ Z \end{bmatrix}_{\mathrm{display}}. \tag{16.8}$$

Recall that the ICC specification requests that these transforms are defined for moving into the PCS and that they must be invertible. As such, we would actually use the inverse of Equations (16.8) and (16.7), applied in that order. We would then have to apply the inverse of the display gamma, essentially $RGB^{\frac{1}{1.8}}$.

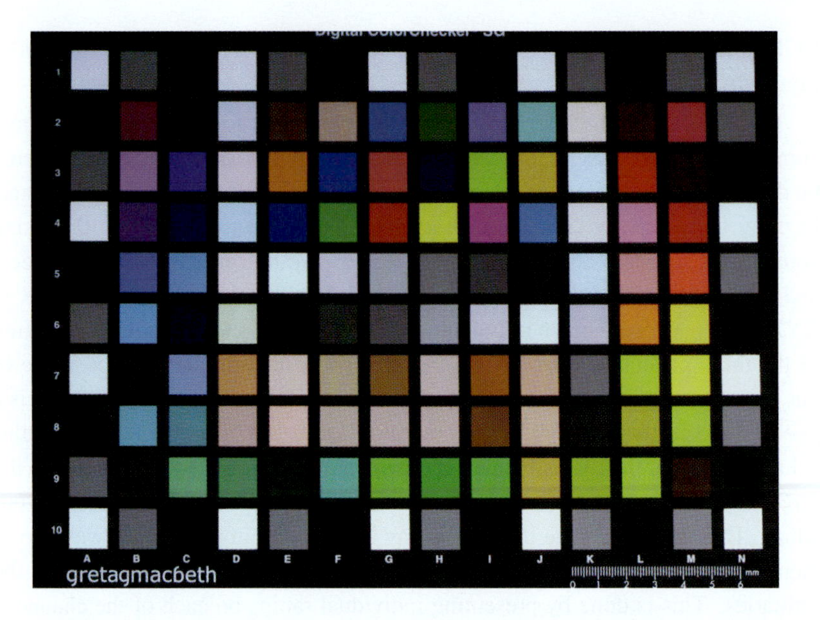

Figure 16.19. An sRGB image transformed into a particular monitor display color space.

After that, the image is ready to be displayed. An example of the image that would be displayed is shown in Figure 16.19. Remember, with the ICC workflow the underlying encoded image bits do not change, the only thing that changes is how we interpret those bits. We can see that in order to display an sRGB image correctly on this particular display, we must make the image more blue and with a higher contrast.

16.3.2 More Details on Trilinear and Tetrahedral Interpolation

The two most common forms of interpolation for color look-up tables are probably trilinear and tetrahedral. An example of calculating values based upon these interpolation techniques is given here. Recall our three-dimensional cube bounded by eight nodes, or vertices, in the device color space. The cube is shown again here in Figure 16.20. Hypothetically, we have the measured CIELAB or CIE XYZ values at each of the corner nodes, represented by points $p_{000} - p_{111}$, and we are interested in knowing the PCS values of point p, with the coordinates in a generic device space of x, y, z.

In the trilinear interpolation, we simply take the three one-dimensional interpolations along the device coordinate axes. Combined, this not so simple equa-

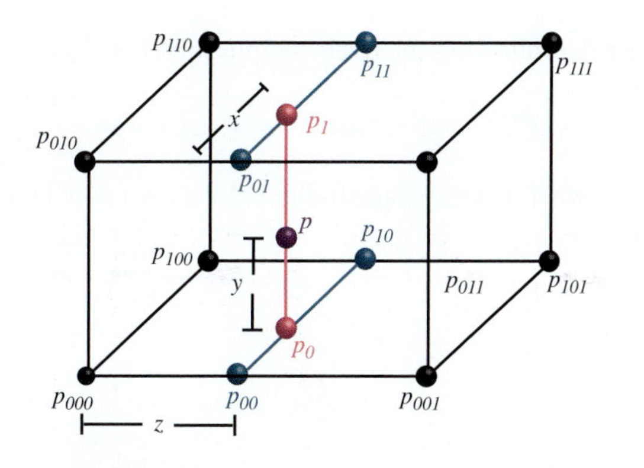

Figure 16.20. A bounding cube in our device color space used in a color look-up table.

tion, based upon [1035], is

$$\begin{aligned} PCS(x,y,z) = & p_{000} + (p_{001} - p_{000})z + (p_{010} + (p_{011} - p_{010})z \\ & - (p_{000} + (p_{001} - p_{000})z))y + \{p_{100} + (p_{101} - p_{100})z \\ & + (p_{110} + (p_{111} - p_{110})z - (p_{100} + (p_{101} - p_{100})z))y \\ & - [p_{000} + (p_{001} - p_{000})z + (p_{010} + (p_{011} - p_{010})z \\ & - (p_{000} + (p_{001} - p_{000})z))y]\}x. \end{aligned} \tag{16.9}$$

To calculate the values in the PCS, you replace the p_{***} values in the equation with the measured CIELAB or CIE XYZ values at those nodes.

For tetrahedral interpolation, as shown in Figure 16.14, we first must figure out in which of the six tetrahedra our point lies. This is done using a series of inequality tests on the (x,y,z)-coordinates of our color space. The interpolation is then calculated based upon the four specific vertices on that tetrahedron. Note, that to maintain the neutral axis as best we can, all the tetrahedra share the p_{000} vertex. Table 16.3 shows the inequality test and the corresponding weights used for the interpolation. Once we have those weights, we can interpolate our values

Test	W_z	W_y	W_x
$x < z < y$	$p_{110} - p_{010}$	$p_{010} - p_{000}$	$p_{111} - p_{110}$
$z < y < x$	$p_{111} - p_{011}$	$p_{011} - p_{001}$	$p_{001} - p_{000}$
$z \leq x \leq y$	$p_{111} - p_{011}$	$p_{010} - p_{000}$	$p_{011} - p_{010}$
$y \leq z < x$	$p_{101} - p_{001}$	$p_{111} - p_{101}$	$p_{001} - p_{000}$
$y < x \leq z$	$p_{100} - p_{000}$	$p_{110} - p_{100}$	$p_{111} - p_{110}$
$x \leq y \leq z$	$p_{100} - p_{000}$	$p_{111} - p_{101}$	$p_{101} - p_{100}$

Table 16.3. The test cases and vertex weights for tetrahedral interpolation.

in the PCS using the following relatively simple equation,

$$PCS\,(xyz) = p_{000} + w_z z + w_y y + w_x x. \tag{16.10}$$

This simplicity is what makes tetrahedral interpolation a popular choice for color look-up tables.

Chapter 17
Dynamic Range Reduction

Real-world environments typically contain a range of illumination much larger than can be represented by conventional 8-bit images. For instance, sunlight at noon may be as much as 100 million times brighter than starlight [312,952,1075]. The human visual system is able to detect 4 or 5 log units of illumination simultaneously and can adapt to a range of around ten orders of magnitude over time [312].

On the other hand, conventional 8-bit images with values between 0 and 255, have a useful dynamic range of around two orders of magnitude. Such images are represented typically by one byte per pixel for each of the red, green and blue channels. The limited dynamic range afforded by 8-bit images is well-matched to the display capabilities of cathode-ray tubes (CRTs), and therefore to all displays that are designed to be backwards compatible with CRTs. Their range, while being larger than two orders of magnitude, lies partially in the dark end of the range, where the human visual system has trouble discerning very small differences under normal viewing circumstances. Hence, CRTs have a useful dynamic range of 2 log units of magnitude.

Of course, the inability to store and display the full range of illumination means that, in practice, compromises have to be sought. For instance, professional photographers tend to select scenes with a dynamic range that can be captured with currently available hardware. Examples of professional photographs taken in "good light" are shown in Figure 17.1. Shooting into the light creates a much higher dynamic range that is sometimes exploited for artistic effect, as shown in Figure 17.2.

Although the limitations of photography can be exploited to create a desired atmosphere, for many applications it would be preferable to capture the full dynamic range of the scene, independent of the lighting conditions. If necessary, the image can then be processed for artistic effect later.

Figure 17.1. Photographs taken in "good light" contain no under- or over-exposed areas. This can be achieved by keeping the sunlight behind the camera. (Images courtesy of Kirt Witte, Savannah School of Art and Design, http://www.TheOtherSavannah.com).

In addition, capturing the full dynamic range of a scene implies that, in many instances, the resulting high dynamic range image cannot be directly displayed, since its range is likely to exceed the two orders of magnitude range afforded by conventional display devices. Examples of images that are typically difficult to capture with conventional photography are shown in Figure 17.3.

For such images, the photographer normally chooses whether the indoor or the outdoor part of the scene is properly exposed. Sometimes, neither can be achieved within a single image, as shown in Figure 17.4 (left). These decisions may be avoided by using high dynamic range imaging (Figure 17.4 (right)).

Landscapes tend to have a bright sky and a much darker foreground; such environments also benefit from high dynamic range imaging. For this particular type of photography, it is also possible to resort to graduated neutral density filters, whereby the top half of the filter attenuates more than the bottom half. Positioning

Figure 17.2. Photographs shot into the light. The stark contrast and under-exposed foregrounds are created for dramatic effect. (Images courtesy of Kirt Witte, Savannah School of Art and Design (http://www.TheOtherSavannah.com).)

Figure 17.3. Some environments contain a larger range of luminance values than can be captured with conventional techniques. From left to right, top to bottom: Harbor of Konstanz, June 2005; Liebfrauenkirche, Trier, Germany, May 2006; the Abbey of Mont St. Michel, France, June 2005, hut near Schloß Dagstuhl, Germany, May 2006.

Figure 17.4. With conventional photography, some parts of the scene may be under- or over-exposed (left). Capturing this scene with nine exposures, and assembling these into one high dynamic range image, affords the result shown on the right; refectory in Mont St. Michel, France, June 2005.

Figure 17.5. Linear scaling of high dynamic range images to fit a given display device may cause significant detail to be lost (left and middle). The left image is linearly scaled. In the middle image, high values are clamped. For comparison, the right image is tone-mapped, allowing details in both bright and dark regions to be visible; Mont St. Michel, France, June 2005.

such a filter in front of the lens allows the photographer to hold detail in both sky and foreground. Nonetheless, a more convenient and widely applicable solution is to capture high dynamic range data.

There are two strategies to display high dynamic range images. First, we may develop display devices that can directly accommodate high dynamic range imagery [1020, 1021]. Second, high dynamic range images may be prepared for display on low dynamic range display devices by applying a *tone-reproduction operator*. The purpose of tone reproduction is therefore to reduce the dynamic range of an image such that it may be displayed on a display device.

We can use a simple compressive function to normalize an image (see Figure 17.5 (left)). This constitutes a linear scaling, which is sufficient only if the dynamic range of the image is slightly higher than the dynamic range of the display. For images with a higher dynamic range, small intensity differences will be quantized to the same display value and visible details are lost. In Figure 17.5 (middle), all large pixel values are clamped. This makes the normalization less dependent on noisy outliers, but here we lose information in the bright areas of the image. In comparison, the right image in Figure 17.5 is tone-mapped, showing detail in both the light and dark regions.

In general linear scaling will not be appropriate for tone reproduction. The key issue in tone reproduction is to compress an image, while at the same time preserving one or more attributes of the image. Different tone-reproduction algorithms focus on different attributes such as contrast, visible detail, brightness, or appearance.

Ideally, displaying a tone-mapped image on a low dynamic range display device would create the same visual response in the observer as the original scene. Given the limitations of display devices, this is, in general, not achievable, although we may approximate this goal as closely as possible.

An orthogonal issue is that of choosing parameters. Dependent on the chosen tone-reproduction operator, different parameters may have to be chosen by the user. Some operators do not require any parameter adjustment, whereas most operators have a few parameter values that help control the final appearance. At the other end of the scale, more advanced user interfaces can be designed to guide the compression algorithm. Such interfaces may consist of scribble tools to roughly indicate which areas are to be affected [686]. In this chapter, however, we focus predominantly on automatic and semi-automatic operators.

17.1 Spatial Operators

In the following sections we discuss tone-reproduction operators that apply compression directly on pixels. Often global and local operators are distinguished. Tone-reproduction operators in the former class change each pixel's luminance values according to a compressive function that is the same for each pixel [264, 310, 780, 1007, 1152, 1211, 1213]. The term *global* stems from the fact that many such functions need to be anchored to some values determined by analyzing the full image. In practice, most operators use the geometric average $\bar{L}_v$ to steer the compression:

$$\bar{L}_v = \exp\left(\frac{1}{N}\sum_{x,y}\log\left(\delta + L_v(x,y)\right)\right). \tag{17.1}$$

The small constant δ is introduced to prevent the average from becoming zero in the presence of black pixels. Note that absolute black tends to occur mostly in computer-generated images. In the real world, it is difficult to find environments with a complete absence of photons. The geometric average is normally mapped to a predefined display value. The effect of mapping the geometric average to different display values is shown in Figure 17.6.

Alternatively, sometimes the minimum or maximum luminance found in the image is used. This approach tends to suffer from the same problems as dynamic range measures that are based on the minimum and maximum luminance: they are sensitive to outliers.

The main challenge faced in the design of a global operator lies in the choice of compressive function. Many functions are possible, for instance based on the image's histogram (Section 17.7) [1211] or on data gathered from psychophysics (Section 17.2).

On the other hand, local operators compress each pixel according to a specific compression function that is modulated by information derived from a selection of neighboring pixels, rather than from the full image [47, 174, 178, 270, 291, 552,

Figure 17.6. Spatial tone mapping applied after mapping the geometric average to different display values (left: 0.09, right: 0.18.); Schloß Dagstuhl, Germany, May 2006.

856, 883, 884, 931, 932, 950]. The rationale is that a bright pixel in a bright neighborhood may be perceived differently than a bright pixel in a dim neighborhood. Design challenges for local operators involve choosing the compressive function, the size of the local neighborhood for each pixel, and the manner in which local pixel values are used. In general, local operators are able to achieve better compression than global operators (Figure 17.7), albeit at a higher computational cost.

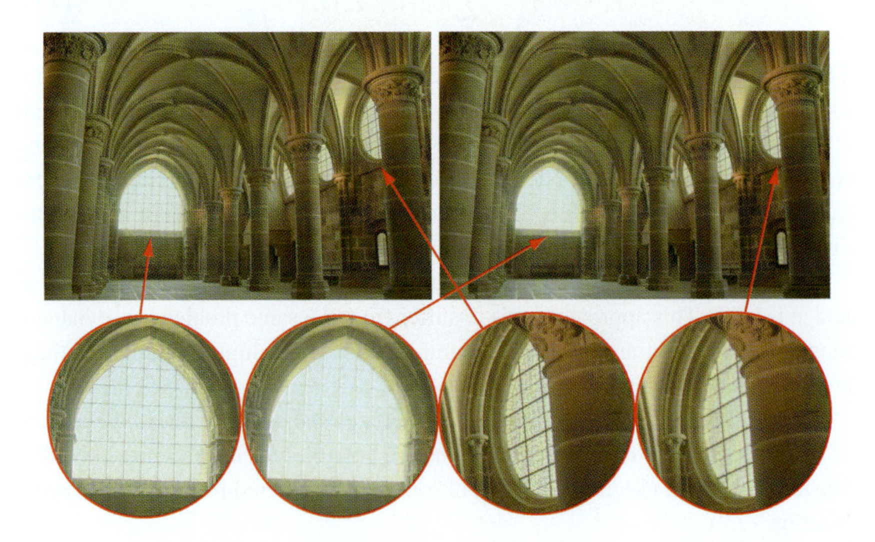

Figure 17.7. A local tone-reproduction operator (left) and a global tone-reproduction operator (right) [950]. The local operator shows more detail, as for instance seen in the insets; Mont St. Michel, France, June 2005.

Both global and local operators are often inspired by the human visual system. Most operators employ one of two distinct compressive functions; this is orthogonal to the distinction between local and global operators. Display values $L_d(x,y)$ are most commonly derived from image luminance $L_v(x,y)$ by the following two functional forms:

$$L_d(x,y) = \frac{L_v(x,y)}{f(x,y)}, \tag{17.2a}$$

$$L_d(x,y) = \frac{L_v^n(x,y)}{L_v^n(x,y) + g^n(x,y)}. \tag{17.2b}$$

In these equations, $f(x,y)$ and $g(x,y)$ may either be constant or a function that varies per pixel. In the former case, we have a global operator, whereas a spatially varying function results in a local operator. The exponent n is a constant that is either fixed, or set differently per image.

Equation (17.2a) divides each pixel's luminance by a value derived from either the full image or a local neighborhood. As an example, the substitution $f(x,y) = L^{\max}/255$ in (17.2a) yields a linear scaling such that values may be di-

Figure 17.8. Halos are artifacts commonly associated with local tone-reproduction operators. Chiu's operator is used here without smoothing iterations to demonstrate the effect of division (left). Several mechanisms exist to minimize halos, for instance by using the scale-selection mechanism used in photographic tone mapping (right); St. Barnabas Monastery, North Cyprus, July 2006.

rectly quantized into a byte, and they can therefore be displayed. A different approach is to substitute $f(x,y) = L^{\mathrm{blur}}(x,y)$, i.e., divide each pixel by a weighted local average, perhaps obtained by applying a Gaussian filter to the image [174]. While this local operator yields a displayable image, it highlights a classical problem whereby areas near bright spots are reproduced too dark. This is often seen as halos, as demonstrated in Figure 17.8.

The cause of halos stems from the fact that Gaussian filters blur across sharp contrast edges in the same way that they blur small details. If there is a large contrast gradient in the neighborhood of the pixel under consideration, this causes the Gaussian blurred pixel to be significantly different from the pixel itself. By using a very large filter kernel in a division-based approach, such large contrasts are averaged out, and the occurrence of halos can be minimized. However, very large filter kernels tend to compute a local average that is not substantially different from the global average. In the limit that the size of the filter kernel tends to infinity, the local average becomes identical to the global average and, therefore, limits the compressive power of the operator to be no better than a global operator. Thus, the size of the filter kernel in division-based operators presents a trade-off between the ability to reduce the dynamic range and the visibility of artifacts.

17.2 Sigmoidal Compression

Equation 17.2b has an S-shaped curve on a log-linear plot and is called a sigmoid for that reason. This functional form fits data obtained from measuring the electrical response of photoreceptors to flashes of light in various species [811]. It has also provided a good fit to other electro-physiological and psychophysical measurements of human visual function [484, 485, 594].

Sigmoids have several desirable properties. For very small luminance values, the mapping is approximately linear, so that contrast is preserved in dark areas of the image. The function has an asymptote at 1, which means that the output mapping is always bounded between 0 and 1. A further advantage of this function is that for intermediate values, the function affords an approximately logarithmic compression. This can be seen for instance in Figure 17.9, where the middle section of the curve is approximately linear on a log-linear plot. To illustrate, both $L_d = L_w/(L_w + 1)$ and $L_d = 0.25 \log(L_w) + 0.5$ are plotted in this figure,[1] showing that these functions are very similar over a range centered around 1.

In Equation 17.2b, the function $g(x,y)$ may be computed as a global constant, or as a spatially varying function. Following common practice in electro-

[1] The constants 0.25 and 0.5 were determined by equating the values as well as the derivatives of both functions for $x = 1$.

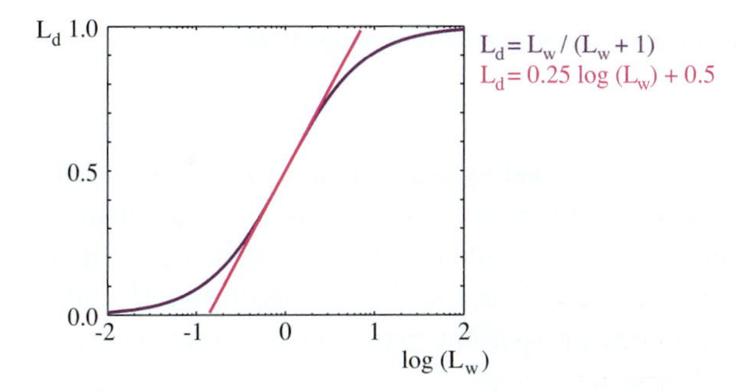

Figure 17.9. Over the middle-range values, sigmoidal compression is approximately logarithmic.

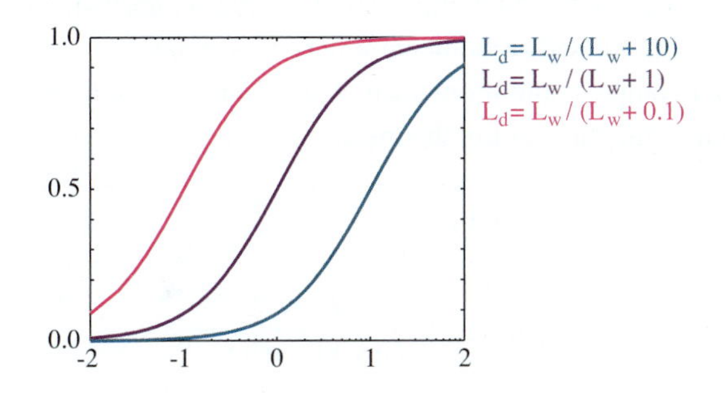

Figure 17.10. The choice of semi-saturation constant determines how input values are mapped to display values.

physiology, we call $g(x,y)$ the "semi-saturation" constant. Its value determines which values in the input image are optimally visible after tone mapping. In particular, if we assume that the exponent n equals 1, then luminance values equal to the semi-saturation constant will be mapped to 0.5, as plotted in Figure 17.9. The effect of choosing different semi-saturation constants is shown in Figure 17.10.

The function $g(x,y)$ may be computed in several different ways. In its simplest form, $g(x,y)$ is set to $\bar{L}_v/k$, so that the geometric average is mapped to user parameter k [950]. In this case, a good initial value for k is 0.18, which conforms to the photographic equivalent of middle gray. For particularly bright or dark scenes, this value may be raised or lowered. Alternatively, its value may be estimated from the image itself [954]. The exponent n in Equation 17.2b may be set to 1; other choices for n are discussed below.

In this approach, the semi-saturation constant is a function of the geometric average, and the operator is therefore global. A variation of this global operator computes the semi-saturation constant by linearly interpolating between the geometric average and each pixel's luminance [948]:

$$g(x,y) = a L_v(x,y) + (1-a)\bar{L}_v. \tag{17.3}$$

The interpolation is governed by user parameter $a \in [0,1]$ which has the effect of varying the amount of contrast in the displayable image (Figure 17.11). More contrast means less visible detail in the light and dark areas and vice versa. This interpolation may be viewed as a half-way house between a fully global and a fully local operator by interpolating between the two extremes without resorting to expensive blurring operations.

Although operators typically compress luminance values, this particular operator may be extended to include a simple form of chromatic adaptation. It thus presents an opportunity to adjust the level of saturation normally associated with tone mapping, as discussed at the beginning of this chapter.

Rather than compress the luminance channel only, sigmoidal compression is applied to each of the three color channels:

$$I_{r,d}(x,y) = \frac{I_r(x,y)}{I_r(x,y) + g^n(x,y)}, \tag{17.4a}$$

$$I_{g,d}(x,y) = \frac{I_g(x,y)}{I_g(x,y) + g^n(x,y)}, \tag{17.4b}$$

$$I_{b,d}(x,y) = \frac{I_b(x,y)}{I_b(x,y) + g^n(x,y)}. \tag{17.4c}$$

The computation of $g(x,y)$ is also modified to bilinearly interpolate between the geometric average luminance and pixel luminance, and between each independent color channel and the pixel's luminance value. We therefore compute the geometric average luminance value $\bar{L}_v$, as well as the geometric average of the red, green, and blue channels ($\bar{I}_r$, $\bar{I}_g$ and $\bar{I}_b$). From these values, we compute $g(x,y)$ for each pixel and for each color channel independently. The equation for the red channel ($g_r(x,y)$) is then

$$G_r(x,y) = c I_r(x,y) + (1-c) L_v(x,y), \tag{17.5a}$$

$$\bar{G}_r(x,y) = c \bar{I}_r + (1-c)\bar{L}_v, \tag{17.5b}$$

$$g_r(x,y) = a G_r(x,y) + (1-a)\bar{G}_r(x,y). \tag{17.5c}$$

The green and blue channels are computed similarly. The interpolation parameter a steers the amount of contrast as before, and the new interpolation parameter $c \in [0,1]$ allows a simple form of color correction, as shown in Figure 17.11.

So far we have not discussed the value of the exponent n in Equation (17.2b). Studies in electro-physiology report values between $n = 0.2$ and $n = 0.9$ [485].

Figure 17.11. Linear interpolation of a varies contrast in the tone-mapped image: this is set to 0.0 (left) and to 1.0 (right). Linear interpolation according to c yields varying amounts of color correction: c was varied between 0.0 (top) and 1.0 (bottom); Rennes, France, June 2005.

While the exponent may be user-specified, for a wide variety of images we may estimate a reasonable value from the geometric average luminance $\bar{L}_v$ and the minimum and maximum luminance in the image ($L_{\min}$ and $L_{\max}$) with the following empirical equation:

$$n = 0.3 + 0.7 \left(\frac{L_{\max} - \bar{L}_v}{L_{\max} - L_{\min}} \right)^{1.4}. \tag{17.6}$$

The different variants of sigmoidal compression shown above are all global in nature. This has the advantage that they are fast to compute, and they are very suitable for medium to high dynamic range images. Their simplicity makes these operators suitable for implementation on graphics hardware as well. For very high dynamic range images, however, it may be necessary to resort to a local operator since this may give somewhat better compression.

A straightforward method to extend sigmoidal compression replaces the global semi-saturation constant by a spatially varying function that also can be computed in several different ways. Thus, $g(x,y)$ then becomes a function of a spatially localized average. Perhaps the simplest way to accomplish this is to again use a Gaussian blurred image. Each pixel in a blurred image represents a locally averaged value that may be viewed as a suitable choice for the semi-saturation constant.[2]

As with division-based operators discussed in the previous section, we have to consider haloing artifacts. If sigmoids are used with a spatially varying semi-saturation constant, the Gaussian filter kernel is typically chosen to be very small in order to minimize artifacts. In practice, filter kernels of only a few pixels wide are sufficient to suppress significant artifacts while at the same time producing more local contrast in the tone-mapped images. Such small filter kernels can be conveniently computed in the spatial domain without losing too much performance. There are, however, several different approaches to compute a local average; these are discussed in the following section.

17.3 Local Neighborhoods

For local schemes, the compression of each pixel is affected by a local neighborhood of pixels. Whether used within a division-based approach, or in sigmoidal compression, the size of the neighborhood is important. For division-based approaches, larger kernels are typically used to minimize artifacts. For sigmoidal techniques, small filter kernels produce the least objectionable artifacts.

In either case, halo artifacts occur when the local average is computed over a region that contains sharp contrasts with respect to the pixel under consideration.

[2]Although $g(x,y)$ is now no longer a constant, we continue to refer to it as the semi-saturation *constant*.

It is therefore important that the local average is computed over pixel values that are not significantly different from the pixel that is being filtered.

This suggests a strategy whereby an image is filtered such that no blurring over such edges occurs. A simple, but computationally expensive, way is to compute a stack of Gaussian blurred images with different kernel sizes, i.e., an *image pyramid*. For each pixel, we may choose the largest Gaussian that does not overlap with a significant gradient. The scale at which this happens can be computed as follows.

In a relatively uniform neighborhood, the value of a Gaussian blurred pixel should be the same regardless of the filter-kernel size. Thus, in this case, the difference between a pixel filtered with two different Gaussians should be around zero. This difference will only change significantly if the wider filter kernel overlaps with a neighborhood containing a sharp contrast step, whereas the smaller filter kernel does not. A difference of Gaussians (DoG) signal $L_i^{\mathrm{DoG}}(x,y)$ at scale i can be computed as follows:

$$L_i^{\mathrm{DoG}}(x,y) = R_{i\sigma}(x,y) - R_{2i\sigma}(x,y). \tag{17.7}$$

It is now possible to find the largest neighborhood around a pixel that does not contain sharp edges by examining differences of Gaussians at different kernel sizes i [950]:

$$\left| \frac{L_i^{\mathrm{DoG}}(x,y)}{R_{i\sigma}(x,y) + \alpha} \right| > t \qquad i = 1 \ldots 8. \tag{17.8}$$

Here, the difference-of-Gaussians filter is divided by one of the Gaussians to normalize the result and, thus, enable comparison against the constant threshold t that determines if the neighborhood at scale i is considered to have significant detail. The constant α is added to avoid division by zero.

For the image shown in Figure 17.12 (left), the scale selected for each pixel is shown in Figure 17.12 (right). Such a scale selection mechanism is employed by the photographic tone-reproduction operator [950] as well as in Ashikhmin's operator [47].

Once the appropriate neighborhood for each pixel is known, the Gaussian blurred average L_{blur} for this neighborhood may be used to steer the semi-saturation constant, such as for instance employed by the photographic tone-reproduction operator:

$$L_{\mathrm{d}} = \frac{L_{\mathrm{w}}}{1 + L_{\mathrm{blur}}}. \tag{17.9}$$

It is instructive to compare the result of this operator with its global equivalent, defined as

$$L_{\mathrm{d}} = \frac{L_{\mathrm{w}}}{1 + L_{\mathrm{w}}}. \tag{17.10}$$

Figure 17.12. Scale selection mechanism. The left image shows the tone-mapped result. The image on the right encodes the selected scale for each pixel as a gray value—the darker the pixel, the smaller the scale. A total of eight different scales were used to compute this image; Clifton Suspension Bridge, Bristol, UK, April 2006.

Images tone mapped with both forms are shown in Figure 17.13. In addition, this figure shows the color difference as computed with the CIE94 color difference metric, presented in Section 8.8.2. This image shows that the main differences occur near (but not precisely at) high-frequency high-contrast edges, predominantly

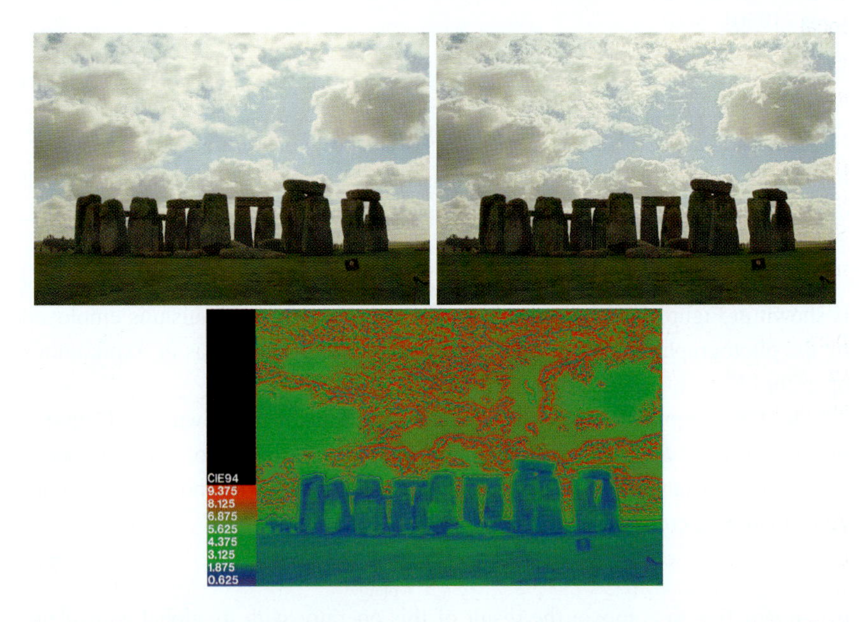

Figure 17.13. This image was tone mapped with both global and local versions of the photographic tone-reproduction operator (top left, and top right, respectively). Below, for each pixel, the CIE ΔE^*_{ab} color difference is shown; Stonehenge, UK, October 2006.

seen in the clouds. These are the regions where more detail is produced by the local operator.

Image pyramids may also be used to great effect to tone map an image directly. Careful design of the filter stack may yield results that are essentially free of halos, as discussed in the following section.

An alternative approach includes the use of edge-preserving smoothing operators that are designed specifically for removing small details while keeping sharp contrasts intact. Such filters have the advantage that sharp discontinuities in the filtered result coincide with the same sharp discontinuities in the input image, and they may therefore help to prevent halos [253]. Several such filters, for example, the bilateral filter, trilateral filter, Susan filter, the LCIS algorithm, and the mean shift algorithm are suitable [178, 204, 270, 883, 1153], although some of them are expensive to compute. Edge-preserving smoothing operators are discussed in Section 17.5.

17.4 Sub-Band Systems

In the preceding section, an image pyramid was used to determine the size of a local area around each pixel within which no sharp edges cross. The weighted average of the pixel values in this area is considered to be representative of a local adaptation value, and can be used as such.

Image pyramids can also be used directly for the purpose of tone reproduction, provided the filter bank is designed carefully [681]. Here, a signal is decomposed into a set of signals that can then be summed to reconstruct the original signal. Such algorithms are known as sub-band systems, or alternatively as wavelet techniques, multi-scale techniques or image pyramids.

An image consisting of luminance signals $L_v(x,y)$ is split into a stack of band-pass signals $L_i^{\mathrm{DoG}}(x,y)$ using (17.7) whereby the n scales, i, increase by factors of two. The original signal L_v can then be reconstructed by simply summing the sub-bands:

$$L_v(x,y) = \sum_{i=1}^{n} L_i^{\mathrm{DoG}}(x,y). \tag{17.11}$$

Assuming that L_v is a high dynamic range image, a tone-mapped image can be created by first applying a non-linearity to the band-pass signals. In its simplest form, the non-linearity applied to each $L_i^{\mathrm{DoG}}(x,y)$ is a sigmoid [253], such as given in (17.2b). However, as argued by Li et al., summing the filtered sub-bands then leads to distortions in the reconstructed signal [681]. To limit such distortions, either the filter bank may be modified or the non-linearity may be redesigned.

Although sigmoids are smooth functions, their application to a (sub-band) signal with arbitrarily sharp discontinuities will yield signals with potentially high frequencies. The high-frequency content of sub-band signals will cause distortions in the reconstruction of the tone-mapped signal L_v [681]. The effective gain $G(x,y)$ applied to each sub-band as a result of applying a sigmoid can be expressed as a pixel-wise multiplier:

$$L_i^{\mathrm{DoG}'}(x,y) = L_i^{\mathrm{DoG}}(x,y)\; G(x,y). \tag{17.12}$$

To avoid distortions in the reconstructed image, the effective gain should have frequencies no higher than the frequencies present in the sub-band signal. This can be achieved by blurring the effective gain map $G(x,y)$ before applying it to the sub-band signal. This approach leads to a significant reduction in artifacts, and it is an important tool in the prevention of halos.

The filter bank itself may also be adjusted to limit distortions in the reconstructed signal. In particular, to remove undesired frequencies in each of the sub-bands caused by applying a non-linear function, a second bank of filters may be applied before summing the sub-bands to yield the reconstructed signal. If the first filter bank that splits the signal into sub-bands is called the analysis filter bank, then the second bank is called the synthesis filter bank. The non-linearity described above can then be applied in between the two filter banks. Each of the synthesis filters should be tuned to the same frequencies as the corresponding analysis filters.

An efficient implementation of this approach, which produces excellent artifact-free results, is described by Li et al. [681], and an example image is shown in Figure 17.14. The source code for this operator is available at http://www.mit.edu/~yzli/hdr_companding.htm. The image benefits from clamping the bottom 2% and the top 1% of the pixels; this is discussed further in Section 17.10.2.

Although this method produces excellent artifact-free images, it has the tendency to over-saturate the image. This effect was ameliorated in Figure 17.14 by desaturating the image using the technique described in Section 17.10.1 (with a value of 0.7, which is the default value used for the sub-band approach). However, even after desaturating the image, its color fidelity remained a little too saturated. Further research would be required to determine the exact cause of this effect, which is shared with gradient-domain compression (Section 17.6). Finally, it should be noted that the scene depicted in Figure 17.14 is a particularly challenging image for tone reproduction. The effects described here would be less pronounced for many high dynamic range photographs.

Figure 17.14. Tone reproduction using a sub-band architecture, computed here using a Haar filter [681]; Gymnasium of Salamis, North Cyprus, July 2006.

17.5 Edge-Preserving Smoothing Operators

An edge-preserving smoothing operator attempts to remove details from the image without removing high-contrast edges. An example is the bilateral filter [871, 1131, 1231], which is a spatial Gaussian filter multiplied with a second Gaussian operating in the intensity domain. With $L_v(x,y)$ representing the luminance at pixel (x,y), and using (15.19), the bilateral filter $L_B(x,y)$ is defined as

$$L_B(x,y) = \frac{\sum_u \sum_v R_{\sigma_1}(x-u, y-v)\, R_{\sigma_2}(L_v(x,y), L_v(u,v))\, L_v(u,v)}{\sum_u \sum_v R_{\sigma_1}(x-u, y-v)\, R_{\sigma_2}(L_v(x,y), L_v(u,v))}. \tag{17.13}$$

Here, σ_1 is the kernel size used for the Gaussian operating in the spatial domain, and σ_2 is the kernel size of the intensity domain Gaussian filter. The bilateral filter can be used to separate an image into "base" and "detail" layers [270]. Applying the bilateral filter to an image results in a blurred image in which sharp edges remain present (Figure 17.15 (left)). Such an image is normally called a base layer. This layer has a dynamic range similar to the input image. The high dynamic range image may be divided pixel-wise by the base layer, obtaining a detail layer $L_D(x,y)$, which contains all the high frequency detail, but typically does not have a very high dynamic range (Figure 17.15 (right)):

$$L_D(x,y) = L_v(x,y)/L_B(x,y). \tag{17.14}$$

By compressing the base layer before recombining into a compressed image, a displayable low dynamic range image may be created (Figure 17.16). Compres-

Figure 17.15. Bilateral filtering removes small details, but preserves sharp gradients (left). The associated detail layer is shown on the right; Bristol, UK, August 2005.

Figure 17.16. An image tone mapped using bilateral filtering. The base and detail layers shown in Figure 17.15 are recombined after compressing the base layer; Bristol, UK, August 2005.

sion of the base layer may be achieved by linear scaling. Tone reproduction on the basis of bilateral filtering is executed in the logarithmic domain.

Aside from bilateral filtering, other operators may be applicable, such as the mean-shift filter [204], the trilateral filter [178], anisotropic diffusion [890], an adaptation of a Retinex-like filter [775], or even the median filter (Section 15.2.3).

Edge-preserving smoothing operators may also be used to compute a local adaptation level for each pixel, to be applied in a spatially varying or local tone-reproduction operator. A local operator based on sigmoidal compression can, for instance, be created by substituting $L_{\text{blur}}(x,y) = L_B(x,y)$ in (17.9).

Alternatively, the semi-saturation constant $g(x,y)$ in (17.4) may be seen as a local adaptation constant, and can therefore be locally approximated [658, 777], for instance, with $L_B(x,y)$ or with any of the other filters mentioned above. As shown in Figure 17.10, the choice of semi-saturation constant shifts the curve horizontally such that its middle portion lies over a desirable range of values. In a local operator, this shift is determined by the values of a local neighborhood of pixels and is thus different for each pixel. This leads to a potentially better compression mechanism than a constant value could afford.

17.6 Gradient-Domain Operators

Local adaptation provides a measure of how different a pixel is from its immediate neighborhood. If a pixel is very different from its neighborhood, it typically needs to be more attenuated. Such a difference may also be expressed in terms of contrast, which can be represented with image gradients (in log space):

$$\nabla L = (L(x+1,y) - L(x,y), L(x,y+1) - L(y)). \tag{17.15}$$

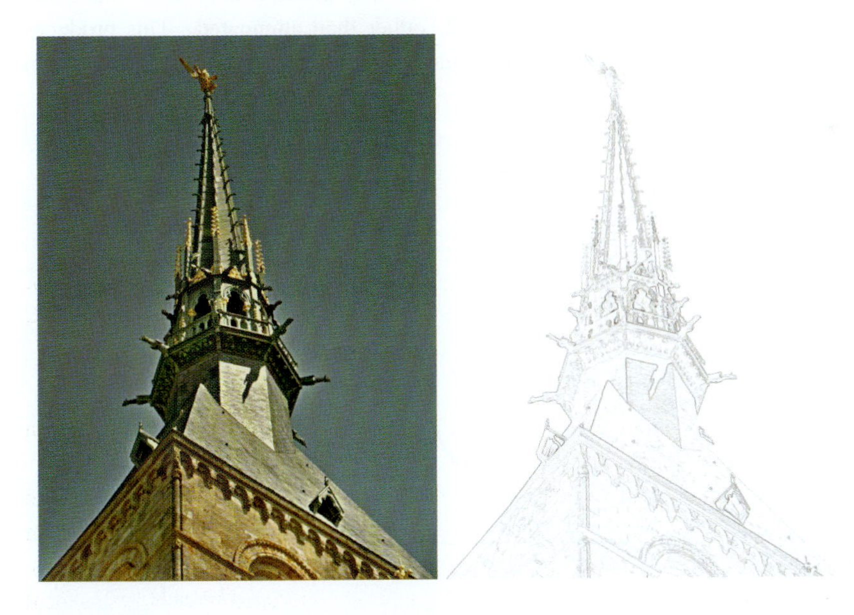

Figure 17.17. The image on the left is tone mapped using gradient domain compression. The magnitude of the gradients $\|\nabla L\|$ is mapped to a greyscale in the right image (white is a gradient of 0; black is the maximum gradient in the image); Mont St. Michel, June 2005.

Here, ∇L is a vector-valued gradient field. By attenuating large gradients more than small gradients, a tone-reproduction operator may be constructed [306]. Afterwards, an image can be reconstructed by integrating the gradient field to form a tone-mapped image. Such integration must be approximated by numerical techniques, by solving a Poisson equation using the Full Multigrid Method [926]. The resulting image then needs to be linearly scaled to fit the range of the target display device. An example of gradient domain compression is shown in Figure 17.17.

17.7 Histogram Adjustment

A simple but effective approach to tone reproduction is to derive a mapping from input luminance to display luminance using the histogram of the input image. Histogram equalization simply adjusts the luminance so that the probability that each display value occurs in the output image is equal. Such a mapping is created by computing the image's histogram, and then integrating this histogram to produce a cumulative histogram. This function can be used directly to map input luminance to display values.

However, dependent on the shape of the histogram, it is possible that some contrasts in the image are exaggerated, rather than attenuated. This produces unnatural results, which may be overcome by restricting the cumulative histogram

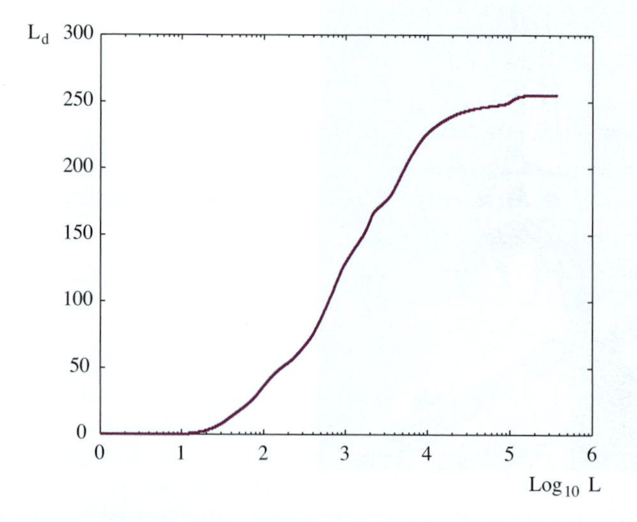

Figure 17.18. Tone-mapping function created by reshaping the cumulative histogram of the image shown in Figure 17.19.

Figure 17.19. Image tone mapped using the histogram-adjustment technique; St. Barnabas monastery, North Cyprus, July 2006.

to never attain a slope that is too large. The magnitude of the slope that can be maximally attained is determined by a threshold-versus-intensity curve (TVI) curve (See Section 5.5.2). The method is then called histogram adjustment, rather than histogram equalization [1211]. An example of a display mapping generated by this method is shown in Figure 17.18. This mapping is derived from the image shown in Figure 17.19.

This method can be extended to include the observer's state of adaptation and a time course of adaptation [529]. Such an operator, optimized to preserve visibility, is then able to simulate what a visually-impaired person would see. The TVI curve that is used in this algorithm is extended to account for adaptation, which in turn is governed by a temporal factor modeling the amount of bleaching occurring in the retina.

17.8 Lightness Perception

The theory of lightness perception, discussed in Chapter 5, provides a model for the perception of surface reflectance (recall that lightness is defined as perceived surface reflectance). To cast this into a computational model, the image needs to be automatically decomposed into frameworks, i.e., regions of common illumination. The influence of each framework on the total lightness needs to be estimated, and the anchors within each framework must be computed [618–620].

It is desirable to assign a probability of belonging to a particular framework to each pixel. This leaves the possibility of a pixel having non-zero participation in multiple frameworks, which is somewhat different from standard segmentation algorithms that assign a pixel to at most one segment. To compute frameworks and probabilities for each pixel, a standard K-means clustering algorithm may be applied. This algorithm results in a set of n centroids dotted around the image. For a centroid with a given luminance value L_C, the probability of a pixel belonging to the framework defined by this centroid is given by

$$P_C(x,y) = \exp\left(\frac{-\log_{10}(L_C/L_v(x,y))^2}{2\sigma^2}\right), \tag{17.16}$$

where σ is the maximum difference between adjacent centroids. If no pixels belong to a centroid with a probability higher than 0.95, this centroid is merged with a neighbor.

The probability map thus obtained is the first step towards defined frameworks. The concept of proximity, however, is not yet accounted for. This feature may be implemented by filtering the probability map with a bilateral filter (17.13), with σ_1 set to half of the smaller of the image width and height and σ_2 set to 0.4.

Once frameworks are computed, their impact on the image's total lightness can be assessed. The strength of a framework depends on both its articulation and its relative size (see Section 5.8.5). Articulation can be estimated from the minimum and maximum luminance within each framework:

$$A_C = 1 - \exp\left(\frac{-\log_{10}\left(L_C^{\max}/L_C^{\min}\right)^2}{2\sigma^2}\right), \tag{17.17}$$

where $\sigma = 1/3$. The size factor of the relative framework X_C is computed from the number of pixels in the framework N_C as follows:

$$X_C = 1 - \exp\left(\frac{-N_C^2}{2\sigma^2}\right), \tag{17.18}$$

where $\sigma = 0.1$. The probability maps P_C are multiplied by their respective articulation and size factors to obtain the final decomposition into frameworks:

$$P'_C(x,y) = P_C(x,y)\, A_C\, X_C. \tag{17.19}$$

Afterwards, these probabilities are normalized so that, for each pixel, they sum to 1.

For each framework, the highest luminance rule may now be applied to find an anchor. However, direct application of this rule may result in the selection of a luminance value of a patch that is perceived as self-luminous. As the anchor should be the highest luminance value that is not perceived as self-luminous, selection of the highest luminance value should be preceded by filtering the area of the local framework with a large Gaussian filter.

Assuming that W_C is the anchor selected for local framework C, the net lightness of the whole image can be computed as follows:

$$L_{\text{lightness}}(x,y) = 0.7 \log_{10}(L_v(x,y)/W_0) + 0.3 \sum_C P'_C(x,y) \; \log_{10}(L_v(x,y)/W_C). \tag{17.20}$$

Here, W_0 is the anchor of the global framework. This then constitutes a computational model of lightness perception that can be extended for the purpose of tone reproduction. To achieve dynamic range reduction, the above equation can be augmented with individual scale factors f_C that scale the output of each of the frameworks independently:

$$L_d(x,y) = 0.7 \log_{10}(L_v(x,y)/W_0) + 0.3 \sum_C P'_C(x,y) \; \frac{\log_{10}(L_v(x,y)/W_C)}{f_C}. \tag{17.21}$$

The factors f_C are chosen such that each framework fits within the display capabilities of the target display device.

One of the strengths of using a computational model of lightness perception for the purpose of tone reproduction, is that traditionally difficult phenomena such as the Gelb effect, introduced in Section 5.8.5, can be handled correctly. To

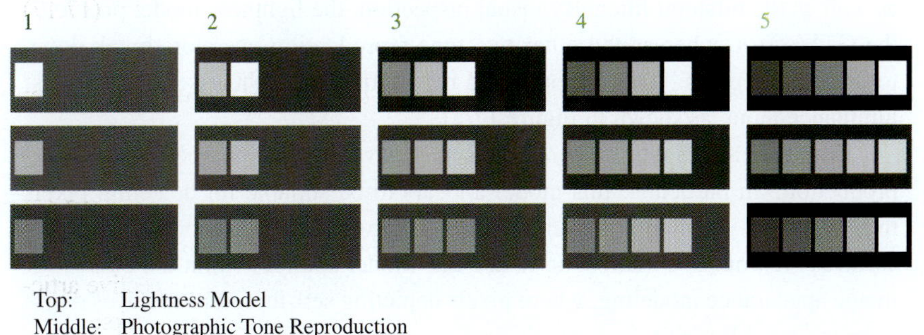

Top: Lightness Model
Middle: Photographic Tone Reproduction
Bottom: Bilateral Filtering

Figure 17.20. Gelb patches as reproduced by three different tone-reproduction operators. (Figure courtesy of Grzegorz Krawczyk, Karol Myszkowski and Hans-Peter Seidel, Max Planck Institute for Informatics, Saarbrücken, Germany.)

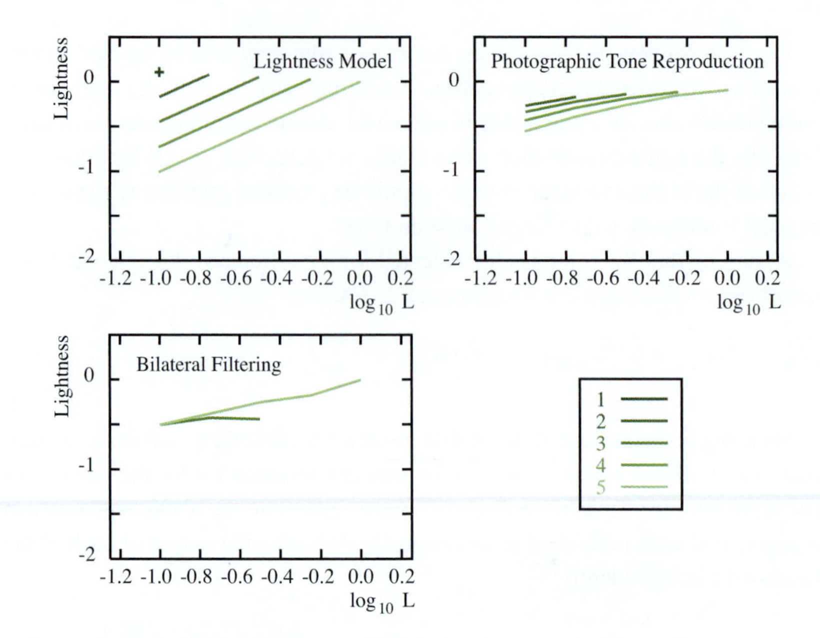

Figure 17.21. Comparison of performance for three different tone-reproduction algorithms with respect to reproducing the Gelb effect. The numbers refer to the test images shown in Figure 17.20. (Figure courtesy of Grzegorz Krawczyk, Karol Myszkowski and Hans-Peter Seidel, Max Planck Institute for Informatics, Saarbrücken, Germany.)

demonstrate, Figure 17.20 shows the output of the lightness model for a set of stimuli and compares them against the photographic tone-reproduction operator as well as the bilateral filter. By visual inspection, the lightness model preserves the Gelb effect, whereas the other two tone-reproduction operators break down to different degrees. This is confirmed by plotting the lightness values against luminance input, as shown in Figure 17.21.

This model of lightness perception may have applications beyond tone reproduction. In particular, this model appears to be suitable for detecting pixels that belong to self-luminous surfaces versus pixels that depict bright reflective surfaces. It would be interesting to see this model applied within the context of image appearance modeling, where pixels depicting self-luminous surfaces ought to be treated differently from pixels depicting bright surfaces.

Further, this model may be used for the purpose of white balancing images of scenes that have different areas illuminated by different light sources. For instance, shops typically illuminate their displays throughout the day, even when daylight enters through the shop window. Images taken under these circumstances may have areas illuminated by tungsten light and other areas illuminated by daylight. The approach discussed here would partition these areas into different frameworks and treat them separately.

17.9 Counter Shading

The design of local tone-reproduction operators often involves a trade-off between the amount of achievable range reduction and the minimization of artifacts. As discussed in this chapter, halos are often a by-product of local operators, unless they are specifically designed out of the system.

This desire to remove haloing artifacts is inspired by early results as shown in Figure 17.8. This image shows particularly large and obtrusive halos (exaggerated in this figure for demonstration purposes), and it is clear that these types of artifacts are best avoided. Recent tone-reproduction operators do an excellent job of avoiding halos.

However, at the same time, artists have used halos and related hatchings and shadings on purpose to suggest increased contrast in drawings and paintings. In those cases, the effect is called counter shading. Further, the existence of the Cornsweet-Craik-O'Brien illusion (Section 5.4.4) suggests that small contrasts can be enhanced by applying a halo-like gradient on either side of an edge.

As a result, it appears that it should be possible to apply counter shading in a controlled fashion within a tone-reproduction operator to increase the apparent contrast, without the results being perceived as artifacts. A first attempt at deliberately applying counter shading for the purpose of enhancing contrast has

Figure 17.22. An example of algorithmic counter shading applied to chromatic channels (left), compared with the same image without counter shading (right); Strasbourg. (Images courtesy of Kaleigh Smith, Grzegorz Krawczyk, Karol Myszkowski, and Hans-Peter Seidel; Max Planck Institute for Informatics, Saarbrücken, Germany.)

recently appeared [1058]. Here, the image is segmented using K-means segmentation [98,724]. At the edge of each segment, contrast is enhanced with a function that models contrasts as found in the Cornsweet illusion:

$$m(x,y) = 1 + a \begin{cases} \exp\left(-d^2/\sigma^2\right) & (x,y) \in \text{segment A}, \\ 1 - \exp\left(-d^2/\sigma^2\right) & (x,y) \in \text{segment B}. \end{cases} \tag{17.22}$$

Here, d is the shortest distance to the edge between segment A and segment B, and σ is a user parameter to specify the width of the enhanced area. It is assumed that the average luminance in segment A is greater than in segment B. Further, a is a measure of the amplitude of the scaling and, thus, specifies how strong the effect is. The value of m is applied as a multiplier to enhance the u* and v* channels in the CIE $L^*u^*v^*$ color space. Applying counter shading only to the chromatic channels yields a subtler effect than using the luminance channel. An example result is shown in Figure 17.22. Note that in this image, halo artifacts are not obvious, although the overall impression is that of a more saturated image.

17.10 Post-Processing

After tone reproduction, it is possible to apply several post-processing steps to either improve the appearance of the image, adjust its saturation, or correct for the gamma of the display device. Here, we discuss two frequently used techniques that have a relatively large impact on the overall appearance of the tone-mapped results—a technique to desaturate the results and a technique to clamp a percentage of the lightest and darkest pixels.

17.10.1 Color in Tone Reproduction

Tone-reproduction operators normally compress luminance values, rather than work directly on the red, green, and blue components of a color image. After these luminance values have been compressed into display values $L_d(x,y)$, a color image may be reconstructed by keeping the ratios between color channels the same as before compression (using $s = 1$) [1007]:

$$I_{r,d}(x,y) = \left(\frac{I_r(x,y)}{L_v(x,y)}\right)^s L_d(x,y), \tag{17.23a}$$

$$I_{g,d}(x,y) = \left(\frac{I_g(x,y)}{L_v(x,y)}\right)^s L_d(x,y), \tag{17.23b}$$

$$I_{b,d}(x,y) = \left(\frac{I_b(x,y)}{L_v(x,y)}\right)^s L_d(x,y). \tag{17.23c}$$

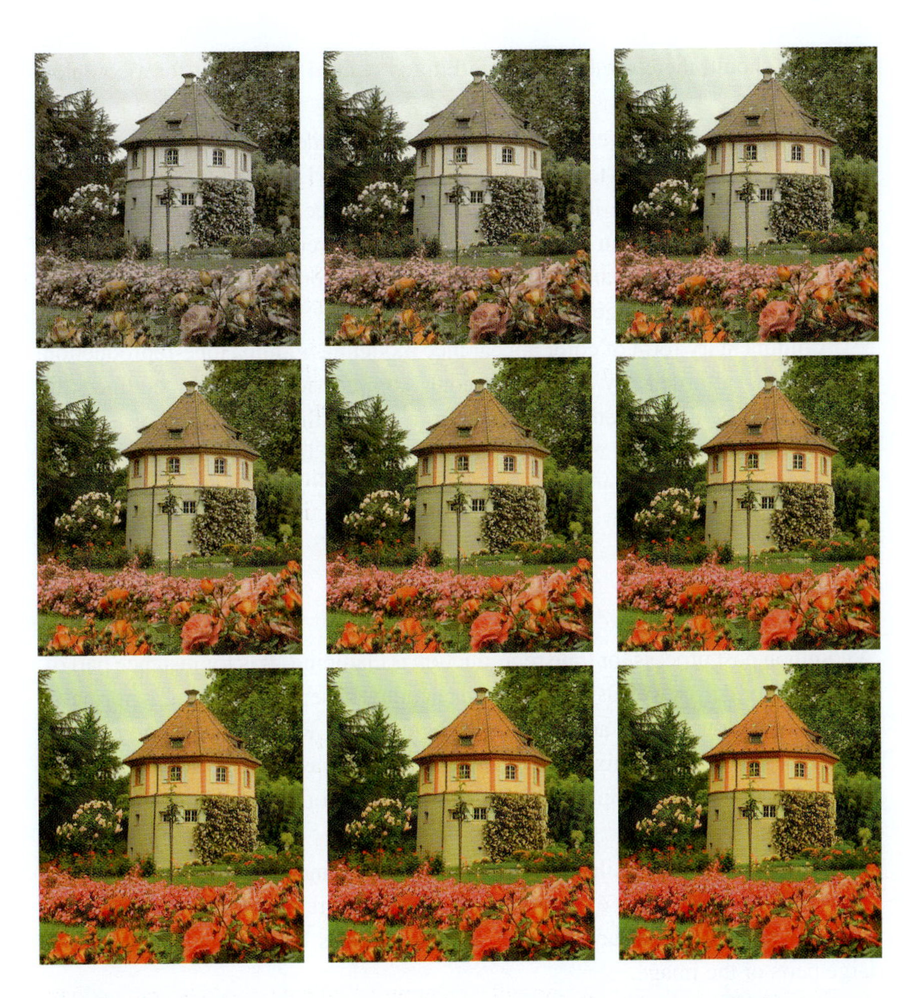

Figure 17.23. Per-channel gamma correction may desaturate the image. The images were desaturated with values of $s = 0.1$ through 0.9 with increments of 0.1; Mainau, Germany, June 2005.

Alternatively, the saturation constant s may be chosen smaller than 1. Such per-channel gamma correction may desaturate the results to an appropriate level, as shown in Figure 17.23 [306].

The results of tone reproduction may appear unnatural, because human color perception is non-linear with respect to overall luminance level. This means that if we view an image of a bright outdoor scene on a monitor in a dim environment, we are adapted to the dim environment rather than the outdoor lighting. By keeping color ratios constant, we do not take this effect into account. The above approach should therefore be seen as a limited control to account for a complex phenomenon.

A more comprehensive solution is to incorporate ideas from the field of color appearance modeling into tone-reproduction operators [292,884,948]. The iCAM image appearance model is the first color appearance model operating on images [291,292,791]. It can also be used as a tone-reproduction operator. It therefore constitutes an important trend towards the incorporation of color appearance modeling in dynamic range reduction and vice versa.

A rudimentary form of color appearance modeling within a tone-reproduction operator is afforded by the sigmoidal compression scheme outlined in Section 17.2 [948]. Nonetheless, we believe that further integration of tone reproduction and color appearance modeling is desirable for the purpose of properly accounting for the differences in adaptation between scene and viewing environments.

17.10.2 Clamping

A common post process for tone reproduction is clamping. It is for instance part of the iCAM model, as well as the sub-band encoding scheme. Clamping is normally applied to both very dark as well as very light pixels. Rather than specify a hard threshold beyond which pixels are clamped, a better way is to specify a percentile of pixels that will be clamped. This gives better control over the final appearance of the image.

By selecting a percentile of pixels to be clamped, inevitably detail will be lost in the dark and light areas of the image. However, the remainder of the luminance values is spread over a larger range, and this creates better detail visibility for large parts of the image.

The percentage of pixels clamped varies usually between 1% and 5%, depending on the image. The effect of clamping is shown in Figure 17.24. All images are treated with the photographic tone-reproduction technique. The images on the left show the results without clamping, and therefore all pixels are within the display range. In the image on the top right, the darkest 7% of the pixels are clamped, as well as the lightest 2% of the pixels. This has resulted in an image that has reduced visible detail in the steps of the amphitheater, as well as in the wall and the bushes in the background. However, the overall appearance of the image has improved, and the clamped image conveys the atmosphere of the environment better than the directly tone-mapped image.

The bottom-right image of Figure 17.24 also conveys the atmosphere of the scene better. This photograph was taken during a bright day. In this case, the

Figure 17.24. Examples of clamping. All images were tone mapped using photographic tone reproduction. The left images were not clamped, whereas 7% of the darkest pixels and 2% of the lightest pixels were clamped in the top-right image. Top: Salamis amphitheater, North Cyprus, July 2006. Bottom: Lala Mustafa Pasha Mosque, Famagusta, North Cyprus, July 2006.

clamping has allowed more detail to become visible in the back wall around the window.

To preserve the appearance of the environment, the disadvantage of losing detail in the lightest and darkest areas is frequently outweighed by a better overall impression of brightness. The photographs shown in Figure 17.24 were taken during a very bright day, and this is not conveyed well in the unclamped images.

Finally, the effect of clamping has a relatively large effect on the results. For typical applications, it is an attractive proposition to add this technique to any tone-reproduction operator. However, as only a few tone-reproduction operators incorporate this feature as standard, it also clouds the ability to assess the quality of tone-reproduction operators. The difference between operators appears to be of similar magnitude as the effect of clamping.

17.11 Validation and Comparison

A recent topic of interest in tone reproduction is validation. It is generally recognized that visually comparing results obtained with different operators is a haphazard way of assessing their quality. A problem with this approach is that different observers apply different criteria to decide which operator they appreciate best. The sample size, i.e., the number of images that would have to be used to draw conclusions, is usually too large to allow meaningful discussion in research papers that are almost always limited in length. With small numbers of images involved in a visual comparison, there is a distinct danger that the choice of image has a more profound impact on the outcome than the choice of tone-reproduction operator.

Other problems with the visual assessment of tone reproduction operators is that it is sometimes unclear which qualities of a tone reproduction are desirable to achieve a specific task. Hence, the influence of the task for which a tone-reproduction operator is chosen is non-negligible. Consider, for instance, two disparate tasks such as assessment of medical imagery for the purpose of diagnosis versus the use of tone-reproduction operators in entertainment (film, video, games). For the latter, a visually pleasing tone-mapped result may well be all that is required. Simple visual inspection may be sufficient to decide between the many tone-reproduction operators currently available.

However, the same could not be said in the case of medical imaging. Here, the tone-reproduction operator does not have to produce visually appealing results, but should enable the task—diagnosis of patients—to be carried out with maximum confidence. Simple visual inspection is likely to be inadequate for this task.

Nonetheless, visual comparison does allow expert observers to infer certain behaviors of tone-reproduction operators. In addition, it is possible to include more tone-reproduction operators in such a comparison than could be realistically included in a validation study. As such, there is still value in presenting a side-by-side comparison of a set of tone-reproduction operators. We therefore present and discuss a selection of tone-mapped images in Section 17.11.2.

A less subjective approach is to define specific image features that are preferentially preserved in tone-mapped images. It is then possible to measure a high dynamic range image and its tone-mapped counterpart and apply the metric to the ratio. If this ratio is close to 1, then the particular image feature is said to be preserved, whereas smaller values indicate deviation from the ideal. This approach, discussed in Section 17.11.3, is less prone to interpretation than visual inspection, yet is not as involved as a full-blown user study.

As mentioned, for many application areas, the performance of a tone-reproduction operator is intimately linked to the task for which it is designed. The notion that a tone-reproduction operator is likely to be good for some specific tasks, but not all tasks, is currently dominant. It should therefore be possible to identify real-world tasks for which tone-reproduction operators may be used and to derive laboratory experiments from these tasks.

The task performance thus recorded under laboratory conditions is then an indication of how well the real-world task may be performed. Unfortunately, such validation studies are time consuming, and this is why only recently a few such studies have appeared. These are discussed in Section 17.11.4.

17.11.1 Operators

For each of the studies discussed in the following, we refer to the original papers for detailed descriptions of the experimental set-up and analysis. We aim to provide an overview as well as the main conclusions of each of the studies. The algorithms included are summarized as follows:

Linear scaling. A global operator, whereby the image is linearly scaled to fit into the range of the display.

Histogram adjustment. A global operator that derives luminance-mapping functions from the cumulative histogram of the high dynamic range input [1211].

Time-dependent operator. A global operator that takes the time course of adaptation into account [885].

Adaptive logarithmic operator. A global tone-reproduction operator that essentially takes the logarithm for each pixel, but adjusts the base of the logarithm on a per-pixel basis for enhanced performance [264].

Photographic tone reproduction. A tone-reproduction operator based on photographic practice. It exists in both local and global variants and is discussed in Section 17.3 [950].

Ashikhmin operator. A local tone-reproduction operator inspired by human visual perception [47]. It uses a scale-selection mechanism similar to the local photographic operator.

Bilateral filtering. A local tone-reproduction operator, discussed in Section 17.5 [270].

Trilateral filtering. A refinement of bilateral filtering, it involves the application of the bilateral filter twice, but in different domains [178].

LCIS. A general image-processing technique to optimize contrast in images. Similar to anisotropic diffusion, it can be used for the purpose of tone reproduction [1153].

iCAM. A local tone-reproduction operator and color appearance model, discussed in Chapter 11 [291, 292].

Sub-band encoding. A local tone-reproduction operator, discussed in Section 17.4 [681].

Gradient-domain compression. Tone reproduction based on compressing gradient fields, discussed in Section 17.6. [306].

Photoreceptor-based compression. Sigmoidal compression, inspired by photoreceptors, as discussed in Section 17.2 [948].

Tumblin-Rushmeier operator. A global tone-reproduction operator, based on preserving brightness [1152].

Local eye adaptation. Sigmoidal compression using local adaptation computed with a bilateral filter, as discussed in Section 17.5. It includes a model of time-dependent adaptation as well [658].

17.11.2 Comparison

In this section, we present side-by-side comparisons of several recent tone-reproduction operators. As discussed above, the successful visual comparison depends on several parameters:

- *Correctness of the implementation of each operator.* Where possible, we have used implementations provided by the authors. However, in most cases the implementations are taken from the supplemental material of *High Dynamic Range Imaging: Acquisition, Display and Image-Based Lighting* [952]. These implementations are not known to be faulty, although correctness cannot be guaranteed.

- *Choice of user parameters.* Most tone-reproduction operators provide user parameters that need to be set appropriately to obtain meaningful results. Only a few operators are either fully automatic, or provide an automatic mode of operation. Where user parameters need to be specified, they were chosen so as to produce a flattering result. However, it is impossible to guarantee in every case that a different combination of parameter settings would not lead to a better image.

- *Choice of test images.* The image used in this comparison is only one of many that we could have chosen. It is well known that some images suit certain operators better than others. An exhaustive comparison, or even a visual comparison with a reasonable number of test images, is beyond the scope of this book. We have therefore tried to select a reasonable test image, but note that for specific applications, a better test image could have been chosen.

- *Experience of the observer.* Even the experts are sometimes unclear as to what it is they are looking for in a good tone-mapped image. There appears to be a correlation between images that are most liked and images that look natural, but this correlation may be weaker than expected. In addition, if a tone-mapped image could be compared against the real environment from which it was derived, the choice of which is the most natural-looking result may change [44].

With these caveats in mind, images of several tone-reproduction operators are shown in Figure 17.25. The first five images are created with local operators, followed by linear scaling. The bottom row consists of global operators. For these particular images, and with the parameter settings chosen by the author, the local operators all preserve details in the dark areas very well. Details are also

Figure 17.25. Visual comparison between several different tone-reproduction operators. First row: sub-band encoding, bilateral filtering, gradient domain compression; Second row: Ashikhmin, photographic, linear scaling; Third row: histogram adjustment, photoreceptor-based, Tumblin-Rushmeier.

well preserved in the bright areas, except by photographic tone reproduction. This operator was designed to let bright areas burn out, to maintain a good impression of brightness. Hence, this operator affords a different trade-off between detail visibility and overall perception of brightness.

We applied clamping to the linearly scaled operator with 10% of the bottom pixels and 5% of the top pixels clamped. This has resulted in significant burn-out in the windows and loss of detail in the dark areas. However, this operator has created a well exposed mid-range, presenting the walls and the floor well. Without clamping, this image would appear to be largely black, with the windows still burnt-out. The linear operator is effectively a baseline result, the quality of which all other operators will have to exceed to justify their computational expense.

Both the histogram-adjustment operator and the photoreceptor-inspired operator present sufficient details in the dark areas; however, the histogram-adjustment operator let the bright area burn out. The bright windows are reasonably well represented by the photoreceptor-based operator. The overall impression that this window is very bright in the real environment is well conveyed by both operators.

The Tumblin-Rushmeier operator trades detail visibility for global contrast. This operator would benefit from additional clamping to improve detail visibility of the mid-range of luminance values. However, we applied clamping only if this was part of the original operator, which for Figure 17.25 was only the sub-band encoding scheme and the linear scaling. Similarly, gamma correction was applied to all images, except the sub-band encoding technique, as this operator was designed to handle gamma issues directly.

17.11.3 Evaluation Metrics

In addition to visual comparison, it is possible to derive metrics of desirable image features. Two such metrics are described here: global contrast and detail visibility [1058].

The impression of global contrast is best evaluated in the brightness domain, which itself can be approximated by taking the logarithm (see also Section 6.4). In Section 17.4, an effective gain map $G(x,y)$ was introduced for each pixel. Tone mapping can thus be seen as multiplying the HDR image by G(x,y). Alternatively, it is possible to approximate the effective gain map by a constant and view tone reproduction as a linear function in the log domain:

$$\log_{10}(L_d(x,y)) \approx f(\log_{10}(L_v(x,y))) = G_1 \log_{10}(L_v(x,y)) + G_2, \tag{17.24}$$

with G_1 and G_2 coefficients estimated through linear regression for all tone-mapped pixel values. Here, f is the estimated tone-reproduction function (linear in the log domain) that is clamped to the minimum and maximum display

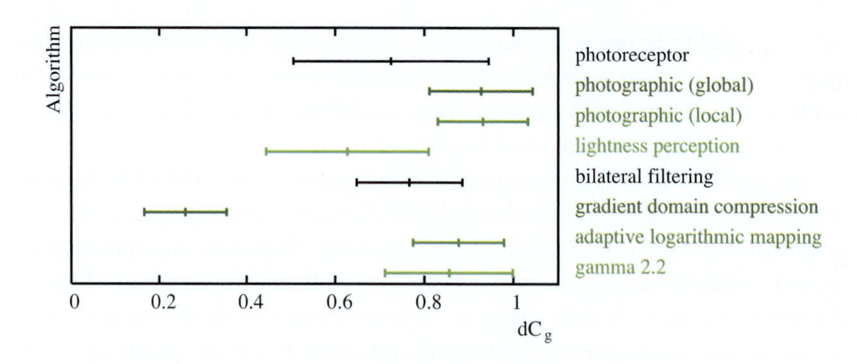

Figure 17.26. Change in global contrast dC_g for different tone-reproduction operators (after Smith, Krawczyk, Myszkowski, and Seidel [1058]).

luminance. The global contrast $\Delta \log_{10}(L_d)$ in the brightness domain can then be derived using the minimum and maximum brightness in the tone-mapped image:

$$\Delta \log_{10}(L_d) = f(\max(\log_{10}(L_v))) - f(\min(\log_{10}(L_v))). \tag{17.25}$$

The change in global contrast dC_g as a result of applying the tone-reproduction operator (approximated by applying the function f) is then given by

$$dC_g = \frac{f(\max(\log_{10}(L_v))) - f(\min(\log_{10}(L_v)))}{\max(\log_{10}(L_v)) - \min(\log_{10}(L_v))}. \tag{17.26}$$

Note that dC_g is equal to G_1 as a result of the linear regression applied above, i.e., it represents the slope of the line obtained through linear regression.

Thus, the change in global contrast can be interpreted as the ratio of the brightness range in the tone-mapped image and the input HDR image. If this ratio is close to 1, then global contrast is preserved. For values less than 1, global contrast is reduced. It is therefore desirable for tone-reproduction operators to show as little change in global contrast as possible. In practice, some operators preserve this feature better than others, as shown in Figure 17.26.

Figure 17.26 shows that many of the operators do a reasonable job of preserving global contrast, with the photographic operator being closest to the ideal value of 1. On the other end of the scale is the gradient-domain operator that can only give a very limited impression of global contrast. It makes a clear trade-off in favor of local detail visibility, as we will show next.

By compressing the global range of values, more headroom is left for expressing local details. Both global contrast and detail visibility are desirable, and a trade-off between the two appears to be unavoidable. A metric to compute detail visibility may be devised as follows.

In Section 17.5 we argued that a bilateral filter could be used to compute a local adaptation luminance for each pixel. The output of the bilateral filter $L_B(x,y)$ can be used to compute a measure of local contrast C_l:

$$C_l(L,x,y) = \log_{10} \frac{\max\left(L(x,y), L_B(x,y)\right)}{\min\left(L(x,y), L_B(x,y)\right)}. \tag{17.27}$$

A model of the response of the human visual system to a given contrast can be approximated with the following transducer function $T(C_l)$ [733]:

$$T(C_l(L,x,y)) = 54.09288\, C_l(L,x,y)^{0.41850}. \tag{17.28}$$

The constants in this equation are chosen such that $T(0) = 0$ and, for an appropriately chosen threshold contrast C_t, we have $T(C_t) = 1$. The threshold contrast was chosen to be 1%, so that $C_t = \log_{10}(1.01)$. For high dynamic range images, this assumption may be inaccurate. To correct for the unique conditions posed by high dynamic range imaging, a correction factor may be applied to $T(C_l)$, leading to the following model of response to local contrast:

$$C_{\text{resp}}(x,y) = T(C_l(L,x,y)) \frac{C_t}{\log_{10}\left(\dfrac{L_B(x,y) + \text{tvi}\left(L_B(x,y)\right)}{L_B(x,y)}\right)}. \tag{17.29}$$

This response function depends on a threshold-versus-intensity function, abbreviated as TVI, that models the smallest change in intensity that a human can observe as a function of the background intensity (see Section 5.5.2).

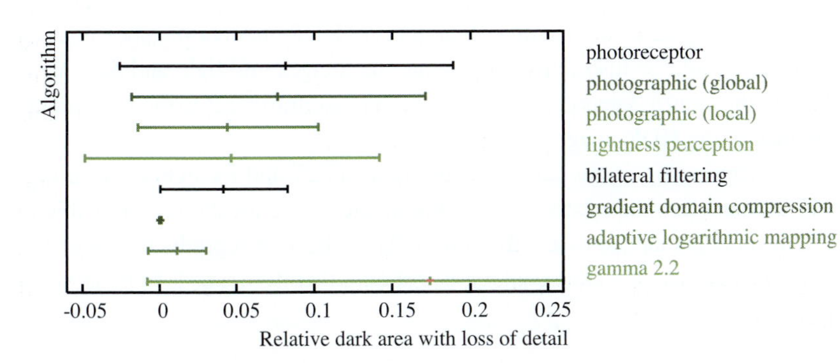

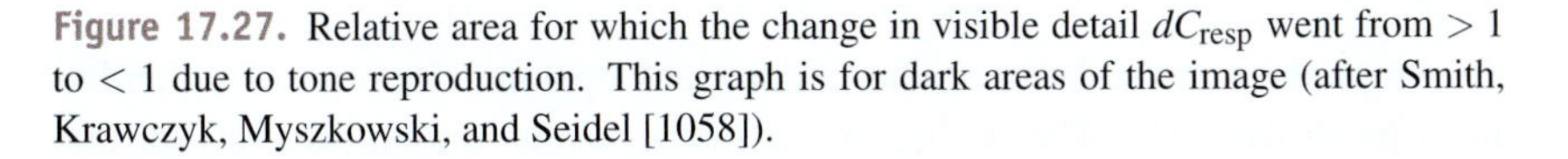

Figure 17.27. Relative area for which the change in visible detail dC_{resp} went from > 1 to < 1 due to tone reproduction. This graph is for dark areas of the image (after Smith, Krawczyk, Myszkowski, and Seidel [1058]).

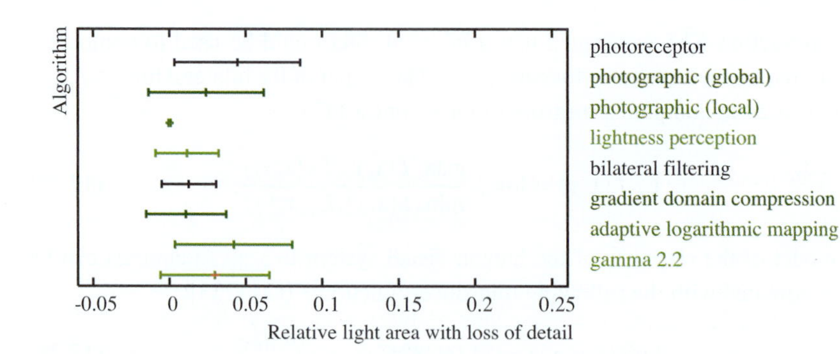

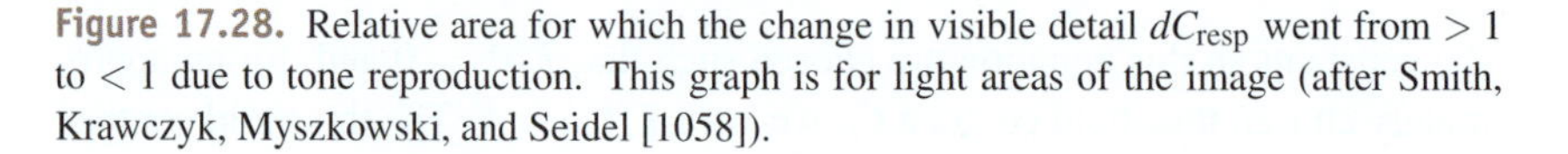

Figure 17.28. Relative area for which the change in visible detail dC_{resp} went from > 1 to < 1 due to tone reproduction. This graph is for light areas of the image (after Smith, Krawczyk, Myszkowski, and Seidel [1058]).

The change in hypothetical response that occurs when a high dynamic range image is tone mapped is thus a relevant metric to evaluate how well local details are preserved. As the unit of measurement of C_{resp} is in just noticeable differences (JNDs), values less than 1 are imperceptible and can therefore be set to 0. Also, if tone reproduction causes the contrast to be reduced from > 1 JND to < 1 JND, the change is set to 1. In all other cases, the change in detail visibility is simply the difference in contrast response:

$$dC_{\text{resp}}(x,y) = \begin{cases} 1 & \text{if } \; C_{\text{resp}}(L_v,x,y) > 1 > C_{\text{resp}}(L_d,x,y), \\ 0 & \text{if } \; \left|C_{\text{resp}}(L_v,x,y) - C_{\text{resp}}(L_d,x,y)\right| < 1, \\ C_{\text{resp}}(L_v,x,y) - C_{\text{resp}}(L_d,x,y) & \text{otherwise.} \end{cases} \tag{17.30}$$

This metric is typically applied directly to pairs of HDR and tone-mapped images. However, it would be instructive to separate the images into light and dark areas and apply this metric separately. The threshold is arbitrarily set to the log average luminance of the HDR image.

The analysis, shown for dark areas in Figure 17.27 and for light areas in Figure 17.28 measures the relative area of the image where local contrast visible in the HDR input is mapped to less than one JND by the tone-reproduction operator. This is the area where visible contrast is lost and, therefore, needs to be as small as possible.

17.11.4 Validation

Tone reproduction validation studies aim to rank operators in terms of suitability for a particular task. The design of such experiments thus involve a task that observers are asked to perform. The accuracy with which the task is performed is then said to be a measure for how suitable a tone reproduction is for that particular task. It is therefore important to select a sensible task, so that practicality of tone-reproduction operators can be inferred from task performance.

This particular requirement causes several difficulties. For instance, the game industry is currently moving to high dynamic range imaging. This means that rendered content needs to be tone mapped first before it can be displayed. The real-world task is then "entertainment." Maybe the aim of the game designer is to achieve a high level of involvement of the player; or perhaps the goal is to enable players to achieve high scores. In any case, it will be difficult to devise a laboratory experiment that measures entertainment, involvement, or engagement.

For other tasks—medical visualization was mentioned earlier—the validation study may be easier to define, but such application-specific validation studies are currently in their infancy. A notable exception is Park and Montag's study, which evaluates tone reproduction operators for their ability to represent non-pictorial data [872]. As a result, validation studies tend to address issues related to preference and/or similarity.

Preference ratings, which measure whether one image is liked better than another, have various advantages [263, 625, 627]. It is, for instance, not necessary to compare tone-mapped images to real scenes, as only pair-wise comparisons between tone-mapped images are required. On the downside, not having seen the original environment can be viewed as a serious limitation [44]. The most preferred image may still deviate significantly from the real environment from which it was derived. Preference scalings will not be able to signal such deviations.

On the other hand, similarity ratings are designed to determine which tone-mapped result looks most natural, i.e., most like the real environment that it depicts [627, 659, 1278]. Such studies benefit from having the real environment present; it has been shown that including the real environment in the study may change observers' opinions [44].

The first validation study that assessed tone-reproduction operators appeared in 2002 [263] and measured image preference and, in a second experiment, naturalness. They found that photographic tone reproduction consistently scored highest, followed by the histogram-adjustment method; LCIS scored low.

In a separate experiment, a set of seven tone-reproduction operators were compared against two real scenes [1278]. A total of 14 observers rated the tone-mapped images for overall brightness, overall contrast, visible detail in dark regions, visible detail in light regions, and naturalness. Results of this experiment are reproduced in Figure 17.29.

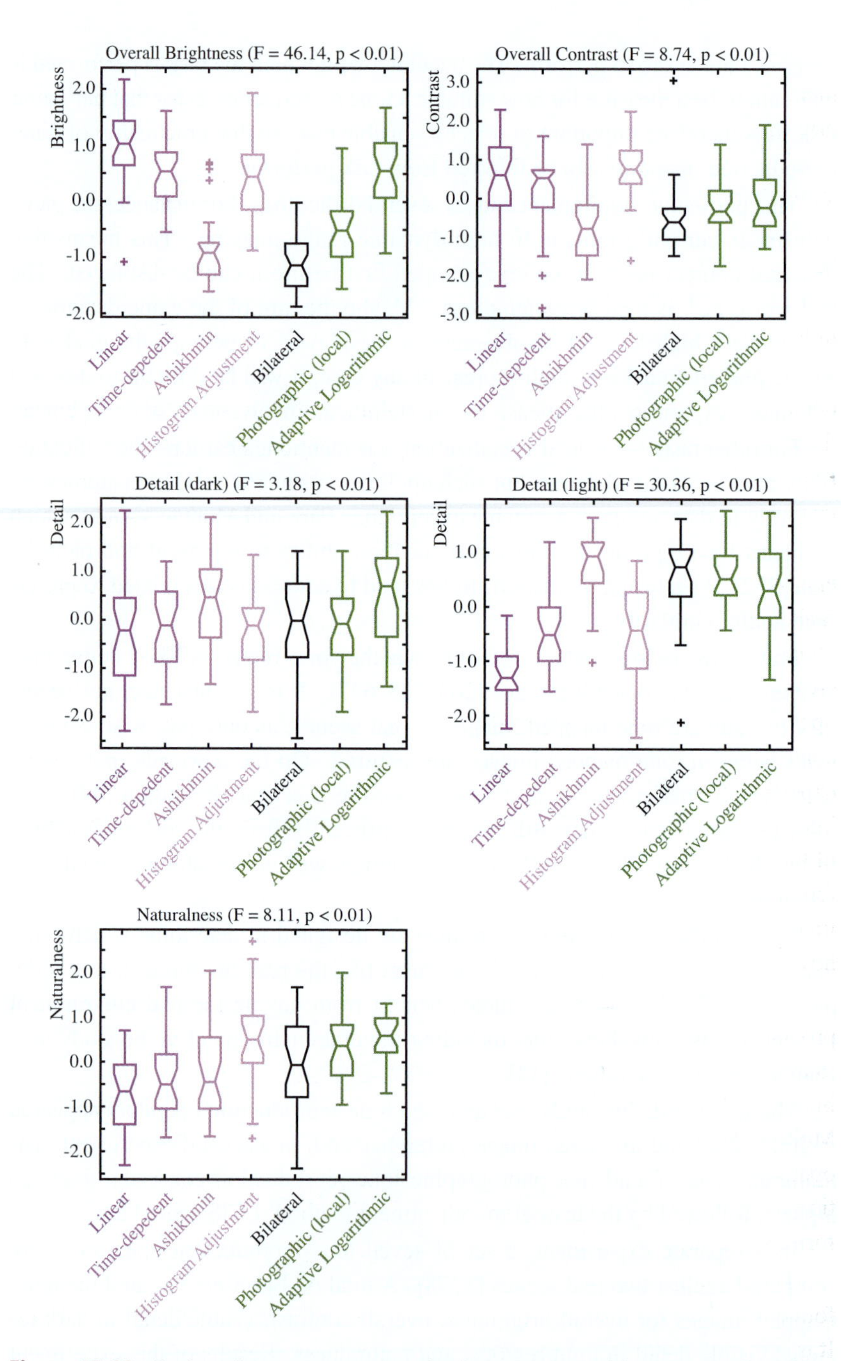

Figure 17.29. Ratings for seven tone-reproduction operators with respect to five different image attributes. The green colors correspond to the operators also included in Figures 17.26, 17.27, and 17.28 (after Yoshida, Blanz, Myszkowski, and Seidel [1278]).

This experiment reveals that the global operators (linear, histogram adjustment, time-dependent and adaptive logarithmic) are perceived to be overall brighter than the local operators (bilateral filtering, Ashikhmin, and photographic). In addition, the global operators have higher overall contrast than the local operators. Details in dark regions are best preserved by the Ashikhmin and adaptive logarithmic operators according to this experiment. Details in light regions are best preserved by the local operators. The histogram-adjustment, photographic and adaptive-logarithmic operators are perceived to be the most natural reproductions of the real scenes.

Correlations between each of brightness, contrast, and detail visibility on the one hand, and naturalness on the other were computed. It was found that none of these attributes alone can sufficiently explain the naturalness results. It is therefore hypothesized that a combination of these attributes, and possibly others, are responsible for naturalness.

A second, large-scale validation study also asked the question of which operators produces images with the most natural appearance [659]. Images were ranked in pairs with respect to a reference image, which was displayed on a high dynamic range display device. Thus, for each trial, the participant viewed a reference image as well as two tone-mapped images that were created with different operators. Preference for one of the two tone-mapped images was recorded in the context of a specific image attribute that varied between experiments. In the first experiment, overall similarity was tested using 23 scenes. In the next two experiments, visibility and reproduction of detail was tested in light and dark regions of the image using 10 scenes each. The experiments were carried out for six different tone-reproduction operators, and the analysis produced a ranking for each image. In Figure 17.30, we summarize the results by showing, for each operator, how many times it was ranked first, second, and so on.

The overall similarity ranking in this figure indicates that the iCAM model produces images that were perceived to be closest to the high dynamic range images, followed by photographic tone reproduction. In this experiment, the bilateral-filter and the adaptive-logarithmic mapping did not perform as well. Multiple comparison scores were computed on the data, which revealed that all operators perform differently in a statistically significant manner, with the exception of histogram adjustment and the local eye-adaptation model which together form one group.

The overall significance test was repeated for gray-scale images. Here it was found that the photographic operator and the iCAM model perform equivalently. It is speculated that this change in ranking may result from iCAM's roots in color appearance modeling, as it produces particularly natural looking color images.

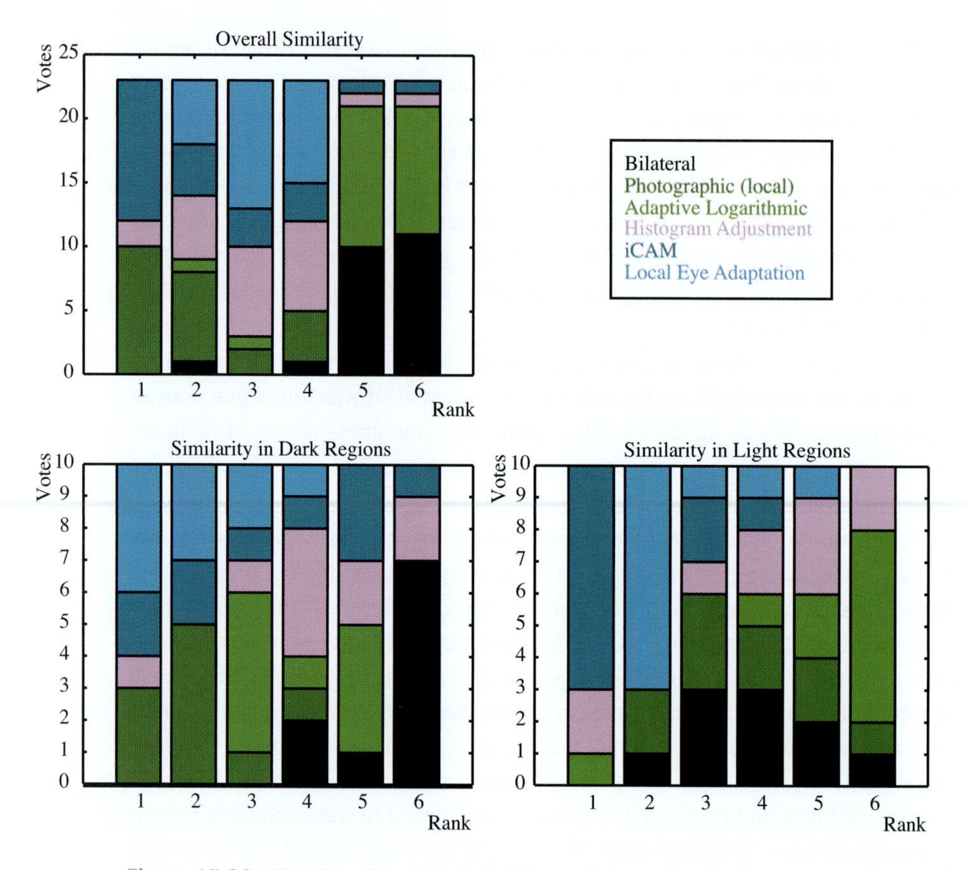

Figure 17.30. Number of images ranked for each tone-reproduction operator in experiments performed by Ledda et al. [659].

In bright regions, the iCAM model still outperforms the other operators, but the photographic operator does not reproduce details in dark regions as well, being an average performer in this experiment and performing better in dark regions. Each of Ledda's experiments provides a ranking, which is summarized in Figure 17.31, showing that in these experiments, iCAM and the photographic operator tend to produce images that are perceived to be closer to the corresponding high dynamic range images than can be achieved with other operators.

A further user study asking observers to rate images in terms of naturalness was recently presented [44]. This study assessed in terms of preference (Experiment I). Operators were also compared for naturalness in the absence of a real scene (Experiment II) and in the presence of a real scene (Experiment III). The

	Overal Similarity (Color)		Overal Similarity (Grayscale)
1	iCAM	1	Photographic (local)
2	Photographic (local)	2	iCAM
3	Histogram Adjustment		Local Eye Adaptation
	Local Eye Adaptation	3	Histogram Adjustment
4	Adaptive Logarithmic	4	Adaptive Logarithmic
5	Bilateral	5	Bilateral

	Detail (Light)		Detail (Dark)
1	iCAM	1	Photographic (local)
2	Local Eye Adaptation		Local Eye Adaptation
3	Photographic (local)	2	iCAM
	Histogram Adjustment	3	Adaptive Logarithmic
4	Bilateral		Histogram Adjustment
5	Adaptive Logarithmic	4	Bilateral

Figure 17.31. Ranking of operators in Ledda's experiments [659]. The boxes around pairs of operators indicate that these operators were not different within statistical significance.

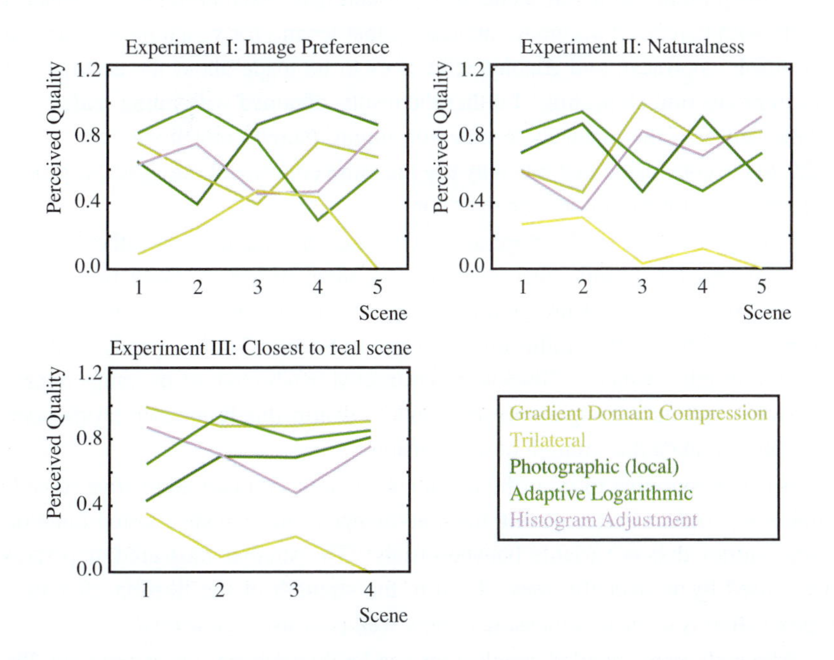

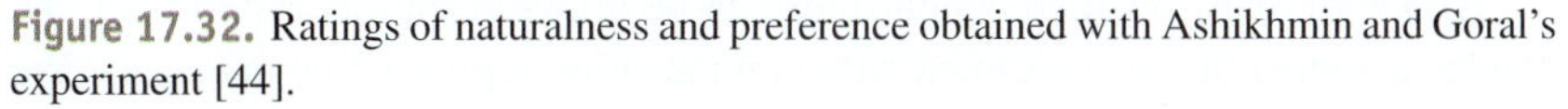

Figure 17.32. Ratings of naturalness and preference obtained with Ashikhmin and Goral's experiment [44].

results for each of these experiments are reproduced in Figure 17.32 for five different tone-reproduction operators.

As can be seen, in this particular experiment, the absence of a real scene caused operators to perform inconsistently across scenes, suggesting that the choice of scene has a non-negligible influence on the outcome. However, the results are more consistent if the images could be compared against the real environment.

When compared directly against real environments, in Experiment III the gradient-domain compression technique performs well, followed by adaptive-logarithmic mapping. The trilateral filter does not produce images that are close to the real environments in this particular experiment.

For the purpose of assessing different tone reproduction operators, as well as for the development of an evaluation protocol for testing such operators, Kuang et al developed a series of experiments centered around visual preference and rendering accuracy in the context of a digital photography task [627].In all three of their experiments the bilateral filter performed well, with the photographic operator scoring similarly in two of the three experiments.

In the presence of a real scene, it was found that visual preference correlates highly with perceived accuracy, suggesting that testing for visual preference is a reasonable approach, and enables inferences to be made about the accuracy of tone reproduction algorithms. Further the results obtained with rating scales correlate well with the much more time-consuming paired comparison tests. This leads to the conclusion that one may use the more convenient rating scales procedure to assess tone reproduction operators.

Finally, a recent study proposed not to use real images, but rather a carefully constructed stimulus that enables assessment of tone-reproduction operators in terms of contrast preservation [25]. It is argued that the strength of the Cornsweet-Craik-O'Brien illusion, discussed in Section 5.4.4, depends on its luminance profile. If such an illusion is constructed in a high dynamic range image, then tone-reproduction operators are likely to distort this luminance profile and, therefore, weaken the strength of this illusion.

It is now possible to infer the illusion's strength in a two-alternative forced-choice experiment by comparing the tone-mapped image against a step function with contrast that is variable between trials. The smaller measured step sizes are caused by weaker illusions. In turn, the strength of the illusion after tone reproduction is taken as a measure of how well contrast is preserved.

Although many detailed conclusions can be drawn from this experiment, the results are generally in agreement with previous tone-reproduction operators. We infer that the preservation of contrast therefore correlates with people's subjective preference. The reproduction of contrast should therefore be an important design criterion in the development of tone-reproduction operators, although it cannot be the only criterion.

17.11.5 User Strategies

Using a high dynamic range display device, it is possible to artificially limit the dynamic range of the display. This affords the opportunity to assess how users adjust generic image features, such as overall brightness and contrast, for different display conditions. Such an experiment has revealed the general strategy that users apply to adjust an image for display [1279].

In general it was found that parameter adjustments proceed in a similar way irrespective of the target dynamic range of the display device. This suggests that tone-reproduction operators do not need to be tailored to specific display devices. In particular, when high dynamic range display devices become available, it may not be necessary to adopt new algorithms to prepare images for such displays when accounting for user preference. Of course, issues relating to visual adaptation to the display device itself, as well as color appearance issues, do still need to be taken into account.

The strategy adopted by most users is to first adjust the image such that an image-dependent percentage of pixels is burnt out. Typically, image areas that are perceived as self-luminous are mapped to the maximum display value. Hence, for images with a large directly visible light source, users chose a high percentage of pixels to burn out. For images with small light sources, such as decorated Christmas trees, a lower percentage of pixels is typically chosen.

Then, contrast is adjusted according to which task is set. If the goal is to produce a good-looking image, the contrast tends to be exaggerated by up to four times the contrast of the original high dynamic range image, even if this means that a large percentage of the darkest pixels needs to be clipped. If the task is to produce a realistic-looking image, as close as possible to a reference image, then users avoid clipping light and dark regions, and contrast is not exaggerated. However, irrespective of task, users tend to over-saturate pixels that belong to self-luminous objects.

This result indicates that users tend to have a strong preference for both bright and high-contrast images. This favors algorithms that achieve good detail reproduction in both dark and light areas of the image. Tone-reproduction operators tailored to follow user preferences should be able to achieve this. However, tone-reproduction operators intended for high fidelity reproduction should not clamp such large numbers of pixels, especially not in the dark areas.

A note of caution is warranted here, though. Dependent on circumstances, different users may have completely different preferences, so that averaging responses across participants in subjective experiments may not always be appropriate. For instance, some users prefer losing details in dark regions rather than light ones. Given that participants are relatively consistent in setting the reference white, their contrast and brightness preferences may be different. As a result, this would alter their attitude toward losing details in dark regions.

17.11.6 Discussion

Now that more validation studies are becoming available, some careful overall conclusions may be drawn. It is clear that operators that perform well in one validation study are not guaranteed to perform well in all validation studies. This hints at the possibility that these operators are not general purpose, but are suitable only for certain specific tasks.

On the other hand, it could also be argued that the choice of task, types of images, as well as further particulars of the experimental designs, have an impact on the outcome. Ashikhmin's study has found that having a real scene included in the experimental paradigm, for instance, significantly changes the result.

Further, there are many different operators, and each validation study necessarily includes only a relatively small subset of these. There are a few operators that have been included in most of the user studies to date. Of these, the photographic operator appears to perform consistently well, except in Ashikhmin's study. The adaptive-logarithmic mapping performed well in Yoshida's study as well as in Ashikhmin's study, but performed poorly in Ledda's study.

Finally, Ashikhmin's study, as well as Ledda's study, did not find a clear distinction between the included local and global operators. On the other hand, Yoshida's methodology did allow local operators to be distinguished from global operators. This suggests that perhaps for some combinations of tasks and images, a computationally expensive local operator would not yield better results than a cheaper global operator. However, this does not mean that there does not exist a class of particularly challenging images for which a local operator would bring significant benefits. However, such scenes would also be challenging to include in validation studies.

17.12 Further Reading

High dynamic range imaging is a relatively new field, with so far only two books available on the topic. Reinhard et al.'s book presents the high dynamic range imaging pipeline, with an emphasis on capture, storage, and display algorithms. It also includes a chapter on image-based lighting [952]. Hoefflinger's book focuses on high dynamic range imaging hardware, including cameras, sensors, lenses, and display hardware [475], but also includes chapters on tone reproduction and HDR data compression.

The DVD accompanying this book contains a large collection of high dynamic range images. Further links to image resources are provided on the DVD as well. Of particular importance is the HDR Photographic Survey [303], which contains a large set of calibrated images, each annotated with appearance and GPS data (Available at http://www.cis.rit.edu/fairchild/HDR.html).

Part IV
Appendices

Appendix A
Vectors and Matrices

While we assume basic vector algebra, Maxwell's equations in differential form, as presented in Chapter 2, are defined on vector fields. In this appendix, we give a brief account of the basic gradient, divergence, and curl operators, as well as vector integration and differentiation. An introduction of vector calculus in the context of Maxwell's equations can be found in *Div, Grad, Curl, and All That* [1004].

All operations assume a Cartesian coordinate system. In the following, we indicate unit vectors along each of the x-, y- and z-axes with $\mathbf{e}_x$, $\mathbf{e}_y$ and $\mathbf{e}_z$.

A.1 Cross and Dot Product

The vector dot product between two vectors **A** and **B** is defined as follows:

$$\mathbf{A} \cdot \mathbf{B} = A_x B_x + A_y B_y + A_z B_z. \tag{A.1}$$

The dot product is therefore a scalar quantity. The geometric interpretation is that the dot product represents the cosine of the angle between two vectors times the length of both vectors (see Figure A.1):

$$\mathbf{A} \cdot \mathbf{B} = \|\mathbf{A}\| \, \|\mathbf{B}\| \cos(\alpha). \tag{A.2}$$

This implies that the dot product will be zero for perpendicular vectors. Vectors pointing in the same direction have a dot product with a value equal to the square of their length:

$$\mathbf{A} \cdot \mathbf{A} = \|\mathbf{A}\|^2. \tag{A.3}$$

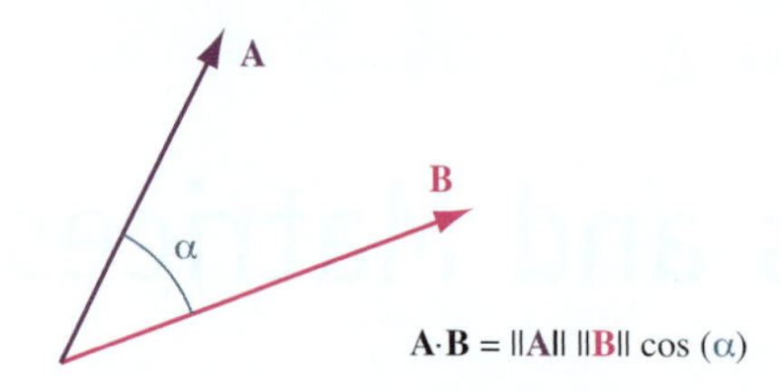

Figure A.1. Dot product between vectors **A** and **B**.

Another implication is that to find the angle between two vectors, we may normalize both vectors and take the dot product:

$$\cos(\alpha) = \frac{\mathbf{A}}{\|\mathbf{A}\|} \cdot \frac{\mathbf{B}}{\|\mathbf{B}\|}. \tag{A.4}$$

The cross product is defined as

$$\mathbf{A} \times \mathbf{B} = \begin{pmatrix} A_z B_y - A_y B_z \\ A_x B_z - A_z B_x \\ A_y B_x - A_x B_y \end{pmatrix} \tag{A.5}$$

and results in a new vector which is perpendicular to both **A** and **B**, which may be expressed as

$$\mathbf{A} \cdot (\mathbf{A} \times \mathbf{B}) = 0, \tag{A.6a}$$
$$\mathbf{B} \cdot (\mathbf{A} \times \mathbf{B}) = 0. \tag{A.6b}$$

As shown in Figure A.2, the right-hand rule applies so that $\mathbf{A} \times \mathbf{B} = -\mathbf{B} \times \mathbf{A}$. Finally, the length of the resulting vector may be interpreted as the sine of the

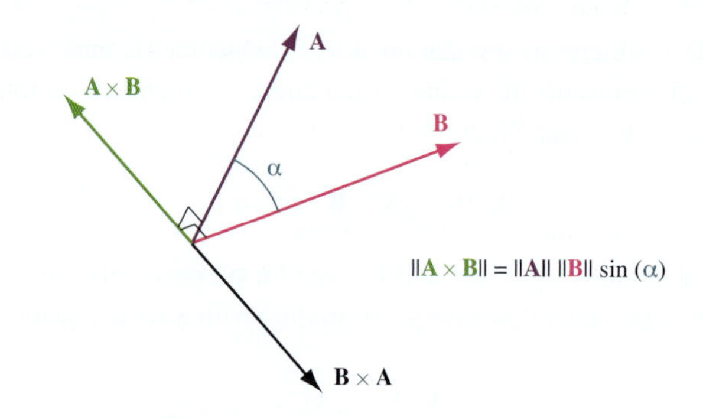

Figure A.2. Cross product between vectors **A** and **B**.

angle between both vectors times the length of the vectors:

$$\|\mathbf{A} \times \mathbf{B}\| = \|\mathbf{A}\| \, \|\mathbf{B}\| \sin(\alpha). \tag{A.7}$$

While this identity could be used to compute the angle between two normalized vectors, it is cheaper to use the dot product as indicated above in (A.4).

A.2 Vector Differentiation

A vector $\mathbf{A}$ may be defined as a function of an independent variable t. By making a change in the value of t, the magnitude and direction of $\mathbf{A}$ will generally change. By increasing or decreasing the value of t, the vector $\mathbf{A}$ will draw a contour in three-dimensional space.[1]

The differential change dt of $\mathbf{A}$ with respect to variable t is then defined as

$$\frac{d\mathbf{A}}{dt} = \lim_{\Delta t \to 0} \frac{\mathbf{A}(t + \Delta t) - \mathbf{A}(t)}{\Delta t}. \tag{A.8}$$

The resulting vector $\frac{d\mathbf{A}}{dt}$ is tangential to the contour drawn by $\mathbf{A}(t)$.

A.3 Gradient of a Scalar Function

A scalar function may be defined in three-dimensional space. This function will return a scalar value for each point in space. Let T be such a scalar function. We are interested in the value of this function at two nearby points (x, y, z) and $(x + \Delta x, y + \Delta y, z + \Delta z)$. Assuming that T is continuous and that the increments Δx, Δy, and Δz are small, the values T_1 at position $P_1 = (x, y, z)$ and T_2 at position $P_2 = (x + \Delta x, y + \Delta y, z + \Delta z)$ are related by a Taylor series expansion:

$$T_2 = T_1 + \left.\frac{\partial T}{\partial x}\right|_{P_1} \Delta x + \left.\frac{\partial T}{\partial y}\right|_{P_1} \Delta y + \left.\frac{\partial T}{\partial z}\right|_{P_1} \Delta z + \text{higher-order terms}. \tag{A.9}$$

As the two points are close to one another, the higher-order terms can be neglected, so that the difference ΔT between T_1 and T_2 is given by

$$\Delta T = T_2 - T_1 = \left.\frac{\partial T}{\partial x}\right|_{P_1} \Delta x + \left.\frac{\partial T}{\partial y}\right|_{P_1} \Delta y + \left.\frac{\partial T}{\partial z}\right|_{P_1} \Delta z. \tag{A.10}$$

[1] We assume throughout that we are dealing with three dimensions.

In the limit that the distance between the two points goes to zero, this becomes

$$\lim_{\substack{\Delta x \to 0 \\ \Delta y \to 0 \\ \Delta z \to 0}} \Delta T = dT = \frac{\partial T}{\partial x} dx + \frac{\partial T}{\partial y} dy + \frac{\partial T}{\partial z} dz. \tag{A.11}$$

We may write this in vector notation as a dot product:

$$dT = \left(\frac{\partial T}{\partial x} \mathbf{e}_x + \frac{\partial T}{\partial y} \mathbf{e}_y + \frac{\partial T}{\partial z} \mathbf{e}_z \right) \cdot (dx\,\mathbf{e}_x + dy\,\mathbf{e}_y + dz\,\mathbf{e}_z). \tag{A.12}$$

With the position vector given as $\mathbf{r}$, the differential change in position can be written as

$$d\mathbf{r} = dx\,\mathbf{e}_x + dy\,\mathbf{e}_y + dz\,\mathbf{e}_z. \tag{A.13}$$

At the same time, the first term of the dot product in (A.12) represents the gradient of T. The gradient operator (or vector differential operator) ∇ may then be defined as

$$\nabla = \text{grad} = \frac{\partial}{\partial x} \mathbf{e}_x + \frac{\partial}{\partial y} \mathbf{e}_y + \frac{\partial}{\partial z} \mathbf{e}_z. \tag{A.14}$$

With this definition, (A.12) becomes simply $dT = \nabla T \cdot d\mathbf{r}$. The interpretation given to the gradient operator is that it indicates rate of change. The gradient operator takes a scalar function and returns a vector.

A.4 Divergence

Considering a small volume Δv of space bounded by a surface Δs, the influence of a vector field $\mathbf{F}$ may cause a flux, i.e., a measurable quantity flows into the volume (in-flux) or out of the volume (out-flux). If the volume goes to zero, the differential out-flux is represented by the divergence of a vector field. The divergence $\nabla \cdot$ of a vector $\mathbf{F}$ is then defined as

$$\nabla \cdot \mathbf{F} = \text{div } \mathbf{F} = \frac{\partial F_x}{\partial x} + \frac{\partial F_y}{\partial y} + \frac{\partial F_z}{\partial z} \tag{A.15}$$

provided that $\mathbf{F}$ and its first partial derivatives are continuous within Δv. The divergence operator applied to a vector, returns a vector.

A.5 Gauss' Theorem

Consider a vector field acting upon a volume v bounded by a closed surface s. At each point on the surface, the vector field may be decomposed into a component

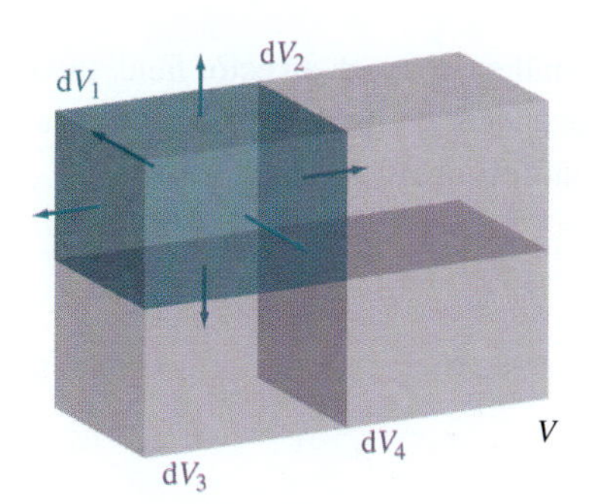

Figure A.3. A volume V is split into four subvolumes dV_1 through dV_4. The flux emanating from dV_1 flows partially into neighboring subvolumes and partially out of volume V. Only subvolumes at the boundary contribute to flux emanating from the volume.

normal to the surface and a component tangential to the surface. The surface integral of the component normal to the surface, taken over the surface s, may be related to the volume integral of the divergence of the vector field, taken over the volume v:

$$\int_s \mathbf{F} \cdot \mathbf{n} \, ds = \int_v \nabla \cdot \mathbf{F} \, dv. \tag{A.16}$$

This relation is known as the *divergence theorem* or *Gauss' theorem*. Its use lies in the fact that the triple integral over the volume v may be replaced by a double integral over the surface s that bounds the volume.

If a volume V is broken up into a set of small subvolumes, then the flux emanating from a given subvolume will flow into its neighbors, as illustrated in Figure A.3. Thus for subvolumes located in the interior of V, the net flux is zero. Only subvolumes located on the boundary of V will contribute to the flux associated with the volume. This contribution is related to the normal component of the vector field at the surface. Hence, Gauss' theorem, as stated above.

A.6 Curl

Suppose water flows through a river. The drag of the banks of the river cause the water to flow slower near the banks than in the middle of the river. If an object is released in the river, the water will flow faster along one side of the object than the opposite side, causing the object to rotate; the water flow on either side of the object is non-uniform. The term *curl* is a measure of non-uniformity, i.e., the ability to cause rotation.

Modeling the flow in the river with a vector field, there will be variation in the strength of the field dependent on location. At each point in a vector field $\mathbf{F}$, the curl measures its non-uniformity. Curl is defined as

$$\nabla \times \mathbf{F} = \text{curl}\, \mathbf{F} = \left(\frac{\partial F_z}{\partial y} - \frac{\partial F_y}{\partial z} \right) \mathbf{e}_x + \left(\frac{\partial F_x}{\partial z} - \frac{\partial F_z}{\partial x} \right) \mathbf{e}_y + \left(\frac{\partial F_y}{\partial x} - \frac{\partial F_x}{\partial y} \right) \mathbf{e}_z. \tag{A.17}$$

The curl operator takes a vector and produces a new vector.

A.7 Stokes' Theorem

The divergence theorem outlined above relates a volume integral to a surface integral over a closed surface. A similar theorem, called Stokes' Theorem, links the line integral over a closed contour c to the surface integral over an open surface s enclosed by a contour c. Examples of contours and open surfaces are shown in Figure A.4.

Figure A.4. Examples of open (or capped) surfaces with associated contours.

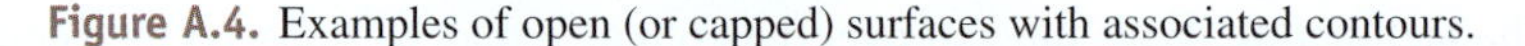

When an open surface is subdivided into small differential surface areas, then the integral of a vector quantity normal to each surface area may be approximated with a line integral over the contour. The contribution of each contour cancels out for adjacent surface areas, as shown in Figure A.5. Thus, only the contour segments that coincide with the contour of the entire surface do not cancel out. This leads to Stokes' Theorem:

$$\int_s \nabla \times \mathbf{F} \cdot \mathbf{n}\, ds = \int_c \mathbf{F} \cdot \mathbf{n}\, dc. \tag{A.18}$$

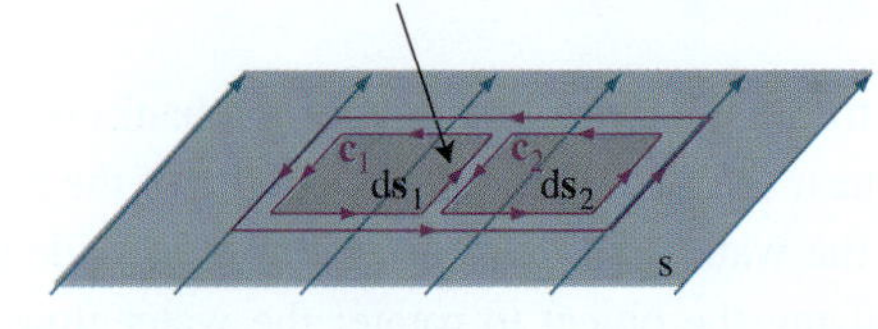

Figure A.5. Neighboring differential surface areas with associated contours. Integrals over contours c_1 and c_2 cancel where they are adjacent and are oriented in opposite directions.

A.8 Laplacian

The Laplacian operator on a scalar function T is given by

$$\nabla^2 T = \frac{\partial^2 T}{\partial x^2} + \frac{\partial^2 T}{\partial y^2} + \frac{\partial^2 T}{\partial z^2}. \tag{A.19}$$

For a vector-valued function **F**, this operator is defined by applying the above operator on each of the three x, y, and z components independently, yielding the Laplacian of vector **F**.

$$\nabla^2 \mathbf{F}_x = \frac{\partial^2 \mathbf{F}_x}{\partial x^2} + \frac{\partial^2 \mathbf{F}_x}{\partial y^2} + \frac{\partial^2 \mathbf{F}_x}{\partial z^2}, \tag{A.20a}$$

$$\nabla^2 \mathbf{F}_y = \frac{\partial^2 \mathbf{F}_y}{\partial x^2} + \frac{\partial^2 \mathbf{F}_y}{\partial y^2} + \frac{\partial^2 \mathbf{F}_y}{\partial z^2}, \tag{A.20b}$$

$$\nabla^2 \mathbf{F}_z = \frac{\partial^2 \mathbf{F}_z}{\partial x^2} + \frac{\partial^2 \mathbf{F}_z}{\partial y^2} + \frac{\partial^2 \mathbf{F}_z}{\partial z^2}. \tag{A.20c}$$

This Laplacian operator thus takes a vector and produces a new vector.

A.9 Vector Identities

For convenience, we present several identities that hold for a given scalar t and vector **F**:

$$\nabla \cdot (t\mathbf{F}) = t \nabla \cdot \mathbf{F} + \mathbf{F} \cdot \nabla t, \tag{A.21}$$

$$\nabla \times (t\mathbf{F}) = t \nabla \times \mathbf{F} + (\nabla t) \times \mathbf{F}, \tag{A.22}$$

$$\nabla \times \nabla \times \mathbf{F} = \nabla (\nabla \cdot \mathbf{F}) - \nabla^2 \mathbf{F}. \tag{A.23}$$

For clarity, we repeat these three equations in alternate notation:

$$\text{div } t\mathbf{F} = t \text{ div } \mathbf{F} + \mathbf{F} \cdot \text{grad } t, \tag{A.24}$$

$$\text{curl } t\mathbf{F} = t \text{ curl } \mathbf{F} + (\text{grad } t) \times \mathbf{F}, \tag{A.25}$$

$$\text{curl curl } \mathbf{F} = \text{grad div } \mathbf{F} - \nabla^2 \mathbf{F}. \tag{A.26}$$

A.10 Homogeneous Coordinates

Affine transformations of vectors can be seen as a linear transform followed by a translation:

$$\mathbf{x} \mapsto \mathbf{Ax} + \mathbf{b}. \tag{A.27}$$

Such transformations can be conveniently represented by a single matrix if the vectors and matrices are augmented by an extra dimension. Thus, a three-dimensional vector $\mathbf{x} = (x\ y\ z)^T$ can be represented with $\mathbf{x}' = (x\ y\ z\ 0)^T$. An affine transformation can then be represented by

$$\mathbf{x}' \mapsto \mathbf{M}\mathbf{x}', \tag{A.28}$$

where $\mathbf{M}$ is a 4×4 matrix. The elements of the four-vector $\mathbf{x}'$ are then called *homogeneous coordinates.*

It should be noted that in homogeneous coordinates, vectors, points, and surface normals are each treated differently. For instance, under translation, a point will be mapped to a different point in space, whereas a vector (indicating a direction) will continue to point in the same direction. Hence, a vector is represented in homogeneous coordinates with

$$\mathbf{v} = (x\ y\ z\ 0)^T, \tag{A.29}$$

whereas a point in space is represented in homogeneous coordinates with

$$\mathbf{p} = (x\ y\ z\ 1)^T. \tag{A.30}$$

If an object were to undergo an affine transformation and the surface normals were represented as vectors, after the transformation the surface normals would no longer point away from the surface as they should. To apply an affine transformation to a surface normal using homogeneous coordinates, the surface normal is represented as a vector, i.e., $\mathbf{n} = (x, y, z, 0)^T$, but instead of using the transformation matrix $\mathbf{M}$, the transpose of its inverse should be used [1154]:

$$\mathbf{n} \mapsto \mathbf{M}^{-1T}\mathbf{n}. \tag{A.31}$$

To convert a vector from its homogeneous representation back to a conventional (three-dimensional) vector, each component is divided by its fourth component. A vector $\mathbf{v}' = (x'\ y'\ z'\ w')^T$ is thus converted to $\mathbf{v} = (x\ y\ z)^T$ using

$$\begin{bmatrix} x \\ y \\ z \end{bmatrix} = \begin{bmatrix} x'/w' \\ y'/w' \\ z'/w' \end{bmatrix}. \tag{A.32}$$

This procedure is known as the *homogeneous divide.*

Appendix B
Trigonometry

In this appendix, we show some of the trigonometric relations that are used elsewhere in this book. While we assume basic familiarity with trigonometry, we review some of the lesser-known relations.

B.1 Sum and Difference Formulae

Starting with two unit vectors $\mathbf{P}_u$ and $\mathbf{P}_v$, the dot product between these two vectors will equal the cosine of the angle between them. If these vectors make an angle of u and v with the positive x-axis, then we have

$$\mathbf{P}_u \cdot \mathbf{P}_v = \cos(u - v). \tag{B.1}$$

However, by the definition of the dot product, we also have

$$\mathbf{P}_u \cdot \mathbf{P}_v = \cos(u)\cos(v) + \sin(u)\sin(v). \tag{B.2}$$

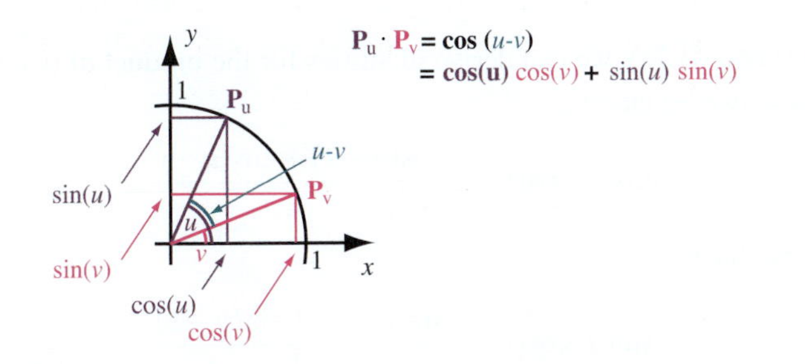

Figure B.1. The cosine of the difference between two angles may be found by computing the dot product between two unit vectors.

By equating these two expressions, we find the first difference formula (as shown in Figure B.1):

$$\cos(u-v) = \cos(u)\cos(v) + \sin(u)\sin(v). \tag{B.3}$$

A similar expression may be derived for the sine of the difference of two angles. Here, we use the definition of the cross product and define two three-dimensional vectors as

$$\mathbf{P}_u = \begin{pmatrix} \cos(u) \\ \sin(u) \\ 0 \end{pmatrix}, \qquad \mathbf{P}_v = \begin{pmatrix} \cos(v) \\ \sin(v) \\ 0. \end{pmatrix} \tag{B.4}$$

The length of $\mathbf{P}_u \times \mathbf{P}_v$ is then equal to $\sin(u-v)$ (see (A.7)). At the same time, this vector is given by

$$\mathbf{P}_u \times \mathbf{P}_v = \begin{pmatrix} 0 \\ 0 \\ \sin(u)\cos(v) - \cos(u)\sin(v) \end{pmatrix}. \tag{B.5}$$

The length of this vector is equal to its z-component. We thus have the following identity:

$$\sin(u-v) = \sin(u)\cos(v) - \cos(u)\sin(v). \tag{B.6}$$

The above two difference formulas may be extended for the cosine and sine of the sum of two angles. Here, we use the identities $\cos(-v) = cos(v)$ and $\sin(-v) = -sin(v)$:

$$\cos(u+v) = \cos(u)\cos(v) - \sin(u)\sin(v), \tag{B.7a}$$

$$\sin(u+v) = \sin(u)\cos(v) + \cos(u)\sin(v). \tag{B.7b}$$

B.2 Product Identities

From (B.3) and (B.7a), we can derive identities for the product of two cosines by adding these two equations:

$$\cos(u)\cos(v) = \frac{\cos(u+v) + \cos(u-v)}{2}. \tag{B.8a}$$

Similarly we have

$$\sin(u)\cos(v) = \frac{\sin(u+v) + \sin(u-v)}{2}, \tag{B.8b}$$

$$\sin(u)\sin(v) = \frac{\cos(u-v) + \cos(u-v)}{2}. \tag{B.8c}$$

B.3 Double-Angle Formulae

The last two equations may be used to derive double-angle formulae for $\cos(2u)$ and $\sin(2u)$ by noting that $2u = u + u$:

$$\cos(2u) = \cos(u)\cos(u) - \sin(u)\sin(u) \tag{B.9a}$$
$$= \cos^2(u) - \sin^2(u) \tag{B.9b}$$
$$= 2\cos^2(u) - 1 \tag{B.9c}$$
$$= 1 - 2\sin^2(u). \tag{B.9d}$$
$$\sin(2u) = \sin(u)\cos(u) + \cos(u)\sin(u) \tag{B.9e}$$
$$= 2\sin(u)cos(u). \tag{B.9f}$$

B.4 Half-Angle Formulae

Two derive formulae for $\sin(u/2)$ and $\cos(u/2)$, we substitute $u/2$ in (B.9), and solve for $\cos(u/2)$ and $\sin(u/2)$, respectively:

$$\cos(u/2) = \sqrt{\frac{1+\cos(u)}{2}}, \tag{B.10a}$$
$$\sin(u/2) = \sqrt{\frac{1-\cos(u)}{2}}. \tag{B.10b}$$

B.5 Sum Identities

The derivation of the sum of two sines is more involved. We begin by using the identity:

$$\sin^2(u) + cos^2(u) = 1 \tag{B.11}$$

leading to:

$$\sin(u) + \sin(v) = \sqrt{1 - \cos^2(u)} + \sqrt{1 - \cos^2(v)}. \tag{B.12a}$$

Rewriting yields

$$\sin(u) + \sin(v) = \sqrt{1-\cos(u)}\sqrt{1+\cos(u)} + \sqrt{1-\cos(v)}\sqrt{1+\cos(v)}. \tag{B.12b}$$

We then introduce the constant 2:

$$\sin(u)+\sin(v)=2\left(\sqrt{\frac{1-\cos(u)}{2}}\sqrt{\frac{1+\cos(u)}{2}}+\sqrt{\frac{1-\cos(v)}{2}}\sqrt{\frac{1+\cos(v)}{2}}\right) \tag{B.12c}$$

and then apply the half-angle rules:

$$\sin(u)+\sin(v)=2\,\sin(u/2)\cos(u/2)+2\,\sin(v/2)\cos(v/2). \tag{B.12d}$$

This expression can be expanded using $\sin^2(u/2)+cos^2(u/2)=1$ once more:

$$\sin(u)+\sin(v)=2\,\sin(u/2)\cos(u/2)\left(\sin^2(u/2)+cos^2(u/2)\right) \tag{B.12e}$$
$$+2\,\sin(v/2)\cos(v/2)\left(\sin^2(v/2)+cos^2(v/2)\right) \tag{B.12f}$$

By rearranging terms we find

$$\sin(u)+\sin(v)=2\left(\sin(u/2)\cos(v/2)+\cos(u/2)\sin(v/2)\right) \tag{B.12g}$$
$$\times\left(\cos(u/2)\cos(v/2)+\sin(u/2)\sin(u/2)\right). \tag{B.12h}$$

We can then apply sum and difference formulae to find the desired expression:

$$\sin(u)+\sin(v)=2\sin\left(\frac{u+v}{2}\right)\cos\left(\frac{u-v}{2}\right). \tag{B.13a}$$

Similarly, the following relations may be derived:

$$\sin(u)-\sin(v)=2\cos\left(\frac{u+v}{2}\right)\sin\left(\frac{u-v}{2}\right), \tag{B.13b}$$
$$\cos(u)+\cos(v)=2\cos\left(\frac{u+v}{2}\right)\cos\left(\frac{u-v}{2}\right), \tag{B.13c}$$
$$\cos(u)-\cos(v)=-2\sin\left(\frac{u+v}{2}\right)\sin\left(\frac{u-v}{2}\right). \tag{B.13d}$$

B.6 Solid Angle

Solid angles can be seen as the 3D equivalents of 2D angles. The solid angle spanned by a surface around a point is equal to the area of the surface projected on a unit sphere surrounding the point as shown in Figure B.2. Solid angles are measured in *steradians* (sr), and their largest value can be 4π which is the surface area of the full sphere.

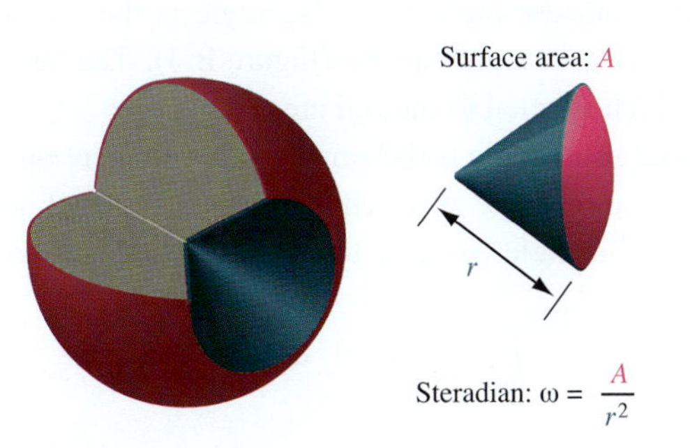

Figure B.2. The definition of solid angle ω is a conical segment of a sphere. One steradian is the angle subtended at the center of this sphere by an area of surface equal to the square of the radius.

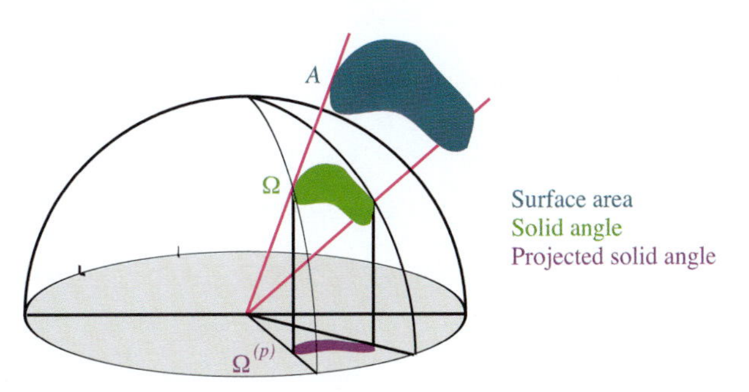

Figure B.3. The solid angle spanned by the surface A is equal to Ω. The projected solid angle is shown by $\Omega^{(p)}$.

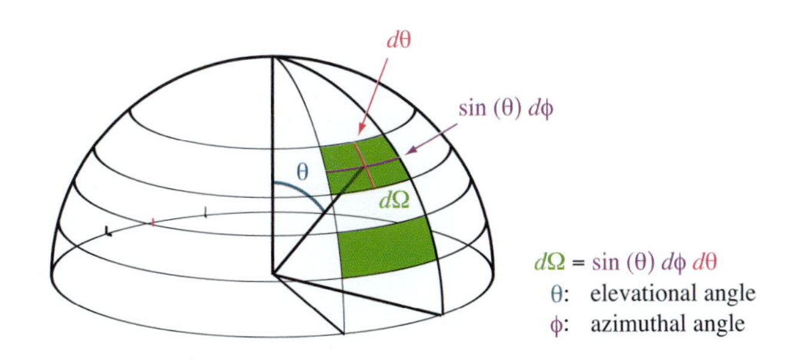

Figure B.4. Two differential solid angles are shown on a unit sphere. Note that the size of the differential solid angle changes based on its vertical position on the sphere. The relationship is given by $d\Omega = \sin(\theta)d\phi d\theta$.

On the other hand, the projected solid angle is the area of the solid angle projected onto the base of a unit sphere (Figure B.3). The largest projected solid angle can be π which is equal to the full area of the base.

The differential solid angle is the smallest area element on a sphere. Its value depends on the elevation angle at which it is computed (see Figure B.4). The sum of all differential solid angles is equal to the surface area of the sphere:

$$\int_0^{2\pi} \int_0^{\pi} \sin(\theta)\, d\phi\, d\theta = 4\pi. \tag{B.14}$$

Appendix C
Complex Numbers

In this appendix, we review some basic formulae and theorems relating to complex numbers.

C.1 Definition

A complex number z is defined as

$$z = x + iy, \tag{C.1}$$

where x and y are real numbers, and i is the imaginary number ($i = \sqrt{-1}$). Complex numbers may be represented by points on a complex plane, as shown in Figure C.1. Here, the point z may either be represented in Cartesian coordinates (x, y) or in polar coordinates (r, φ). To convert a complex number from Cartesian

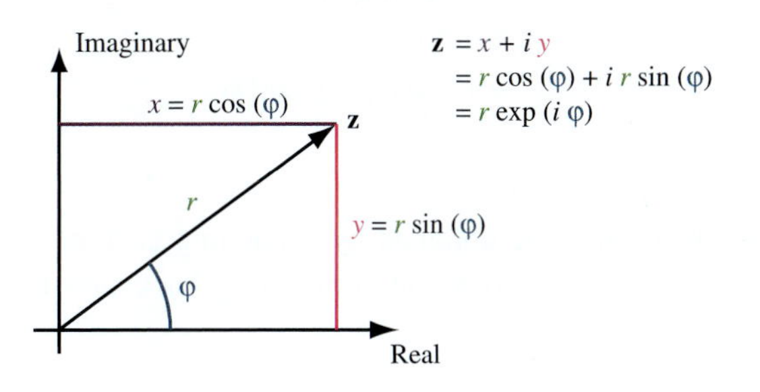

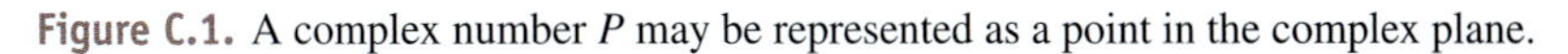
Figure C.1. A complex number P may be represented as a point in the complex plane.

coordinates to polar coordinates:

$$r = \sqrt{x^2 + y^2}, \tag{C.2a}$$

$$\varphi = \tan^{-1}(y/x). \tag{C.2b}$$

For a complex number z, the real part is indicated with $x = \mathrm{Re}\{z\}$ and the imaginary part is given by $y = \mathrm{Im}\{z\}$. From (C.1), it can be seen that the operators $\mathrm{Re}\{z\}$ and $\mathrm{Im}\{z\}$ are defined as follows:

$$\mathrm{Re}\{z\} = \frac{z + z^*}{2}, \tag{C.3}$$

$$\mathrm{Im}\{z\} = \frac{z - z^*}{2i}, \tag{C.4}$$

where z^* indicates the complex conjugate of z, i.e., $z = x - iy = \mathrm{Re}\{z\} - i\mathrm{Im}\{z\}$. The magnitude $|z|$ of a complex number is given by

$$|z| = \sqrt{zz^*} = \sqrt{\mathrm{Re}\{z\}^2 + \mathrm{Im}\{z\}^2} \tag{C.5}$$

An frequently occurring quantity is the square of the magnitude of a complex number, which can be computed by omitting the square root in the above equation:

$$|z|^2 = zz^* = \mathrm{Re}\{z\}^2 + \mathrm{Im}\{z\}^2 \tag{C.6}$$

C.2 Euler's Formula

If a complex number is represented in polar coordinates, we may apply Euler's formula and rewrite the equation as a complex exponential function:

$$z = x + iy \tag{C.7a}$$

$$= r\cos(\varphi) + ir\sin(\varphi) \tag{C.7b}$$

$$= re^{i\varphi}. \tag{C.7c}$$

A proof of this result may be obtained by expanding each of the exp, cos and sin functions into a Taylor series and substituting back into Equations (C.7b) and (C.7c).

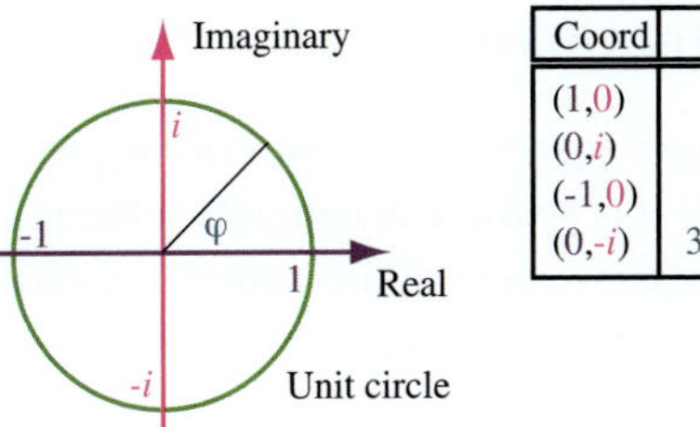

Coord	φ	exp (iφ)
(1,0)	0	1
(0,i)	π/2	i
(-1,0)	π	-1
(0,-i)	3π/2	-i

Figure C.2. Values of $e^{i\varphi}$ may be deduced for specific values of φ by observing the unit circle in the complex plane.

C.3 Theorems

By drawing a unit circle around the origin in the complex plane, we obtain several useful expressions for complex exponential functions (see Figure C.2):

$$e^0 = 1, \tag{C.8a}$$

$$e^{i\pi/2} = i, \tag{C.8b}$$

$$e^{i\pi} = -1, \tag{C.8c}$$

$$e^{3i\pi/2} = -i. \tag{C.8d}$$

Further identities are as follows:

$$e^{-i\varphi} = \cos(\varphi) - i\sin(\varphi) = z^* \tag{C.9a}$$

$$\cos(\varphi) = \frac{1}{2}\left(e^{i\varphi} + e^{-i\varphi}\right), \tag{C.9b}$$

$$\sin(\varphi) = \frac{1}{2i}\left(e^{i\varphi} - e^{-i\varphi}\right), \tag{C.9c}$$

$$r_1 e^{i\varphi_1}\, r_2 e^{i\varphi_2} = r_1\, r_2 e^{i(\varphi_1+\varphi_2)}, \tag{C.9d}$$

$$\frac{r_1 e^{i\varphi_1}}{r_2 e^{i\varphi_2}} = r_1\, r_2 e^{i(\varphi_1-\varphi_2)}. \tag{C.9e}$$

Finally, an important property of the exponential function e is that for all values of φ, the function $e^{i\varphi}$ is a measure of the rate of change of $e^{i\varphi}$, i.e., its derivative:

$$\frac{d}{d\varphi} e^{i\varphi} = i\varphi\, e^{i\varphi}. \tag{C.10}$$

C.4 Time-Harmonic Quantities

One of the solutions to the wave equations discussed in Chapter 2 is in the form of sinusoidal waveforms. While it is possible to compute solutions in terms of sines and cosines, it is often more convenient to use complex notation. A sinusoidal waveform $f(t)$ may be defined as

$$f(t) = a\cos(\omega t + \varphi). \tag{C.11}$$

Here, a is the amplitude and $\omega t + \varphi$ is the phase. In most cases, we view t as the time variable and φ is an arbitrary phase constant. We may complement this function by its complex counterpart to find the complex function $\underline{f}(t)$:

$$\underline{f}(t) = a\left(\cos(\omega t + \varphi) + i\sin(\omega t + \varphi)\right). \tag{C.12}$$

By applying Euler's formula we get

$$\underline{f}(t) = a\,e^{i(\omega t+\varphi)} \tag{C.13}$$
$$= a\,e^{i\varphi}\,e^{i\omega t} \tag{C.14}$$
$$= \underline{A}\,e^{i\omega t}. \tag{C.15}$$

Thus, the complex function $\underline{f}(t)$ may be written as the product of the complex phasor $\underline{A} = a\,e^{i\varphi}$ and the time factor $e^{i\omega t}$. This is an important result that allows the time dependency to be eliminated from partial differential equations. It thereby reduces partial differential equations to ordinary differential equations. These results are used in finding solutions to the wave equations for plane waves in time-harmonic fields in Section 2.2.3.

To reconstitute a solution, it suffices to take the real part:

$$f(t) = \mathrm{Re}\{\underline{f}(t)\}. \tag{C.16}$$

Appendix D

Units and Constants

In this appendix, we enumerate the constants, symbols and units used in this book. Also given are conversions between commonly encountered units, insofar as they deviate from SI units. The basic quantities are given in Table D.1. Some relevant units outside the SI system are given in Table D.2. The 22 derived SI units are enumerated in Table D.3. Contants (in SI units) are given in Table D.4. For further information on SI units, see the NIST website (http://physics.nist.gov/cuu/Units/index.html.)

Unit	Symbol	Physical quantity
Meter	m	Length
Kilogram	kg	Mass
Second	s	Time
Ampere	A	Electric current
Kelvin	K	Thermodynamic temperature
Mole	mol	Amount of substance
Candela	cd	luminous intensity

Table D.1. Fundamental SI units.

Name	Symbol	SI equivalent
minute (time)	min	60 s
hour	h	3600 s
day	d	86,400 s
degree (angle)	°	$(\pi/180)$ rad
minute (angle)	'	$(\pi/10,800)$ rad
second (angle)	"	$(\pi/648,000)$ rad
liter	L	10^{-3} m^3
metric ton	t	10^3 kg
neper	Np	1
bel	B	$(1/2)\ln(10Np)$
electronvolt	eV	$1.602\,18 \times 10^{-19}$ J
unified atomic mass unit	u	$1.660\,54 \times 10^{-27}$ kg
astonomical unit	ua	$1.495\,98 \times 10^{11}$ m

Table D.2. Units outside the SI system.

Unit	Symbol	Definition	Physical quantity
radian	rad	$m\ m^{-1} = 1$	plane angle
steradian	sr	$m^2\ m^{-2} = 1$	solid angle
hertz	Hz	s^{-1}	frequency
newton	N	$kg\ m\ s^{-2}$	force
pascal	Pa	$N\ m^{-2}$	pressure
joule	J	N m	energy
watt	W	$J\ s^{-1}$	power
coulomb	C	s A	charge
volt	V	$W\ A^{-1}$	potential
farad	F	$C\ V^{-1}$	capacitance
ohm	Ω	$V\ A^{-1}$	electric resistance
siemens	S	$A\ V^{-1}$	electric conductance
weber	Wb	V s	magnetic flux
tesla	T	$Wb\ m^{-1}$	magnetic flux density
henry	H	$Wb\ A^{-1}$	inductance
degree Celsius	°C	K	Celsius temperature
lumen	lm	cd sr	luminous flux
lux	lx	$lm\ m^{-2}$	illuminance
becquerel	Bq	s^{-1}	activity
gray	Gy	$J\ kg^{-1}$	absorbed dose
sievert	Sv	$J\ kg^{-1}$	dose equivalent
katal	kat	$mol\ s^{-1}$	catalytic activity

Table D.3. Derived SI units used in eletromagnetic theory.

Constant	Symbol	Value	Unit
Permittivity of vacuum	ε_0	$\frac{1}{36\pi} \times 10^{-9}$	$F\ m^{-1}$
Permeability of vacuum	μ_0	$4\pi \cdot 10^{-7}$	$H\ m^{-1}$
Stefan-Boltzmann constant	σ	5.670×10^{-8}	$W\ m^{-2}\ K^{-4}$
Speed of light in vacuum	c	2.998×10^{8}	$m\ s^{-1}$
Planck's constant	h	6.626×10^{-34}	J s
Boltzmann's constant	k_B	1.380×10^{-23}	$J\ K^{-1}$

Table D.4. Constants.

Appendix E
The CIE Luminous Efficiency Functions

λ (nm)	+ 0 nm	+ 1 nm	+ 2nm	+ 3 nm	+ 4 nm	+ 5 nm	+ 6 nm	+ 7nm	+ 8 nm	+ 9 nm
360	3.9e-06	4.4e-06	4.9e-06	5.5e-06	6.2e-06	7.0e-06	7.8e-06	8.8e-06	9.8e-06	1.1e-05
370	1.2e-05	1.4e-05	1.6e-05	1.7e-05	2.0e-05	2.2e-05	2.5e-05	2.8e-05	3.2e-05	3.5e-05
380	3.9e-05	4.3e-05	4.7e-05	5.2e-05	5.7e-05	6.4e-05	7.2e-05	8.2e-05	9.4e-05	1.1e-04
390	1.2e-04	1.3e-04	1.5e-04	1.7e-04	1.9e-04	2.2e-04	2.5e-04	2.8e-04	3.2e-04	3.6e-04
400	4.0e-04	4.3e-04	4.7e-04	5.2e-04	5.7e-04	6.4e-04	7.2e-04	8.3e-04	9.4e-04	1.1e-03
410	1.2e-03	1.4e-03	1.5e-03	1.7e-03	1.9e-03	2.2e-03	2.5e-03	2.8e-03	3.1e-03	3.5e-03
420	4.0e-03	4.5e-03	5.2e-03	5.8e-03	6.5e-03	7.3e-03	8.1e-03	8.9e-03	9.8e-03	1.1e-02
430	1.2e-02	1.3e-02	1.4e-02	1.5e-02	1.6e-02	1.7e-02	1.8e-02	1.9e-02	2.0e-02	2.2e-02
440	2.3e-02	2.4e-02	2.6e-02	2.7e-02	2.8e-02	3.0e-02	3.1e-02	3.3e-02	3.5e-02	3.6e-02
450	3.8e-02	4.0e-02	4.2e-02	4.4e-02	4.6e-02	4.8e-02	5.0e-02	5.3e-02	5.5e-02	5.7e-02
460	6.0e-02	6.3e-02	6.5e-02	6.8e-02	7.1e-02	7.4e-02	7.7e-02	8.0e-02	8.4e-02	8.7e-02
470	9.1e-02	9.5e-02	9.9e-02	1.0e-01	1.1e-01	1.1e-01	1.2e-01	1.2e-01	1.3e-01	1.3e-01
480	1.4e-01	1.4e-01	1.5e-01	1.6e-01	1.6e-01	1.7e-01	1.8e-01	1.8e-01	1.9e-01	2.0e-01
490	2.1e-01	2.2e-01	2.3e-01	2.4e-01	2.5e-01	2.6e-01	2.7e-01	2.8e-01	3.0e-01	3.1e-01
500	3.2e-01	3.4e-01	3.5e-01	3.7e-01	3.9e-01	4.1e-01	4.3e-01	4.4e-01	4.6e-01	4.8e-01
510	5.0e-01	5.2e-01	5.4e-01	5.7e-01	5.9e-01	6.1e-01	6.3e-01	6.5e-01	6.7e-01	6.9e-01
520	7.1e-01	7.3e-01	7.5e-01	7.6e-01	7.8e-01	7.9e-01	8.1e-01	8.2e-01	8.4e-01	8.5e-01
530	8.6e-01	8.7e-01	8.8e-01	9.0e-01	9.1e-01	9.1e-01	9.2e-01	9.3e-01	9.4e-01	9.5e-01
540	9.5e-01	9.6e-01	9.7e-01	9.7e-01	9.8e-01	9.8e-01	9.8e-01	9.9e-01	9.9e-01	9.9e-01
550	9.9e-01	1.0e+00	1.0e+00	1.0e+00	1.0e+00	1.0e+00	1.0e+00	1.0e+00	1.0e+00	1.0e+00
560	9.9e-01	9.9e-01	9.9e-01	9.9e-01	9.8e-01	9.8e-01	9.7e-01	9.7e-01	9.6e-01	9.6e-01
570	9.5e-01	9.5e-01	9.4e-01	9.3e-01	9.2e-01	9.2e-01	9.1e-01	9.0e-01	8.9e-01	8.8e-01
580	8.7e-01	8.6e-01	8.5e-01	8.4e-01	8.3e-01	8.2e-01	8.0e-01	7.9e-01	7.8e-01	7.7e-01
590	7.6e-01	7.4e-01	7.3e-01	7.2e-01	7.1e-01	6.9e-01	6.8e-01	6.7e-01	6.6e-01	6.4e-01
600	6.3e-01	6.2e-01	6.1e-01	5.9e-01	5.8e-01	5.7e-01	5.5e-01	5.4e-01	5.3e-01	5.2e-01
610	5.0e-01	4.9e-01	4.8e-01	4.7e-01	4.5e-01	4.4e-01	4.3e-01	4.2e-01	4.1e-01	3.9e-01

Table E.1. The CIE 1924 Photopic luminous efficiency function, $V(\lambda)$.

λ (nm)	+ 0 nm	+ 1 nm	+ 2nm	+ 3 nm	+ 4 nm	+ 5 nm	+ 6 nm	+ 7nm	+ 8 nm	+ 9 nm
620	3.8e-01	3.7e-01	3.6e-01	3.4e-01	3.3e-01	3.2e-01	3.1e-01	3.0e-01	2.9e-01	2.8e-01
630	2.7e-01	2.5e-01	2.4e-01	2.4e-01	2.3e-01	2.2e-01	2.1e-01	2.0e-01	1.9e-01	1.8e-01
640	1.7e-01	1.7e-01	1.6e-01	1.5e-01	1.5e-01	1.4e-01	1.3e-01	1.3e-01	1.2e-01	1.1e-01
650	1.1e-01	1.0e-01	9.6e-02	9.1e-02	8.6e-02	8.2e-02	7.7e-02	7.3e-02	6.9e-02	6.5e-02
660	6.1e-02	5.7e-02	5.4e-02	5.1e-02	4.8e-02	4.5e-02	4.2e-02	3.9e-02	3.7e-02	3.4e-02
670	3.2e-02	3.0e-02	2.8e-02	2.6e-02	2.5e-02	2.3e-02	2.2e-02	2.1e-02	1.9e-02	1.8e-02
680	1.7e-02	1.6e-02	1.5e-02	1.4e-02	1.3e-02	1.2e-02	1.1e-02	1.0e-02	9.5e-03	8.8e-03
690	8.2e-03	7.6e-03	7.1e-03	6.6e-03	6.1e-03	5.7e-03	5.3e-03	5.0e-03	4.7e-03	4.4e-03
700	4.1e-03	3.8e-03	3.6e-03	3.4e-03	3.1e-03	2.9e-03	2.7e-03	2.6e-03	2.4e-03	2.2e-03
710	2.1e-03	2.0e-03	1.8e-03	1.7e-03	1.6e-03	1.5e-03	1.4e-03	1.3e-03	1.2e-03	1.1e-03
720	1.0e-03	9.8e-04	9.1e-04	8.5e-04	7.9e-04	7.4e-04	6.9e-04	6.4e-04	6.0e-04	5.6e-04
730	5.2e-04	4.8e-04	4.5e-04	4.2e-04	3.9e-04	3.6e-04	3.4e-04	3.1e-04	2.9e-04	2.7e-04
740	2.5e-04	2.3e-04	2.1e-04	2.0e-04	1.9e-04	1.7e-04	1.6e-04	1.5e-04	1.4e-04	1.3e-04
750	1.2e-04	1.1e-04	1.0e-04	9.7e-05	9.1e-05	8.5e-05	7.9e-05	7.4e-05	6.9e-05	6.4e-05
760	6.0e-05	5.6e-05	5.2e-05	4.9e-05	4.5e-05	4.2e-05	4.0e-05	3.7e-05	3.4e-05	3.2e-05
770	3.0e-05	2.8e-05	2.6e-05	2.4e-05	2.3e-05	2.1e-05	2.0e-05	1.8e-05	1.7e-05	1.6e-05
780	1.5e-05	1.4e-05	1.3e-05	1.2e-05	1.1e-05	1.1e-05	9.9e-06	9.2e-06	8.6e-06	8.0e-06
790	7.5e-06	7.0e-06	6.5e-06	6.0e-06	5.6e-06	5.3e-06	4.9e-06	4.6e-06	4.3e-06	4.0e-06
800	3.7e-06	3.5e-06	3.2e-06	3.0e-06	2.8e-06	2.6e-06	2.4e-06	2.3e-06	2.1e-06	2.0e-06
810	1.8e-06	1.7e-06	1.6e-06	1.5e-06	1.4e-06	1.3e-06	1.2e-06	1.1e-06	1.0e-06	9.8e-07
820	9.1e-07	8.5e-07	7.9e-07	7.4e-07	6.9e-07	6.4e-07	6.0e-07	5.6e-07	5.2e-07	4.8e-07
830	4.5e-07									

Table E.1. (continued) The CIE 1924 Photopic luminous efficiency function, $V(\lambda)$.

λ (nm)	+ 0 nm	+ 1 nm	+ 2nm	+ 3 nm	+ 4 nm	+ 5 nm	+ 6 nm	+ 7nm	+ 8 nm	+ 9 nm
380	5.9e-04	6.7e-04	7.5e-04	8.5e-04	9.7e-04	1.1e-03	1.3e-03	1.5e-03	1.7e-03	1.9e-03
390	2.2e-03	2.5e-03	2.9e-03	3.4e-03	3.9e-03	4.5e-03	5.2e-03	6.0e-03	7.0e-03	8.1e-03
400	9.3e-03	1.1e-02	1.2e-02	1.4e-02	1.6e-02	1.9e-02	2.1e-02	2.4e-02	2.7e-02	3.1e-02
410	3.5e-02	3.9e-02	4.4e-02	4.9e-02	5.4e-02	6.0e-02	6.7e-02	7.4e-02	8.1e-02	8.8e-02
420	9.7e-02	1.1e-01	1.1e-01	1.2e-01	1.3e-01	1.4e-01	1.5e-01	1.7e-01	1.8e-01	1.9e-01
430	2.0e-01	2.1e-01	2.2e-01	2.4e-01	2.5e-01	2.6e-01	2.8e-01	2.9e-01	3.0e-01	3.1e-01
440	3.3e-01	3.4e-01	3.5e-01	3.7e-01	3.8e-01	3.9e-01	4.1e-01	4.2e-01	4.3e-01	4.4e-01
450	4.6e-01	4.7e-01	4.8e-01	4.9e-01	5.0e-01	5.1e-01	5.2e-01	5.4e-01	5.5e-01	5.6e-01
460	5.7e-01	5.8e-01	5.9e-01	6.0e-01	6.1e-01	6.2e-01	6.3e-01	6.4e-01	6.5e-01	6.6e-01
470	6.8e-01	6.9e-01	7.0e-01	7.1e-01	7.2e-01	7.3e-01	7.4e-01	7.6e-01	7.7e-01	7.8e-01
480	7.9e-01	8.1e-01	8.2e-01	8.3e-01	8.4e-01	8.5e-01	8.6e-01	8.7e-01	8.8e-01	8.9e-01
490	9.0e-01	9.1e-01	9.2e-01	9.3e-01	9.4e-01	9.5e-01	9.6e-01	9.6e-01	9.7e-01	9.8e-01
500	9.8e-01	9.9e-01	9.9e-01	9.9e-01	1.0e+00	1.0e+00	1.0e+00	1.0e+00	1.0e+00	1.0e+00
510	1.0e+00	9.9e-01	9.9e-01	9.9e-01	9.8e-01	9.7e-01	9.7e-01	9.6e-01	9.5e-01	9.4e-01
520	9.4e-01	9.3e-01	9.2e-01	9.0e-01	8.9e-01	8.8e-01	8.7e-01	8.5e-01	8.4e-01	8.3e-01
530	8.1e-01	8.0e-01	7.8e-01	7.7e-01	7.5e-01	7.3e-01	7.2e-01	7.0e-01	6.8e-01	6.7e-01
540	6.5e-01	6.3e-01	6.2e-01	6.0e-01	5.8e-01	5.6e-01	5.5e-01	5.3e-01	5.1e-01	5.0e-01
550	4.8e-01	4.7e-01	4.5e-01	4.3e-01	4.2e-01	4.0e-01	3.9e-01	3.7e-01	3.6e-01	3.4e-01
560	3.3e-01	3.2e-01	3.0e-01	2.9e-01	2.8e-01	2.6e-01	2.5e-01	2.4e-01	2.3e-01	2.2e-01
570	2.1e-01	2.0e-01	1.9e-01	1.8e-01	1.7e-01	1.6e-01	1.5e-01	1.4e-01	1.4e-01	1.3e-01
580	1.2e-01	1.1e-01	1.1e-01	1.0e-01	9.6e-02	9.0e-02	8.5e-02	7.9e-02	7.4e-02	7.0e-02
590	6.6e-02	6.1e-02	5.7e-02	5.4e-02	5.0e-02	4.7e-02	4.4e-02	4.1e-02	3.8e-02	3.6e-02

Table E.2. The CIE 1951 Scotopic luminous efficiency function, $V'(\lambda)$.

λ (nm)	+ 0 nm	+ 1 nm	+ 2nm	+ 3 nm	+ 4 nm	+ 5 nm	+ 6 nm	+ 7nm	+ 8 nm	+ 9 nm
600	3.3e-02	3.1e-02	2.9e-02	2.7e-02	2.5e-02	2.3e-02	2.1e-02	2.0e-02	1.9e-02	1.7e-02
610	1.6e-02	1.5e-02	1.4e-02	1.3e-02	1.2e-02	1.1e-02	1.0e-02	9.3e-03	8.6e-03	8.0e-03
620	7.4e-03	6.8e-03	6.3e-03	5.8e-03	5.4e-03	5.0e-03	4.6e-03	4.2e-03	3.9e-03	3.6e-03
630	3.3e-03	3.1e-03	2.8e-03	2.6e-03	2.4e-03	2.2e-03	2.1e-03	1.9e-03	1.8e-03	1.6e-03
640	1.5e-03	1.4e-03	1.3e-03	1.2e-03	1.1e-03	1.0e-03	9.3e-04	8.6e-04	7.9e-04	7.3e-04
650	6.8e-04	6.3e-04	5.8e-04	5.4e-04	5.0e-04	4.6e-04	4.2e-04	3.9e-04	3.6e-04	3.4e-04
660	3.1e-04	2.9e-04	2.7e-04	2.5e-04	2.3e-04	2.1e-04	2.0e-04	1.8e-04	1.7e-04	1.6e-04
670	1.5e-04	1.4e-04	1.3e-04	1.2e-04	1.1e-04	1.0e-04	9.5e-05	8.9e-05	8.3e-05	7.7e-05
680	7.2e-05	6.7e-05	6.2e-05	5.8e-05	5.4e-05	5.0e-05	4.7e-05	4.4e-05	4.1e-05	3.8e-05
690	3.5e-05	3.3e-05	3.1e-05	2.9e-05	2.7e-05	2.5e-05	2.3e-05	2.2e-05	2.0e-05	1.9e-05
700	1.8e-05	1.7e-05	1.6e-05	1.5e-05	1.4e-05	1.3e-05	1.2e-05	1.1e-05	1.0e-05	9.8e-06
710	9.1e-06	8.6e-06	8.0e-06	7.5e-06	7.0e-06	6.6e-06	6.2e-06	5.8e-06	5.4e-06	5.1e-06
720	4.8e-06	4.5e-06	4.2e-06	4.0e-06	3.7e-06	3.5e-06	3.3e-06	3.1e-06	2.9e-06	2.7e-06
730	2.5e-06	2.4e-06	2.3e-06	2.1e-06	2.0e-06	1.9e-06	1.8e-06	1.7e-06	1.6e-06	1.5e-06
740	1.4e-06	1.3e-06	1.2e-06	1.2e-06	1.1e-06	1.0e-06	9.6e-07	9.1e-07	8.5e-07	8.1e-07
750	7.6e-07	7.2e-07	6.7e-07	6.4e-07	6.0e-07	5.7e-07	5.3e-07	5.1e-07	4.8e-07	4.5e-07
760	4.3e-07	4.0e-07	3.8e-07	3.6e-07	3.4e-07	3.2e-07	3.0e-07	2.9e-07	2.7e-07	2.6e-07
770	2.4e-07	2.3e-07	2.2e-07	2.0e-07	1.9e-07	1.8e-07	1.7e-07	1.6e-07	1.6e-07	1.5e-07
780	1.4e-07									

Table E.2. (continued) The CIE 1951 Scotopic luminous efficiency function, $V'(\lambda)$.

λ (nm)	+ 0 nm	+ 1 nm	+ 2nm	+ 3 nm	+ 4 nm	+ 5 nm	+ 6 nm	+ 7nm	+ 8 nm	+ 9 nm
380	2.0e-04	2.3e-04	2.6e-04	3.0e-04	3.4e-04	4.0e-04	4.6e-04	5.2e-04	6.0e-04	7.0e-04
390	8.0e-04	9.2e-04	1.0e-03	1.2e-03	1.4e-03	1.5e-03	1.8e-03	1.9e-03	2.2e-03	2.5e-03
400	2.8e-03	3.1e-03	3.5e-03	3.8e-03	4.2e-03	4.7e-03	5.1e-03	5.6e-03	6.2e-03	6.8e-03
410	7.4e-03	8.1e-03	9.0e-03	9.8e-03	1.1e-02	1.2e-02	1.3e-02	1.4e-02	1.5e-02	1.6e-02
420	1.8e-02	1.9e-02	2.0e-02	2.1e-02	2.2e-02	2.3e-02	2.4e-02	2.5e-02	2.5e-02	2.6e-02
430	2.7e-02	2.8e-02	2.9e-02	3.0e-02	3.2e-02	3.3e-02	3.4e-02	3.5e-02	3.6e-02	3.7e-02
440	3.8e-02	3.9e-02	4.0e-02	4.1e-02	4.2e-02	4.2e-02	4.3e-02	4.4e-02	4.5e-02	4.6e-02
450	4.7e-02	4.8e-02	4.9e-02	5.0e-02	5.1e-02	5.2e-02	5.3e-02	5.5e-02	5.6e-02	5.8e-02
460	6.0e-02	6.3e-02	6.5e-02	6.8e-02	7.1e-02	7.4e-02	7.7e-02	8.0e-02	8.4e-02	8.7e-02
470	9.1e-02	9.5e-02	9.9e-02	1.0e-01	1.1e-01	1.1e-01	1.2e-01	1.2e-01	1.3e-01	1.3e-01
480	1.4e-01	1.4e-01	1.5e-01	1.6e-01	1.6e-01	1.7e-01	1.8e-01	1.8e-01	1.9e-01	2.0e-01
490	2.1e-01	2.2e-01	2.3e-01	2.4e-01	2.5e-01	2.6e-01	2.7e-01	2.8e-01	3.0e-01	3.1e-01
500	3.2e-01	3.4e-01	3.5e-01	3.7e-01	3.9e-01	4.1e-01	4.3e-01	4.4e-01	4.6e-01	4.8e-01
510	5.0e-01	5.2e-01	5.4e-01	5.7e-01	5.9e-01	6.1e-01	6.3e-01	6.5e-01	6.7e-01	6.9e-01
520	7.1e-01	7.3e-01	7.5e-01	7.6e-01	7.8e-01	7.9e-01	8.1e-01	8.2e-01	8.4e-01	8.5e-01
530	8.6e-01	8.7e-01	8.8e-01	9.0e-01	9.1e-01	9.1e-01	9.2e-01	9.3e-01	9.4e-01	9.5e-01
540	9.5e-01	9.6e-01	9.7e-01	9.7e-01	9.8e-01	9.8e-01	9.8e-01	9.9e-01	9.9e-01	9.9e-01
550	9.9e-01	1.0e+00	1.0e+00	1.0e+00	1.0e+00	1.0e+00	1.0e+00	1.0e+00	1.0e+00	1.0e+00
560	9.9e-01	9.9e-01	9.9e-01	9.9e-01	9.8e-01	9.8e-01	9.7e-01	9.7e-01	9.6e-01	9.6e-01
570	9.5e-01	9.5e-01	9.4e-01	9.3e-01	9.2e-01	9.2e-01	9.1e-01	9.0e-01	8.9e-01	8.8e-01
580	8.7e-01	8.6e-01	8.5e-01	8.4e-01	8.3e-01	8.2e-01	8.0e-01	7.9e-01	7.8e-01	7.7e-01
590	7.6e-01	7.4e-01	7.3e-01	7.2e-01	7.1e-01	6.9e-01	6.8e-01	6.7e-01	6.6e-01	6.4e-01

Table E.3. The CIE 1988 Photopic luminous efficiency function, $V_M(\lambda)$

λ (nm)	+ 0 nm	+ 1 nm	+ 2nm	+ 3 nm	+ 4 nm	+ 5 nm	+ 6 nm	+ 7nm	+ 8 nm	+ 9 nm
600	6.3e-01	6.2e-01	6.1e-01	5.9e-01	5.8e-01	5.7e-01	5.5e-01	5.4e-01	5.3e-01	5.2e-01
610	5.0e-01	4.9e-01	4.8e-01	4.7e-01	4.5e-01	4.4e-01	4.3e-01	4.2e-01	4.1e-01	3.9e-01
620	3.8e-01	3.7e-01	3.6e-01	3.4e-01	3.3e-01	3.2e-01	3.1e-01	3.0e-01	2.9e-01	2.8e-01
630	2.7e-01	2.5e-01	2.4e-01	2.4e-01	2.3e-01	2.2e-01	2.1e-01	2.0e-01	1.9e-01	1.8e-01
640	1.7e-01	1.7e-01	1.6e-01	1.5e-01	1.5e-01	1.4e-01	1.3e-01	1.3e-01	1.2e-01	1.1e-01
650	1.1e-01	1.0e-01	9.6e-02	9.1e-02	8.6e-02	8.2e-02	7.7e-02	7.3e-02	6.9e-02	6.5e-02
660	6.1e-02	5.7e-02	5.4e-02	5.1e-02	4.8e-02	4.5e-02	4.2e-02	3.9e-02	3.7e-02	3.4e-02
670	3.2e-02	3.0e-02	2.8e-02	2.6e-02	2.5e-02	2.3e-02	2.2e-02	2.1e-02	1.9e-02	1.8e-02
680	1.7e-02	1.6e-02	1.5e-02	1.4e-02	1.3e-02	1.2e-02	1.1e-02	1.0e-02	9.5e-03	8.8e-03
690	8.2e-03	7.6e-03	7.1e-03	6.6e-03	6.1e-03	5.7e-03	5.3e-03	5.0e-03	4.7e-03	4.4e-03
700	4.1e-03	3.8e-03	3.6e-03	3.4e-03	3.1e-03	2.9e-03	2.7e-03	2.6e-03	2.4e-03	2.2e-03
710	2.1e-03	2.0e-03	1.8e-03	1.7e-03	1.6e-03	1.5e-03	1.4e-03	1.3e-03	1.2e-03	1.1e-03
720	1.0e-03	9.8e-04	9.1e-04	8.5e-04	7.9e-04	7.4e-04	6.9e-04	6.4e-04	6.0e-04	5.6e-04
730	5.2e-04	4.8e-04	4.5e-04	4.2e-04	3.9e-04	3.6e-04	3.4e-04	3.1e-04	2.9e-04	2.7e-04
740	2.5e-04	2.3e-04	2.1e-04	2.0e-04	1.9e-04	1.7e-04	1.6e-04	1.5e-04	1.4e-04	1.3e-04
750	1.2e-04	1.1e-04	1.0e-04	9.7e-05	9.1e-05	8.5e-05	7.9e-05	7.4e-05	6.9e-05	6.4e-05
760	6.0e-05	5.6e-05	5.2e-05	4.9e-05	4.5e-05	4.2e-05	4.0e-05	3.7e-05	3.4e-05	3.2e-05
770	3.0e-05	2.8e-05	2.6e-05	2.4e-05	2.3e-05	2.1e-05	2.0e-05	1.8e-05	1.7e-05	1.6e-05
780	1.5e-05									

Table E.3. (continued) The CIE 1988 Photopic luminous efficiency function, $V_M(\lambda)$.

Appendix F
CIE Illuminants

λ(nm)	$S_0(\lambda)$	$S_1(\lambda)$	$S_2(\lambda)$	D_{50}	D_{55}	D_{65}	D_{75}
300	0.04	0.02	0.00	0.02	0.02	0.03	0.04
305	3.02	2.26	1.00	1.04	1.05	1.66	2.59
310	6.00	4.50	2.00	2.05	2.07	3.29	5.13
315	17.80	13.45	3.00	4.92	6.65	11.77	17.47
320	29.60	22.40	4.00	7.78	11.23	20.24	29.82
325	42.45	32.20	6.25	11.27	15.94	28.64	42.38
330	55.30	42.00	8.50	14.76	20.66	37.05	54.94
335	56.30	41.30	8.15	16.36	22.27	38.50	56.11
340	57.30	40.60	7.80	17.96	23.89	39.95	57.27
345	59.55	41.10	7.25	19.49	25.86	42.43	60.01
350	61.80	41.60	6.70	21.02	27.83	44.91	62.75
355	61.65	39.80	6.00	22.49	29.23	45.77	62.87
360	61.50	38.00	5.30	23.95	30.63	46.64	62.99
365	65.15	40.20	5.70	25.46	32.47	49.36	66.66
370	68.80	42.40	6.10	26.97	34.32	52.09	70.33
375	66.10	40.45	4.55	25.74	33.45	51.03	68.52
380	63.40	38.50	3.00	24.50	32.59	49.98	66.71
385	64.60	36.75	2.10	27.19	35.34	52.31	68.34
390	65.80	35.00	1.20	29.88	38.09	54.65	69.97
395	80.30	39.20	0.05	39.60	49.53	68.70	85.96
400	94.80	43.40	-1.10	49.32	60.96	82.75	101.94
405	99.80	44.85	-0.80	52.93	64.76	87.12	106.92
410	104.80	46.30	-0.50	56.53	68.56	91.49	111.91
415	105.35	45.10	-0.60	58.29	70.07	92.46	112.36
420	105.90	43.90	-0.70	60.05	71.59	93.43	112.81
425	101.35	40.50	-0.95	58.94	69.75	90.06	107.96
430	96.80	37.10	-1.20	57.83	67.92	86.68	103.10
435	105.35	36.90	-1.90	66.33	76.77	95.77	112.15
440	113.90	36.70	-2.60	74.84	85.61	104.86	121.21

Table F.1. Components $S_0(\lambda)$, $S_1(\lambda)$, and $S_2(\lambda)$ used to compute CIE illuminants D.

λ(nm)	$S_0(\lambda)$	$S_1(\lambda)$	$S_2(\lambda)$	D_{50}	D_{55}	D_{65}	D_{75}
445	119.75	36.30	-2.75	81.05	91.81	110.94	127.11
450	125.60	35.90	-2.90	87.26	98.00	117.01	133.02
455	125.55	34.25	-2.85	88.94	99.24	117.41	132.69
460	125.50	32.60	-2.80	90.62	100.47	117.81	132.36
465	123.40	30.25	-2.70	91.00	100.19	116.34	129.85
470	121.30	27.90	-2.60	91.38	99.92	114.86	127.33
475	121.30	26.10	-2.60	93.25	101.33	115.39	127.07
480	121.30	24.30	-2.60	95.12	102.74	115.92	126.81
485	117.40	22.20	-2.20	93.54	100.41	112.37	122.30
490	113.50	20.10	-1.80	91.97	98.08	108.81	117.79
495	113.30	18.15	-1.65	93.85	99.38	109.08	117.19
500	113.10	16.20	-1.50	95.73	100.68	109.35	116.59
505	111.95	14.70	-1.40	96.17	100.69	108.58	115.15
510	110.80	13.20	-1.30	96.62	100.70	107.80	113.71
515	108.65	10.90	-1.25	96.88	100.34	106.30	111.18
520	106.50	8.60	-1.20	97.13	99.99	104.79	108.66
525	107.65	7.35	-1.10	99.62	102.10	106.24	109.55
530	108.80	6.10	-1.00	102.10	104.21	107.69	110.45
535	107.05	5.15	-0.75	101.43	103.16	106.05	108.37
540	105.30	4.20	-0.50	100.76	102.10	104.41	106.29
545	104.85	3.05	-0.40	101.54	102.54	104.22	105.60
550	104.40	1.90	-0.30	102.32	102.97	104.05	104.90
555	102.20	0.95	-0.15	101.16	101.48	102.02	102.45
560	100.00	0.00	0.00	100.00	100.00	100.00	100.00
565	98.00	-0.80	0.10	98.87	98.61	98.17	97.81
570	96.00	-1.60	0.20	97.73	97.22	96.33	95.62
575	95.55	-2.55	0.35	98.33	97.48	96.06	94.91
580	95.10	-3.50	0.50	98.92	97.75	95.79	94.21
585	92.10	-3.50	1.30	96.21	94.59	92.24	90.60
590	89.10	-3.50	2.10	93.50	91.43	88.69	87.00
595	89.80	-4.65	2.65	95.59	92.92	89.35	87.11
600	90.50	-5.80	3.20	97.68	94.42	90.01	87.23
605	90.40	-6.50	3.65	98.48	94.78	89.80	86.68
610	90.30	-7.20	4.10	99.27	95.14	89.60	86.14
615	89.35	-7.90	4.40	99.15	94.68	88.65	84.86
620	88.40	-8.60	4.70	99.04	94.22	87.70	83.58
625	86.20	-9.05	4.90	97.38	92.33	85.49	81.16
630	84.00	-9.50	5.10	95.72	90.45	83.29	78.75
635	84.55	-10.20	5.90	97.28	91.39	83.49	78.59
640	85.10	-10.90	6.70	98.85	92.33	83.70	78.43
645	83.50	-10.80	7.00	97.26	90.59	81.86	76.61
650	81.90	-10.70	7.30	95.66	88.85	80.03	74.80
655	82.25	-11.35	7.95	96.92	89.58	80.12	74.56
660	82.60	-12.00	8.60	98.18	90.31	80.21	74.32
665	83.75	-13.00	9.20	100.59	92.13	81.25	74.87
670	84.90	-14.00	9.80	103.00	93.95	82.28	75.42

Table F.1. (continued) Components $S_0(\lambda)$, $S_1(\lambda)$, and $S_2(\lambda)$ used to compute CIE illuminants D.

λ(nm)	$S_0(\lambda)$	$S_1(\lambda)$	$S_2(\lambda)$	D_{50}	D_{55}	D_{65}	D_{75}
675	83.10	-13.80	10.00	101.06	91.95	80.28	73.50
680	81.30	-13.60	10.20	99.13	89.95	78.28	71.57
685	76.60	-12.80	9.25	93.25	84.81	74.00	67.71
690	71.90	-12.00	8.30	87.37	79.67	69.72	63.85
695	73.10	-12.65	8.95	89.48	81.26	70.67	64.46
700	74.30	-13.30	9.60	91.60	82.84	71.61	65.07
705	75.35	-13.10	9.05	92.24	83.84	72.98	66.57
710	76.40	-12.90	8.50	92.88	84.84	74.35	68.07
715	69.85	-11.75	7.75	84.86	77.54	67.98	62.25
720	63.30	-10.60	7.00	76.85	70.23	61.60	56.44
725	67.50	-11.10	7.30	81.68	74.77	65.74	60.34
730	71.70	-11.60	7.60	86.50	79.30	69.89	64.24
735	74.35	-11.90	7.80	89.54	82.14	72.49	66.69
740	77.00	-12.20	8.00	92.57	84.99	75.09	69.15
745	71.10	-11.20	7.35	85.40	78.43	69.34	63.89
750	65.20	-10.20	6.70	78.22	71.88	63.59	58.63
755	56.45	-9.00	5.95	67.96	62.34	55.01	50.62
760	47.70	-7.80	5.20	57.69	52.79	46.42	42.62
765	58.15	-9.50	6.30	70.30	64.36	56.61	51.98
770	68.60	-11.20	7.40	82.92	75.92	66.81	61.35
775	66.80	-10.80	7.10	80.59	73.87	65.09	59.84
780	65.00	-10.40	6.80	78.27	71.82	63.38	58.32
785	65.50	-10.50	6.90	78.91	72.37	63.84	58.73
790	66.00	-10.60	7.00	79.55	72.93	64.30	59.14
795	63.50	-10.15	6.70	76.47	70.14	61.88	56.93
800	61.00	-9.70	6.40	73.40	67.35	59.45	54.73
805	57.15	-9.00	5.95	68.66	63.04	55.71	51.32
810	53.30	-8.30	5.50	63.92	58.72	51.96	47.92
815	56.10	-8.80	5.80	67.34	61.86	54.70	50.41
820	58.90	-9.30	6.10	70.77	64.99	57.44	52.91
825	60.40	-9.55	6.30	72.60	66.65	58.88	54.23
830	61.90	-9.80	6.50	74.44	68.30	60.31	55.54

Table F.1. (continued) Components $S_0(\lambda)$, $S_1(\lambda)$, and $S_2(\lambda)$ used to compute CIE illuminants D.

Appendix G
Chromaticity Coordinates of Paints

In this appendix, we list a set of chromaticity coordinates for paints. Table G.1 is after Barnes, who measured the reflectances of both organic and inorganic pigments used for painting [68]. This table is useful for specifying colors in rendering equations. It should be noted that these values were obtained under standard illuminant C.

Pigment	Y	x	y	Pigment	Y	x	y
Red lead	32.76	0.5321	0.3695	Burnt umber	5.08	0.4000	0.3540
English vermillion	22.26	0.5197	0.3309	Malachite	41.77	0.2924	0.3493
Cadmium red	20.78	0.5375	0.3402	Chrome green (medium)	16.04	0.3133	0.4410
Madder lake	33.55	0.3985	0.2756	Verdigris	16.67	0.1696	0.2843
Alizarin crimson	6.61	0.5361	0.3038	Emerald green	39.12	0.2446	0.4215
Carmine lake	5.04	0.4929	0.3107	Terre verte	29.04	0.3092	0.3510
Dragon's blood	4.94	0.4460	0.3228	Cobalt green	19.99	0.2339	0.3346
Realgar	32.35	0.4868	0.3734	Viridian	9.85	0.2167	0.3635
Venetian red	13.12	0.4672	0.3462	Cobalt blue	16.81	0.1798	0.1641
Indian red	10.34	0.3797	0.3194	Genuine ultramarine	18.64	0.2126	0.2016
Cadmium orange (medium)	42.18	0.5245	0.4260	Cerulean blue	18.19	0.1931	0.2096
Camboge	29.87	0.4955	0.4335	Azurite	9.26	0.2062	0.2008
Yellow ochre	41.15	0.4303	0.4045	Manganese violet	26.99	0.3073	0.2612
Zinc yellow	82.57	0.4486	0.4746	Cobalt violet	9.34	0.2817	0.1821
Indian yellow	40.01	0.4902	0.4519	Smalt	8.25	0.1898	0.1306
Saffron	33.45	0.4470	0.4345	Blue verditer	16.46	0.1899	0.1880
Yellow lake	27.24	0.4410	0.4217	French ultramarine blue	7.84	0.1747	0.1151
Strontium lemon yellow	84.36	0.4157	0.4732	Indigo	3.62	0.3180	0.2728
Hansa yellow 5 G	75.09	0.4447	0.5020	Prussian blue	1.30	0.2883	0.2453
Chrome yellow (medium)	63.07	0.4843	0.4470	Titanium white A	95.75	0.3124	0.3199
Cadmium yellow (light)	76.66	0.4500	0.4819	Titanium white B	95.66	0.3122	0.3196
Orpiment	64.45	0.4477	0.4402	Zinc white	94.88	0.3122	0.3200
Cobalt yellow	50.48	0.4472	0.4652	White lead	87.32	0.3176	0.3242
Raw sienna	20.03	0.4507	0.4043	Lamp black	4.08	0.3081	0.3157
Burnt sienna	7.55	0.4347	0.3414	Ivory black	2.22	0.3055	0.3176
Raw umber	7.44	0.3802	0.3693				

Table G.1. Yxy coordinates for a selection of organic and inorganic paints.

Bibliography

[1] E. Aas and J. Bogen. "Colors of Glacier Water." *Water Resources Research* 24:4 (1988), 561–565.

[2] P. M. Acosta-Serafini, I. Masaki, and C. G. Sodini. "A 1/3" VGA Linear Wide Dynamic Range CMOS Image Sensor Implementing a Predictive Multiple Sampling Algorithm with Overlapping Integration Intervals." *IEEE Journal of Solid-State Circuits* 39:9 (2004), 1487–1496.

[3] A. Adams and G. Haegerstrom-Portnoy. "Color Deficiency." In *Diagnosis and Management in Vision Care*. Boston: Butterworth, 1987.

[4] D. L. Adams and S. Zeki. "Functional Organization of Macaque V3 for Stereoscopic Depth." *Journal of Neurophysiology* 86:5 (2001), 2195–2203.

[5] J. Adams, K. Parulski, and K. Spaulding. "Color Processing in Digital Cameras." *IEEE Micro* 18:6 (1998), 20–30.

[6] M. M. Adams, P. R. Hof, R. Gattass, M. J. Webster, and L. G. Ungerleider. "Visual Cortical Projections and Chemoarchitecture of Macaque Monkey Pulvinar." *Journal of Comparative Neurology* 419:3 (2000), 377–393.

[7] J. E. Adams Jr. "Interactions between Color Plane Interpolation and Other Image Processing Functions in Electronic Photography." In *Proceedings of the SPIE 2416*, pp. 144–155. Bellingham, WA: SPIE, 1995.

[8] A. Adams. *The camera*. The Ansel Adams Photography series, Little, Brown and Company, 1980.

[9] A. Adams. *The negative*. The Ansel Adams Photography series, Little, Brown and Company, 1981.

[10] A. Adams. *The print*. The Ansel Adams Photography series, Little, Brown and Company, 1983.

[11] E. H. Adelson and J. R. Bergen. "The Plenoptic Function and the Elements of Early Vision." In *Computational Models of Visual Processing*, pp. 3–20. Cambridge, MA: MIT Press, 1991.

[12] E. H. Adelson and J. Y. A. Wang. "The Plenoptic Camera." *IEEE Transactions on Pattern Analysis and Machine Intelligence* 14:2 (1992), 99–106.

[13] E. A. Adelson. "Lightness Perception and Lightness Illusions." In *The New Cognitive Neurosciences*, edited by M. Gazzaniga, 2nd edition, pp. 339–351. Cambridge, MA: MIT Press, 2000.

[14] M. Aggarwal and N. Ahuja. "Split Aperture Imaging for High Dynamic Range." *International Journal of Computer Vision* 58:1 (2004), 7–17.

[15] M. Aggarwal, H. Hua, and N. Ahuja. "On Cosine-Fourth and Vignetting Effects in Real Lenses." In *Proceedings of the Eighth IEEE International Conference on Computer Vision (ICCV)*, 1, 1, pp. 472–479, 2001.

[16] M. Aguilar and W. S. Stiles. "Saturation of the Rod Mechanism of the Retina at High Levels of Stimulation." *Optica Acta* 1:1 (1954), 59–65.

[17] W. E. Ahearn and O. Sahni. "The Dependence of the Spectral and Electrical Properties of AC Plasma Panels on the Choice and Purity of the Gas Mixture." *SID International Symposium Digest* 7 (1978), 44–45.

[18] L. von Ahn, M. Blum, N. J. Hopper, and J. Langford. "Captcha: Telling Humans and Computers Apart Automatically." In *Advances in Cryptology - Proceedings of Eurocrypt 2003, Lecture Notes in Computer Science 2656*, pp. 294–311. Berlin: Springer, 2003.

[19] L. von Ahn, M. Blum, N. J. Hopper, and J. Langford. "Telling Humans and Computers Apart Automatically." *Communications of the ACM* 47:2 (2004), 57–60.

[20] T. Ajito, T. Obi, M. Yamaguchi, and N. Ohyama. "Multiprimary Color Display for Liquid Crystal Display Projectors using Diffraction Grating." *Optical Engineering* 38:11 (1999), 1883–1888.

[21] T. Ajito, T. Obi, M. Yamaguchi, and N. Ohyama. "Expanded Color Gamut Reproduced by Six-Primary Projection Display." In *Proceedings of the SPIE 3954*, edited by M. H. Wu, pp. 130–137. Bellingham, WA: SPIE, 2000.

[22] N. Akahane, S. Sugawa, S. Adachi, K. Mori, T. Ishiuchi, and K. Mizobuchi. "A Sensitivity and Linearity Improvement of a 100-dB Dynamic Range CMOS Image Sensor using a Lateral Overflow Integration Capacitor." *IEEE Journal of Solid-State Circuits* 41:4 (2006), 851–858.

[23] T. Akenine-Möller and E. Haines. *Real-Time Rendering*, 2nd edition. Natick, MA: A K Peters, 2002.

[24] A. O. Akyüz and E. Reinhard. "Noise Reduction in High Dynamic Range Imaging." *Journal of Visual Communication and Image Representation* 18:5 (2007), 366–376.

[25] A. O. Akyüz and E. Reinhard. "Perceptual Evaluation of Tone Reproduction Operators using the Cornsweet-Craik-O'Brien Illusion." *ACM Transactions on Applied Perception* 4:4 (2007), 20–1 – 20–29.

[26] A. O. Akyüz, R. Fleming, B. Riecke, E. Reinhard, and H. Bülthoff. "Do HDR Displays Support LDR Content? A Psychophysical Investigation." *ACM Transactions on Graphics* 26:3 (2007), 38–1 – 38–7.

[27] E. Allen. "Analytical Color Matching." *Journal of Paint Technology* 39:509 (1967), 368–376.

[28] E. Allen. "Basic Equations used in Computer Color matching II. Tristimulus Match, Two-Constant Theory." *Journal of the Optical Society of America* 64:7 (1974), 991–993.

[29] D. H. Alman. "CIE Technical Committee 1-29, Industrial Color Difference Evaluation Progress Report." *Colour Research and Application* 18 (1993), 137–139.

[30] M. Alpern and J. Moeller. "The Red and Green Cone Visual Pigments of Deuteranomalous Trichromacy." *Journal of Physiology* 266 (1977), 647–675.

[31] M. Alpern and S. Torii. "The Luminosity Curve of the Deuteranomalous Fovea." *Journal of General Physiology* 52:5 (1968), 738–749.

[32] M. Alpern and S. Torii. "The Luminosity Curve of the Protanomalous Fovea." *Journal of General Physiology* 52:5 (1968), 717–737.

[33] M. Alpern and T. Wake. "Cone Pigment in Human Deutan Color Vision Defects." *Journal of Physiology* 266 (1977), 595–612.

[34] M. Alpern. "The Stiles-Crawford Effect of the Second Kind (SCII): A Review." *Perception* 15:6 (1986), 785–799.

[35] B. Anderson and J. Winawer. "Image Segmentation and Lightness Perception." *Nature* 434 (2005), 79–83.

[36] R. Anderson. "Matrix Description of Radiometric Quantities." *Applied Optics* 30:7 (1991), 858–867.

[37] B. L. Anderson. "Perceptual Organization and White's Illusion." *Perception* 32 (2003), 269–284.

[38] A. Angelucci, J. B. Levitt, E. J. S. Walton, J.-M. Hupé, J. Bullier, and J. S. Lund. "Circuits for Local and Global Signal Integration in Primary Visual Cortex." *Journal of Neuroscience* 22:19 (2002), 8633–8646.

[39] A. Angström. "The Albedo of Various Surfaces of Ground." *Geografiska Annalen* 7 (1925), 323–342.

[40] R. A. Applegate and V. Lakshminarayanan. "Parametric Representation of Stiles-Crawford Functions: Normal Variation of Peak Location and Directionality." *Journal of the Optical Society of America A* 10:7 (1993), 1611–1623.

[41] G. B. Arden. "The Importance of Measuring Contrast Sensitivity in Cases of Visual Disturbances." *British Journal of Ophthalmology* 62 (1978), 198–209.

[42] L. Arend and A. Reeves. "Simultaneous Color Constancy." *Journal of the Optical Society of America A* 3:10 (1986), 1743–1751.

[43] P. Artal, E. Berrio, A. Guirao, and P. Piers. "Contribution of the Cornea and Internal Surfaces to the Change of Ocular Aberrations with Age." *Journal of the Optical Society of America A* 19:1 (2002), 137–143.

[44] M. Ashikhmin and J. Goral. "A Reality Check for Tone Mapping Operators." *ACM Transactions on Applied Perception* 3:4 (2006), 399–411.

[45] M. Ashikhmin and P. Shirley. "An Anisotropic Phong BRDF Model." *journal of graphics tools* 5:2 (2002), 25–32.

[46] M. Ashikhmin, S. Premože, and P. Shirley. "A Microfacet-Based BRDF Generator." In *Proceedings of ACM SIGGRAPH 2000, Computer Graphics Proceedings, Annual Conference Series*, edited by K. Akeley, pp. 65–74. Reading, MA: Addison-Wesley, 2000.

[47] M. Ashikhmin. "A Tone Mapping Algorithm for High Contrast Images." In *Proceedings of the 13th Eurographics Workshop on Rendering*, pp. 145–155. Aire-la-Ville, Switzerland: Eurographics Association, 2002.

[48] ASTM. "Standard Tables for Reference Solar Spectral Irradiances: Direct Normal and Hemispherical on 37° Tilted Surface." Technical Report G173-03, American Society for Testing and Materials, West Conshohocken, PA, 2003.

[49] ASTM. "Standard Terminology of Appearance." Technical Report E 284, American Society for Testing and Materials, West Conshohocken, PA, 2006.

[50] D. A. Atchison and G. Smith. "Chromatic Dispersions of the Ocular Media of Human Eyes." *Journal of the Optical Society of America A* 22:1 (2005), 29–37.

[51] D. A. Atchison, G. Smith, and N. Efron. "The Effect of Pupil Size on Visual Acuity in Uncorrected and Corrected Myopia." *American Journal of Optometry and Physiological Optics* 56:5 (1979), 315–323.

[52] J. J. Atick and N. A. Redlich. "What Does the Retina Know about Natural Scenes?" *Neural Computation* 4 (1992), 196–210.

[53] H. Aubert. *Physiologie der Netzhaut*. Breslau, Germany: E Morgenstern, 1865.

[54] V. A. Babenko, L. G. Astafyeva, and V. N. Kuzmin. *Electromagnetic Scattering in Disperse Media: Inhomogeneous and Anisotropic Particles*. Berlin: Springer-Verlag, 2003.

[55] R. J. Baddeley and P. J. B. Hancock. "A Statistical Analysis of Natural Images Matches Psychophysically Derived Orientation Tuning Curves." *Proceedings of the Royal Society of London B* 246 (1991), 219–223.

[56] R. Bajcsy, S. W. Lee, and A. Leonardis. "Color Image Segmentation with Detection of Highlights and Local Illumination Induced by Interreflection." In *Proceedings of the International Conference on Pattern Recognition*, pp. 785–790. Washington, DC: IEEE, 1990.

[57] R. Bajcsy, S. W. Lee, and A. Leonardis. "Detection of Diffuse and Specular Interface Reflections and Inter-reflections by Color Image Segmentation." *International Journal of Computer Vision* 17:3 (1996), 241–272.

[58] R. M. Balboa, C. W. Tyler, and N. M. Grzywacz. "Occlusions Contribute to Scaling in Natural Images." *Vision Research* 41:7 (2001), 955–964.

[59] G. V. G. Baranoski, P. Shirley, J. G. Rokne, T. Trondsen, and R. Bastos. "Simulating the Aurora Borealis." In *Eighth Pacific Conference on Computer Graphics and Applications*, pp. 2–14. Los Alamitos, CA: IEEE Computer Society Press, 2000.

[60] H. B. Barlow, R. M. Hill, and W. R. Levick. "Retinal Ganglion Cells Responding Selectively to Direction and Speed of Image Motion in the Rabbit." *Journal of Physiology (London)* 173 (1964), 377–407.

[61] H. B. Barlow. "The Size of Ommatidia in Apposition Eyes." *Journal of Experimental Biology* 29:4 (1952), 667–674.

[62] H. B. Barlow. "Summation and Inhibition in the Frog's Retina." *Journal of Physiology (London)* 119 (1953), 69–88.

[63] H. B. Barlow. "Dark and Light Adaptation: Psychophysics." In *Handbook of Sensory Physiology VII/4*, edited by D. Jameson and L. M. Hurvich, pp. 1–28. Berlin: Springer-Verlag, 1972.

[64] K. Barnard and B. Funt. "Investigations into Multi-Scale Retinex (MSR)." In *Color Imaging: Vision and Technology*, edited by L. W. MacDonald and M. R. Luo, pp. 9–17. New York: John Wiley and Sons, 1999.

[65] K. Barnard and B. Funt. "Camera Characterization for Color Research." *Color Research and Application* 27 (2002), 152–163.

[66] K. Barnard, L. Martin, A. Coath, and B. Funt. "A Comparison of Computational Color Constancy Algorithms — Part II: Experiments with Image Cata." *IEEE Transactions on Image Processing* 11:9 (2002), 985–996.

[67] C. Barnes, J. Wei, and S. K. Shevell. "Chromatic Induction with Remote Chromatic Contrast Varied in Magnitude, Spatial Frequency, and Chromaticity." *Vision Research* 39 (1999), 3561–3574.

[68] N. F. Barnes. "Color Characteristics of Artists' Paints." *Journal of the Optical Society of America* 29:5 (1939), 208–214.

[69] P. Barone, A. Batardiere, K. Knoblauch, and H. Kennedy. "Laminar Distribution of Neurons in Extrastriate Areas Projecting to Visual Areas V1 and V4 Correlates with the Hierarchical Rank and Indicates the Operation of a Distance Rule." *Journal of Neuroscience* 20:9 (2000), 3263–3281.

[70] B. Barsky, L. Chu, and S. Klein. "Cylindrical Coordinate Representations for Modeling Surfaces of the Cornea and Contact Lenses." In *Proceedings of the International Conference on Shape Modeling and Applications (SMI-99)*, edited by B. Werner, pp. 98–115. Los Alamitos, CA: IEEE Computer Society, 1999.

[71] A. Bartels and S. Zeki. "The Architecture of the Colour Centre in the Human Visual Brain: New Results and a Review." *European Journal of Neuroscience* 12:1 (2000), 172–193.

[72] C. Bartleson and E. Breneman. "Brightness Perception in Complex Fields." *Journal of the Optical Society of America* 57 (1967), 953–957.

[73] M. Bass, E. W. van Stryland, D. R. Williams, and W. L. Wolfe, editors. *Handbook of Optics: Fundamentals, Techniques and Design*, Second edition. New York: McGraw-Hill, 1995.

[74] B. Bastani, W. Cressman, and B. Funt. "Calibrated Color Mapping Between LCD and CRT Displays: A Case Study." *Colour Research and Application* 30:6 (2005), 438–447.

[75] B. Baxter, H. Ravindra, and R. A. Normann. "Changes in Lesion Detectability Caused by Light Adaptation in Retinal Photoreceptors." *Investigative Radiology* 17 (1982), 394–401.

[76] B. E. Bayer. "Color Imaging Array.", 1976. U.S. Patent 3,971,065.

[77] D. A. Baylor, B. J. Nunn, and J. L. Schnapf. "The Photocurrent, Noise and Spectral Sensitivity of Rods of the Monkey *Macaca fascicularis*." *Journal of Physiology* 357:1 (1984), 576–607.

[78] D. A. Baylor, B. J. Nunn, and J. L. Schnapf. "Spectral Sensitivity of Cones of the Monkey *Macaca fascicularis*." *Journal of Physiology* 390:1 (1987), 145–160.

[79] H. Becker, H. Vestweber, A. Gerhard, P. Stoessel, and R. Fortte. "Novel Host Materials for Efficient and Stable Phosphorescent OLED Devices." In *Proceedings of the International Display Manufacturing Conference*, pp. 329–330. San Jose, CA: Society for Information Display, 2005.

[80] R. E. Bedford and G. Wyszecki. "Axial Chromatic Aberration of the Human Eye." *Journal of the Optical Society of America* 47 (1957), 564–565.

[81] P. R. Bélanger. "Linear-Programming Approach to Color-Recipe Formulations." *Journal of the Optical Society of America* 64:11 (1974), 1541–1544.

[82] A. J. Bell and T. J. Sejnowski. "Edges Are the 'Independent Components' of Natural Scenes." In *Advances in Neural Information Processing Systems*, pp. 831–837. Cambridge, MA: MIT Press, 1996.

[83] A. J. Bell and T. J. Sejnowski. "The Independent Components of Natural Scenes Are Edge Filters." *Vision Research* 37 (1997), 3327–3338.

[84] G. Berbecel. *Digital Image Display: Algorithms and Implementation.* Chichester: John Wiley and Sons, 2003.

[85] T. T. J. M. Berendschot, J. van de Kraats, and D. van Norren. "Wavelength Dependence of the Stiles-Crawford Effect Explained by Perception of Backscattered Light from the Choroid." *Journal of the Optical Society of America A* 18:7 (2001), 1445–1451.

[86] L. D. Bergman, B. E. Rogowitz, and L. A. Treinish. "A Rule-Based Tool for Assisting Colormap Selection." In *Proceedings of IEEE Visualization*, pp. 118–125. Los Alamitos, CA: IEEE Computer Society, 1995.

[87] B. Berlin and P. Kay. *Basic Color Terms: Their Universality and Evolution.* Berkeley, CA: University of California Press, 1969.

[88] B. J. Berne and R. Pecora. *Dynamic Light Scattering (with Applications to Chemistry, Biology and Physics).* Mineola, NY: Dover Publications, 2000.

[89] R. S. Berns and M. J. Shiyu. "Colorimetric Characterization of a Desk-Top Drum Scanner using a Spectral Model." *Journal of Electronic Imaging* 4 (1995), 360–372.

[90] R. S. Berns, D. H. Alman, L. Reniff, G. D. Snyder, and M. R. Balonon-Rosen. "Visual Determination of Suprathreshold Color-Difference Tolerances using Probit Analysis." *Colour Research and Application* 16 (1991), 297–316.

[91] R. S. Berns, M. E. Gorzynski, and R. J. Motta. "CRT Colorimetry – Part II: Metrology." *Color Research and Application* 18:5 (1993), 315–325.

[92] R. S. Berns. "Methods for Characterizing CRT Displays." *Displays* 16:4 (1996), 173–182.

[93] R. S. Berns. *Billmeyer and Saltzman's Principles of Color Technology*, Third edition. New York: John Wiley and Sons, 2000.

[94] I. Biederman and P. Kalocsai. "Neural and Psychological Analysis of Object and Face Recognition." In *Face Recognition: From Theory to Applications*, pp. 3–25. Berlin: Springer-Verlag, 1998.

[95] M. Bigas, E. Cabruja, J. Forest, and J. Salvi. "Review of CMOS Image Sensors." *Microelectronics Journal* 37:5 (2006), 433–451.

[96] J. Birnstock, J. Blässing, A. Hunze, M. Scheffel, M. Stößel, K. Heuser, G. Wittmann, J. Wörle, and A. Winnacker. "Screen-Printed Passive Matrix Displays Based on Light Emitting Polymers." *Applied Physics Letters* 78:24 (2001), 3905–3907.

[97] J. Birnstock, J. Blässing, A. Hunze, M. Scheffel, M. Stößel, K. Heuser, J. Wörle, G. Wittmann, and A. Winnacker. "Screen-Printed Passive Matrix Displays and Multicolor Devices." In *Proceedings of the SPIE 4464*, pp. 68–75. Bellingham, WA: SPIE, 2002.

[98] C. M. Bishop. *Neural Networks for Pattern Recognition*. Oxford: Oxford University Press, 1995.

[99] D. L. Bitzer and H. G. Slottow. "The Plasma Display Panel — A Digitally Addressable Display with Inherent Memory." In *Proceedings of the AFIPS Conference 29*, pp. 541–547. American Federation of Information Processing Societies, 1966.

[100] H. R. Blackwell. "Luminance difference thresholds." In *Handbook of Sensory Physiology, VII/4*, edited by D. Jameson and L. M. Hurvich, pp. 78–101. Berlin: Springer-Verlag, 1972.

[101] C. Blakemore and F. Campbell. "On the Existence of Neurons in the Human Visual System Electively Sensitive to Orientation and Size of Retinal Images." *Journal of Physiology* 203 (1969), 237–260.

[102] B. Blakeslee and M. E. McCourt. "A Multi-Scale Spatial Filtering Account of the White Effect, Simultaneous Brightness Contrast and Grating Induction." *Vision Research* 39 (1999), 4361–4377.

[103] N. Blanc. "CCD versus CMOS — Has CCD Imaging Come to an End?" In *Photogrammetric Week 01*, edited by D. Fritsch and R. Spiller, pp. 131–137. Heidelberg: Wichmann Verlag, 2001.

[104] G. G. Blasdel and J. S. Lund. "Termination of Afferent Axons in Macaque Striate Cortex." *Journal of Neuroscience* 3:7 (1983), 1389–1413.

[105] J. F. Blinn. "Models of Light Reflection for Computer Synthesized Pictures." *Proc. SIGGRAPH '77, Computer Graphics* 11 (1977), 192–198.

[106] J. F. Blinn. "Return of the Jaggy." *IEEE Computer Graphics and Applications* 9:2 (1989), 82–89.

[107] J. F. Blinn. "What We Need Around Here Is More Aliasing." *IEEE Computer Graphics and Applications* 9:1 (1989), 75–79.

[108] C. F. Bohren. "Colors of the Sea." *Weatherwise* 35:5 (1982), 256–260.

[109] G. F. Bohren. "The Green Flash." *Weatherwise* 35:6 (1982), 271–275.

[110] C. F. Bohren. "More About Colors of the Sea." *Weatherwise* 36:6 (1983), 311–316.

[111] C. F. Bohren. "Scattering by Particles." In *Handbook of Optics: Fundamentals, Techniques and Design, Volume 1*, edited by M. Bass, E. W. van Stryland, D. R. Williams, and W. L. Wolfe, Second edition. New York: McGraw-Hill, 1995.

[112] M. Born and E. Wolf. *Principles of Optics*, Seventh edition. Cambridge, UK: Cambridge University Press, 1999.

[113] T. Bossomaier and A. W. Snyder. "Why Spatial Frequency Processing in the Visual Cortex?" *Vision Research* 26 (1986), 1307–1309.

[114] P. Bouguer. *Traité d'optique sur la gradation de la lumiere: Ouvrage posthume*. Paris: M l'Abbé de la Caille, 1760.

[115] D. Boussaoud, L. G. Ungerleider, and R. Desimone. "Pathways for Motion Analysis: Cortical Connections of the Medial Superior Temporal and Fundus of the Superior Temporal Visual Areas in the Macaque." *Journal of Comparative Neurology* 296:3 (1990), 462–495.

[116] A. Bovik, editor. *Handbook of Image and Video Processing*. San Diego, CA: Academic Press, 2000.

[117] R. M. Boynton and D. N. Whitten. "Visual Adaptation in Monkey Cones: Recordings of Late Receptor Potentials." *Science* 170 (1970), 1423–1426.

[118] R. M. Boynton. *Human Color Vision*. New York: Holt, Rinehart and Winston, 1979.

[119] D. H. Brainard and W. T. Freeman. "Bayesean Color Constancy." *Journal of the Optical Society of America A* 14:7 (1997), 1393–1411.

[120] D. H. Brainard, W. A. Brunt, and J. M. Speigle. "Color Constancy in the Nearly Natural Image. I. Assymetric Matches." *Journal of the Optical Society of America A* 14:9 (1997), 2091–2110.

[121] D. H. Brainard, A. Roorda, Y. Yamauchi, J. B. Calderone, A. Metha, M. Neitz, J. Neitz, D. R. Williams, and G. H. Jacobs. "Functional Consequences of the Relative Numbers of L and M Cones." *Journal of the Optical Society of America A* 17:3 (2000), 607–614.

[122] D. H. Brainard, D. G. Pelli, and T. Robson. "Display Characterization." In *Encylopedia of Imaging Science and Technology*, edited by J. Hornak, pp. 172–188. New York: John Wiley and Sons, 2002.

[123] D. H. Brainard. "Color Constancy in the Nearly Natural Image. II. Achromatic Loci." *Journal of the Optical Society of America A* 15 (1998), 307–325.

[124] G. J. Braun and M. D. Fairchild. "Techniques for Gamut Surface Definition and Visualization." In *Proceedings of the 5th IS&T/SID Color Imaging Conference*, pp. 147–152. Springfield, VA: IS&T, 1997.

[125] G. Brelstaff and F. Chessa. "Practical Application of Visual Illusions: Errare Humanum Est." In *Proceedings of the 2nd Symposium on Applied Perception in Graphics and Visualization (APGV)*, pp. 161–161. New York: ACM, 2005.

[126] E. Breneman. "Corresponding Chromaticities for Different States of Adaptation to Complex Visual Fields." *Journal of the Optical Society of America, A* 4 (1987), 1115–1129.

[127] P. Bressan, E. Mingolla, L. Spillmann, and T. Watanabe. "Neon Color Spreading: A Review." *Perception* 26:11 (1997), 1353–1366.

[128] H. Brettel and F. Viénot. "Web Design for the Colour-Blind User." In *Colour Imaging: Vision and Technology*, edited by L. W. MacDonald and M. R. Luo, pp. 55–71. Chichester, UK: John Wiley and Sons, Ltd., 1999.

[129] H. Brettel, F. Viénot, and J. D. Mollon. "Computerized Simulation of Color Appearance for Dichromats." *Journal of the Optical Society of America A* 14:10 (1997), 2647–2655.

[130] D. Brewster. "A Note Explaining the Cause of an Optical Phenomenon Observed by the Rev. W. Selwyn." In *Report of the Fourteenth Meeting of the British Association for the Advancement of Science*, edited by J. Murray. London, 1844.

[131] M. H. Brill and J. Larimer. "Avoiding On-Screen Metamerism in N-Primary Displays." *Journal of the Society for Information Display* 13:6 (2005), 509–516.

[132] M. H. Brill. "Image Segmentation by Object Color: A Unifying Framework and Connection to Color Constancy." *Journal of the Optical Society of America A* 7:10 (1986), 2041–2047.

[133] G. S. Brindley. "The Discrimination of After-Images." *Journal of Physiology* 147:1 (1959), 194–203.

[134] R. W. Brislin. "The Ponzo Illusion: Additional Cues, Age, Orientation, and Culture." *Journal of Cross-Cultural Psychology* 5:2 (1974), 139–161.

[135] K. H. Britten, M. N. Shadlen, W. T. Newsome, and J. A. Movshon. "The Analysis of Visual Motion: A Comparison of Neuronal and Psychophysical Performance." *Journal of Neuroscience* 12:12 (1992), 4745–4765.

[136] A. Brockes. "Vergleich der Metamerie-Indizes bei Lichtartwechsel von Tageslicht zur Glühlampe and zu verschiedenen Leuchtstofflampen." *Die Farbe* 18:223.

[137] A. Brockes. "Vergleich von berechneten Metamerie-Indizes mit Abmusterungsergebnissen." *Die Farbe* 19:135 (1970), 1–10.

[138] K. Brodmann. *Vergleichende Lokalisationslehre der Großhirnrhinde*. Leipzig, Germany: Barth Verlag, 1909. Translated by L J Garey, *Localisation in the Cerebral Cortex* (Smith-Gordon, London, 1994).

[139] J. Broerse, T. Vladusich, and R. P. O'Shea. "Colour at Edges and Colour Spreading in McCollough Effects." *Vision Research* 39 (1999), 1305–1320.

[140] N. Bruno, P. Bernardis, and J. Schirillo. "Lightness, Equivalent Backgrounds and Anchoring." *Perception and Psychophysics* 59:5 (1997), 643–654.

[141] G. Buchsbaum. "A Spatial Processor Model for Object Color Perception." *Journal of the Franklin Institute* 310 (1980), 1–26.

[142] Bureau International des Poids et Mesures. "Le Système International d'Unités (The International System of Units).", 2006.

[143] J. H. Burroughes, D. D. C. Bradley, A. R. Brown, R. N. Marks, K. Mackay, R. H. Friend, P. L. Burns, and A. B. Holmes. "Light-Emitting Diodes based on Conjugated Polymers." *Nature* 347:6293 (1990), 539–541.

[144] G. J. Burton and I. R. Moorhead. "Color and Spatial Structure in Natural Scenes." *Applied Optics* 26:1 (1987), 157–170.

[145] A. Calabria and M. D. Fairchild. "Herding CATs: A Comparison of Linear Chromatic Adaptation Transforms for CIECAM97s." In *IS&T/SID 9th Color Imaging Conference*, pp. 174–178. Springfield, VA: Society for Imaging Science and Technology, 2001.

[146] A. J. Calabria and M. D. Fairchild. "Perceived Image Contrast and Observer Preference I: The Effects of Lightness, Chroma, and Sharpness Manipulations on Contrast Perception." *Journal of Imaging Science and Technology* 47 (2003), 479–493.

[147] A. Calabria and M. D. Fairchild. "Perceived Image Contrast and Observer Preference II: Empirical Modeling of Perceived Image Contrast and Observer Preference Data." *Journal of Imaging Science and Technology* 47 (2003), 494–508.

[148] E. M. Callaway and A. K. Wiser. "Contributions of Individual Layer 2-5 Spiny Neurons to Local Circuits in Macaque Primary Visual Cortex." *Visual Neuroscience* 13:5 (1996), 907–922.

[149] E. Camalini, editor. *Optical and Acoustical Holography*. New York: Plenum Press, 1972.

[150] F. W. Campbell and A. H. Gregory. "Effect of Pupil Size on Acuity." *Nature (London)* 187:4743 (1960), 1121–1123.

[151] F. W. Campbell and R. W. Gubisch. "Optical Quality of the Human Eye." *Journal of Physiology* 186 (1966), 558–578.

[152] F. W. Campbell and J. G. Robson. "Application of Fourier Analysis to the Visibility of Gratings." *Journal of Physiology (London)* 197 (1968), 551–566.

[153] V. C. Cardei and B. Funt. "Color Correcting Uncalibrated Digital Camera." *Journal of Imaging Science and Technology* 44:4 (2000), 288–294.

[154] B. S. Carlson. "Comparison of Modern CCD and CMOS Image Sensor Technologies and Systems for Low Resolution Imaging." In *Proceedings of IEEE Sensors*, 1, 1, pp. 171–176, 2002.

[155] J. E. Carnes and W. F. Kosonocky. "Noise Sources in Charge-Coupled Devices." *RCA Review* 33 (1972), 327–343.

[156] J. E. Carnes and W. F. Kosonocky. "Noise Sources in Charge-Coupled Devices." In *SPIE Milestone Series 177*, pp. 101–109. Bellingham, WA: SPIE, 2003.

[157] V. A. Casagrande. "A Third Parallel Pathway to the Primate Retina V1." *Trends in Neuroscience* 17 (1994), 305–310.

[158] J. Cataliotti and A. Gilchrist. "Local and Global Processes in Surface Lightness Perception." *Perception and Psychophysics* 57:2 (1995), 125–135.

[159] P. Cavanagh and S. Anstis. "The Contribution of Color and Motion in Normal and Color-Deficient Observers." *Vision Research* 31:12 (1991), 2109–2148.

[160] P. Cavanagh and Y. G. Leclerc. "Shape from Shadows." *Journal of Experimental Psychology* 15:1 (1989), 3–27.

[161] A. Chalmers, T. Davis, and E. Reinhard, editors. *Practical Parallel Rendering*. Natick, MA: A K Peters, 2002.

[162] Y.-C. Chang and J. F. Reid. "RGB Calibration for Color Image Analysis in Machine Vision." *IEEE Transactions on Image Processing* 5:10 (1996), 1414–1422.

[163] G. Charles and G. van Loan. *Matrix Computations*, Third edition. Baltimore, MD: The John Hopkins University Press, 1996.

[164] W. N. Charman. "The Flawed Retinal Image. Part I." *Optician* 190:5020 (1985), 31–34.

[165] W. N. Charman. "The Flawed Retinal Image. Part II." *Optician* 190:5022 (1985), 16–20.

[166] W. N. Charman. "Optics of the Eye." In *Handbook of Optics: Fundamentals, Techniques and Design, Volume 1*, edited by M. Bass, E. W. van Stryland, D. R. Williams, and W. L. Wolfe, Second edition. New York: McGraw-Hill, 1995.

[167] B. Chen and W. Makous. "Light Capture by Human Cones." *Journal of Physiology* 414 (1989), 89–109.

[168] C.-J. Chen, A. Lien, and M. I. Nathan. "4 $\times$ 4 and 2 $\times$ 2 Matrix Formulations for the Optics in Stratified and Biaxial Media." *Journal of the Optical Society of America A* 14:11 (1997), 3125–3134.

[169] Y. Chen, J. Au, P. Kazlas, A. Ritenour, H. Gates, and J. Goodman. "Ultra-Thin, High-Resolution, Flexible Electronic Ink Displays Addressed by a-Si Active-Matrix TFT Backplanes on Stainless Steel Foil." In *Electron Devices Meeting, IEDM '02*, pp. 389–392, 2002.

[170] W.-C. Cheng and M. Pedram. "Power Minimization in a Backlit TFT=LCD Display by Concurrent Brightness and Contrast Scaling." *IEEE Transactions on Consumer Electronics* 50:1 (2004), 25–32.

[171] M. E. Chevreul. *De la loi du contraste simultane des couleurs*. Paris: Pitois Levreault, 1839.

[172] C.-C. Chiao, T. W. Cronin, and D. Osorio. "Color Signals in Natural Scenes: Characteristics of Reflectance Spectra and the Effects of Natural Illuminants." *Journal of the Optical Society of America A* 17:2 (2000), 218–224.

[173] C. D. Child. "Discharge from Hot CaO." *Physical Review I* 32:5 (1911), 492–511.

[174] K. Chiu, M. Herf, P. Shirley, S. Swamy, C. Wang, and K. Zimmerman. "Spatially Nonuniform Scaling Functions for High Contrast Images." In *Proceedings of Graphics Interface '93*, pp. 245–253. Toronto, Canada: Canadian Information Processing Society, 1993.

[175] I. Choi, H. Shim, and N. Chang. "Low Power Color TFT-LCD Display for Hand-Held Embedded Systems." In *Proceedings of the Symposium on Low Power Electronics and Design*, pp. 112–117, 2002.

[176] T. J. Cholewo and S. Love. "Gamut Boundary Determination using Alpha-Shapes." In *Proceedings of the 7th IS&T/SID Color Imaging Conference*, pp. 200–204. Springfield, VA: IS&T, 1999.

[177] B. Choubey, S. Aoyoma, S. Otim, D. Joseph, and S. Collins. "An Electronic Calibration Scheme for Logarithmic CMOS Pixels." *IEEE Sensors Journal* 6:4 (2006), 950–056.

[178] P. Choudhury and J. Tumblin. "The Trilateral Filter for High Contrast Images and Meshes." In *EGRW '03: Proceedings of the 14th Eurographics Workshop on Rendering*, pp. 186–196. Aire-la-Ville, Switzerland: Eurographics Association, 2003.

[179] R. E. Christoffersen. *Basic Principles and Techniques of Molecular Quantum Mechanics*. New York: Springer-Verlag, 1989.

[180] C. Chubb, G. Sperling, and J. A. Solomon. "Texture Interactions Determine Perceived Contrast." *Proceedings of the National Academy of Sciences of the United States of America* 86:23 (1989), 9631–9635.

[181] J. W. Chung, H. R. Guo, C. T. Wu, K. C. Wang, W. J. Hsieh, T. M. Wu, and C. T. Chung. "Long-Operating Lifetime of Green Phosphorescence Top-Emitting Organic Light Emitting Devices." In *Proceedings of the International Display Manufacturing Conference*, pp. 278–280. San Jose, CA: Society for Information Display, 2005.

[182] CIE. "CIE Proceedings 1951." Technical Report Vol. 1, Sec 4; Vol 3, p. 37, Commision Internationale De L'Eclairage, Vienna, 1951.

[183] CIE. "Light as a True Visual Quantity: Principles of Measurement." Technical Report Publ. No. 41 (TC-1.4), Commision Internationale De L'Eclairage, Vienna, 1978.

[184] CIE. "A Method For Assessing the Quality of Daylight Simulators for Colorimetry." Technical Report CIE 51-1981, Commision Internationale De L'Eclairage, Vienna, 1981.

[185] CIE. "The Basis of Physical Photometry." Technical Report Publ. No. 18.2 (TC-1.2), Commision Internationale De L'Eclairage, Vienna, 1983.

[186] CIE. "Colorimetry, Second Edition." Technical Report CIE 15.2, Commision Internationale De L'Eclairage, Vienna, 1986.

[187] CIE. "International Lighting Vocabulary." Technical Report 17.4, Commision Internationale De L'Eclairage, Vienna, 1987.

[188] CIE. "Special Metamerism Index: Change in Observer." Technical Report CIE 80-1989, Commision Internationale De L'Eclairage, Vienna, 1989.

[189] CIE. "CIE 1988 2° Spectral Luminous Efficiency Function for Photopic Vision." Technical Report Publ. No. 86, Commision Internationale De L'Eclairage, Vienna, 1990.

[190] CIE. "Method of Measuring and Specifying Colour Rendering Properties of Light Sources." Technical Report CIE 13.3-1995, Commision Internationale De L'Eclairage, Vienna, 1995.

[191] CIE. "The CIE 1997 Interim Colour Appearance Model (Simple Version), CIECAM97s." Technical Report CIE 131-1998, Commision Internationale De L'Eclairage, Vienna, 1998.

[192] CIE. "Virtual Metamers for Assessing the Quality of Simulators of CIE Illuminant D50." Technical Report Supplement 1-1999 to CIE 51-1981, Commision Internationale De L'Eclairage, Vienna, 1999.

[193] CIE. "CIE Standard Illuminants for Colorimetry, Standard CIE S005/E-1998." Technical Report Standard CIE S005/E-1998, Commision Internationale De L'Eclairage, Vienna, 2004. Published also as ISO 10526/CIE S 005/E-1999.

[194] CIE. "Colorimetry, Third Edition." Technical Report CIE Publ. No. 15:2004, Commision Internationale De L'Eclairage, Vienna, 2004.

[195] CIE. "A Colour Appearance Model for Colour Management Systems: CIECAM02." Technical Report CIE 159:2004, Commision Internationale De L'Eclairage, Vienna, 2004.

[196] F. R. Clapper and J. A. C. Yule. "The Effect of Multiple Internal Reflections on the Densities of Halftone Prints on Paper." *Journal of the Optical Society of America* 43 (1953), 600–603.

[197] F. R. Clapper and J. A. C. Yule. "Reproduction of Color with Halftone Images." In *Proceedings of the Technical Association of Graphic Arts*, pp. 1–12. Chicago, IL: Technical Assn. of the Graphic Arts, 1955.

[198] F. J. J. Clark, R. McDonald, and B. Rigg. "Modifications to the JPC79 Colour-Difference Formula." *Journal of the Society of Dyers and Colourists* 100 (1984), 128–132, (Errata: 281–282).

[199] S. Coe, W.-K. Woo, M. Bawendi, and V. Bulovic. "Electroluminescence from Single Monolayers of Nanocrystals in Molecular Organic Devices." *Nature* 420:6917 (2002), 800–803.

[200] M. F. Cohen and J. R. Wallace. *Radiosity and Realistic Image Synthesis*. Cambridge, MA: Academic Press, Inc., 1993.

[201] J. Cohen. "Dependency of the Spectral Reflectance Curves of the Munsell Color Chips." *Psychonomic Science* 1 (1964), 369–370.

[202] M. Colbert, E. Reinhard, and C. E. Hughes. "Painting in High Dynamic Range." *Journal of Visual Communication and Image Representation* 18:5 (2007), 387–396.

[203] M. Collet. "Solid-State Image Sensors." *Sensors and Actuators* 10 (1986), 287–302.

[204] D. Comaniciu and P. Meer. "Mean Shift: A Robust Approach toward Feature Space Analysis." *IEEE Transactions on Pattern Analysis and Machine Intelligence* 24:5 (2002), 603–619.

[205] B. Comiskey, J. D. Albert, H. Yoshizawa, and J. Jacobson. "An Electrophoretic Ink for All-Printed Reflective Electronic Displays." *Nature* 394:6690 (1998), 253–255.

[206] M. Corbetta, F. M. Miezin, S. Dobmeyer, G. L. Shulman, and S. E. Petersen. "Selective and Divided Attention during Visual Discriminations of Shape, Color and Speed: Functional Anatomy by Positron Emission Tomography." *Journal of Neuroscience* 11:8 (1991), 2383–2402.

[207] T. N. Cornsweet and D. Teller. "Relation of Increment Thresholds to Brightness and Luminance." *Journal of the Optical Society of America* 55 (1965), 1303–1308.

[208] T. N. Cornsweet. *Visual Perception*. New York: Academic Press, 1970.

[209] N. P. Cottaris and R. L. de Valois. "Temporal Dynamics of Chromatic Tuning in Macaque Primary Visual Cortex." *Nature* 395:6705 (1998), 896–900.

[210] W. B. Cowan and N. Rowell. "On the Gun Independence and Phosphor Constancy of Color Video Monitors." *Color Research and Application* 11 (1986), Supplement 34–38.

[211] W. B. Cowan. "An Inexpensive Scheme for Calibration of a Colour Monitor in Terms of CIE Standard Coordinates." *Proc. SIGGRAPH '83 Computer Graphics* 17:3 (1983), 315–321.

[212] B. H. Crawford. "The Scotopic Visibility Function." *Proceedings of the Physical Society of London* 62:5 (1949), 321–334.

[213] G. P. Crawford, editor. *Flexible Flat Panel Displays*. Chichester, UK: John Wiley and Sons, 2005.

[214] W. Crookes. "The Bakerian lecture: On the Illumination of Lines of Electrical Pressure, and the Trajectory of Molecules." *Philosophical Transactions of the Royal Society of London* 170 (1879), 135–164.

[215] M. E. Crovella and M. S. Taqqu. "Estimating the Heavy Tail Index from Scaling Properties." *Methodology and Computing in Applied Probability* 1:1 (1999), 55–79.

[216] R. Cruz-Coke. *Colour Blindness — An Evolutionary Approach*. Springfield, IL: Charles C Springfield, 1970.

[217] R. Cucchiara, C. Grana, M. Piccardi, A. Prati, and S. Sirotti. "Improving Shadow Suppression in Moving Object Detection with HSV Color Information." In *IEEE Intelligent Transportation Systems*, pp. 334–339. IEEE Press, 2001.

[218] J. A. Curcio and C. C. Petty. "The Near-Infrared Absorption Spectrum of Liquid Water." *Journal of the Optical Society of America* 41:5 (1951), 302–304.

[219] C. A. Curcio, K. R. Sloan, R. E. Kalina, and A. E. Hendrickson. "Human Photoreceptor Topography." *Journal of Comparative Neurology* 292:4 (1990), 497–523.

[220] C. A. Curcio, K. A. Allen, K. R. Sloan, C. L. Lerea, J. B. Hurley, I. B. Klock, and A. H. Milam. "Distribution and Morphology of Human Cone Photoreceptors Stained with Anti-Blue Opsin." *Journal of Comparative Neurology* 312:4 (1991), 610–624.

[221] D. M. Dacey and B. B. Lee. "The 'Blue-On' Opponent Pathways in Primate Retina Originates from a Distinct Bistratified Ganglion Cell." *Nature* 367:6465 (1994), 731–735.

[222] D. M. Dacey and B. B. Lee. "Cone Inputs to the Receptive Field of Midget Ganglion Cells in the Periphery of the Macaque Retina." *Investigative Ophthalmology and Visual Science, Supplement* 38 (1997), S708.

[223] D. M. Dacey and B. B. Lee. "Functional Architecture of Cone Signal Pathways in the Primate Retina." In *Color Vision: From Genes to Perception*, edited by K. R. Gegenfurtner and L. Sharpe, pp. 181–202. Cambridge, UK: Cambridge University Press, 1999.

[224] D. M. Dacey and M. R. Petersen. "Dendritic Field Size and Morphology of Midget and Parasol Ganglion Cells of the Human Retina." *Proceedings of the National Academy of Sciences of the United States of America* 89:20 (1992), 9666–9670.

[225] D. M. Dacey. "Morphology of a Small-Field Bistratified Ganglion Cell Type in the Macaque and Human Retina." *Visual Neuroscience* 10 (1993), 1081–1098.

[226] D. M. Dacey. "The Mosaic of Midget Ganglion Cells in the Human Retina." *Journal of Neuroscience* 13:12 (1993), 5334–5355.

[227] D. M. Dacey. "Parallel Pathways for Spectral Coding in Primate Retina." *Annual Review of Neuroscience* 23 (2000), 743–775.

[228] S. C. Dakin and P. J. Bex. "Natural Image Statistics Mediate Brightness Filling In." *Proceedings of the Royal Society of London, B* 270 (2003), 2341–2348.

[229] S. Daly. "The Visible Differences Predictor: An Algorithm for the Assessment of Image Fidelity." In *Digital Images and Human Vision*, edited by A. Watson, pp. 179–206. Cambridge, MA: MIT Press, 1993.

[230] A. Damasio, T. Yamada, H. Damasio, J. Corbett, and J. McKnee. "Central Achromatopsia: Behavioral, Anatomic, and Physiologic Aspects." *Neurology* 30:10 (1980), 1064–1071.

[231] K. J. Dana, B. van Ginneken, S. K. Nayar, and J. J. Koenderink. "Reflectance and Texture of Real World Surfaces." *ACM Transactions on Computer Graphics* 18:1 (1999), 1–34.

[232] K. J. Dana. "BRDF/BTF Measurement Device." In *Proceedings of the Eighth IEEE International Conference on Computer Vision (ICCV)*, pp. 460–466. Washington, DC: IEEE, 2001.

[233] J. L. Dannemiller. "Spectral Reflectance of Natural Objects: How Many Basis Functions Are Necessary?" *Journal of the Optical Society of America A* 9:4 (1992), 507–515.

[234] J. Daugman. "High Confidence Visual Recognition of Persons by a Test of Statistical Independence." *IEEE Transactions on Pattern Analysis and Machine Intelligence* 15:11 (1993), 1148–1161.

[235] R. Davis and K. S. Gibson. *Filters for the Reproduction of Sunlight and Daylight and the Determination of Color Temperature*. Bureau of Standards, Miscellaneous. Publication 114, Washington, DC, 1931.

[236] H. Davson. *Physiology of the Eye*, Fifth edition. Pergamon Press, 1990.

[237] E. A. Day, L. Taplin, and R. S. Berns. "Colorimetric Characterization of a Computer-Controlled Liquid Crystal Display." *Color Research and Application* 29:5 (2004), 365–373.

[238] P. Debevec and J. Malik. "Recovering High Dynamic Range Radiance Maps from Photographs." In *Proceedings SIGGRAPH '97, Computer Graphics Proceedings, Annual Conference Series*, pp. 369–378. Reading, MA: Addison-Wesley, 1997.

[239] P. E. Debevec. "Rendering Synthetic Objects into Real Scenes: Bridging Traditional and Image-Based Graphics with Illumination and High Dynamic Range Photography." In *Proceedings of SIGGRAPH '98, Computer Graphics Proceedings, Annual Conference Series*, pp. 45–50. Reading, MA: Addison-Wesley, 1998.

[240] P. E. Debevec. "A Tutorial on Image-Based Lighting." *IEEE Computer Graphics and Applications* 22:2 (2002), 26–34.

[241] S. Decker, R. McGrath, K. Brehmer, and C. Sodini. "A 256 $\times$ 256 CMOS Imaging Array with Wide Dynamic Range Pixels and Column-Parallel Digital Output." In *IEEE International Solid-State Circuits Conference (ISSCC)*, pp. 176–177, 1998.

[242] C. DeCusatis, editor. *Handbook of Applied Photometry*. Philadelphia: American Institute of Physics, 1997.

[243] M. F. Deering. "A Photon Accurate Model of the Human Eye." *ACM Transactions on Graphics* 24:3 (2005), 649–658.

[244] P. Delahunt and D. Brainard. "Control of Chromatic Adaptation Signals from Separate Cone Classes Interact." *Vision Research* 40 (2000), 2885–2903.

[245] F. Delamare and B. Guineau. *Colors: The Story of Dyes and Pigments*. New York: Harry N Abrams, 2000.

[246] E. Demichel. *Le Procédé* 26 (1924), 17–21.

[247] E. Demichel. *Le Procédé* 26 (1924), 26–27.

[248] E. J. Denton. "The Contributions of the Orientated Photosensitive and Other Molecules to the Absorption of the Whole Retina." *Proceedings of the Royal Society of London, B* 150:938 (1959), 78–94.

[249] G. Derra, H. Moensch, E. Fischer, H. Giese, U. Hechtfischer, G. Heusler, A. Koerber, U. Niemann, F.-C. Noertemann, P. Pekarski, J. Pollmann-Retsch, A. Ritz, and U. Weichmann. "UHP Lamp Systems for Projection Applications." *Journal of Physics D: Applied Physics* 38 (2005), 2995–3010.

[250] A. M. Derrington, J. Krauskopf, and P. Lennie. "Chromatic Mechanisms in Lateral Geniculate Nucleus of Macaque." *Journal of Physiology (London)* 357 (1984), 241–265.

[251] K. Devlin, A. Chalmers, and E. Reinhard. "Visual Self-Calibration and Correction for Ambient Illumination." *ACM Transactions on Applied Perception* 3:4 (2006), 429–452.

[252] D. S. Dewald, S. M. Penn, and M. Davis. "Sequential Color Recapture and Dynamic Filtering: A Method of Scrolling Color." *SID Symposium Digest of Technical Papers* 32:1 (2001), 1076–1079.

[253] J. M. DiCarlo and B. A. Wandell. "Rendering High Dynamic Range Images." In *Proceedings of the SPIE 3965 (Electronic Imaging 2000 Conference)*, pp. 392–401. Bellingham, WA: SPIE, 2000.

[254] J. Dillon, R. H. Wang, and S. J. Atherton. "Photochemical and Photophysical Studies on Human Lens Constituents." *Photochemistry and Photobiology* 52 (1990), 849–854.

[255] J. Dillon. "Photochemical Mechanisms in the Lens." In *The Ocular Lens*, edited by H. Maisel, pp. 349–366. New York: Marcel Dekker, 1985.

[256] R. W. Ditchburn. *Eye Movements and Visual Perception*. Oxford, UK: Clarendon Press, 1973.

[257] K. R. Dobkins. "Moving Colors in the Lime Light." *Neuron* 25 (2000), 15–18.

[258] D. Doherty and G. Hewlett. "Pulse Width Modulation Control in DLP Projectors." *Texas Instruments Technical Journal*, pp. 115–121.

[259] D. W. Dong and J. J. Atick. "Statistics of Natural Time-Varying Images." *Network: Computation in Neural Systems* 6:3 (1995), 345–358.

[260] R. W. Doty. "Nongeniculate Afferents to Striate Cortex in Macaques." *Journal of Comparative Neurology* 218:2 (1983), 159–173.

[261] R. F. Dougherty and A. R. Wade. Available online (http://www.vischeck.com/daltonize).

[262] J. E. Dowling. *The Retina: An Approachable Part of the Brain.* Cambridge, MA: Belknap Press, 1987.

[263] F. Drago, W. L. Martens, K. Myszkowski, and H.-P. Seidel. "Perceptual Evaluation of Tone Mapping Operators with Regard to Similarity and Preference." Technical Report MPI-I-2002-4-002, Max Plank Institut für Informatik, Saarbrücken, Germany, 2002.

[264] F. Drago, K. Myszkowski, T. Annen, and N. Chiba. "Adaptive Logarithmic Mapping for Displaying High Contrast Scenes." *Computer Graphics Forum* 22:3 (2003), 419–426.

[265] M. Drew and B. Funt. "Natural Metamers." *Computer Vision, Graphics and Image Processing* 56:2 (1992), 139–151.

[266] M. S. Drew, J. Wei, and Z. N. Li. "Illumination-Invariant Image Retrieval and Video Segmentation." *Pattern Recognition* 32:8 (1999), 1369–1388.

[267] D. D. Dudley. *Holography: A Survey.* Washington, DC: Technology Utilization Office, National Aeronautics and Space Administration, 1973.

[268] R. O. Duncan and G. M. Boynton. "Cortical Magnification within Human Primary Visual Cortex Correlates with Acuity Thresholds." *Neuron* 38:4 (2003), 659–671.

[269] A. Dür. "An Improved normalization for the Ward Reflectance Model." *journal of graphics tools* 11:1 (2006), 51–59.

[270] F. Durand and J. Dorsey. "Fast Bilateral Filtering for the Display of High-Dynamic-Range Images." *ACM Transactions on Graphics* 21:3 (2002), 257–266.

[271] P. Dutré, P. Bekaert, and K. Bala. *Advanced Globel Illumination.* Natick, MA: A K Peters, 2003.

[272] D. M. Eagleman. "Visual Illusions and Neurobiology." *Nature Reviews Neuroscience* 2:12 (2001), 920–926.

[273] F. Ebner and M. D. Fairchild. "Constant Hue Surfaces in Color Space." In *Proceedings of the SPIE 3300*, pp. 107–117. Bellingham, WA: SPIE, 1998.

[274] F. Ebner and M. D. Fairchild. "Development and Testing of a Color Space (IPT) with Improved Hue Uniformity." In *IS&T/SID Sixth Color Imaging Conference: Color Science, Systems and Applications*, pp. 8–13. Springfield, VA: Society for Imaging Science & Technology, 1998.

[275] M. Ebner. *Color Constancy.* Chichester, UK: John Wiley and Sons, Ltd., 2007.

[276] J. G. Eden. "Information Display Early in the 21st Century: Overview of Selected Emissive Display Technologies." *Proceedings of the IEEE* 94:3 (2006), 567–574.

[277] W. Ehrenstein. "Über Abwandlungen der L Hermannschen Heiligkeitserscheinung." *Zeitschrift für Psychologie* 150 (1941), 83–91.

[278] W. Ehrenstein. "Modifications of the Brightness Phenomenon of L Hermann." In *The Perception of Illusory Contours*, edited by S. Petry and G. E. Meyer. New York: Springer-Verlag, 1987. Translated by A Hogg from [277].

[279] A. Einstein. "Ist die Trägheit eines Körpers von seinem Energieinhalt abhängig?" *Annalen der Physik* 18:13 (1905), 639–641.

[280] J. H. Elder. "Are Edges Incomplete?" *International Journal of Computer Vision* 34:2/3 (1999), 97–122.

[281] H. D. Ellis. "Introduction to Aspects of Face Processing: Ten questions in Need of Answers." In *Aspects of Face Processing*, edited by H. Ellis, M. Jeeves, F. Newcombe, and A. Young, pp. 3–13. Dordrecht, The Netherlands: Nijhoff, 1986.

[282] S. A. Engel, X. Zhang, and B. A. Wandell. "Color Tuning in Human Visual Cortex Measured using Functional Magnetic Resonance Imaging." *Nature* 388:6637 (1997), 68–71.

[283] P. G. Engeldrum. "Computing Color Gamuts of Ink-Jet Printing Systems." *Proceedings of the Society for Information Display* 27 (1986), 25–30.

[284] C. Enroth-Cugell and J. G. Robson. "The Contrast Sensitivity of Retinal Ganglion Cells of the Cat." *Journal of Physiology (London)* 187 (1966), 517–552.

[285] C. J. Erkerlers, J. van der Steen, R. M. Steinman, and H. Collewijn. "Ocular Vergence under Natural Conditions I: Continuous Changes of Target Distance along the Median Plane." *Proceedings of the Royal Society of London* B326 (1989), 417–440.

[286] C. J. Erkerlers, R. M. Steinman, and H. Collewijn. "Ocular Vergence under Natural Conditions II: Gaze Shifts between Real Targets Differing in Distance and Direction." *Proceedings of the Royal Society of London* B326 (1989), 441–465.

[287] D. C. van Essen and J. L. Gallant. "Neural Mechanisms of Form and Motion Processing in the Primate Visual System." *Neuron* 13:1 (1994), 1–10.

[288] R. Evans. *The Perception of Color.* New York: John Wiley & Sons, 1974.

[289] G. L. Fain, H. R. Matthews, M. C. Cornwall, and Y. Koutalos. "Adaptation in Vertebrate Photoreceptors." *Physiological Review* 81 (2001), 117–151.

[290] M. D. Fairchild and G. M. Johnson. "On Contrast Sensitivity in an Image Difference Model." In *IS&T PICS Conference*, pp. 18–23. Springfield, VA: Society for Imaging Science and Technology, 2001.

[291] M. D. Fairchild and G. Johnson. "Meet iCAM: A Next-Generation Color Appearance Model." In *Proceedings of the IS&T/SID 10th Color Imaging Conference*, pp. 33–38. Springfield, VA: Society for Imaging Science and Technology, 2002.

[292] M. D. Fairchild and G. M. Johnson. "The iCAM Framework for Image Appearance, Image Differences, and Image Quality." *Journal of Electronic Imaging* 13 (2004), 126–138.

[293] M. D. Fairchild and G. M. Johnson. "METACOW: A Public-Domain, High-Resolution, Fully-Digital, Extended-Dynamic-Range, Spectral Test Target for Imaging System Analysis and Simulation." In *IS&T/SID 12th Color Imaging Conference*, pp. 239–245. Springfield, VA: Society for Imaging Science and Technology, 2004.

[294] M. D. Fairchild and G. M. Johnson. "On the Salience of Novel Stimuli: Adaptation and Image Noise." In *IS&T/SID 13th Color Imaging Conference*, pp. 333–338. Springfield, VA: Society for Imaging Science and Technology, 2005.

[295] M. D. Fairchild and E. Pirrotta. "Predicting the Lightness of Chromatic Object Colors using CIELAB." *Color Research and Application* 16 (1991), 385–393.

[296] M. D. Fairchild and L. Reniff. "Time-Course of Chromatic Adaptation for Color-Appearance Judgements." *Journal of the Optical Society of America* 12 (1995), 824–833.

[297] M. D. Fairchild and D. R. Wyble. "Colorimetric Characterization of the Apple Studio Display (Flat Panel LCD)." Technical report, Munsell Color Science Laboratory, Rochester, NY, 1998.

[298] M. D. Fairchild, E. Pirrotta, and T. G. Kim. "Successive-Ganzfeld Haploscopic Viewing Technique for Color-Appearance Research." *Color Research and Application* 19 (1994), 214–221.

[299] M. D. Fairchild. "Formulation and Testing of an Incomplete-Chromatic-Adaptation Model." *Color Research and Application* 16 (1991), 243–250.

[300] M. D. Fairchild. "A Model of Incomplete Chromatic Adaptation." In *Proceedings 22nd Session of the CIE*, pp. 33–34. Vienna: Commission Internationale de l'Eclairage, 1991.

[301] M. D. Fairchild. "A Revision of CIECAM97s for Practical Applications." *Color Research and Application* 26 (2001), 418–427.

[302] M. D. Fairchild. *Color Appearance Models*, Second edition. Chichester, UK: John Wiley and Sons, Ltd., 2005.

[303] M. D. Fairchild. "The HDR Photographic Survey." In *Proceedings of the 15th Color Imaging Conference: Color Science and Engineering Systems, Technologies, and Applications*, pp. 233–238. Albuquerque, NM, 2007.

[304] H. Fairman, M. Brill, and H. Hemmendinger. "How the CIE 1931 Color-Matching Functions Were Derived from Wright-Guild Data." *Color Research and Application* 22 (1997), 11–23.

[305] R. J. Farrell and J. M. Booth. *Design Handbook for Imagery Interpretation Equipment, Section 3.2.* Seattle: Boeing Aerospace Co., 1984.

[306] R. Fattal, D. Lischinski, and M. Werman. "Gradient Domain High Dynamic Range Compression." *ACM Transactions on Graphics* 21:3 (2002), 249–256.

[307] R. Fedkiw, J. Stam, and H. W. Jensen. "Visual Simulation of Smoke." In *Proceedings of SIGGRAPH 2001, Computer Graphics Proceedings, Annual Conference Series*, edited by E. Fiume, pp. 15–22. Reading, MA: Addison-Wesley, 2001.

[308] D. J. Felleman and D. C. van Essen. "Receptive Field Properties of Neurons in Area V3 of Macaque Monkey Extrastriate Cortex." *Journal of Neurophysiology* 57:4 (1987), 889–920.

[309] R. Fergus, B. Singh, A. Hertzmann, S. T. Roweis, and W. T. Freeman. "Removing Camera Shake from a Single Photograph." *ACM Transactions on Graphics* 25:3 (2006), 787–794.

[310] J. A. Ferwerda, S. Pattanaik, P. Shirley, and D. P. Greenberg. "A Model of Visual Adaptation for Realistic Image Synthesis." In *Proceedings of SIGGRAPH 96, Computer Graphics Proceedings, Annual Conference Series*, edited by H. Rushmeier, pp. 249–258. Reading, MA: Addison-Wesley, 1996.

[311] J. A. Ferwerda, S. N. Pattanaik, P. Shirley, and D. P. Greenberg. "A Model of Visual Masking for Computer Graphics." In *Proceedings of SIGGRAPH '97, Computer Graphics Proceedings, Annual Conference Series*, pp. 143–152. Reading, MA: Addison-Wesley, 1997.

[312] J. A. Ferwerda. "Elements of Early Vision for Computer Graphics." *IEEE Computer Graphics and Applications* 21:5 (2001), 22–33.

[313] D. J. Field and N. Brady. "Visual Sensitivity, Blur and the Sources of Variability in the Amplitude Spectra of Natural Scenes." *Vision Research* 37:23 (1997), 3367–3383.

[314] D. J. Field. "Relations between the Statistics of Natural Images and the Response Properties of Cortical Cells." *Journal of the Optical Society of America A* 4:12 (1987), 2379–2394.

[315] D. J. Field. "Scale-Invariance and Self-Similar 'Wavelet' Transforms: An Analysis of Natural Scenes and Mammalian Visual Systems." In *Wavelets, Fractals and Fourier Transforms*, edited by M. Farge, J. C. R. Hunt, and J. C. Vassilicos, pp. 151–193. Oxford, UK: Clarendon Press, 1993.

[316] E. F. Fincham. "Defects of the Colour-Sense Mechanism as Indicated by the Accommodation Reflect." *Journal of Physiology* 121 (1953), 570–580.

[317] G. D. Finlayson and S. D. Hordley. "Color Constancy at a Pixel." *Journal of the Optical Society of America A* 18:2 (2001), 253–264.

[318] G. D. Finlayson and S. Süsstrunk. "Color Ratios and Chromatic Adaptation." In *Proceedings of IS&T CGIV, First European Conference on Color Graphics, Imaging and Vision*, pp. 7–10. Springfield, VA: Society for Imaging Science and Technology, 2002.

[319] G. D. Finlayson, S. D. Hordley, and P. M. Hubel. "Color by Correlation: A Simple, Unifying Framework for Color Constancy." *IEEE Transactions on Patterns Analysis and Machine Intelligence* 23:11 (2001), 1209–1221.

[320] G. D. Finlayson, S. D. Hordley, and M. S. Drew. "Removing Shadows from Images using Retinex." In *Proceedings IS&T Color Imaging Conference*, pp. 73–79. Springfield, VA: Society for Imaging Science and Technology, 2002.

[321] A. Fiorentini and A. M. Ercoles. "Vision of Oscillating Visual Fields." *Optica Acta* 4 (1957), 370.

[322] D. Fitzpatrick, K. Itoh, and I. T. Diamond. "The Laminar Organization of the Lateral Geniculate Body and the Striate Cortex in the Squirrel Monkey (*Saimiri sciureus*)." *Journal of Neuroscience* 3:4 (1983), 673–702.

[323] D. Fitzpatrick, W. M. Usrey, B. R. Schofield, and G. Einstein. "The Sublaminar Organization of Corticogeniculate Neurons in Layer 6 of Macaque Striate Cortex." *Visual Neuroscience* 11 (1994), 307–315.

[324] P. D. Floyd, D. Heald, B. Arbuckle, A. Lewis, M. Kothari, B. J. Gally, B. Cummings, B. R. Natarajan, L. Palmateer, J. Bos, D. Chang, J. Chiang, D. Chu, L.-M. Wang, E. Pao, F. Su, V. Huang, W.-J. Lin, W.-C. Tang, J.-J. Yeh, C.-C. Chan, F.-A. Shu, and Y.-D. Ju. "IMOD Display Manufacturing." *SID Symposium Digest of Technical Papers* 37:1 (2006), 1980–1983.

[325] J. M. Foley and M. E. McCourt. "Visual Grating Induction." *Journal of the Optical Society of America A* 2 (1985), 1220–1230.

[326] J. Foley, A. van Dam, S. Feiner, and J. Hughes. *Computer Graphics Principles and Practice*, Second edition. Reading, MA: Addison-Wesley, 1990.

[327] A. Ford and A. Roberts. "Colour Space Conversions.", 1998. Available online (http://www.poynton.com/PFDs/coloureq.pdf).

[328] S. Forrest, P. Burrows, and M. Thompson. "The Dawn of Organic Electronics." *IEEE Spectrum* 37:8 (2000), 29–34.

[329] J. Forrester, A. Dick, P. McMenamin, and W. Lee. *The Eye: Basic Sciences in Practice*. London: W. B. Saunders Company Ltd., 2001.

[330] E. R. Fossum. "CMOS Image Sensors: Electronics Camera-On-A-Chip." *IEEE Transactions on Electron Devices* 44:10 (1997), 1689–1698.

[331] D. H. Foster and S. M. C. Nascimento. "Relational Color Constancy from Invariant Cone-Excitation Ratios." *Proceedings of the Royal Society: Biological Sciences* 257:1349 (1994), 115–121.

[332] J. Frankle and J. McCann. "Method and Apparatus for Lightness Imaging.", 1983. U.S. patent #4,384,336.

[333] W. Fries and H. Distel. "Large Layer V1 Neurons of Monkey Striate Cortex (Meynert Cells) Project to the Superior Colliculus." *Proceedings of the Royal Society of London B, Biological Sciences* 219 (1983), 53–59.

[334] W. Fries. "Pontine Projection from Striate and Prestriate Visual Cortex in the Macaque Monkey: An Anterograde Study." *Visual Neuroscience* 4 (1990), 205–216.

[335] K. Fritsche. *Faults in Photography*. London: Focal Press, 1968.

[336] G. A. Fry and M. Alpern. "The Effect of a Peripheral Glare Source upon the Apparent Brightness of an Object." *Journal of the Optical Society of America* 43:3 (1953), 189–195.

[337] G. D. Funka-Lea. "The Visual Recognition of Shadows by an Active Observer." Ph.D. thesis, University of Pennsylvania, 1994.

[338] B. V. Funt and G. Finlayson. "Color Constant Color Indexing." *IEEE Transactions on Pattern Analysis and Machine Intelligence* 17 (1995), 522–529.

[339] B. V. Funt, M. S. Drew, and J. Ho. "Color Constancy from Mutual Reflection." *International Journal of Computer Vision* 6 (1991), 5–24.

[340] B. Funt, F. Ciurea, and J. McCann. "Retinex in Matlab." In *Proceedings of the IS&T/SID Eighth Color Imaging Conference: Color Science, Systems and Applications*, pp. 112–121. Springfield, VA: Society for Imaging Science and Technology, 2000.

[341] M. Furuta, Y. Nishikawa, T. Inoue, and S. Kawahito. "A High-Speed, High-Sensitivity Digital CMOS Image Sensor with a Global Shutter and 12-Bit Column-Parallel Cyclic A/D Converters." *IEEE Journal of Solid-State Circuits* 42:4 (2007), 766–774.

[342] D. Gabor. "A New Microscopic Principle." *Nature* 161 (1948), 777–778.

[343] D. Gabor. "Wavefront Reconstruction." In *Optical and Acoustical Holography*, edited by E. Camatini, pp. 15–21. New York: Plenum Press, 1972.

[344] E. R. Gaillard, L. Zheng, J. C. Merriam, and J. Dillon. "Age-Related Changes in the Absorption Characteristics of the Primate Lens." *Investigative Ophthalmology and Visual Science* 41:6 (2000), 1454–1459.

[345] L. Gall. "Computer Color Matching." In *Colour 73: Second Congress International Colour Association*, pp. 153–178. London: Adam Hilger, 1973.

[346] L. Ganglof, E. Minoux, K. B. K. Teo, P. Vincent, V. T. Semet, V. T. Binh, M. H. Yang, I. Y. Y. Bu, R. G. Lacerda, G. Pirio, J. P. Schnell, D. Pribat, D. G. Hasko, G. A. J. Amaratunga, W. Milne, and P. Legagneux. "Self-Aligned, Gated Arrays of Individual Nanotube and Nanowire Emitters." *Nano Letters* 4 (2004), 1575–1579.

[347] M. J. Gastinger, J. J. O'Brien, J. N. B. Larsen, and D. W. Marshak. "Histamine Immunoreactive Axons in the Macaque Retina." *Investigative Ophthalmology and Visual Science* 40:2 (1999), 487–495.

[348] H. Gates, R. Zehner, H. Doshi, and J. Au. "A5 Sized Electronic Paper Display for Document Viewing." *SID Symposium Digest of Technical Papers* 36:1 (2005), 1214–1217.

[349] F. Gatti, A. Acquaviva, L. Benini, and B. Ricco. "Low Power Control Techniques for TFT-LCD Displays." In *Proceedings of the International Conference on Compilers, Architecture, and Synthesis for Embedded Systems*, pp. 218–224, 2002.

[350] I. Gauthier, M. J. Tarr, A. W. Anderson, P. Skudlarski, and J. C. Gore. "Activation of the Middle Fusiform 'Face Area' Increases with Expertise in Recognizing Novel Objects." *Nature Neuroscience* 2:6 (1999), 568–573.

[351] W. J. Geeraets and E. R. Berry. "Ocular Spectral Characteristics as Related to Hazards from Lasers and Other Sources." *American Journal of Ophthalmology* 66:1 (1968), 15–20.

[352] K. Gegenfurtner and D. C. Kiper. "Color Vision." *Annual Review of Neuroscience* 26 (2003), 181–206.

[353] K. R. Gegenfurtner and L. Sharpe, editors. *Color Vision: From Genes to Perception*. Cambridge, UK: Cambridge University Press, 1999.

[354] J. Geier, L. Sera, and L. Bernath. "Stopping the Hermann Grid Illusion by Simple Sine Distortion." *Perception, ECVP 2004 supplement* 33 (2004), 53.

[355] W. S. Geisler and M. S. Banks. "Visual Performance." In *Handbook of Optics: Fundamentals, Techniques and Design, Volume 1*, edited by M. Bass, E. W. van Stryland, D. R. Williams, and W. L. Wolfe, Second edition. New York: McGraw-Hill, 1995.

[356] N. George, R. J. Dolan, G. R. Fink, G. C. Baylis, C. Russell, and J. Driver. "Contrast Polarity and Face Recognition in the Human Fusiform Gyrus." *Nature Neuroscience* 2:6 (1999), 574–580.

[357] T. Gevers and H. Stokman. "Classifying Color Edges in Video into Shadow-Geometry, Highlight, or Material Transitions." *IEEE Transactions on Multimedia* 5:2 (2003), 237–243.

[358] K. S. Gibson. "The Relative Visibility Function." CIE Compte Rendu des Séances, Sixième Session, Genève, 1924.

[359] J. E. Gibson and M. D. Fairchild. "Colorimetric Characterization of Three Computer Displays (LCD and CRT)." Technical report, Munsell Color Science Laboratory, Rochester, NY, 2000.

[360] K. S. Gibson and E. P. T. Tyndall. "Visibility of Radiant Energy." *Scientific Papers of the Bureau of Standards* 19 (1923), 131–191.

[361] I. M. Gibson. "Visual Mechanisms in a Cone Monochromat." *Journal of Physiology* 161 (1962), 10–11.

[362] A. Gilchrist and J. Cataliotti. "Anchoring of Surface Lightness with Multiple Illumination Levels." *Investigative Ophthalmology and Visual Science* 35:4 (1994), 2165–2165.

[363] A. Gilchrist, C. Kossyfidis, F. Bonato, T. Agostini, J. Cataliotti, X. Li, B. Spehar, V. Annan, and E. Economou. "An Anchoring Theory of Lightness Perception." *Psychological Review* 106:4 (1999), 795–834.

[364] A. L. Gilchrist. "Lightness Contrast and Failures of Constancy: A Common Explanation." *Perception and Psychophysics* 43:5 (1988), 415–424.

[365] A. L. Gilchrist. "The Importance of Errors in Perception." In *Colour Perception: Mind and the Physical World*, edited by R. M. D. Heyer, pp. 437–452. Oxford, UK: Oxford University Press, 2003.

[366] A. L. Gilchrist. "Lightness Perception: Seeing One Color through Another." *Current Biology* 15:9 (2005), R330–R332.

[367] A. Glasser and M. C. W. Campbell. "Presbyopia and the Optical Changes in the Human Crystalline Lens with Age." *Vision Research* 38 (1998), 209–229.

[368] J. Glasser. "Principles of Display Measurement and Calibration." In *Display Systems: Design and Applications*, edited by L. W. MacDonald and A. C. Lowe. Chichester: John Wiley and Sons, 1997.

[369] A. S. Glassner. "How to Derive a Spectrum from an RGB Triplet." *IEEE Computer Graphics and Applications* 9:4 (1989), 95–99.

[370] A. S. Glassner. *Principles of Digital Image Synthesis*. San Fransisco, CA: Morgan Kaufmann, 1995.

[371] A. Glassner. "Computer-Generated Solar Halos and Sun Dogs." *IEEE Computer Graphics and Applications* 16:2 (1996), 77–81.

[372] A. Glassner. "Solar Halos and Sun Dogs." *IEEE Computer Graphics and Applications* 16:1 (1996), 83–87.

[373] J. W. von Goethe. *Zur Farbenlehre*. Tübingen, Germany: Freies Geistesleben, 1810.

[374] N. Goldberg. *Camera Technology: The Dark Side of the Lens*. Boston: Academic Press, Inc., 1992.

[375] S. Gong, J. Kanicki, G. Xu, and J. Z. Z. Zhong. "A Novel Structure to Improve the Viewing Angle Characteristics of Twisted-Nematic Liquid Crystal Displays." *Japanese Journal of Applied Physics* 48 (1999), 4110–4116.

[376] R. C. Gonzales and R. E. Woods. *Digital Image Processing*, Second edition. Upper Saddle River, NJ: Prentice-Hall, 2002.

[377] J.-C. Gonzato and S. Marchand. "Photo-Realistic Simulation and Rendering of Halos." In *Winter School of Computer Graphics (WSCG '01) Proceedings*. Plzen, Czech Republic: University of West Bohemia, 2001.

[378] A. Gooch, S. C. Olsen, J. Tumblin, and B. Gooch. "Color2Gray: Salience-Preserving Color Removal." *ACM Transactions on Graphics* 24:3 (2005), 634–639.

[379] R. H. Good, Jr and T. J. Nelson. *Classical Theory of Electric and Magnetic Fields*. New York: Academic Press, 1971.

[380] C. M. Goral, K. E. Torrance, D. P. Greenberg, and B. Battaile. "Modeling the Interaction of Light Between Diffuse Surfaces." *Computer Graphics (SIGGRAPH '84 Proceedings)* 18:3 (1984), 213–222.

[381] S. J. Gortler, R. Grzeszczuk, R. Szeliski, and M. F. Cohen. "The Lumigraph." In *Proceedings SIGGRAPH '96, Computer Graphics Proceedings, Annual Conference Series*, pp. 43 –54. Reading, MA: Addison-Wesley, 1996.

[382] P. Gouras. "Identification of Cone Mechanisms in Monkey Ganglion Cells." *Journal of Physiology (London)* 199 (1968), 533–547.

[383] C. F. O. Graeff, G. B. Silva, F. Nüesch, and L. Zuppiroli. "Transport and Recombination in Organic Light-Emitting Diodes Studied by Electrically Detected Magnetic Resonance." *European Physical Journal E* 18 (2005), 21–28.

[384] J. Graham. "Some Topographical Connections of the Striate Cortex with Subcortical Structures in Macaca Fascicularis." *Experimental Brain Research* 47 (1982), 1–14.

[385] T. Granlund, L. A. A. Petterson, M. R. Anderson, and O. Inganäs. "Interference Phenomenon Determines the Color in an Organic Light Emitting Diode." *Journal of Applied Physics* 81:12 (1997), 8097–8104.

[386] H. Grassmann. "Zur Theorie der Farbernmischung." *Ann. Phys. Chem.* 89 (1853), 69–84.

[387] R. Greenler. *Rainbows, Halos, and Glories*. Cambridge, UK: Cambridge University Press, 1980.

[388] R. L. Gregory and P. F. Heard. "Border Locking and the Café Wall Illusion." *Perception* 8:4 (1979), 365–380.

[389] R. L. Gregory and P. F. Heard. "Visual Dissociations of Movement, Position, and Stereo Depth: Some Phenomenal Phenomena." *Quarterly Journal of Experimental Psychology* 35A (1983), 217–237.

[390] J. Greivenkamp. "Color Dependent Optical Filter for the Suppression of Aliasing Artifacts." *Applied Optics* 29:5 (1990), 676–684.

[391] L. D. Griffin. "Optimality of the Basic Color Categories for Classification." *Journal of the Royal Society, Interface* 3:6 (2006), 71–85.

[392] K. Grill-Spector and R. Malach. "The Human Visual Cortex." *Annual Review of Neuroscience* 27 (2004), 647–677.

[393] K. Grill-Spector, T. Kushnir, S. Edelman, Y. Itzchak, and R. Malach. "Cue-Invariant Activation in Object-Related Areas of the Human Occipital Lobe." *Neuron* 21:1 (1998), 191–202.

[394] K. Grill-Spector, T. Kushnir, T. Hendler, S. Edelman, Y. Itzchak, and R. Malach. "A Sequence of Object-Processing Stages Revealed by fMRI in the Human Occipital Lobe." *Human Brain Mapping* 6:4 (1998), 316–328.

[395] S. Grossberg and E. Mingolla. "Neural Dynamics of Form Perception: Boundary Adaptation, Illusory Figures, and Neon Color Spreading." *Psychological Review* 92 (1985), 173–211.

[396] S. Grossberg and J. Todorović. "Neural Dynamics of 1-D and 2-D Brightness Perception: A Unified Model of Classical and Recent Phenomena." *Perception and Psychophysics* 43 (1988), 241–277.

[397] F. Grum and R. J. Becherer. *Optical Radiation Measurements*. New York: Academic Press, Inc., 1979.

[398] C. Gu and P. Yeh. "Extended Jones Matrix Method. II." *Journal of the Optical Society of America A* 10:5 (1993), 966–973.

[399] G. Gu, V. Boluvić, P. E. Burrows, S. R. Forrest, and M. E. Thompson. "Transparent Organic Light Emitting Devices." *Applied Physics Letters* 68:19 (1996), 2606–2608.

[400] G. Gu, P. E. Burrows, S. Venkatesh, S. R. Forrest, and M. E. Thompson. "Vacuum-Deposited, Nonpolymeric Flexible Organic Light-Emitting Devices." *Optics Letters* 22:3 (1997), 172–174.

[401] G. Gu, G. Parthasarathy, P. E. Burrows, P. Tian, I. G. Hill, A. Kahn, and S. R. Forrest. "Transparent Stacked Organic Light Emitting Devices. I. Design Principles and Transparent Compound Electrodes." *Journal of Applied Physics* 86:8 (1999), 4067–4075.

[402] R. W. Gubbisch. "Optical Performance of the Eye." *Journal of Physiology* 186 (1966), 558–578.

[403] C. Gueymard, D. Myers, and K. Emery. "Proposed Reference Irradiance Spectra for Solar Energy Systems Testing." *Solar Energy* 73:6 (2002), 443–467.

[404] C. Gueymard. "The Sun's Total and Spectral Irradiance for Solar Energy Applications and Solar Radiation Models." *Solar Energy* 76:4 (2004), 423–453.

[405] J. Guild. "The Colorimetric Properties of the Spectrum." *Phil. Trans. Roy. Soc. A* 230 (1931), 149–187.

[406] R. W. Guillery and M. Colonnier. "Synaptic Patterns in the Dorsal Lateral Geniculate Nucleus of Cat and Monkey: A Brief Review." *Zeitschrift für Zellforschung und mikroskopische Anatomie* 103 (1970), 90–108.

[407] B. Gulyas, C. A. Heywood, D. A. Popplewell, P. E. Roland, and A. Cowey. "Visual Form Discrimination from Color or Motion Cues: Functional Anatomy by Positron Emission Tomography." *Proceedings of the National Academy of Sciences of the United States of America* 91:21 (1994), 9965–9969.

[408] M. Gur, I. Kagan, and D. M. Snodderly. "Orientation and Direction Selectivity of Neurons in V1 of Alert Monkeys: Functional Relationships and Laminar Distributions." *Cerebral Cortex* 15:8 (2005), 1207–1221.

[409] C. Gutierrez and C. G. Cusick. "Area V1 in Macaque Monkeys Projects to Multiple Histochemically Defined Subdivisions of the Inferior Pulvinar Complex." *Brain Research* 765 (1997), 349–356.

[410] D. Gutierrez, F. J. Seron, A. Munoz, and O. Anson. "Chasing the Green Flash: A Global Illumination Solution for Inhomogeneous Media." In *Proceedings of the Spring Conference on Computer Graphics*, pp. 95–103. New York: ACM Press, 2004.

[411] D. Gutierrez, A. Munoz, O. Anson, and F. J. Seron. "Non-Linear Volume Photon Mapping." In *Rendering Techniques 2005: Eurographics Symposium on Rendering*, edited by K. Bala and P. Dutré, pp. 291–300. Aire-la-Ville, Switzerland: Eurographics Association, 2005.

[412] C. S. Haase and G. W. Meyer. "Modeling Pigmented Materials for Realistic Image Synthesis." *ACM Transactions on Graphics* 11:4 (1992), 305–335.

[413] J. Haber, M. Magnor, and H.-P. Seidel. "Physically-Based Simulation of Twilight Phenomena." *ACM Transactions on Graphics* 24:4 (2005), 1353–1373.

[414] R. A. Hall and D. P. Greenberg. "A Testbed for Realistic Image Synthesis." *IEEE Computer Graphics and Applications* 3:6 (1983), 10–20.

[415] R. A. Hall. *Illumination and Color in Computer Generated Imagery*. Monographs in Visual Communication, New York: Springer-Verlag, 1989.

[416] R. A. Hall. "Comparing Spectral Color Computation Methods." *IEEE Computer Graphics and Applications* 19:4 (1999), 36–45.

[417] P. E. Hallet. "The Variations in Visual Threshold Measurement." *Journal of Physiology* 202 (1969), 403–419.

[418] M. Halstead, B. Barsky, S. Klein, and R. Mandell. "Reconstructing Curved Surfaces from Specular Reflection Patterns using Spline Surface Fitting of Normals." In *Proceedings of SIGGRAPH 96, Computer Graphics Proceedings, Annual Conference Series*, edited by H. Rushmeier, pp. 335–342. Reading, MA: Addison-Wesley, 1996.

[419] A. Hanazawa, H. Komatsu, and I. Murakami. "Neural Selectivity for Hue and Saturation of Colour in the Primary Visual Cortex of the Monkey." *European Journal of Neuroscience* 12:5 (2000), 1753–1763.

[420] P. J. B. Hancock, R. J. Baddeley, and L. S. Smith. "The Principle Components of Natural Images." *Network* 3 (1992), 61–70.

[421] J. A. Hanley and B. J. McNeil. "The Meaning and Use of the Area under the Receiver Operating Characteristic (ROC) Curve." *Radiology* 143:1 (1982), 29–36.

[422] J. H. Hannay. "The Clausius-Mossotti Equation: An Alternative Derivation." *European Journal of Physics* 4 (1983), 141–143.

[423] A. Hård and L. Sivik. "NCS — Natural Color System: A Swedish Standard for Color Notation." *Colour Research and Application* 6 (1981), 129–138.

[424] A. Hård, L. Sivik, and G. Tonnquist. "NCS, Natural Color System — From Concepts to Research and Applications. Part I." *Colour Research and Application* 21 (1996), 180–205.

[425] A. Hård, L. Sivik, and G. Tonnquist. "NCS, Natural Color System — From Concepts to Research and Applications. Part II." *Colour Research and Application* 21 (1996), 206–220.

[426] A. C. Hardy and F. L. Wurzburg, Jr. "Color Correction in Color Printing." *Journal of the Optical Society of America* 38:4 (1948), 300–307.

[427] P. Hariharan. *Basics of Holography*. Cambridge, UK: Cambridge University Press, 2002.

[428] H. F. Harmuth, R. N. Boules, and M. G. M. Hussain. *Electromagnetic Dignals: Reflection, Focussing, Distortion, and their Practical Applications*. New York: Kluwer Academic / Plenum Publishers, 1999.

[429] F. J. Harris. "On the Use of Windows for Harmonic Analysis with the Discrete Fourier Transform." *Proceedings of the IEEE* 66:1 (1978), 51–84.

[430] H. K. Hartline. "The Response of Single Optic Nerve Fibers of the Vertebrate Eye to Illumination of the Retina." *American Journal of Physiology* 121 (1938), 400–415.

[431] E. I. Haskal, M. Buechel, J. F. Dijksman, P. C. Duineveld, E. A. Meulenkamp, C. A. H. A. Mutsaers, A. Sempel, P. Snijder, S. I. E. Vulto, P. van de Weijer, and S. H. P. M. de Winter. "Ink jet Printing of Passive Matrix Polymer Light-Emitting Displays." *SID Symposium Digest of Technical Papers* 33:1 (2002), 776–779.

[432] U. Hasson, M. Harel, I. Levy, and R. Malach. "Large-Scale Mirror-Symmetry Organization of Human Occipito-Temporal Object Areas." *Neuron* 37:6 (2003), 1027–1041.

[433] J. A. van Hateren and A. van der Schaaf. "Independent Component Filters of Natural Images Compared with Simple Cells in Primary Visual Cortex." *Proceedings of the Royal Society of London B* 265 (1998), 359–366.

[434] J. H. van Hateren. "A Cellular and Molecular Model of Response Kinetics and Adaptation in Primate Cones and Horizontal Cells." *Journal of Vision* 5 (2005), 331–347.

[435] J. H. van Hateren. "Encoding of High Dynamic Range Video with a Model of Human Cones." *ACM Transactions on Graphics* 25:4 (2006), 1380–1399.

[436] W. Hauser. *Introduction to the Principles of Electromagnetism.* Reading: Addison-Wesley, 1971.

[437] V. Havran. "Heuristic Ray Shooting Algorithms." Ph.D. thesis, Department of Computer Science and Engineering, Faculty of Electrical Engineering, Czech Technical University in Prague, 2000.

[438] M. J. Hawken, A. J. Parker, and J. S. Lund. "Laminar Organization and Contrast Sensitivity of Direction-Selective Cells in the Striate Cortex of the Old World Monkey." *Journal of Neuroscience* 8:10 (1988), 3541–3548.

[439] G. E. Healey and R. Kondepudy. "Radiometric CCD Camera Calibration and Noise Estimation." *IEEE Transactions on Pattern Analysis and Machine Intelligence* 16:3 (1994), 267–276.

[440] G. Healey and D. Slater. "Global Color Constancy: Recognition of Objects by Use of Illumination-Invariant Properties of Color Distributions." *Journal of the Optical Society of America A* 11:11 (1994), 3003–3010.

[441] G. Healey. "Estimating Spectral Reflectance using Highlights." *Image Vision Computing* 9:5 (1991), 333–337.

[442] G. Healey. "Segmenting Images using Normalized Color." *IEEE Transactions on Systems, Man, and Cybernetics* 22 (1992), 64–73.

[443] S. Hecht and J. Mandelbaum. "The Relation between Vitamin A and Dark Adaptation." *Journal of the American Medical Association* 112 (1939), 1910–1916.

[444] S. Hecht, C. Haig, and G. Wald. "The Dark Adaptation of Retinal Fields of Different Size and Location." *Journal of General Physiology* 19:2 (1935), 321–327.

[445] S. Hecht, C. Haig, and A. M. Chase. "The Influence of Light Adaptation on Subsequent Dark Adaptation of the Eye." *Journal of General Physiology* 20:6 (1937), 831–850.

[446] S. Hecht, S. Schlaer, and M. H. Pirenne. "Energy, Quanta and Vision." *Journal of General Physiology* 25 (1942), 819–840.

[447] E. Hecht. *Optics*, Fourth edition. San Francisco: Addison-Wesley, 2002.

[448] R. Heckaman. "Talking about Color...Brilliance." *Color Research and Application* 30 (2005), 250–151.

[449] E. R. Heider and D. C. Olivier. "The Structure of the Color Space in Naming and Memory for Two Languages." *Cognitive Psychology* 3 (1972), 337–354.

[450] E. Heider. "Universals in Color Naming and Memory." *Journal of Experimental Psychology* 93:1 (1972), 10–20.

[451] W. Helfrich and W. G. Schneider. "Recombination Radiation in Anthracene Crystals." *Physical Review Letters* 14:7 (1965), 229–231.

[452] W. Helfrich and W. G. Schneider. "Transients of vVolume-cControlled Current and of Recombination Radiation in Anthracene." *The Journal of Chemical Physics* 44:8 (1966), 2902–2909.

[453] H. L. F. von Helmholtz. *Handbuch der physiologischen Optik.* Leipzig, Germany: Leopold Voss, 1867.

[454] H. Helson and D. B. Judd. "An Experimental and Theoretical Study of Changes in Surface Colors under Changing Illuminations." *Psychological Bulletin* 33 (1936), 740–741.

[455] H. Helson. "Fundamental Problmes in Color vision I. The Principle Governing Changes in Hue, Saturation, and Lightness of Non-Selective Samples in Chromatic Illuminantion." *Journal of Experimental Psychology* 23 (1938), 439–477.

[456] H. Helson. *Adaptation-Level Theory.* New York: Harper and Row, 1964.

[457] S. H. C. Hendry and D. J. Calkins. "Neuronal Chemistry and Functional Organization in the Primate Visual System." *Trends in Neuroscience* 21:8 (1998), 344–349.

[458] S. H. C. Hendry and R. C. Reid. "The Koniocellular Pathway in Primate Vision." *Annual Review of Neuroscience* 23 (2000), 127–153.

[459] S. H. Hendry and T. Yoshioka. "A Neurochemically Distinct Third Channel in the Macaque Dorsal Lateral Geniculate Nucleus." *Science* 264 (1994), 575–577.

[460] L. Henyey and J. Greenstein. "Diffuse Radiation in the Galaxy." *Astrophysics Journal* 93 (1941), 70–83.

[461] E. Hering. "Zur Lehre vom Lichtsinne. IV. Über die sogenannte Intensität der Lichtempfindung und über die Empfindung des Schwarzen." *Sitzungsberichte / Akademie der Wissenschaften in Wien, Mathematisch-Naturwissenschaftliche Klasse Abteilung III, Anatomie und Physiology des Menschen und der Tiere sowie theoretische Medizin* 69 (1874), 85–104.

[462] E. Hering. *Outlines of a Theory of the Light Sense (*Translation from German*: Zur Lehre vom Lichtsinne, 1878).* Cambridge, MA: Harvard University Press, 1920.

[463] L. Hermann. "Eine Erscheinung simultanen Contrastes." *Pflügers Archiv für die gesamte Physiologie des Menschen und Tiere* 3 (1870), 13–15.

[464] E. W. Herold. "History and Development of the Color Picture Tube." *Proceedings of the Society for Information Display* 15:4 (1974), 141–149.

[465] R. D. Hersch, F. Collaud, and P. Emmel. "Reproducing Color Images with Embedded Metallic Patterns." *ACM Transactions on Graphics* 22:3 (2003), 427–434.

[466] M. Herzberger. *Modern Geometrical Optics.* New York: Interscience, 1958.

[467] P. G. Herzog. "Analytical Color Gamut Representations." *Journal for Imaging Science and Technology* 40 (1996), 516–521.

[468] P. G. Herzog. "Further Developments of the Analytical Color Gamut Representations." In *Proceedings of the SPIE 3300*, pp. 118–128. Bellingham, WA: SPIE, 1998.

[469] M. Hess and M. Wiegner. "COP: A Data Library of Optical Properties of Hexagonal Ice Crystals." *Applied Optics* 33 (1994), 7740–7749.

[470] S. Hesselgren. *Hesselgrens färgatla med kortfattad färglära.* Stockholm, Sweden: T Palmer, AB, 1952.

[471] D. Hideaki, H. Y. K. Yukio, and S. Masataka. "Image Data Processing Apparatus for Processing Combined Image Signals in order to Extend Dynamic Range." In *Image Sensor. 8223491, Japanese Patent*, 1996.

[472] B. M. Hill. "A Simple General Approach to Inference about the Tail of a Distribution." *The Annals of Statistics* 3:5 (1975), 1163–1174.

[473] J. W. Hittorf. "Über die Electricitätsleitung der Gase. Erste Mitteilungen." *Annalen der Physik und Chemie* 136 (1869), 1–31, 197–234.

[474] J. Ho, B. V. Funt, and M. S. Drew. "Separating a Color Signal into Illumination and Surface Reflectance Components: Theory and Applications." *IEEE Transactions on Patterns Analysis and Machine Intelligence* 12:10 (1990), 966–977.

[475] B. Hoefflinger, editor. *High-Dynamic-Range (HDR) Vision: Microelectronics, Image Processing, Computer Graphics.* Springer Series in Advanced Microelectronics, Berlin: Springer, 2007.

[476] H. Hofer, B. Singer, and D. R. Williams. "Different Sensations from Cones with the Same Photopigment." *Journal of Vision* 5 (2005), 444–454.

[477] A. Hohmann and C. von der Malsburg. "McCollough Effect and Eye Optics." *Perception* 7 (1978), 551–555.

[478] G. Hollemann, B. Braun, P. Heist, J. Symanowski, U. Krause, J. Kränert, and C. Deter. "High-Power Laser Projection Displays." In *Proceedings of the SPIE 4294*, edited by M. H. Wu, pp. 36–46. Bellingham, WA: SPIE, 2001.

[479] N. Holonyak Jr. and S. F. Bevaqua. "Coherent (Visible) Light Emission from $GaAs_{1-\infty}P_{\infty}$ Junctions." *Applied Physics Letters* 1 (1962), 82–83.

[480] G. Hong, M. R. Luo, and P. A. Rhodes. "A Study of Digital Camera Colorimetric Characterization Based on Polynomial Modeling." *Color Research and Application* 26:1 (2001), 76–84.

[481] Q. Hong, T. X. Wu, X. Zhu, R. Lu, and S.-T. Wu. "Extraordinarily High-Contrast and Wide-View Liquid-Crystal Displays." *Applied Physics Letters* 86 (2005), 121107–1–121107–3.

[482] F. M. Honrubia and J. H. Elliott. "Efferent Innervation of the Retina I: Morphologic Study of the Human Retina." *Archives of Ophthalmology* 80 (1968), 98–103.

[483] D. C. Hood and D. G. Birch. "A Quantitative Measure of the Electrical Activity of Human Rod Photoreceptors using Electroretinography." *Visual Neuroscience* 5 (1990), 379–387.

[484] D. C. Hood and M. A. Finkelstein. "Comparison of Changes in Sensitivity and Sensation: Implications for the Response-Intensity Function of the Human Photopic System." *Journal of Experimental Psychology: Human Perceptual Performance* 5:3 (1979), 391–405.

[485] D. C. Hood, M. A. Finkelstein, and E. Buckingham. "Psychophysical Tests of Models of the Response Function." *Vision Research* 19:4 (1979), 401–406.

[486] S. J. Hook. "ASTER Spectral Library.", 1999. Available online (http://speclib.jpl.nasa.gov).

[487] S. D. Hordley and G. D. Finlayson. "Reevaluation of Color Constancy Algorithm Performance." *Journal of the Optical Society of America A* 23:5 (2006), 1008–1020.

[488] B. K. P. Horn. *Robot Vision*. Cambridge, MA: MIT Press, 1986.

[489] L. J. Hornbeck. "From Cathode Rays to Digital Micromirrors: A History of Electronic Projection Display Technology." *Texas Instruments Technical Journal* 15:3 (1998), 7–46.

[490] J. C. Horton and L. C. Sincich. "How Specific is V1 Input to V2 Thin Stripes?" *Society for Neuroscience Abstracts* 34 (2004), 18.1.

[491] G. D. Horwitz, E. J. Chichilnisky, and T. D. Albright. "Spatial Opponency and Color Tuning Dynamics in Macaque V1." *Society for Neuroscience Abstracts* 34 (2004), 370.9.

[492] C. Hou, M. W. Pettet, V. Sampath, T. R. Candy, and A. M. Norcia. "Development of the Spatial Organization and Dynamics of Lateral Interactions in the Human Visual System." *Journal of Neuroscience* 23:25 (2003), 8630–8640.

[493] D. H. Hubel and T. N. Wiesel. "Receptive Fields, Binocular Interaction and Functional Architecture in the Cat's Visual Cortex." *Journal of Physiology* 160 (1962), 106–154.

[494] D. H. Hubel and T. N. Wiesel. "Ferrier Lecture: Functional Architecture of Macaque Monkey Visual Cortex." *Proceedings of the Royal Society of London B* 198:1130 (1977), 1–59.

[495] P. M. Hubel, J. Holm, and G. Finlayson. "Illuminant Estimation and Color Correction." In *Color Imaging*, pp. 73–95. New York: Wiley, 1999.

[496] D. H. Hubel. *Eye, Brain, and Vision*, Reprint edition, see also http://neuro.med.harvard.edu/site/dh/ edition. New York: W. H. Freeman and Company, 1995.

[497] A. J. Hughes. "Controlled Illumination for Birefringent Colour LCDs." *Displays* 8:3 (1987), 139–141.

[498] A. C. Huk, D. Ress, and D. J. Heeger. "Neuronal Basis of the Motion Aftereffect Reconsidered." *Neuron* 32:1 (2001), 161–172.

[499] H. C. van de Hulst. *Light Scattering by Small Particles*. Mineola, NY: Dover Publications, 1981.

[500] P.-C. Hung and R. S. Berns. "Determination of Constant Hue Loci for a CRT Gamut and Their Predictions using Color Appearance Spaces." *Color Research and Application* 20 (1995), 285–295.

[501] P.-C. Hung. "Colorimetric Calibration for Scanners and Media." In *Proceedings of the SPIE 1498*, edited by W. Chang and J. R. Milch, pp. 164–174. Bellingham, WA: SPIE, 1991.

[502] P.-C. Hung. "Colorimetric Calibration in Electronic Imaging Devices using a Look-Up Table Model and Interpolations." *Journal of Electronic Imaging* 2 (1993), 53–61.

[503] P.-C. Hung. "Color Theory and its Application to Digital Still Cameras." In *Image Sensors and Signal Processing for Digital Still Cameras*, edited by J. Nakamura, pp. 205–221. Boca Raton, FL: Taylor and Francis, 2006.

[504] R. W. G. Hunt. "Light and Dark Adaptation and the Perception of Color." *Journal of the Optical Society of America* 42 (1952), 190–199.

[505] R. W. G. Hunt. "The Strange Journey from Retina to Brain." *Journal of the Royal Television Society* 11 (1967), 220–229.

[506] R. W. G. Hunt. "Hue Shifts in Unrelated and Related Colors." *Color Research and Application* 14 (1989), 235–239.

[507] R. W. G. Hunt. "Why is Black and White So Important in Colour?" In *Colour Imaging: Vision and Technology*, edited by L. W. MacDonald and M. R. Luo, pp. 3–15. Chichester, UK: John Wiley and Sons, Ltd., 1999.

[508] R. W. G. Hunt. "Saturation: Superfluous or Superior?" In *IS&T/SID 9th Color Imaging Conference*, pp. 1–5. Springfield, VA: Society for Imaging Science and Technology, 2001.

[509] R. W. G. Hunt. *The Reproduction of Colour*, Sixth edition. Chichester, UK: John Wiley and Sons Ltd., 2004.

[510] J.-M. Hupé, A. C. James, B. R. Payne, S. G. Lomber, P. Girard, and J. Bullier. "Cortical Feedback Improves Discrimination between Figure and Ground." *Nature* 394:6695 (1998), 784–787.

[511] J.-M. Hupé, A. C. James, P. Girard, and J. Bullier. "Response Modulations by Static Texture Surround in Area V1 of the Macaque Monkey Do Not Depend on Feedback Connections from V2." *Journal of Neurophysiology* 85:1 (2001), 146–163.

[512] B. S. Hur and M. G. Kang. "High Definition Color Interpolation Scheme for Progressive Scan CCD Image Sensor." *IEEE Transactions on Consumer Electronics* 47:1 (2001), 179–186.

[513] A. Hurlbert. "Formal Connections between Lightness Algorithms." *Journal of the Optical Society of America A* 3 (1986), 1684–1693.

[514] J. Hurri, A. Hyvärinen, and E. Oja. "Wavelets and Natural Image Statistics." In *Proceedings of the 10th Scandinavian Conference on Image Analysis*, pp. 13–18. International Association for Pattern Recognition, 1997.

[515] L. M. Hurvich and D. Jameson. "Some Quantitative Aspects of Opponent-Colors Theory. IV. A Psychological Color Specification System." *Journal of the Optical Society of America* 46:6 (1956), 416–421.

[516] L. M. Hurvich and D. Jameson. "The Opponent Process Theory of Color Vision." *Psychological Review* 64 (1957), 384–404.

[517] L. M. Hurvich. *Color Vision*. Sunderland, MA: Sinauer Associates, 1981.

[518] J. B. Hutchings. "Colour and Appearance in Nature, Part 1." *Colour Research and Application* 11 (1986), 107–111.

[519] J. B. Hutchings. "Color in Anthopology and Folklore." In *Color for Science, Art and Technology*, edited by K. Nassau. Amsterdam: Elsevier Science B.V., 1998.

[520] J. B. Hutchings. "Color in Plants, Animals and Man." In *Color for Science, Art and Technology*, edited by K. Nassau. Amsterdam: Elsevier Science B.V., 1998.

[521] J. Hynecek. "CDS Noise Reduction of Partially Reset Charge-Detection Nodes." *IEEE Transactions on Circuits and Systems — I: Fundamental Theory and Applications* 49:3 (2002), 276–280.

[522] A. Hyvärinen. "Survey on Independent Components Analysis." *Neural Computing Surveys* 2 (1999), 94–128.

[523] ICC. "Image Technology Colour Management - Architecture, Profile Format, and Data Structure." Technical Report ICC.1:2004-10, International Color Consortium, 2004. Available online (http://www.color.org).

[524] IEC. "Part 2-1: Colour Management - Default RGB Colour Space - sRGB." Technical Report 61966, International Electrotechnical Commission, Geneva, Switzerland, 1999.

[525] M. Ikebe and K. Saito. "A Wide-Dynamic-Range Compression Image Sensor with Negative-Feedback Resetting." *IEEE Sensors Journal* 7:5 (2007), 897–904.

[526] E. Ikeda. "Image Data Processing Apparatus for Processing Combined Image Signals in order to Extend Dynamic Range." In *5801773, United States Patent*, 2003.

[527] Q. M. T. Inc. "Interferometric Modulator (IMOD) Technology Overview.", 2007. White paper.

[528] M. Inui. "Fast Algorithm for Computing Color Gamuts." *Color Research and Application* 18 (1993), 341–348.

[529] P. Irawan, J. A. Ferwerda, and S. R. Marshner. "Perceptually Based Tone Mapping of High Dynamic Range Image Streams." In *Eurographics Symposium on Rendering*, pp. 231–242. Aire-la-Ville, Switzerland: Eurographics Association, 2005.

[530] S. Ishida, Y. Yamashita, T. Matsuishi, M. Ohshima, T. Ohshima, K. Kato, and H. Maeda. "Photosensitive Sseizures Provoked while Viewing "Pocket Monster", a Made-for-Television Animation Program in Japan." *Epilepsia* 39 (1998), 1340–1344.

[531] S. Ishihara. *Tests for Colour-Blindness*. Tokyo: Hongo Harukicho, 1917.

[532] M. F. Iskander. *Electromagnetic Fields and Waves*. Upper Saddle River, NJ: Prentice Hall, 1992.

[533] ISO. "Graphic technology - Spectral measurement and colorimetric computation for graphic arts images." Technical Report ISO 13655:1996, International Organization for Standardization, 1996.

[534] ISO. "Photography – Digital Still Cameras – Determination of Exposure Index, ISO Speed Ratings, Standard Output Sensitivity, and Recommended Exposure Index." 12232:2006.

[535] S. Itoh and M. Tanaka. "Current Status of Field-Emission Displays." *Proceedings of the IEEE* 90:4 (2002), 514–520.

[536] ITU-R. "Parameter Values for HDTV Standards for Production and International Programme Exchange,." Technical report, BT.709-2, International Telecommunication Union, Geneva, Switzerland, 1995.

[537] A. V. Ivashchenko. *Dichroic Dyes for Liquid Crystal Displays*. Boca Raton, Florida: CRC Press, 1994.

[538] H. Ives. "The Transformation of Color-Mixture Equations from One System to Another." *Journal of Franklin Inst.* 16 (1915), 673–701.

[539] H. E. Ives. "The Resolution of Mixed Colors by Differential Visual Activity." *Philosophical Magazine* 35 (1918), 413–421.

[540] J. Iwao. "Errors in Color Calculations Due to Fluorescence when using the Neugebauer Equations." In *Proceedings of the Technical Association of the Graphical Arts*, pp. 254–266. Sewickley, PA: Technical Association of the Graphic Arts, 1973.

[541] D. Jackèl and B. Walter. "Modeling and Rendering of the Atmosphere using Mie Scattering." *Computer Graphics Forum* 16:4 (1997), 201–210.

[542] D. Jackèl and B. Walter. "Simulation and Visualization of Halos." In *Proceedings of ANIGRAPH '97*, 1997.

[543] A. K. Jain. *Fundamentals of Digital Image Processing*. Upper Saddle River, NJ: Prentice-Hall, 1989.

[544] D. Jameson and L. M. Hurvich. "Color Adaptation: Sensitivity, Contrast and After-Images." In *Handbook of Sensory Physiology*, VII/4, edited by D. Jameson and L. M. Hurvich, VII/4, pp. 568–581. Berlin: Springer-Verlag, 1972.

[545] T. H. Jamieson. "Thin-Lens Theory of Zoom Systems." *Optica Acta* 17:8 (1970), 565–584.

[546] J. Janesick, T. Elliott, S. Collins, M. Blouke, and J. Freeman. "Scientific Charge-Coupled Devices." *Optical Engineering* 26:8 (1987), 692–715.

[547] J.-S. Jang and B. Javidi. "Improvement of Viewing Angle in Integral Imaging by Use of Moving Lenslet Arrays with Low Fill Factor." *Applied Optics* 42:11 (2003), 1996–2002.

[548] H. W. Jensen, J. Legakis, and J. Dorsey. "Rendering of Wet Materials." In *Proceedings of the* 10th *Eurographics Symposium on Rendering*, edited by D. Lischinski and G. W. Larson, pp. 273–281. Vienna: Springer-Verlag, 1999.

[549] H. W. Jensen. *Realistic Image Synthesis using Photon Mapping*. Natick, MA: A K Peters, 2001.

[550] E. Jin, X. Feng, and J. Newell. "The Development of a Color Visual Difference Model (CVDM)." In *IS&T PICS Conference*, pp. 154–158. Springfield, VA: Society for Imaging Science and Technology, 1998.

[551] L. Jing and K. Urahama. "Image Recoloring by Eigenvector Mapping." In *Proceedings of the International Workshop on Advanced Image Technology*, 2006.

[552] D. J. Jobson, Z. Rahman, and G. A. Woodell. "Retinex Image Processing: Improved Fidelity to Direct Visual Observation." In *Proceedings of the IS&T Fourth Color Imaging Conference: Color Science, Systems, and Applications*, pp. 124–125. Springfield, VA: Society for Imaging Science and Technology, 1995.

[553] D. J. Jobson, Z. Rahman, and G. A. Woodell. "A Multi-Scale Retinex for Bridging the Gap between Color Images and Human Observation of Scenes." *IEEE Transactions on Image Processing* 6:7 (1997), 965–976.

[554] T. Johansson. *Färg*. Stockholm, Sweden: Lindfors Bokförlag, AB, 1937.

[555] G. M. Johnson and M. D. Fairchild. "A Top Down Description of S-CIELAB and CIEDE2000." *Color Research and Application* 28 (2003), 425–435.

[556] G. M. Johnson and F. Mark D. "The Effect of Opponent Noise on Image Quality." In *SPIE Proceedings 5668 (Electronic Imaging Conference)*, pp. 82–89. Bellingham, WA: SPIE, 2005.

[557] P. D. Jones and D. H. Holding. "Extremely Long-Term Persistence of the McCollough Effect." *Journal of Experimental Psychology* 1:4 (1975), 323–327.

[558] D. B. Judd and G. Wyszecki. *Color in business, science and industry*, Third edition. New York: Wiley Interscience, 1975.

[559] D. B. Judd, D. L. MacAdam, and G. Wyszecki. "Spectral Distribution of Typical Light as a Function of Correlated Color Temperature." *Journal of the Optical Society of America* 54:8 (1964), 1031–1040.

[560] D. B. Judd. "Chromaticity Sensibility to Stimulus Differences." *Journal of the Optical Society of America* 22 (1932), 72–108.

[561] D. B. Judd. "Sensibility to Color-Temperature Change as a Function of Temperature." *Journal of the Optical Society of America* 23 (1933), 7.

[562] T. B. Jung, J. C. Kim, and S. H. Lee. "Wide-Viewing-Angle Transflective Display Associated with a Fring-Field Driven Homogeneously Aligned Nematic Liquid Crystal Display." *Japanese Journal of Applied Physics* 42:5A (2003), L464–L467.

[563] J. H. Kaas and D. C. Lyon. "Visual Cortex Organization in Primates: Theories of V3 and Adjoining Visual Areas." *Progress in Brain Research* 134 (2001), 285–295.

[564] J. H. Kaas, M. F. Huerta, J. T. Weber, and J. K. Harting. "Patterns of Retinal Terminations and Laminar Organization of the Lateral Geniculate Nucleus of Primates." *Journal of Comparative Neurology* 182:3 (1978), 517–554.

[565] F. Kainz, R. Bogart, and D. Hess. "The OpenEXR Image File Format." In *SIGGRAPH '03 Technical Sketches*, 2003. See also: http://www.openexr.com.

[566] P. K. Kaiser and R. M. Boynton, editors. *Human Color Vision*, Second edition. Washington, DC: Optical Society of America, 1996.

[567] P. K. Kaiser. "Photopic and Mesopic Photometry: Yesterday, Today, and Tomorrow." In *Golden Jubilee of Colour in the CIE : Proceedings of a Symposium held by The Colour Group (Great Britain) at Imperial College, London*. Bradford, UK: The Society of Dyers and Colourists, 1981.

[568] J. T. Kajiya. "The Rendering Equation." *Computer Graphics (SIGGRAPH '86 Proceedings)* 20:4 (1986), 143–150.

[569] K. Kakinuma. "Technology of Wide Color Gamut Backlight with Light-Emitting Diode for Liquid Crystal Display Television." *Japanese Journal of Applied Physics* 45:5B (2006), 4330–4334.

[570] M. Kalloniatis and C. Luu. "Psychophysics of Vision.", 2006. http://webvision.med.utah.edu/KallSpatial.html.

[571] E. Kaneko. *Liquid crystal TV displays: Principles and applications of liquid crystal displays*. Tokyo: KTK Scientific Publishers, 1987. In co-publication with D Reidel Publishing Company, Dordrecht, Holland.

[572] H. R. Kang and P. G. Anderson. "Neural Network Applications to the Colour Scanner and Printer Calibrations." *Journal of Electronic Imaging* 1 (1992), 125–134.

[573] S. B. Kang, M. Uttendaele, S. Winder, and R. Szeliski. "High Dynamic Range Video." *ACM Transactions on Graphics* 22:3 (2003), 319–325.

[574] H. R. Kang. "Colour Scanner Calibration." *Journal of Imaging Science and Technology* 36 (1992), 162–170.

[575] H. R. Kang. *Color Technology for Electronic Imaging Devices*. Belingham, Washington: SPIE Optical Engineering Press, 1997.

[576] H. R. Kang. "Computational Accuracy of RGB Encoding Standards." In *IS&Ts NIP16: International Conference on Digital Printing Technologies*, pp. 661–664. Springfield, VA: Society for Imaging Science and Technology, 2000.

[577] H. R. Kang. *Computational Color Technology*. Bellingham, WA: SPIE Press, 2006.

[578] R. Kanthack. *Tables of Refractive Indices*, II. London: Hilger, 1921.

[579] K. Karhunen. "Über lineare Methoden in der Warscheinlichkeitsrechnung." *Annales Academiae Scientiarium Fennicae* 37 (1947), 3–79.

[580] J. E. Kasper and S. A. Feller. *The Complete Book of Holograms: How They Work and How to Make Them*. New York: John Wiley and Sons, Inc., 1987.

[581] N. Katoh and K. Nakabayashi. "Applying Mixed Adaptation to Various Chromatic Adaptation Transformation (CAT) Models." In *IS&T PICS Conference*, pp. 299–305. Springfield, VA: Society for Imaging Science and Technology, 2001.

[582] J. Kautz and M. McCool. "Interactive Rendering with Arbitrary BRDFs using Separable Approximations." In *Proceedings of the* 10th *Eurographics Symposium on Rendering*, edited by D. Lischinski and G. W. Larson, pp. 247–260. Vienna: Springer-Verlag, 1999.

[583] R. Kawakami, J. Takamatsu, and K. Ikeuchi. "Color Constancy from Blackbody Illumination." *Journal of the Optical Society of America A*. (to appear), 2007.

[584] P. Kay and C. K. McDaniel. "The Linguistic Significance of the Meanings of Basic Color Terms." *Language* 54:3 (1978), 610–646.

[585] E. A. Khan and E. Reinhard. "Evaluation of Color Spaces for Edge Classification in Outdoor Scenes." In *IEEE International Conference on Image Processing*, pp. 952–955. Washington, DC: IEEE Press, 2005.

[586] E. A. Khan, A. O. Akyüz, and E. Reinhard. "Ghost Removal in High Dynamic Range Images." In *IEEE International Conference on Image Processing*, pp. 2005–2008. Washington, DC: IEEE Press, 2006.

[587] E. A. Khan, E. Reinhard, R. Fleming, and H. Bülthoff. "Image-Based Material Editing." *ACM Transactions on Graphics* 25:3 (2006), 654–663.

[588] J.-H. Kim and J. P. Allebach. "Color Filters for CRT-Based Rear Projection Television." *IEEE Transactions on Consumer Electronics* 42:4 (1996), 1050–1054.

[589] H. Kim, J.-Y. Kim, S. H. Hwang, I.-C. Park, and C.-M. Kyung. "Digital Signal Processor with Efficient RGB Interpolation and Histogram Accumulation." *IEEE Transactions on Consumer Electronics* 44:4 (1998), 1389–1395.

[590] R. Kimmel. "Demosaicing: Image Reconstruction from Color CCD Samples." *IEEE Transactions on Image Processing* 8:9 (1999), 1221–1228.

[591] G. Kindlmann, E. Reinhard, and S. Creem. "Face-Based Luminance Matching for Perceptual Colormap Generation." In *Proceedings of IEEE Visualization*, pp. 309–406. Washington, DC: IEEE, 2002.

[592] D. C. Kiper, J. B. Levitt, and K. R. Gegenfurtner. "Chromatic Signals in Extrastriate Areas V2 and V3." In *Color Vision: From Genes to Perception*, edited by K. R. Gegenfurtner and L. Sharpe, pp. 249–268. Cambridge, UK: Cambridge University Press, 1999.

[593] A. Kitaoka, J. Gyoba, H. Kawabata, and K. Sakurai. "Two Competing Mechanisms Underlying Neon Color Spreading, Visual Phantoms and Grating Induction." *Vision Research* 41:18 (2001), 2347–2354.

[594] J. Kleinschmidt and J. E. Dowling. "Intracellular Recordings from Gecko Photoreceptors during Light and Dark Adaptation." *Journal of General Physiology* 66:5 (1975), 617–648.

[595] G. J. Klinker, S. A. Shafer, and T. Kanade. "Image Segmentation and Reflectance Analysis through Color." In *Proceedings of the SPIE 937 (Application of Artificial Intelligence VI)*, pp. 229–244. Bellingham, WA: SPIE, 1988.

[596] G. J. Klinker, S. A. Shafer, and T. Kanade. "A Physical Approach to Color Image Understanding." *International Journal of Computer Vision* 4:1 (1990), 7–38.

[597] K. Klug, N. Tiv, Y. Tsukamoto, P. Sterling, and S. Schein. "Blue Cones Contact OFF-Midget Bipolar Cells." *Society for Neuroscience Abstracts* 18 (1992), 838.

[598] K. Klug, Y. Tsukamoto, P. Sterling, and S. Schein. "Blue Cone OFF-Midget Ganglion Cells in Macaque." *Investigative Ophthalmology and Visual Science, Supplement* 34 (1993), 986–986.

[599] R. Knight, S. L. Buck, G. A. Fowler, and A. Nguyen. "Rods Affect S-Cone Discrimination on the Farnsworth-Munsell 100-Hue Test." *Vision Research* 38:21 (1998), 3477–3481.

[600] W. E. Kock. *Lasers and Holography*, Second edition. New York: Dover Publications, Inc, 1981.

[601] J. J. Koenderink, A. van Doorn, and M. Stavridi. "Bidirectional Reflection Distribution Function Expressed in Terms of Surface Scattering Modes." In *European Conference on Computer Vision*, pp. 28–39. London: Springer-Verlag, 1996.

[602] K. Koffka. *Principles of Gestallt Psychology*. New York: Harcourt, Brace, and World, 1935.

[603] A. J. F. Kok. "Ray Tracing and Radiosity Algorithms for Photorealistic Image Synthesis." Ph.D. thesis, Delft University of Technology, 1994.

[604] H. Kolb and L. Dekorver. "Midget Ganglion Cells of the Parafovea of the Human Retina: A Study by Electron Microscopy and Serial Section Reconstructions." *Journal of Comparative Neurology* 303:4 (1991), 617–636.

[605] H. Kolb and E. V. Famiglietti. "Rod and Cone Pathways in the Inner Plexiform Layer of the Cat Retina." *Science* 186 (1974), 47–49.

[606] H. Kolb, K. A. Linberg, and S. K. Fisher. "Neurons of the Human Retina: A Golgi Study." *Journal of Comparative Neurology* 318:2 (1992), 147–187.

[607] H. Kolb. "The Inner Plexiform Layer in the Retina of the Cat: Electron Microscopic Observations." *Journal of Neurocytology* 8 (1979), 295–329.

[608] H. Kolb. "Anatomical Pathways for Color Vision in the Human Retina." *Visual Neuroscience* 7 (1991), 61–74.

[609] H. Kolb. "How the Retina Works." *American Scientist* 91 (2003), 28–35.

[610] A. König. "Über den Menschlichen Sehpurpur und seine Bedeutung für das Sehen." *Sitzungsberichte der Akademie der Wissenschaften, Berlin*, pp. 577–598.

[611] A. König. *Gesammelte Abhandlungen zur Physiologischen Optik*. Leipzig: Barth, 1903.

[612] N. Kouyama and D. W. Marshak. "Bipolar Cells Specific for Blue Cones in the Macaque Retina." *Journal of Neuroscience* 12:4 (1992), 1233–1252.

[613] T. Koyama. "Optics in Digital Still Cameras." In *Image Sensors and Signal Processing for Digital Still Cameras*, edited by J. Nakamura, pp. 21–51. Boca Raton, FL: Taylor and Francis, 2006.

[614] J. Krauskopf, Q. Zaidi, and M. B. Mandler. "Mechanisms of Simultaneous Color Induction." *Journal of the Optical Society of America A* 3:10 (1986), 1752–1757.

[615] J. Krauskopf. "Light Distribution in Human Retinal Images." *Journal of the Optical Society of America* 52:9 (1962), 1046–1050.

[616] J. Krauskopf. "Effect of Retinal Stabilization on the Appearance of Heterochromatic Targets." *Journal of the Optical Society of America* 53 (1963), 741–744.

[617] Y. A. Kravtsov and Y. I. Orlov. *Geometrical Optics of Inhomogeneous Media*. Springer Series on Wave Phenomena, Berlin: Springer-Verlag, 1990.

[618] G. Krawczyk, R. Mantiuk, K. Myszkowski, and H.-P. Seidel. "Lightness Perception Inspired Tone Mapping." In *APGV '04: Proceedings of the 1st Symposium on Applied Perception in Graphics and Visualization*, pp. 172–172. New York: ACM Press, 2004.

[619] G. Krawczyk, K. Myszkowski, and H.-P. Seidel. "Lightness Perception in Tone Reproduction for High Dynamic Range Images." *Computer Graphics Forum* 24:3 (2005), 635–645.

[620] G. Krawczyk, K. Myszkowski, and H.-P. Seidel. "Computational Model of Lightness Perception in High Dynamic Range Imaging." In *Proceedings of SPIE 6057 (Human Vision and Electronic Imaging)*. Bellingham, WA: SPIE, 2006.

[621] W. Kress and M. Stevens. "Derivation of 3-Dimensional Gamut Descriptors for Graphic Arts Output Devices." In *Proceedings of the Technical Association of Graphical Arts*, pp. 199–214. Sewickley, PA: Technical Assocation of the Graphics Arts, 1994.

[622] J. von Kries. "Die Gesichtsempfindungen." In *Handbuch der Physiologie des Menschen*, edited by W. Nagel, pp. 109–282. Braunschweig, Germany: Vieweg, 1905.

[623] J. von Kries. "Chromatic Adaptation." In *Sources of Color Vision*, edited by D. L. MacAdam. Cambridge, MA: MIT Press, 1970. Originally published in *Festschrift der Albrecht-Ludwig-Universitat*, 1902.

[624] J. Kruger and P. Gouras. "Spectral Selectivity of Cells and Its Dependence on Slit Length in Monkey Visual Cortex." *Journal of Neurophysiology* 43 (1979), 1055–1069.

[625] J. Kuang, H. Yamaguchi, G. M. Johnson, and M. D. Fairchild. "Testing HDR Image Rendering Algorithms." In *Proceedings of IS&T/SID 12th Color Imaging Conference*, pp. 315–320. Springfield, VA: Society for Imaging Science and Technology, 2004.

[626] J. Kuang, G. M. Johnson, and M. D. Fairchild. "Image Preference Scaling for HDR Image Rendering." In *IS&T/SID 13th Color Imaging Conference*, pp. 8–13. Springfield, VA: Society for Imaging Science and Technology, 2005.

[627] J. Kuang, H. Yamaguchi, C. Liu, G. M. Johnson, and M. D. Fairchild. "Evaluating HDR Rendering Algorithms." *ACM Transactions on Applied Perception* 4:2 (2007), 9–1 – 9–27.

[628] P. Kubelka and F. Munk. "Ein Beitrag zur Optik der Farbanstriche." *Zeitschrift für technische Physik* 12 (1931), 593–601.

[629] R. G. Kuehni. *Computer Colorant Formulation*. Lexington, MA: Lexington Books, 1975.

[630] R. G. Kuehni. *Color: An Introduction to Practice and Principles*. New York: John Wiley & Sons, 1997.

[631] R. G. Kuehni. *Color Space and Its Divisions: Color Order from Antiquity to the Present*. New York: Wiley-Interscience, 2003.

[632] S. W. Kuffler. "Discharge Patterns and Functional Organization of Mammalian Retina." *Journal of Neurophysiology* 16 (1953), 37–68.

[633] T. Kuno, H. Sugiura, and N. Matoba. "New Interpolation Method using Discriminated Color Correction for Digital Still Cameras." *IEEE Transactions on Consumer Electronics* 45:1 (1999), 259–267.

[634] C. J. Kuo and M. H. Tsai, editors. *Three-Dimensional Holographic Imaging*. New York: John Wiley and Sons, Inc., 2002.

[635] E. A. Lachica and V. A. Casagrande. "Direct W-Like Geniculate Projections to the Cytochrome-Oxydase (CO) Blobs in Primate Visual Cortex: Axon Morphology." *Journal of Comparative Neurology* 319:1 (1992), 141–158.

[636] E. A. Lachica, P. D. Beck, and V. A. Casagrande. "Parallel Pathways in Macaque Monkey Striate Cortex: Anatomically Defined Columns in Layer III." *Proceedings of the National Academy of Sciences of the United States of America* 89:8 (1992), 3566–3560.

[637] E. P. F. Lafortune, S.-C. Foo, K. E. Torrance, and D. P. Greenberg. "Non-Linear Approximation of Reflectance Functions." In *Proceedings of SIGGRAPH'97, Computer Graphics Proceedings, Annual Conference Series*, pp. 117–126. Reading, MA: Addison-Wesley, 1997.

[638] R. Lakowski. "Colorimetric and Photometric Data for the 10th Edition of the Ishihara Plates." *The British Journal of Physiological Optics* 22:4 (1965), 195–207.

[639] P. Lalonde and A. Fournier. "Filtered Local Shading in the Wavelet Domain." In *Proceedings of the 8th Eurographics Symposium on Rendering*, pp. 163–174. Vienna: Springer-Verlag, 1997.

[640] P. Lalonde and A. Fournier. "A Wavelet Representation of Reflectance Functions." *IEEE Transactions on Visualization and Computer Graphics* 3:4 (1997), 329–336.

[641] K. Lam. "Metamerism and Colour Constancy." Ph.D. thesis, University of Bradford, 1985.

[642] T. D. Lamb and J. Bourriau, editors. *Colour: Art and Science*. Cambridge, UK: Cambridge University Press, 1995.

[643] T. D. Lamb. "Photoreceptor Spectral Sensitivities: Common Shape in the Long Wavelength Region." *Vision Research* 35:22 (1995), 3083–3091.

[644] T. D. Lamb. "Photopigments and the Biophysics of Transduction in Cone Photoreceptors." In *Color Vision: From Genes to Perception*, edited by K. R. Gegenfurtner and L. Sharpe, pp. 89–101. Cambridge, UK: Cambridge University Press, 1999.

[645] J. H. Lambert. "Photometria sive de mensura et gradibus luminis, colorum et umbrae.", 1760.

[646] D. Lamming. "Contrast Sensitivity." In *Vision and Visual Dysfunction*, edited by J. R. Cronley-Dillon. London: Macmillan Press, 1991.

[647] E. Land and J. McCann. "Lightness and Retinex Theory." *Journal of the Optical Society of America* 61:1 (1971), 1–11.

[648] E. Land. "The Retinex Theory of Color Vision." *Scientific American* 237:6 (1977), 108–128.

[649] E. Land. "Recent Advances in Retinex Theory." *Vision Research* 26:1 (1986), 7–21.

[650] L. D. Landau and E. M. Lifshitz. *Fluid Mechanics*, Course of Theoretical Physics, 6, Second edition. Oxford: Pergamon, 1987.

[651] M. S. Langer. "Large-Scale Failures of $f^{-\alpha}$ Scaling in Natural Image Spectra." *Journal of the Optical Society of America A* 17:1 (2000), 28–33.

[652] I. Langmuir. "The Effect of Space Charge and Residual Gases on Thermionic Currents in High Vacuum." *Physical Review* 2:6 (1913), 450–486.

[653] R. D. Larrabee. "Spectral Emissivity of Tungsten." *Journal of the Optical Society of America* 49:6 (1959), 619–625.

[654] P. Laven. "Mieplot." Available online (http://www.philiplaven.com).

[655] J. Lawrence, S. Rusinkiewicz, and R. Ramamoorthi. "Efficient BRDF Importance Sampling using a Factored Representation." *ACM Transactions on Graphics* 23:3 (2004), 496–505.

[656] Y. Le Grand. *Light, Colour and Vision*, Second edition. Somerset, N.J.: Halsted Press, 1968. Translated by R. W. G. Hunt, J. W. T. Walsh and F. R. W. Hunt.

[657] S. J. Leat, R. V. North, and H. Bryson. "Do Long Wavelength Pass Filters Improve Low Vision Performance?" *Ophthalmic and Physiological Optics* 10:3 (1991), 219–224.

[658] P. Ledda, L.-P. Santos, and A. Chalmers. "A Local Model of Eye Adaptation for High Dynamic Range Images." In *Proceedings of ACM Afrigraph*, pp. 151–160. New York: ACM, 2004.

[659] P. Ledda, A. Chalmers, T. Troscianko, and H. Seetzen. "Evaluation of Tone Mapping Operators using a High Dynamic Range Display." *ACM Transactions on Graphics* 24:3 (2005), 640–648.

[660] B. B. Lee, P. R. Martin, and A. Valberg. "The Physiological Basis of Heterochromatic Flicker Photometry Demonstrated in the Ganglion Cells of the Macaque Retina." *Journal of Physiology* 404 (1988), 323–347.

[661] B. B. Lee, J. Kremers, and T. Yeh. "Receptive Fields of Primate Retinal Ganglion Cells Studied with a Novel Technique." *Visual Neuroscience* 15:1 (1998), 161–175.

[662] C. H. Lee, B. J. Moon, H. Y. Lee, E. Y. Chung, and Y. H. Ha. "Estimation of Spectral Distribution of Scene Illumination from a Single Image." *Journal of Imaging Science and Technology* 44:4 (2000), 308–320.

[663] T. Lee, J. Zaumseil, Z. Bao, J. W. P. Hsu, and J. A. Rogers. "Organic Light-Emitting Diodes Formed by Soft Contact Lamination." *Proceedings of the National Academy of Sciences of the United States of America* 101:2 (2004), 429–433.

[664] H.-C. Lee. "Method for Computing the Scene-Illuminant Chromaticity from Specular Highlight." *Journal of the Optical Society of America A* 3:10 (1986), 1694–1699.

[665] H.-C. Lee. "Internet Color Imaging." In *Proceedings of the SPIE 3080*, pp. 122–135. Bellingham, WA: SPIE, 2000.

[666] H.-C. Lee. *Introduction to Color Imaging Science*. Cambridge, UK: Cambridge University Press, 2005.

[667] A. Lefohn, R. Caruso, E. Reinhard, B. Budge, and P. Shirley. "An Ocularist's Approach to Human Iris Synthesis." *IEEE Computer Graphics and Applications* 23:6 (2003), 70–75.

[668] J. Lekner and M. C. Dorf. "Why Some Things Are Darker when Wet." *Applied Optics* 27:7 (1988), 1278–1280.

[669] P. Lennie, J. Krauskopf, and G. Sclar. "Chromatic Mechanisms in Striate Cortex of Macaque." *Journal of Neuroscience* 10:2 (1990), 649–669.

[670] P. Lennie, J. Pokorny, and V. C. Smith. "Luminance." *Journal of the Optical Society of America A* 10:6 (1993), 1283–1293.

[671] H. Lensch, J. Kautz, M. Goessele, W. Heidrich, and H. Seidel. "Image-Based Reconstruction of Spatially Varying Materials." In *Proceedings of the 12th Eurographics Workshop on Rendering*, edited by S. J. Gortler and K. Myszkowski, pp. 104–115. Aire-la-Ville, Switzerland: Eurographics Association, 2001.

[672] S. Lerman, J. M. Megaw, and M. N. Moran. "Further Studies on the Effects of UV Radiation on the Human Lens." *Ophthalmic Research* 17:6 (1985), 354–361.

[673] A. G. Leventhal, R. W. Rodieck, and B. Dreher. "Retinal Ganglion Cell Classes in the Old World Monkey: Morphology and Central Projections." *Science* 213 (1981), 1139–1142.

[674] A. G. Leventhal, K. G. Thompson, D. Liu, Y. Zhou, and S. J. Ault. "Concomitant Sensitivity to Orientation, Direction, and Color of Cells in Layers 2, 3, and 4 of Monkey Striate Cortex." *Journal of Neuroscience* 15:3 (1995), 1808–1818.

[675] H. W. Leverenz. *An Introduction to Luminescence of Solids*. Mineola, NY: Dover Publications, 1968.

[676] M. Levoy and P. Hanrahan. "Light Field Rendering." In *Proceedings SIGGRAPH '96, Computer Graphics Proceedings, Annual Conference Series*, pp. 31–42. Reading, MA: Addison-Wesley, 1996.

[677] I. Levy, U. Hasson, G. Avidan, T. Hendler, and R. Malach. "Center-Periphery Organization of Human Object Areas." *Nature Neuroscience* 4:5 (2001), 533–539.

[678] P. A. Lewis. "Colorants: Organic and Inorganic Pigments." In *Color for Science, Art and Technology*, edited by K. Nassau. Amsterdam: Elsevier Science B.V., 1998.

[679] X. Li and A. Gilchrist. "Relative Area and Relative Luminance Combine to Anchor Surface Lightness Values." *Perception and Psychophysics* 61 (1999), 771–785.

[680] M. Li and J.-M. Lavest. "Some Aspects of Zoom Lens Camera Calibration." *IEEE Transactions on Pattern Analysis and Machine Intelligence* 18:11 (1996), 1105–1110.

[681] Y. Li, L. Sharan, and E. H. Adelson. "Compressing and Companding High Dynamic Range Images with Subband Architectures." *ACM Transactions on Graphics* 24:3 (2005), 836–844.

[682] Z. Li-Tolt, R. L. Fink, and Z. Yaniv. "Electron Emission from Patterned Diamond Flat Cathodes." *J. Vac. Sci. Tech. B* 16:3 (1998), 1197–1198.

[683] A. Liberles. *Introduction to Molecular-Orbital Theory*. New York: Holt, Rinehart and Winston, Inc., 1966.

[684] A. Lien. "A Detailed Derivation of Extended Jones Matrix Representation for Twisted Nematic Liquid Crystal Displays." *Liquid Crystals* 22:2 (1997), 171–175.

[685] B. J. Lindbloom, 2005. Available online (http://www.brucelindbloom.com).

[686] D. Lischinski, Z. Farbman, M. Uyttendaele, and R. Szeliski. "Interactive Local Adjustment of Tonal Values." *ACM Transactions on Graphics* 25:3 (2006), 646–653.

[687] D. Litwiller. "CCD vs. CMOS: Facts and Fiction." *Photonics Spectra* 35:1 (2001), 151–154.

[688] D. Litwiller. "CCD vs. CMOS: Maturing Technologies, Maturing Markets." *Photonics Spectra* 39:8 (2005), 54–58.

[689] C. Liu and M. D. Fairchild. "Measuring the Relationship between Pereived Image Contrast and Surround Illuminantion." In *IS&T/SID 12th Color Imaging Conference*, pp. 282–288. Springfield, VA: Society for Imaging Science and Technology, 2004.

[690] Y. Liu, J. Shigley, E. Fritsch, and S. Hemphill. "Abnormal Hue-Angle Change of the Gemstone Tanzanite between CIE Illuminants D65 and A in CIELAB Color Space." *Color Research and Application* 20 (1995), 245–250.

[691] C. Liu, P. R. Jonas, and C. P. R. Saunders. "Pyramidal Ice Crystal Scattering Phase Functions and Concentric Halos." *Annales Geophysicae* 14 (1996), 1192–1197.

[692] M. S. Livingstone and D. H. Hubel. "Anatomy and Physiology of a Color System in the Primate Visual Cortex." *Journal of Neuroscience* 4:1 (1984), 309–356.

[693] M. S. Livingstone and D. H. Hubel. "Connections between Layer 4B of Area 17 and the Thick Cytochrome Oxidase Stripes of Area 18 in the Squirrel Monkey." *Journal of Neuroscience* 7:11 (1987), 3371–3377.

[694] M. S. Livingstone and D. H. Hubel. "Segregation of Form, Color, Movement, and Depth: Anatomy, Physiology, and Perception." *Science* 240:4853 (1988), 740–749.

[695] M. S. Livingstone. *Vision and Art: The Biology of Seeing*. New York: Harry N Abrams, 2002.

[696] M. M. Loève. *Probability Theory*. Princeton: Van Nostrand, 1955.

[697] B. London and J. Upton. *Photography*, Sixth edition. London: Longman, 1998.

[698] A. V. Loughren. "Recommendations of the National Television System Committee for a Color Television Signal." *Journal of the SMPTE* 60 (1953), 321–326, 596.

[699] O. Lowenstein and I. E. Loewenfeld. "The Pupil." In *The Eye, Volume 3*, edited by H. Davson, pp. 255–337. New York: Academic Press, 1969.

[700] J. Lu and D. M. Healey. "Contrast Enhancement via Multiscale Gradient Transformation." In *Proceedings of the* 16th *IEEE International Conference on Image Processing, Volume II*, pp. 482–486. Washington, DC: IEEE, 1992.

[701] R. Lu, J. J. Koenderink, and A. M. L. Kappers. "Optical Properties (Bidirectional Reflectance Distribution Functions) of Velvet." *Applied Optics* 37:25 (1998), 5974–5984.

[702] R. Lu, J. J. Koenderink, and A. M. L. Kappers. "Specularities on Surfaces with Tangential Hairs or Grooves." *Computer Vision and Image Understanding* 78 (1999), 320–335.

[703] B. D. Lucas and T. Kanade. "An Iterative Image Registration Technique with an Application to Stereo Vision." In *Proceedings of the 1981 DARPA Image Understanding Workshop*, pp. 121–130, 1981.

[704] M. P. Lucassen and J. Walraven. "Evaluation of a Simple Method for Color Monitor Recalibration." *Color Research and Application* 15:6 (1990), 321–326.

[705] C. J. Lueck, S. Zeki, K. J. Friston, M.-P. Deiber, P. Cope, V. J. Cunningham, A. A. Lammertsma, C. Kennard, and R. S. J. Frackowiak. "The Colour Centre in the Cerebral Cortex of Man." *Nature* 340:6232 (1989), 386–389.

[706] E. Lueder. *Liquid Crystal Displays: Addressing Schemes and Electro-Optical Effects*. Chichester, UK: John Wiley and Sons, Ltd., 2001.

[707] T. Lulé, B. Schneider, and M. Böhm. "Design and Fabrication of a High-Dynamic-Range Image Sensor in TFA Technology." *IEEE Journal of Solid-State Circuits* 34:5 (1999), 704–711.

[708] T. Lulé, M. Wagner, M. Verhoeven, H. Keller, and M. Böhm. "100000-Pixel, 120-dB Imager in TFA Technology." *IEEE Journal of Solid-State Circuits* 35:5 (2000), 732–739.

[709] J. S. Lund, R. D. Lund, A. E. Hendrickson, A. H. Bunt, and A. F. Fuchs. "The Origin of Efferent Pathways from the Primary Visual Cortex, Area 17, of the Macaque monkey as Shown by Retrogade Transport of Horseradish Peroxidase." *Journal of Comparative Neurology* 164:3 (1975), 287–303.

[710] J. S. Lund. "Organization of Neurons in the Visual Cortex, Area 17, of the Monkey (Macaca mulatta)." *Journal of Comparative Neurology* 147:4 (1973), 455–496.

[711] M. Luo, A. Clark, P. Rhodes, A. Schappo, S. Scrivner, and C. Tait. "Quantifying Colour Appearance. Part I. LUTCHI Colour Appearance Data." *Color Research and Application* 16 (1991), 166–180.

[712] M. Luo, A. Clark, P. Rhodes, A. Schappo, S. Scrivner, and C. Tait. "Quantifying Colour Appearance. Part II. Testing Colour Models Performance using LUTCHI Color Appearance Data." *Color Research and Application* 16 (1991), 181–197.

[713] M. Luo, M.-C. Lo, and W.-G. Kuo. "The LLAB(l:c) Colour Model." *Color Research and Application* 21 (1996), 412–429.

[714] M. R. Luo, G. Cui, and B. Rigg. "The Development of CIE2000 Colour-Difference Formula: CIEDE2000." *Color Research and Application* 26:5 (2001), 340–350.

[715] R. Luther. "Aus dem Gebiet der Farbreizmetrik." *Zeitschrift für Technische Physik* 8 (1927), 540–558.

[716] D. K. Lynch and W. Livingstone. *Color and Light in Nature*, Second edition. Cambridge, UK: Cambridge University Press, 2001.

[717] D. L. MacAdam. *Sources of Color Science*. Cambridge, MA: MIT Press, 1970.

[718] D. L. MacAdam. "Uniform Color Scales." *Journal of the Optical Society of America* 64 (1974), 1619–1702.

[719] D. L. MacAdam. "Colorimetric Data for Samples of OSA Uniform Color Scales." *Journal of the Optical Society of America* 68 (1978), 121–130.

[720] D. L. MacAdam. "Introduction." In *Selected Papers on Colorimetry — Fundamentals*, edited by D. L. MacAdam. Bellingham, Washington: SPIE Optical Engineering Press, 1993.

[721] L. W. MacDonald. "Gamut Mapping in Perceptual Color Space." In *Proceedings of the First IS&T/SID Color Imaging Conference*, pp. 193–196. Spriongfield, VA: Society for Imaging Science and Technology, 1993.

[722] D. I. A. MacLeod and R. M. Boynton. "Chromaticity Diagram Showing Cone Excitation by Stimuli of Equal Luminance." *Journal of the Optical Society of America* 69:8 (1979), 1183–1186.

[723] M. A. MacNeil and R. H. Masland. "Extreme Diversity among Amacrine Cells: Implications for Function." *Neuron* 20 (1998), 971–982.

[724] J. B. MacQueen. "Some Methods for Classification and Analysis of Multivariate Observations." In *Proceedings of the 5th Berkeley Symposium on Mathematical Statistics and Probability*, pp. 281–297. Berkeley: University of California Press, 1967.

[725] M. Mahy. "Calculation of Color Gamuts Based on the Neugebauer Model." *Color Research and Application* 22 (1997), 365–374.

[726] R. Malach, R. B. Tootell, and D. Malonek. "Relationship between Orientation Domains, Cytochrome Oxydase Strips, and Intrinsic Horizontal Connections in Squirrel Monkey Area V2." *Cerebral Cortex* 4:2 (1994), 151–165.

[727] R. Malladi and J. A. Sethian. "Image Processing via Level Set Curvature Flow." *Proceedings of the National Academy of Science* 92:15 (1995), 7046–7050.

[728] L. T. Maloney and B. A. Wandell. "Color Constancy: A Method for Recovering Surface Spectral Reflectance." *Journal of the Optical Society of America A* 3:1 (1986), 29–33.

[729] L. T. Maloney. "Evaluation of Linear Models of Surface Spectral Reflectance with Small Number of Parameters." *Journal of the Optical Society of America A* 3 (1986), 1673–1683.

[730] Étienne. L. Malus. "Sur la mesure du pouvoir réfringent des corps opaques." *Journal de l'École Polytechnique* 15:8 (1809), 219–228.

[731] S. Mann and R. W. Picard. "On Being 'Undigital' with Digital Cameras: Extending Dynamic Range by Combining Differently Exposed Pictures." In *IS&T's 48th Annual Conference*, pp. 422–428. Springfield, VA: Society for Imaging Science and Technology, 1995.

[732] R. J. W. Mansfield. "Primate Photopigments and Cone Mechanisms." In *The Visual System*, edited by A. Fein and J. S. Levine, pp. 89–106. New York: Alan R Liss, 1985.

[733] R. Mantiuk, A. Efremov, K. Myszkowski, and H.-P. Seidel. "Backward Compatible High Dynamic Range MPEG Video Compression." *ACM Transactions on Graphics* 25:3 (2006), 713–723.

[734] E. W. Marchand. "Gradient Index Lenses." In *Progress in Optics*, XI, edited by E. Wolf, XI, pp. 305–337. Amsterdam: North Holland, 1973.

[735] R. F. Marcinik and K. M. Revak. "Effects of New Cool White Fluorescent Lamps on Viewing and Measuring Color." *Textile Chemist and Colorist* 30:1 (1998), 14–16.

[736] D. Marcuse. *Light Transmission Optics*, Second edition. New York: Van Nostrand Reinhold Company, 1982.

[737] D. Marini and A. Rizzi. "A Computational Approach to Color Adaptation Effects." *Image and Vision Computing* 18:13 (2000), 1005–1014.

[738] D. Marini, A. Rizzi, and C. Carati. "Color Constancy Effects Measurement of the Retinex Theory." In *Proceedings of the IS&T/SPIE Conference on Electronic Imaging*, pp. 23–29. Springfield, VA: Society for Imaging Science and Technology, 1999.

[739] K. Z. Markov. "Elementary Micromechanisms of Heterogeneous Media." In *Heterogeneous Media: Modelling and Simulation*, edited by K. Z. Markov and L. Preziosi, pp. 1–162. Boston: Birkhauser, 1999.

[740] T. J. Marks, J. G. C. Veinot, J. Cui, H. Yan, A. Wang, N. L. Edlerman, J. Ni, Q. Huang, P. Lee, and N. R. Armstrong. "Progress in High Work Function TCO OLED Anode Alternatives and OLED Nanopixelation." *Synthetic Metals* 127 (2002), 29–35.

[741] G. H. Markstein. *Nonsteady Flame Propagation*. Oxford, UK: Pergamon, 1964.

[742] D. Marr. *Vision, A Computational Investigation into the Human Representation and Processing of Visual Information*. San Fransisco: W. H. Freeman and Company, 1982.

[743] J. S. Marsh and D. A. Plant. "Instrumental Colour Match Prediction using Organic Pigments." *Journal of the Oil and Colour Chemists Association* 47 (1964), 554–575.

[744] S. R. Marshner, S. H. Westin, E. P. F. Lafortune, K. E. Torrance, and D. P. Greenberg. "Image-Based BRDF Measurement Including Human Skin." In *Proceedings of the 10th Eurographics Workshop on Rendering*, pp. 139–152. Aire-la-Ville, Switzerland: Eurographics Association, 1999.

[745] S. R. Marshner, S. H. Westin, E. P. F. Lafortune, and K. Torrance. "Image-Based Measurement of the Bidirectional Reflectance Distribution Function." *Applied Optics* 39:16 (2000), 2592–2600.

[746] P. R. Martin, B. B. Lee, A. J. R. White, S. G. Solomon, and L. Rüttiger. "Chromatic Sensitivity of Ganglion Cells in the Peripheral Primate Retina." *Nature* 410:6831 (2001), 933–936.

[747] K. A. C. Martin. "Microcircuits in Visual Cortex." *Current Opinion in Neurobiology* 12:4 (2002), 418–425.

[748] M. Mase, S. Kawahito, M. Sasaki, and Y. Wakamori. "A 19.5b Dynamic Range CMOS Image Sensor with 12b Column-Parallel Cyclic A/D Converters." In *IEEE International Solid-State Circuits Conference (ISSCC)*, pp. 350–351, 2005.

[749] R. H. Masland. "Processing and Encoding of Visual Information in the Retina." *Current Opinion in Neurobiology* 6 (1996), 467–474.

[750] M. Massimi. *Pauli's Exclusion Principle: The Origin and Validation of a Scientific Principle*. Cambridge, UK: Cambridge University Press, 2005.

[751] W. Matusik, H. Pfister, M. Brand, and L. McMillan. "A Data-Driven Reflectance Model." *ACM Transactions on Graphics* 22:3 (2003), 759–769.

[752] W. Matusik, H. Pfister, M. Brand, and L. McMillan. "Efficient Isotropic BRDF Measurement." In *Proceedings of the* 14th *Eurographics Symposium on Rendering*, pp. 241–316. Aire-la-Ville, Switzerland: Eurographics Association, 2003.

[753] J. H. R. Maunsell and D. C. van Essen. "The Connections of the Middle Temporal Visual Area (MT) and Their Relationship to a Cortical Hierarchy in the Macaque Monkey." *Journal of Neuroscience* 3:12 (1983), 2563–2586.

[754] J. H. R. Maunsell and W. T. Newsome. "Visual Processing in Monkey Extrastriate Cortex." *Annual Review of Neuroscience* 10 (1987), 363–401.

[755] D. M. Maurice. "The Structure and Transparency of the Cornea." *Journal of Physiology* 136:2 (1957), 263–286.

[756] J. C. Maxwell. "Experiments on Color, as Perceived by the Eye, with Remarks on Color-Blindness." *Transactions of the Royal Society of Edinburgh* 21 (1855), 275–298.

[757] C. S. McCamy, H. Marcus, and J. G. Davidson. "A Color Rendition Chart." *Journal of Applied Photographic Engineering* 2:3 (1976), 95–99.

[758] J. J. McCann and A. Rizzi. "Veiling Glare: The Dynamic Range Limit of HDR Images." In *Human Vision and Electronic Imaging XII*, 6492, 6492. SPIE, 2007.

[759] J. McCann. "Lessons Learned from Mondrians Applied to Real Images and Color Gamuts." In *Proceedings of the IS&T Seventh Color Imaging Conference*, pp. 1–8. Springfield, VA: Society for Imaging Science and Technology, 1999.

[760] E. J. McCartney. *Optics of the Atmosphere: Scattering by Molecules and Particles*. New York: John Wiley and Sons, Inc., 1976.

[761] W. R. McCluney. *Introduction to Radiometry and Photometry*. Boston: Artech House Publishers, 1994.

[762] C. McCollough. "Color Adaptation of Edge Detectors in the Human Visual System." *Science* 149:3688 (1965), 1115–1116.

[763] C. McCollough. "Do McCollough Effects Provide Evidence for Global Pattern Processing?" *Perception and Psychophysics* 62:2 (2000), 350–362.

[764] M. McCool, J. Ang, and A. Ahmad. "Homomorphic Factorization of BRDFs for High-Performance Rendering." In *Proceedings of SIGGRAPH 2001, Computer Graphics Proceedings, Annual Conference Series*, pp. 171–178. Resding, MA: Addison-Wesley, 2001.

[765] R. McDonald. "Acceptibility and Perceptibility Decisions using the CMC Colour Difference Formula." *Textile Chemist and Colorist* 20:6 (1988), 31–31.

[766] D. McIntyre. *Colour Blindness: Causes and Effects*. Austin, TX: Dalton Publishing, 2002.

[767] K. McLaren. *The Colour Science of Dyes and Pigments*, Second edition. Bristol, UK: Adam Hilger Ltd., 1986.

[768] D. McQuarrie. *Quantum Chemistry*. Mill Valley, CA: University Science Books, 1983.

[769] H. McTavish. "A Demonstration of Photosynthetic State Transitions in Nature." *Photosynthesis Research* 17 (1988), 247–247.

[770] C. A. Mead. "Operation of Tunnel-Emission Devices." *Journal of Applied Physics* 32:4 (1961), 646–652.

[771] J. A. Medeiros. *Cone Shape and Color Vision: Unification of Structure and Perception*. Blountsville, AL: Fifth Estate Publishers, 2006.

[772] J. Mellerio. "Yellowing of the Human Lens: Nuclear and Cortical Contributions." *Vision Research* 27:9 (1987), 1581–1587.

[773] C. E. Metz. "Basic Principles of ROC Analysis." *Seminars in Nuclear Medicine* 8 (1978), 283–298.

[774] L. Meylan and S. Süsstrunk. "Color Image Enhancement using a Retinex-Based Adaptive Filter." In *Proceedings of the 2nd IS&T European Conference on Color in Graphics, Image, and Vision (CGIV 2004)*, pp. 359–363. Springfield, VA: Society for Imaging Science and Technology, 2004.

[775] L. Meylan and S. Süsstrunk. "High Dynamic Range Image Rendering with a Retinex-Based Adaptive Filter." *IEEE Transactions on Image Processing* 15:9 (2006), 2820–2930.

[776] L. Meylan, S. Daly, and S. Süsstrunk. "The Reproduction of Specular Highlights on High Dynamic Range Displays." In *Proceedings of the 14th IS&T/SID Color Imaging Conference*. Springfield, VA: Society for Imaging Science and Technology, 2006.

[777] L. Meylan, D. Alleysson, and S. Süsstrunk. "A Model of Retinal Local Adaptation for the Tone Mapping of CFA Images." *Journal of the Optical Society of America A*.

[778] L. Meylan, S. Daly, and S. Süsstrunk. "Tone Mapping for High Dynamic Range Displays." In *Proceedings of SPIE 6492 (Human Vision and Electronic Imaging XII)*. Belingham, WA: SPIE, 2007.

[779] G. Mie. "Beiträge zur Optik trüber Medien, speziell kolloidaler Metallösungen." *Annalen der Physik* 25:4 (1908), 377–377.

[780] N. J. Miller, P. Y. Ngai, and D. D. Miller. "The Application of Computer Graphics in Lighting Design." *Journal of the IES* 14 (1984), 6–26.

[781] J. Millman. "Electronic Energy Bands in Metallic Lithium." *Physical Review* 47:4 (1935), 286–290.

[782] W. I. Milne and K. B. K. Teo. "Growth and Characterization of Carbon Nanotubes for Field Emission Display Applications." *Journal of the Society for Information Display* 12 (2004), 289–292.

[783] M. G. J. Minnaert. *Light and Color in the Outdoors*. New York: Springer-Verlag, 1993.

[784] T. Mitsunaga and S. K. Nayar. "Radiometric Self Calibration." In *International Conference on Computer Vision and Pattern Recognition*, pp. 374–380. Washington, DC: IEEE Press, 1999.

[785] J. D. Mollon and J. K. Bowmaker. "The Spatial Arrangement of Cones in the Primate Retina." *Nature* 360:6405 (1992), 677–679.

[786] J. D. Mollon and G. Estévez. "Tyndall's Paradox of Hue Discrimination." *Journal of the Optical Society of America A* 5 (1988), 151–159.

[787] J. D. Mollon. "'Tho' She Kneel'd in that Place where They Grew...', The Uses and Origins of Primate Colour Vision." *Journal of Experimental Biology* 146 (1989), 21–38.

[788] F. M. De Monasterio and P. Gouras. "Functional Properties of Ganglion Cells of the Rhesus Monkey Retina." *Journal of Physiology (London)* 251 (1975), 167–195.

[789] P. Moon and D. E. Spencer. "On the Stiles-Crawford effect." *Journal of the Optical Society of America* 34:6 (1944), 319–329.

[790] G. Mori and J. Malik. "Recognizing Objects in Adversarial Clutter: Breaking a Visual Captcha." In *Proceedings of CVPR*, pp. 134–141. Los Alamitos, CA: IEEE Computer Society, 2003.

[791] N. Moroney and I. Tastl. "A Comparison of Retinex and iCAM for Scene Rendering." *Journal of Electronic Imaging* 13:1 (2004), 139–145.

[792] N. Moroney. "Chroma Scaling and Crispening." In *IS&T/SID 9th Color Imaging Conference*, pp. 97–101. Springfield, VA: Society for Imaging Science and Technology, 2001.

[793] N. Moroney. "A Hypothesis Regarding the Poor Blue Constancy of CIELAB." *Color Research and Application* 28 (2003), 371–378.

[794] J. Morovic and M. R. Luo. "Gamut Mapping Algorithms Based on Psychophysical Experiment." In *Proceedings of the* 5th *IS&T/SID Color Imaging Conference*, pp. 44–49. Springfield, VA: Society for Imaging Science and Technology, 1997.

[795] J. Morovic and M. R. Luo. "The Fundamentals of Gamut Mapping: A Survey." *Journal of Imaging Science and Technology* 45:3 (2001), 283–290.

[796] J. Morovic and P. L. Sun. "The Influence of Image Gamuts on Cross-Media Colour Image Reproduction." In *Proceedings of the* 8th *IS&T/SID Color Imaging Conference*, pp. 324–329. Springfield, VA: Society for Imaging Science and Technology, 2000.

[797] J. Morovic. "To Develop a Universal Gamut Mapping Algorithm." Ph.D. thesis, University of Derby, Derby, UK, 1998.

[798] J. Morovic. "Gamut Mapping." In *Digital Color Imaging Handbook*, edited by G. Sharma, Chapter 10, pp. 639–685. Boca Raton, FL: CRC Press, 2003.

[799] B. Moulden and F. A. A. Kingdom. "White's Effect: A Dual Mechanism." *Vision Research* 29 (1989), 1245–1236.

[800] K. Moutoussis and S. Zeki. "Responses of Spectrally Selective Cells in Macaque Area V2 to Wavelengths and Colors." *Journal of Neurophysiology* 87:4 (2002), 2104–2112.

[801] K. T. Mullen and F. A. A. Kingdom. "Losses in Periferal Colour Sensitivity Predicted from "Hit and Miss" Post-Receptoral Cone Connections." *Vision Research* 36:13 (1996), 1995–2000.

[802] C. D. Müller, A. Falcou, N. Reckefuss, M. Rojahn, V. Wiederhirn, P. Rudati, H. Frohne, O. Nuyken, H. Becker, and K. Meerholtz. "Multi-Colour Organic Light Emitting Displays by Solution Processing." *Nature* 421 (2003), 829–833.

[803] H. Müller. "Über die entoptische Wahrnehmung der Netzhautgefässe, insbesondere als Beweismittel für die Lichtperzeption durch die nach hinten gelegenen Netzhautelemente." *Verhandlungen der Physikalisch Medizinische Gesellschaft in Würzburg* 5 (1854), 411–411.

[804] J. C. Mullikin, L. J. van Vliet, H. Netten, F. R. Boddeke, G. van der Feltz, and I. T. Young. "Methods for CCD Camera Characterization." In *Image Acquisition and Scientific Imaging Systems*, SPIE vol. 2173, edited by H. C. Titus and A. Waks, SPIE vol. 2173, pp. 73–84, 1994.

[805] A. H. Munsell. *A Color Notation*, First edition. Baltimore, MD: Munsell Color Company, 1905.

[806] A. H. Munsell. *Atlas of the Munsell Color System.* Malden, MA: Wadsworth-Howland and Company, 1915.

[807] A. H. Munsell. *Munsell Book for Color.* Baltimore, MD: Munsell Color Company, 1929.

[808] A. H. Munsell. *A Color Notation: An Illustrated System Defining All Colors and Their Relations by Measured Scales of Hue, Value, and Chroma*, Ninth edition. Baltimore, MD: Munsell Color Company, 1941.

[809] Y. Muramatsu, S. Kurosawa, M. Furumiya, H. Ohkubo, and Y. Nakashiba. "A Signal-Processing CMOS Image Sensor using a Simple Analog Operation." In *IEEE International Solid-State Circuits Conference (ISSCC)*, pp. 98–99, 2001.

[810] S. Musa. "Active-Matrix Liquid-Crystal Displays." *Scientific American* November.

[811] K. I. Naka and W. A. H. Rushton. "S-Potentials from Luminosity Units in the Retina of Fish (Cyprinidae)." *Journal of Physiology (London)* 185:3 (1966), 587–599.

[812] J. Nakamura. "Basics of Image Sensors." In *Image Sensors and Signal Processing for Digital Still Cameras*, edited by J. Nakamura, pp. 53–93. Boca Raton, FL: Taylor and Francis, 2006.

[813] J. Nakamura, editor. *Image Sensors and Signal Processing for Digital Still Cameras.* Boca Raton, FL: Taylor and Francis, 2006.

[814] K. Nassau. "Fundamentals of Color Science." In *Color for Science, Art and Technology*, edited by K. Nassau. Amsterdam: Elsevier Science B.V., 1998.

[815] K. Nassau. *The Physics and Chemistry of Color: The Fifteen Causes of Color.* New York: John Wiley and Sons, 2001.

[816] S. K. Nayar and V. Branzoi. "Adaptive Dynamic Range Imaging: Optical Control of Pixel Exposures over Space and Time." In *International Conference on Computer Vision, Volume 2*, pp. 1168–1175. Washington, DC: IEEE Press, 2003.

[817] S. K. Nayar and T. Mitsunaga. "High Dynamic Range Imaging: Spatially Varying Pixel Exposures." In *International Conference on Computer Vision and Pattern Recognition, Volume 1*, pp. 472–479. Washington, DC: IEEE Press, 2000.

[818] S. K. Nayar, G. Krishnan, M. D. Grossberg, and R. Raskar. "Fast Separation of Direct and Global Components of a Scene using High Frequency Illumination." *ACM Transactions on Graphics* 25:3 (2006), 935–944.

[819] Y. Nayatani, K. Takahama, and H. Sobagaki. "Formulation of a Nonlinear Model of Chromatic Adaptation." *Color Research and Application* 6 (1981), 161–171.

[820] Y. Nayatani, K. Takahama, H. Sobagaki, and J. Hirono. "On Exponents of a Nonlinear Model of Chromatic Adaptation." *Color Research and Application* 7 (1982), 34–45.

[821] Y. Nayatani, T. Mori, K. Hashimoto, and H. Sobagaki. "Comparison of Color Appearance Models." *Color Research and Application* 15 (1990), 272–284.

[822] Y. Nayatani. "Proposal of an Opponent Colors System Based on Color Appearance and Color Vision Studies." *Color Research and Application* 29 (2004), 135–150.

[823] M. Neitz, J. Neitz, and G. H. Jacobs. "Spectral Tuning of Pigments Underlying Red-Green Color Vision." *Science* 252 (1991), 971–974.

[824] J. Neitz, J. Carroll, Y. Yamauchi, M. Neitz, and D. R. Williams. "Color Perception Is Mediated by a Plastic Neural Mechanism that Is Adjustable in Adults." *Neuron* 35:4 (2002), 783–792.

[825] R. Nelson and H. Kolb. "A17: A Broad-Field Amacrine Cell of the Rod System in the Retina of the Cat." *Joural of Neurophysiology* 54 (1985), 592–614.

[826] R. Nelson. "Visual Responses of Ganglion Cells.", 2006. Webvision Article. Available online (http://webvision.med.utah.edu).

[827] F. L. van Ness and M. A. Bouman. "Spatial Modulation Transfer in the Human Eye." *Journal of the Optical Society of America* 57 (1967), 401–406.

[828] H. E. J. Neugebauer. "Die theoretischen Grundlagen des Mehrfarbenbuchdrucks." *Zeitschrift für wissenschaftliche Photographie, Photophysik und Photochemie* 36:4 (1937), 73–89.

[829] H. J. Neugebauer. "Clausius-Mosotti Equation for Anisotropic Crystals." *Physical Review* 88:5 (1952), 1210–1210.

[830] H. Neumann, L. Pessoa, and T. Hansen. "Visual Filling-In for Computing Perceptual Surface Properties." *Biological Cybernetics* 85 (2001), 355–369.

[831] S. Newhall. "Preliminary Report of the O.S.A. Subcommittee on the Spacing of the Munsell Colors." *Journal of the Optical Society of America* 30 (1940), 617–645.

[832] R. G. Newton. *Scattering Theory of Waves and Particles*, Second edition. Mineola, NY: Dover Publications, 2002.

[833] M. Nezamabadi and R. S. Berns. "The Effect of Image Size on the Color Appearance of Image Reproduction using Colorimetrically Calibrated LCD and DLP Displays." *Journal of the Society for Information Display* 14:9 (2006), 773–783.

[834] R. Ng, M. Levoy, M. Bredif, G. Duval, M. Horowitz, and P. Hanrahan. "Light Field Photography with a Hand-Held Plenoptic Camera.", 2005.

[835] D. Q. Nguyen, R. Fedkiw, and H. W. Jensen. "Physically Based Modeling and Animation of Fire." *ACM Transactions on Graphics* 21:3 (2002), 721–728.

[836] D. Nickerson. "Munsell Renotations for Samples of OSA Uniform Color Scales." *Journal of the Optical Society of America* 68 (1978), 1343–1347.

[837] D. Nickerson. "OSA Uniform Color Scales Samples — A Unique Set." *Colour Research and Application* 6 (1981), 7–33.

[838] F. E. Nicodemus, J. C. Richmond, J. J. Hsia, I. W. Ginsberg, and T. Limperis. "Geometrical Considerations and Nomenclature for Reflectance." Technical Report Monograph 160, National Bureau of Standards (US), Gaithersburg, MD, 1977.

[839] F. E. Nicodemus, editor. *Self-Study Manual on Optical Radiation Measurements*. National Bureau of Standards, 1976. Technical Note 910-1.

[840] C. L. Nikias and A. P. Petropulu. *Higher-Order Spectra Analysis*. Signal Processing Series, Englewood Cliffs, NJ: Prentice Hall, 1993.

[841] M. Nonako, T. Kashmina, and Y. Knodo. "A 5-Meter Integrating Sphere." *Applied Optics* 6:4 (1967), 757–771.

[842] V. O'Brien. "Contour Perception, Illusion and Reality." *Journal of the Optical Society of America* 49 (1958), 112–119.

[843] S. Ogata, J. Ishida, and T. Sasano. "Optical Sensor Array in an Artificial Compound Eye." *Optical Engineering* 33:11 (1994), 3649–3655.

[844] M. P. Ogren and A. E. Hendrickson. "Pathways between Striate Cortex and Subcortical Regions in *Macaca mulatta* and *Saimiri sciureus*: Evidence for a Reciprocal Pulvinar Connection." *Journal of Comparative Neurology* 53 (1976), 780–800.

[845] T. Oguchi, E. Yamaguchi, K. Sazaki, K. Suzuki, S. Uzawa, and K. Hatanaka. "A 36 Inch Surface-Conduction Electron-Emitter Display (SED)." In *Technical Digest of the Society for Information Display (SID)*, pp. 1929–1931. San Jose, CA: Society for Information Display, 2005.

[846] K. Ohki, S. Chung, Y. H. Ch'ng, P. Kara, and R. C. Reid. "Functional Imaging with Cellular Resolution Reveals Precise Micro-Architecture in Visual Cortex." *Nature* 433:7026 (2005), 597–603.

[847] K. Ohno and T. Kusunoki. "The Effect of Ultrafine Pigment Color Filters on Cathode Ray Tube Brightness, Contrast, and Color Purity." *Journal of the Electrochemical Society* 143:3 (1996), 1063–1067.

[848] Y. Ohno. "Spectral Design Considerations for White LED Color Rendering." *Optical Engineering* 44:11 (2005), 111302–1 – 111302–9.

[849] N. Ohta and A. R. Robertson. *Colorimetry: Fundamentals and Applications*. Chichester, UK: John Wiley and Sons, 2005.

[850] R. Oi and K. Aizawa. "Wide Dynamic Range Imaging by Sensitivity Adjustable CMOS Image Sensor." In *Proceedings of the IEEE International Conference on Image Processing (ICIP)*, pp. 583–586, 2003.

[851] H. W. Ok, S. D. Lee, W. H. Choe, D. S. Park, and C. Y. Kim. "Color Processing for Multi-Primary Display Devices." In *IEEE International Conference on Image Processing 2005*, pp. 980–983. Washington, DC: IEEE, 2005.

[852] K. Okinaka, A. Saitoh, T. Kawai, A. Senoo, and K. Ueno. "High Performable Green-Emitting OLEDs." *SID Symposium Digest of Technical Papers* 35:1 (2004), 686–689.

[853] B. A. Olshausen and D. J. Field. "Emergence of Simple-Cell Receptive Field Properties by Learning a Sparse Code for Natural Images." *Nature* 381 (1996), 607–609.

[854] B. A. Olshausen and D. J. Field. "How Close Are We to Understanding V1?" *Neural Computation* 17:8 (2005), 1665–1699.

[855] W. O'Mara. *Liquid Crystal Flat Panel Displays: Manufacturing Science and Technology*. New York: Van Nostrand Reinhold, 1993.

[856] A. V. Oppenheim, R. Schafer, and T. Stockham. "Nonlinear Filtering of Multiplied and Convolved Signals." *Proceedings of the IEEE* 56:8 (1968), 1264–1291.

[857] A. V. Opppenheim and R. Schafer. *Digital Signal Processing*. Englewood Cliffs, NJ: Prentice Hall, 1975.

[858] A. V. Opppenheim, R. Schafer, and J. R. Buck. *Discrete-Time Signal Processing*, Second edition. Englewood Cliffs, NJ: Prentice-Hall, 1999.

[859] M. Oren and S. K. Nayar. "Generalization of Lambert's Reflectance Model." In *Proceedings of SIGGRAPH '94, Computer Graphics Proceedings, Annual Conference Series*, edited by A. Glassner, pp. 239–246. New York: ACM Press, 1994.

[860] OSA. *The Science of Color*. Washington: Optical Society of America, 1963.

[861] D. Osorio, D. L. Ruderman, and T. W. Cronin. "Estimation of Errors in Luminance Signals Encoded by Primate Retina Resulting from Sampling of Natural Images with Red and Green Cones." *Journal of the Optical Society of America A* 15:1 (1998), 16–22.

[862] G. Østerberg. "Topography of the Layer of Rods and Cones in the Human Retina." *Acta Ophthalmologica, Supplement* 6 (1935), 1–103.

[863] V. Ostromoukhov, C. Donohue, and P. Jodoin. "Fast Hierarchical Importance Sampling with Blue Noise Properties." *ACM Transactions on Graphics* 23:3 (2004), 488–495.

[864] V. Ostromoukhov. "Chromaticity Gamut Enhancement by Heptatone Multi-Color Printing." In *Proceedings IS&T/SPIE 1909, 1993 International Symposium on Electronic Imaging: Science & Technology*, pp. 139–151. Bellingham, WA: SPIE, 1993.

[865] C. A. Padgham and J. E. Saunders. *The Perception of Light and Color*. New York: Academic Press, 1975.

[866] R. H. Page, K. I. Schaffers, P. A. Waide, J. B. Tassano, S. A. Payne, W. F. Krupke, and W. K. Bischel. "Upconversion-Pumped Luminescence Efficiency of Rare-Earth-Doped Hosts Sensitized with Trivalent Ytterbium." *Journal of the Optical Society of America B* 15:3 (1998), 996–1008.

[867] G. Palmer. *Theory of Colors and Vision*. London: Leacroft, 1777. Reprinted in David L MacAdam (ed.), *Selected Papers on Colorimetry — Fundamentals*, SPIE Optical Engineering Press, 1993.

[907] I. Pobbaravski. "Methods of Computing Inks Amount to Produce a Scale of Neutrals for Photomechanical Reproduction." In *Proceedings of the Technical Association of the Graphical Arts*, pp. 10–33. Sewickley, PA: Technical Association of the Graphic Arts, 1966.

[908] A. B. Poirson and B. A. Wandell. "The Appearance of Colored Patterns: Pattern-Color Separability." *Journal of the Optical Society of America A* 10:12 (1993), 2458–2471.

[909] A. Poirson and B. Wandell. "Pattern-Color Separable Pathways Predicts Sensitivity to Simple Colored Patterns." *Vision Research* 35:2 (1996), 239–254.

[910] J. Pokorny, V. C. Smith, G. Verriest, and A. J. L. G. Pinkers. *Congenital and Acquired Color Vision Defects*. New York: Grune and Stratton, 1979.

[911] J. Pokorny, V. C. Smith, and M. Wesner. "Variability in Cone Populations and Implications." In *From Pigments to Perception*, edited by A. Valberg and B. B. Lee, pp. 23–34. New York: Plenum, 1991.

[912] F. Pollak. "Masking for Halftone." *Journal of Photographic Science* 3 (1955), 180–188.

[913] F. Pollak. "The Relationship between the Densities and Dot Size of Halftone Multicolor Images." *Journal of Photographic Science* 3 (1955), 112–116.

[914] F. Pollak. "New Thoughts on Halftone Color Masking." *Penrose Annual* 50 (1956), 106–110.

[915] S. L. Polyak. *The Vertebrate Visual System*. Chicago, IL: Chicago University Press, 1957.

[916] M. Ponzo. "Rapports entre quelque illusions visuelles de contraste angulaire et l'appréciation de grandeur des astres à l'horizon." *Archives Italiennes de Biologie* 58 (1913), 327–329.

[917] R. M. Pope and E. S. Fry. "Absorption Spectrum (380–700 nm) of Pure Water. II. Integrating Cavity Measurements." *Applied Optics* 36:33 (1997), 8710–8723.

[918] D. Pope. "The Elusive Blue Laser." *The Industrial Physicist* 3:3 (1997), 16.

[919] J. Portilla and E. P. Simoncelli. "A Parametric Texture Model based on Joint Statistics of Complex Wavelet Coefficients." *International Journal of Computer Vision* 40:1 (2000), 49–71. Available online ({www.cns.nyu.edu/~eero/publications.html}).

[920] D. L. Post and C. S. Calhoun. "An eEvaluation of Methods for Producing Desired Colors on CRT Monitors." *Color Research and Application* 14:4 (1989), 172–186.

[921] P. Poulin and A. Fournier. "A Model for Anisotropic Reflection." *Computer Graphics (Proceedings of SIGGRAPH '90)* 24:4 (1990), 273–282.

[922] C. Poynton. "Color FAQ.", 2002. Available online (http://www.poynton.com/ColorFAQ.html).

[923] C. Poynton. *Digital Video and HDTV: Algorithms and Interfaces*. San Francisco: Morgan Kaufmann Publishers, 2003.

[924] O. Prache. "Active Matrix Molecular OLED Microdisplays." *Displays* 22 (2001), 49–56.

[925] A. J. Preetham, P. Shirley, and B. Smits. "A Practical Analytic Model for Daylight." In *Proceedings of SIGGRAPH '99, Computer Graphics Proceedings, Annual Conference Series*, pp. 91–100. Reading, MA: Addison-Wesley Longman, 1999.

[926] W. H. Press, S. A. Teukolsky, W. T. Vetterling, and B. P. Flannery. *Numerical Recipes in C: The Art of Scientific Computing*, Second edition. Cambridge, UK: Cambridge University Press, 1992.

[927] I. G. Priest. "A Proposed Scale for Use in Specifying the Chromaticity of Incandescent Illuminants and Various Phases of Daylight." *Journal of the Optical Society of America* 23 (1933), 141–141.

[928] E. N. Pugh Jr. and T. D. Lamb. "Phototransduction in Vertebrate Rods and Cones: Molecular Mechanisms of Amplification, Recovery and Light Adaptation." In *Handbook of Biological Physics, Volume 3*, edited by D. G. S. et al., pp. 183–254. Amsterdam: Elsevier, 2000.

[929] M. D. Purdy. "Spectral Hue as a Fucntion of Intensity." *American Journal of Psychology* 43 (1931), 541–559.

[930] J. E. Purkinje. *Neue Beiträge zur Kenntnis des Sehens in subjectiver Hinsicht.* Berlin: G Reimer, 1825.

[931] Z. Rahman, D. J. Jobson, and G. A. Woodell. "A Multiscale Retinex for Color Rendition and Dynamic Range Compression." In *Proceedings of the SPIE 2847 (Applications of Digital Image Processing XIX)*. Bellingham, WA: SPIE, 1996.

[932] Z. Rahman, G. A. Woodell, and D. J. Jobson. "A Comparison of the Multiscale Retinex with Other Image Enhancement Techniques." In *IS&T's 50th Annual Conference: A Celebration of All Imaging*, pp. 426–431. Springfield, VA: Society for Imaging Science and Technology, 1997.

[933] S. J. M. Rainville and F. A. A. Kingdom. "Spatial-Scale Contribution to the Detection of Mirror Symmetry in Fractal Noise." *Journal of the Optical Society of America A* 16:9 (1999), 2112–2123.

[934] V. S. Ramachandran. "Perception of Shape from Shading." *Nature* 331:6152 (1988), 163–166.

[935] S. Rámon y Cajal. "La rétine des vertébrés." *Cellule* 9 (1893), 119–255.

[936] A. Rapaport, J. Milliez, M. Bass, A. Cassanho, and H. Jenssen. "Review of the Properties of Up-Conversion Phosphors for New Emissive Displays." *SID Symposium Digest of Technical Papers* 35:1 (2004), 1237–1239.

[937] A. Rapaport, J. Milliez, M. Bass, A. Cassanho, and H. Jenssen. "Review of the Properties of Up-Conversion Phosphors for New Emissive Displays." *IEEE/OSA Journal of Display Technology* 2:1 (2006), 68–78.

[938] K. Rasche, R. Geist, and J. Westall. "Detail Preserving Reproduction of Color Images for Monochromats and Dichromats." *IEEE Computer Graphics and Applications* 25:3 (2005), 22–30.

[939] K. Rasche, R. Geist, and J. Westall. "Re-Coloring Images for Gamuts of Lower Dimension." *Eurographic Forum* 24:3 (2005), 423–432.

[940] D. D. Rathman, D. M. O'Mara, M. K. Rose, R. M. Osgood III, and R. K. Reich. "Dynamic Response of an Electronically Shuttered CCD Imager." *IEEE Transactions on Electron Devices* 51:6.

[941] M. S. Rea, editor. *The IESNA Lighting Handbook: Reference and Application*, Ninth edition. New York: The Illuminating Engineering Society of North America, 2000.

[942] C. Redies and L. Spillmann. "The Neon Color Effect in the Ehrenstein Illusion." *Perception* 10:6 (1981), 667–681.

[943] C. Redies, L. Spillmann, and K. Kunz. "Colored Neon Flanks and Line Gap Enhancement." *Vision Research* 24:10 (1984), 1301–1309.

[944] B. C. Regan, C. Julliot, B. Simmen, F. Viénot, P. Charles-Dominique, and J. D. Mollon. "Fruits, Foliage and the Evolution of Primate Colour Vision." *Philosophical Transactions of the Royal Society of London, B* 356:1407 (2001), 229–283.

[945] T. Regier, P. Kay, and N. Khetarpal. "Color Naming Reflects Optimal Partitions of Color Space." *Proceedings of the National Academy of Sciences* 104:4 (2007), 1436–1441.

[946] R. C. Reid and R. M. Shapley. "Spatial Structure of Cone Inputs to Receptive Fields in Primate Lateral Geniculate Nucleus." *Nature* 356:6371 (1992), 716–718.

[947] R. C. Reid and R. M. Shapley. "Space and Time Maps of Cone Photoreceptor Signals in Macaque Lateral Geniculate Nucleus." *Journal of Neuroscience* 22:14 (2002), 6158–6175.

[948] E. Reinhard and K. Devlin. "Dynamic Range Reduction Inspired by Photoreceptor Physiology." *IEEE Transactions on Visualization and Computer Graphics* 11:1 (2005), 13–24.

[949] E. Reinhard, M. Ashikhmin, B. Gooch, and P. Shirley. "Color Transfer between Images." *IEEE Computer Graphics and Applications* 21 (2001), 34–41.

[950] E. Reinhard, M. Stark, P. Shirley, and J. Ferwerda. "Photographic Tone Reproduction for Digital Images." *ACM Transactions on Graphics* 21:3 (2002), 267–276.

[951] E. Reinhard, A. O. Akyüz, M. Colbert, C. E. Hughes, and M. O'Connor. "Real-Time Color Blending of Rendered and Captured Video." In *Proceedings of the Interservice/Industry Training, Simulation and Education Conference*. Arlington, VA: National Defense Industrial Association, 2004.

[952] E. Reinhard, G. Ward, S. Pattanaik, and P. Debevec. *High Dynamic Range Imaging: Acquisition, Display and Image-Based Lighting*. San Francisco: Morgan Kaufmann Publishers, 2005.

[953] E. Reinhard, P. Debevec, G. Ward, K. Myszkowski, H. Seetzen, D. Hess, G. McTaggart, and H. Zargarpour. "High Dynamic Range Imaging: Theory and Practice.", 2006. (SIGGRAPH 2006 Course).

[954] E. Reinhard. "Parameter Estimation for Photographic Tone Reproduction." *journal of graphics tools* 7:1 (2003), 45–51.

[955] A. G. Rempel, M. Trentacoste, H. Seetzen, H. D. Young, W. Heidrich, L. Whitehead, and G. Ward. "Ldr2Hdr: On-the-fly Reverse Tone Mapping of Legacy Video and Photographs." *ACM Transactions on Graphics* 26:3 (2007), 39–1 – 39–6.

[956] E. Renner. *Pinhole Photography: Rediscovering a Historic Technique*, Third edition. Focal Press, 2004.

[957] M. Rezak and L. A. Benevento. "A Comparison of the Organization of the Projections of the Dorsal Lateral Geniculate Nucleus, the Inferior Pulvinar and Adjacent Lateral Pulvinar to Primary Visual Cortex (Area 17) in the Macaque Monkey." *Brain Research* 167 (1979), 19–40.

[958] P. Rheingans. "Color, Change, and Control for Quantitative Data Display." In *Proceedings of IEEE Visualization*, pp. 252–258. Los Alamitos, CA: IEEE Press, 1999.

[959] P. Rheingans. "Task-Based Color Scale Design." In *SPIE Proceedings (Applied Image and Pattern Recognition)*, pp. 33–43. Bellingham, WA: SPIE, 1999.

[960] D. C. Rich. "Light Sources and Illuminants: How to Standardize Retail Lighting." *Textile Chemist and Colorist* 30:1 (1998), 8–12.

[961] D. W. Rickman and N. C. Brecha. "Morphologies of Somatostatin-Immunoreactive Neurons in the Rabbit Retina." In *Neurobiology of the Inner Retina*, H31, edited by R. Weiler and N. N. Osborne, H31, pp. 461–468. Berlin: Springer-Verlag, 1989.

[962] L. A. Riggs. "Visual Acuity." In *Vision and Visual Perception*, edited by C. H. Graham. New York: John Wiley and Sons, Inc., 1965.

[963] K. Riley, D. S. Ebert, M. Kraus, J. Tessendorf, and C. Hansen. "Efficient Rendering of Atmospheric Phenomena." In *Eurographics Symposium on Rendering*, edited by H. W. Jensen and A. Keller, pp. 375–386. Aire-la-Ville, Switzerland: Eurographics Association, 2004.

[964] D. L. Ringach, R. M. Shapley, and M. J. Hawken. "Orientation Selectivity in Macaque V1: Diversity and Laminar Dependence." *Journal of Neuroscience* 22:13 (2002), 5639–5651.

[965] A. Rizzi, C. Gatta, and D. Marini. "From Retinex to Automatic Color Equalization: Issues in Developing a New Algorithm for Unsupervised Color Equalization." *SPIE Journal of Electronic Imaging* 13:1 (2004), 75–84.

[966] P. K. Robertson and J. F. O'Callaghan. "The Generation of Color Sequences for Univariate and Bivariate Mapping." *IEEE Computer Graphics and Applications* 6:2 (1986), 24–32.

[967] M. Robertson, S. Borman, and R. Stevenson. "Estimation-Theoretic Approach to Dynamic Range Enhancement using Multiple Exposures." *Journal of Electronic Imaging* 12:2 (2003), 219–228.

[968] A. H. Robins. *Perspectives on Human Pigmentation.* Cambridge, UK: Cambridge University Press, 1991.

[969] D. A. Robinson. "The Mechanics of Human Saccadic Eye Movements." *Journal of Physiology* 174 (1964), 245–264.

[970] K. S. Rockland, K. S. Saleem, and K. Tanaka. "Divergent Feedback Connections from Areas V4 and TEO in the Macaque." *Visual Neuroscience* 11 (1994), 579–600.

[971] K. S. Rockland. "Elements of Cortical Architecture Revisited." In *Cerebral Cortex*, edited by K. S. Rockland, J. H. Kaas, and A. Peters. New York: Plenum, 1997.

[972] R. W. Rodieck. "The Primate Retina." *Comparative Primate Biology* 4 (1988), 203–278.

[973] H. R. Rodman, K. M. Sorenson, A. J. Shim, and D. P. Hexter. "Calbindin Immunoreactivity in the Geniculo-Extrastriate System of the Macaque: Implications for Heterogeneity in the Koniocellular Pathway and Recovery from Cortical Damage." *Journal of Comparative Neurology* 431:2 (2001), 168–181.

[974] B. E. Rogowitz and A. D. Kalvin. "The 'Which Blair Project:' A Quick Visual Method for Evaluating Perceptual Color Maps." In *Proceedings of IEEE Visualization*, pp. 183–190. Los Alamitos, CA: IEEE Computer Society, 2001.

[975] J. Roizen and R. Lipkin. "SECAM: An End to Crosstalk." *Electronics* 38 (1965), 104–105.

[976] A. K. Romney and R. G. D'Andrade. "Modeling Lateral Geniculate Nucleus Cell Response Spectra and Munsell Reflectance Spectra with Cone Sensitivity Curves." *Proceedings of the National Academy of Sciences of the United States of America* 102:45 (2005), 16512–16517.

[977] A. Roorda and D. R. Williams. "The Arrangement of the Three Cone Classes in the Living Human Eye." *Nature* 397:6719 (1999), 520–522.

[978] E. Rosche. "Natural Categories." *Cognitive Psychology* 4:3 (1973), 328–350.

[979] A. Rose. "The Relative Sensitivities of Television Pick-Up Tubes, Photographics Film, and the Human Eye." In *Proceedings of the Institute of Radio Engineers*, 293–300, 293–300. Washington, DC: IEEE Press, 1942.

[980] M. Rosenblatt. "Remarks on Some Nonparametric Estimates of Density Functions." *Annals of Mathematical Statistics* 27:3 (1956), 832–837.

[981] S. Roth and W. Caldwell. "Four Primary Color Projection Display." *SID Symposium Digest of Technical Papers* 36:1 (2005), 1818–1821.

[982] J. M. Rubin and W. A. Richards. "Color Vision and Image Intensities: When Are Changes Material?" *Biological Cybernetics* 45:3 (1982), 215–226.

[983] D. L. Ruderman and W. Bialek. "Statistics of Natural Images: Scaling in the Woods." *Physical Review Letters* 73:6 (1994), 814–817.

[984] D. L. Ruderman, T. W. Cronin, and C.-C. Chiao. "Statistics of Cone Responses to Natural Images: Implications for Visual Coding." *Journal of the Optical Society of America A* 15:8 (1998), 2036–2045.

[985] D. L. Ruderman. "Origins of Scaling in Natural Images." *Vision Research* 37 (1997), 3385–3398.

[986] D. L. Ruderman. "The Statistics of Natural Images." *Network: Computation in Neural Systems* 5:4 (1997), 517–548.

[987] W. A. H. Rushton. "Visual Pigments in Man." In *Handbook of Sensory Physiology, VII/1*, edited by H. J. A. Dartnal. New York: Springer-Verlag, 1972.

[988] S. M. Sagar. "Somatostatin-Like Immunoreactive Material in the Rabbit Retina: Immunohistochemical staining using monoclonal antibodies." *Journal of Comparative Neurology* 266:2 (1987), 291–299.

[989] R. Saito and H. Kotera. "Extraction of Image Gamut Surface and Calculations of Its Volume." In *Proceedings of the 8th IS&T/SID Color Imaging Conference*, pp. 330–334. Springfield, VA: Society for Imaging Science and Technology, 2000.

[990] E. Salvador, A. Cavallaro, and T. Ebrahimi. "Shadow Identification and Classification using Invariant Color Models." In *International Conference on Acoustics, Speech, and Signal Processing (ICASSP), Volume 3*, pp. 1545–1548. Washington, DC: IEEE Press, 2001.

[991] C. D. Salzman, C. M. Murasugi, K. H. Britten, and W. T. Newsome. "Microstimulation in Visual Area MT: Effects on Direction Discrimination Performance." *Journal of Neuroscience* 12:6 (1992), 2331–2355.

[992] F. Samadzadegan, M. Hahn, M. Sarpulaki, and N. Mostofi. "Geometric and Radiometric Evaluation of the Potential of a High Resolution CMOS-Camera." In *Proceedings of the International Society for Photogrammetry and Remote Sensing (ISPRS)*, XXXV, B3, XXXV, B3, pp. 488–493. Istanbul, Turkey, 2004.

[993] J. H. Sandell and P. H. Schiller. "Effect of Cooling Area 18 on Striate Cortex Cells in the Squirrel Monkey." *Journal of Neurophysiology* 48:1 (1982), 38–48.

[994] G. Sapiro. "Color and Illuminant Voting." *IEEE Transactions on Pattern Analysis and Machine Intelligence* 21 (1999), 1210–1215.

[995] M. Sasaki, M. Mase, S. Kawahito, and Y. Tadokoro. "A Wide-Dynamic-Range CMOS Image Sensor based on Multiple Short Exposure-Time Readout with Multiple-Resolution Column Parallel ADC." *IEEE Sensors Journal* 7:1 (2007), 151–158.

[996] K. Sassen, N. C. Knight, Y. Takano, and A. J. Heymsfield. "Effects of Ice-Crystal Structure on Halo Formation: Cirrus Cloud Experimental and Ray-Tracing Modeling Studies." *Applied Optics* 33:21 (1994), 4590–4601.

[997] K. Sassen. "Halos in Cirrus Clouds: Why Are Classic Displays So Rare?" *Applied Optics* 44:27 (2005), 5684–5687.

[998] Y. Sato, M. D. Wheeler, and K. Ikeuchi. "Object Shape and Reflectance Modeling from Observation." In *Proceedings of SIGGRAPH'97, Computer Graphics Proceedings, Annual Conference Series*, pp. 379–387. Reading, MA: Addison-Wesley, 1997.

[999] Y. Sato, K. Amemiya, and M. Uchidoi. "Recent Progress in Device Performance and Picture Quality of Color Plasma Displays." *Journal of the Society for Information Display* 10 (2002), 17–23.

[1000] K. Sato. "Image-Processing Algorithms." In *Image Sensors and Signal Processing for Digital Still Cameras*, edited by J. Nakamura, pp. 223–253. Boca Raton, FL: Taylor and Francis, 2006.

[1001] A. van der Schaaf. "Natural Image Statistics and Visual Processing." Ph.D. thesis, Rijksuniversiteit Groningen, The Netherlands, 1998. Available online ({http://www.ub.rug.nl/eldoc/dis/science/a.v.d.schaaf/}).

[1002] J. Schanda. "Current CIE Work to Achieve Physiologically-Correct Color Metrics." In *Color Vision: Perspecitves from Different Disciplines*, edited by W. Backhaus, R. Kliegl, and J. S. Werner, pp. 307–318. Berlin: Walter de Gruyter, 1998.

[1003] R. Schettini, G. Ciocca, and I. Gagliardi. "Content-Based Color Image Retrieval with Relevance Feedback." In *Proceedings of the International Conference on Image Processing*, pp. 27AS2.8–27AS2.8. Los Alamitos, CA: IEEE Computer Society, 1999.

[1004] H. M. Schey. *Div, Grad, Curl, and All That.* New York: W. W. Norton and Company, 1973.

[1005] C. Schlick. "A Customizable Reflectance Model for Everyday Rendering." In *Fourth Eurographics Workshop on Rendering*, pp. 73–84. Aire-la-Ville, Switzerland: Eurographics Association, 1993.

[1006] C. Schlick. "Fast Alternatives to Perlin's Bias and Gain Functions." In *Graphics Gems IV*, pp. 401–403. San Diego: Academic Press, 1994.

[1007] C. Schlick. "Quantization Techniques for the Visualization of High Dynamic Range Pictures." In *Photorealistic Rendering Techniques*, edited by P. Shirley, G. Sakas, and S. Müller, pp. 7–20. Berlin: Springer-Verlag, 1994.

[1008] M. Schmolesky. "The Primary Visual Cortex.", 2006. Available online (http://webvision.med.utah.edu/VisualCortex.html).

[1009] J. L. Schnapf, T. W. Kraft, and D. A. Baylor. "Spectral Sensitivity of Human Cone Photoreceptors." *Nature* 325:6103 (1987), 439–441.

[1010] J. L. Schnapf, T. W. Kraft, B. J. Nunn, and D. A. Baylor. "Spectral Sensitivity of Primate Photoreceptors." *Visual Neuroscience* 1 (1988), 255–261.

[1011] M. Schrauf, B. Lingelbach, E. Lingelbach, and E. R. Wist. "The Hermann Grid and the Scintillation Effect." *Perception* 24, supplement (1995), 88–89.

[1012] M. Schrauf, B. Lingelbach, and E. R. Wist. "The Scintillating Grid Illusion." *Vision Research* 37:8 (1997), 1033–1038.

[1013] P. Schröder and W. Sweldens. "Spherical Wavelets: Efficiently Representing Functions on the Sphere." In *Proceedings of SIGGRAPH '95, Computer Graphics Proceedings, Annual Conference Series*, pp. 161–172. Reading, MA: Addison-Wesley, 1995.

[1014] E. F. Schubert and J. K. Kim. "Solid-State Light Sources Getting Smart." *Science* 308 (2005), 1274–1278.

[1015] M. Schultze. "Zur Anatomie und Physiologie der Retina." *Archiv für mikroskopische Anatomie und Entwicklungsmechanik* 2 (1866), 165–286.

[1016] A. J. Schwab. *Field Theory Concepts: Electromagnetic Fields, Maxwell's Equations, Grad, Curl, Div, etc.* Berlin: Springer-Verlag, 1988.

[1017] E. Schwartz, R. B. Tootell, M. S. Silverman, E. Switkes, and R. L. de Valois. "On the Mathematical Structure of the Visuotopic Mapping of Macaque Striate Cortex." *Science* 227 (1985), 1065–1066.

[1018] E. A. Schwartz. "Voltage Noise Observed in Rods of the Turtle Retina." *Journal of Physiology* 272 (1977), 217–246.

[1019] S. H. Schwartz. *Visual Perception: A Clinical Orientation.* New York: McGraw-Hill, 2004.

[1020] H. Seetzen, L. A. Whitehead, and G. Ward. "A High Dynamic Range Display using Low and High Resolution Modulators." *SID Symposium Digest of Technical Papers* 34:1 (2003), 1450–1453.

[1021] H. Seetzen, W. Heidrich, W. Stuerzlinger, G. Ward, L. Whitehead, M. Trentacoste, A. Ghosh, and A. Vorozcovs. "High Dynamic Range Display Systems." *ACM Transactions on Graphics* 23:3 (2004), 760–768.

[1022] H. Seetzen, H. Li, G. Ward, L. Whitehead, and W. Heidrich. "Guidelines for Contrast, Brightness, and Amplitude Resolution of Displays." In *Society for Information Display (SID) Digest*, pp. 1229–1233, 2006.

[1023] H. Seetzen, S. Makki, H. Ip, T. Wan, V. Kwong, G. Ward, W. Heidrich, and L. Whitehead. "Self-Calibrating Wide Color Gamut High Dynamic Range Display." In *Human Vision and Electronic Imaging XII*, 6492, edited by B. E. Rogowitz, T. N. Pappas, and S. J. Daly, 6492, p. 64920Z. SPIE, 2007.

[1024] R. G. Seippel. *Optoelectronics for Technicians and Engineering.* Englewood Cliffs, NJ: Prentice Hall, 1989.

[1025] C. C. Semmelroth. "Prediction of Lightness and Brightness on Different Backgrounds." *Journal of the Optical Society of America* 60 (1970), 1685–1689.

[1026] A. Sempel and M. Büchel. "Design Aspects of Low Power Polymer/OLED Passive-Matrix Displays." *Organic Electronics* 3 (2002), 89–92.

[1027] J. Sethian. "A Fast Marching Level Set Method for Monotonically Advancing Fronts." *Proceedings of the National Academy of Sciences of the United States of America* 93 (1996), 1591–1595.

[1028] S. A. Shafer. "Using Color to Separate Reflection Components." *Color Research and Application* 10:4 (1985), 210–218.

[1029] S. E. Shaheen, G. E. Jabbour, B. Kippelen, N. Peyghambarian, J. D. Anderson, S. R. Marder, N. R. Armstrong, E. Bellmann, and R. H. Grubbs. "Organic Light-Emitting Diode with 20 lm/W Efficiency using a Triphenyldiamine Side-Group Polymer as the Hole Transport Layer." *Applied Physics Letters* 74:21 (1999), 3212–3214.

[1030] R. Shapley and M. Hawken. "Neural Mechanisms for Color Perception in the Primary Visual Cortex." *Current Opinion in Neurobiology* 12:4 (2002), 426–432.

[1031] R. Shapley and J. Victor. "Hyperacuity in Cat Retinal Ganglion Cells." *Science* 231 (1986), 999–1002.

[1032] R. Shapley. "Visual Sensitivity and Parallel Retinocortical Channels." *Annual Review of Psychology* 41 (1990), 635–658.

[1033] R. Shapley. "A New View of the Primary Visual Cortex." *Neural Networks* 17:5–6 (2004), 615–623.

[1034] G. Sharma and H. J. Trussell. "Digital Color Imaging." *IEEE Transactions on Image Processing* 6:7 (1997), 901–932.

[1035] G. Sharma. "LCDs Versus CRTs: Color Calibration and Gamut Considerations." *Proceedings of the IEEE: Special Issue on Flat Panel Display Technologies* 90:4 (2002), 605–622.

[1036] L. T. Sharpe, H. J. Andrew Stockman, and J. Nathans. "Opsin Genes, Cone Photopigments, Color Vision and Color Blindness." In *Color Vision: From Genes to Perception*, edited by K. R. Gegenfurtner and L. T. Sharpe, pp. 3–51. Cambridge, UK: Cambridge University Press, 1999.

[1037] J. H. Shaw. "Solar Radiation." *The Ohio Journal of Science* 53:5 (1953), 258–271.

[1038] S. L. Sherwood, editor. *The Nature of Psychology: A Selection of Papers, Essays and Other Writings by Kenneth J W Craik*. Cambridge: Cambridge University Press, 1966.

[1039] S. K. Shevell and J. Wei. "Chromatic Induction: Border Contrast or Adaptation to Surrounding Light." *Vision Research* 38 (1998), 1561–1566.

[1040] J. Shinar, editor. *Organic Light-Emitting Devices: A Survey*. New York: Springer-Verlag, 2003.

[1041] S. Shipp and S. Zeki. "Segregation of Pathways Leading from Area V2 to Areas V4 and V5 of Macaque Monkey Visual Cortex." *Nature* 315:6017 (1985), 322–324.

[1042] P. Shirley and R. K. Morley. *Realistic Ray Tracing*, Second edition. Wellesley, Massachusetts: A K Peters, 2003.

[1043] P. Shirley, M. Ashikhmin, S. R. Marschner, E. Reinhard, K. Sung, W. B. Thompson, and P. Willemsen. *Fundamentals of Computer Graphics*, Second edition. Wellesley, Massachusetts: A K Peters, 2005.

[1044] A. Shmuel, M. Korman, A. Sterkin, M. Harel, S. Ullman, R. Malach, and A. Grinvald. "Retinotopic Axis Specificity and Selective Clustering of Feedback Projections from V2 to V1 in the Owl Monkey." *Journal of Neuroscience* 25:8 (2005), 2117–2131.

[1045] A. E. Siegman. *Lasers*. Sausalito, CA: University Science Books, 1986.

[1046] F. X. Sillion and C. Puech. *Radiosity and Global Illumination*. San Francisco, CA: Morgan Kaufmann Publishers, Inc., 1994.

[1047] E. P. Simoncelli and J. Portilla. "Texture Characterization via Joint Statistics of Wavelet Coefficient Magnitudes." In *Fifth IEEE Int'l Conf on Image Processing, Volume I*, pp. 62–66. Washington, DC: IEEE Press, 1998.

[1048] E. P. Simoncelli. "Bayesian Denoising of Visual Images in the Wavelet Domain." In *Bayesian Inference in Wavelet Based Models, Lecture Notes in Statistics 141*, edited by P. Müller and B. Vidakovic, Chapter 18, pp. 291–308. New York: Springer-Verlag, 1999.

[1049] E. P. Simoncelli. "Modelling the Joint Statistics of Images in the Wavelet Domain." In *Proceedings of the SPIE 3818 (44th SPIE Annual Meeting)*, pp. 188–195. Bellingham, WA: SPIE, 1999.

[1050] L. C. Sincich and J. C. Horton. "Divided by Cytochrome Oxydase: A Map of the Projections from V1 to V2 in Macaques." *Science* 295:5560 (2002), 1734–1737.

[1051] L. C. Sincich and J. C. Horton. "The Circuitry of V1 and V2: Integration of Color, Form and Motion." *Annual Review of Neuroscience* 28 (2005), 303–326.

[1052] L. C. Sincich, K. F. Park, M. J. Wohlgemuth, and J. C. Horton. "Bypassing V1: A Direct Geniculate Input to Area MT." *Nature Neuroscience* 7:10 (2004), 1123–1128.

[1053] L. Sivik. "Systems for Descriptive Colour Notations — Implications of Definitions and Methodology." *Farbe* 40 (1994), 37–49.

[1054] G. Smith and D. A. Atchison. *The Eye and Visual Optical Instruments*. New York: Cambridge University Press, 1997.

[1055] R. C. Smith and K. A. Baker. "Optical Properties of the Clearest Natural Waters (200-800 nm)." *Applied Optics* 20:2 (1981), 177–184.

[1056] R. C. Smith and J. E. Tyler. "Optical Properties of Clear Natural Water." *Journal of the Optical Society of America* 57:5 (1967), 589–601.

[1057] V. C. Smith, J. Pokorny, B. B. Lee, and D. M. Dacey. "Primate Horizontal Cell Dynamics: An Analysis of Sensitivity Regulation in the Outer Retina." *Journal of Neurophysiology* 85 (2001), 545–558.

[1058] K. Smith, G. Krawczyk, K. Myszkowski, and H.-P. Seidel. "Beyond Tone Mapping: Enhanced Depiction of Tone Mapped HDR Images." *Computer Graphics Forum* 25:3 (2006), 427–438.

[1059] R. Smith-Gillespie. "Design Considerations for LED Backlights in Large Format Color LCDs." In *LEDs in Displays — SID Technical Symposium*, 2006.

[1060] H. M. Smith. *Principles of Holography*, Second edition. New York, NY: John Wiley and Sons,, Inc., 1975.

[1061] W. J. Smith. *Modern Optical Engineering*, Third edition. New York: McGraw-Hill, 2000.

[1062] W. J. Smith. *Modern Lens Design*, Second edition. New York: McGraw Hill, 2005.

[1063] B. Smits. "An RGB to Spectrum Conversion for Reflectances." *journal of graphics tools* 4:4 (1999), 11–22.

[1064] E. Snitzer. "Cylindrical Dielectric Waveguide Modes." *Journal of the Optical Society of America* 51:5 (1961), 491–498.

[1065] D. M. Snodderly, J. D. Auran, and F. C. Delori. "The Macular Pigment II: Spatial Distribution in Primate Retinas." *Investigative Ophthalmology and Visual Science* 25:6 (1984), 674–685.

[1066] A. Sobel. "Television's Bright New Technology." *Scientific American*.

[1067] Society of Dyers and Colourists. *Colour Index.* Bradford, UK: Society of Dyers and Colourists, 1971. Available online (http://www.colour-index.org).

[1068] F. M. Sogandares and E. S. Fry. "Absorption Spectrum (340–640 nm) of Pure Water. II. Photothermal Measurements." *Applied Optics* 36:33 (1997), 8699–8709.

[1069] J. A. Solomon, G. Sperling, and C. Chubb. "The Lateral Inhibition of Perceived Contrast Is Indifferent to On-Center/Off-Center Segregation, but Specific to Orientation." *Vision Research* 33:18 (1993), 2671–2683.

[1070] X. Song, G. M. Johnson, and M. D. Fairchild. "Minimizing the Perception of Chromatic Noise in Digital Images." In *IS&T/SID 12th Color Imaging Conference*, pp. 340–346. Springfield, VA: Society for Imaging Science and Technology, 2004.

[1071] M. Sonka, V. Hlavac, and R. Boyle. *Image Processing, Analysis, and Machine Vision*, Second edition. Pacific Grove, CA: PWS Publishing, 1999.

[1072] K. E. Spaulding, E. J. Giorgianni, and G. Woolfe. "Reference Input/Output Medium Metric RGB Color Encoding (RIMM-ROMM RGB)." In *PICS2000: Image Processing, Image Quality, Image Capture, Systems Conference*, pp. 155–163. Springfield, VA: Society for Imaging Science and Technology, 2000.

[1073] K. E. Spauling, G. J. Woolfe, and R. L. Joshi. "Extending the Color Gamut and Dynamic Range of an sRGB Image using a Residual Image." *Color Research and Application* 25:4 (2003), 251–266.

[1074] N. Speranskaya. "Determination of Spectrum Color Co-ordinates for Twenty-Seven Normal Observers." *Optics Spectrosc.* 7 (1959), 424–428.

[1075] L. Spillmann and J. S. Werner, editors. *Visual Perception: The Neurological Foundations.* San Diego: Academic Press, 1990.

[1076] L. Spillmann. "The Hermann Grid Illusion: A Tool for Studying Human Receptive Field Organization." *Perception* 23 (1994), 691–708.

[1077] B. Stabell and U. Stabell. "Chromatic Rod-Cone Interaction during Dark Adaptation." *Journal of the Optical Society of America A* 15:11 (1998), 2809–2815.

[1078] B. Stabell and U. Stabell. "Effects of Rod Activity on Color Perception with Light Adaptation." *Journal of the Optical Society of America A* 19:7 (2002), 1248–1258.

[1079] J. Stam. "Stable Fluids." In *Proceedings SIGGRAPH '99, Computer Graphics Proceedings,Annual Conference Series*, pp. 121–128. Reading, MA: Addison-Wesley Longman, 1999.

[1080] J. Starck, E. Pantin, and F. Murtagh. "Deconvolution in Astronomy: A Review." *Publications of the Astronomical Society of the Pacific* 114 (2002), 1051–1069.

[1081] O. N. Stavroudis. "Orthomic Systems of Rays in Inhomogeneous Isotropic Media." *Applied Optics* 10:2 (1971), 260–263.

[1082] O. N. Stavroudis. *The Optics of Rays, Wavefronts and Caustics.* New York: Academic Press, 1972.

[1083] D. A. Steigerwald, J. C. Bhat, D. Collins, R. M. Fletcher, M. O. Holcomb, M. J. Ludowise, P. S. Martin, and S. L. Rudaz. "Illumination with Solid State Lighting Technology." *IEEE Journal on Selected Topics in Quantum Electronics* 8:2 (2002), 310–320.

[1084] J. Steinhoff and D. Underhill. "Modifications of the Euler Equations for "Vorticity Confinement": Application to the Computation of Interacting Vortex Rings." *Physics of Fluids* 6:8 (1994), 2738–2744.

[1085] J. C. Stevens and S. S. Stevens. "Brightness fFunctions: Effects of Adaptation." *Journal of the Optical Society of America* 53 (1963), 375–385.

[1086] S. S. Stevens. "To Honor Fechner and Repeal His Law." *Science* 133:3446 (1961), 80–86.

[1087] W. Stiles and J. Burch. "N.P.L Colour-Matching Investigation: Final Report." *Optica Acta* 6 (1959), 1–26.

[1088] W. S. Stiles and B. H. Crawford. "The Luminous Efficiency of Rays Entering the Eye Pupil at Different Points." *Proceedings of the Royal Society of London, B* 112:778 (1933), 428–450.

[1089] W. S. Stiles. "The Luminous Sensitivity of Monochromatic Rays Entering the Eye Pupil at Different Points and a New Colour Effect." *Proceedings of the Royal Society of London, B* 123:830 (1937), 64–118.

[1090] A. Stimson. *Photometry and Radiometry for Engineers*. New York: John Wiley and Sons, Inc., 1974.

[1091] A. Stockman and L. T. Sharpe. "The Spectral Sensitivities of the Middle- and Long-Wavelength-Sensitive Cones Derived from Measurements on Observers of Known Genotype." *Vision Research* 40:13 (2000), 1711–1737.

[1092] A. Stockman, D. I. MacLeod, and N. E. Johnson. "Spectral Sensitivities of the Human Cones." *Journal of the Optical Society of America A* 10:12 (1993), 2491–2521.

[1093] E. J. Stollnitz, V. Ostromoukhov, and D. H. Salesin. "Reproducing Color Images using Custom Inks." In *Proceedings SIGGRAPH '98, Computer Graphics Proceedings, Annual Conference Series*, pp. 267–274. Reading, MA: Addison-Wesley, 1998.

[1094] R. I. Stolyarevskaya. "The 25th Meeting of CIE and Measurements on LED Optical Parameters." *Measurement Techniques* 47:3 (2004), 316–320.

[1095] M. C. Stone, W. B. Cowan, and J. C. Beatty. "Color Gamut Mapping and the Printing of Digital Color Images." *ACM Transactions on Graphics* 7:4 (1988), 249–292.

[1096] M. C. Stone. "Color Balancing Experimental Projection Displays." In *Proceedings of the 9th Color Imaging Conference: Color Science and Engineering: Systems, Technologies, Applications*, pp. 342–347. Springfield, VA: Society for Imaging Science and Technology, 2001.

[1097] M. C. Stone. *A Field Guide to Digital Color*. Natick, MA: A K Peters, 2003.

[1098] M. Störring, E. Granum, and H. J. Andersen. "Estimation of the Illuminant Colour using Highlights from Human Skin." In *International Conference on Color in Graphics and Image Processing*, pp. 45–50. Paris: Lavoisier, 2000.

[1099] A. Streitwieser Jr, C. H. Heathcock, and E. M. Kosower. *Introduction to Organic Chemistry*, Fourth edition. New York: Macmillan, 1992.

[1100] C. F. Stromeyer. "Form-Color After Effects in Human Vision." In *Handbook of Sensory Physiology: Perception, VIII*, edited by R. Held, H. W. Leibowitz, and H. L. Teuber, pp. 97–142. New York: Springer-Verlag, 1978.

[1101] S. Subramanian and I. Biederman. "Does Contrast Reversal Affect Object Identification?" *Investigative Ophthalmology and Visual Science* 38:998.

[1102] K. Suffern. *Ray Tracing from the Ground Up*. Wellesley, MA: A K Peters, 2007.

[1103] A. Sugimoto, H. Ochi, S. Fujimura, A. Yoshida, T. Miyadera, and M. Tsuchida. "Flexible OLED Displays using Plastic Substrates." *IEEE Journal of Selected Topics in Quantum Electronics* 10:1 (2004), 107–114.

[1104] H. Sugiura, K. Sakawa, and J. Fujimo. "False Color Signal Reduction Method for Single-Chip Color Video Cameras." *IEEE Transactions on Consumer Electronics* 40:2 (1994), 100–106.

[1105] W. E. Sumpner. "The Diffusion of Light." *Proceedings of the Physical Society of London* 12 (1892), 10–29.

[1106] Y. Sun, D. Fracchia, T. W. Calvert, and M. S. Drew. "Deriving Spectra from Colors and Rendering Light Interference." *IEEE Computer Graphics and Applications* 19:4 (1999), 61–67.

[1107] Y. Sun. "Rendering Biological Iridescences with RGB-Based Renderers." *ACM Transactions on Graphics* 25:1 (2006), 100–129.

[1108] W. Suzuki, K. S. Saleem, and K. Tanaka. "Divergent Backward Projections from the Anterior Part of the Inferotemporal Cortex (Area TE) in the Macaque." *Journal of Comparative Neurology* 422:2 (2000), 206–228.

[1109] H. Suzuki. *Electronic Absorption Spectra and Geometry of Organic Molecules: An Application of Molecular Orbital Theory*. New York: Academic Press, 1967.

[1110] M. J. Swain and D. H. Ballard. "Color Indexing." *International Journal of Computer Vision* 7:1 (1991), 11–32.

[1111] Swedish Standards Institute. *Colour Notation System*, Second edition. Stockholm: Swedish Standards Institute, 1990.

[1112] K. Takatoh, M. Hasegawa, M. Koden, N. Itoh, R. Hasegawa, and M. Sakamoto. *Alignment Technologies and Applications of Liquid Crystal Devices*. London: Taylor and Francis, 2005.

[1113] H. Takeshi, H. Onoe, H. Shizuno, E. Yoshikawa, N. Sadato, H. Tsukada, and Y. Watanabe. "Mapping of Cortical Areas Involved in Color Vision in Non-Human Primates." *Neuroscience Letters* 230:1 (1997), 17–20.

[1114] A. A. Talin, K. A. Dean, and J. E. Jaskie. "Field Emission Displays: A Critical Review." *Solid-State Electronics* 45 (2001), 963–976.

[1115] E.-V. Talvala, A. Adams, M. Horowitz, and M. Levoy. "Veiling Glare in High Dynamic Range Imaging." *ACM Transactions on Graphics* 26:3 (2007), 37–1 – 37–9.

[1116] N. Tamura, N. Tsumura, and Y. Miyake. "Masking Model for Accurate Colorimetric Characterization of LCD." In *Proceedings of the IS&T/SID Tenth Color Imaging Conference*, pp. 312–316. Springfield, VA: Society of Imaging Science and Technology, 2002.

[1117] C. W. Tang and S. A. VanSlyke. "Organic Electroluminescent Diode." *Applied Physics Letters* 51:12 (1987), 913–915.

[1118] J. Tanida and K. Yamada. "TOMBO: Thin Observation by Bound Optics." *Applied Optics* 40:11 (2001), 1806–1813.

[1119] J. Tanida, R. Shogenji, Y. Kitamura, M. Miyamoto, and S. Miyatake. "Color Imaging with an Integrated Compound Imaging System." *Journal of the Optical Society of America A* 11:18 (2003), 2109–2117.

[1120] L. E. Tannas, editor. *Flat-Panel Displays and CRTs*. New York: Van Nostrand Reinhold Company, 1985.

[1121] A. H. Taylor and E. B. Rosa. "Theory, Construction, and Use of the Photometric Integrating Sphere." *Scientific Papers of the Bureau of Standards* 18 (1922), 280–325.

[1122] H. Terstiege. "Chromatic Adaptation: A State-of-the-Art Report." *Journal of Color and Appearance* 1 (1972), 19–23.

[1123] A. J. P. Theuwissen. "CCD or CMOS Image Sensors for Consumer Digital Still Photography." In *IEEE International Symposium on VLSI Technology, Systems, and Applications*, pp. 168–171, 2001.

[1124] W. B. Thompson, P. Shirley, and J. Ferwerda. "A Spatial Post-Processing Algorithm for Images of Night Scenes." *journal of graphics tools* 7:1 (2002), 1–12.

[1125] M. G. A. Thomson. "Higher-Order Structure in Natural Scenes." *Journal of the Optical Society of America A* 16:7 (1999), 1549–1553.

[1126] M. G. A. Thomson. "Visual Coding and the Phase Structure of Natural Scenes." *Network: Computation in Neural Systems* 10:2 (1999), 123–132.

[1127] L. G. Thorell, R. L. de Valois, and D. G. Albrecht. "Spatial Mapping of Monkey V1 Cells with Pure Color and Luminance Stimuli." *Vision Research* 24:7 (1984), 751–769.

[1128] R. Tilley. *Colour and the Optical Properties of Materials*. Chichester, UK: John Wiley and Sons, 1999.

[1129] D. Todorović. "Lightness and Junctions." *Perception* 26 (1997), 379–395.

[1130] D. J. Tolhurst, Y. Tadmor, and T. Chiao. "Amplitude Spectra of Natural Images." *Ophthalmic and Physiological Optics* 12 (1992), 229–232.

[1131] C. Tomasi and R. Manduchi. "Bilateral Filtering for Gray and Color Images." In *Proceedings of the IEEE International Conference on Computer Vision*, pp. 836–846. Washington, DC: IEEE Press, 1998.

[1132] S. Tominaga and B. A. Wandell. "The Standard Surface Reflectance Model and Illuminant Estimation." *Journal of the Optical Society of America A* 6:4 (1989), 576–584.

[1133] S. Tominaga and B. A. Wandell. "Natural Scene-Illuminant Estimation using the Sensor Correlation." *Proceedings of the IEEE* 90:1 (2002), 42–56.

[1134] S. Tominaga, S. Ebisui, and B. A. Wandell. "Scene Illuminant Classification: Brighter is Better." *Journal of the Optical Society of America A* 18:1 (2001), 55–64.

[1135] S. Tominaga. "Surface Identification using the Dichromatic Reflection Model." *IEEE Transactions on Pattern Analysis and Machine Intelligence* 13 (1991), 658–670.

[1136] S. Tominaga. "Dichromatic Reflection Models for a Variety of Materials." *Color Research and Application* 19:4 (1994), 277–285.

[1137] S. Tominaga. "Multichannel Vision System for Estimation Surface and Illumination Functions." *Journal of the Optical Society of America A* 13:11 (1996), 2163–2173.

[1138] R. B. H. Tootell, J. B. Reppas, A. M. Dale, R. B. Look, M. I. Sereno, R. Malach, T. J. Brady, and B. R. Rosen. "Visual Motion Aftereffect in Human Cortical Area MT Revealed by Functional Magnetic Resonance Imaging." *Nature* 375:6527 (1995), 139–141.

[1139] R. B. H. Tootell, K. Nelissen, W. Vanduffel, and G. A. Orban. "Search for Color 'Center(s)' in Macaque Visual Cortex." *Cerebral Cortex* 14:4 (2004), 353–363.

[1140] K. E. Torrance and E. M. Sparrow. "Theory for Off-Specular Reflection from Roughened Surfaces." *Journal of the Optical Society of America* 57:9 (1967), 1105–1114.

[1141] K. Toyoda. "Digital Still Cameras at a Glance." In *Image Sensors and Signal Processing for Digital Still Cameras*, edited by J. Nakamura, pp. 1–19. Boca Raton, FL: Taylor and Francis, 2006.

[1142] C. van Trigt. "Color Rendering, A Reassessment." *Color Research and Application* 24:3 (1999), 197–206.

[1143] T. Troscianko and I. Low. "A Technique for Presenting Isoluminant Stimuli using a Microcomputer." *Spatial Vision* 1 (1985), 197–202.

[1144] T. Troscianko, R. Baddeley, C. Párraga, U. Leonards, and J. Troscianko. "Visual Encoding of Green Leaves in Primate Vision." *Journal of Vision* 3 (2003), 137a.

[1145] T. Troscianko. "A Given Visual Field Location Has a Wide Range of Receptive Field Sizes." *Vision Research* 22 (1982), 1363–1369.

[1146] J. Tsitsiklis. "Efficient Algorithms for Globally Optimal Trajectories." *IEEE Transactions on Automatic Control* 40 (1995), 1528–1538.

[1147] D. Y. Ts'o and C. D. Gilbert. "The Organization of Chromatic and Spatial Interactions in the Primate Striate Cortex." *Journal of Neuroscience* 8:5 (1988), 1712–1727.

[1148] T. Tsuji, S. Kawami, S. Miyaguchi, T. Naijo, T. Yuki, S. Matsuo, and H. Miyazaki. "Red-Phosphorescent OLEDs Employing bis-(8-Quinolinolato)-Phenolato-Aluminum (III) Complexes as Emission-Layer Hosts." *SID Symposium Digest of Technical Papers* 35:1 (2004), 900–903.

[1149] M. Tsukada and Y. Ohta. "An Approach to Color Constancy from Mutual Reflection." In *Proceedings of the 3rd International Conference on Computer Vision*, pp. 385–389. Washington, DC: IEEE Press, 1990.

[1150] T. Tsutsui. "A Light-Emitting Sandwich Filling." *Nature* 420:6917 (2002), 752–755.

[1151] H. F. J. M. van Tuijl. "A New Visual Illusion: Neonlike Color Spreading and Complementary Color Induction between Subjective Contours." *Acta Psychologica* 39 (1975), 441–445.

[1152] J. Tumblin and H. Rushmeier. "Tone Reproduction for Computer Generated Images." *IEEE Computer Graphics and Applications* 13:6 (1993), 42–48.

[1153] J. Tumblin and G. Turk. "LCIS: A Boundary Hierarchy for Detail-Preserving Contrast Reduction." In *Proceedings SIGGRAPH '99, Computer Graphics Proceedings, Annual Conference Series*, edited by A. Rockwood, pp. 83–90. Reading, MA: Addison-Wesley Longman, 1999.

[1154] K. Turkowski. "Transformations of Surface Normal Vectors with Applications to Three Dimensional Computer Graphics." Technical Report 22, Apple Computer, Inc., 1990.

[1155] S. R. Turns. *An Introduction to Combustion: Concepts and Applications*, Second edition. New York: McGraw-Hill, 1996.

[1156] S. A. Twomey, C. F. Bohren, and J. L. Mergenthaler. "Reflectance and Albedo Differences between Wet and Dry Surfaces." *Applied Optics* 25:3 (1986), 431–437.

[1157] C. W. Tyler. "Analysis of Human Receptor Density." In *Basic and Clinical Applications of Vision Science*, edited by V. Lakshminarayanan, pp. 63–71. Norwood, MA: Kluwer Academic Publishers, 1997.

[1158] E. P. T. Tyndall. "Chromaticity Sensibility to Wavelength Difference as a Function of Purity." *Journal of the Optical Society of America* 23:1 (1933), 15–22.

[1159] H. Uchiike and T. Hirakawa. "Color Plasma Displays." *Proceedings of the IEEE* 90:4 (2002), 533–539.

[1160] H. Uchiike, K. Miura, N. Nakayama, and T. Shinoda. "Secondary Electron Emission Characteristics of Dielectric Materials in AC-Operated Plasma Display Panels." *IEEE Transactions on Electron Devices* ED-23 (1976), 1211–1217.

[1161] Z. Ulanowski. "Ice Analog Halos." *Applied Optics* 44:27 (2005), 5754–5758.

[1162] R. Ulbricht. "Die Bestimmung der mittleren raumlichen Lichtintensitat durch nur eine Messung." *Elektrotechnische Zeitschrift* 21 (1900), 595–597.

[1163] I. Underwood. "A Review of Microdisplay Technologies." In *The Annual Conference of the UK and Ireland Chapter of the Society for Information Display*. San Jose, CA: Society for Information Display, 2000.

[1164] L. G. Ungerleider and R. Desimone. "Cortical Connections of Visual Area MT in the Macaque." *Journal of Comparative Neurology* 248:2 (1986), 190–222.

[1165] L. G. Ungerleider and R. Desimone. "Projections to the Superior Temporal Sulcus from the Central and Peripheral Field Representations of V1 and V2." *Journal of Comparative Neurology* 248:2 (1986), 147–163.

[1166] United States Committee on Extension to the Standard Atmosphere. *U.S. Standard Atmosphere, 1976*. Washington, D.C.: National Oceanic and Atmospheric Administration, National Aeronautics and Space Administration, United States Air Force, 1976.

[1167] M. Uyttendaele, A. Eden, and R. Szeliski. "Eliminating Ghosting and Exposure Artifacts in Image Mosaics." In *IEEE Computer Society Conference on Computer Vision and Pattern Recognition, Volume 2*, p. 509. Washington, DC: IEEE Press, 2001.

[1168] K. M. Vaeth. "OLED-display technology." *Information Display* 19:6 (2003), 12–17.

[1169] J. M. Valeton and D. van Norren. "Light-Adaptation of Primate Cones: An Analysis Based on Extracellular Data." *Vision Research* 23:12 (1982), 1539–1547.

[1170] R. L. de Valois and K. K. de Valois. "Neural Coding of Color." In *Handbook of Perception V*, edited by E. C. Carterette and M. P. Friedman, pp. 117–166. New York: Academic Press, 1975.

[1171] R. L. de Valois and K. K. de Valois. "A Multi-Stage Color Model." *Vision Research* 33 (1993), 1053–1065.

[1172] R. L. de Valois, I. Abramov, and G. H. Jacobs. "Analysis of Response Patterns of LGN Cells." *Journal of the Optical Society of America* 56:7 (1966), 966–977.

[1173] R. L. de Valois, D. M. Snodderly Jr, E. W. Yund, and N. K. Hepler. "Responses of Macaque Lateral Geniculate Cells to Luminance and Color Figures." *Sensory Processes* 1:3 (1977), 244–259.

[1174] R. L. de Valois, K. K. de Valois, E. Switkes, and L. E. Mahon. "Hue Scaling of Isoluminant and Cone-Specific Lights." *Vision Research* 37:7 (1997), 885–897.

[1175] R. L. de Valois, N. P. Cottaris, S. D. Elfar, L. E. Mahon, and J. A. Wilson. "Some Transformations of Color Information from Lateral Geniculate Nucleus to Striate Cortex." *Proceedings of the National Academy of Sciences of the United States of America* 97:9 (2000), 4997–5002.

[1176] R. L. de Valois, K. K. de Valois, and L. E. Mahon. "Contribution of S Opponent Cells to Color Appearance." *Proceedings of the National Academy of Sciences of the United States of America* 97:1 (2000), 512–517.

[1177] D. I. Vaney. "The Mosaic of Amacrine Cells in the Mammalian Retina." *Progress in Retinal and Eye Research* 9 (1990), 49–100.

[1178] R. VanRullen and T. Dong. "Attention and Scintillation." *Vision Research* 43:21 (2003), 2191–2196.

[1179] D. Varin. "Fenomeni di contrasto e diffusione cromatica nell organizzazione spaziale del campo percettivo." *Rivista di Psicologia* 65 (1971), 101–128.

[1180] R. G. Vautin and B. M. Dow. "Color Cell Groups in Foveal Striate Cortex of the Behaving Macaque." *Journal of Neurophysiology* 54:2 (1985), 273–292.

[1181] T. R. Vidyasagar, J. J. Kulikowski, D. M. Lipnicki, and B. Dreher. "Convergence of Parvocellular and Magnocellular Information Channels in the Primary Visual Cortex of the Macaque." *European Journal of Neuroscience* 16:5 (2002), 945–956.

[1182] J. A. S. Viggiano. "Modeling the Color of Multi-Colored Halftones." In *Proceedings of the Technical Association of Graphical Arts*, pp. 44–62. Sewickley, PA: Technical Association of the Graphic Arts, 1990.

[1183] B. Vohnsen. "Photoreceptor Waveguides and Effective Retinal Image Quality." *Journal of the Optical Society of America A* 24:3 (2007), 597–607.

[1184] E. Völkl, L. F. Allard, and D. C. Joy, editors. *Introduction to Electron Holography*. New York: Kluwer Academic / Plenum Publishers, 1999.

[1185] J. J. Vos, J. Walraven, and A. van Meeteren. "Light Profiles of the Foveal Image of a Point Source." *Vision Research* 16:2 (1976), 215–219.

[1186] J. Vos. "Disability Glare—A State of the Art Report." *CIE Journal* 3:2 (1984), 39–53.

[1187] M. J. Vrhel and H. J. Trussell. "Color Device Calibration: A Mathematical Formulation." *IEEE Transactions on Image Processing* 8:12 (1999), 1796–1806.

[1188] H. L. de Vries. "The Quantum Character of Light and Its Bearing upon the Threshold of Vision, the Differential Sensitivity and Acuity of the Eye." *Physica* 10:7 (1943), 553–564.

[1189] T. Wachtler, T. J. Sejnowski, and T. D. Albright. "Representation of Color Stimuli in Awake Macaque Primary Visual Cortex." *Neuron* 37:4 (2003), 681–691.

[1190] H. G. Wagner, E. F. MacNichol, and M. L. Wolbarsht. "Three Response Properties of Single Ganglion Cells in the Goldfish Retina." *Journal of General Physiology* 43 (1960), 45–62.

[1191] G. Wald and D. R. Griffin. "The Change in Refractive Power of the Human Eye in Dim and Bright Light." *Journal of the Optical Society of America* 37 (1946), 321–336.

[1192] I. Wald and P. Slusallek. "State of the Art in Interactive Ray Tracing." In *Eurographics STAR - State of the Art Reports*, pp. 21–42. Aire-la-Ville, Switzerland: Eurographics Association, 2001.

[1193] I. Wald, P. Slusallek, and C. Benthin. "Interactive Distributed Ray Tracing of Highly Complex Models." In *Proceedings of the* 12th *Eurographics Workshop on Rendering*, edited by S. J. Gortler and K. Myszkowski, pp. 274–285. Aire-la-Ville, Switzerland: Eurographics Association, 2001.

[1194] G. Wald. "Human Vision and the Spectrum." *Science 101*, pp. 653–658.

[1195] M. M. Waldrop. "Brilliant Displays." *Scientific American* 297:3 (2007), 94–97.

[1196] H. Wallach. "Brightness Constancy and the Nature of Achromatic Colors." *Journal of Experimental Psychology* 38 (1948), 310–324.

[1197] G. Walls. "The Filling-In Process." *American Journal of Optometry* 31 (1954), 329–340.

[1198] E. Walowit, C. J. McCarthy, and R. S. Berns. "Spectrophotometric Color Matching Based on Two-Constant Kubelka-Munk Theory." *Color Research and Application* 13:6 (1988), 358–362.

[1199] J. W. T. Walsh. *Photometry*, Third edition. New York: Dover Publications, Inc., 1965.

[1200] B. Walter. "Notes on the Ward BRDF." Technical Report PCG-05-06, Cornell Program of Computer Graphics, Ithaca, NY, 2005.

[1201] B. A. Wandell and J. E. Farrell. "Water into Wine: Converting Scanner RGB to Tristimulus XYZ [1909-11]." In *Proceedings of SPIE 1909*, pp. 92–101. Bellingham, WA: SPIE, 1993.

[1202] B. A. Wandell, H. A. Baseler, A. B. Poirson, G. M. Boynton, and S. A. Engel. "Computational Neuroimaging: Color Tuning in Two Human Cortical Areas Measured using fMRI." In *Color Vision: From Genes to Perception*, edited by K. R. Gegenfurtner and L. Sharpe, pp. 269–282. Cambridge, UK: Cambridge University Press, 1999.

[1203] B. A. Wandell. "Color Rendering of Color Camera Data." *Color Research and Application* 11 (1986), S30–S33.

[1204] B. A. Wandell. "The Synthesis and Analysis of Color Images." *IEEE Transactions on Patterns Analysis and Machine Intelligence* 9:1 (1987), 2–13.

[1205] B. A. Wandell. *Foundations of Vision*. Sunderland, Massachusetts: Sinauer Associates, inc., 1995.

[1206] J. Y. A. Wang and E. H. Adelson. "Representing Moving Images with Layers." *IEEE Transactions on Image Processing* 3:5 (1994), 625–638.

[1207] L. Wang, L.-Y. Wei, K. Zhou, B. Guo, and H.-Y. Shum. "High Dynamic Range Image Hallucination." In *Rendering Techniques '07 (Proceedings of the Eurographics Symposium on Rendering)*, 2007.

[1208] G. Ward and E. Eydelberg-Vileshin. "Picture Perfect RGB Rendering using Spectral Prefiltering and Sharp Color Primaries." In *Thirteenth Eurographics Workshop on Rendering*, edited by P. Debevec and S. Gibson, pp. 123–130. Aire-la-Ville, Switzerland: Eurographics Association, 2002.

[1209] G. Ward and M. Simmons. "Subband Encoding of High Dynamic Range Imagery." In *APGV '04: Proceedings of the 1st Symposium on Applied Perception in Graphics and Visualization*, pp. 83–90. New York: ACM Press, 2004.

[1210] G. Ward and M. Simmons. "JPEG-HDR: A Backwards Compatible, High Dynamic Range Extension to JPEG." In *Proceedings of the Thirteenth Color Imaging Conference*, pp. 283–290. Springfield, VA: Society for Imaging Science and Technology, 2005.

[1211] G. Ward, H. Rushmeier, and C. Piatko. "A Visibility Matching Tone Reproduction Operator for High Dynamic Range Scenes." *IEEE Transactions on Visualization and Computer Graphics* 3:4 (1997), 291–306.

[1212] G. Ward. "Measuring and Modeling Anisotropic Reflection." *Proc. SIGGRAPH '92, Computer Graphics* 26:2 (1992), 265–272.

[1213] G. Ward. "A Contrast-Based Scalefactor for Luminance Display." In *Graphics Gems IV*, edited by P. Heckbert, pp. 415–421. Boston: Academic Press, 1994.

[1214] G. Ward. "Fast, Robust Image Registration for Compositing High Dynamic Range Photographs from Hand-Held Exposures." *journal of graphics tools* 8:2 (2003), 17–30.

[1215] C. Ware. "Color Sequences for Univariate Maps: Theory, Experiments, and Principles." *IEEE Computer Graphics and Applications* 8:5 (1988), 41–49.

[1216] C. Ware. *Information Visualization: Perception for Design.* Morgan Kaufmann, 2000.

[1217] H. Wässle and B. B. Boycott. "Functional Architecture of the Mammalian Retina." *Physiological Reviews* 71:2 (1991), 447–480.

[1218] H. Wässle, U. Grünert, M.-H. Chun, and B. B. Boycott. "The Rod Pathway of the Macaque Monkey Retina: Identification of AII-Amacrine Cells with Antibodies against Calretinin." *Journal of Comparative Neurology* 361:3 (1995), 537–551.

[1219] A. Watson and A. Ahumada. "A Standard Model for Foveal Detection of Spatial Contrast." *Journal of Vision* 5:9 (2005), 717–740.

[1220] A. Watson and J. Solomon. "A Model of Visual Contrast Gain Control and Pattern Masking." *Journal of the Optical Society of America* 14 (1997), 2378–2390.

[1221] A. Watson. "The Spatial Standard Observer: A Human Vision Model for Display Inspection." *SID Symposium Digest of Tehcnical Papers* 37 (2006), 1312–1315.

[1222] R. A. Weale. "Human Lenticular Fluorescence and Transmissivity, and Their Effects on Vision." *Experimental Eye Research* 41:4 (1985), 457–473.

[1223] R. A. Weale. "Age and Transmittance of the Human Crystalline Lens." *Journal of Physiology* 395 (1988), 577–587.

[1224] M. S. Weaver, V. Adamovich, M. Hack, R. Kwong, and J. J. Brown. "High Efficiency and Long Lifetime Phosphorescent OLEDs." In *Proceedings of the International Conference on Electroluminescence of Molecular Materials and Related Phenomena*, pp. 10–35. Washington, DC: IEEE, 2003.

[1225] E. H. Weber. "De pulsu, resorptione, auditu et tactu." In *Annotationes Anatomicae et Physiologicae*. Lipsiae: C F Köhler, 1834.

[1226] E. H. Weber. "On the Sense of Touch and Common Sensibility." In *E H Weber On the Tactile Senses*, edited by H. E. Ross and D. J. Murray, Second edition. Hove, UK: Erlbaum, 1996.

[1227] M. A. Webster and J. D. Mollon. "Color Constancy Influenced by Contrast Adaptation." *Nature* 373:6516 (1995), 694–698.

[1228] M. Webster. "Light Adaptation, Contrast Adaptation, and Human Colour Vision." In *Colour Perception: Mind and the Physical World*, edited by R. Mausfeld and D. Heyer, pp. 67–110. Oxford, UK: Oxford University Press, 2003.

[1229] M. Webster. "Pattern-Selective Adaptation in Color and Form Perception." In *The Visual Neurosciences, Volume 2*, edited by L. Chalupa and J. Werner, pp. 936–947. Cambridge, MA: MIT Press, 2004.

[1230] A. R. Weeks. *Fundamentals of Image Processing*. New York: Wiley-IEEE Press, 1996.

[1231] B. Weiss. "Fast Median and Bilateral Filtering." *ACM Transactions on Graphics* 25:3 (2006), 519–526.

[1232] E. W. Weisstein. "Full Width at Half Maximum." In *MathWorld — A Wolfram Web Resource*. Wolfram Research, 1999. Available online (http://mathworld.wolfram.com/).

[1233] E. W. Weisstein. "Mie Scattering." In *MathWorld — A Wolfram Web Resource*. Wolfram Research, 1999. Available online (http://mathworld.wolfram.com/).

[1234] S.-W. Wen, M.-T. Lee, and C. H. Chen. "Recent Development of Blue Fluorescent OLED Materials and Devices." *IEEE/OSA Journal of Display Technology* 1:1 (2005), 90–99.

[1235] J. Weng, P. Cohen, and M. Herniou. "Camera Calibration with Distortion Models and Accuracy Evaluation." *IEEE Transactions on Pattern Analysis and Machine Intelligence* 14:10 (1994), 965–980.

[1236] S. Wesolkowski, S. Tominaga, and R. D. Dony. "Shading and Highlight Invariant Color Image Segmentation using the MPC Algorithm." In *Proceedings of the SPIE (Color Imaging: Device-Independent Color, Color Hard Copy and Graphic Arts VI)*, pp. 229–240. Bellingham, WA: SPIE, 2001.

[1237] R. S. West, H. Konijn, W. Sillevis-Smitt, S. Kuppens, N. Pfeffer, Y. Martynov, Y. Takaaki, S. Eberle, G. Harbers, T. W. Tan, and C. E. Chan. "High Brightness Direct LED Backlight for LCD-TV." *SID Symposium Digest of Technical Papers* 34:1 (2003), 1262–1265.

[1238] G. Westheimer. "Visual Acuity." *Annual Review of Psychology* 16 (1965), 359–380.

[1239] G. Westheimer. "Visual Acuity." In *Adler's Physiology of the Eye*, edited by R. A. Moses and W. M. Hart. St. Louis: The C V Mosby Company, 1987.

[1240] S. Westin, J. Arvo, and K. Torrance. "Predicting Reflectance Functions from Complex Surfaces." In *Proceedings SIGGRAPH '99, Computer Graphics Proceedings, Annual Conference Series*, pp. 255–264. Reading, MA: Addison-Wesley Longman, 1999.

[1241] M. White. "A New Effect on Perceived Lightness." *Perception* 8:4 (1979), 413–416.

[1242] T. Whitted. "An Improved Illumination Model for Shaded Display." *Communications of the ACM* 23:6 (1980), 343–349.

[1243] H. Widdel, D. L. Grossman, D. L. Post, and J. Walraven, editors. *Color in Electronic Displays*. Dordrecht, Netherlands: Kluwer Academic / Plenum Publishers, 1992.

[1244] T. N. Wiesel and D. H. Hubel. "Spatial and Chromatic Interactions in the Lateral Geniculate Body of the Rhesus Monkey." *Journal of Neurophysiology* 29 (1966), 1115–1156.

[1245] D. R. Williams and A. Roorda. "The Trichromatic Cone Mosaic in the Human Eye." In *Color Vision: From Genes to Perception*, edited by K. R. Gegenfurtner and L. T. Sharpe, pp. 113–122. Cambridge, UK: Cambridge University Press, 1999.

[1246] T. L. Williams. *The Optical Transfer Function of Imaging Systems*. Bristol, UK: Institute of Physics Publishing, 1999.

[1247] S. J. Williamson and H. Z. Cummins. *Light and Color in Nature and Art*. New York: John Wiley and Sons, 1983.

[1248] E. N. Wilmer and W. D. Wright. "Colour Sensitivity of the Fovea Centralis." *Nature* 156 (1945), 119–121.

[1249] K. Witt. "Modified CIELAB Formula Tested using a Textile Pass/Fail Data Set." *Colour Research and Application* 19 (1994), 273–285.

[1250] L. B. Wolff. "Polarization-Based Material Classification from Specular Reflection." *IEEE Transactions on Patterns Analysis and Machine Intelligence* 12:11 (1990), 1059–1071.

[1251] M. Wong-Riley. "Columnar Cortico-Cortical Interconnections within the Visual System of the Squirrel and Macaque Monkeys." *Brain Research* 162:2 (1979), 201–217.

[1252] H.-S. P. Wong. "CMOS Imge Sensors — Recent Advances and Device Scaling Considerations." In *International Electron Devices Meeting Technical Digest*, pp. 201–204, 1997.

[1253] W. Wright. "A Red-Determination of the Trichromatic Coefficients of the Spectral Colours." *Trans. Opt. Soc.* 30 (1928–1929), 141–161.

[1254] W. D. Wright. "The Characteristics of Tritanopia." *Journal of the Optical Society of America* 42 (1952), 509–520.

[1255] W. Wright. "Why and How Chromatic Adaptation Has Been Studied." *Color Research and Application* 6 (1981), 147–152.

[1256] W. Wu and J. P. Allebach. "Imaging Colorimetry using a Digital Camera." *Journal of Imaging Science and Technology* 44:4 (2000), 267–279.

[1257] S. Wurmfeld. "Color in Abstract Painting." In *Color for Science, Art and Technology*, edited by K. Nassau. Amsterdam: Elsevier Science B.V., 1998.

[1258] D. R. Wyble and R. Berns. "A Critical Review of Spectral Models Applied to Binary Color Printing." *Color Research and Application* 25 (2000), 4–19.

[1259] D. R. Wyble and M. R. Rosen. "Color Management of DLP Projectors." In *Proceedings of the* 12th *IS&T/SID Color Imaging Conference*, pp. 228–232. Springfield, VA: Society for Imaging Science and Technology, 2004.

[1260] D. R. Wyble and H. Zhang. "Colorimetric Characterization Model for DLP Projectors." In *Proceedings of the* 11th *IS&T/SID Color Imaging Conference*, pp. 346–350. Springfield, VA: Society for Imaging Science and Technology, 2003.

[1261] G. Wyszecki and W. S. Stiles. *Color Science: Concepts and Methods, Quantitative Data and Formulae*. New York: John Wiley & Sons, 1982.

[1262] G. Wyszecki and W. S. Stiles. *Color science: Concepts and methods, quantitative data and formulae*, Second edition. John Wiley and Sons, Ltd, 2000.

[1263] K. Xiao, C. Li, M. Luo, and C. Taylor. "Color Appearance for Dissimilar Sizes." In *IS&T CGIV, 2nd European Conference on Color in Graphics, Imaging, and Vision*, pp. 12–16. Springfield, VA: Society for Imaging Science and Technology, 2004.

[1264] C. Xu, W. Zhang, W.-H. Ki, and M. Chan. "A 1.0-V V_{DD} CMOS Active-Pixel Sensor with Complementary Pixel Architecture and Pulsewidth Modulation Fabricated with a 0.25 μm CMOS Process." *IEEE Journal of Solid-State Circuits* 37:12 (2002), 1853–1859.

[1265] N. H. Yabuta and E. M. Callaway. "Functional Streams and Local Connections of Layer 4C Neurons in Primary Visual Cortex of the Macaque Monkey." *Journal of Neuroscience* 18:22 (1998), 9489–9499.

[1266] O. Yadid-Pecht and E. R. Fossum. "Wide Intrascene Dynamic Range CMOS APS using Dual Sampling." *IEEE Transactions on Electron Devices* 44:10 (1997), 1721–1723.

[1267] H. Yamaguchi and M. D. Fairchild. "A Study of Simultaneous Lightness Perception for Stimuli with Multiple Illumination Levels." In *IS&T/SID 12th Color Imaging Conference*, pp. 22–28. Springfield, VA: Society for Imaging Science and Technology, 2004.

[1268] M. Yamaguchi, T. Ajito, and N. Ohyama. "Multiprimary Color Display using Holographic Optical Element." In *Proceedings of the SPIE 3293*, edited by S. A. Benton, pp. 70–77. Bellingham, WA: SPIE, 1998.

[1269] M. Yamaguchi, H. Haneishi, H. Fukuda, J. Kishimoto, H. Kanazawa, M. Tsuchida, R. Iwama, and N. Ohyama. "High-Fidelity Video and Still-Image Communication based on Spectral Information: Natural Vision System and Its Applications." In *Proceedings of SPIE 6062*, edited by M. R. Rosen, F. H. Imai, and S. Tominaga, pp. 60620G–1–60620G–12. Bellingham, WA: SPIE, 2006.

[1270] M. Yamauchi. "Jones-Matrix Models for Twisted-Nematic Liquid-Crystal Devices." *Applied Optics* 44:21 (2005), 4484–4493.

[1271] J. N. Yang and L. T. Maloney. "Illuminant Cues in Surface Color Perception: Tests of Three Candidate Cues." *Vision Research* 41 (2001), 2581–2600.

[1272] D. X. D. Yang, A. E. Gamal, B. Fowler, and H. Tian. "A 640 × 512 CMOS Image Sensor with Ultrawide Dynamic Range Floating-Point Pixel-Level ADC." *IEEE Journal of Solid-State Circuits* 34:12 (1999), 1821–1834.

[1273] T. Yasuda, T. Hamamoto, and K. Aizawa. "Adaptive-Integration-Time Sensor with Real-Time Reconstruction Function." *IEEE Transactions on Electron Devices* 50:1 (2003), 111–120.

[1274] P. Yeh. "Extended Jones Matrix Method." *Journal of the Optical Society of America* 72:4 (1982), 507–513.

[1275] H. Yersin. "Triplet emitters for OLED applications. Mechanisms of exciton trapping and control of emission properties." In *Transition Metal and Rare Earth Compounds, Topics in Current Chemistry 241*, edited by H. Yersin. Berlin: Springer-Verlag, 2004.

[1276] M.-S. Yoo, J.-N. Kim, D.-J. Woo, S.-H. Yim, and Y.-H. Cho. "Development of Performance and Cost-Oriented HD PDP TV with High Picture Quality using a High Efficiency Hexagonal Array Structure." *Journal of the Society for Information Display* 12 (2004), 251–256.

[1277] Y. Yoshida and Y. Yamamoto. "Color Calibration of LCDs." In *Proceedings of the IS&T/SID Tenth Color Imaging Conference*, pp. 305–311. Springfield, VA: Society of Imaging Science and Technology, 2002.

[1278] A. Yoshida, V. Blanz, K. Myszkowski, and H.-P. Seidel. "Perceptual Evaluation of Tone Mapping Operators with Real-World Scenes." In *Proceedings of SPIE 5666 (Human Vision and Electronic Imaging X)*, pp. 192–203. Bellingham, WA: SPIE, 2005.

[1279] A. Yoshida, R. Mantiuk, K. Myszkowski, and H.-P. Seidel. "Analysis of Reproducing Real-World Appearance on Displays of Varying Dynamic Range." *Computer Graphics Forum* 25:3 (2006), 415–426.

[1280] K. Yoshikawa, Y. Kanazawa, M. Wakitani, T. Shinoda, and A. Ohtsuka. "A Full Color AC Plasma Display with 256 Gray Scale." In *Proceedings of Asia Display*, pp. 605–608. San Jose, CA: Society for Information Display, 1992.

[1281] T. Yoshioka, J. B. Levitt, and J. S. Lund. "Independence and Merger of Thalamocortical Channels within Macaque Monkey Primary Visual Ccortex: Anatomy of Interlaminar Projections." *Visual Neuroscience* 11:3 (1994), 467–489.

[1282] I. T. Young, J. J. Gerbrands, and L. J. van Vliet. "Fundamentals of Image Processing (Online Image Processing Course at www.ph.tn.tudelft.nl/Courses/FIP/).", 2006.

[1283] T. Young. "On the Theory of Light and Colors." *Philosophical Transactions of the Royal Society of London* 92 (1802), 20–21. Reprinted in David L MacAdam (ed.), *Selected Papers on Colorimetry — Fundamentals*, SPIE Optical Engineering Press, 1993.

[1284] F. H. Yu and H. S. Kwok. "Comparison of Extended Jones Matrices for Twisted Nematic Liquid-Crystal Displays at Oblique Angles of Incidence." *Journal of the Optical Society of America A* 16:11 (1999), 2772–2780.

[1285] Y. Yu, P. Debevec, J. Malik, and T. Hawkins. "Inverse Global Illumination: Recovering Reflectance Models of Real Scenes from Photographs." In *Proceedings SIGGRAPH '99, Computer Graphics Proceedings, Annual Conference Series*, pp. 215–224. Reading, MA: Addison-Wesley Longman, 1999.

[1286] M. Yukie and E. Iwai. "Direct Projection from the Dorsal Lateral Geniculate Nucleus to the Prestriate Cortex in Macaque Monkeys." *Journal of Comparative Neurology* 201:1 (1981), 81–97.

[1287] J. A. C. Yule. *Principles of Color Reproduction.* New York: Wiley, 1967.

[1288] Q. Zaidi, B. Spehar, and M. Shy. "Induced Effects of Background and Foregrounds in Three-Dimensional Configurations: The Role of T-Junctions." *Perception* 26 (1997), 395–408.

[1289] S. A. Zeki. "A Century of Cerebral Achromatopsia." *Brain* 113:6 (1990), 1721–1777.

[1290] S. A. Zeki. *Vision of the Brain.* Oxford, UK: Blackwell Publishers, 1993.

[1291] S. A. Zeki. *Inner Vision: An Exploration of Art and the Brain.* Oxford, UK: Oxford University Press, 2000.

[1292] X. M. Zhang and B. Wandell. "A Spatial Extension to CIELAB for Digital Color Image Reproductions." *Proceedings of the SID Symposium Digest* 27 (1996), 731–734.

[1293] R. Zhang, P. Tsai, J. Cryer, and M. Shah. "Shape from Shading: A Survey." *IEEE Transactions on Pattern Analysis and Machine Intelligence* 21:8 (1999), 690–706.

[1294] X. Zhu and S.-T. Wu. "Overview on Transflective Liquid Crystal Displays." In *The* 16th *Annual Meeting of the IEEE Lasers and Electro-Optical Society, Volume 2*, pp. 953–954. Washington, DC: IEEE Press, 2003.

[1295] C. Ziegaus and E. W. Lang. "Statistics of Natural and Urban Images." In *Proceedings of the* 7th *International Conference on Artificial Neural Networks, Lecture Notes in Computer Science 1327*, pp. 219–224. Berlin: Springer-Verlag, 1997.

[1296] C. Ziegaus and E. W. Lang. "Statistical Invariances in Artificial, Natural and Urban Images." *Zeitschrift für Naturforschung A* 53a:12 (1998), 1009–1021.

[1297] S. Zigman. "Vision Enhancement using a Short Wavelength Light Absorbing Filter." *Optometry and Vision Science* 67:2 (1990), 100–104.

[1298] M. D'Zmura and G. Iverson. "Color Constancy. I. Basic Theory of Two-Stage Linear Recovery of Spectral Description for Lights and Surfaces." *Journal of the Optical Society of America A* 10:10 (1993), 2148–2165.

[1299] M. D'Zmura and G. Iverson. "Color Constancy. II. Results for Two-Stage Linear Recovery of Spectral Description for Lights and Surfaces." *Journal of the Optical Society of America A* 10:10 (1993), 2166–2180.

[1300] M. D'Zmura and P. Lennie. "Mechanisms of Color Constancy." *Journal of the Optical Society of America A* 3:10 (1986), 1662–1673.

[1301] M. D'Zmura, G. Iverson, and B. Singer. "Probabilistic Color Constancy." In *Geometric Representations of Perceptual Phenomena*, pp. 187–202. Mahwah, NJ: Lawrence Erlbaum, 1995.

[1302] J. A. Zuclich, R. D. Glickman, and A. R. Menendez. "In Situ Measurements of Lens Fluorescence and Its Interference With Visual Function." *Investigative Ophthalmology and Visual Science* 33:2 (1992), 410–415.

Index